D1523059

Stormwater Effects Handbook

A Toolbox for
Watershed Managers,
Scientists, and Engineers

Stormwater Effects Handbook

A Toolbox for
Watershed Managers,
Scientists, and Engineers

G. Allen Burton, Jr., Ph.D.
Robert E. Pitt, Ph.D., P.E.

LEWIS PUBLISHERS

A CRC Press Company
Boca Raton   London   New York   Washington, D.C.

Library of Congress Cataloging-in-Publication Data

Burton, G. Allen
    Stormwater effects handbook : a toolbox for watershed managers, scientists, and engineers / by G. Allen Burton, Jr. and Robert Pitt.
       p. cm.
    Includes bibliographical references and index.
    ISBN 0-87371-924-7 (alk. paper)
    1. Runoff—Management—Handbooks, manuals, etc. 2. Runoff—Environmental aspects—Handbooks, manuals, etc. 3. Water quality—Measurement—Handbooks, manuals, etc. 4. Water quality biological assessment—Handbooks, manuals, etc. I. Pitt, Robert. II. Title.

TD657 .B86 2001
628.1'68—dc21
                                             2001029906
                                                  CIP

Visit the CRC Press Web site at www.crcpress.com

© 2002 by CRC Press LLC
Lewis Publishers is an imprint of CRC Press LLC

No claim to original U.S. Government works
International Standard Book Number 0-87371-924-7
Library of Congress Card Number 2001029906
Printed in the United States of America 1 2 3 4 5 6 7 8 9 0
Printed on acid-free paper

Dedication

This book is dedicated to those who were instrumental in guiding and supporting our development as scientists and engineers and our appreciation of the outdoors.

Preface

This handbook is intended to be a working document which assists scientists, engineers, consultants, regulators, citizen groups, and environmental managers in determining if stormwater runoff is causing adverse effects and beneficial-use impairments in local receiving waters. This includes adverse effects on aquatic life and human health and considers exposures to multiple stressors such as pathogens, chemicals, and habitat alteration. Given the complicated nature of the problem, where diffuse inputs contain multiple stressors which vary in intensity with time (and often in areas which are simultaneously impacted by point source discharges or other development activities, e.g., channelization), it is difficult to define and separate stormwater effects from these other factors. To accomplish this task requires an integrated watershed-based assessment approach which focuses on sampling before, during, and after storm events.

This handbook provides a logical approach for an experimental design that can be tailored to address a wide range of environmental concerns, such as ecological and human health risk assessments, determining water quality or biological criteria exceedances, use impairment, source identification, trend analysis, determination of best management practice (BMP) effectiveness, stormwater quality monitoring for NPDES Phase I and II permits and applications, and total maximum daily load (TMDL) assessments. Despite the complexity of stormwater, successful and accurate assessments of its impact are possible by following the logical integrated approaches described in this handbook.

New methods and technologies are rapidly being developed, so this should be considered a "living" document which will be updated as the science warrants. We welcome your input on ways to improve future editions.

Allen Burton
Bob Pitt
May 2001

Disclaimer: The views presented within this document do not necessarily represent those of the U.S. Environmental Protection Agency.

Acknowledgments

We are indebted to our professional colleagues whose prior contributions enabled us to produce this book. In addition, the long productive hours of a host of graduate and undergraduate students at Wright State and the University of Alabama at Birmingham are acknowledged for their essential research contributions.

We greatly appreciate the word processing of Nancy Pestian and Amy Ray. We also thank the production staff and editors at Lewis Publishers/CRC Press for their hard work and patience.

The support of the U.S. EPA, especially Richard Field, is also appreciated, not only for help in the preparation of this current work, but also for the prior support given to many of the research projects described in this book.

Special thanks are also due to our families, who provided never-ending support during the preparation of this book.

About the Authors

G. Allen Burton, Jr., is the Brage Golding Distinguished Professor of Research and Director of the Institute for Environmental Quality at Wright State University. He obtained a Ph.D. degree in Environmental Science from the University of Texas at Dallas in 1984. From 1980 until 1985 he was a Life Scientist with the U.S. Environmental Protection Agency. He was a Postdoctoral Fellow at the National Oceanic and Atmospheric Administration's Cooperative Institute for Research in Environmental Sciences at the University of Colorado. Since then, he has had positions as a NATO Senior Research Fellow in Portugal and Visiting Senior Scientist in Italy and New Zealand.

Dr. Burton's research during the past 20 years has focused on developing effective methods for identifying significant effects and stressors in aquatic systems where sediment and stormwater contamination is a concern. His ecosystem risk assessments have evaluated multiple levels of biological organization, ranging from microbial to amphibian effects. He has been active in the development and standardization of toxicity methods for the U.S. EPA, American Society for Testing and Materials (ASTM), Environment Canada, and the Organization of Economic Cooperation and Development (OECD). Dr. Burton has served on numerous national and international scientific committees and review panels, and written more than 100 publications dealing with aquatic systems.

Robert Pitt is currently a Professor in the Department of Civil and Environmental Engineering at the University of Alabama. Bob had previously served on the School of Engineering faculty at the University of Alabama at Birmingham since 1987. Prior to that, he was a Senior Engineer for 16 years in industry and government, and continues to consult to many municipalities and engineering firms. He received his Ph.D. in Civil and Environmental Engineering from the University of Wisconsin–Madison, his M.S.C.E. in Environmental Engineering/Hydraulic Engineering from San Jose State University, CA, and his B.S. in Engineering Science, from Humboldt State University, Arcata, CA. He is a registered professional engineer (WI) and a Diplomate of the American Academy of Environmental Engineers.

During the past 30 years, Bob has been the project manager and principal investigator for many water resources research projects conducted for the U.S. EPA, Environment Canada, Ontario Ministry of the Environment, and state and local governments concerning the effects, sources, and control of urban runoff. Some are used as case studies in this book. His major area of interest is in stormwater management, especially the integration of drainage and water quality objectives. He currently teaches classes in water supply and drainage design, hydrology, hydraulics, experimental design, and field sampling, plus a series on stormwater management. Bob has published more than 100 chapters, books, journal articles, and major research reports. He is a member of the American Society of Civil Engineers, the Water Environment Federation, the North American Lake Management Society, the American Water Resources Association, and the Society for Environmental Toxicology and Chemistry.

Contents

Unit 1: The Problem of Stormwater Runoff

Chapter 1   Introduction

Overview: The Problem of Stormwater Runoff 3
Sources of NPS Pollution...4
Regulatory Program...8
Applications of the Handbook 10
References 13

Chapter 2   Receiving Water Uses, Impairments, and Sources of Stormwater Pollutants

Introduction..15
Beneficial Use Impairments 22
Likely Causes of Receiving Water Use Impairments 30
Major Urban Runoff Sources 31
Summary..42
References 43

Chapter 3   Stressor Categories and Their Effects on Humans and Ecosystems

Effects of Runoff on Receiving Waters 47
Stressor Categories and Their Effects..63
Receiving Water Effect Summary 90
References 92

Unit 2: Components of the Assessment

Chapter 4   Overview of Assessment Problem Formulation

Introduction..102
Watershed Indicators of Biological Receiving Water Problems 103
Summary of Assessment Tools 107
Study Design Overview..107
Beginning the Assessment...108
Example Outline of a Comprehensive Runoff Effect Study 119
Case Studies of Previous Receiving Water Evaluations 123
Summary: Typical Recommended Study Plans..213
References 218

Chapter 5   Sampling Effort and Collection Methods

Introduction..224
Experimental Design: Sampling Number and Frequency 224
Data Quality Objectives (DQO) and Associated QA/QC Requirements...................247
General Considerations for Sample Collection 254
Receiving Water, Point Source Discharge, and Source Area Sampling....................278
Sediment and Pore Water Sampling 313
Summary: Basic Sample Collection Methods 336
References 338

Chapter 6   Ecosystem Component Characterization
Overview..346
Flow and Rainfall Monitoring..349
Soil Evaluations..388
Aesthetics, Litter, and Safety 398
Habitat...400
Water and Sediment Analytes and Methods.....................................423
Microorganisms in Stormwater and Urban Receiving Waters 485
Benthos Sampling and Evaluation in Urban Streams 491
Zooplankton Sampling 502
Fish Sampling...502
Toxicity and Bioaccumulation 507
Summary..546
References 550

Chapter 7   Statistical Analyses of Receiving Water Data
Selection of Appropriate Statistical Analysis Tools and Procedures 575
Comments on Selected Statistical Analyses Frequently Applied to Receiving Water
Data..582
Summary of Statistical Elements of Concern When Conducting a Receiving Water
Investigation..605
References 606

Chapter 8   Data Interpretation
Is There a Problem? 609
Evaluating Biological Stream Impairments Using the Weight-of-Evidence Approach......611
Evaluating Human Health Impairments Using a Risk Assessment Approach....................619
Identifying and Prioritizing Critical Stormwater Sources 626
Summary..636
References 637

Unit 3: Toolbox of Assessment Methods

Appendix A   Habitat Characterization
The Qualitative Habitat Evaluation Index (QHEI) 643
The USEPA Habitat Assessment for the Rapid Bioassessment Protocols.........................652
References..662

Appendix B   Benthic Community Assessment
Rapid Bioassessment Protocol: Benthic Macroinvertebrates 665
The Ohio EPA Invertebrate Community Index Approach 681
A Partial Listing of Agencies that Have Developed Tolerance Classifications and/or
Biotic Indices..687
References 690

Appendix C   Fish Community Assessment
Rapid Bioassessment Protocol V — Fish 693
References 707

Appendix D   Toxicity and Bioaccumulation Testing

General Toxicity Testing Methods 710

Methods for Conducting Long-Term Sediment Toxicity Tests with *Hyalella azteca* 710

Methods for Conducting Long-Term Sediment Toxicity Tests with *Chironomus tentans* 718

In Situ Testing Using Confined Organisms 724

Toxicity Identification Evaluations 729

Toxicity — Microtox Screening Test..730

References 733

Appendix E   Laboratory Safety, Waste Disposal, and Chemical Analyses Methods

Introduction..736

Fundamentals of Laboratory Safety..737

Basic Rules and Procedures for Working with Chemicals.......................738

Use and Storage of Chemicals in the Laboratory 743

Procedures for Specific Classes of Hazardous Materials 748

Emergency Procedures 758

Chemical Waste Disposal Program..760

Material Safety Data Sheets (MSDS) 763

Summary of Field Test Kits 767

Special Comments Pertaining to Heavy Metal Analyses 774

Stormwater Sample Extractions for EPA Methods 608 and 625............779

Calibration and Deployment Setup Procedure for YSI 6000upg Water Quality Monitoring Sonde...782

References 785

Appendix F   Sampling Requirements for Paired Tests

Charts 787

Appendix G   Water Quality Criteria

Introduction..798

EPA's Water Quality Criteria and Standards Plan — Priorities for the Future 798

Compilation of Recommended Water Quality Criteria and EPA's Process for Deriving New and Revised Criteria 799

Ammonia 813

Bacteria 816

Chloride, Conductivity, and Total Dissolved Solids.................................822

Chromium 823

Copper..824

Hardness 825

Hydrocarbons...826

Lead 827

Nitrate and Nitrite 828

Phosphate 830

pH 832

Suspended Solids and Turbidity..834

Zinc 835

Sediment Guidelines...836

References 839

Appendix H   Watershed and Receiving Water Modeling
    Introduction..843
    Modeling Stormwater Effects and the Need for Local Data for Calibration and
    Verification..845
    Summary...860
References 866

Appendix I   Glossary...867

Appendix J   Vendors of Supplies and Equipment Used in Receiving Water Monitoring
    General Field and Laboratory Equipment 871
    Automatic Samplers 872
    Basic Field Test Kits 873
    Specialized Field Test Kits..873
    Parts and Supplies for Custom Equipment...873
    Toxicity Test Organisms..874
    Laboratory Chemical Supplies (and other equipment)........................874

Index 875

The Problem of Stormwater Runoff

## CHAPTER 1

Introduction

"A stench from its inky surface putrescent with the oxidizing processes to which the shadows of the over-reaching trees add stygian blackness and the suggestion of some mythological river of death. With this burden of filth the purifying agencies of the stream are prostrated; it lodges against obstructions in the stream and rots, becoming hatcheries of mosquitoes and malaria. A thing of beauty is thus transformed into one of hideous danger."

Texas Department of Health 1925

CONTENTS

Overview: The Problem of Stormwater Runoff 3
Sources of NPS Pollution 4
Regulatory Program 8
Applications of the Handbook 10
    Stormwater Management Planning (Local Problem Evaluations and Source
    Identifications) 10
    Risk Assessments 11
    Total Maximum Daily Load (TMDL) Evaluations 11
    Model Calibration and Validation 11
    Effectiveness of Control Programs 12
    Compliance with Standards and Regulations 13
References 13

OVERVIEW: THE PROBLEM OF STORMWATER RUNOFF

The vivid description, above, of the Trinity River as it flowed through Fort Worth and Dallas, TX, in 1925 is no longer appropriate. The acute pollution problems that occurred in the Trinity River and throughout the United States before the 1970s have been visibly and dramatically improved. The creation of the U.S. Environmental Protection Agency (EPA) and the passage of the Clean Water Act (CWA) in 1972 resulted in improved treatment of municipal and industrial wastewaters, new and more stringent water quality criteria and standards, and an increased public awareness of water quality issues. During the first 18 years of the CWA, regulatory efforts, aimed at pollution control, focused almost entirely on point source, end-of-pipe, wastewater discharges. However, during this same period, widespread water quality monitoring programs and special studies conducted by state and federal agencies and other institutions implicated nonpoint sources

(NPS) as a major pollutant category, affecting most degraded waters around the country. For example, in Ohio 51% of the streams assessed were thought to be adversely impacted by NPS pollution. Nonpoint source pollution presents a challenge from both a regulatory and an assessment perspective. Unlike many point source discharges, pollution inputs are not constant, do not reoccur in a consistent pattern (i.e., discharge volume and period), often occur over a diffuse area, and originate from watersheds whose characteristics and pollutant loadings vary through time. Given this extreme heterogeneity, simple solutions to NPS pollution control and the assessment of eco-system degradation are unlikely. Fortunately, methods do exist to accomplish both control and accurate assessments quite effectively. To accomplish this, however, one must have a clear under-standing of the nature of the problem, the pollutant sources, the receiving ecosystem, the strengths and weaknesses of the assessment tools, and proper quality assurance (QA) and quality control (QC) practices. This handbook will discuss these issues as they pertain to assessing stormwater runoff effects on freshwater ecosystems.

SOURCES OF NPS POLLUTION

A wide variety of activities and media comprise NPS pollution in waters of the United States (Table 1.1). The major categories of sources include agriculture, silviculture, resource extraction, hydro-modification, urban areas, land disposal, and contaminated sediments. The contribution of each category is, of course, a site-specific issue. In Ohio, as in many midwestern and southern states, agriculture is the principal source of NPS stressors, as shown in Table 1.2 (ODNR 1989).

These stressors include habitat destruction (e.g., channelization, removal of stream canopy and riparian zone, loss of sheltered areas, turbidity, siltation) and agrichemicals (e.g., pesticides and nutrients). In urban areas, stream and lake impairment is also due to habitat destruction; but, in addition, physical and chemical contaminant loadings come from runoff from impervious areas (e.g., parking lots, streets) of construction sites, and industrial, commercial, and residential areas. Numerous studies (such as May 1996) have examined the extent of urbanization in relation to decaying receiving water conditions (Figure 1.1). Other contaminant sources that have been doc-

Table 1.1 Nonpoint Source Pollution Categories and Subcategories

Category: Agriculture	Category: Hydromodification
General agriculture	General hydromodification
Crop production	Channelization
Livestock production	Dredging
Pasture	Dam construction
Specialty crop production	Stream bank modification
Category: Silviculture	Bridge construction
General silviculture	**Category: Urban**
Harvesting, reforestation	General urban
Residue management	Storm sewers
Road construction	Sanitary sewers
Forest management	Construction sites
Category: Resource Extraction	Surface runoff
General resource extraction	**Category: Land Disposal**
Surface coal mining	General land disposal
Subsurface coal mining	Sludge disposal
Oil/Gas production	Wastewater
Category: In-place (Sediment) Pollutants	Sanitary landfills
	Industrial land treatment
	On-site wastewater treatment

From EPA. *Results of the Nationwide Urban Runoff Program.* Water Planning Division, PB 84-185552, Washington, D.C. December 1983.

Table 1.2 Major Categories of Nonpoint Source Pollution
Impacting Surface Water Quality in Ohio

Major Categories of Nonpoint Source Pollution	Stream Miles Affected	Percentage of Miles Affected
Agriculture	5300	44
Resource extraction	2000	17
Land disposal	1600	13
Hydromodification	1500	13
Urban	1100	9
Silviculture	400	3
In-place pollutants	100	1
Total stream miles affected	12,000	

From ODNR (Ohio Department of Natural Resources). Ohio Nonpoint
Source Management Program. Ohio Department of Natural Resources,
Columbus, OH. 1989.

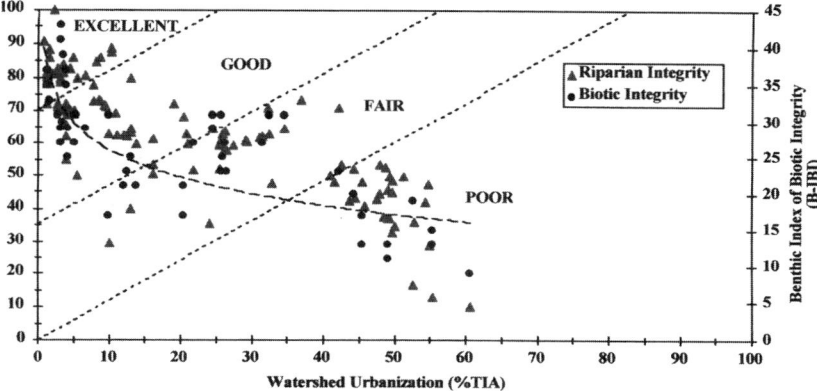

Figure 1.1 Relationship between basin development, riparian buffer width, and biological integrity in Puget Sound lowland streams. (From May, C.W. *Assessment of the Cumulative Effects of Urbanization on Small Streams in the Puget Sound Lowland Ecoregion: Implications for Salmonid Resource Management.* Ph.D. dissertation, University of Washington, Seattle. 1996. With permission.)

umented, but are even more difficult to assess, include accidental spills, unintended discharges, and atmospheric deposition.

The pollutants present in stormwater runoff vary with each watershed; however, certain pollutants are associated with specific activities (e.g., soybean farming, automobile service areas) and with area uses (e.g., parking lots, construction). By analyzing the land use patterns, watershed characteristics, and meteorological and hydrological conditions, an NPS assessment program can be focused and streamlined.

A number of studies have linked specific pollutants in stormwater runoff with their sources (Table 1.3). Pitt et al. (1995) reviewed the literature on stormwater pollutant sources and effects and also measured pollutants and sample toxicity from a variety of urban source categories of an impervious and pervious nature. The highest concentrations of synthetic organics were in roof runoff, urban creeks, and combined sewer overflows (CSOs). Zinc was highest from roof runoff (galvanized gutters). Nickel was highest in runoff from parking areas. Vehicle service areas produced the highest cadmium and lead concentrations, while copper was highest in urban creeks (Pitt et al. 1995). Most metals in stormwater runoff originate from streets (Table 1.4, FWHA 1987) and parking areas. Other metal sources include wood preservatives, algicides, metal corrosion, road salt, batteries, paint, and industrial electroplating waste. One large survey (EPA 1983) found only 13 organics occurring in at least 10% of the samples. The most common were 1,3-dichlorobenzene

Table 1.3   Potential Sources of Stormwater Toxicants

	Automobile Use	Pesticide Use	Industrial/Other
Halogenated Aliphatics			
Methylene chloride		Fumigant[a]	Plastics, paint remover, solvent
Methyl chloride	Leaded gas[a]	Fumigant[a]	Refrigerant, solvent
Phthalate Esters			
Di-N-butyl phthalate		Insecticide	Plasticizer[a], printing inks, paper, stain, adhesive
Bis (2-ethyhexyl) phthalate			Plasticizer[a]
Butylbenzyl phthalate			Plasticizer[a]
Polycyclic Aromatic Hydrocarbons			
Chrysene	Gasoline[a], oil/grease		
Phenanthrene	Gasoline		Wood/coal combustion[a]
Pyrene	Gasoline, oil, asphalt	Wood preservatives	Wood/coal combustion[a]
Volatiles			
Benzene	Gasoline[a]		Solvent formed from salt, gasoline and asphalt
Chloroform		Insecticide	Solvent, formed from chlorination[a]
Toluene	Gasoline[a], asphalt		Solvent
Heavy Metals			
Chromium	Metal corrosion[a]		Paint, metal corrosion, electroplating waste[a]
Copper	Metal corrosion, brake linings	Algicide	Paint, metal corrosion, electroplating waste[a]
Lead	Gasoline, batteries		Paint
Zinc	Metal corrosion, road salt, rubber[a]	Wood preservative	Paint, metal corrosion[a]
Organochlorides and Pesticides			
Lindane		Mosquito control[a] Seed pretreatment	
Chlordane		Termite control[a]	
Pentachlorophenol		Wood preservative	Paint Wood processing
PCBs			Electrical, insulation, paper adhesives
Dieldrin			
Diazinon			
Chlorpyrifos			
Atrazine			

[a] Most significant sources.

Modified from Callahan, M.A., et al., *Water Related Environmental Fates of 129 Priority Pollutants.* U.S. Environmental Protection Agency, Monitoring and Data Support Division, EPA-4-79-029a and b. Washington D.C. 1979; Verschueren, K. *Handbook of Environmental Data on Organic Chemicals,* 2nd edition. Van Nostrand Reinhold Co., New York. 1983.

Table 1.4   Highway Runoff Constituents and Their Primary Sources

Constituents	Primary Sources
Particulates	Pavement wear, vehicles, atmosphere, maintenance
Nitrogen, phosphorus	Atmosphere, roadside fertilizer application
Lead	Leaded gasoline (auto exhaust), tire wear (lead oxide filler material, lubricating oil and grease, bearing wear)
Zinc	Tire wear (filler materials), motor oil (stabilizing additive), grease
Iron	Auto body rust, steel highway structures (guard rails, etc.), moving engine parts
Copper	Metal plating, bearing and bushing wear, moving engine parts, brake lining wear, fungicides and insecticides
Cadmium	Tire wear (filler material), insecticide application
Chromium	Metal plating, moving engine parts, break lining wear
Nickel	Diesel fuel and gasoline (exhaust), lubricating oil, metal plating, bushing wear, brake lining wear, asphalt paving
Manganese	Moving engine parts
Cyanide	Anticake compound (ferric ferrocyanide, sodium ferrocyanide, yellow prussiate of soda) used to keep deicing salt granular
Sodium, calcium, chloride	Deicing salts
Sulfate	Roadway beds, fuel, deicing salts
Petroleum	Spills, leaks, or blow-by of motor lubricants, antifreeze and hydraulic fluids, asphalt surface leachate
PCB	Spraying of highway rights-of-way, background atmospheric deposition, PCB catalyst in synthetic tires

From U.S. DOT, FHWA, Report No. FHWA/RD-84/056-060, June 1987.

and fluoranthene (23% of the samples). These 13 compounds were similar to those reported in most areas. The most common organic toxicants have been from automobile usage (polycyclic aromatic hydrocarbons, or PAHs), combustion of wood and coal (PAHs), industrial and home use solvents (halogenated aliphatics and other volatiles), wood preservatives (PAHs, creosote, pentachlorophenol), and a variety of agricultural, municipal, and highway compounds, and pesticides.

The major urban pollution sources are construction sites, on-site sewage disposal systems, households, roadways, golf courses, parks, service stations, and parking areas (Pitt et al. 1995). The primary pollutant from construction is eroded soils (suspended and bedload sediments, dissolved solids, turbidity), followed by hydrocarbons, metals, and fertilizers.

Silviculture is a major source of nonpoint pollution in many areas of the country. The primary pollutant is eroded soils, which result in elevated turbidity, silted substrates, altered habitat, higher dissolved solids, and altered ion ratios in the streams and lakes of the watershed. Water temperatures increase as tree canopies are removed and stream flow slows. Fertilizers and pesticides may also be used which are transported to the streams via surface runoff, groundwater, and drift.

Agricultural activities contribute a wide variety of stormwater pollutants, depending on the production focus and ecoregion. Major pollutants include eroded soils, fertilizers, pesticides, hydrocarbons (equipment-related), animal wastes, and soil salts.

The hydromodification category of NPS includes dredging, channelization, bank stabilization, and impoundments. Stormwaters obviously do not "run off" any of these sources, but stormwater (high flow) does degrade waters associated with these sources. Water quality parameters which may be affected by these sources during stormwater events include turbidity, sediment loading (habitat alteration), dissolved solids, temperature, nutrients, metals, synthetic organics, dissolved oxygen, pathogens, and toxicity.

Of a more site-specific nature, resource extraction, land waste disposal, and contaminated sediments are sources of pollutants during stormwater events. Activities such as sand and gravel, metal, coal, and oil and gas extraction from or near receiving waters may contribute to habitat alteration and increased turbidity, siltation, metals, hydrocarbons, and salt during storm events. Land waste disposal sources consist of sludge farm runoff, landfill and lagoon runoff and leachate, and on-site septic system (leachfield) overflows. These sources may contribute a variety of pollutants to receiving waters such as nutrients, solids (dissolved and suspended), pathogens, metals, and synthetic organics. Contaminated sediments occur in numerous areas throughout the United States (EPA 1994). Many nutrients and toxic metals, metalloids, and synthetic organics readily sorb to particulates (organic or inorganic) which accumulate as bedded sediments. During storm events, these sediments may be resuspended and then become more biologically active by pollutant desorption, transformation, or particle uptake by organism ingestion.

The specific stormwater pollutants vary dramatically in their fate and effect characteristics. In most assessments of NPS pollution, there are many unknowns, such as:

- What are the pollutants of concern?
- What are the pollutant sources?
- What are the pollutant loadings?

These common unknowns provide the rationale for use of an integrated assessment strategy (see Unit 2) which incorporates several essential components of runoff-receiving water systems.

REGULATORY PROGRAM

In February 1987, amendments to the federal Clean Water Act (CWA) were passed by Congress and required states (Sections 101 and 319) to assess NPS pollution and develop management programs. These programs are to be tailored on a watershed-specific basis, although they are structured along political jurisdictions. There are also NPS requirements under Section 6217 of the Coastal Zone Act Reauthorization Amendments of 1990. The EPA published the Phase 1 stormwater discharge regulations for the CWA in the *Federal Register* on November 16, 1990. The regulations confirm stormwater as a point source that must be regulated through permits issued under the National Pollutant Discharge Elimination System (NPDES). Certain specified industrial facilities and large municipalities (>100,000 population) fell under the Phase 1 regulations. The Phase 2 regulations were enacted in October 1999, requiring municipalities of 10,000 and greater to comply with stormwater control guidelines.

Monitoring activities must be part of the Phase 1 NPDES stormwater permit requirements. One monitoring element is a field screening program to investigate inappropriate discharges to the storm drainage system (Pitt et al. 1993). The Phase 1 requirements also specified outfall monitoring during wet weather to characterize discharges from different land uses. Specified industries are also required to periodically monitor their stormwater discharges. Much of the local municipal effort associated with the Phase 1 permit requirements involved describing the drainage areas and outfalls. Large construction sites are also supposed to be controlled, but enforcement has been very spotty. Local governments have been encouraged by the EPA to develop local stormwater utilities to pay for the review and enforcement activities required by this regulation. The Phase 2 permit requirements are likely to have reduced required monitoring efforts for small communities and remaining industries.

The Stormwater Phase 2 Rule was published in early November 1999 in the *Federal Register*. The purpose of the rule is to designate additional sources of stormwater that need to be regulated to protect water quality. Two new classes of facilities are designated for automatic coverage on a nationwide basis:

1. Small municipal separate storm sewer systems located in urbanized areas (about 3500 municipalities) [Phase 1 included medium and large municipalities, having populations greater than 100,000]
2. Construction activities that disturb between 1 and 5 acres of land (about 110,000 sites a year) [Phase 1 included construction sites larger than 5 acres]

There is also a new "no exposure" incentive for Phase 1 sites having industrial activities. It is expected that this will exclude about 70,000 facilities nationwide from the stormwater regulations. The NPDES permitting authority would need to issue permits (most likely general permits) by May 31, 2002.

Proposed construction site regulations in the Phase 2 rule include:

1. Ensure control of other wastes at construction sites (discarded building materials, concrete truck washout, sanitary wastes, etc.)
2. Implement appropriate best management practices (such as silt fences, temporary detention ponds, etc.)
3. Require preconstruction reviews of site management plans
4. Receive and consider public information
5. Require regular inspections during construction
6. Have penalties to ensure compliance

If local regulations incorporate the following principles and elements into the stormwater program, they would be considered as "qualifying" programs that meet the federal requirements:

Five Principles
1. Good site planning
2. Minimize soil movement
3. Capture sediment
4. Good housekeeping practices
5. Mitigation of post-construction stormwater discharges

Eight Elements
1. Program description
2. Coordination mechanism
3. Requirements for nonstructural and structural BMPs
4. Priorities for site inspections
5. Education and training
6. Exemption of some activities due to limited impacts
7. Incentives, awards, and streamlining mechanisms
8. Description of staff and resources

Unfortunately, many common stormwater parameters which cause acute and chronic toxicity or habitat problems are not included in typical monitoring programs conducted under the NPDES stormwater permit program. Therefore, stormwater discharges that are degrading receiving waters may not be identified as significant outfalls from these monitoring efforts. Conversely, these data may suggest significant pollution is adversely affecting receiving waters, when in fact it is not. As discussed later in this book, the recent promotion and adoption of integrated assessment approaches which utilize stream biological community indices, toxicity, and habitat characterization of receiving waters provide much more reliable data on stormwater discharge effects and water quality.

Section 304 of the CWA directs EPA to develop and publish information on methods for measuring water quality and establishing water quality criteria for toxic pollutants. These other approaches include biological monitoring and assessment methods which assess the effects of pollutants on aquatic communities and factors necessary to restore and maintain the chemical, physical, and biological integrity of all waters. These "toolboxes" are intended to enable local users to make more efficient use of their limited monitoring resources. Of course, a primary purpose of this book is also to provide guidance to this user community. As such, it is hoped that this book can be considered a "super" toolbox, especially with its large number of references for additional information and its detailed case studies.

APPLICATIONS OF THE HANDBOOK

The first aspect of designing a monitoring program is describing how the data are to be used. This may include future uses of the data and must also include the necessary quality of the data (allowable errors). Many uses of the data may be envisioned, as shown in the following brief discussion. Data may be used in the evaluation of local stormwater problems (risk assessments) and identification of pollutant sources to support a comprehensive stormwater management program, compliance monitoring required by regulations, model calibration and verification for TMDL (total maximum daily load) evaluations, evaluation of the performance of control practices, screening analyses to identify sources of pollutants, etc. It is critical that an integrated assessment approach (designed on a site-specific basis) be used to improve the validity of the assessment and its resulting conclusions. Critical aspects of this are discussed below.

Stormwater Management Planning (Local Problem Evaluations and Source Identifications)

Stormwater management planning encompasses a wide range of site-specific issues. The local issues that affect stormwater management decisions include understanding local problems and the sources of pollutants or flows that affect these problems. Local monitoring therefore plays an important role in identifying local problems and sources.

The main purpose of treating stormwater is to reduce its adverse impacts on receiving water beneficial uses. Therefore, it is important in any stormwater runoff study to assess the detrimental effects that runoff is actually having on a receiving water. Receiving waters may have many beneficial use goals, including:

- Stormwater conveyance (flood prevention)
- Biological uses (warm water fishery, biological integrity, etc.)
- Noncontact recreation (linear parks, aesthetics, boating, etc.)
- Contact recreation (swimming)
- Water supply

As discussed in Chapter 2, it is unlikely that any of these uses can be fully obtained with full development in a watershed and with no stormwater controls. However, the magnitude of these effects varies greatly for different conditions. Obviously, local monitoring and evaluation of data are needed to describe specific local problems, especially through the use of an integrated monitoring approach that considers physical, chemical, and biological observations collectively. As described throughout this book, relying only on a single aspect of receiving water conditions, or applying general criteria to local data, can be very misleading, and ultimately expensive and ineffective.

After local receiving problems are identified, it is necessary to understand what is causing the problems. Again, this can be most effectively determined through local monitoring. Runoff is comprised of many separate source area flow components and phases that are discharged through the storm drainage system and includes warm weather stormwater, snowmelt, baseflows, and inappropriate discharges to the storm drainage ("dry-weather" flows). It may be important to consider all of these potential urban flow discharges when evaluating alternative stormwater management options.

It may be adequate to consider the combined outfall conditions alone when evaluating the long-term, area-wide effects of many separate outfall discharges to a receiving water. However, if better predictions of outfall characteristics (or the effects of source area controls) are needed, then the separate source area components must be characterized. The discharge at an outfall is made up of a mixture of contributions from different source areas. The "mix" depends on the characteristics of the drainage area and the specific rain event. The effectiveness of source area controls is therefore highly site and storm specific.

Risk Assessments

Risk assessments contain four major components (NRC 1983):

- Hazard identification
- Effects characterization
- Exposure characterization
- Risk characterization

Hazard identification includes quantifying pollutant discharges, plus modeling the fate of the discharged contaminants. Obviously, substantial site-specific data are needed to prepare the selected model for this important aspect of a risk assessment. Knowledge about the mass and concentration discharges of a contaminant is needed so the transport and fate evaluations of the contaminant can be quantified. Knowledge of the variations of these discharges with time and flow conditions is needed to determine the critical dose–response characteristics for the contaminants of concern. A suitable model, supported by adequate data, is necessary to produce the likely dose–stressor response characteristics. Exposure assessment is related to knowledge of the users of receiving waters and contaminated components (such as contaminated fish that are eaten, contaminated drinking water being consumed, children exposed to contaminated swimming by playing in urban creeks, etc.). Finally, the risk is quantified based on this information, including the effects of all of the possible exposure pathways. Obviously, many types of receiving water and discharge data are needed to make an appropriate risk assessment associated with exposure to stormwater, especially related to discharge characteristics, fate of contaminants, and verification of contaminated components. The use of calibrated and validated discharge and fate models is therefore necessary when conducting risk assessments.

Total Maximum Daily Load (TMDL) Evaluations

The total maximum daily load (TMDL) for a stream is the estimated maximum discharge that can enter a water body without affecting its designated uses. TMDLs can be used to allocate discharges from multiple sources and to define the level of control that may be needed. Historically, assimilative capacities of many receiving waters were based on expected dissolved oxygen conditions using in-stream models. Point source discharges of BOD were then allocated based on the predicted assimilative capacity. Allowed discharges of toxic pollutants can be determined in a similar manner. Existing background toxicant concentrations are compared to water quality criteria under critical conditions. The margin in the pollutant concentration (difference between the existing and critical concentrations) is multiplied by the stream flow to estimate the maximum allowable increased discharge, before the critical criteria would likely be exceeded. There has always been concern about margins of safety and other pollutant sources in the simple application of assimilative capacity analyses. The TMDL process is a more comprehensive approach that attempts to examine and consider all likely pollutant sources in the watershed. The EPA periodically publishes guidance manuals describing resources available for conducting TMDL analyses (Shoemaker et al. 1997, for example).

Model Calibration and Validation

A typical use of stormwater monitoring data is to calibrate and validate models that can be used to examine many questions associated with urbanization, especially related to the design of control programs to reduce problem discharges effectively. All models need to be calibrated for local conditions. Local rain patterns and development characteristics, for example, all affect runoff characteristics. Calibration usually involves the collection of an initial set of data that is used to modify the model for these local characteristics. Validation is an independent check to ensure that the calibrated model produces predictions within an acceptable error range. Unfortunately, many models are used to predict future conditions that are not well represented in available data sets, or the likely future conditions are not available in areas that could be monitored. These problems, plus many other aspects of modeling, require someone with good skill and support to ensure successful model use.

Model calibration and validation involves several steps that are similar for most stormwater modeling processes. The best scenario may be to collect all calibration information from one watershed and then validate the calibrated model using independent observations from another watershed. Another common approach is to collect calibration information for a series of events from one watershed, and then validate the calibrated model using additional data from other storms from the same watershed. Numerous individual rainfall-runoff events may need to be sampled to cover the range of conditions of interest. For most stormwater models, detailed watershed information is also needed. Jewell et al. (1978) presented one of the first papers describing the problems and approaches needed for calibrating and validating nonpoint source watershed scale models. Most models have descriptions of recommended calibration and validation procedures. Models that have been used for many years (such as SWMM and HSPF) also have many publications available describing the sensitivity of model components and the need for adequate calibration.

It is very important that adequate QA/QC procedures be used to ensure the accuracy and suitability of the data. Common problems during the most important rainfall-runoff monitoring activities are associated with unrepresentative rainfall data (using too few rain gauges and locating them incorrectly in the watershed), incorrect rain gauge calibrations, poor flow-monitoring conditions (surcharged flows, relying on Manning's equation for V and Q, poor conditions at the monitoring location), etc. The use of a calibrated flume is preferred, for example. Other common errors are associated with inaccurate descriptions of the watershed (incorrect area, amount of impervious area, understanding of drainage efficiency, soil characteristics, etc.). Few people appreciate the inherent errors associated with measuring rainfall and runoff. Most monitoring programs are probably no more than ±25% accurate for each event. It is very demanding to obtain rainfall and runoff data that is only 10% in error. This is most evident when highly paved areas (such as shopping centers or strip commercial areas) are monitored and the volumetric runoff coefficients are examined. For these areas, it is not uncommon for many of the events to have volumetric runoff coefficient (Rv) values greater than 1.0 (implying more runoff than rainfall). Similar errors occur with other sites but are not as obvious.

Data from several watersheds are available for the calibration and validation process. If so, start with data from the simplest area (mostly directly connected paved areas and roofs, with little unpaved areas). This area probably represents commercial roofs and parking/storage areas alone. These areas should be calibrated first, before moving on to more complex areas. The most complex areas, such as typical residential areas having large expanses of landscaped areas and with most of the roofs being disconnected from the drainage areas, should be examined last.

Effectiveness of Control Programs

Effective stormwater management programs include a wide variety of control options that can be utilized to reduce receiving water problems. With time and experience, some of these will be found to be more effective than others. In order to identify which controls are most cost-effective for a specific area, local performance evaluations should be conducted. In many cases, straightforward effectiveness monitoring (comparing influent with effluent concentrations for a stormwater filter, for example) can be utilized, while other program elements (such as public education or street cleaning) can be much more difficult to evaluate. Therefore, this book presents monitoring approaches that can be utilized for a broad range of control programs. These monitoring activities may appear to be expensive. However, the true cost of not knowing how well currently utilized controls function under local conditions can be much more costly than obtaining accurate local data and making appropriate changes in design methods.

The first concern when investigating alternative treatment methods is determining the needed level of stormwater control. This determination has a great effect on the cost of the stormwater management program and needs to be made carefully. Problems that need to be addressed range from sewerage maintenance issues to protecting many receiving water uses. As an example, Laplace et al. (1992) recommends that all particles greater than about 1 to 2 mm in diameter be removed from stormwater in order to prevent deposition in sewerage. The specific value is dependent on the energy gradient of the flowing water in the drainage system and the hydraulic radius of the sewerage. This treatment objective can be easily achieved using a number of cost-effective source area and inlet treatment practices. In contrast, much greater levels of stormwater control are likely needed to prevent excessive receiving water degradation. Typical treatment goals usually specify about 80% reductions in suspended solids concentrations. For most stormwaters, this would require the removal of most particulates greater than about 10 μm in diameter, about 1% of the 1 mm size noted above to prevent sewerage deposition problems. Obviously, the selection of a treatment goal must be done with great care.

There are many stormwater control practices, but not all are suitable in every situation. It is important to understand which controls are suitable for the site conditions and can also achieve the required goals. This will assist in the realistic evaluation for each practice of the technical feasibility, implementation costs, and long-term maintenance requirements and costs. It is also important to appreciate that the reliability and performance of many of these controls have not been well established, with most still in the development stage. This is not to say that emerging controls cannot be effective; however, there is not a large amount of historical data on which to base designs or to provide confidence that performance criteria will be met under the local conditions. Local monitoring can be used to identify the most effective controls based on the sources of the identified problem pollutants, and monitoring can be utilized to measure how well in-place controls are functioning over the long term. These important data can be used to modify recommendations for the use of specific controls, design approaches, and sizing requirements.

Compliance with Standards and Regulations

The receiving water (and associated) monitoring tools described in this book can also be used to measure compliance with standards and regulations. Numerous state and local agencies have established regulatory programs for moderate and large-sized communities due to the EPA's NPDES (National Pollutant Discharge Elimination System) stormwater permit program. The recently enacted Phase 2 regulations will extend some stormwater regulations to small communities through-out the United States. In addition, the increasing interest in TMDL evaluations in critical watersheds also emphasizes the need for local receiving water and discharge information. These regulatory programs all require certain monitoring, modeling, and evaluation efforts that can be conducted using procedures and methods described in this book.

REFERENCES

Callahan, M.A., M.W. Slimak, N.W. Gabel, I.P. May, C.F. Fowler, J.R. Freed, P. Jennings, R.L. Durfee, F.C. Whitmore, B. Maestri, W.R. Mabey, B.R. Holt, and C. Gould, *Water Related Environmental Fates of 129 Priority Pollutants*. U.S. Environmental Protection Agency, Monitoring and Data Support Division, EPA-4-79-029a and b. Washington D.C. 1979.

EPA. *Results of the Nationwide Urban Runoff Program.* Water Planning Division, PB 84-185552, Washington, D.C. December 1983.

EPA. *Procedures for Assessing the Toxicity and Bioaccumulation of Sediment-Associated Contaminants with Freshwater Invertebrates,* EPA 600/R-94/024, U.S. Environmental Protection Agency, Duluth, MN. 1994.

Jewell, T.K., T.J. Nunno, and D.D. Adrian. Methodology for calibrating stormwater models. *J. Environ. Eng. Div.* 104: 485. 1978.

Laplace, D., A. Bachoc, Y. Sanchez, and D. Dartus. Truck sewer clogging development — description and solutions. *Water Sci. Technol.* 25(8): 91–100. 1992.

May, C.W. *Assessment of the Cumulative Effects of Urbanization on Small Streams in the Puget Sound Lowland Ecoregion: Implications for Salmonid Resource Management.* Ph.D. dissertation, University of Washington, Seattle. 1996.

NRC (National Research Council). *Risk Assessment in the Federal Government: Managing the Process.* National Academy Press. Washington, D.C. 1983.

ODNR (Ohio Department of Natural Resources). Ohio Nonpoint Source Management Program. Ohio Department of Natural Resources, Columbus, OH. 1989.

Pitt, R.E., R.I. Field, M.M. Lalor, D.D. Adrian, D. Barbé, *Investigation of Inappropriate Pollutant Entries into Storm Drainage Systems: A User's Guide.* Rep. No. EPA/600/R-92/238, NTIS Rep. No. PB93-131472/AS, U.S. EPA, Storm and Combined Sewer Pollution Control Program, Edison, NJ. Risk Reduction Engineering Lab., Cincinnati, OH. 1993.

Pitt, R., R. Field, M. Lalor, and M. Brown. Urban stormwater toxic pollutants: assessment, sources and treatability. *Water Environ. Res.* 67(3): 260–275. May/June 1995.

Shoemaker, L., M. Lahlou, M. Bryer, D. Kumar, and K. Kratt. *Compendium of Tools for Watershed Assessment and TMDL Development.* EPA 841-B-97-006. U.S. Environmental Protection Agency. Office of Water. Washington, D.C. May 1997.

U.S. Department of Transportation (DOT). FHWA/RD-84/056-060, June 1987.

Verschueren, K. *Handbook of Environmental Data on Organic Chemicals,* 2nd edition. Van Nostrand Reinhold, New York. 1983.

## CHAPTER 2

Receiving Water Uses, Impairments, and Sources of Stormwater Pollutants

"Bathing in sewage-polluted seawater carries only a negligible risk to health, even on beaches that are aesthetically very unsatisfactory."

Committee on Bathing Beach Contamination
Public Health Laboratory Service of the U.K.
1959

CONTENTS

Introduction 15
Beneficial Use Impairments...22
    Recognized Value of Human-Dominated Waterways..22
    Stormwater Conveyance (Flood Prevention) 26
    Recreation (Non-Water Contact) Uses..26
    Biological Uses (Warm-Water Fishery, Aquatic Life Use, Biological Integrity, etc.)..........27
    Human Health-Related Uses (Swimming, Fishing, and Water Supply) 28
Likely Causes of Receiving Water Use Impairments 30
Major Urban Runoff Sources 31
    Construction Site Erosion Characterization...32
    Urban Runoff Contaminants 34
Summary 42
References 43

INTRODUCTION

Wet-weather flow impacts on receiving waters have been historically misunderstood and de-emphasized, especially in times and areas of poorly treated municipal and industrial discharges. The above 1959 quote from the Committee on Bathing Beach Contamination of the Public Health Laboratory Service of the U.K. demonstrates the assumption that periodic combined sewer over-flows (CSOs), or even raw sewage discharges, produced negligible human health risks. Is it any wonder then that the much less dramatically contaminated stormwater discharges have commonly been considered "clear" water by many regulators?

The EPA reported that only 57% of the rivers and streams in the United States fully support their beneficial uses (Figure 2.1). A wide variety of pollutants and sources are the cause of impaired

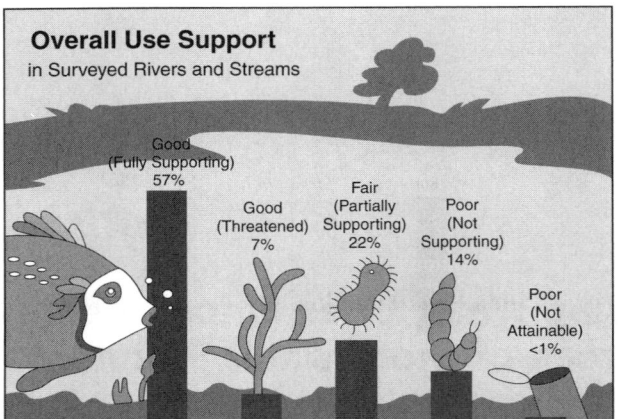

Figure 2.1 U.S. rivers and streams meeting designated beneficial uses. Note: Percentages do not add to 100% because more than one pollutant or source may impair a segment of ocean shoreline. (From U.S. Environmental Protection Agency. *National Water Quality Inventory. 1994 Report to Congress.* Office of Water. EPA 841-R-95-005. Washington, D.C. December 1995.)

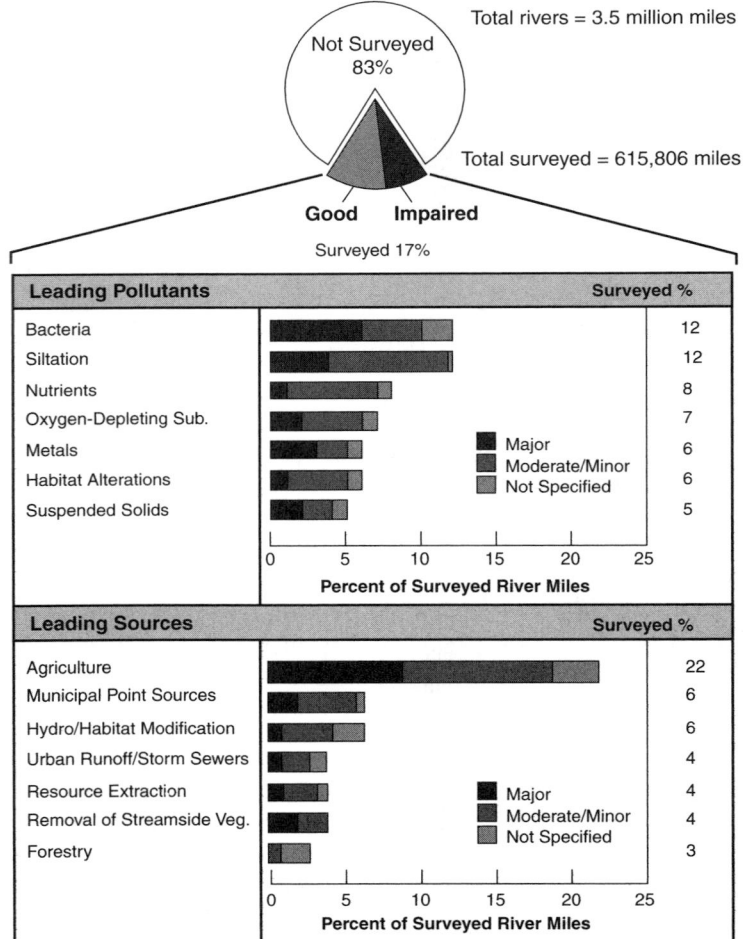

Figure 2.2 Pollutants and sources impairing U.S. rivers. Note: Percentages do not add to 100% because more than one pollutant or source may impair a segment of ocean shoreline. (From U.S. Environmental Protection Agency. *National Water Quality Inventory. 1994 Report to Congress.* Office of Water. EPA 841-R-95-005. Washington, D.C. December 1995.)

uses (Figures 2.2 through 2.6) but runoff from urban and agricultural sources dominate. This book contains discussions of instances of beneficial use impairments associated with stormwater runoff and the possible sources of the stressors of these effects. However, stormwater effects on receiving waters are not always clear and obvious. As will be evident to the reader, most stormwater runoff

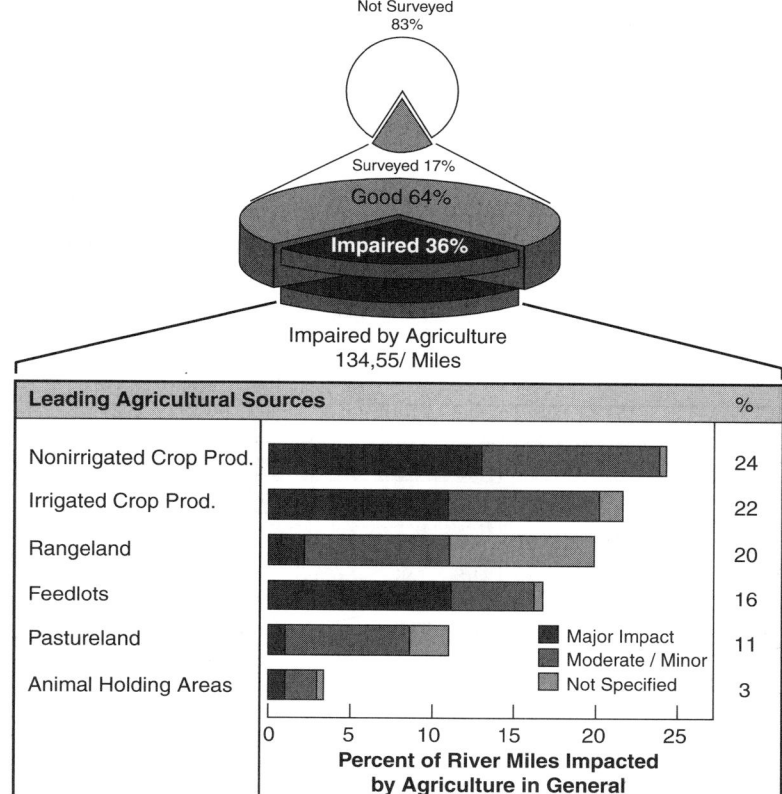

Figure 2.3 Agricultural activities affecting U.S. rivers and streams. Note: Percentages do not add to 100% because more than one pollutant or source may impair a segment of ocean shoreline. (From U.S. Environmental Protection Agency. *National Water Quality Inventory. 1994 Report to Congress.* Office of Water. EPA 841-R-95-005. Washington, D.C. December 1995.)

assessments have been conducted in urban waterways, with fewer examples for agricultural systems. However, many of the approaches, methods, and receiving water effects are similar in both urban and agriculturally dominated waterways. In completely urbanized watersheds, the small urban streams are commonly severely degraded, but they typically have no official beneficial uses or monitoring programs (and may be intermittent in flow), and are therefore unrecognized as being impacted or important. Unfortunately, these streams receive substantial recreational use by neighborhood children. Besides the obvious safety concerns and potential drowning fears, the water quality of urban streams can present significant risks. In older cities, stream sediments downstream from historical industrial areas can be heavily contaminated by heavy metals and organic compounds. Even in nonindustrialized areas, metallic and organic contamination can be high. Unfortunately, bacteria concentrations, especially near outfalls during and soon after rains, are always very high in these small streams, although the health risks are poorly understood. Sediment bacteria conditions are also always high, as the sediments appear to be an excellent sink for bacteria. Children, and others, playing in and near the streams therefore are exposed to potentially hazardous conditions. In addition, inner-city residents sometimes rely on nearby urban waterways for fishing opportunities, both for recreation and to supplement food supplies.

In contrast to the above obvious conditions associated with small streams in completely urbanized watersheds, wet-weather flows from relatively large cities discharging into large waterways may not be associated with obvious in-stream detrimental conditions. In one example, frequent CSO discharges from Nashville, TN, into the Cumberland River were not found to produce any significant dissolved oxygen (DO) or fecal coliform problems (Cardozo et al. 1994). However, Nashville is currently investigating sources of high bacteria levels in the small urban streams draining heavily urbanized city watersheds. A series of studies of airport deicing compound runoff

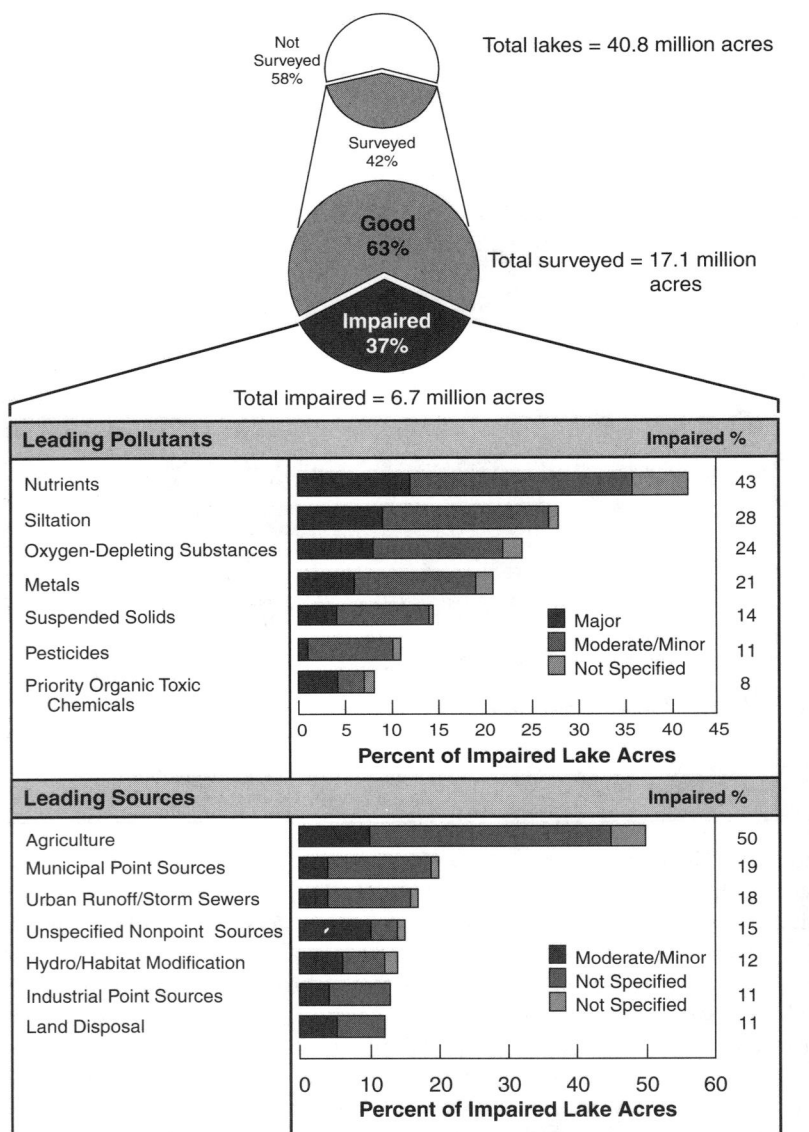

Figure 2.4 Pollutants and sources affecting U.S. lakes. Note: Percentages do not add to 100% because more than one pollutant or source may impair a segment of ocean shoreline. (From U.S. Environmental Protection Agency. *National Water Quality Inventory. 1994 Report to Congress.* Office of Water. EPA 841-R-95-005. Washington, D.C. December 1995.)

at Milwaukee's Mitchell Field is another example that demonstrates unique site-specific conditions affecting receiving water impacts. This study, conducted by the USGS and the Wisconsin Department of Natural Resources, found that the extremely high BOD concentrations (several thousand mg/L) associated with the deicing runoff had negligible effects on the DO levels in the small streams draining the airport area to Lake Michigan. They concluded that the cold temperatures occurring during the times of deicing runoff significantly reduced the BOD decomposition rate, and that the small streams had short travel times before discharging into Lake Michigan, where it was well mixed. Under laboratory conditions, the BOD rate would be much faster, and would be expected to produce dramatically low DO conditions for almost any condition in these small streams.

Other obvious receiving water problems, such as fish kills, are also rarely associated with stormwater discharges, as described in Chapter 3. Stormwater discharges occur frequently, and normally do not create acute toxicity problems (or extremely low DO conditions). Fish surviving

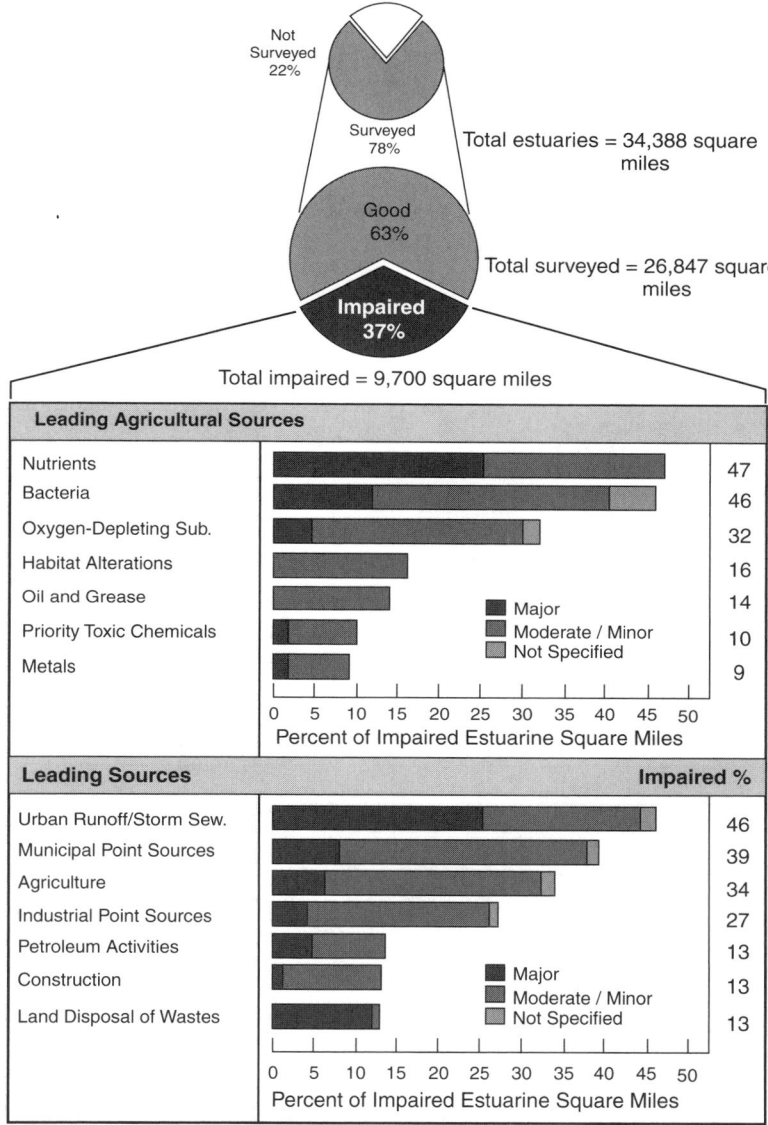

Figure 2.5 Pollutants and sources affecting U.S. estuaries. Note: Percentages do not add to 100% because more than one pollutant or source may impair a segment of ocean shoreline. (From U.S. Environmental Protection Agency. *National Water Quality Inventory. 1994 Report to Congress.* Office of Water. EPA 841-R-95-005. Washington, D.C. December 1995.)

in urban streams are tolerant species, with most of the intolerant organisms long since gone. It is therefore unusual for fish kills to occur, unless severe inappropriate discharges infrequently occur (such as those associated with industrial accidents, runoff from fire fighting, or improper waste disposal activities). However, chronic toxicity, mostly associated with contaminated sediments or suspended solids, is associated with stormwater. The effects of this chronic toxicity, plus habitat problems, are the likely causes of the commonly observed significant shifts in the in-stream biological community from naturally diverse (mostly intolerant) species to a much less diverse assemblage of introduced tolerant species. There is increasing evidence that stormwaters in urban and agriculturally dominated watersheds are often toxic (see Chapter 6). However, traditional toxicity approaches often do not detect problems associated with pulse exposures and or particulate-associated toxicity. More recently, both laboratory and in-stream (*in situ*) toxicity tests, especially associated with moderate to long-term exposures to contaminated sediments and particulates, have shown significant stormwater toxicity.

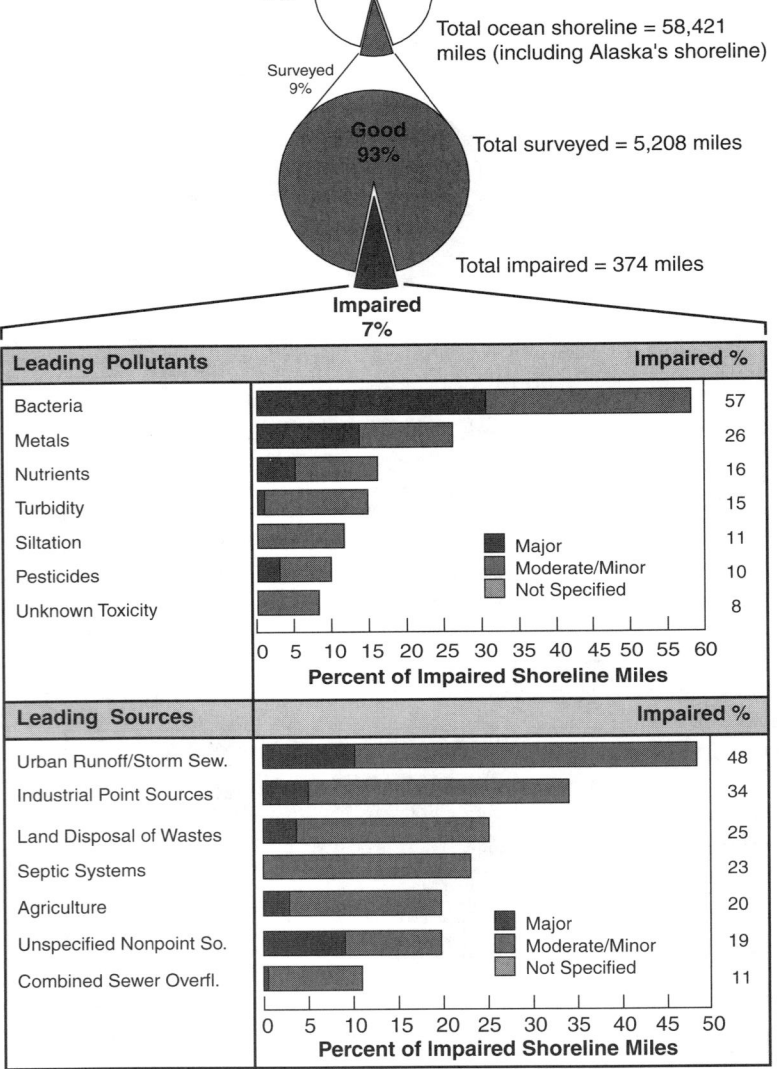

Figure 2.6 Pollutants and sources affecting U.S. ocean shorelines. Note: Percentages do not add to 100% because more than one pollutant or source may impair a segment of ocean shoreline. (From U.S. Environmental Protection Agency. *National Water Quality Inventory. 1994 Report to Congress.* Office of Water. EPA 841-R-95-005. Washington, D.C. December 1995.)

The discharges of stormwater are also periodic, causing different types of effects than the better-regulated continuous point source discharges. Stormwater causes episodic disturbances in aquatic ecosystems (Minshall 1988) whose patterns of occurrence are chaotic in nature (Pool 1989) and characteristics are unique to each event. The sciences of aquatic ecology and aquatic toxicology have progressed to the point where the effects of continuous levels of single stressors (e.g., dissolved oxygen, temperature, copper, DDT, diazinon, chlorpyrifos) on a wide variety of common aquatic species are known. The effects that the single stressors have, or may have, in stormwater are therefore known with reasonable certainty. However, as is shown in Table 2.1, nonpoint sources, including stormwater, contain multiple stressors that are applied intermittently, and science currently has a poor understanding of stressor interactions and effects.

The attributes of each stormwater event are a result of previous meteorological conditions (e.g., dry deposition, air patterns, humidity), land use patterns (e.g., traffic and parking patterns, construction and landscaping activities), storm intensity and duration, and other watershed character-

Table 2.1 Potential Effects of Some Sources of Alteration on Stream Parameters

Stream Parameter	Acid Mine Drainage or Acid Precipitation	Sewage Treatment Plant Discharge	Agriculture Runoff (pasture or cropland)	Urban Runoff	Channelization	Pulp and Paper	Textile	Metal Finishing and Electroplating	Petroleum	Iron and Steel	Paint and Ink	Dairy and Meat Products	Fertilizer Production and Lime Crushing	Plastics and Synthetics
pH	D					C	C	C		D	C		D,I	C
Alkalinity	D						I						D,I	
Hardness	I	I	I				I					I		I
Chlorides		I												
Sulfates	I	I	I	I					I					
TDS	I	I	I	I			I		I			I		
TKN		I	I	I				I	I	I		I	I	I
NH3-N		I	I	I					I			I	I	I
Total-P		I	I				I	I					I	
Ortho-P		I	I											
BOD5		I	I	I		I			I	I	I	I		I
COD	I	I	I	I		I	I	I	I	I	I	I	I	I
TOC		I	I	I		I	I				I			I
COD/BOD						D		D	D	D	D			
D.O.		D				D						D		I
Volatile compounds						I			I					
Fluoride			I					I						I
Cr	I			I			I	I	I	I	I	I		I
Cu	I			I						I	I			
Pb	I			I					I					I
Zn	I			I					I					I
Cd	I			I										I
Fe	I	I	I	I						I		I	I	
Arsenic				I						I				
Mercury				I						I				I
Cyanide				I										I
Oil and grease				I			I	D	I		I			
Coliforms	D	I	I	I		I		D		D		I		
Chlorophyll	D	I				D		D	D	D	D			D
Diversity	D	D		D	D	D	D	D	D	D	D		D	D
Biomass	D	D			I	I	I		D	D	D		I	D
Riparian factors				C	C									
Temperature				I	I									
TSS			I	I	I								I	I
VSS														
Color		I				I	I						I	I
Conductivity	I			C								I		
Channel factors				C	C									

D = decrease, I = Increase, C = change.
From EPA (U.S. Environmental Protection Agency). *Results of the Nationwide Urban Runoff Program.* Water Planning Division, PB 84-185552, Washington, D.C. December 1983.

istics. Because of the potentials for extreme heterogeneity in stormwater and its associated quality, predicting effects to receiving waters is difficult and crude at best. Stormwaters often contain a large number of potential stressors to aquatic ecosystems. These stressors include oxygen demand, suspended solids, dissolved solids (including salts), altered ion ratios, nutrients, pathogens, metals, natural and synthetic organics, pH, and temperature. These stressors may interact to varying degrees in an antagonistic, additive, or synergistic fashion, affecting organisms in the receiving water.

There are numerous receiving water problems associated with stormwater that interfere with beneficial uses. The most obvious is the substantial increase in runoff causing increases in the frequency and magnitude of flooding along urban streams. Increases in stream flows also cause significant habitat problems in urban streams by attempting to enlarge the stream cross sections, causing significant channel erosion and unstable conditions. Stream-side residents also dramatically affect habitat by removing riparian vegetation and large organic debris from the streams. Another significant and obvious effect is the increase in sediment associated with poorly controlled construction site runoff. This sediment smothers coarse stream sediments that are needed by many spawning fish, and fills in stream pool areas. Another obvious receiving water problem associated with stormwater is the large amount of floating trash and litter (some hazardous) that is discharged by stormwater and that accumulates along urban waterways. This creates unsightly and potentially hazardous conditions interfering with noncontact recreational uses of the stream corridors.

The degree of impact on an exposed organism is dependent on numerous factors, such as the organism's sensitivity, life stage, feeding habits, frequency of exposure, and magnitude and duration of exposure. The organism or community affected by stormwater induces changes in other components of their ecosystem including habitat, food sources, predator–prey relationships, competition, and other behavior patterns. It is clear that there is no simple method by which to detect an effect of stormwaters on the receiving water ecosystem. Human health and safety concerns associated with stormwater discharges are also highly variable depending on many site conditions. Chapters 3 and 4 discuss ways in which effects can be assessed effectively, despite the complex, heterogeneous nature of the system, while Chapters 5 and 6 describe how specific monitoring activities can be carried out. Chapters 7 and 8 outline ways to evaluate the collected data to accomplish the study goals, outlined in Chapter 4.

The main purpose of treating stormwater is to reduce its adverse impacts on receiving water beneficial uses. Therefore, it is important in any stormwater runoff study to assess the detrimental effects that runoff is actually having on a receiving water. Below are discussions of the basic receiving water beneficial uses that need to be considered in all cases.

BENEFICIAL USE IMPAIRMENTS

Recognized Value of Human-Dominated Waterways

With full development in a watershed and with no stormwater controls, it is unlikely that any of the basic beneficial uses can be achieved. With less development, and with the application of stormwater controls, some uses may be possible. However, it is important that unreasonable expectations not be placed on urban or agricultural waters, as the cost to obtain these uses may be prohibitive. With full-scale development and lack of adequate stormwater controls, severely degraded streams will be common. In all cases, stormwater conveyance and aesthetics should be the basic beneficial use goals for all human-dominated waters. Biological uses should also be a goal, but with the realization that the natural stream ecosystem will be severely modified with urbanization and agricultural activities. Certain basic stormwater controls, installed at the time of development, plus protection of stream habitat, may enable partial to full use of some of these basic goals. Careful planning and optimal utilization of stormwater controls are necessary to obtain these basic goals in most watersheds. Water contact recreation, consumptive fisheries, and water

Figure 2.7   Original section of Riverwalk in San Antonio, TX.

Figure 2.8   New section of Riverwalk in San Antonio, TX.

supplies are not appropriate goals for most heavily developed watersheds. However, these higher uses may be possible in urban areas where the receiving waters are large and drain mostly undeveloped areas.

There are many examples throughout the world where local citizens have recognized the added value that aesthetically pleasing waters contribute to cities. With this recognition comes a local pride in these waters and a genuine desire to improve their condition. In many cases, water has played an important part in the economic renewal of an inner city area. Dreiseitl (1998) states that "stormwater is a valuable resource and opportunity to provide an aesthetic experience for the city dweller while furthering environmental awareness and citizen interest and involvement." He found that water flow patterns observed in nature can be duplicated in the urban environment to provide healthy water systems of potentially great beauty. Without reducing safety, urban drainage elements can utilize water's refractive characteristics and natural flow patterns to create very pleasing urban areas. Successful stormwater management in Germany has been best achieved by using several measures together. Small open drainage channels placed across streets have been constructed of cobbles. These collect and direct the runoff, plus slow automobile traffic and provide dividing lines for diverse urban landscaping elements. The use of rooftop retention and evaporation areas reduce peak flows. Dreiseitl has found that infiltration and retention ponds can also be used to great advantage by providing a visible and enjoyable design element in urban landscapes.

Probably the most famous U.S. example of the economic benefits that water has contributed in an older part of a city is Riverwalk in San Antonio, TX. Many cities would like to emulate Riverwalk, with the great economic benefit that it has provided to San Antonio (Figures 2.7 through 2.9). Riverwalk was conceived and constructed many decades ago, but only in recent years has its full value been realized. Bellingham, WA (Figure 2.10), Austin, TX (Figure 2.11), and Denver, CO (Figures 2.12 through 2.14) are some of the other U.S. cities that have long enjoyed central city urban creek corridors.

Dreiseitl (1998) described the use of stormwater as an important component of the Potsdamer Platz in the center of Berlin. Roof runoff will be stored in large underground cisterns, with some filtered and used for toilet flushing and irrigation. The rest of the roof runoff will flow into a 1.4-ha (3.8-acre) concrete-lined lake in the center of the project area. The small lake provides an important natural element in the center of this massive development and regulates the stormwater discharge rate to the receiving water (Landwehrkanal). The project is also characterized by numerous fountains, including some located in underground parking garages.

Figure 2.9 Litter control along Riverwalk, San Antonio, TX.

Figure 2.10 Bike and walking trail along Watcom Creek, Bellingham, WA.

Figure 2.11 Barton Springs swimming area, Austin, TX.

Figure 2.12 Cherry Creek walkway, downtown Denver, CO.

Figure 2.13 Cherry Creek walk in Denver, CO.

Figure 2.14 Cherry Creek and Platte River junction in Denver, CO.

Göransson (1998) also described the aesthetic use of stormwater in Swedish urban areas. The main emphasis was to retain the stormwater in surface drainages instead of rapidly diverting it to underground conveyances. Small, sculpted rainwater channels are used to convey roof runoff downspouts to the drainage system. Some of these channels are spiral in form and provide much visual interest in areas dominated by the typically harsh urban environment. Some of these spirals are also formed in infiltration areas and are barely noticeable during dry weather. During rains, increasing water depths extenuate the patterns. Glazed tile, small channels with perforated covers, and geometrically placed bricks with large gaps to provide water passage slightly below the surface help urban dwellers better appreciate the beauty of flowing water.

Tokyo has instituted major efforts to restore historical urban rivers that have been badly polluted, buried, or have had all of their flows diverted. Fujita (1998) describes how Tokyo residents place great value on surface waterways: "Waterfront areas provide urban citizens with comfort and joy as a place to observe nature and to enjoy the landscape." Unfortunately, the extensive urbanization that has taken place in Tokyo over the past several decades has resulted in severe stream degradation, including the disappearance of streams altogether. However, there has recently been a growing demand for the restoration of polluted urban watercourses in Tokyo. This has been accomplished in many areas by improved treatment of sanitary sewage, reductions in combined sewer overflows, and by infiltration of stormwater.

Fujita (1998) repeatedly states the great importance the Japanese place on nature, especially flowing water and the associated landscaping and attracted animals. They are therefore willing to perform what seems to be extraordinary efforts in urban stream recovery programs in one of the world's largest cities. The stream recovery program is but one element of the local efforts to provide a reasonably balanced urban water program. Water reuse and conservation are also important elements in their efforts. Stormwater infiltration to recharge groundwaters and the use of treated wastewaters for beneficial uses (including stream restoration, plus landscaping irrigation, train washing, sewer flushing, fire fighting, etc.) are all important elements of these efforts, although this reuse currently only amounts to about 7% of the total annual water use in Tokyo.

At many U.S. wet detention pond project sites, the stormwater treatment pond is used to increase the value of the property. Figures 2.15 and 2.16 show two examples (in Austin, TX, and in Lake Oswego, OR, respectively). Many people live near wet detention ponds because of the close presence of the wetlands, and their property values are typically greater than lots farther from the ponds (Marsalek et al. 1982). They also reported that small (well-maintained) wet detention ponds are less subject to controversy than larger ponds (that are more commonly neglected). Debo and Ruby (1982) summarized a survey conducted in Atlanta, GA, of residents living near and downstream of 15 small detention ponds and found that almost half the people surveyed who lived in the immediate areas of the ponds did not even know that they existed. Wiegand et al. (1986) found that wet detention ponds, when properly maintained, are preferred by residents over any other urban runoff control practice.

Figure 2.15 Advertising the benefits of a stormwater pond (Austin, TX).

Figure 2.16 Stormwater pond adding value to apartment complex (Lake Oswego, OR).

Emmerling-DiNovo (1995) reported on a survey of homeowners in the Champaign-Urbana, IL, area living in seven subdivisions having either dry or wet detention ponds. She reported that past studies have recognized that developers are well aware that proximity to water increases the appeal of a development. Detention ponds can create a sense of identity, distinguishing one development from another, and can be prominent design elements. Increased value is important because the added cost of the detention facility, including loss of developable land, must be recovered by increasing the housing costs. Others have also found that the higher costs of developments having stormwater detention facilities can also be offset by being able to sell the housing faster. In a survey in Columbia, MD, 73% of the respondents were found to be willing to pay more for property located in an area having a wet detention pond if designed to enhance fish and wildlife use. Although the residents were concerned about nuisances and hazards, they felt that the benefits outweighed these concerns. In her survey, Emmerling-DiNovo (1995) received 143 completed surveys. Respondents reported that the overall attractiveness of the neighborhood was the most important factor in their decision to purchase their home. Resale value was the second most important factor, while proximity to water was slightly important. More than 74% of the respondents believed that wet detention ponds contributed positively to the image of the neighborhood and that they were a positive factor in choosing that subdivision. In contrast, the respondents living in the subdivisions with dry ponds felt that the dry ponds were not a positive factor for locating in their subdivision. Respondents living adjacent to wet ponds felt that the presence of the pond was very positive in the selection of their specific lot. The lots adjacent to the wet ponds were reported to be worth about 22% more than lots that were not adjacent to the wet ponds. Lots adjacent to the dry ponds were actually worth less (by about 10%) than other lots in two of the three dry basin subdivisions studied. The respondents favored living adjacent to wet ponds even more than next to golf courses. Living adjacent to dry ponds was the least preferred location.

Stormwater Conveyance (Flood Prevention)

This is a basic beneficial use of streams and storm drainage systems that must be considered. Problems are caused by increases in peak runoff flow rates that are associated with large increases in runoff volume and decreases in the drainage time of concentration. Because of high flows during wet weather, it is common for urban streams to have much lower flows during dry weather due to lack of recharge from shallow groundwaters (Color Figure 2.1).* Debris and obstructions in the receiving waters, which assist aquatic life uses, typically degrade flooding and drainage uses and are often cleared to provide better drainage. Other common conflicts are associated with the desire to have homogeneous channels (smooth bottoms and straight alignments) for drainage (Figure 2.17), while aquatic life requires diversity in the channel characteristics. These conflicts must be resolved through comprehensive planning, including source controls and drainage controls that have minimal effects on aquatic life. The best solutions would provide for the necessary flooding and drainage benefits while also providing suitable biological habitat (including improved channel stability, decreased bank erosion, artificial pools and riffle areas, overstory shading, gravel linings, low flow meandering channel alignments, and other refuge areas).

Figure 2.17 Channelized urban stream, Nor-X-Way, Menomonee Falls, WI.

Recreation (Non-water Contact) Uses

This basic beneficial use is concerned with odors, trash, beauty, access, and rapidly fluctuating flows. Safety is an important issue in urban

Figure 2.18 Degraded stream banks along New York City shoreline.

Figure 2.19 Debris in riparian area, New York City.

Figure 2.20 Algal mats and other floating debris, Orlando, FL.

Figure 2.21 Litter controlled behind floating booms, New York City.

areas where children frequently play near small streams. Bank stability and rapidly fluctuating flows are, therefore, of prime importance (Figures 2.18 and 2.19). Many communities have also established linear parks along urban streams as part of their flood control and parks programs. In these cases, aesthetics (trash, odor, and beauty), access (paths and bridges), and the above safety issues are also important. Excessive algal growths, with attendant odors and unsightly conditions, may also occur along stressed urban waterways (Figures 2.20 and Color Figure 2.2). Some simple controls have been instituted in some areas to reduce aesthetic impacts (Figure 2.21). Human health may be an issue if water contact (especially by wading children) or if consumptive fishing occurs. These human health uses will be very difficult to maintain in urban areas.

Biological Uses (Warm-Water Fishery, Aquatic Life Use, Biological Integrity, etc.)

This basic beneficial use is also important, but it is defined differently by different people. It is unreasonable to expect natural receiving water conditions in agricultural or urbanized streams. Some degradation is inevitable. The goal is to have an acceptable diversity of aquatic life and an absence of episodic fish kills, at a minimum. It is unfortunate if sensitive and important species exist in an agricultural or urbanized stream and need special protection, as it is probably unrealistic to believe that it is possible to maintain these species in the absence of dramatic and extensive stormwater controls (which are not likely to occur). The most significant impairments to aquatic life beneficial uses are likely: habitat destruction (including channel and bank instability, sedimentation, and loss of refuge areas and vegetative overstory/canopy), highly fluctuating flow rates, inappropriate dry-weather contaminated discharges (toxicants and pathogens), polluted sediment (toxicants and oxygen-demanding materials), and possibly wet weather water quality degradation. Decreases in groundwater recharge and increased peak flows during periods of storm events are obviously associated with decreased flows during dry periods. Aquatic life undergoes additional stress during periods of low flow due to associated increased water temperatures, decreased pollutant mixing and transport, and simple decreased mobility and forage opportunities.

It may be possible to obtain significant short-term biological beneficial use improvements in a degraded stream with improvements in habitat conditions alone. Longer-term benefits would likely require sediment removal and control, plus the control of inappropriate dry-weather toxic discharges. It is unlikely that large improvements in wet weather water quality would be possible in heavily developed watersheds, nor may it be needed to obtain acceptable (but degraded) biological uses. The retrofitting of stormwater controls to improve wet-weather runoff quality in an urban area is very costly and is limited in effectiveness. However, the basic use of construction site erosion controls and biofiltration/infiltration and sedimentation stormwater controls in newly developing areas should be mandatory to decrease the further degradation of biological conditions in receiving waters.

Human Health-Related Uses (Swimming, Fishing, and Water Supply)

In many areas of the country, urban and agricultural runoff drains into public water supplies, swimming areas, or fisheries. In these cases, additional concerns need to be considered, especially relating to toxicants and pathogens. Public water supplies are frequently affected by upstream wastewater discharges (both point and nonpoint sources) and are designed to reduce and monitor constituents of concern. As upstream discharges increase, water treatment becomes more difficult and costly, with increased probabilities of waterborne disease outbreaks and increased (but "legal") taste and odor problems. Swimming areas in urban receiving waters (large rivers and lakes) have also been more frequently closed to the public because of high bacteria counts for extended periods after rains, and because of other unsafe conditions (Figures 2.22 through 2.25 and Color Figure 2.3). In addition, although fishing in urban and agricultural areas is relatively common (Figures 2.26 and 2.27), many communities are posting fishing advisories to discourage this practice (Figure 2.28).

Figure 2.22 Swimming restriction in urban lake, San Francisco, CA.

Figure 2.23 Swimming near stormwater outfall, Navesink River, NJ.

Figure 2.24 Children playing in Lincoln Creek, Milwaukee, WI. (Courtesy of Wisconsin Department of Natural Resources.)

Figure 2.25 Floatable trash from CSO and stormwater discharges, New York City.

Figure 2.26 Fishing in urban stream, Birmingham, AL.

Figure 2.27 Urban fishing in Neva River, St. Petersburg, Russia.

Unfortunately, pathogen levels in stormwater may be high. Fecal coliform levels can be very high, but fecal coliform levels are not thought to be a good indicator of pathogens in stormwater (see also Chapter 4). Direct pathogen monitoring in stormwater has shown very large numbers of some specific pathogens, however, requiring careful consideration for human health issues. In addition, sediments may contain elevated levels of pathogens which live for extended periods following high flow events (Burton et al. 1987). It is very difficult to reduce the high levels using typical stormwater controls. Common disinfection controls are also very costly and may create additional problems associated with trihalomethane production. The consumption of fish or shellfish in waters receiving agricultural and urban runoff is also a cause of concern because of pathogens and toxicants. This has been shown with the recent outbreaks of *Pfiesteria* in nutrient-laden waters of the East Coast. Many of the toxic compounds found in stormwater may readily bioaccumulate in aquatic organisms, and pathogens can also contaminate the aquatic organisms. All of these human health issues require careful study by epidemiologists and public health professionals.

Figure 2.28 Fish advisory for Village Creek, Jefferson Co., AL.

LIKELY CAUSES OF RECEIVING WATER USE IMPAIRMENTS

In general, monitoring of urban and agricultural stormwater runoff has indicated that the biological beneficial uses of receiving waters are most likely affected by habitat destruction and long-term pollutant exposures (especially to macroinvertebrates via contaminated sediment). Pulse exposures to suspended solids and toxicants and contaminated sediments have also been shown to be common in urban and agricultural waterways (see Chapter 6; also review by Burton et al. 2000). Mancini and Plummer (1986) have long been advocates of numeric water quality standards for stormwater that reflect the partitioning of the toxicants and the short periods of exposure during rains. Unfortunately, this approach attempts to isolate individual runoff events and does not consider the accumulative adverse effects caused by the frequent exposures of receiving water organisms to stormwater (Davies 1995; Herricks et al. 1996a,b). Recent investigations have identified acute toxicity problems associated with intermediate-term (about 10 to 20 days) exposures to adverse toxicant concentrations in urban receiving streams (Crunkilton et al. 1996). The most severe receiving water problems may be associated with chronic exposures to contaminated sediment and to habitat destruction.

Heaney et al. (1980) conducted a comprehensive evaluation of the early literature pertaining to urban runoff effects on receiving waters. They found that well-documented cases of receiving water detrimental effects were scarce. Through their review of many reports, they found several reasons to question the implied cause-and-effect relationships between urban runoff and receiving water conditions. Impacts that were attributed to urban runoff were probably caused, in many cases, by other water pollution sources (such as combined sewer overflows, agricultural nonpoint sources, etc.). One of the major difficulties encountered in their study was the definition of "problem" that had been used in the reviewed projects. They found that very little substantive data had been collected to document beneficial use impairments. In addition, urban runoff impacts are most likely to be associated with small receiving waters, while most of the existing urban water quality monitoring information exists for larger bodies of water. It was also very difficult for many researchers to isolate urban runoff effects from other water pollutant sources, such as municipal and industrial wastes. This was especially important in areas that had combined sewers that overflowed during wet weather, contributing to the receiving water impacts during wet-weather conditions.

Claytor (1996a) summarized the approach developed by the Center for Watershed Protection as part of their EPA-sponsored research on stormwater indicators (Claytor and Brown 1996). The 26 stormwater indicators used for assessing receiving water conditions were divided into six broad categories: water quality, physical/hydrological, biological, social, programmatic, and site. These were presented as tools to measure stress (impacting receiving waters), to assess the resource itself, and to indicate stormwater control program implementation effectiveness. The biological communities in Delaware's Piedmont streams have been severely impacted by stormwater, after the extent of imperviousness in the watersheds exceeded about 8 to 15%, according to a review article by Claytor (1996b). If just conventional water quality measures are used, almost all (87%) of the state's nontidal streams supported their designated biological uses. However, when biological assessments are included, only 13% of the streams were satisfactory.

MAJOR URBAN RUNOFF SOURCES

Soil erosion from construction sites and increased stormwater runoff generated from newly established urban areas cause significant economic, social, and environmental problems. These problems may result from all land development activities such as subdivision development, individual homesite construction, large-scale construction projects such as shopping centers and industrial sites, highway construction, and public utility construction projects. Problems caused by construction site erosion and stormwater runoff include sediment that destroys fish habitat and fills in lakes; urban runoff volumes and flow rates that increase flooding; nutrient discharges that produce nuisance algae growths; toxic heavy metal and organic discharges that result in inedible fish, undrinkable water, and shifts in aquatic life to more pollution-tolerant species; and pathogenic bacteria discharges that necessitate swimming beach closures.

Erosion losses and downstream sedimentation peak during construction, when soil exposure is greatest, and decline after construction is completed. Thus, while the impacts of erosion and sedimentation may be severe, they are relatively short term in nature for any specific construction site.

Stormwater runoff and pollutant discharges, on the other hand, increase steadily as development progresses and remain at an elevated level for the lifetime of the development. This happens because impervious surfaces such as roads, sidewalks, driveways, rooftops, etc., permanently reduce infiltration of rainfall and runoff into the ground.

Accelerated stormwater runoff rates also occur with development and can significantly increase the water's ability to detach sediment and associated pollutants, to carry them off site, and to deposit them downstream. Increased runoff rates may also cause stream bank and channel erosion. Increased stormwater runoff volumes and flow rates also increase urban flooding and the resultant loss of human life and property.

Urbanization may also affect groundwater adversely. In some cases, polluted stormwater contaminates groundwater. More frequently, impervious surfaces block infiltration of rainfall and runoff that otherwise would recharge groundwater supplies. Reduced infiltration affects not only groundwater levels but also the amount of groundwater-derived stream flow available during low flow periods. From a water quality standpoint, low flow periods are critical because the amount of water available to dilute stream pollutants is at a minimum at those times. Reduced flows during extended dry periods also adversely affect aquatic life.

Urban runoff, which includes stormwater, construction site runoff, snowmelt, and contaminated baseflows, has been found to cause significant receiving water impacts on aquatic life. The effects are obviously most severe for small receiving waters draining heavily urbanized and rapidly developing watersheds. However, some studies have shown important aquatic life impacts for streams in watersheds that are less than 10% urbanized.

In order to best identify and understand these impacts, it is necessary to include biological monitoring (using a variety of techniques) and sediment quality analyses in a monitoring program. Water column testing alone has been shown to be very misleading. Most aquatic life impacts associated with urbanization are probably related to chronic long-term problems caused by habitat destruction, polluted sediments, and food web disruption. Transient water column quality conditions associated with urban runoff probably rarely cause significant direct aquatic life acute impacts.

The underlying theme of many researchers is that an adequate analysis of receiving water biological impacts must include investigations of a number of biological organism groups (fish, benthic macroinvertebrates, algae, rooted macrophytes, etc.) in addition to studies of water and sediment quality. Simple studies of water quality alone, even with possible comparisons with water quality criteria for the protection of aquatic life, are usually inadequate to predict biological impacts associated with urban runoff.

Duda et al. (1982) presented a discussion on why traditional approaches for assessing water quality, and selecting control options, in urban areas have failed. The main difficulties of traditional approaches when applied to urban runoff are the complexity of pollutant sources, wet weather monitoring problems, and limitations when using water quality standards to evaluate the severity of wet weather receiving water problems. They also discuss the difficulty of meeting water quality goals (that were promulgated in the Water Pollution Control Act of 1972) in urban areas.

Relationships between observed receiving water biological effects and possible causes have been especially difficult to identify, let alone quantify. The studies reported in this chapter have identified a wide variety of possible causative agents, including sediment contamination, poor water quality (low dissolved oxygen, high toxicants, etc.), and factors affecting the physical habitat of the stream (high flows, unstable stream beds, absence of refuge areas, etc.). It is expected that all of these factors are problems, but their relative importance varies greatly depending on the watershed and receiving water conditions. Horner (1991), as an example, notes that many watershed, site, and organism-specific factors must be determined before the best combination of runoff control practices to protect aquatic life can be determined.

Construction Site Erosion Characterization

Sediment is, by weight, the greatest pollutant of water resources. Willett (1980) estimated that approximately 5 billion tons of sediment reach U.S. surface waters annually, of which 30% is generated by natural processes and 70% by human activities. Half of this 70% is attributed to eroding croplands. Although urban construction accounts for only 10%, this amount equals the combined contributions of forestry, mining, industrial, and commercial activities (Willett 1980; Virginia 1980).

Construction accounts for a much greater proportion of the sediment load in urban areas — sometimes more than 50% — than it does in the nation as a whole. Urban areas experience large sediment loads from construction site erosion because construction sites usually have extremely high erosion rates and because urban construction sites are efficiently drained by stormwater drainage systems. Construction sites at most U.S. locations have an erosion rate of approximately 20 to 200 tons per acre per year, a rate that is about 3 to 100 times that of croplands. Construction site erosion losses vary greatly depending on local rain, soil, topographic, and management conditions. As an example, the Birmingham, AL, area may have some of the highest erosion rates in the nation because of its combination of very high-energy rains, moderately erosive soils, and steep topography. The typically high erosion rates mean that even a small construction project may have a significant detrimental effect on local water bodies. While construction occurs on only about 0.007% of U.S. land, it accounts for about 10% of the sediment load to U.S. surface waters (Willett 1980).

Data from the highly urbanized Menomonee River watershed in southeastern Wisconsin illustrate the impact of construction on water quality. These data indicate that construction sites have much greater potential for generating sediment and phosphorus than do areas in other land uses (Chesters et al. 1979). For example, construction sites can generate approximately 8 times more sediment and 18 times more phosphorus than industrial sites, the land use that contributes the second highest amount of these pollutants, and 25 times more sediment and phosphorus than row crops. In fact, construction contributes more sediment and phosphorus to the river than any other land use. In 1979, construction comprised only 3.3% of the watershed's total land area, but it contributed about 50% of the suspended sediment and total phosphorus loading at the river mouth (Novotny et al. 1979).

Similar conclusions were reported by the Southeastern Wisconsin Regional Planning Commission in a 1978 modeling study of the relative pollutant contributions of 17 categories of point and nonpoint pollution sources to 14 watersheds in the southeast Wisconsin regional planning area (SEWRPC 1978). This study revealed construction as the first or second largest contributor of sediment and phosphorus in 12 of the 14 watersheds. Although construction occupied only 2% of the region's total land area in 1978, it contributed approximately 36% of the sediment and 28% of the total phosphorus load to inland waters, making construction the region's second largest source of sediment and phosphorus. The largest source of sediment was estimated to be cropland; livestock operations were estimated to be the largest source of phosphorus. By comparison, cropland comprised 72% of the region's land area and contributed about 45% of the sediment and only 11% of the phosphorus to regional watersheds. This study again points out the high pollution-generating ability of construction sites and the significant water quality impact a small amount of construction may have on a watershed.

A monitoring study of construction site runoff water quality in the Village of Germantown (Washington County, WI) yielded similar results (Madison et al. 1979). Several large subdivisions being developed with single and multifamily residences were selected for runoff monitoring. All utility construction, including the storm drainage system and streets, was completed before monitoring began.

Analysis of the monitoring data showed that sediment leaving the developing subdivisions averaged about 25 to 30 tons per acre per year (Madison et al. 1979). Construction practices identified as contributing to these high yields were removing surface vegetation; stripping and stockpiling topsoil; placing large, highly erodible mounds of excavated soil on and near the streets; pumping water from flooded basement excavations; and tracking of mud into the streets by construction vehicles. If the amount of sediment leaving the sites during utility development had been added in, the total amount of eroded sediment leaving the site would have been substantially greater.

Analysis of the Germantown data also showed that the amount of sediment leaving areas undergoing development is a function of the extent and duration of development and is independent of the type of development. In other words, there is no difference in the per acre sediment loads produced by single-family or multifamily construction. This finding is significant because local and state regulatory programs sometimes exempt single-family home construction from erosion control requirements.

Almost all eroded sediment from the Germantown construction areas entered the receiving waters. The delivery of sediment to the receiving waters was found to be nearly 100% when 10% or more of the watershed was experiencing development. The smallest delivery value obtained during the Germantown monitoring was 50%, observed when only 5% of the watershed was undergoing development. These high delivery values occurred (even during periods with small amounts of development) because storm drainage systems, which efficiently transport water and its sediment load, had been installed during an early stage of development.

Local Birmingham, AL, erosion rates from construction sites can be 10 times the erosion rates from row crops and 100 times the erosion rates from forests or pastures (Nelson 1996). The site-specific factors affecting construction site erosion include:

- Rainfall energy (Alabama has the highest in the nation)
- Soil erodibility (northern part of the state has fine-grained, highly erosive soils)
- Site topography (northeastern part of the state has steep hills under development)
- Surface cover (usually totally removed during initial site grading)

The rain energy is directly related to rainfall intensity, and the rainfall erosion index varies from 250 to 550+ for Alabama (most of the state is about 350), which is the highest in the United States. The months having the greatest erosion potential are February and March, while September through November have the lowest erosion potential. Nelson (1996) monitored sediment quantity and particle size from 70 construction site runoff samples from the Birmingham area. He measured suspended solids concentrations ranging from 100 to more than 25,000 mg/L (overall median about 4000 mg/L), while the turbidity values ranged from about 300 to >50,000 NTU (average of about 4000 NTU). About 90% of the particles (by mass) were smaller than about 20 μm (0.02 mm) in diameter, and the median size was about 5 μm (0.005 mm). The local construction site erosion discharges were estimated to be about 100 tons/acre/year. Table 2.2 summarizes the measured suspended solids and median particle sizes as a function of rain intensity. High-intensity rains were found to have the most severe erosion problems, as expected, with much greater suspended solids

Table 2.2 Birmingham (AL) Construction Site Erosion Runoff Characteristics

	Low-Intensity Rains (<0.25 in/hr)	Moderate-Intensity Rains (about 0.25 in/hr)	High-Intensity Rains (>1 in/hr)
Suspended solids, mg/L	400	2000	25,000
Particle size (median), μm	3.5	5	8.5

Data from Nelson, J. *Characterizing Erosion Processes and Sediment Yields on Construction Sites.* M.S.C.E. thesis. Department of Civil and Environmental Engineering, University of Alabama at Birmingham. 94 pp. 1996.

concentrations. Typical small particle sizes of erosion particulates make it very difficult to remove these particulates after they have been eroded from the site. The extreme turbidity values also cause very high in-stream turbidity conditions in local receiving waters for great distances downstream of eroding sites.

Urban Runoff Contaminants

Urban runoff is comprised of many different flow phases. These may include dry-weather base flows, stormwater runoff, combined sewer overflows (CSOs), and snowmelt. The relative magnitudes of these discharges vary considerably, based on a number of factors. Season (such as cold vs. warm weather, or dry vs. wet weather) and land use have been identified as important factors affecting baseflow and stormwater runoff quality.

Land development increases stormwater runoff volumes and pollutant concentrations. Impervious surfaces, such as rooftops, driveways, and roads, reduce infiltration of rainfall and runoff into the ground and degrade runoff quality. The most important hydraulic factors affecting urban runoff volume (and therefore the amount of water available for groundwater infiltration) are the quantity of rain and the extent of impervious surfaces directly connected to a stream or drainage system. Directly connected impervious areas include paved streets, driveways, and parking areas draining to curb and gutter drainage systems, and roofs draining directly to a storm or combined sewer pipe. Table 2.3 presents older stormwater quality data (APWA 1969), while Table 2.4 is a summary of the Nationwide Urban Runoff Program (NURP) stormwater data collected from about 1979 through 1982 (EPA 1983). The NURP data are the most comprehensive stormwater data available from throughout the nation. The recently collected data for the stormwater NPDES permits is a potentially large and important database of information, but it has not been made conveniently available. Land use and source areas (parking areas, rooftops, streets, landscaped areas, etc.) all have important effects on stormwater runoff quality. BOD_5 bacteria and nutrient concentrations in stormwater are lower than in raw sanitary wastewater. However, urban stormwater still has relatively high concentrations of bacteria, along with high concentrations of many metallic and some organic toxicants.

NURP found that stormwater pollutant concentrations, runoff volumes, and therefore annual pollutant yields often vary with land use. Although inconsistencies in local development practices within a single land use category make land use a less than perfect indicator of urban runoff characteristics, land use must serve as a surrogate for more appropriate indicators because development data are typically reported in land use categories. The amount of directly connected impervious area is a very good indicator of an area's runoff volume. The extent of "effective" impervious surfaces, however, is a function of local development customs (lot sizes, use of swale drainages, single or multilevel buildings, type of landscaping, etc.), which can vary significantly within a single land use category (such as medium-density residential). Development characteristics are not uniform throughout a region, and they may also vary by age of development or location within a single city.

Bannerman et al. (1979) found a high correlation between pollutant loading values and percent connected-imperviousness during monitoring of seven subwatersheds of the Menomonee River basin: pollutant loading to the river increased as the extent of impervious areas directly connected to the storm drainage system increased. Although larger amounts of runoff and pollutants were

Table 2.3   Characteristics of Stormwater Runoff from Early Studies

City	BOD$_5$ (mg/L)	Total Solids (mg/L)	Suspended Solids (mg/L)	Chlorides (mg/L)	COD (mg/L)
East Bay Sanitary District: Oakland, California					
Minimum	3	726	16	300	
Maximum	7700		4400	10,260	
Average	87	1401	613	5100	
Cincinnati, Ohio					
Maximum Seasonal Means	12	260			110
Average	17		227		111
Los Angeles County					
Average 1962–63	161	2909		199	
Washington, D.C.					
Catch-basin samples during storm	6		26	11	
Minimum	625		36,250	160	
Maximum	126		2100	42	
Average					
Seattle, Washington	10				
Oxney, England	100[a]	2045			
Moscow, Russia	186–285	1000–3500[a]			
Leningrad, Russia	36	14,541			
Stockholm, Sweden	17–80	30–8000			18–3100
Pretoria, South Africa					
Residential	30				29
Business	34				28
Detroit, Michigan	96–234	310–914	102–213[b]		

[a] Maximum.
[b] Mean.

From APWA (American Public Works Association). *Water Pollution Aspects of Urban Runoff.* Water Pollution Control Research Series WP-20-15, Federal Water Pollution Control Administration. January 1969.

Table 2.4   Median Stormwater Pollutant Concentrations for All Sites by Land Use

Constituent	Residential Median	COV[a]	Mixed Land Use Median	COV	Commercial Median	COV	Open/Non-urban Median	COV
BOD$_5$, mg/L	10	0.41	7.8	0.52	9.3	0.31	—	—
COD, mg/L	73	0.55	65	0.58	57	0.39	40	0.78
TSS, mg/L	101	0.96	67	1.14	69	0.85	70	2.92
Total Kjeldahl nitrogen, µg/L	1900	0.73	1288.8	0.50	1179	0.43	965	1.00
NO$_2$ + NO$_3$ (as N) µg/L	736	0.83	558	0.67	572	0.48	543	0.91
Total P, µg/L	383	0.69	263	0.75	201	0.67	121	1.66
Soluble P, µg/L	143	0.46	56	0.75	80	0.71	26	2.11
Total lead, µg/L	144	0.75	114	1.35	104	0.68	30	1.52
Total copper, µg/L	33	0.99	27	1.32	29	0.81	—	—
Total zinc, µg/L	135	0.84	154	0.78	226	1.07	195	0.66

[a] COV: coefficient of variation = standard deviation/mean.

From EPA (U.S. Environmental Protection Agency). *Results of the Nationwide Urban Runoff Program.* Water Planning Division, PB 84-185552, Washington, D.C. December 1983.

generated in low-density residential areas, compared to undisturbed areas, runoff and pollutant delivery from the source areas to streams was still low due to the use of grass-lined roadside drainage channels. Soil and vegetation have a greater chance to reduce runoff water and pollutants in areas drained by grass-lined drainage channels than in similar areas drained by conventional curb-and-gutter drainage systems.

Table 2.5 presents estimates of typical urban area pollutant yields from several separate studies. Local conditions and development characteristics significantly affect these estimates. The most significant factor is the drainage efficiency of the areas, specifically if the areas are drained by grass swales. The low-density residential area values shown on this table reflect grass swale drained areas. If conventional curbs and gutters were used instead of grass swales, the yields would be about 10 times greater. Other important development characteristics affecting runoff yields include roof drainage connections and the presence of alleyways. Increased drainage efficiency invariably leads to increased pollutant discharges.

A number of urban runoff monitoring projects (such as EPA 1983; Pitt and McLean 1986) have found inorganic and organic hazardous and toxic substances in urban runoff. The NURP data, collected from mostly residential areas throughout the United States, did not indicate any regional differences in the substances detected, or in their concentrations. However, residential and industrial data obtained by Pitt and McLean (1986) in Toronto found significant concentration and yield differences for these two land uses and for dry weather and wet weather urban runoff flows.

Tables 2.6 and 2.7 list the toxic and hazardous organic substances that have been found in greater than 10% of industrial and residential urban runoff samples. NURP data do not reveal toxic urban runoff conditions significantly different for different geographical areas throughout North America (EPA 1983). The pesticides shown were mostly found in urban runoff from residential areas, while other hazardous materials were much more prevalent in industrial areas. Urban runoff dry weather baseflows may also be important contributors of hazardous and toxic pollutants.

Urban Runoff Pollutant Sources

Sources of the toxic and hazardous substances found in urban runoff vary widely. Table 1.3 listed the major expected sources of these substances. Automobile use contributes significantly to many of these materials. Polycyclic aromatic hydrocarbons (PAHs), the most commonly detected toxic organic compounds found in urban runoff, are mostly from fossil fuel combustion. Phthalate esters, another group of relatively common toxic organic compounds, are derived from plastics. Pentachlorophenol, also frequently found, comes from preserved wood. Such compounds are very hard to control at their sources, and, unfortunately, their control by typical stormwater management practices is little understood.

Urban runoff includes warm and cold weather baseflows, stormwater runoff, and snowmelt. Table 2.8 shows median concentrations of some of the pollutants monitored in a mixed residential and commercial catchment and from an industrial area in Toronto, Ontario, for these different flow phases (Pitt and McLean 1986). Samples were obtained from baseflow discharges, stormwater runoff, and snowmelt. The baseflows had surprisingly high concentrations of several pollutants, especially dissolved solids (filtrate residue) and fecal coliforms from the residential catchment. The concentrations of some constituents in the stormwater from the industrial watershed were typically much greater than the concentrations of the same constituents in the residential stormwater. The industrial warm weather baseflows were also much closer in quality to the industrial stormwater quality than the residential baseflows were to the residential stormwater quality. The data collected for pesticides and PCBs indicate that the industrial stormwater and baseflows typically contained much greater concentrations of these pollutants than the residential waters. Similarly, the more commonly analyzed heavy metals were also more prevalent in the industrial stormwater. However, herbicides were only detected in residential urban runoff, especially the baseflows.

During cold weather, the increases in filtrate residue were quite apparent for both study catchments and for both baseflows and snowmelt. These increases were probably caused by high chlorides from road salt applications. In contrast, bacteria populations were noticeably lower in all outfall discharges during cold weather. Few changes were noted in concentrations of nutrients and heavy metals at the outfall, between cold- and warm-weather periods.

Table 2.5 Typical Urban Area Pollutant Yields (lb/acre/year or kg/ha/yr)[a]

Land Use	Total Solids	Suspended Solids	Chloride	Total Phosphorus	TKN	NH$_3$	NO$_3$ plus NO$_2$	BOD$_5$
Commercial	2100	1000	420	1.5	6.7	1.9	3.1	62
Parking lot	1300	400	300	0.7	5.1	2.0	2.9	47
High-density residential	670	420	54	1.0	4.2	0.8	2.0	27
Medium-density residential	450	250	30	0.3	2.5	0.5	1.4	13
Low-density residential[b]	65	10	9	0.04	0.3	0.02	0.1	1
Freeways	1700	880	470	0.9	7.9	1.5	4.2	NA[b]
Industrial	670	500	25	1.3	3.4	0.2	1.3	NA
Parks	NA[c]	3	NA	0.03	NA	NA	NA	NA
Shopping center	720	440	36	0.5	3.1	0.5	1.7	NA

Land Use	COD	Lead[d]	Zinc	Chromium	Copper	Cadmium	Arsenic
Commercial	420	2.7	2.1	0.15	0.4	0.03	0.02
Parking lot	270	0.8	0.8	NA	0.06	0.01	NA
High-density residential	170	0.8	0.7	NA	0.03	0.01	NA
Medium-density residential	50	0.05	0.1	0.02	0.03	0.01	0.01
Low-density residential[e]	7	0.01	0.04	0.002	0.01	0.001	0.001
Freeways	NA	4.5	2.1	0.09	0.37	0.02	0.02
Industrial	200	0.2	0.4	0.6	0.10	0.05	0.04
Parks	NA	0.005	NA	NA	NA	NA	NA
Shopping center	NA	1.1	0.6	0.04	0.09	0.01	0.02

[a] The difference between lb/acre/year and kg/ha/yr is less than 15%, and the accuracy of the values shown in this table cannot differentiate between such close values.

[b] The monitored low-density residential areas were drained by grass swales.

[c] NA = Not available.

[d] The lead unit area loadings shown on this table are currently expected to be significantly less than shown on this table, as these values are from periods when leaded gasoline adversely affected stormwater lead quality.

[e] The monitored low-density residential areas were drained by grass swales.

Data from Bannerman et al. (1979, 1983); Madison et al. (1979); EPA (1983); Pitt and McLean (1986).

Table 2.6   Hazardous Substances Observed in Urban Runoff

Hazardous Substances	Residential Areas	Industrial Areas
Benzene	5 µg/L	5 µg/L
Chlordane	17 ng/L	—
Chloroform	—	5 µg/L
Dieldrin	2 to 6 ng/L	—
Endrin	44 ng/L	—
Methoxychlor	20 ng/L	—
Pentachlorophenol	70 to 280 ng/L	50 to 710 ng/L
Phenol	1 µg/L	4 µg/L
Phosphorus	0.1 mg/L	0.5 µg/L
Toluene	—	5 µg/L

Data from EPA 1983; Pitt and McLean 1986 (Toronto); and Pitt et al. 1996 (Birmingham).

Table 2.7   Other Toxic Substances Observed in Urban Runoff

GC/MS Volatiles	Residential Areas	Industrial Areas
1,2-Dichloroethane	—	6 µg/L
Methylene chloride	—	5 µg/L
Tetrachloroethylene	—	High in some source areas
GC/MS Base/Neutrals		
Bis (2-ethylene) phthalate	8 µg/L	18 µg/L
Butyl benzyl phthalate	5 µg/L	58 µg/L
Diethyl phthalate	—	20 µg/L
Di-N-butyl phthalate	3 µg/L	4 µg/L
Isophorone	2 µg/L	—
N-Nitrosodimethylamine	—	3 µg/L
Phenanthrene	—	High in some source areas
Pyrene	—	High in some source areas
GC/MS Pesticides		
BHC	up to 20 ng/L	—
Chlordane	up to 15 ng/L	—
Dieldrin	up to 6 ng/L	—
Endosulfan sulfate	up to 10 ng/L	—
Endrin	up to 45 ng/L	—
PCB-1254	—	up to 630 ng/L
PCB-1260	—	up to 440 ng/L

Data from EPA 1983; Pitt and McLean 1986 (Toronto); and Pitt et al. 1996 (Birmingham).

   Table 2.9 compares the estimated annual discharges from the residential and industrial catchments during the different runoff periods. The unit area annual yields for many of the heavy metals and nutrients are greater from the industrial catchment. Industrial catchments contribute most of the chromium to the local receiving waters, and approximately equal amounts with the residential and commercial catchments for phosphorus, chemical oxygen demand, copper, and zinc. This table also shows the great importance of warm weather baseflow discharges to the annual urban runoff pollutant yields, especially for industrial areas. Cold weather bacteria discharges are insignificant when compared to the warm weather bacteria discharges, but chloride (and filtrate residue) loadings are much more important during cold weather.

   Table 2.10 shows the fraction of the annual estimated yields for different warm and cold periods (warm weather baseflow, stormwater flows, cold weather baseflow, and snowmelt). Typical storm-

Table 2.8   Median Urban Runoff Pollutant Concentrations

Constituent	Warm-Weather Baseflow		Warm-Weather Stormwater	
	Residential	Industrial	Residential	Industrial
Total residue	979	554	256	371
Filterable residue	973	454	230	208
Particulate residue	<5	43	22	117
Total phosphorus	0.09	0.73	0.28	0.75
Total Kjeldahl N	0.9	2.4	2.5	2.0
Phenolics (μg/L)	<1.5	2.0	1.2	5.1
COD	22	108	55	106
Fecal coliforms (no./100 mL)	33,000	7000	40,000	49,000
Fecal streptococci (no./100 mL)	2300	8800	20,000	39,000
Chromium	<0.06	0.42	<0.06	0.32
Copper	0.02	0.045	0.03	0.06
Lead	<0.04	<0.04	<0.06	0.08
Zinc	0.04	0.18	0.06	0.19

Constituent	Cold-Weather Baseflow		Cold-Weather Melting Periods	
	Residential	Industrial	Residential	Industrial
Total residue	2230	1080	1580	1340
Filterable residue	2210	1020	1530	1240
Particulate residue	21	50	30	95
Total phosphorus	0.18	0.34	0.23	0.50
Total Kjeldahl N	1.4	2.0	1.7	2.5
Phenolics (μg/L)	2.0	7.3	2.5	15.0
COD	48	68	40	94
Fecal coliforms (no./100 mL)	9800	400	2320	300
Fecal streptococci (no./100 mL)	1400	2400	1900	2500
Chromium	<0.01	0.24	<0.01	0.35
Copper	0.015	0.04	0.04	0.07
Lead	<0.06	<0.04	0.09	0.08
Zinc	0.065	0.15	0.12	0.31

From Pitt, R. and J. McLean. *Humber River Pilot Watershed Project*, Ontario Ministry of the Environment, Toronto, Canada. 483 pp. June 1986.

water flow contributions from these separate stormwater outfalls were only about 20 to 30% of the total annual discharges (by volume). Baseflows contributed the majority of flows. Many constituents were also contributed mostly by snowmelt and baseflows, with the stormwater contributions being less than 50% of the total annual yields. The ratios of expected annual pollutant yields from the industrial catchment divided by the yields from the residential/commercial catchment can be high, as summarized below.

Ratios of Industrial to Mixed Residential/Commercial Unit Area Yields

Particulate residue (suspended solids)	4.4
Phosphorus	3.0
Phosphates	5.1
Chemical oxygen demand	2.0
Fecal streptococci bacteria	2.6
Chromium	53.0
Zinc	2.5

The only constituents with annual unit area yields that were lower in the industrial catchment than in the mixed residential/commercial catchment were chloride and filtrate residue (dissolved

Table 2.9 Monitored Annual Pollutant Discharges for Toronto's Humber River Watershed Test Sites

Constituent	Units	Thistledowns (Residential/Commercial)					Emery (Industrial)					Approx. Indus. to Resid. Total Yield Ratios	Weighted Indus. to Resid. Total Yield Ratios[a]
		Warm		Cold			Warm		Cold				
		Base-flow	Storm-water	Base-flow	Melt-water	Approx. Total	Base-flow	Storm-water	Base-flow	Melt-water	Approx. Total		
Runoff	m³/ha	1700	950	1100	1800	5600	2100	1500	660	830	5100	0.9	0.3
Total residue	kg/ha	1700	240	2400	1700	6100	1100	670	710	1500	4000	0.7	0.2
Chlorides	kg/ha	480	33	1200	720	2400	160	26	310	700	1200	0.5	0.2
Total P	g/ha	150	290	200	570	1200	1500	1300	220	540	3600	3.0	1.0
Total Kjeldahl N	g/ha	1500	2800	1500	3500	9300	4900	3400	1300	2800	12,000	1.3	0.4
Phenolics	g/ha	<2.6	1.2	2.3	23	26	4.1	8.1	4.8	14	31	1.2	0.4
COD	kg/ha	38	51	52	130	270	220	170	45	91	530	2.0	0.7
Chromium	g/ha	<100	21	<10	15	36	860	600	160	290	1900	50	18
Copper	g/ha	35	30	16	77	160	92	120	26	76	310	1.9	0.7
Lead	g/ha	<70	41	<70	170	210	<75	170	<25	150	320	1.5	0.5
Zinc	g/ha	70	74	70	270	480	370	430	100	350	1200	2.5	0.8
Fecal coliform	10⁹ org/ha	560	480	110	62	1200	144	760	3	6	910	0.8	0.3

"Warm weather" is for the period from about March 15 through December 15, while "cold weather" is for the period from about December 15 through March 15.

a The Humber River basin is about 25% industrial and 75% residential and commercial.

From Pitt, R. and J. McLean. Humber River Pilot Watershed Project, Ontario Ministry of the Environment, Toronto, Canada. 483 pp. June 1986.

Table 2.10 Major Concentration Periods by Parameter

	Runoff Volume		Total Residue		Filtrate Residue		Particulate Residue		Chlorides	
	Residential	Industrial	Residential	Industrial	Residential	Industrial	Residential	Industrial	Residential	Industrial
Warm baseflow	31%	41%	28%	28%	28%	30%	4%	16%	20%	13%
Stormwater	17	29	4	17	4	10	18	53	1	2
Cold baseflow	20	13	40	18	40	18	14	5	49	26
Meltwater	33	16	29	38	27	41	63	26	29	58

	Phosphorus		Phosphate		Total Kjeldahl Nitrogen		Ammonia Nitrogen		Pseudomonas aeruginosa	
	Residential	Industrial	Residential	Industrial	Residential	Industrial	Residential	Industrial	Residential	Industrial
Warm baseflow	12	42	—	35	16	39	—	—	53	41
Stormwater	24	36	24	51	30	27	21	24	46	58
Cold baseflow	16	6	—	—	16	10	—	—	1	—
Meltwater	47	15	76	14	38	23	78	76	—	1

	Phenolics		COD		Fecal Coliform		Fecal Streptococci	
	Residential	Industrial	Residential	Industrial	Residential	Industrial	Residential	Industrial
Warm baseflow	—	13	14	42	46	16	12	20
Stormwater	5	27	19	32	40	84	61	73
Cold baseflow	9	16	19	9	9	—	4	2
Meltwater	87	45	48	17	5	—	22	4

	Chromium		Copper		Lead		Zinc	
	Residential	Industrial	Residential	Industrial	Residential	Industrial	Residential	Industrial
Warm baseflow	—	45	22	29	—	—	14	30
Stormwater	59	31	19	38	19	54	15	35
Cold baseflow	—	8	10	8	—	—	14	8
Meltwater	41	16	49	24	81	46	56	27

Warm period included samples from Thistledowns from July 28 through Nov. 15, 1983, and from Emery from May 14 through Nov.15, 1983. Cold period samples from Thistledowns were from Feb. 2 through March 25, 1984, and from Emery from Jan. through March 22, 1984.

From Pitt, R. and J. McLean. *Humber River Pilot Watershed Project*, Ontario Ministry of the Environment, Toronto, Canada. 483 pp. June 1986.

solids). The annual unit area yields from the residential/commercial catchment were approximately twice the annual unit area yields from the industrial catchment for these constituents.

If only warm weather stormwater runoff is considered (and not baseflows and snowmelts), then significant yield and control measure selection errors are probable. Residential/commercial unit area annual yields for total residue (total solids) for stormwater alone are approximately 240 kg/ha, compared with approximately 670 kg/ha for the industrial catchment. These yields are similar to yields reported elsewhere for total annual total residue unit area yields. However, these warm weather stormwater runoff yields only contributed approximately 5 to 20% of the total annual total residue yields for these study catchments. Annual yields of several constituents were dominated by cold weather processes irrespective of the land use monitored. These constituents include total residue, filtrate residue, chlorides, ammonia nitrogen, and phenolics. The only constituents for which the annual yields were dominated by warm weather processes, irrespective of land use, were bacteria (fecal coliforms, fecal streptococci, and *Pseudomonas aeruginosa*), and chromium. Lead and zinc were both dominated by either stormwater or snowmelt runoff, with lower yields of these heavy metals originating from baseflows.

Warm weather stormwater runoff alone was the most significant contributor to the annual yields for a number of constituents from the industrial catchment. These constituents included particulate residue, phosphorus, phosphates, the three bacteria types, copper, lead, and zinc. In the residential/commercial catchment, only fecal streptococcus bacteria and chromium were contributed by warm weather stormwater runoff more than by the other three sources of water shown. Either warm or cold weather baseflows were most responsible for the yields of many constituents from the industrial catchment. These constituents included runoff volume, phosphorus, total Kjeldahl nitrogen, chemical oxygen demand, and chromium. Important constituents that have high yields in the baseflow from the residential/commercial catchment included total residue, filtrate residue, chlorides, and fecal coliform and *P. aeruginosa* bacteria. More recently, agricultural pesticides have been detected in urban rainfall and urban pesticides in agricultural rainfall and have also been detected in receiving waters.

SUMMARY

This chapter reviewed some of the major receiving water use impairments that have been associated with urban stormwater discharges. The problems associated with urban stormwater discharges can be many, but varied, depending on the specific site conditions. It is therefore important that local objectives and conditions be considered when evaluating local receiving water problems. There has been a great deal of experience in receiving water assessments over the past decade, especially focusing on urban nonpoint source problems. The main purpose of this book is to provide techniques and direction that can be applied to local waters to assess problems based on actual successful field activities. Of course, monitoring and evaluation techniques are constantly changing and improving, and this book also periodically presents short summaries of emerging techniques that hold promise, but may require additional development to be easily used by most people.

Generally, receiving water problems are not readily recognized or understood if one relies on only a limited set of tools. It is critical that conventional water quality measurements be supplemented with habitat evaluations and biological studies, for example. In many cases, receiving water problems caused by urbanization may be mostly associated with habitat destruction, contaminated sediment, and inappropriate discharges, all of which would be poorly indicated by relying only on conventional water quality measurements. In contrast, eliminating water quality measurements from an assessment and relying only on less expensive indicators, such as the currently popular citizen monitoring of benthic conditions, is also problematic, especially from a human health perspective.

A well-balanced assessment approach is therefore needed to understand the local problems of most concern and is the focus of this book.

This chapter also summarized stormwater characteristics. Runoff from established urban areas may not be the major source of some of the problem pollutants in urban areas. Obviously, construction site runoff is typically the major source of sediment in many areas, but snowmelt contributions of sediment (and many other constituents) is also very important in northern areas. Dry weather flows in separate storm drainage systems can be contaminated with inappropriate discharges from commercial and industrial establishments and sewage. Obviously, these inappropriate discharges need to be identified and corrected.

The rest of this book establishes an approach for investigating receiving water use impairments and in identifying the likely causes for these problems. When this information is known, it is possible to begin to develop an effective stormwater management program.

REFERENCES

APWA (American Public Works Assoc.). *Water Pollution Aspects of Urban Runoff.* Water Pollution Control Research Series WP-20-15, Federal Water Pollution Control Administration. January 1969.

Bannerman, R., J. Konrad, D. Becker, G.V. Simsiman, G. Chesters, J. Goodrich-Mahoney, and B. Abrams. *The IJC Menomonee River Watershed Study — Surface Water Monitoring Data.* EPA-905/4-79-029. U.S. Environmental Protection Agency, Chicago, IL. 1979.

Bannerman, R., K. Baun, M. Bohn, P.E. Hughes, and D.A. Graczyk. *Evaluation of Urban Nonpoint Source Pollution Management in Milwaukee County, Wisconsin*, Vol. I. PB 84-114164. U.S. Environmental Protection Agency, Water Planning Division. November 1983.

Burton, G.A., Jr., D. Gunnison, and G.R. Lanza. Survival of pathogenic bacteria in various freshwater sediments. *Appl. Environ. Microbiol.*, 53: 633–638. 1987.

Burton, G.A., Jr., R. Pitt, and S. Clark. The role of whole effluent toxicity test methods in assessing stormwater and sediment contamination. *CRC Crit. Rev. Environ. Sci. Technol.*, 30: 413–447. 2000.

Cardozo, R.J., W.R. Adams, and E.L. Thackston. CSO's real impact on water quality: the Nashville experience. *A Global Perspective for Reducing CSOs: Balancing Technologies, Costs, and Water Quality.* July 10–13, 1994. Louisville, KY. Water Environment Federation. Alexandria, VA. 1994.

Chesters, G., J. Konrad, and G. Simsiman. *Menomonee River Pilot Watershed Study — Summary and Recommendations,* EPA-905/4-79-029. U.S. Environmental Protection Agency, Chicago, IL. 1979.

Claytor, R. Multiple indicators used to evaluate degrading conditions in Milwaukee County. *Watershed Prot. Techn.*, 2(2): 348–351. Spring 1996a.

Claytor, R.A. An introduction to stormwater indicators: an urban runoff assessment tool. *Watershed Prot. Techn.*, 2(2): 321–328. Spring 1996b.

Claytor, R.A. and W. Brown. *Environmental Indicators to Assess the Effectiveness of Municipal and Industrial Stormwater Control Programs.* Prepared for the U.S. EPA, Office of Wastewater Management. Center for Watershed Protection, Silver Spring, MD. 210 pp. 1996.

Crunkilton, R., J. Kleist, J. Ramcheck, B. DeVita, and D. Villeneuve. Assessment of the response of aquatic organisms to long-term *in situ* exposures to urban runoff. Engineering Foundation Conference: *Effects of Watershed Development & Management on Aquatic Ecosystems,* Snowbird, UT. Published by the American Society of Civil Engineers, New York. August 1996.

Davies, P.H. Factors in controlling nonpoint source impacts. In *Stormwater Runoff and Receiving Systems: Impact, Monitoring, and Assessment.* Edited by E.E. Herricks, CRC/Lewis Publishers, Boca Raton, FL. pp. 53–64. 1995.

Debo, D.N. and H. Ruby. Detention basins — an urban experience. *Public Works*, 113(1): 42. January 1982.

Dreiseitl, Herbert. The role of water in cities. Presented at the Engineering Foundation/ASCE sponsored symposium on *Sustaining Urban Water Resources in the 21st Century*, Malmo, Sweden. Edited by A.C. Rowney, P. Stahre, and L.A. Roesner. September 7–12, 1997. ASCE/Engineering Foundation, New York. 1998.

Duda, A.M., D.R. Lenat, and D. Penrose. Water quality in urban streams — what we can expect. *J. Water Pollut. Control Fed.*, 54(7): 1139–1147. July 1982.

Emmerling-DiNovo, C. Stormwater detention basins and residential locational decisions. *Water Resour. Bull.*, 31(3): 515–521. June 1995.

EPA (U.S. Environmental Protection Agency). *Results of the Nationwide Urban Runoff Program.* Water Planning Division, PB 84-185552, Washington, D.C. December 1983.

Fujita, S. Restoration of polluted urban watercourses in Tokyo for community use. Presented at the Engineering Foundation/ASCE sponsored symposium on *Sustaining Urban Water Resources in the 21st Century.* Malmo, Sweden, Edited by A.C. Rowney, P. Stahre, and L.A. Roesner. September 7–12, 1997. Malmo, Sweden. ASCE/Engineering Foundation, New York. 1998.

Göransson, C. Aesthetic aspects of stormwater management in an urban environment. Presented at the Engineering Foundation/ASCE sponsored symposium on *Sustaining Urban Water Resources in the 21st Century*, Malmo, Sweden, Edited by A.C. Rowney, P. Stahre, and L.A. Roesner. September 7–12, 1997. ASCE/Engineering Foundation, New York. 1998.

Heaney, J.P., W.C. Huber, and M.E. Lehman. *Nationwide Assessment of Receiving Water Impacts from Urban Storm Water Pollution.* U.S. Environmental Protection Agency, Cincinnati, OH. April 1980.

Herricks, E.E, I. Milne, and I. Johnson. A protocol for wet weather discharge toxicity assessment. Volume 4, pg. 13–24. *WEFTEC'96: Proceedings of the 69th Annual Conference & Exposition.* Dallas, TX. 1996a.

Herricks, E.E., R. Brent, I. Milne, and I. Johnson. Assessing the response of aquatic organisms to short-term exposures to urban runoff. Engineering Foundation Conference: *Effects of Watershed Development & Management on Aquatic Ecosystems,* Snowbird, UT. Published by the American Society of Civil Engineers, New York. August 1996b.

Horner, R.R. Toward ecologically based urban runoff management. Engineering Foundation Conference: *Effects of Urban Runoff on Receiving Systems: An Interdisciplinary Analysis of Impact, Monitoring, and Management.* Mt. Crested Butte, CO. Published by the American Society of Civil Engineers, New York. 1991.

Madison, F., J. Arts, S. Berkowitz, E. Salmon, and B. Hagman. *Washington County Project.* EPA 905/9-80-003, U.S. Environmental Protection Agency, Chicago, IL. 1979.

Mancini, J. and A. Plummer. Urban runoff and water quality criteria. Engineering Foundation Conference: *Urban Runoff Quality — Impact and Quality Enhancement Technology.* Henniker, NH. Edited by B. Urbonas and L.A. Roesner. Published by the American Society of Civil Engineers, New York. pp. 133–149. June 1986.

Marsalek, J., D. Weatherbe, and G. Zukovs. Institutional aspects of stormwater detention. Proceedings of the Conference: *Stormwater Detention Facilities, Planning, Design, Operation, and Maintenance.* Henniker, NH. Edited by W. DeGroot. Published by the American Society of Civil Engineers, New York. August 1982.

Minshall, G.W. Stream ecosystem theory: a global perspective, *J. North Am. Benthol. Soc.,* 7: 263–288. 1988.

Nelson, J. *Characterizing Erosion Processes and Sediment Yields on Construction Sites.* MSCE thesis. Department of Civil and Environmental Engineering, University of Alabama, Birmingham. 94 pp. 1996.

Novotny, V., D. Balsiger, R. Bannerman, J. Konrad, D. Cherkauer, G. Simsiman, and G. Chesters. *The IJC Menomonee River Watershed Study — Simulation of Pollutant Loadings and Runoff Quality.* EPA-905/4-79-029. U.S. Environmental Protection Agency, Chicago, IL. 1979.

Pitt, R. and J. McLean. *Humber River Pilot Watershed Project*, Ontario Ministry of the Environment, Toronto, Canada. 483 pp. June 1986.

Pitt, R., R. Field, M. Lalor, and M. Brown. Urban stormwater toxic pollutants: assessment, sources and treatability. *Water Environ. Res.,* 67(3): 260–275. May/June 1995. Discussion and closure in 68(4): 953–955. July/August 1996.

Pool, R. Is it chaos, or is it just noise? *Science,* 243: 25–27. 1989.

SEWRPC (Southeastern Wisconsin Planning Commission). *Sources of Water Pollution in Southeastern Wisconsin: 1975.* Technical Report No. 21. Waukesha, WI. 1978.

Virginia. *Erosion and Sediment Control Handbook. 2nd edition.* Division of Soil and Water Conservation. Virginia Dept. of Conservation and Historic Resources, Richmond, VA. 1980.

Wiegand, C., T. Schueler, W. Chittenden, and D. Jellick. Comparative costs and cost effectiveness of urban best management practices. Engineering Foundation Conference: *Urban Runoff Quality — Impact and Quality Enhancement Technology*. Henniker, NH, Edited by B. Urbonas and L.A. Roesner. Published by the American Society of Civil Engineers, New York. June 1986.

Willett, G. Urban erosion, in *National Conference on Urban Erosion and Sediment Control: Institutions and Technology*. EPA 905/9-80-002. U.S. Environmental Protection Agency, 1980.

## CHAPTER 3

Stressor Categories and Their Effects on Humans and Ecosystems

"As for Paris, within the last few years, it has been necessary to move most of the mouths of the sewers down stream below the last bridge."

Victor Hugo, 1862

CONTENTS

Effects of Runoff on Receiving Waters 47
    Indicators of Receiving Water Biological Effects and Analysis Methodologies 48
    Fish Kills and Advisories 49
    Adverse Aquatic Life Effects Caused by Runoff 50
    Observed Habitat Problems Caused by Runoff 54
    Groundwater Impacts from Stormwater Infiltration 56
Stressor Categories and Their Effects 63
    Stream Flow Effects and Associated Habitat Modifications 63
    Safety Concerns with Stormwater 66
    Aesthetics, Litter/Floatables, and Other Debris Associated with Stormwater 68
    Solids (Suspended, Bedded, and Dissolved) 71
    Dissolved Oxygen 73
    Temperature 75
    Nutrients 76
    Toxicants 76
    Pathogens 78
Receiving Water Effect Summary 90
References 92

EFFECTS OF RUNOFF ON RECEIVING WATERS

Many studies have shown the severe detrimental effects of urban and agricultural runoff on receiving waters. These studies have generally examined receiving water conditions above and below a city, by comparing two parallel streams, or by comparing to an ecoregion reference. However, only a few studies have examined direct cause-and-effect relationships of runoff for receiving water aquatic organisms (Heaney and Huber 1984; Burton and Moore 1999; Werner et al. 2000; Vlaming et al. 2000; Bailey et al. 2000; Wenholz and Crunkilton 1995). Chapter 4 presents several case studies representing the major approaches to assessing receiving water problems, while this chapter presents a review of the major stressor categories and summarizes their observed effects.

Indicators of Receiving Water Biological Effects and Analysis Methodologies

There are a number of useful, well-proven tools that can detect adverse biological effects in receiving waters (see also Chapter 6). When these tools are used correctly and combined in the proper framework, they can be used to identify runoff-related problems. Kuehne (1975) studied the usefulness of aquatic organisms as indicators of pollution. He found that invertebrate responses are indicative of pollution for some time after an event, but they may not give an accurate indication of the nature of the pollutants. In-stream fish studies were not employed as biological indicators much before 1975, but they are comparable in many ways to invertebrates as quality indicators and can be more easily identified. However, because of better information pertaining to invertebrates and due to their limited mobility, certain invertebrate species may be sensitive to minor changes in water quality. Fish can be highly mobile and cover large sections of a stream, as long as their passage is not totally blocked by adverse conditions. Fish disease surveys were also used during the Bellevue, WA, urban runoff studies as an indicator of water quality problems (Scott et al. 1982; Pitt and Bissonnette 1984). McHardy et al. (1985) examined heavy metal uptake in green algae (*Cladophora glomerata*) from urban runoff for use as a biological monitor of specific metals.

It is necessary to use a range of measurement endpoints to characterize ecosystem quality in systems that receive multiple stressors (Marcy and Gerritsen 1996; Baird and Burton 2001). Dyer and White (1996) examined the problem of multiple stressors affecting toxicity assessments. They felt that field surveys can rarely be used to verify simple single parameter laboratory experiments. They developed a watershed approach integrating numerous databases in conjunction with *in situ* biological observations to help examine the effects of many possible causative factors (see also Chapter 6).

The interactions of stressors such as suspended solids and chemicals can be confounding and easily overlooked. Ireland et al. (1996) found that exposure to UV radiation (natural sunlight) increased the toxicity of PAH-contaminated sediments to *C. dubia*. The toxicity was removed when the UV wavelengths did not penetrate the water column to the exposed organisms. Toxicity was also reduced significantly in the presence of UV when the organic fraction of the stormwater was removed. Photo-induced toxicity occurred frequently during low flow conditions and wet-weather runoff and was reduced during turbid conditions.

Johnson et al. (1996) and Herricks et al. (1996a,b) describe a structured tier testing protocol to assess both short-term and long-term wet-weather discharge toxicity that they developed and tested. The protocol recognizes that the test systems must be appropriate to the time-scale of exposure during the discharge. Therefore, three time-scale protocols were developed, for intra-event, event, and long-term exposures. The use of standard whole effluent toxicity (WET) tests were found to overestimate the potential toxicity of stormwater discharges.

The effects of stormwater on Lincoln Creek, near Milwaukee, WI, were described by Crunkilton et al. (1996). Lincoln Creek drains a heavily urbanized watershed of 19 mi^2 that is about 9 miles long. On-site toxicity testing was conducted with side-stream flow-through aquaria using fathead minnows, plus in-stream biological assessments, along with water and sediment chemical measurements. In the basic tests, Lincoln Creek water was continuously pumped through the test tanks, reflecting the natural changes in water quality during both dry and wet-weather conditions. The continuous flow-through mortality tests indicated no toxicity until after about the 14th day of exposure, with more than 80% mortality after about 25 days, indicating that short-term toxicity tests likely underestimate stormwater toxicity. The biological and physical habitat assessments supported a definitive relationship between degraded stream ecology and urban runoff.

Rainbow (1996) presented a detailed overview of heavy metals in aquatic invertebrates. He concluded that the presence of a metal in an organism cannot tell us directly whether that metal is poisoning the organism. However, if compared to concentrations in a suite of well-researched biomonitors, it may be possible to determine if the accumulated concentrations are atypically high, with a possibility that toxic effects may be present. The user should be cautious, however, when attempting to relate tissue concentrations to effects or with bioconcentration factors. Many metals are essential and/or regulated by organisms and their internal concentrations might bear no relationship to the concentrations in surrounding waters or sediments.

A battery of laboratory and *in situ* bioassay tests are most useful when determining aquatic biota problems (Burton and Stemmer 1988; Burton et al. 1996; Chapter 6). The test series may include microbial activity tests, along with exposures of zooplankton, amphipods, aquatic insects, bivalves, and fish. Indigenous microbial activity responses correlated well with *in situ* biological and chemical profiles. Bascombe et al. (1990) also reported on the use of *in situ* biological tests, using an amphipod exposed for 5 to 6 weeks in urban streams, to examine urban runoff receiving water effects. Ellis et al. (1992) examined bioassay procedures for evaluating urban runoff effects on receiving water biota. They concluded that an acceptable criteria for protecting receiving water organisms should not only provide information on concentration and exposure relationships for *in situ* bioassays, but also consider body burdens, recovery rates, and sediment-related effects.

During the Coyote Creek, San Jose, CA, receiving water study, 41 stations were studied in both urban and non-urban perennial flow stretches of the creek. Short- and long-term sampling techniques were used to evaluate the effects of urban runoff on water quality, sediment properties, fish, macroinvertebrates, attached algae, and rooted aquatic vegetation (Pitt and Bozeman 1982).

Fish Kills and Advisories

Runoff impacts are sometimes difficult for many people to appreciate in urban and agricultural areas. Fish kills are the most obvious indication of water quality problems for many people. However, because receiving water quality is often so poor, the aquatic life in typical urban and agricultural receiving waters is usually limited in abundance and diversity, and quite resistant to poor water quality. Sensitive native organisms have typically been displaced, or killed, long ago, and it usually requires an unusual event to cause a fish fill (Figure 3.1). Ray and White (1979) stated that one of the complicating factors in determining fish kills related to heavy metals is that the fish mortality may lag behind the first toxic exposure by several days and is usually detected many miles downstream from the discharge location. The actual concentrations of the water quality constituents that may have caused the kill could then be diluted beyond detection limits, making probable sources of the toxic materials impossible to determine in many cases.

Heaney et al. (1980) reviewed fish kill information reported to government agencies from 1970 to 1979. They found that less than 3% of the reported 10,000 fish kills was identified as having been caused by urban runoff. This is fewer than 30 fish kills per year nationwide. However, the cause of most of these 10,000 fish kills could not be identified. It is expected that many of these fish kills could have been caused by runoff, or a combination of problems that could have been worsened by runoff. For example, elevated nutrient loading causes eutrophication that may lead to dissolved oxygen deficits and subsequent fish kills. These events are exacerbated by natural stressors such as low flow conditions. More recent surveys have found nearly 30% of fish kills is attributable to runoff (Figure 3.2; EPA 1995).

During the Bellevue, WA, receiving water studies, some fish kills were noted in the unusually clean urban streams (Pitt and Bissonnette 1984). The fish kills were usually associated with inappropriate discharges to the storm drainage system (such as cleaning materials and industrial chemical spills) and not from "typical" urban runoff. However, as noted later, the composition of the fish in the Bellevue urban stream was quite different, as compared to the control stream (Scott et al. 1986).

Figure 3.1   Fish kill in Village Creek, Birmingham, AL, due to Dursban entering storm drainage during warehouse fire.

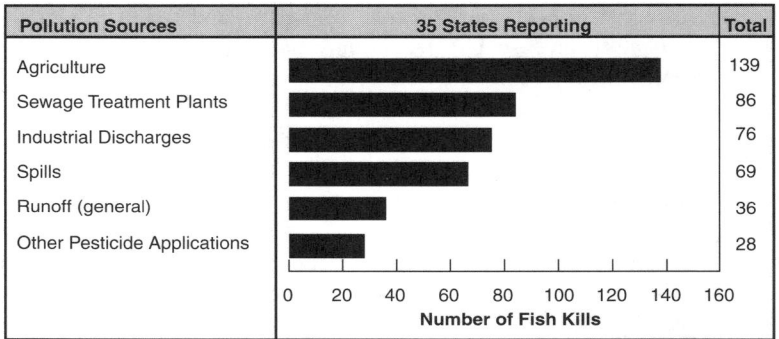

Pollution Sources	35 States Reporting	Total
Agriculture		139
Sewage Treatment Plants		86
Industrial Discharges		76
Spills		69
Runoff (general)		36
Other Pesticide Applications		28

Number of Fish Kills

Figure 3.2   Sources associated with fish kills. (From U.S. Environmental Protection Agency. *National Water Quality Inventory. 1994 Report to Congress*. Office of Water. EPA 841-R-95-005. Washington, D.C. December 1995.)

Fish kill data have, therefore, not been a good indicator for identifying stressor categories or types. However, the composition of the fisheries and other aquatic life taxonomic information are sensitive indicators of receiving water problems in streams.

In addition to fish kills, a significant concern is the increasing number of fish advisories being issued by states across the nation (Figure 3.3; EPA 1995). The causes of fish contamination and fish kills vary, but runoff is a primary contributor.

Adverse Aquatic Life Effects Caused by Runoff

Aquatic organisms are sensitive indicators of water quality. There have been many studies that describe aquatic life impairments that may result from exposure to contaminated runoff and/or habitat degradation. The following section summarizes some of these studies, which are typical of urban and agricultural watersheds.

Klein (1979) studied 27 small watersheds having similar characteristics, but having varying land uses, in the Piedmont region of Maryland. During an initial phase of the study, definite relationships were found between water quality and land use. Subsequent study phases examined aquatic life relationships in the watersheds. The principal finding was that stream aquatic life problems were first identified with watersheds having imperviousness areas comprising at least 12% of the watershed. Severe problems were noted after the imperviousness quantities reached 30%.

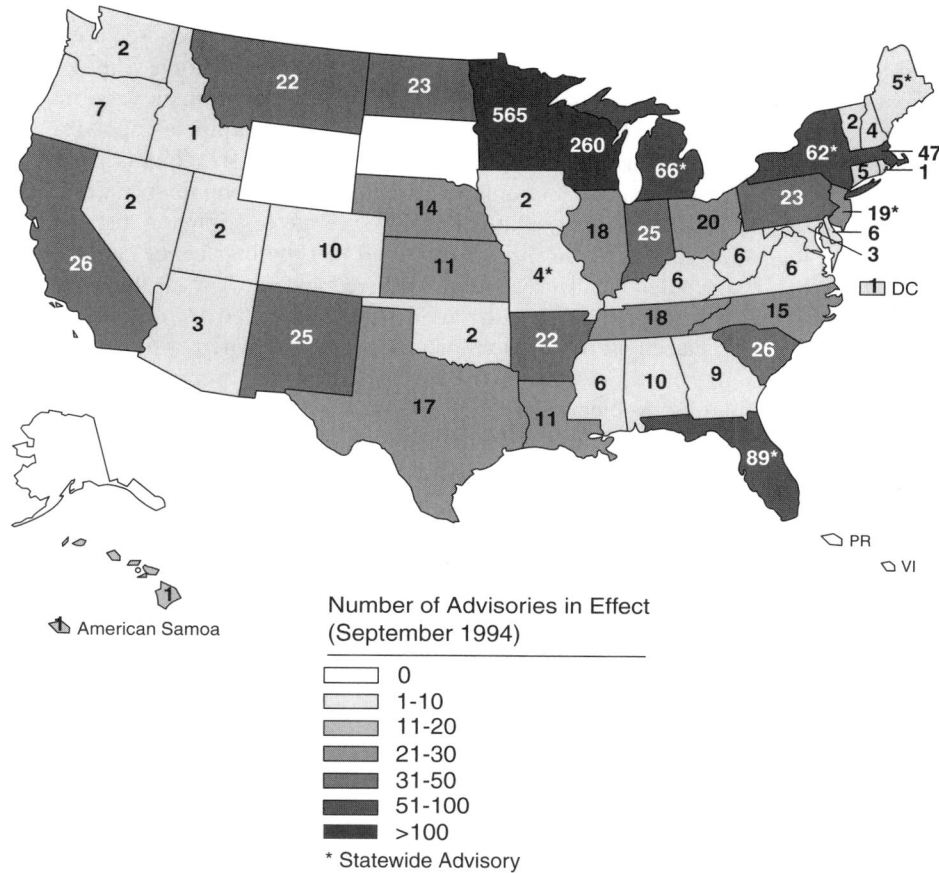

Figure 3.3   U.S. fish consumption advisories. **Note:** States that perform routine fish tissue analysis (such as Great Lake States) will detect more cases of fish contamination and issue more advisories than states with less rigorous fish sampling programs. In many cases, the states with the most fish advisories support the best monitoring programs for measuring toxic contamination in fish, and their water quality is no worse than the water quality in other states. (From U.S. Environmental Protection Agency. *National Water Quality Inventory. 1994 Report to Congress.* Office of Water. EPA 841-R-95-005. Washington, D.C. December 1995.)

Receiving water impact studies were also conducted in North Carolina by Lenat et al. (1979), Lenat and Eagleson (1981), and Lenat et al. (1981). The benthic fauna occurred mainly on rocks. As sedimentation increased, the amount of exposed rocks decreased, with a decreasing density of benthic macroinvertebrates. Data from 1978 and 1979 in five cities showed that urban streams were grossly polluted by a combination of toxicants and sediment. Chemical analyses, without biological analyses, would have underestimated the severity of the problems because the water column quality varied rapidly, while the major problems were associated with sediment quality and effects on macroinvertebrates. Macroinvertebrate diversities were severely reduced in the urban streams, compared to the control streams. The biotic indices indicated "very poor" conditions for all urban streams. Occasionally, high populations of pollutant-tolerant organisms were found in the urban streams, but would abruptly disappear before subsequent sampling efforts. This was probably caused by intermittent discharges of spills or illegal dumping of toxicants. Although the cities studied were located in different geographic areas of North Carolina, the results were remarkably uniform.

A major nonpoint runoff receiving water impact research program was conducted in Georgia (Cook et al. 1983). Several groups of researchers examined streams in major areas of the state.

Benke et al. (1981) studied 21 stream ecosystems near Atlanta having watersheds of 1 to 3 square miles each and land uses ranging from 0 to 98% urbanization. They measured stream water quality but found little relationship between water quality and degree of urbanization. The water quality parameters also did not identify a major degree of pollution. In contrast, there were major correlations between urbanization and the number of species. They had problems applying diversity indices to their study because the individual organisms varied greatly in size (biomass). CTA (1983) also examined receiving water aquatic biota impacts associated with nonpoint sources in Georgia. They studied habitat composition, water quality, macroinvertebrates, periphyton, fish, and toxicant concentrations in the water, sediment, and fish. They found that the impacts of land use were the greatest in the urban basins. Beneficial uses were impaired or denied in all three urban basins studied. Fish were absent in two of the basins and severely restricted in the third. The native macroinvertebrates were replaced with pollution-tolerant organisms. The periphyton in the urban streams were very different from those found in the control streams and were dominated by species known to create taste and odor problems.

Pratt et al. (1981) used basket artificial substrates to compare benthic population trends along urban and nonurban areas of the Green River in Massachusetts. The benthic community became increasingly disrupted as urbanization increased. The problems were not only associated with times of heavy rain, but seemed to be affected at all times. The stress was greatest during summer low flow periods and was probably localized near the stream bed. They concluded that the high degree of correspondence between the known sources of urban runoff and the observed effects on the benthic community was a forceful argument that urban runoff was the causal agent of the disruption observed.

Cedar swamps in the New Jersey Pine Barrens were studied by Ehrenfeld and Schneider (1983). They examined 19 swamps subjected to varying amounts of urbanization. Typical plant species were lost and replaced by weeds and exotic plants in urban runoff-affected swamps. Increased uptakes of phosphorus and lead in the plants were found. It was concluded that the presence of stormwater runoff to the cedar swamps caused marked changes in community structure, vegetation dynamics, and plant tissue element concentrations.

Medeiros and Coler (1982) and Medeiros et al. (1984) used a combination of laboratory and field studies to investigate the effects of urban runoff on fathead minnows. Hatchability, survival, and growth were assessed in the laboratory in flow-through and static bioassay tests. Growth was reduced to one half of the control growth rates at 60% dilutions of urban runoff. The observed effects were believed to be associated with a combination of toxicants.

The benthos in the upper reaches of Coyote Creek (San Jose, CA) consisted primarily of amphipods and a diverse assemblage of aquatic insects (Pitt and Bozeman 1982). Together those groups comprised two thirds of the benthos collected from the non-urban portion of the creek. Clean water forms were abundant and included amphipods (*Hyaella azteca*) and various genera of mayflies, caddisflies, black flies, crane flies, alderflies, and riffle beetles. In contrast, the benthos of the urban reaches of the creek consisted almost exclusively of pollution-tolerant oligochaete worms (tubificids). Tubificids accounted for 97% of the benthos collected from the lower portion of Coyote Creek.

There were significant differences in the numbers and types of benthic organisms found during the Bellevue Urban Runoff Program (Pederson 1981; Perkins 1982; Richey et al. 1981; Richey 1982; Scott et al. 1982). Mayflies, stoneflies, caddisflies, and beetles were rarely observed in urbanized Kelsey Creek, but were quite abundant in rural Bear Creek. These organisms are commonly regarded as sensitive indicators to environmental degradation. As an example of a degraded aquatic habitat, a species of clams (*Unionidae*) was not found in Kelsey Creek, but was found in Bear Creek. These clams are very sensitive to heavy siltation and unstable sediments. Empty clam shells, however, were found buried in the Kelsey Creek sediments indicating their previous presence in the creek and their inability to adjust to the changing conditions. The benthic organism composition in Kelsey Creek varied radically with time and place, while the organisms were much more stable in Bear Creek.

Introduced fishes often cause radical changes in the nature of the fish fauna present in a given water body. In many cases, they become the dominant fishes because they are able to outcompete the native fishes for food or space, or they may possess greater tolerance to environmental stress. In general, introduced species are most abundant in aquatic habitats modified by man, while native fishes tend to persist mostly in undisturbed areas. Such is apparently the case within Coyote Creek, San Jose, CA (Pitt and Bozeman 1982).

Samples from the non-urban portion of the study area were dominated by an assemblage of native fish species such as hitch, three spine stickleback, Sacramento sucker, and prickly sculpin. Rainbow trout, riffle sculpin, and Sacramento squawfish were captured only in the headwater reaches and tributary streams of Coyote Creek. Collectively, native species comprised 89% of the number and 79% of the biomass of the 2379 fishes collected from the upper reaches of the study area. In contrast, native species accounted for only 7% of the number and 31% of the biomass of the 2899 fishes collected from the urban reach of the study area.

Hitch was the most numerous native fish species present. Hitch generally exhibit a preference for quiet water habitat and are characteristic of warm, low elevation lakes, sloughs, sluggish rivers, and ponds. Mosquitofish dominated the collections from the urbanized section of the creek and accounted for over two thirds of the total number of fish collected from the area. This fish is particularly well adapted to withstand extreme environmental conditions, including those imposed by stagnant waters with low dissolved oxygen concentrations and elevated temperatures. The second most abundant fish species in the urbanized reach of Coyote Creek, the fathead minnow, is equally well suited to tolerate extreme environmental conditions. The species can withstand low dissolved oxygen, high temperature, high organic pollution, and high alkalinity. Often thriving in unstable environments such as intermittent streams, the fathead minnow can survive in a wide variety of habitats.

The University of Washington (Pederson 1981; Perkins 1982; Richey et al. 1981; Richey 1982; Scott et al. 1982) conducted a series of studies to contrast the biological and chemical conditions in urban Kelsey Creek with rural Bear Creek. The urban creek was significantly degraded when compared to the rural creek, but still supported a productive but limited and unhealthy salmonid fishery. Many of the fish in the urban creek, however, had respiratory anomalies. The urban creek was not grossly polluted, but flooding from urban developments has increased dramatically in recent years. These increased flows have dramatically changed the urban stream's channel, by causing unstable conditions with increased stream bed movement, and by altering the availability of food for the aquatic organisms. The aquatic organisms are very dependent on the few relatively undisturbed reaches. Dissolved oxygen concentrations in the sediments depressed embryo salmon survival in the urban creek. Various organic and metallic priority pollutants were discharged to the urban creek, but most of them were apparently carried through the creek system by the high storm flows to Lake Washington. The urbanized Kelsey Creek also had higher water temperatures (probably due to reduced shading) than Bear Creek. This probably caused the faster fish growth in Kelsey Creek.

The fish population in Kelsey Creek had adapted to its degrading environment by shifting the species composition from coho salmon to less sensitive cutthroat trout and by making extensive use of less-disturbed refuge areas (Figure 4.22). Studies of damaged gills found that up to three fourths of the fish in Kelsey Creek were affected with respiratory anomalies, while no cutthroat trout and only two of the coho salmon sampled in Bear Creek had damaged gills. Massive fish kills in Kelsey Creek and its tributaries were observed on several occasions during the project due to the dumping of toxic materials down the storm drains.

Urban runoff impact studies were conducted in the Hillsborough River near Tampa Bay, FL, as part of the NURP program (Mote Marine Laboratory 1984). Plants, animals, sediment, and water quality were all studied in the field and supplemented by laboratory bioassay tests. Effects of saltwater intrusion and urban runoff were both measured because of the estuarine environment. During wet weather, freshwater species were found closer to the bay than during dry weather. In coastal areas, these additional natural factors make it even more difficult to identify the

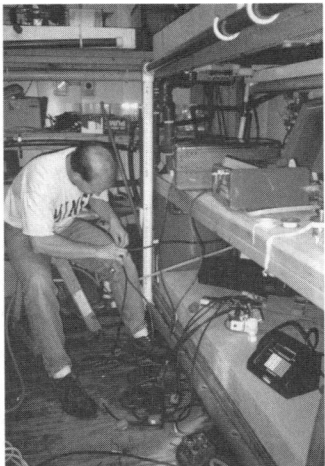

Figure 3.4 Installation of side-stream fish bioassay test facilities at Lincoln Creek, Milwaukee, WI.

Figure 3.5 Lincoln Creek side-stream fish bioassay test facilities nearing completion.

cause-and-effect relationships for aquatic life problems. During another NURP project, Striegl (1985) found that the effects of accumulated pollutants in Lake Ellyn (Glen Ellyn, IL) inhibited desirable benthic invertebrates and fish and increased undesirable phytoplankton blooms. LaRoe (1985) summarized the off-site effects of construction sediment on fish and wildlife. He noted that physical, chemical, and biological processes all affect receiving water aquatic life.

The number of benthic organism taxa in Shabakunk Creek in Mercer County, NJ, declined from 13 in relatively undeveloped areas to 4 below heavily urbanized areas (Garie and McIntosh 1986, 1990). Periphyton samples were also analyzed for heavy metals, with significantly higher metal concentrations found below the heavily urbanized area than above.

The Wisconsin Department of Natural Resources, in conjunction with the USGS and the University of Wisconsin, conducted side-stream fish bioassay tests in Lincoln Creek in Milwaukee (Figures 3.4 and 3.5) (Crunkilton et al. 1996). They identified significant acute toxicity problems associated with intermediate-term (about 10 to 20 day) exposures to adverse toxicant concentrations in urban receiving streams, with no indication of toxicity for shorter exposures. These toxicity effects were substantially (but not completely) reduced through the removal of stormwater particulates using a typical wet detention pond designed to remove most of the particles larger than 5 μm.

Observed Habitat Problems Caused by Runoff

Some of the most serious effects of urban and agricultural runoff are on the aquatic habitat of the receiving waters. These habitat effects are in addition to the pollutant concentration effects. The major effects of sediment on the aquatic habitat include silting of spawning and food production areas and unstable bed conditions (Cordone and Kelley 1961). Other major habitat destruction problems include rapidly changing flows and the absence of refuge areas to protect the biota during these flow changes. Removal of riparian vegetation can increase water temperatures and eliminate a major source of debris, which provides important refuge areas. The major source of these habitat problems is the increased discharge volumes and flow rates associated with stormwater in developing areas that cause significant enlargements and unstable banks of small and moderate sized streams (Figures 3.6 and 3.7). Other habitat problems are caused by attempts to "correct" these problems by construction of lined channels (Figures 3.8 and 3.9) or small drop structures which hinder migration of aquatic life and create areas for the accumulation of contaminated silt (Figure 3.10).

Figure 3.6 Creek blowout after initial significant spring rains in newly developed area. (Courtesy of Wisconsin Department of Natural Resources.)

Figure 3.7 Unstable banks and trash along Five-Mile Creek, Birmingham, AL.

Figure 3.8 Lined embankment along Waller Creek, Austin, TX.

Figure 3.9 Lined channel in Milwaukee, WI.

Schueler (1996) stated that channel geometry stability can be a good indicator of the effectiveness of stormwater control practices. He also found that once a watershed area has more than about 10 to 15% effective impervious cover, noticeable changes in channel morphology occur, along with quantifiable impacts on water quality and biological conditions. Stephenson (1996) studied changes in streamflow volumes in South Africa during urbanization. He found increased stormwater runoff, decreases in the groundwater table, and dramatically decreased times of concentration. The peak flow rates increased by about twofold, about half caused by increased pavement (in an area having only about 5% effective impervious cover), with the remainder caused by decreased times of concentration.

Figure 3.10 Small drop structure obstruction in Lincoln Creek, Milwaukee, WI.

Brookes (1988) has documented many cases in the United States and Great Britain of stream morphological changes associated with urbanization. These changes are mostly responsible for habitat destruction which is usually the most significant detriment to aquatic life. In many cases, water quality improvement would result in very little aquatic life benefit if the physical habitat is grossly modified. The most obvious habitat problems are associated with stream "improvement" projects, ranging from removal of debris, to straightening streams, to channelization projects. Brookes (1988, 1991) presents a number of ways to minimize habitat problems associated with stream channel projects, including stream restoration.

Wolman and Schick (1967) observed deposition of channel bars, erosion of channel banks, obstruction of flows, increased flooding, shifting of channel bottoms, along with concurrent changes in the aquatic life, in Maryland streams affected by urban construction activities. Robinson (1976) studied eight streams in watersheds undergoing urbanization and found that the increased magnitudes and frequencies of flooding, along with the increased sediment yields, had considerable impact on stream morphology (and therefore aquatic life habitat).

The aquatic organism differences found during the Bellevue Urban Runoff Program were probably most associated with the increased peak flows in Kelsey Creek caused by urbanization and the resultant increase in sediment-carrying capacity and channel instability of the creek (Pederson 1981; Perkins 1982; Richey et al. 1981; Richey 1982; Scott et al. 1982). Developed Kelsey Creek had much lower flows than rural Bear Creek during periods between storms. About 30% less water was available in Kelsey Creek during the summers. These low flows may also have significantly affected the aquatic habitat and the ability of the urban creek to flush toxic spills or other dry-weather pollutants from the creek system (Ebbert et al. 1983; Prych and Ebbert undated). Kelsey Creek had extreme hydrologic responses to storm. Flooding substantially increased in Kelsey Creek during the period of urban development; the peak annual discharges have almost doubled in the last 30 years, and the flooding frequency has also increased due to urbanization (Ebbert et al. 1983; Prych and Ebbert undated). These increased flows in urbanized Kelsey Creek resulted in greatly increased sediment transport and channel instability. The Bellevue studies (summarized by Pitt and Bissonnette 1984) indicated very significant interrelationships between the physical, biological, and chemical characteristics of the urbanized Kelsey Creek system. The aquatic life beneficial uses were found to be impaired, and stormwater conveyance was most likely associated with increased flows from the impervious areas in the urban area. Changes in the flow characteristics could radically alter the ability of the stream to carry the polluted sediments into the other receiving waters. If the stream power (directly related to sediment-carrying capacity) of Kelsey Creek were reduced, these toxic materials could be expected to be settled into its sediment, with increased effects on the stream's aquatic life. Reducing peak flows would also reduce the flushing of smaller fish and other aquatic organisms from the system.

Many recent studies on urban stream habitats and restoration efforts have been conducted, especially in the Pacific Northwest. In one example, May et al. (1999) found that maintaining natural land cover offers the best protection for maintaining stream ecological integrity and that best management practices have generally been limited in their ability to preserve appropriate conditions for lowland salmon spawning and rearing streams. They found that Puget Sound watersheds having a 10% impervious cover (likely resulting in marginal in-stream conditions) maintained at least 50% forested cover.

Groundwater Impacts from Stormwater Infiltration

There have been some nationwide studies that have shown virtually every agricultural and urban watershed contains elevated levels of nutrients, pesticides, and other organic chemicals in surface and groundwaters, sediments, and fish tissues (e.g., USGS 1999). Since groundwaters are widely used as a drinking water and irrigation source and recharge many surface water bodies, the implications of chemical contamination are quite serious.

Prior to urbanization, groundwater recharge resulted from infiltration of precipitation through pervious surfaces, including grasslands and woods. This infiltrating water was relatively uncontam-

Figure 3.11 Groundwater recharge basin in Long Island, NY, using stormwater. (Courtesy of New York Department of USGS).

Figure 3.12 Karst geology at an Austin, TX, roadcut showing major channeling opportunities for surface water to enter the Edwards Aquifer.

Figure 3.13 Public education roadside sign in Austin, TX, warning about sensitive recharge zone.

Figure 3.14 Paver blocks for on-site infiltration in Essen, Germany.

inated. Urbanization reduced the permeable soil surface area through which recharge by infiltration could occur. This resulted in much less groundwater recharge and greatly increased surface runoff. In addition, the waters available for recharge generally carried increased quantities of pollutants. With urbanization, new sources of groundwater recharge also occurred, including recharge from domestic septic tanks, percolation basins (Figure 3.11), and industrial waste injection wells, and from agricultural and residential irrigation. Special groundwater contamination problems may occur in areas having Karst geology where surface waters can be easily and quickly directed to the subsurface (Figures 3.12 and 3.13). Of course, there are many less dramatic opportunities for stormwater to enter the groundwater, including areas of porous paver blocks (Figures 3.14 through 3.16), grass swales (Figures 3.17 and 3.18), infiltration trenches (Figure 3.19), biofiltration areas (Figure 3.20), and simply from runoff flowing across grass (Figure 3.21). Many of these infiltration practices are done to reduce surface water impacts associated with stormwater discharges. If the infiltration is conducted through surface soils (such as for grass swales and grass landscaped areas), groundwater contamination problems are significantly reduced. However, if subsurface infiltration is used (especially through the use of injection wells), the risk of groundwater contamination for many stormwater pollutants substantially increases (Pitt et al. 1994, 1996).

Figure 3.15 Paver blocks for emergency and utility vehicle access, Madison, WI (under construction).

Figure 3.16 Paver blocks for occasional access road, Seattle Science Center, WA.

Figure 3.17 Grass swale in residential area, Milwaukee, WI.

Figure 3.18 Grass swale in office park area, Milwaukee, WI.

The Technical University of Denmark (Mikkelsen et al. 1996a,b) has been involved in a series of tests to examine the effects of stormwater infiltration on soil and groundwater quality. It found that heavy metals and PAHs present little groundwater contamination threat if surface infiltration systems are used. However, it expresses concern about pesticides, which are much more mobile. Squillace et al. (1996) along with Zogorski et al. (1996) presented information concerning stormwater and its potential as a source of groundwater MTBE contamination. Mull (1996) stated that traffic areas are the third most important source of groundwater contamination in Germany (after abandoned industrial sites and leaky sewers). The most important contaminants are chlorinated hydrocarbons, sulfate, organic compounds, and nitrates. Heavy metals are generally not an important groundwater contaminant because of their affinity for soils. Trauth and Xanthopoulus (1996) examined the long-term trends in groundwater quality at Karlsruhe, Germany. They found that urban land use is having a long-term influence on the groundwater quality. The concentration of many pollutants has increased by about 30 to 40% over 20 years. Hütter and Remmler (1996)

Figure 3.19 Stormwater infiltration through infiltration trench, office park, Lake Oswego, OR.

Figure 3.20 Biofiltration in parking area (Photo used with permission of Center for Watershed Protection.)

describe a groundwater monitoring plan, including monitoring wells that were established during the construction of an infiltration trench for stormwater disposal, in Dortmund, Germany. The worst problem expected is with zinc if the infiltration water has a pH value of 4.

The following paragraphs (summarized from Pitt et al. 1994, 1996) describe the stormwater pollutants that have the greatest potential of adversely affecting groundwater quality during inadvertent or intentional stormwater infiltration, along with suggestions on how to minimize these potential problems.

Nutrients

Groundwater contamination with phosphorus has not been as widespread, or as severe, as with nitrogen compounds. Nitrates are one of the most

Figure 3.21 Infiltration through grassed areas.

frequently encountered contaminants in groundwater. Whenever nitrogen-containing compounds come into contact with soil, a potential for nitrate leaching into groundwater exists, especially in rapid-infiltration wastewater basins, stormwater infiltration devices, and in agricultural areas. Nitrate has leached from fertilizers and affected groundwaters under various turf grasses in urban areas, including golf courses, parks, and home lawns. Significant leaching of nitrates occurs during the cool, wet seasons. Cool temperatures reduce denitrification and ammonia volatilization, and limit microbial nitrogen immobilization and plant uptake. The use of slow-release fertilizers is recommended in areas having potential groundwater nitrate problems. The slow-release fertilizers include urea formaldehyde (UF), methylene urea, isobutyldiene diurea (IBDU), and sulfur-coated urea. Residual nitrate concentrations are highly variable in soil due to soil texture, mineralization, rainfall and irrigation patterns, organic matter content, crop yield, nitrogen fertilizer/sludge rate, denitrification, and soil compaction. Nitrate is highly soluble (>1 kg/L) and will stay in solution in the percolation water, after leaving the root zone, until it reaches the groundwater.

Nitrate has a low to moderate groundwater contamination potential for both surface percolation and subsurface infiltration/injection practices because of its relatively low concentrations found in most stormwaters. However, if the stormwater nitrate concentration were high, then the groundwater contamination potential would also likely be high.

Pesticides

Pesticide contamination of groundwater can result from agricultural, municipal, and homeowner use of pesticides for pest control and their subsequent collection in stormwater runoff. A wide range of pesticides and their metabolites have been found in watersheds, which include typical urban pesticides in agricultural areas, and vice versa. This cross-contamination of pesticides into areas in which they are not being used is attributed to atmospheric transport. Heavy repetitive use of mobile pesticides on irrigated and sandy soils likely contaminates groundwater. Some insecticides, fungicides, and nematocides must be mobile in order to reach the target pest and, hence, they generally have the highest contamination potential. Pesticide leaching depends on patterns of use, soil texture, total organic carbon content of the soil, pesticide persistence, and depth to the water table.

The greatest pesticide mobility occurs in areas with coarse-grained or sandy soils without a hardpan layer, having low clay and organic matter content and high permeability. Structural voids, which are generally found in the surface layer of finer-textured soils rich in clay, can transmit pesticides rapidly when the voids are filled with water and the adsorbing surfaces of the soil matrix are bypassed. In general, pesticides with low water solubilities, high octanol-water partitioning coefficients, and high carbon partitioning coefficients are less mobile. The slower-moving pesticides have been recommended in areas of groundwater contamination concern. These include the fungicides iprodione and triadimefon, the insecticides isofenphos and chlorpyrifos, and the herbicide glyphosate. The most mobile pesticides include 2,4-D, acenaphthylene, alachlor, atrazine, cyanazine, dacthal, diazinon, dicamba, malathion, and metolachlor.

Pesticides decompose in soil and water, but the total decomposition time can range from days to years. Literature half-lives for pesticides generally apply to surface soils and do not account for the reduced microbial activity found deep in the vadose zone. Pesticides with a 30-day half-life can show considerable leaching. An order-of-magnitude difference in half-life results in a five- to tenfold difference in percolation loss. Organophosphate pesticides are less persistent than organochlorine pesticides, but they also are not strongly adsorbed by the sediment and are likely to leach into the vadose zone and the groundwater. Perhaps a greater concern that has recently emerged is the widespread prevalence of toxic pesticide metabolites (breakdown products) that are not routinely analyzed. The ecological and human health significance of this is not known at present, but will be a future topic of investigation.

Lindane and chlordane have moderate groundwater contamination potentials for surface percolation practices (with no pretreatment) and for subsurface injection (with minimal pretreatment). The groundwater contamination potentials for both of these compounds would likely be substantially reduced with adequate sedimentation pretreatment. Pesticides have mostly been found in urban runoff from residential areas, especially in dry-weather flows associated with landscaping irrigation runoff.

Other Organics

The most commonly occurring organic compounds that have been found in urban groundwaters include phthalate esters (especially bis(2-ethylhexyl)phthalate) and phenolic compounds. Other organics more rarely found, possibly due to losses during sample collection, have included the volatiles: benzene, chloroform, methylene chloride, trichloroethylene, tetrachloroethylene, toluene, and xylene. PAHs (especially benzo(a)anthracene, chrysene, anthracene, and benzo(b)fluoroan-thenene) have also been found in groundwaters near industrial sites.

Groundwater contamination from organics, as from other pollutants, occurs more readily in areas with sandy soils and where the water table is near the land surface. Removal of organics from the soil and recharge water can occur by one of three methods: volatilization, sorption, and degradation. Volatilization can significantly reduce the concentrations of the most volatile com-pounds in groundwater, but the rate of gas transfer from the soil to the air is usually limited by the presence of soil water. Hydrophobic sorption onto soil organic matter limits the mobility of less soluble base/neutral and acid extractable compounds through organic soils and the vadose zone. Sorption is not always a permanent removal mechanism, however. Organic resolubilization can occur during wet periods following dry periods. Many organics can be at least partially degraded by microorganisms, but others cannot. Temperature, pH, moisture content, ion-exchange capacity of soil, and air availability may limit the microbial degradation potential for even the most degrad-able organic.

1,3-Dichlorobenzene may have a high groundwater contamination potential for subsurface infiltration/injection (with minimal pretreatment). However, it would likely have a lower ground-water contamination potential for most surface percolation practices because of its relatively strong sorption to vadose zone soils. Both pyrene and fluoranthene would also likely have high ground-water contamination potentials for subsurface infiltration/injection practices, but lower contami-nation potentials for surface percolation practices because of their more limited mobility through the unsaturated zone (vadose zone). Others (including benzo(a)anthracene, bis(2-ethylhexyl) phtha-late, pentachlorophenol, and phenanthrene) may also have moderate groundwater contamination potentials if surface percolation with no pretreatment or subsurface injection/infiltration is used. These compounds would have low groundwater contamination potentials if surface infiltration was used with sedimentation pretreatment. Volatile organic compounds (VOCs) may also have high groundwater contamination potentials if present in the stormwater (likely for some industrial and commercial facilities and vehicle service establishments). The other organics, especially the vol-atiles, are mostly found in industrial areas. The phthalates are found in all areas. The PAHs are also found in runoff from all areas, but they are in higher concentrations and occur more frequently in industrial areas.

Pathogenic Microorganisms

Viruses have been detected in groundwater where stormwater recharge basins are located short distances above the aquifer. Enteric viruses are more resistant to environmental factors than enteric bacteria and they exhibit longer survival times in natural waters. They can occur in potable and marine waters in the absence of fecal coliforms. Enteroviruses are also more resistant to commonly used disinfectants than are indicator bacteria, and can occur in groundwater in the absence of indicator bacteria.

The factors that affect the survival of enteric bacteria and viruses in the soil include pH, antagonism from soil microflora, moisture content, temperature, sunlight, and organic matter. The two most important attributes of viruses that permit their long-term survival in the environment are their structure and very small size. These characteristics permit virus occlusion and protection within colloid-size particles. Viral adsorption is promoted by increasing cation concentration, decreasing pH, and decreasing soluble organics. Since the movement of viruses through soil to groundwater occurs in the liquid phase and involves water movement and associated suspended virus particles, the distribution of viruses between the adsorbed and liquid phases determines the viral mass available for movement. Once the virus reaches the groundwater, it can travel laterally through the aquifer until it is either adsorbed or inactivated.

The major bacterial removal mechanisms in soil are straining at the soil surface and at intergrain contacts, sedimentation, sorption by soil particles, and inactivation. Because they are larger than viruses, most bacteria are retained near the soil surface due to this straining effect. In general, enteric bacteria survive in soil for 2 to 3 months, although survival times up to 5 years have been documented.

Enteroviruses likely have a high groundwater contamination potential for all percolation practices and subsurface infiltration/injection practices, depending on their presence in stormwater (likely, if contaminated with sanitary sewage). Other pathogens, including *Shigella*, *Pseudomonas aeruginosa*, and various protozoa, would also have high groundwater contamination potentials if subsurface infiltration/injection practices are used without disinfection. If disinfection (especially by chlorine or ozone) is used, then disinfection by-products (such as trihalomethanes or ozonated bromides) would have high groundwater contamination potentials. Pathogens are most likely associated with sanitary sewage contamination of storm drainage systems, but several bacterial pathogens are commonly found in surface runoff in residential areas.

Heavy Metals and Other Inorganic Compounds

The heavy metals and other inorganic compounds in stormwater of most environmental concern, from a groundwater pollution standpoint, are chromium, copper, lead, nickel, and zinc. However, the majority of metals, with the consistent exception of zinc, are mostly found associated with the particulate solids in stormwaters and are thus relatively easily removed through sedimentation practices. Filterable forms of the metals may also be removed by either sediment adsorption or organically complexing with other particulates.

In general, studies of recharge basins receiving large metal loads found that most of the heavy metals are removed either in the basin sediment or in the vadose zone. Dissolved metal ions are removed from stormwater during infiltration mostly by adsorption onto the near-surface particles in the vadose zone, while the particulate metals are filtered out near the soil surface. Studies at recharge basins found that lead, zinc, cadmium, and copper accumulated at the soil surface with little downward movement over many years. However, nickel, chromium, and zinc concentrations have exceeded regulatory limits in the soils below a recharge area at a commercial site. Elevated groundwater heavy metal concentrations of aluminum, cadmium, copper, chromium, lead, and zinc have been found below stormwater infiltration devices where the groundwater pH has been acidic. Allowing percolation ponds to go dry between storms can be counterproductive to the removal of lead from the water during recharge. Apparently, the adsorption bonds between the sediment and the metals can be weakened during the drying period.

Similarities in water quality between runoff water and groundwater have shown that there is significant downward movement of copper and iron in sandy and loamy soils. However, arsenic, nickel, and lead did not significantly move downward through the soil to the groundwater. The exception to this was some downward movement of lead with the percolation water in sandy soils beneath stormwater recharge basins. Zinc, which is more soluble than iron, has been found in higher concentrations in groundwater than has iron. The order of attenuation in the vadose zone from infiltrating stormwater is zinc (most mobile) > lead > cadmium > manganese > copper > iron > chromium > nickel > aluminum (least mobile).

Nickel and zinc would likely have high groundwater contamination potentials if subsurface infiltration/injection were used. Chromium and lead would have moderate groundwater contamination potentials for subsurface infiltration/injection practices. All metals would likely have low groundwater contamination potentials if surface infiltration were used with sedimentation pretreatment.

Salts

Salt applications for winter traffic safety is a common practice in many northern areas, and the sodium and chloride, which are collected in the snowmelt, travel down through the vadose zone to the groundwater with little attenuation. Soil is not very effective at removing salts. Salts that are still in the percolation water after it travels through the vadose zone will contaminate the groundwater. Infiltration of stormwater has led to increases in sodium and chloride groundwater concentrations above background concentrations. Fertilizer and pesticide salts also accumulate in urban areas and can leach through the soil to the groundwater.

Studies of depth of pollutant penetration in soil have shown that sulfate and potassium concentrations decrease with depth, while sodium, calcium, bicarbonate, and chloride concentrations increase with depth. Once contamination with salts begins, the movement of salts into the groundwater can be rapid. The salt concentration may not decrease until the source of the salts is removed.

Chloride would likely have a high groundwater contamination potential in northern areas where road salts are used for traffic safety, irrespective of the pretreatment, infiltration, or percolation practice used. Salts are at their greatest concentrations in snowmelt and in early spring runoff in northern areas.

STRESSOR CATEGORIES AND THEIR EFFECTS

There are several ways in which stormwater stressors may be grouped. Overlap between these categories will occur since the ecosystem is comprised of interrelated, interactive components. Attempts at studying single stressors or single categories represents a "reductionist" approach as opposed to a more realistic "holistic" ecosystem approach (Chapman et al. 1992). However, for one to understand the whole system and its response to stormwater stressors, there must first be a basic understanding of single component effects and patterns (see also Chapters 3 through 6). The adverse effect of stormwater runoff has been mainly documented indirectly in NPS effect studies in urban and agricultural watersheds. The aquatic ecosystems in these environments typically show a loss of sensitive species, loss of species numbers (diversity and richness), and increases in numbers of pollution-tolerant organisms (e.g., Schueler 1987; EPA 1987a; Pitt and Bozeman 1982; Pitt 1995). These trends are observed at all levels of biological organization including fish, insects, zooplankton, phytoplankton, benthic invertebrates, protozoa, bacteria, and macrophytes. These alterations tend to change an aquatic ecosystem from a stable system to an unstable one, and from a complex system to an overly simplistic one. As disturbances (e.g., toxic stormwater discharges) increase in frequency and severity, the recovery phase will increase and the ability to cope with a disturbance will decrease. The following categories are but a generalized summary of commonly observed characteristics and effects in previous stormwater and ecotoxicological studies.

Stream Flow Effects and Associated Habitat Modifications

Some of the most serious effects of urban and agricultural runoff are on the aquatic habitat of the receiving waters. A major threat to habitat comes from the rapidly changing flows and the absence of refuge areas to protect the biota during these flow changes. The natural changes in stream hydrology will change naturally at a slow, relatively nondetectable rate in most areas of the United States where stream banks are stabilized by riparian vegetation. In other areas, however, natural erosion and bank slumping will occur in response to high flow events. This "natural" contribution to stream solids is accelerated by hydromodifications, such as increases in stream power due to upstream channelization, installation of impervious drainage networks, increased impervious areas in the watershed (roof tops, roadways, parking areas), and removal of trees and vegetation. All of these increase the runoff volume and stream power, and decrease the time period for stream peak discharge.

In moderately developed watersheds, peak discharges are two to five times those of predevelopment levels (Leopold 1968; Anderson 1970). These storm events may have 50% greater volume, which may result in flooding. The quicker runoff periods reduce infiltration; thus, interflows and baseflows into the stream from groundwater during drought periods are reduced, as are groundwater levels. As stream power increases, channel morphology will change with an initial widening of the channel to as much as two to four times its original size (Robinson 1976; Hammer 1972). Flood-plains increase in size, stream banks are undercut, and riparian vegetation lost. The increased sediment loading from erosion moves through the watershed as bedload, covering sand, gravel, and cobble substrates.

The aquatic organism differences found during the Bellevue Urban Runoff Program were probably most associated with the increased peak flows in Kelsey Creek caused by urbanization and the resultant increase in sediment-carrying capacity and channel instability of the creek (Ped-erson 1981; Perkins 1982; Richey et al. 1981; Richey 1982; Scott et al. 1982). Kelsey Creek had much lower flows than Bear Creek during periods between storms. About 30% less water was available in Kelsey Creek during the summers. These low flows may also have significantly affected the aquatic habitat and the ability of the urban creek to flush toxic spills or other dry-weather pollutants from the creek system (Ebbert et al. 1983; Prych and Ebbert undated). Kelsey Creek had extreme hydrologic responses to storms. Flooding substantially increased in Kelsey Creek during the period of urban development; the peak annual discharges have almost doubled in the last 30 years, and the flooding frequency has also increased due to urbanization (Ebbert et al. 1983; Prych and Ebbert undated). These increased flows in urbanized Kelsey Creek resulted in greatly increased sediment transport and channel instability.

The Bellevue studies (Pitt and Bissonnette 1984) indicated very significant interrelationships among the physical, biological, and chemical characteristics of the urbanized Kelsey Creek system. The aquatic life beneficial uses were found to be impaired, and stormwater conveyance was most likely associated with increased flows from the impervious areas in the urban area. Changes in the flow characteristics could radically alter the ability of the stream to carry the polluted sediments into the other receiving waters.

Stephenson (1996) studied changes in stream flow volumes in South Africa during urbanization. He found increased stormwater runoff, decreases in the groundwater table, and dramatically decreased times of concentration. The peak flow rates increased by about twofold, about half caused by increased pavement (in an area having only about 5% effective impervious cover), with the remainder caused by decreased times of concentration.

Bhaduri et al. (1997) quantified the changes in stream flow and decreases in groundwater recharge associated with urbanization. They point out that the most widely addressed hydrologic effect of urbanization is the peak discharge increases that cause local flooding. However, the increase in surface runoff volume also represents a net loss in groundwater recharge. They point out that urbanization is linked to increased variability in volume of water available for wetlands and small streams, causing "flashy" or "flood-and-drought" conditions. In northern Ohio, urbanization at a study area was found to have caused a 195% increase in the annual volume of runoff, while the expected increase in the peak flow for the local 100-year event was 26% for the same site. Although any increase in severe flooding is problematic and cause for concern, the much larger increase in annual runoff volume, and associated decrease in groundwater recharge, likely has a much greater effect on in-stream biological conditions.

A number of presentations concerning aquatic habitat effects from urbanization were made at the *Effects of Watershed Development and Management on Aquatic Ecosystems* conference held in Snowbird, UT, in August of 1996, and sponsored by the Engineering Foundation and the ASCE. MacRae (1997) presented a review of the development of the common zero runoff increase (ZRI) discharge criterion, referring to peak discharges before and after development. This criterion is commonly met using detention ponds for the 2-year storm. MacRae shows how this criterion has not effectively protected the receiving water habitat. He found that stream bed and bank erosion is controlled by the frequency and duration of the mid-depth flows (generally occurring more often than once a year), not the bank-full condition (approximated by the 2-year event). During monitoring

Table 3.1 Hours of Exceedance of Developed Conditions with Zero Runoff Increase (ZRI) Controls Compared to Predevelopment Conditions

Recurrence Interval (yrs)	Existing Flow Rate (m³/s)	Exceedance for Predevelopment Conditions (hrs per 5 yrs)	Exceedance for Existing Development Conditions, with ZRI Controls (hrs per 5 yrs)	Exceedance for Ultimate Development Conditions, with ZRI Controls (hrs per 5 yrs)
1.01 (critical mid-bank-full conditions)	1.24	90	380	900
1.5 (bank-full conditions)	2.1	30	34	120

near Toronto, he found that the duration of the geomorphically significant predevelopment mid-bank-full flows increased by a factor of 4.2 times, after 34% of the basin had been urbanized, compared to flow conditions before development. The channel had responded by increasing in cross-sectional area by as much as three times in some areas, and was still expanding. Table 3.1 shows the modeled durations of critical discharges for predevelopment conditions, compared to current and ultimate levels of development with "zero runoff increase" controls in place. At full development and even with full ZRI compliance in this watershed, the hours exceeding the critical mid-bank-full conditions will increase by a factor of 10, with significant effects on channel stability and the physical habitat.

MacRae (1997) also reported other studies that found channel cross-sectional areas began to enlarge after about 20 to 25% of the watershed was developed, corresponding to about a 5% impervious cover in the watershed. When the watersheds are completely developed, the channel enlargements were about five to seven times the original cross-sectional areas. Changes from stable stream bed conditions to unstable conditions appear to occur with basin imperviousness of about 10%, similar to the value reported for serious biological degradation. He also summarized a study conducted in British Columbia that examined 30 stream reaches in natural areas, in urbanized areas having peak flow attenuation ponds, and in urbanized areas not having any stormwater controls. The channel widths in the uncontrolled urban streams were about 1.7 times the widths of the natural streams. The streams having the ponds also showed widening, but at a reduced amount compared to the uncontrolled urban streams. He concluded that an effective criterion to protect stream stability (a major component of habitat protection) must address mid-bank-full events, especially by requiring similar durations and frequencies of stream power (the product of shear stress and flow velocity, not just flow velocity alone) at these depths, compared to satisfactory reference conditions.

Urbanization radically affects many natural stream characteristics. Pitt and Bissonnette (1984) reported that the coho and cutthroat were affected by the increased nutrients and elevated temperatures of the urbanized streams in Bellevue, as studied by the University of Washington as part of the EPA NURP project (EPA 1983). These conditions were probably responsible for accelerated growth of the fry, which were observed to migrate to Puget Sound and the Pacific Ocean sooner than their counterparts in the control forested watershed that was also studied. However, the degradation of sediments, mainly the decreased particle sizes, adversely affected their spawning areas in streams that had become urbanized. Sovern and Washington (1997) reported that, in Western Washington, frequent high flow rates can be 10 to 100 times the predevelopment flows in urbanized areas, but that the low flows in the urban streams are commonly lower than the predevelopment low flows. They have concluded that the effects of urbanization on western Washington streams are dramatic, in most cases permanently changing the stream hydrologic balance, by increasing the annual water volume in the stream, increasing the volume and rate of storm flows, decreasing the low flows during dry periods, and increasing the sediment and pollutant discharges from the watershed. With urbanization, the streams increase in cross-sectional area to accommodate these increased flows, and headwater downcutting occurs to decrease the channel gradient. The gradients of stable urban streams are often only about 1 to 2%, compared to 2 to 10% gradients in natural areas. These changes in width and the downcutting result in very different and changing stream conditions. For example, the common pool/drop habitats are generally replaced by pool/riffle habitats, and the stream bed material is comprised of much finer material. Along urban streams, fewer than 50 aquatic plant and animal species are usually found. Researchers have concluded that once urbanization begins, the effects on stream shape are not completely reversible. Developing and maintaining quality aquatic life habitat, however, is possible under urban conditions, but it requires human intervention and it will not be the same as for forested watersheds.

Increased flows due to urban and agricultural modification obviously cause aquatic life impacts due to destroyed habitat (unstable channel linings, scour of sediments, enlarging stream cross sections, changes in stream gradient, collapsing of riparian stands of mature vegetation, siltation, embeddedness, etc.) plus physical flushing of aquatic life from refuge areas downstream. The increases in peak flows, annual runoff amounts, and associated decreases in groundwater recharge obviously cause decreased dry-weather flows in receiving streams. Many small and moderate-sized streams become intermittent after urbanization, causing extreme aquatic life impacts. Even with less severe decreased flows, aquatic life impacts can be significant. Lower flows are associated with increased temperatures, increased pollutant concentrations (due to decreased mixing and transport), and decreased mobility and forage opportunities.

Safety Concerns with Stormwater

There are many aspects of safety associated with urban and agricultural waters, including:

- Exposure to pathogens and toxicants
- Flows (rapidly changing and common high flows)
- Steep banks/cut banks/muddy/slippery banks
- Mucky sediments
- Debris (sharps and strainers)
- Habitat for nuisance organisms (e.g., mosquitoes, rats, snakes)

Most urban receiving waters having direct storm drainage outfalls are quite small and have no formally designated beneficial uses. Larger receiving waters typically have basic uses established, but few urban receiving waters have water contact recreation as a designated beneficial use. Unfortunately, these small waters typically attract local children who may be exposed to some of the hazards associated with stormwater, as noted above. Conditions associated with pathogens and toxicants are likely a serious problem, but the other hazards listed are also very serious. Obviously, drowning should be a concern to all and is often a topic of heated discussion at public meetings where wet detention ponds for stormwater treatment are proposed. However, drowning hazards may be more common in typical urban streams than in well-designed wet detention ponds. These hazards are related to rapidly changing water flows, high flow rates, steep and muddy stream banks, and mucky stream deposits. These hazards are all increased with stormwater discharges and are typically much worse than in predevelopment times when the streams were much more stable. This can be especially critical in newly developing areas where the local streams are thought to be relatively safe from prior experience, but rapidly degrade with increased development and associated stormwater discharges. Other potentially serious hazards are related to debris thrown into streams or trash dumped along stream banks. In unstable urban streams, banks are often continuously cut away, with debris (bankside trees, small buildings, trash piles, and even automobiles) falling into the waterway.

Many people also see untidy urban stream corridors as habitat for snakes and other undesirable creatures and like to clearcut the riparian vegetation and plant grass to the water's edge. Others see creeks as convenient dumping grounds and throw all manner of junk (yard wastes, old appliances, etc.) over their back fences or off bridges into stream corridors. Both of these approaches greatly hinder the use of streams. In contrast, residents of Bellevue, WA, have long accepted the value of small urban streams as habitat for fish. As an example, they have placed large amounts of gravel into streams to provide suitable spawning habitat. In other Northwest area streams, large woody debris is carefully placed into urban streams (using large street-side cranes, and sometimes even helicopters) to improve the aquatic habitat. In these areas, local residents are paying a great deal of money to improve the habitat along the streams and are obviously much more careful about creating hazards associated with trash and other inappropriate debris or discharges.

Drowning Hazards

Marcy and Flack (1981) state that drownings in general most often occur because of slips and falls into water, unexpected depths, cold water temperatures, and fast currents. Four methods to minimize these problems include eliminating or minimizing the hazard, keeping people away, making the onset of the hazard gradual, and providing escape routes.

Jones and Jones (1982) consider safety and landscaping together because landscaping should be used as an effective safety element. They feel that appropriate slope grading and landscaping near the water's edge can provide a more desirable approach than widespread fencing around wet detention ponds. Fences are expensive to install and maintain and usually produce unsightly pond edges. They collect trash and litter, challenge some individuals who like to defy barriers, and impede emergency access if needed. Marcy and Flack (1981) state that limited fencing may be appropriate in special areas. When the side slopes of a wet detention pond cannot be made gradual (such as when against a railroad right-of-way or close to a roadway), steep sides with submerged retaining walls may be needed. A chain-link fence located directly on the top of the retaining wall very close to the water's edge may be needed (to prevent human occupancy of the narrow ledge on the water side of the fence). Another area where fencing may be needed is at the inlet or outlet structures of wet detention ponds. However, fencing usually gives a false sense of security, because most can be easily crossed (Eccher 1991).

Common recommendations to maximize safety near wet detention ponds include suggestions that the pond side slopes be gradual near the water's edge, with a submerged ledge close to shore. Aquatic plants on the ledge would decrease the chance of continued movement to deeper water, and thick vegetation on shore near the water's edge would discourage access to the water and decrease the possibility of falling accidentally. Pathways should not be located close to the water's edge, or turn abruptly near the water. Marcy and Flack (1981) also encourage the placement of escape routes in the water whenever possible. These could be floats on cables, ladders, hand-holds, safety nets, or ramps. They should not be placed to encourage entering the water.

The use of inlet and outlet trash racks and antivortex baffles is also needed to prevent access to locations with dangerous water velocities. Several types are recommended by the NRCS (SCS 1982). Racks need to have openings smaller than about 6 in, to prevent people from passing through them, and they need to be placed where water velocities are less than 3 ft/s, to allow people to escape (Marcy and Flack 1981). Besides maintaining safe conditions, racks also help keep trash from interfering with the operation of the outlet structure.

Eccher (1991) lists the following pond attributes to ensure maximum safety, while having good ecological control:

1. There should be no major abrupt changes in water depth in areas of uncontrolled access.
2. Slopes should be controlled to ensure good footing.

3. All slope areas should be designed and constructed to prevent or restrict weed and insect growth (generally requiring some form of hardened surface on the slopes).
4. Shoreline erosion needs to be controlled.

Obviously, many of these suggestions to improve safety near wet detention ponds may also be applicable to urban stream corridors. Of course, streams can periodically have high water velocities, and steep banks may be natural. However, landscaping and trail placement along urban stream corridors can be carefully done to minimize exposure to the hazardous areas.

Aesthetics, Litter/Floatables, and Other Debris Associated with Stormwater

One of the major problems with the aesthetic degradation of receiving waters in urban areas is a general lack of respect for the local water bodies. In areas where stormwater is considered a beneficial component of the urban water system, these problems are not as severe, and inhabitants and visitors enjoy the local waterscape. The following list indicates the types of aesthetic problems that are common for neglected waters:

- Low flows
- Mucky sediments
- Trash from illegal dumping
- Floatables from discharges of litter
- Unnatural riparian areas
- Unnatural channel modifications
- Odiferous water and sediment
- Rotting vegetation and dead fish
- Objectionable sanitary wastes from CSOs and SSOs

The above list indicates the most obvious aesthetic problems in receiving waters. Many of these problems are directly associated with poor water quality (such as degraded sediments, eutrophication, and fish kills). Other direct problems associated with runoff include massive modifications of the hydrologic cycle with more severe and longer durations of low flow periods due to reduced infiltration of rainwater. Many of the other problems on the above list are related to indirect activities of the inhabitants of the watershed, namely, illegal dumping of trash into streams, littering in the drainage area, and improper modifications. In many areas, separate sewer overflows (SSOs) and combined sewer overflows (CSOs) also contribute unsightly and hazardous debris to urban receiving waters.

Floatable Litter Associated with Wet-Weather Flows

As previously indicated, aesthetics is one of the most important beneficial uses recognized for urban waterways. Floatable litter significantly degrades the aesthetic enjoyment of receiving waters. The control of floatables has therefore long been a goal of most communities.

In coastal areas, land-based sources of beach debris and floatable material have generally been found to originate from wet-weather discharges from the land, and not from marine sources (such as shipping). Of course, in areas where solid wastes (garbage or sewage sludge, for example) have been (or are still being) dumped in the sea, these sources may also be significant beach litter sources. In CSO areas, items of sanitary origin are found in the receiving waters and along the beaches, but stormwater discharges are responsible for most of the bulk litter material, including much of the hazardous materials. In inland areas, marine contributions are obviously not an issue. Therefore, with such direct linkages to the drainage areas, much of the floatable material control efforts have focused on watershed sources and controls (including being part of the "nine minimum" controls

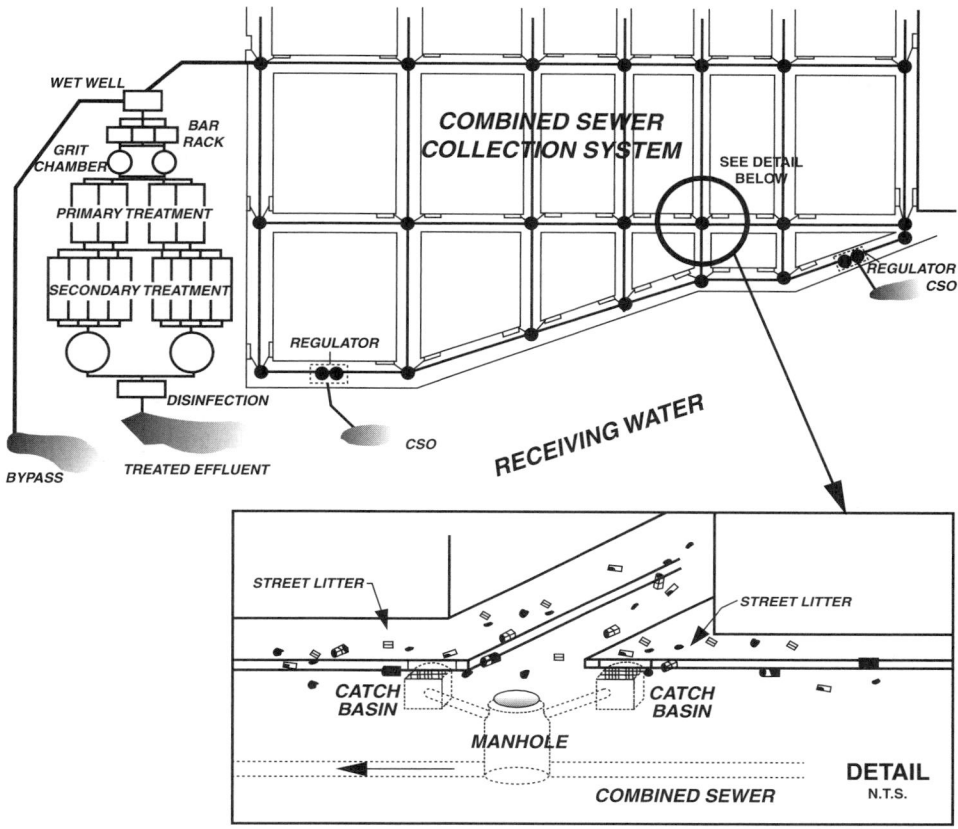

Figure 3.22 Schematic of transport of street and sidewalk litter into receiving waters. (From HydroQual, Inc. *Floatables Pilot Program Final Report: Evaluation of Non-Structural Methods to Control Combined and Storm Sewer Floatable Materials.* City-Wide Floatables Study, Contract II. Prepared for New York City, Department of Environmental Protection, Bureau of Environmental Engineering, Division of Water Quality Improvement. NYDP2000. December 1995.)

for CSOs required by the EPA). Figure 3.22 shows a schematic of how street and sidewalk litter enter the receiving waters (HydroQual 1995).

An example of an investigation of beach litter sources was conducted by Williams and Simmons (1997) along the Bristol Channel in the U.K. They concluded that most of the litter accumulating on the beaches originated from river discharges, and not from litter being deposited directly on the beaches by visitors or from shipping or other oceanic sources. The sources of the litter into the major rivers were the many combined sewer overflows in the area. About 3000 CSOs exist in Wales, and 86 of the 126 CSOs discharging into the study area receive no treatment. They summarized previous studies that have concluded that about half of Britain's coastline is contaminated, with an average of 22 plastic bottles, 17 cans, and 20 sanitary items occurring per km of coast. In some areas, the beach litter can exceed 100 items per category per kilometer. Their survey found that low energy (relatively flat) sandy beaches collected the most debris. Winter litter loadings were generally higher than during the summer, further indicating that storm-related sources were more important than visitor-related sources. They concluded that the linear strip development in South Wales' valleys had led to rivers being used as open sewers and as general dumping grounds.

One of the largest and most comprehensive beach litter and floatable control investigations and control efforts in the United States has been conducted by New York City. At the beginning of their description of this floatable control program, Grey and Olivieri (1998) stated that "one of the major

Figure 3.23 Trash boom, New York City.

Figure 3.24 New York booms and skimmers for the control of floatable discharges.

issues of urban wet-weather pollution is the control of floatable pollution." The comprehensive New York City program included investigations of the sources of the litter contributing to the floatable discharges (mostly street and sidewalk litter) and the effectiveness of many floatable control practices (including public education, enhanced street cleaning, catchbasin hoods, floatable capture nets, and booming and skimmer boats) (Figures 3.23 through 3.26).

New York City used in-line net boxes installed below catchbasin inlets to capture the discharge of floatables for identification and quantification. Much of the work was directed at the capture efficiency of the floatable material in catchbasins. It was found that it was critical that hoods (covers over the catchbasin outlets that extended below the standing water) be used in the catchbasins to help retain the captured material. The hoods increased the capture of the floatables by 70 to 85%. Unhooded catchbasins were found to discharge about 11 g/100 ft of curb length per day, while

Figure 3.25 TrashTrap™ at Fresh Creek, Brooklyn, NY.

Figure 3.26 New York City's use of end-of-pipe TrashTrap systems.

Table 3.2   Floatable Litter Characteristics Found on New York City Streets

	No. of Items (%)	Weight of Items (%)	Density of Items (lb/ft³)
Plastic	57.2	44.3	2.8
Metal	18.9	12.0	3.8
Paper (coated/waxed)	5.9	4.0	2.0
Wood	5.9	5.3	7.7
Polystyrene	5.4	1.3	0.7
Cloth/fabric	2.5	12.5	8.3
Sensitive items	1.7	0.4	na
Rubber	1.1	1.1	10.5
Misc.	1.0	3.6	9.8
Glass	0.4	15.6	13.8

From HydroQual, Inc. *Floatables Pilot Program Final Report: Evaluation of Non-Structural Methods to Control Combined and Storm Sewer Floatable Materials.* City-Wide Floatables Study, Contract II. Prepared for New York City, Department of Environmental Protection, Bureau of Environmental Engineering, Division of Water Quality Improvement. NYDP2000. December 1995.

hooded catchbasins reduced this discharge to about 3.3 g/100 ft of curb length per day. It was also found that the hoods greatly extended the period of time between cleanings and the depth of accumulated litter that could be captured in the catchbasins without degraded capture performance.

There are about 130,000 stormwater inlet structures in New York City's 190,000 acres served by combined and separate sewers, or about 1.5 acres served by each inlet. They are surveying all of these inlet structures, replacing damaged or missing hoods, and accurately measuring their dimensions and indicating their exact locations for a citywide GIS system. Catchbasin cleaning costs are about $170 per inlet, while the inspection and mapping costs are about $45 per inlet. Replacement hoods cost about $45 per inlet.

Litter surveys conducted by the New York City Department of Sanitation (DOS) in 1984 and 1986 found that 70% of the street litter items consisted of food and beverage wrappers and containers (60%) and the paper and plastic bags (10%) used to carry these items. The early studies also found that litter levels on the streets and sidewalks were about 20 to 25% higher in the afternoon than in the morning. The DOS conducted similar surveys in 1993 at 90 blockfaces throughout the city (HydroQual 1995). Each litter monitoring site was monitored several times simultaneously when the surveys were conducted with the floatable litter separated into 13 basic categories. They found that twice as much floatable litter was located on the sidewalks compared to the streets (especially glass) and that land use had little effect on the litter loadings (except in the special business districts where enhanced street cleaning/litter control was utilized, resulting in cleaner conditions). Their baseline monitoring program determined that an average of 2.3 floatable litter items were discharged through the catchbasin inlets per day per 100 ft of curb. This amount was equivalent to about 6.2 in² and 0.013 lb (8.5 g) of material. The total litter load discharged was about twice this floatable amount. Table 3.2 summarizes the characteristics of the floatable litter found on the streets.

Solids (Suspended, Bedded, and Dissolved)

The detrimental effects of elevated suspended and dissolved solids and increases in siltation and fine-grained bedded sediments have been well documented (EPA 1987b). The sources of these solids are primarily from dry deposition, roadways, construction, and channel alteration and have significant effects on receiving-system habitats. Solids concentrations are directly related to watershed use characteristics and watershed hydrology.

In the United States, 64% of the land is dominated by agriculture and silviculture from which the major pollutant is sediment (approximately 1.8 billion metric tons per year) (EPA 1977). The suspended sediments transport toxicants, nutrients, and lower the aesthetic value of the waterways

Table 3.3   Classification of Suspended and Dissolved Solids and Their Probable Major Impacts on Freshwater Ecosystems

	Chemical and Physical Effects	Biochemical and Biological Effects
Suspended Solids		
Clays, silts, sand	Sedimentation, erosion, and abrasion turbidity (light reduction), habitat change	Respiratory interference habitat restriction, light limitation
Natural organic matter	Sedimentation, DO utilization	Food sources, DO effects
Wastewater organic particles	Sedimentation, DO utilization	DO effects, eutrophication, nutrient source
Toxicants sorbed to particles	All of the above	Toxicity
Dissolved Solids		
Major inorganic salts	Salinity, buffering, precipitation, element ratios	Nutrient availability, succession, salt effects
Important nutrients		Eutrophication, DO production
Natural organic matter		DO effects and utilization
Wastewater organic matter		DO effects and utilization
Toxicants		Toxicity and effects on DO

From EPA (U.S. Environmental Protection Agency). *Suspended and Dissolved Solids Effects on Freshwater Biota: A Review,* Environmental Research Laboratory, U.S. Environmental Protection Agency, Corvallis, OR, EPA 600/3-77/042. 1977.

(EPA 1977). Suspended sediments decrease light penetration and photosynthesis, clog gills and filtering systems of aquatic organisms, reduce prey capture, reduce spawning, reduce survival of sensitive species, and carry adsorbed pollutants (Tables 3.3 through 3.5). Acute effects of suspended solids are commonly observed at 80,000 mg/L with death at 200,000 mg/L. Recovery is quick at lower exposures (EPA 1977). As the suspended sediments settle, they cover silt-free spawning substrates, suffocating embryos, and alter the sediment environment. Suspended solids reduce primary productivity and alter temperatures, thus affecting summer stratification. Solids should not reduce photosynthesis by more than 10% of the seasonal average, using the "light–dark" bottle method (APHA 1992). Reduced productivity may then reduce zooplankton populations. Desirable benthic species may be smothered, and tolerant species, such as oligochaetes, will increase in numbers. The sediment environment plays a major role in aquatic ecosystem functioning and overlying water quality (Wetzel 1975). These new bedded sediments may possess different chemical, physical, and biological characteristics from pre-impact sediments. So any alteration to the micro-, meio-, and macrobenthic communities, sorption and desorption dynamics of essential and toxic chemical species, and organic matter and nutrient cycling processes may profoundly influence the aquatic ecosystem (Power and Chapman 1992). As the rate of bedload sediment movement increases and the frequency of occurrence of bedload movement increases, the stress to the system increases.

Dissolved solids concentrations can often be very high in stormwaters and baseflows. The associated dissolved constituents consist primarily of road salts and salts from exposed soils. Though the major cations and anions are nontoxic to most species in relatively high concentrations, stormwaters may exceed threshold levels (EPA 1977) and alter ion ratios, which may cause chronic toxicity effects. In addition, toxic trace metal-metalloids such as selenium may be dissolved from natural soil matrices (as dramatically demonstrated in the San Joaquin Valley's Kesterson Reservoir of California), or dissolved zinc may be discharged from roof runoff components of urban runoff. Long-term and repeated exposures result as the dissolved species accumulate in interstitial water, bacteria, macrophytes, phytoplankton, and other food chain components (Burton et al. 1987; EPA 1977) and result in increased mortality, teratogenicity and other adverse effects (EPA 1977).

Table 3.4   Summary of Suspended Solids Effects on Aquatic Macroinvertebrates

Organisms	Effect	Suspended Solid Concentration	Source of Suspended Solids	Comment
Mixed populations	Lower summer populations		Mining area	
	Reduced populations to 25%	261–390 NTU (turbidity)	Log dragging	
	Densities 11% of normal	1000–6000 mg/L		Normal populations at 60 mg/L
	No organisms in the zone of setting	>5000 mg/L	Glass manufacturing	Effect noted 13 miles downstream
Chironomus and Tubificidae	Normal fauna replaced by species selection		Colliery	Reduction in light-reduced submerged plants
Chematopsyche (net spinners)	Number reduced	(High concentrations)	Limestone quarry	Suspended solids as high as 250 mg/L
Tricorythoides	Number increased		Limestone quarry	Due to preference for mud or silt
Mixed populations	90% increase in drift	80 mg/L	Limestone quarry	
	Reduction in numbers	40–200 NTU	Manganese strip mine	Also caused changes in density and diversity
Chironomidae	Increased drift with suspended sediment		Experimental sediment addition	
Ephemoptera, Simuliidae, Hydracarina	Inconsistent drift response to added sediment		Experimental sediment addition	

From EPA (U.S. Environmental Protection Agency). *Suspended and Dissolved Solids Effects on Freshwater Biota: A Review,* Environmental Research Laboratory, U.S. Environmental Protection Agency, Corvallis, OR, EPA 600/3-77/042. 1977.

Dissolved Oxygen

Historically, dissolved oxygen has received much attention when researchers investigate biological receiving water effects of pollutant discharges. Therefore, the earliest efforts to evaluate the potential problems caused by urban runoff included investigations of dissolved oxygen conditions in urban receiving waters.

Bacteria respond rapidly (within minutes) in temperate streams and lakes to their surrounding environment. Due to the low level of nutrients normally present, most of the indigenous bacteria are dominant. During a storm event, however, micro- to submicrogram levels of organic nutrients (e.g., carbon, nitrogen, phosphorus, and sulfur-containing compounds) suddenly increase by orders of magnitude. Consequently, bacterial reproduction and respiration rates increase dramatically; thus exerting biochemical oxygen demand (BOD). Oxygen depletion problems may occur during the high flow event, but it is likely more serious days later when associated with organic material affecting the sediment oxygen demand (Pitt 1979). BOD_5 levels may exceed 20 mg/L during storm events, which may result in anoxia in downstream receiving waters (Schueler 1987). Predicting this problem is complicated by toxicants that may be present and interfere with the BOD test (OWML 1982). Sediment resuspension contributes to both BOD and chemical oxygen demand (COD). BOD_5 values were elevated tenfold (10 to 20 days after a storm event) related to sediment oxygen demand (SOD). Stormwater dissolved oxygen (DO) levels less than 5 mg/L are common (Keefer et al. 1979).

Aquatic macrofauna are cold-blooded and sensitive to temperature changes. In cold water systems, sustained temperatures in excess of 21°C are stressful to resident biota. Many agricultural and urban watersheds contribute to thermal pollution by removing shade canopies over streams, and runoff temperatures increase rapidly as water flows over impervious surfaces (Schueler 1987).

Table 3.5 Summary of Suspended Solids Effects on Fish[a]

Fish (Special)	Effect	Concentration of Suspended Solids (mg/L)	Source of Suspended Materials
Rainbow trout	Survived 1 day	80,000	Gravel washing
(*Salmo gairdneri*)	Killed in 1 day	160,000	Gravel washing
	50% Mortality in 3½ weeks	4250	Gypsum
	Killed in 20 days	1000–2500	Natural sediment
	50% mortality in 16 weeks	200	
	1/5 mortality in 37 days	1000	Spruce fiber
	No deaths in 4 weeks	553	Cellulose fiber
	No deaths in 9–10 weeks	200	Gypsum
	20% mortality in 2–6 months	90	Coal washery waste
	No deaths in 8 months	100	Kaolin and diatomaceous earth
	No deaths in 8 months	50	Spruce fiber
	No increased mortality	30	Coal washery waste
	Reduced growth	50	Kaolin or diatomaceous earth
	Reduced growth	50	Wood fiber
	Fair growth	200	Coal washery waste
	"Fin-rot" disease	270	Coal washery waste
	"Fin-rot" disease	100	Diatomaceous earth
	No "fin-rot"	50	Wood fiber
	Reduced egg survival	(Siltation)	Wood fiber
	Total egg mortality in 6 days	1000–2500	
	Reduced survival of eggs	(Silting)	Wood fiber
	Supports populations	(Heavy loads)	Mining operations
	Avoid during migration	(Muddy waters)	Glacial silt
Brown trout	Do not dig redds	(Sediment in gravel)	
(*Salmo trutta*)	Reduced populations to 1/7 of clean streams	1000–6000	China-clay waste
Cutthroat trout	Abandon redds	(If silt is encountered)	
(*Salmo clarkii*)	Sought cover and stopped feeding	35	
Brook trout (*Salvelinus fontinalis*)	No effect on movement	(Turbidity)	
Golden shiner	Reaction	20,000–50,000	
(*Notemigonus crysoleucas*)	Death	50,000–100,000	
Carp	Reaction	20,000	
(*Cyrinus carpio*)	Death	175,000–250,000	
Largemouth black bass	Reaction	20,000	
(*Micropterus salmoides*)	Death	101,000 (average)	
Smallmouth bass (*Micropterus dolomieu*)	Successful nesting, spawning, hatching	(Sporadic periods of high turbidity)	

[a] See EPA 1977 for additional species-specific effect information.

From EPA (U.S. Environmental Protection Agency). *Suspended and Dissolved Solids Effects on Freshwater Biota: A Review*, Environmental Research Laboratory, U.S. Environmental Protection Agency, Corvallis, OR, EPA 600/3-77/042. 1977.

Acid precipitation and acid mine drainage cause NPS pollution problems in some parts of the United States which are, at times, aggravated by storm events. During the spring in areas where snows have accumulated, rain events intensify the snowmelt process. This results in pulses of low pH runoff and snowmelts which may be stressful or lethal to aquatic macrofauna, particularly the sensitive life stages of fish occurring during the spring spawning period.

Keefer et al. (1979) examined the data from 104 water quality monitoring sites near urban areas throughout the country for DO conditions. These stations were selected from more than 1000 nationwide monitoring stations operated by various federal and state agencies. They conducted analyses of daily DO data for 83 of these sites. About one half of the monitoring stations examined showed a 60% or greater probability of a higher than average dissolved oxygen deficit occurring at times of higher than average stream flow, or on days with rainfall. This result was based on daily data for entire water years; not all years at any given location exhibited this 60% probability condition. They found that the DO levels fell to less than 75% saturation at most of the stations that had this 60% or greater probability condition. They also found that DO concentrations of less than 5 mg/L were common. Keefer et al. (1979) examined hourly DO data at 22 nationwide sites to find correlations between flows and DO deficit. They found that for periods of steady low flows, the DO fluctuated widely on a daily cycle, ranging from 1 to 7 mg/L. During rain periods, however, the flow increased, of course, but the diurnal cycle of this DO fluctuation disappeared. The minimum DO dropped from 1 to 1.5 mg/L below the minimum values observed during steady flows, and remained constant for periods ranging from 1 to 5 days. They also reported that as the high flow conditions ended, the DO levels resumed diurnal cyclic behavior. About 50% of the stations examined in detail on an hour-by-hour basis would not meet a 5 mg/L DO standard, and about 25% of these stations would not even meet a 2.0 mg/L standard for 4-hour averages. The frequency of these violations was estimated to be up to five times a year per station.

Ketchum (1978) conducted another study in Indiana that examined DO depletion on a regional basis. Sampling was conducted at nine cities, and the project was designed to detect significant DO deficits in streams during periods of rainfall and runoff. The results of this study indicated that wet-weather DO levels generally appeared to be similar or higher than those observed during dry-weather conditions in the same streams. They found that significant wet-weather DO depletions were not observed, and due to the screening nature of the sampling program, more subtle impacts could not be measured.

Heaney et al. (1980), during their review of studies that examined continuous DO stations downstream from urbanized areas, indicated that the worst DO levels occurred after the storms in about one third of the cases studied. This lowered DO could be due to urban runoff moving downstream, combined sewer overflows, and/or resuspension of benthic deposits. Resuspended benthic deposits could have been previously settled urban runoff solids.

Pitt (1979) found that the BOD of urban runoff, after a 10- to 20-day incubation period, can be more than five to ten times the BOD of a 1- to 5-day incubation period (Figure 3.27). Therefore, urban runoff effects on DO may occur at times substantially different from the actual storm period and be associated with interaction between sediment and the overlying water column. It is especially important to use acclimated microorganisms for the BOD test seed for stormwater BOD analyses. The standard activated sludge seed may require substantial acclimation periods. Even in natural waters, several-day acclimation periods may be needed (see Lalor and Pitt 1998; P/R *in situ* test descriptions in Chapter 6).

Temperature

In-stream temperature increases have been noted in many studies as being adversely affected by urbanization. Rainwater flowing across heated pavement can significantly elevate stormwater temperatures. This temperature increase can be very detrimental in steams having sensitive cold-water fisheries. Removal of riparian vegetation can also increase in-stream water temperatures. Higher water temperatures increase the toxicity of ammonia and also affect the survival of pathogens. The temperature increases in urban streams are most important during the hot summer months when the natural stream temperatures may already be nearing critical conditions and when the stream flows are lowest. Pavement is also the hottest at this time and stormwater temperature increases are therefore the highest. Much of the habitat recovery efforts in urban streams focus on restoring an overstory for the streams to provide shading, refuge areas, and bank stability. Wet detention ponds in urban areas have also been shown to cause significant temperature increases. Grass-lined channels, however, provide some relief, compared to rock-lined or asphalt-lined drain-

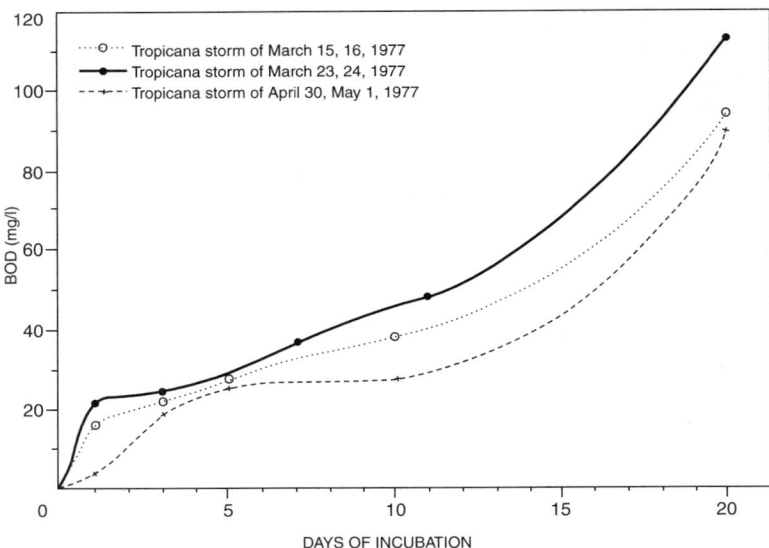

Figure 3.27 BOD rate curve for stormwater, showing dramatic increase after 10 days of incubation. (From Pitt, R. *Demonstration of Nonpoint Pollution Abatement through Improved Street Cleaning Practices,* EPA-600/2-79-161, U.S. Environmental Protection Agency, Cincinnati, OH. 270 pp. 1979.)

age channels. Since temperature is simple to monitor and is a critical stressor for many aquatic organics, it should be included in most monitoring efforts.

Nutrients

In general, urban stormwater is relatively low in organic matter and nutrients and high in toxicants. However, the nutrient levels in stormwaters can periodically be high and produce large mass discharges of nitrogen and phosphorus compounds (e.g., EPA 1977, 1983; Schueler 1987). Single spring storm events have been shown to contribute 90% of the annual phosphorus input into receiving impoundments. However, urban and agricultural runoff may contain nutrient concentrations which exceed the normal (predevelopment) ranges, and result in adverse responses such as cyanobacterial (blue-green algae) and green algal blooms. Many of the nutrients present in urban runoff are soluble and thus readily assimilated by planktonic organisms (Schueler 1987). Sources include rain, dry deposition, soils, fertilizers, and animal wastes. Impoundments receiving contaminated runoff, with retention times of 2 weeks or longer, may develop symptoms of eutrophication. Blue-green algal blooms can produce hepato- and neurotoxins implicated in cattle deaths, human liver cancer, and allergic responses (Zhang et al. 1991). As algal blooms eventually decompose, bacterial respiration may result in DO sags and anoxia, with associated fish kills.

A large amount of the nutrients enter receiving waters adsorbed to suspended solids (Lin 1972; Middlebrooks 1974; Carlile et al. 1974). These fractions will largely end up as bedded sediments which may or may not be subsequently released to overlying waters. The sediment nutrients may stimulate bacterial activity, ammonia production, and rooted macrophyte growth.

Toxicants

Heavy Metals

Stormwater runoff commonly contains elevated levels of metals and metalloids, particularly in urban areas (EPA 1983; Pitt et al. 1995; Schueler 1987). Some of these constituents are very toxic at relatively low concentrations (Table 3.6). The metals of principal concern that often occur in

Table 3.6 U.S. EPA Trace Metal Criteria for Human Health and Aquatic Life Beneficial Uses

Trace Metal Contaminant	Water Hardness (mg/L as CaCO₃)	Humanᵃ Ingestion (food/drink) (μg/L)	Ambient Life Criteria for Intermittent Exposure (μg/L)ᵇ	
			Thresholdᶜ Effect	Significantᵈ Mortality
Copper	50	—	20	50–90
	100	—	35	90–150
	200	—	80	120–350
Cadmium	50	10	3	7–160
	100	10	6.6	15–350
	300	10	20	45–1070
Lead	50	50	150	350–3200
	100	50	360	820–7500
	200	50	850	1950–17850
Zinc	50	—	380	870–3200
	100	—	680	1550–4500
	200	—	1200	2750–8000
Nickel	—	13.4	—	—

ᵃ Derived from EPA drinking water criteria.
ᵇ EPA estimate of toxicity under intermittent, short-duration exposure (several hours once every several days).
ᶜ Concentration causing mortality to the most sensitive individual of the most sensitive species.
ᵈ Significant mortality shown as a range: 50% mortality in the most sensitive species, and mortality of the most sensitive individual in the species in the 25th percentile of sensitivity.

From EPA (U.S. Environmental Protection Agency). *Quality Criteria for Water.* EPA 440/5-86-001. U.S. Environmental Protection Agency, Washington, D.C. May 1986.

urban runoff are arsenic, cadmium, copper, lead, mercury, and zinc (EPA 1983). Metal bioavailability is reduced in waters of higher hardness (Table 3.6) by sorption to solids and by stormwater dilution. However, acute and chronic effects have been attributed to stormwater metals (Ray and White 1979; Ellis 1992). The highest metal concentrations are not always associated with the "first flush," but are better correlated with the peak flow period (Heaney 1978). Most metals are bound to street and parking area particulates and subsequently deposited in stream and lake sediments (Pitt et al. 1995). Sediment metal concentrations are dependent on particle size (Wilber and Hunter 1980). Wilber and Hunter (1980) suggest that larger particle sizes are better indicators of urban inputs since they are less affected by scouring. Zinc and copper are often present in runoff as soluble forms (Schueler 1987; Pitt et al. 1995).

Predicting detrimental effects from water or sediment metal concentration or loading data is difficult due to the myriad of processes which control bioavailability and fate. Speciation, availability, and toxicity are affected by pH, redox potential, temperature, hardness, alkalinity, solids, iron and manganese oxyhydroxides, sulfide fractions, and other organic-inorganic chelators. These constituents and conditions are often rapidly changing during a storm event and processes which increase and decrease bioavailability (e.g., loss of sulfide complexes and formation of oxyhydroxide complexes) may occur simultaneously. This makes accurate modeling of toxicity difficult, if not impossible.

Episodic exposures of organisms to stormwaters laden with metals can produce stress and lethality (see also Chapter 6). Ray and White (1976) observed fish death days after exposure and miles downstream after metals were diluted to nondetectable levels. Ellis et al. (1992) showed amphipods bioaccumulated zinc from episodic, *in situ* exposures. Repeated exposures increased their sensitivity, and mortality was observed 3 weeks after the storm event.

Toxic Organic Compounds

The types and concentrations of toxic organic compounds that are in stormwaters are driven primarily by land use patterns and automobile activity in the watershed. Most nonpesticide organic compounds originate as washoff from impervious areas in commercial areas having large numbers of automobile startups and/or other high levels of vehicle activities, including vehicle maintenance operations and heavily traveled roads. The compounds of most interest are the polycyclic aromatic hydrocarbons (PAHs). Other organics include phthalate esters (plasticizers) and aliphatic hydrocarbons. Other compounds frequently detected in residential and agricultural areas are cresol constituents (and other wood preservatives), herbicides, and insecticides. Many of these organic compounds are strongly associated with the particulate fraction of stormwater. Volatile organic compounds (VOCs) are rarely found in urban runoff. While most organics are not detected or are detected at low µg/L concentrations, some are acutely toxic, including freshly applied pesticides and photoactivated PAHs (Skalski 1991; Oris and Giesy 1986). The extent of detrimental impact from these constituents has not been well documented, but likely is significant in some areas.

Environmental Fates of Runoff Toxicants

The fate of runoff toxicants after discharge significantly determines their associated biological effects. If the pollutants are discharged in a soluble form and remain in solution, they may have significant acute toxicity effects on fish, for example. However, if discharged soluble pollutants form insoluble complexes or sorb onto particulates, chronic toxicity effects associated with contaminated sediments are more likely. For many of the metallic and organic toxicants discharged in urban runoff, the particulate fractions are much greater than the soluble fractions (Pitt et al. 1995). Particulate forms of pollutants may remain in suspension, if their settling rates are low and the receiving water is sufficiently turbulent. However, polluted sediments are common in many urban and agricultural streams, indicating significant accumulations of runoff particulate pollutants (Pitt 1995).

Tables 3.7 through 3.9 summarize the importance of various environmental processes for the aquatic fates of some runoff heavy metals and organic priority pollutants, as described by Callahan et al. (1979). Photolysis (the breakdown of the compounds in the presence of sunlight) and volatilization (the transfer of the materials from the water into the air as a gas or vapor) are not nearly as important as the other mechanisms for heavy metals. Chemical speciation (the formation of chemical compounds) is very important in determining the solubilities of the specific metals. Sorption (adsorption is the attachment of the material onto the outside of a solid, and absorption is the attachment of the material within a solid) is very important for all of the heavy metals shown. Sorption can typically be the controlling mechanism affecting the mobility and the precipitation of most heavy metals. Bioaccumulation (the uptake of the material into organic tissue) can occur for all of the heavy metals shown. Biotransformation (the change of chemical form of the metal by organic processes) is very important for some of the metals, especially mercury, arsenic, and lead. In many cases, mercury, arsenic, or lead compounds discharged in forms that are unavailable can be accumulated in aquatic sediments. They are then exposed to various benthic organisms that can biotransform the material through metabolization to methylated forms, which can be highly toxic and soluble.

Tables 3.8 and 3.9 also summarize various environmental fates for some of the toxic organic pollutants found in typical runoff from human-modified watersheds, mainly various phenols, polycyclic aromatic hydrocarbons (PAHs), and phthalate esters. Photolysis may be an important fate process for phenols and PAHs but is probably not important for the phthalate esters. Oxidation or hydrolysis may be important for some phenols. Volatilization may be important for some phenols and PAHs. Sorption is an important fate process for most of the materials, except for phenols. Bioaccumulation, biotransformation, and biodegradation are important for many of these organic materials.

Pathogens

Water Environment & Technology (1996) reported that the latest National Water Quality Inventory released by the EPA only showed a slight improvement in the attainment of beneficial uses in

Table 3.7 Importance of Environmental Processes on the Aquatic Fates of Selected Urban Runoff Heavy Metals

Environmental Process	Arsenic	Cadmium	Copper	Mercury	Lead	Zinc
Photolysis	Not important	Not important	Not important	May be important in some aquatic environments	Determines the form of lead entering the aquatic system	Not important
Chemical speciation	Important in determining distribution and mobility[a]	Complexation with organics; most important in polluted waters	Complexation with organics; most important in polluted waters	Conversion to complex species; HgS will precipitate in reducing sediments	Determines which solid phase controls solubility	Complexation predominates in polluted waters
Volatilization	Important when biological activity or highly reducing conditions produce AsH_3 or methyl-arsenic	Not important	Not important	Important	Not important	Not important
Sorption	Sorption onto clays, oxides, and organic material important	Sorption onto organic materials, clays, hydrous iron and manganese oxides most important	Can reduce Cu mobility and enrich suspended and bed sediments; sorption onto organics in polluted waters, clay minerals or hydrous iron and manganese oxides	Strongest onto organic material, results in partitioning of mercury into suspended and bed sediments	Adsorption to inorganic solids, organic materials and hydrous iron and manganese oxides control mobility of lead	Strong affinity for hydrous metal oxides, clays, and organic matter; adsorption increases with pH
Bioaccumulation	Most important at lower trophic levels; toxicity limits bioaccumulation	Biota strongly bioaccumulate cadmium	Biota strongly bioaccumulate copper	Occurs by many mechanisms, most connected to methylated forms of mercury	Biota strongly bioaccumulates lead	Zinc is strongly bioaccumulated
Biotransformation	Arsenic can be metabolized to organic arsenicals	Not methylized biologically, organic ligands may affect solubility and adsorption	Source Cu complexes may be metabolized; organic ligands are important in sorption and complexation processes	Can be metabolized by bacteria to methyl and dimethyl forms which are quite mobile	Biomethylation of lead in sediments can remobilize lead	Not evident; organic ligands of biological origin may affect solubility and adsorption

[a] Conversion of As^{3+} and As^{5+} and organic complexation most important.

From Callahan, M.A. et al. *Water Related Environmental Fates of 129 Priority Pollutants.* U.S. Environmental Protection Agency, Monitoring and Data Support Division, EPA-4-79-029a and b. Washington, D.C. 1979.

Table 3.8 Importance of Environmental Processes on the Aquatic Fates of Various Polycyclic Aromatic Hydrocarbons and Phthalate Esters

Environmental Process[a]	Anthracene	Fluoranthene	Pheranthrene	Diethyl Phthalate (DEP)	Di-n-Butyl Phthalate (DBP)	Bis (2-Ethyl-hexyl) Phthalate (DEHP)	Butyl Benzyl Phthalate (BBP)
Photolysis	Dissolved portion may undergo rapid photolysis	Dissolved portion may undergo rapid photolysis	Dissolved portion may undergo rapid photolysis	Not important	Not important	Not important	Not important
Volatilization	May be competitive with adsorption	May be competitive with adsorption	May be competitive with adsorption	Not important	Not important	Not important	Not important
Sorption	Adsorbs onto suspended solids; movement by suspended solids is important transport process	Adsorbs onto suspended solids; movement by suspended solids is important transport process	Adsorbs onto suspended solids; movement by suspended solids is important transport process	Sorbed onto suspended solids and biota; complexation with humic substances most important transport process	Sorbed onto suspended solids and biota; complexation with humic substances most important transport process	Sorbed onto suspended solids and biota; complexation with humic substances most important transport process	Sorbed onto suspended solids and biota; complexation with humic substances most important transport process
Bioaccumulation	Short-term process; is readily metabolized	Short-term process; is readily metabolized	Short-term process; is readily metabolized	Variety of organisms accumulate phthalates (lipophilic)	Variety of organisms accumulate phthalates (lipophilic)	Variety of organisms accumulate phthalates (lipophilic)	Variety of organisms accumulate phthalates (lipophilic)
Biotransformation	Readily metabolized by organisms and biodegradation, probably ultimate fate mechanisms	Readily metabolized by organisms and biodegradation, probably ultimate fate mechanism	Readily metabolized by organisms and biodegradation, probably ultimate fate mechanisms	Can be metabolized	Can be metabolized	Can be metabolized	Can be metabolized

[a] Oxidation and hydrolysis are not important fate mechanisms for any of these compounds.

From Callahan, M.A. et al. *Water Related Environmental Fates of 129 Priority Pollutants*. U.S. Environmental Protection Agency, Monitoring and Data Support Division, EPA-4-79-029a and b. Washington, D.C. 1979.

Table 3.9 Importance of Environmental Processes on the Aquatic Fates of Various Phenols and Pyrene

Environmental Process	Phenol	Pentachlorophenol (PCP)	2,4,6-Trichlorophenol	2,4-Dimethyl Phenol (2,4-Xylenol)	Pyrene
Photolysis	Photooxidation may be important degradation process in aerated, clear, surface waters	Reported to occur in natural waters; important near water surface	Reported, but importance is uncertain	May be important degradation process in clear aerated surface waters	Dissolved portion may undergo rapid photolysis
Oxidation	Metal-catalyzed oxidation may be important in aerated surface waters	Not important	Not important	Metal-catalyzed oxidation may be important in aerated surface waters	Not important
Volatilization	Possibility of some phenol passing into the atmosphere	Not important	Not important	Not important	Not as important as adsorption
Sorption	Not important	Sorbed by organic litter in soil and sediments	Potentially important for organic material; not important for clays	Not important	Adsorption onto suspended solids important; movement by suspended solids important
Bioaccumulation	Not important	Bioaccumulates in numerous aquatic organisms	Not important	Not important	Short-term process not significant; metabolized over long term
Biotransformation	Very significant	Can be metabolized to other phenol forms	Reported in soil and sewage sludge; uncertain for natural surface waters	Inconclusive information	Readily metabolized; biodegradation probably ultimate fate process

From Callahan, M.A. et al. *Water Related Environmental Fates of 129 Priority Pollutants.* U.S. Environmental Protection Agency, Monitoring and Data Support Division, EPA-4-79-029a and b. Washington, D.C. 1979.

the nation's waters. Urban runoff was cited as the leading source of problems in estuaries, with nutrients and bacteria as the primary problems. Problems in rivers and lakes were mostly caused by agricultural runoff, with urban runoff the third ranked source for lakes and the fourth ranked source for rivers. Bacteria, siltation, and nutrients were the leading problems in the nation's rivers and lakes.

Pathogens in stormwater are a significant concern potentially affecting human health. The use of indicator bacteria is controversial for stormwater, as is the assumed time of typical exposure of swimmers to contaminated receiving waters. However, recent epidemiological studies have shown significant health effects associated with stormwater-contaminated marine swimming areas. Protozoan pathogens, especially associated with likely sewage-contaminated stormwater, are also a public health concern.

Fecal indicators (i.e., fecal coliforms, fecal streptococci, *Escherichia coli*, and enterococci) are usually found in elevated concentrations in stormwater runoff, greatly exceeding water quality criteria and standards for primary and secondary contact (MWCOG 1984). This suggests that fecal pathogen levels are also elevated, though significant correlations with fecal coliforms are tenuous (EPA 1986). Die-off of fecal organisms in receiving waters during summer months is relatively rapid, with 99% dying within 24 to 48 hours (Burton 1985). However, fecal microorganisms also accumulate in sediments where survival is extended for weeks to months (Burton et al. 1987). Recent sediment bacteriological analyses conducted by UAB in local Birmingham (AL) area urban lakes have found elevated pore water concentrations (several hundred to several thousand organisms/100 mL) of *E. coli* and enterococci extending to at least 0.1 m into the sediments. Also, when gently disturbed, the water layer over the sediments is also found to significantly increase in microorganism concentrations. *In situ* die-off studies also indicated that bacteria sedimentation may be a more important fate mechanism of stormwater bacteria than die-off (Easton 2000).

Good correlations between the incidence of gastroenteritis in swimmers and *E. coli* and enterococci concentrations in water have resulted in new recreational water criteria (EPA 1986). High fecal microorganism concentrations in stormwaters originate from wastes of wildlife, pets, livestock, septic systems, and combined sewer overflows (CSOs). The ecological effects of these inputs of fecal organisms are unknown; however, public health is at risk in swimming areas that receive stormwaters.

Urban Bacteria Sources

The Regional Municipality of Ottawa–Carleton (1972) recognized the importance of rooftop, street surface, and field runoff in contributing bacteria contaminants to surface waters in the Ottawa area. Gore & Storrie/Proctor and Redfern (1981) also investigated various urban bacteria sources affecting the Rideau River. They examined dry-weather continuous coliform sources, the resuspension of contaminated river bottom sediments, exfiltration from sanitary sewers, and bird feces. These sources were all considered in an attempt to explain the relatively high dry-weather coliform bacteria concentrations found in the river. They concluded, however, that stormwater runoff is the most probable source for the wet-weather and continuing dry-weather bacteria concentrations in the Rideau River. The slow travel time of the river water usually does not allow the river to recover completely from one rainstorm before another begins.

The Regional Municipality of Ottawa–Carleton (1972) noted the early Ottawa activities in correcting stormwater and sanitary sewage cross-connections. Since that time, many combined sewer overflows have also been eliminated from the Rideau River. Loijens (1981) stated that, as a result of sewer separation activities, only one overflow remained active by 1981 (Clegg Street). During river surveys in 1978 and 1979 in the vicinity of this outfall, increased bacteria levels were not found. Gore & Storrie/Proctor and Redfern (1981) stated that there was no evidence that combined sewer overflows are causing the elevated fecal coliform bacteria levels in the river. Environment Canada (1980), however, stated that high dry-weather bacteria density levels, especially when considering the fecal coliform to fecal streptococci ratio, constitutes presumptive evidence of low-volume sporadic inputs of sanitary sewage from diverse sources into the downstream Rideau River sectors.

Street surfaces have been identified as potential major sources of urban runoff bacteria. Pitt and Bozeman (1982) found that parking lots, street surfaces, and sidewalks were the major contributors of indicator bacteria in the Coyote Creek watershed in California. Gupta et al. (1981) found high concentrations of fecal coliforms at a highway runoff site in Milwaukee. This site was entirely impervious and located on an elevated bridge deck. The only likely sources of fecal coliforms at this site were atmospheric deposition, bird droppings, and possibly feces debris falling from livestock trucks or other vehicles.

Several studies have found that the bacteria in stormwater in residential and light commercial areas were from predominantly nonhuman origins. Geldreich and Kenner (1969) stated that the fecal coliforms in stormwater are from dogs, cats, and rodents in city areas, and from farm animals and wildlife in rural areas. Qureshi and Dutka (1979) found that there may be an initial flush of animal feces when runoff first develops. The most important source, however, may be feces bacteria that are distributed in the soil and not the fresh feces washing off the impervious surfaces.

Some studies have investigated vegetation sources of coliform bacteria. For example, Geldreich (1965) found that the washoff of bacteria from vegetation does not contribute significant bacteria to the runoff. They also found that most of the bacteria on vegetation is of insect origin. Geldreich et al. (1980) found that recreation activities in water bodies also increase the fecal coliform and fecal streptococci concentrations. These organisms of intestinal origin will concentrate in areas near the shore or in areas of stratification. Fennell et al. (1974) found that open dumps containing domestic refuse can be a reservoir of *Salmonella* bacteria that can be spread to nearby water bodies by foraging animals and birds.

When a drainage basin has much of its surface paved, the urban runoff bacteria concentrations can be expected to peak near the beginning of the rainfall event and then decrease as the event continues. Initial high levels of bacteria may be associated with direct flushing of feces from paved surfaces. These feces are from dogs defecating on parking lots and street areas and from birds roosting on rooftops. When a drainage area has a lot of landscaped areas or open land, relatively high bacteria concentrations in the urban runoff may occur throughout the rain event associated with continuous erosion of contaminated soils.

Fecal Coliform to Fecal Streptococci Bacteria Ratios

Geldreich (1965) found that the ratio of fecal coliform to fecal streptococci bacteria concentrations may be indicative of the probable fecal source. In fresh human fecal material and domestic wastes, he found that the fecal coliform densities were more than four times the fecal streptococcal densities. However, this ratio for livestock, poultry, dogs, cats, and rodents was found to be less than 0.6. These ratios must be applied carefully because of the effects of travel time and various chemical changes (especially pH) on the die-off rates of the component bacteria. This can result in the ratio changing, as the fecal coliform organisms tend to die faster than the fecal streptococcal bacteria. As a generality, he stated that fecal coliform to fecal streptococci ratios greater than 4 indicate that the bacteria pollution is from domestic wastes, which are composed mostly of human fecal material, laundry wastes, and food refuse. If the ratio is less than 0.6, the bacteria are probably from livestock or poultry in agricultural areas or from stormwater runoff in urban areas. He found that agricultural and stormwater runoff can be differentiated by studying the types of fecal streptococci bacteria found in the water samples. Geldreich and Kenner (1969) further stressed the importance of using this ratio carefully. They stressed that samples must be taken at the wastewater outfalls. At these locations, domestic waste, meat packing wastes, stormwater discharges, and feedlot drainage contain large numbers of fecal organisms recently discharged from warm-blooded animals. Once these organisms are diffused into the receiving stream, however, water temperature,

Table 3.10 Fecal Coliform to Fecal Streptococci Bacteria Population Ratios in Study Area

Source Areas	FC/FS Ratio
Rooftop runoff	0.5
Vacant land sheetflow	0.3
Parking lot sheetflow	0.2
Gutter flows	0.2
Average of source area values	0.3
Rideau River segment	
A	1.2
B	0.6
C	0.5
D	0.5
E	1.0
Average of river segment values	0.7
River swimming beaches	
Strathcona	2.8
Brantwood	2.3
Brighton	2.1
Mooney's Bay	1.7
Average of swimming beach values	2.2

From Pitt, R. *Urban Bacteria Sources and Control by Street Cleaning in the Lower Rideau River Watershed.* Rideau River Stormwater Management Study Technical Report. Prepared for the Ontario Ministry of the Environment, Environment Canada, Regional Municipality of Ottawa-Carleton, City of Ottawa, and Nepean. 1983.

organic nutrients, toxic metals, and adverse pH values may alter the relationship between the indicator organisms. This ratio should only be applied within 24 hours following the discharge of the bacteria.

Feachem (1975) examined how these ratios could be used with bacteria observations taken over a period of time. Because the fecal coliform and fecal streptococci bacteria die-off rates are not the same, the ratio gradually changes with time. He found that bacteria are predominantly from human sources if the FC/FS ratios are initially high (greater than 4) and then decrease with time. Nonhuman bacteria sources would result in initially low FC/FS ratios (less than 0.7), which then rise with time.

Pitt (1983) examined the FC/FS bacteria population ratios observed in the Rideau River study area in Ottawa, as shown in Table 3.10. These ratios were divided into groups corresponding to source area samples, Rideau River water samples, and water samples collected at the swimming beaches farther downstream. The source area sheet-flow samples contained the most recent pollution, while the river segment and beach samples contained "older" bacteria. The initial source area samples all had ratios of less than 0.7. However, the river averages ranged from 0.5 to 1.2, and the beach samples (which may be "older" than the river samples) ranged from 1.7 to 2.8. These ratios are seen to start with values less than 0.7 and increase with time. Based on Feachem's (1975) work, this would indicate that the major bacteria sources in the Rideau River are from nonhuman sources. Periodic high bacteria ratios in the river and at the beaches could be caused by the greater die-off ratio of fecal streptococci as compared to fecal coliform. The observed periodic high Rideau River FC/FS ratios (which can be greater than 4) may therefore be from old, nonhuman fecal discharges and not from fresh human fecal discharges.

Human Health Effects of Stormwater

There are several mechanisms whereby stormwater exposure can cause potential human health problems. These include exposure to stormwater contaminants at swimming areas affected by stormwater discharges, drinking water supplies contaminated by stormwater discharges, and the consumption of fish and shellfish that have been contaminated by stormwater pollutants. Understanding the risks associated with these exposure mechanisms is difficult and not very clear. Receiving waters where human uses are evident are usually very large, and the receiving waters are affected by many sanitary sewage and industrial point discharges, along with upstream agricultural nonpoint discharges, in addition to the local stormwater discharges. In receiving waters having only stormwater discharges, it is well known that inappropriate sanitary and other wastewaters are also discharging through the storm drainage system. These "interferences" make it especially difficult to identify specific cause-and-effect relationships associated with stormwater discharges alone, in contrast to the many receiving water studies that have investigated ecological problems that can more easily study streams affected by stormwater alone. Therefore, much of the human risk assessment associated with stormwater exposure must use theoretical evaluations relying on stormwater characteristics and laboratory studies in lieu of actual population studies. However, some site investigations, especially related to swimming beach problems associated with nearby stormwater discharges, have been conducted and are summarized (from Lalor and Pitt 1998) in the following discussion.

Contact recreation in pathogen-contaminated waters has been studied at many locations. The sources of the pathogens are typically assumed to be sanitary sewage effluent, or periodic industrial discharges from certain food preparation industries (especially meat packing and fish and shellfish processing). However, several studies have investigated pathogen problems associated with stormwater discharges. It has generally been assumed that the source of pathogens in stormwater are from inappropriate sanitary connections. However, stormwater unaffected by these inappropriate sources still contains high counts of pathogens that are also found in surface runoff samples from many urban surfaces. Needless to say, sewage contamination of urban streams is an important issue that needs attention during a receiving water investigation.

Inappropriate Sanitary Sewage Discharges into Urban Streams

Urban stormwater runoff includes waters from many other sources that find their way into storm drainage systems, besides from precipitation. There are cases where pollutant levels in storm drainage are much higher than they would otherwise be because of excessive amounts of contaminants that are introduced into the storm drainage system by various non-stormwater discharges. Additionally, baseflows (during dry weather) are also common in storm drainage systems. Dry-weather flows and wet-weather flows have been monitored during numerous urban runoff studies. These studies have found that discharges observed at outfalls during dry weather were significantly different from wet-weather discharges and may account for the majority of the annual discharges for some pollutants of concern from the storm drainage system.

In many cases, sanitary sewage was an important component (although not necessarily the only component) of the dry-weather discharges from the storm drainage systems. From a human health perspective (associated with pathogens), it may not require much raw or poorly treated sewage to cause a receiving water problem. However, at low discharge rates, the DO receiving water levels may be minimally affected. The effects these discharges have on receiving waters is therefore highly dependent on many site-specific factors, including frequency and quantity of sewage discharges and the creek flows. In many urban areas, the receiving waters are small creeks in completely developed watersheds. These creeks are the most at risk from these discharges as dry baseflows may be predominantly dry-weather flows from the drainage systems. In Tokyo (Fujita 1998), for example, numerous instances were found where correcting inappropriate sanitary sewage discharges resulted in the urban streams losing all of their flow. In cities adjacent to large receiving waters, these discharges likely have little impact (such as DO impacts from Nashville, TN, CSO discharges on the Cumberland River, as studied by Cardozo et al. 1994). The presence of pathogens from raw or poorly treated sewage in urban streams, however, obviously presents a potentially serious public health threat. Even if the receiving waters are not designated as water contact recreation, children are often seen playing in small city streams.

There have been a few epidemiology studies describing the increased health risks associated with contaminated dry-weather flows affecting public swimming beaches. The following discussion presents an overview of the development of water quality criteria for water contact recreation, plus the results of a recent epidemiological study that specifically examined human health problems associated with swimming in water affected by stormwater. In most cases, the levels of indicator organisms and pathogens causing increased illness were well within the range found in urban streams.

Runoff Pathogens and Their Sanitary Significance

The occurrence of *Salmonella* biotypes is typically low, and their reported density is less than one organism/100 mL in stormwater. *Pseudomonas aeruginosa* are frequently encountered at densities greater than 10 organisms/100 mL, but only after rains. The observed ranges of concentrations and percent isolations of bacterial biotypes vary significantly from site to site and at the same location for different times. Many potentially pathogenic bacteria biotypes may be present in urban runoff. Because of the low probability of ingestion of urban runoff, many of the potential human diseases associated with these biotypes are not likely to occur. The pathogenic organisms of most concern in urban runoff are usually associated with skin infections and body contact. The most important biotype causing skin infections would be *P. aeruginosa*. This biotype has been detected frequently in most urban runoff studies in concentrations that may cause infections. However, there is little information associating the cause and effect of increased *P.* concentrations with increased infections. *Shigella* may be present in urban runoff and receiving waters. This pathogen, when ingested in low numbers, can cause dysentery.

Salmonella

Salmonella has been reported in some, but not all, urban stormwaters. Qureshi and Dutka (1979) frequently detected *Salmonella* in southern Ontario stormwaters. They did not find any predictable patterns of *Salmonella* isolations; they were found throughout the various sampling periods. Olivieri et al. (1977a) found *Salmonella* frequently in Baltimore runoff, but at relatively low concentrations. Typical concentrations were from 5 to 300 *Salmonella* organisms/10 L. The concentrations of *Salmonella* were about ten times higher in the stormwater samples than in the urban stream receiving the runoff. The researchers also did not find any marked seasonal variations in *Salmonella* concentrations. Almost all of the stormwater samples that had fecal coliform concentrations greater that 2000 organisms/100 mL had detectable *Salmonella* concentrations, while about 275 of the samples having fecal coliform concentrations less than 200 organisms/100 mL had detectable *Salmonella*.

Quite a few urban runoff studies have not detected *Salmonella*. Schillinger and Stuart (1978) found that *Salmonella* isolations were not common in a Montana subdivision runoff study and that the isolations did not correlate well with fecal coliform concentrations. Environment Canada (1980) stated that *Salmonella* were virtually absent from Ottawa storm drainage samples in 1979. It concluded that *Salmonella* are seldom present in significant numbers in Ottawa urban runoff. The types of *Salmonella* found in southern Ontario were *S. thompson* and *S. typhimurium* var. *copenhagen* (Qureshi and Dutka 1979).

Olivieri et al. (1977b) stated that the primary human enteric disease producing *Salmonella* biotypes associated with the ingestion of water include *S. typhi* (typhoid fever), *S. paratyphi* (paratyphoid fever), and *Salmonella* species (salmonellosis). These biotypes are all rare except for

Salmonella sp. The dose of *Salmonella* sp. required to produce an infection is quite large (approximately 10^5 organisms). The salmonellosis health hazard associated with water contact in urban streams is believed to be small because of this relatively large infective dose. If 2 L of stormwater having typical *Salmonella* concentrations (10 *Salmonella* organisms/10 L) is ingested, less than 0.001 of the required infective dose would be ingested. If a worst-case *Salmonella* stormwater concentration of 10,000 organisms/10 L occurred, the ingestion of 20 L of stormwater would be necessary for an infective dose. They stated that the low concentrations of *Salmonella*, coupled with the unlikely event of consuming enough stormwater, make the *Salmonella* health hazard associated with urban runoff small.

Staphylococcus

Staphylococcus aureus is an important human pathogen it can cause boils, carbuncles, abscesses, and impetigo on skin on contact. Olivieri et al. (1977b) stated that the typical concentrations of *Staphylococci* are not very high in urban streams. They also noted that there was little information available relating the degree of risk of staph infections with water concentrations. They concluded that *Staphylococcus aureus* appears to be the most potentially hazardous pathogen associated with urban runoff, but there is no evidence available that skin, eye, or ear infections can be caused by the presence of this organism in recreational waters. They concluded that there is little reason for extensive public health concern over recreational waters receiving urban storm runoff containing staph organisms.

Shigella

Olivieri et al. (1977b) stated that there is circumstantial evidence that *Shigella* is present in urban runoff and receiving waters and could present a significant health hazard. *Shigella* species causing bacillary dysentery are one of the primary human enteric disease-producing bacteria agents present in water. The infective dose of *Shigella* necessary to cause dysentery is quite low (10 to 100 organisms). Because of this low required infective dose and the assumed presence of *Shigella* in urban waters, it may be a significant health hazard associated with urban runoff.

Streptococcus

Streptococcus faecalis and atypical *S. faecalis* are of limited sanitary significance (Geldreich 1976). *Streptococcus* determinations on urban runoff are most useful for identifying the presence of *S. bovis* and *S. equinus,* which are specific indicators of nonhuman, warm-blooded animal pollution. However, it is difficult to interpret fecal streptococcal data when their concentrations are lower than 100 organisms/100 mL because of the ubiquitous occurrence of *S. faecalis* var. *liquifaciens.* This biotype is generally the predominant streptococcal biotype occurring at low fecal streptococcal concentrations.

Pseudomonas aeruginosa

Pseudomonas is reported to be the most abundant pathogenic bacteria in urban runoff and streams (Olivieri et al. 1977b). This pathogen is associated with eye and ear infections and is resistant to antibiotics. Oliveri et al. also stated that past studies have failed to show any relationships between *P. aeruginosa* concentrations in bathing waters and ear infections. However, *Pseudomonas* concentrations in urban runoff are significantly higher (about 100 times) than the values associated with past bathing beach studies. Cabelli et al. (1976) stated that *P. aeruginosa* is indigenous in about 15% of the human population. Swimmer's ear or other *Pseudomonas* infections may, therefore, be caused by trauma to the ear canals associated with swimming and diving, and not exposure to *Pseudomonas* in the bathing water.

Environment Canada (1980) stated that there is preliminary evidence of the direct relationship between very low levels of *P. aeruginosa* and an increase in incidents of ear infections in swimmers. It stated that a control level for this *Pseudomonas* biotype of between 23 and 30 organisms/100 mL was considered. Cabelli et al. (1976) stated that *P. aeruginosa* densities greater than 10 organisms/100 mL were frequently associated with fecal coliform levels considerably less than 200 organisms/100 mL. *Pseudomonas aeruginosa* densities were sometimes very low when the fecal coliform levels were greater than 200 organisms/100 mL. An average estimated *P. aeruginosa* density associated with a fecal coliform concentration of 200 organisms/100 mL is about 12/100 mL. It further stated that *P. aeruginosa* by itself cannot be used as a basis for water standards for the prevention of enteric diseases during recreational uses of surface waters. The determinations of this biotype should be used in conjunction with fecal coliform or other indicator organism concentrations for a specific location. It recommended that bathing beaches that are subject to urban runoff be temporarily closed until the *P. aeruginosa* concentrations return to a baseline concentration.

Campylobacter

Koenraad et al. (1997) investigated the contamination of surface waters by *Campylobacter* and its associated human health risks. They reported that campylobacteriosis is one of the most frequently occurring acute gastroenteritis diseases in humans. Typical investigations have focused on the consumption of poultry, raw milk, and untreated water as the major sources of this bacterial illness. Koenraad et al. (1997) found that human exposures to *Campylobacter*-contaminated surface waters is likely a more important risk factor than previously considered. In fact, they felt that *Campylobacter* infections may be more common than *Salmonella* infections. The incidence of campylobacteriosis due to exposure to contaminated recreational waters has been estimated to be between 1.2 to 170 per 100,000 individuals. The natural habitat of *Campylobacter* is the intestinal tract of warm-blooded animals (including poultry, pigs, cattle, gulls, geese, pigeons, magpies, rodents, shellfish, and even flies). It does not seem to multiply outside of its host, but it can survive fairly well in aquatic environments. It can remain culturable and infective for more than 2 months under ideal environmental conditions. Besides runoff, treated wastewater effluent is also a major source of *Campylobacter* in surface waters. Sanitary wastewater may contain up to 50,000 MPN of *Campylobacter* per 100 mL, with 90 to 99% reductions occurring during typical wastewater treatment.

Cryptosporidium, Giardia, *and* Pfiesteria

Protozoa became an important public issue with the 1993 *Cryptosporidium*-caused disease outbreak in Milwaukee when about 400,000 people become ill from drinking contaminated water. Mac Kenzie et al. (1994) prepared an overview of the outbreak, describing the investigation of the causes of the illness and the number of people affected. They point out that *Cryptosporidium*-caused disease in humans was first documented in 1976, but had received little attention and no routine monitoring. *Cryptosporidium* is now being monitored routinely in many areas and is the subject of much research concerning its sources and pathways. At the time of the Milwaukee outbreak, both of the city's water treatment plants (using water from Lake Michigan) were operating within acceptable limits, based on required monitoring. However, at one of the plants (which delivered water to most of the infected people), at the time of the outbreak the treated water underwent a large increase in turbidity (from about 0.3 NTU to about 1.5 NTU) that was not being well monitored (the continuous monitoring equipment was not functioning, and values were obtained only every 8 hours). More than half of the residents receiving water from this plant became ill. The plant had recently changed its coagulant from polyaluminum chloride to alum, and equipment to assist in determining the correct chemical dosages was not being used. The finished water had apparently relatively high levels of *Cryptosporidium* because some individuals became ill after drinking less than 1 L of water.

Cryptosporidium oocysts have often been found in untreated surface waters, and it was thought that *Cryptosporidium* oocysts entered the water treatment supply before the increase in turbidity was apparent. MacKenzie et al. (1994) point out that monitoring in the United Kingdom has uncovered sudden, irregular, community-wide increases in cryptosporidiosis that were likely caused by waterborne transmission. They also stated that the source of the *Cryptosporidium* oocysts was speculative, but could have included cattle feces contamination in the Milwaukee and Menomonee Rivers, slaughterhouse wastes, and human sewage. The rivers were also swelled by high spring rains and snowmelt runoff that may have aided the transport of upstream *Cryptosporidium* oocysts into the lake near the water intakes.

The *Journal of the American Water Works Association* has published numerous articles on protozoa contamination of drinking water supplies. Crockett and Haas (1997) describe a watershed investigation to identify sources of *Giardia* and *Cryptosporidium* in the Philadelphia watershed. They describe the difficulties associated with monitoring *Cryptosporidium* and *Giardia* in surface waters because of low analytical recoveries and the cost of analyses. Large variations in observed protozoa concentrations made it difficult to identify major sources during the preliminary stages of their investigations. They do expect that wastewater treatment plant discharges are a major local source, although animals (especially calves and lambs) are likely significant contributors. Combined sewer overflows had *Giardia* levels similar to raw sewage, but the CSOs had much less *Cryptosporidium* than the raw sewage. LeChevallier et al. (1997) investigated *Giardia* and *Cryptosporidium* in open reservoirs storing finished drinking water. This gave them an opportunity to observe small increases in oocyst concentrations associated from nonpoint sources of contamination from the highly controlled surrounding area. They observed significantly larger oocyst concentrations at the effluent (median values of 6.0 *Giardia*/100 L and 14 *Cryptosporidium*/100 L) in the reservoirs than in the influents (median values of 1.6 *Giardia*/100 L and 1.0 *Cryptosporidium*/100 L). No human wastes could influence any of the tested reservoirs, and the increases were therefore likely caused by wastes from indigenous animals or birds, either directly contaminating the water or through runoff from the adjacent wooded areas.

A Management Training Audioconference Seminar on *Cryptosporidium* and Water (MTA 1997) was broadcast in May of 1997 to familiarize state and local agencies about possible *Cryptosporidium* problems that may be evident as a result of the EPA's Information Collection Rule which began in July of 1997. This regulation requires all communities serving more than 100,000 people to monitor their source water for *Cryptosporidium* oocysts. If the source water has more than 10 *Cryptosporidium* oocysts/L, the finished water must also be monitored. It is likely that many source waters will be found to be affected by *Cryptosporidium*. The researchers reviewed one study that found the percentage of positive samples of *Cryptosporidium* in lakes, rivers, and springs was about 50 to 60% and about 5% in wells. In contrast, the percentage of samples testing positive for *Giardia* was about 10 to 20% in lakes and rivers, and very low in springs and wells.

Special human health concerns have also been recently expressed about *Pfiesteria piscicida*, a marine dinoflagellate that is apparently associated with coastal eutrophication caused by runoff nutrients (Maguire and Walker 1997). Dramatic blooms and resulting fish kills have been associated with increased nutrient loading from manure-laden runoff from large livestock feedlot operations. This organism has garnered much attention in the popular press, usually called the "cell from hell" (Zimmerman 1998). It has been implicated as causing symptoms of nausea, fatigue, memory loss, and skin infections in south Atlantic coastal bay watermen. *Pfiesteria* and *Pfiesteria*-like organisms have also been implicated as the primary cause of many major fish kills and fish disease events in Virginia, Maryland, North Carolina, and Delaware. In August 1997, hundreds of dead and dying fish were found in the Pocomoke River, near Shelltown, MD, in the Chesapeake Bay, prompting the closure of a portion of the river. Subsequent fish kills and confirmed occurrences of *Pfiesteria* led to further closures of the Manokin and Chicamacomico Rivers. The Maryland Department of Health and Mental Hygiene also presented preliminary evidence that adverse public health effects could result from exposure to the toxins released by *Pfiesteria* and *Pfiesteria*-like organisms. The increasing numbers of fish kills of Atlantic menhaden (an oily, non-game fish) motivated Maryland's governor to appoint a Citizens *Pfiesteria* Action Commission. The commission convened a forum of noted scientists to examine the existing information on *Pfiesteria*. The results of the State of Maryland's *Pfiesteria* monitoring program are available on the Maryland Department of Natural Resources' Web site: http://www.dnr.state.md.us/pfiesteria/.

Pfiesteria has a complex life cycle, including at lease 24 flagellated, amoeboid, and encysted stages. Only a few of these stages appear to be toxic, but their complex nature makes them difficult to identify by non-experts (Maguire and Walker 1997). *Pfiesteria* spends much of its life span in a nontoxic predatory form, feeding on bacteria and algae, or as encysted dormant cells in muddy sediment. Large schools of oily fish (such as the Atlantic menhaden) trigger the encysted cells to emerge and excrete toxins. These toxins make the fish lethargic, so the fish remain in the area where the toxins attack the fish skin, causing open sores to develop. The *Pfiesteria* then feed on the sloughing fish tissue. Unfortunately, people working in the water during these toxin releases may also be affected (Zimmerman 1998).

Researchers suggest that excessive nutrients (causing eutrophication) increase the algae and other organic matter that the *Pfiesteria* and Atlantic menhaden use for food. The increased concentrations of *Pfiesteria* above natural background levels increase the likelihood of toxic problems. Maguire and Walker (1997) state that other factors are also apparently involved, including stream hydraulics, water temperature, and salinity. They feel that *Pfiesteria* is only one example of the increasing threats affecting coastal ecosystems that are experiencing increased nutrient levels. Most of the resulting algal blooms only present nuisance conditions, but a small number can result in human health problems (mostly as shellfish poisonings). The increased nutrient discharges are mostly associated with agricultural operations, especially animal wastes from large poultry and swine operations. In the Pocomoke River watershed, the Maryland Department of Natural Resources estimates that about 80% of the phosphorus and 75% of the nitrogen load is from agricultural sources. Urban runoff may also be a causative factor of eutrophication in coastal communities, especially those having small enclosed coastal lagoons or embayments, or in rapidly growing urban areas. Zimmerman (1998) points out that the Chesapeake Bay area is one of the country's most rapidly growing areas, with the population expected to increase by 12% by the year 2010.

Viruses

It is believed that approximately half of all waterborne diseases are of viral origin. Unfortunately, it is very difficult and time-consuming to identify viruses from either environmental samples or sick individuals. When the EPA conducted its extensive epidemiological investigations of freshwater and marine swimming beaches in the 1980s, two viruses common to human gastrointestinal tracts (coliphage and enterovirus) were evaluated as potential pathogen indicators. These two indicators did not show good correlations between their presence and the incidence of gastroenteritis. Viruses tend to survive for slightly longer periods in natural waters than do Gram-negative bacteria. It is believed that the high correlation observed between gastroenteritis and the presence of enterococci may be because the Gram-positive enterococci's longer survival more closely mimics viral survival. Therefore, enterococci may serve as a good recreational water indicator for the presence of viral pathogens.

RECEIVING WATER EFFECT SUMMARY

Recent studies have combined chemical-physical characterizations of water and sediment with biosurveys and laboratory/*in situ* toxicity surveys (low and high flow) to effectively characterized major water column and sediment stressors (Burton and Rowland 1999; Burton et al. 1998; Dyer and White 1996; Burton and Moore 1999). Suspended solids, ammonia, sediments, temperature,

PAHs, and/or stormwater runoff were observed to be primary stressors in these test systems. These primary stressors could not have been identified without low and high flow and sediment quality assessments both in the laboratory and field. It is apparent that to determine the role of chemicals as stressors in the receiving waters, the role of other stressors (both natural and anthropogenic) must be assessed (see also Chapters 6 and 8).

Johnson et al. (1996) and Herricks et al. (1996a,b) describe a structured tier testing protocol to assess both short-term and long-term wet-weather discharge toxicity. The protocol recognizes that the test systems must be appropriate to the time-scale of exposure during the discharge. Therefore, three time-scale protocols were developed, for intra-event, event, and long-term exposures.

There is a natural tendency in the popular "weight-of-evidence" or "sediment quality triad" approaches to look for "validation" of one assessment tool with another (see also Chapters 6 and 8). For example, matching a toxic response in a WET test with that of an impaired community gives a greater weight of evidence. This does not, however, necessarily "validate" the results (or invalidate, if there are differences) (Chapman 1995). Natural temporal changes in aquatic populations at different sites within a study system need not be the same (Power et al. 1988; Resh 1988; Underwood 1993); therefore, predictions of effect or no-effect from WET testing of reference sites may be in error. Each monitoring tool (i.e., chemical, physical, and indigenous biota characterizations, laboratory and field toxicity, and bioaccumulation) provides unique and often essential information (Burton 1995; Chapman et al. 1992; Burton et al. 1996; Baird and Burton 2001). If the responses of each of the biological tools disagree, it is likely due to species differences or a differing stressor exposure dynamic/interaction. These critical exposures issues can be characterized through a systematic process of separating stressors and their respective dynamics into low and high flow and sediment compartments using both laboratory and field exposures. Then, a more efficient and focused assessment can identify critical stressors and determine their ecological significance with less uncertainty than the more commonly used approaches. The chronic degradation potential of complex ecosystems receiving multiple stressors cannot be adequately evaluated without a comprehensive assessment that characterizes water, sediment, and biological dynamics and their interactions.

Because most sites have multiple stressors (physical, chemical, and biological), it is essential that the relative contributions of these stressors be defined to design effective corrective measures. The integrated laboratory and field approach rigorously defines the exposures of organisms (media of exposure and contaminant concentration), separating it into overlying water, surficial sediment, historical sediment, and interstitial water. The degree of contaminant-associated toxicity can best be assessed using a combination of laboratory and field screening methods which separate stressors (i.e., a Stressor Identification Evaluation (SIE) approach) (Burton et al. 1996), into different, major stressor categories, including metals, nonpolar organics, photoinduced toxicity from PAHs, ammonia, suspended solids, predators, dissolved oxygen, and flow. There is much research to be done to refine these approaches, but the tools are there to make ecologically relevant assessments of aquatic ecosystem contamination with reasonable certainty.

The effects of urban runoff on receiving water aquatic organisms or other beneficial uses is also very site specific. Different land development practices may create substantially different runoff flows. Different rain patterns cause different particulate washoff, transport, and dilution conditions. Local attitudes also define specific beneficial uses and desired controls. There are also a wide variety of water types receiving urban and agricultural runoff, and these waters all have watersheds that are urbanized to various degrees. Therefore, it is not surprising that runoff effects, though generally dramatic, are also quite variable and site specific.

Previous attempts to identify runoff problems using existing data have not generally been conclusive because of differences in sampling procedures and the common practice of pooling data from various sites or conditions. It is therefore necessary to carefully design comprehensive, long-term studies to investigate runoff problems on a site-specific basis. Sediment transport, deposition, and chemistry play key roles in receiving waters and need additional research. Receiving water aquatic biological conditions, especially compared to unaffected receiving waters, should be studied in preference to laboratory bioassays.

These specific studies need to examine beneficial uses directly, and not rely on published water quality criteria and water column measurements alone. Published criteria are usually not applicable to urban runoff because of the sluggish nature of runoff and the unique chemical speciation of its components.

The long-term aquatic life effects of runoff are probably more important than short-term effects associated with specific events. The long-term effects are probably related to the deposition and accumulation of toxic sediments, or the inability of the aquatic organisms to adjust to repeated exposures to high concentrations of toxic materials or high flow rates.

REFERENCES

Anderson, D.G. *Effects of Urban Development of Floods in Northern Virginia,* U.S. Geological Survey, Water Supply Paper 2001-C, Washington, D.C. 1970.

APHA (American Public Health Association). *Standard Methods for the Examination of Water and Wastewater, 18th edition.* American Public Health Association, American Water Works Association, Water Pollution Control Federation. 1992.

Bailey, H.C., L. Deanovic, E. Reyes, T. Kimball, K. Larson, K. Cortright, V. Connor, and D.E. Hinton. Diazinon and chlorpyrifos in urban waterways in northern California, USA. *Environ. Toxicol. Chem.*, 19: 82–87. 2000.

Baird, D. and G.A. Burton, Jr. (Eds.) *Ecosystem Complexity: New Directions for Assessing Responses to Stress.* Pellston Workshop Series. SETAC Press. To be published in 2001.

Bascombe, A.D., J.B. Ellis, D.M. Revitt, and R.B.E. Shutes. Development of ecotoxicological criteria in urban catchments. *Water Sci. Techn.*, 22(10/1) 173–179. 1990.

Benke, A.C., G.E. Willeke, F.K. Parrish, and D.L. Stites. *Effects of Urbanization on Stream Ecosystems.* School of Biology, Environmental Resources Center, Report No. ERC 07-81, Georgia Institute of Technology, Atlanta, GA. 1981.

Bhaduri, B., M. Grove, C. Lowry, and J. Harbor. Assessing long-term hydrologic effects of land use change. *J. AWWA,* 89(11): 94–106. Nov. 1997.

Brookes, A. *Channelized Rivers: Perspectives for Environmental Management.* John Wiley & Sons, New York. 1988.

Brookes, A. Design issues. Engineering Foundation Conference: *Effects of Urban Runoff on Receiving Systems: An Interdisciplinary Analysis of Impact, Monitoring, and Management.* Mt. Crested Butte, CO. Publisher by the American Society of Civil Engineers, New York. 1991.

Burton G.A., C.G. Ingersoll, L.C. Burnett, M. Henry, M.L. Hinman, S.J. Klaine, P.F. Landrum, P. Ross, and M. Tuchman. A comparison of sediment toxicity test methods at three Great Lakes Areas of Concern. *J. Great Lakes Res.,* 22: 495–511. 1996.

Burton G.A., Jr. Quality assurance issues in assessing receiving water, in *Stormwater Runoff and Receiving Systems.* Edited by E.E. Herricks. CRC/Lewis, Boca Raton, FL. pp. 275–284. 1995.

Burton, G.A., Jr., C. Rowland, K. Kroeger, M. Greenberg, D. Lavoie, and J. Brooker. Determining the effect of ammonia at complex sites: laboratory and in situ approaches. *Abstr. Annu. Meet. Soc. Environ. Toxicol. Chem.,* 563: 120, Pensacola, FL. 1998.

Burton, G.A., Jr., and C. Rowland. *Continuous in Situ Toxicity Monitoring and Thermal Effect Characterization.* Final Report to Upper Illinois Waterway Task Force. Commonwealth Edison Corp. Chicago, IL. 1999.

Burton, G.A., Jr. Microbiological water quality, in *Microbial Processes in Reservoirs,* Edited by D. Gunnison, Junk Publishers. pp. 79–97. 1985.

Burton, G. A., Jr., T. Giddings, P. DeBrine, and R. Fall. A high incidence of selenite-resistant bacteria from a site polluted with selenium, *Appl. Environ. Microbiol.,* 53: 185–188. 1987.

Burton, G.A., Jr. and B.L. Stemmer. Evaluation of surrogate tests in toxicant impact assessments. *Toxicity Assess. Int. J.,* 3: 255–269. 1988.

Burton, G.A., Jr., and L. Moore. *An Assessment of Stormwater Runoff Effects in Wolf Creek, Dayton, OH.* Final Report. City of Dayton, OH. 1999.

Cabelli, V.J., H. Kennedy, and M.A. Levin. *Pseudomonas aeruginosa*-fecal coliform relationships in estuarine and fresh recreational waters. *J. WPCF,* 48(2): 367–376. Feb. 1976.

Callahan, M.A., M.W. Slimak, N.W. Gabel, I.P. May, C.F. Fowler, J.R. Freed, P. Jennings, R.L. Durfee, F.C. Whitmore, B. Maestri, W.R. Mabey, B.R. Holt, and C. Gould. *Water Related Environmental Fates of 129 Priority Pollutants.* U.S. Environmental Protection Agency, Monitoring and Data Support Division, EPA-4-79-029a and b. Washington, D.C. 1979.

Cardozo, R.J., W.R. Adams, and E.L. Thackston. CSO's real impact on water quality: the Nashville experience. *A Global Perspective for Reducing CSOs: Balancing Technologies, Costs, and Water Quality.* July 10–13, 1994. Louisville, KY. Water Environment Federation. Alexandria, VA. 1994.

Carlile, B., B. McNeal, J. Kittrick, L. Johnson, and H. Cheng. *Characterization of Suspended Sediments in Water from Selected Watersheds as Related to Control Processes, Nutrient Contents and Lake Eutrophication,* PB232-167. National Technical Information Service, Springfield, VA. 1974.

Chapman, P.M. Extrapolating laboratory toxicity results to the field. *Environ. Toxicol. Chem.,* 14: 927–930. 1995.

Chapman, P.M., E.A. Power, and G.A. Burton, Jr. Integrative assessments in aquatic ecosystems, in *Sediment Toxicity Assessment.* Edited by G.A. Burton, Jr., Lewis Publishers, Boca Raton, FL. 1992.

Cook, W.L., F. Parrish, J.D. Satterfield, W.G. Nolan, and P.E. Gaffney. *Biological and Chemical Assessment of Nonpoint Source Pollution in Georgia: Ridge-Valley and Sea Island Streams,* Department of Biology, Georgia State University, Atlanta, GA. 1983.

Cordone, A.J. and D.W. Kelley. Influences of inorganic sediments on aquatic life of streams. *Cal. Fish Game,* 47: 189–228. 1961.

Crockett, C.S. and C.N. Hass. Understanding protozoa in your watershed. *J. Am. Water Works Assoc.,* 89(9): 62–73. Sept. 1997.

Crunkilton, R., J. Kleist, J. Ramcheck, B. DeVita, and D. Villeneuve. Assessment of the response of aquatic organisms to long-term *in situ* exposures to urban runoff. Engineering Foundation Conference: *Effects of Watershed Development & Management on Aquatic Ecosystems,* Snowbird, UT. Published by the American Society of Civil Engineers, New York. August 1996.

CTA, Inc. *Georgia Nonpoint Source Impact Assessment Study: Blue Ridge/Upland Georgia Cluster, Piedmont Cluster, and Gulf Coastal Plain Cluster,* Georgia Environmental Protection Division, Department of Natural Resources, Atlanta, GA. 1983.

Dyer, S.D. and C.E. White. A watershed approach to assess mixture toxicity via integration of public and private databases. Abstract Book: *SETAC 17th Annual Meeting.* pp. 96. Washington, D.C., Nov. 17–21, 1996.

Easton, J. *The Development of Pathogen Fate and Transport Parameters for Use in Assessing Health Risks Associated with Sewage Contamination.* Ph.D. dissertation, Dept. of Civil and Environmental Engineering, University of Alabama at Birmingham. 2000.

Ebbert, J.C., J.E. Poole, and K.L. Payne. *Data Collected by the U.S. Geological Survey during a Study of Urban Runoff in Bellevue, Washington, 1979–82.* Preliminary U.S. Geological Survey Open-File Report, Tacoma, WA. 1983.

Eccher. C.J. Thoughtful design is prime factor in water safety. *Lake Line,* pp. 4–8. May 1991.

Ehrenfeld, J.G. and J.P. Schneider. *The Sensitivity of Cedar Swamps to the Effects of Non-Point Pollution Associated with Suburbanization in the New Jersey Pine Barrens,* PB8-4-136779, U.S. Environmental Protection Agency, Office of Water Policy, Washington, D.C. 1983.

Ellis, J., R. Shutes, and D. Revitt. Ecotoxicological approaches and criteria for the assessment of urban runoff impacts on receiving waters, in *Proc. of the Effects of Urban Runoff on Receiving Systems: An Interdisciplinary Analysis of Impact, Monitoring and Management,* Edited by E. Herricks, J. Jones, and B. Urbonas. Engineering Foundation, New York. 1992.

Ellis, J.B., R. Hamilton, and A.H. Roberts. Composition of Suspended solids in urban stormwater. *Second International Conference on Urban Storm Drainage,* Urbana, IL. June 1981.

Environment Canada. *Rideau River Water Quality and Stormwater Monitor Study, 1979.* MS Rept. No. OR-29. Feb. 1980.

EPA (U.S. Environmental Protection Agency). *Suspended and Dissolved Solids Effects on Freshwater Biota: A Review,* Environmental Research Laboratory, U.S. Environmental Protection Agency, Corvallis, OR, EPA 600/3-77/042. 1977.

EPA (U.S. Environmental Protection Agency). *Results of the Nationwide Urban Runoff Program.* Water Planning Division, PB 84-185552, Washington, D.C. December 1983.

EPA (U.S. Environmental Protection Agency). *Quality Criteria for Water.* EPA 440/5-86-001. U.S. Environmental Protection Agency, Washington, D.C. May 1986.

EPA. *Biomonitoring to Achieve Control of Toxic Effluents,* Office of Water, U.S. Environmental Protection Agency, Washington, D.C., EPA 625/8-87/013. 1987a.

EPA. *An Overview of Sediment Quality in the United States,* Office of Water Regulations and Standards, U.S. Environmental Protection Agency, Washington, D.C. and Region 5, Chicago, IL, EPA 905/9-88/002. 1987b.

EPA. *Short-Term Methods for Estimating the Chronic Toxicity of Effluents and Receiving Water to Freshwater Organisms.* Research and Development, Cincinnati, OH. EPA/600/4-91/002. 1995.

EPA. *National Water Quality Inventory. 1994 Report to Congress.* Office of Water. EPA 841-R-95-005. Washington, D.C. December 1995.

Feachem, R. An improved role for faecal coliform to faecal streptococci ratios in the differentiation between human and non-human pollution sources. *Water Res.,* 9: 689–690. 1975.

Fennell, H., D.B. James, and J. Morris. Pollution of a storage reservoir by roosting gulls. *J. Soc. Water Treatment Exam.,* 23(1): 5–24. 1974.

Fujita, S. Restoration of polluted urban watercourses in Tokyo for community use. Engineering Foundation Conference: *Sustaining Urban Water Resources in the 21st Century.* September 7–12, 1997. Malmo, Sweden. Published by the American Society of Civil Engineers, New York. 1998.

Garie, H.L. and A. McIntosh. Distribution of benthic macroinvertebrates in a stream exposed to urban runoff, *Water Res. Bull.,* 22: 447. 1986.

Garie, H.L. and A. McIntosh. Distribution of benthic macroinvertebrates in a stream exposed to urban runoff. *Water Sci. Technol.,* 22: 10/11. 1990.

Geldreich, E.E. Origins of microbial pollution in streams. In *Transmission of Viruses by the Water Route.* Edited by G. Berg, Interscience Publishers, New York. 1965.

Geldreich, E.E. and Kenner, B.A. Concepts of fecal streptococci in stream pollution. *J. Water Pollut. Control Fed.,* 41(8): R336–R352. 1969.

Geldreich, E.E. "Fecal coliform and fecal streptococcus density relationships in waste discharges and receiving waters." *Crit. Rev. Environ. Control,* 6(4): 349. Oct. 1976.

Geldreich, E.E., H.D. Nash, D.F. Spino, and D.J. Reasoner. Bacterial dynamics in a water supply reservoir: a case study. *J. AWWA,* Jan. 1980.

Gore & Storrie Ltd./Proctor & Redfern Ltd. *Executive Summary Report on Rideau River Stormwater Management Study, Phase I.* Rideau River Stormwater Management Study, Ottawa, and the Ontario Ministry of the Environment, Kingston, Ontario. 1981.

Grey, G. and F. Oliveri. Catch basins — effective floatables control devices. Presented at the *Advances in Urban Wet Weather Pollution Reduction* conference. Cleveland, OH, June 28–July 1, 1998. Water Environment Federation, Alexandria, VA. 1998.

Gupta, M.K., R.W. Agnew, D. Gruber, and W. Kreutzberger. *Constituents of Highway Runoff. Vol. IV, Characteristics of Highway Runoff from Operating Highways.* FHWA Rept. No. fhwa/rd-81/045. Feb. 1981.

Hammer, T. R. Stream channel enlargement due to urbanization, *Water Resources Res.,* 8:453-471. 1972.

Heaney, J.P. *Nationwide Assessment of Receiving Water Impacts from Urban Storm Water Pollution: First Quarterly Progress Report.* Environmental Engineering Sciences, University of Florida, Gainesville, FL. 1978.

Heaney, J.P., W.C. Huber, and M.E. Lehman. *Nationwide Assessment of Receiving Water Impacts from Urban Storm Water Pollution.* U.S. Environmental Protection Agency, Cincinnati, OH. April 1980.

Heaney, J.P. and W.C. Huber. Nationwide assessment of urban runoff impact on receiving water quality. *Water Res. Bull.,* 20(1): 35–42. February 1984.

Herricks, E.E, I. Milne, and I. Johnson. A protocol for wet weather discharge toxicity assessment. Vol. 4, pp. 13–24. *WEFTEC'96: Proceedings of the 69th Annual Conference & Exposition,* Dallas, TX. 1996a.

Herricks, E.E., R. Brent, I. Milne, and I. Johnson. Assessing the response of aquatic organisms to short-term exposures to urban runoff. Engineering Foundation Conference: *Effects of Watershed Development & Management on Aquatic Ecosystems,* Snowbird, UT. Published by the American Society of Civil Engineers, New York. August 1996b.

Hütter, U. and F. Remmler. Stormwater infiltration at a site with critical subsoil conditions: investigations of soil, seepage water, and groundwater. *Proceedings: 7th International Conference on Urban Storm Drainage.* Hannover, Germany. Sept. 9–13, 1996. Edited by F. Sieker and H-R. Verworn. IAHR/IAWQ. SuG Verlagsgesellschaft. Hannover, Germany. pp. 713–718. 1996.

HydroQual, Inc. *Floatables Pilot Program Final Report: Evaluation of Non-Structural Methods to Control Combined and Storm Sewer Floatable Materials.* City-Wide Floatables Study, Contract II. Prepared for New York City, Department of Environmental Protection, Bureau of Environmental Engineering, Division of Water Quality Improvement. NYDP2000. December 1995.

Ireland, D.S., G.A. Burton, Jr., and G.G. Hess. In-situ toxicity evaluations of turbidity and photoinduction of polycyclic aromatic hydrocarbons. *Environ. Toxicol. Chem.*, 15(4): 574–581. April 1996.

Johnson, I., E.E. Herricks, and I. Milne. Application of a test battery for wet weather discharge toxicity analyses. Vol. 4, pp. 219–229. *WEFTEC'96: Proceedings of the 69th Annual Conference & Exposition.* Dallas, TX. 1996.

Jones, J.E. and D.E. Jones, Interfacing considerations in urban detention ponding. *Proceedings of the Conference on Stormwater Detention Facilities, Planning, Design, Operation, and Maintenance*, Henniker, NH, Edited by W. DeGroot, published by the American Society of Civil Engineers, New York. August 1982.

Keefer, T.N., R.K. Simons, and R.S. McQuivey. Dissolved Oxygen Impact from Urban Storm Runoff. EPA-600/2-79-150, U.S. Environmental Protection Agency, Cincinnati, OH. March 1979.

Ketchum, L.H., Jr. *Dissolved Oxygen Measurements in Indiana Streams during Urban Runoff. EPA-600/2-78-135.* U.S. Environmental Protection Agency, Cincinnati, OH. August 1978.

Klein, R.D. Urbanization and stream quality impairment. *Water Resour. Bull.*, 15(4). August 1979.

Koenradd, P.M.F.J., F.M. Rombouts, and S.H.W. Notermans. Epidemiological aspects of thermophilic *Campylobacter* in water-related environments: a review. *Water Environ. Res.*, 69(1): 52–63. Jan–Feb 1997.

Kuehne, R.A. *Evaluation of Recovery in a Polluted Creek after Installment of New Sewage Treatment Procedures.* University of Kentucky Water Resources Research Institute, Lexington, Kentucky. May 1975.

Lalor, M. and R. Pitt. *Assessment Strategy for Evaluating the Environmental and Health Effects of Sanitary Sewer Overflows from Separate Sewer Systems.* First Year Report. Wet-Weather Flow Management Program, National Risk Management Research Laboratory, U.S. Environmental Protection Agency, Cincinnati, OH. January 1998.

LaRoe, E.T. *Instream Impacts of Soil Erosion on Fish and Wildlife.* Division of Biological Services, U.S. Fish and Wildlife Service. May 1985.

LeChevallier, M.W., W.D. Norton, and T.B. Atherholt. Protozoa in open reservoirs. *J. Am.Water Works Assoc.*, 89(9): 84–96. Sept. 1997.

Lenat, D.R., D.L. Penrose, and K. Eagleson. *Biological Evaluation of Non-Point Sources of Pollutants in North Carolina Streams and Rivers*, North Carolina Division of Environmental Management, Biological Series #102, North Carolina Dept. of Natural Resources and Community Development, Raleigh, NC. 1979.

Lenat, D.R., D.L. Penrose, and K.W. Eagleson. Variable effects of sediment addition on stream benthos. *Hydrobiologia*, 79: 187–194. 1981.

Lenat, D. and K. Eagleson. *Ecological Effects of Urban Runoff on North Carolina Streams. North Carolina Division of Environmental Management, Biological Series #104.* North Carolina Dept. of Natural Resources and Community Development, Raleigh, NC. 1981.

Leopold, L.B. *Hydrology for Urban Land Planning: A Guidebook on the Hydrologic Effects of Land Use,* U.S. Geological Survey, Washington, D.C., Circular 554. 1968.

Lin, S. *Nonpoint Rural Sources of Water Pollution.* Circular III, Illinois State Water Survey, Urbana, IL, 1972.

Loijens, H.S. *Status Report on the Rideau River Stormwater Management Study.* Rideau River Stormwater Management Study, Ottawa, and the Ontario Ministry of the Environment, Kingston, Ontario. June 1981.

Mac Kenzie, W.R., N.J. Hoxie, M.E. Proctor. M.S. Gradus, K.A. Blair, D.E. Peterson, J.J. Kazmierczak, D.G. Addiss, K.R. Fox, J.B. Rose, and J.P. Davis. A massive outbreak in Milwaukee of cryptosporidium infection transmitted through the public water supply. *N. Engl. J. Med.*, 331(3): 161–167. July 21, 1994.

MacRae, C.R. Experience from morphological research on Canadian streams: Is control of the two-year frequency runoff event the best basis for stream channel protection? Presented at the *Effects of Watershed Developments and Management on Aquatic Ecosystems* conference. Snowbird, UT, August 4–9, 1996. Edited by L.A. Roesner. ASCE, New York, 1997.

Maguire, S. and D. Walker. *Pfiesteria piscicida* implicated in fish kills in Chesapeake Bay tributaries and other mid-Atlantic estuaries. *WSTB. A Newsletter from the Water Science and Technology Board*. National Research Council. Washington, D.C. 14(4): 1–3. October/November 1997.

Marcy, S.J. and J.E. Flack. Safety considerations in urban storm drainage design. *Second International Conference on Urban Storm Drainage*. Urbana, IL. June 1981.

Marcy, S. and J. Gerritsen. Developing diverse assessment endpoints to address multiple stressors in watershed ecological risk assessment. Abstract Book: *SETAC 17th Annual Meeting*. p. 96. Washington, D.C. Nov. 17–21, 1996.

May, C., R.R. Horner, J.R. Karr, B.W. Mar, and E.B. Welch. *The Cumulative Effects of Urbanization on Small Streams in the Puget Sound Lowland Ecoregion*. University of Washington, Seattle. 1999.

McHardy, B.M., J.J. George, and J. Salanki (Ed.). The uptake of selected heavy metals by the green algae *Cladophora glomerata*. In *Proceedings of Symposium on Heavy Metals in Water Organisms*. Tihany, Hungary. Akademiai Kiado, Budapest, Hungary. *Symp. Biol. Hung.*, 29: 3–20. 1985.

Medeiros, C. and R.A. Coler. *A Laboratory/Field Investigation into the Biological Effects of Urban Runoff*, Water Resources Research Center, University of Massachusetts, Amherst, MA. July 1982.

Medeiros, C., R.A. Coler, and E.J. Calabrese. A laboratory assessment of the toxicity of urban runoff on the fathead minnow (*Pimephales promelas*). *J. Environ. Sci. Health*, A19(7): 847–861. 1984.

Middlebrooks, E.J. Review paper: animal waste management and characterization, *Water Res.*, 8: 697–712. 1974.

Mikkelsen, P.S., K. Arngjerg-Nielsen, and P. Harremoës. Consequences for established design practice from geographical variation of historical rainfall data. *Proceedings: 7th International Conference on Urban Storm Drainage*. Hannover, Germany. Sept. 9–13, 1996a.

Mikkelsen, P.S., M. Häfliger, M. Ochs, J.C. Tjell, M. Jacobsen, and M. Boller. Experimental assessment of soil and groundwater contamination from two old infiltration systems for road run-off in Switzerland. *Sci. Total Environ.*, 1996b.

Moore, L. and G.A. Burton, Jr. *Assessment Of Stormwater Runoff Effects on Wolf Creek, Dayton, Ohio*. Final Report. City of Dayton, Dayton OH. 1999.

Mote Marine Laboratory. *Biological and Chemical Studies on the Impact of Stormwater Runoff upon the Biological Community of the Hillsborough River, Tampa, Florida*, Stormwater Management Division, Dept. of Public Works, Tampa. 1984.

MTA (Management Training Audioconferences). *Participant Program Guide: Cryptosporidium and Water*. 1997 Management Training Audioconference Seminars. Public Health Foundation. National Center for Infectious Diseases, CDC. Atlanta, GA. 1997.

Mull, R. Water exchange between leaky sewers and aquifers. *7th International Conference on Urban Storm Drainage*. Hannover, Germany. Sept. 9–13. Edited by F. Sieker and H-R. Verworn. IAHR/IAWQ. SuG-Verlagsgesellschaft. Hannover, Germany. pp. 695–700. 1996.

MWCOG (Metropolitan Washington Council of Governments). *Potomac River Water Quality: Conditions and Trends in Metropolitan Washington*, Water Resources Planning Board, Washington, D.C. 1984.

Olivieri, V.P., C.W. Kruse, and K. Kawata. Selected pathogenic microorganisms contributed from urban watersheds. In *Watershed Research in Eastern North America*, Vol. II. Edited by D.L. Correll. NTIS No. PB-279 920/3SL. 1977a.

Olivieri, V.P., C.W. Kruse, and K. Kawata. *Microorganisms in Urban Stormwater*. U.S. EPA Rept. No. EPA-600/2-77-087. July 1977b.

Oris, J.T. and J.P. Giesy. The photoinduced toxicity of polycyclic aromatic hydrocarbons to larvae of the fathead minnow, *Pimephales promelas, Chemosphere,* 16: 1396–1404. 1987.

OWML (Occoquan Watershed Monitoring Lab). *Evaluation of Management Tools in the Occoquan Watershed*, Final Report, Virginia State Water Control Board, Richmond, VA. 1982.

Pedersen, E.R. *The Use of Benthic Invertebrate Data for Evaluating Impacts of Urban Stormwater Runoff*, Masters thesis submitted to the College of Engineering, University of Washington, Seattle. 1981.

Perkins, M.A. *An Evaluation of Instream Ecological Effects Associated with Urban Runoff to a Lowland Stream in Western Washington*. U.S. Environmental Protection Agency, Corvallis Environmental Research Laboratory, Corvallis, OR, July, 1982.

Pitt, R. *Demonstration of Nonpoint Pollution Abatement through Improved Street Cleaning Practices*, EPA-600/2-79-161, U.S. Environmental Protection Agency, Cincinnati, OH. 270 pp. 1979.

Pitt, R. and Bozeman, M. *Sources of Urban Runoff Pollution and Its Effects on an Urban Creek*, EPA-600/S2-82-090, U.S. Environmental Protection Agency, Cincinnati, OH. 1982.

Pitt, R. *Urban Bacteria Sources and Control by Street Cleaning in the Lower Rideau River Watershed.* Rideau River Stormwater Management Study Technical Report. Prepared for the Ontario Ministry of the Environment, Environment Canada, Regional Municipality of Ottawa –Carleton, City of Ottawa, and Nepean. 1983.

Pitt, R. and P. Bissonnette. *Bellevue Urban Runoff Program, Summary Report.* PB84 237213, Water Planning Division, U.S. Environmental Protection Agency and the Storm and Surface Water Utility, Bellevue, WA. 1984.

Pitt, R., S. Clark, and K. Parmer. *Protection of Groundwater from Intentional and Nonintentional Stormwater Infiltration.* U.S. Environmental Protection Agency, EPA/600/SR-94/051. PB94-165354AS, Storm and Combined Sewer Program, Cincinnati, OH. 187 pp. May 1994.

Pitt, R. "Effects of Urban Runoff on Aquatic Biota." In *Handbook of Ecotoxicology.* Edited by D.J. Hoffman, B.A. Rattner, G.A. Burton, Jr. and J.Cairns, Jr. Lewis Publishers/CRC Press, Boca Raton, FL. pp. 609–630. 1995.

Pitt, R., R. Field, M. Lalor, and M. Brown. Urban stormwater toxic pollutants: assessment, sources and treatability. *Water Environ. Res.,* 67(3): 260–275. May/June 1995.

Pitt, R., S. Clark, K. Parmer, and R. Field. *Groundwater Contamination from Stormwater.* Ann Arbor Press, Chelsea, MI. 219 pp. 1996.

Power, M.E., S.J. Stout, C.E. Cushing, P.P. Harper, F.R. Hauer, W.J. Matthews, P.B. Moyle, B. Statzner, and I.R. Wars de Badgen. Biotic and abiotic controls in river and stream communities. *J. North Am. Benthol. Soc.,* 7: 456–479. 1988.

Power, E.A., and P.M. Chapman. Assessing sediment quality, in *Sediment Toxicity Assessment*, Edited by G.A. Burton, Jr., Lewis Publishers, Boca Raton, FL. 1992.

Pratt, J.M., R.A. Coler, and P.J. Godfrey. Ecological effects of urban stormwater runoff on benthic macroin-vertebrates inhabiting the Green River, Massachusetts. *Hydrobiologia*, 83: 29. 1981.

Prych, E.A. and J.C. Ebbert. *Quantity and Quality of Storm Runoff from Three Urban Catchments in Bellevue, Washington*, Preliminary U.S. Geological Survey Water Resources Investigations Report, Tacoma, WA. Undated.

Qureshi, A.A. and B.J. Dutka. Microbiological studies on the quality of urban stormwater runoff in southern Ontario, Canada. *Water Res.,* 13: 977–985. 1979.

Rainbow, P.S. Chapter 18: Heavy metals in aquatic invertebrates. In *Environmental Contaminants in Wildlife; Interpreting Tissue Concentrations.* Edited by W.N. Beyer, G.H. Heinz, and A.W. Redmon-Norwood. CRC/Lewis Press, Boca Raton, FL. pp. 405–425. 1996.

Ray, S. and W. White. Selected aquatic plants as indicator species for heavy metal pollution. *J. Environ. Sci. Health*, A11: 717. 1976.

Regional Municipality of Ottawa-Carleton. *Report on Pollution Investigation and Abatement.* Ottawa, Ontario. Sept. 1972.

Resh, V.H. Sampling variability and life history features: basic considerations in the design of aquatic insect studies, *J. Fish. Res. Bd. Can.,* 36: 290–311.

Richey, J.S., M.A. Perkins, and K.W. Malueg. The effects of urbanization and stormwater runoff on the food quality in two salmonid streams. *Verh. Int. Werein. Limnol.,* 21: 812–818, Stuttgart, October 1981.

Richey, J. S. *Effects of Urbanization on a Lowland Stream in Western Washington*, Doctor of philosophy dissertation, University of Washington, Seattle. 1982.

Robinson, A.M. The effects of urbanization on stream channel morphology. *Proceedings of the National Symposium on Urban Hydrology, Hydraulics, and Sediment Control.* University of Kentucky. Lexington. 1976.

Schillinger, J.E. and D.G. Stuart. *Quantification of Non-Point Water Pollutants from Logging, Cattle Grazing, Mining, and Subdivision Activities.* NTIS No. PB 80-174063. 1978.

Schueler, T.R. *Controlling Urban Runoff: A Practical Manual for Planning and Designing Urban BMPs.* Department of Environmental Programs. Metropolitan Washington Council of Governments. Water Resources Planning Board. 1987.

Schueler, T. (Ed.). Stream channel geometry used to assess land use impacts in the Pacific Northwest. *Watershed Prot. Techn.,* 2(2): 345–348. Spring 1996.

Scott, J.B., C.R. Steward, and Q.J. Stober. *Impacts of Urban Runoff on Fish Populations in Kelsey Creek, Washington*, Contract No. R806387020, U.S. Environmental Protection Agency, Corvallis Environmental Research Laboratory, Corvallis, OR. 1982.

Scott, J.B., C.R. Steward, and Q.J. Stober. Effects of urban development on fish population dynamics in Kelsey Creek, Washington. *Trans. Am. Fish. Soc.,* 115(4): 555–567. July 1986.

SCS (now NRCS) (U.S. Soil Conservation Service). *Ponds — Planning, Design, Construction*. Agriculture Handbook Number 590, U.S. Department of Agriculture, 1982.

Skalski, C. *Laboratory and in* Situ *Sediment Toxicity Evaluations with Early Life Stages of* Pimephales promelas, M.S. thesis, Wright State University, Dayton, OH. 1991.

Sovern, D.T. and P.M. Washington. Effects of urban growth on stream habitat. Presented at the *Effects of Watershed Developments and Management on Aquatic Ecosystems* conference. Snowbird, UT, August 4–9, 1996. Edited by L.A. Roesner. pp. 163–177. ASCE, New York. 1997.

Squillace, P.J., J.S. Zogorski, W.G. Wilber, and C.V. Price. Preliminary assessment of the occurrence and possible sources of MTBE in groundwater in the United States, 1993–94. *Environ. Sci. Technol.,* 30(5): 1721–1730. May 1996.

Stephenson, D. Evaluation of effects of urbanization on storm runoff. *7th International Conference on Urban Storm Drainage*. Hannover, Germany. Sept. 9–13, 1996. Edited by F. Sieker and H.-R. Verworn. IAHR/IAWQ. SuG-Verlagsgesellschaft. Hannover, Germany. pp. 31–36. 1996.

Striegl, R.G. *Effects of Stormwater Runoff on an Urban Lake, Lake Ellyn at Glen Ellyn, Illinois*, USGS Open File Report 84-603, U.S. Geological Survey, Lakewood, CO. 1985.

Trauth, R. and C. Xanthopoulos. Non-point pollution of groundwater in urban areas. *7th International Conference on Urban Storm Drainage*. Hannover, Germany. Sept. 9–13, 1996. Edited by F. Sieker and H.-R. Verworn. IAHR/IAWQ. SuG-Verlagsgesellschaft. Hannover, Germany. pp. 701–706. 1996.

Underwood, A.J. The mechanics of spatially replicated sampling programmes to detect environmental impacts in a variable world. *Aust. J. Ecol.,* 18: 99–116. 1993.

USGS. *The Quality of Our Nation's Waters — Nutrients and Pesticides*. U.S. Geological Survey Circular 1225. Denver, CO. 1999.

Vlaming, V. de, V. Connor, C. DiGiorgio, H.C. Bailey, L.A. Deanovic, Edited by D.E. Hinton. Application of whole effluent toxicity test procedures to ambient water quality assessment. *Environ. Toxicol Chem.,* 19: 42–62. 2000.

Water Environ. Technol. Research notes: beachgoers at risk from urban runoff. 8(11): 65. Nov. 1996.

Wenholz, M., and R. Crunkilton. Use of toxicity identification evaluation procedures in the assessment of sediment pore water toxicity from an urban stormwater retention pond in Madison Wisconsin. *Bull. Environ. Contam. Toxicol.,* 54: 676–682. 1995.

Werner, I., L.A. Deanovic, V. Connor, V. de Vlaming, H.C. Bailey, and D.E. Hinton. Insecticide-caused toxicity to *Ceriodaphnia dubia* (Cladocera) in the Sacramento–San Joaquin River Delta, California, USA. *Environ. Toxicol. Chem.,* 19: 215–227. 2000.

Wetzel, R. G. *Limnology*. Saunders Publishing, Philadelphia, PA. 1975.

Wilber, W.G., and J.V. Hunter. *The Influence of Urbanization on the Transport of Heavy Metals in New Jersey Streams*, Water Resources Research Institute, Rutgers University, New Brunswick, NJ. 1980.

Williams, A.T. and S.L. Simmons. Estuarine litter at the river/beach interface in the Bristol Channel, United Kingdom. *J. Coastal Res.,* 13(4): 1159–1165. Fall 1997.

Wissmar, R.C. and F.J. Swanson. Landscape disturbances and lotic ecotones, in *The Ecology and Management of Aquatic Terrestrial Ecotones*, Vol. 4. Edited by R.J. Naiman, and H. Decamps. UNESCO and Parthenon Publishing Group, Paris, France. pp. 65–89. 1990.

Wolman, M.G. and A.P. Schick. Effects of construction on fluvial sediment, urban and suburban areas of Maryland. *Water Resour. Res.,* 3(2): 451–464. 1967.

Zhang, Q.-X., W. Carmichael, M.-J. Yu, and S.-H. Li. Cyclic peptide hepatotoxins from freshwater cyanobacterial (blue-green algae) waterblooms collected in central China, *Environ. Toxicol. Chem.,* 10: 313–321. 1991.

Zimmerman, T. How to revive the Chesapeake Bay: Filter it with billions and billions of oysters. *U.S. News & World Report*. p. 63. December 29, 1997/January 5, 1998.

Zogorski, J.S., A.B. Morduchowitz, A.L. Baehr, B.J. Bauman, D.L. Conrad, R.T. Drew, N.E. Korte, W.W. Lapham, J.F. Pankow, and E.R. Washington. *Fuel Oxygenates and Water Quality: Current Understanding of Sources, Occurrence in Natural Waters, Environmental Behavior, Fate, and Significance*. Office of Science and Technology, Washington, D.C. 91 pp. 1996.

Components of the Assessment

## CHAPTER 4

Overview of Assessment Problem Formulation

"If the Lord Almighty had consulted me before embarking on the Creation, I would have recommended something simpler."

Alfonso X of Castile (Alfonso the Wise), 1221–1284

CONTENTS

Introduction 102
    Rationale for an Integrated Approach to Assessing Receiving Water Problems 102
Watershed Indicators of Biological Receiving Water Problems 103
Summary of Assessment Tools..107
Study Design Overview 107
Beginning the Assessment 108
    Specific Study Objectives and Goals 110
    Initial Site Assessment and Problem Identification 110
    Review of Historical Site Data 112
    Formulation of a Conceptual Framework 113
    Selecting Optimal Assessment Parameters (Endpoints) 113
    Data Quality Objectives and Quality Assurance Issues 118
Example Outline of a Comprehensive Runoff Effect Study...119
    Step 1. What's the Question?..119
    Step 2. Decide on Problem Formulation 119
    Step 3. Project Design...120
    Step 4. Project Implementation (Routine Initial Semiquantitative Survey)..................121
    Step 5. Data Evaluation..122
    Step 6. Confirmatory Assessment (Optional Tier 2 Testing) 122
    Step 7. Project Conclusions 123
Case Studies of Previous Receiving Water Evaluations 123
    Example of a Longitudinal Experimental Design — Coyote Creek, San Jose, CA,
    Receiving Water Study 124
    Example of Parallel Creeks Experimental Design — Kelsey and Bear Creeks, Bellevue,
    WA, Receiving Water Study..139
    Example of Long-Term Trend Experimental Design — Lake Rönningesjön, Sweden,
    Receiving Water Study 169
    Case Studies of Current, Ongoing, Stormwater Projects 181
    Outlines of Hypothetical Case Studies 205

Summary: Typical Recommended Study Plans 213
        Components of Typical Receiving Water Investigations 213
        Example Receiving Water Investigations 213
References 218

INTRODUCTION

This chapter summarizes various approaches that have been used and recommended for evaluating receiving water effects. It outlines a reasonable method that allows the study designer to consider many factors that may affect the outcome of the project. Major study approaches are presented with extensive case study examples. The chapters and appendices in this book complement this material by providing guidance for developing an experimental design, methods for the collection of samples and their analysis, various other field evaluation efforts, and the statistical analysis of the data.

Rationale for an Integrated Approach to Assessing Receiving Water Problems

During the past decade, it has become apparent from numerous water and sediment quality assessment studies that no one single approach (e.g., chemical-specific criteria) can be routinely used to accurately determine or predict ecosystem health and beneficial use impairment. In Ohio, evaluation of indigenous biota showed that many of the impaired stream segments could not be detected using chemical criteria alone (EPA 1990b). In an intensive survey, 431 sites in Ohio were assessed using in-stream chemical and biological surveys. In 36% of the cases, chemical evaluations implied no impairment, but the biological survey evaluations did show impairment. In 58% of the cases the chemical and biological assessments agreed. Of these, 17% identified waters with no impairment, while 41% identified waters which were considered impaired. Realization of the inadequacy of nationwide criteria prompted the EPA to look for other site-specific criteria modifications. Numerous studies of bulk sediment contaminant concentrations failed to show significant correlations with toxic effects to test species (Burton 1991).

Each assessment approach or component has associated strengths and weaknesses (Table 4.1). The ultimate objective of the CWA (Sec. 101(a)) is "to restore and maintain the chemical, physical,

Table 4.1 Components of an Integrated Approach to Assess Receiving Water Quality

Control Approach	What It Provides	What It Doesn't Provide:
Chemical specific	Human health protection	All toxics present
	Complete toxicology	Bioavailability
	Straightforward treatability	Interactions of mixtures (e.g., additivity)
	Familiarity with control	Poor trend analysis
	Persistency coverage	Accurate toxicology (false assumptions)
	Regulatory ease	Actual and direct evaluations of receiving water beneficial use impairments
Toxicity	Aggregate toxicity	Human health protection
	All toxicants present	Complete toxicology (few species may be tested)
	Bioavailability	Simple treatability
	Accurate toxicology	Persistency coverage
	Good trend analysis	
	Lab or *in situ* testing	
Bioassessments	Actual receiving water effects	Critical flow effects
	Trend analysis	Straightforward interpretation of results
	Severity of impact	Cause of impact
	Total effect of all sources	Differentiation of sources
		Habitat and site variation influence

Modified from EPA. Wisconsin legislature establishes a nonpoint pollution committee. *Nonpoint Source EPA News-Notes.* #8. October 1990a.

and biological integrity of the Nation's waters." These three components define the overall ecological integrity of an aquatic ecosystem (EPA 1990a). Pollutant loadings into receiving waters from point and nonpoint sources vary in magnitude, frequency, duration, and type. They are also strongly influenced by meteorological and hydrologic conditions, terrestrial processes, and land use activities.

A myriad of potential stressor combinations are possible in waters that are in human-dominated watersheds. In the laboratory, it would be impossible to evaluate even a small number of the possible stressor combinations, varying the magnitude, frequency, and duration of each stressor. Traditional bioassay methods simply look at one simple exposure scenario. Chemical criteria provide a bench-mark from which to evaluate the significance of contaminant concentrations and direct further monitoring resources. Biological assessments indicate if the aquatic community is of a pollution-and/or habitat-tolerant or sensitive nature by showing the effect of long-term exposures. By con-sidering habitat influence and comparing to reference sites, evaluations of ecological integrity (health) can be made. Habitat (physical) evaluations are essential to separate point source and nonpoint source toxicity effects from physical effects. As an example, some NPS pollution effects from stormwater may be of a physical nature, such as habitat alteration and destruction from increased stream flow, increased suspended and bedload sediments, or elevated water temperatures. In addition, a fourth major assessment component (toxicity) is needed beyond the three components of chemical, physical, and biological integrity (EPA 1990a). Biosurvey data may not detect subtle, short-term, or recent toxic effects due to the natural variation (spatial and temporal) that occurs in aquatic communities. Toxicity testing also removes the effects of habitat problems relatively well, focusing on the availability of chemical contaminants alone. The EPA (1990a) states that when any assessment approach (i.e., chemical-specific, toxicity, or biosurvey) shows water quality standards not being achieved, regulatory action should be taken.

The complexity of ecosystems dictates that these assessment tools be used in an integrated fashion. Scientists in any of the traditional disciplines (such as chemistry, microbiology, ecology, limnology, oceanography, hydrology, agronomy) are quick to point out the multitude of ecosystem complexities associated with their science. Many of these complexities influence chemical fate and effects and, more importantly, affect natural and anthropogenic stressor fate and effects. For example, it is well documented that many natural factors may act as significant stressors to organisms in aquatic systems, including light, temperature, flow, dissolved oxygen, sediment particle size, sus-pended solids, habitat quality, ammonia, salinity, food quality and quantity, predators, parasites, and pathogens. In addition, ecotoxicologists have long been aware of the differences between species and their life stages in regard to toxicant sensitivity. Unfortunately, toxicity information exists only for a fraction of the 1.5 to 100 million species (Wilson 1992; May 1994) and 7 million chemicals (U.S. General Accounting Office 1994) in the world. This reality makes extrapolations between species and chemicals tenuous at best. Despite these many and often interacting complexities, some excellent and proven tools exist for conducting ecologically relevant assessments of contamination.

The necessity of using each of the above assessment components and the degree to which each is utilized is a site-specific issue. At sites of extensive chemical pollution, extreme habitat destruc-tion, or absence of desirable aquatic organisms, the impact can be clearly established with only one or two components, or simply qualitative measures. However, at most study sites, there will be "gray" areas where the ecosystem's integrity (quality) is less clear and should be measured via multiple components, using a weight-of-evidence approach to evaluate adverse effects.

WATERSHED INDICATORS OF BIOLOGICAL RECEIVING WATER PROBLEMS

The EPA (1996) published a list of 18 indicators to track the health of the nation's aquatic ecosystems. These indicators are intended to supplement conventional water quality analyses in compliance-monitoring activities. The use of broader indicators of environmental health is increas-ing. As an example, by 1996, 12 states were using biological indicators and 27 states were developing local biological indicators, according to Pelley (1996). Because of the broad nature of the nation's potential receiving water problems, this list is more general than typically used for any one specific discharge type (such as stormwater, municipal wastewaters, or industrial wastewaters). These 18 indicators are (EPA 1996):

1. Population served by drinking water systems violating health-based requirements
2. Population served by unfiltered surface water systems at risk from microbiological contamination
3. Population served by community drinking water systems exceeding lead action levels
4. Drinking water systems with source water protection programs
5. Fish consumption advisories
6. Shellfish-growing waters approved for harvest for human consumption
7. Biological integrity of rivers and estuaries
8. Species at risk of extinction
9. Rate of wetland acreage loss
10. Designated uses: drinking water supply, fish, and shellfish consumption, recreation, aquatic life
11. Groundwater pollutants (nitrates)
12. Surface water pollutants
13. Selected coastal surface water pollutants in shellfish
14. Estuarine eutrophication conditions
15. Contaminated sediments
16. Selected point source loadings to surface water and groundwater
17. Nonpoint source sediment loadings from cropland
18. Marine debris

In one example of the use of watershed indicators, Claytor (1996, 1997) summarized the approach developed by the Center for Watershed Protection as part of its EPA-sponsored research for assessing the effectiveness of stormwater management programs (Claytor and Brown 1996). The indicators selected are direct or indirect measurements of conditions or elements that indicate trends or responses of watershed conditions to stormwater management activities. Categories of these environmental indicators are shown in Table 4.2, ranging from conventional water quality measurements to citizen surveys. Biological and habitat categories are also represented. Table 4.3 lists 26 indicators, by category. It was recommended that appropriate indicators be selected from each category for a specific area under study. This will enable a better understanding of the linkage of what is done on the land, how the sources are regulated or managed, and the associated receiving water problems. The indicators were selected to (1) measure stress or the activities that lead to

Table 4.2 Stormwater Indicator Categories

Category	Description	Principal Element Being Assessed
Water quality	Specific water quality characteristics	Receiving water quality
Physical/hydrologic	Measure changes to, or impacts on, the physical environment	Receiving water quality
Biological	Use of biological communities to measure changes to, or impacts on, biological parameters	Receiving water quality
Social	Responses to surveys or questionnaires to assess social concerns	Human activity on the land surface
Programmatic	Quantify various nonaquatic parameters for measuring program activities	Regulatory compliance or program initiatives
Site	Indicators adapted for assessing specific conditions at the site level	Human activity on the land surface

From Claytor, R.A. An introduction to stormwater indicators: urban runoff assessment tools. Presented at the *Assessing the Cumulative Impacts of Watershed Development on Aquatic Ecosystems and Water Quality* conference. March 20–21, 1996. Northeastern Illinois Planning Commission. pp. 217–224. Chicago, IL. April 1997.

Table 4.3 Environmental Indicators

Indicator Category	Indicator Name
Water quality indicators	Water quality pollutant constituent monitoring
	Toxicity testing
	Nonpoint source loadings
	Exceedance frequencies of water quality standards
	Sediment contamination
	Human health criteria
Physical and hydrologic indicators	Stream widening/downcutting
	Physical habitat monitoring
	Impacted dry-weather flows
	Increased flooding frequency
	Stream temperature monitoring
Biological indicators	Fish assemblage
	Macroinvertebrate assemblage
	Single species indicator
	Composite indicators
	Other biological indicators
Social indicators	Public attitude surveys
	Industrial/commercial pollution prevention
	Public involvement and monitoring
	User perception
Programmatic indicators	Illicit connections identified/corrected
	BMPs installed, inspected, and maintained
	Permitting and compliance
	Growth and development
Site indicators	BMP performance monitoring
	Industrial site compliance monitoring

From Claytor, R.A. An introduction to stormwater indicators: urban runoff assessment tools. Presented at the *Assessing the Cumulative Impacts of Watershed Development on Aquatic Ecosystems and Water Quality* conference. March 20–21, 1996. Northeastern Illinois Planning Commission. pp. 217–224. Chicago, IL. April 1997.

impacts on receiving waters, (2) assess the resource itself, and (3) measure the regulatory compliance or program initiatives. Claytor (1997) presented a framework for using stormwater indicators that is similar to many others recommended in hazard and risk assessment, as shown below:

Level 1 (Problem Identification):
1. Establish management sphere (who is responsible, other regulatory agencies involved, etc.).
2. Gather and review historical data.
3. Identify local uses that may be impacted by stormwater (flooding/drainage, biological integrity, noncontact recreation, drinking water supply, contact recreation, and aquaculture).
4. Inventory resources and identify constraints (time frame, expertise, funding and labor limitations).
5. Assess baseline conditions (use rapid assessment methods).

Obviously, the selection of the indicators to assess the baseline conditions should be based on the local uses of concern. Most of the anticipated important uses are shown to require indicators selected for each of the categories. However, the indicator selection process requires more than just a beneficial use consideration. Additional issues, such as the questions being asked, regulatory and societal concerns, the characteristics of the ecoregion, sensitive and threatened indigenous species, resource availability, and time constraints, are also important considerations.

Claytor (1997) also recommends a Level 2 assessment strategy for examining the local management program as outlined below:

Level 2:
    1. State goals for program (based on baseline conditions, resources, and constraints)
    2. Inventory prior and ongoing efforts (including evaluating the success of ongoing efforts)
    3. Develop and implement management program
    4. Develop and implement monitoring program (more quantitative indicators than typically used for the Level 1 evaluations above)
    5. Assess indicator results (does the stormwater indicator monitoring program measure the overall watershed health?)
    6. Reevaluate management program (update and revise management program based on measured successes and failures)

While the approach and recommendations of Claytor (1997) have merit and provide a good overall framework, they may not adequately consider all the important study design issues for every specific area. Most important, their indicator guidance for determining receiving water effects from stormwater runoff may not provide a characterization of all the important stressors. For example, short-term pulses of polycyclic aromatic hydrocarbons from roadways and parking lots may be creating photoinduced toxicity problems not detected by traditional bioassessment approaches.

Another example of the effective use of environmental indicators is in the Detroit, MI, area. Cave (1998) described how they are being used to summarize the massive amounts of data being generated by the Rouge River National Wet Weather Demonstration Project in Wayne County. This large project is examining existing receiving water problems, the performance of stormwater and CSO management practices, and receiving water responses in a 438 mi^2 watershed having more than 1.5 million people in 48 separate communities. The baseline monitoring program has now more than 4 years of continuous monitoring of flow, pH, temperature, conductivity, and DO, supplemented by automatic sampling for other water quality constituents, at 18 river stations. More than 60 projects are examining the effectiveness of stormwater management practices, and 20 projects are examining the effectiveness of CSO controls, each also generating large amounts of data. Toxicants are also being monitored in sediment, water, fish tissue, and with semipermeable membranes to help evaluate human health and aquatic life effects. Habitat surveys were conducted at 83 locations along more than 200 miles of waterway. Algal diversity and benthic macroinvertebrate assessments were also conducted at these survey locations. Electrofishing surveys were conducted at 36 locations along the main river and in tributaries. Several computer models were also used to predict sources, loadings, and wet-weather flow management options for the receiving waters and for the drainage systems. A geographic information system was used to manage and provide spatial analyses of the massive amounts of data collected. However, there was still a great need to simplify the presentation of the data and findings, especially for public presentations. Cave described how they developed a short list of 35 indicators, based on the list of 18 from EPA and on discussions with state and national regulatory personnel. They then developed seven indices that could be color-coded and placed on maps to indicate areas of existing problems and projected conditions based on alternative management scenarios. These indices are described as follows:

Condition Quality Indicators:
    1. Dissolved oxygen. Concentration and % saturation values (ecologically important)
    2. Fish consumption index. Based on advisories from the Michigan Department of Public Health
    3. River flow. Significant for aquatic habitat and fish communities
    4. Bacteria count. *E. coli* counts based on Michigan Water Quality Standards, distinguished for wet and dry conditions

Multifactor Indices:
    1. Aquatic biology index. Composite index based on fish and macroinvertebrate community assessments (populations and individuals)

2. Aquatic habitat index. Habitat suitability index, based on substrate, cover, channel morphology, riparian/bank condition, and water quality
3. Aesthetic index. Based on water clarity, color, odor, and visible debris

These seven indicators represent 30 physical, chemical, and biological conditions that directly impact the local receiving water uses (water contact recreation, warm water fishery, and general aesthetics). Cave presented specific descriptions for each of the indices and gave examples of how they are color-coded for map presentation. These data presentations have clearly demonstrated how the Rouge River is degraded in specific areas and show the relationships of these critical river areas with adjacent watershed activities.

SUMMARY OF ASSESSMENT TOOLS

Almost all states using bioassessment tools have relied on the EPA reference documents as the basis for their programs. Common components of these bioassessment programs (in general order of popularity) include:

- Macroinvertebrate surveys (almost all programs, but with varying identification and sampling efforts)
- Habitat surveys (almost all programs)
- Some simple water quality analyses
- Some watershed characterizations
- Few fish surveys
- Limited sediment quality analyses
- Limited stream flow analyses
- Hardly any toxicity testing
- Hardly any comprehensive water quality analyses

Normally, numerous metrics are used, typically only based on macroinvertebrate survey results, which are then assembled into a composite index. Many researchers have identified correlations between these composite index values and habitat conditions. Water quality analyses in many of these assessments are seldom comprehensive, a possible overreaction to conventional, very costly programs that have typically resulted in minimally worthwhile information. This book recommends a more balanced assessment approach, using toxicity testing and carefully selected water and sediment analyses to supplement the needed biological and habitat monitoring activities. A multi-component assessment enables a more complete evaluation of causative factors and potential mitigation approaches.

STUDY DESIGN OVERVIEW

The study design must be developed based on the study objectives, preliminary site-problem assessments, regulatory mandates, and available resources. This chapter includes detailed information for developing the experimental design aspects of the study design. Many of the typical monitoring subcomponents of each approach are listed in Table 4.4. All of these parameters cannot realistically be evaluated in routine water quality assessments. The amount and type of monitoring hinges not only on the above issues but the degree of confidence and accuracy expected from the results. This issue falls under the Data Quality Objectives process and is also discussed in later chapters.

The most commonly used test hypotheses in assessing receiving water impacts is that the designated use or integrity of the water body is not impaired (null hypothesis), or the alternative hypotheses that it or some component is impaired or some specific factor (e.g., stormwater) is

Table 4.4 Summary of Recommended Aquatic Ecosystem Assessment Parameters

Physical Evaluations	Chemical Evaluations	Indigenous Biota Evaluations	Toxicity Evaluations
In-stream characteristics	Dissolved oxygen (W)	Biological inventory	Acute/Short-term
Size (mean	Toxicants (WS)	(Existing Use Analysis):	Chronic
width/depth)	Nutrients (W)	Fish	Responses(WS):
Flow/velocity	Nitrogen	Macroinvertebrates	Fish (*Pimephales*
Total volume	Phosphorus	Microinvertebrates	*promelas*)
Reaeration rates	Biochemical oxygen	Phytoplankton	Zooplankton
Gradient/pools/riffles	demand (W)	Macrophytes	(*Ceriodaphnia dubia*)
Temperature	Sediment oxygen	Biological	Benthic
Suspended solids	demand (S)	Condition/Health	macroinvertebrates
Sedimentation	Conductivity/salinity(W)	Analysis:	(*Selenastrum*
Channel modifications	Hardness (W)	Diversity indices	*capricornutum*)
Channel stability	Alkalinity (W)	HIS models	Other (microbial,
Substrate composition	pH (WS)	Tissue analysis	protozoan,
and characteristics	Temperature (W)	Recovery index	macrophytes,
Particle size distribution	Dissolved solids (W)	Intolerant species	amphibian, or
Sediment dry weight	Total organic carbon (S)	Omnivore-carnivore	indigenous species)
Channel debris	Acid volatile sulfides (S)	analysis	
Sludge deposits	Ammonia (WS)	Biological potential	
Riparian characteristics		analysis	
Downstream		Reference reach	
characteristics		comparison	

W = Water
S = Sediment

causing impairment. To detect differences between ambient and/or reference (nonimpacted) conditions in an aquatic system and the test system, it is important to establish the appropriate level of sensitivity. A 5% difference in condition or integrity is more difficult to detect than a 50% difference. The level of detection needs to be predetermined to establish the sample size (see Chapter 5).

A thorough assessment of ecosystem impact, hazard, or risk may follow the general approach proposed by EPA for ecological risk assessments. The toxicity assessment process consists of identifying the stressors (hazards), using various measurement endpoints to determine concentration (exposure)–response gradients, and then characterizing the stressor–effect level (threshold) and degree of impact, hazard, or risk that exists so that management decisions regarding remediation (corrective action) can be made. The impact characterization step is the most difficult given the many natural and anthropogenic unknowns, such as spatial and temporal variation; chemical fate, effects, and interactions through time and food webs; and biotic and abiotic patch interactions. For these reasons, the weight-of-evidence approach is the most reliable, as discussed in Chapter 8. The most effective use of resources in routine stormwater assessments is via a tiered monitoring approach (see also Chapter 8).

BEGINNING THE ASSESSMENT

Designing and implementing an assessment study requires careful and methodical planning to ensure that the study objectives will be accomplished. The preceding section described the watershed indicator approach recommended by Claytor (1996, 1997) and the EPA. The following sections in this chapter will provide additional critical considerations, approach details, and method options for conducting receiving water impact assessments.

The main objectives of most environmental monitoring studies may be divided into two general categories: characterization and/or comparisons. Characterization pertains to quantifying a few simple attributes of the parameter of interest. As an example, the concentration of copper in the sediment near an outfall may be of concern. The important question would be, "What is the most likely concentration of the copper?" Other questions of interest include changes in the copper concentrations between surface deposits and buried deposits, or in upstream vs. downstream locations. These additional questions are considered in the second category, namely, comparisons. Other comparison questions may relate to comparing the observed copper concentrations with criteria or standards. Finally, many researchers would also be interested in quantifying trends in the copper concentrations. This extends beyond the above comparison category, as trends usually consider more than just two locations or conditions. Examples of trend analyses would examine copper gradients along the receiving stream, or trends of copper concentrations with time. Another type of analysis related to comparisons is the identification of hot spots, where the gradient of concentrations in an area is used to identify areas having unusually high concentrations.

An adequate experimental design enables a researcher to efficiently investigate a study hypothesis. The results of the experiments will theoretically either prove or disprove the hypothesis. In reality, the experiments will tend to shed some light on the real problem and will probably result in many more questions that need addressing. In many cases, the real question may not have even been recognized initially. Therefore, even though it is very important to have a study hypothesis and appropriate experimental design, it may be important to reserve enough study resources to enable additional unanticipated experiments. In this discussion, sampling plans and specific statistical tools will be briefly examined.

Experimental design covers several aspects of a monitoring program. The most important aspect of an experimental design is being able to write down the study objectives and why the data are needed. The quality of the data (accuracy of the measurements) must also be known. Allowable errors need to be identified based on how the information will change a conclusion. Specifically, how sensitive are the data that are to be collected in defining the needed answer? A logical experimental process that can be used to set up an assessment of receiving waters consists of several steps:

1. Establish clear study objectives and goals (hypothesis to be tested, calibration of equation or model to be used, etc.).
2. Assess initial site assessment and identify preliminary problem.
3. Review historical site data. Collect information on the physical conditions of the system to be studied (watershed characteristics, etc.), estimate the time and space variabilities of the parameters of interest (assumed, based on prior knowledge, or other methods).
4. Formulate a conceptual framework (e.g., the EPA ecological risk framework) and model.
5. Determine optimal assessment parameters. Determine the sampling plan (strata and relationships that need to be defined), including the number of samples needed (when and where, within budget restraints).
6. Establish data quality objectives (DQO) and procedures needed for QA/QC during sample collection, processing, analysis, data management, and data analyses.
7. Locate sampling sites.
8. Establish field procedures, including the sampling specifics (volumes, bottle types, preservatives, samplers to be used, etc.).
9. Review QA/QC issues.
10. Construct data analysis plan by determining the statistical procedures that will be used to analyze the data (including field data sheets and laboratory QA/QC plan).
11. Implement the study.

Preliminary project data obtained at the beginning of the project should be analyzed to verify assumptions used in the experimental design process. However, one needs to be cautious and not make major changes until sufficient data have been collected to verify new assumptions. After the data have been analyzed and evaluated, it is likely that follow-up monitoring should be conducted to address new concerns uncovered during the project.

Table 4.5   Principles for Designing Successful Environmental Studies

1. State concisely to someone what question you are asking. Your results will be as coherent and as comprehensible as your initial conception of the problem.
2. Take replicate samples within each combination of time, location, and any other controlled variable. Differences between groups can only be demonstrated by comparison to differences within groups.
3. To test whether a condition has an effect, collect samples both where the condition is present and where the condition is absent (reference site) but all else is the same. An effect can only be demonstrated by comparison with a control.
4. Carry out some preliminary sampling to provide a basis for evaluation of sampling design and statistical analysis options. Deleting this step to save time usually results in losing time.
5. Verify that the sampling device or method is sampling the population it should be sampling, and with equal and adequate efficiency over the entire range of sampling conditions to be encountered. Variation in efficiency of sampling from area to area biases among-area comparisons.
6. If the area to be sampled has a large-scale environmental pattern, break the area up into relatively homogeneous subareas and allocate samples to each in proportion to the size of the subarea. If it is an estimate of total abundance over the entire area that is desired, make the allocation proportional to the number of organisms in the subarea.
7. Verify that the sample unit size is appropriate to the size, densities, and spatial distributions of the organisms being sampled. Then estimate the number of replicate samples required to obtain the needed precision.
8. Test the data to determine whether the error variation is homogeneous, normally distributed, and independent of the mean. If it is not, as will be the case for most field data, then (a) appropriately transform the data, (b) use a distribution-free (nonparametric procedure, (c) use an appropriate sequential sampling design, or (d) test against simulated H_0 data.
9. Having chosen the best statistical method to test the hypothesis, stick with the result. An unexpected or undesired result is *not* a valid reason for rejecting the method and searching for a "better" one.

Green, R.H. *Sampling Design and Statistical Methods for Environmental Biologists*. John Wiley & Sons, New York. 1979.

Most of the first six of these elements are described in this chapter, while the remaining ones are included in the later chapters. If any of these process components are inadequately addressed, the study outputs may not achieve the necessary study goals and/or may lead to erroneous conclusions. An early paper by Green (1979) lists principles (Table 4.5) that are still valid for preparing environmental study designs.

Specific Study Objectives and Goals

The study objectives and goals should be clearly defined, addressing ecosystem characterization and protection concerns and also the role of the assessment in the decision-making process for managing the particular problem. There are four primary reasons for an assessment program: planning, research or design, control and process optimization, and corrective action/regulation. The overall scope of planning studies is often general, while the other program types are more specific in nature. Study goals may range from establishing trends or background levels to optimizing control design or even enforcement actions. Once the objectives are defined, the needed sensitivity of the evaluation can be determined in the DQO process.

Initial Site Assessment and Problem Identification

It is essential that a reconnaissance survey be conducted or an individual who has previously studied the site be included in the design process. A substantial degree of qualitative site characterization information is gained through this process and cannot be acquired through reading report descriptions. These preliminary studies should be conducted by personnel with expertise in evaluating pollution effects on aquatic ecosystems. The preliminary survey should focus on several watershed characteristics (Table 4.6) that will need to be addressed in the study design and final assessment. Most of these factors are interwoven in a cause–effect relationship, but will often affect the study design and field methods as separate, influencing components. As an example, the most

Table 4.6 Stream Assessment Factors for Nonpoint Source-Affected Streams

Watershed development factor	Imperviousness of contributing watershed and drainage efficiency of land use. Watershed area. Age of development. Nature of upstream land use. Percent forest cover. Pollutant (NPS and PS) input locations and dynamics.
Best management practice	Proportion of contributing watershed effectively controlled by a proposed BMP or retrofit. Type and performance of BMP.
Hydrologic change factor	Drainage efficiency (such as pre- vs. post-development runoff coefficients and times of concentrations). Dry-weather flow rate in modified vs. reference watershed. Frequent return period flows and associated channel dimensions.
Channel form/stability factor	Natural, eroded, open, lined, protected or enclosed channel form. Dry-weather wetted perimeter vs. reference watershed. Evidence of widening or downcutting. Bedrock controlled channel. Consolidated or unconsolidated banks. Channel gradient.
Substrate quality factor	Median diameter or bed sediment. Degree of embeddedness. Reference substrate in undeveloped stream. Existing and future disturbed areas. Evidence of shifting sand bars, discolored cobbles.
Water quality factor	Summer maximum temperature. Benthic algal growth. Organic slime on rocks. Silt and sand deposits in stream. Presence/absence of point source discharge or pipes along stream. Type and height of debris jams. Discolored or black rocks upon turning. Dry-weather water velocity.
Stream community factor	Reference macroinvertebrate and fish species expected. Evidence of benthic algae or leaf processing. Rock turning or kick sampling. Cold, cool, or warm water community.
Refugia factor	Presence of refuge habitats allowing species escape and reintroduction.
Riparian cover factor	Presence or absence of riparian canopy cover over stream. Width of buffer $2\frac{1}{2}$ H max. Is vegetation stabilizing banks?
Stream reach factor	Presence or absence of pool and riffle structure. Minimum dry-weather flow. Sinuosity of channel. Open or closed to fish migration. Creation of linear barrier across stream.
Contiguous wetland factor	Presence or absence of nontidal wetlands in riparian, floodplain, or BMP zone. Quality, area, and function of wetlands present. Downstream wetlands to be affected?
Floodplain change factor	Constrained or unconstrained floodplain. Extent of ultimate flood plain. Property in floodplain.
Receiving water target factor	Are there any unique watershed water quality targets in a downstream river, lake, or estuary?

Modified from Schueler, T.R. *Controlling Urban Runoff: A Practical Manual for Planning and Designing Urban BMPs*. Department of Environmental Programs. Metropolitan Washington Council of Governments. Water Resources Planning Board. 1987.

important factors at the root of most nonpoint source pollution-related problems include watershed development characteristics whether of an urban, agricultural, or silviculture nature. Therefore, the preliminary problem identification process should begin with observations on the type, number, size, and location of point source discharges, stormwater inputs, upstream land use drainage patterns, and combined sewer overflows (CSOs).

A reference watershed should be located in the same type of ecoregion, but which has an undeveloped (unimpacted) watershed of a similar size with a stream (or lake) of a similar size. It is not practical to expect to find a completely natural and totally unimpacted watershed that can be used as a reference. The amount of allowable impact in the reference watershed will depend on the frequency and degree of exposure, persistence of the stressors, substrate composition, habitat and riparian quality, ecoregion and species sensitivity, and the range in water quality conditions.

The use of reference sites is common to most bioassessment approaches. Reference sites are typically selected to represent natural conditions as nearly as possible. However, it is not possible to identify such pristine locations representing varied habitat conditions in most areas of the country. Schueler (1997) points out that in many cases, a completely natural forested area is not a suitable benchmark for current conditions before urbanization. In many areas of the country, land that has long been in agricultural use is being converted to urban land, and the in-stream changes expected should therefore be more reasonably compared to agricultural conditions.

The Ohio EPA has been recognized for having one of the more advanced biological assessments in place, especially in its efforts to incorporate biological criteria as part of the regulatory program. It relies heavily on a large network of reference sites representing the various ecological conditions throughout the state. Many of the states waterways were channelized decades ago. This severe habitat disruption prevents them from ever attaining as high a quality as a similar unchannelized waterway. Therefore, Ohio EPA established "modified" warm water habitat designations with appropriate modified reference sites. Few of these reference sites are completely unimpacted by modifications or human activity in the watersheds. Yoder and Rankin (1997) reported that biological monitoring of small streams in Ohio has indicated a general lowering of biological index scores with increasing urbanization, especially in areas having CSOs and industrial discharges. Of 110 sampling sites, only 23% had good to exceptional biological resources. Poor or very poor scores were evident in 85% of the urbanized areas. They also found that more than 40% of the suburban, urbanizing sites were impaired, due to increasing residential and commercial developments. An earlier Ohio study found that biological impairments were evident in about half the locations where no impairments were indicated, based on chemical ambient monitoring data alone. They have, therefore, come to rely on biological monitoring, such as expressed in the Index of Biotic Integrity (IBI) and the Invertebrate Community Index (ICI), as a less expensive and more accurate overall indication of receiving water problems than conventional chemical water pollutant monitoring.

Crawford and Lenat (1989) examined the differences between streams located in forested, agricultural, and urban watersheds in North Carolina. The USGS study found that the stream impacted by agricultural operations was intermediate in quality, with higher nutrient and worse substrate conditions than the urban stream, but better macroinvertebrate and fish conditions. The forested watershed had the best conditions (good conditions for all categories), except for somewhat higher heavy metal concentrations in sediment than expected. Even though the agricultural watershed had little impervious area, it had high sediment and nutrient discharges, plus some impacted stream corridors. The urban stream had poor macroinvertebrate and fish conditions, poor sediment and temperature conditions, and fair substrate and nutrient conditions.

Review of Historical Site Data

As in any environmental assessment process, historical site data should be reviewed initially. Municipal, county, regional, state, and federal information sources of public information may be available concerning:

1. Predevelopment water quality, fisheries, and flow conditions (e.g., state and EPA STORET database)
2. Annual hydrological conditions vs. development area (e.g., USGS)
3. Business and industrial categories (e.g., municipality)
4. Historical hazardous spills, large quantity toxicant releases and storage (e.g., fire department, state EPA, and EPA's Toxics Release Inventory), and hazardous waste and sanitary landfill locations (e.g., state and EPA)

The initial information search should review land use patterns from a chronological approach and attempt to correlate development with hydrological data and previous water quality surveys. Unfortunately, these data are often nonexistent for the small and more heavily impacted urban streams (headwaters). If the contaminants (stressors) of concern are known, site or area stream quality survey data can be used to determine the likely background levels in water, sediment, soil, and fish. Also, one should determine what the effects and threshold levels are likely to be, and whether any rare, threatened, or endangered species are indigenous to the area. Sources of the above information may include state environmental and natural resource agencies; state game and fish agencies; conservation agencies; societies; citizens' and sportsman's groups; state agricultural agencies; relevant university departments; museums; park officials; local water and wastewater utilities; and regional offices of federal agencies (i.e., U.S. Fish and Wildlife Service, U.S. Environmental Protection Agency, U.S. Department of Agriculture, and Natural Resources Conservation Service). From this information, it is possible to determine which species are most likely to be present and what problems may exist in an area.

Formulation of a Conceptual Framework

A conceptual framework is similar to logistical critical-path control schedules, where the major components of the study (i.e., investigation of pollutant sources, hydrologic analyses, and stream and ecosystem monitoring) are blended to describe source movement, distribution, and interaction with the receiving water ecosystem. Once the previous steps are completed, it should be possible to formulate a suitable assessment problem formulation. This process is improved if there are adequate knowledge and expertise to address the key issues of pollutant types expected, predicted pollutant fate and effects, beneficial use designations, stream hydrological characteristics, meteorological characteristics, reference and test stream water quality, and key indicator aquatic organisms present at the reference and test locations. This design stage leads directly to the next step of defining measurement endpoints.

This process should be tailored toward addressing the study objectives. If the study is to be an "endangerment," "hazard," or "risk" assessment to meet EPA regulatory requirements (e.g., RCRA, CERCLA), it would be best to follow their assessment paradigm:

1. Hazard identification: qualitative stress (e.g., lead) and receptor (e.g., trout) identification
2. Exposure assessment: contaminant (stress) dynamics vs. receptor patterns and characteristics
3. Toxicity assessment: stress–response relationship quantified
4. Hazard or risk characterization: combine above information to predict or assign adverse effects vs. source exposure

The specifics of these approaches are currently still under development by the EPA. This book could possibly be used to support any program directive which includes assessing the effects of stormwater runoff on receiving water ecosystems.

Selecting Optimal Assessment Parameters (Endpoints)

Characterization of the ecosystem should allow for differentiation of its present "natural" status from its present condition caused by polluted discharges and/or other anthropogenic stressors. This requires that a number of chemical, biological, and physical parameters be monitored, including flow and habitat. There are a wide variety of potentially useful study parameters which vary in importance with the study objectives and program needs, as shown in Table 4.7. Many of the chemical endpoints would be specifically selected based on the likely pollutant sources in the watershed. Those shown in Table 4.7 are a general list.

The selection of the specific endpoints for monitoring should be based on expected/known receiving water problems. The parameters being monitored should confirm if these uses are being impaired. If they are, then more detailed investigations can be conducted to understand the discharges of the problem pollutants, or the other factors, causing the documented problems. Finally, control programs can be designed, implemented, and monitored for success. Therefore, any receiving water investigation should proceed in stages if at all possible. It is much more cost-effective

Table 4.7   Useful Receiving Water Assessment Parameters

Chemical	Physical	Biological
Oxygen	Habitat quantification[a]	*Escherichia coli*
Dissolved	Flow, velocity	Enterococci
Biochemical demand	Temperature	Fecal coliforms
Carbonaceous	Conductivity, salinity	Benthic macroinvertebrate indices[a]
Nitrogenous	Suspended solids	Fish community indices[a]
Ultimate	Dissolved solids	Blue-green algal (cyanobacteria) blooms
Chemical demand	Reach lengths	Toxicity tests[b]
Sediment demand	Channel morphology	*Pimephales promelas* early-life stage
Nutrients	Tributary loadings	*Ceriodaphnia* or *Daphnia* sp.
Nitrogen: Total, Organic, Nitrate,	Point source loadings	*Selenastrum capricornutum*
Nitrite, Ammonia (total, un-ionized)	Nonpoint source loadings	Microtox
Phosphorus	Particle size distributions	*Hyalella azteca*
Total, Organic	Bedload	*Chironomus tentans*
Carbon	Precipitation	Tissue contaminants[b]
Total, Dissolved		Fish or bivalve tissue residues
pH		Bioaccumulation testing with *Lumbriculus*
Alkalinity		*variegatus*, bivalves, or fish
Hardness		Uptake in semipermeable membrane
Metals: Cd, Cu, Zn, Pb		devices (SPMD)
Organics: Polycyclic aromatic		
hydrocarbons (PAHs)		
Aliphatic hydrocarbons		
Pesticides (chlorinated and new age)		
Oil and grease		

[a] Comprised of multiple endpoints (see EPA 1989 and OEPA 1989 and Chapter 5).
[b] Water, whole sediment, and effluent exposures (see Chapter 5 for specific effect endpoints).

to begin with a relatively simple and inexpensive monitoring program to document the problems that may exist in a receiving water than it is to conduct a large and comprehensive monitoring program with little prior knowledge. Without having information on the potential existing problems, the initial list of parameters to be monitored has to be based on best judgment. Chapter 3 contains a review of the potential problems caused by stormwater in urban streams. The parameters to be monitored can be taken from Table 4.7 and grouped into general categories depending on expected beneficial use impairments, as follows:

- Flooding and drainage: debris and obstructions affecting flow conveyance are parameters of concern.
- Biological integrity: habitat destruction, high/low flows, inappropriate discharges, polluted sediment (SOD and toxicants), benthic macroinvertebrate and fish species impairment (toxicity and bioaccumulation of contaminants), and wet-weather quality (toxicants, nutrients, DO) are key parameters.
- Noncontact recreation: odors, trash, high/low flows, aesthetics, and public access are the key parameters.
- Swimming and other contact recreation: pathogens and above-listed noncontact parameters are key parameters.
- Water supply: water quality standards (especially pathogens and toxicants) are key parameters.
- Shellfish harvesting and other consumptive fishing: pathogens, toxicants, and those listed under biological integrity are key parameters.

Point source discharges, stormwater runoff, snowmelt, baseflows in receiving waters, sediments, and biological specimens may all need to be sampled and analyzed to obtain a complete understanding of receiving water effects from pollutant discharges.

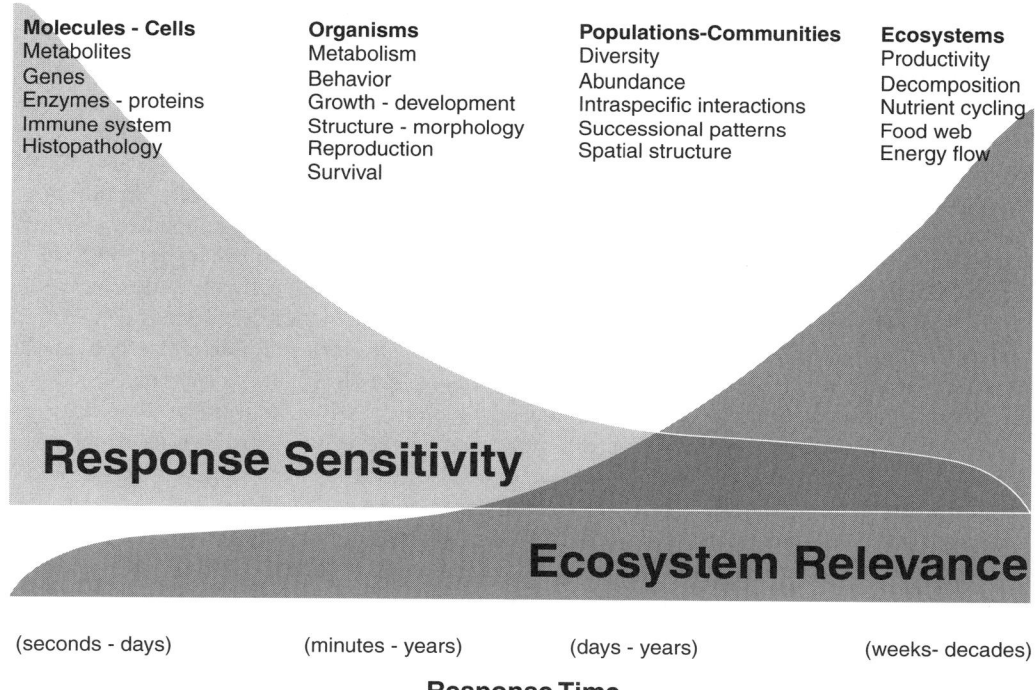

Molecules - Cells	Organisms	Populations-Communities	Ecosystems
Metabolites	Metabolism	Diversity	Productivity
Genes	Behavior	Abundance	Decomposition
Enzymes - proteins	Growth - development	Intraspecific interactions	Nutrient cycling
Immune system	Structure - morphology	Successional patterns	Food web
Histopathology	Reproduction	Spatial structure	Energy flow
	Survival		

Response Sensitivity

Ecosystem Relevance

(seconds - days)   (minutes - years)   (days - years)   (weeks- decades)

Response Time

Figure 4.1   Ecotoxicological endpoints: sensitivity and relevance. (Reprinted with permission from Burton, G.A., Jr. Assessing freshwater sediment toxicity. *Environ. Toxicol. Chem.*, 10: 1585–1627, 1991. © SETAC, Pensacola, FL, U.S.A.)

Selection of Biological Endpoints for Monitoring

The optimal assessment parameters which should be included depend on the project objectives. These parameters can be defined as measured characteristics, responses, or endpoints. For example, if the affected stream is classified as a high quality water and cold water fishery, then possible assessment or measured responses (endpoints) could include trout survival and hatchability, population and community indices (e.g., species richness), spawning area quantity and quality, dissolved oxygen, suspended solids, and water temperature. Endpoints vary dramatically in their sensitivity to pollutants and ecological relevance (Figure 4.1). The endpoints that are more sensitive are often more variable or respond to natural "nonpollutant" factors, so that adverse effects (stressors) are more difficult to classify with certainty. The most commonly and successfully used biotic indicators and endpoints are discussed in subsequent sections.

Aquatic ecosystems are quite complex, consisting of a wide variety of organisms. These organisms have their own unique function in the ecosystem and are directly or indirectly linked with other organisms. For example, bacteria, fungi, insects, and other invertebrates that inhabit the bottom of the waterways each need the others to assist in the decomposition of organic matter (such as leaves) so that they may consume it as food. If any one of these groups of organisms is lost or reduced, then the others will also be adversely affected. If the invertebrates are lost, their fish predators will be impacted. These groups are made up of a number of species with varying tolerance levels to stressors, and each possesses unique or overlapping functional characteristics (e.g., organic matter processing, nitrogen cycling). By carefully selecting the biological monitoring parameters, a broad range of relevant and sensitive indicator organisms can be used to efficiently assess ecosystem quality.

The most commonly used biological groups in aquatic assessments are fish, benthic macroinvertebrates, zooplankton, and algae. In lotic (flowing water) systems, fish and benthic macroinvertebrates are often chosen as monitoring tools. Benthic refers to sediment or bottom surfaces (organic
and inorganic). Macroinvertebrates are typically classified as those organisms which are retained
in sieves larger than 0.3 to 0.5 mm. They include a wide range of invertebrates, such as worms,
insect larvae, snails, and bivalves. They are excellent indicators of water quality because they are
relatively sedentary and do not move between different parts of a stream or lake. In addition, a
great deal is known about their life histories and pollution sensitivity. Algae, zooplankton, and fish
are used more in lentic (lake) environments. Of these, fish are most often used (both in lotic and
lentic habitats). Fish are transient, moving between sites, so it is more difficult to determine their
source of exposure to stressors; however, they are excellent indicators of water quality and provide
a direct link to human health and wildlife consumption advisories. Rooted macrophytes and
terrestrial plant species are good wetland health indicators, but are used less frequently.

In order to effectively and accurately evaluate ecosystem integrity, biosurveys should use two
to three types of organisms which have different roles in the ecosystem, such as decomposers
(bacteria), producers, primary to tertiary consumers (EPA 1990b). This same approach should be
used in toxicity testing (Burton et al. 1989, 1996; Burton 1991). This increases the power of the
assessment, providing greater certainty that if there is a type of organism(s) (species, population,
or community) in the ecosystem being adversely affected, either directly or indirectly, it will be
detected. This also allows for better predictions of effects, such as in food chain bioaccumulation
with subsequent risk to fish-eating organisms (e.g., birds, wildlife, humans). A large database exists
for many useful indicator species concerning their life history, distribution, abundance in specific
habitats or ecoregions, ecological function, and pollutant (stressor) sensitivity.

In the monitoring of fish and benthic macroinvertebrate communities, a wide variety of approaches
have been used. A particularly popular approach recommended by the U.S. EPA, Ohio EPA, state
volunteer monitoring programs, and other agencies is a multimetric approach, as summarized previously. The multimetric approach uses the basic data of which organisms are present at the site and
analyzes the data using a number of different metrics, such as richness (number of species present),
abundance (number of individuals present), and groups types of pollution-sensitive and resistant
species. The various metrics provide unique and sometimes overlapping information on the quality
of the aquatic community. Structural metrics describe the composition of a community, that is, the
number and abundance of different species, with associated tolerance rankings. Functional metrics
may measure photosynthesis, respiration, enzymatic activity, nutrient cycling, or proportions of feeding groups, such as omnivores, herbivores, insectivores, shredders, collectors, and grazers. The U.S.
EPA and Ohio EPA approaches are described in more detail in Chapter 6 and Appendices A, B, and C.

The Microtox™ (from Azur) toxicity screening test has been successfully used in numerous
studies to indicate the sources and variability of toxicant discharges. However, these tests have not
been standardized by the U.S. EPA or state environmental agencies but have been in Europe. More
typically, whole effluent toxicity test methods are employed (see Chapter 6, and also review by
Burton et al. 2000). These tests may miss toxicant pulses and do not reflect real-world exposure
dynamics. Many of the *in situ* toxicity tests, especially in conjunction with biological surveys (at
least habitat and benthic macroinvertebrate evaluations) and sediment chemical analyses, can
provide more useful information to document actual receiving water toxicity problems than relying
on water analyses alone. If a water body is shown to have toxicant problems, it is best to conduct
a toxicity identification evaluation (TIE) to attempt to isolate the specific problematic compounds
(or groups of compounds) before long lists of toxicants are routinely analyzed.

Selection of Chemical Endpoints for Monitoring

An initial monitoring program must include parameters associated with the above beneficial
uses. However, as the receiving water study progresses, it is likely that many locations and some

beneficial uses may not be found to be problematic. This would enable a reduction in the list of parameters to be routinely monitored. Similarly, additional problems may also become evident with time, possibly requiring an expansion of the monitoring program. The following paragraphs briefly describe the main chemical monitoring parameters that could be included for the beneficial use impact categories listed previously for a receiving water only affected by stormwater. However, it might be a good idea to periodically conduct a more-detailed analysis as a screening tool to observe less obvious, but persistent problems. If industrial or municipal point discharges or other nonpoint discharges (such as from agriculture, forestry, or mining activities) also affect the receiving water under study, additional constituents might need to be added to this list.

Obviously, chemical analyses can be very expensive. Therefore, care should be taken to select an appropriate list of parameters for monitoring. However, the appropriate number of samples must be collected (see Chapter 5) to ensure reliable conclusions. Chemical analyses of sediments may be more informative of many receiving water problems (especially related to toxicants) than chemical analyses of water samples. This is fortunate because sediment chemical characteristics do not change much with time, so generally fewer sediment samples need to be analyzed during a study period, compared to water samples. In addition, the concentrations of many of the constituents are much higher in sediment samples than in water samples, requiring less expensive methods for analyses. Unfortunately, sediment sample preparation (especially extractions for organic toxicant analyses and digestions for heavy metal analyses) can be much more difficult for sediments than for water.

Sediment Chemical Analyses

The basic list for chemical analyses for sediment samples, depending on beneficial use impairments, includes toxicants and sediment oxygen demand. The toxicants should include heavy metals (likely routine analyses for copper, zinc, lead, and cadmium, in addition to periodic ICP analyses for a broad list of metals). Acid volatile sulfides (AVS) are also sometimes analyzed to better understand the availability of the sediment heavy metals. Other sediment toxicant analyses may include PAHs and pesticides. Particle size analyses should also be routinely conducted on the sediment samples. Sediment oxygen demand analyses, in addition to an indication of sediment organic content (preferably particulate organic carbon, or at least COD and volatile solids), and nutrient analyses are important in areas having nutrient enrichment or oxygen depletion problems. Microorganisms (*Escherichia coli*, enterococci, and fecal coliforms) should also be evaluated in sediments in areas having likely pathogen problems (all urban areas). Interstitial water may also need to be periodically sampled and analyzed at important locations for the above constituents.

Water Chemical Analyses

The basic list for chemical analyses for water samples, depending on beneficial use impairments, includes toxicants, nutrients, solids, dissolved oxygen, and pathogens.

The list of specific toxicants is similar to that for the sediments (copper, zinc, lead, and cadmium, plus PAHs and pesticides). However, because of the generally lower concentrations of the constituents in the sample extracts for these analyses, more difficult analytical methods are generally needed, but the extraction and digestion processes are usually less complex than for sediments. In addition, because of the high variability of the constituent concentrations with time, many water samples are usually required to be analyzed for acceptable error levels. Therefore, less costly screening methods should be stressed for indicating toxicants in water. Because of the their strong associations with particulates, the toxicants should also be periodically analyzed in both their total and filterable forms. This increases the laboratory costs, but is necessary to understand the fates and controllability of the toxicant discharges. Typical chemical analyses for stormwater toxicants may include:

- Metals (lead, copper, cadmium, and zinc using graphite furnace atomic adsorption spectrophotometry, or other methods having comparable detection limits), periodic total and filtered sample analyses
- Organics (PAHs, phenols, and phthalate esters using GC/MSD with SIM, or HPLC), pesticides (using GC/ECD, or immunoassays), periodic total and filtered sample analyses

Pesticides in urban stormwater have recently started to receive more attention (USGS 1999). The USGS's National Water Quality Assessment (NAWQA) program has extensively sampled urban and rural waters throughout the nation. Herbicides commonly detected in urban water samples include simazine, prometon, 2,4-D, diuron, and tebuthiuron. These herbicides are extensively used in urban areas. However, other herbicides frequently found in urban waters are used in agricultural areas almost exclusively (and likely drift in to urban lands from adjacent farm lands) and include atrazine, metolachlor, deethylatrazine, alachlor, cyanezine, and EPTC. Insecticides commonly detected in urban waters include diazinon, carbaryl, chlorpyrifos, and malathion.

Nutrient analyses are also important when evaluating several beneficial uses. These analyses are not as complex as the toxicants listed above and are therefore much less expensive. However, relatively large numbers of analyses are still required. Water analyses may include the following typical nutrients: total phosphorus, inorganic phosphates (and, by difference, organic phosphates), ammonia, Kjeldahl nitrogen (or the new HACH total nitrogen), nitrate plus nitrite, and TOC. Periodic analyses for total and filtered forms of the phosphorus and TOC should also be conducted.

Dissolved oxygen is a basic water quality parameter and is important for several beneficial uses. Historical discharge limits have typically been set based on expected DO conditions in the receiving water. The typical approach is to use a portable DO meter for grab analyses of DO. Continuous *in situ* monitors, described in Chapter 6, are much more useful, especially the new units that have much more stable DO monitoring capabilities and can also frequently record temperature, specific conductance, turbidity, pH, and ORP. These long-term analyses are especially useful when evaluating diurnal variations or storm-induced discharges.

Pathogens should be monitored frequently in most receiving waters. Both urban and rural streams are apparently much more contaminated by problematic pathogenic conditions than has previously been assumed. Historically monitored organisms (such as fecal coliforms), in addition to *E. coli* and enterococci which are now more commonly monitored, can be present at very high levels and be persistent in urban streams. Specific pathogens (such as *Pseudomonas aeruginosa* and *Shigella*) can also be more easily monitored now than in the past. Most monitoring efforts should probably focus on fecal coliforms, *E. coli*, and enterococci.

Additional conventional parameters affecting fates and effects of pollutants in receiving waters should also be routinely monitored, including hardness, alkalinity, pH, specific conductivity, COD, turbidity, suspended solids (SS), volatile suspended solids (VSS), and total dissolved solids (TDS).

Selection of Additional Endpoints Needed for Monitoring

Several other stream parameters also need to be evaluated when investigating beneficial uses. These may include debris and flow obstructions, high/low flow variations, inappropriate discharges, aesthetics (odors and trash), and public access.

Data Quality Objectives and Quality Assurance Issues

For each study parameter, the precision and accuracy needed to meet the project objectives should be defined. After this is accomplished, the procedures for monitoring and controlling data quality must be specific and incorporated within all aspects of the assessment, including sample collection, processing, analysis, data management, and statistical procedures (see also Chapter 7).

When designing a plan one should look at the study objectives and ask:

- How will the data be used to arrive at conclusions?
- What will the resulting actions be?
- What are the allowable errors?

This process establishes the Data Quality Objectives (DQOs), which determine the level of uncertainty that the manager is willing to accept in the results. DQOs, in theory, require the study designers (decision makers and technical staff) to decide what are allowable probabilities for Type I and II errors (false-positive and false-negative errors) and issues such as what difference in replicate means is significant. The DQO process is a pragmatic approach to environmental studies, where limited resources prevent the collection of data not essential to the decision-making process. Uncertainty in ecological impact assessments is natural due to variability and unknowns, sampling measurement errors, and data interpretation errors. Determining the degree of uncertainty in any of these areas can be difficult or impractical. Yet an understanding of these uncertainties and their relative magnitudes is critical to the QA objectives of producing meaningful, reliable, and representative data. The more traditional practices of QA/QC should be expanded to encompass these objectives and thus help achieve valid conclusions on the test ecosystem's health (Burton 1992).

The first stage in developing DQOs requires the decision makers to determine what information is needed, reasons for the need, how it will be used, and to specify time and resource limits. During the second stage, the problem is clarified and constraints on data collection identified. The third stage develops alternative approaches to data selection, selecting the optimal approach, and establishing the DQOs (EPA 1984, 1986). Chapter 5 includes detailed information concerning the required sampling efforts to achieve the necessary DQOs, based on measured or estimated parameter variabilities and the uncertainty goals.

EXAMPLE OUTLINE OF A COMPREHENSIVE RUNOFF EFFECT STUDY

The following is an outline of the specific steps that generally need to be followed when designing and conducting a receiving water investigation. This outline includes the topics that are described in detail in later chapters of this book.

Step 1. What's the Question?

For example: Does site runoff degrade the quality of the receiving-stream ecosystem? Chapter 3 is a summary of documented receiving water problems associated with urban stormwater, for example. That chapter will enable the investigator to identify the likely problems that may be occurring in local receiving waters, and to identify the likely causes.

Step 2. Decide on Problem Formulation

Candidate experimental designs can be organized in one of the following basic patterns:

1. Parallel watersheds (developed and undeveloped)
2. Upstream and downstream of a city
3. Long-term trend
4. Preferably, most elements of all of the above approaches combined in a staged approach

Examples of these problem formulations are included at the end of this chapter, while Chapter 5 describes basic study designs, such as stratified random sampling, cluster sampling, and search sampling.

Another important issue is determining the appropriate study duration. In most cases, at least 1 year should be planned in order to examine seasonal variations, but a longer duration may be needed if unusual or dynamic conditions are present. As shown in Chapter 7, trend analyses can require many years. In addition, variations in the parameters being investigated will require specific numbers of observations in order to obtain the necessary levels of errors in the program (as described in Chapter 5). If the numbers of observations relate to events (such as runoff events), the study will need to last for the duration necessary to observe and monitor the required number of events.

Step 3. Project Design

1. Qualitative watershed characterization
    A. Establish degree of residential, commercial, and industrial area to predict potential stressors. Typically, elevated solids, flows, and temperatures are stressors common to all urban land uses. The following lists typical problem pollutants that may be associated with each of these land uses:
        1. Residential: nutrients, pesticides, fecal pathogens, PAHs, and metals
        2. Commercial: petroleum compounds, metals
        3. Industrial: petroleum compounds, other organics, metals
        4. Construction: suspended solids
    Topographical maps are used to determine watershed areas and drainage patterns.
2. Stream characterization
    A. Identify potential upstream stressor sources and potential stressors
        1. Photograph and describe sites.
    B. Survey upstream and downstream (from outfall to 1 km minimum) quality. Record observations on physical characteristics, including channel morphology (pools, riffles, runs, modification), flow levels, habitat (for fish and benthos), riparian zone, sediment type, organic matter, oil sheens, and odors. Record observations on biological communities, such as waterfowl, fish-eating birds or mammals, fish, benthic invertebrates, algal blooms, benthic algae, and filamentous bacteria.
    C. Identify appropriate reference site upstream and/or in a similar sized watershed with same ecoregion.
    D. Collect historical data on water quality and flows.
3. Select monitoring parameters
    A. Habitat evaluation. Should be conducted at project initiation and termination. Includes Quantitative Habitat Evaluation Index (QHEI), bed instability survey (bed lining materials and channel cross-sectional area changes), aesthetic/litter survey, inappropriate discharges (field screening), etc.
    B. Stressors and their indicators:
        1. Physical: flow, temperature, turbidity. Determine at intervals throughout base to high flow conditions.
        2. Chemical: conductivity, dissolved oxygen, hardness, alkalinity, pH, nutrients (nitrates, ammonia, orthophosphates), metals (cadmium, copper, lead, and zinc), and immunoassays (pesticides and polycyclic aromatic hydrocarbons) and/or toxicity screening (Microtox). The necessity of testing nutrients, metals, and organics will depend on the watershed characteristics. Determine at intervals throughout base to high flow conditions.
        3. Biological: benthic community structure (e.g., RBP), fish community structure, and tissue residues (confirmatory studies only). Benthic structure should be determined at the end of the project. Sediment bioaccumulation potential can be determined using the benthic invertebrate *Lumbriculus variegatus*.
        4. Toxicity: short-term chronic toxicity assays of stream water, outfalls, and sediment. Sediment should be sampled during baseflow conditions and tested before and after a high flow event. Water samples should be collected during baseflow and during pre-crest levels. Test species selection is discussed in Chapter 6 and in Appendix D. Expose test chambers with and without sunlight-simulating light (containing ultraviolet light wavelengths) to detect PAH toxicity. *In situ* toxicity assays should be deployed in the stream for confirmatory studies during base and high flow periods.

4. Data quality objectives. Determine the kinds of data needed and the levels of accuracy and precision necessary to meet the project objectives. These decisions must consider that there is typically a large amount of spatial and temporal variation associated with runoff study parameters. Chapter 5 relates sampling efforts associated with actual variability and accuracy and precision goals. This requires additional resources for adequate quantification.

5. Triggers and tiered testing. Establish the trigger levels or criteria that will be used to determine when there is a significant effect, when the objective has been answered, and/or when additional testing is required. Appropriate trigger levels may include:
   A. An arbitrary 20% difference in the test site sample, as compared to the reference site, might constitute a significant effect. (However, as noted in Chapter 5, a difference this small for many parameters may be difficult and therefore expensive to detect because of the natural variability.)
   B. An exceedance of the 95% statistical confidence intervals as compared to the reference sample.
   C. High toxicity in the test site sample, measured as Toxic Units (TUs) (e.g., 1/LC50).
   D. Exceedance of biotic integrity, sediment, or water quality criteria/guidelines/standards at the test site
   E. Exceedance of a hazard quotient of 1 (e.g., site concentration/environmental effect or background concentration).

A tiered or a phased testing approach is most cost effective, if time permits. A qualitative or semiquantitative study may include a greater number of indicator or screening parameters, such as turbidity, temperature, DO, specific conductivity, and pH using a continuous recording water quality sonde, plus artificial substrate macroinvertebrate colonization tests, and "quick" sediment toxicity tests. If possible, Microtox screening toxicity tests, immunoassay tests for pesticides and PAHs, and sediment metal analyses should also be added to this initial effort. These simple tests can be conducted with more widespread sampling to better focus later tiers on quantifying appropriate stressors in critical sampling areas and times. Final project tiers can identify specific stressors, their contribution to the problem, their sources, or simply confirm the ecological significance of the observed effects.

6. Sampling station selection. Select the study sites, such as upstream reference sites, outfall(s), and downstream impacted sites. In the selection of the upstream/reference and downstream sites, consider flow dynamics, stressor sources, and reference habitat similarities.

7. Quality assurance project plans (QAPP). It is essential that the quality of the project be ensured with adequate quality assurance and quality control measures. This will include routine laboratory and field documentation of operator and instrumentation performance, chain-of-custody procedures, adequate sample replication, QA/QC samples (blanks and spikes, etc.), performance criteria, and ensuring data validity. Appropriate experimental design (study design and sampling efforts) is also a critical component of a QAPP.

Step 4. Project Implementation (Routine Initial Semiquantitative Survey)

1. Baseflow conditions
   A. Habitat survey (e.g., Qualitative Habitat Evaluation Index)
   B. Benthic RBP
   C. Test water and sediment from all test sites for short-term chronic toxicity with two species.
   D. Establish spatial and diurnal variation (YSI 6000 for several weeks, plus grab samples or time composites).
   E. Set up automatic stream samplers/monitors, stream depth gauges, and rain gauges.
   F. Establish local contacts to oversee field equipment and provide rain event notification.
   G. Conduct field screening survey at outfalls to identify sources of dry-weather flows.
2. High flow conditions
   A. Confirm that the samplers and monitors are operational. Collect grab samples if necessary (for microbiological and VOC analyses, for example).
   B. Deploy *in situ* toxicity test assays.
   C. Measure flow and note staff gauge depth, using manual or automatic samplers and flow recorders. Repeat flow measurements at intervals of 0.5- to 1.0-ft stream depth intervals as the stream rises, noting time and depth. Focus on first flush to crest period.

    D. Measure DO, temperature, turbidity, conductivity, and stage at each station following each flow measurement. Establish spatial variance. May use continuous recording water quality sondes.

    E. Collect flow-weighted composited (or combine many discrete) samples for other analyses.

3. Sample analyses

    A. Filter, preserve, and chill samples, as required.

    B. Deliver samples to analytical laboratories with chain-of-custody forms.

    C. Initiate toxicity testing and other chemical and microbiological analyses within required time period since sample collection.

    D. Document QA/QC.

4. Follow-up (post-event) monitoring

    A. Check *in situ* assay chambers at 24 and 48 hours and at 7 and 14 days if deployed.

    B. Conduct benthic RBP.

    C. Conduct QHEI, noting bedload movement.

    D. Collect fish for tissue residue analyses.

Step 5. Data Evaluation

1. Plot flow vs. physical and chemical analysis results.
2. Statistically compare responses/loadings during base, first flush, and post-crest conditions. This will provide a characterization of flow dynamics and its effect on stressor profiles.
3. Statistically compare stations (instantaneous, mean periods) for significant differences and correlations.
4. Calculate and compare physical, chemical, and toxicity (using Toxicity Units) loadings. This will show the relative load contribution of stressors from reference (upstream) vs. impacted (downstream) reach.
5. Identify magnitude and duration of trigger exceedances.
6. Identify sources of uncertainty.
7. Identify potential sources of pollutants and stressors.
8. Determine literature value thresholds for key stressors on key indigenous species.

Step 6. Confirmatory Assessment (Optional Tier 2 Testing)

1. Repeat Steps 2 and 3 using Tier 1 information to select fewer test parameters with increased sampling frequency and/or select more descriptive methods. Increased sampling will better quantify the magnitude and duration of stressor dynamics. Expanded sampling will better document the quality of the receiving water. More definitive testing could include:

    A. Short-term chronic toxicity testing with additional species (lab and *in situ*)

    B. Increased testing of toxicants

    C. Characterizing fish, plankton, periphyton, or mussel populations

    D. Measuring assimilative capacity via long-term BOD and SOD testing

    E. Measuring productivity with light/dark bottle BOD *in situ* tests

2. Conduct toxicity identification evaluation (TIE) study of water, outfalls, and/or sediment to determine contribution of each stressor to total toxicity. This information can better determine which stressors are important to control and can also identify sources of toxicity.

3. Conduct bioaccumulation testing of site sediments. Some pollutants, such as highly chlorinated organic compounds (e.g., chlordane, DDT, PCBs, dioxins) are readily bioaccumulated, yet may not be detected using the above study design. The EPA has a benthic invertebrate 28-day assay to measure sediment bioaccumulation potential. Also SPMDs may be used.

4. Indigenous biological community characterization and tissue analysis. More in-depth quantification of benthic and/or fish community structure on a seasonal basis will better identify significant ecological effects. Tissue sampling of fish for contaminants will provide information on bioaccumulative pollutants and potential food web or human health effects from consumption.

Table 4.8   Watershed Study Complexity Matrix

Situation: Complexity Scale (Simple to Complex)	Primary Considerations
Single outfall	
Small stream (small watershed)	Focus on loading of site stressors from site and from upstream. Reference upstream.
Large stream (larger watershed)	Determine if upstream inputs are degrading water quality. Upstream and separate ideal reference sites.
Pristine estuary	Focus on outfall quality and mixing zone. Deploy *in situ* monitors. Use far-field reference.
Multiple outfalls	
River (multi-watersheds)	Multistation network with habitat, benthos, and select toxicity evaluations of water and sediments. Tiered study with TIE, outfall, and *in situ* studies to find major problem sources. Use upstream and adjacent watershed references. Focus on tributary mouths for initial sampling and use SPMDs.
Coastal harbor	Focus on outfall quality and near-field mixing zones. Deploy *in situ* monitors. Use far-field, adjacent watershed references.

Step 7. Project Conclusions

1. List probable stressors.
2. Document trigger exceedances.
3. Discuss relative contribution of stressors(s) to ecosystem degradation. Support documentation may include:
    A. Literature threshold values
    B. Criteria exceedances
    C. Toxicity observed (from TIE, photoactivation, or *in situ* assays)
    D. Bioaccumulation factors and potential for food web contamination
4. Provide recommendations for stressor reduction and ecosystem enhancement.
5. Include suggestions on habitat improvement, flow reduction, turbidity removal, and reduced siltation.

Table 4.8 summarizes the primary considerations that should be examined for different levels of receiving water complexity. Obviously, increasingly complex situations require more complex study designs and elements. However, this table briefly outlines the major issues that should be considered.

CASE STUDIES OF PREVIOUS RECEIVING WATER EVALUATIONS

This section presents several case studies that have been conducted to investigate receiving water problems associated with runoff. These case studies illustrate the major approaches used to identify a potentially affected area through comparisons with a control area. The basic experimental designs are:

- Above/below longitudinal study where a stream is studied as it flows from above a city through a city. Obviously, the upstream control reach must be in a relatively undisturbed portion of the watershed and only wet-weather flows of interest affect any of the test reaches.
- Parallel stream study where two (or more) streams are studied. One of the streams is a control stream in a relatively undisturbed area, while the other stream is in an urbanized area.
- Trend analyses with time in a single stream to investigate changes that may occur with time as a watershed becomes urbanized, or with the application of stormwater controls.

The selection of suitable test areas is critical. As noted, the control water body should be minimally affected by urbanization, while the urban test water body should be affected only by urban runoff (and not municipal or industrial discharges, for example) if possible. In addition, the test and control water bodies must be otherwise very similar (especially as watershed area, topography, habitat potential, etc., are concerned). In a longitudinal study, the watershed area obviously increases in a downstream (urbanized) direction. In addition, the urban water body has a substantially different flow regime than an undisturbed water body. These differences should be the result of urbanization and not other factors. A successful receiving water study usually requires several years of study at many locations in each stream segment. As noted throughout this book, the selection of monitoring parameters is also critical. In most cases, varied and complementary analyses should be conducted, covering a range of biological, physical, and chemical parameters. However, carefully designed investigations can be more successfully focused on limited project objectives.

The first three case studies are examples of these three basic experimental designs for conducting a receiving water investigation and include both test and control conditions. Most of the receiving water studies reported in the literature only focus on potentially impacted water bodies, without any adequate control sites. This may be suitable in an area where the receiving water potential is well understood through extensive prior studies (such as in Ohio). However, it is very problematic to rely solely on various criteria to identify the magnitude of receiving water problems, without extensive local expertise on relatively natural conditions.

The identification of a "problem" is also highly dependent on desired beneficial uses. The local perception of use is critical. Obviously, human health considerations associated with potentially contaminated water supplies, consumptive fisheries, or contact recreation areas must be stringently addressed. Biological uses may be more open to local interpretation, however. It is unreasonable to expect completely natural receiving water conditions in an urban area. There are unavoidable impacts that will prevent the best natural conditions from occurring in an urbanized watershed. Obviously, general biological uses can still be met by providing suitable habitat and somewhat degraded conditions that would allow a reasonable assemblage of aquatic organisms to exist in an area. Noncontact recreational uses (especially the aesthetic factors of odors and trash) should also be provided in urban receiving waters. Test and control receiving water investigations are very useful in that they enable contrasting of existing degraded conditions with less impacted conditions. Perhaps the control reference sites should include not natural conditions, but acceptable degraded conditions associated with partial urbanization. This is possible with a longitudinal study where a receiving water is studied as it flows through an urban area, becoming more degraded in the downstream direction. Parallel stream studies can also include partially degraded, but acceptable, sites. In addition, trend analyses with time will indicate when unacceptable degradation occurs.

Example of a Longitudinal Experimental Design — Coyote Creek, San Jose, CA, Receiving Water Study

The Coyote Creek study is an example of an investigation of the effects of stormwater on the biological conditions in an urban creek as it passed through the City of San Jose, CA. This was an early comprehensive receiving water study that examined many attributes of the creek above and within the city.

This research project included many different biological, chemical, and physical parameters to quantify biological effects. The project was conducted by Pitt and Bozeman (1982) from 1977 through 1982, with funding from the Storm and Combined Sewer Section of the U.S. Environmental Protection Agency. The objective of this 3-year field monitoring study was to evaluate the sources and impacts of urban runoff on water quality and biological conditions in Coyote Creek. In many cases, very pronounced gradients of water and biological quality indicators were observed. Cause-

and-effect relationships cannot be conclusively proven in a study such as this; the degradation of conditions in Coyote Creek may be due to several factors, including urban runoff, stream flows (both associated and not associated with urban runoff), and natural conditions (e.g., drought, stream gradient, groundwater infiltration, etc.). Information collected during this study implied that the effects of various urban runoff constituents, especially organics and heavy metals in the water and in the polluted sediment, may be responsible for many of the adverse biological conditions observed.

The beginning of the project followed 2 years of severe drought. The first major rains occurred the previous November (1977), and seasonal rains that occurred during the study period were considered normal. Typical rainfall averaged 33 cm (13 in) per year in the area below Lake Anderson, and 50 to 71 cm (20 to 28 in) per year in the watershed above Lake Anderson. During the drought, which preceded this study, rainfall was only about one half of these amounts.

Step 1. What's the Question?

The major questions that were to be addressed during the Coyote Creek study were:

1. Identify and describe important sources of urban runoff pollutants.
2. Describe the effects of those pollutants on water quality, sediment quality, aquatic organisms, and the creek's associated beneficial uses.
3. Assess potential measures for controlling the problem pollutants in urban runoff.

Step 2. Decide on Problem Formulation

This project was designed to examine the changes in conditions in Coyote Creek as it passed through San Jose, CA. It was therefore a longitudinal study. The several-year duration of the study also enabled year-to-year variations to be compared to the differences in locations.

Step 3. Project Design

Qualitative Watershed Characterization

Figure 4.2 is a map of the San Francisco Bay area showing the location of the Coyote Creek watershed, while Figure 4.3 is a detailed map of the Coyote Creek watershed. The watershed itself is about 70 km (45 miles) long, 15 km (10 miles) wide, and contains about 80,000 ha (200,000 acres). Nearly 15% of the watershed consisted of developed urban areas during the study period. Most of the urban development is located in the northwest portion of the watershed.

Stream Characterization

For much of its length, Coyote Creek flows northwesterly along the western edge of the watershed. Elevations in the watershed range from sea level to nearly 920 m (3000 ft). Figure 4.4 shows the elevations of the various major sampling locations. Near the San Jose urban area, the watershed can be characterized as a broad plain with rolling foothills to the east. A portion of the watershed (i.e., the narrow strip between Lake Anderson and the urban area) is used for light but productive agriculture. The upper reaches and the headwaters of Coyote Creek are in extremely rugged terrain, with slopes commonly exceeding 30%. These upper areas can be characterized as chaparral-covered hills and gullies in a fairly natural state; they receive little use by man. Much of this land is within the Henry Coe State Park; non-park land is used primarily for low-density cattle grazing. Even though the watershed is very large and has upstream dams, the flow variations are extreme. Figure 4.5 shows the creek during a wet-weather period where the flows are overtopping

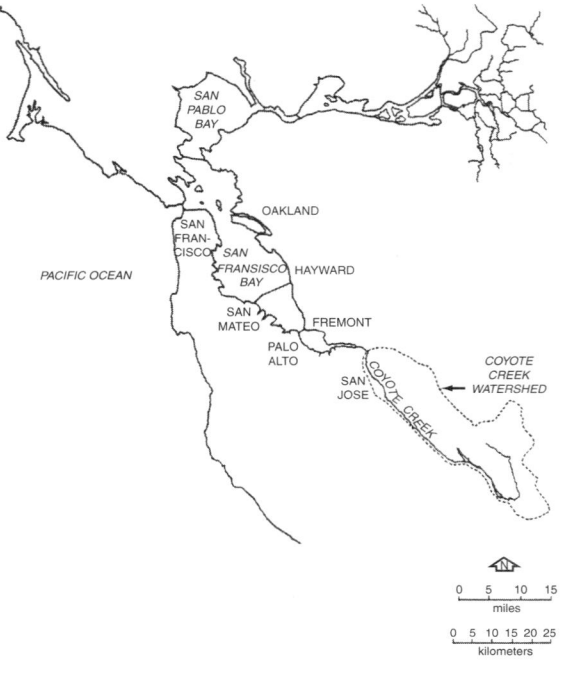

Figure 4.2 San Francisco Bay area and the location of the Coyote Creek watershed. (From Pitt, R. and M. Bozeman. *Sources of Urban Runoff Pollution and Its Effects on an Urban Creek*, EPA-600/S2-82-090, U.S. Environmental Protection Agency, Cincinnati, OH. 1982.)

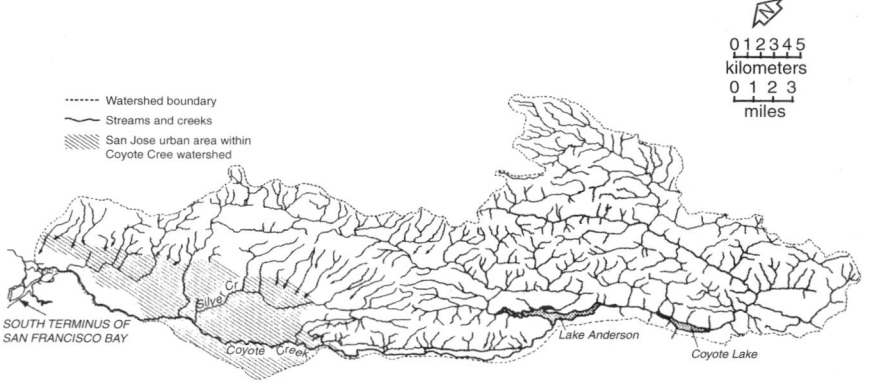

Figure 4.3 Detailed map of the Coyote Creek watershed. (From Pitt, R. and M. Bozeman. *Sources of Urban Runoff Pollution and Its Effects on an Urban Creek*, EPA-600/S2-82-090, U.S. Environmental Protection Agency, Cincinnati, OH. 1982.)

a road culvert, while Figure 4.6 shows the creek during a typical dry period (commonly lasting for 100 days without rain during summer months).

Several major facilities have been built on Coyote Creek to provide flood control and groundwater recharge. The largest are the dams, which contain man-made reservoirs: Lake Anderson and Coyote Lake. Discharges from these lakes are controlled by the Santa Clara Valley Water District. The major study area was located between the farthest downstream dam (Lake Anderson) and the first major confluence (where Coyote Creek meets Silver Creek, within the City of San Jose). Within this 39-km (24-mile) study area, approximately 16 km (10 miles) are urban and 23 km (14 miles) are non-urban. Sampling stations were located in both the urban and non-urban reaches of the stream for comparison.

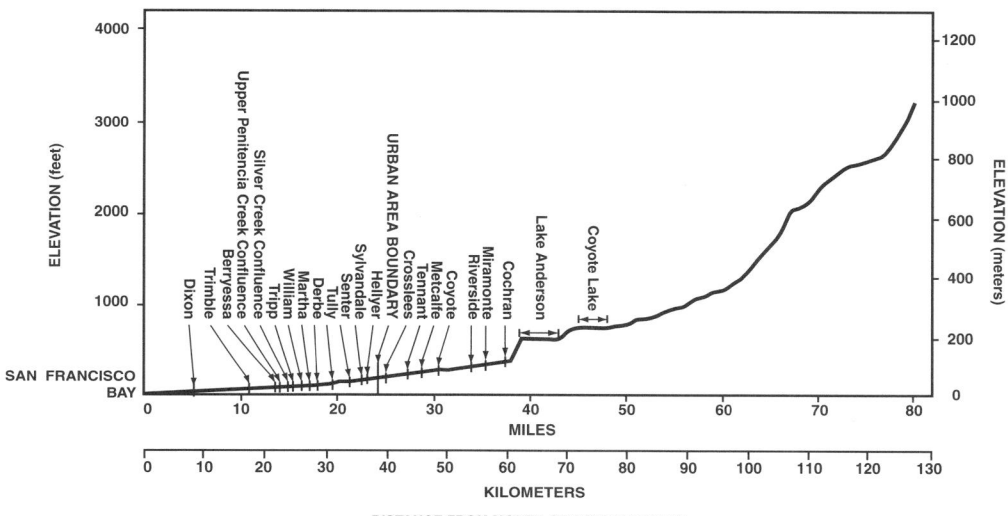

Figure 4.4   Elevations of the major sampling locations. (From Pitt, R. and M. Bozeman. *Sources of Urban Runoff Pollution and Its Effects on an Urban Creek*, EPA-600/S2-82-090, U.S. Environmental Protection Agency, Cincinnati, OH. 1982.)

Figure 4.5   High flows in Coyote Creek overtopping road culvert.

Figure 4.6   Low flows in Coyote Creek during typically extended summer dry period.

Average daily flows in the northern part of the creek during dry weather were typically less than 1.5 m³/s (50 cfs). Major storm flows, however, approach 30 m³/s (1000 cfs). The flows in the northern part of the creek were controlled largely by the discharges from Lake Anderson and Coyote Lake.

Coyote Creek is an important element of the Santa Clara Valley Water District's groundwater recharge program. Several recharge basins have been established adjacent to the stream channel within the study area. Diversion channels withdraw water from Coyote Creek, route it into these large basins, and return it back to the creek, depending upon such factors as season, stream flow, and groundwater level.

There is an average of 0.6 to 3 storm drain outfalls per kilometer (1 to 5 per mile) along the urban reach of Coyote Creek that was studied. The outfalls ranged from 20 to 180 cm (8 to 70 in) in diameter, but most are about 75 cm (30 in) in diameter. The drainage area per outfall

Table 4.9   Coyote Creek Drainage Areas above Each Monitoring Station

Sampling Station	Total Area (hectares)	Urban Area (hectares)	Non-urban Area (hectares)	Percent Urban
Cochran	49,510	<5	49,510	<0.01
Miramonte	50,260	<5	50,260	<0.01
Metcalfe	52,360	<50	54,360	<0.1
Crosslees	54,030	50	53,980	0.1
Hellyer	54,400	350	54,050	0.6
Sylvandale	54,720	450	54,320	0.7
Senter	55,300	800	50,500	1.5
Derbe	56,300	1740	54,560	3.2
William	56,920	2150	54,770	3.9
Tripp	57,260	2460	54,800	4.5

From Pitt, R. and M. Bozeman. *Sources of Urban Runoff Pollution and Its Effects on an Urban Creek*, EPA-600/S2-82-090, U.S. Environmental Protection Agency, Cincinnati, OH. 1982.

ranged from 2 to 320 ha (5 to 800 acres), but most of the outfalls drained areas smaller than 40 ha (100 acres).

Table 4.9 describes the drainage areas which cumulatively contribute runoff flows to selected monitoring stations. The urban area stations had about 3 to 5% (1700 to 2500 ha or 4000 to 6000 acres) of their total drainage areas urbanized, whereas the non-urban area stations had less than 0.1% of their drainage areas urbanized. The three stations designated as Hellyer, Sylvandale, and Senter were transition stations (about 0.6 to 1.5% of their drainage areas were urbanized).

Select Monitoring Parameters

The project involved conducting field measurements, observations, sampling, and other studies of Coyote Creek from March 1977 through August 1980. The study focused on the urban reaches of Coyote Creek, extending from Lake Anderson to the confluence with Silver Creek. In this reach of Coyote Creek, there are no known flow or pollutant contributions other than urban runoff. The sampling areas were selected such that each included a stretch of stream several hundred meters long, which met prescribed criteria for physical, biological, and chemical homogeneity.

The following parameters were typically examined at each sampling location:

- Basic hydrologic conditions
- Water quality
- Sediment properties
- General habitat characteristics
- Fish
- Benthic organisms (e.g., aquatic insects, crustaceans, mollusks)
- Attached algae
- Rooted aquatic vegetation (e.g., cattails)

Step 4. Project Implementation (Routine Initial Semiquantitative Survey)

Sampling took place during all months during the complete project period. As an example, the biological sampling stressed the spring and summer seasons of all project years, while the water column and sediment samples were conducted approximately monthly.

All water and sediment sampling was conducted manually using either plastic (HDPE) or glass wide-mouth bottles. Sediment core samples were obtained using a liquid carbon dioxide freezing core sampler. All water and sediment samples were comprised of at least six subsamples from the sampling location reach that were composited before analysis. The samples were then appropriately preserved and delivered to a commercial analytical laboratory for EPA-approved analyses.

Biological samples for lead and zinc bioaccumulation measurements (e.g., mosquito fish, filamentous algae, crayfish, cattail plant segments) were obtained at selected sampling stations during the routine fish sampling activities.

Fish were collected by seining and electroshocking representative pool and riffle habitats at 40 locations within the Coyote Creek system. Most of the collection efforts (conducted during the spring and summer of the project years) were focused on the portion of Coyote Creek between Lake Anderson and the confluence of Silver Creek. However, to further define the species composition and distribution of fishes, additional samples were obtained from both the upper and lower reaches of Coyote Creek, as well as from several locations within major tributaries. Captured fishes were identified and counted. The total length and weight were recorded for each specimen. Where numerous individuals of a particular species were encountered, only length range and aggregate weight were recorded, along with any abnormalities.

Quantitative collections of benthic macroinvertebrates were made at nine locations in Coyote Creek. Benthic macroinvertebrate samples were collected from natural substrates (e.g., cobbles, gravel, sand) in both pool and riffle habitats by means of an Ekman dredge (sample area of 0.023 m^2) or a Surber sampler (sample area of 0.093 m^2). Additionally, artificial substrates were used at six sampling locations. These consisted of pairs of Hester-Dendy multiplate samplers constructed of multiple, parallel plates of tempered hardboard (sample area of 0.120 m^2). The Hester-Dendy samplers were left in riffle sections of the stream for 8 weeks and then removed and examined in the laboratory.

Qualitative benthic collections were also made with the use of a D-frame sweep net at all biological monitoring stations. The benthic samples were washed through a sieve having a mesh size of 500 mm. Organisms retained on the screen were removed and preserved in 10% formalin, transferred to 70% ethanol, identified to the lowest practicable taxon, and enumerated.

Attached algae samples were obtained from both natural and artificial substrates throughout the various reaches of Coyote Creek. Qualitative samples of attached algae were collected by scraping uniform areas of natural substrates such as logs and rocks. Quantitative collections of attached algae were made with the use of artificial substrates consisting of diatometers equipped with glass slides. These were suspended in the water column at six locations within the study area for 8 weeks, then removed and examined in the laboratory.

Rooted aquatic plants were sampled qualitatively whenever they were encountered in the study area. Plant specimens were collected, pressed or preserved, and identified.

Step 5. Data Evaluation and Step 6. Confirmatory Assessment

Observed Conditions in Coyote Creek

Water Quality — The purpose of the water quality monitoring program in Coyote Creek was to define receiving water conditions in the urban and non-urban areas during dry-weather conditions. Data on wet-weather Coyote Creek water quality conditions were also obtained from other sources for comparison (Pitt 1979; Metcalf and Eddy 1978; Pitt and Shawley 1982; SCVWD 1978; USDA 1978). Table 4.10 summarizes Coyote Creek water quality data for the wet- and dry-weather conditions and for both the urban and non-urban creek reaches. Dry-weather concentrations of many constituents exceeded corresponding wet-weather concentrations by factors of two to five times. For example, during dry weather, many of the major constituents (e.g., major ions, hardness, alkalinity, total solids, total dissolved solids, specific conductance, ammonia nitrogen, and ortho-phosphate) were significantly greater in both the urban and non-urban reaches. These constituents were all found at substantially lower concentrations in the urban runoff affecting Coyote Creek (Pitt 1979). Temperature, pH, dissolved oxygen, nitrate nitrogen, and arsenic were found to be about the same for wet and dry weather, for both the urban and non-urban areas. Within the urban area, several constituents were found in greater concentrations during wet weather than during dry

Table 4.10   Typical Coyote Water Quality Condition by Location and Season (mg/L unless otherwise noted)

	Urban Area		Non-Urban Area	
	Wet Weather	Dry Weather	Wet Weather	Dry Weather
Common Parameters and Major Ions				
pH	7	8		8
Temperature	16	17	—	16
Calcium — dissolved	20	100	40	100
Magnesium — dissolved	6	70	20	60
Sodium — dissolved	0.01	—	—	20
Potassium — dissolved	2	4	2	2
Bicarbonate	50	150	—	200
Sulfate	20	60	—	40
Chloride	10	60	—	20
Total hardness	70	500	200	600
Total alkalinity	50	300	150	300
Residuals				
Total solids	350	1000	600	1000
Total dissolved solids	150	1000	300	1000
Suspended solids	300	4	600	20
Volatile suspended solids	60	2	90	10
Turbidity (NTU)	50	15	—	20
Specific conductance (μmhos/cm)	200	500	—	400
Organics and Oxygen Demand Material				
Dissolved oxygen (DO)	8	7	—	9
Biochemical oxygen demand (5-day) (BOD_5)	25	—	5	—
Chemical oxygen demand (COD)	100	40	90	30
Total organic carbon (TOC)	110	—	—	0.6
Nutrients				
Total Kjeldahl nitrogen (TKN)	7	0.5	2	<0.3
Nitrate (as N)	0.7	0.8	—	1.2
Nitrite (as N)	—	0.02	—	<0.002
Ammonia (as N)	0.1	0.8	0.1	0.3
Orthophosphate	0.2	0.5	0.1	0.4
Heavy Metals				
Lead (μg/L)	2000	40	200	2
Zinc (μg/L)	400	30	200	20
Copper (μg/L)	20	10	50	5
Chromium (μg/L)	20	10	5	5
Cadmium (μg/L)	5	<1	5	<1
Mercury (μg/L)	1	0.2	1	0.2
Arsenic (μg/L)	4	3	5	2
Iron (μg/L)	10,000	1000	20,000	2000
Nickel (μg/L)	40	<1	80	<1

From Pitt, R. and M. Bozeman. *Sources of Urban Runoff Pollution and Its Effects on an Urban Creek*, EPA-600/S2-82-090, U.S. Environmental Protection Agency, Cincinnati, OH. 1982.

weather (e.g., suspended solids, volatile suspended solids, and turbidity). COD and organic nitrogen were also present in the urban area in greater abundance during wet weather than dry, as were heavy metals (e.g., lead, zinc, copper, cadmium, mercury, iron, and nickel).

Water quality upstream of the urbanized area was fairly consistent from site to site, but the quality changed markedly as the creek passed through the urbanized area. The water quality within the urbanized reach was generally poorer than at the stations upstream. Similar differences between wet and dry weather were also noted for the non-urban area. However, the wet-weather concentrations were typically much higher in the urban area than in the non-urban area. Several other constituents were also found in higher concentrations in the urban area than in the non-urban area during wet weather. Lead concentrations were more than seven times greater in the urban reach than in the non-urban reach during dry weather. Nitrite concentrations were almost seven times greater in the urban area. Ammonia nitrogen values in the urban area were 2.8 times greater than in the non-urban area. Other significant increases in urban area concentrations included chloride, nitrate, orthophosphate, COD, specific conductance, sulfate, and zinc. Conversely, the dissolved oxygen measurements were about 20% less in the urban reach than in the non-urban reach of the creek.

Selected water and sediment samples from the urban area reaches of Coyote Creek were analyzed as part of a nationwide screening effort to assess priority pollutant concentrations in urban runoff and urban receiving waters. Three samples were collected in January 1979, during a major storm. These included a runoff sample and samples of sediment and water from Coyote Creek. The sampling was conducted in and near the Martha Street outfall, which is located in a heavily urbanized area. Only 18 of the approximately 120 priority pollutants analyzed were detected (base-neutrals: fluoranthene, diethyl phthalate, di-n-butyl phthalate, bis(2-ethyl hexyl)phthalate, anthracene, phenanthrene, and pyrene; the phenols: 2,4,6-tricholorphenol, 2,4-dimethylphenol, pentachlorophenol, and phenol; and heavy metals: arsenic, cadmium, copper, lead, mercury, and zinc). These priority pollutants are generally the same as those found in most other urban runoff and receiving water samples collected nationwide (EPA 1983, Pitt et al. 1995).

Sediment Quality — Sediment samples were collected at the major sampling locations three times during the study. Table 4.11 summarizes all of the Coyote Creek sediment quality measurements obtained during the entire project. Orthophosphates, TOC, BOD_5, sulfates, sulfur, and lead were all found in higher concentrations in the sediments from the urban area stations, as compared with those from the upstream, non-urban area stations. The median sediment particle sizes were also found to be significantly smaller at the urban area stations, reflecting a higher silt content. Sulfur, lead, and arsenic were found in substantially greater concentrations (4 to 60 times greater) for the urban area sediments compared to the non-urban area sediments.

When all of the sediment data from the three monitoring periods were combined, very few differences were found between the urban and non-urban area values for COD, total phosphate, arsenic, and median particle size. However, seasonal variations were found to be important. When the data from just one sampling period were considered alone, greater and more significant variations in constituent concentrations between the two reaches were observed.

Lead concentrations in the urban area sediments were markedly greater than those from the non-urban area, by a factor of about six times (which is the widest margin for any constituent monitored). Large differences were also found between the urban and non-urban area data for both sulfate and phosphate. Average zinc concentrations in the sediments were found to increase by only about 1.5 times, but with a high degree of confidence.

The largest difference between urban and non-urban area sediment (mg/kg) to water (mg/L) concentration ratios (S/W) was for lead, where the S/W ratio was over 3000 for the urban area and only about 400 for the non-urban area. The total Kjeldahl nitrogen S/W ratio was about 5500 for the urban area but exceeded 22,000 for the non-urban area. For the other constituents studied, the differences between the urban and non-urban area S/W ratios were much less. Lead, zinc, arsenic,

Table 4.11 Coyote Creek Sediment Quality

All Units Are mg/kg Total Solids, Except for Particle Size	Non-Urban Area Stations below Anderson Dam						Urban Area Stations above Silver Creek						Urban and Non-Urban Differences	
	No. of Obs.	Mean	Min	Max	St. Dev.	COV	No. of Obs.	Mean	Min	Max	St. Dev.	COV	Ratio of Means	Confidence that Urban/Non-Urban Values
Chemical oxygen demand	7	35,500	7400	98,000	34,800	0.98	13	39,300	4600	131,000	41,000	1.0	1.1	<60%
Total phosphate	4	148	7.5	344	168	1.1	10	168	14	406	161	0.96	1.1	<60%
Orthophosphate	3	1.2	0.46	1.7	0.65	0.54	3	3.6	1.2	6.6	2.8	0.78	3.0	85%
Total Kjeldahl nitrogen	7	6500	138	29,000	10,500	1.6	13	2490	146	14,000	4100	1.7	0.4	85%
Sulfate	7	136	<200	478	229	1.7	13	430	<200	3670	1010	2.4	3.2	80%
Arsenic	7	11.1	<1.0	28	11.1	1.0	13	13.0	1.5	45	10.3	0.79	1.2	65%
Lead	7	18.8	6.7	37	10.2	0.54	13	114	20	400	132	1.2	6.1	96%
Zinc	7	64	14	90	25	0.39	13	96	30	170	37	0.39	1.5	97%
Median particle size (μm)	7	4350	210	8760	4085	0.94	13	4480	70	8600	3650	0.81	1.0	<60%

From Pitt, R. and M. Bozeman. *Sources of Urban Runoff Pollution and Its Effects on an Urban Creek*, EPA-600/S2-82-090, U.S. Environmental Protection Agency, Cincinnati, OH. 1982.

and total Kjeldahl nitrogen all had S/W ratios of between 2000 and 5000 in the urban area. COD and total phosphate had S/W ratios of 1300 and 670, respectively, while orthophosphate and sulfate had S/W ratios of only about 20 and 6, respectively.

Because of these high observed sediment pollutant concentrations, it is likely that urban runoff-affected sediment is an important factor in the general decline in biological quality as Coyote Creek passes through the San Jose urban area. Other natural factors (e.g., stream gradient, temperature, and velocity changes) also probably contribute to this decline. For example, relatively flat creek gradients in the urban reach lead to low velocities which, in turn, encourage sedimentation of polluted particulates and allow temperatures to rise. Decreased flows in the urban area (due to diversions and infiltration) are an additional cause for changes in flow regime, water quality, and biological conditions.

Bioaccumulation of Lead and Zinc — Biological samples were collected from six stations in Coyote Creek and were analyzed to determine the lead and zinc they had accumulated while living in the creek. This sampling program was restricted to a single collection of organisms, with representative samples obtained from throughout the urban and non-urban stretches of the creek. Fish (*Gambusia affinis*), filamentous algae (*Cladophora* sp.), crayfish (*Procambarus clarkii*), and cattail plant segments (*Typha* sp.) were collected for analysis. An effort was made to collect similar specimens of the same species from each sampling location. All samples were rinsed to remove adhering sediment and were then chemically digested and analyzed for total lead and zinc content.

Some evidence of bioaccumulation of lead and zinc was found in many of the samples of algae, crayfish, and cattails. The measured concentrations of these metals in organisms (mg/kg) exceeded concentrations in the sediments (mg/kg) by up to a maximum factor of about 6. Concentrations of lead and zinc in the organisms exceeded water column concentrations by factors of 100 to 500 times, depending on the organism. Lead concentrations in urban area samples of algae, crayfish, and cattails were found to be two to three times as high as in non-urban area samples (Table 4.12), whereas zinc concentrations in urban area algae and cattail samples were about three times as high as the concentrations in the samples from the non-urban areas (Table 4.13). Lead and zinc concentrations in fish tissue were not significantly different between the urban and non-urban area samples.

Several early studies examined metal bioaccumulations in urban aquatic environments (Wilber and Hunter 1980; Neff et al. 1978; Phillips and Russo 1978; Ray and While 1976; Rolfe et al. 1977; Spehan et al. 1978). The lead concentrations in Coyote Creek waters are probably lower than the critical levels necessary to cause significant bioaccumulation in most aquatic organisms. The whole-body concentrations of zinc for the fish and crayfish were greater than many of the whole-body concentrations reported in the literature. The zinc concentrations in the Coyote Creek plants, however, were smaller than concentrations reported elsewhere for polluted waters.

Table 4.12 Lead Concentrations (mg lead/kg dry tissue) in Biological Samples[a]

	Non-Urbanized Area Stations			Urbanized Area Stations		
	Cochran	Miramonte	Metcalfe	Derbe	William	Tripp
Fish	<40	NS	NS	<30	<40	<50
Attached algae	<20	<30	<30	200	170	70
Crayfish	14	NS	<30	29	<36	40
Higher aquatics	<20	<30	<30	<30	<50	60
Sediment	28	37	16	37	370	400

[a] During storm events, lead concentrations in the urban reaches of Coyote Creek averaged about 2 mg/L. Dry weather, lead concentrations averaged about 0.04 mg/L in the urban reach. Non-urbanized reaches had lead water concentrations about 1/10 these values.

NS = No sample collected.

From Pitt, R. and M. Bozeman. *Sources of Urban Runoff Pollution and Its Effects on an Urban Creek*, EPA-600/S2-82-090, U.S. Environmental Protection Agency, Cincinnati, OH. 1982.

Table 4.13 Zinc Concentrations (mg zinc/kg dry tissue) in Biological Samples[a]

	Non-Urbanized Area Stations			Urbanized Area Stations		
	Cochran	Miramonte	Metcalfe	Derbe	William	Tripp
Fish	135	NS	NS	100	120	130
Attached algae	6.5	24	17	160	135	69
Crayfish	80	NS	90	89	140	62
Higher aquatics	9	78	26	40	150	210
Sediment	70	70	14	30	120	70

[a] During storm events, zinc concentration in the urban reaches of Coyote Creek averaged about 0.4 mg/L. Dry-weather zinc concentration in the urban reaches averaged about 0.03 mg/L. Non-urban reach water sample zinc concentrations were about half of these values.
NS = No sample collected.

From Pitt, R. and M. Bozeman. *Sources of Urban Runoff Pollution and Its Effects on an Urban Creek*, EPA-600/S2-82-090, U.S. Environmental Protection Agency, Cincinnati, OH. 1982.

Aquatic Biota Conditions

Fish — The fish fauna known to exist in the Coyote Creek drainage system at the time of the study was comprised of 27 species, 11 of which are native California fishes. The remainder were introduced through stocking by the California Department of Fish and Game and by the activities of bait dealers, fisherman, farm pond owners, and others. Although a relatively large variety of fish species was present in the Coyote Creek drainage, the existing distribution of some species was not widespread. Both Lake Anderson and Coyote Lake reservoirs sustained warm-water sport fisheries, and several of the fish species reported from the drainage were apparently confined to the specific habitat provided by those reservoirs. This included brown bullhead, channel catfish, Mississippi silverside, pumpkinseed, and redear sunfish. Of the remaining 22 species of fish known in Coyote Creek, 21 were encountered during this study, in which a total of 7198 fish were collected from 40 locations throughout the drainage. Rainbow trout and riffle sculpin were captured only in the headwater reaches and tributary streams of Coyote Creek. Likewise, Sacramento squawfish were found only in the upper reaches of the creek and reportedly have not been encountered downstream of Lake Anderson since 1960 (Scoppettone and Smith 1978). Seventeen fish species were collected from the major study area between Lake Anderson and the confluence of Silver Creek. Speckled dace, a native species previously reported to occur in the study area, was not encountered. Pacific lamprey, an anadromous species which moves into fresh water to spawn, was found only in and around the mouth of Upper Penitencia Creek, a tributary that enters the lower reaches of Coyote Creek.

Introduced fishes often cause radical changes in the nature of the fish fauna present in a given water body or drainage system. In many cases, they become the dominant fishes because they are able to outcompete the native fish for food or space, or they may possess greater tolerance to environmental stress. In general, introduced species are most abundant in aquatic habitats modified by man, while native fish tend to persist mostly in undisturbed areas (Moyle and Nichols 1973). Such was apparently the case within Coyote Creek. As seen in Table 4.14, samples from the non-urban portion of the study area were dominated by an assemblage of native fish species such as hitch, threespine stickleback, Sacramento sucker, and prickly sculpin. Collectively, native species comprised 89% of the number and 79% of the biomass of the 2379 fish collected from the upper reaches of the study area. In contrast, native species accounted for only 7% of the number and 31% of the biomass of the 2899 fish collected from the urban reach of the study area.

Hitch was the most numerous native fish species present. Hitch generally exhibit a preference for quiet water habitat and are characteristic of warm, low elevation lakes, sloughs, sluggish rivers, and ponds (Calhoun 1966; Moyle and Nichols 1976). In streams of the San Joaquin River system in the Sierra Nevada foothills of central California, Moyle and Nichols (1973) found hitch to be

Table 4.14 Relative Abundance of Fish in Coyote Creek

	Urban Reach	Rural Reach
Native Fish		
Hitch	4.9%	34.8%
Threespine stickleback	0.8	27.3
Sacramento sucker	0.1	12.6
Prickly sculpin	<0.1	8.2
Introduced Fish		
Mosquitofish	66.9	5.6
Fathead minnow	20.6	0.6
Threadfin shad	2.4	nd
Green sunfish	1.2	<0.1
Bluegill	1.0	0.2

From Pitt, R. and M. Bozeman. *Sources of Urban Runoff Pollution and Its Effects on an Urban Creek*, EPA-600/S2-82-090, U.S. Environmental Protection Agency, Cincinnati, OH. 1982.

most abundant in warm, sandy-bottomed streams with large pools, where introduced species such as green sunfish, largemouth bass, and mosquitofish were common. Likewise, during this Coyote Creek study, hitch were found to be associated with green sunfish, fathead minnows, and mosquitofish in the lower portions of Coyote Creek. However, mosquitofish dominated the collections from the urbanized section of the creek and accounted for over two thirds of the total number of fish collected from that area. In foothill streams of the Sierra Nevada, Moyle and Nichols (1973) found mosquitofish to be most abundant in disturbed portions of the intermittent streams, especially in warm, turbid pools. The fish is particularly well adapted to withstand extreme environmental conditions, including those imposed by stagnant waters with low dissolved oxygen concentrations and elevated temperature. The second most abundant fish species in the urbanized reach of Coyote Creek, the fathead minnow, is equally well suited to tolerate extreme environmental conditions. The species can withstand low dissolved oxygen, high temperature, high organic pollution, and high alkalinities. Often thriving in unstable environments such as intermittent streams, the fathead minnow can survive in a wide variety of habitats. However, the species seems to do best in pools of small, muddy streams and in ponds (Moyle and Nichols 1976).

Benthic Macroinvertebrates — The taxonomic composition and relative abundance of benthic macroinvertebrates were collected from both natural and artificial substrates in Coyote Creek (Figures 4.7 through 4.9). The abundance and diversity of benthic taxa were greatest in the non-urbanized sections of the stream. Figure 4.10 shows the trend of the overall decrease in the total number of benthic taxa encountered in the urbanized sections of the study area during 1978 and 1979. An overall increase in number and diversity of benthic organisms was encountered in 1979, compared to 1978 collections. This may be attributed to further recovery from the drought conditions that preceded this study. The benthos in the upper reaches of Coyote Creek consisted primarily of amphipods and a diverse assemblage of aquatic insects. Together those groups comprised two thirds of the benthos collected from the non-urban portion of the creek. Clean-water forms were abundant and included amphipods (*Hyalella azteca*) and various genera of mayflies, caddisflies, black flies, crane flies, alderflies, and riffle beetles. In contrast, the benthos of the urban reaches of the creek consisted almost exclusively of pollution-tolerant oligochaete worms (tubificids). Tubificids accounted for 97% of the benthos collected from the lower portion of Coyote Creek.

Crayfish were present throughout the study area and were collected in conjunction with the fish sampling effort. Two species of crayfish were encountered in Coyote Creek waters — *Pacifastacus leniusculus* and *Procambarus clarkii*. Neither species is native to California waters. *Pacifastacus*

Figure 4.7 Natural substrate sampling using a Surber sampler in Coyote Creek.

Figure 4.8 Removing benthic macroinvertebrate samples from Surber sampler.

Figure 4.9 Artificial substrate sampling using a Hester-Dendy multiplate sampler in Coyote Creek.

leniusculus was collected in the non-urbanized section of the study area. It is typically found in a wide variety of habitats including large rivers, swift or sluggish streams, lakes, and, occasionally, muddy sloughs. *Procambarus clarkii* was collected in both the urbanized and non-urbanized sections of the stream. The species prefers sloughs where the water is relatively warm and vegetation plentiful; however, it is also found in large streams. Because of its burrowing activities *P. clarkii* often becomes a nuisance by damaging irrigation ditches and earthen dams.

Attached Algae — Qualitative samples from natural substrates indicated that the filamentous alga *Cladophora* sp. was found throughout the study area. However, its growth reached greatest proportions in the upper sections of the stream. Table 4.15 presents the taxonomic composition and relative abundance of diatoms collected from artificial substrates (Figure 4.11) placed at selected sample locations. The periphyton of the non-urban reaches of the stream was dominated by the genera *Cocconeis* and *Achnanthes*. The genera *Nitzschia* and *Navicula*, generally accepted to be more pollution-tolerant forms, dominated the periphyton of the urbanized reaches of Coyote Creek.

Rooted Aquatic Vegetation — Rooted aquatic plants were not greatly abundant in the Coyote Creek study area. Submerged macrophytes were restricted entirely to the upper reaches of the study area and consisted of occasional stands of sago pondweed (*Potamogeton pectinatus*) and curly-leaf pondweed (*P. crispus*). Emergent forms consisted of water primrose (*Jussiaea* sp.), confined to several areas in the non-urban reach of the stream, and numerous small stands of cattails (*Typha* sp.) sparsely distributed throughout the length of the study area.

Step 7. Project Conclusions

The biological investigations in Coyote Creek indicated distinct differences in the taxonomic composition and relative abundance of the aquatic biota present in Coyote Creek. The non-urban

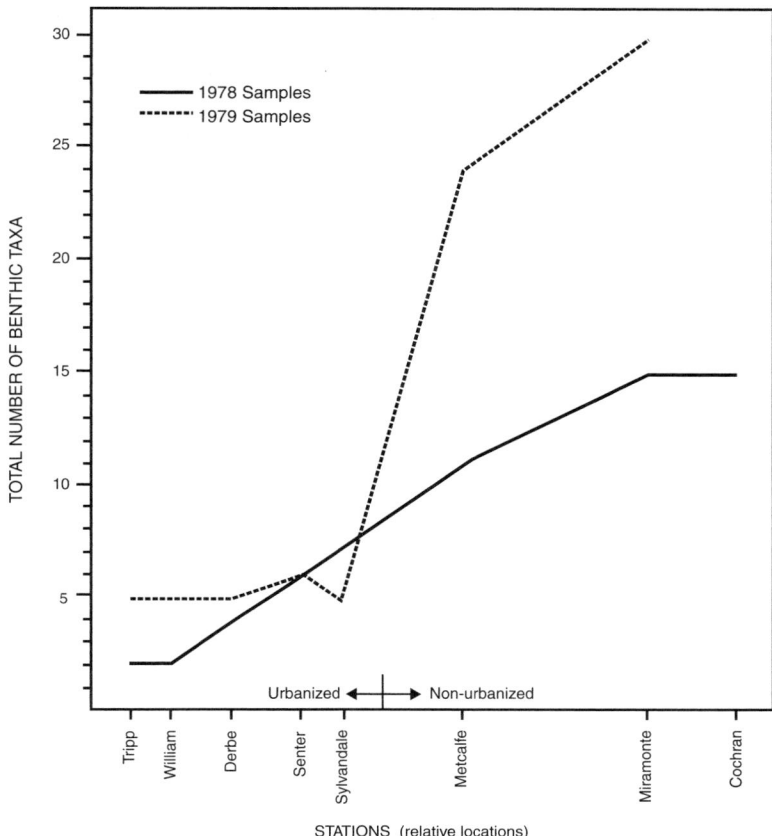

Figure 4.10 Trend of total number of benthic taxa observed during 1978 and 1979 (From Pitt, R. and M. Bozeman. *Sources of Urban Runoff Pollution and Its Effects on an Urban Creek*, EPA-600/S2-82-090, U.S. Environmental Protection Agency, Cincinnati, OH. 1982.)

sections of the creek supported a comparatively diverse assemblage of aquatic organisms, including an abundance of native fishes and numerous benthic macroinvertebrate taxa. In contrast, however, the urban portions of the creek comprised an aquatic community generally lacking in diversity and was dominated by pollution-tolerant organisms such as mosquitofish and tubificid worms.

Although certain differences in physical habitat occurred in the downstream reaches of the study area (e.g., a decrease in stream gradient, shorter riffles, wider, deeper pools, etc.), such differences were not thought to be responsible for the magnitude of change noted in the aquatic biota of the urban reach of Coyote Creek.

Urban runoff monitoring during this project showed that stormwater was the significant contributor to the high levels of many toxic materials in the receiving water and sediments of the stream. In addition, changes in the nature of the stream substrate occurred as a result of the deposition of silt and debris, which largely originate from urban runoff. Such changes were likely the primary reason for the decline in species abundance and diversity observed in the urban reaches of Coyote Creek.

Critique of the Longitudinal Analyses in Coyote Creek

The Coyote Creek study was very comprehensive, and therefore costly. This was probably the earliest large-scale receiving water study conducted to investigate urban runoff effects on in-stream

Table 4.15 Taxonomic Composition and Relative Abundance of Diatoms Collected on Glass Slides in Coyote Creek during the Spring of 1978

| | Relative Abundance (%) of each Taxon within the Sample | | | | | |
| | Non-Urban Area Stations | | | Urban Area Stations | | |
Taxon	Cochran	Miramonte	Metcalfe	Derbe	Williams	Tripp
Centrales						
Coscinodiscaceae						
Melosira sp.	0.4	—	—	—	1.2	0.8
Pennales						
Diatomaceae						
Diatoma vulgare	0.4	—	1.5	—	—	—
Fragilariaceae	—	—	—	0.8	0.9	0.4
Synedra sp.						
Achnanthaceae						
Achanthes lanceolata	20.6	37.8	56.1	49.8	0.9	1.6
Rhoicosphenia curvata	0.4	—	—	1.2	—	—
Cocconeis pediculus	15.0	18.2	0.4	—	—	—
Cocconeis placentula	62.4	44.0	41.2	—	—	—
Naviculaceae						
Navicula spp.	—	—	—	—	10.5	23.8
Diploneis sp.	—	—	—	—	2.4	—
Frustulia rhomboides	—	—	—	—	0.4	—
Gyrosigma sp.	—	—	—	—	—	0.4
Gomphonenataceae						
Gomphonema sp.	—	—	—	2.8	6.9	0.8
Cybellaceae						
Cymbella sp.	0.8	—	—	—	2.0	0.4
Rhopalodia spp.	—	—	—	—	—	0.4
Nitzschiaceae						
Nitzachia sp.	—	—	0.8	43.4	67.5	70.6
Denticula elegans	—	—	—	—	2.4	0.4
Surirellaceae						
Cymatopleura solea	—	—	—	—	0.9	—
Surirella sp.	—	—	—	2.0	4.0	0.4
Total Number Frustules/mm²	5545	4950	1874	4488	1189	4575

From Pitt, R. and M. Bozeman. *Sources of Urban Runoff Pollution and Its Effects on an Urban Creek*, EPA-600/S2-82-090, U.S. Environmental Protection Agency, Cincinnati, OH. 1982.

Figure 4.11 Artificial substrate diatometer sampler being loaded with glass microscope slides in Coyote Creek.

biological conditions. As such, many elements were considered in the site investigation. The project included field sampling over a period of 3 years, and more than 40 sampling sites were periodically visited. A broad list of biological, chemical, and physical measurements was obtained. Even though the project was comprehensive, several omissions seem obvious. The most notable is the lack of toxicity testing. Some limited laboratory fathead minnow 96-hour exposure tests were conducted as part of the study, but were inconclusive and therefore not reported. The project was also conducted before effective and less costly *in situ* toxicity tests were developed. Another element that was missing was comprehensive habitat surveys. Formalized habitat survey procedures detailed in this book (Chapter 6 and Appendix A) would have been very useful during the Coyote Creek study. Finally, because the study design did not have any precedence, it was probably inefficient in that it obtained more information than was actually needed, and at more locations than necessary.

The longitudinal study design is very helpful in that gradients of conditions can be examined. The Coyote Creek study examined a very large number of locations along the creek in an attempt to identify locations that were partially degraded, but still in acceptable condition. When these locations are identified, watershed modeling can be used to calculate the assimilative capacity of the stream, which can then be used to determine necessary stormwater controls to provide these conditions farther downstream. Unfortunately, Coyote Creek was found to degrade very rapidly at the edge of development. Additional monitoring locations were therefore added in an attempt to isolate the degradation gradient. The highly variable conditions in the creek at the edge of urbanization were likely due to major flow changes seasonally and from year-to-year, preventing identification of an acceptably degraded site.

In many cases, a longitudinal study design can be combined with the other two major types of designs (parallel and trend studies) to obtain additional information. The trend case study presented is for a trend with time, but a trend with distance can also be evaluated using similar statistical procedures described in Chapter 7.

Example of Parallel Creeks Experimental Design — Kelsey and Bear Creeks, Bellevue, WA, Receiving Water Study

Several separate urban stormwater projects (as part of the U.S. EPA's Nationwide Urban Runoff Program, or NURP) were conducted in Bellevue, WA, to address the three major phases in designing an urban runoff control program (quantifying the specific local urban runoff receiving water problems, determining the sources of the problem pollutants, and selecting the most appropriate control measures). These projects were conducted from 1977 through 1982 and constitute one of the most comprehensive urban runoff/receiving water impact research programs ever conducted.

The U.S. Geological Survey (USGS) through its Tacoma, WA, office, conducted one of the projects, which was funded by the USGS and the Water Planning Division of EPA. The USGS (Ebbert et al. 1983; Prych and Ebbert undated) intensively monitored urban runoff quality and quantity from three residential areas in Bellevue and evaluated the effectiveness of a detention facility. Wet and dry atmospheric sources were also monitored by the USGS.

The University of Washington's Civil Engineering Department and the College of Fisheries Research Institute prepared five reports based on their studies, which were funded by the Corvallis Environmental Research Laboratory of EPA (Pedersen 1981; Perkins 1982; Richey et al. 1981; Richey 1982; Scott et al. 1982). Generally, the University of Washington's projects evaluated the receiving water conditions for direct impairments of beneficial uses.

The Municipality of Metropolitan Seattle (METRO) research was funded by Region X of EPA and was prepared by Galvin and Moore (1982). METRO analyzed many source area, urban runoff, and creek samples for metallic and organic priority pollutants.

The City of Bellevue also conducted a study, which was funded by the Storm and Combined Sewer Section of EPA and the City of Bellevue. The Bellevue report was prepared by Pitt (1985) and Pitt and Bissonnette (1984). The City of Bellevue collected and analyzed urban runoff and baseflow samples using flow-weighted techniques for more than 300 storms from two residential areas, in addition to extensively evaluating street and sewerage cleaning as stormwater management practices.

Bellevue's moderate climate has a mean annual precipitation of about 1.1 m (44 in) which occurs mostly as rainfall from October through May. Most of the rainfall results from frontal storms formed over the Pacific Ocean. During fall and winter months, low to moderate rainfall intensities are common. Even though the runoff quality was found to be much cleaner than in other locations in the United States, the urban creek was significantly degraded when compared to the rural creek, but still supported a productive, but limited and unhealthy salmonid fishery. Many of the fish in the urban creek, however, had respiratory anomalies. The urban creek was not grossly polluted, but flooding from urban developments has increased dramatically in recent years. These increased

Figure 4.12 Rural Bear Creek, Bellevue, WA.

Figure 4.13 Rural Bear Creek, Bellevue, WA, in undeveloped area.

Figure 4.14 Rural Bear Creek, Bellevue, WA, passing through trailer park.

Figure 4.15 Urbanized Kelsey Creek, Bellevue, WA, in low-density residential area.

flows have dramatically changed the urban stream's channel, by causing unstable conditions with increased stream bed movement, and by altering the availability of food for the aquatic organisms. The aquatic organisms are very dependent on the few relatively undisturbed reaches. Dissolved oxygen concentrations in the sediments depressed embryo salmon survival in the urban creek. Various organic and metallic priority pollutants were discharged to the urban creek, but most of them were apparently carried through the creek system by the high storm flows to Lake Washington.

The in-stream studies were conducted in Bear Creek (Figures 4.12 through 4.14), a relatively undisturbed natural stream, and in Kelsey Creek (Figures 4.15 through 4.17), a heavily urbanized stream. The watershed studies were conducted in the Lake Hills and Surrey Downs neighborhoods (Figure 4.18).

Step 1. What's the Question?

Does urban runoff significantly affect Bellevue's receiving water uses; what are the sources of the urban runoff problem pollutants; and can public works practices (street cleaning and catchbasin cleaning) reduce the magnitude of these problems?

Figure 4.16 Urbanized Kelsey Creek, Bellevue, WA, in commercial area.

Figure 4.17 Kelsey Creek, Bellevue, WA, street crossing with sign.

Bellevue area waters have five designated beneficial uses:

1. Preservation of habitat suitable for aquatic organisms
2. Flood prevention by the conveyance of stormwater
3. Open space and resource preservation
4. Recreational uses (swimming and boating)
5. Aesthetics

The Bellevue research projects (especially those conducted by the University of Washington team) investigated the potential impairments of these uses in the urbanized Kelsey Creek, compared to Bear Creek, the control stream.

Figure 4.18 Typical residential neighborhood in monitored Lake Hills and Surrey Downs watersheds, Bellevue, WA.

Step 2. Decide on Problem Formulation

The basic problem formulation was to investigate parallel watersheds. Kelsey Creek is completely urbanized, while Bear Creek had only minor development and was used as a control stream. In addition, the street cleaning portions of the study compared parallel portions of the urban area (the Lake Hills and Surrey Downs catchments), with rotating street cleaning operations and outfall monitoring.

Step 3. Project Design, Step 4. Project Implementation, Step 5. Data Evaluation, and Step 6. Confirmatory Assessment

1. Qualitative Watershed Characterization

The Surrey Downs and Lake Hills test catchments are about 5 km apart and are each about 40 ha in size. They are both fully developed, mostly as single-family residential areas. The 148th Avenue dry detention basin study area is about 10 ha in area and is primarily a street arterial with adjacent landscaping. The Surrey Downs area was developed in the late 1950s. Most of the slopes in the basin are moderate with some steeper slopes on the west side of the area. About 60% of the Surrey Downs area is pervious. Back and front yards make up most of the land surface area, while the streets make up 10%. There is relatively little automobile traffic in the Surrey Downs area and the on-street parking density is low. The storm drainage system discharges into an artificial pond located in an adjacent development. This pond discharges into Mercer Slough, which eventually drains to Lake Washington and Puget Sound. The Surrey Downs catchment ranges in elevation from about 3 to 55 m.

The Lake Hills catchment is about 41 ha in size and contains the St. Louise parish church and school in addition to single-family homes. These homes were also developed in the 1950s. Lake Hills has a slightly larger percentage of pervious area than Surrey Downs, but a slightly smaller typical lot size. The slopes in Lake Hills are also more moderate (with a few exceptions) than those found in Surrey Downs. Most of the streets in Lake Hills also carry low volumes of traffic and have low parking densities, except for two busy roads which cross through the area. The Lake Hills storm drainage system discharges into a short open channel which joins Kelsey Creek just downstream from Larsen Lake. Kelsey Creek also discharges into Mercer Slough and finally into Lake Washington and Puget Sound. The elevation of the Lake Hills study catchment ranges from 80 to 125 m.

The 148th Avenue S.E. catchment was used to investigate the effects of a dry detention facility on stormwater quality. The drainage area is about 10 ha. Slightly more than one fourth of this area is the actual street surface of 148th Avenue S.E., a divided, four-lane arterial. Other impervious areas include sidewalks, parking lots, office buildings, and parts of Robins Wood Elementary School.

The soils in all three of these test catchments are mostly the Arents-Alderwood variety, having 6 to 15% slopes. The surface soils are made up of gravelly sand loams with an estimated natural permeability of between 50 and 150 mm/hour. The total water capacity of this soil horizon is about 20 mm.

A demographic survey was conducted in the test catchments by the City of Bellevue at the beginning of the project (in 1977). Slightly more than three people per household were reported in both basins, while the population density per hectare was about 30 in Lake Hills and about 23 in Surrey Downs. More than half of the people in both basins had no dogs or cats, with the remainder of the households having one or more of each. Slightly more than two cars per household were reported, with about 10% of the households in each basin reporting four or more cars. Most of the automobile oil was disposed of properly in the household garbage or recycled, but between 5 and 10% of the households used oil to treat fence posts, dumped it onto the ground, or into the storm sewers. Most of the people carried their grass and leaves to the dump, or put them into the garbage, and about one third composted the organic debris on their lots.

2. Stream Characterization

Kelsey Creek flows through the City of Bellevue, while Bear Creek is about 30 km farther east. Kelsey Creek drains a watershed about 3200 ha in area, which is predominantly urban. About 54% of the Kelsey Creek watershed has single- and multiple-family residences, 24% has commercial or light industrial uses, and 22% has parks and undeveloped areas. A main channel of Kelsey Creek starts at Larson Lake and flows about 12 km through the City of Bellevue before discharging into Lake Washington. The USGS has continuously monitored Kelsey Creek flows since 1959 at a location about 2.5 km upstream from Lake Washington. Kelsey Creek is a relatively narrow stream with a mild slope. The mean channel slope is about 1.5% and the bank full width ranges from about 3.5 to 6.5 m in the study area. Along much of its length, Kelsey Creek appears disturbed. Channelization, riprapping, storm drain outfalls, scoured and eroded banks, and culverts are common. The stream bank (riparian) vegetation is mostly composed of low growing alder and vine maple with scattered big leaf maples and western red cedar trees. The understory is dominated by blackberry bushes.

Bear Creek starts at Paradise Lake and drains into Cottage Lake Creek. Its drainage area is about 3400 ha and is mostly rural in character. About 85% of the Bear Creek drainage is in pasture or woodlands with about 15% developed in single-family residences. Bear Creek also has a mild slope (about 0.6%) and is slightly wider than Kelsey Creek with a bank-full width ranging from about 7 to 11 m. Bear Creek has the appearance of a relatively undisturbed stream, especially when compared to Kelsey Creek. The vegetation along some reaches in Bear Creek has been modified, and there is some riprapping for bank stabilization. Most of these disturbances are quite small. Throughout most of the Bear Creek study reach, the creek is composed of alternating series of pools and riffles, frequent debris dams, side channels, and sloughs. The riparian vegetation along Bear Creek is mostly old growth alder, western red cedar, and douglas fir, with an understory of vine maple and salmonberry. Richey (1982) states that while Bear Creek receives no point source discharges, it is not pristine. Drainage from septic tanks, fertilizers, and livestock wastes has enriched the stream. Many homeowners have cut or modified the bank vegetation, installed small diversions, and created small waterfalls. These activities appear to have generated an increase in sediment transport. Building activity has also increased in the upper parts of the watershed since 1981. Much of the creek, however, remains in a natural condition and is typical of many of the gravel-bottomed streams in the Pacific Northwest.

3. Select Monitoring Parameters

The Bellevue city project included monitoring of the quality and quantity of stormwater runoff from two urban areas in the City of Bellevue. Street surface particulate samples were collected in these two basins along with storm drainage sediment samples. The City of Bellevue conducted various street cleaning operations in the two test basins and evaluated the effectiveness of various types of street cleaning programs and catchbasin cleaning activities in improving the quality of urban runoff. The USGS also monitored stormwater runoff quality and quantity in these two test basins. The USGS used different sampling techniques to monitor fewer storms but in much greater detail. The USGS monitored rainfall and dustfall quality and quantity along with the performance of a series of detention basins at a third Bellevue test site. The University of Washington's projects investigated urban runoff receiving water conditions and conditions in a control stream much less affected by urban runoff. The University's projects studied physical, chemical, and biological conditions to identify impacts associated with urban development on receiving water quality. The Seattle METRO project involved conducting trace metal and organic pollutant analyses for samples collected from these three other projects. The following list summarizes the major components of the Bellevue investigations:

- In-stream effects from urban stormwater (comparing test and control stream conditions over a 2-year period)
  - In-stream water quality (wet and dry weather observations) for conventional and nutrient constituents, plus some toxicants
  - Interstitial water quality in test and control streams for dissolved oxygen, nutrients, and metallic toxicants
  - Continuous stream flow rates
  - Aquatic organism food availability and utilization studies
  - Riparian vegetation, algae, benthic organisms, and fish
  - Creek sediment quality for conventional and toxic pollutants
  - Creek bank stability and stream bed erosion, and creek sedimentation and sediment transport
- Sources of urban runoff pollutants in two test catchments for 2-year period
  - Atmospheric particulate and rainfall contributions
  - Runoff monitoring from about 400 rain effects (91 to 99% of annual flow monitored during 2 years)

- Stormwater quality from more than 200 events for conventional, nutrient, and toxic constituents (200 to 1000 analyses per constituent)
- Baseflow quality from about 25 sampling periods for conventional, nutrient, and toxic constituents
- Street dirt characteristics from about 600 samples (loading, particle size, washoff, and chemical quality)
- Sewerage and catchbasin sediment accumulations over 2-year period (accumulation and quality) from about 200 inlets
- Effectiveness of urban runoff controls
  - Monitored street dirt loadings and runoff characteristics at two test catchments over 2-year period, comparing none with three times a week street cleaning effort
  - Measured changes in catchbasin sump accumulations of pollutants in about 200 inlets over 2 years in two catchments
  - Monitored influent and effluent from a dry detention pond for the 20 storms during the 2-year period when flows were sufficient to enter the pond system

Observations

Effects of Urban Runoff on Bellevue's Stream's Beneficial Uses — Richey (1982) summarizes some of the beneficial use impairments that the University of Washington study team addressed. Urbanization and stormwater runoff discharges to streams can have a wide variety of effects on these receiving waters. These include increased runoff, decreased surface storage, decreased transpiration, decreased infiltration, and a degradation in water quality. These effects may be either long term or intermittent. Changes in channel geomorphology caused by channelization in the clearing of stream bank vegetation may cause permanent stresses to the stream. Changes in the stream flows during runoff events, such as rapidly rising and falling hydrographs and increased total flows and peak discharges, are intermittent stress factors. The discharge and transport of pollutants can act as an intermittent stress factor, but the storage of these pollutants in the stream system (in the sediments or bioaccumulation) can act as a long-term or chronic stress factor. Therefore, it is necessary to identify not only the causative factor in impairing receiving water quality but also the times when these effects occur. Elevated concentrations of toxic materials in the runoff may affect receiving water organisms during a runoff event. However, they may also accumulate in the sediments and not affect the receiving water aquatic life until some time after they were discharged.

Richey points out the difficulty in identifying problem pollutants or their causes based upon their different destructive powers. She presents a hypothetical example where the gradual introduction of toxic pollutants in the receiving water results in a decline of fish species diversity and system productivity. Because the watershed has been urbanizing, increased flows have also occurred. If it is assumed that the increased flows causing flooding and scouring in the water body were the most important element restricting the fish populations, an abatement program incorporating detention facilities to reduce these flooding problems may be implemented. However, the input of toxic substances may not be reduced and significant improvements in the beneficial use may not occur. She concludes that it is very important to study all effects on a receiving body including hydrology, geomorphology, and pollutant inputs.

Scott et al. (1982) state that factors contributing to the instability of the physical receiving water system are relatively easy to identify but that their combined effect on the receiving water aquatic life is difficult to measure. They also mention that the Resource Planning Section of the King County Planning Division analyzed available data for 15 local streams in an attempt to establish a cause-and-effect relationship between urban development and stream degradation. They examined watershed variables such as the magnitude of the impervious areas, peak flows, water quality, aquatic insects, and salmonid escapement to rank the streams. Bear Creek ranked 12th in impervious surfaces, lowest in peak flow, 5th in water quality, 6th in aquatic insects, and 2nd in salmonid escapement. Kelsey Creek ranked 2nd in impervious areas, 6th in peak flow, 15th in water quality, 50th in aquatic insects, and 8th in salmonid escapement.

1. Bellevue Receiving Water Beneficial Uses

Kelsey Creek, the urban receiving water studied during this project, has three primary functions: conveyance of stormwater from Bellevue to Lake Washington, providing a scenic resource for the area, and providing a habitat for fish. The most important beneficial use of Kelsey Creek is the conveyance of stormwater out of the city. The City of Bellevue, in its Storm Drainage Utility and support of projects such as these, has a commitment to provide the other beneficial uses. Richey (1982) states that Kelsey Creek can physically provide for all of these beneficial uses. The creek has been developed for the conveyance of stormwaters, but there are also areas in its lower reaches where the canopy cover is relatively intact and the stream banks and morphology are still quite natural. Dense growth of shrubbery and blackberry vines also provides cover and shade for stream aquatic life. The riprap allows the development of deep pools which can be a good habitat for fish. Perkins (1982) states that some of the upstream reaches and tributaries of Kelsey Creek are less disturbed and serve as a potential refuge area for aquatic life. The downstream reaches of Kelsey Creek, however, are less supportive of aquatic life due to channel instability and erosion, along with flashy flows and increasing floods.

2. Bear Creek and Kelsey Creek Water Quality

The University of Washington project monitored Kelsey Creek and Bear Creek water quality from May 1979 through April 1981. Table 4.16 (Richey 1982) summarizes these creek water quality observations. The values for the constituent concentrations were obtained during stable flow periods only when the creeks were not rising or falling rapidly. The major ion types are similar for both Bear and Kelsey Creeks: calcium/magnesium bicarbonate. The concentrations of these ions were typically lower in Bear Creek. Richey found that during the study period the average nutrient levels in Kelsey Creek were greater than those found in Bear Creek. Total phosphorus and soluble reactive phosphorus in Kelsey Creek were about 2.5 times higher than those found in Bear Creek. Both streams have ample supplies of both nitrogen and phosphorus for the aquatic organisms, and the nitrate plus nitrite concentrations had a distinct seasonal trend in Bear Creek, while they were essentially random in Kelsey Creek. High winter and low summer concentrations of nitrate plus nitrite have been observed in other rural streams and are thought to be controlled largely by the seasonal nitrogen uptake of terrestrial vegetation. Bear Creek has much more riparian vegetation than does Kelsey Creek. The high nitrogen concentrations in Bear Creek may also be caused by in-stream nitrification. In addition, the maximum ammonium concentrations in Bear Creek occurred during the autumn when there was decomposition of sockeye salmon bodies in the creek.

The observed low dissolved lead concentrations in Kelsey Creek and Bear Creek are not expected to exert a major impact on the aquatic life. However, other possible toxic compounds which may be washing into the stream system were not continuously monitored. Pedersen (1981) notes that massive fish kills in Kelsey Creek or its tributaries were observed on several occasions due to the dumping of toxic materials down storm drains. The resultant impact of this toxic material on the benthic organisms from these dumps was found to be substantial, but no permanent impact over long time periods was observed. The 5-day biochemical oxygen demand (BOD_5) concentrations were low in both streams. They found that the greatest differences in constituents between the two streams occurred in constituents that were in particulate forms.

Scott et al. (1982) listed the most important water quality differences between these two creeks:

- Kelsey Creek had higher nutrient concentrations than Bear Creek.
- Kelsey Creek had one to two times the suspended particulate loads of Bear Creek.

Table 4.16 Surface Water Quality (monthly average concentrations from May 1979 through April 1981)

	Units	Kelsey Creek				Bear Creek				Ratio of Kelsey Creek Mean Values to Bear Creek Mean Values
		Mean	SD*	Minimum	Maximum	Mean	SD	Minimum	Maximum	
Drainage area	ha	3109				3600				0.9
Instantaneous discharge	m³/s	36.7	6.8	0.20	8.68	27.5	4.9	0.13	6.31	1.5/1.4 (min/max ratios)
Substrate size	mm									1.3
Summer temperature	°C				23.0				23.0	1.0 (ratio of max.)
Winter temperature	°C			5.0				3.2		1.6 (ratio of min.)
Total suspended solids	mg/L	11.0	7.4	2.5	32.9	4.7	3.0	0.8	11.9	2.3
Fine particulate organic carbon	mg C/L	0.87	0.53	0.10	2.51	0.75	0.06	0.32	1.51	1.2
Dissolved organic carbon	mg C/L	7.5	3.4	3.8	14.8	6.4	3.3	3.0	16.8	1.2
Total phosphorus	µg P/L	116	32	72	193	43	16	15	79	2.7
Soluble reactive phosphorus	µg P/L	82	27	54	167	24	16	8	63	3.4
Nitrate plus nitrite nitrogen	µg N/L	743	137	468	962	508	540	59	2350	1.5
Ammonia nitrogen	µg N/L	36	14	12	66	30	26	9	114	1.2
BOD$_5$	mg O$_2$/L	2.26	1.27	0.86	5.3	1.63	1.8	0.03	3.59	1.4
Dissolved lead	µg Pb/L	5	2	2	11	<4	—	<4	<4	>1.3

* SD = standard deviation.

Data from Richey, J. S. *Effects of Urbanization on a Lowland Stream in Western Washington*, Ph.D. dissertation, University of Washington, Seattle. 1982. With permission.

Table 4.17 Annual Kelsey and Bear Creek Discharges (June 1979 through May 1980, kg/ha/year)

Constituent	Kelsey Creek	Bear Creek	Ratio of Kelsey to Bear Creek Discharges
Total suspended solids	300	78	3.8
Fine particulate organic carbon (FPOC)	33	12	2.8
Dissolved organic carbon (DOC)	53	55	1.0
Soluble reactive phosphorus	0.56	0.17	3.3
Total phosphorus	0.87	0.33	2.6
Nitrate plus nitrite nitrogen	4.3	7.1	0.6

Data from Richey, J.S., et al. The effects of urbanization and stormwater runoff on the food quality in two salmonid streams. *Verh. Internat. Werein. Limnol.*, Vol. 21, pp. 812–818, Stuttgart. October 1981.

- Inorganic silt was the dominant fraction of the suspended particulate load in Kelsey Creek.
- The concentrations of potentially toxic materials in both study streams were quite low and possibly negligible.

Observed problems in Kelsey Creek included high water temperatures and elevated fecal coliform counts. The fecal coliform counts, however, varied considerably throughout the Kelsey Creek drainage system. Bear Creek also had high fecal coliform counts along with high inorganic nitrogen and total phosphorus concentrations.

The annual creek discharges of various water quality constituents are shown in Table 4.17 (Richey et al. 1981). The total solids concentrations were highest during the periods of high flows (late fall, winter, and early spring). Therefore, most of the solid material was transported during only a few months of the year. Thirty-three percent of the solids were transported out of Kelsey Creek and 35% out of Bear Creek during the high flow month of December alone. The annual yields of both particulate and soluble phosphorus were about three times greater in Kelsey Creek than in Bear Creek. The total suspended solids transport in Kelsey Creek was almost four times greater than Bear Creek. While the fine particulate organic matter in Kelsey Creek was almost three times more than in Bear Creek on an annual basis, the dissolved organic carbon transport was about the same. High phosphorus concentrations in the fall in Bear Creek may also be caused by decomposing sockeye salmon. Scott reported more than 1000 sockeye carcasses in the stream channel during the fall of 1979 and 1980.

Richey (1982) states that Kelsey Creek is surprisingly clean for a heavily urbanized stream. This might be because of the in-stream dilution of the contaminants, because some of the watershed is still relatively protected, or possibly the result of differences in the occurrence of the urban contaminants. She further states that Kelsey Creek is enriched but does not appear to be polluted in the classic sense. The rapid transport of water and materials appears to protect the stream by removing many of the potentially hazardous pollutants to downstream locations. In addition, the rapid transport of water also helps to maintain high levels of dissolved oxygen.

The City of Bellevue project (Pitt 1985) evaluated water quality with beneficial use criteria. Potential long-term problem pollutants are settleable solids, lead, and zinc. These long-term problems are caused by settled organic and inorganic debris and particulates. This material may silt up salmon spawning beds in the Bellevue streams and introduce high concentrations of potentially toxic materials directly to the sediments. Oxygen depletion caused by organic sediments may also occur under certain conditions, and the lead and zinc concentrations in the sediments may affect the benthic organisms. The discharge of particulate heavy metals, which settle out in the sediments, may be converted to more soluble forms through chemical or biological processes.

3. Creek Interstitial Water Quality

The University of Washington and the Seattle METRO project teams analyzed interstitial water for various constituents. These samples were obtained by inserting perforated aluminum standpipes into the creek sediment. This water is most affected by the sediment quality and in turn affects the benthic organisms much more than the creek water column. Scott et al. (1982) found that the interstitial water pH ranged from 6.5 to 7.6 and did not significantly differ between the two streams but did tend to decrease during the spring months. The lower fall temperatures and pH levels contributed to reductions in ammonium concentrations. The total ammonia and ammonium concentrations were significantly greater in Kelsey Creek than in Bear Creek. They also found that the interstitial dissolved oxygen concentrations in Kelsey Creek were much below concentrations considered normal for undisturbed watersheds. These decreased interstitial oxygen concentrations were much less than the water column concentrations and indicated the possible impact of urban development. The dissolved oxygen concentrations in the interstitial waters and Bear Creek were also lower than expected, potentially suggesting deteriorating fish spawning conditions. During the winter and spring months, the interstitial oxygen concentrations appeared to be intermediate between those characteristic of disturbed and undisturbed watersheds.

The University of Washington (Richey 1982) also analyzed heavy metals in the interstitial waters. They found that copper and chromium concentrations were very low or undetectable, while lead and zinc were higher. Kelsey Creek interstitial water also had concentrations approximately twice those found in the Bear Creek interstitial water. They expect that most of the metals were loosely bound to fine sediment particles. Most of the lead was associated with the particulates and very little soluble lead was found in the interstitial waters. The interstitial samples taken from the standpipes were full of sediment particles that could be expected to release lead into solution following the mild acid digestion for exchangeable lead analyses. They also found that the metal concentrations in Kelsey Creek interstitial water decreased in a downstream direction. They felt that this might be caused by stream scouring of the benthic material in that part of the creek. The downstream Kelsey Creek sites were more prone to erosion and channel scouring, while the most upstream station was relatively stable.

Seattle METRO (Galvin and Moore 1982) also monitored heavy metals in the interstitial waters in Kelsey and Bear Creeks. They found large variations in heavy metal concentrations depending upon whether the sample was obtained during the wet or the dry season. During storm periods, the interstitial water and creek water heavy metal concentrations approached the stormwater values (200 µg/L for lead). During nonstorm periods, the interstitial lead concentrations were typically only about 1 µg/L. They also analyzed priority pollutant organics in interstitial waters. Only benzene was found and only in the urban stream. The observed benzene concentrations in two Kelsey Creek samples were 22 and 24 µg/L, while the reported concentrations were less than 1 µg/L in all other interstitial water samples analyzed for benzene.

4. Increased Kelsey Creek Water Flows

The increasing population of the City of Bellevue and the observed peak annual discharges have been studied by the University of Washington (Richey 1982). Bellevue was initially settled in 1883 but it grew slowly, reaching a population of only 400 by 1900. The Bellevue population density continued to be low until the 1940s. During this time, almost the entire Kelsey Creek drainage basin was undeveloped. In the late 1940s, the City of Bellevue's population was stimulated by the construction of the Lake Washington floating bridge connecting Bellevue to Seattle. From 1950 to 1970, low-density residential housing progressed rapidly, and the population of the greater Bellevue area increased by nearly 600%. By 1959, residential housing occupied a substantial portion of the Kelsey Creek watershed. The Bellevue population slowed during the 1970s due to the depressed local economy and the saturation of land development. In 1976, the population of the City of Bellevue was estimated to be 67,000 people. The peak annual discharges of Kelsey Creek almost doubled between the 1950s and the late 1970s. The frequency of flooding during this period of time also increased. Floods that used to return every 10 years in the early 1950s returned at least every other year during the late 1970s. The increase in the rate of runoff has also had a measurable effect on the channel stability in Kelsey Creek.

Figure 4.19 Stilling well at Bellevue flow monitoring station.

Figure 4.20 Level recorder at Bellevue flow monitoring station.

The University of Washington, in conjunction with the USGS, monitored flows from June 1979 through May 1980 (Perkins 1982; Richey et al. 1981) (Figures 4.19 and 4.20). The frequency of floods and the observed high flows have increased substantially in recent times. The peak flow for the same recurrence intervals have approximately doubled for recurrence intervals greater than 2 years. During the early period, a discharge of 7 m³/s had a 10-year recurrence interval, while it had only a 1- to 2-year recurrence interval during the more recent period. Also, a 100-year recurrence interval storm had a peak flow of 8.4 m³/s during the earlier period and was almost doubled to 16.7 m³/s during the latter period.

The responses of the two streams during individual storms were also significantly different. Figure 4.21 shows how Kelsey Creek responded much more dramatically during two storms than did Bear Creek. The response of Kelsey Creek to these two example rains showed a very rapidly rising hydrograph, while Bear Creek responded relatively slowly. After peaking, the flows in Kelsey Creek typically returned to baseflow rates in less than 24 hours, while 48 hours or more were required in Bear Creek. The maximum annual discharges in Kelsey Creek during the study period were much greater than in Bear Creek (4.6 vs. 1.9 L/ha). The total annual runoff yields in both watersheds were similar; therefore, much more of the total runoff occurs during storms in Kelsey than in Bear Creek, while baseflows are much less in Kelsey than in Bear Creek.

Because of these increased flow rates, much of Kelsey Creek is characterized by unstable banks with much erosion and deposition of sediment. The amount of stream power available in Kelsey Creek is greater than in Bear Creek despite the slightly greater slope of Bear Creek. During peak flows, Kelsey Creek has more than twice the available power of Bear Creek. Kelsey Creek can therefore move and erode sediments much more effectively than Bear Creek.

Richey (1982) also summarized low flows observed in Kelsey and Bear Creeks. On a unit area basis, about 30% more water was flowing in Bear Creek during the summer of 1981 than in Kelsey Creek. The low flow summer discharge in Kelsey Creek was about 250 L/hour/ha while the Bear Creek flows were about 350 L/hour/ha.

5. Aquatic Organism Food Availability and Utilization

The University of Washington studied primary productivity and the availability of food in the two streams. Richey (1982) also examined primary productivity in both Kelsey and Bear Creeks. She found that on an annual basis, primary productivity per unit area (measured as carbon fixation)

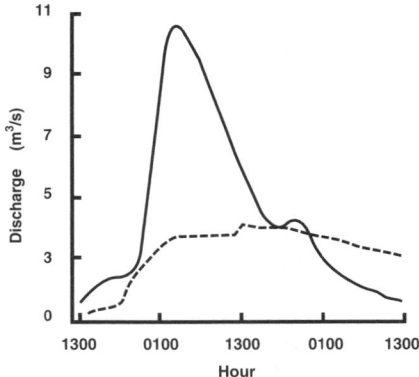 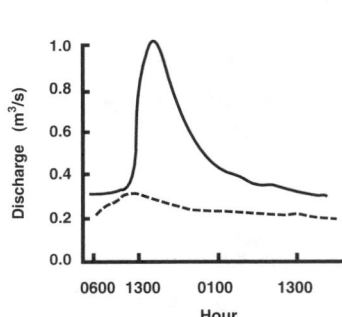

Figure 4.21 Hydrographs during winter and summer storms (December 14–16, 1979, and July 12–13, 1979) Note: solid line = urban Kelsey Creek; dashed line = rural Bear Creek. (From Richey, J.S. *Effects of Urbanization on a Lowland Stream in Western Washington*, Ph.D. dissertation, University of Washington, Seattle. 1982.)

was almost twice as large in Kelsey Creek (56 g C/m^2) than it was in Bear Creek (34 g C/m^2). She concluded that the scouring of the biomass during periods of high flows in Kelsey Creek limited the amount of primary production, even though there were sufficient nutrients available. The low levels of primary productivity measured in Bear Creek during October may have been the result of high turbidity, limiting the infiltration of sunlight in the water. Richey (1982) also examined the consumption of large organic material by grazing macroinvertebrates and microbes. The loss of leaf litter in both streams occurred at approximately equal rates. The causes for the loss of the leaf litter, however, were quite different. The microbial degradation and consumption by leaf shredding organisms are more important in Bear Creek while downstream transport of the leaf material in Kelsey Creek was most important. There was some macroinvertebrate consumption of leaf material in some of the Kelsey Creek locations, but this consumption occurred at a slower rate than in Bear Creek.

Richey (1982) also conducted experiments examining the toxicity of the periphyton in Kelsey Creek using mayflies. The adults emerged successfully in equal numbers, and the surviving larvae were indistinguishable in terms of activity levels from both Kelsey and Bear Creek periphyton.

The University of Washington's projects also examined the availability and quality of particulate organic matter as food in both creeks. They found no differences in the amount of particulate organic matter measured in the two creeks (about 100 g/m^2). There was significantly more particulate organic matter in Kelsey Creek during August and significantly less during November than in Bear Creek. The surface accumulations of material in Kelsey Creek had much more fine silts associated with them and had a lower carbohydrate content. They also analyzed the protein content of a particulate organic matter but with varying results.

Refuge areas seem to play an important role in Kelsey Creek. The more stable areas in Kelsey creek had aquatic life populations comparable to those found in Bear Creek. These refuge areas did not balance the lack of diversity observed in Kelsey Creek. The Kelsey Creek biota are relatively inefficient in utilizing food resources. The efficiency of utilization was only 3% in Kelsey Creek and about 20% in Bear Creek when the throughput of dissolved organic carbon was excluded (Perkins 1982).

6. Riparian Vegetation

Richey (1982) states that modifications to the vegetative cover have been very significant in Kelsey Creek. The riparian vegetation was relatively intact throughout the entire length of Bear Creek, while only the upper 800 m of Kelsey Creek had a significant amount of intact riparian vegetation. Most of the riparian vegetation along Kelsey Creek was new growth alders less than 150 mm in diameter, vine maple, and blackberry vines. The riparian vegetation along most of Bear Creek was old growth fir, cedar, and alder, which are greater than 300 mm in diameter with an understory of salmonberry and vine maple. Riparian cover in the stream channel in both streams was common, however. Many sections of Kelsey Creek were overhung with dense blackberry vines, which did provide some shade and in-stream cover. Pedersen (1981) states that the vegetation along each watershed was possibly the major factor affecting species composition.

Scott et al. (1982) state that the most beneficial effect of stream alteration is the increase in solar energy reaching the stream surface as the result of the removal of a significant portion of the overhanging canopy. The current riparian vegetation along the middle and lower reaches of Kelsey Creek are only a small fraction of its former growth. The removal of this stream side cover, however, has not resulted in excessive water temperatures and appears to have indirectly benefited the trout populations in the urban stream. Bear Creek, which is heavily canopied along most of its length, can be considered light-limited. Maximum fish growth in Bear Creek occurs in the fall months after leaf fall when sunlight can reach the water. This is different from Kelsey Creek where fish growth is stimulated during the spring and early summer months when the periphyton and probably the benthic productions are greatest. Regardless of the relative production of the benthic invertebrates in each stream, it was found that the salmonids grew more rapidly in Kelsey Creek than in Bear Creek. The size of an age I migrant cutthroat trout from Kelsey Creek was near the length of age II outmigrants from Bear Creek.

7. Algae

University of Washington studies (Richey et al. 1981) found that periphyton algae were the predominant ingredient in the organic accumulation of material in Kelsey Creek. Algae was not nearly as important in Bear Creek. Richey (1982) conducted some algae bioassays with interstitial water, stormwater, and direct runoff water from the urban stream and its watershed. Only very low levels of inhibition to growth were found, and there were few instances where there were growth differences from samples taken from the two different streams. These tests indicated that the particulate-bound metals were mostly not available to the algae. She found that the stream interstitial water caused slight growth inhibition during the laboratory algal tests but that the indigenous algal cells were much less affected. Similar results were found with the stormwater and the runoff waters. She concludes that there is a potential for some toxic impacts of the stormwaters on the algae in Kelsey Creek, but it did not appear to be a dominant factor in limiting algae survival.

8. Benthic Organisms

Pedersen examined the benthic organisms in Kelsey and Bear Creeks as part of the University of Washington's project. He studied the relative occurrence of these bottom organisms in the two streams from about 350 samples. The variety of the organisms found was striking. Insects such as mayflies, stoneflies, caddisflies, and beetles were observed only rarely in Kelsey Creek and were usually of the same few families. Baetids, however, were found in large numbers in certain regions of Kelsey Creek (relatively undisturbed channel sections with riparian vegetation intact). Bear Creek demonstrated a much more diverse distribution of benthic organisms and usually showed more than one dominant family in each major grouping. However, the overall abundance of benthic organisms based on the average number of organisms per sample was not significantly different in Kelsey and Bear Creeks. Kelsey Creek had a mean abundance of about 53 organisms per sample, while Bear Creek had a value of about 48. A total of 179 samples were obtained at Kelsey Creek, while 127 samples were obtained from Bear Creek.

The worm category in Kelsey Creek was dominated by oligochaetes, which represented about 50% of benthic biota in Kelsey Creek. Amphipods, and occasional crayfish, made up about 36%

of the total benthos population. In Bear Creek, the worm category counted for only about 12% of the total benthos, while the amphipod and crayfish group accounted for less than 15% of the total. Chironomids showed up at about 10% in Kelsey Creek, demonstrating a fairly stable population over time except in late July when the population jumped to nearly 30% of the total benthos. The chironomids in Bear Creek made up closer to 20% of the total benthos population. In summary, the benthic life-forms dominating Kelsey Creek were of the collector-gatherer feeding types, which have a greater potential to survive in disturbed systems.

The benthos in Kelsey Creek generally showed a constantly changing composition with large variations in total numbers while the composition in Bear Creek did not change as much. The Bear Creek benthic organisms were also much more evenly distributed among the different taxa. Several of the Kelsey Creek stations can be considered polluted with some marginally unpolluted, while most of the Bear Creek stations were considered to be unpolluted.

The lack of the different representatives of the herbivores in Kelsey Creek (such as stoneflies or caddisflies) which were found in Bear Creek was probably due to the sensitive nature of Hemouridae and most trichoptera to environmental stress (Pederson 1981). Mayflies such as the baetids are more adaptable to minor disturbances. The lack of other herbivores could have allowed the baetids to increase their numbers due to a lack of competition and predators.

The violent flows and increased sedimentation in Kelsey Creek could be a problem for most benthic organisms, except those such as oligochaetes and chironomids, which are burrowers and filter feeders, and amphipods, which can burrow or swim and filter feed. Generally, filter feeders prefer areas of little sediment accumulation where they are exposed to maximum current. The fact that the chironomids maintain relatively stable populations in Kelsey Creek through storms and possible extreme water quality conditions as compared to other groups of insects could be due to their relatively short generation time and high recovery potential. Not all chironomids or oligochaetes, however, are limited to strictly polluted conditions; they can have dense populations where other insects are also found.

Richey (1982) found frequent dense beds of large clams (Unionidae) in Bear Creek, while they were not found in Kelsey Creek. The clams found in Bear Creek were large, indicating a stable and old population. These clams are very sensitive to heavy siltation and bed instability. They depend upon fine particulates carried in the water column for their diet. Therefore, it is not surprising that they were not found in Kelsey Creek. The high inorganic content in the suspended solids in Kelsey Creek and the unstable nature of the channel bed probably prevents their survival in Kelsey Creek. However, empty shells were found buried in the Kelsey Creek stream bed and no live organisms were observed. Therefore, they had probably existed in Kelsey Creek but have been gradually excluded by a shifting habitat and a gradual decrease in the quality of the available food and problems associated with channel instability.

9. Fish

Scott et al. (1982) reviewed two earlier studies that examined the fish populations in Bear and Kelsey Creeks. They stated that Kelsey Creek was a major producer of coho salmon and also supported significant numbers of cutthroat trout and kokanee salmon at one time. A 1956 survey, however, indicated that the Kelsey Creek salmon population was already in jeopardy due to increased urban development. Another study in 1972 found that the cutthroat were more abundant than the coho. Kokanee populations are noted to have declined throughout the Lake Washington drainage area because of the successful introduction of sockeye salmon in major tributaries. This 1972 study also observed occasional chinook salmon in Kelsey Creek. Food availability was determined not to be a limiting factor in the fish populations at that time. This earlier study did, however, find that a new culvert at the lower end of Kelsey Creek did block upstream fish passage under certain flow conditions. This problem was then corrected and the major factor impairing salmon reproduction in the urban streams was thought to be siltation resulting from construction activities.

The University of Washington (Scott et al. 1982) examined fish life in Kelsey and Bear Creeks for 3 years ending in 1981. Figure 4.22 summarizes the fish biomass observed at these two creeks for the different species during an example month (August 1981). Coho was found to comprise only a small fraction of the salmon found in Kelsey Creek, but they frequently exceeded 50% of the total salmon population of Bear Creek. There was also a limited number of cutthroat trout older than age II inhabiting Kelsey Creek. Cutthroat of up to age III were found in Bear Creek, although in limited numbers. The Kelsey Creek salmon were reduced substantially in 1980 relative to both 1979 and 1981. The maximum salmon density in Kelsey Creek in 1981 was about 1 fish/3 m³, which was less than 30% of what was observed in 1979 and 1981. The salmonid population of Bear Creek during this 3-year period was also unstable, as the density of salmon increased in each succeeding year.

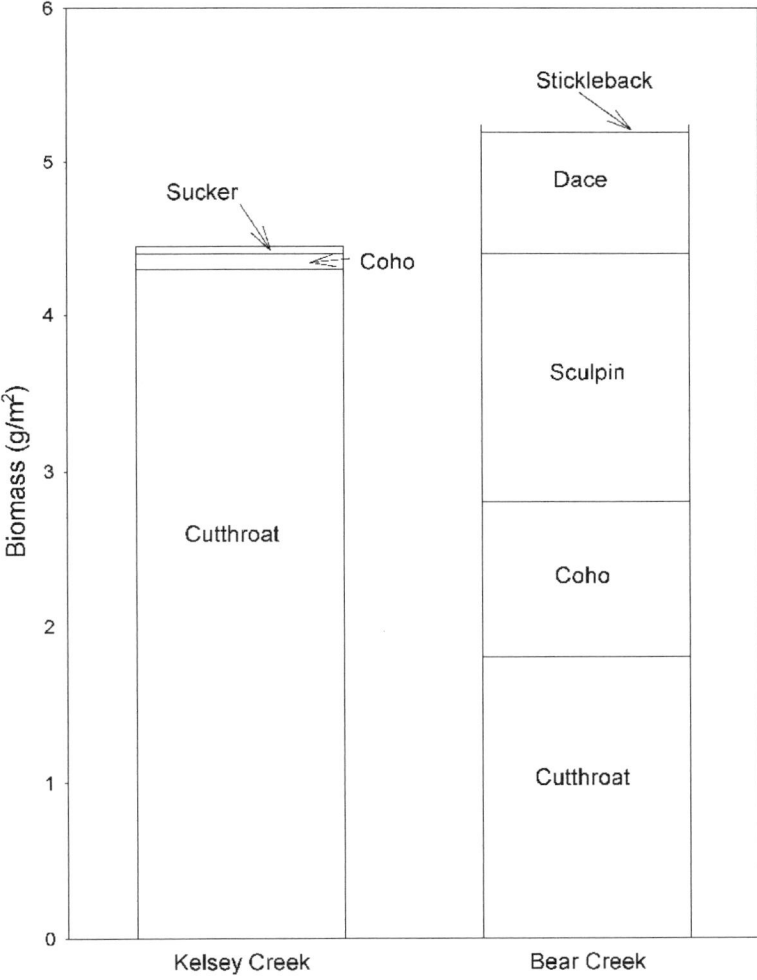

Figure 4.22 Average biomass of fish at sample sites in Bear (reference) and Kelsey (urbanized) Creeks, August 1981. (From Scott, J.B. et al. *Impacts of Urban Runoff on Fish Populations in Kelsey Creek, Washington*, Contract No. R806387020, U.S. Environmental Protection Agency, Corvallis Environmental Research Laboratory, Corvallis, OR. 1982.)

The dominant seasonal trends of fish biomass in Kelsey Creek showed a rapid buildup of biomass in the late winter and early spring followed by a sharp decline in early summer. The generally increasing trend of salmonid biomass in Bear Creek ended with a maximum of 3.7 g/m^2 in May of 1981. The maximum biomass in Kelsey Creek was about 6.5 g/m^2 in comparison. Non-salmon fish species were also quite abundant in Bear Creek, made up mostly of various species of sculpins and dace. Non-salmonids in Kelsey Creek were not very important, with only a few large-scale sucker found. Some dace stickleback and sculpin were also found in Kelsey Creek but in very small numbers. When all fish species were considered, it was found that Bear Creek supported only a slightly greater quantity of total fish biomass (5.2 g/m^2) compared to Kelsey Creek (4.5 g/m^2). Also, no single grouping of fish accounted for more than about 35% of the total fish biomass in Bear Creek. However, the salmonid biomass in Kelsey Creek was greater than the salmonid biomass in Bear Creek, with cutthroat trout comprising almost all of the salmon species found in Kelsey Creek, while large populations of coho salmon were found in Bear Creek along with cutthroat trout. In comparison to some standards, the salmonid production in Bear Creek is low, the direct consequence of a depressed standing crop.

Scott et al. (1982) state that perhaps the best measure of the relative health of a stream in the Pacific Northwest is the number of smolts it produces. The number of smolts in Kelsey Creek is approximately 40% less than that observed in other area creeks. The relative abundance of the cutthroat trout may explain the apparently poor salmonid smolt production of the Kelsey Creek watershed. Cutthroat trout require a larger territory than the typical coho smolt. Therefore, because of the large cutthroat population in Kelsey Creek, the smaller than normal smolt production may be expected.

The Kelsey Creek cutthroat appeared to grow considerably more rapidly than cutthroat observed previously in other streams. The average length of an age I cutthroat smolt in Kelsey Creek was close to the average length of an average age II cutthroat smolt in other streams. The Kelsey Creek age II smolts were typical of the lengths for other age III smolts. Also, most of the fish in Kelsey Creek outmigrated at age I. Typically, cutthroat smolts from other streams generally outmigrate from ages II through IV. It is believed that the cutthroat migrating from Kelsey Creek spend an additional year in Lake Washington before entering Puget Sound.

Scott et al. (1982) summarized the potential effects of sedimentation on stream-living salmon. These include the clogging and abrasion of gills, abrasion or adherence of sediment to the egg chorion, increasing susceptibility to diseases, modification of behavior, blocking emergence of alevins, reducing spawning habitat, changing intragravel permeability with reduced dissolved oxygen concentrations, introducing potentially toxic materials associated with the suspended material, and altering the structure and productivity of the food resources available to the fish. They studied the incidence of damaged gills on the fish in Kelsey and Bear Creeks (Scott et al. 1982). They found that from 0 to 77% of the fish sampled in Kelsey Creek were afflicted with respiratory anomalies. The season and location along the channel, as well as the age and species of the fish, affected these anomalies. Cutthroat, as an example, had afflictions that rapidly increased after mid-May. Older cutthroat also had less incidence of gill damage. Small coho salmon in Kelsey Creek had little gill damage. They also note that the incidence of damage to gills in the cutthroat trout in Kelsey Creek generally decreased in a downstream direction. No cutthroat trout and only two of the coho salmon sampled in Bear Creek had damaged gills.

In-stream embryo bioassays indicated that coho embryo salmon survival was significantly greater in Bear Creek but that no difference was found when using rainbow trout embryos. Streamside bioassays, however, indicated that the surface waters of Kelsey Creek did not significantly reduce the survival of the salmon embryos. The survival of the embryos during the winter bioassays was significantly greater in Bear Creek than in Kelsey Creek, but no difference in survival was noted during the spring bioassay tests. While the laboratory and field bioassays tended to indicate minimal toxic influences, other field observations suggested a stronger possibility of toxic problems. Coho salmon were absent in the more heavily developed areas, and the incidence of cutthroat trout with gill damage increased in those areas. Higher levels of toxic pollutants, such as lead, were occurring with the increases of sediment transport in these more developed areas and may have contributed to the observed increase in gill damage.

Creek flows may also significantly affect the salmon fisheries. Scott et al. (1982) state that high creek flows may increase the sweeping of poorly swimming fish from the creeks. The highest flows where migration of fish from the creek were monitored was a little over 4 m³/s which was less than one third of the peak flow recorded during the study period on Kelsey Creek. At these monitored flows, the species with relatively poor swimming ability were swept from the system, while the salmon were better able to withstand these increased flows. They estimated that a flood with a recurrence interval of about 5 years in Kelsey Creek having a peak instantaneous discharge of about 11 m³/s may be expected to increase the coho embryo mortality by about 20%. This would increase the scour mortality during a 5-year flood to 10% or less. The lower summer flows may also limit the survival of some salmon populations (especially coho salmon) (Richey 1982).

Pedersen (1981) states that the salmon in Kelsey Creek seem to be adjusting their feeding to invertebrates that are present based upon fish stomach contents analyses. Their growth did not appear to be limited by the type of diet available in Kelsey Creek. The salmon fishery in Kelsey Creek seems to be surviving; the City of Bellevue and the Municipality of Metropolitan Seattle have supported the installation and maintenance of stream-side boxes for the incubation of sockeye salmon eggs. This program has provided direct involvement for the local school children and, therefore, also serves the educational aspects of the beneficial uses for these urban streams.

10. Creek Sediment Quality

Several of the University of Washington projects and the Seattle METRO project investigated physical and chemical characteristics of the Kelsey and Bear Creek sediments. Perkins (1982) stated that the size and composition of the sediments near the water interface tended to be more variable and of a larger median size in Kelsey Creek than in Bear Creek. These particle sizes varied in both streams on an annual cycle in response to runoff events. Larger particle sizes were more common during the winter months when the larger flows were probably more efficient in flushing through the finer materials. Pedersen (1981) also states that Kelsey Creek demonstrated a much greater accumulation of sandy sediments in the early spring. This decreases the suitability of the stream substrates for benthic colonization. Scott et al. (1982) state that the level of fines in the sediment samples appears to be a more sensitive measure of substrate quality than the geometric mean of the particle size distribution. Fines were defined as all material less than about 840 µm in diameter.

METRO (Galvin and Moore 1982) also analyzed organic priority pollutants in 17 creek sediments including several in Kelsey and Bear Creeks. Very few organic compounds were detected in either stream, with the most notable trend being the much more common occurrence of various PAHs in Kelsey Creek while none was detected in Bear Creek.

Scott et al. (1982) state that stream bed substrate quality can be an important factor in the survival of salmonid embryos. Richey (1982) describes sediment bioassay tests which were performed using Kelsey and Bear Creek sediments. She found that during the 4-day bioassay experiment, no mortalities or loss of activities were observed in any of the tests. She concluded that the chemical constituents in the sediment were not acutely toxic to the test organism. However, the chronic and/or low level toxicities of these materials were not tested.

11. Creek Bank and Stream Bed Erosion

Richey (1982) made some observations about bank stabilities in Kelsey and Bear Creeks. She notes that the Kelsey Creek channel width has been constrained during urban development. Thirty-five percent of the Kelsey Creek channel mapped during these projects was modified by the addition of some type of stabilization structure. Only 8% of Bear Creek's length was stabilized.

Most of the stabilization structures in Bear Creek were low walls in disrepair, while more than half the structures observed along Kelsey Creek were large riprap or concrete retention walls. The need for the stabilization structures was evident from the extent and severity of erosion cuts and the number of deposition bars observed along the Kelsey Creek stream banks. Bridges and culverts were also frequently found along Kelsey Creek; these structures further act to constrict the channel. As discharges increase and the channel width is constrained, the velocity increases, causing increases in erosion and sediment transport.

The use of heavy riprapping along the creek seems to worsen the flood problems. Storm flows are unable to spread out onto a floodplain, and the increased velocities are evident downstream along with increased sediment loads. This rapidly moving water has enough energy to erode unprotected banks downstream of riprap. Many erosion cuts along Kelsey Creek downstream of these riprap structures were found. Similar erosion of the banks did not occur in Bear Creek. Much of the Bear Creek channel had a wide floodplain with many side sloughs and back eddies. High flows in Bear Creek could spread onto the floodplains and drop much of their sediment load as the water velocities decreased.

12. Creek Sedimentation and Sediment Transport

The University of Washington studies also examined sediment transport in Kelsey and Bear Creeks. Richey (1982) found that the relative lack of debris dams and off-channel storage areas and sloughs in Kelsey Creek contributed to the rapid downstream transit of water and materials. The small size of the riparian vegetation and the increased stream power probably both contributed to the lack of debris in the channel. It is also possible that the channel debris may have been cleared from the stream to facilitate rapid drainage. The high flows from high velocities caused the sediments to be relatively coarse. The finer materials were more easily transported downstream. Larger boulders were also found in the sediment but were probably from failed riprap or gabion structures.

The effects of erosion and sediment deposition in Kelsey Creek were more severe than those found in Bear Creek. Kelsey Creek's channel was scoured to deeper depth, there was much more channel instability in Kelsey Creek, and the numbers of erosion cuts and deposition bars were much more frequent in Kelsey Creek. Richey (1982) reported that the sediment transport in Bear Creek during December 1979 was 27 kg/ha, while 98 kg/ha left Kelsey Creek. The suspended solids transport was almost exponentially related to discharge. On an annual basis, Kelsey Creek discharged almost four times as much suspended solids as did Bear Creek, but most of this material passed through the stream in a few hours or days. Richey (1982) found that much of the solids transport in Kelsey Creek occurred during the rapid rise of the hydrograph when the energy to move sediment material was increasing. The silts and associated pollutants were rapidly transported through the system during these periods. The scouring of the channel appeared to remove temporarily stored silts and the associated pollutants. The higher levels of particulate transport in Kelsey Creek are probably due to increased stream power rather than increased sources of sediment material in the watershed. However, there were substantial amounts of in-stream sources of sediment material in Kelsey Creek to augment the runoff discharged sediment. Because of the lack of debris dams in the downstream sections of Kelsey Creek, the transported materials are carried significant distances before deposition. The high stream power available to transport the materials and the erodable nature of the stream banks in the watershed areas along with the lack of storage sites along the stream all contributed to high particulate yields from Kelsey Creek. Because much of the suspended particulate material in Kelsey Creek was from the relatively unpolluted bank materials, the sediments and suspended loads in Kelsey Creek had much lower concentrations of many of the typical urban pollutants compared to the urban runoff that was discharged to the creek.

Sources of Urban Runoff Pollutants

1. Atmospheric Particulate and Rainfall Deposition of Pollutants

The USGS (Ebbert et al. 1983; Prych and Ebbert undated) studied dustfall quantity and quality along with rain quality at each of three locations in the test watersheds. Seattle METRO also examined the metallic and organic priority pollutant quality of atmospheric particulates. These data indicated that the airborne PAHs are combustion products, while the street dirt PAHs are from petroleum product spills. In August of 1980, ash from the eruption of Mt. St. Helens fell in the study area and substantially increased the dustfall measurements. These increased dustfall values were typically two to four times the average monthly values. During most months, dry atmospheric dustfall was much greater than the total solids associated with the rainwater.

2. Runoff Water Sources

The City of Bellevue study (Pitt 1985) monitored runoff and rainfall characteristics at the two main study locations (Surrey Downs and Lake Hills) during the 2 years of the project. Bellevue receives about 1 m of rain each year. Dry periods of more than a week are quite rare. Rains come on an average about once every 2 or 3 days throughout the year. Slightly more than 100 rains may occur each year, but the amount of rain associated with each is quite small. Most of the rains are less than 6 mm. The largest rains monitored during this project were about 100 mm.

The Lake Hills rain depths were about 12% more than the comparable Surrey Downs rains. The average duration of the Lake Hills rains was also about 10% longer than the Surrey Downs rains. The Lake Hills rains also started about a half hour before the rains in Surrey Downs began. Most of the rain events had less than 6 mm of rain, and less than 10% of the rain events had depths greater than 25 mm. Most of the rainfall quantities were associated with rain events greater than about 15 mm. The much more common small rains did not add up to much total depth. The rains that were smaller than 6 mm accounted for less than 25% of the total rainfall depth, while about 30% of the total rainfall depth was associated with rains greater than 25 mm.

Almost 400 runoff events were monitored at the Surrey Downs and Lake Hills monitoring stations during the 2-year study period. Almost 99% of the rains that occurred in Surrey Downs and 91% of the Lake Hills rains were monitored. The baseflow in the Surrey Downs basin accounted for about 23% of the total annual flow, while the baseflow was only about 13% of the total annual flow in Lake Hills. The stormwater flows in Lake Hills were about 35% greater than in Surrey Downs. Overall, the base plus stormwater urban flows from Lake Hills were about 18% greater than Surrey Downs on an equal area basis.

For both study years and test basins, only about 25% of the rain that fell in the test basins left the areas as runoff. The small rains typically had the smallest runoff factors, while the large rains had the largest factors. For very small rains, no runoff is expected to occur from the pervious areas nor from the impervious areas that drain to these pervious areas. Starting at about 2.5 mm of rain, however, the volumetric runoff coefficients (Rv) are about 0.3 to 0.5 times the maximum values that they are likely to obtain. The dry season runoff coefficients are less than the wet season values due to different soil moisture conditions. For all rains greater than about 2.5 mm, impervious surfaces contribute more than 60% of the total urban runoff flows. The remainder of the flows are approximately evenly divided between front and back yards, while vacant lots and parks contribute very little flow due to their limited presence in the area. Street surfaces contribute about 25% of the total urban flows for most rains causing runoff.

3. Stormwater and Baseflow Urban Runoff Quality and Pollutant Source Areas

Collecting stormwater runoff quality data was a major aspect of the City of Bellevue's and the USGS projects. In addition, Seattle METRO analyzed some of the samples collected by the City of Bellevue for metallic and organic priority pollutants. Most of the analytical effort was associated with a core list of important constituents. Tables 4.18 and 4.19 summarize USGS and City of Bellevue stormwater quality data for these core constituents. The USGS obtained many discrete

Table 4.18   Urban Runoff Quality Reported by the USGS (many discrete samples for a limited number of storms)

Constituent (mg/L, unless otherwise noted)	Maximum	Minimum	Approx. Median	No. of Discrete Samples Analyzed
Temperature, °C	14.8	2.6	8.0	49
Specific conductance, μmhos/cm	1480	12	41	1299
pH, pH units	7.9	3.4	6.7	1093
COD	780	8	60	681
BOD_5	40	<0.1	6.6	321
BOD ultimate	115	3.5	20	138
Particulate organic carbon	40	<0.1	2.1	638
Dissolved organic carbon	120	0.2	7.5	681
Fecal coliforms, No./100 mL	66,000	1	980	326
Suspended solids	2740	1	50	1180
Dissolved solids	788	8	35	241
Nitrate plus nitrite nitrogen	4.5	<0.01	0.21	691
Ammonia nitrogen	7.2	<0.01	0.14	689
Total Kjeldahl nitrogen	45	0.21	1.1	687
Dissolved Kjeldahl nitrogen	33	<0.01	0.63	686
Total phosphorus	9.2	0.01	0.15	686
Dissolved phosphorus	7.2	<0.01	0.06	685
Lead	1.8	0.004	0.14	693
Oil and grease	10	<1	2.5	16

Data from Ebbert, J.C. et al. *Data Collected by the U.S. Geological Survey During a Study of Urban Runoff in Bellevue, Washington, 1979–82.* Preliminary U.S. Geological Survey Open-File Report, Tacoma, WA. 1983.

Table 4.19   Urban Runoff Quality Reported by the City of Bellevue (total storm, flow-weighted composite samples for most runoff events, Surrey Downs and Lake Hills observations combined, 2/80–1/82) (mg/L, unless otherwise noted)

Constituent	Maximum	Minimum	Average	No. of Flow-Weighted Total Storm Samples Analyzed
Specific conductance, μmhos/cm	300	16	41	204
pH, pH units	7.4	5.2	6.3	204
Turbidity, NTU	150	4	19	204
Total solids	620	24	109	208
Total Kjeldahl nitrogen	5.9	<0.5	1.0	208
COD	150	13	46	208
Total phosphorus	3.6	0.002	0.26	208
Lead	0.82	<0.1	0.17	208
Zinc	0.37	0.03	0.12	208

From Pitt, R. *Characterizing and Controlling Urban Runoff through Street and Sewerage Cleaning.* U.S. Environmental Protection Agency, Storm and Combined Sewer Program, Risk Reduction Engineering Laboratory. EPA/600/S2-85/038. PB 85-186500. Cincinnati, OH. 467 pp. June 1985.

samples throughout individual storms but only analyzed data from a small percentage of the total runoff events that occurred during the study period. The City of Bellevue's sampling procedures involved collecting total storm flow-weighted composite samples throughout most of the events that occurred during the sampling period at the Surrey Downs and Lake Hills sites.

The USGS (Ebbert et al. 1983) found that when the stormwater runoff discharge was high, the concentrations of the constituents in particulate forms tended to be high, and the concentrations of the constituents in dissolved forms tended to be low. During periods of low discharge, particulate concentrations were low, and the dissolved concentrations were high. There was very little variation in most of the constituent concentrations for each of the three sites for most rains. The hardness of the stormwater was generally very low. About two thirds of the total solids and phosphorus loads, and one third of the total Kjeldahl nitrogen, total nitrogen, and organic carbon loads were associated with particulates. They also found that about 15% of the total nitrogen load was in the form of dissolved nitrate plus nitrite nitrogen, about 10% is as dissolved ammonia, 40% as dissolved organic nitrogen, and 35% was particulate Kjeldahl nitrogen.

Seattle METRO (Galvin and Moore 1982) analyzed about 21 of the total storm flow-weighted composite samples from Bellevue for 14 metallic priority pollutants. The stormwater metal concentrations were very low when compared to other urban runoff metal data for other locations (except for arsenic). They also found that the stormwater metal concentrations did not vary significantly between the study areas. METRO also analyzed many of the samples for dissolved concentrations of the different metals in addition to the total concentrations. Only copper and zinc showed significant dissolved concentrations, while the other metals were almost completely associated with the particulates in the stormwater. None of the organic priority pollutants detected by METRO was found in more than 25% of the samples submitted. Of the 111 organic priority pollutants, only 19 were detected at least once in the METRO stormwater sample analyses. Except for one value (a pentachlorophenol value of 115 µg/L), they were all very close to the detection limits.

The USGS also analyzed about 16 of their discrete samples for a long list of insecticides and herbicides. Lindane, Diazinon, Malathion, Dieldrin, and 2,4-D were detected in more than half the samples. Endosulfan, Silvex, and 2,4,5-T were found in about one third of the samples submitted. Many of the insecticides and herbicides analyzed were not detected in any of the samples.

The USGS (Prych and Ebbert undated) also examined stormwater-suspended sediment size distributions in four to seven samples. These analyses showed that 64% of the particulate material in stormwater was associated with particle sizes smaller than 62 µm. Only about 10% of the stormwater particles had sizes greater than 250 µm.

The City of Bellevue study (Pitt 1983) also examined the baseflow quality at Surrey Downs and Lake Hills. The runoff water quality at Bellevue was much better compared to most other locations. The baseflow quality, on the other hand, was found to be worse than expected. This was probably because the study basins were completely urbanized and the baseflows were percolated urban sheet flow waters from previous storms that were draining out of the surface soils. In basins with undeveloped upstream areas, the baseflow would originate mostly from the non-urbanized upper reaches and would have much better quality. The data shown in Table 4.20 were collected from 26 composite samples collected over 24-hour periods from both Surrey Downs and Lake Hills.

Table 4.21 shows the measured annual baseflow and stormwater runoff yields for the two test catchments. There was an apparent increase in storm runoff discharges at Lake Hills, while Surrey Downs had larger baseflow contributions. The baseflow contributions were much less than the storm-generated flows, but the phosphorus and TKN baseflow discharges comprised about 25 to 30% of the total Surrey Downs discharges.

Pitt (1985) made estimates of the pollutant contributions from the different source areas. Table 4.22 summarizes these estimates. During very small rains, most of the runoff, and therefore pollutant discharges, was associated with the directly connected impervious areas. As the rain total increased (greater than about 2.5 mm), the pervious areas became much more important. These

Table 4.20 Baseflow Water Quality Reported by the City of Bellevue (Surrey Downs and Lake Hills data combined) (mg/L, unless otherwise noted)

Constituent	Maximum	Minimum	Average	No. of 24-hr Composite Baseflow Samples Analyzed
Specific conductance, μmhos/cm	430	138	260	18
Total solids	326	108	202	26
COD	67	6.8	23	26
Total Kjeldahl nitrogen	2.4	0.20	0.8	26
Total phosphorus	1.2	0.027	0.16	26
Lead	0.1	<0.1	<0.1	26
Zinc	0.47	0.026	0.09	26

From Pitt, R. *Characterizing and Controlling Urban Runoff through Street and Sewerage Cleaning.* U.S. Environmental Protection Agency, Storm and Combined Sewer Program, Risk Reduction Engineering Laboratory. EPA/600/S2-85/038. PB 85-186500. Cincinnati, OH. 467 pp. June 1985.

Table 4.21 Annual Baseflow and Stormwater Runoff Mass Yields Reported by the City of Bellevue (kg/ha/yr)

Constituent	Surrey Downs			Lake Hills		
	Baseflow	Storm Runoff	Total	Baseflow	Storm Runoff	Total
Total solids	110	205	315	76	280	360
COD	11	90	100	9.9	110	120
Total Kjeldahl nitrogen	0.60	1.8	2.4	0.20	2.7	2.9
Total phosphorus	0.11	0.40	0.51	0.04	0.69	0.73
Lead	0.03	0.26	0.29	0.02	0.45	0.47
Zinc	0.060	0.24	0.30	0.027	0.31	0.34

From Pitt, R. *Characterizing and Controlling Urban Runoff through Street and Sewerage Cleaning.* U.S. Environmental Protection Agency, Storm and Combined Sewer Program, Risk Reduction Engineering Laboratory. EPA/600/S2-85/038. PB 85-186500. Cincinnati, OH. 467 pp. June 1985.

Table 4.22 Source Area Contributions for Runoff Pollutants from Bellevue Residential Areas (for 2.5 to 65 mm rains) (% contributions from source areas)

Source Area	Total Solids	COD	Phosphates	Total Kjeldahl Nitrogen	Lead	Zinc
Streets	9	45	32	31	60	44
Driveways and parking lots	6	27	21	20	37	28
Rooftops	<1	3	5	10	<1	24
Front yards	44	13	22	19	<1	2
Back yards	39	12	20	20	<1	2
Vacant lots and parks	2	<1	<1	<1	<1	<1

From Pitt, R. *Characterizing and Controlling Urban Runoff through Street and Sewerage Cleaning.* U.S. Environmental Protection Agency, Storm and Combined Sewer Program, Risk Reduction Engineering Laboratory. EPA/600/S2-85/038. PB 85-186500. Cincinnati, OH. 467 pp. June 1985.

patterns varied significantly for different areas depending on the rain characteristics and land uses. It was estimated that for most rain events, total solids originated mostly from the back and front yards in the test areas, and street surfaces contributed only a small fraction of the total solids urban runoff discharge. Street surfaces, however, were expected to make up most of the lead, zinc, and COD concentrations in urban runoff. Phosphates and total Kjeldahl nitrogen were mostly contributed from street surfaces, driveways, and parking lots combined. Front and back yards made up slightly less than half of these nutrient contributions to the outfall. It was noted that zinc contributions from rooftops made up about one fourth of the total zinc discharges. These zinc rooftop sources were expected to be associated with galvanized metal rain gutters and downspouts.

4. Street Dirt Contributions to Urban Runoff Discharges

The City of Bellevue examined street dirt loadings in the three urban runoff test areas during the 2-year period of study (Pitt 1985). By the end of January 1982, about 600 street surface accumulation samples were collected from the test areas in Bellevue. Each of these 600 street surface samples was separated into eight different particle sizes. The smallest particle sizes account for only a small fraction of the total material. This was especially true during the wet season when the rains were most effective in removing the smallest particles. During the dry season, the larger particle sizes accounted for relatively small fractions of the total solids weight. Most of the street surface particulates were associated with particles in the size range of 125 to 1000 μm.

The Bellevue street surfaces were relatively clean when compared to other locations throughout the country. This difference is expected to be mostly due to the frequent rains that occur in Bellevue. The initial accumulation rates (assumed to be equal to the deposition rates) in the test areas were estimated to vary between 1 and 6 (with an average of about 3) g/curb-meter/day. This is comparable to accumulation rates observed in other locations for smooth streets in good condition. However, the Bellevue streets never have an opportunity to become extremely dirty due to the relatively frequent rains.

The Bellevue study (Pitt 1985) also examined the chemical characteristics associated with the particulates in different size ranges. The chemical characteristics were not unusual when compared to other locations throughout the United States. The Seattle METRO project (Galvin and Moore 1982) also examined heavy metals in the street surface particulate samples collected by the Bellevue sampling team. All of the inorganic priority pollutants, except selenium, were detected in the street dirt. The most abundant metals were lead, zinc, chromium, copper, nickel, arsenic, cadmium, and beryllium. METRO did not find any clear differences between metal concentrations in the two residential basins nor when these residential basin street dirt characteristics were compared with commercial and industrial samples collected in Seattle. They also found that the concentrations of metals were greatest in the finer size particles, but these fine particles accounted for only a small portion of the total solids loadings on the street surfaces. When these metallic priority pollutant analyses were compared with similar analyses conducted elsewhere in the United States, the Bellevue concentrations tended to be quite low (except for arsenic).

Seattle METRO (Galvin and Moore 1982) also analyzed street dirt samples for organic priority pollutants. Of the 111 organic priority pollutants, only about 30 were detected in the street dirt samples. Two of the PAHs (fluoranthene and phenanthrene) were found in all of the street dirt samples. Several of the compounds had concentrations greater than 1 mg constituent/kg total solids, while one phthalate was recorded as great as 35 mg constituent/kg total solids. It was also noted that most of the organic priority pollutants were associated with the finest particle size fractions. The halogenated aliphatics, monocyclic aromatics, phenolics, and phthalate esters were very common in the residential samples but were only infrequently found in the other samples. The industrial sample, however, periodically had very high concentrations of some of the organic constituents.

Most of the material that washed off the street surfaces during rains occurred in particle sizes less than about 125 μm. Only about 10% of the washoff material was greater than about 500 μm in size. The largest street surface particulates were notably absent in the runoff water. For all of the sites combined, only about 14% of the total solids were removed by rains observed during the test period. The washoff percentage is substantially greater for lead (about 21%) because of the greater abundance of lead found in the smaller particle sizes.

5. Sewerage and Catchbasin Sediment Accumulations

Sewerage system sediment loadings were periodically observed in the Surrey Downs and Lake Hills study areas during the City of Bellevue project (Pitt 1985). The storm drainage system was cleaned before the start of the project and the accumulating sediment volumes in inlets and catchbasins were observed nine times during the 2 years. During the second year of observations, the amount of accumulated material remained relatively constant. Typically, there was about twice as much sediment in the storm drainage systems at any one time as there was on the streets. Table 4.23 shows the calculated sewerage accumulation rates in inlets and catchbasins in Surrey Downs and Lake Hills. These accumulation values were the rates observed after the initial cleaning and before the stable Year 2 volumes were obtained. During the second year (October 1981) a very large storm (about 100 mm) occurred. However, the loading observations before and after this event were not significantly different, indicating very little net removal due to flushing. The chemical quality of the catchbasin and inlet sump material was very similar to the street dirt materials, for similar particle sizes.

A survey of the pipe dimensions and slopes throughout each of the study areas was made during the early months of the project by the City of Bellevue (Pitt 1985). Very few pipes in either Surrey Downs or Lake Hills had slopes less than 1%, the slope assumed to be critical for sediment accumulation. Frequent observations of sediment accumulations in the pipes throughout the two study areas were also made. Generally, very small amounts of sediment were found in the sewerage in Lake Hills and Surrey Downs. The pipes that had significant quantities of sediment were sloped less than 1.5% and/or located close to a source of sediment. The characteristics of the sewerage sediment were also similar to the characteristics of the sediment in the close-by manholes and catchbasins and the street surface materials. The volume of sediment accumulated in the Lake Hills

Table 4.23   Stormwater Inlet Sediment Volumes and Accumulation Rates

	Total Inlets	Inlets per ha	Sediment per ha (L/month)	Sediment per Inlet (L/month)	Approximate Months Needed to Reach Steady-State Volume	Steady-State Volume per ha (L)	Steady-State Volume per Inlet (L)
Surrey Downs (38.0 ha)							
Catchbasins	43	1.1	5.3	4.8	13	68	62
Inlets	27	0.7	2.0	2.8	20	40	57
Manholes	6	0.2	0.8	4.0	19	15	76
Average	76 (total)	2.0 (total)	8.1	4.2	15	123	62
Lake Hills (40.7 ha)							
Catchbasins	71	1.7	2.4	1.4	18	43	25
Inlets	45	1.1	1.5	1.4	14	22	20
Manholes	15	0.4	1.6	4.0	23	36	90
Average	131 (total)	3.2 (total)	5.5	1.7	18	100	31

From Pitt, R. *Characterizing and Controlling Urban Runoff through Street and Sewerage Cleaning.* U.S. Environmental Protection Agency, Storm and Combined Sewer Program, Risk Reduction Engineering Laboratory. EPA/600/S2-85/038. PB 85-186500. Cincinnati, OH. 467 pp. June 1985.

pipes was about 0.04 m³/ha (about 70 kg/ha). In Surrey Downs, the pipe sediment volume was estimated to be more than 0.5 m³/ha (about 1000 kg/ha). Most of the sediment in Surrey Downs was located in silted-up pipes along 108th Street and Westwood Homes Road, which were not swept and had nearby major sediment sources. The pipe sediment volume estimated to be available for runoff transport in Surrey Downs was about 0.01 m³/ha (about 15 kg/ha).

Urban Runoff Controls

1. The Effects of Street Cleaning in Controlling Urban Runoff Pollutant Discharges

The coordination of the street surface sampling, street cleaning operations, and runoff monitoring activities during the City of Bellevue project allowed many different data analysis procedures to be used to investigate possible effects of street cleaning on runoff water quality. The use of two test basins and the rotation of the street cleaning operations also allowed one basin to be compared against the other along with internal basin comparisons.

The design of an effective street cleaning program requires not only a determination of the accumulation rates, but also an assessment of the performance of specific street cleaning equipment for the actual conditions encountered. The street cleaning tests conducted by the City of Bellevue (Pitt 1985) utilized two different street cleaning frequencies. These two frequencies included no cleaning and intensive three times a week cleaning. Each cleaning frequency was employed in both the Surrey Downs and the Lake Hills test catchments for a several-month period and were then rotated. There was also a several-month period when no street cleaning was conducted in either test catchment. Runoff was simultaneously monitored for the two catchments during these varying street cleaning programs.

During the entire project period, street dirt loadings were about 115 g/curb-meter (with an extreme value of about 350) during the period of no street cleaning. The loadings were reduced to about 60 g/curb-meter shortly after the start of street cleaning. Median particle sizes decreased with the start of street cleaning because of the selective removal of the large particle sizes by street cleaners. The rain periods all reduced the street surface loadings appreciably, except for the largest rain observed during the study. The rains also increased the median particle sizes because they were most effective in removing the finer material. The largest rain had little effect on the net loading change, probably because of substantial erosion material carried onto the street during this major storm and the relative cleanliness of the street surface before the storm occurred.

Street loadings responded rapidly to initiation of street cleaning. Changes from periods of street cleaning to no street cleaning were not as rapid. The Bellevue study collected many street surface particulate samples in the two test basins immediately before and immediately after the streets were cleaned. Street cleaning equipment cannot remove particulates from the street surface unless the loadings are greater than a certain amount. This value was about 85 g/curb-meter in the test basins for the mechanical broom street cleaners and about 30 g/curb-meter for the regenerative air street cleaner. If the initial street surface loading values were smaller than this, the residual loadings typically were equal to the initial loadings.

Statistical analysis showed that the frequent rains in Bellevue were probably more effective than the street cleaning in keeping Bellevue streets clean. The street surface loadings after rains were usually about 50 g/curb-meter, and the mechanical street cleaning equipment could only remove the street surface particulates down to about 85 g/curb-meter. It was also found that typical mechanical street cleaning equipment is quite ineffective in removing the small particle sizes that are removed by rains. However, a modified street cleaner resulted in an almost constant residual loading value in the cleaning width after cleaning, irrespective of the initial loading. This indicates a very important advantage in the cleaning effectiveness for this street cleaner.

Much data analysis effort during the Bellevue City project was directed toward attempting to identify differences in runoff concentrations and yields caused by street cleaning operations (Pitt

1985). No significant differences in runoff yields or concentrations during periods of intensive street cleaning vs. no street cleaning were observed. Street surfaces contributed less than 25% of the runoff yield for most storms. Therefore, street cleaning would have to be extremely effective to cause stormwater yield improvements approaching 25%. For very small rains, street surface washoff is estimated to contribute more than 60% of most of the constituents to the runoff yield. For larger rains, however, the importance of street washoff diminishes. With intensive street cleaning, only the larger particle sizes are significantly reduced, while particle sizes most subject to washoff by rains are not effectively reduced. This may result in less than a 6% expected improvement in runoff water quality for intensive street cleaning. The modified regenerative air street cleaner is expected to have only slightly better effectiveness in reducing runoff yields. The modified street cleaner may reduce the runoff yields by as much as 10%.

2. Sewerage Inlet Cleaning Effects in Reducing Urban Runoff Yields

The City of Bellevue's project (Pitt 1985) also studied the potential benefits of cleaning sewerage inlet structures in controlling urban runoff discharges. The rains preferentially removed the finer, more heavily polluted, and more available materials during washoff. The sediments in the catchbasins and the sewerage were mostly the largest particles that were washed off the street. Catchbasin sump sediments can be relatively conveniently removed to eliminate this potential source of urban runoff pollutants. Because the catchbasin sediment accumulation rate is quite low, frequent cleaning of catchbasins is not necessary.

Only about 60% of the available sump volumes in the inlets were used for detention of particulates. The structures with large sump volumes required less frequent cleaning and held larger volumes of sediments. It is expected that cleaning these inlet sumps about twice a year could reduce the lead and total solids urban runoff discharges by between 10 and 25%. COD, total Kjeldahl nitrogen, total phosphorus, and zinc may be controlled by between 5 and 10% with semiannual catchbasin cleaning. Cleaning less frequently than this would reduce these expected improvements. If the catchbasin sumps are left full, the potential exists for dramatically increased runoff yields during rare events that may flush captured material. Some pollutants may also be chemically changed by oxidation-reduction reactions or other chemical or biological changes in the catchbasins.

3. The Use of Dry Detention Basins in Controlling Urban Runoff Discharges

The USGS (Ebbert et al. 1983) tested the effectiveness of a dry detention facility in the 148th Avenue S.E. test catchment. The detention basin system consisted of five normally dry grass-lined swales which were contoured into a small park adjacent to the road. The swales were about 300 m long and 30 m wide. There were five control structures used to regulate the flow and the storage along the 27-in trunk line running under the park. The original design of the detention system permitted the flow and storage to be regulated by weirs and valves. Runoff from low-intensity storms was originally allowed to pass through the system with little detention, while discharge from higher intensity storms was detained behind the weirs in the 27-in trunk line. During extreme events, the higher flows ran over the weirs when the detention basins were full.

During the study, the USGS (Ebbert et al. 1983) modified the control structures to permit the slow release of water stored in the detention basin, which was then monitored with a recorder installed behind the weir. Water was therefore stored during much smaller rains than in the original configuration. The detention time was about 30 min or less, which was sufficient time for settling of sand and some coarse silt. Much of the finer material, however, was probably transported directly through the detention system. Earlier data indicated that most of the suspended sediment in the storm runoff at this site was finer than 62 μm. The results of the monitoring (Prych and Ebbert undated) indicated that the detention of the storm runoff had little effect on the concentrations of the runoff constituents. The performance of the detention basins on the four to seven storms that were tested seemed to depend mostly on the distribution of the constituents between the suspended and dissolved phases. The volume of the storm sewer behind the weir used to control the flow was adequate to store the runoff during about 70% of the storms that occurred during that phase of the study. For the other 30% of the storms, the volume of the sewer was insufficient to store all the detained water and some was backed up into the grass-lined depressed area. When the grassy area was inspected after a storm, only a trace of fine residual material was noted on the blades of grass.

Over the entire detention phase of the study, there were about 20 storms (about 10% of all storms) large enough to cause detention in the grassy swale. At the end of the study, only a small amount of suspended sediment was seen on the grass. It was estimated that less than one tenth of the total amount transported through the system was detained. The USGS (Prych and Ebbert undated) also examined the ability of the detention facility to affect the discharge rate of storms. The average ratio of peak discharge rates without detention to detention was 0.63.

Step 7. Project Conclusions

Degradation of Habitat and Biological Communities

- The urbanized Kelsey Creek environmental quality was much better than expected, but was degraded when compared to the less urbanized Bear Creek. Kelsey Creek apparently lacked gross contamination by pollutants. The direct toxic effects of pollutants during storms appeared to be small; the stream did support a small, unhealthy salmonid population. Kelsey Creek salmon did grow faster than Bear Creek salmon, however.
- The fish population in Kelsey Creek had adapted to its degrading environment by shifting the species composition from coho salmon to less sensitive cutthroat trout and by making extensive use of less disturbed refuge areas.
- Studies of damaged gills found that up to three fourths of the fish in Kelsey Creek were affected by respiratory anomalies, while no cutthroat trout and only two of the coho salmon sampled in Bear Creek had damaged gills.
- Massive fish kills in Kelsey Creek and its tributaries were observed on several occasions during the project due to the dumping of toxic materials into storm drains.
- There were significant differences in the numbers and types of benthic organisms found. Mayflies, stoneflies, caddisflies, and beetles were rarely observed in Kelsey Creek but were quite abundant in Bear Creek. These organisms are commonly regarded as sensitive indicators of environmental degradation. By comparison, Kelsey Creek fauna was dominated by oligochaetes, chironomids, and amphipods, commonly regarded as species more tolerant to environmental degradation.
- As an example of a degraded aquatic habitat in Kelsey Creek, a species of clams (Unionidae) was not found in Kelsey Creek, but was found in Bear Creek. These clams are very sensitive to heavy siltation and unstable sediments. Empty clam shells, however, were found buried in the Kelsey Creek sediments, indicating their previous presence in the creek and their inability to adjust to the changing conditions.
- The benthic organism composition in Kelsey Creek varied radically with time and place while the organisms were much more stable in Bear Creek.

Degradation of Habitat and Biological Conditions, Possible Causes

- These aquatic organism differences were probably mostly associated with the increased peak flows in Kelsey Creek caused by urbanization and the resultant increase in sediment-carrying capacity and channel instability of the creek.
- There was also the potential for accumulation of toxic materials in the stream system affecting aquatic organisms, but only low concentrations of toxic materials were found in the receiving waters.
- The concentrations of dissolved oxygen in the urban creek's gravel waters were quite low and may have decreased the survival of salmon embryos. In-stream embryo bioassays indicated that

coho embryo salmon survival was significantly greater in Bear Creek than in Kelsey Creek, but no difference was found when using rainbow trout embryos.

- Direct receiving water effects from urban runoff may not have been significant for most storms. Potential long-term problems, however, may be associated with settleable solids, lead, and zinc. These settled materials may have silted up spawning beds and introduced high concentrations of potentially toxic materials directly to the sediments. The oxygen depletion observed in the interstitial waters was probably caused by organic sediment buildup from runoff events.
- Kelsey Creek had much lower flows than Bear Creek during periods between storms. About 30% less water was available in Kelsey Creek during the summers, even though both creeks have drainage basins of similar size, rainfall characteristics, and soils. These low flows may also have significantly affected the aquatic habitat and the ability of the urban creek to flush toxic spills or other dry-weather pollutants from the creek system.
- Kelsey Creek had higher water temperatures (probably due to reduced shading) than Bear Creek. This probably caused the faster fish growth in Kelsey Creek.

Conveyance of Stormwater

- Kelsey Creek had extreme hydrologic responses to storms. Flooding substantially increased in Kelsey Creek during the period of urban development; the peak annual discharges have almost doubled in the last 30 years, and the flooding frequency has also increased due to urbanization.
- These increased flows in urbanized Kelsey Creek resulted in greatly increased sediment transport and channel instability.

Open Space and Resource Preservation Beneficial Uses

- The lack of adequate buffer zones and natural creek banks along much of the urban reaches of Kelsey Creek is balanced by extensive park system developments along selected reaches. Natural creek reaches are very important for the aquatic organisms in Kelsey Creek.
- Creek bank-side homeowners have made extensive channel and riparian vegetative changes, which significantly reduced the ability of the creek to support aquatic life.

Recreational Beneficial Uses

- The natural small size of Kelsey Creek restricts its usefulness for most water contact-related activities, although swimming does occur in the lower reaches of Kelsey Creek during the summer.
- The fecal coliform bacteria counts in Kelsey Creek were high and variable. These organisms indicate the potential presence of pathogenic bacteria and commonly exceeded water contact numeric criteria.

Aesthetics Beneficial Uses

- This use is related to most of the above uses; unsightly creeks are not utilized in educational field trips or as swimming areas, or desired as amenities to property.
- Dead fish from periodic toxic material spills significantly degrade this use.
- Debris and unstable channels also adversely affect the aesthetic quality of Kelsey Creek.

Sources of Increased Flows and Pollutants

- For all rains greater than about 2.5 mm (0.1 in), the impervious surfaces (streets, sidewalks, driveways, parking lots, and rooftops) were found to contribute more than 60% of the total urban runoff flows. The remainder of the flows were approximately evenly divided between front and back yards, while vacant lots and parks contributed very little to the flows due to their limited presence in the test areas. For most of the rain events monitored, the street surfaces contributed about 25% of the total urban runoff flows.

- Most of the total solids in urban runoff originated from front and back yards in the test areas. The street surfaces contributed only a small fraction to the total solids of urban runoff discharges. Lead, zinc, and COD, however, were mostly contributed from street surfaces. Nutrients (phosphorus and total Kjeldahl nitrogen) were found to originate mostly from street surfaces, driveways, and parking lots combined.
- Pesticides were only found in the residential street dirt samples, and not in the arterial, commercial, or industrial street dirt samples. The arterial street dirt samples had much higher concentrations of lead, most likely due to increased automobile activity.
- Many organic priority pollutants were detected in the soil samples. The most important organics found were the polycyclic aromatic hydrocarbons (PAHs), which were frequently detected in the street dirt samples and the Kelsey Creek sediment samples.
- Motor vehicle activity was expected to be the primary contributor of most of the toxic organic and inorganic priority pollutants. Gasoline and diesel fuel combustion products, lubricant and fuel leakages, and wear of the vehicles affected the street dirt material most significantly.
- Almost as much of the street dirt was lost to the air, as suspended particulates, as was washed off during rain events.
- Only a small fraction of the total particulate loadings on the impervious surfaces was removed by the rains (about 15%). Large particles were not effectively removed, while about one half of the smallest particles (less than 50 μm) were washed off during rains. These small particles were not very abundant, but had very high heavy metal and nutrient concentrations.
- Most of the settled particulate material in the storm drainage inlets and sewerage pipes was not removed by the observed storms.

Control of Urban Runoff by Street and Storm Drainage Inlet Cleaning and by Dry Detention Ponds

- Intensive street cleaning (three times a week) resulted in rapid and significant decreases in street surface loadings; from about 110 g/curb-meter down to about 55 g/curb-meter. The median particle sizes also decreased significantly with intensive street cleaning. A regenerative air street cleaner showed substantially better performance in removing the finer street surface materials than the regular mechanical street cleaner.
- Extensive data analysis did not show any significant improvements in runoff water quality during periods of intensive street cleaning. The street cleaning operations tested are only expected to improve runoff quality by a maximum of about 10%. The street cleaning equipment preferentially removed the larger particle sizes, while the rain events preferentially removed the finer materials. Street cleaning was not very effective in removing the particulates available for washoff.
- Mechanical broom street cleaning was effective in removing the larger litter from the streets.
- Infrequent street cleaning may result in significant increases in fugitive dust losses to the atmosphere.
- After an initial cleaning, it required almost a full year for sediment to reach a stable volume in the inlet structures. Only about 60% of the total available sump volumes in inlets and catchbasins was used for detention of particulates. Cleaning the inlets and catchbasin sumps about twice a year was expected to reduce the lead and total solids urban runoff concentrations by between 10 and 25%. COD, the nutrients, and zinc might be controlled between 5 and 10%.
- The small detention basin tested (detention time of 30 min or less) did not have any significant effect on urban runoff quality.
- The small detention basin did have a significant effect on the peak flow rates. The peak flow rates were reduced by about 60%.

Summary

The Bellevue studies indicated the very significant interrelationships between the physical, biological, and chemical characteristics of the urbanized Kelsey Creek system. The aquatic life beneficial uses were found to be impaired and stormwater conveyance was found to be significantly stressed by urbanization. These degradations were most likely associated with increased flows from the impervious areas in the urban area. Changes in the flow characteristics could radically alter the ability of the stream to carry the polluted sediments into the other receiving waters. If the stream power of Kelsey Creek was reduced, then these toxic materials could be expected to be settled into its sediment, with increased effects on the stream's aquatic life. Reducing peak flows would also reduce the flushing of smaller fish and other aquatic organisms from the system.

If detention basins were used to control peak flows, they would have to be carefully located and designed so that increased flow rates did not occur in downstream areas. The placement of flow-modifying structures throughout the watershed could significantly affect the response time of the watershed to rain events, with possible resultant increases in downstream peak flows.

It was found that substantial quantities of water originated from the impervious areas in the developed areas. More careful planning to increase the perviousness of these areas should also be considered.

Another recommendation is to preserve any of the refuge areas in Kelsey Creek and to carefully design any channelization project to include refuge areas for the aquatic life. Because of the larger potential for sedimentation of toxic pollutants in Kelsey Creek, increased awareness of the beneficial uses and undesirable discharges to the drainage system will be more important. The large assimilative capacity of the water bodies that currently receive most of these pollutants are currently masking this concern.

Many recommendations concerning the public works practices in the Bellevue area can also be made based on this project. However, their effects on improving the urban runoff quality would probably be quite small. If intensive street cleaning was implemented, along with semiannual catchbasin sediment cleaning, urban runoff discharges for most pollutants would be reduced by about 10%, while some of the heavy metal discharges may be reduced by as much as 25%. Even though these reductions are quite small, they may be important to reduce the accumulation of these highly polluted sediments in the smaller creek systems, especially if peak flushing flows are reduced.

Critique of Parallel Stream Analyses in Bellevue

The Bellevue, WA, NURP project included many in-stream measurements to compare the test Kelsey Creek with the control Bear Creek. The study included numerous physical and biological measurements. In addition, in-stream toxicity tests were conducted. This large research program included numerous components. As for the Coyote Creek study, this program was likely much larger than needed. Newer tools and the use of efficient indicators could have reduced the sampling and analytical effort. The very large number of storms evaluated and the long-term stream studies were extremely enlightening, but similar conclusions could have been obtained through less expensive means. Again, this was one of the first comprehensive receiving water studies conducted, and there was little guidance to indicate what to expect.

The numerous researchers and different institutions conducting this research program indicated numerous communication and coordination problems, especially concerning preliminary conclusions. Most of the researchers were reluctant to share their results with the other groups until they had completed their thorough evaluations. If better communications were practiced, efficient modifications to the field activities would have been possible. However, the many experts involved in this research program resulted in a very important multidisciplinary study that would not have been possible with a smaller team of researchers.

In general, parallel stream investigations can be expanded well beyond a two-stream comparison by including numerous streams having variable levels of development. This has been a common experimental design for recent receiving water investigations. However, it is still important to conduct the study over a long duration and in numerous locations to best understand the dynamics of the systems. In many cases, in-stream variations can easily mask differences between streams.

Figure 4.23 Drawing showing underwater features of an FBM facility. (Used with permission of Fresh Creek Technologies, Inc.)

Figure 4.24 FBM installation located at Lake Trehormingen, Sweden. (Used with permission of Fresh Creek Technologies, Inc.)

Example of Long-Term Trend Experimental Design — Lake Rönningesjön, Sweden, Receiving Water Study

An example showing the use of trend analyses for investigating receiving water effects of stormwater is presented here, using a Swedish lake example that has undergone stormwater treatment (Pitt 1995a). The significant beneficial use impairment issue is related to decreasing transparency due to eutrophication. The nutrient enrichment was thought to have been aggravated by stormwater discharges of phosphorus. Stormwater treatment was shown to decrease the phosphorus discharges in the lake, with an associated increase in transparency. The data available include nutrient, chlorophyll *a*, transparency, and algal evaluations conducted over a 20- to 30-year period, plus treatment plant performance information for 10 years of operation. This trend evaluation was conducted by Pitt (1995a) using data collected by Swedish researchers, especially Enell and Henriksson-Fejes (1989–1992).

A full-scale plant, using the Karl Dunkers' system for treatment of separate stormwater (the Flow Balancing Method, or FBM) and lake water, has been operating since 1981 in Lake Rönningesjön, Taby (near Stockholm), Sweden. The FBM and the associated treatment system significantly improved lake water quality through direct treatment of stormwater and by pumping lake water through the treatment system during dry weather. Figure 4.23 is an illustration of an idealized FBM system showing how inflowing stormwater is routed though a series of interconnected compartments, before being discharged to the lake. A pump can also be used to withdraw water from the first compartment to a treatment facility. Figure 4.24 is a photograph of an FBM installation located at Lake Trehormingen, Sweden. Figure 4.25 shows wetland vegetation growing in one of the compartments of the FBM at Lake Rönningesjön, while Figure 4.26 shows the building containing the chemical treatment facility at the Lake Rönningesjön facility.

The annual average removal of phosphorus from stormwater and lake water by the ferric chloride precipitation and clarification treatment system was 66%, while the annual average total lake

Figure 4.25 Wetland vegetation growing in FBM cell at Lake Rönningesjön, Sweden. (Used with permission of Fresh Creek Technologies, Inc.)

Figure 4.26 Chemical treatment facility at FBM installation at Lake Rönningesjön, Sweden. (Used with permission of Fresh Creek Technologies, Inc.)

phosphorus concentration reductions averaged about 36%. Excess flows are temporarily stored in the FBM before treatment. Stormwater is pumped to the treatment facility during rains, with excess flows stored inside in-lake flow-balancing tanks. The treatment system consists of a chemical treatment system designed for the removal of phosphorus and uses ferric chloride precipitation and crossflow lamella clarifiers. The stormwater is pumped from the flow-balancing storage tanks to the treatment facility. Lake water is also pumped to the treatment facility during dry periods, after any excess stormwater is treated.

Step 1. What's the Question?

The specific question to be addressed by this research was whether controlling phosphorus in stormwater discharges to a lake would result in improved lake water quality. Secondly, this evaluation was made to determine if the treatment system was designed and operated satisfactorily.

Step 2. Decide on Problem Formulation

The problem formulation employed for this project was a long-term trend analysis. Up to 30 years of data were available for some water quality parameters, including about 10 years of observations before the treatment system was implemented. Data were available for two sampling locations in the lake, plus at the stormwater discharge location. In addition, mass balance data were available for the treatment operation.

Monitored water quality in Lake Rönningesjön, near Stockholm, Sweden, was evaluated to determine the changes in transparency and nutrient concentrations associated with retrofitted stormwater controls. Statistical trend analyses were used to evaluate these changes. Several publications have excellent descriptions of statistical trend analyses for water quality data. In addition to containing detailed descriptions and examples of experimental design methods to determine required sampling effort, Gilbert (1987) devotes a large portion of his book to detecting trends in water quality data and includes the code for a comprehensive computer program for trend analysis. That information and other experimental design issues on conducting a trend investigation are briefly reviewed in Chapter 7 of this book.

Step 3. Project Design

Qualitative Watershed and Lake Characterization

Lake Rönningesjön is located in Taby, Sweden, near Stockholm. Figure 4.27 shows the lake location, the watershed, and the surrounding urban areas. The watershed area is 650 ha, including Lake Rönningesjön itself (about 60 ha) and the urban area that has its stormwater drainage bypassing the lake (about 175 ha). The effective total drainage area (including the lake surface) is therefore about 475 ha. Table 4.24 summarizes the land use of the lake watershed area. About one half of the drainage area (including the lake itself) is treated by the treatment and storage operation.

The lake volume is about 2,000,000 m³ and the lake has an annual outflow of about 950,000 m³. The estimated mean lake resident time is therefore slightly longer than 2 years. The average lake depth is 3.3 m. It is estimated that rain falling directly on the lake surface contributes about one half of the total lake outflow.

The treatment process consists of an in-lake flow-balancing storage tank system (the Flow Balancing Method, or FBM) to contain excess stormwater flows which are pumped to a treatment facility during dry weather. The treatment facility uses ferric chloride and polymer precipitation and crossflow lamella clarifiers.[*] Figure 4.28 shows the cross section of the FBM in the lake. It is made of plastic curtains forming the cell walls, supported by floating pontoons and anchored to the lake bottom with weights.

Figure 4.29 shows that the FBM provides storage of contaminated water by displacing clean lake water that enters the storage facility during dry weather as the FBM water is pumped to the

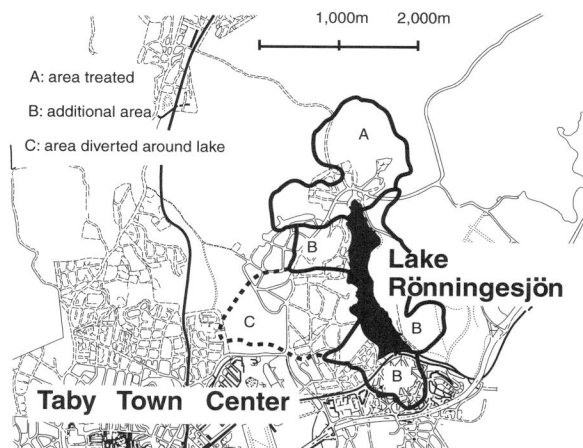

Figure 4.27 Lake Rönningesjön watershed in Taby, Sweden. (From Pitt 1995a. Used with permission of Fresh Creek Technologies, Inc.)

Table 4.24 Lake Rönningesjön Watershed Characteristics

	Area Treated, ha	Additional Area, ha	Total Area, ha
Urban	50	100	150 (32%)
Forest	75	80	155 (32%)
Agriculture	65	45	110 (23%)
Lake surface	60	0	60 (13%)
Total drainage	250	225	475 (100%)

From Pitt 1995a.

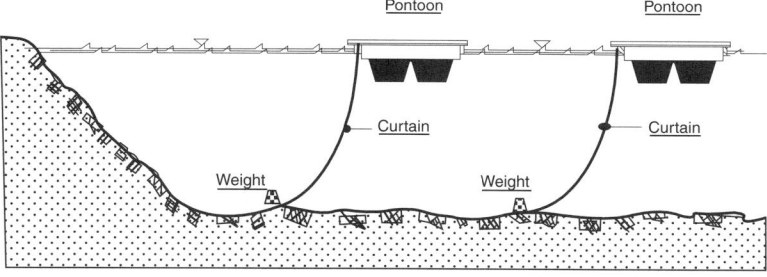

Figure 4.28 Cross section of FBM in-lake tanks. (From Pitt 1995a. Used with permission of Fresh Creek Technologies, Inc.)

treatment system. All stormwater enters the FBM directly (into cell A). The pump continuously pumps water from cell A to the chemical treatment area. If the stormwater enters cell A faster than the pump can remove it, portions of the stormwater flows through curtain openings (as a slug flow) into cells B, C, D, and finally E, displacing lake water (hence the term *flow balancing*). As the pump continues to operate, stormwater is drawn back into cell A and then to the treatment facility. The FBM is designed to capture the entire runoff volume of most storms. The Lake Rönningesjön treatment system is designed to treat water at a higher rate than normal to enable lake water to be pumped through the treatment system after all the runoff is treated.

The FBM is mainly intended to be a storage device, but it also operates as a wet detention pond, resulting in sedimentation of particulate pollutants within the storage device. The first two cells of the FBM facility at Lake Rönningesjön were dredged in 1991, after 10 years of operation, to remove about 1 m of polluted sediment.

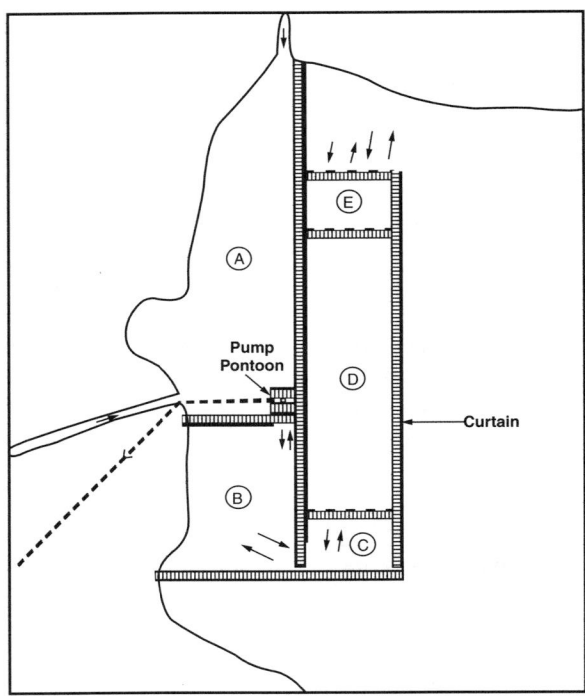

Figure 4.29 Flow pattern in FBM. (From Pitt 1995a. Used with permission of Fresh Creek Technologies, Inc.)

Table 4.25 Stormwater Treatment System Operating Cost Breakdown

Chemicals	26%
Electricity	8
Sludge transport	3
Labor	41
Sampling and analyses	22

From Pitt 1995a.

The treatment flow rate is 60 m³/hour (about 0.4 MGD). The ferric chloride feed rate is about 20 to 35 g/m³ of water. About 30 m³ of thickened sludge is produced per day for co-disposal with sludge produced at the regional sanitary wastewater treatment facility. The annual operating costs are about $28,000 per year (or about $0.03 per 100 gallons of water treated), as shown in Table 4.25.

From 1981 through 1987, the FBM operated an average of about 5500 hours per year (about 7.6 months per year), treating an average of about 0.33 million m³ per year. The treatment period ranged from 28 to 36 weeks (generally from April through November). The FBM treatment system treated stormwater about 40% of its operating time and lake water about 60% of its operating time. The FBM treatment system directly treated about one half of the waters flowing into the lake (at a level of about 70% phosphorus removal).

Lake Rönningesjön and Treatment System Phosphorus Budgets

Two tributaries flow directly to the treatment facility. Excess flows (exceeding the treatment plant flow capacity) are directed to the FBM in the lake. As the flows in the tributaries fall below the treatment plant capacity, pumps in the FBM deliver stored stormwater runoff for treatment. When all of the stormwater is pumped from the FBM, the pumps deliver lake water for treatment. Tables 4.26 and 4.27 summarize the runoff and lake volumes treated and phosphorus removals during the period of treatment.

Table 4.26 Water Balance for Treatment System (m³)

	From Trib. A	From Trib. B	Total Stormwater	From Lake	Total Treated and Discharged	Stormwater, % of Total Treated
1981	185,100	101,100	286,200	121,600	407,700	70
1982	112,700	41,000	153,700	238,700	391,900	39
1983	14,400	6400	20,800	250,000	271,000	8
1984	122,000	53,000	175,000	95,000	270,000	65
1985	96,600	46,500	143,100	149,000	292,400	49
1986	216,000	86,000	302,000	48,000	350,000	86
1987	243,000	97,000	340,000	13,000	353,000	96
1988	26,200	19,300	45,500	186,300	231,800	20
1989	24,900	19,900	44,800	267,700	312,500	14
1990	12,160	8,330	20,490	201,270	221,760	9
1991	11,610	7780	19,390	121,730	141,120	14

From Pitt 1995a.

Table 4.27 Phosphorus Treatment Mass Balance (kg)

	From Trib. A	From Trib. B	From Lake	Total to Treatment	P Discharged to Lake	P Removal	% Removal
1981	20.3	16.8	10.2	47.3	13.6	33.7	71.2
1982	8.0	8.0	18.0	34.0	12.8	21.2	62.4
1983	1.5	2.5	20.0	24.0	11.0	13.0	54.2
1984	10.0	9.5	3.0	22.5	10.0	12.5	55.6
1985	7.1	5.9	2.1	15.1	4.3	10.8	71.5
1986	15.2	21.4	3.7	40.3	5.1	35.2	87.3
1987	18.6	7.5	1.7	27.8	4.3	23.5	84.5
1988	1.7	2.3	9.2	13.2	6.1	7.1	53.8
1989	1.7	1.4	14.1	17.2	7.6	9.6	55.8
1990	1.3	0.3	10.5	12.1	3.7	8.4	69.4
1991	7.7	9.8	5.6	23.1	8.9	14.2	61.5

From Pitt 1995a.

There have been highly variable levels of phosphorus treatment from stormwater during the period of operation. The years from 1988 through 1990 had low phosphorus removals. These years had relatively mild winters with substantial stormwater runoff occurring during the winter months when the treatment system was not operating. Normally, substantial phosphorus removal occurred with spring snowmelt during the early weeks of the treatment plant operation each year. The greatest phosphorus improvements in the lake occurred during the years when the largest amounts of stormwater were treated.

The overall phosphorus removal rate for the 11 years from 1981 through 1991 was about 17 kg/year. About 40% of the phosphorus removal occurred in the FBM from sedimentation processes, while the remainder occurred in the chemical treatment facility. This phosphorus removal would theoretically cause a reduction in phosphorus concentrations of about 10 μg/L per year in the lake, or a total phosphorus reduction of about 100 μg/L during the data period since the treatment system began operation. About 70% of this phosphorus removal was associated with the treatment of stormwater, while about 30% was associated with the treatment of lake water.

Select Monitoring Parameters

Lake Rönningesjön water quality has been monitored since 1967 by the Institute for Water and Air Pollution Research (IVL); the University of Technology, Stockholm; the Limnological Institute at the University of Uppsala; and by Hydroconsult Corp. Surface and subsurface samples were obtained at one or two lake locations about five times per year. In addition, the tributaries being treated, incoming lake water, and discharged water were all monitored on all weekdays of treatment plant operation. The creek tributary flow rates were also monitored using overflow weirs. Phosphorus, nitrogen, chlorophyll *a*, and Secchi disk transparency were all monitored at the lake stations.

Step 4. Project Implementation, Step 5. Data Evaluation, and Step 6. Confirmatory Assessment

Observed Long-Term Lake Rönningesjön Water Quality Trends

The FBM started operation in 1981. Based on the hydraulic detention time of the lake, several years would be required before a new water quality equilibrium condition would be established. A new water quality equilibrium will eventually be reached after existing pollutants are reduced from the lake water and sediments. The new water quality conditions would be dependent on the lake flushing rate (or detention time, estimated to be about 2.1 years), and the new (reduced) pollutant discharge levels to the lake. Without lake water treatment, the equilibrium water quality would be worse and would take longer to obtain.

Figure 4.30 is a plot of all chlorophyll *a* data collected at both the south and north sampling stations. Very little trend is obvious, but the wide swings in chlorophyll *a* values appeared to have been reduced after the start of stormwater treatment. Figure 4.31 is a three-dimensional plot of

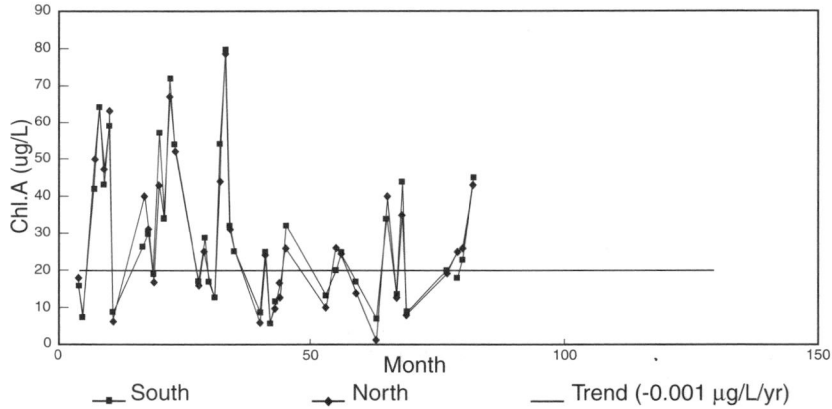

Figure 4.30 Chlorophyll *a* observations with time (μg/L). (From Pitt 1995a.)

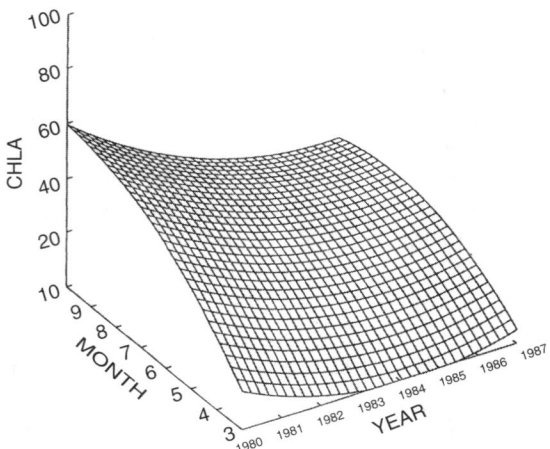

Figure 4.31 Chlorophyll *a* trends by season and year (μg/L). (From Pitt 1995a.)

smoothed chlorophyll *a* data, indicating significant trends by season. The values started out relatively low each early spring and dramatically increased as the summer progressed. This was expected and was a function of algal growth. Homogeneity, seasonal Kendall, and Mann–Kendall statistical tests (Gilbert 1987) were conducted using the chlorophyll *a* data. The homogeneity test was used to determine if any trends found at the north and south sampling stations were different. The probabilities that the trends at these two stations were the same were calculated as follows:

	χ^2	Probability
Season	14.19	0.223
Station	0.00001	1.000
Station–season	0.458	1.000
Trend	21.64	0.000

This test shows that the trend was very significant (P < 0.001) and was the same at both sampling stations (P = 1.000). The seasonal trend tests only compared data obtained for each season, such as comparing trends for June observations alone. The station-season interaction term shows that the chlorophyll *a* concentration trends at the two stations were also very similar for all months (P = 1.000). Therefore, the sampling data from both stations were combined for further analyses.

The seasonal Kendall test calculated the chlorophyll *a* concentration trends and determined the probabilities that they were not zero, for all months separately. This test and the Mann–Kendall tests found that both the north and south sampling locations had slight decreasing (but very significant) overall trends in concentrations with increasing years (P ≤ 0.001). However, individual monthly trends were not very significant (P ≥ 0.05). The trends do show an important decrease in the peak concentrations of chlorophyll *a* that occurred during the fall months during the years of the FBM operation. The 1980 peak values were about 60 µg/L, while the 1987 peak values were lower, at about 40 µg/L.

Swedish engineers (Söderlund 1981; Lundkvist and Söderlund 1988) summarized major changes in the algal species present and in the algal biomass in Lake Rönningesjön, corroborating the chlorophyll *a* and phosphorus-limiting nutrient observations. From 1977 through 1983, the lake was dominated by a stable population of thread-shaped blue-green algae species (especially *Oscillatoria* sp. and *Aphanizomenon flos aquae f. gracile*). Since 1985, the algae population has been unstable, with only a small amount of varying blue-green (*Gomphosphaeria*), silicon (*Melosira*, *Asterionella*, and *Synedra*), and gold (*Chrysochromulina*) algae species. They also found a substantial decrease in the algal biomass in the lake. From 1978 through 1981, the biomass concentration was commonly greater than 10 mg/L. The observed maximum was about 20 mg/L, with common annual maximums of 15 mg/L in July and August of each year. From 1982 through 1986, the algal biomass was usually less than 10 mg/L. The observed maximum was 14 mg/L and the typical annual maximum was about 6 mg/L each late summer. The lake showed an improvement in its eutrophication level since the start of stormwater treatment, going from hypotrophic to eutrophic.

Figure 4.32 is a plot of all Secchi disk transparency data obtained during the project period. A very large improvement in transparency is apparent from this plot, but large variations were observed in most years. A large improvement may have occurred in the first 5 years of stormwater treatment and then the trend may have decreased. The smoothed plot in Figure 4.33 shows significant improvement in Secchi disk transparency since 1980. This three-dimensional plot shows that the early years started off with clearer water (as high as 1 m transparency) in the spring and then degraded as the seasons progressed, with transparency levels decreasing to less than 0.5 m in the fall. The later years indicated a significant improvement, especially in the later months of the year.

Homogeneity, seasonal Kendall, and Mann–Kendall statistical tests (Gilbert 1987) were conducted using the Secchi disk transparency data. The homogeneity test was used to determine if any

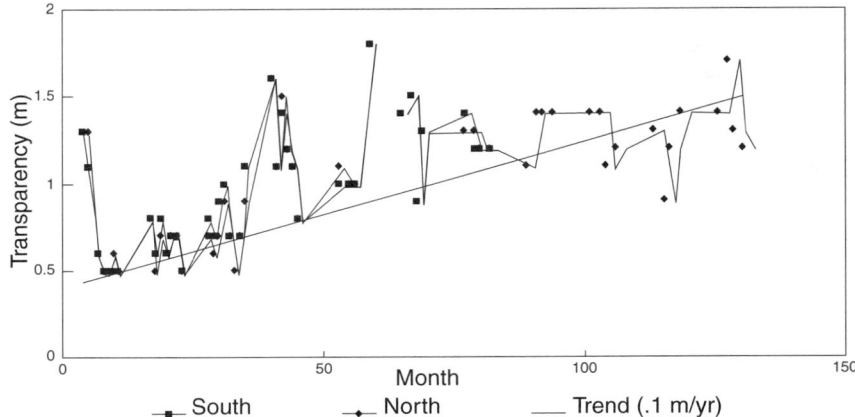

Figure 4.32 Secchi disk transparency observations with time (m). (From Pitt 1995a.)

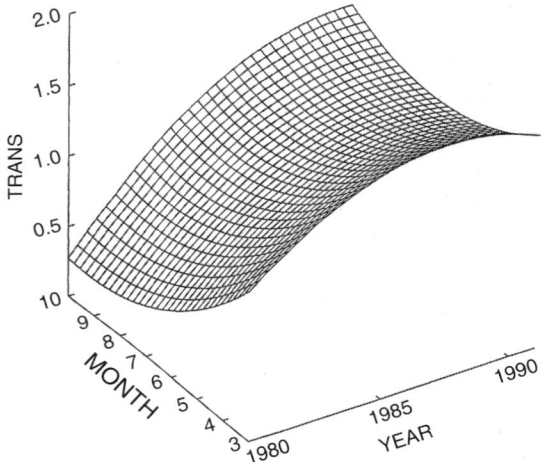

Figure 4.33 Secchi disk trends by season and year (m). (From Pitt 1995a.)

trends found at the north and south sampling stations were different. The probabilities that the trends at these two stations were the same were calculated as follows:

	χ^2	**Probability**
Season	17.15	0.103
Station	0.012	0.913
Station–season	3.03	0.990
Trend	29.44	0.000

These statistics show that the observed trend was very significant (P < 0.001) and was the same at both stations. The seasonal Kendall and Mann–Kendall tests found that both the north and south sampling locations had increasing transparency values (the average trend was about 0.11 m per year) with increasing years (P < 0.001). The trend in later years was found to be less than in the early years. The transparency has remained relatively stable since about 1987 (ranging from about 1 to 1.5 m), with less seasonal variation.

Figure 4.34 plots observed phosphorus concentrations with time, while Figure 4.35 is a smoothed plot showing seasonal and annual variations together. The initial steep decreases in

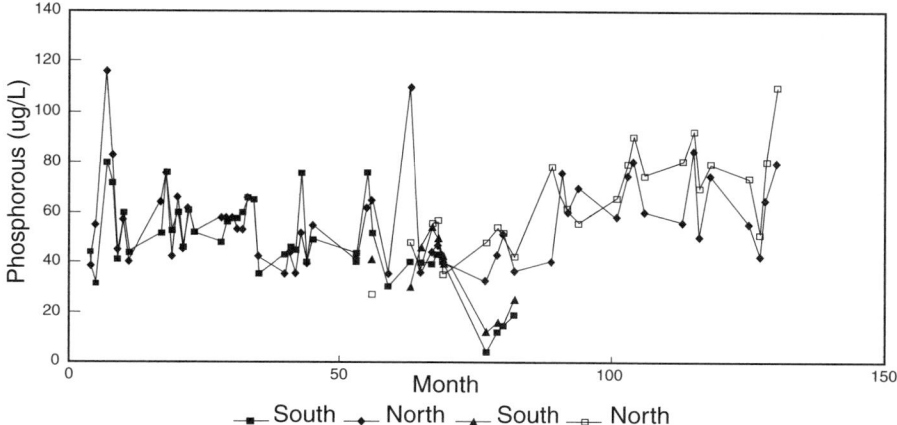

Figure 4.34 Total phosphorus observations with time (µg/L). (From Pitt 1995a.)

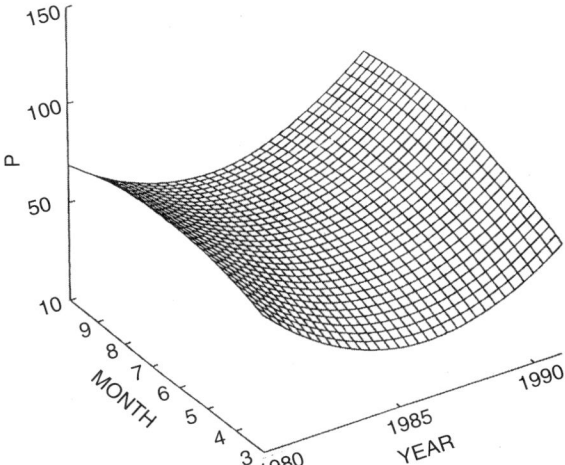

Figure 4.35 Total phosphorus trends by season and year (µg/L). (From Pitt 1995a.)

phosphorus concentration in the early years of the FBM operation were followed by a sharp increase during later years. The increase was likely associated with the decreased levels of stormwater treatment during the mild winters of 1988 through 1990 when the treatment system was not operating; large amounts of untreated stormwater were discharged into the lake instead of being tied up as snow to be treated in the spring as snowmelt runoff.

Individual year phosphorus concentrations leveled off in the summer (about July). These seasonal phosphorus trends were found to be very significant ($P \leq 0.002$), but were very small, using the seasonal Kendall test (Gilbert 1987). Homogeneity tests found no significant differences between lake sample phosphorus concentrations obtained at the different sampling locations, or depths, irrespective of season:

	χ^2	**Probability**
Season	15.38	0.166
Station	0.0033	0.954
Station–season	1.64	0.999
Trend	12.43	0.000

The overall lake phosphorus concentrations ranged from about 15 to 130 µg/L, with an average of about 65 µg/L. The monitored stormwater, before treatment, had phosphorus concentrations ranging from 40 to >1000 µg/L, with an average of about 200 µg/L.

An increase in nitrogen concentrations also occurred from the beginning of each year to the fall months. The overall annual trend decreased during the first few years of the FBM operation, but it then subsequently increased. These total nitrogen concentration variations were similar to the total phosphorus concentration variations. However, homogeneity, seasonal Kendall, and Mann–Kendall statistical tests (Gilbert 1987) conducted using the nitrogen data found that neither the north nor south sampling locations had significant concentration trends with increasing years ($P > 0.2$). However, lake Kjeldahl nitrogen concentration reductions were found to occur during years when the FBM system was treating the largest amounts of stormwater.

Lake Water Quality Model

A simple water quality model was used with the Lake Rönningesjön data to determine the total annual net phosphorus discharges into the lake and to estimate the relative magnitude of various in-lake phosphorus-controlling processes (associated with algal growth and sediment interactions, for example). These estimated total phosphorus discharges were compared to the phosphorus removed by the treatment system. The benefits of the treatment system on the lake water quality were then estimated by comparing the expected lake phosphorus concentrations (as if the treatment system was not operating) to the observed phosphorus concentrations.

Thomann and Mueller (1987) presented the following equation to estimate the resulting water pollutant concentrations associated with varying input loadings for a well-mixed lake:

$$S_t = (M/V) \exp(-T/Td) \tag{4.1}$$

where S_t = concentration associated with a step input at time t
M = mass discharge per time-step interval (kg)
V = volume of lake (2,000,000 m^3)
T = time since input (years)
Td = hydraulic residence time, or lake volume/lake outflow (2.1 years)

This equation was used to calculate the yearly total mass discharges of phosphorus to Lake Rönningesjön, based on observed lake concentrations and lake hydraulic flushing rates. It was assumed that the varying concentrations observed were mostly caused by varying mass discharges and much less by variations in the hydraulic flushing rate. The flushing rate was likely to vary, but by relatively small amounts. The lake volume was quite constant, and the outflow rate was expected to vary by less than 20% because of the relatively constant rainfall that occurred during the years of observation (average rainfall of about 600 mm, with a coefficient of variation of about 0.15).

The total mass of phosphorus discharged into the lake each year from 1972 to 1991 was calculated using the following equation (an expansion of Equation 4.1), solving for the M_{n-x} terms:

$$S_n = M_n\left[\exp(-T_n/Td)/V\right] + M_{n-1}\left[\exp(-T_{n-1}/Td)/V\right] + M_{n-2}\left[\exp(-T_{n-2}/Td)/V\right]$$
$$+ M_{n-3}\left[\exp(-T_{n-3}/Td)/V\right] + \cdots \tag{4.2}$$

where S_n is the annual average phosphorus concentration during the current year, M_n is the net phosphorus mass discharged into the lake during the current year, M_{n-1} is the phosphorus mass discharged during the previous year, M_{n-2} is the phosphorus mass that was discharged 2 years previously, etc.

The effects of discharges into the lake many years earlier have little effect on the current year's observations. Similarly, more recent discharges have greater effects on the lake's concentrations. The magnitude of effect that each year's step discharge has on a more recent concentration observation is dependent on the $exp(-T_n/Td)$ factors shown in Equation 4.2. A current year's discharge affects that year's concentration observations by about 40% of the steady-state theoretical value (M/V), and a discharge from 5 years earlier would affect the current year's concentration observations by less than 10% of the theoretical value for Lake Rönningesjön. Similarly, a new steady-state discharge would require about 4 years before 90% of its equilibrium concentration would be obtained. It would therefore require several years before the effects of a decrease in pollutant discharges would have a major effect on the lake pollutant concentrations.

The annual control of phosphorus ranged from about 10 to 50%, with an average lake-wide level of control of about 36%, during the years of treatment plant operation. It is estimated that there would have been about a 1.6 times increase in phosphorus discharges into Lake Rönningesjön if the treatment system was not operating. There was a substantial variation in the year-to-year phosphorus discharges, but several trends were evident. If there was no treatment, the phosphorus discharges would have increased over the 20-year period from about 50 to 75 kg/year, associated with increasing amounts of contaminated stormwater, in turn associated with increasing urbanization in the watershed. With treatment, the discharges were held relatively constant at about 50 kg/year (as evidenced by the lack of any observed phosphorus concentration trend in the lake). During 1984 through 1987, the phosphorus discharges were quite low compared to other years, but increased substantially in 1988 and 1989 because of the lack of stormwater treatment during the unusually mild winters.

Figure 4.36 is a plot of the annual average lake phosphorus concentrations with time. If there had been no treatment, the phosphorus concentrations in the lake would have shown a relatively

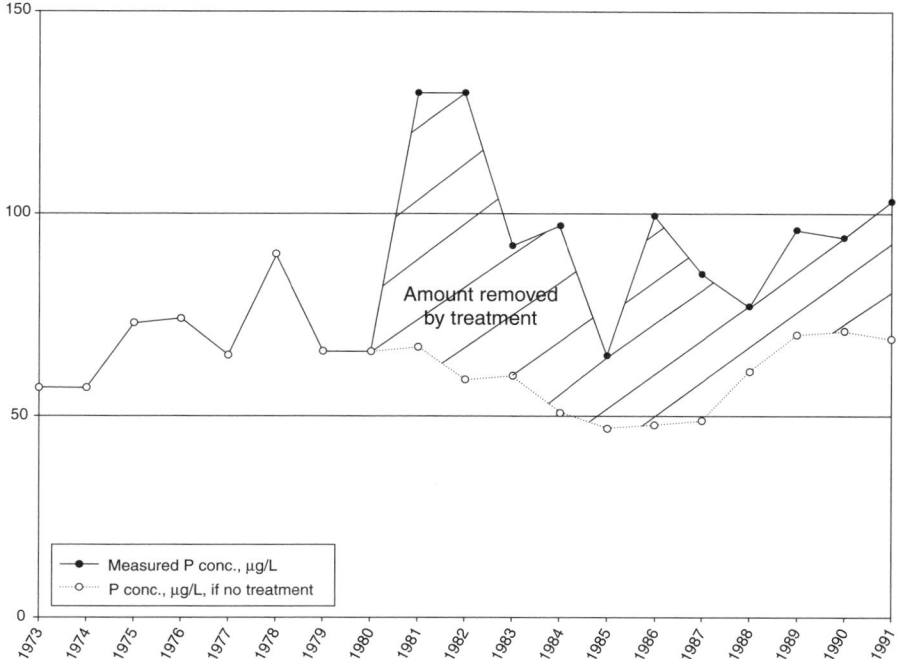

Figure 4.36 Effects of treatment on Lake Rönningesjön total phosphorus concentrations (µg/L). (From Pitt 1995a.)

steady increase from about 50 to about 100 µg/L over the 20-year period. With treatment, the lake phosphorus concentrations were held within a relatively narrower range (from about 50 to 75 µg/L). The lake phosphorus concentration improvements averaged about 50 µg/L over this period of time, compared to an expected theoretical improvement of about 100 µg/L. Therefore, only about one half of the theoretical improvement occurred, probably because of sediment-water interchange of phosphorus, or other unmeasured phosphorus sources.

Step 7. Project Conclusions

The in-lake flow-balancing method (FBM) for storage of excess stormwater during periods of high flows allowed for lower treatment flow rates, while still enabling a large fraction of the stormwater to be treated for phosphorus removal. The treatment system also enabled lake water to be treated during periods of low (or no) stormwater flow. The treatment of the stormwater before lake discharge accounted for about 70% of the total observed phosphorus discharge reductions, while the lake water treatment was responsible for the remaining 30% of the discharge reductions. The lake water was treated during 60% of the operating time, but resulted in less phosphorus removal, compared to stormwater treatment. The increased efficiency of phosphorus removal from stormwater compared to lake water was likely due to the more abundant particulate forms of phosphorus that were removed in the FBM by sedimentation and by the stormwater's higher dissolved phosphorus concentrations that were more efficiently removed during the chemical treatment process.

Lake transparency improved with treatment. Secchi disk transparencies were about 0.5 m before treatment began and improved to about 1 to 1.5 m after treatment. The total phosphorus concentrations ranged from about 65 to 90 µg/L during periods of low levels of stormwater treatment, to about 40 to 60 µg/L during periods of high levels of stormwater treatment.

The annual average removal of phosphorus by the ferric chloride precipitation and clarification treatment system was 66%, with a maximum of 87%. The observed phosphorus concentration improvements in the lake were strongly dependent on the fraction of the annual stormwater flow that was treated. The annual average total lake phosphorus discharge and concentration reductions averaged about 36%, or about one half the maximum expected benefit.

Critique of the Trend Analyses at Lake Rönningesjön

The water sampling for this project was irregular. Only a relatively few samples were obtained in any one year, but up to 30 years of data were obtained. In addition, no winter data were available due to icing of the lake. In general, statistically based trend analyses are more powerful with evenly spaced data over the entire period of time. However, this is typically unrealistic in environmental investigations because of an inability to control other important factors. If all samples were taken on the 15th of each month, for example, the samples would be taken under highly variable weather conditions. Weather is a significant factor in urban runoff studies, obviously, and this statistical methodology requirement would have severely confounded the results. The trend analyses presented by Gilbert (1987) enable a more reasonable sample collection effort, with some missing data. However, the procedure does require relatively complete data collected over an extended period of time. It would have been very difficult to conduct this analysis with only a few years of data, for example. The seasonal patterns were very obvious when multiple years of before and after treatment were monitored. In addition, the many years of data enabled unusual weather conditions (such as the years with unusually mild winters) to stand out from the more typical weather conditions.

The analytical effort only focused on a few parameters. This is acceptable for a well-designed and executed project, but prohibits further insights that a more expansive effort may obtain. Since this project was specifically investigating transparency-associated eutrophication, the parameters evaluated enabled the basic project objectives to be effectively evaluated. However, the cost of labor for the sampling effort is a major component of an investigation like this one, and some additional supportive analyses may not have added much to the overall project cost while adding potentially valuable additional information.

In general, trend analyses require a large amount of data, typically obtained over a long period of time. These requirements cause potential problems. Experimental designs for a several-year (or several-decade) monitoring effort are difficult to carry out. Many uncontrolled changes may occur during a long period, such as changes in laboratory analysis methods. Laboratory method changes can affect the specific chemical species being measured, or at least have differing detection limit capabilities. This study examined basic measurements that have not undergone major historical changes, and very few "non-detectable" values were reported. In contrast, examining historical heavy metal data is very difficult because of changes in instrumentation and associated detection limits. The need for a typically long-duration study also requires a long period before statistically relevant conclusions can be obtained. Budget reductions in the future always threaten long-term efforts. In addition, personnel changes lead to inconsistent sampling and may also possibly lead to other errors. Basically, adequate trend analyses require a large amount of resources (including time) to be successful. The use of historical data not collected for a specific trend analysis objective is obvious and should be investigated to supplement an anticipated project. However, great care must be expended to ensure the quality of the data. In most cases, incorrect sampling locations and dates, let alone obvious errors in reported concentrations, will be found in historical data files. These problems, in conjunction with problems associated with changing laboratory methods during the monitoring period, require special attention and effort.

Case Studies of Current, Ongoing, Stormwater Projects

Los Angeles County Stormwater Monitoring Program to Support Its Stormwater Discharge Permit

Step 1. What's the Question?

Los Angeles County is currently conducting a comprehensive stormwater monitoring program in conjunction with its stormwater discharge permit. The Los Angeles region of the California Regional Water Quality Control Board (RWQCB) oversees the enforcement of the NPDES stormwater discharge permit for the Los Angeles area. The County of Los Angeles is the principal permittee of the municipal permit and is the permit coordinator responsible for administration for the 80 co-permittees (Rashedi and Liu 1996). The municipal permit had partitioned Los Angeles County and adjacent areas into five regional drainage basins: Santa Monica Bay, Upstream Los Angeles River, Upper San Gabriel River, Lower Los Angeles River, Lower San Gabriel River, and Santa Clarita Valley.

The originally proposed monitoring program was thought to be insufficient by local environmental groups and a suit was filed by the NRDC (*Natural Resources Defense Council v. County of Los Angeles,* CV 94-5978, C.D. Cal). After lengthy discussions between experts representing Los Angeles County and the NRDC, a settlement was reached between NRDC and Los Angeles County (with the approval of the California RWQCB) which specified the scope of work for the monitoring program needed to support the stormwater discharge permit. This program is described in the following paragraphs. Because of the importance and magnitude of the work involved, it is likely that changes to this program will be needed as information is collected and reviewed. Like all monitoring programs, it is necessary to retain a certain degree of flexibility and make slight changes in the monitoring program based on periodic comprehensive data reviews. In this case study, for example, certain monitoring parameters may be eliminated from the basic monitoring program if they are infrequently observed. However, they should still be periodically monitored on a less frequent schedule in case their initial absence was due to seasonal or unusual weather-related factors.

Figure 4.37 Santa Monica Bay/Beach.

Figure 4.38 Downtown Los Angeles.

Figure 4.39 Los Angeles River and roadway cross-ings.

Figure 4.40 Los Angeles River showing small central pilot channel containing perennial flow.

This monitoring program is multifaceted and will last for several years. The information to be obtained will enable the county to fulfill its permit obligations by conducting a stormwater management program based on local data and conditions. Without this local information, decisions that would have been made and stormwater management activities to be conducted would likely result in inadequate stormwater control and be very expensive for the benefits received. The comprehensive monitoring program being conducted will enable cost-effective management decisions to be made in the future. Figure 4.37 shows one of the major receiving waters addressed in the Los Angeles County stormwater management program (Santa Monica Bay), while Figure 4.38 shows the characteristics of the intensively developed ultra-urban area affecting local receiving waters. Figures 4.39 and 4.40 show the massive concrete-lined Los Angeles River draining much of the Los Angeles basin (discharges to Long Beach, not to the Santa Monica Bay).

Step 2. Decide on Problem Formulation

The Los Angeles County activities address the three main topics necessary in a comprehensive stormwater monitoring program: (1) measurements of the effects of stormwater on local receiving water beneficial uses, (2) identification of the sources of the problem pollutants responsible for these problems, and (3) local evaluations of candidate stormwater control practices to reduce the discharge of these problem pollutants and conditions.

This is a large effort and will include components of many of the sampling strategies available (such as comparing stormwater characteristics from multiple land use areas and evaluating trends in receiving water quality over time). Most of the monitoring activities will be conducted over a

3- to 5-year period and will include sampling during all seasons. Long-term evaluations are especially important in southern California because of the tremendous variability in precipitation from year to year. Some years have very little rain, while others, like the 1997–98 rain year affected by El Niño, are characterized by massive flooding. Under these conditions, it is very difficult to define what is "typical" and to design a comprehensive and effective stormwater management program without a monitoring program extending over several years and including many events.

Step 3. Project Design

The Los Angeles County stormwater permit (CA0061654) required the implementation of a monitoring program to control and eliminate the sources of stormwater pollution being discharged from the separate municipal stormwater drainage system. The California Regional Water Quality Control Board (Board Order No. 90-079) required the following actions in the monitoring program:

1. Initiate a monitoring network of initially nine stations to establish long-term trends in stormwater quality in the Santa Monica Drainage Basin.
2. Use a stormwater model in conjunction with the monitoring program to refine annual estimates of pollutant loads to Santa Monica Bay.
3. Implement targeted monitoring to identify sources of specific toxic pollutants in the local stormwater.
4. Implement a monitoring program to evaluate the effectiveness of specific stormwater controls.
5. Implement monitoring to identify locations of illegal practices and to eliminate pollutant sources.
6. Develop and implement a program to evaluate stormwater impacts on selected receiving waters including conducting toxicity studies in the Santa Monica Bay Drainage Basin.

The nine initial sampling locations were first separated into four "mass emission" stations to examine long-term water quality trends, and five land use stations that were relatively homogeneous to obtain unit area loadings and typical effluent concentrations. Critical source area locations will also be monitored to characterize stormwater from locations expected to contribute especially high loadings of toxicants. Thirteen "baseline" stormwater management practices will also be selected for evaluation. Public education (inlet sign painting, billboards, and radio messages) are of special interest.

1. Qualitative Watershed Characterization — The four mass emission sites currently being monitored are in large watersheds and are as follows (LACDPW 1995):

- Ballona Creek. 89 mi^2, representing much of the 127 mi^2 watershed that is not tidally influenced. The overall level of imperviousness is about 53%, and the land uses are approximately as follows: 19% open space, 30% single-family residential, 32% multiple-family residential, 14% commercial, and 4% industrial. The gauging/sampling station location is in a concrete-lined trapezoid channel, about 100 ft wide with a maximum depth of about 25 ft.
- Malibu Creek. 105 mi^2, representing almost all of the 110 mi^2 watershed. The overall imperviousness is about 13%, and the land uses are approximately as follows: 54% open space, 36% single-family residential, 5% multiple-family residential, and about 5% commercial and industrial combined. The monitoring station is located in a natural section of the creek, about 200 ft wide.
- Los Angeles River at Wardlow Rd. 815 mi^2, the largest watershed discharging into the Pacific Ocean in Los Angeles County. This site has been an active gauging station since 1931. The channel is concrete-lined and 400 ft wide. The maximum depth is 22 ft, while a shallow 28-ft-wide pilot channel carries dry-weather flows. This very large watershed contains all of the Los Angeles County land uses. Stream diversions, dams, and spreading areas are common in the watershed, all affecting the flows, especially from the upper foothill areas.
- San Gabriel River. 460 mi^2, also at an existing gauging station. Numerous flow regulation facilities also exist in this large watershed. The river is partially stabilized with concrete at the monitoring station and is 200 ft wide. The maximum depth is from 11 to 14 ft.

These stations represent the four major drainage points for the watersheds that discharge into the ocean from Los Angeles County. Up to 10 storms per year will be monitored at each of these locations. The purpose of monitoring at these drainages is to observe trends in stormwater quality over the period of monitoring. The data will also be useful in confirming the models calibrated from the land use specific monitoring stations. However, the large number of flow modification structures in the large watersheds will hinder some of the comparisons.

Besides the initial mass emission drainage monitoring stations listed above, initial land use monitoring stations were also established. These drainages represent relatively homogeneous (or simple combined) land uses and are as follows:

- Trancas Canyon. 7.45 mi^2, 97% open space (mostly in the Santa Monica Mountains National Recreation Area), and 3% low-density residential, with 1% imperviousness
- Palos Verdes Estates. 1.7 mi^2, 81% single-family residential, and 19% open space, with 40% imperviousness
- Manhattan Beach. 200 acres, 98% single-family residential and 2% commercial, with 42% imperviousness
- Downtown Los Angeles drain. 150 acres, 51% industrial and 49% commercial, with 91% imperviousness
- City of Santa Monica drain. 50 acres, 96% commercial (Santa Monica Mall) and 4% multifamily residential, with 92% imperviousness

A marginal benefit analysis was conducted by Woodward Clyde Consultants (WCC) and Psomas (1996), using the procedures described in Chapter 5, to identify additional land use monitoring sites to best represent the wide range of land uses in Los Angeles County. Table 4.28 lists the general land use categories for Los Angeles County, showing the percentage of each in the area covered by the NPDES stormwater discharge permit, plus the percentage of the total area total suspended solids (TSS) and copper loadings. Site surveys were conducted for the 12 most important land uses shown on this table (excluding vacant land). These 12 land uses comprised about 75% of the area of all land uses, excluding the vacant land. Seven to eight homogeneous areas representing each of these land use areas were surveyed during a 5-week period in the summer of 1996. Site survey information included detailed descriptions of the land use and age of the area, the nature and character of the buildings, the routing of on-site drainage (roof drainage and paved area drainage), the condition of the streets and other impervious areas, gutter types, the nature of the landscaping adjacent to the road, the presence of treated wood near the streets, and landscaping practices. In addition, measurements from maps and aerial photographs were made to determine the areas of each element of the development (roofs, streets, sidewalks, gutters, driveways, parking/storage areas, paved playgrounds, other paved areas, landscaped areas, and other pervious areas). Figure 4.41 shows box plots of the site-measured directly connected impervious areas for each of these 12 major land use areas.

The individual land use categories are also ranked in Table 4.28 according to their total area contributions of these attributes. The estimated contributions for each land use category were based on measured site characteristics (especially imperviousness) of the most important land uses, plus the best estimates of runoff characteristics for these land uses. Analyses using other expected critical pollutants (especially bacteria) would have been informative, but preliminary data were not available. Similar analyses using runoff volume, COD, and P were also conducted, with very similar results: the same land uses were always included in the group of the most important land uses.

Figure 4.42 is the plot from the marginal benefit analysis of all Los Angeles County land use areas, showing the decreasing marginal benefits associated with monitoring an increasing number of land use monitoring sites. From this analysis, a total of seven land uses were identified: high-density single-family residential, vacant land, light industrial, transportation, retail and commercial, multifamily residential, and educational facilities. Multifamily residential and educational facilities were therefore added to the five land use areas previously selected for monitoring. It must be noted that heavy industrial land use data are being collected by the industrial component of the NPDES

Table 4.28 Land Uses in Los Angeles County and Estimated Pollutant Discharge Rankings

Land Use Category	% of Area	Rank Based on Area	% of TSS Load	Rank Based on TSS Load	% of Copper Load	Rank Based on Copper Load
Vacant land	56.0	1	19.5	2	13.3	3
High-density single-family residential	18.6	2	22.9	1	32.5	1
Light industry	3.2	3	14.8	3	17.1	2
Multifamily residential	2.8	4	4.9	6	6.9	4
Retail and commercial	2.5	5	9.5	4	4.6	6
Transportation	1.7	6	5.6	5	6.5	5
Low-density SFR	1.6	7	1.6	11	2.2	8
Educational facilities	1.6	8	3.6	7	1.7	11
Receiving waters	1.4	9	0.0	34	0.0	34
Open space/recreation	1.2	10	1.6	13	0.54	19
Mixed residential	1.1	11	1.5	14	2.1	10
Utility facilities	1.1	12	1.2	15	0.69	16
Natural resources extraction	0.73	13	2.1	8	2.4	7
Institutions	0.66	14	1.6	12	0.76	14
Urban vacant	0.64	15	0.26	24	0.14	26
Golf courses	0.64	16	0.46	21	0.16	25
Rural residential	0.62	17	0.29	23	0.40	22
Floodways and structures	0.62	18	0.85	17	0.29	23
Heavy industry	0.51	19	1.9	9	2.2	9
General office use	0.49	20	1.8	10	0.86	12
Agriculture	0.45	21	0.21	25	0.11	29
Under construction	0.41	22	0.56	19	0.65	17
Other commercial	0.33	23	1.2	16	0.58	18
Nurseries and vineyards	0.33	24	0.10	29	0.27	24
Mobile homes and trailer parks	0.25	25	0.50	20	0.71	15
Mixed transportation and utility	0.14	26	0.66	18	0.77	13
Animal husbandry	0.11	27	0.09	30	0.09	31
Military installations	0.10	28	0.12	27	0.13	27
Maintenance yards	0.08	29	0.38	22	0.44	21
Mixed commercial and industrial	0.04	30	0.07	31	0.09	30
Harbor facilities	0.04	31	0.12	26	0.52	20
Marina facilities	0.03	32	0.03	33	0.07	32
Mixed urban	0.03	33	0.05	32	0.06	33
Communication facilities	0.02	34	0.11	28	0.13	28

program, and construction sites were not deemed an appropriate source to be included in this program by the county.

Further analyses were conducted to select smaller watershed areas for monitoring critical sources (WCC and Psomas Assoc. 1996). A list of industrial categories (by SIC codes), along with their ranking by their pollution potential and the number of the facilities, is shown in Table 4.29. The pollution potential rank was determined based on the number of sources in the area, the relative size of the paved areas at each source, the likelihood of specific toxic pollutants, and the exposure potential of the on-site sources. From this analysis, the following critical light industrial and commercial sources were selected for potential monitoring:

- Wholesale trade (including scrap yards and auto dismantlers)
- Automotive repair/parking (intend to stress repair facilities over parking areas in the monitoring program)
- Fabricated metal products (including electroplating)
- Motor freight (including trucking)
- Chemical manufacturing

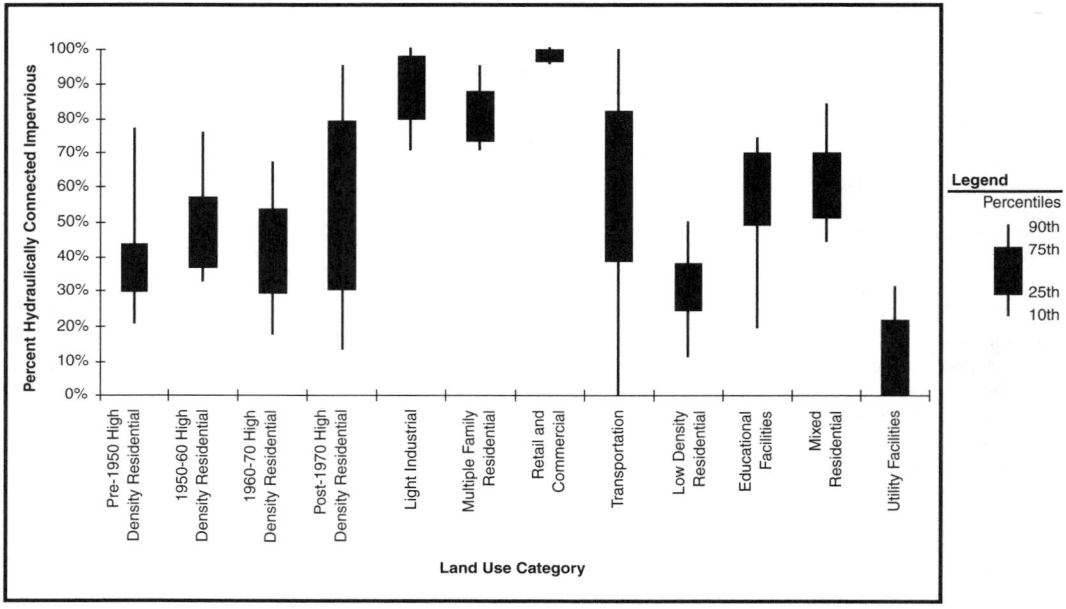

Figure 4.41 Box plots of hydraulically connected impervious areas of the most important Los Angeles County land use areas. (From Woodward Clyde Consultants and Psomas and Associates. *Evaluation of Land Use Monitoring Stations.* Prepared for the Los Angeles County Department of Public Works. August 1996.)

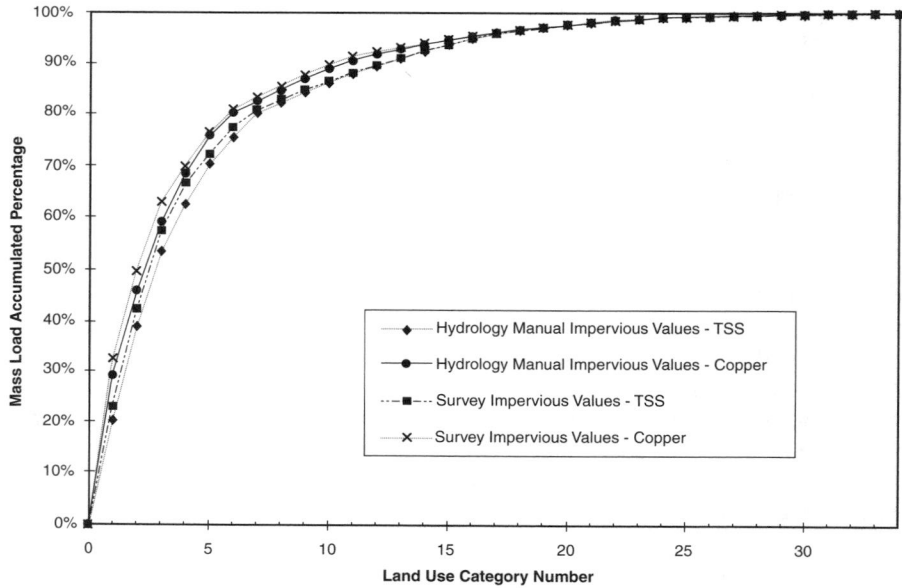

Figure 4.42 Marginal benefit analysis of all Los Angeles County land use areas. (From Woodward Clyde Consultants and Psomas and Associates. *Evaluation of Land Use Monitoring Stations.* Prepared for the Los Angeles County Department of Public Works. August 1996.)

Table 4.29 Ranking of Candidate Critical Sources in Los Angeles County

Industrial Category	SIC Code	No. of Facilities in Los Angeles County Study Area	Ranking Based on Pollution Potential
Wholesale trade (scrap, auto dismantling)	50	587	1
Automotive repair/parking	75	6067	2
Fabricated metal products	34	3283	3
Motor freight	42	872	4
Chemical manufacturing	28	1069	5
Automotive dealers/gas stations	55	2744	6
Primary metals products	33	703	7
Electric/gas/sanitary	49	2001	8
Air transportation	45	431	9
Rubbers/miscellaneous plastics	30	1034	10
Local/suburban transit	41	336	11
Railroad transportation	40	319	12
Oil and gas extraction	13	327	13
Lumber/wood products	24	905	14
Machinery manufacturing	35	4223	15
Transportation equipment	37	1838	16
Stone, clay, glass, concrete	32	733	17
Leather/leather products	31	163	18
Miscellaneous manufacturing	39	1144	19
Food and kindred products	20	1249	20
Petroleum refining	29	231	21
Mining of nonmetallic minerals	14	39	22
Printing and publishing	27	2432	23
Electric/electronic	36	1636	24
Paper and allied products	26	451	25
Furniture and fixtures	25	1368	26
Personal services (laundries)	72	2515	27
Instruments	38	1029	28
Textile mills products	22	440	29
Apparel	23	1900	30

From WCC and Psomas 1996.

These source categories were found to be poorly represented in past stormwater studies, with very little characterization data already available. Therefore, all of these categories were selected for further monitoring.

2. Receiving Water Characterization — The near-shore Pacific Ocean, local ocean beaches, and the large streams and major rivers are the receiving waters examined during this monitoring effort. As an example of the characteristics of the receiving waters, the Los Angeles River has a watershed of 827 mi^2, draining portions of the San Gabriel Mountains, the San Fernando Valley, and a large part of the metropolitan area of the city of Los Angeles. Lowe and Rashedi (1996) reviewed the historical flows in the Los Angeles River and reported an average runoff flow of about 235 million m^3/year, corresponding to about 4.4 in of runoff (a volumetric runoff coefficient of about 1/3, typical for large urban areas). The Los Angeles River also has a relatively small base flow, of about 14 million m^3/year, which is primarily treated wastewater discharged from upstream treatment facilities. Seasonal variations of flows are very large. Lowe and Rashedi (1996) reported that about 80% of the rainfall occurs in the winter, between November and March, with about 84% of the annual runoff also occurring during these months. January typically has the greatest flows and only about 2% of the annual runoff occurs in June through August. There is also a great variation in flows from year to year. They found about a 15 times difference in annual flows between the 10th percentile year and the 90th percentile year. These flow variations reported for the Los Angeles

River are likely similar to the variations that may be found in other urbanized rivers and streams of Los Angeles County. The physical nature of the Los Angeles River is greatly modified. It is completely channelized and concrete-lined for most of its length through the urban area toward the ocean. The river is very wide (about 400 ft) and relatively shallow (about 20 ft) in the downstream reaches. It has a shallow low-flow pilot channel about 25 ft wide and 2 ft deep. Many of the other major receiving waters in the county are also greatly modified, although all are smaller than the Los Angeles River.

A receiving waters study is also planned as part of the Los Angeles County monitoring program. This will be a joint effort between USC, UCSB, and the Southern California Coastal Water Research Project. An ongoing toxicity study conducted by UCLA will also be supported by the Los Angeles County Department of Public Works (LACDPW). The receiving water studies include a plume study to investigate the dispersion of stormwater flows and pollutants into the ocean from Malibu and Ballona Creeks. Marine benthic conditions near the outfalls of these two large creeks will also be investigated. The toxicity studies will investigate the stormwater flows from these two creeks, plus the affected sediments. The plume study will investigate discharges over 2 years from these creeks into Santa Monica Bay following strong winter storms. The spatial and temporal nature of the stormwater plumes will be mapped, and the interaction between the stormwater and the ocean water will be determined. The suspended particulate matter and dissolved organic material discharges will be of special interest. The benthic study will investigate water quality (DO, salinity, density, temperature, light transmissivity, and pH), sediment characteristics (grain size, organic and other constituent concentrations), and the structure of the benthic invertebrate community. The toxicity study will examine water column toxicity by using sea urchin fertilization tests and toxicity identification examinations (TIE). Sediment toxicity tests will include amphipod survival tests, sea urchin growth tests, chemical analyses of sea urchin tissue, and TIE tests. Two stormwater and one dry-weather flow sample will also be tested for toxicity (using sea urchin fertilization tests) at the Los Angeles River and the San Gabriel River monitoring stations in each of 2 years.

3. Select Monitoring Parameters and Magnitude of Sampling — The nine initial monitoring stations were instrumented with refrigerated automatic water samplers. Since the mass emission sampling locations required lifts greater than 15 ft and very long sample line lengths, auxiliary pumps were located in the stream channels that delivered a continuous flow of water close to the automatic samplers. The stormwater samples are being collected on a flow-proportionate basis, using existing flow monitoring facilities if available, or installing flow monitoring equipment, if needed. The samples were collected as discrete samples and then manually composited for analyses. Certain parameters (bacteria and VOCs) required manual sampling. The dry-weather sampling uses the same automatic samplers, but the samplers are reprogrammed to obtain samples on a time-weighted basis. At least one rain gauge capable of measuring rain intensity was also installed in the upper watersheds. The LACDPW operates many rain gauges throughout the Santa Monica Drainage Basin, and these were used to supplement the installed gauges.

Table 4.30 lists the priorities for the monitored constituents and the associated sample volumes needed to conduct the selected constituents. The total sample volume needed for the complete list of analyses to be collected from the automatically collected stormwater samples is about 8 L. As shown in Chapter 6, many of these analyses may be conducted using procedures requiring much smaller sample volumes. However, the use of alternative (but acceptable) methods can be more costly, especially if the laboratory needs to develop new methods. Only 40 mL of water is needed for the VOC analyses, but the samples must be manually collected because specialized automatic VOC samplers are not being used. Other analyses to be conducted on manually collected grab samples include total coliforms, fecal coliforms, fecal streptococcus, oil and grease, total phenols, cyanide, pH, and temperature. About 2.5 L of water is needed for these additional analyses.

Table 4.30   Analyses Priority and Sample Volumes Needed for Automatically Collected Stormwater Samples

Priority	Constituent	Method	Sample Volume Needed (mL)
1	Heavy metals (total and dissolved)	EPA[a] 200	500
2	Total petroleum hydrocarbons (TPH)	EPA 418.1	1000
3	Semivolatile organic compounds	EPA 8250	1000
4	Pesticides and PCBs	EPA 8250 or 608	1000
5	Total suspended solids (TSS)	EPA 160.1	100
6	Volatile suspended solids (VSS)	EPA 160.1	100
7	Total organic carbon (TOC)	EPA 415.1	25
8	Chemical oxygen demand (COD)	EPA 410.4	500
9	Specific conductance	EPA 120.1	100
10	Total dissolved solids (TDS)	EPA 160.1	100
11	Turbidity	EPA 180.1	100
12	Biochemical oxygen demand (BOD_5)	EPA 405.1	1000
13	Dissolved phosphorus	EPA 300	50
14	Total phosphorus	EPA 300	50
15	Total ammonia nitrogen	EPA 350.2	500
16	Total Kjeldahl nitrogen	EPA 351.3	100
17	Nitrate and nitrite nitrogen	SM[b] 4110	100
18	Alkalinity	EPA 310.1	100
19	Chloride	SM 4110	50
20	Fluoride	SM 4110	300
21	Sulfate	SM 4110	50
22	Herbicides	EPA 619	1000

[a] EPA published method.
[b] *Standard Methods for the Examination of Water and Wastewater.*

Sampling at the land use monitoring locations will include the complete list of constituents, unless the constituent is frequently not detected. If the constituent is not found at the method detection limit (MDL) in at least 25% of the samples, it will be eliminated from the list for routine analyses. However, the constituent will be analyzed at least once a year. In addition, once sufficient storms at a specific location have been sampled to allow the event mean concentration (EMC) of a constituent to be determined with an error rate of 25%, or less, that constituent will also be removed from the list of analyses to be conducted at that location. The land use station will remain in operation until the following constituent EMCs are determined at the 25% error level:

Total PAHs
Chlordane
Cd, Cu, Ni, Pb, Cr, Ag, Zn
TSS
Total nitrogen
Total phosphorus

A chain-of-custody record was prepared specifically for this project by the LACDPW. The sampling program also included routine QA/QC field activities, such as the use of field blanks for manual VOC sampling and field duplicates for all events. Before the sampling program began, a sampling instruction manual was prepared, detailing such things as specific sampling equipment features, sample handling, and field equipment lists. The *Quality Assurance Manual* from the local laboratory being used (Environmental Toxicology Laboratory of the County of Los Angeles Office of Agricultural Commissioner/Weights and Measures) was also included in the initial proposed stormwater monitoring program description prepared by the LACDPW.

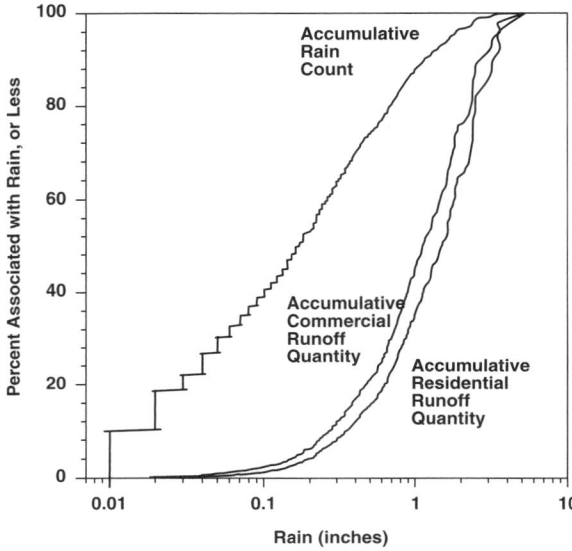

Figure 4.43 Probability plots of rain depth and runoff depths for 1969–1993 LAX conditions. (From Pitt et al. 1999.)

The initial monitoring design was to program the automatic samplers to obtain the needed sample volume for a 0.4-inch storm, with a maximum rain depth of 1.7 inches capable of filling the samplers. During the 45-year period from 1948 to 1993, about 1350 rains occurred at LAX (assuming a conventional 6-hour inter-event dry period), or about 30 rain events per year. Figure 4.43 shows a probability plot of rain event depths and estimated runoff depths for residential and commercial sites in the Los Angeles area for 1969 through 1993 rains. The median rain depth (by count and considering all rains) was about 0.2 inches, and about 70% of all recorded rains at LAX were less than 0.4 inches. About 5% of the rain events were greater than 1.7 inches in rain depth. Therefore, only about 25% of all rains (by occurrence) were in the range of 0.4 to 1.7 inches in depth. The 0.4-inch rain depth needed for complete analyses was therefore found to be relatively large, resulting in a significant number of events that would not be represented in the monitoring program. A special monitoring test was therefore conducted to determine the minimum rain event size that would produce significant runoff that could also be adequately sampled. The results of this special test indicated that the samplers could be programmed to capture runoff from at least a 0.25-inch rain, resulting in about 90% of the annual runoff volume being represented in the monitoring program.

Experimental design calculations also indicated the need for very large paired data sets to observe statistically significant differences in stormwater runoff quality from most public education and public works practices. With a coefficient of variation of 1 (common for most stormwater concentration data), plus a 20% likelihood of false negatives and 95% confidence, about 200 paired observations would be needed if the control program produces a change of about 25% in stormwater characteristics. If the change is about 50%, then about 50 paired observations would be needed. If the control program produced about 95% differences in stormwater characteristics (only possible for the most effective stormwater controls, such as well-designed and operated wet detention ponds or grass swales), then only 15 pairs of data would be needed. In an area having relatively few rain events per year, it could take many years to obtain adequate data for important decisions.

The sampling plan for the critical source areas includes monitoring at six sites in each of the five categories (WCC and Psomas Assoc. 1996). These monitoring activities will also include evaluations of site stormwater controls. The first year will include monitoring of the sites without controls, while the second year of monitoring will include the use of site controls at three of the sites in each category. These paired tests will enable site and rainfall differences to be identified to enable more accurate stormwater control evaluations. Five rain events will be monitored using manual grab sampling during the first year, and ten will be monitored during the second year. The

Table 4.31 Constituents to Be Monitored as Part of the Critical Source Area Monitoring Program

Constituent	Wholesale Trade	Automotive Repair/Parking	Fabricated Metal Products	Motor Freight	Chemical Manufacturing
pH	X	X	X	X	X
Specific conductance	X	X	X	X	X
Oil and grease	X	X	X	X	
Semivolatile organics	X	X	X	X	X
Total petroleum hydrocarbons (TPH)		X		X	
Chemical oxygen demand (COD)	X	X	X	X	X
Total suspended solids (TSS)	X	X	X	X	X
Total dissolved solids (TDS)	X	X	X	X	X
Total organic carbon (TOC)	X	X	X	X	X
MBAS (detergents)		X			
Heavy metals (Al, Cd, Cr, Cu, Fe, Pb, Ni, and Zn)	X	X	X	X	
Other (based on chemicals handled at facility)			X		X

From WCC and Psomas Assoc. 1996.

samples will be composited before analysis into test and control samples for each source area category. The samples will be analyzed for the constituents shown on Table 4.31.

The stormwater controls to be investigated will be selected from the following ranked listing:

Infiltration
Media filtration (sand filters and similar devices)
Oil/water separators
Water quality inlets (oil spill containment)
Biofiltration (vegetated swales or filter strips)
Wet or extended detention dry ponds
Constructed wetlands
Runoff quantity control ponds and vaults
Multiple systems

In addition, industrial and commercial source controls will also be considered, including preventive maintenance, spill containment, material handling, litter control, etc.

Step 4. Project Implementation (Routine Initial Semiquantitative Survey)

An important initial step in any monitoring program is to collect and review any existing data and information. LACDPW has been actively monitoring surface water quality since the late 1960s (Rashedi and Liu 1996). Since the mid-1980s, 28 sampling sites have been routinely monitored during both dry weather (monthly observations) and wet weather (three to four storms per year). Table 4.32 lists the constituents that have been included in these monitoring activities.

The available data were reported by LACDPW as part of its evaluation of existing stormwater quality monitoring data (task 5.2, *Report of Waste Discharge*, volume 8). This report included some of the stormwater data (TDS, chloride, pH, sulfate, nitrite, lead, fecal coliforms, enterococcus, and total coliforms) for several storms a year. The bacteria were generally high, as is typical for stormwater. Fecal coliforms averaged from 10,000 to 100,000 organisms per 100 mL, and the enterococci were only slightly lower. Similar monitoring was also conducted at these locations during dry weather. The dry weather fecal coliform observations were much lower, being about 1000 to 10,000 organisms per 100 mL, while the TDS and chlorides were higher. The "Basin Plan Objective" for fecal coliforms is only 200 organisms per 100 mL, with most observations greatly exceeding this value.

Table 4.32   Constituents Monitored at 28 Surface Water Sampling Locations since the Late 1980s

Constituent	Dry Weather	Wet Weather
Minerals	X	X
Pesticides	X	X
Total petroleum hydrocarbons	X	
Heavy metals	X	Total and filtered
Bacteria (total and fecal coliforms, streptococci, and enterococci)	X	X
Suspended solids (total and volatile)		X
Oil and grease		X
Biochemical oxygen demand		X
Total organic carbon		X
Volatile organic compounds	Semiannually	X

Rashedi and Liu (1996) reported that the top ten compounds with the highest numbers of exceedances of the water quality objectives were: fecal coliforms, enterococcus, TDS, ammonia, chloride, nitrite, pH, sulfate, total coliforms, and lead. The available data indicated very high variabilities in concentrations, with no obvious and consistent trends observed. However, most of the lower basin monitoring data showed higher concentrations of chloride, sulfate, lead, and TDS than the corresponding upper basin areas. Lead concentrations sharply decreased after 1990, and the most recent data were mostly below the water quality objective limits. The dry-weather flow lead concentrations were generally higher than the storm-generated flows in the Los Angeles River (Lowe and Rashedi 1996).

Rashedi and Liu (1996) also evaluated the available data for different land uses. They found higher concentrations of total and fecal coliforms, lead, TDS, chloride, and sulfate in drainages having large industrial areas. Higher chloride, sulfate, TDS, nitrate, ammonia, total coliforms, and lead concentrations were found in watersheds that were heavily urbanized.

Because of the observed high variability (typical for stormwater quality), a large number of samples (probably at least 50) will be needed to obtain event mean concentration values having errors of 25%, or less. If only five storms can be monitored per year at each of the monitoring locations, it may require at least a decade before enough data are collected for the necessary statistical analyses to satisfy the project objectives.

Several special studies were also conducted to investigate potential local monitoring problems. One included an investigation of reducing the smallest storm size that could be monitored, and another investigated problems associated with monitoring in very wide and shallow channels. As noted previously, the samplers were programmed to sample storms as small as 0.25 inches, reduced from the initial design of 0.4 inches. This reduction in the small storm size that could be sampled should increase the capture of the annual runoff significantly. About 15 to 20% of the annual runoff is associated with rains less than 0.4 inches, while less than 10% of the annual runoff is expected from storms less than 0.25 inches in depth (using a conventional interevent dry period of 6 hours and for the LAX rain history from 1969 to 1993). The larger range of storms to be monitored will enable the collection of most storms that occur and will allow analyses of concentration variations associated with rain depth. The design of many less expensive stormwater controls is based on the assumption that higher concentrations of pollutants occur with small rains, or with the first portion of rains. Therefore, this monitoring effort will enable this important characterization aspect to be investigated. The number of events associated with these small storms is also very large and is therefore important in relation to water quality objectives (especially bacteria). Characterizing these smaller events will therefore enable better evaluations of exceedance frequency and durations of water quality objectives.

A study was conducted at the monitoring station at Ballona Creek to investigate whether the single midstream sampling location was reasonably representative of the 100×25 ft channel (WCC and CDM 1996). Four surface samples (collected from locations evenly spaced along the width of the channel) were compared to the single midchannel automatic sampling location at the channel bottom during three storms. Samples were obtained at 3-hour intervals during the storm durations and were analyzed for temperature, pH, specific conductivity, turbidity, TDS, TSS, copper (total and dissolved), zinc (total and dissolved), and nitrate. The three storms monitored were 1.8, 3.1, and 2.2 inches in depth, all quite large, but sufficient to create enough depth in the channel to enable sampling over a wide area. The flows were confined in a channel about 50 to 100 ft wide and from 2 to 8 ft deep, and the water velocities ranged from 0.2 to 0.3 ft/s during this study. The differences in constituent concentrations for the different sampling locations for any storm were found to be much less than the differences in concentrations between storms. As an example, the middle bottom sample was from 5 to 25% different from the overall average, with no clear bias, for suspended solids. Calculations were also made by LACDPW (1998) to determine the flow distances required for complete mixing in the channel during these events (to achieve less than a 10% variation in water quality). It may require from 600 to 2500 ft of channel length from a discharge to achieve this level of mixing for these storms. At the Ballona Creek monitoring station, three upstream outfalls are within 2500 ft. However, these outfalls only represent about 2% of the complete drainage area. The required flow distances for complete mixing at the other wide channel sites (200 to 400 ft in width) would likely be substantially longer, depending on the expected flow rates and water depths. However, problems associated with automating a multilocation sampling system are difficult, requiring multiple sampling pumps spread across the channel, instead of the single unit used here.

An important aspect of any monitoring program is the health and safety of the project personnel. The LACDPW requires all employees to identify the likely hazards that may be encountered on their jobs. For this project, these hazards included hazardous weather conditions, working in confined spaces, hazards associated with chemicals, snakes, poison ivy, traffic, falling, drowning, etc. The county requires field sampling personnel to undergo a minimum of 40 hours of Hazardous Materials Awareness training and other training to enable the personnel to evaluate potentially hazardous situations and safety concerns.

Step 5. Data Evaluation

This case study describes the development of a workplan for a large and comprehensive stormwater management program. Only preliminary data are currently available, as described above, which were used to modify and refine the initial workplan.

Step 6. Confirmatory Assessment (Optional Tier 2 Testing)

There are several additional stormwater monitoring programs being conducted in southern California that can be very useful for Los Angeles County. One of the most interesting is a unique epidemiological study conducted at Santa Monica Bay beaches to examine human health risks associated with swimming in water contaminated by stormwater. It is summarized in the following paragraphs and tables. This study was the first large-scale epidemiological study in the United States to investigate possible adverse health effects associated with swimming in ocean waters affected by discharges from separate storm drains (*Water Environment & Technology* 1996a,b; *Environmental Science & Technology* 1996; Haile et al. 1996).

During a 4-month period in the summer of 1995, about 15,000 ocean swimmers were interviewed on the beach and by telephone 1 to 2 weeks later. They were queried concerning illnesses since their beach outing. The incidence of illness (such as fever, chills, ear discharge, vomiting, coughing with phlegm, and credible gastrointestinal illness) was significantly greater (from 44 to 127% increased incidence) for oceangoers who swam directly off the outfalls, compared to those who swam 400 yards away, as shown on Table 4.33. As an example, the rate ratio (RR) for fever was 1.6, while it was 2.3 for ear discharges, and 2.2 for highly credible gastrointestinal illness

Table 4.33 Comparative Health Outcomes for Swimming in Front of Storm Drain Outfalls, Compared to Swimming at Least 400 Yards Away

Health Outcome	Relative Risk, %	Rate Ratio	Estimated Association	Estimated No. of Excess Cases per 10,000 Swimmers (rate difference)
Fever	57	1.57	Moderate	259
Chills	58	1.58	Moderate	138
Ear discharge	127	2.27	Moderate	88
Vomiting	61	1.61	Moderate	115
Coughing with phlegm	59	1.59	Moderate	175
Any of the above symptoms	44	1.44	Weak	373
HCGI-2	111	2.11	Moderate	95
SRD (significant respiratory disease)	66	1.66	Moderate	303
HCGI-2 or SRD	53	1.53	Moderate	314

From SMBRP (Santa Monica Bay Restoration Project). *A Health Effects Study of Swimmers in Santa Monica Bay.* Santa Monica Bay Restoration Project. Monterey Park, CA. October 1996.

comprised of vomiting and fever (HCGI). Disease incidence dropped significantly with distance from the storm drain. At 400 yards, and beyond, upcoast or downcoast, elevated disease risks were not found. The results did not change when adjusted for age, beach, gender, race, socioeconomic status, or worry about health risks associated with swimming at the beach.

These interviews were supplemented with indicator and pathogenic bacteria and virus analyses in the waters. The greatest health problems were associated with times of highest concentrations (*E. coli* > 320 cfu/100 mL, enterococcus > 106 cfu/100 mL, total coliforms >10,000 cfu/100 mL, and fecal coliforms > 400 cfu/100 mL). Bacteria populations greater than these are common in urban runoff and in urban receiving waters. Symptoms were found to be associated with swimming in areas where bacterial indicator levels were greater than these critical counts. Table 4.34 shows the health outcomes associated with swimming in areas having bacterial counts greater than these critical values. The association for enterococcus with bloody diarrhea was strong, and the association of total coliforms with skin rash was moderate, but nearly strong.

The ratio of total coliform to fecal coliform was found to be one of the better indicators for predicting health risks when swimming close to a storm drain. When the total coliforms were greater than 1000 cfu/100 mL, the strongest effects were generally observed when the total to fecal coliform ratio was 2. The risks decreased as the ratio increased. In addition, illnesses were more common on days when enteric viruses were found in the water.

The percentage of survey days exceeding the critical bacterial counts was high, especially when closest to the storm drains, as shown on Table 4.35. High densities of *E. coli*, fecal coliforms, and enterococcus were observed on more than 25% of the days; however, there was a significant amount

Table 4.34 Health Outcomes Associated with Swimming in Areas Having High Bacterial Counts

Indicator (and critical cutoff count)	Health Outcome	Increased Risk, %	Risk Ratio	Estimated Association	Excess Cases per 10,000 Swimmers
E. coli (>320 cfu/100 mL)	Ear ache and nasal congestion	46 24	1.46 1.24	Weak Weak	149 211
Enterococcus (>106 cfu/100 mL)	Diarrhea w/blood and HCGI-1	323 44	4.23 1.44	Strong Weak	27 130
Total coliform bacteria (>10,000 cfu/100 mL)	Skin rash	200	3.00	Moderate	165
Fecal coliform bacteria (>400 cfu/100 mL)	Skin rash	88	1.88	Moderate	74

From SMBRP (Santa Monica Bay Restoration Project). *A Health Effects Study of Swimmers in Santa Monica Bay.* Santa Monica Bay Restoration Project. Monterey Park, CA. October 1996.

Table 4.35 Percentages of Days When Samples Exceeded Critical Levels

Bacterial Indicator	0 yards	1 to 100 yards Upcoast	1 to 100 yards Downcoast	400+ yards Upcoast
E. coli (>320 cfu/100 mL)	25.0	3.5	6.7	0.6
Total coliforms (>10,000 cfu/100 mL)	8.6	0.4	0.9	0.0
Fecal coliforms (>400 cfu/100 mL)	29.7	3.0	8.6	0.9
Enterococcus (>106 cfu/100 mL)	28.7	6.0	9.6	1.3
Total/Fecal coliform ratio ≤5 (and total coliforms >1000 cfu/100 mL)	12.0	0.5	3.9	0.4

From SMBRP (Santa Monica Bay Restoration Project). *A Health Effects Study of Swimmers in Santa Monica Bay.* Santa Monica Bay Restoration Project. Monterey Park, CA. October 1996.

of variability in observed counts in the water samples obtained directly in front of the drains. The variability and the frequency of high counts dropped considerably with distance from the storm drains. Upcoast bacteria densities were less than downcoast densities probably because of prevailing near-shore currents.

The SMBRP (1996) concluded that less than 2 miles of Santa Monica Bay's 50-mile coastline had problematic health concerns due to the storm drains flowing into the bay. They also concluded that the bacterial indicators currently being monitored do help predict risk. In addition, the total to fecal coliform ratio was found to be a useful additional indicator of illness. As an outcome of this study, the Los Angeles County Department of Health Services will post new warning signs advising against swimming near the outfalls ("Warning! Storm drain water may cause illness. No swimming"). These signs will be posted on both sides of all flowing storm drains in Los Angeles County. In addition, county lifeguards will attempt to warn and advise swimmers to stay away from areas directly in front of storm drain outlets, especially in ponded areas. The county is also accelerating its studies on sources of pathogens in stormwater.

Step 7. Project Conclusions

It was necessary to modify the original workplan for conducting this large and comprehensive stormwater management study in support of the local stormwater discharge permit. Los Angeles County is probably the largest and most complex urban area that has ever attempted to conduct such a comprehensive study needed for the permit and to direct its future stormwater management decisions. In addition to its unique complexity and size, highly variable and sometimes violent rain conditions also occur. These have all contributed to produce a study that is examining many scales of the stormwater problem. Even though there will still exist some deficiencies in this project (such as not examining beneficial use problems in the smaller urban drainages that have informal human contact recreation), the results of this work will be very important for many years to come.

Birmingham Separate Sewer Overflow Program Monitoring

The Department of Civil and Environmental Engineering at the University of Alabama at Birmingham (Lalor and Pitt 1998) participated in a multiyear research project funded by the U.S. Environmental Protection Agency to develop a protocol to enable municipalities to assess local problems associated with sanitary sewer overflows (SSOs). SSOs and receiving waters are highly variable, resulting in highly variable conclusions pertaining to local problems. If SSOs occur frequently and affect small streams having substantial human contact, the problem is likely serious. However, if the receiving water is relatively large, the SSOs infrequent, and human contact rare, the problems associated with these discharges may be insignificant. This project therefore developed and demonstrated a preliminary protocol to enable municipalities to understand their specific local SSO-related problems and to plan better for their control.

Step 1. What's the Question?

Identify and quantify the human and environmental risks associated with SSOs in urban streams. Need to quantify the sources, fates, and exposure mechanisms of pathogens and toxicants in SSOs. Human exposure ranges from informal human contact associated with children playing in urban receiving waters to consumption of water and fish contaminated by upstream SSOs.

Step 2. Decide on Problem Formulation

As in most environmental research projects, this project was designed as a series of overlapping individual experiments, some of short duration and some long, some examining specific individual processes and some examining many processes interacting together. The conventional stream monitoring activities associated with this project involve longitudinal "above" and "below" monitoring following the stream path as it flows past several known SSO locations. The project test sites have different characteristics to test the sensitivity of the monitoring program in identifying the known SSO discharges and to determine if the SSO discharges were causing measurable beneficial use impairments. Initial monitoring during the first project phase only included specific tracer analyses that were thought to be the most sensitive in detecting SSO discharges. Later project phases could include more comprehensive chemical and biological monitoring at the locations along the streams that were found to have a variety of SSO effects. From this sequence of tests, the ability of these different parameters to detect SSO discharges and their effects for different stream conditions will be determined. The initial test locations include:

- A local hillside where a low-volume, but constant SSO is occurring, flowing into a moderate-sized stream
- A moderate-sized stream (Five-Mile Creek), having a watershed area of about 100 mi^2 with a large intermittent SSO and a small continuous SSO
- A small, completely urbanized stream (Griffin Brook), having a watershed area of about 10 mi^2 with numerous small SSOs

A sampling strategy examining the individual streams as they flowed past the SSO locations (longitudinal sampling along the flow path) was used for most of the field studies. The variable conditions that these test sites provide enabled us to investigate a range of discharge and receiving water conditions, and different resulting problems. The hillside site was used to investigate changes in the SSO's characteristics as it flowed toward the creek. The moderate- and small-sized receiving waters also used longitudinal sampling, with samples collected above and below the known discharge locations, and for an extended distance downstream. The moderate-sized stream also included small-scale up- and downgradient analyses of sediment conditions. The field studies were also conducted during different seasons and flow patterns, contrasting wet- and dry-weather conditions and warm and cold weather.

Another important aspect of this research was to determine suitable risk assessment approaches and tools to enable municipalities to determine the magnitude of local SSO-related problems. Therefore, various experiments were conducted to enable receiving water models to be calibrated for expected local SSO characteristics. The experiments conducted and planned include:

- *In situ* bacteria and other pathogen die-off tests
- Photosynthesis and respiration (P/R) of sewage-contaminated waters
- Interaction of water column pollutants and contaminated sediments and interstitial waters
- Interstitial water measurements
- Measurement of frequency, duration, and magnitude of WWF events
- Sediment oxygen demand (SOD) and sediment P/R tests
- Settleability of SSO-related bacteria and toxicants

Step 3. Project Design

Qualitative Watershed Characterization and Stream Characterization — There are several sites where samples were taken. The sites were located in and along two urban streams in the Birmingham, AL, area. These sites were chosen to allow for overland, upstream, in-stream, and downstream samples near known SSO locations.

Five-Mile Creek — The Five-Mile Creek area has ten sampling sites along an approximately 3-mile reach from Five-Mile Creek Road to Highway 79. Five-Mile Creek is located in the northern part of Birmingham and is surrounded by industrial and suburban development. This series of sampling locations includes sites from 500 ft upstream to 1000 ft downstream from known SSO discharge points.

Overland Flow Sampling Site — The small-volume, overland flow/continuous discharge SSO site is located on Five-Mile Creek, and in-stream sampling points are above and below its location. In order to evaluate the effects of overland flow on SSO characteristics (especially pathogen die-off and particulate toxicant settling), several hillside locations were sampled as the discharge flowed overland toward the stream.

Griffin Brook — Griffin Brook is within a small, fully developed watershed, and is a first-order stream. Griffin Brook is located within Homewood, a suburb located in the southern Birmingham area, and discharges into Shades Creek. The Griffin Brook test reach is approximately 2.5 miles in length, bracketing several known small SSO discharges.

Select Monitoring Parameters — The stream sampling locations were tested during the first project phase using a brief set of chemical and microbiological parameters. These parameters were thought to be the most sensitive to enable the identification of SSO discharges. These parameters (mostly based on earlier work on identifying inappropriate discharges into storm drainage systems; Pitt et al. 1993; Lalor 1994) were:

- Indicators of sewage (detergents, ammonia, potassium, fluoride, color, and odor)
- Other conventional parameters (pH, turbidity, and conductivity)
- Rapid microbiological analyses for *E. coli.*, enterococci, and total coliforms (using IDEXX Quantitrays)

The later phase of the project could involve more comprehensive analyses at the sites found to have detectable SSO discharges. These analyses will be used to quantify the receiving water effects of SSOs on beneficial uses (contact and noncontact recreation, water supply, consumptive fishing, and aquatic life uses). These analyses may include the following parameters:

Primary list (for routine analysis of most samples):
- Pathogens, including protozoa (*Giardia* and *Cryptosporidium*), *Pseudomonas aeruginosa*, and *Shigella*, along with *E. coli*. Viruses, if possible, will also be investigated.
- Trash and other debris along the streams.
- Toxicants, including partitioned metals (lead, copper, cadmium, and zinc, using graphite furnace atomic adsorption spectrophotometer, or other methods having comparable detection limits), partitioned organics (PAHs, phenols, and phthalate esters using GC/MSD with SIM, or HPLC), herbicides, and insecticides (using GC/ECD or immunoassays); suggest routinely using toxicant screening method, such as Azur's Microtox™, for possible guidance in modifying specific list of toxicants.
- Nutrients, including phosphates, total phosphorus, ammonia, total Kjeldahl nitrogen, nitrate plus nitrite, and partitioned TOC (or at least COD).

- Additional conventional parameters affecting fates and effects of pollutants in receiving waters, including hardness, alkalinity, pH, specific conductivity, particle size analyses, turbidity, suspended solids (SS), volatile suspended solids (VSS), and dissolved solids (TDS).

Secondary list (in addition to the above-listed analyses at selected critical locations at least seasonally):

- Selected additional metallic toxicants (such as arsenic and mercury and possible screening using mass spec/mass spec) and selected additional organic toxicants (such as VOCs)
- Long-term NBOD and CBOD (for k rates and ultimate BOD)
- Particulate organic carbon (POC)
- Major cations and anions
- Continuous pH, ORP, specific conductivity, temperature, and turbidity should also be conducted using an *in situ* water quality sonde.

Sediment analyses (seasonal analyses):

- Particle size distributions of sediment
- Acid volatile sulfides (AVS) in sediments
- Toxicants and nutrients by particle size
- BOD and COD (and possibly POC) by particle size
- Interstitial water analyses for key parameters, especially pathogens, nutrients, pH, and ORP, plus others, volume permitting

Numerous seasonal biological attributes should also be included at each sampling reach, including:

- Benthic macroinvertebrates (natural and artificial substrates)
- Algae (natural and artificial substrates) and macrophytes
- *In situ* toxicity test assays

Partitioned analyses of the toxicants in runoff and in the receiving water is very important, as the form of the pollutants will have great effects on their fate and treatability. Conventional assumptions that only filterable toxicants have a toxic effect on receiving water organisms is not always correct.

The sampling requirements will vary for each primary parameter, based on the concentration variations observed. In most cases, 1 year of data (including about 15 to 35 events) will likely be sufficient. For most parameters (assuming a COV of 0.75 to 1.0), this number of samples will result in an event-mean concentration (EMC) value estimate with about 25% levels of error, and will enable effective comparisons to be made between paired upstream and downstream locations. The secondary parameters will only be analyzed about four times (seasonally) and at fewer locations. The likely errors in their EMCs will therefore be quite large. However, the purpose of these measurements is for screening: to identify the presence of additional significant parameters. The seasonal sediment and biological analyses should be sufficient because their variability is much less than for the water parameters.

An important aspect of this research project is to develop an approach useful for municipalities to determine the local risks and the role that SSOs play in TMDL calculations. As such, this project will develop several alternative field program recommendations that should result in different levels of confidence. The above list of parameters will therefore be narrowed considerably for these alternative approaches.

Step 4. Project Implementation (Routine Initial Semiquantitative Survey) and Step 5, Data Evaluation

A series of initial tests was conducted during the first project period to investigate methods to measure the fates of the critical pathogens and toxicants associated with SSO events. This initial effort includes the following experiments:

Initial Steam Surveys in Five-Mile Creek and in Griffin Brook — A number of SSO discharge points were observed along Five-Mile Creek. Figure 4.44 shows a large, intermittent, SSO discharge

Figure 4.44 Five-Mile Creek SSO discharge during large flow.

Figure 4.45 Five-Mile Creek under normal flow conditions.

Figure 4.46 Typical SSO discharge point along banks of Five-Mile Creek.

Figure 4.47 Unusual continuous SSO discharge from surcharged/broken sewerage along Five-Mile Creek.

during a large rain event, Figure 4.45 shows Five-Mile Creek under normal flow conditions, while Color Figure 4.1* shows this discharge mixing with the creek during this large overflow. Figure 4.46 shows another intermittent SSO discharge location at a poorly sealed sanitary sewer manhole in the creek right-of-way. Moderate rains causing surcharging conditions in the sewerage would obviously cause a large SSO at this location. Figure 4.47 shows an unusual continuous (but relatively low volume) SSO discharge that was caused by a leaking sewer on a hillside discharging to Five-Mile Creek.

The initial stream surveys in Five-Mile Creek found no significant SSO discharge effects in the stream during wet or dry weather in the proximity of the small continuous hillside discharge shown in Figure 4.47, except within a few feet of the discharge location. No samples were obtained during high creek flows when the large intermittent SSO was discharging. However, visual observations were obtained during one large discharge event, indicating very large amounts of SSO being discharged into Five-Mile Creek (Figures 4.44 and Color Figure 4.1). During this event, the SSO discharge was likely about 10% of the creek flow and was visually obvious for several hundred feet downstream of the discharge location. This SSO discharge is scheduled to be corrected by Jefferson County in the near future.

The stream surveys in Griffin Brook indicated significant effects from continuous SSO discharges during dry weather, but no noticeable SSO effects during wet weather. The numerous SSOs were all individually quite small, but were responsible for a significant portion of the dry-weather

* Color figures follow page 370.

Figure 4.48 Griffin Brook during wet weather conditions. (Courtesy of Robin Chapman.)

flow in the stream during the summer. During rains, the much higher flows and the moderate to high concentrations of most pollutants in the urban runoff masked the continuous SSO discharges, effectively diluting the SSOs below detection (Figure 4.48).

In Situ Bacteria and Other Pathogen Die-off Tests — Dialysis bags were initially used to measure *in situ* die-off of pathogens (Figures 4.49 and 4.50). *In situ* die-off tests are more accurate indicators of pathogen die-off compared to laboratory tests, as actual environmental conditions are allowed to affect the test organisms. The dialysis bags allow water, nutrients, and gases to enter the bags, but restrain the test organisms. Samples of raw sewage collected from known SSO discharge locations were diluted with stream water and placed in sealed bags. The bags were fitted into large-diameter plastic pipes (with coarse screening on the ends) for protection and anchored in the streams. Bags were then periodically removed and the pathogen populations determined and compared to the initial conditions. In later, extended tests lasting several weeks, we found that the dialysis bag material decomposed, allowing substantial leakage. We have since replaced these initial chamber designs with ones using plastic tubing with membrane filter ports. These new designs and test results are described in Chapter 6.

Photosynthesis and Respiration of Sewage-Contaminated Waters — The aim of this experiment was to examine the acclimation period of the effects of a sewage discharge to a receiving water's dissolved oxygen, and to measure the photosynthesis and respiration (P/R) rates for several mixtures of sewage and receiving waters. The P/R discussion in Chapter 6 describes the test results and summarizes the specific procedures used. The acclimation period of an intermittent discharge into a receiving water may be relatively long, requiring extended observations to obtain an understanding of the likely dissolved oxygen effects. The use of continuously recording water quality sondes enables the collection of water quality data over an extended period (14 days during this

Figure 4.49 Placement of *in situ* pathogen die-off test chambers in Five-Mile Creek. (Courtesy of John Easton.)

Figure 4.50 In place pathogen die-off test chambers. (Courtesy of John Easton.)

field study). Traditional measurements of P/R rates are performed using light and dark bottles over a short period of time, usually several hours, and with little replication. These short period data are then used to construct a dissolved oxygen curve for a 1-day cycle, for the light and dark bottles, from which P/R calculations are made. With the continuously recording sondes, several curves can be constructed over multiple days having variable weather, providing far more useful results than the traditional method. In addition, the acclimation period can be accurately determined and considered in DO calculations.

The net effect of the P/R processes is that the dissolved oxygen level in the water rises during the daylight and falls at night. In addition, the pH of typical receiving waters is governed by the carbonic acid/bicarbonate/carbonate buffering system. Increases in the dissolved CO_2 concentration cause corresponding decreases in pH, and vice versa. Therefore, the pH increases during the daytime hours because CO_2 is being fixed by photosynthetic organisms and is thereby removed from the water. Then, at night, pH drops because atmospheric CO_2 and CO_2 being produced by respiration increase the concentration of CO_2 in the water. The DO and pH sonde probes measured these changes directly. In addition, changes in temperature, ORP, and specific conductance were also observed.

The site for this experiment was a small lake on private property located in Shelby County, AL, to ensure security for the sondes. This lake rarely, if ever, received sanitary sewage, producing a likely worst case for acclimation. YSI 6000 sondes were used to measure the following parameters during these experiments: depth, specific conductance, dissolved oxygen, turbidity, pH, oxidation-reduction potential, and temperature. The sondes were programmed to acquire data in unattended mode for 2 weeks at 15-min intervals. Raw sewage was obtained at the Riverview Sewage Treatment Plant. Lake water was used for diluting the sewage in the following ratios: 0/100%, 33/67%, 67/33%, and 100/0% (sewage/lake water). The test chambers were 5-gallon clear plastic bags containing 15 L of the test water mixtures. The measurement ends of the sondes were placed into the test chamber bags and sealed with tape after as much air as possible was removed. The test chambers and sondes were placed on the lake bottom in approximately 1 to 2 ft of water near the shore and in full sun.

The 0% sewage test chamber indicated a 5-day biochemical oxygen demand, BOD_5, of approximately 2.5 mg/L. The 33% sewage chamber had initial anoxic conditions, but after acclimating for approximately 5 days, there was a diurnal photosynthesis/respiration variation observed: the DO levels in this chamber were supersaturated during the daylight hours. When this chamber was pulled at the experiment's end, there was a large amount of green biomass present, indicating large amounts of photosynthesizing material. The 67% and the 100% sewage test chambers stayed at anoxic DO levels throughout the test period.

Plots of DO were then created using the 0 and 33% sewage results for the last 5-day period in order to calculate the P/R rates, corrected for the experimental photoperiod. The net photosynthesis rates for the 33% sewage were very high, ranging from 12 to 30 mg/L/day for the 5 days of useful data, indicating variations associated with different cloud cover. The net photosynthesis rates for the 0% sewage/100% lake water mixture were typical for local lake waters, being approximately 1 to 2 mg/L/day.

The use of the YSI 6000 sonde, with the rapid-pulse DO sensor, allowed these simple experiments to be conducted. Conventional P/R measurements using light and dark bottles would not be sensitive to the relatively long acclimation period noted for raw sewage discharges into waters that rarely receive SSOs. In areas having more consistent SSOs, the acclimation period would not be as long. In addition, the long-duration experiment enabled us to observe variations in the P/R rates corresponding to different weather conditions and other factors. The use of only a single random P/R value (which would be obtained using conventional *in situ* light/dark bottle tests) could result in large errors.

Interaction of Water Column Pollutants and Contaminated Sediments and Interstitial Waters — There are five processes that affect the pollutant exchange between the water column and the sediment interstitial water and that affect the fates of SSO discharged pollutants: (1)

hydrodynamics, currents, and wave action; (2) resuspension/erosion of sediments; (3) flocculation, settling speeds, and deposition; (4) sorption of chemicals to sediments; and (5) flux/diffusion of chemicals from the water column to interstitial water, and vice versa. The most important processes, or those that contribute most to short-period chemical exchange, in a stream such as Five-Mile Creek, are those that promote turbulent mixing of the water column and the interstitial water. Therefore, experiments were conducted to measure the relative exchange rates between the water column and interstitial water for coarse and fine stream bed sediments. Results of these tests are presented in Chapter 6.

This study examined the exchange of water and the degradation of interstitial water due to poor water quality flowing over its surface. It was expected that differences in sediment particle size between the monitored sites will impact exchange, i.e., sites having larger, well-graded sediment particles will allow more rapid and complete exchange between the interstitial water and the stream water than will smaller sediment particle sizes.

The test locations for this experiment on Five-Mile Creek were near a site of a continuous SSO. At this site, raw sewage, at a rate of several liters per minute, flows over about 300 ft of ground before discharging into the creek. The flow in the creek ranged from approximately 2 to 10 m^3/s during the experiment. Four sondes were deployed: two were located upstream and two were located downstream of the SSO discharge point. At each upstream and downstream site, one sonde was located on the creek bottom and the second sonde was buried under approximately 6 in of sediment. The sondes were protected from large particles by placing them inside 75 μm aperture nylon mesh bags.

The YSI 6000 sondes enabled direct measurements of the lag time and magnitude response from the surface to the interstitial water for several parameters. There were no detectable differences between the upstream and downstream water quality data, in relation to the continuous SSO location. The background levels of pollutants in the creek masked the smaller SSO discharge effects. The differences in the flow rates of the SSO discharge and the creek were high, causing great dilution. However, the data from the buried sondes were used to compare interstitial water characteristics at the two sites based upon different sediment characteristics.

At the fine sediment site, the temperature plots indicated a definite lag time between changes in the water column and the sediment interstitial water of approximately 6 hours from peak to peak at the fine sediment site and approximately 2 hours at the coarse sediment site. The data at the coarse sediment site showed a much closer correlation between the water and the interstitial water than for the fine sediment site. The interstitial water at the coarse sediment site changed with the water column, albeit at a reduced magnitude, while the interstitial water at the fine sediment site showed no change.

Specific conductance was selected as the best parameter for monitoring chemical exchange between the water column and sediment interstitial water. The rate of relative chemical exchange was much higher and more variable in the coarse sediment than in the fine sediment. In the coarse sediment, the much more rapid process of turbulent mixing was occurring, as opposed to the slower process of diffusion, which is the driving force in the fine sediment.

The use of the continuously recording sondes, especially with the rapid-pulse DO sensors, enabled real-time interstitial water quality changes to be made. These measurements are especially important for sensitive parameters that are not possible to accurately measure in collected samples (especially ORP). The continuous measurements showed that interstitial water within fine sediments was basically isolated from the overlying water column, and the quality of the interstitial water was therefore affected by sediment quality. The coarse sediments, however, allowed a relatively free exchange of water between the overlying water and the interstitial water, with much less of an influence of sediment quality on interstitial water quality.

Interstitial Water Measurements — Peepers (described in Chapter 5) were used to contrast interstitial water conditions in sediments having different textures and levels of contamination. The

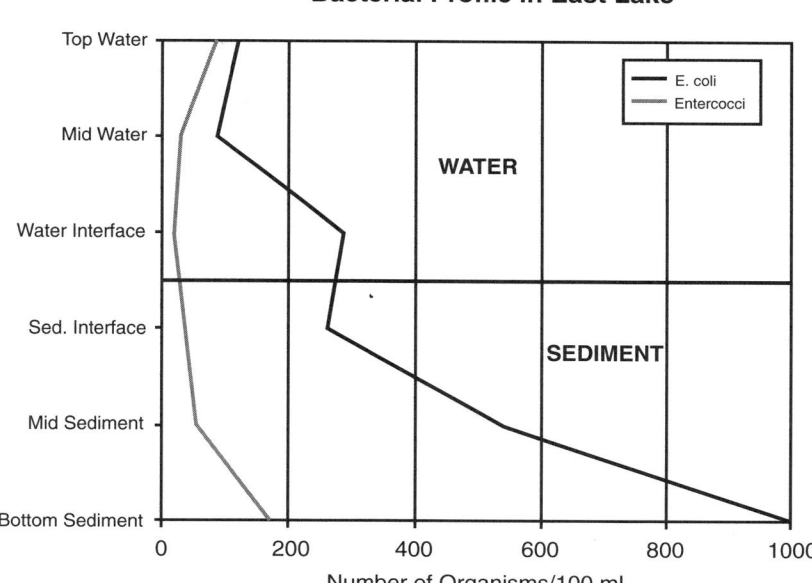

Figure 4.51 Interstitial water bacteria populations contrasted to overlying water conditions.

fine vertical spatial resolution enables measurements close to the sediment–water interface and at deeper depths. Initial experiments were conducted to examine bacteria population variations with depth. Figure 4.51 presents bacteria observations using the peepers. Very high bacteria populations were observed in the sediments, much greater than the overlying water column observations. These data indicate that the deposition of particulates, with associated bacteria, is likely an important fate mechanism for wet-weather flow bacteria. These bacteria may also be readily scoured during periods of high flows, as shown during monitoring on the Rideau River in Ottawa (Pitt 1983b).

Ten peepers were constructed for monitoring vertical variations in interstitial water quality. The peepers are machined from Delrin™ and have 46 (8 mL) cells, 1 cm apart. For use, the cells are covered with a 74-μm nylon screen, which will hold water, but allow diffusion of most pollutants, bacteria, and silts. The peepers are washed with concentrated nitric acid, rinsed with deionized water, and all cells are filled with Reverse Osmosis quality water (18 Mohms resistivity). The cells are then sealed with the nylon screen membrane, and the slotted covers are bolted on. Special stainless steel covers slide over the peepers, protecting the membranes during transport and placement. The prepared peepers are then brought to the field (keeping them horizontal to minimize water loss) and carefully pushed into the soft deposits of the stream bed, leaving at least a few of the uppermost cells above the sediment surface. After installation, the stainless steel covers are then carefully removed by sliding them off, leaving the membranes relatively unimpacted by sediments.

The array of cells allows investigations of the effects of depth on interstitial water chemistry and microbiology. The peeper is placed in the sediment and allowed to equilibrate for a period of time, usually at least 2 hours with the relatively coarse screen. After this period, the stainless steel covers are pushed over the peepers, and the units are removed from the sediment when they are carefully rinsed with clean water to remove any superficial sediment from the cell coverings. In order to extract the water samples from the cells, a small hole is made in the mesh covering with a sharp object, allowing a 10-mL plastic syringe to withdraw the sample water. The water is then transferred to a small storage vial and sealed and brought to the laboratory for analysis. pH and conductivity are measured on site using a micro probe.

Measurement of Frequency, Duration, and Magnitude of WWF Events — This experiment was conducted to examine the utility of the continuous recording YSI 6000 sondes as a tool for monitoring the duration, frequency, and magnitude of wet-weather flow events. Short-term, or runoff-induced, pollution effects can be studied in detail using these instruments. The long deployment time and continuous monitoring capability of the YSI 6000 enables acquisition of data for multiple events, i.e., as many as occur during the time of deployment. The sonde can be programmed to record stream depth, turbidity, and specific conductivity, all found to be all good indicators of wet-weather flows. Chapter 6 contains illustrations of the data obtained during these experiments.

Depth and turbidity values both increased, and the specific conductivity values decreased simultaneously at the beginning of a WWF event. The rise period for all of the parameters was very rapid, and the peaks occurred very early in the runoff event. They then returned to the previous levels within 1 to 2 days, depending upon the parameter. The data set acquired for water depth is obviously the parameter that best correlates to the runoff hydrographs.

The flow in Five-Mile Creek rapidly changes with rain conditions, especially considering that the watershed is relatively large (many square miles). However, the water quality remained degraded long after the water levels decreased to baseflow conditions. The turbidity remained elevated for about 30 hours, and the specific conductivity remained depressed for about 40 hours, although the hydrograph response was completed in about 12 hours. Because of the common rains in Alabama (rains occurring about every 3 to 5 days, and moderate rains similar to that which was monitored occurring about every 10 to 15 days), the degraded water quality associated with the WWF could affect the creek about 10 to 20% of the time. In addition, several days of exposure to degraded conditions may be common, instead of the several hours of exposure to degraded conditions typically assumed for WWF effects.

Continuously recording sondes, especially those capable of long-term monitoring of depth, turbidity, and specific conductivity, are therefore very useful in indicating the frequency, magnitude, and duration of WWF degradation on in-stream water quality. If located upstream and downstream from a major SSO discharge point, these devices can also continuously measure the magnitude of the SSO flows in relation to the receiving water flow. The SSO location where the sondes were located for this demonstration did not cause any measurable difference in the sonde parameters (DO, temperature, specific conductivity, pH, ORP, turbidity, or water depth) because of its relatively small flow in relation to the large creek flow.

Additional Tests for Sediment Oxygen Demand (SOD), Sediment P/R, and Settleability of Bacteria and Toxicants — A series of tests is also planned to more fully examine the role that sediments play with SSO pathogens, oxygen-demanding material, and toxicants. These tests will be necessary to calibrate receiving water models and estimate the fates and risks associated with SSO discharges. Four clear plastic bottomless boxes are being constructed as SOD chambers. A flange opening on one side of the boxes will hold the YSI 6000 continuously by recording sondes. During short-term use, two of the test chambers will be covered with opaque material (such as aluminum foil) to act as a dark chamber for respiration analyses, while two will remain clear for respiration plus photosynthesis measurements. During extended tests, the chambers will remain clear, measuring respiration during the night and photosynthesis plus respiration during the day. The chambers will also have temporary bottoms available for background water P/R analyses. This will enable the SOD to be directly measured over a period of several days, as in the previously described *in situ* water column P/R tests. Deployment of the test chambers over a several-day period above and in the vicinity of an SSO discharge will result in sufficient data to indicate SSO-impacted SOD under various weather conditions.

It is possible that much of the reported die-off of bacteria in natural waters is actually associated with settling. Very high bacteria populations have been noted near the sediment–water interface and these can be easily resuspended during periods of high flow or other turbulence (Pitt 1983b). These settling experiments will therefore supplement the *in situ* peeper tests and the *in situ* die-off tests to distinguish settling and die-off of bacteria and biochemical changes of the pollutants. Conventional laboratory settling columns (30 cm in diameter and 1.3 m in height) will be used to measure the settling rate of SSO pollutants, especially bacteria and toxicants. Control tests (using a continuously stirred chamber) will indicate die-off of the bacteria and biochemical modifications of the chemicals.

Step 6. Confirmatory Assessment

Being a multiyear project, numerous project elements should be conducted during later project phases. An important element of this later work will be confirmation of the recommended approach developed during the earlier phases, based on actual receiving water beneficial use impairment measurements. The recommended approach will present several options, having increasingly complex and expensive activities, but with increasing confidence in the conclusions. It is expected that a moderate level of activity will be the most cost-effective approach. However, the costs associated with correcting SSOs in an area are extremely high and additional information and associated higher confidence in the assessment studies will result in a greater degree of success of the control program.

Step 7. Project Conclusions

The preliminary results confirmed several obvious hypotheses: small SSOs discharging into large receiving waters produce little measurable effects, while large intermittent SSOs discharging into smaller water bodies can be serious. However, many small, continuous SSOs in small urbanized waterways can dominate dry-weather conditions, producing hazardous situations, while they are completely obscured during most wet-weather events by the larger flows and pollutants associated with urban runoff.

The small experiments demonstrated useful tools needed for calibrating receiving water models used for estimating fates and exposures of SSO pollutants. Without site and SSO specific tests, modeling estimates could be very misleading.

It is expected that the extensive list of chemical and biological parameters being investigated during this project can be effectively reduced to result in cost-effective investigations of local SSO problems, especially considering the very high cost of reducing SSO discharges. The information obtained can also be used in a TMDL evaluation to determine the role of SSOs in relation to other discharges in a watershed.

Outlines of Hypothetical Case Studies

The following hypothetical case studies represent commonly encountered situations where the effects of stormwater runoff may need to be determined. These brief examples are based on similar studies and reflect integrated, weight-of-evidence study designs (as described previously). As always, available resources will determine how comprehensive a design is feasible. The following designs assume relatively limited resources, yet address the essential components that allow for reliable weight-of-evidence-based conclusions and decision making. Additional resources are needed for Tier 2 level "confirmatory" assessments that identify specific stressors, their relative contribution to degradation, and their sources. These test designs can easily fit into the EPA Ecological Risk Assessment paradigm or Stressor Identification Evaluation Process. For additional information on useful multistressor assessment methods see Baird and Burton (2001).

Effect of Outfall on Algal Growth

Case Situation: A permitted industrial effluent contains low levels of nitrogen and phosphorus and is discharged into a small urban stream. The upstream watershed is predominantly older residential neighborhoods. Stormwater runoff is discharged directly into the stream. Upstream of the outfall the stream is intermittent, with occasional no-flow conditions occurring during dry, summer periods. However, the industry provides continual flow from its outfall, providing aquatic habitat downstream throughout the year. The receiving stream has excessive algal growth downstream of the outfall. The state environmental agency is concerned over the role of the effluent on the algal growth and suggests additional wastewater treatment should be added to reduce nutrient levels.

Step 1. What's the Question?

Does the outfall degrade water quality and cause excessive algal growth in the receiving stream?

Step 2. Problem Formulation

There are basically two separate issues that should be addressed. First, if there were no industrial outfall, what would be the quality of the downstream reach? Given the intermittent nature of the stream, it is likely that both the benthic macroinvertebrate and fish populations would be very limited and only of a brief seasonal nature. However, the environmental agency may argue that since the outfall does occur, it must be free of excess nutrients. The second issue is whether nutrients from the outfall are degrading downstream conditions. These two issues dictate that upstream and downstream sampling be conducted during low and high flow conditions, monitoring the relationship of flow and outfall loadings with both biological communities and nutrient concentrations.

Steps 3 and 4. Project Design and Implementation

A site reconnaissance found >90% of the upstream watershed was an older, middle-income residential neighborhood with no septic systems. There were no continual discharges or combined sewers evident; however, stormwater discharges emptied directly to the stream. This suggests that runoff would include nutrients (from lawn fertilizers and small mammal feces), pathogens (from small mammal feces), pesticides (from lawn/garden care chemicals and agrichemicals in rainfall), and some metals and petroleum products (from automobiles and roadways). The stream habitat was relatively good throughout, with a good riparian zone, some stream canopy, and sand to cobble substrates with little siltation or embeddedness. However, there were no pools of depths greater than 1 ft, indicating a susceptibility to drought conditions. Excessive algal growth occurred near the outfall, but decreased downstream. Various fish species and benthic macroinvertebrates were observed downstream, but not upstream of the outfall.

A weight-of-evidence, multicomponent assessment design was used. This included physico-chemical monitoring of key parameters (ammonia, nitrate, nitrite, total and orthophosphorus, turbidity, temperature, pH, conductivity, flow) during a low and high flow event. The outfall was sampled with an ISCO automatic sampler during each event. Composited samples were collected during low flow by grab sampling and during high flow with a flow-activated ISCO automatic sampler. ISCO samples were separated into 15-min intervals. Flow was measured using a Marsh-McBirney flow meter. Effluent flow was monitored continuously by the plant and did not vary during the low and high flow sampling events. During high flow, flow was measured during pre-crest and post-crest for comparisons to ISCO samples and stage graphs. In addition, the benthic macroinvertebrate and fish communities were assessed at two sites upstream and downstream of the outfall during low flow conditions in late summer using the EPA's Rapid Bioassessment Protocol I (EPA 1987, Appendices B and C). This process includes a Qualitative Habitat Evaluation Index assessment at the same sites (Appendix A). Finally, the EPA algal 96-hour growth test using *Selenastrum capricornutum* (Appendix D) was conducted during the low and high flow events on three samples (upstream, outfall, and downstream).

Steps 5 and 6. Data Evaluation and Confirmatory Assessment

The data showed significant water quality differences between high and low flow conditions in the stream. The outfall contributed nondetectable levels of phosphorus, nitrite, and ammonia and approximately 1 mg/L nitrate. The habitat downstream was better than the upstream habitat. The upstream reach was only isolated pools during late summer. The downstream habitat had flowing water and greater canopy cover. During low flow, nitrogen and phosphorus were nondetectable in both upstream and downstream water samples. During high flow conditions nutrient levels were highly elevated and did not differ significantly between upstream and downstream samples. A conversion to mass loading based on flow conditions showed the outfall contribution less than 1% of the nitrogen to the stream on an annual basis, as compared to one high flow event. No fish or benthic macroinvertebrates were recovered from the upstream isolated pools during the summer low flow sampling. Several pollution-tolerant species were recovered downstream. The algal growth test showed increased growth in the outfall sample. The upstream and downstream samples showed similar low levels of growth. No confirmatory assessment was deemed necessary.

Step 7. Conclusions

The weight of evidence clearly established that while the outfall does contribute nitrogen to the stream, it is insignificant in comparison to the nutrient loading during high flow conditions from the upstream residential area. The pollution-tolerant species found downstream of the outfall are typical of an urban waterway and likely reflect the stressor loadings from the upstream watershed. Stormwater controls should be installed to yield the greatest improvement to water quality.

Effect of On-Site Runoff from an Industry

Case Situation: A manufacturer has site runoff discharging into a drain which empties directly into a small stream. The manufacturer has a large amount of on-site vehicular traffic and uses a variety of inorganics (e.g., caustics, metals) and petroleum products in the production process. The upstream watershed is mixed urban and agricultural. As part of the stormwater permitting process, the company must determine whether its runoff is contaminated.

Step 1. What's the Question?

Does the on-site runoff degrade receiving water quality?

Step 2. Problem Formulation

A potential for stormwater contamination exists since there is a large amount of impervious area being drained that is susceptible to spills from industrial processes, chemical accidents, and diesel-gasoline-powered vehicles. The watershed upstream of the stormwater outfall is approximately 50% commercial and industrial sites and 50% agriculture (crops and pasture). The brief survey of the stream showed primarily pollution-tolerant species with occasional sensitive species both upstream and downstream of the outfall.

Steps 3 and 4. Project Design and Implementation

The stormwater from the test site had the potential to be contaminated with a wide range of compounds, which may or may not have water quality standards. Given the changing nature of the stormwater quality and the sporadic discharges, it is unlikely that any chemical data could be logically interpreted using water quality standards. The uncertain and changing exposures that organisms would undergo in the stream would not allow for reliable predictions of ecological effects using chemical data only. To achieve an adequate database would require extensive inorganic and organic monitoring during many runoff events. Therefore, to improve data interpretation in a cost-effective manner, a tiered approach was chosen, whereby biological effects were first monitored to determine if detrimental impacts were occurring.

Tier 1 of the study involved a stream survey of benthic macroinvertebrates upstream and downstream of the stormwater outfall using the Ohio EPA's Invertebrate Community Index approach (Appendix B). This was conducted for 30 days during the summer, during which two storm events occurred. During those storm events, flow-activated ISCO samplers collected samples from the outfall, upstream and downstream. Short-term chronic toxicity testing was conducted on the water samples using *Ceriodaphnia dubia* (Appendix D). In addition, toxicity testing was conducted on upstream and downstream samples during low flow conditions.

In the event that toxicity or biological impairment was suspected due to the outfall, a Tier 2 study was designed that focused on identification of the stressor. This involved both laboratory and field testing, using EPA's Toxicity Identification Evaluation (TIE) procedure and *in situ* exposures of caged organisms (Appendix D). The Phase 1 TIE was conducted on a fresh composited outfall sample using *C. dubia* acute exposures. The *in situ* exposures were conducted during low and high flow events (4 days each), upstream and downstream of the outfall. Two species were used: *Daphnia magna* (a zooplankton similar to *C. dubia*) and *Hyalella azteca* (a benthic macroinvertebrate recommended by EPA for sediment toxicity testing). These organisms were exposed in different treatments to better identify potential stressors: (1) light vs. dark cages to identify whether photo-induced toxicity from polycyclic aromatic hydrocarbons (PAHs) exists, and (2) small vs. large mesh cages to identify whether suspended solids contribute to mortality. Basic water quality measures monitored during the exposures were DO, pH, temperature, conductivity, turbidity, ammonia, alkalinity, and hardness.

Steps 5 and 6. Data Evaluation and Confirmatory Assessment

Tier 1 testing found the benthic invertebrate populations upstream and downstream of the outfall were of fair quality; however, those downstream scored lower. This suggested that the outfall may be contributing stressors to the stream; however, given the variable nature of benthic invertebrate communities and stormwater, these results were not conclusive. The toxicity testing results were mixed as shown in Table 4.36.

These data suggest that toxicity from the outfall is variable, but does exist. Its effect on the receiving water is uncertain, as the upstream and downstream samples were not significantly different statistically. It is also apparent that storm events are toxic in the stream, but baseflow conditions are not. The results of the ICI showed both upstream and downstream communities were of poor quality.

Table 4.36   *C. dubia* Survival and Reproduction at Manufacturing Site

	Upstream	Outfall	Downstream
Storm event no. 1	60%	70%	62%
	15 neonates	13 neonates	10 neonates
Storm event no. 2	75%	20%	65%
	20 neonates	0 neonates	10 neonates
Baseflow event	90%	NA	95%
	28 neonates		32 neonates

Tier 2 testing was then initiated. The laboratory-based TIE Phase 1 suggested metals as a primary toxicant and nonpolar organics as a possible toxicant. The in-stream *in situ* exposures showed no significant differences between upstream and downstream, at low or high flows. High turbidity existed during high flow events, and hardness values, while lower during high flow events, were still >300 mg/L $CaCO_3$. The suspended solids exposure treatment during high flow showed relatively high survival when solids were removed. During low flows there was greater toxicity in the light treatments suggesting PAH-photoinduced toxicity exists.

Step 7. Conclusions

These results show the outfall is toxic, primarily due to metals. The concentrations of metals found, however, are not at a level that is likely to cause toxicity in the receiving water due to its high hardness. The nonpolar organic toxicity observed in the effluent may be contributing to the photoinduced toxicity observed during baseflows. However, since these effects were also noted upstream, there are likely additional sources of PAHs upstream. The high levels of suspended solids appear to be contributing to the poor benthic community quality also and will require watershed-based controls to mitigate the problem. These studies did not ascertain whether or not chemicals associated with the suspended solids are contributing to mortality, nor did they rule out other stressors in the receiving stream, such as pesticides. The conclusion is that the outfall does contribute some toxicity to the receiving water, but not at a significant level that could be detected in the stream.

Effect of a Dry Detention Pond

Case Situation: A shopping center has many acres of property that drain into a dry detention pond. The detention pond outfall empties into a stream. A local citizens group expresses concern that water quality is poor downstream of the outfall. A study is initiated to determine whether the dry detention pond drainage is contributing to stream degradation.

Step 1. What's the Question?

Does the dry detention pond outfall degrade water quality in the stream?

Step 2. Problem Formulation

Four different situations are likely to be encountered in urban watersheds where dry detention ponds are used that will affect the study design. First, the outfall discharges into the headwaters of a stream so that the upstream–downstream sampling design is not possible. In this case, a nearby ecoregion reference site may be used that has a similar sized drainage area and the habitat is similar. If habitat modification is a possible cause of impairment (stress), the reference site should have a reasonably good habitat that is unmodified. Since this is a headwater area, fish and benthic communities are likely to be limited by stream size, available habitat, and food availability. Therefore, monitoring should focus on toxicity and loadings of pollutants (chemical and physical) to downstream areas.

The second situation often encountered is that the upstream reach is also degraded, so the upstream–downstream sampling design is somewhat problematic. Again, a nearby reference site is useful, but mainly as a control site to ensure method validity. The key approach in this situation is to assess the outfall quality and its loading of pollutants to the stream during high flow conditions.

An upstream–downstream sampling approach may show increased toxicity and contamination downstream or dilution of upstream contamination.

The third situation encountered is that the upstream area is relatively unimpacted, so traditional upstream–downstream sampling designs as described above may be used.

Finally, the fourth situation is the use of "side-stream" detention ponds where the detention pond is located adjacent to the stream or drainage and captures water only during unusually high flow periods (possibly only a couple of times a year). In small drainages, a dry detention pond may have a lined channel passing through the excavated area that carries the stormwater. Only when the stormwater flow exceeds the capacity of a downstream culvert does the water back up into the adjacent area (like an artificial floodplain). Side-stream dry ponds can also be located adjacent to larger receiving waters, and can fill with excessive flows when the stream stage exceeds a side overflow weir. In many cases, these larger side-stream dry ponds are used as recreation areas. It is difficult to monitor the benefits of these ponds during events where the pond is in operation, as their operation is commonly so intermittent that they rarely divert water.

The primary benefit of a dry detention pond is the reduction in peak stormwater runoff flow rates and associated energy. The increased flow and energy resulting from greater runoff across impervious areas and loss of infiltration basins can cause flooding and/or destroy stream habitat, resulting in beneficial use impairments. Unfortunately, many of the detention ponds in use do not reduce flow enough, still resulting in habitat alteration. In addition, monitoring dry detention ponds rarely has shown significant and important pollutant concentration and mass yield reductions. Some dry ponds partially may act as percolation ponds where some of the runoff is infiltrated.

Steps 3 and 4. Project Design and Implementation

Since many detention pond outfalls discharge into small headwaters or tributaries, the first situation described in Step 2 will be addressed. A site reconnaissance showed that the watershed that drains into the dry detention pond is >90% impervious parking lots. This suggests that runoff may contain suspended soils, salt (during periods of snowmelt and possibly for a few additional months, depending on the levels of deicing salt applications), petroleum products and metals (from automobiles), and perhaps low levels of pesticides associated with precipitation events. The stream into which the pond discharges is a first-order tributary and is intermittent in flow; however, it joins a small, high-quality, perennial stream approximately 200 yards from the pond.

As in the previous case study examples, this site should be studied at both low and high flow conditions. There should be a minimum of four stations, two on the tributary (near outfall and near mouth) and two on the perennial stream just upstream and downstream of the tributary confluence. In addition, it would be useful to have a similar ecoregion reference site for comparison. At each site, qualitative habitat evaluation indices (Appendix A) would be evaluated, along with rapid bioassessments of the benthic macroinvertebrate communities (Appendix B) on one occasion during the summer. Toxicity testing (*Pimephales promelas* 7-day survival and growth assay, Appendix D) was conducted on grab water samples collected during first flush conditions and at low flow. In addition, toxicity of depositional sediments (*Hyalella azteca* 10-day assay, Appendix D) was conducted at three sites (near mouth of tributary, and upstream and downstream of confluence at the first depositional sites). General water quality measures were also made during low and high flow collection periods.

If toxicity was observed, confirmatory assessments would consist of *in situ* toxicity exposures on the tributary and two sites on the perennial stream. These exposures would include treatments to evaluate whether toxicity was associated with water or sediments, suspended solid or dissolved fractions, and whether PAH-photoinduced toxicity was a stressor (as described in the preceding Case Study Example). Extensive chemical analyses were not warranted as the only source was a parking lot. If advanced treatment was recommended, then identification of the dominant chemical stressors might be needed.

Steps 5 and 6. Data Evaluation and Confirmatory Assessment

It was apparent that the tributary had received substantial loadings of eroded soils during the construction of the shopping center, as the natural large-grained sediments were embedded with clays and silts. The habitat quality of the two tributary sites was very different due to the change in gradient, which precluded comparisons of station impairment. The perennial stream habitats did not vary appreciably from each other.

The laboratory toxicity tests showed growth impairment in both of the tributary high flow samples, but not in any other water samples (high or low flow). The amphipod *H. azteca* had poor survival in the tributary and downstream perennial stream sediments. The benthic community results showed only a fair community in the intermittent tributary, but a good community in the perennial stream.

Confirmatory Tier 2 studies revealed that most of the toxicity was associated with the suspended solids; however, some toxicity was also observed in the small mesh (50 μm) chambers. No water column treatment difference were observed in the light–dark treatments. However, the sediment light treatments showed increased toxicity during baseflow conditions in the tributary and downstream samples.

Step 7. Conclusions

The dry detention pond outfall was toxic during the first flush of the events. Since the drainage area was mostly a large paved area, with simple drainage, high concentrations are more common near the beginning of storms than later. However, if short periods of high rain intensity occur later in the storm, an additional surge of high concentrations would likely occur due to the increased storm energy. If the drainage area was a typical mixed urban area, the drainage system would be more complex and the different surfaces would cause flows coming from different areas to be much more mixed, significantly reducing any first-flush effect.

Most of the toxicity was associated with suspended solids and likely contributed to the toxic sediments observed downstream. It is uncertain whether this toxicity from the pond is significantly impacting the perennial stream without more extensive studies. Improved reduction of suspended solids, possibly by retrofitting the pond to an extended detention pond or a wet pond, would likely result in improved downstream aquatic communities.

Effect of a Wet Detention Pond

Case Situation: A wet detention pond is located on-line, in a creek that drains a developing watershed of approximately 3 mi^2. The pond was created by constructing a small dam across the creek. The creek begins in farmland and drains into the residential development containing expensive homes before reaching the detention pond. The detention pond water quality has degraded, with eutrophic conditions such as algal blooms and occasional fish kills. The state environmental protection agency suspects additional downstream problems may be due to the pond and conducts an assessment.

Step 1. What's the Question?

Is the wet detention pond impairing water quality downstream?

Step 2. Problem Formulation

Wet detention ponds typically are located on or off a stream. On-line ponds are constructed in the existing waterway and capture all upstream flows. Adjacent ponds are located next to the stream, before the outfall, and only treat water originating from the smaller drainage, and not the complete receiving watershed. The advantage of these ponds is as for dry detention ponds, in that they can reduce the power associated with high flow events, thereby reducing habitat destruction and loss of aquatic organisms. If large enough, they can also capture appreciable amounts of the stormwater particulates and associated pollutants. Since on-line ponds may treat much larger areas, they need to be correspondingly larger for similar levels of treatment. In addition, the low head dams across the stream result in a loss of flowing stream reach, block fish migration, degrade the habitat needed for more pollution-sensitive species, and allow accumulation of depositional sediments that contain toxicants. This study will not focus on the water quality of the pond, but whether the outflow from the pond degrades downstream beneficial uses.

Steps 3 and 4. Project Design and Implementation

Water quality was evaluated during both low and high flow conditions. There were three stations, two downstream of the pond and one upstream. An ecoregion reference site was also selected with which to compare fish and benthic community results. At each site, qualitative habitat evaluation indices (Appendix A) were evaluated, along with rapid bioassessments of the fish and benthic macroinvertebrate communities (Appendices B and C). Toxicity was assessed using *in situ* exposures of caged organisms (Appendix D). The *in situ* exposures were conducted during low and high flow events (4 days each), upstream and downstream of the outfall. Two species were used: *Daphnia magna* and *H. azteca*. Contrary to earlier case studies, PAH-photoinduced toxicity was not suspected as a potential stressor in this watershed. So *in situ* treatments were limited to water and sediment exposures. Basic water quality measures monitored during the exposures were DO, pH, temperature, conductivity, turbidity, ammonia, alkalinity, and hardness. Testing was conducted during the spring and late summer to investigate critical time periods of pesticide application, fish spawning, and low flow conditions.

If toxicity was observed in the water column during high flow, a TIE would be conducted as described above. This would help identify the source of the toxicity. If sediment toxicity or community impairment was observed, confirmatory assessments would consist of additional sediment toxicity testing and bioaccumulation testing. Toxicity of depositional sediments (*H. azteca* and *Chironomus tentans* 10-day assay, Appendix D) would be conducted at all sites where depositional sediments occurred. Since pesticides were suspected from both the farming and residential areas, bioaccumulation of organochlorines (such as DDT, chlordane) was also investigated by looking at fish tissue samples. If upper trophic level fish could not be captured, then semipermeable membrane devices (SPMDs) would be used to collect bioaccumulable substances (see Chapter 6).

Steps 5 and 6. Data Evaluation and Confirmatory Assessment

High levels of turbidity were observed during high flow events. The majority of this turbidity appeared to originate from upstream farmland and erodable stream banks. Toxicity was observed during high flow conditions in the water column. Slight toxicity was observed in stream sediment exposures. Habitat conditions did not vary appreciably among sites. The benthic communities were of fair quality at all sites and were not significantly different. The fish community was poor upstream of the detention pond and fair to poor below. Ammonia was found at elevated levels during the late summer period at all sites.

Follow-up confirmatory assessments showed significant sediment toxicity in laboratory exposures. A TIE evaluation suggested pesticides may be present during the spring high flow periods. Fish tissue residues showed detectable levels of chlordane and DDE.

Step 7. Conclusions

The wet detention pond affected downstream water quality appreciably. The upstream and downstream portions appeared to be impacted by elevated levels of pesticides and nutrients from the farming and residential drainage. The poor water quality observed in the pond was likely due to the buildup of nutrients in the sediments and water, allowing for excessive productivity and occasional anoxia. The widespread toxicity and detection of pesticides in the fish suggest upstream stormwater controls are needed.

SUMMARY: TYPICAL RECOMMENDED STUDY PLANS

Components of Typical Receiving Water Investigations

The specifics for any receiving monitoring program would be determined by the study objectives and the site conditions. As an example, Table 4.37 summarizes some general parameters that should be included in an urban water use evaluation study, depending on the specific beneficial uses of interest. Of course, the final parameters selected for study would vary for specific site conditions and historical information. As expected, an investigation of drainage uses (the primary use for an urban waterway) would be relatively straightforward compared to studies of other use impairments. However, investigations of drainage problems can be expensive and time-consuming. When the other uses are added to the list of potential objectives, the necessary data collection effort can become very comprehensive and expensive. Therefore, a staged approach is usually recommended, with a fairly simple initial effort used to obtain basic information. This information can then be used to develop specific experimental designs for later study stages.

Example Receiving Water Investigations

The following scenarios are brief examples of simple to complex receiving water investigations that incorporate many of the elements shown in Table 4.37. The first example, budgeted in Table 4.38, is the least expensive and would be appropriate for a single monitoring condition, such as a small lake or pond, or a short segment of a relatively small and homogeneous stream, having a single stormwater outfall. The proposed sampling effort is:

Water quality:	1 location × 1 season × 2 phases × 5 events/periods = 10 samples for analyses
Bacteria:	With above water samples, lab to analyze (*E. coli* and enterococci)
YSI sondes:	Rental for first/single deployment, $1000 per month
Inappropriate discharge screens:	1 outfall × 2 replicates = 2 samples
Habitat:	1 season × 2 phases × 2 locations = 4 station tests
Rapid bioassessment (RBP):	1 season × 2 locations × 3 replicates = 6 site visits
Toxicity:	1 season × 2 phases × 2 locations = 4 station tests

Twenty sets of outfall water samples during both wet- and dry-weather phases would be needed to obtain an allowable error of 40% for typical levels of variation (as described in Chapter 5). However, since this is a single season sampling effort, not many wet-weather events are likely to occur. Therefore, it is assumed that five wet-weather events would be monitored during about a 1- to 3-month period, and the error in estimating the event mean concentration (EMC) could therefore be larger than 40%. A laboratory budget of $225 per sample should cover both *E. coli* and enterococci bacteria analyses, and selected total heavy metals and nutrients, plus COD and suspended solids

Table 4.37 Parameters of Concern When Evaluating Different Receiving Water Uses

	Drainage	Biological Life and Integrity	Noncontact Recreation	Swimming and Other Contact Recreation	Water Supply	Shellfish Harvesting and Other Consumptive Fishing Uses
Debris and obstructions (channel conveyance capacity)	X					
Habitat destruction (channel stability, sediment scour, and deposition)		X				X
High/low flows (rates and durations)		X	X	X		X
Aesthetics, odors, and trash			X	X		
Safety (bank condition, garbage)			X	X		
Public access			X	X		
Inappropriate discharges		X	X	X	X	X
Benthic macroinvertebrate species present		X				X
Fish species present		X				X
Polluted sediment (SOD and toxicants[a])		X				X
Toxicity and bioaccumulation of toxicants[a]		X				X
Health-related water quality standards (especially microorganisms[b] and toxicants[a])				X	X	X
Wet-weather quality (toxicants[a], nutrients[c], DO, temperature, alkalinity, and hardness)		X				X

Primary constituents are indicated in bold/underlined and should be analyzed for most all samples. Others can be analyzed less often as screening tests. In all cases, the common constituents should also be analyzed for all samples.

[a] Toxicants (organic toxicants such as pesticides, herbicides, and PAHs; metallic toxicants such as **zinc**, **copper**, **lead**, cadmium, arsenic, and mercury) and toxicity tests (such as **Microtox screening test**, plus other *in situ* and laboratory toxicity tests).

[b] Microorganisms (indicator bacteria and selected pathogens such as: **fecal coliforms**, *E. coli*, **enterococci**, and *Pseudomonas aeruginosa*).

[c] Nutrients (**ammonia**, TKN, **nitrates**, TP, **phosphates**).

Common constituents, added to all water quality investigations (**pH**, **conductivity**, **turbidity**, **suspended solids**, **COD**).

analyses. These data would be supplemented with field screening in the drainage system during two dry-weather flow periods (assuming water was found during both visits) to identify inappropriate sources of wet-weather flows. It is recommended that a YSI 6000 probe be rented for a 1 month to measure flow (depth values) and water quality variations (DO, temperature, conductivity, turbidity, and pH) during several runoff events and periods of dry weather in the receiving water. This would indicate the duration and severity of the runoff events and the associated recovery periods. Diurnal DO and temperature fluctuations would also be measured. This water quality data would be supplemented with habitat, rapid bioassessment (RBP), and limited *in situ* and laboratory toxicity testing above and below the outfall (two locations). This collective information should give a good indication of the presence of receiving water problems at the site. Of course, because it is a single season analysis, an appropriate sampling schedule needs to be carefully selected, probably based on critical biological conditions in the receiving water (likely early spring or late summer, depending on the expected organisms present and the local weather patterns). Besides being a minimum sampling

Table 4.38 Monitoring Cost Estimate for Single Outfall in a Single Receiving Water Segment of Interest

	Equipment Cost			Analytical Cost			Labor Cost					
	Unit Cost	No. Needed	Total Cost	Unit Cost	No. Needed	Total Cost	Labor (hrs)	No. Needed	Total hrs	Costs at $35/hr	Total Costs	
Field screening				$50	2	$100	1.5	2	3	$105	$205	
Habitat				na	4	na	0.35	4	1.4	49	49	
Toxicity				500	4	2000	na	na	na	na	2000	
RBP				na	na	na	2	6	12	420	420	
YSI probe (rental)	$1000	1	$1000	na	na	na	8	1	8	280	1280	
Water and bacteria (lab)				225	10	2250	na	na	na	na	2250	
Site costs			1000						100	3500	4500	
Total			$2000			$4350			124.4	$4354	$10,704	

effort incorporating all recommended phases of a monitoring program, this scheme could be used as the initial effort for a longer-duration and more complex study.

The next scenario is for a more complex situation where there are 25 outfalls in a moderately sized (first-order) receiving water about 2 miles long in a completely urbanized watershed, 3 mi^2 in area. This is also presented as a first step in a possible recurring effort to cover more seasons or several years. The main purpose of this program is to identify possible serious receiving-water problems that would warrant more extensive evaluations. This scenario could be repeated at other similarly sized receiving waters in an area. In many ways, this scenario is very similar to the previously described program, except that the water sampling for bacteriological and chemical analyses would be conducted in the receiving water with some outfall samples. Outfall screening (using purchased test kits) during dry weather would also be conducted to identify inappropriate discharges. Table 4.39 shows the estimated costs, and the following lists the proposed effort for this program:

Water quality:	1 location × 1 season × 2 phases × 20 sets = 40
Bacteria:	With above water samples, lab to analyze (*E. coli* and enterococci)
YSI sondes:	Rental for first/single deployment, $1000 per month
Inappropriate discharge screens:	25 outfalls × 2 replicates = 50 samples
Habitat:	1 season × 2 phases × 25 locations = 50 station tests
RBP:	1 season × 4 locations × 3 replicates = 12 site visits
Toxicity:	1 season × 2 phases × 4 locations = 8 station tests

The last option shown is a relatively complete approach, covering all seasons, and is reasonably comprehensive and, therefore, relatively expensive. Again, the components are similar to the above programs, but the number of samples is greatly increased to cover the two critical seasons (RBP and sondes during four seasons) and to collect both outfall and receiving water samples. Because of the study duration, it would likely be more economical to purchase the YSI 6000 sondes and the bacteriological test equipment. The other water quality analyses would be conducted by a commercial laboratory. It may be appropriate to add selected immunoassay tests for pesticides and PAHs for some of the water samples (at about $25 each). Much greater site costs are shown because flow monitoring and rainfall monitoring will also be conducted during this effort. The sampling effort is shown below, while the estimated cost is shown in Table 4.40:

Water quality:	4 locations × 2 seasons × 2 phases × 20 sets = 320
Bacteria:	4 locations × 2 seasons × 2 phases × 20 sets = 320
YSI sondes:	4 locations × 4 seasons = 16 deployments
Inappropriate discharge screens:	25 outfalls × 2 seasons × 3 replicates = 150 samples
Habitat:	4 seasons × 2 phases × 25 locations = 200 station tests
RBP:	2 seasons × 4 locations × 5 replicates = 40 site visits
Toxicity:	4 seasons × 2 phases × 4 locations = 32 station tests

In all cases, major modifications are expected to be made to the above scenarios for real situations. In addition, the initial analyses will provide information that should be used to reexamine the complete workplan. Obviously, the above costs are only crude approximations, depending on local labor costs, site access, the availability of equipment, etc.

This chapter outlined an approach for designing appropriate multicomponent assessment projects for various conditions and objectives. As will be stressed throughout this book, it is critical that potential problems be examined using complementary and supportive procedures. It is inefficient, and subject to significant evaluation errors, to rely on simplistic single parameter/media approaches. Typical urban receiving waters are likely most affected by habitat degradation, frequent

Table 4.39 First Evaluation for 2-Mile Stream Segment Having 25 Outfalls

	Equipment Cost			Analytical Cost			Labor Cost				Total Costs
	Unit Cost	No. Needed	Total Cost	Unit Cost	No. Needed	Total Cost	Labor (hrs)	No. Needed	Total hrs	Costs at $35/hr	
Field screening	$1600	1	$1600	20	50	$1000	1.5	50	75	$2625	$5225
Habitat				na	50	na	0.35	50	17.5	612	612
Toxicity				500	8	4000	na	na	na	na	4000
RBP				na	na	na	2	12	24	840	840
YSI probe (rental)	1000	1	1000	na	na	na	8	1	8	280	1280
Water and bacteria (lab)				225	40	9000	na	na	na	na	9000
Site costs			2500						200	7000	9500
Total			$5100			$14,000			324	$11,357	$30,457

Table 4.40 Annual Sampling Effort for a Moderately Sized, Completely Urbanized Watershed Having 25 Outfalls

	Equipment Cost			Analytical Cost			Labor Cost				Total Costs
	Unit Cost	No. Needed	Total Cost	Unit Cost	No. Needed	Total Cost	Labor (hrs)	No. Needed	Total hrs	Costs at $35/hr	
Field screening	$1600	1	$1600	20	150	$3000	1.5	150	225	$7875	$12,475
Habitat				na	200	na	0.35	200	70	2450	2450
Toxicity				500	32	16,000	na	na	na	na	16,000
RBP				na	na	na	2	40	80	2800	2800
Bacteria	3000	1	3000	15	320	4800	0.05	320	16	560	8360
YSI probe	7000	4	28,000	na	na	na	8	16	128	4480	32,480
Water quality				175	320	56,000	na	na	na	na	56,000
Site costs			15,000						640	22,400	37,400
Total			$47,600			$79,800			1159	$40,565	$167,965

high flows, and contaminated sediment. While water and sediment chemical analyses can be expensive, they should not necessarily be rejected outright. Some of these more expensive analyses may be critical when evaluating biological and habitat information, for example. The number of needed data observations (as discussed in Chapter 5) and the sampling methods (described in Chapters 5 and 6) are critical for a successful assessment, in addition to the selection of the most appropriate assessment endpoints and overall assessment strategy.

REFERENCES

Baird, D. and G.A. Burton, Jr. (Eds.) *Ecosystem Complexity: New Directions for Assessing Responses to Stress.* Pellston Workshop Series. SETAC Press. To be published in 2001.

Burton, G.A., B.L. Stemmer, K.L. Winks, P.E. Ross, and L.C. Burnett. A multitrophic level evaluation of sediment toxicity in Waukegan and Indiana harbors, *Environ. Toxicol. Chem.,* 8:1057–1066. 1989.

Burton G.A., Jr. Assessing freshwater sediment toxicity, *Environ. Toxicol. Chem.,* 10:1585–1627. 1991.

Burton, G.A., Jr. Quality assurance issues in assessing receiving waters, in *Proc. of the Conf. on Effects of Urban Runoff on Receiving Systems.* Edited by J. Saxena, New York, Engineering Foundation Publ. 1992.

Burton G.A., Jr., C. Hickey, T. DeWitt, D. Morrison, D. Roper, and M. Nipper. *In situ* toxicity testing: teasing out the environmental stressors. *SETAC News,* 16(5):20–22. 1996.

Burton, G.A., Jr., R. Pitt, and S. Clark. The role of whole effluent toxicity test methods in assessing stormwater and sediment contamination. *CRC Crit. Rev. Environ. Sci. Technol.,* 30:413–447. 2000.

Calhoun, A.C. (Ed.) *Inland Fisheries Management,* California Department of Fish and Game, Sacramento. 1966.

Cave, K.A. Receiving water quality indicators for judging stream improvement. In *Sustaining Urban Water Resources in the 21st Century.* Proceedings of an Engineering Foundation Conference. Edited by A.C. Rowney, P. Stahre, and L.A. Roesner. Malmo, Sweden. Sept. 7–12, 1997. Published by ACSE, New York. 1998.

Claytor, R. Multiple indicators used to evaluate degrading conditions in Milwaukee County. *Watershed Prot. Techn.,* 2(2):348–351. Spring 1996.

Claytor, R.A. An introduction to stormwater indicators: urban runoff assessment tools. Presented at the *Assessing the Cumulative Impacts of Watershed Development on Aquatic Ecosystems and Water Quality* conference. March 20–21, 1996. Northeastern Illinois Planning Commission. pp. 217–224. Chicago, IL. April 1997.

Claytor, R.A. and W. Brown. *Environmental Indicators to Assess the Effectiveness of Municipal and Industrial Stormwater Control Programs.* Prepared for the U.S. EPA, Office of Wastewater Management. Center for Watershed Protection, Silver Spring, MD. 210 pp. 1996.

Crawford, J.K. and D.R. Lenat. *Effects of Land Use on the Water Quality and Biota of Three Streams in the Piedmont Province of North Carolina.* U.S. Geological Survey. Water Resources Investigation Report 89-4007. Raleigh, NC. 67 pp. 1989.

Ebbert, J.C., J.E. Poole, and K.L. Payne. *Data Collected by the U.S. Geological Survey during a Study of Urban Runoff in Bellevue, Washington, 1979–82.* Preliminary U.S. Geological Survey Open-File Report, Tacoma, WA. 1983.

Enell, M. and J. Henriksson-Fejes. *Dagvattenreningsverket vid Rönningesjön, Täby Kommun. Undersokning-sresultat* [Investigation Results of Water Purification Works near Täby Municipality], in Swedish. Institutet for Vatten-och Luftvardsforskning (IVL). Stockholm, Sweden. 1989–1992.

Environ. Sci. Technol. News briefs, 30(7):290a. July 1996b.

EPA. *Results of the Nationwide Urban Runoff Program.* Water Planning Division, PB 84-185552, Washington, D.C. December 1983.

EPA. *Sampling Guidance Manual for the National Dioxin Manual,* Draft, Office of Water Regulations and Standards, Monitoring and Data Support Division, U.S. Environmental Protection Agency/Corps of Engineers, Washington, D.C. 1984.

EPA. *The Development of Data Quality Objectives,* Office of Research and Development, U.S. Environmental Protection Agency, Washington, D.C. 1986.

EPA. *Biomonitoring to Achieve Control of Toxic Effluents*, Office of Water, U.S. Environmental Protection Agency, Washington, D.C., EPA 625/8-87/013. 1987.

EPA. *Rapid Bioassessment Protocols for Use in Streams and Rivers: Benthic Macroinvertebrates and Fish*, Office of Water, U.S. Environmental Protection Agency, Washington, D.C., EPA 444/4-89/001. 1989a.

EPA. Wisconsin legislature establishes a nonpoint pollution committee. *Nonpoint Source EPA News-Notes*, #8. October 1990a.

EPA. Milwaukee River South declared a priority watershed in Wisconsin. *Nonpoint Source EPA News-Notes*, #9. December 1990b.

EPA. *Environmental Indicators of Water Quality in the United States*. Office of Water, U.S. Environmental Protection Agency. EPA 841-F-96-002. Washington, D.C. June 1996.

Galvin, D.V. and R.K. Moore. *Toxicants in Urban Runoff*. Toxicant Control Planning Section, Municipality of Metropolitan Seattle. Contract #P-16101. U.S. Environmental Protection Agency, Lacy, WA. December 1982.

Gilbert, R.O. *Statistical Methods for Environmental Pollution Monitoring*. Van Nostrand Reinhold, New York. 1987.

Green, R.H. *Sampling Design and Statistical Methods for Environmental Biologists*. John Wiley & Sons, New York. 1979.

Haile, R.W., J. Alamillo, K. Barrett, R. Cressey, J. Dermond, C. Ervin, A. Glasser, N. Harawa, P. Harmon, J. Harper, C. McGee, R.C. Millikan, M. Nides, and J.S. Witte. *An Epidemiological Study of Possible Health Effects of Swimming in Santa Monica Bay*. Santa Monica Bay Restoration Project. Monterey Park, CA. May 1996.

LACDPW (Los Angeles County Department of Public Works, Environmental Programs Division). *Wide Channel Analysis*. January 1998.

Lalor, M. *An Assessment of Non-Stormwater Discharges to Storm Drainage Systems in Residential and Commercial Land Use Areas*. Ph.D. dissertation. Department of Civil and Environmental Engineering. Vanderbilt University. 1994.

Lalor, M. and R. Pitt. *Assessment Strategy for Evaluating the Environmental and Health Effects of Sanitary Sewer Overflows from Separate Sewer Systems*. First Year Report. Wet-Weather Flow Management Program, National Risk Management Research Laboratory, U.S. Environmental Protection Agency, Cincinnati, OH. January 1998.

Lowe, P. and N. Rashedi. Los Angeles River as a water source for a freshwater reservoir. *North American Water and Environment Congress*. American Society of Civil Engineers, New York. 1996.

Lundkvist, S. and H. Söderlund. Rönningesjöns Tillfrisknande. Resultat Efter Dag-och Sjövattenbehandling Åren 1981–1987. [Recovery of the Lake Rönningesjön in Täby, Sweden. Results of Storm and Lake Water Treatment over the Years 1981–1987], in Swedish. *Vatten*, 44(4):305–312. 1988.

May, R.M. Biological diversity: differences between land and sea. *Philos. Trans. R. Soc. London B*, 343:105–111. 1994.

Metcalf and Eddy, Inc. *Surface Runoff Management Plan for Santa Clara County*, Santa Clara Valley Water District, Palo Alto, CA. December 1978.

Moyle, P.B. and B.D. Nichols. Ecology of some native and introduced fishes of the Sierra Nevada foothills in central California. *Copeia*, 3:478. 1973.

Neff, J.W., R.S. Foster, and J.F. Slowey. *Availability of Sediment-Absorbed Heavy Metals to Benthos with Particular Emphasis on Deposit-Feeding in Fauna*. Technical Report D-78-42, Office, Chief of Engineers, U.S. Army, Washington, D.C. 311 pp. August 1978.

OEPA (Ohio Environmental Protection Agency). *The Qualitative Habitat Evaluation Index (QHEI): Rationale, Methods, and Application*, Ecological Assessment Section, Ohio Environmental Protection Agency, Columbus, OH. 1989.

Pedersen, ER. *The Use of Benthic Invertebrate Data for Evaluating Impacts of Urban Stormwater Runoff*. Masters thesis. College of Engineering. University of Washington, Seattle. 1981.

Pelley, J. National "environmental indicators" issued by EPA to track health of U.S. waters. *Environ. Sci. Technol.*, 31(9):381a. Sept. 1996.

Perkins, M.A. *An Evaluation of Instream Ecological Effects Associated with Urban Runoff to a Lowland Stream in Western Washington*. U.S. Environmental Protection Agency, Corvallis Environmental Research Laboratory, Corvallis, OR. July, 1982.

Phillips, G.R., and R.C. Russo. *Metal Bioaccumulation in Fishes and Aquatic Invertebrates: A Literature Review,* EPA-600/3-78-103, U.S. Environmental Protection Agency, Duluth, MN. 1978.

Pitt, R. *Demonstration of Nonpoint Pollution Abatement through Improved Street Cleaning Practices,* EPA-600/2-79-161, U.S. Environmental Protection Agency, Cincinnati, OH. 270 pp. 1979.

Pitt, R. and G. Shawley. *A Demonstration of Non-Point Source Pollution Management on Castro Valley Creek.* Alameda County Flood Control and Water Conservation District and the U.S. Environmental Protection Agency, Water Planning Division (Nationwide Urban Runoff Program). Washington, D.C. June 1982.

Pitt, R. and M. Bozeman. *Sources of Urban Runoff Pollution and Its Effects on an Urban Creek,* EPA-600/S2-82-090, U.S. Environmental Protection Agency, Cincinnati, OH. 1982.

Pitt, R. *Urban Bacteria Sources and Control by Street Cleaning in the Lower Rideau River Watershed.* Rideau River Stormwater Management Study Technical Report. Prepared for the Ontario Ministry of the Environment, Environment Canada, Regional Municipality of Ottawa–Carleton, City of Ottawa, and Nepean. 1983.

Pitt, R. and P. Bissonnette. *Bellevue Urban Runoff Program, Summary Report.* PB84 237213, Water Planning Division, U.S. Environmental Protection Agency and the Storm and Surface Water Utility, Bellevue, WA. 1984.

Pitt, R. *Characterizing and Controlling Urban Runoff through Street and Sewerage Cleaning.* U.S. Environmental Protection Agency, Storm and Combined Sewer Program, Risk Reduction Engineering Laboratory. EPA/600/S2-85/038. PB 85-186500. Cincinnati, OH. 467 pp. June 1985.

Pitt, R., M. Lalor, R. Field, D.D. Adrian, and D. Barbé. *A User's Guide for the Assessment of Non-Stormwater Discharges into Separate Storm Drainage Systems.* Jointly published by the Center of Environmental Research Information, U.S. EPA, and the Urban Waste Management & Research Center (UWM&RC). EPA/600/R-92/238. PB93-131472. Cincinnati, OH. January 1993.

Pitt, R., R. Field, M. Lalor, and M. Brown. Urban stormwater toxic pollutants: assessment, sources and treatability. *Water Environment Research,* Vol. 67, No. 3, pp. 260-275. May/June 1995.

Pitt, R. Water quality trends from stormwater controls. In *Stormwater NPDES Related Monitoring Needs.* Edited by H.C. Torno. Engineering Foundation and ASCE. New York. pp. 413–434. 1995a.

Pitt, R. Preliminary investigation of EquaFlow™ system for Town Lake, Austin, TX. Unpublished report prepared for Loomis and Associates and the City of Austin, TX, October 1995b.

Pitt, R., M. Lilburn, S. Nix, S.R. Durrans, S. Burian, J. Voorhees, and J. Martinson. *Guidance Manual for Integrated Wet Weather Flow (WWF) Collection and Treatment Systems for Newly Urbanized Areas (New WWF Systems).* U.S. Environmental Protection Agency. 612 pp. 1999.

Prych, E. A. and J. C. Ebbert. *Quantity and Quality of Storm Runoff from Three Urban Catchments in Bellevue, Washington,* Preliminary U.S. Geological Survey Water Resources Investigations Report, Tacoma, WA. Undated.

Rashedi, N. and D. Liu. Los Angeles County Department of Public Works Storm Water Quality Assessments. *North American Water and Environment Congress.* American Society of Civil Engineers. New York. 1996.

Ray, S. and W. White. Selected aquatic plants as indicator species for heavy metal pollution. *Journal of Environmental Science and Health,* A11, 717, 1976.

Richey, J. S. *Effects of Urbanization on a Lowland Stream in Western Washington,* Ph.D. dissertation, University of Washington, Seattle. 1982.

Richey, Joanne Sloane, Michael A. Perkins, and Kenneth W. Malueg. The effects of urbanization and stormwater runoff on the food quality in two salmonid streams. *Verh. Internat. Werein. Limnol.,* Vol. 21, pp. 812–818, Stuttgart. October 1981.

Rolfe, G. L., A. Haney, and K. A. Reinbold. *Environmental Contamination by Lead and Other Heavy Metals. Vol. II: Ecosystem Analysis.* Institute for Environmental Studies, University of Illinois, Urbana-Champaign, IL. 1977.

Rolfe, G.L. and K.A. Reinhold. *Vol. I: Introduction and Summary. Environmental Contamination by Lead and Other Heavy Metals.* Institute for Environmental Studies, University of Illinois, Champaign-Urbana, IL, July 1977.

Schueler, Thomas R. *Controlling Urban Runoff: A Practical Manual for Planning and Designing Urban BMPs.* Department of Environmental Programs. Metropolitan Washington Council of Governments. Water Resources Planning Board. 1987.

Schueler, T. (Ed.). Comparison of forest, urban and agricultural streams in North Carolina. *Watershed Protection Techniques*. Vol. 2, No. 4, pp. 503–506. June 1997.

Scoppettone, G. G. and J. J. Smith. Additional records on the distribution and status of native fishes in Alameda and Coyote Creeks, California. *California Department of Fish and Game*, 64, 61, 1978.

Scott, J. B., C. R. Steward, and Q. J. Stober. *Impacts of Urban Runoff on Fish Populations in Kelsey Creek, Washington*, Contract No. R806387020, U.S. Environmental Protection Agency, Corvallis Environmental Research Laboratory, Corvallis, OR. 1982.

SCVWD (Santa Clara Valley Water District), *Surface Water Data: 1976–77 Season*, Santa Clara Valley Water District, CA. April 1978.

SMBRP (Santa Monica Bay Restoration Project). *A Health Effects Study of Swimmers in Santa Monica Bay*. Santa Monica Bay Restoration Project. Monterey Park, CA. October 1996.

Söderlund, H. Dag-och Sjövattenbehandling med Utjämning i Flytbassänger Samt Kemisk Fällning med Tvärlamellsedimentering" [Treatment of Storm- and Lakewater with Compensation in Floating Basins and Chemical Precipitation with Crossflow Lamella Clarifier], in Swedish. *Vatten*, 37(2):166–175. 1981.

Spehan, R.L., R.L. Anderson, and J.T. Fiandt. Toxicity and bioaccumulation of cadmium and lead in aquatic invertebrates. *Environ. Pollut.*, 15:195, 1978.

Thomann, R.V. and J.A. Mueller. *Principles of Surface Water Quality Modeling and Control*. Harper & Row. New York. 1987.

USDA (U.S. Department of the Interior), *Geological Survey. 1973–1977 Water Resources Data for California: Part 1. Surface Water Records, and Part 2. Water Quality Records*, U.S. Department of the Interior. 1978.

U.S. General Accounting Office. *Toxic Substances Control Act: Legislative Changes Could Make the Act More Effective*. GAO/RCED-94-103. Washington, D.C. 1994.

USGS. *The Quality of Our Nation's Waters — Nutrients and Pesticides*. U.S. Geological Survey Circular 1225. Denver, CO. 1999.

Water Environ. Technol. News watch: sewer separation lowers fecal coliform levels in the Mississippi River. 8(11):21–22. Nov. 1996a.

Water Environ. Technol. Research notes: beachgoers at risk from urban runoff. 8(11):65. Nov. 1996b.

Wilber, W.G. and J.V. Hunter. *The Influence of Urbanization on the Transport of Heavy Metals in New Jersey Streams*, Water Resources Research Institute, Rutgers University, New Brunswick, NJ. 1980.

Wilson, E.O. *The Diversity of Life*. Harvard University Press, Cambridge, MA. 1992.

Woodward Clyde Consultants and CDM (WCC and CDM). *Monitoring Plan for 1996–1997 Wide Channel Pilot Study*. Prepared for the Los Angeles County Department of Public Works. November 1996.

Woodward Clyde Consultants and Psomas and Associates (WCC and Psomas). *Evaluation of Land Use Monitoring Stations*. Prepared for the Los Angeles County Department of Public Works. August 1996.

Yoder, C.O. and E.T. Rankin. Assessing the condition and status of aquatic life designated uses in urban and suburban watersheds. Presented at the *Effects of Watershed Developments and Management on Aquatic Ecosystems* conference. Snowbird, UT, August 4–9, 1996. Edited by L.A. Roesner. ASCE, New York, pp. 201–227. 1997.

## CHAPTER 5

Sampling Effort and Collection Methods

"A little experience often upsets a lot of theory."

Cadman

CONTENTS

Introduction 224
Experimental Design: Sampling Number and Frequency..224
    Sampling Plans 225
    Factorial Experimental Designs 227
    Number of Samples Needed to Characterize Conditions......................................231
    Determining the Number of Samples Needed to Identify Unusual Conditions 243
    Number of Samples Needed for Comparisons between Different Sites or Times 244
    Need for Probability Information and Confidence Intervals 245
Data Quality Objectives (DQO) and Associated QA/QC Requirements......................247
    Quality Control and Quality Assurance to Identify Sampling and Analysis Problems......247
    Identifying the Needed Detection Limits and Selecting the Appropriate Analytical
    Method 252
General Considerations for Sample Collection 254
    Basic Safety Considerations When Sampling 255
    Selecting the Sampling Locations...256
    Sampler and Other Test Apparatus Materials 260
    Volumes to Be Collected, Container Types, Preservatives to Be Used, and Shipping
    of Samples 263
    Personnel Requirements 275
Receiving Water, Point Source Discharge, and Source Area Sampling 278
    Automatic Water Sampling Equipment 278
    Manual Sampling Procedures 289
    Source Area Sampling...297
Sediment and Pore Water Sampling 313
    Sediment Sampling Procedures...313
    Interstitial Water and Hyporheic Zone Sampling 326
Summary: Basic Sample Collection Methods...336
References 338

INTRODUCTION

This chapter begins by describing experimental design methods enabling the user to determine the sampling effort needed to accomplish project objectives. The statistical basis for this approach is required to justify the allocation of scarce resources. In many cases, certain elements of a multifaceted study program, as required for practically all receiving water studies, require much more time and money than other elements of the program. The approach and tools given in this chapter enable one to balance project resources and scope with expected outcomes. It can be devastating to project conclusions if needed numbers of samples are not obtained at the appropriate time. The tools in this chapter enable one to better plan and conduct a sampling program to minimize this possibility. Of course, all projects conclude with some unresolved issues that were not considered at the outset. This can only be minimized with increased experience and subject knowledge, and by retaining some flexibility during project execution.

The tools presented here assume some prior knowledge of the situation (especially expected variation in a variable to be measured) in order to determine the sampling effort. This is initially obtained through professional judgment (based on one's experience in similar situations and from the literature), and is generally followed up with a multistaged sampling effort where an initial experimental design sampling effort is conducted to obtain a better estimate of parameter variability. That estimate can then be used to help foresee and estimate the needed sampling effort during later sampling periods. In all cases, the tools presented here enable one to obtain a level of confidence concerning the significance of the project conclusions. As an example, if it is necessary to compare two sampling location conditions (a very common objective), the sampling effort will determine the sensitivity of the study. Depending on the variability of the parameter of interest, a few samples collected may be useful to identify only very large differences in conditions between two sampling locations. Of course, the objective of the study may be only to confirm large differences (such as between reference and grossly contaminated sites, or between influent and effluent conditions for a stormwater measure known to be very effective). Unfortunately, in most cases involving nonpoint source discharges, the differences are likely to be much more subtle, requiring numerous samples and careful allocations of project resources. The tools presented in this chapter enable one to predict the statistical sensitivity of different sampling schemes, allowing informed decisions and sound budget requests to be made.

The other elements of this chapter involve specific options for collecting samples from the many ecosystem components of interest. Quality control/quality assurance (QA/QC) sampling requirements are described along with basic considerations for safe sample collection (selecting sampling locations, preventing sample contamination, sample volumes needed, sample shipping, personnel requirements, etc.). Water sampling (manual sampling, automatic samplers, sampler setup options, sampler modifications, bedload samples, suspended sediment samples, floatable material sampling, source area sheetflow sampling, etc.) are also described and discussed. This chapter also includes important considerations pertaining to sediment sampling and interstitial (pore water) sampling. The material included in this chapter, therefore, describes how to collect basic water and sediment samples for receiving water studies. Chapter 6, in turn, discusses measurement methods, including the collection of biological samples.

EXPERIMENTAL DESIGN: SAMPLING NUMBER AND FREQUENCY

The first task in any study is to formulate the questions being addressed. The expected statistical analysis tools (described in Chapter 7) that are expected to be used for evaluating the data should also be an early part of the experimental design. Alternative study plans can then be examined, and finally, the sampling effort can be estimated.

Sampling Plans

All sampling plans attempt to obtain certain information (usually average values, totals, ranges, etc.) about a large population by sampling and analyzing a much smaller sample. The first step in this process is to select the sampling plan and then to determine the number of samples needed. Many sampling plans have been well described in the environmental literature. The following are the four main categories, plus subcategories, of sampling plans (Gilbert 1987):

- Haphazard sampling. Samples are taken in a haphazard (not random) manner, usually at the convenience of the sampler when time permits. Especially common when the weather is pleasant. This is only possible with a very homogeneous condition over time and space; otherwise biases are introduced in the measured population parameters. It is therefore not recommended because of the difficulty in verifying the homogeneous assumption. This is the most common sampling strategy when volunteers are used for sampling, unless the grateful agency is able to spend sufficient time to educate the volunteer samplers about the problems of this type of sampling and to specify a more appropriate strategy.
- Judgment sampling. This strategy is used when only a specific subset of the total population is to be evaluated, with no desire to obtain "universal" characteristics. The target population must be clearly defined (such as during wet-weather conditions only) and sampling is conducted appropriately. This could be the first stage of later, more comprehensive sampling of other target population groups (multistage sampling).
- Probability sampling. Several subcategories of probability sampling have been described:
  - Simple random sampling. Samples are taken randomly from the complete population. This usually results in total population information, but it is usually inefficient as a greater sampling effort may be required than if the population was subdivided into distinct groups. Simple random sampling doesn't allow information to be obtained for trends or patterns in the population. This method is used when there is no reason to believe that the sample variation is dependent on any known or measurable factor.
  - Stratified random sampling. This may be the most appropriate sampling strategy for most receiving water studies, especially if combined with an initial limited field effort as part of a multistage sampling effort. The goal is to define strata that result in little variation within any one strata, and great variation between different strata. Samples are randomly obtained from several population groups that are assumed to be internally more homogeneous than the population as a whole, such as separating an annual sampling effort by season, lake depth, site location, habitat category, rainfall depth, land use, etc. This results in the individual groups having smaller variations in the characteristics of interest than in the population as a whole. Therefore, sample efforts within each group will vary, depending on the variability of characteristics for each group, and the total sum of the sampling effort may be less than if the complete population was sampled as a whole. Also, much additional useful information is likely if the groups are shown to actually be different.
  - Multistage sampling. One type of multistage sampling commonly used is associated with the required subsampling of samples obtained in the field and brought to the laboratory for subsequent splitting for several different analyses. Another type of multistage sampling is when an initial sampling effort is used to examine major categories of the population that may be divided into separate clusters during later sampling activities. This is especially useful when reasonable estimates of variability within a potential cluster are needed for the determination of the sampling effort for composite sampling. These variability measurements may need to be periodically reverified during the monitoring program.
  - Cluster sampling. Gilbert (1987) illustrates this sampling plan by specifically targeting specific population units that cluster together, such as a school of fish or clump of plants. Every unit in each randomly selected cluster can then be monitored.
  - Systematic sampling. This approach is most useful for basic trend analyses, where evenly spaced samples are collected for an extended time. Evenly spaced sampling is also most efficient when trying to find localized hot spots that randomly occur over an area. Gilbert (1987) presents

guidelines for spacing of sampling locations for specific project objectives relating to the size of the hot spot to be found. Spatial gradient sampling is a systematic sampling strategy that may be worthy of consideration when historical information implies an aerial variation of conditions in a river or other receiving water. One example would be to examine the effects of a point source discharge on receiving-sediment quality. A grid would be described in the receiving water in the discharge vicinity whose spacing would be determined by preliminary investigations.
- Search sampling. This sampling plan is used to find specific conditions where prior knowledge is available, such as the location of a historical (but now absent) waste discharger affecting a receiving water. Therefore, the sampling pattern is not systematic or random over an area, but stresses areas thought to have a greater probability of success.

Box et al. (1978) contains much information concerning sampling strategies, specifically addressing problems associated with randomizing the experiments and blocking the sampling experiments. Blocking (such as in paired analyses to determine the effectiveness of a control device, or to compare upstream and downstream locations) eliminates unwanted sources of variability. Another way of blocking is to conduct repeated analyses (such as for different seasons) at the same locations. Most of the above probability sampling strategies should include randomization and blocking within the final sampling plans (as demonstrated in the following example and in the use of factorial experiments).

Albert and Horwitz (1988) warn that the user of statistics should be critical and alert in making decisions based on sample estimates, and they list the following as essential aspects of statistical sampling:

- Sampling should not be undertaken until the questions have been determined and properly framed. The expense of conducting a survey can only be justified if the questions answered have a value. Vague or unstructured exploratory surveys are wasteful.
- The individuals included in the sample must be chosen at random, specifically from a population that is well defined.

Example Use of Stratified Random Sampling Plan

Street dirt samples were collected in San Jose, CA, during an early EPA project to identify sources of urban runoff pollutants (Pitt 1979). The samples were collected from narrow strips, from curb to curb, using an industrial vacuum. Many of these strips were to be collected in each area and combined to determine the dust and dirt loadings and their associated characteristics (particle size and pollutant concentrations). Each area (stratum) was to be sampled frequently to determine the changes in loadings with time and to measure the effects of street cleaning and rains in reducing the loadings. The analytical procedure used to determine the number of subsamples needed for each composite sample involved weighing individual subsamples in each study area to calculate the coefficient of variation (COV = standard deviation/mean) of the street surface loading. The number of subsamples necessary (N), depending on the allowable error (L), was then determined. An allowable error value of about 25%, or less, was needed to keep the precision and sampling effort at reasonable levels. The formula used (after Cochran 1963) was:

$$N = 4\sigma^2/L^2$$

With 95% confidence, this equation estimates the number of subsamples necessary to determine the true mean value for the loading within a range of ±L. As will be shown in the following discussions, more samples are required for a specific allowable error as the COV increases. Similarly, as the allowable error decreases for a specific COV, more samples are also required. Therefore, with an allowable error of 25%, the required number of subsamples for a study area with a COV of 0.8 would be 36.

Initially, individual samples were taken at 49 locations in the three study areas to determine the loading variabilities. The loadings averaged about 2700 lb/curb-mile in the Downtown and Keyes Street areas, but were found to vary greatly within these two areas. The Tropicana area loadings were not as high, and averaged 310 lb/curb-mile. The Cochran (1963) equation was then used to determine the required number of subsamples in each test area. The data were then examined to determine if the study areas should be divided into meaningful test area groups.

The purpose of these divisions was to identify a small number of meaningful test area-groupings (strata) that would require a reasonable number of subsamples and to increase the usefulness of the test data by identifying important groupings. Five different strata were identified for this research: two of the areas were divided by street texture conditions into two separate strata each (good vs. poor), while the other area was left undivided. The total number of individual subsamples for all five areas combined was 111, and the number of subsamples per strata ranged from 10 to 35. In contrast, 150 subsamples would have been needed if the individual areas were not subdivided. Subdividing the main sampling areas into separate strata not only resulted in a savings of about 25% in the sampling effort, but also resulted in much more useful information concerning the factors affecting the values measured. The loading variations in each strata were reexamined seasonally, and the sampling effort was readjusted accordingly.

Factorial Experimental Designs

Factorial experiments are described in Box et al. (1978) and in Berthouex and Brown (1994). Both of these books include many alternative experimental designs and examples of this method. Berthouex and Brown (1994) state that "experiments are done to:

1. Screen a set of factors (independent variables) and learn which produce an effect
2. Estimate the magnitude of effects produced by experimental factors
3. Develop an empirical model
4. Develop a mechanistic model."

They concluded that factorial experiments are efficient tools in meeting the first two objectives and are also excellent for meeting the third objective in many cases. Information obtained during the experiments can also be very helpful in planning the strategy for developing mechanistic models. The main feature of factorial experimental designs is that they enable a large number of possible factors that may influence the experimental outcome to be simultaneously evaluated.

Box et al. (1978) presents a comprehensive description of many variations of factorial experimental designs. A simple 2^3 design (three factors: temperature, catalyst, and concentrations at two levels each) is shown in Figure 5.1 (Box et al. 1978). All possible combinations of these three factors are tested, representing each corner of the cube. The experimental results are placed at the appropriate corners. Significant main effects can usually be easily seen by comparing the values on opposite faces of the cube. If the values on one face are consistently larger than on the opposite face, then the experimental factor separating the faces likely has a significant effect on the outcome of the experiments. Figure 5.2 (Box et al. 1978) shows how these main effects are represented, along with all possible two-factor interactions and the one three-factor interaction. The analysis of the results to identify the significant factors is straightforward.

One of the major advantages of factorial experimental designs is that the main effect of each factor, plus the effects of all possible interactions of all of the factors can be examined with relatively few experiments. The initial experiments are usually conducted with each factor tested at two levels (a high and a low level). All possible combinations of these factors are then tested. Table 5.1 shows an experimental design for testing four factors. This experiment therefore requires 2^4 (=16) separate experiments to examine the main effects and all possible interactions of these four factors. The signs signify the experimental conditions for each main factor during each of the 16 experiments.

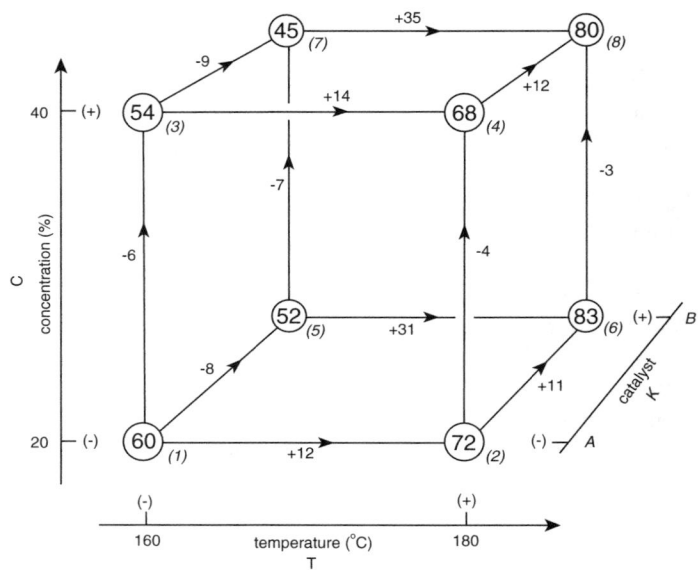

Figure 5.1   Basic cubic design of 2^3 factorial test. (From Box, G.E.P., W.G. Hunter, and J.S. Hunter. *Statistics for Experimenters*. Copyright 1978. This material used by permission of John Wiley & Sons, Inc., New York.)

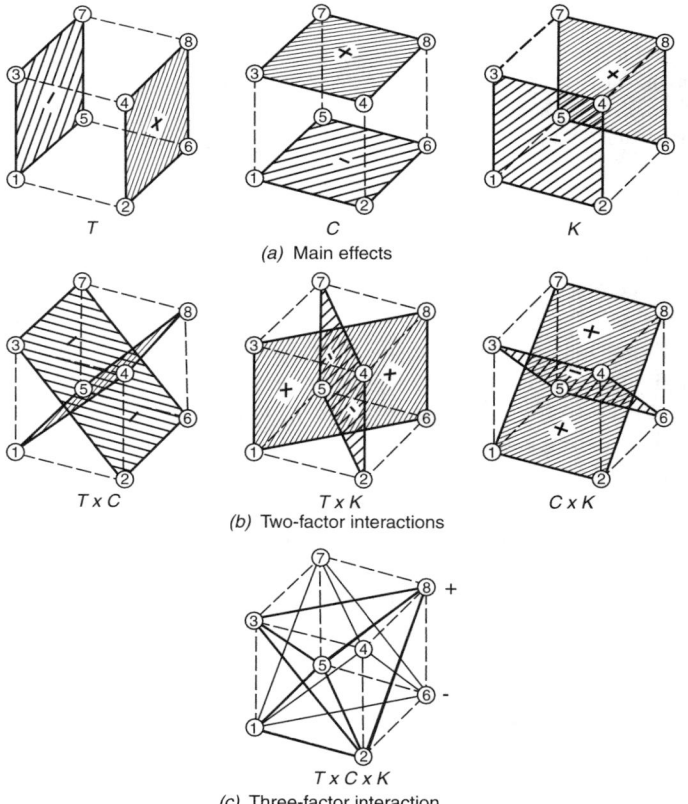

Figure 5.2   Main effects and interactions for 2^3 factorial test. (From Box, G.E.P., W.G. Hunter, and J.S. Hunter. *Statistics for Experimenters*. Copyright 1978. This material used by permission of John Wiley & Sons, Inc., New York.)

Table 5.1 Factorial Experimental Design for Four Factors and 16 Experiments

Experiment No.	A	B	C	D	AB	AC	AD	BC	BD	CD	ABC	ABD	BCD	ABCD
1	+	+	+	+	+	+	+	+	+	+	+	+	+	+
2	−	+	+	+	−	−	−	+	+	+	−	−	+	−
3	+	−	+	+	−	+	+	−	−	+	−	−	−	−
4	−	−	+	+	+	−	−	−	−	+	+	+	−	+
5	+	+	−	+	+	−	+	−	+	−	−	+	−	−
6	−	+	−	+	−	+	−	−	+	−	+	−	−	+
7	+	−	−	+	−	−	+	+	−	−	+	−	+	+
8	−	−	−	+	+	+	−	+	−	−	−	+	+	−
9	+	+	+	−	+	+	−	+	−	−	+	−	−	−
10	−	+	+	−	−	−	+	+	−	−	−	+	−	+
11	+	−	+	−	−	+	−	−	+	−	−	+	+	+
12	−	−	+	−	+	−	+	−	+	−	+	−	+	−
13	+	+	−	−	+	−	−	−	−	+	−	−	+	+
14	−	+	−	−	−	+	+	−	−	+	+	+	+	−
15	+	−	−	−	−	−	−	+	+	+	+	+	−	−
16	−	−	−	−	+	+	+	+	+	+	−	−	−	+

The shaded main factors are the experimental conditions, while the other columns specify the data reduction procedures for the other interactions. A plus sign shows when the factor is to be held at the high level, while a minus sign shows when the factor is to be held at the low level for the main experimental conditions (A through D). This table also shows all possible two-way, three-way, and four-way interactions, in addition to the main factors. Simple analysis of the experimental results allows the significance of each of these factors and interactions to be determined. As an example, the following list shows the four factors and the associated levels for tests conducted to identify factors affecting runoff quality:

A: Season (plus: winter; minus: summer)
B: Land use (plus: industrial; minus: residential)
C: Age of development (plus: old; minus: new)
D: Rain depth (plus: >1 in; minus: <1 in)

These factors would require the selection of four sampling locations:

1. Old industrial area
2. New industrial area
3. Old residential area
4. New residential area

The above experiments are designed to collect stormwater runoff data from four test locations. Obviously, both winter and summer seasons must be monitored, and rainfall events of varying depths will be sampled. Rains both less than 1 inch and greater than 1 inch will need to be sampled at all monitoring stations in both seasons in order to obtain the needed information.

Even though factorial experiments are best suited in controlled laboratory settings, they have been very useful in organizing environmental data for analysis. Table 5.2 shows an example where environmental data were organized using a simple factorial design. The design called for a 2^3 experiment to investigate the effects of soil moisture, soil texture, and soil compaction on observed soil infiltration rates (Pitt et al. 1999a). This table shows the calculations from 152 double-ring infiltration tests for the Horton (1939) equation final infiltration rate coefficient (f_c).

Replicate observations enhance the data analysis efforts, and grouped standard error values can be calculated (Box et al. 1978) to identify the significant factors affecting runoff quality. In Table 5.2, at least 12 replicates were conducted for each test condition to improve the statistical basis

Table 5.2 Example Factorial Experiment Analysis for Field Project Investigating Infiltration into Disturbed Urban Soils

Moisture (Wet = +/Dry = −)	Texture (Clay = +/Sand = −)	Compacted (Yes = +/No = −)	Factorial Group	Average	Standard Error	Number
+	+	+	1	0.23	0.13	18
+	+	−	2	0.43	0.50	27
+	−	+	3	1.31	1.13	18
+	−	−	4	16.49	1.40	12
−	+	+	5	0.59	0.35	15
−	+	−	6	7.78	4.00	17
−	−	+	7	2.25	0.98	21
−	−	−	8	13.08	2.78	24

Overall average 5.27
Calculated polled S.E. 1.90

$$f_c = 5.27 \pm (T/2) \pm (C/2)$$
$$f_c = 5.27 \pm (-6.02/2) \pm (-8.35/2)$$

T	C	Calculated Values
−	+	−1.92
+	−	6.43
+	+	4.10
−	−	12.45

Factorial Group	Prob.	Rank
1	7.14	1
2	21.43	2
3	35.71	3
4	50.00	4
5	64.29	5
6	78.57	6
7	92.86	7

Factorial Group	Effects
C	−8.35
T	−6.02
MT	−2.55
M	−1.31
MC	0.66
MTC	2.83
TC	4.66

Probability of Effects for f_c

Probability of Residuals for f_c

From Pitt, R., J. Lantrip, R. Harrison, C. Henry, and D. Hue. *Infiltration through Disturbed Urban Soils and Compost-Amended Soil Effects on Runoff Quality and Quantity.* U.S. Environmental Protection Agency, Water Supply and Water Resources Division, National Risk Management Research Laboratory. EPA 600/R-00/016. Cincinnati, OH. 231 pp. December 1999a.

for the conclusions. These unusually large numbers of replicates were needed because of the inherently large variability within each test category. If the variability was less, then the number of required replicates could have been much less (as described later in this chapter). In addition, the site test conditions were not known with certainty when the field tests were run, as some field estimates required confirmation with later laboratory tests that resulted in the reclassification of some of the data.

If observations are not available for some of the needed conditions (such as the monitoring equipment failing during the only large event that occurred at the old industrial site during the summer), then a fractional factorial design can still be used to organize the data and calculate the effects for all of the main factors, and for most of the interactions (as noted in the above experiment). Once the initial experiments are completed, follow-up experiments can be efficiently designed to examine the linearity of the effects of the significant factors by conducting response surface experimental designs. In addition, further experiments can be conducted and merged with these initial experiments to examine other factors that were not considered in the first experiments. Because of the usefulness and adaptability of factorial experimental designs, Berthouex and Brown (1994) recommend that they "should be the backbone of an experimenter's design strategy."

Number of Samples Needed to Characterize Conditions

An important aspect of any research is the assurance that the samples collected represent the conditions to be tested and that the number of samples to be collected is sufficient to provide statistically relevant conclusions. Unfortunately, sample numbers are most often not based on a statistical process and follow traditional "best professional judgments," or are resource driven. The sample numbers should be equal between sampling locations if comparing station data (EPA 1983b) and paired sampling should be conducted, if at all possible (the samples at the two comparison sites should be collected at the "same" time, for example), allowing for much more powerful paired statistical comparison tests (see Chapter 7). In addition, replicate subsamples must also be collected and then combined to provide a single sample for analysis for many types of ecosystem sampling. Cairns and Dickson (1971) observed from many years of experience that at least three artificial substrate samplers, 3 to 10 dredge hauls, and three Surber square foot samples were the minimum number of samples required to describe benthic macroinvertebrates at a given station. These are then combined (to reduce analysis expenses) or kept as separate samples (more costly, but provides a legitimate measure of variation/precision).

Receiving water studies frequently include objectives to characterize various chemical, biological, and physical parameters of the water body itself, or influencing features (meteorological, discharges, watershed, etc.). An experimental design process can be used that estimates the number of needed samples based on the allowable error, the variance of the observations, and the degree of confidence and power needed for each parameter. A basic equation that can be used is as follows:

$$n = [COV(Z_{1-\alpha} + Z_{1-\beta})/(error)]^2$$

where
$n =$ number of samples needed
$\alpha =$ false positive rate ($1 - \alpha$ is the degree of confidence. A value of α of 0.05 is usually considered statistically significant, corresponding to a $1 - \alpha$ degree of confidence of 0.95, or 95%)
$\beta =$ false negative rate ($1 - \beta$ is the power. If used, a value of β of 0.2 is common, but it is frequently ignored, corresponding to a β of 0.5)
$Z_{1-\alpha} =$ Z score (associated with area under normal curve) corresponding to $1 - \alpha$. If α is 0.05 (95% degree of confidence), then the corresponding $Z_{1-\alpha}$ score is 1.645 (from

standard statistical tables).

$Z_{1-\beta}$ = Z score corresponding to $1 - \beta$ value. If β is 0.2 (power of 80%), then the corresponding $Z_{1-\beta}$ score is 0.85 (from standard statistical tables). However, if power is ignored and β is 0.5, then the corresponding $Z_{1-\beta}$ score is 0.

error = allowable error, as a fraction of the true value of the mean

COV = coefficient of variation (sometimes noted as CV), the standard deviation divided by the mean. (Data set assumed to be normally distributed.)

This equation is only approximate, as it requires that the data set be normally distributed. However, if the coefficient of variation (COV) values are low (less than about 0.4), then there is probably no significant difference in the predicted sampling effort. This equation is only appropriate as an approximation in many cases, as normal distributions are rare (log-normal distributions are appropriate for most water quality parameters) and the COV values are typically relatively large (closer to 1). The presentation of the results and the statistical procedures used to evaluate the data, however, should calculate the exact degree of confidence of the measured values.

Figure 5.3 (Pitt and Parmer 1995) is a plot of this equation, showing the approximate number of samples needed for an α of 0.05 (degree of confidence of 95%), and a β of 0.2 (power of 80%). As an example, if an allowable error of about 25% is desired and the COV is estimated to be 0.4, then about 20 samples would have to be analyzed. The samples could be composited and a single analysis conducted, but this would not allow the COV assumption to be confirmed, or the actual confidence range of the concentration to be determined. The use of stratified random sampling can usually be used to advantage by significantly reducing the COV of the subpopulation in the strata, requiring fewer samples for characterization, as illustrated above.

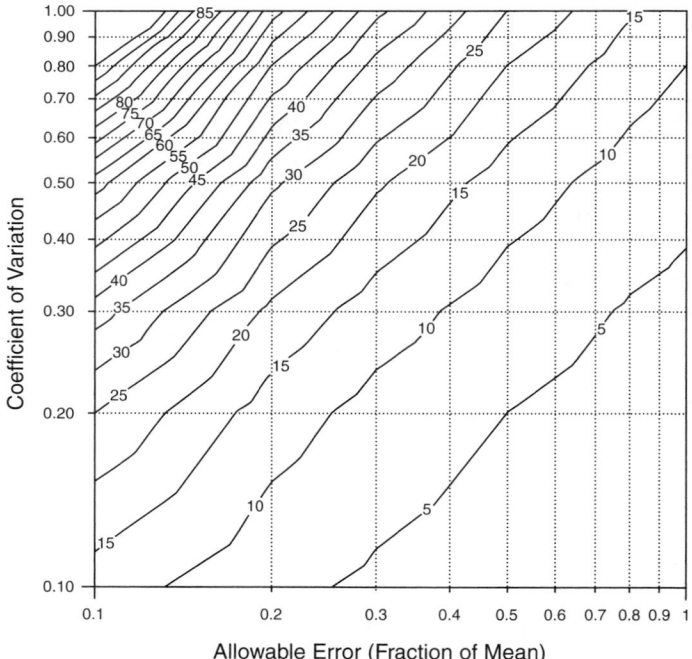

Figure 5.3   Sample requirements for confidence of 95% (α = 0.05) and power of 80% (β = 0.20). (From Pitt, R. and K. Parmer. *Quality Assurance Project Plan: Effects, Sources, and Treatability of Stormwater Toxicants.* Contract No. CR819573. U.S. Environmental Protection Agency, Storm and Combined Sewer Program, Risk Reduction Engineering Laboratory. Cincinnati, OH. February 1995.)

Gilbert (1987) presents variations of this basic equation that consider the number of samples needed to determine the probability of occurrence within a specified range (such as to calculate the frequency of standard violations). He also presents equations that consider correlated data, such as when the observations are not truly independent, as when very high pollutant concentrations affect values in close spatial or temporal proximity. As expected, correlated data necessitate more samples than indicated from the basic equations. Additional sample size equations are presented in experimental design texts and in listings from government agencies (such as Table 5.3 from Environment Canada 1994).

Types of Errors Associated with Sampling

Unfortunately, there are many errors associated with a receiving water study. Errors associated with too few (or too many) samples for a parameter of interest is only one category. Sampling and analytical errors may also be significant and could add to these other errors. Hopefully, the collective sum of all errors is known (through QA/QC activities and adequate experimental design) and manageable. An important aspect of a monitoring program is recognizing the levels of errors and considering the uncertainties in developing recommendations and conclusions.

Generally, errors can be divided into precision and bias problems. Both of these errors, either together or separately, have dramatic effects on the final conclusions of a study. Figure 5.4 (Gilbert 1987) shows the effects of these errors. Bias is a measure of how close the measured median value is to the true median value, while precision is a measure of how "fuzzy" the median estimate is (the repeatability of the analyses; used to determine the confidence of the measurements).

Errors in decision making are usually divided into Type 1 (α: alpha) and Type 2 (β: beta) errors:

α (alpha) (Type 1 error) — a false positive, or assuming something is true when it is actually false. An example would be concluding that a tested water was adversely contaminated, when it actually was clean. The most common value of α is 0.05 (accepting a 5% risk of having a Type 1 error). Confidence is $1 - \alpha$, or the confidence of not having a false positive.

β (beta) (Type 2 error) — a false negative, or assuming something is false when it is actually true. An example would be concluding that a tested water was clean when it actually was contaminated. If this was an effluent, it would therefore be an illegal discharge with the possible imposition of severe penalties from the regulatory agency. In most statistical tests, β is usually ignored (if ignored, β is 0.5). If it is considered, a typical value is 0.2, implying accepting a 20% risk of having a Type 2 error. Power is $1 - \beta$, or the certainty of not having a false negative.

It is important that power and confidence be balanced for an effective monitoring program. Most studies ignore power, while providing a high value (typically 95%) for the level of confidence. This is an unrealistic approach because both false negatives and false positives are important. In many environmental programs, power (false negative problems) may actually be more critical than confidence. If a tested water had a Type 2 error (false negative), inappropriate discharges would occur. Typical fines imposed by regulatory agencies are $10,000 per day for nonpermitted discharges. Future liability for wastes discharged due to an error in measurement or negligence can easily reach into millions of dollars for cleanup and mitigation of health effects. Clearly, one wants to minimize costs, yet have the assurance that the correct decision is being made. However, errors will always be present in any analysis, and some uncertainty in the conclusions must be accepted. Obviously, it can become prohibitively expensive to attempt to reduce monitoring errors to extremely low levels, especially when the monitoring program is affected by uncontrollable environmental factors.

Chapter 7 describes statistical analysis procedures that can be used for data analyses. It is always important to report the statistical significance (and importance) of the test results. The "importance" of the test results relates to the magnitude of the difference between two alternatives, for example,

Table 5.3 Typical Listing of Sample Size Equations That Are Useful for Environmental Research

Objective	Formula	Ref.
To determine the sample size required to detect an effect in an impacted area vs. a control area over time: a) Resampling same sites before and after impact and testing if the mean change in the control area is the same as that in the impacted area	$n = 2(t_\alpha + t_\beta)^2 \left(\dfrac{S}{\Delta}\right)^2$	Green 1989
b) Sampling different sites before and after impact and testing if the mean change in the control area is the same as that in impacted area	$n = 4(t_\alpha + t_\beta)^2 \left(\dfrac{S}{\Delta}\right)^2$	Green 1989

where:
n = number of samples for each of the control and impact areas
S = standard deviation
Δ = magnitude of change required to be a real effect with specified power $(1 - b)$
t_α = t statistic given a Type I error probability
t_β = t statistic given a Type II error probability

Objective	Formula	Ref.
To determine if the mean value for an impacted area differs significantly from a standard value (e.g., sediment quality criterion)	$n \geq \dfrac{(Z_\alpha + Z_\beta)^2}{d^2} + 0.5Z_\alpha^2$	Alldredge 1987

where:
n = sample size
Z_α = Z statistic for Type I error probability (e.g., x = 0.05)
Z_β = Z statistical for Type II error probability (e.g., B − 0.90)
d = magnitude of the difference to be detected (i.e., effect level)

Objective	Formula	Ref.
To determine if the mean value for an impacted area differs significantly from the mean of a control site	$n \geq \dfrac{2(Z_\alpha + Z_\beta)^2}{d^2} + 0.25Z_\alpha^2$	Alldredge 1987

where:
n = sample size
Z_α = Z statistic for Type I error probability (e.g. x = 0.05)
Z_β = Z statistical for Type II error probability (e.g., B − 0.90)
d = magnitude of the difference to be detected (i.e., effect level)

Objective	Formula	Ref.
To determine the number of samples that would be required to determine a mean value (representative of the area) with a given statistical certainty	$y\bar{x} = t_c \left[\dfrac{S_x}{(N-1)^{0.5}}\right]$	Håkanson 1984

where:
y = accepted error in the percent of the mean value (e.g., y = 10%)
$\bar{x}$ = mean value of x_i (i = 1...n)
S_x = standard deviation
t_c = confidence coefficient (e.g., 90% or $t_{0.95}$)
N = number of samples

Table 5.3 Typical Listing of Sample Size Equations That Are Useful for Environmental Research (Continued)

Objective	Formula	Ref.
To determine the number of samples required to give a result with a specific confidence limit	$N = \dfrac{(t_1 + t_2)^2}{d^2} \, S$	Gad and Weil 1988

where:
t_1 = one-tailed t value with $N - 1$ d.f. corresponding to a level of confidence
t_2 = one-tailed t value with $N - 1$ degrees of freedom corresponding to the probability that the sample size will be adequate to achieve the desired precision
S = sample standard deviation
d = the acceptable range of variation for the variable being measured

Objective	Formula	Ref.
To determine the number of samples required to achieve a maximum acceptable error	$n = \dfrac{Z^2 \sigma^2}{E^2}$	Gilbert 1981

where:
n = number of samples
Z = Z statistic
E = maximum acceptable error

Objective	Formula	Ref.
To determine the number of samples required to estimate a mean	$n = \dfrac{(Z_{\alpha/2})\sigma^2}{d^2}$	Milton et al. 1986

where:
n = number of samples
Z = Z statistic (standard normal curve)
σ^2 = variance
$\alpha/2$ = probability of a 95% confidence level
d = the distance between the center of the lower confidence and the upper confidence bound

Objective	Formula	Ref.
To determine the number of samples required for a particular power for: a) A normal distribution (i.e., $x > s^2$) b) A Poisson distribution (i.e., $x - S^2$) c) A negative binomial distribution (i.e., $s < S^2$)	a) $N = \dfrac{10^4 (t^2 s^2)}{(R^2 \bar{x}^2)}$ b) $N = \dfrac{10^4 t^2}{(R^2 \bar{x}^2)}$ c) $N = 10^4 \left(\dfrac{t^2}{R^2} \right) \left[\left(\dfrac{1}{x} \right) + \left(\dfrac{1}{K} \right) \right]$	Kratochvil and Taylor 1981

where:
N = number of samples
t = t statistic for a desired confidence level
$\bar{x}$ = mean value from preliminary sampling or historical data
s = standard deviation of mean
R^2 = percentage coefficient of variation
K = index of clumping

Data from EC (Environment Canada). *Guidance Document on Collection and Preparation of Sediments for Physicochemical Characterization and Biological Testing.* Environmental Protection Series Report, EPS 1/RM/29. Ottawa, Canada. pp. 111–113, December, 1994.)

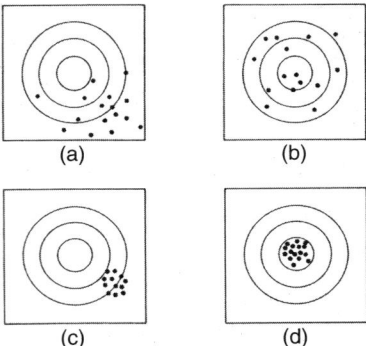

(a)   (b)

(c)   (d)

Figure 5.4 Accuracy definitions: (a) low precision, large bias, (b) low precision, small bias, (c) high precision, large bias, and (d) high precision, small bias (the only "accurate" case). (From Gilbert, R.O. *Statistical Methods for Environmental Pollution Monitoring.* Van Nostrand Reinhold. New York. Copyright 1987. This material is used by permission of John Wiley & Sons.)

and determines if a decision should be changed. In some cases, statistically significant results may occur simultaneously with small data differences (usually if low variations and/or large data sets are available). In this case, it may not be worthwhile, or feasible, to change a process or make other major changes.

Determining Sample Concentration Variations

An important requirement for using the above sampling effort equation is estimating the COV of the parameter of interest. In many cases, the approximate range of likely concentrations can be estimated for a parameter of interest. Figure 5.5 (Pitt and Lalor 2001) can be used to estimate the COV value for a parameter by knowing the 10th and 90th percentile ratios (the "range ratio"), assuming a log-normal distribution. Extreme values are usually not well known, but the approximate 10th and 90th percentile values can be estimated with better confidence. As an example, assume that the 10th and 90th percentile values of a water quality constituent of interest was estimated to be about 0.7 and 1.5 mg/L, respectively. The resulting range ratio is therefore 1.5/0.7 = 2.1 and the estimated COV value is 0.25.

Also shown in Figure 5.5 is an indication of the median value, compared to the 10th percentile value and the range ratio, assuming a log-normal distribution. As the range ratio decreases, the median comes close to the midpoint between the 10th and 90th percentile values. Therefore, at low COV values, the differences between normal distributions and log-normal distributions diminish, as stated previously. As the COV values increase, the mean values are located much closer to the 10th percentile value. In log-normal distributions, no negative concentration values are

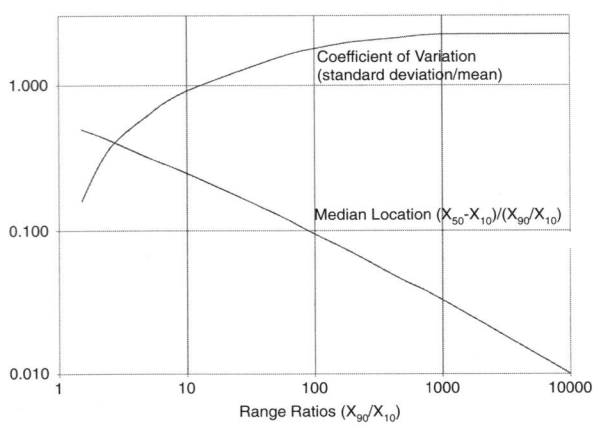

Figure 5.5 Determination of coefficient of variation from range of observations (Pitt, R. and M. Lalor. *Identification and Control of Non-Stormwater Discharges into Separate Storm Drainage Systems. Development of Methodology for a Manual of Practice.* U.S. Environmental Protection Agency, Water Supply and Water Resources Division, National Risk Management Research Laboratory, Cincinnati, OH. 451 pp. To be published in 2001.)

allowed, but very large positive "outliers" can occur. In the previous example, the median location is about 0.4 for the range ratio of 2.1. The following calculation shows how the median value can be estimated using this "median location" value:

median location = 0.4 = $(X_{50} - X_{10})/(X_{90} - X_{10})$
therefore $X_{50} - X_{10} = 0.4(X_{90} - X_{10})$.
$(X_{90} - X_{10}) = 1.5$ mg/L $- 0.7$ mg/L $= 0.8$ mg/L.
Therefore $X_{50} - X_{10} = 0.4 (0.8) = 0.32$ mg/L,
and $X_{10} = 0.7$ mg/L, $X_{50} = 0.32$ mg/L $+ 0.7$ mg/L $= 1.0$ mg/L.

For comparison, the average of the 10th and 90th percentile values is 1.1 mg/L. Therefore, the concentration distribution is likely close to being normally distributed and the equation shown previously can be used to estimate the required number of samples needed because these two values are within about 10% of each other. The following paragraphs (from Pitt and Lalor 2001) show how log transformations of real-space data descriptors (COV and median) can be used in modifications of these equations.

Example of Log₁₀ Transformations for Experimental Design Calculations

For relatively large COV values, it may be necessary to transform the data from known log-normal distributions (checked using log-normal probability paper, for example) before calculating the actual error associated with the collected data. Much urban receiving water quality data from the 10th to 90th percentile can typically be described as a normal probability distribution, after $\log_{10}$ transformations of the data. However, values less than the 10th percentile value are usually less than predicted from the log-normal probability plot, while values greater than the 90th percentile value are usually greater than predicted from the log-normal probability plot. Nontransformed water quality data do not typically fit normal probability distributions very well, except for pH (which are log transformed, by definition).

Figure 5.6 (Pitt and Lalor 2001) presents a relationship between the COV value in real space (nontransformed) and the standard deviation of $\log_{10}$ transformed data. Knowing the $\log_{10}$ transformed standard deviation values enables certain statistical experimental design features to be determined. The most significant feature is determining the number of observations needed to enable the data to be described with a specific error level. It can also be used to calculate the error associated with any observation, based on the assumed population distribution characteristics and the number of observations. As an example, consider a pollutant having a COV of 0.23 and a median value of 0.14. The resulting $\log_{10}$ transformed standard deviation would be about 0.12. One

Figure 5.6 Relationship between COV (real space) and standard deviation ($\log_{10}$ space) (From Pitt, R. and M. Lalor. *Identification and Control of Non-Stormwater Discharges into Separate Storm Drainage Systems*. Development of Methodology for a Manual of Practice. U.S. Environmental Protection Agency, Water Supply and Water Resources Division, National Risk Management Research Laboratory, Cincinnati, OH. 451 pp. To be published in 2001.)

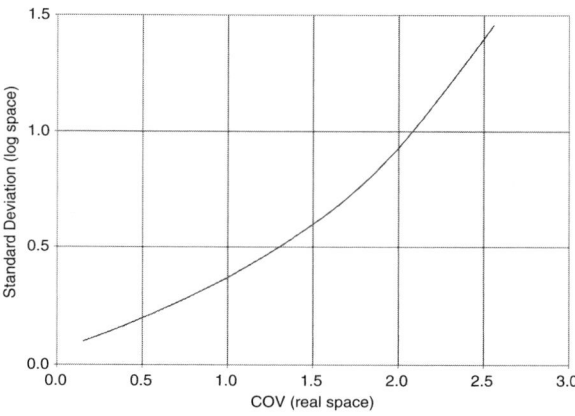

equation that has been historically used to calculate the number of analyses needed, based on the allowable error is (Cochran 1963):

$$\text{Number of samples} = 4(\text{standard deviation})^2/(\text{allowable error})^2$$

With an approximate 95% level of confidence ($1.96^2 \doteq 4$), this relationship determines the number of samples needed to obtain a value within the range of the sample mean, plus and minus the error. This equation can be rearranged to obtain the error, based on the number of samples obtained and the standard deviation. As an example, for 10 samples and the above standard deviation (0.12), the resulting approximate 95% confidence range (ignoring false negatives) of the median observation (0.14 mg/L) is:

$$\text{Error} = 2(0.12)/(10)^{0.5} = 0.076 \text{ in } \log_{10} \text{ space}$$

The confidence interval is therefore $\log_{10}(0.14) \pm 0.076$, which is -0.778 to -0.930 in $\log_{10}$ space. This results in an approximate 95% confidence range of $10^{-0.930}$ ($= 0.12$) to $10^{-0.778}$ ($= 0.17$). The absolute value for the error in the estimate of the median value is therefore between 14% (100 × (0.14 – 0.12)/0.14) and 21% (100 × (0.17 – 0.14)/0.14) for 10 samples. If the original untransformed data were used, the error associated with 10 samples is about 15%, within the range of the estimate after log transformations. These results are close because of the low COV value (0.23). If the COV value is large (>0.4), the need for log transformations increases.

Example Showing Improvement of Mean Concentrations with Increasing Sampling Effort

Many stormwater discharge samples were obtained from two study areas during the Bellevue, WA, Urban Runoff Program (Pitt 1985). The runoff from each drainage area was affected by different public works stormwater control practices, and the outfall data were compared to identify if any runoff quality improvements were associated with this effort. These data offer an opportunity to examine how increasing numbers of outfall data decreased the uncertainty of the overall average concentrations of the stormwater pollutants. Table 5.4 shows how the accumulative average of the observed concentrations eventually becomes reasonable steady, but only after a significant sampling effort. As an example: the average on the first three observations results in an EMC (event-mean concentration) that is in error by about 40%. It would require more than 15 samples before the average value would be consistently less than 10% from the seasonal average value, which only had a total population of 25 storm events, even with the relatively small COV value of 0.65.

Albert and Horwitz (1988) point out that taking averages leads to a tighter distribution. As shown above, the extreme values have little effect on the overall average, even with a relatively few observations (for a Gaussian distribution). The reduction in the standard deviation is proportional to $1/n^{0.5}$, for n observations. Even if the population is not Gaussian, the averages tend to be Gaussian-like. In addition, the larger the sample size, the more Gaussian-like is the population of averages.

Determining the Number of Sampling Locations (or Land Uses) Needed to Be Represented in a Monitoring Program

The above example for characterizing a parameter briefly examined a method to determine the appropriate number of samples to be collected and analyzed at a specific location. However, another aspect of sample design is determining how many components (specifically sampling locations) need to be characterized. The following example uses a marginal benefit analysis to help identify a basic characterization monitoring program. The sampling effort procedure discussed previously applies to the number of samples needed for each sampling location, while this analysis identifies

Table 5.4   Event-Mean Concentrations for Series of Storm Samples in Bellevue, WA

Storm No.	Lead Concentration (mg/L)	Moving Average Concentration (EMC)	Error from Seasonal Average (percent)
1	0.53	0.53	119
2	0.10	0.32	30
3	0.38	0.34	39
4	0.15	0.29	20
5	0.12	0.26	6
6	0.12	0.23	−3
7	0.56	0.28	16
8	0.19	0.27	11
9	0.38	0.28	16
10	0.23	0.28	14
11	0.20	0.27	11
12	0.39	0.28	16
13	0.53	0.30	24
14	0.05	0.28	16
15	0.26	0.28	16
16	0.05	0.27	10
17	0.05	0.25	5
18	0.39	0.26	8
19	0.28	0.26	8
20	0.10	0.25	5
21	0.29	0.25	6
22	0.18	0.25	4
23	0.31	0.25	5
24	0.10	0.25	2
25	0.10	0.24	0

From Pitt, R. *Characterizing and Controlling Urban Runoff through Street and Sewerage Cleaning.* U.S. Environmental Protection Agency, Storm and Combined Sewer Program, Risk Reduction Engineering Laboratory. EPA/600/S2-85/038. PB 85-186500. Cincinnati, OH. 467 pp. June 1985.

the number of sampling locations that should be monitored. This example specifically examines which land use categories should be included in a city-wide monitoring program when the total city's stormwater discharges need to be quantified with a reasonable error.

Land Use Monitoring for Wet-Weather Discharge Characteristics

The following paragraphs outline the steps needed to select the specific land uses that need to be included in a monitoring program to characterize stormwater runoff from an urban area to a specific receiving water. This method was also shown earlier in Chapter 4 for the Los Angeles County monitoring effort case study. The following example is loosely based on analyses of data for the Waller Creek drainage in Austin, TX.

Step 1 — This step identifies the land use categories that exist in the area of study. The information compiled during site selection activities will enable effective monitoring sites to be selected. In addition, this information will be very useful in extrapolating the monitoring results across the whole drainage area (by understanding the locations of similar areas represented by the land use-specific monitoring stations) in helping to identify the retrofit control programs that may be suitable for these types of areas, and in understanding the benefits of the most cost-effective controls for new development.

The initial list of land use areas to be considered for monitoring should be based on available land use maps, but they will have to be modified by overlaying additional information that should have an obvious effect on stormwater quality and quantity. The most obvious overlays would be the age of development (an "easy" surrogate for directly connected imperviousness, maturity of vegetation, width of streets, conditions of streets, etc., that all affect runoff conditions and control measure applications) and the presence of grass swale drainage (which has a major effect on mass emissions and runoff frequency). Some of these areas may not be important (very small area represented in study area, especially with known very low concentrations or runoff mass) and may be eliminated at this step. After this initial list (with overlays) is developed, locations that are representative of each potential category need to be identified for preliminary surveys. About 10 representative neighborhoods in each category that reflect the full range of development conditions for each category should be identified. The 10 locations in each land use would be relatively small areas, such as a square block for residential areas, a single school or church, a few blocks of strip commercial, etc. The 10 sites would be selected over a wide geographical area of the study area to include topographical effects, distance from ocean, etc.

Step 2 — This step includes preliminary surveys of the land uses identified above. For each of the 10 neighborhoods identified in each category, simple field sheets are filled out with information that may affect runoff quality or quantity, including type of roof connections, type of drainage, age of development, housing density, socioeconomic conditions, quantity and maintenance of landscaping, condition of pavement, soils, inspections of storm drainage to ensure no inappropriate discharges, and existing stormwater control practices. These are simple field surveys that can be completed by a team of two people at the rate of about 10 locations a day, depending on navigation problems, traffic, and how spread out the sites are. Several photographs can also be made of each site and be archived with the field sheets for future reference.

Step 3 — In this step, measurements of important surface area components are made for each of the neighborhoods surveyed above. These measurements are made using aerial photographs of each of the 10 areas in each land use category. Measurements will include areas of rooftops, streets, driveways, sidewalks, parking areas, storage areas, front grass strips, sidewalks and streets, playgrounds, backyards, front yards, large turf areas, undeveloped areas, decks and sheds, pools, railroad rows, alleyways, and other paved and nonpaved areas. This step requires the use of good aerial photography in order to resolve the elements of interest for measurement. Print scales of about 100 ft per 1 inch are probably adequate, if the photographs are sharp. Photographic prints for each of the homogeneous neighborhoods examined on the ground in step 2 are needed. The actual measurements require about an hour per site.

Step 4 — In this step, the site survey and measurement information are used to confirm the groupings of the individual examples for each land use category. This step finalizes the categories to be examined, based on the actual measured values. As an example, some of the sites selected for field measurement may actually belong in another category (based on actual housing density, for example) and would then be reassigned before the final data evaluation. More important, the development characteristics (especially drainage paths) and areas of important elements (especially directly connected pavement) may indicate greater variability within an initial category than between other categories in the same land use (such as for differently aged residential areas, or high-density residential and duplex home areas). A simple ANOVA test would indicate if differences exist, and additional statistical tests can be used to identify the specific areas that are similar. If there is no other reason to suspect differences that would affect drainage quality or quantity (such as landscaping maintenance for golf courses vs. undeveloped areas), these areas could be combined to reduce the total number of individual land use categories/subcategories used in subsequent evaluations.

Step 5 — This step includes the ranking of the selected land use categories according to their predominance and pollutant generation. A marginal benefit analysis can be used to identify which land use categories should be monitored. Each land use category has a known area in the drainage area and an estimated pollutant mass discharge. This step involves estimating the total annual mass discharges associated with each land use category for the complete study area. These sums are then ranked, from largest to smallest, and an accumulated percentage contribution is produced. These accumulated percentage values are plotted against the number of land use categories. The curve will be relatively steep initially and then level off as it approaches 100%. A marginal benefit analysis can then be used to select the most effective number of land uses that should be monitored.

The following is an example of this marginal benefit analysis to help select the most appropriate number of land uses to monitor. The numbers and categories are based on the Waller Creek, Austin, TX, watershed. Table 5.5 shows 16 initial land use categories, their land cover (as a percentage), and the estimated unit area loadings for each category for a critical pollutant. These loading numbers will have to be obtained using best judgment and prior knowledge. This table then shows the relative masses of the pollutant for each land use category (simply the % area times the unit area loading). The land uses are shown ranked by their relative mass discharges and a summed total is shown. This sum is then used to calculate the percentage of the pollutant associated with each land use category. These are then accumulated. The "straight-line model" is the straight line from 0 mass at 0 stations to 100% of the mass at 16 stations. The final column is the difference between these two lines (the marginal benefit).

Figure 5.7 is a marginal benefit plot of these values. The most effective monitoring strategy is to monitor seven land uses in this example. After this number, the marginal benefit starts to decrease. Seven (out of 16) land uses will also account for about 75% of the total annual emissions from these land uses in this area. A basic examination of the plot shows a strong leveling of the curve at 12 land uses, where the marginal benefit dramatically decreases and where there is little doubt of additional benefit for additional effort. The interpretation of these data should include the following issues that may expand the basic monitoring effort:

- The marginal benefit (as shown to include 7 of the 16 land uses for monitoring in this example)
- Land uses that have expected high unit area mass discharges that may not be included in the above list because of relatively low abundance, such as shopping malls in this example
- Land uses that are expected to become a significant component (such as the new medium-density residential area in this example)
- Land uses that have special conditions, such as a grass swale site in this example, that may need to be demonstrated/evaluated.

Step 6 — Final selection of monitoring locations. The top-ranked land uses will then be selected for monitoring. In most cases, a maximum of about 10 sites would be initiated each year. The remaining top-ranked land uses will then be monitored starting in future years because of the time needed to establish monitoring stations. In selecting sites for monitoring, sites draining homogeneous areas need to be found. In addition, monitoring locations will need to be selected that have sampling access, no safety problems, etc. To save laboratory resources, three categories of land uses can be identified. The top group would have the most comprehensive monitoring efforts (including most of the critical source area monitoring activities), while the lowest group may only have flow monitoring (with possibly some manual sampling). The middle group would have a shorter list of constituents routinely monitored, with periodic checks for all constituents being investigated.

Step 7 — The monitoring facilities need to be installed. The monitoring equipment should be comprised of automatic water samplers and flow sensors (velocity and depth of flow in areas expected to have surcharging flow problems), plus a tipping bucket rain gauge. The samples should all be obtained as flow-weighted composites, requiring only one sample to be analyzed per event at each monitoring station.

The sampler should initiate sampling after three tips (about 0.03 inches of rain) of the tipping bucket rain gauge at the sampling site. Another sample initiation method is to use an offset of the flow

Table 5.5 Example Marginal Benefit Analysis

	Land Use (ranked by % mass per category)	% of Area	Critical Unit Area Loading	Relative Mass	% Mass per Category	Accum. (% mass)	Straight-line Model	Marginal Benefit
1	Older medium-density residential	24	200	4800	22.8	22.8	6.25	16.5
2	High-density residential	7	300	2100	10.0	32.7	12.5	20.2
3	Office	7	300	2100	10.0	42.7	18.8	24.0
4	Strip commercial	8	250	2000	9.5	52.2	25.0	27.2
5	Multiple-family	8	200	1600	7.6	59.8	31.3	28.5
6	Manufacturing industrial	3	500	1500	7.1	66.9	37.5	29.4
7	Warehousing	5	300	1500	7.1	74.0	43.8	30.3
8	New medium-density residential	5	250	1250	5.9	80.0	50.0	30.0
9	Light industrial	5	200	1000	4.7	84.7	56.3	28.4
10	Major roadways	5	200	1000	4.7	89.4	62.5	26.9
11	Civic/educational	10	100	1000	4.7	94.2	68.8	25.4
12	Shopping malls	3	250	750	3.6	97.7	75.0	22.7
13	Utilities	1	150	150	0.7	98.5	81.3	17.2
14	Low-density residential with swales	5	25	125	0.6	99.1	87.5	11.6
15	Vacant	2	50	100	0.5	99.5	93.8	5.8
16	Park	2	50	100	0.5	100.0	100.0	0.0
	Total	100		21,075	100			

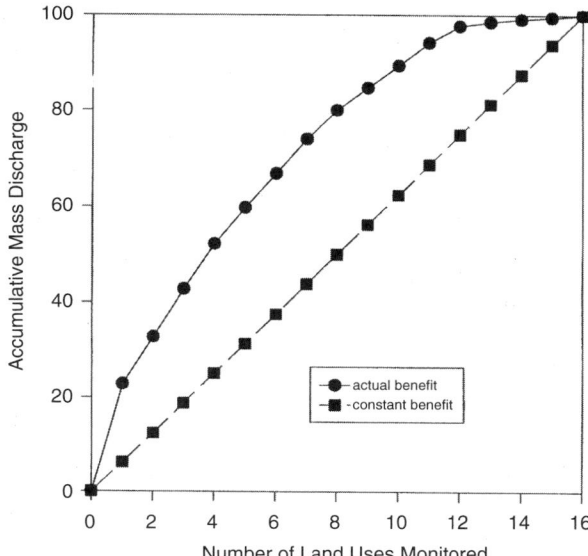

Figure 5.7 Marginal benefit associated with increasing sampling effort.

stage recorder to cause the sampler to begin sampling after a predetermined rise in flow conditions. False starts are then possible, caused by inappropriate discharges in the watershed above the sampling station. Frequent querying of sampler, flow, and rain conditions (using a data logger with phone connections) will detect this condition to enable retrieval of these dry-weather samples for analyses and to clean and reset the sampler. Both methods can be used simultaneously to ensure that only wet-weather samples are obtained. Of course, periodic (on random days about a month apart) dry-weather sampling (on a time composite basis over 24 hours) is also likely to be needed.

The base of the automatic sampler will need to be modified for a larger sample bottle (as much as a 100 L Teflon®-lined drum, with a 10 L glass bottle suspended for small events) in order to automatically sample a wide range of rain conditions without problems. A refrigerated base may also be needed, depending on ambient air conditions and sample holding requirements. The large drum will need to be located in a small freezer, with a hole in the lid where the sample line from the automatic sampler passes through.

Each sampler should also be connected to a telephone so the sampler status (including the temperature of the sample) and rainfall and flow conditions can be observed remotely. This significantly reduces personnel time and enables sampler problems to be identified quickly. Each sampler site will also need to be visited periodically (about weekly) to ensure that everything is ready to sample.

Step 8 — The monitoring initiation should continue down the list of ranked land use categories and repeat steps 6 and 7 for each category. At some point the marginal benefit from monitoring an additional land use category will not be sufficient to justify the additional cost.

While it is difficult to state how long this eight-step process should take, as a very rough estimate, it could take the following times to complete each step for a large city: Steps 1 to 3, 1 month each; Steps 4 and 5, 1 month combined; Step 6, 3 months; Step 7, 3 months; Step 8, continuous, for a total of about 10 months. This process was totally completed by Los Angeles County, for the unincorporated areas, in just a few months (see Chapter 4 case study).

Determining the Number of Samples Needed to Identify Unusual Conditions

An important aspect of receiving water effects studies is investigating unusual conditions. The methods presented by Gilbert (1987) ("Locating Hot Spots") can be used to select sampling

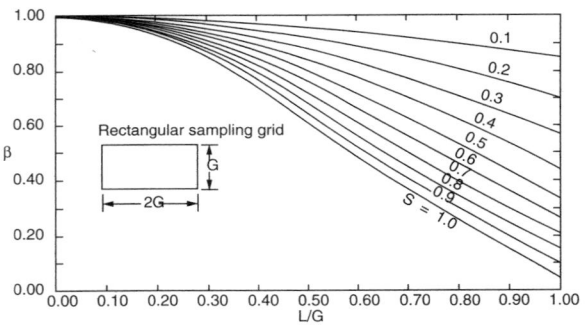

Figure 5.8 Sample spacing needed to identify unusual conditions. (From Gilbert, R.O. *Statistical Methods for Environmental Pollution Monitoring.* Van Nostrand Reinhold, New York. Copyright 1987. This material used with permission of John Wiley & Sons.)

locations that have acceptable probabilities of locating these unusual conditions. These methods are probably most applicable for lake or large stream sediment investigations in two dimensions. One-dimensional (longitudinal) studies can also be designed using a similar approach. Gilbert concluded that the use of a regular spacing of samples over an area was more effective when the contamination pattern was irregular, and an irregular pattern was best if the contamination existed in a repeating pattern. In almost all cases, unusual contamination has an irregular pattern and a regular grid is recommended. Gilbert presents square, rectangular, and triangular grid patterns to help locate sampling locations over an area. The sampling locations are located at the nodes of the resulting grids. Figure 5.8 (Gilbert 1987) is for the rectangular grid pattern, where the grid has a 2-to-1 aspect ratio. The figure relates the ratio of the size of a circular hot spot to the rectangular grid dimensions (sampling spacing) to the probability of detection. β is the probability of not finding the spot, while S is the shape factor for the hot spot (S = 1 for a circular spot; S = 0.5 for an elliptical spot). For example, if a semi-elliptical spot was to be targeted (S = 0.7) and the probability of not finding the spot was set at 25% (β = 0.25), the required L/G ratio would be 0.95+, with the rectangular width (G) about equal to the minor radius of the target.

Number of Samples Needed for Comparisons between Different Sites or Times

The comparison of paired data sets is commonly used when evaluating the differences between two situations (locations, times, practices, etc.). An equation related to the one given previously can be used to estimate the needed samples for a paired comparison:

$$n = 2\,[(Z_{1-\alpha} + Z_{1-\beta})/(\mu_1 - \mu_2)]^2\sigma^2$$

where α = false positive rate ($1 - \alpha$ is the degree of confidence. A value of α of 0.05 is usually considered statistically significant, corresponding to a $1 - \alpha$ degree of confidence of 0.95, or 95%)

β = false negative rate ($1 - \beta$ is the power. If used, a value of β of 0.2 is common, but it is frequently ignored, corresponding to a β of 0.5)

$Z_{1-\alpha}$ = Z score (associated with area under normal curve) corresponding to $1 - \alpha$

$Z_{1-\beta}$ = Z score corresponding to $1 - \beta$ value

μ_1 = mean of data set one

μ_2 = mean of data set two

σ = standard deviation (same for both data sets, same units as μ; both data sets are assumed to be normally distributed)

This equation is also only approximate, as it requires that the two data sets be normally distributed and have the same standard deviations. As noted previously, many parameters of interest in receiving water studies are likely closer to being log-normally distributed. Again, if the coefficient of variation

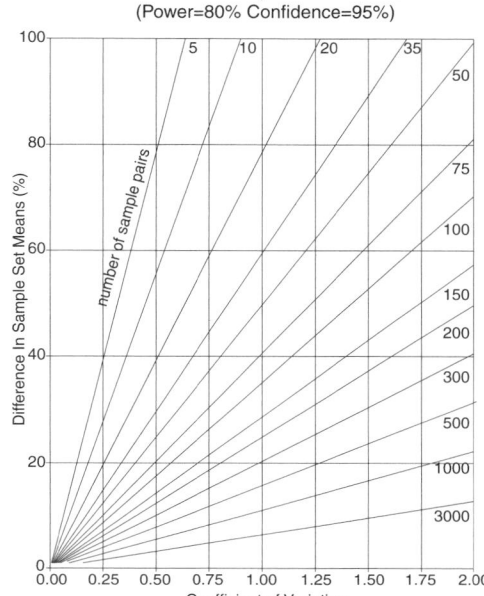

Figure 5.9   Sample effort needed for paired testing (power of 80% and confidence of 95%). (From Pitt, R. and K. Parmer. *Quality Assurance Project Plan: Effects, Sources, and Treatability of Stormwater Toxicants.* Contract No. CR819573. U.S. Environmental Protection Agency, Storm and Combined Sewer Program, Risk Reduction Engineering Laboratory. Cincinnati, OH. February 1995.)

(COV) values are low (less than about 0.4), then there is probably no real difference in the predicted sampling effort. Also, variations after treatment are commonly lower than before treatment.

Figure 5.9 (Pitt and Parmer 1995) is a plot of this equation (normalized using COV and differences of sample means) showing the approximate number of sample pairs needed for an α of 0.05 (degree of confidence of 95%), and a β of 0.2 (power of 80%). As an example, 12 sample pairs will be sufficient to detect significant differences (with at least a 50% difference in the parameter value) for two locations, if the coefficient of variation is no more than about 0.5. Appendix A (Pitt and Parmer 1995) contains similar plots for many combinations of other levels of power, confidence, and expected differences.

Need for Probability Information and Confidence Intervals

The above discussions have presented information mostly pertaining to a simple characteristic of the population being sampled: the "central tendency," usually presented as the average, or mean, of the observations. However, much greater information is typically needed, especially when conducting statistical analyses of the information. Information concerning the probability distribution of the data (especially variance) was used previously as it affected the sampling effort. However, many more uses of the probability distributions exist. Albert and Horwitz (1988) state that the researcher must be aware of how misleading an average value alone can be, because the average tells nothing about the underlying spread of values. Berthouex and Brown (1994) also point out the importance of knowing the confidence interval (and the probability) of a statistical conclusion. It can be misleading to state simply that the results of an analysis are significant (implying that the null hypothesis, the difference between the means of two sets of data is zero, is rejected at the 0.05 level), for example, when the difference may not be very important. It is much more informative to present the 95% confidence interval of the difference between the means of the two sets of data.

One important example of how probability affects decisions concerns the selection of critical and infrequent conditions. In hydrology analyses, the selection of a "design" rainfall dramatically affects the design of a drainage system. Similarly, the likelihood of extreme events is also important for receiving water analyses (such as the frequency of high flushing flows vs. needed recovery periods). The probability that a high flow rate in a stream (or any other factor of interest having a recurrence interval of "T" years) will occur during "n" years is:

$$P = 1 - (1 - 1/T)^n$$

As an example, the probability of a 5-year rain occurring at least once in a 5-year period is not 1, but is:

$$P = 1 - (1 - 1/5)^5 = 1 - (0.8)^5 = 1 - 0.328 = 0.67 \text{ (or 67\%)}$$

In another example, a flow having a recurrence interval of 20 years is assumed to cause substantial damage to critical biological species in a stream. That flow is likely to have the following probability of occurrence during a 100-year period:

$$P = 1 - (1 - 1/20)^{100} = 1 - (0.95)^{100} = 1 - 0.0059 = 0.994 \text{ (99.4\%)}$$

but only the following probability of occurrence during a 5-year period:

$$P = 1 - (1 - 1/20)^5 = 1 - (0.95)^5 = 1 - 0.774 = 0.227 \text{ (22.7\%)}$$

Figure 5.10 (McGee 1991) illustrates this equation. If a construction site is undergoing development for 2 years and the erosion control practices had to be certain of survival at least at the 95% level, then a 40-year design storm condition must be used! Similarly, a 1000-year design flow (one having only a 0.1% chance of occurring in any 1 year) would be needed if one needed to be 90% certain that it would not be exceeded during a 100-year period.

An entertaining example presented by Albert and Horwitz (1988) illustrates an interesting case concerning the upper limits of a confidence interval. In their example, an investigator wishes to

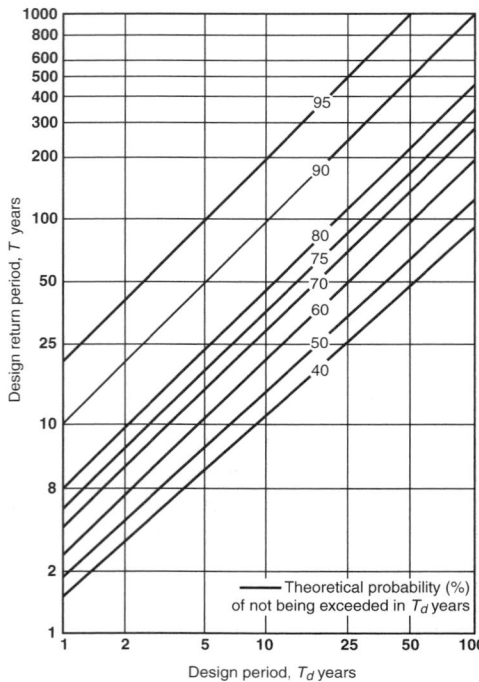

Figure 5.10 Design period and return period. (From McGee, T.J. *Water Supply and Sewerage*. McGraw-Hill, Inc., New York. 1991. With permission.)

determine if purple cows really exist. While traveling through a farming district, 20 cows are spotted, but none is purple. What is the actual percentage of cows that are purple (at a 95% confidence level), based on this sampling? The following formula can be used to calculate the upper limit of the 95% confidence interval:

$$(1 - 0)^n - (1 - x)^n = 0.95$$

or

$$1 - (1 - x)^n = 0.95$$

where n is the number of absolute negative observations and x is the upper limit of the 95% confidence interval. Therefore, for a sampling of 20 cows (n = 20), the actual percentage of cows that are purple is between 0.0% and 13.9% (x = 0.139). If the sample was extended to 40 cows (n = 40), the actual percentage of cows that are purple would be between 0.0% and 7.2% (x = 0.072). The upper limit of both of these cases is well above zero and, for most people, these results generally conflict with common sense. Obviously, the main problem with the above purple cow example is the violation of the need for random sampling throughout the whole population.

DATA QUALITY OBJECTIVES (DQO) AND ASSOCIATED QA/QC REQUIREMENTS

As noted in Chapter 4, the precision and accuracy necessary to meet the project objectives should be defined. After this is accomplished, the procedures for monitoring and controlling data quality must be specific and incorporated within all aspects of the assessment, including sample collection, processing, analysis, data management, and statistical procedures.

- When designing a plan, one should look at the study objectives and ask:
  - How will the data be used to arrive at conclusions?
  - What will the resulting actions be?
  - What are the allowable errors?

The first stage in developing DQOs requires the decision makers to determine what information in needed, reasons for the need, how the information will be used, and to specify time and resource limits. During the second stage, the problem is clarified and constraints on data collection identified. The third stage develops alternative approaches to data selection, selecting the optimal approach, and establishing the DQOs (EPA 1984, 1986).

Quality Control and Quality Assurance to Identify Sampling and Analysis Problems

Quality assurance and quality control (QA/QC) have been used in laboratories for many years to ensure the accuracy of analytical results. Unfortunately, similar formal QA/QC programs have been lacking in field collection and field analysis programs. Without carefully planned and executed sample collection activities, the best laboratory results are meaningless. Previous sections of this chapter have discussed the necessary experimental design aspects that enable the magnitude of the sampling effort to be determined. They specifically showed how the sample collection and data analysis efforts need to be balanced with experimental objectives. These sections stressed the need for a well-conceived experimental design to enable the questions at hand to be answered. This section presents additional information for conducting a water sampling program. These two discussions therefore contain information pertaining to "good practice" in conducting a field investigation and are therefore fundamental components of a QA/QC program for field activities.

This section reviews some of the aspects of conventional laboratory QA/QC programs that must also be used in field investigations of receiving water problems. This is not a comprehensive presentation of these topics suitable for conventional laboratory use. It is intended only as a description of many of the components that should be used in field or screening analyses. It is also suitable as a description of the QA/QC efforts that supporting analytical laboratories should be using and can help the scientist or engineer interpret the analytical reports.

Use of Blanks to Minimize and Identify Errors

Blanks are the most effective tools for assessing and controlling contamination, which is a common source of error in environmental measurements. Contamination can occur from many sources, including during sample collection, sample transport and storage, sample preparation, and sample analysis. Proper cleaning of sampling equipment and sample containers, as previously described, is critical in reducing contamination. The use of appropriate materials that contact the sample (sampling equipment and sample containers especially) was also previously noted as being critical in reducing sample contamination. Field handling of samples (such as adding preservatives) may also cause sample contamination. During the Castro Valley urban runoff study, Pitt and Shawley (1982) found very high, but inconsistent, concentrations of lead in the samples. This was especially critical because the several months' delay between sending the samples to the laboratory and receiving the results prevented repeating the collection and analysis of the suspect samples. After many months of investigation, the use of trip blanks identified the source of contamination. The glass vials containing the HNO_3 used for sample preservation were color-coded with a painted strip. The paint apparently had a high heavy metal content. When the acid was poured into the sample container in the field, some of it flowed across the paint strip, leaching lead into the sample. About 1 year of runoff data for heavy metals had to be discarded.

There are many types of blanks that should be used in monitoring programs. The following are typical blanks and their purpose:

- Instrument blank (system blank). Used to establish the baseline response of an instrument in the absence of the analyte. This is a blank analysis using only the minimal reagents needed for instrument operation (doesn't include reagents needed to prepare the sample); could be only ultrapure water.
- Calibration blank (solvent blank). Used to detect and measure solvent impurities. Similar to the above blank but only contains the solvent used to dilute the sample. This typically is the zero concentration in a calibration series.
- Method blank (reagent blank). Used to detect and measure contamination from all of the reagents used in sample preparation. A blank sample (using ultrapure water) with all reagents needed in sample preparation is processed and analyzed. This value is commonly subtracted from the analytical results for the samples prepared in the same way during the same analytical run. This blank is carried through the complete sample preparation procedures, in contrast to the calibration blank which doesn't require any preparation, but is injected directly into the instrument.
- Trip blank (sampling media blank). Used to detect contamination associated with field filtration apparatus and sample bottles. A known water (similar to sample) is carried from the laboratory and processed in the field in an identical manner as a sample.
- Equipment blank. Used to detect contamination associated with the sampling equipment. Also used to verify the effectiveness of cleaning the sampling equipment. A known water (similar to sample) is pumped through the sampling equipment and analyzed. Rinse water (or solvent) after the final equipment cleaning can also be collected and analyzed for comparison with a sample of the fluid before rinsing.

Quality Control

Standard Methods for the Examination of Water and Wastewater (1995) lists seven elements of a good quality control program: certification of operator competence, recovery of known additions, analysis of externally supplied standards, analysis of reagent blanks, calibration with standards, analysis of duplicates, and the use of control charts. These elements are briefly described below.

Certification of Operators

Adequate training and suitable experience of analysts are necessary for good laboratory work. Periodic tests of analytical skill are needed. A test proposed by *Standard Methods* (1995) is to use at least four replicate analyses of a check sample that is between 5 and 50 times the method detection limit (MDL) of the procedure. The precision of the results should be within the values shown in Table 5.6.

Recovery of Known Additions

The use of known additions should be a standard component of regular laboratory procedures. A known concentration is added to periodic samples before sample processing. This increase should be detected compared to a split of the same sample that did not receive the known addition. Matrix interferences are detected if the concentration increase is outside the tolerance limit, as shown in Table 5.6. The known addition concentration should be between 5 and 50 times the MDL (or 1 to 10 times the expected sample concentration). Care should be taken to ensure that the total concentration is within the linear response of the method. *Standard Methods* (1995) suggests that known additions be added to 10% of the samples analyzed.

Analysis of External Standards

These standards are periodically analyzed to check the performance of the instrument and the calibration procedure. The concentrations should be between 5 and 50 times the MDL, or close to the sample concentrations (whichever is greater). *Standard Methods* (1995) prefers the use of certified standards, which are traceable to National Institute of Standards and Technology (NIST) standard reference materials, at least once a day. Do not confuse these external standards with the standards used to calibrate the instrument.

Table 5.6 Acceptance Limits for Replicate Samples and Known Additions

Parameter	Recovery of Known Additions (%)	Precision of Low-Level (<20 × MDL) Duplicates (±%)	Precision of High-Level (>20 × MDL) Duplicates (±%)
Metals, anions, nutrients, other inorganics, and TOC	80–120	25	10
Volatile and base/neutral organics	70–130	40	20
Acid extractable organics	60–140	40	20
Herbicides	40–160	40	20
Organochlorine pesticides	50–140	40	20
Organophosphate pesticides	50–200	40	20
Carbamate pesticides	50–150	40	20

Data from *Standard Methods for the Examination of Water and Wastewater. 19th edition.* Water Environment Federation. Washington, D.C. 1995.

Analysis of Reagent Blanks

Reagent blanks must also be analyzed periodically. *Standard Methods* (1995) suggests that at least 5% of the total analytical effort be reagent blanks. These blanks should be randomly spaced between samples in the analytical run order, and after samples having very high concentrations. These samples will measure sample carry-over, baseline drift of the instrument, and impurity of the reagents.

Calibration with Standards

Obviously, the instrument must be calibrated with known standards according to specific guidelines for the instrument and the method. However, at least three known concentrations of the parameter should be analyzed at the beginning of the instrument run, according to *Standard Methods* (1995). It is also preferable to repeat these analyses at least at the end of the analytical run to check for instrument drift.

Analysis of Duplicates

Standard Methods (1995) suggests that at least 5% of the samples have duplicate analyses, including those used for matrix interferences (known additions), while other guidance may suggest more duplicate analyses. Table 5.6 presents the acceptable limits of the precision of the duplicate analyses for different parameters.

Control Charts

The use of control charts enables rapid and visual indications of QA/QC problems, which can then be corrected in a timely manner, especially while it may still be possible to reanalyze samples. However, many laboratories are slow to upgrade the charts, losing their main benefit. Most automated instrument procedures and laboratory information management systems (LIMs) have control charting capabilities built in. *Standard Methods* (1995) describes a "means" chart for standards, blanks, and recoveries. A means chart is simply a display of the results of analyses in run order, with the ±2 (warning level) and ±3 (control level) standard deviation limits shown. At least five means charts should be prepared (and kept updated) for each analyte: one for each of the three standards analyzed at the beginning (and at least at the end) of each analytical run, one for the blank samples, and one for the recoveries. Figure 5.11 is an example of a means chart. The pattern of observations should be random and most within the warning limits. Drift, or sudden change, should also be cause for concern, needing immediate investigation. Of course, if the warning levels are at the 95% confidence limit (approximate ±2 standard deviations), then approximately 1 out of 20 samples will exceed the limits, on average. Only 1 out of 100 should exceed the control limits (if at the 99% confidence limit, or approximate ±3 standard deviations).

Standard Methods (1995) suggests that if one measurement exceeds the control limits, the sample should be immediately reanalyzed. If the repeat is within acceptable limits, then continue. If the repeat analysis is again outside the control limits, the analyses must be discontinued and the problem identified and corrected. If two out of three successive analyses exceed the warning limits, another replicate analysis is made. If the replicate is within the warning limits, then continue. However, if the third analysis is also outside the warning limits, the analyses must be discontinued and the problem identified and corrected. If four out of five successive analyses are greater than ±1 standard deviation of the expected value, or are in decreasing or increasing order, another sample is to be analyzed. If the trend continues, or if the sample is still greater than ±1 standard deviation of the expected value, then the analyses must be discontinued and the problem identified and corrected. If six successive samples are all on one side of the average concentration line, and the

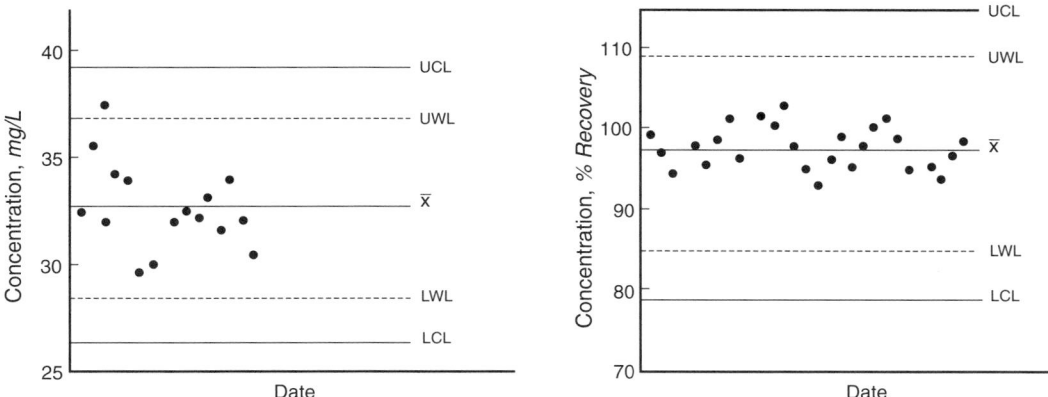

Figure 5.11 Means quality control chart (From *Standard Methods for the Examination of Water and Waste-water. 20th edition.* Water Environment Federation. Washington, D.C. Copyright 1998. APHA. With permission.)

next is also on the same side as the others, the analyses must be discontinued and the problem identified and corrected. After correcting the problem, *Standard Methods* (1995) recommends that at least half the samples analyzed between the last in-control measurement and the out-of-control measurement be reanalyzed.

Standard Methods (1995) also points out that another major function of control charts is to identify changes in detection limits. Recalculate the warning and control limits (based on the standard deviations of the results) for every 20 samples. Running averages of these limits can be used to easily detect trends in precision (and therefore detection limits).

Carrying out a QA/QC program in the laboratory is not inexpensive. It can significantly add to the analytical effort. ASTM (1995) summarizes these typical extra sample analyses:

- Three or more standards to develop or check a calibration curve per run
- One method blank per run
- One field blank per set of samples
- At least one duplicate analysis for precision calculations for every 20 samples
- One standard sample to check the calibration for every 20 samples
- One spiked sample for matrix interference analyses for every 20 samples.

This can total at least eight additional analyses for every run of up to 20 samples.

Checking Results

Good sense is very important and should be used in reviewing analytical results. Extreme values should be questioned, for example, not routinely discarded. With a complete QA/QC program, including laboratory and field blanks, there should be little question if a problem has occurred and what the source of the problem may be. Unfortunately, few monitoring efforts actually carry out adequate or complete QA/QC programs. Especially lacking is timely updating of control charts and other tools that can easily detect problems. The reasons for this may be cost, ignorance, or insufficient time. However, the cost of discarded results may be very high, such as for resampling. In many cases, resampling is not possible, and much associated data may be worth much less without necessary supporting analytical information. In all cases, unusual analytical results should be reported to the field sampling crew and other personnel as soon as possible to solicit their assistance in verifying that the results are valid and not associated with labeling or sampling error.

Standard Methods (1995) presents several ways to check analytical results for basic measurements, based on a paper by Rossum (1975). The total dissolved solids concentration can be estimated using the following calculation:

$$TDS \cong 0.6 \ (alkalinity) + Na + K + Ca + Mg + Cl + SO_4 + SiO_3 + NO_3 + F$$

where the ions are measured in mg/L (alkalinity as $CaCO_3$, SO_4 as SO_4, and NO_3 as NO_3). The measured TDS should be higher than the calculated value because of likely missing important components in the calculation. If the measured value is smaller than the calculated TDS value, the sample should be reanalyzed. If the measured TDS is more than 20% higher than the calculated value, the sample should also be reanalyzed.

The anion–cation balance should also be checked. The milliequivalents per liter (meq/L) sums of the anions and the cations should be close to 1.0. The percentage difference is calculated by (*Standard Methods* 1995):

$$\% \text{ difference} = 100 \ (\Sigma \text{ cations} - \Sigma \text{ anions}) \ / \ (\Sigma \text{ cations} + \Sigma \text{ anions})$$

with the following acceptance criteria:

Anion Sum (meq/L)	Acceptable Difference
0 to 3.0	±0.2 meq/L
3.1 to 10.0	±2%
10.1 to 800	±2 to 5%

In addition, *Standard Methods* (1995) states that both the anion and cation sums (in meq/L) should be 1/100 of the measured electrical conductivity value (measured as µS/cm). If either of the sums is more than 10% different from this criterion, the sample should be reanalyzed. The ratio of the measured TDS (in mg/L) and measured electrical conductivity (as µS/cm) values should also be within the range of 0.55 to 0.70.

Identifying the Needed Detection Limits and Selecting the Appropriate Analytical Method

The selection of the analytical procedure depends on a number of factors, including (in order of general importance):

- Appropriate detection limits
- Freedom from interferences
- Good analytical precision (repeatability)
- Minimal cost
- Reasonable operator training and needed expertise

One of the most critical and obvious determinants used for selecting an appropriate analytical method is the identification of the needed analytical detection limit. It is possible to select available analytical methods that have extremely low detection limits. Unfortunately, these very sensitive methods are typically costly and difficult to utilize. However, in many cases, these extremely sensitive methods are not needed. The basic method of selecting an appropriate analytical method is to ensure that it can identify samples that exceed appropriate criteria for the parameter being measured. If detection limits are smaller than a critical water quality criterion or standard, then analytical results that may indicate interference with a beneficial use can be selected directly. Appendix G presents water quality criteria for many constituents of concern in receiving water studies, while Chapter 6 and Appendix E describe typical levels of performance for different analytical methods.

There are several different detection limits that are used in laboratory analyses. *Standard Methods* (1995) states that the common definition of a detection limit is that it is the smallest concentration that can be detected above background noise, using a specific procedure and with a specific confidence. The instrument detection limit (IDL) is the concentration that produces a signal that is three standard deviations of the noise level. This would result in about a 99% confidence that the signal was different from background noise. This is the simplest measure of detection and is solely a function of the instrument and is not dependent on sample preparation. The MDL accounts for sample preparation in addition to the instrument sensitivity. The MDL is about four times greater than the IDL because sample preparation increases the variability in the analytical results. Automated methods have MDLs much closer to the IDLs than manual sample preparation methods. An MDL is determined by spiking reagent water with a known concentration of the analyte of interest at a concentration close to the expected MDL. Seven portions of this solution are then analyzed (with complete sample preparation) and the standard deviation is calculated. The MDL is 3.14 times this measured standard deviation (at the 99% confidence level). The practical quantification limit (PQL) is a more conservative detection limit and considers the variability between laboratories using the same methods on a routine basis. The PQL is estimated in *Standard Methods* to be about five times the MDL.

A quick estimate of the needed detection limit can be made by assuming the likely concentration of the compound necessary for detection and the associated coefficient of variation (the COV, or the standard deviation divided by the mean) of the distribution of the analytical results, and applying a multiplier. If an estimated COV is not available, an alternative is to use the expected ratio of the 90th and 10th percentile values (the "range ratio") of the data and using Figure 5.5, assuming a log-normal probability distribution of the data (Pitt and Lalor 2001). Log-normal probability distributions are commonly used to describe the concentration distributions of water quality data, including stormwater data (EPA 1983a,b). The data ranging from the 10th to the 90th percentile can typically be suitably described as a log-normal probability distribution. However, values less than the 10th percentile value are usually less than predicted from the log-normal probability plot, while values greater than the 90th percentile value are usually greater than predicted from the log-normal probability plot. The range ratio can generally be selected easily based on the expected concentrations to be encountered, ignoring the most extreme values. As the range ratio increases, the COV also increases, up to a maximum value of about 2.5 for the set of conditions studied by Pitt and Lalor 2001.

Pitt and Lalor (2001) conducted numerous Monte Carlo analyses using mixtures having broad ranges of concentrations. Using these data, they developed guidelines for estimating the needed detection limits to characterize water samples. If the analyte has an expected narrow range of concentrations (a low COV), then the detection limit can be greater than if the analyte has a wider range of expected concentrations (a high COV). These guidelines are as follows:

- If the analyte has a low level of variation (a 90th to 10th percentile range ratio of 1.5, or a COV of <0.5), then the estimated required detection limit is about 0.8 times the expected median concentration.
- If the analyte has a medium level of variation (a 90th to 10th percentile range ratio of 10, or a COV of about 0.5 to 1.25), then the estimated required detection limit is about 0.23 times the expected median concentration.
- Finally, if the analyte has a high level of variation (a 90th to 10th percentile range ratio of 100, or a COV of about >1.25), then the estimated required detection limit is about 0.12 times the expected median concentration.

Reporting Results Affected by Detection Limits

Reporting chemical analysis results should be clear, based on the measured detection limits and QA/QC program. Concentrations below the IDL are not present with sufficient confidence to detect them as significantly different from the baseline random noise of the instrument. These should be reported as not detected (generally given a "U" qualifier in organic compound analytical reports). Concentrations of a parameter above the IDL, but below the MDL, are present, but the confidence in the concentration value is less than 99% (can be given a "J" qualifier in organic analytical reports). Concentrations above the MDL indicate that the parameter is present in the sample and that the reported concentration is certain, at the 99% confidence level, or greater. Many other conditions may be present that degrade the confidence of the analytical results. These should all be carefully noted in the analytical report.

As noted in Chapter 7, nondetected ("left-censored") values present special problems in analyzing data. If only a few (or most) of the observations are below the detection limit, these problems are not very serious. However, if the detection limit available results in many left-censored data (say, between 25 and 75% of the observations), statistical analyses are severely limited. It may not be possible to statistically evaluate the effectiveness of a treatment process completely, for example, if many of the effluent concentrations of a critical pollutant are below the detection limit, even if the influent concentrations are well above the MDL. The removal of the pollutant is obviously important and effective, but it is not possible to calculate the significance of the differences in the observed concentrations. From a statistical (and engineering) viewpoint, it would be better if all concentrations determined by the analytical procedure be reported, even if they are below the designated "formal" detection limit, set using (extreme) 99% confidence limits. The use of the qualifiers (such as U and J as used in reporting GC/MS data) along with the numeric values and obvious reporting of the MDL should serve as a warning for the limited use of these values. However, analytical chemists are justifiably concerned about the misuse of "nondetected" values, and the availability of these values for statistical analyses will likely remain elusive. Unfortunately, nondetected values can be legally reported as "zero" in NPDES discharge reports, likely skewing mass calculations needed for TMDL, and other, evaluations.

GENERAL CONSIDERATIONS FOR SAMPLE COLLECTION

Sample collection and processing methods are dictated in part by the study objectives, regulatory requirements/recommendations, and proper QA/QC practice. The typical stormwater effects assessment will be comprised of in-stream water, sediment, and benthic invertebrate sampling. More intensive surveys may also sample other biological communities (e.g., fish, periphyton, zooplankton, phytoplankton, rooted macrophytes), watershed soils, interstitial sediment pore waters, dry- and wet-weather outfall effluents, and possibly sheet flows during rains. A number of publications have reviewed sampling methods which are applicable to stormwater assessments (Håkanson and Jansson 1983; EPA 1982, 1990c; ASTM 1991a).

It is important when sampling dynamic ecosystem components that there be an understanding that once the sample is collected and removed from the ecosystem, it no longer is a part of that ecosystem. It no longer will interact with the other ecosystem components spatially and temporally. A new ecosystem (the sample container) is created with different microenvironments, patch dynamics, and chemical transformations. For many sample constituents and parameters of concern, such as pesticides, suspended solids, and conductivity, the sampling process may do little to alter their levels from those present *in situ*. However, for other sample constituents and parameters, such as dissolved oxygen, un-ionized ammonia, metal speciation/solubility, microbial activity, pathogen survival, acid volatile sulfides, contaminant bioavailability, and toxicity, changes in the sample may be significant after sample collection. These changes cannot be predicted and are sample specific. Since the laboratory results of sample analyses are extrapolated to field conditions, these changes can potentially lead to erroneous conclusions on receiving water effects. Despite this bleak reality, accurate and precise studies have and can be conducted, provided proper sampling and processing practices are followed and there is an understanding of method limitations, procedurally induced artifacts, and constituents interactions. There is no one optimal method by which to sample all streams and lakes. The major types of sampling activities are discussed in this chapter.

The discussion of the selection of analytical methods in Chapter 6 also includes information on field determinations. These may lessen these sample disturbance problems, but the typically less precise and less sensitive field methods may not offer a great advantage over the generally superior laboratory methods. Combinations or replicates of methods are therefore usually used (such as conducting both field and lab pH determinations and toxicity surveys), along with special tests to examine the effects of sample storage, to quantify possible sample modifications that may affect the analyte concentrations.

Discrete samples are needed for defining minimum and maximum values, for statistical analyses of point-in-time using replicates rather than composite samples, when constituents are labile, or when spatial variance at a site is to be measured. Continuous *in situ* monitors (discussed in Chapter 6) are also available to indicate real-time variations for key parameters (such as DO, temperature, conductivity, turbidity, pH, and ORP). These can be used to supplement composite analyses for a cost-effective solution compared to conducting only discrete analyses. Composites provide an estimate of the mean of the constituent (population) from which the individual samples are drawn. They should only be collected on an individual event or subevent basis, or for a defined time interval. Variance of the mean and precision cannot be obtained from a composite. Proper QA/QC requires that accuracy and precision be determined, which is usually not possible with compositing. Compositing reduces maximum and increases minimum values and thus is a better indicator of chronic, long-term exposure values (EPA 1990a). Coefficients of variation and errors can be based on EMCs (event mean concentrations) (EPA 1983a,b). There are much greater variations observed between different events than within events for most in-stream or outfall chemical conditions. Collecting discrete samples greatly increases the laboratory analytical costs, reducing the number of events represented. Clearly, the best sampling plan must be carefully selected based on the specific study requirements and usually includes components of several different basic approaches.

Samplers should be constructed of inert, nonreactive materials and capable of collecting the necessary sample volume. They must also be capable of programming to meet the specific sampling schedule and protocol needed for the specific study. There are many automatic water samplers that are relatively inexpensive and have a great deal of flexibility to meet many different project needs. However, some modifications may be needed, as described later in this chapter. Metal, low-density polyethylene, or polyvinyl chloride (PVC) samplers may slightly contaminate water samples with metals and organics, respectively. Sampler material is not as critical when sampling sediments because the quantity of contaminant contributed to the edge of the sample is not significant.

Basic Safety Considerations When Sampling

The most important factor when conducting a field monitoring program is personnel safety. If an adequate program cannot be carried out in a reasonably safe manner, an alternative to the monitoring program must be used. Similarly, an inadequate monitoring program would be hard to justify. Most of the hazards reflect site selection and sampling times. The use of automatic samplers and well-trained crews (more than one) will reduce many of the hazards.

Water and sediment sampling may expose field personnel to hazardous conditions. Obviously, water hazards (high flows, deep pools, soft sediments, etc.) are usually of initial concern. In many stormwater assessment studies, sampling during rainy weather in streams that may undergo rapid velocity and depth changes is necessary. Great care must be taken when approaching a stream in wet weather, as steep and slippery banks may cause one to slide into the water. Always sample in pairs and have adequate safety equipment available. At a minimum, this will include:

- Throw rope
- Inflatable life vests
- Nylon-covered neoprene waders (that offer some flotation, even when swamped)
- 2-way radio or cellular phone
- Weather radio

If the conditions warrant (such as with steep and slippery stream banks), the sampler personnel should be tied together, with an attachment to a rigid shore object. In all cases, only go into the stream if absolutely necessary. Try to collect all samples from shore, especially during heavy rains. Be extremely cautious of changing weather and stream conditions and cancel sampling when hazardous conditions warrant. Never enter a stream where your footing is unstable or if the water is too deep (probably more than 2 ft deep) or fast (probably more than 2.5 ft/s). Always enter the water cautiously and be prepared to make an efficient retreat if you feel insecure.

Other hazardous conditions may also occur when working near urban streams. Sharp debris in the water and along the banks require that protective waders be worn at all times while in the stream. No one should enter the water barefooted. Poison ivy, poison oak, and ticks thrive along many stream banks, requiring long pants and shirts. When in the field during sunny weather, sun screen and a hat are necessities. In many parts of the country, especially in the South, special caution is also required concerning snakes. Water moccasins are very common, and coral snakes and copperheads may also be present along streams. Again, waders offer some protection, but be careful when moving through thick underbrush where visibility is limited.

These cautions are necessary and are basically common sense. However, the greatest dangers associated with field sampling, especially in urban areas, are likely associated with dogs running loose, odd people, automobiles/trucks, and eating greasy fast food (dangers which are not restricted to stream sampling).

Selecting the Sampling Locations

Specific sampling locations are determined based on the objectives of the study and site-specific conditions. Obviously, safety is a prime consideration, along with statistical requirements expressed in the experimental design. In all cases, the sample must represent the conditions being characterized.

The process of selecting a sampling site is often given minimal thought when designing an assessment study. Site selections are driven by two basic criteria: accessibility/safety and upstream–downstream locations of pollutant discharges. However, given the ecosystem complexities and statistical concerns, the importance of this process in achieving representative samples and one's study objectives cannot be overemphasized. Stormwater runoff effects may not be detected unless the proper samples are obtained from the affected site during the critical time periods and compared to baseline conditions.

As described earlier in this chapter, random or nonrandom sampling plans are used to determine *within*-site sampling locations. Few studies follow a random selection process, but it is the preferred method allowing for quantitative analyses which meet statistical assumptions (EPA 1990c). Only by knowing the probability (from random selection) of selecting a specific sample can one extrapolate from the sample to the population in an objective way. Only by using a grid-random number approach may one consciously select sample locations without subconscious bias (EPA 1990c). This process only occurs after the measurements, station locations, and number of samples have been determined. (See Gilbert 1987 and EPA guidance for grid sampling and stratified random sampling for hot spots, as summarized earlier in this chapter.)

Because benthic community spatial distributions are related to habitat conditions, a simple random approach is not optimal. Rather, it is best to stratify the habitat types based on known physical differences and then select subsampling units in which randomization is used. See Ford and Turina (1985). Sampling increases precision and most likely accuracy. Strata which may be

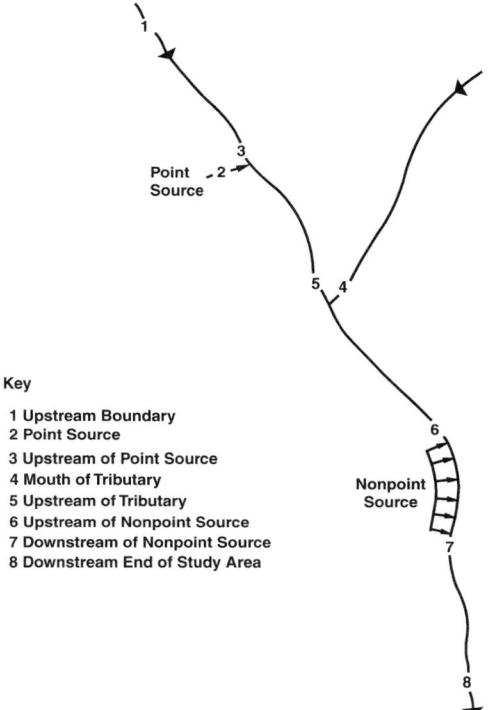

Key

1 Upstream Boundary
2 Point Source
3 Upstream of Point Source
4 Mouth of Tributary
5 Upstream of Tributary
6 Upstream of Nonpoint Source
7 Downstream of Nonpoint Source
8 Downstream End of Study Area

Figure 5.12 Recommended station locations for a minimal sampling program. (From EPA. *Handbook for Stream Sampling for Waste Load Allocation Applications*, Office of Research and Development, U.S. Environmental Protection Agency, Washington, D.C. EPA/625/6-86/013. 1986.)

used to define sampling units include habitat (pool vs. riffle), flow, temperature, sediment types, and others whose presence or effects may correlate to the parameter of interest. When locating sampling subunits in a nonrandom manner, one must consider samples semiquantitative for data extrapolation purposes (EPA 1990c).

Systematic sampling is often used in reconnaissance surveys and produces qualitative data. Samples are usually collected at key locations (e.g., a river bend) or at discrete intervals along a transect. This allows one to revisit fixed stations but ignores physical changes and disallows probability analyses. Kriging and other contaminant mapping techniques may be used when lake samples are collected using a systematic grid approach.

It is often more efficient and precise to have varying types of random sampling approaches for different parameters, such as: plankton — grid; macrophytes — shoreline transect; periphyton — shoreline transect. In small streams, fish and benthic macrobenthic sampling may be nonrandom, encompassing a total sub-reach section with true replication being impossible. This, of course, will violate some statistical assumptions.

Sites for sampling in a typical stream assessment are shown in Figure 5.12. Basic guidance for site location is as follows (modified from Cairns and Dickson 1971):

1. Two upstream reference stations are preferred, one immediately upstream of stressor inputs and one in upper reaches unimpacted by any anthropogenic influence. In addition, a nearby reference stream in the same ecoregion, which has similar watershed, flow, and habitat characteristics, is useful (EPA 1989).
2. Sample principal impact station, immediately below stressor inputs.
3. Note mixing patterns for point source inputs during subsampling.
4. Locate subsequent downstream stations based on pollutant loading, stream flow, sensitive areas, and suspected recovery–impact gradient. The maximum flow travel time between stations for conservative pollutants should be less than 2 days, and 5 to 8 km for reactive toxicants (EPA 1986). Sample station intervals are often about 0.5 day time-of-passage below a pollutant input for the

first 3 days, and 1 day thereafter (Kittrell 1969). In many urban streams, the sample locations are much more closely spaced, possibly only a few hundred meters apart, because of the large number of outfalls and frequent stream character variations due to artificial stream modifications. If a sample design is investigating the effects of a reach containing numerous outfalls on downstream waters, or possibly even an entire community, instead of a single discharge, wider spaced sampling locations below these areas would be needed.

5. Sample above and below tributaries.
6. Stations should have similar habitat and flow conditions, which typify the stream reach.
7. Samples should be replicated and collected in 1 day. Time of sampling must be noted, as many constituents have obvious natural diurnal cycles, e.g., dissolved oxygen (DO) and temperature. Sampling of indigenous communities such as periphyton, benthic invertebrates, and fish should occur as near as possible to the time that water quality samples are collected. In addition, weather conditions (air and water temperature, cloud cover, precipitation) during the sampling effort also should be noted. Riparian vegetation condition (especially seasonal growth) may also affect in-stream observations and also needs to be routinely noted.
8. Sampling should occur during each annual season in long-term studies to observe temporal cycles, seasonal stresses, and different organism groups and life stages.
9. Sampling should occur during a wide range of flow conditions.

Channel, flow, and stratification characteristics are particularly important when locating sample sites in streams, rivers, lakes, and reservoirs. Sampling near shore is seldom satisfactory except in small, upper reach streams. Whether using a random or systematic approach, one should carefully note the channel, flow, or stratification (lakes and reservoirs) conditions. In reservoirs, it is common for the principal flow to follow the old river channel and at a depth similar to the temperature (density) of the feeder stream. This area thus often contains the highest pollutant concentrations (e.g., suspended solids, fecal pathogens). Depositional zones, such as river bends and mouths, pools, and impoundment structures, should be sampled for sediment contamination and toxicity. For additional guidance on factors to consider in selecting station locations see below and Håkanson and Jansson (1983), and EPA (1983b, 1985, 1986, 1988, 1990 a,b,c).

As noted in Chapter 7, paired analyses are the most efficient sampling strategy. This can be simply sampling the influent and effluent of a control structure, outfalls of test and control watersheds, comparable stream habitats in test and control streams, or even the same stream sampling location, but at different seasons. Paired sampling can eliminate much variability, as many influencing factors are assumed to remain constant, enabling effects to be more easily seen. Obviously, if the differences between the two elements in the pair are expected to be large, and the background random variability is small, many fewer sampling pairs are needed to identify a statistically significant difference in the observations. Great care must be taken to select correct pairs, as the random variability can easily be greater than expected. Earlier sections of this chapter presented methods to determine the sampling effort for paired testing.

One example of likely inefficient paired sampling is sampling above and below an outfall in a stream. In almost all cases, the stream pollutant loads and flows are much greater than a single outfall discharge. Therefore, the differences expected in stream water quality upstream vs. downstream of an outfall would be very small and very difficult to detect. Exceptions may occur with large point source outfalls discharging during very low flow conditions. Otherwise, one large number is basically subtracted from another large number (with both having uncertainty) to determine the effects of a relatively small discharge. If this sampling strategy needs to be employed, make sure that the outfall discharge is also well characterized.

If loadings or stormwater concentrations of runoff from different land uses in a watershed are needed, then a sufficient number of examples need to be monitored. Many watersheds have several distinct land uses in their drainage area. It is important that a sufficient number of the land uses be adequately monitored in order to make an adequate mass balance. Examples of marginal benefits for increasing sampling locations was given earlier in this chapter and in Chapter 4.

The actual location of sampling is somewhat dependent on the type of sampler to be used. However, in all cases, the sample taken must be representative of the flow to be characterized. Permanently mounted automatic or semiautomatic samplers are most restricted in their placement, as security and better access is needed with them than with manual grab sampling. With manual sampling, less equipment is generally being carried to the sampling location (some type of manual dipper sampler, plus sample bottles, for example), while automatic samplers require a relatively large sample container, a multi-bottle sampler base, and batteries and other maintenance and cleaning supplies to be periodically carried to the sampler. Weekly visits to automatic samplers, at least, are needed for maintenance. In all cases, access during rains must be provided to all stormwater sampler locations. Manual stormwater sampling takes place during rains, of course, while automatic samplers may need to have their bottles switched during rains, or other checks made. Therefore, dangerous locations, such as those requiring steep ascents down clayey stream banks obviously must be avoided.

Permanently mounted samplers must have their intakes located to represent flow conditions. This is much easier with relatively small urban streams or outfalls compared to larger receiving waters. Wide, shallow, and fast-flowing streams are the most difficult to sample adequately. Great distances may be required before flows from individual discharges are completely mixed in these situations. Thomann and Mueller (1987) present the following USGS equation that can be used to estimate the distance needed before complete mixing occurs (for a side-stream discharge):

$$L_m = (2.6 \, UB^2)/H$$

where   U = the stream velocity in ft/s
        B = the average stream width in feet
        H = average stream width in feet

As an example, about 2000 m (6700 ft) may be required before complete mixing occurs for a stream that is 12 m (40 ft) wide, 1.5 m (5 ft) deep, and flowing at 2.4 m/s (8 ft/s). For a more typical urban stream with a 3 m (10 ft) width, 0.6 m (2 ft) depth, and flowing at 0.9 m/s (3 ft/s), the mixing length would be about 120 m (390 ft). Half of these distances would be needed if the discharge is located at the centerline of the stream (such as may occur for a diffuser for an industrial outfall). ASTM (1995) in standard D 3370 states that a distance of 1 to 3 miles below a tributary is usually sufficient to obtain complete mixing. It also suggests that samples be taken at least one half mile below dams or waterfalls to allow entrained air to escape.

These distances may be too great for many practical reasons, including the typical presence of numerous and fairly closely spaced outfalls along an urban creek (every several hundred feet). If it is not possible to site the sampler intake where the water will be well mixed, several sample intakes may be needed to obtain a composite sample across the stream. This can be accomplished by using several submerged pumps at different locations feeding a central large container located near the samplers. Automatic samplers are also restricted to a vertical height from the water surface to the sampler pump of about 7 m (since most use a peristaltic pump located on the sampler and therefore pull the water sample using vacuum suction). If the sampler height is greater than this critical height, a submerged pump can also be used to solve this problem. The automatic sampler would then sample from the large container that the submerged pumps are discharging into. In most cases, the submerged pumps would run continuously (needing on-site AC power or solar-charged batteries) and the flow-weighted sampler would be programmed to appropriately sample from the composite container, based on measured flows in the stream. The excess flow from the multiple pumps would overflow the composite container. Chapter 4 presented a case study for Los Angeles County, where this was an important consideration. The sample velocity in the sampler lines must be at least 100 cm/s to minimize particulate settling in the sampling lines. Care must also be taken to select a pump and sampler line that will not contaminate the samples (require stainless steel, Teflon, or appropriate plastic) and be easy to clean in the field. Manual pump samplers, discussed later, may be suitable when sampling wide or deep streams or rivers from a bridge or boat.

Obviously, care must be taken to locate the sampler intakes to minimize induced scour of sediments and to prevent clogging from debris. All submerged pumps can quickly fail if the pump draws coarse particles into the pump, but doesn't have enough velocity in the sample line to discharge most of them completely through the sample line. If the intake is located on a creek bottom, the water entering the sampler intake will likely scour sediment from the surrounding area. Locating the sampler intake on top of a small anchored concrete slab in the creek minimizes scour. Elevating the sampler intake above the creek bottom also minimizes scour, but presents an obstruction to flows and catches debris easily. Elevating the intake slightly is important in obtaining a better sample if the flow is vertically stratified. In some cases, sampler intakes can be successfully located on the downstream side of a bridge piling or pier. Do not locate the sample intake near any treated wood structure if heavy metals or organics are to be sampled. Bedload sampling is discussed later.

Locating a sampler intake in an outfall pipe presents other problems. Because the pipe is likely to be smaller than a receiving water, horizontal differences in water quality should not be a problem. However, vertical differences may occur. The sampler intake also presents a greater obstruction to the pipe flow and therefore has a greater tendency to catch debris. To ensure a well-mixed water sample, the intake can be placed in an area that has turbulent flow. This may decrease volatile components in the water sample, but typical automatic samplers are inappropriate for collecting samples for volatile analyses anyway. Locating the intake on the downstream side of a flow monitoring flume would help obtain a mixed sample. In addition, added obstructions (bricks and concrete blocks) can be cemented to the pipe above the sampling location to induce well-mixed conditions during low to moderate flows, being careful not to cause pooling of water and sedimentation. Obviously, flow measurements would not be taken where obstructions are used to mix the flow.

Manual sampling is much more flexible and can be modified to better represent the flow conditions at the time of sampling. Obviously, multiple dips across a stream, and at multiple depths, will result in a better representation of the stream than a single sampling location. Special manual samplers (described later) are needed to collect depth-integrated samples that may be needed for sediment transport studies.

The advantages of manual sampling compared to automatic sampling are offset by the time frame that is represented in the sample. A grab sample taken at a single time will not be as representative of a storm event as an automatic sampler taking subsamples from many time periods during the event, even considering multiple vs. single sampling points. A single sampling location will be subjected to varying conditions during the storm, including horizontal and vertical variations. However, if a single sampling location is consistently biased compared to the cross section of the stream, that needs to be recognized and corrected. Therefore, it is necessary to observe conditions in the stream during the sampling times as much as possible to detect any potential bias. A bias may be caused by currents or nearby discharges, for example, and may be visually observed if colored or turbid water is indicating current conditions near the sampler. A hand-held *in situ* probe that can measure turbidity (such as sold by YSI, Solomat, or Horiba) is extremely helpful in checking flow variations near the sampler intake. These probes can also be very helpful during manual grab sampling to measure the likely flow variabilities during the time of sampling. Other parameters are usually available on these probes (such as conductivity, temperature, DO, pH, and specific ions) that would also be helpful in these field checks.

Sampler and Other Test Apparatus Materials

A major concern when samples are analyzed for trace contaminants is the need to use sampling equipment that will have minimal effect on the sample characteristics. Most modern automatic water samplers have been continuously improved over the years, and current models are designed

Table 5.7 Potential Sample Contamination from Sampler Material

Material	Contaminant
PVC – threaded joints	Chloroform
PVC – cemented joints	Methylethyl ketone, toluene, acetone, methylene chloride, benzene, ethyl acetate, tetrahydrofuran, cyclohexanone, organic tin compounds, and vinyl chloride
Teflon	Nothing
Polypropylene and polyethylene	Plasticizers and phthalates
Fiberglass-reinforced epoxy material (FRE)	Nothing
Stainless steel	Chromium, iron, nickel, and molybdenum
Glass	Boron and silica

Data from Cowgill, U.M. Sampling waters, the impact of sample variability on planning and confidence levels, in *Principles of Environmental Sampling*. Edited by L.H. Keith. ACS Professional Reference Book. American Chemical Society. pp. 171–189. 1988.

to have little effect on sample quality. Teflon-lined sample tubing, special silicon peristaltic pump tubing, and glass sample bottles are all that contact the sample for automatic water samplers designed for monitoring toxicants and most other stormwater pollutants.

Careful selection of materials for manual samplers is just as important as for automatic samplers. Sediment samplers made with stainless steel are available to minimize sample contamination. Cole Parmer includes an extensive table in its standard catalog that lists chemical compatibility with different materials, including many plastics, elastomers, metals, and nonmetals. The effects listed include "no effect," "minor effect," "moderate effect," and "severe effect, not recommended." This guidance is mostly for material degradation and high concentrations of the chemicals, but it is useful when considering potential contamination problems.

Table 5.7 lists potential contaminants from some sampler materials (Cowgill 1988). It was found that extensive steam cleaning (at least five washings using steam produced from distilled water) practically eliminated all contamination problems. Cemented materials should probably be avoided, as is evident from Table 5.7. Threaded or bolted-together sampler components are preferable. ASTM (1995), in standard E 1391, recommends preconditioning samplers (plus test chambers and sample containers) before their first use. ASTM summarized research that found that all plastics (including Teflon) leached elements, but that this could be minimized with a 7-day leaching using a 1:1 solution of HCl and deionized water and then another 7 days in a 1:1 solution of HNO_3 in deionized water. Overnight soaking in these solutions was found to be adequate for glassware. Care should be taken, however, when soaking material for long periods in relatively strong acids. We have destroyed some plastic sampler components (including Delrin) after several days. Therefore, always conduct a soaking test to ensure compatibility and use the least aggressive cleaning method suitable.

Pitt et al. (1999) tested leaching potentials for many other materials that may be used in sampling apparatus and also pilot-scale treatment units (Table 5.8). The most serious problems occurred with plywood, including untreated wood. Attempting to seal the wood with Formica™ and caulking was partially successful, but toxicants were still leached. Lining large wooden boxes with cleaned plastic sheeting is probably more suitable than using the Formica lining. Fiberglass screening material, especially before cleaning, also causes a potential problem with plasticizers and other organics. PVC and aluminum may be acceptable sampling apparatus material, if phthalate esters and aluminum contamination can be tolerated. Pitt et al. (1999) used aggressive water (18 megohm water, prepared using ion exchange) when conducting their leaching tests. They were also conducted over a 3-day period (for worst-case conditions during treatability tests). The much shorter contact times associated with sampling (especially after the sampler has been rigorously cleaned) should result in minimal contamination problems when using sampling equipment that has been reasonably selected to avoid contamination of compounds of major interest.

These tables indicate that care must be taken when selecting and cleaning sampling equipment. The use of Teflon reduces most of the problems, but it is quite expensive. Delrin is almost

Table 5.8   Potential Sample Contamination from Materials Used in Sampler and Pilot-Scale Treatability Test Apparatus

Material	Contaminant
Untreated plywood	Toxicity, chloride, sulfate, sodium, potassium, calcium, 2,4-dimethylphenol, benzylbutyl phthalate, bis(2-ethylhexyl) phthalate, phenol, N-nitro-so-di-n-propylamine, 4-chloro-3-methylphenol, 2,4-dinitrotoluene, 4-nitrophenol, alpha BHC, gamma BHC, 4,4′-DDE, endosulfan II, methoxychlor, and endrin ketone
Treated plywood (CCA)	Toxicity, chloride, sulfate, sodium, potassium, hexachloroethane, 2,4-dimethylphenol, bis(2-chloroethoxy) methane, 2,4-dichlorophenol, benzylbutyl phthalate, bis(2-ethylhexyl) phthalate, phenol, 4-chloro-3-methylphenol, acenaphthene, 2,4-dinitrotoluene, 4-nitrophenol, alpha BHC, gamma BHC, beta BHC, 4,4′-DDE, 4,4′-DDD, endosulfan II, endosulfan sulfate, methoxychlor, endrin ketone, and copper (likely), chromium (likely), arsenic (likely)
Treated plywood (CCA) and Formica	Toxicity, chloride, sulfate, sodium, potassium, bis(2-chloroethyl) ether,* diethylphthalate, phenanthrene, anthracene, benzylbutyl phthalate, bis(2-ethylhexyl) phthalate, phenol,* N-nitro-so-di-n-propylamine, 4-chloro-3-methylphenol,* 4-nitrophenol, pentachlorophenol, alpha BHC, 4,4′-DDE, endosulfan II, methoxychlor, endrin ketone, and copper (likely), chromium (likely), arsenic (likely)
Treated plywood (CCA), Formica, and silica caulk	Lowered pH, toxicity, bis(2-chloroethyl) ether,* hexachlorocyclopentadiene, diethylphthalate, bis(2-ethylhexyl) phthalate, phenol,* N-nitro-so-di-n-propylamine, 4-chloro-3-methylphenol,* alpha BHC, heptachlor epoxide, 4,4′-DDE, endosulfan II, and copper (likely), chromium (likely), arsenic (likely)
Formica and silica caulk	Lowered pH, toxicity, 4-chloro-3-methylphenol, aldrin, and endosulfan 1
Silica caulk	Lowered pH, toxicity, and heptachlor epoxide
PVC pipe	N-nitrosodiphenylamine, and 2,4-dinitrotoluene
PVC pipe with cemented joint	Bis(2-ethylhexyl) phthalate,* acenaphthene, and endosulfan sulfate
Plexiglas and Plexiglas cement	Naphthalene, benzylbutyl phthalate, bis(2-ethylhexyl) phthalate, and endosulfan II
Aluminum	Toxicity and aluminum (likely)
Plastic aeration balls	2,6-Dinitrotoluene
Filter fabric material	Acenaphthylene, diethylphthalate, benzylbutyl phthalate, bis(2-ethylhexyl) phthalate, and pentachlorophenol
Sorbent pillows	Diethylphthalate and bis(2-ethylhexyl) phthalate
Black plastic fittings	Pentachlorophenol
Reinforced PVC tubing	Diethylphthalate, and benzylbutyl phthalate
Fiberglass window screening	Toxicity, dimethylphthalate, diethylphthalate,* bis(2-ethylhexyl) phthalate, di-n-octyl phthalate, phenol, 4-nitrophenol, pentachlorophenol, and 4,4′-DDD
Delrin	Benzylbutyl phthalate
Teflon	Nothing (likely)
Glass	Zinc (likely)

* Signifies that the observed concentrations in the leaching solution were very large compared to the other materials. Not all of the heavy metals had been verified.

From Pitt, R. et al. *Stormwater Treatment at Critical Areas: The Multi-Chambered Treatment Train (MCTT).* U.S. Environmental Protection Agency, Wet Weather Flow Management Program, National Risk Management Research Laboratory. EPA/600/R-99/017. Cincinnati, OH. 505 pp. March 1999.

as effective, is somewhat less expensive, and is much easier to machine when manufacturing custom equipment. Both of these materials are fragile and cannot withstand rough handling. They are therefore not appropriate for sediment sampling, but can be used to advantage in water samplers. Glass is not usable for most sampling equipment, but is commonly used in bench-scale tests and when storing and preparing samples. Glass presents a problem with heavy metals attaching to the glass walls, and zinc leaching out of the glass. It is a necessary material when analyzing organics, however. Stainless steel is preferred for most sediment samplers and for hardware for water samplers. Plastics should not be used if contamination by phthalate esters is to be avoided. Many adequate and inexpensive sampler apparatus can be made of plastics, especially if cements are not used. In all cases, careful cleaning and preconditioning has been shown to significantly reduce the concentrations of the contaminants in the leach water, stressing the need to thoroughly clean and condition the sampling equipment.

Volumes to Be Collected, Container Types, Preservatives to Be Used, and Shipping of Samples

The specific sample volume, bottle type, and preservative requirements should be specified by the analytical laboratory used. *Standard Methods* (1995) lists the basic container requirements, minimum sample sizes, required preservative, and the maximum storage period before the analyses need to be conducted. Table 5.9 shows these guidelines for water samples, while Table 5.10 lists the guidelines for sediment and pore water samples. Care must be taken to handle the samples properly to ensure the best analytical results. Numerous losses, transformations, and increases in pollutant concentrations may occur if these guidelines are not followed. Some analyses should be conducted as soon as possible (within a few hours of sample collection, or preferably on-site or *in situ*). These include CO_2, chlorine residual, DO (unless fixed), iodine, nitrite, ozone, pH, and temperature. ORP (oxidation-reduction potential) is also in this category of required on-site analyses, even though not included in this table. Parameters that need to be analyzed within 24 hours of sample collection (same day) include acidity, alkalinity, BOD, cyanide, chromium VI (and other specific ionic forms of metals), taste and odor, and turbidity. Microorganisms are not shown on this table either, and need to be analyzed within 24 hours of sample collection. Most of the nutrients need to be analyzed within 2 days. Many parameters can be stored for long periods of time, after preservation, specifically total forms of most heavy metals (6 months) and extracted organic compounds (30 days). In some cases, it may be possible to deviate from these guidelines if site-specific testing is conducted to demonstrate acceptable pollutant stability. The most important guidelines are the bottle type and preservative. Some parameters may be able to undergo longer storage periods, but this must be tested for specific conditions. The required sample volumes are all much greater than needed for most modern laboratory procedures and may be reduced (with permission from the laboratory) if shipping costs or sample storage facilities are a concern. Make sure that extra sample is available to redo critical analyses if problems develop, however. Be sure to verify these guidelines with the newest version of *Standard Methods*.

Sample Volumes

The volume of water or sediment needed depends on the types of toxicity assays, physical and chemical analyses, and level of precision (replicate numbers) needed. Usually 1 to 2 L is adequate for physical and chemical analyses. For static (daily) renewal toxicity assays, the quantities needed vary with the assay (Table 5.11). Volumes listed for sediments may be excessive if the sediment contains little interstitial water, such as found in sand, gravel, or compacted sediments, and few interstitial water chemical analyses are to be conducted. It is recommended that un-ionized ammonia generally be determined on interstitial water of sediments. If using the ion-selective electrode method, about 100 mL of aqueous solution is needed.

The following example for determining the water volume needed for laboratory analyses is based on the requirements of the UAB Environmental Engineering Laboratory. We have developed analytical modifications that require minimal amounts of sample in order to decrease shipping costs and storage problems, plus enabling small-scale treatability tests. Obviously, it is critical that the laboratory specify the sample volume requirements to ensure enough sample is available. Table 5.12 summarizes the sample quantities collected for each set of analysis. Also shown in this table is whether the sample is filtered or unfiltered (for constituent partitioning analyses). As an example, the metallic and organic toxicants are analyzed in both unfiltered and filtered sample portions in order to determine the amount of the pollutants associated with particulates and the amount that are considered "soluble." Filtering is through 0.45 μm membrane filters (using all-glass filtering apparatus and membrane filters that are found to have minimal effects

Table 5.9 Summary of Special Sampling and Handling Requirements for Water and Wastewater Samples[a]

Determination	Container[b]	Minimum Sample Size (mL)	Sample Type[c]	Preservation[d]	Maximum Storage Recommended/ Regulatory[e]
Acidity	P, G(B)	100	g	Refrigerate	24h/14d
Alkalinity	P, G	200	g	Refrigerate	24h/14d
BOD	P, G	1000	g, c	Refrigerate	6h/48h
Boron	P (PTFE) or quartz	100	g, c	None required	28d/6months
Bromide	P, G	100	g, c	None required	28d/28d
Carbon, organic, total	G	100	g, c	Analyze immediately; or refrigerate and add H_3PO_4 or H_2SO_4 to pH<2	7d/28d
Carbon dioxide	P, G	100	g	Analyze immediately	0.25h/N.S.
COD	P, G	100	g, c	Analyze as soon as possible, or add H_2SO_4 to pH<2; refrigerate	7d/28d
Chloride	P, G	50	g, c	None required	28d
Chlorine, total, residual	P, G	500	g	Analyze immediately	0.25h/0.25h
Chlorine, dioxide	P, G	500	g	Analyze immediately	0.5 h/N.S.
Chlorophyll	P, G	500	g, c	Unfiltered, dark, 4°C Filtered, dark, −20°C (Do not store in frost-free refrigerator)	28d/–
Color	P, G	500	g, c	Refrigerate	48h/48h
Conductivity	P, G	500	g, c	Refrigerate	28d/28d
Cyanide: Total	P, G	1000	g, c	Add NaOH to pH>12, refrigerate in dark	24h/14d;24h if sulfide present
Fluoride	P	100	g, c	None required	28d/28d
Hardness	P, G	100	g, c	Add HNO_3 to pH<2	6 months/6months
Iodine	P, G	500	g, c	Analyze immediately	0.5h/N.S.
Metals, general	P(A), G(A)	1000	g, c	For dissolved metals filter immediately, add HNO_3 to pH<2	6months/6months
Chromium VI	P(A), G(A)	1000	g	Refrigerate	24h/24h
Mercury	P(A), G(A)	1000	g, c	Add HNO_3 to pH<2, 4°C, refrigerate	28d/28d
Nitrogen:					
Ammonia	P, G	500	g, c	Analyze as soon as possible or add H_2SO_4 to pH<2, refrigerate	7d/28d
Nitrate	P, G	100	g, c	Analyze as soon as possible or refrigerate	48h/48h (28d for chlorinated samples)
Nitrate + nitrite	P, G	200	g, c	Add H_2SO_4 to pH<2, refrigerate	1–2d/28d
Nitrite	P, G	100	g, c	Analyze as soon as possible refrigerate	None /48h
Organic, Kjeldahl	P, G	500	g, c	Refrigerate; add H_2SO_4 to pH<2	7d/28d
Oil and grease	G,wide-mouth calibrated	1000	g, c	Add HCl to pH<2, refrigerate	28d/28d
Organic compounds:		200			
MBAS	P, G	250	g, c	Refrigerate	48h/N.S.
Pesticides	G(S), PTFE-lined cap	1000	g, c	Refrigerate; add 1000 mg ascorbic acid/L if residual chlorine present	7d/7d until extraction 40d after extraction
Phenols	P, G PTFE-lined cap	500	g, c	Refrigerate add H_2SO_4 to pH<2	*/28d until extraction

Table 5.9 Summary of Special Sampling and Handling Requirements for Water and Wastewater Samples[a] (Continued)

Determination	Container[b]	Minimum Sample Size (mL)	Sample Type[c]	Preservation[d]	Maximum Storage Recommended/ Regulatory[e]
Purgeables* by purge and trap	G, PTFE-lined cap	2×40	g	Refrigerate; add HCl to pH<2; add 1000 mg ascorbic acid/L if residual chlorine present	7d/14d
Base/neutrals and acids	G (S), amber	1000	g, c	Refrigerate	7d/7d until extraction; 40d after extraction
Oxygen, dissolved: Electrode	G, BOD bottle	300	g	Analyze immediately	0.25h/0.25h
Winkler				Titration may be delayed after acidification	8h/8h
Ozone	G	1000	g	Analyze immediately	0.25h/N.S.
pH	P, G	50	g	Analyze immediately	0.25h/0.25h
Phosphate	G(A)	100	g	For dissolved phosphate filter immediately; refrigerate	48h/N.S.
Phosphorus, total	P, G	100	g, c	Add H_2SO_4 to pH<2 and refrigerate	28d/–
Salinity	G, wax seal	240	g	Analyze immediately or use wax seal	6 months/N.S.
Silica	P (PTFE) or quartz	200	g, c	Refrigerate, do not freeze	28d/28d
Solids	P, G	200	g, c	Refrigerate	7d/2-7d
Sulfate	P, G	100	g, c	Refrigerate	28 /28d
Sulfide	P, G	100	g, c	Refrigerate; add 4 drops 2N zinc acetate/100 mL; add NaOH to pH>9	28d/7d
Temperature	P, G	—	g	Analyze immediately	0.25h
Turbidity	P, G	100	g, c	Analyze same day; store in dark up to 24 h, refrigerate	24/h48h

[a] See *Standard Methods* for additional details. For determination not listed, use glass or plastic containers; preferably refrigerate during storage and analyze as soon as possible.
[b] P = plastic (polyethylene or equivalent); G = glass; G (A) or P(A) = rinsed with 1 + 1 HNO; G(B) = glass, borosilicate; G(S) = glass, rinsed with organic solvents or baked.
[c] g = grab; c = composite
[d] Refrigerate = storage at 4°C ± 2 °C, in the dark; analyze immediately = analyze usually within 15 min of sample collection.
[e] Environmental Protection Agency, Rules and Regulation, 40 CFR Parts 100-149, July 1, 1992. See this citation for possible differences regarding container and preservation requirements.
Note: N.S. = not stated in cited reference; stat = no storage allowed; analyze immediately.

From *Standard Methods for the Examination of Water and Wastewater. 20th edition.* Water Environment Federation. Washington, D.C. Copyright 1998. APHA. With permission.

on constituent concentrations). The sample volumes that need to be delivered to the laboratory (where further filtering, splitting, and chemical preservation will be performed) and the required containers are as follows:

- Three 500 mL amber glass containers with Teflon-lined screw caps
- Three 500 mL HDPE (high-density polyethylene) plastic containers with screw caps

A total of 3 L of each water sample is therefore needed for comprehensive analyses. In addition to the water samples, collected sediment must be shipped in the following sample bottles:

- One 500 mL amber glass wide-mouth container with Teflon-lined screw cap
- One 500 mL HDPE (high-density polyethylene) wide-mouth plastic container

Table 5.10 Type of Container and Conditions Recommended for Storing Samples of Sediment or Pore Water

End Use	Container Type	Wet Weight or Volume of Sample	Temperature	Holding Time
Sediment				
Particle size distribution	1 Teflon 2 Glass 3 High-density polyethylene containers or bags	250 g	4 to 40°C Do not freeze	<6 mo
Major ions and elements: Al, C, Ca, Cl, Cr, Fe, Fl, H, K, Mn, Na, P, S, Si, Ti (oxides and total)	1 Teflon 2 High-density polyethylene containers or bags	250 g	<2°C	<2 wk
Nutrients: NH_4-N, NO_2-N, NO_3-N, TKN, TC, TOC	1 Teflon 2 Glass with Teflon or polyethylene-lined cap	100 g	<2°C	<48 h
Trace elements: Ag, Ba, Be, Cd, Co, Cr, Cu, Hg, Li, Mn, Mo, Ni, Pb, Sb, Sr, Va, Zn	1 Teflon 2 High-density polyethylene containers or bags	250 to 500 g	<2°C or −20°C	<2 wk <6 mo
Organic contaminants	1 Stainless steel canisters 2 Aluminum canisters 3 Amber glass with aluminum-lined cap	250 to 500 g	<2°C or −20°C	<2 wk <6 mo
Sediments for toxicity tests where the suspected contaminants are metals	1 Teflon 2 Glass 3 High-density polyethylene bags or containers	1 to 3 L	<2°C	<8 wk preferably <2 wk
Sediments for toxicity tests where the suspected contaminants are organic(s)	1 Glass with Al- or polyethylene-lined caps 2 Teflon 3 Stainless steel 4 High-density polyethylene bags or containers	1 to 3 L	<2°C	<8 wk preferably <2 wk
Control and reference sediment for toxicity tests	1 Teflon 2 Glass 3 High-density polyethylene bags or containers	>15 L	<2°C	<12 mo[a]
Pore Water				
Major ions and elements: Ca, Mg, Cl, Si, Fl, Na, SO_4, K, Al, Fe, acidity, alkalinity	1 Teflon 2 Amber glass with Teflon-lined lids 3 High-density polyethylene containers	40 mL	−20°C	<6 wk
Nutrients in pore water: NH_4-N, NO_2-N, NO_3-N, C (total organic), P (soluble reactive), DIC, DOC	4 Amber glass with Teflon-lined lids	40 mL	−20°C	<6 mo
P (total)	1 Amber glass with Teflon-lined lids	40 mL	−20°C or <2°C with 1 mL of 30% H_2SO_4 per 100 mL	<6 wk <2 wk
Trace elements (total) in pore water: Ba, Be, Cd, Cr, Cu, Co, Li, Mn, Mo, Ni, Pb, Sb, Sr, Va, Zn	1 Teflon 2 Polyethylene	10 to 250 g	−20°C or <2°C with 2 mL of 1 M HNO_3 per 1000 mL pore water	<6 mo <6 wk

Table 5.10 Type of Container and Conditions Recommended for Storing Samples of Sediment or Pore Water *(Continued)*

End Use	Container Type	Wet Weight or Volume of Sample	Temperature	Holding Time
Ag	1 Amber Polyethylene	250 mL	<2°C with 1 g Na_2 EDTA per 250 mL pore water	<6 wk
Hg	1 Teflon 2 Glass (Soviral/Wheaton)	100 mL	<2°C with 1 mL H_2SO_4 per 100 mL of pore water	<6 wk
Organic contaminants in pore water[b]	1 Amber glass with Al-lined caps 2 Amber glass with Teflon-lined caps	1000 mL	−20°C or <2°C acidified with H_2SO_4 or with the addition of 10 g Na_2SO_4 per L of pore water	<6 mo <6 wk
Organochlorine and PCBs	1 Amber glass with Al-lined caps 2 Amber glass with Teflon-lined caps	1000 mL	−20°C or <2°C	<6 mo <6 wk
Organophosphates	1 Amber glass with Al-lined caps 2 Amber glass with Teflon-lined caps	1000 mL	−20°C or <2°C acidified with HCl to pH 4.4	<6 mo <6 wk
PCP	1 Amber glass with Al-lined caps 2 Amber glass with Teflon-lined caps	1000 mL	−20°C or <2°C acidified with H_2SO_4 to pH <4 or preserved with 0.5 g $CuSO_4$ per liter or pore water	<6 mo <6 wk
Phenoxy acid herbicides	1 Amber glass with Al-lined caps 2 Amber glass with Teflon-lined caps	1000 mL	−20°C or <2°C with acidification to pH <2 with H_2SO_4	<6 mo <6 wk
PAHs	1 Amber glass with Al-lined caps 2 Amber glass with Teflon-lined caps	1000 mL	−20°C or <2°C	<6 mo <6 wk
Pore water[c] or elutriate for toxicity tests	1 Amber glass with Teflon-lined caps	1 to 3 L	2°C	<72 h

[a] These sediments should be monitored over this period of time to ensure that changes that might occur to the physicochemical characteristics are acceptable.

[b] It is very difficult to collect sufficient pore water for analyses of volatile organic compounds and aromatic organic compounds.

[c] It is very difficult to collect sufficient pore water for standard toxicity testing; however, smaller quantities will suffice if the experimental design of the test accommodates extraction of successive samples of sediment and/or compositing of within-station replicate samples. It should be recognized that once pore water that has been collected in situ is exposed to oxygen (e.g., air) it becomes geochemically distinct (Mudroch 1992). The Microtox toxicity test only requires a few mL of sample and could be used as an indicator of pore water toxicity.

Table 5.11 Sample Volumes Needed for Toxicity Testing[a]

Assay	Aqueous Phase[b] (L)		Solid Phase[c] (g wet weight)	
	Acute	Short-Term Chronic[d]	Acute	Short-Term Chronic
Fish	2.5	2.5	400	600
Zooplankton				
Daphnia magna or *pulex*	0.2	0.3	200	100
Ceriodaphnia dubia	0.2	0.3	200	100
Amphipod				
Hyalella azteca	2.5	—	1000[d]	1500
Midge				
Chironomus tentans or *C. riparius*	2.5	—	1000[d]	1500
Phytoplankton				
Selenastrum capricornutum	—	0.4	—	—
Microtox[e]	0.1	—	—	—
Chemical analyses[f]	2.0		1000	

[a] Screening only. Definitive assays to produce effect levels (e.g., LC50, NOEL) require testing of five concentrations (e.g., 100%, 50%, 25%, 12.5%, 6.25%).

[b] Surface or interstitial waters, elutriates, or effluents.

[c] Whole sediment or soil, overlain with site, reference, or reconstituted water.

[d] Exposure periods of 10 days.

[e] Definitive test.

[f] Routine chemical analyses of alkalinity, hardness, conductivity, pH, turbidity, temperature, and dissolved oxygen. For sediment samples, interstitial waters may be used for most analyses. Volume of sediment needed will depend on sediments water content. Ammonia and particle size measurements recommended when testing sediments.

Table 5.12 Example Water Volume Requirements for Different Analytes When Using Special Low-Volume Analytical Methods

Constituent	Volume (mL)	Filtered?	Unfiltered?
Total solids	100		Yes
Dissolved solids	100	Yes	
Turbidity	30	Yes	Yes
Particle size (by Coulter Counter MultiSizer IIe)	20		Yes
Conductivity	70		Yes
pH (also on-site or *in situ*)	25		Yes
Color	25		Yes
Hardness	100		Yes
Alkalinity	50		Yes
Anions (F^-, Cl^-, NO_2^-, NO_3^-, SO_4^{2-}, and PO_4^{2-})	25	Yes	
Cations (Li^+, Na^+, NH_4^+, K^+, Ca^{2+}, and Mg^{2+})	25	Yes	
COD	10	Yes	Yes
Metals (Pb, Cr, Cd, Cu, and Zn)	70	Yes	Yes
Semivolatile compounds (by GC/MSD)	315	Yes	Yes
Pesticides (by GC/ECD)	315	Yes	Yes
Microtox toxicity screen	10	Yes	Yes

The following list shows the amounts of sediment sample generally required for different chemical and physical analyses:

Inorganic chemicals	90–1000 mL
Organic chemicals	50–2000 mL
TOC, moisture	100–300 mL
Particle size	230–500 mL

Petroleum hydrocarbons	250–1000 mL
Acute toxicity tests	1–3 L
Bioaccumulation tests	3–4 L
Pore water extraction	2 L (sediment and assay dependent)
Elutriate preparation	1 L (assay dependent)

Sample Containers

Aqueous samples for toxicity testing may be collected and shipped in plastic containers, e.g., Cubitainers®. Dark borosilicate glass with Teflon-lined caps is recommended for samples to be used for organics analyses. High-density polyethylene containers are needed when metals are to be analyzed. Metals can sorb to glass, and new glassware may have zinc contaminants. Polyethylene is not recommended when samples are contaminated with oil, grease, or creosote.

All containers have been shown to adsorb various organic contaminants (Batley 1989; Batley and Gardner 1977; Schults et al. 1992). Polytetrafluoroethylene (PTF), e.g., Teflon, glass, and stainless steel have been shown to adsorb metals and organic compounds, acting as ion exchangers. However, sediments have many more binding sites than the container walls, and likely decrease the significance of container-associated loss for short-term exposures.

Wide-mouth containers made of either Teflon or high-density polyethylene, with Teflon-lined or polypropylene screw caps, are available in a variety of sizes from any scientific supply company and are considered the optimal all-purpose choice for sediment samples collected for both chemical and toxicity testing. Wide-mouth, screw-capped containers made of clear or amber borosilicate glass are also suitable for most types of analyses, with the notable exception of sediment metals, where polyethylene or Teflon is preferred. In addition, if a sediment or pore water sample is to be analyzed for organic contaminants, amber glass bottles are recommended over plastic. It should be noted that glass containers have several disadvantages, such as greater weight and volume and susceptibility to breakage, particularly when they are filled with sediment and frozen. Plastic bags made of high-density polyethylene can also be used for storing wet or dry sediment samples for certain end uses. Generally, when the end use of the sample is known, Tables 5.9 and 5.10 (and the primary references) should be consulted for specific recommendations regarding type of container, volume, and storage times.

Precleaned sample containers can be obtained from I-Chem (through Fisher Scientific at 800-766-7000) or Eagle Picher (at 800-331-7425). Fisher's catalog numbers and prices are as follows:

I-Chem #	Fisher #	Approx. Cost	Description
241-0500	05-719-74	$35/case of 12	Wide-mouth amber 0.5 L glass jars with Teflon-lined lids and labels
311-0500	05-719-242	$68/case of 24	Wide-mouth 0.5 L HDPE jars with Teflon-lined lids and labels

Eagle Picher sample containers are as follows:

122-16A	$25/case of 12	Wide-mouth amber 0.5 L glass jars with Teflon-lined lids
151500WWM	$46/case of 24	Wide-mouth 0.5 L HDPE jar with Teflon-lined lids

Cleaning Sample Bottles

ASTM (1995) has listed bottle cleaning/conditioning requirements in standard D 3370. New glass bottles (unless purchased precleaned) must be preconditioned before use by filling them with water for several days. This conditioning time can be shortened by using a dilute solution of HCl. ASTM also points out that polyethylene is the only suitable material for sample containers when low concentrations of hardness, silica, sodium, or potassium are to be determined (in conflict with the above recommendation that warned of using polyethylene for samples containing creosote, oils, or greases). All sample containers must also be sealed with Teflon (preferred) or aluminum-lined caps. The bottles must be washed using a protocol similar to that described below for sampling equipment. ASTM (1995), in standard E 1391, also recommended more stringent preconditioning of sample containers before their first use in critical toxicological testing, as noted above (7-day leaching using a 1:1 solution of HCl and deionized water and then another 7 days in a 1:1 solution of HNO_3 in deionized water for plastics. Overnight soaking in these solutions was found to be adequate for glassware. Again, take care, and test for damage before soaking equipment in strong acid solutions).

Minimum cleaning includes cleaning the samplers, including sampling lines, with domestic tap water immediately after sample retrieval. Components that can be taken to the laboratory (such as the containers in the automatic samplers) are washed using warm tap water and laboratory detergent (phosphate free), rinsed with tap water, then distilled water, and finally laboratory grade (18 megohm) water.

ASTM (1995) presents standard D 5088-90 covering the cleaning of sampling equipment and sample bottles. This guidance varies from the above ASTM standard. It recommends a series of washings, depending on the analyses to be performed. The first wash is with a phosphate-free detergent solution (with a scrub brush, if possible), followed by a rinse of clean (known characteristics) water, such as tap water. If inorganic analyses are to be performed (especially trace heavy metals), then the sample-contacting components of the equipment and the sample bottles need to be rinsed with a 10% solution of reagent grade nitric or hydrochloric acid and deionized water. The equipment is rinsed again. If organic analyses are to be performed (especially trace organic compounds by GC/MSD), then the sample-contacting components of the equipment and sample bottles must be rinsed with pesticide-grade isopropanol alcohol, acetone, or methanol. The equipment and bottles are then rinsed with deionized water and allowed to air dry. The cleaned equipment needs to be wrapped with suitable inert material (such as aluminum foil or plastic wrap) for storage and transport. If sample components, such as tubing, cannot be reached with a brush, the cleaning solutions need to be recirculated through the equipment. Be careful of potentially explosive conditions when using alcohol or acetone. Intrinsically safe sampling equipment that does not produce sparks with electronic contacts or from motors, or friction heat, should be used whenever possible. Obviously, work in a well-ventilated area and wear protective garments, including eye protection, when cleaning the sampling equipment with the acid or solvents.

ASTM also recommends that the equipment components that do not contact the sample be cleaned with a portable power washer or steam-cleaning machine. If these are not available, a hand brush must be used with the detergent solution.

Containers can be a potential source of contamination and must be cleaned before receiving a field sample of sediment or pore water. New glass and most plastics should be cleaned to remove residues and/or leachable compounds, and to minimize potential sites of adsorption (Environment Canada 1994). A recommended sequence of cleaning activities for sediment samples is detailed in Table 5.13. It should be noted that precleaned containers for water and sediment samples are commercially available and are used with increasing frequency in many sampling programs.

Different general cleaning procedures are recommended for inorganic vs. organic analyses of sediment and pore water samples (Table 5.13). However, it should be noted that there is no universal procedure for all projects; a specific cleaning method can be very effective for one element, but not sufficient for another (Mudroch and Azcue 1995). Special attention must be paid in cases where sediment samples are collected in one type of container and subsequently analyzed for different types of organic and inorganic compounds. In such cases, the cleaning procedure can be a source of contamination for some of the parameters of interest. For example, contamination problems have been reported in the determination of chromium when sodium dichromate solution was used to

Table 5.13 Cleaning Procedures for Containers Destined to Hold Sediment Samples

For determination of inorganic constituents in the sediment samples:

1. Scrub containers with phosphate-free soap and hot water
2. Wash in high-pressure tap water
3. Degrease with Versa Clean (Fisher) or similar soap bath for 24 hours
4. Soak in a 72-hour acid bath with reagent grade 6 M nitric acid; drain off acid and rinse with hot water
5. Rinse with double-distilled water and allow to dry in a particle-free environment
6. Place containers in heavy polyethylene bags

For determination of organic constituents in the sediment samples:

1. Scrub containers with phosphate-free soap and hot water
2. Wash with high-pressure tap water
3. Clean with detergent such as Versa Clean (Fisher) or similar
4. Rinse three times with organic-free water
5. Rinse twice with methyl alcohol
6. Rinse twice with dichloromethane
7. Dry in an oven at 360°C for at least 6 hours

clean glass containers, or nitrate contamination was introduced by washing the containers with nitric acid, and phosphate contamination was introduced by washing the containers with phosphate-containing detergents (Mudroch and Azcue 1995). In these situations, it is usually advisable to use separate containers made of appropriate material and cleaned following applicable procedures for the different types of analyses to be performed. Finally, the rigorous cleaning procedures outlined in Table 5.13 may not always be necessary, especially if the chemicals of interest in the samples are expected to be present at high concentrations. Thus, the choice of cleaning procedure often must be left to the professional judgment of principal scientists based on study objectives and expected levels of the parameters of interest.

Field Processing of Samples and Preparation for Shipping

Water Samples

If the samples are to be analyzed locally, the field collection bottles (such as the automatic sampler base with bottles) can be delivered directly to the laboratory for processing. We generally conduct all filtering and preservation in the laboratory if at all possible, as this lessens the severe problems associated with field filtration and acid handling. Critical parameters (pH, DO, ORP, temperature) are analyzed *in situ* or on-site. If samples cannot be delivered to the laboratory quickly, field filtration and preservation will be necessary. Samples need to be split and individually preserved, as described in *Standard Methods*. A commercial sample splitter is available from Markson Scientific (800-858-2243) (catalog # 6614K1455 at about $265 for a 14 L polyethylene churn sample splitter, with 4 and 8 L splitters also available, Figure 5.13). Cone splitters are much more effective than churn splitters when suspended solids and particle size analyses are critical. A sample splitter is also useful if numerous individual sampler bottles are to be combined as a composite. The appropriate sample volumes are poured into the splitter from the individual bottles; the composite sample is then agitated and drained into individual bottles for shipping or further processing.

Personnel should wear latex gloves and safety glasses when handling the samples. Sample containers should be filled with no remaining headspace to reduce the loss of volatile components. Samples collected for microbiological analyses or suspended solids, however, should have air space to allow for sample mixing prior to testing. The caps must be screwed on securely and taped shut to reduce the possibility of losing some of the sample. The chain-of-custody seal can then be applied over the sealing tape. The paper chain-of-custody seals are not adequate to seal the lids on the jars. Do not let the water samples freeze.

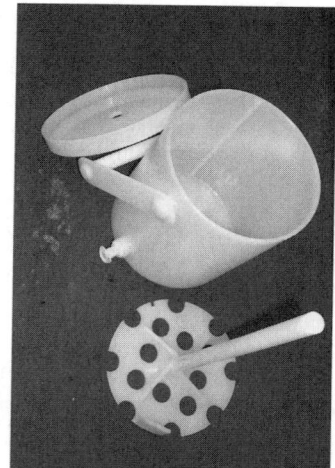

Figure 5.13 Churn splitter used to divide sample into individual bottles for separate preservative treatments and storage conditions, plus for preparing QA/QC split samples for independent analyses.

Sediment Samples

In the field, sediment samples can be stored temporarily in refrigerated units on board the sampling vessel, placed into insulated containers containing ice or frozen ice packs, or taken immediately to a local storage facility where they can be placed either in a freezer or a refrigerator. Dry ice can be used to freeze samples for temporary storage or transport, as long as its efficacy is known and the user is aware of the regulations regarding the transportation of samples stored in this manner.

Sediment samples for toxicity or particle size testing must not be frozen. While in transit to a storage facility or laboratory, frozen samples must not be thawed. Samples that have a recommended storage temperature of 4°C should be cooled to that temperature using ice or refrigeration prior to placement in the transport container. The transport container should be refrigerated to 4°C or contain sufficient ice or frozen gel packs to keep the samples at 4 (±3)°C during transport to the laboratory. Depending on the logistics of the operation, field personnel may either transport samples to the laboratory themselves or utilize an overnight courier service. Samples must not freeze during transport, and light should be excluded from the transport container.

If a container with a sediment sample is to be frozen, it should be filled to only two thirds of its volume. For studies in which it is critical to maintain the collected sediment under anoxic conditions, the headspace in the container should be purged with an inert gas (e.g., nitrogen) before capping tightly. If samples are to be stored at 4°C, containers can be filled to the rim and air excluded during capping. Clear glass containers are often wrapped tightly with an opaque material (e.g., clean aluminum foil) to eliminate light and reduce accidental breakage (Environment Canada 1994).

Shipping Samples

Once the samples are split/divided into the appropriate shipping bottles (and preserved, if needed), the sample container label should be filled out completely and logged onto a shipping list for each shipping container. Shipping containers are usually plastic coolers. There needs to be adequate packing (preferably as many "ice" packs as can fit, plus bubble wrap) inside the shipping container to ensure that the sample bottles do not rub or bang against each other en route. Newspapers (flat, not wadded) can be placed on top of the samples and ice packs, directly under the lid, to further fill up any extra volume. Do not use packing peanuts (especially the water-soluble type) to fill up space. Wrap glass bottles with bubble wrap. Use sufficient "blue ice" or other cooling packs to ensure the coolers stay cool during shipment. Do not use water ice. The coolers must also be securely taped shut (seal the seams) to minimize leakage if a bottle breaks during shipment.

The samples should be sent via overnight courier so they arrive while laboratory personnel are present and sufficient time is available to initiate the critical analyses immediately (unless special arrangements have been made with the laboratory). Always call to schedule a sample shipment and fax a confirmation of the sample shipping information. Always keep a copy of any sample identification sheets and send the originals (by mail, not in the coolers). Include a shipping list (and copy of appropriate sampling forms) in an envelope taped to the outside of the cooler.

Chain-of-Custody and Other Documentation

When the sample is collected, the bottle labels and chain-of-custody forms must be filled out. In many cases, additional field sheets containing site or sample information are also completed. Documentation of collection and analysis of samples requires all the information necessary to: (1) trace a sample from the field to the final result of analysis; (2) describe the sampling and analytical methodology; and (3) describe the QA/QC program (Mudroch and Azcue 1995; Keith et al. 1983).

Correct and complete field notes are absolutely necessary in any sampling program. Poor or incomplete documentation of sample collection can make analytical results impossible to interpret. The following items should be recorded at the time of sediment sampling (Mudroch and Azcue 1995):

1. Project or client number
2. Name of sampling site and sample number
3. Time and date of sample collection
4. Weather conditions (particularly wind strength and direction, air and water temperature)
5. Sample collection information
6. Type of vessel used (size, power, engine type)
7. Type of sampler used (grab, corer, automatic, etc.) and any modifications made to the sampler during sampling
8. Names of sampling personnel
9. Notes on any unusual events that occurred during sampling (e.g., problems with recovered samples or sampling equipment, observations of possible contamination)
10. Sample physical description including texture and consistency, color, odor, estimate of quantity of recovered samples by a grab sampler, length and appearance of recovered sediment cores
11. Notes on further processing of samples in the field, particularly subsampling methods, type of containers, and temperature used for sample storage
12. Record any measurements made in the field, such as pH and ORP

Bound notebooks are preferred to the loose-leaf type and should be kept in a room or container that will protect against fire or water damage. Whenever legal or regulatory objectives are involved, notebook data should be entered in ink, each page should be signed and witnessed, and all errors or changes should be struck through one time and initialed (Keith 1991).

When samples are transported to a laboratory, an inventory list of each individual sample should be included in the shipment, and a separate copy sent to the laboratory. The inventory list should indicate the required analyses for each enclosed sample. The transport container should be labeled properly, including a description of the contents, the destination, any special handling instructions, and phone numbers to call on arrival or in case of an emergency. It is highly recommended that laboratories receiving samples be alerted to their impending arrival, particularly if samples will arrive on a weekend or holiday, so that appropriate arrangements can be made for their receipt.

Samples collected for legal purposes typically require the use of strict chain-of-custody procedures during handling and transport. This includes preparing detailed documentation regarding sample collection, preparation, and handling. All transport containers must remain locked during transport to and from the sampling site. The name and signature of the person who collected the sample should be placed on each sample container and witnessed, and the label should be securely fastened to the container after the sample has been placed in it and the lid tightly secured.

Appropriate chain-of-custody forms must be filled out for each transport container, including a complete listing and description of the enclosed samples. Each transport should be locked during pickup, transit, and delivery and should have a tape seal to demonstrate that it has not been opened during transport. The chain-of-custody documentation must accompany the transport container, and every time the package changes hands, the transfer of responsibilities must be documented with names and signatures. A file of all documentation (e.g., signed package slips, waybills, chain-of-custody forms) should be established, and all samples must be kept in a locked area of the laboratory with restricted access. All documentation of the analytical procedures and results should be kept on file and in control of the laboratory and/or project QA/QC officer (EC 1994).

The typical information provided on a chain-of-custody form includes:

- The sampling location
- The sample identification number
- The type of test or analytical procedure
- The name of the person who relinquishes the samples
- The date and time of sample collection
- The date and time when samples are relinquished
- The name of the person who should receive the sampling results

Sample Preservation and Storage at the Laboratory

Once the samples arrive in the laboratory, they must be logged in, sorted for further processing, and filtered and preserved, as needed. In addition, the sample temperatures and the presence of ice in the coolers should be checked upon arrival in the laboratory to verify that the samples were kept below critical temperatures during shipping. A reading of pH and temperature is conducted as soon as the samples arrive, and bacteria analyses need to be started as soon as possible.

Within a day, chilled samples must be filtered. Glass filters used for suspended solids analyses typically contain large amounts of zinc that easily contaminates samples, therefore, membrane filters need to be used for filtered (dissolved) metal analyses. The filtered and unfiltered sample portions are then divided and preserved. The following is an example from the UAB environmental engineering laboratories:

- Unfiltered sample in two 250 mL amber glass bottles (Teflon-lined lids) (no preservatives) for total forms of toxicity, COD, and GC analyses (using MSD and ECD detectors)
- Filtered sample in one 250 mL amber glass bottle (Teflon-lined lids) (no preservative) for filtered forms of toxicity, COD, and GC analyses (using MSD and ECD detectors)
- Unfiltered sample in one 250 mL high-density polyethylene (no preservatives) for solids, turbidity, color, particle size, and conductivity
- Filtered sample in one 250 mL high-density polyethylene (no preservatives) for anion and cation analyses (using ion chromatography), hardness, dissolved solids, and alkalinity
- Unfiltered sample in one 250 mL high-density polyethylene (HNO_3 preservative to pH < 2) for total forms of heavy metal, using the graphite furnace atomic adsorption spectrophotometer
- Filtered sample in one 125 mL high-density polyethylene (HNO_3 preservative to pH < 2) for filtered forms of heavy metal, using the graphite furnace atomic adsorption spectrophotometer

All samples are chilled on ice or in a refrigerator at 4°C (except for the HNO_3-preserved samples for heavy metal analyses) and analyzed within the holding times shown below:

- Immediately after sample collection or upon arrival in the laboratory: pH and microorganisms
- Within 24 hours: toxicity, ions, color, and turbidity
- Within 7 days: GC extractions, solids, and conductivity
- Within 40 days: GC analyses
- Within 6 months: heavy metal digestions and analyses

Drying, freezing, and storage temperature all affect toxicity (ASTM 1991a). Significant changes in metal toxicity to cladocerans and microbial activity have been observed in stored sediments (Stemmer et al. 1990b). Recommended limits for storage of metal-spiked sediments have ranged from less than 2 to 5 days (Swartz et al. 1985), less than 2 weeks (ASTM 1991a; Nebeker et al. 1984), to 2 to 8 weeks (EPA 2000). Cadmium toxicity in sediments has been shown to be related to acid volatile sulfide (AVS) complexation (DiToro et al. 1991). AVS is a reactive solid phase sulfide pool that apparently binds some metals, thus reducing toxicity (DiToro et al. 1991). When anoxic sediments were exposed to air, AVS was volatilized. If a study intends to investigate metal toxicity and the sediment environment is anoxic, then exposure to air might reduce or increase toxicity due to oxidation and precipitation of the metal species or loss of AVS complexation. It is generally agreed that sediments used for toxicity testing should not be frozen (Schuytema et al. 1989; ASTM 1991), should be stored at 4°C with no air space or under nitrogen, and analyzed as soon as possible (Reynoldson 1987).

Samples should be handled and manipulated as little as possible to reduce artifact formation and constituent alteration. It is sometimes necessary to remove debris and predatory organisms from samples to be used for toxicity testing. As large a filter pore size as possible should be used to prevent removal of suspended solids, which affect toxicity. Dredge (grab) collected sediment samples (for toxicity testing) should be placed in wide-mouth containers which allow the sample to be gently stirred. The sediment should be stirred until it is a slurry or any overlying water is mixed into the sediment matrix. If necessary, the sample may be sieved to remove large debris and homogenize the particle size distribution. It may not be possible to remove all predatory or nontest organisms from whole sediment toxicity assays. Caution should be exercised when sieved samples are used for testing, as the particle size distribution, redox gradients, and other alterations have occurred which may affect toxicity responses and the accuracy of lab-to-field extrapolations. Sieving is recommended for macroinvertebrate analyses because it increases counting efficiency (see EPA 1990c for additional information).

Elutriate testing was developed by the U.S. Army Corps of Engineers to simulate a condition that occurs during a dredging operation. When dredging effects are a study objective, elutriate analysis should be included in the test design. Elutriate samples are prepared by mixing (shaking) a 1 to 4 ratio of sediment to water for 30 minutes. The mixture is allowed to settle for 1 hour, and the supernatant is used for testing. There are modified methods which mix for longer periods, mix by aeration, or filter the supernatant. It is important that the method used be consistent because any modification may alter the elutriate's characteristics. TCLP tests are also sometimes conducted to determine the leaching potential of sediments under more severe conditions.

Personnel Requirements

Personnel needed to carry out an effective monitoring program fall into several classifications. Obviously, project directors need to design the program to fulfill the project objectives while staying within the available resources. In many cases, a calculated monitoring program may be impossible to carry out because of insufficient monitoring opportunities (necessary length of monitoring period available, number of rain events expected, etc.). Obviously, the project personnel therefore need to understand the local conditions. The project directors also need a varied understanding of many components of the ecosystem being investigated (hydrology, biology, chemistry, land use, etc.). Project field staff must be able to collect samples in an efficient and safe manner and be capable of working under changing and uncomfortable conditions. In all cases, at least two people need to go into the field together. Selection of laboratory personnel depends on the analyses to be conducted, and candidates will likely need to have substantial wet-weather sample analysis experience. Statistical experts are also needed to assist in the project design and to help analyze the data. Some of this effort could be handled by volunteers, but most comprehensive monitoring programs will also require a substantial effort by highly trained technical personnel. Obviously, volunteer support can be very successful from an economical and educational viewpoint. This is especially important in nonpoint source/watershed studies where local residents need to have a greater role in decision making and in taking responsibility for the watershed.

Uses of Monitoring Data and the Appropriate Use of Volunteers in Monitoring Programs

An increasingly common method to obtain water quality data in receiving waters affected by stormwater is through the use of volunteer programs. Typically, a group of interested people is recruited by a local environmental organization. These people are trained in the use of relatively simple field test kits and carry out relatively broad-based observations. Usually, these people obtain relatively frequent data from local waters that supplement regulatory agency monitoring efforts. Historically, the most common volunteer efforts have been conducted mostly by lakeshore property owners who take Secchi disk readings of lake water transparency. However, with decreasing budgets for regulatory agencies and decreasing formal monitoring efforts conducted by state agencies, volunteer monitoring programs are increasing. The objectives for the use of these data must still define the parameters to be measured and other aspects of the experimental designs (sampling locations, frequencies, etc.). All too often, volunteer monitoring programs are relatively unstructured and are restricted to parameters that are relatively simple to measure. They therefore cannot truly replace most professional monitoring programs, but can be good supplements. Recent evaluations of simple field test kits have also identified their limitations, along with their advantages (Day 1996).

Volunteer monitoring programs are currently being conducted by several hundred groups throughout the U.S. The following list shows the number of volunteer monitoring programs having specific objectives for the use of the data (EPA 1994):

Education	439
Problem identification	333
Local decisions	288
Research	226
Nonpoint source assessment	225
Watershed planning	213
Habitat restoration	160
Water classification and standards	127
Enforcement	120
Legislation	84
305b compliance	53

Most of these uses require accurate information, because the data may have profound effects on regulatory agency decisions. In many states, however, water quality monitoring data collected by anyone who is not an employee of the state regulatory agency is not admissible as evidence in court. The lack of adequate quality assurance and quality control plus legal chain-of-custody procedures (including proof that samples or observations were obtained where claimed) are the most obvious problems with volunteer collected data.

The users of volunteer-collected data are also varied. The following list indicates the numbers of volunteer monitoring programs collecting data used by various groups (EPA 1994):

State governments	319
Local governments	315
Advocacy groups	288
Federal government	156
University scientists	142

The types of data being collected by volunteer monitoring groups have greatly expanded since the early days of Secchi disk surveys. The following list shows the number of volunteer monitoring programs that are collecting specific information/data (EPA 1994):

Water temperature	377
pH	313
Dissolved oxygen	296
Macroinvertebrates	259
Debris cleanups	218
Habitat assessments	211
Nitrogen	205
Phosphorus	202
Turbidity	192
Coliform bacteria	184
Secchi disk transparency	177
Aquatic vegetation	173
Flow	157
Birds and wildlife	152
Fish	150
Watershed mapping	138
Rainfall	131
Photographic surveys	129
Salinity	101
Sediment assessments	100
Alkalinity	98
Pipe surveys	96
TSS/TDS	91
Construction site inspections	81
BOD	75
Hardness	71
Chlorides	62
Chlorophyll a	60
Metals	56
Pesticides	24
Other bacteria	24
Hydrocarbons	14

Many of these parameters are well suited for trained volunteers. They can conduct relatively low-cost observations, which require minimal sampling or analytical equipment costs, for temperature, salinity, debris cleanup, habitat assessments, Secchi disk transparency, watershed mapping, photographic surveys, pipe surveys, and construction site inspections. Most of the other parameters (including most of the chemical analyses) would require the use of analytical equipment.

Relatively simple field test kits have been marketed in the United States for the past 30 years that can evaluate many of these parameters. However, few of these kits are suitable substitutes for conventional laboratory procedures. With care, good "screening" observations can be obtained from many of these kits. The sample collector, kit user, and data user must be aware of the limitations and hazards associated with many of these kits. The main concerns include:

- Safety (safe and correctly labeled reagents and clear instructions, including disposal guidance)
- Adequate sensitivity for required use of data
- Problems with interferences
- Ease of use and level of training needed
- Cost

Tests recently conducted at the University of Alabama at Birmingham have evaluated numerous field test kits for these criteria (Day 1996). The results are summarized in Chapter 6.

RECEIVING WATER, POINT SOURCE DISCHARGE, AND SOURCE AREA SAMPLING

Samples can be collected by manual grab or automatic samplers, the latter being more expensive but often superior when conditions fluctuate rapidly or sporadically, or when available personnel are lacking. Automatic samplers are essential for the NPDES program when effluents are monitored for permit requirements. Many types of automatic samplers exist (e.g., see EPA 1982) and none is ideal for all situations. The following variables must be considered when selecting a sampler (EPA 1982):

- Water or effluent variation (flow and constituents)
- Suspended solids concentration, dissolved gases, and specific gravity of effluent
- Vertical lift required
- Maintenance

Commonly used water samplers are listed in Table 5.14 and are discussed later in this section.

Automatic Water Sampling Equipment

Automatic water samplers that are commonly used for stormwater monitoring are available from ISCO and American Sigma, among others (Figures 5.14 to 5.22). These manufactures have samplers that have very flexible programming capabilities specifically designed for stormwater sampling and designed for priority pollutant sampling. A simpler automatic sampler is the Masterflex self-contained composite sampler (from Forestry Suppliers, Inc., for about $1500). This sampler is restricted to composite sampling only on a time-increment basis, and there is little control over the sample volumes that can be obtained. However, it may be a worthwhile option for simple sampling needs.

The American Sigma (800-635-4567) samplers are an excellent example of a highly flexible automatic sampler (Figure 5.14). They have an integral flowmeter option and can directly connect to a liquid level actuator or a depth sensor. The depth sensor is placed in the storm drainage upstream of a flow monitoring device (such as a weir or flume, or any calibrated stage-discharge relationship can be used). The flow indicators can control sample initiation and/or sampling frequency. A rain gauge is also available that can be connected directly to the sampler. Rainfall data can therefore be logged by the sampler, along with flow information and sampling history. Rainfall can also be

Table 5.14 The Advantages and Disadvantages of Manual and Automatic Sampling

Type	Advantages	Disadvantages
Manual	Low capital cost Not a composite Point-in-time characterization Compensate for various situations Note unusual conditions No maintenance Can collect extra samples in short time when necessary	Probability of increased variability due to sample handling Inconsistency in collection High cost of labor[a] Repetitious and monotonous task for personnel
Automatic	Consistent samples Probability of decreased variability caused by sample handling Minimal labor requirement for sampling Has capability to collect multiple bottle samples for visual estimate of variability and analysis of individual bottles	Considerable maintenance for batteries and cleaning; susceptible to plugging by solids Restricted in size to the general specifications Inflexibility Sample contamination potential Subject to damage by vandals

[a] High cost of labor assumes that several samples are taken daily, large distances between sampling sites, and labor is used solely for sampling.

From EPA. *Handbook for Sampling and Sample Preservation of Water and Wastewater*, Environmental Monitoring and Support Laboratory, U.S. Environmental Protection Agency, Cincinnati, OH, EPA 600/4-82/029. 1982.

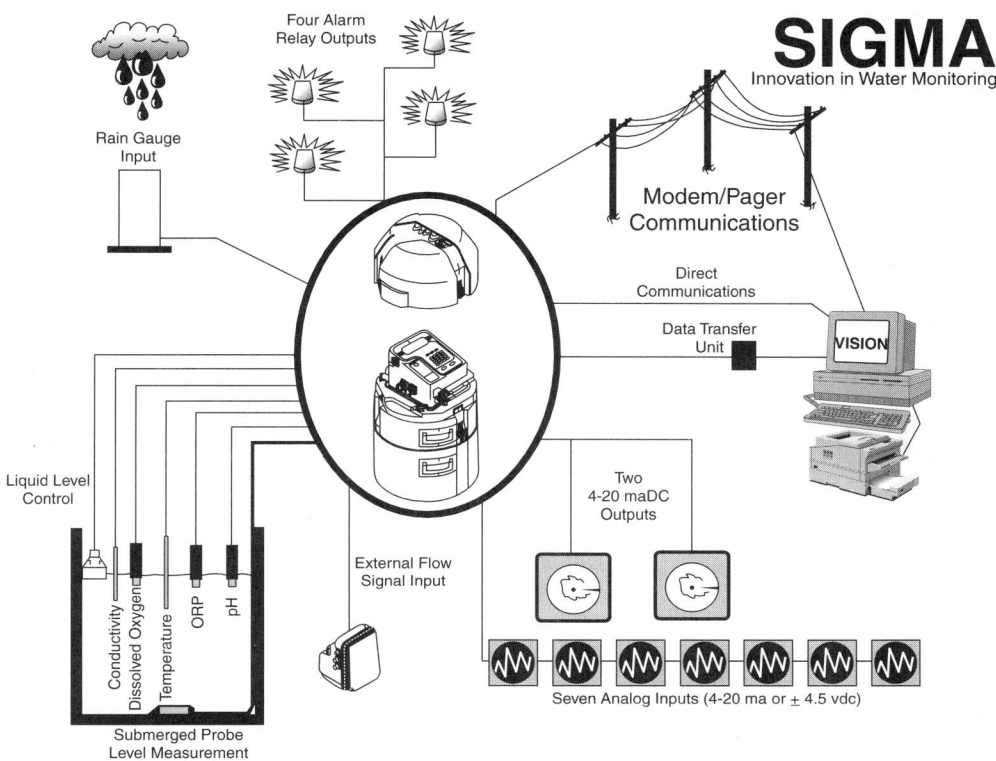

Figure 5.14 American Sigma connection options to ancillary equipment. (Used with permission.)

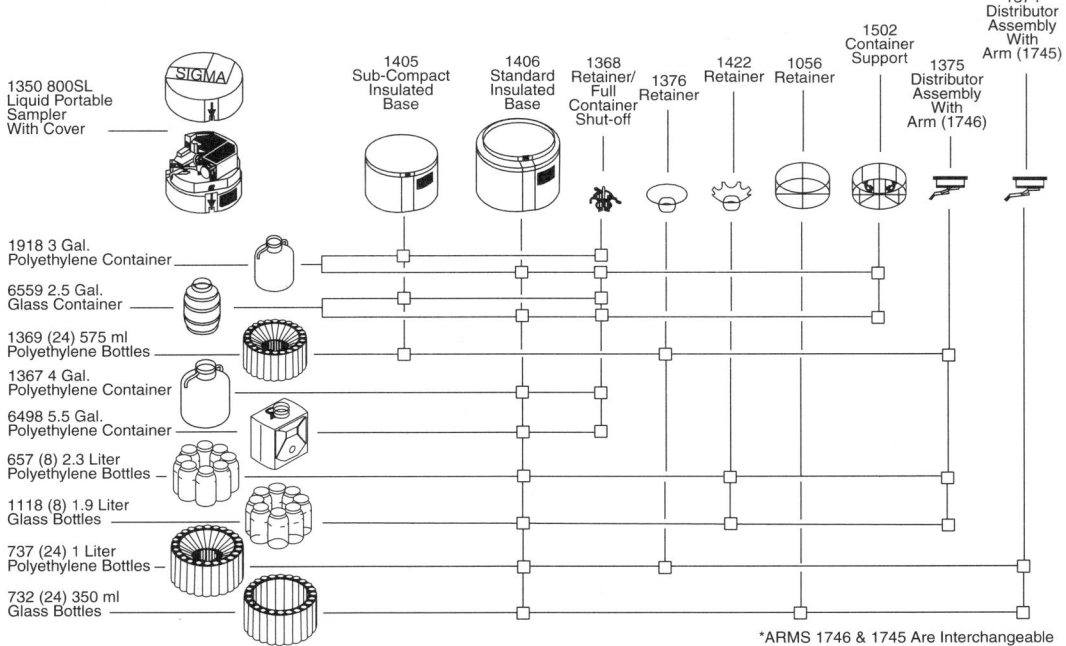

Figure 5.15 American Sigma sample bottle options. (Used with permission.)

Figure 5.16 Automatic ISCO sampler used to monitor snowmelt in Toronto, Ontario, manhole.

Figure 5.17 ISCO sampler used in instrument shelter with flow monitoring and telemetry equipment in Madison, WI.

Figure 5.18 Intermittent stream monitoring in Austin, TX.

Figure 5.19 Refrigerated automatic sampler located at detention pond outfall in Madison, WI.

used to trigger sample initiation. A solar panel is also available to keep the sampler's battery charged. Several sample bases and sample bottle options are also available (Figure 5.15). Single bottle composite sample bases are available having glass or polyethylene bottles from 2.5 to 5.5 gallons in volume. Up to four 1 gallon glass or polyethylene bottles can also be used to obtain composite samples over segments of the runoff event. In addition, several 24 bottle options are also available, with 575 mL or 1 L polyethylene bottles, or 350 mL glass bottles. American Sigma also has several AC-powered samplers that are refrigerated.

ISCO (800-228-4373) also offers a complete line of automatic water samplers that have been used for stormwater sampling for many years. Flowmeter and rain gauge options are available, along with numerous sample base and sample bottle options. ISCO also has several AC-powered refrigerated samplers. The ISCO 6100 sampler (about $8000, with bladder pump and special bottle rack for 40 mL VOC bottles) is especially designed to obtain samples for volatile analyses. Samples are collected directly in capped 40 mL VOC vials in the sampler, with minimal loss of volatile compounds. Very few volatile hydrocarbons have ever been detected in stormwater, so this sampler

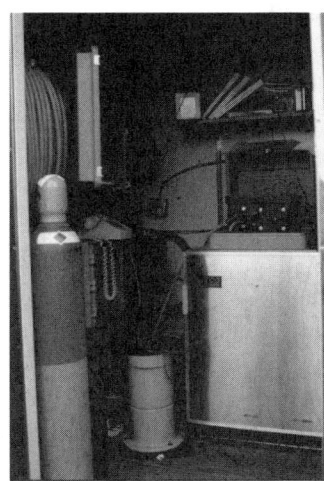

Figure 5.20 Refrigerated automatic sampler in Madison, WI, instrument shelter.

Figure 5.21 Discrete sample bottle base for ISCO automatic sampler.

(and VOC analyses) would probably be used only for specialized studies where VOCs are expected (such as in commercial areas with older dry cleaners or near gasoline stations).

Sigma and ISCO also have new automatic samplers that interface with continuously recording water quality probes that can be used to control sampling during critical periods, irrespective of time or flow. McCrone (1996) describes American Sigma's options for using numerous probes (such as conductivity, DO, temperature, ORP, and pH). The sampler can be programmed to collect a special sample when any of these monitored parameters meets a preset criterion. ISCO has a new sampler series that interfaces with the YSI 6000 water quality probes, allowing specific water quality conditions to also trigger sampling (similar to Sigma's list, plus turbidity).

If a refrigerated sampler cannot be used (due to lack of AC power), ice may be used if sample chilling is needed. Ice is placed in the central cavity surrounded by the sample bottles in the sampler base. The ice must be placed soon before an expected storm event, as it will generally melt within a day. The placement of any sampler in a cool location (such as a manhole) is much preferred over placement in a small shelter that may heat excessively in the summer. In most cases, chilling stormwater during sample collection is not done due to lack of AC power and the inconvenience of using ice. If the sampler is located in a cool location and the samples retrieved soon after the storm has ended, few problems are expected. Bacteria sampling, for example, requires manual sampling to ensure sterile equipment and to minimize storage problems. VOC analyses have previously required manual sampling, but the VOC sampler from ISCO can be used for automatic sample collection. The use of probes to measure pH, ORP, and temperature *in situ* also reduces the need for manual samples for these parameters. Therefore, it is possible to conduct a stormwater sampling program using automatic samplers that do not require AC-powered refrigerated

Figure 5.22 Composite sample bottle from Toronto snowmelt sampler.

samplers, if supplemented with manual sampling for microorganism determinations, and if the samples are retrieved soon after the event has ended. Some analyses may not be available using automatically collected samples, and other options may need to be used to supplement the automatic sampling. In all cases, special storage tests can be used to determine the likely errors associated with long storage in the samplers, with and without chilling.

Required Sample Line Velocities to Minimize Particle Sampling Errors

Typical sample lines are Teflon-lined polyethylene and are 10 mm in diameter. Table 5.15 shows the particle sizes that would be lost in vertical sampling lines at a pumping rate of 30 and 100 cm/s. The water velocity in sample lines is about 100 cm/s, enabling practically all sediment to be transported to the sample containers. A water velocity of 100 cm/s (about 3 ft/s) would result in very little loss of stormwater particles. Particles of 8 to 25 mm would not be lifted in the sample line at all at this velocity, but these particles would not fit through the openings of the intake or even fit in most sample lines. They are also not present in stormwater, but may be a component of bedload in a stream, or gravel in the bottom of a storm drain pipe, requiring special sampling. Very few particles larger than several hundred micrometers occur in stormwater and these should only have a loss rate of 10% at the most. Most particles in stormwater are between 1 and 100 μm in diameter and have a density of between 1.5 and 2.65 g/cm³. Even at 30 cm/s, these particles should experience insignificant losses. A pumping rate of about 100 cm/s would add extra confidence in minimizing particle losses. ASTM (1995) in method D 4411 recommends that the sample velocity in the sampler line be at least 17 times the fall rate of the largest particle of interest. As an example, for the 100 cm/s example above, the ASTM recommended critical fall rate would be about 6 cm/s, enabling a particle of several hundred micrometers in diameter to be sampled with a loss rate of less than 10%. This is certainly adequate for most stormwater sampling needs.

Automatic Sampler Line Flushing

Automatic samplers generally go through three phases when activated to collect a sample. First, the sample line is back-flushed to minimize sample cross-over and to clear debris from the sample intake. Next, the sample is collected. Finally, the sample is back-flushed again before going into a sleep mode to await the next sampling instruction. It can require several minutes to cycle through this process. A volume of 1850 mL of water fills a 10 mm (3/8 in) diameter sample line that is 7.5 m (25 ft) long. If a sample volume of 350 mL is to be collected for each sample interval, the following total volume of water is pumped by the sampler for each sample instruction:

Back-flush line	1850 mL
Fill tube	1850 mL
Collect sample	350 mL
Back-flush line	1850 mL

Table 5.15 Losses of Particles in Sampling Lines

	30 cm/s Flow Rate		100 cm/s Flow Rate	
% Loss	Critical Settling Rate (cm/s)	Size range (μm, for ρ = 1.5 to 2.65 g/cm³)	Critical Settling Rate (cm/s)	Size Range (μm, for ρ = 1.5 to 2.65 g/cm³)
100	30	2000–5000	100	8000–25,000
50	15	800–1500	50	3000–10,000
25	7.5	300–800	25	1500–3000
10	3.7	200–300	10	350–900
1	0.37	50–150	1	100–200

This totals about 6000 mL of water to be pumped. Typical automatic samplers have a pumping rate of about 3500 mL/min for low head conditions (about 1 m). It would therefore require about 1.7 min to pump this water. With pump reversing and slower pumping speeds at typical pumping heads, this could easily extend to 2 min, or more. If the sampler collects 3 L of sample instead of 350 mL, then another minute can be added to this sampling time for one cycle.

This sampler cycle time necessitates various decisions when setting up and programming a sampler, especially for flow-weighted composite sampling. The most important decisions relate to selecting the sampling interval that can accommodate expected peak flows and the sample volume needed for the smallest events to be sampled. Sample storage in the samplers is limited, further complicating the issue. The samplers are generally programmed to sample every 15 min to 1 hour for time-compositing sampling, or for an appropriate sample volume increment for flow-weighted sampling. If each sample increment is 0.25 L, a total of 40 subsamples can accumulate in a 10 L composite sample container.

Time or Flow-Weighted Composite Sampling

Automatic samplers can operate in two sampling modes, based on either time or flow increments. The sample bases can generally hold up to 24 bottles, each 1 L in volume. A single sample bottle of up to about 20 L is generally available for compositing the sample into one container. These bottle choices and the cycle time requirements of automatic samplers restrict the range of rain conditions that can be represented in a single sampler program for flow-weighted sampling. It is important to include samples from small rains (at least as small as 0.1 to 0.2 in) in a stormwater sampling program because they are very frequent and commonly exceed numeric water quality criteria, especially for fecal coliform bacteria and heavy metals. Moderate-sized rains (from about 0.5 to 2 in) are very important because they represent the majority of flow (and pollutant mass) discharges. The largest rains (greater than about 3 in) are important from a drainage design perspective to minimize flooding problems. It is very difficult to collect a wide range of rain depths in an automatic sampler using flow-weighted sampling. Conflicts occur between needing to have enough subsamples during the smallest event desired (including obtaining enough sample volume for the chemical analyses) and the resulting sampling frequency during peak flows for the largest sampling event desired. As an example, consider the following problem:

- Desired minimum rain to be sampled: 0.15 in in depth, 4-hour runoff duration, having a 0.20 Rv (volumetric runoff coefficient)
- Largest rain desired to be sampled: 2.5 in in depth, 12-hour runoff duration, having a 0.50 Rv
- The watershed is 250 acres in size and 3 samples, at least, are needed during the smallest rain

The calculated total runoff is therefore:

- Minimum rain: 0.10 (0.15 in) (250 ac) (ft/12 in) (43,560 ft^2/ac) = 13,600 ft^3
- Maximum rain: 0.50 (2.5 in) (250 ac) (ft/12 in) (43,560 ft^2/ac) = 1,130,000 ft^3

The average runoff flow rates expected are roughly estimated to be:

- Minimum rain: (13,600 ft^3/4 hr) (hr/3600 s) = 0.95 ft^3/s
- Maximum rain: (1,130,000 ft^3/12 hr) (hr/3600 s) = 26 ft^3/s

Using a simple triangular hydrograph, the peak flows are estimated to be about twice these average flow rates:

- Minimum rain: 1.9 ft^3/s
- Maximum rain: 53 ft^3/s

Actual peak flow rates are obviously related to the watershed time of concentration and other factors of the watershed and drainage system, but this triangular hydrograph has been found to roughly estimate high flows during small and moderate rains. It is certainly not an adequate procedure for drainage design, however. As the smallest storm is to be sampled three times during the runoff period, the volume of flow per subsample is simply:

$$13,600 \text{ ft}^3/3 \cong 4500 \text{ ft}^3$$

Therefore, the total number of samples collected during the maximum rain would be:

$$1,130,000 \text{ ft}^3/4500 \text{ ft}^3 \cong 250 \text{ samples}$$

If the minimum sample volume required was 1 L, then each subsample could be as small as 350 mL. This would result in about 1 L of sample during the minimum storm, but result in about 90 L during the maximum storm (obviously much larger than the typical 10 to 20 L container). During the estimated high flow conditions of the largest storm, a subsample would be collected every:

$$4500 \text{ ft}^3 \text{ per sample}/53 \text{ ft}^3/\text{s} \cong 85 \text{ s}$$

If the sampler required 2 min to collect 350 mL, the sampler would not complete its cycle before it was signaled to collect another subsample. This would result in the sampler pump running continuously during this peak time. Since the peak flow period is not expected to have a long duration, this continuous pumping may not be a serious problem, especially considering that about 250 samples are being collected. The biggest problem with this setup is the large volume of sample collected during the large event.

This problem was solved during numerous stormwater monitoring projects (including Pitt and Shawley 1982 during the Castro Valley, CA, NURP project, and Pitt 1985 during the Bellevue, WA, NURP project) by substituting a large container for the standard sample base and installing the sampler in a small shelter. The large container can be a large steel drum (Teflon-lined), a stainless steel drum, or a large Nalgene™ container, depending on the sample bottle requirements. In order to minimize handling the large container during most of the events, a 10 L glass jar can be suspended inside to collect all of the subsamples for the majority of the events. The jar would overflow into the large container for the largest events. Glass bottles are used in the sampler when organics are to be analyzed, with the assumption that the short period of storage in the glass would not adversely affect the metal concentrations. The small shelter should be well vented to minimize extreme temperatures, as it is difficult to ice the large container. Obviously, the sampling stations need to be visited soon after a potential runoff event to verify sample collection, to collect and preserve the collected sample, and to clean the sampler to prepare it for the next event.

Alternatives to using a large sample base (Figure 5.23) in order to accommodate a wide range of runoff events include:

- Use time-compositing instead of flow-weighted sampling
- Use two samplers located at the same location, one optimized for small events, the other optimized for larger events (Figures 5.24 through 5.26)
- Visit the sampling station during the storm and reprogram the sampler, switch out the bottles, or manual sample

The most common option is the last one, which is expensive, uncertain, and somewhat dangerous. Few monitoring stations have ever used multiple samplers, but that may be the best all-around solution, but at an increased cost. The first option above, using time-compositing instead of flow-weighted sampling, should be considered.

Figure 5.23 Automatic sampler with large base for monitoring wide range of flows, with large chest freezer USGS discrete sampler in background, at Bellevue, WA.

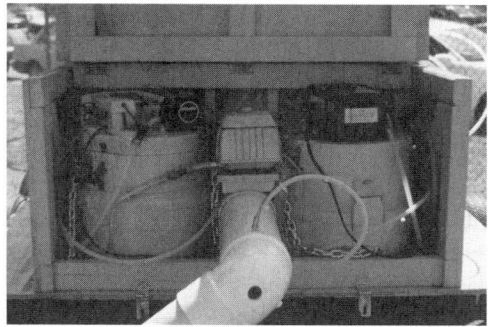

Figure 5.24 Double monitor setup for simultaneously monitoring influent and effluent at small treatment device in Birmingham, AL.

Figure 5.25 Double monitor setup for sampling over a wide range of flow conditions.

Figure 5.26 Multiple flow monitor and sampler setup for simultaneously monitoring influent and effluent over wide range of flow conditions at a small treatment device in Madison, WI.

The Wisconsin Department of Natural Resources conducted a through evaluation of alternative sampling modes for stormwater sampling to determine the average pollutant concentrations for individual events (Roa-Espinosa and Bannerman 1994). Four sampling modes were compared at outfalls at five industrial sites, including flow-weighted composite sampling, time-discrete sampling, time-composite sampling, and "first-flush" sampling during the first 30 min of runoff. Based on many attributes, they concluded that time-composite sampling at outfalls is the best method due to simplicity, low cost, and good comparisons to flow-weighted composite sampling. The time-composite sampling cost was about $1/4$ of the cost of the time discrete and flow-weighted sampling schemes, for example (but was about three times the cost of the first-flush sampling only). The accuracy and reproducibility of the composite samples were all good, while these attributes for the first-flush samples were poor.

It is important to ensure that the time-weighted composite sampling include many subsamples. It would not be unusual to have the automatic samplers take samples every 10 min for the duration of an event. If the minimum sample volume needed is 1 L and the shortest rain to be sampled is 30 min, then each subsample would need to be about 350 mL. The total volume collected would be about 50 L (144 samples) if a storm lasted 24 hours. The sampler would have to have an enlarged container (as in the above flow-weighted example), or the sampler would have to be visited about every 5 hours if a 10 L composite sample container was used.

Another important attribute of time-compositing sampling is that intermittent discharges and other short-term high concentration flows would be more readily detected. Flow-weighted composite sampling may allow very long periods to be unrepresented in the sample, while time-composite sampling can be adjusted to include relatively short sampling periods. Long periods between samplings could allow short-period episodes to be missed. However, sampling periods that are too short may result in almost continuous pumping activity that may exceed the continuous duty cycle of the sampler, resulting in frequent maintenance. Pump tubing should be carefully inspected and frequently replaced in any case, especially considering the gritty nature of stormwater. A new option is the use of *in situ* probes attached to the sampler that can be used to trigger sampling during unusual water quality shifts.

Automatic Sampler Initiation and the Use of Telemetry to Signal or Query Sampler Conditions

Automatic sampling equipment is typically located semipermanently in the field and is set to automatically begin sampling for a predetermined set of conditions. The most common method to start samplers is to use a stage indicator. This simple device, available from most sampler manufactures, may be a float switch (as from American Sigma) or an electronic sensor that shorts out when wet (ISCO). These devices plug into the sampler at the flow sensor connection. If flow monitoring is simultaneously being monitored, a Y connection is available to allow both connections. The stage sensor is typically placed slightly above the baseflow water elevation (in a pipe, open channel, or creek). It is difficult to sample small events that may not cause a large-enough stage elevation increase to trip the indicator. False alarms are also common when the sensor is placed too close to the baseflow water elevation or in areas of high humidity (for the moisture sensor). In addition, the baseflow water stage changes seasonally, requiring constant modifications in the sensor location. If the channel or pipe is normally dry, these problems are significantly reduced, as the sensor can be placed on the bottom of the drainage way or pipe. Flow-weighted sampling schemes can eliminate the use of sensors all together. In this case, some water may collect in the sample container during baseflow conditions, however. Frequent visits to the sampler are needed to empty and clean the sample container.

Another method used to initiate sampling is to trip the sampler using a rain gauge. Pitt and McLean (1986) used a rain gauge to initiate sampling at an industrial site in Toronto, while simultaneously monitoring flow. A tipping bucket rain gauge was used and three trips (about 0.03 in of rain) of the rain gauge within a few hours were usually used to initiate sampling.

In all cases, the use of telemetry (radio, telephone, or cellular phone) is extremely useful in minimizing false trips to a remote sampler by automatically signaling that samples have been collected (Figure 5.27). Campbell Scientific of

Figure 5.27 Telemetry equipment at USGS monitoring site in Madison, WI.

Figure 5.28 In-stream continuous probes at Dort- **Figure 5.29** Automatic sampler connected to contin-
mund, Germany, CSO monitoring site.   uous probes and telemetry at Dortmund,
                                               Germany.

Logan, UT (801-753-2342), supplies many options allowing remote inquiring or automatic signaling to indicate sampler status. It is also possible to phone a monitoring station and immediately determine if a sampler is operating, and to download or observe instantaneous or compiled rain, flow, or continuous *in situ* water quality monitoring information. The use of telemetry is extremely important when many remote systems are being operated by a small group. It should be considered an integral part of all sampling and monitoring programs where high reliability and good quality data are needed. There are potential problems with RF interference between cellular phones and some monitoring equipment, so care must be taken to use an external antenna, to electronically shield the monitoring equipment, and to thoroughly test the setup.

An early example of an automatic stormwater monitoring program using telemetry to excellent advantage was the Champaign/Urbana NURP study conducted in the early 1980s (EPA 1983a). The Universität Gesamthochschule in Essen, Germany, has also used standard telemetry equipment components and specialized software in CSO monitoring in Dortmund, Germany, to inquire about monitoring station and flow status (Wolfgang Geiger, personal communication) (Figures 5.28 and 5.29). Numerous municipalities and state agencies in the United States have also installed telemetry-coupled monitoring stations using relatively inexpensive components, including cellular telephone service and solar-powered battery chargers. This has eliminated most of the concern about the availability of remote utility installations. Cooling collected samples still requires AC-powered chillers, or ice. For remote installations with a small sampling crew, it is impractical to ice the sampler in anticipation of a rain, but that is possible when the samplers are more accessible. It would be more important to recover the samples from the samplers as soon as possible after the event. This is made much more practical, especially with remote samplers, when telemetry is used to inquire about the sampler status.

Siphon Samplers

The USGS recently published a review of siphon samplers, compared to flow-weighted composite samplers for use along small streams (Graczyk et al. 2000). These are inexpensive units that can be utilized in many locations (Figure 5.30). They operate semiautomatically by starting to fill when the water level reaches level B (the top of the loop connected to the intake) in Figure 5.30. The sample

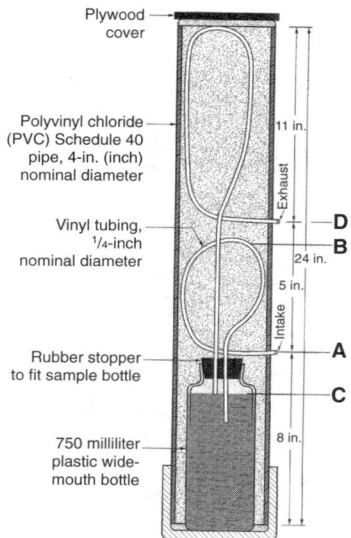

Figure 5.30 Siphon sampler. (From Graczyk, D.J. et al. *Comparison of
Water Quality Samples Collected by Siphon Samplers and
Automatic Samplers in Wisconsin.* USGS Fact Sheet FS-067-
00. U.S. Geological Survey,
Middleton, WI. July 2000.)

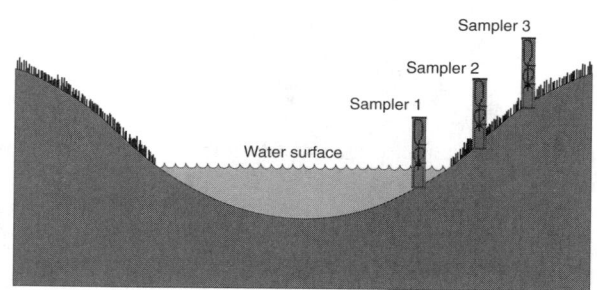

Figure 5.31 Placement of siphon samplers along stream
bank. (From Graczyk, D.J. et al. *Comparison of
Water Quality Samples Collected by Siphon
Samplers and Automatic Samplers in Wisconsin.*
USGS Fact Sheet FS-067-00. U.S. Geological
Survey, Middleton, WI. July 2000.)

bottle fills rapidly due to the hydraulic
head (the elevation of the stream surface
above the discharge end of the intake tube,
level C, in the bottle). After the stream
level reaches level D, an airlock is created
in the top loop, stopping the filling. Therefore, the siphon collects a sample near the
water surface when the stream stage is
between levels B and D, which can be
adjusted. Since they collect samples over
narrow ranges of stream stages, several
can be placed at different heights along a
receiving water, as illustrated in Figure
5.31. Graczyk et al. (2000) compared sets
of three siphon samplers, set at different
elevations, along three streams that also
had flow-weighted automatic samplers
(ISCO) for comparison. They collected 40
to 50 pairs of samples and analyzed them
for suspended solids, ammonia, and total
phosphorus. Figure 5.32 illustrates the
comparison for suspended solids. There
was substantial scatter in the data, but the
differences in the results averaged about
10% for suspended solids and ammonia,

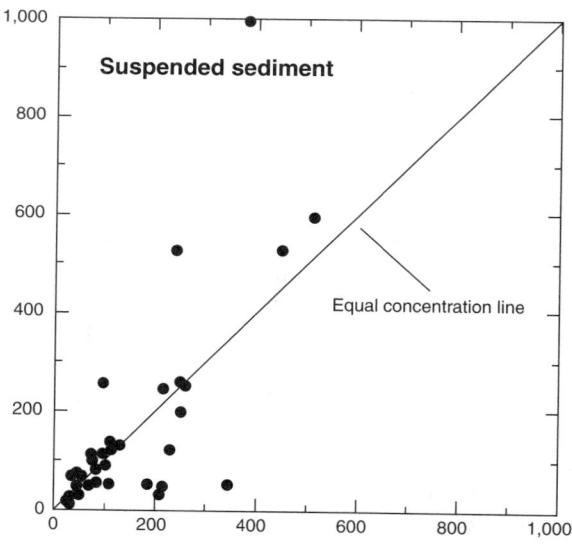

Figure 5.32 Comparison of siphon sampler (y axis) and ISCO
sampler (x axis) suspended solids observations.
(From Graczyk, D.J. et al. *Comparison of Water
Quality Samples Collected by Siphon Samplers
and Automatic Samplers in Wisconsin.* USGS
Fact Sheet FS-067-00. U.S. Geological Survey,
Middleton, WI. July 2000.)

and about 25% for phosphate. However, the differences between individual pairs of samples were much greater. Some of the larger differences may reflect the siphon samplers only collecting samples at specific stage increments, while the automatic samplers collected samples at a single depth over longer periods of time. The siphon samplers may be useful when many samples can be collected and overall conditions are desired, in contrast to more accurate individual results. Their low cost and ability to sample for specific stage conditions makes them an interesting alternative to more expensive automatic samplers, or difficult manual sampling.

Retrieving Samples

Each sampler site will need to be visited soon after the runoff event to retrieve the sample for delivery to the laboratory. The storage time allowed in the sampler before collection should be determined from a special holding-time study conducted in conjunction with the analytical laboratory. Stormwater samples can usually withstand longer holding times than those implied from standard laboratory method descriptions without significant degradation. However, this will need to be verified by local tests. In all cases, the allowable holding times noted in Table 5.10 should be followed except in unusual situations and then only with specific tests. This is especially important when organizing sample deliveries to the laboratory after hours (which can happen frequently).

Manual Sampling Procedures

The following paragraphs summarize the procedures needed for manually collecting water and sediment samples from a creek or small stream.

1. Fill out the sample sheet and take photographs of the surrounding area and the sampling location. Conduct any *in situ* analyses (such as stream flow measurements, along with dissolved oxygen, pH, temperature, conductivity, and turbidity measurements in the water).
2. Use a dipper sampler to reach out into the flow of the stream to collect the sample. Slowly lower the sampler onto the water, gently rolling the top opening into the flow. Be careful not to disturb the bottom sediments. Submerge the sampler lip several inches into the water so floating debris are not collected. Lift out the sampler and pour the water into a compositing container (such as a churn sample splitter). Several samples should be collected in the area of concern and composited. In some cases, it may be useful to sample the water–air interface. This surficial layer is known to trap many types of organic chemicals (e.g., oils and surfactants) and have elevated microbial populations (e.g., pathogens).
3. Each water subsample can be poured into a large clean container during this sampling period. At the end of the sampling period, this composite sample is mixed and poured into the appropriate sample bottles (with preservatives) for delivery to the analytical laboratory.

Microbiological sampling requires special sampling techniques. ASTM (1995) in standard D 3370 describes the grab sampling procedures that must be used for collecting samples that will be analyzed for bacteria. The samples need to be glass and sterile. If the sample contains chlorine, then the sample bottle must contain sodium thiosulfate so any residual disinfection action will be destroyed. The bottle lid is removed and the bottle is placed under flowing water and filled to about $\frac{3}{4}$ of its capacity. Care must be taken when handling the bottle and lid (including not setting them down on any surface and not touching any part of the upper bottle portion) to minimize contamination. Do not rinse the bottle with the sample or submerge it under water.

Sampling sediment can be difficult (see also later discussion). The simplest method is to use a lake bottom sampler. Specifically, a small Ekman dredge sediment sampler, which is typically used for sand, silt, and mud sediments, is usually most useful. Corer samplers are generally not as successful for stream sediments. An exception is the freezing core sampler, where liquid CO_2 is pumped inside a stainless steel tube (with the bottom end sealed with a point) to freeze sediment to the outside of the tube. Again, the sediment would have to be at least several inches deep. In all cases, multiple sediment samples would have to be obtained and composited. Any water samples should be obtained first, as the sediment sampling will create substantial disturbance and resuspension of sediment in the water column. All sampling equipment must also be constructed of noncontaminating materials. Stainless steel, polypropylene, or Teflon are the obvious choices.

Dipper Samplers

The simplest manual sampler is a dipper sampler (Figure 5.33). Markson (telephone: 800-858-2243) sells a dipper sampler that has a 1 L polyethylene beaker on the end of a two-piece, 4-m pole (catalog # MK34438 for about $60). They also sell units on 1- and 2-m poles and with 500 mL capacities. These samplers can only obtain samples from the surface of the water. If subsurface samples are needed, samplers with closure mechanisms need to be used, as described below. A dipper allows sampling of surface waters away from the immediate shoreline and from outfalls or sewerage pipes more conveniently than other types of samplers. Dippers are commonly used to sample small discharges from outfalls, where the flow is allowed to pour directly into the sampler. ASTM (1995) in standard D 5358 describes the correct stream water sampling procedure using a dipper sampler. The dipper needs to be slowly lowered into the water on its side to allow the water to flow into the sampler. The dipper is then rotated to capture the sample and is lifted from the water. Care must be taken to prevent splashing or disturbing the water. The sample is then poured directly into the sample bottles or into a larger container (preferably a churn sampler splitter, as previously described) for compositing several dipped samples.

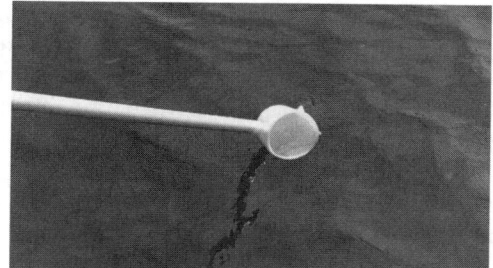

Figure 5.33 Manual dipper sampler.

Submerged Water Samplers with Remotely Operated Closures

There are numerous historical and modern designs of samplers that can take water samples at specific depths. These all have a way to remotely operate closures in a sample container. The sampler capacities usually range from 0.5 to 3 L. Older designs include the Kemmerer and Van Dorn samplers, shown on Figure 5.34 (*Standard Methods* 1995). These samplers have a tube made of metal or plastic and end closures made of plastic or rubber. All Teflon units are available to minimize sample contamination. Newer designs commonly used for small lakes or streams are

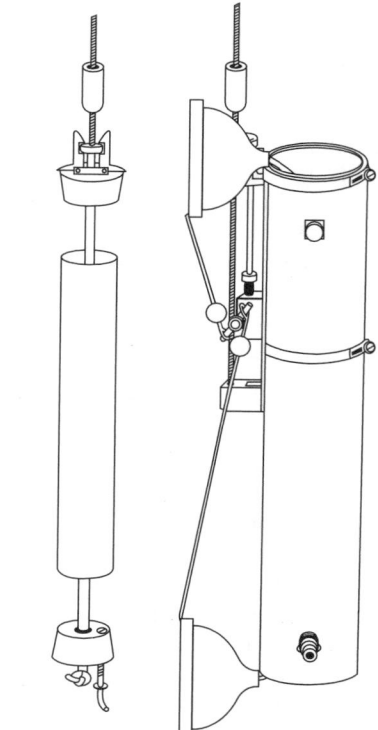

Figure 5.34 Kemmerer and Van Dorn samplers. (From *Standard Methods for the Examination of Water and Wastewater.* 19th edition. Water Environment Federation. Washington, D.C. Copyright 1995 APHA. With permission.)

Figure 5.35 Horizontal water sampler in open position before use.

Figure 5.36 Tripped horizontal water sampler being withdrawn from water with messenger resting on trigger mechanism.

Figure 5.37 Open vertical water sampler being lowered into water, above a horizontal sampler on the same line.

Figure 5.38 Tripped vertical water sampler being withdrawn from water with messenger resting on trigger mechanism.

similar to the Van Dorn design (Figures 5.35 through 5.38). This design allows unhindered flow through the sample container before closure, enabling faster equilibrium with surrounding waters. These samplers are also available in horizontal models (for shallow water) or vertical models. Several of the vertical units can be used on a single line to obtain water samples from various depths simultaneously. A weighted messenger slides down the line that the samplers are attached to, striking a trigger mechanism that closes the end seals. If multiple samplers are used, the trigger releases another messenger that slides down to the next sampler to close that sampler and to release another messenger. A vertical alpha end-closure 2.2-L sampler (polyurethane end seals and transparent acrylic cylinder) is available from Forestry Suppliers, Inc. (800-647-5368) as catalog #77244, with messenger #77285, for a total cost of about $450. Several of these samplers can be installed on a line for simultaneous sampling at various depths. Forestry Suppliers, Inc., also sells a 1.2-L Teflon Kemmerer vertical bottle sampler (catalog #77190) for about $800. A water sample collected with this sampler only contacts Teflon.

Another surface operated design is a sampler that contains a 1-L glass bottle on the end of a long pole (such as catalog #53879 from Forestry Suppliers, Inc. at about $400). A stopper is spring loaded and is attached to a wire extending to the other end of the pole. The bottle end is lowered to the desired sampling depth and the wire is then pulled to fill the bottle. After a short period to allow the bottle to fill, the wire is released, resealing the bottle. This sampler was designed specifically for collecting water samples for Winkler titrations for DO analyses at sewage treatment plants. The bottle is initially full of air before the water enters and aeration may elevate the DO reading. If the bottle is prefilled with clean water, it is difficult to assume that the desired water sample will replace the water in the bottle. However, this sampler type might be useful for collecting subsurface samples for bacteriological analyses that should be collected in glass bottles with minimal handling.

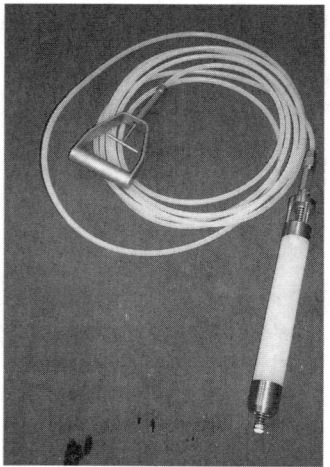

Figure 5.39 Tube sampler.

Figure 5.40 Grundfos Redi-Flo2 pump sampler with controller.

A newer alternative is a Teflon tube sampler that contains a wire-activated sealant mechanism and flow-through design (Figure 5.39). This overcomes the above limitations of the bottle sampler and still allows direct sampling at a specific depth. The AMS Cable Control Liquid Sampler is available from Forestry Suppliers, Inc. (catalog #77623), and costs about $550.

Manual Pump Samplers

A Grundfos Redi-Flo2 (Figure 5.40) pump and converter (designed and commonly used for well sampling) is available with a 300-foot polyurethane hose on a reel that can be used to deliver a water sample to a convenient location, especially useful when sampling wide and swift streams from a bridge. These pumps are available from Forestry Suppliers, Inc. (800-543-4203, catalog #76328 for pump, hose, and reel, and #76333 for voltage converter, for a total cost of about $4500). Hazco (800-332-0435) also sells (and rents) the Redi-Flo2 pump and converter for about $2100 without a hose (catalog #B-L020001 for converter and #B-L020005 for 150 motor lead and pump). A Teflon-lined polyethylene hose is available from Hazco for about $3.25 per foot, with support cable (catalog #A-N010041 and #C-L020009). This pump has an adjustable pumping rate of between 100 mL/min and 9 gal/min and can pump against a head of about 250 ft. However, this pump should be operated at least at 4.5 gal/min to meet the 100 cm/s criterion to minimize particulate settling in the 1 in ID hose. Low pumping rates from a submerged pump can also lead to "sand jamming," in addition to preventing an adequate sample from being obtained.

A less expensive alternative is the XP-100 pump, also available from Forestry Suppliers (#76216 for XP-060 pump and #76230 for control box, for a total cost of about $525). This is an adjustable rate pump and can deliver the needed 100 cm/s pump rate through a $^3/_8$-in tubing against a head of about 30 ft or less. This pump operates from a 12V DC power supply and has a limited service life, compared to the Grundfos pump. It may be useful for temporary installations having limited head, but needing several pumping locations across a stream. It is also useful for continuous sampling at different lake depths.

Depth-Integrated Samplers for Suspended Sediment

Suspended sediment is usually poorly distributed in both flowing and quiescent water bodies. The sediment is usually in greater concentrations near the bottom, as shown in Figure 5.41 (ASTM 1995). Larger and denser particles are also located predominantly in lower depths. Flowing water

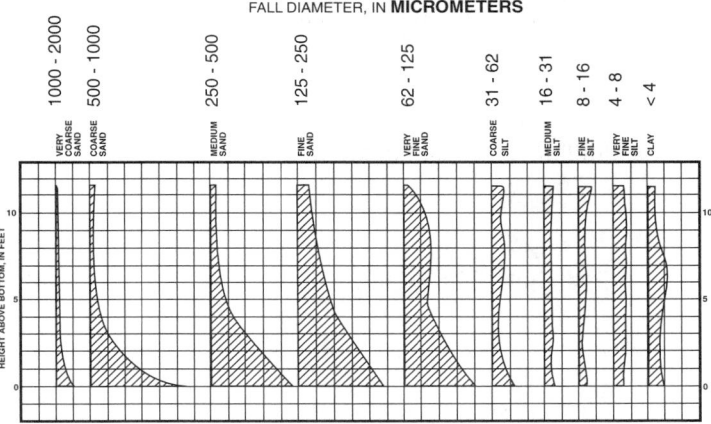

Figure 5.41 Sediment concentrations by depth and particle size, Missouri River, Kansas City, MO. (From American Society for Testing and Materials. *ASTM Standards on Environmental Sampling.* ASTM Pub Code No. 03-418095-38. ASTM, Philadelphia. 1995. Copyright ASTM. Reprinted with permission.)

in a sinuous stream also distributes the suspended sediment horizontally, as shown in Figure 5.42 (ASTM 1995), differently for large and small particles. Collecting representative samples in these situations for sediment analyses is therefore difficult. Because most of the pollutants in stormwater are associated with the particulates, this unequal distribution of sediment also affects the ability to collect representative samples of many pollutants. Depth-integrating sampling is commonly done in small upland streams. Sampling in smaller and more turbulent flows (such as in sewerage or at outfalls during moderate to large storms) is not as severely affected by sediment stratification.

Clay and silt-sized particles are generally well mixed with depth, depending mostly on water mixing conditions near discharges, etc., and not on gravity. ASTM (1995) states that the concentrations of particles smaller than about 60 μm in diameter will be uniform throughout the stream depth (Figure 5.41). However, larger particles will be more affected by gravitational forces and may not be represented well with typical sampling procedures. Conventional water samplers may be used to represent all of the sediment in flowing water (floating material, suspended sediment, and bedload), if the water is very turbulent and capable of mixing the sediment of interest. ASTM refers to these locations as "total-load" stations, allowing the collection of all sediment greater than about 2 mm in diameter. These are generally located at outfalls or other free-falling locations.

Automatic samplers (or any pumped sampler) may disproportionately collect particulates if the intake velocities vary significantly from the water velocity. Isokinetic sampling requires that

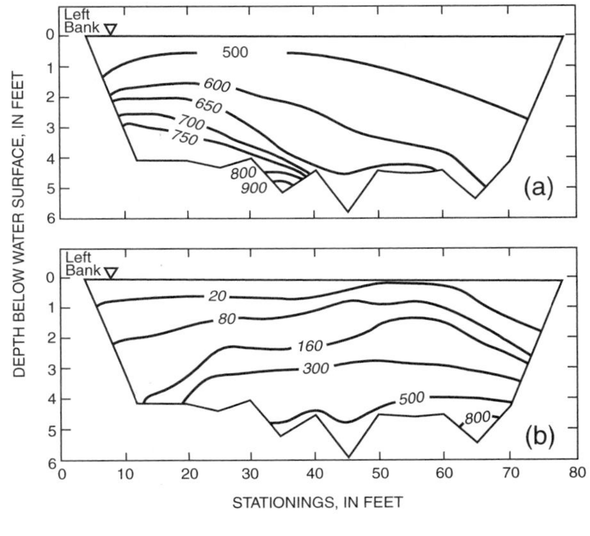

Figure 5.42 Suspended solids concentrations in the Rio Grande River, near Bernardo, NM, for different sediment sizes: (a) material between 62.5 and 125 mm; (b) material between 250 and 500 mm). (From American Society for Testing and Materials). *ASTM Standards on Environmental Sampling.* ASTM Pub Code No. 03-418095-38. ASTM, Philadelphia. 1995. Copyright ASTM. Reprinted with permission.)

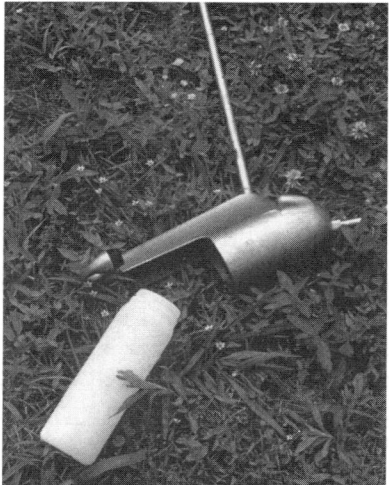

Figure 5.43 Depth-integrated sediment sampler parts.

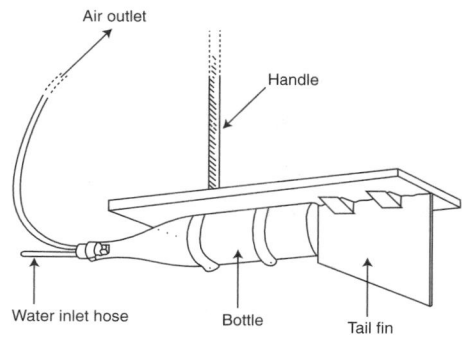

Figure 5.44 Plan for a home-made depth integrated sampler. (Modified from Finlayson 1981.)

Figure 5.45 Depth-integrated sediment sampler being readied for use.

the sampler intake be pointed directly into the flowing water and that the velocity in the intake be the same as the flowing water. The water and sediment streamlines will therefore be parallel in this situation and a sample representative of the flowing water will be obtained. If the sample intake velocity is greater than the water velocity, water will be drawn into the sampler, while heavier particles will tend to flow past. This effect is most evident for heavier particles (larger and denser) than for lighter particles. Berg (1982) reports that particles approaching 100 μm in diameter with densities of 2.65 g/cm^3 have less than a 20% sampling error when the velocities are not matched. Almost all stormwater and stream-suspended particulates are smaller and have a lighter density than this and would therefore generally follow the flow streamlines. These particles would therefore not be significantly affected by this possible problem.

Large-sized (larger than several hundred micrometers in diameter) suspended sediment measurements may be important for receiving water studies, especially in areas having flash flood flows in sandy soil regions (such as the southwest United States). The depth integrated sampler is designed to obtain a sample continuously as the sampler is lowered vertically through the water column at a constant velocity (Figures 5.43 through 5.45). These units vary significantly from commercial grab samplers that have remotely operated valves in that they have air vents to allow the air in the sample bottle to uniformly escape as the sample bottle fills with water. The home-made unit has a narrow-mouthed bottle mounted on a rod with stabilizing fins. The mouth of the bottle is fitted with a two-holed stopper. The top hole has a long flexible tube (which could extend above the water surface for most streams) to act as an air outlet, while the bottom hole has a rigid tube extending at least an inch to act as an intake. The intake nozzle should have a sharp front edge, with a narrow tubing thickness (less than $^1/_{16}$ in) and an inner diameter of 5 to 6 mm ($^3/_{16}$ or $^1/_4$ in) (ASTM 1995, standard D 4411). These are available commercially from Forestry Suppliers, Inc. (800-543-4203) and in Canada from Halltech Environmental, Inc. (519-766-4568), or they can be constructed (Figure 5.44).

When collecting a depth-integrated sample, the sampler needs to stand to the side and downstream of the sampling area to minimize disturbance. The rod is lowered vertically through the water column at a constant rate of about 0.4 times the stream velocity. Detailed vertical sampling rates are presented by ASTM (1995) in standard D 4411 for the series of older depth-integrated samplers. The sampler is lowered at this constant rate from the surface of the stream to the stream bottom, and then reversed and brought back to the surface at the same rate. The sampler does not collect samples within several inches of the stream bottom. Moving sediment near the bottom is usually included in the bedload sample, which requires other sampling methods. The sample bottle should be between $^2/_3$ and $^3/_4$ full after sample collection. If it is full, then the sampler did not represent the complete stream depth and the sample should be discarded and collected again, at a faster vertical rate. If the sampler is less than $^2/_3$ full, another vertical sample pass can be collected. After the sample is collected, the sample is poured from the sampler into a sample bottle. It is possible to mount an appropriate sample bottle directly to the sampler, and sample transfer would therefore not be needed.

Several vertical samples will normally need to be collected across the stream, as the coarser suspended sediment is likely highly variable in both time and space (ASTM 1995). The location and number of sampling verticals required at a sampling site is dependent primarily on the degree of mixing at the cross section.

Settleable Solids Samplers

Sediment traps suspended in the water column can be used to capture settleable solids. Zeng and Vista (1997) describe the use of these samplers off San Diego to capture marine settleable solids for organic compound analyses in the water column at several off-shore locations. The sediment traps were located 1 and 5 m from the seafloor and were retrieved after 30 days. The traps were made of two parts, a glass centrifuge bottle at the bottom and a glass funnel positioned on the bottle through a Teflon-lined silicone rubber seal. When retrieved, the two parts of the traps were separated and water covering the particulates was carefully removed. The centrifuge bottles were then capped with Teflon-lined caps and brought to the laboratory for analysis.

Similar sediment traps were used in the Seattle area to investigate the amount and fate of CSO settleable solids in the receiving waters. These traps were generally similar to those described above but were located much closer to shore and in shallower water. Several were placed vertically on an anchored line in a grid pattern near and surrounding CSO discharge locations being investigated.

Sediment traps were also placed in Fresh Creek, New York City, at the Equi-Flow demonstration facility. These traps were placed within and outside the facility to quantify the amount of settleable material that was captured during the CSO storage operations before being pumped back to the treatment plant. This use of sediment traps was not very successful due to very dynamic flow conditions and the short exposure periods used in an attempt to obtain data during frequently occurring CSO events. Longer exposure periods would have enabled the capture of more measurable material, but would have blended together material from adjacent events.

Sediment traps can be useful sampling devices to capture and measure slowly settling solids *in situ* in the water column. This information is especially important when quantifying the effects of sediment-laden discharges into relatively large water bodies having slow to moderate currents. They may not be suitable for small streams, unless they can be miniaturized. Several traps should be suspended at one location at different depths, and redundant devices should be used to compensate for traps lost during the exposure period. Like the bedload samplers described next, the exposure periods should probably be long (several weeks). The sampler materials also need to be compatible with the constituents intended to be analyzed. A simple framework (made of inert materials) should also be constructed to brace the assembled sediment trap and to allow easy attachment to the anchored line, but it should not extend above the funnel to minimize interference with settling materials.

Bedload Samplers

Bedload is the material that travels in almost continuous contact with the stream bed (ASTM 1995). The bedload material moves when hit by another moving particle, or when water forces overcome its resisting forces. Bedload is sampled by using a trapping sampler located on the stream bottom. The simplest bedload samplers are box or basket samplers which are containers having open ends facing upstream. Bedload material bounces and rolls into the sampler and is trapped. Other types of bedload samplers consist of containers set into the sediment with slot openings about flush with the sediment surface. The bedload material falls through a slot and is trapped. Slot widths and lengths can be varied to represent various fractions of the bedload actually moving in the stream. The errors associated with sampling bedload are greater than with sampling suspended sediment because the larger particles move more irregularly under the influence of gravitational forces and are not well mixed in the water.

Bedload may be important when characterizing stormwater sediment discharges. In northern areas where sands are used for ice control, relatively large amounts of sand can be transported along the drainage system as bedload. At the Monroe St. detention pond site in Madison, WI, the bedload accounted for about 10% of the total annual sediment loading. This fraction was much greater during the spring when most of the sand was flushed from the drainage area.

Conventional water samplers may not adequately collect bedload material. A slot sampler placed in a drilled hole in the bottom of a discharge pipe can effectively collect this material. However, the slot dimensions and placement exposure times must usually be determined by trial and error. In addition, several bedload samplers should be used in close proximity because of the varied nature of bedload transport. Bedload samplers that are full upon retrieval may not represent actual conditions. If full, then the slot widths should be reduced and/or the exposure time should be shortened. The slot length should be as long as possible for the container lid, as bouncing bedload particles may jump over openings that are too short. In addition, the slot widths should be at least $1/4$ in wide, as narrower slots will filter out large materials. Basket samplers are probably most applicable in streams, where the opening width is a small fraction of the stream width. Again, several samplers need to be used in close proximity, and the best exposure period needs to be determined by trial. For grab samples, both hand-held and cable suspended Helley Smith (Geological Survey) bedload samplers are available from Halltech Environmental, Inc. (519-766-4568).

Floatable Litter Sampling

One example of quantifying litter discharges during wet weather was described by Grey and Oliveri (1998). New York City has been involved in a comprehensive litter analysis and capture effectiveness program since the mid-1980s. As part of this investigation, it studied litter discharges from stormwater inlets using baskets that were inserted in manholes below catchbasins (Figure 5.46). The baskets were made of galvanized mesh and were 13 in square and 36 in high. The lower half of the baskets was made of $1/4$-in mesh, while the upper half was of $1/2$-in mesh. The baskets were positioned on a wooden platform just beneath the catchbasin outlet pipe and were held in place with ropes, allowing removal without requiring entry into the manholes. These baskets were installed at 38 locations throughout the city and were in place for 3 to 4 months. Most baskets were removed, emptied, and replaced every 2 weeks, although some were in place for only a week before emptying. The captured material was placed in sample bags, brought to the laboratory, sorted into 13 categories, counted, and weighed. The surface areas of the collected material were also measured.

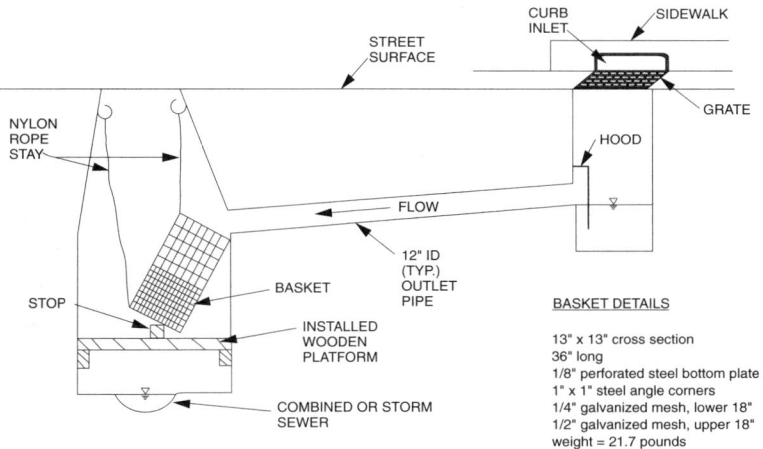

Figure 5.46 New York City catchbasin litter sampling setup. (From HydroQual, Inc. *Floatables Pilot Program Final Report: Evaluation of Non-Structural Methods to Control Combined and Storm Sewer Floatable Materials.* City-Wide Floatables Study, Contract II. Prepared for New York City, Department of Environmental Protection, Bureau of Environmental Engineering, Division of Water Quality Improvement. NYDP2000. December 1995.)

In addition to characterizing the litter discharges, New York City also examined the effectiveness of the catchbasins in capturing this material. Grey and Oliveri (1998) also described these tests. They placed a known amount of litter (10 pieces each of 12 different floatable items, totaling about 1 ft³ in volume of each material), including plastic bags, candy wrappers, straws, bottle caps, juice bottles, hard plastic pieces, glass vials, aluminum cans, polystyrene cups and pieces, cigarette butts, and medical syringes. They then opened a fire hydrant to produce a basic flow rate of about 75 gal/min (corresponding to a rain intensity of about 0.28 in/hour over a 40,000 ft² drainage area). They also ran tests at ⅓ and 2× this flow. The flow was continued until no more items were transported to the sampling basket (usually about 5 to 10 min). The items remaining in the catchbasin were then retrieved and counted. This test was repeated five times for each test, and 10 tests in all were conducted (some with and some without catchbasin hoods).

Source Area Sampling

Much information can be obtained by collecting stormwater samples at source areas. Source areas are where the runoff originates before it is collected in the storm drainage system. Source area sampling also includes rainfall sampling for water quality analyses, conventionally done using a wet/dry-fall sampler. This sampler also collects dust fall during dry periods. This atmospheric contribution can have a significant affect on stormwater quality. However, very little of the dry-fall pollutants occurring over a watershed actually are washed off during rains.

This information can help identify the critical areas in the watershed where most of the problem pollutants may be originating and where control measures should be implemented (Pitt et al. 1995). These areas may include paved industrial storage areas, convenience store parking areas, vehicle maintenance areas, landscaped areas, roof runoff, etc. Conventional automatic samplers may not be efficiently used in these areas because of the small scale of the sampling areas and limited places where the samplers can be located that would only receive runoff from the area of concern. Three sampling methods have been used:

- Manual sheetflow samplers
- Semiautomatic samplers
- Special designs for automatic sample collection

Figure 5.47 Sheetflow sampler operated by hand vacuum pump.

Figure 5.48 Sheetflow sampler being used to sample snowmelt.

Manual Sheetflow Samplers

Manual sheetflow samplers are usually used when collecting grab samples from many different sampling locations. A small team can visit many sampling sites during a single rain to obtain multiple grab samples for statistical comparisons (Figures 5.47 and 5.48). The main drawback is that the samples are not composited during the rain and only represent the conditions during the short sampling period. It is therefore very important to carefully document rain and flow conditions during the sampling period, and for the short time before the sample was obtained. Rain conditions up to the time of sampling can also have a significant effect on measured pollutant concentrations. In many cases, the ability to obtain many samples in a relatively short time is more important than obtaining flow-weighted composite samples. Roa-Espinosa and Bannerman (1994) found that many discrete samples (which could be composited before analysis) are just as useful in obtaining an event-mean concentration (EMC) as are more difficult to obtain flow-weighted composite samples.

Sheetflow samples should be obtained in areas where the sheetflow is originating from a homogeneous area, such as from a parking area, roof runoff, runoff from a landscaped area, etc. Sheetflow samples can be collected by collecting the flow directly into the sample containers, if the flow is deep enough. The flow may be "scooped" using a small container and by pouring the collected samples into the sample container. For shallow sheetflows, a hand-operated vacuum pump can be used to draw the sample into the sample container, as shown in Figure 5.47. A Teflon-lined lid that fits the sample containers can be fitted with two Teflon bulk-head connectors. One of the connectors has a Teflon tubing (about 18 in long and $1/4$ in ID) attached that is used to draw the sample into the container. The other connector has a Tygon™ tube leading to a water trap (another bottle) that is in turn attached to a hand-operated vacuum pump (such as a Nalgene #6132-0020, at about $100). To collect a sample, the Teflon tubing is immersed in the sheetflow and the hand pump draws the water into the sample bottle. The pump should be operated slowly to prevent cavitation at the tubing inlet. The short lengths of Teflon tubing are inexpensive and can be replaced after each sample to prevent cross-contamination. Since the sample is drawn directly into the sample bottle, sample transfer is unnecessary.

An alternative to the hand-operated vacuum pump and water trap arrangement is to use a battery-operated peristaltic pump (such as a Masterflex L/S portable sampling pump, catalog #FE-07570-10, at about $850, with a Teflon tubing pump head, catalog #FE-77390-00, at about $400, available from Cole-Parmer, 800-323-4340). This battery-operated pump can be used to pump directly into the sample containers. The Teflon tubing used in this pump (catalog #FE-77390-60) costs about

$15 each and would therefore not likely be replaced after each sample. The tubing would therefore require field cleaning between each sample. Since the battery is built into this pump, and no water trap is needed, this sampling arrangement is relatively compact.

Semiautomatic Sheetflow Samplers

Source area samplers have been developed to semiautomatically collect composite stormwater samples from small drainages. Samplers (at $250 to $650) from the Vortox Company (909-621-3843) are an attractive option for some studies (Figure 5.49). These 0.8- to 5.5-gallon units (available Teflon lined) are completely passive and operate with a double ball closure system. They are installed in the bottom of intermittent flow paths, requiring a sump for installation. They have a screw closure to adjust the rate of filling. A top ball seals the inlet during dry conditions. When a flow occurs, this ball floats, opening the inlet. An inner ball on the underside of the inlet then seals the inlet when the sampler is full.

Potential problems may occur with sediment clogging the very small inlet and fouling the ball seals. However, this sampler also collects bedload

Figure 5.49 Vortox sampler.

from the flowing stormwater (if the ball valve is opened sufficiently) that is not collected using conventional stormwater samplers. The sampler is somewhat awkward to clean. Another problem is the rapid time (less than 20 minutes for the 0.8-gal unit and less than 2 hours for the 5.5-gal unit) to completely fill the sampler. Sheetflows from homogeneous areas (especially small paved areas where these samplers are likely to be used) usually demonstrate strong "first-flush" conditions. The initial flows have much greater concentrations than the EMC, especially for relatively constant rain intensities. This would result in biased concentrations if only the first 20 min of the flow is represented in the sample.

Because of its low cost and passive operation, this sampler may be attractive in situations where many source areas are to be sampled with a small sampling crew. Again, caution must be expressed in interpreting the results, as the concentrations may be greater than the EMC values for source area flows. At outfalls, in complex drainage ways, or with highly variable rain intensities, the initial samples are not likely to be consistently different from the EMC. Frequent site visits will be necessary when runoff has been expected in order to retrieve samples. It may be desirable to have additional samplers so clean units can be substituted in the field for full samplers. The full samplers can then be brought to the laboratory to be emptied and cleaned.

Automatic Source Area Samplers

Problems associated with the above two sampling methods for source area sheetflows can be largely overcome using automatic samplers. Conventional automatic water samplers discussed earlier are probably the most flexible. However, they are expensive and large. Their size limits where they can be located and the size of flow they can sample. Their cost limits the number of units that can be simultaneously deployed. It is possible to rotate a relatively few samplers randomly between semipermanent sampling locations after every few storms. The samplers would be programmed for time-composite sampling (or time-discrete sampling) and automatically activated with

Figure 5.50 Prototype WI DNR/USGS automatic sheetflow sampler.

flow level sensors, or by rain gauge activity. As noted earlier, telemetry can be used to call the project personnel automatically when the sampler has been activated.

Roa-Espinosa and Bannerman (1994) describe a new automatic source area sheetflow sampler that the Wisconsin Department of Natural Resources and the Madison USGS office have jointly developed (Figure 5.50). Their initial source area sampler was similar to a slot bedload sampler and located in the flow path to be sampled. Like the Vortox unit, it usually filled quickly and did not represent the complete runoff event. This initial sampler consisted of a 10-in ID PVC pipe 12-in long. A 10-in PVC pipe coupling was cut in half and glued to the top of the pipe as a reinforcing collar. This pipe was then cemented in a drilled hole in the pavement (for pavement runoff sampling). A 1-in-thick PVC cap, having a $^5/_8$-in center hole, was fitted snugly in the coupling sleeve of the pipe section cemented in the pavement. The upper surface of this cap was flush with the pavement surface. A sample bottle lid was bolted to the underside of the removable cap, which also had a $^5/_8$-in hole matching the hole in the cap. A 2.5-L glass sample bottle was screwed into this lid and placed in the pipe cemented into the pavement when rain was expected. After the runoff ended, the bottles were retrieved and brought to the laboratory. As noted above, sample bottles commonly were full after the runoff ended, indicating that the samples did not represent the complete event. The sampling holes were reduced to reduce the inflow rate, but clogging was a concern and they still were frequently full. Investigators then developed a new sheetflow sampler that was electronically activated (Figure 5.50). A relatively large sample inlet was used to minimize clogging, but an electronically operated ball valve was added. It is possible to program the sampler to schedule the duration of the open and closed times. This enabled the complete runoff events to be represented in the sample. When commercially available, these samplers are likely to cost about $1000.

Source Area Soil Sampling

Soil sampling in urban areas usually involves collecting material from both paved and unpaved areas. Collecting particulates from paved areas ("street dirt") is described in the following subsection and can be applied to many paved source areas, in addition to streets, the original area of most interest. Soil sampling from nonpaved areas involves more traditional soil sampling procedures and is discussed in any agricultural soils textbook. Generally, small trowels are used to collect surface soil samples for analyses, while small hand coring tools are used to collect subsurface samples down to about 1 ft in depth. Deeper soil samples can be best obtained from the walls of trenches that have been excavated using small backhoes.

If soil characteristics associated with particulates most likely to erode during rain events are of most interest, then care should be taken to emphasize the surface soils during sample collection. In this case, careful "scrapings" of surface dirt by a trowel or stiff brush into a sample container may be most efficient, as only very thin layers of most surface soils are typically eroded. If subsurface soil characteristics are needed, such as observing signs of seasonal high groundwater, then small trenches may be needed. Small soil cores should be used when measuring soil texture when soil infiltration studies are being conducted. Cores (or trenches) are also needed if soil chemical quality is needed for different soil depths.

Street Surface Particulate Sampling Procedures

The street dirt sampling procedures described in this section were developed by Pitt (1979) and were used extensively in many of the EPA's Nationwide Urban Runoff Program (NURP) projects (EPA 1983a) and other street cleaning performance studies and washoff studies (Pitt 1987). These procedures are flexible and more accurate indicators of street dirt loading conditions than previous sampling methods used during earlier studies (such as Sartor and Boyd 1972, for example). The procedures are described here in detail so that they can be used by those wishing to determine loading conditions, accumulation rates, washoff rates, and street cleaning effectiveness for their own locations.

Powerful dry vacuum sampling, as used in this sampling procedure, is capable of removing practically all of the particulates (>99%) from the street surface, compared to wet sampling. It can also remove most of the other major pollutants from the street surface (>80% for COD, phosphates, and metals, for example). Wet sampling, which would better remove some of these other constituents, is restricted to single area sampling, requires long periods of time, requires water (and usually fire hydrants, further restricting sample collection locations to areas that have no parked cars), and basically is poorly representative of the variable conditions present. Dry sampling can be used in many locations throughout an area; it is fast, and it can also be used to isolate specific sampling areas (such as driving lanes, areas with intensive parking, and even airport runways and freeways, if special safety precautions are used). It is especially useful when coupled with appropriate experimental design tools to enable suitable numbers of subsamples to be collected representing subareas, and finally, the collected dry samples can be readily separated into different particle sizes for discrete analyses.

Equipment Description

A small half-ton trailer can be used to carry the generator, two stainless steel industrial vacuum units, vacuum hose and wand, miscellaneous tools, and a fire extinguisher. This equipment can also be fitted in a pickup truck, but much time is then lost with frequent loading and unloading of equipment, especially considering the frequent sampling that is typically used for a study of this nature (sampling at least once a week, and sometimes twice a day before and after street cleaning or rains). A truck with a suitable hitch and signal light connections is needed to pull the trailer. The truck also requires warning lights, including a rooftop flasher unit. The truck is operated with its headlights and warning lights on during the entire period of sample collection. The sampler and hose tender both need to wear orange, high-visibility vests. The trailer also needs to be equipped with a caution sign on its tailgate. In addition, both the truck and the street cleaner used to clean the test area can be equipped with radios (CB radios are adequate), so that the sampling team can contact the street cleaner operator when necessary to verify location and schedule for specific test areas.

Experiments were conducted by Pitt (1979) to determine the most appropriate vacuum and filter bag combination. Two-horsepower (hp) industrial vacuum cleaners with one secondary filter and a primary dacron filter bag are recommended as the best combination. The vacuum units are heavy duty and made of stainless steel to reduce contamination of the samples. Two separate 2-hp vacuums are used together by joining their intakes with a Y connector. This combination extends the useful length of the 1.5-in vacuum hose to 35 ft and increases the suction so that it is adequate to remove all particles of interest from the street surface. Unfortunately, two vacuums need to be cleaned to recover the samples after the subsample collections. A wand and a "gobbler" attachment are also needed. The aluminum gobbler attaches to the end of the wand and is triangular in shape and about 6 in across. Since it was scraped across the street during sample collection, it wears out frequently and must be replaced. The generator needed to power the vacuum units must be of sufficient power to handle the electrical current load drawn by the vacuum units, about 5000 watts for two 2-hp vacuums. Honda water-cooled generators are extremely quiet and reliable for this purpose. Finally, a secure, protected garage is needed to store the trailer and equipment near the study areas when they are not in use.

Sampling Procedure

Because the street surfaces are more likely to be dry during daylight hours (necessary for good sample collection), collection should not begin before sunrise nor continue after sunset. During extremely dry periods, sampling can be conducted during dark hours, but that requires additional personnel for traffic control. Two people are needed for sampling at all times, one acting as the sampler, the other acting as the vacuum hose tender and traffic controller. This lessens individual responsibility and enables both persons to be more aware of traffic conditions.

Before each day of sampling, the equipment is checked to make sure that the generator's oil and gasoline levels are adequate, and that vacuum hose, wand, and gobbler are in good condition. Dragging the vacuum hose across asphalt streets requires periodic hose repairs (usually made using gray duct tape). A check is also made to ensure that the vacuum units are clean, the electrical cords are securely attached to the generator, and the trailer lights and warning lights are operable. The generator requires about 3 to 5 min to warm up before the vacuum units are turned on one at a time (about 5 to 10 s apart to prevent excessive current loading on the generator). The amperage and voltage meters of the generator are also periodically checked. The generator and vacuums are left on during the complete subsampling period to lessen strain associated with multiple shutoffs and startups. Obviously, the sampling end of the vacuum hose needs to be carefully secured between subsamples to prevent contamination.

Figure 5.51 illustrates the general sampling procedure. Each subsample includes all of the street surface material that would be removed during a severe rain (including loose materials and caked-on mud in the gutter and street areas). The location of the subsample strip is carefully selected to ensure that it has no unusual loading conditions (e.g., a subsample should not be collected through the middle of a pile of leaves; rather, it is collected where the leaves are lying on the street in their normal distribution pattern). When possible, wet areas are avoided. If a sample is wet and the particles are caked around the intake nozzle, the caked mud from the gobbler is carefully scraped into the vacuum hose while the vacuum units are running. In addition, the hose needs to be struck against the ground at the end of the sampling period to knock loose any material stuck on the inside of the hose.

Subsamples are collected in a narrow strip about 6 in wide (the width of the gobbler) from one side of the street to the other (curb to curb). In heavily traveled streets where traffic is a problem, some subsamples consist of two separate one-half street strips (curb to crown). Traffic is not stopped for subsample collection; the operators wait for a suitable traffic break. On wide or busy roadways, a subsample is often collected from two strips several feet apart, halfway into the street. On busy roadways with no parking and good street surfaces, most particulates are found within a few feet of the curb, and a good subsample could be collected by vacuuming two strips adjacent to the curb

Figure 5.51 Street dirt subsample collection.

and as far into the traffic lanes as possible. Only a sufficient (and safe) break in traffic allows a subsample to be collected halfway across the street.

Subsamples taken in areas of heavy parking are collected between vehicles along the curb, as necessary. The sampling line across the street does not have to be a continuous line if a parked car blocks the most obvious and easiest subsample strip. A subsample can be collected in shorter (but very close) strips, provided the combined length of the strip is representative of different distances from the curb. Again, in all instances, each subsample must be representative of the overall curb-to-curb loading condition.

When sampling, the leading edge of the gobbler is slightly elevated above the street surface (0.125 in) to permit an adequate air flow and to collect pebbles and large particles. The gobbler is lifted further to accept larger material as necessary. If necessary, leaves in the subsample strip are manually removed and placed in the sample storage container to prevent the hose from clogging. If a noticeable decrease in sampling efficiency is observed, the vacuum hoses are cleaned immediately by disconnecting the hose lengths, cleaning out the connectors (placing the debris into the sample storage container), and reversing the air flows in the hoses (blowing them out by connecting the hose to the vacuum exhaust and directing the dislodged debris into the vacuum inlet). If any mud is caked on the street surface in the subsample strip, the sampler loosens it by scraping a shoe along the subsample path (being certain that street construction material is not removed from the subsample path unless it was very loose). Scraping caked-on mud is done after an initial vacuum pass. After scraping is completed, the strip is revacuumed. A rough street surface is sampled most easily by pulling (not pushing) the wand and gobbler toward the curb. Smooth and busy streets are usually sampled with a pushing action, away from the curb.

An important aspect of the sample collection is the speed at which the gobbler is moved across the street. A very rapid movement significantly decreases the amount of material collected; too slow a movement requires more time than is necessary. The correct movement rate depends on the roughness of the street and the amount of material on it. When sampling a street that has a heavy loading of particulates, or a rough surface, the wand needs to be pulled at a velocity of less than 1 ft/s. In areas of lower loading and smoother streets, the wand can be pushed at a velocity of 2 to 3 ft/s. The best indicators of the correct collection speed are achieved by visually examining how well the street is being cleaned in the sampling strip and by listening to the collected material rattle up the wand and through the vacuum hose. It is quite common to leave a visually cleaner strip on the street where the subsample was collected, even on streets that appeared to be clean before sampling.

In all cases, the hose tender must continuously watch traffic and alert the sampler of potentially hazardous conditions. In addition, the hose tender plays out the hose to the sampler as needed and keeps the hose as straight as possible to prevent kinking. If a kink develops, sampling is stopped until the hose tender straightens the hose. While working near the curb out of the traffic lane (typically an area of high loadings), the sampler visually monitors the performance of the vacuum sampler and periodically checks for vehicles. In the street, the sampler constantly watches traffic and monitors the collection process by listening to particles moving up the wand. A large break in traffic is required to collect dust and dirt from street cracks in the traffic lanes because the sampler has to watch the gobbler to make sure that all of the loose material in the cracks is removed.

When moving from one subsample location to another, the hose, wand, and gobbler need to be securely placed in the trailer. All subsamples are composited in the vacuums for each study area, and the hose must be placed away from the generator's hot muffler to prevent damage. The generator and vacuum units are left on and in the trailer during the entire subsample collection period. This helps dry damp samples and reduces the strain on the vacuum and generator motors.

The length of time it takes to collect all of the subsamples in an area varies with the number of subsamples and the test area road texture and traffic conditions. The number of subsamples required in each area can be determined using the experimental design sample effort equations described earlier in this chapter, with seasonal special sampling efforts to measure the variability of street dirt loadings in each area. The variabilities can be measured using a single, small 1.5-hp industrial vacuum, with a short hose to make sample collection simpler. The vacuum needs to be emptied, the sample collected and placed in individual Ziploc™ baggies, and weighed (later in the lab) for each individual sample to enable the variability in loadings to be measured. As an example, during the first phase of the San Jose, CA, study (Pitt 1979), the test areas required the following sampling effort:

Test Area	No. of Subsamples	Sampling Duration, h
Downtown — poor (rough) asphalt street surface	14	0.5
Downtown — good (smooth) asphalt street surface	35	1
Keyes Street — oil and screens street surface	10	0.5–1
Keyes Street — good asphalt street surface	36	1
Tropicana — good asphalt street surface	16	0.5–1

In the oil and screens test area, the sampling procedure was slightly different because of the relatively large amount of pea gravel (screens) that was removed from the street surface. The gobbler attachment was drawn across the street more slowly (at a rate of about 3 s/ft). Each subsample was collected by a half pass (from the crown to the curb of the street) and therefore contained one half of the normal sample. Two curb-to-curb passes were made for each Tropicana subsample because of the relatively low particulate loadings in this area, as several hundred grams of sample material are needed for the laboratory tests. In addition, an "after" street cleaning subsample is not collected from exactly the same location as the "before" street cleaning subsample (they need to be taken from the same general area, but at least a few feet apart).

A field data record sheet kept for each sample contains:

- Subsample numbers
- Dates and time of the collection period
- Any unusual conditions or sampling techniques

Subsample numbers are crossed off as each subsample is collected. After cleaning, subsample numbers are marked if the street cleaner operated next to the curb at that location. This differentiation enables the effect of parked cars on street cleaning performance to be analyzed. In addition, photographs (and movies) are periodically made to document the methods and street loading conditions.

Sample Transfer

After all subsamples for a test area are collected, the hose and Y connections are cleaned by disconnecting the hose lengths, reversing them, and holding them in front of the vacuum intake. Leaves and rocks that may have become caught are carefully removed and placed in the vacuum can; the generator is then turned off. The vacuums are either emptied at the last station or at a more convenient location (especially in a sheltered location out of the wind and sun).

To empty the vacuums, the top motor units are removed and placed out of the way of traffic. The vacuum units are then disconnected from the trailer and lifted out. The secondary, coarse vacuum filters are removed from the vacuum can and are carefully brushed with a small stiff brush into a large funnel placed in the storage can. The primary dacron filter bags are kept in the vacuum can and shaken carefully to knock off most of the filtered material. The dust inside the can is allowed to settle for a few minutes, then the primary filter is removed and brushed carefully into the sample can with the brush. Any dirt from the top part of the bag where it is bent over the top of the vacuum is also carefully removed and placed into the sample can. Respirators and eye protection are necessary to minimize exposure to the fine dust.

After the filters are removed and cleaned, one person picks up the vacuum can and pours it into the large funnel on top of the sample can, while the other person carefully brushes the inside of the vacuum can with a soft 3- to 4-in paintbrush to remove the collected sample. In order to prevent excessive dust losses, the emptying and brushing is done in areas protected from the wind. To prevent inhaling the sample dust, both the sampler and the hose tender wear mouth and nose dust filters while removing the samples from the vacuums.

To reassemble the vacuum cans, the primary dacron filter bag is inserted into the top of the vacuum can with the filter's elastic edge bent over the top of the can. The secondary, coarse filter is placed into the can and assembled on the trailer. The motor heads are then carefully replaced on the vacuum cans, making sure that the filters are on correctly and the excess electrical cord is wrapped around the handles of the vacuum units. The vacuum hoses and wand are attached so that the unit is ready for the next sample collection.

The sample storage cans are labeled with the date, the test area's name, and an indication of whether the sample was taken before or after the street cleaning test, or if it was an accumulation (or other type) of sample. Finally, the lids of the sample cans are taped shut and transported to the laboratory for logging-in, storage, and analysis.

Measurements of Street Dirt Accumulation

The washoff of street dirt and the effectiveness of street cleaning as a stormwater control practice are highly dependent on the street dirt loading. Street dirt loadings are the result of deposition and removal rates, plus "permanent storage." The permanent storage component is a function of street texture and condition and is the quantity of street dust and dirt that cannot be removed naturally or by street cleaning equipment. It is literally trapped in the texture, or cracks, of the street. The street dirt loading at any time is this initial permanent loading plus the accumulation amount corresponding to the exposure period, minus the resuspended material removed by wind and traffic-induced turbulence. Removal of street dirt can occur naturally by winds and rain, or by human activity (by the turbulence of traffic or by street cleaning equipment). Very little removal occurs by any process when the street dirt loadings are small, but wind removal may be very large with larger loadings, especially for smooth streets (Pitt 1979).

It takes many and frequent samples to ascertain the accumulation characteristics of street dirt. The studies briefly described in the following paragraphs typically involved collecting many hundreds of composite street dirt samples during the course of the 1- to 3-year projects from each study area. With each composite sample made up of about 10 to 35 subsamples, a great number of subsamples were used to obtain the data. Without high resolution (and effective) sampling, it is not possible to identify the variations in loadings and effects of rains and street cleaning.

The most important factors affecting the initial loading and maximum loading values are street pavement texture and street pavement condition. When data from many locations are studied, it is apparent that smooth streets have substantially smaller street dirt loadings at any accumulation period compared to rough streets for the same land use. Very long accumulation periods relative to the rain frequency result in high street dirt loadings. During these conditions, the losses of street dirt to wind (as fugitive dust) may approximate the deposition rate, resulting in relatively constant street dirt loadings. At Bellevue, WA, typical inter-event rain periods average about 3 days. Relatively constant street dirt loadings were observed in Bellevue because the frequent rains kept the loadings low and very close to the initial storage value, with little observed increase in dirt accumulation over time (Pitt 1985). In Castro Valley, CA, the rain inter-event periods were much longer (ranging from about 20 to 100 days) and steady street dirt loadings were only observed after about 30 days when the loadings became very high and fugitive dust losses caused by the winds and traffic turbulence moderated the loadings (Pitt and Shawley 1982).

An example of the type of sampling needed to obtain accumulation rate values was conducted by Pitt and McLean (1986) in Toronto. They measured street dirt accumulation rates and the effects of street cleaning as part of a comprehensive stormwater research project. An industrial street with heavy traffic and a residential street with light traffic were monitored about twice a week for 3 months. At the beginning of this period, intensive street cleaning (one pass per day for each of 3 consecutive days) was conducted to obtain reasonably clean streets. Street dirt loadings were then monitored every few days to measure the accumulation rates of street dirt. The street dirt sampling procedures previously described were used to clean many separate subsample strips across the roads, which were then combined for physical and chemical analyses.

In Toronto, the street dirt particulate loadings were quite high before the initial intensive street cleaning period and were reduced to their lowest observed levels immediately after the last street cleaning. After street cleaning, the loadings on the industrial street increased much faster than on the residential street. Right after intensive cleaning, the street dirt particle sizes were also similar for the two land uses. However, the loadings of larger particles on the industrial street increased at a much faster rate than on the residential street, indicating more erosion or tracking materials were deposited on the industrial street. The residential street dirt measurements did not indicate that any material was lost to the atmosphere as fugitive dust, likely due to the low street dirt accumulation rate and the short periods of time between rains. The street dirt loadings never had the opportunity to reach the high loading values needed before they could be blown from the streets by winds or by traffic-induced turbulence. The industrial street, in contrast, had a much greater street dirt accumulation rate and was able to reach the critical loading values needed for fugitive losses in the relatively short periods between the rains.

A street dirt sampling program must be conducted over a long enough period of time to obtain accumulation information. Infrequent observations hinder the analyses. It requires a continuous period of sampling, possibly with samples collected at least once a week, plus additional sampling close to the beginning and end of rains. Infrequent sampling, especially when interrupted by rains, does not allow changes in loadings to be determined. In addition, seasonal measurement periods are also likely needed because street dirt accumulation rates may change for different periods of the year. Infrequent and few samples may be useful to statistically describe the street dirt loading and to measure pollutant strengths associated with the samples, but they are not suitable for trend analyses. Chapter 7 presents statistical test procedures for identifying trends and should be consulted for different alternative methods to measure street dirt accumulation rates.

Small-Scale Washoff Tests

Washoff tests may be necessary to directly measure the energy available to dislodge and transport street dirt from paved areas to the drainage system. These tests are not usually conducted, as many rely on the process descriptions contained in commonly used stormwater models. Unfortunately, many of the process descriptions are in error due to improper interpretations of the test data. The following discussion therefore briefly describes these tests to encourage watershed researchers to obtain local data for accurate model calibration.

Observations of particulate washoff during controlled tests using actual streets and natural street dirt and debris are affected by street dirt distributions and armoring. The earliest controlled street dirt washoff experiments were conducted by Sartor and Boyd (1972) during the summer of 1970 in Bakersfield, CA. Their data were used in many stormwater models (including SWMM, Huber and Heaney 1981; STORM, COE 1975; and HSPF, Donigian and Crawford 1976) to estimate the percentage of the available particulates on the streets that would wash off during rains of different magnitudes. Sartor and Boyd used a rain simulator having many nozzles and a drop height of $1^{1}/_{2}$ to 2 m in street test areas of about 5 by 10 m. Tests were conducted on concrete, new asphalt, and old asphalt, using simulated rain intensities of about 5 and 20 mm/hour. They collected and analyzed runoff samples every 15 min for about 2 hours for each test. Sartor and Boyd fitted their data to an exponential curve, assuming that the rate of particle removal of a given size is proportional to the street dirt loading and the constant rain intensity:

$$dN/dt = krN$$

where: dN/dt = the change in street dirt loading per unit time
k = proportionality constant
r = rain intensity (in/hour)
N = street dirt loading (lb/curb-mile)

This equation, upon integration, becomes:

$$N = N_o e^{-krt}$$

where: N = residual street dirt load (after the rain)
N_o = initial street dirt load
t = rain duration

Street dirt washoff is therefore equal to N_o minus N. The variable combination rt, or rain intensity (in/h) times rain duration (h), is equal to total rain depth (R), in inches. This equation then further reduces to:

$$N = N_o e^{-kR}$$

Therefore, this equation is only sensitive to the total depth of the rain that has fallen since the beginning of the rain, and not rain intensity. Because of decreasing particulate supplies, the exponential washoff curve also predicts decreasing concentrations of particulates with time since the start of a constant rain (Alley 1980, 1981).

The proportionality constant, k, was found by Sartor and Boyd to be slightly dependent on street texture and condition, but was independent of rain intensity and particle size. The value of this constant is usually taken as 0.18/mm, assuming that 90% of the particulates will be washed from a paved surface in 1 hour during a 13 mm/hour rain. However, Alley (1981) fitted this model to watershed outfall runoff data and found that the constant varied for different storms and pollutants for a single study area. Novotny (as part of Bannerman et al. 1983) also examined "before" and "after" rain event street particulate loading data from the Milwaukee Nationwide Urban Runoff Program (NURP) project and found almost a threefold difference between the constant value of k for fine (<45 μm) and medium-sized particles (100 to 250 μm). The calculated values were 0.026/mm for the fine particles and 0.01/mm for the medium-sized particles, both much less than the "accepted" value of 0.18/mm. Jewell et al. (1980) also found large variations in outfall "fitted" constant values for different rains compared to the typical default value. Either the assumption of the high removal of particulates during the 13 mm/hour storm was incorrect or the equation cannot be fitted to outfall data (most likely, as this would require that all the particulates originate from homogeneous paved surfaces during all storm conditions).

This washoff equation has been used in many stormwater models, along with an expression for an availability factor. An availability factor is needed, as N_o is only the portion of the total street load available for washoff. This availability factor (the fraction of the total street dirt loading available for washoff) is generally used as 1.0 for all rain intensities greater than about 18 mm/hour and reduces to about 0.10 for rains of 1 mm/hour.

The Bellevue, WA, urban runoff project (Pitt 1985) included about 50 pairs of street dirt loading observations close to the beginnings and ends of rains. Very large reductions in street dirt loadings during rains were observed in Bellevue for the smallest particles, but the largest particles actually increased in loadings (due to deposited erosion materials originating from off-street areas). The particles were not source limited, but armor shielding may have been important. Most of the particulates in the runoff were in the fine particle sizes (<63 μm). Very few particles greater than 1000 μm were found in the washoff water. Care must be taken to not confuse street dirt particle size distributions with stormwater runoff particle size distributions. The stormwater particle size distributions are much more biased toward the smaller sizes, as described later.

Washoff tests can be designed to investigate several important factors and interactions that may affect washoff of different sized particulates from impervious areas (Pitt 1987):

- Street texture
- Street dirt loading
- Rain intensity
- Rain duration
- Rain volume

Multiple parameters that may affect a process can be effectively evaluated using factorial tests as described by Box et al. (1978) and earlier in this chapter. As an example, the tests conducted by Pitt (1987) were arranged as an overlapping series of 2^3 factorial tests, one for each particle size and rain total, and were analyzed using factorial test procedures. Nonlinear analyses were also used to identify a set of equations to describe the resulting curve shapes. The differences between available and total loads were also related to the experimental factors. This experimental setup can be effectively repeated elsewhere, with possible adjustments in the levels used in the experiments to reflect local conditions.

All tests were conducted for about 2 hours, with total rain volumes ranging from about 5 to 25 mm. The test code explanations follow:

Test Code	Rain Intensity	Street Dirt Loading	Street Texture
HCR	High	Clean	Rough
HDR	High	Dirty	Rough
LCR	Light	Clean	Rough
LDR	Light	Dirty	Rough
HCS	High	Clean	Smooth
HDS	High	Dirty	Smooth
LCS	Light	Clean	Smooth
LDS	Light	Dirty	Smooth

Unfortunately, the streets during the LDS (light rain intensity; dirty street; smooth texture) test were not as dirty as anticipated and actually replicated the LCS tests. The experimental analyses were modified to indicate these unanticipated duplicate observations.

A simple artificial rain simulator was constructed using 12 lengths of "soaker" hose, suspended on a wooden framework about 1 m above the road surface (Figures 5.52 and 5.53). "Rain" was applied by connecting the hoses to a manifold having individual valves to adjust constant

Figure 5.52 Washoff test site in Toronto.

Figure 5.53 Runoff collection area for Toronto washoff tests.

Figure 5.54 Sprinklers at freeway washoff test site in Austin, TX.

Figure 5.55 Sampler and rain gauge location at Austin freeway washoff test site.

rain intensities for the different areas. The manifold was in turn connected to a fire hydrant. The flow rate needed for each test was calculated based on the desired rain intensity and the area covered. The flow rates were carefully monitored by using a series of ball flow gauges before the manifold. The distributions of the test rains over the study areas were also monitored by placing about 20 small graduated cylinders over the area during the rains. In order to keep the drop sizes representative of sizes found during natural rains, the surface tension of the water drops hanging on the plastic soaker hoses was reduced by applying a light coating of Teflon spray to the hoses.

A different washoff test site is shown in Figures 5.54 through 5.56, where large sprinklers were located along the side of a freeway in Austin, TX. The sprinklers rained water directly onto the freeway during traffic conditions to better represent the combined effect of rain and auto-induced turbulence. Unfortunately, in order to get "rain" over a substantial area of the freeway, the "rain intensity" was extremely high, supplying much more energy than was typical, even for extreme events. In addition, this setup, while useful in obtaining hard-to-get data, may also have imposed an unusually high accident risk to freeway users (although large amounts of publicity, signage, and available alternate routes were all used to reduce this risk). This semipermanent installation was also used to monitor runoff from natural rains for comparison.

It was difficult to obtain even distributions of rain during the light rain tests in Toronto using the manifold, so a single hose was used that was manually moved back and forth over the test area during the smaller rain tests (three people took 30-min shifts). To keep evaporation reasonable for the rain conditions, the test sites were also shaded during sunny days. Blank water samples were also obtained from the manifold for background residue analyses. The filterable residue of the "rain" water (about 185 mg/L) could cause substantial errors when calculating washoff.

Figure 5.56 Sampler and flow monitoring equipment at Austin freeway washoff test site.

The areas studied were about 3 by 7 m each. The street side edges of the test areas were edged with plywood, about 30 cm in height and embedded in thick caulking, to direct the runoff toward the curbs with minimal leakage. All runoff was pumped continuously from downstream sumps (made of caulking and plastic sand bags) to graduated 1000-L Nalgene containers. The washoff samples were obtained from the pumped water going to the containers every 5 to 10 min at the beginning of the tests, and every 30 min near the end of the test. Final complete rinses of the test areas were also conducted (and sampled) at the tests' conclusions to determine total loadings of the monitored constituents.

The samples were analyzed for total residue, filtrate residue, and particulate residue. Runoff samples were also filtered through 0.4-μm filters and microscopically analyzed (using low power polarized light microscopes to differentiate between inorganic and organic debris) to determine particulate residue size distributions from about 1 to 500 μm. The runoff flow quantities were also carefully monitored to determine the magnitude of initial and total rainwater losses on impervious surfaces.

These tests are different from the important early Sartor and Boyd (1972) washoff experiments in the following ways:

- They were organized in overlapping factorial experimental designs to identify the most important main factors and interactions.
- Particle sizes were measured down to about 1 μm (in addition to particulate residue and filterable residue measurements).
- The precipitation intensities were lower in order to better represent actual rain conditions of the upper Midwest.
- Observations were made with more resolution at the beginning of the tests.
- Washoff flow rates were frequently measured.
- Emphasis was placed on total street loading, not just total available loading.
- Bacteria population measurements were also periodically obtained.

Sampling of Atmospheric Contributions

Atmospheric processes affecting urban runoff pollutants include dry dustfall and precipitation quality. These have been monitored in many urban and rural areas. In many instances, however, the samples were combined as a bulk precipitation sample before processing. Automatic precipitation sampling equipment can distinguish between dry periods of fallout and precipitation. These devices cover and uncover appropriate collection jars exposed to the atmosphere. Much of this information has been collected as part of the Nationwide Urban Runoff Program (NURP) and the Atmospheric Deposition Program, both sponsored by the U.S. Environmental Protection Agency (EPA 1983a).

One must be very careful in interpreting this information, however, because of the ability of many polluted dust and dirt particles to be resuspended and then redeposited within the urban area. In many cases, the atmospheric deposition measurements include material that previously resided and was measured in other urban runoff pollutant source areas. Also, only small amounts of the atmospheric deposition material would directly contribute to runoff. Rain is subjected to infiltration and the dry-fall particulates are most likely incorporated with surface soils and only small fractions are then eroded during rains. Therefore, mass balances and determinations of urban runoff deposition and accumulation from different source areas can be highly misleading, unless transfer of material between source areas and the effective yield of this material to the receiving water is considered. Depending on the land use, relatively little of the dustfall in urban areas likely contributes to stormwater discharges. The major exception would be dustfall directly on receiving waters.

Dustfall and precipitation affect all of the major urban runoff source areas in an urban area. Dustfall, is typically not a major pollutant source, but fugitive dust is mostly a mechanism for pollutant transport. Most of the dustfall monitored in an urban area is resuspended particulate matter from street surfaces or wind erosion products from vacant areas (Pitt 1979). Point source pollutant emissions can also significantly contribute to dustfall pollution, especially in industrial areas. Transported dust from regional agricultural activities can also significantly affect urban stormwater.

Wind-transported materials are commonly called "dustfall." Dustfall is normally measured by collecting dry samples, excluding rainfall and snowfall. If rainout and washout are included, one has a measure of total atmospheric fallout. This total atmospheric fallout is sometimes called "bulk precipitation." Rainout removes contaminants from the atmosphere by condensation processes in clouds, while washout is the removal of contaminants by the falling rain. Therefore, precipitation can include natural contamination associated with condensation nuclei in addition to collecting atmospheric pollutants as the rain- or snowfalls. In some areas, the contaminant contribution by dry deposition is small, compared to the contribution by precipitation (Malmquist 1978). However, in heavily urbanized areas, dustfall can contribute more of an annual load than the wet precipitation, especially when dustfall includes resuspended materials.

Much of the monitored atmospheric dustfall and precipitation would not reach the urban runoff receiving waters. The percentage of dry atmospheric deposition retained in a rural watershed was extensively monitored and modeled in Oakridge, TN (Barkdoll et al. 1977). They found that about 98% of the lead in dry atmospheric deposits was retained in the watershed, along with about 95% of the cadmium, 85% of the copper, 60% of the chromium and magnesium, and 75% of the zinc and mercury. Therefore, if the dry deposition rates were added directly to the yields from other urban runoff pollutant sources, the resultant urban runoff loads would be very much overestimated.

Rubin (1976) stated that resuspended urban particulates are returned to the earth's surface and waters in four main ways: gravitational settling, impaction, precipitation, and washout. Gravitational settling, as dry deposition, returns most of the particles. This not only involves the settling of relatively large fly ash and soil particles, but also the settling of smaller particles that collide and coagulate. Rubin stated that particles that are less than 0.1 µm in diameter move randomly in the air and collide often with other particles. These small particles can grow rapidly by this coagulation process. They would soon be totally depleted in the air if they were not constantly replenished. Particles in the 0.1 to 1.0 µm range are also removed primarily by coagulation. These larger particles grow more slowly than the smaller particles because they move less rapidly in the air, are somewhat less numerous, and, therefore, collide less often with other particles. Particles with diameters larger than 1 µm have appreciable settling velocities. Those particles about 10 µm in diameter can settle rapidly, although they can be kept airborne for extended periods and for long distances by atmospheric turbulence.

The second important particulate removal process is impaction. Impaction of particles near the earth's surface can occur on vegetation, rocks, and building surfaces. The third form of particulate removal from the atmosphere is precipitation, in the form of rain and snow. This is caused by the rainout process in which the particulates are removed in the cloud-forming process. The fourth important removal process is washout of the particulates below the clouds during the precipitation event. Therefore, it is easy to see that reentrained particles (especially from street surfaces, other paved surfaces, rooftops, and from soil erosion) in urban areas can be readily redeposited through these various processes, either close to the points of origin, or some distance away.

Pitt (1979) monitored airborne concentrations of particulates near typical urban roads using Climat Particle Counters (Figure 5.57). He found that on a particle count basis, the downwind roadside particulate concentrations were about 10% greater than upwind conditions. About 80% of the concentration increases, by particle count, were associated with particles in the 0.5 to 1.0 µm range. However, about 90% of the particle concentration increases by weight were associated with particles greater than 10 µm. He found that the rate of particulate resuspension from street surfaces increases when the streets are dirty (cleaned infrequently) and varied widely for different street and traffic conditions. The resuspension rates were calculated based upon observed long-term accumulation conditions on street surfaces for many different study area conditions, and varied from about 0.30 to 3.6 kg/curb-km (1 to 12 lb/curb-mile) of street per day.

Figure 5.57 Hi-vol suspended particulate sampler, along with particle counters and wind velocity meters used to measure fugitive dust losses caused by traffic-induced turbulence and dirty roads in San Jose, CA, tests.

Murphy (1975) described a Chicago study in which airborne particulate material within the city was microscopically examined, along with street surface particulates. The particulates from both of these areas were found to be similar (mostly limestone and quartz) indicating that the airborne particulates were most likely resuspended street surface particulates, or were from the same source. PEDCo (1977) found that the reentrained portion of the traffic-related particulate emissions (by weight) is an order of magnitude greater than the direct emissions accounted for by vehicle exhaust and tire wear. They also found that particulate resuspensions from a street are directly proportional to the traffic volume and that the suspended particulate concentrations near the streets are associated with relatively large particle sizes. The medium particle size found, by weight, was about 15 μm, with about 22% of the particulates occurring at sizes greater than 30 μm. These relatively large particle sizes resulted in substantial particulate fallout near the road. They found that about 15% of the resuspended particulates fall out within 10 m, 25% within 20 m, and 35% within 30 m from the street (by weight). In a similar study Cowherd et al. (1977) reported a wind erosion threshold value of about 5.8 m/s (13 mph). At this wind speed, or greater, significant dust and dirt losses from the road surface could result, even in the absence of traffic-induced turbulence. Rolfe and Reinbold (1977) also found that most of the particulate lead from automobile emissions settled out within 100 m of roads. However, the automobile lead does widely disperse over a large area. They found, through multielemental analyses, that the settled outdoor dust collected at or near the curb was contaminated by automobile activity and originated from the streets.

The experimental design and interpretation of atmospheric contributions must therefore be done carefully. Measurements can be obtained using numerous procedures, as summarized below:

- Conventional air pollution monitoring equipment, especially hi-vol samplers for particulates. The captured particulates can be chemically analyzed for pollutants, especially heavy metals.
- Real-time air pollution monitoring equipment, such as nephelometers and particle counters (Figure 5.57). These are especially useful for short-term measurements of resuspended particulates from nearby pavements to indicate turbulence effects from vehicles or natural winds. They are also useful for fugitive dust measurements from construction sites and can also be used to indicate the effects of vehicular traffic and wind losses from construction roads, etc.
- Sticky paper fugitive dust samplers. These are simple upright cylinders about 10 cm in diameter and 20 cm in height that are carefully oriented to enable moderate- or long-term measurements of fugitive dust losses from specific directions. Simple measurements are made by comparing the color and tone of the exposed paper for different exposed directions to standards. The exposed

Figure 5.58 Wet-dry atmospheric deposition sampler in Bellevue, WA.

Figure 5.59 Large surface area used to capture sufficient rain for chemical analyses in early San Jose, CA, tests.

paper can also be examined under a microscope for more specific measurements and identification of particle characteristics.

- Wet- and dry-fall automatic samplers (Figure 5.58). These were commonly used during the EPA's NURP and Atmospheric Deposition Program and allow long-term sampling of dustfall during dry weather and rainwater during wet weather. A lid, connected to a moisture sensor, automatically moves to cover the appropriate sampling bucket. The collected samples are rinsed from the appropriate buckets after the desired exposure periods and chemically analyzed. If a single bucket sampler is used (without the automatic lid), then the dry dustfall and the rainwater samples are combined in one sample for a bulk precipitation analysis. Evaporation of the rainwater sample and obvious chemical transformations occur in these samplers during the typically long-term exposures. These samplers are therefore most useful for evaluations for stable compounds (such as suspended solids and most heavy metals) and are not very suitable for nutrient, bacteria, or organic analyses.
- Precipitation sampler. Because rainwater has little buffer capacity, short-term collections of rainwater are needed for many constituents (especially major ions, pH, and nutrients). However, in order to collect sufficient sample volume in a short period, a large collection area is needed. One simple solution is to construct a large collection area using a plastic tarp supported around its edges (Figure 5.59). The tarp is allowed to sag toward the center, where a weight surrounds a central hole that is located over an appropriate sample bottle. A tarp having about a 10 m^2 surface area can collect several liters of rainwater in a few minutes during a relatively light rainfall. Of course, potential contamination of the sample is possible through the use of the tarp. For a semipermanent installation, it would be possible to construct a relatively large collection area using a piece of glass (being careful of joint materials), or a Teflon-coated surface could be used with fewer interferences than a plastic surface. See the earlier discussion on sample contamination potential from various materials. Many laboratory suppliers sell Teflon-coated sticky paper that is used for covering laboratory benches. It may be possible to use this material to cover a simple seamless rigid platform, having a central trough for rainwater collection.

SEDIMENT AND PORE WATER SAMPLING

Sediment Sampling Procedures

As discussed previously, sediments act as sinks and sources of contaminants and have been implicated as the cause of beneficial use impairments, such as fish consumption advisories, at numerous sites throughout North America. Sediments that should be targeted as potential problem sources during any receiving water assessment are the small-grained, depositional-type sediments in urban, industrial, and agricultural drainages. Stormwater discharges can cause metal and organic chemicals, nutrients, and pathogens to accumulate in depositional sediments. These contaminants then may enter groundwater or reenter surface waters for further transport, or contaminate resident organisms and the overlying food web (see also Chapter 6). Once stormwater flows subside, the influence of contaminated sediments on overlying water persists and even increases during low flow conditions. Even though the short-term BOD of stormwater is not very high (BOD_5 of about 25 mg/L), the long-term BOD (BOD_{90} of about 250 mg/L) is high and resulting accumulations of organic debris in urban streams create anaerobic sediment conditions (Pitt 1979). These depositional sediments will continue to degrade in quality as long as organic and contaminant loadings continue, resulting in replacement with pollution-tolerant benthic macroinvertebrates, such as midges and worms, and also degrade the fish community (Burton and Scott 1992). Assessing the role of sediments in beneficial use attainment and ecosystem health is a necessary aspect of a receiving water investigation. As noted previously, heavy metals and nutrient and organic toxicants are of most interest in urban stream sediments while nutrients and pesticides are of primary concern in agricultural waterways. Pathogens may be a problem in either urban or agricultural watersheds. Contaminated stream sediments likely impart the most important impairments to aquatic life in urban areas (after direct habitat destruction and frequent high flows) and may also in agricultural areas. Collecting and analyzing these sediments and their biota are therefore necessary to establish water quality and the sources of any degradation.

In many ways, sampling and evaluating the quality of sediments is more difficult than water quality sampling. Though sediments vary less than waters on a temporal basis, they exhibit greater variation spatially, in a complex, semisolid, three-dimensional structure. Understanding and preserving this structure has tremendous ramifications in the assessment process. The surficial sediment layers that interface with overlying waters are the most dynamic and recent sediments, subject to resuspension and downstream deposition, oxidation, and rapid changes in quality based on overlying water conditions. As sediment depth increases, the biological communities and chemical conditions may change orders of magnitude over a millimeter to centimeter scale. This has been observed in oxygen-redox vertical gradients (Carlton and Klug 1990) and toxicity (horizontally and vertically) (Stemmer et al. 1990b). In addition to the high degree of heterogeneity often observed, maintaining sediment structure integrity is crucial when attempting to characterize the sample based on physical (e.g., redox potential, percent fines), chemical (e.g., metal speciation, nutrient concentration and speciation, volatile components), biological (e.g., biotransformations, microbial-meiofaunal communities), and toxicity (e.g., contaminant bioavailability) characteristics (ASTM 1991b; Burton 1992b). Maintaining complete sediment integrity is nearly impossible since the very process of sample collection is disruptive (Figure 5.60). There are effective methods, however, by which to reduce this disruption (see also Chapter 6). The importance of maintaining sample integrity depends on the type of problem and the data quality objectives (DQOs) of the study. Several guidance documents exist that address sediment sampling in detail. The most comprehensive and current guidance documents to date include ASTM 1994 and EPA 2001.

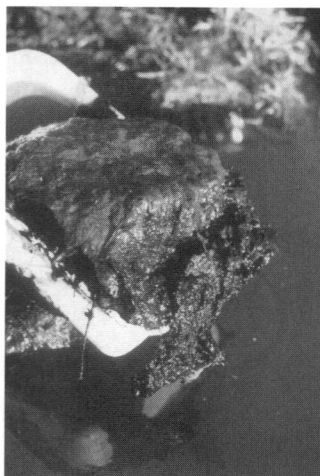

Figure 5.60 The fine-grained and muddy nature of most urban sediments requires specific sediment sampling procedures.

Disrupting the sensitive sediment environment is a major concern when collecting samples for toxicity studies, since the bioavailability and resulting toxicity can change significantly when in-place sediments are disturbed. An additional major concern is that the sediment depth sampled and chemically analyzed matches that being assessed for organism exposure (indigenous organisms and/or toxicity and bioaccumulation using surrogate species). Too often sediment grab samples are collected at unknown sediment depths (0 to 30 cm). The sediments are homogenized and then subsampled for chemical and physical analyses. Contaminant peaks occurring near the surface or deeper in the sediments may be diluted via the mixing process and then compared to biological effects. Resident benthic organisms are likely not being exposed to the same chemicals or concentrations that result from this process. In addition, laboratory toxicity testing will yield results that may bear little resemblance to field conditions. Therefore, it is best to establish whether recent or historical contamination is a concern, sample the appropriate sediment depth, and match the chemical analyses with realistic organism exposures.

A number of sampling-related factors can contribute to loss of the sediment sample's original characteristics, including sampler-induced pressure waves, washout of fine-grained sediments during retrieval, compaction due to sampler wall friction, sampling vessel or person-induced disturbance of surficial layers, disruption during subsampling or transport, oxidation, and temperature alterations. While it is impossible to remove all of these factors from routine assessments, reducing their influence increases the certainty that the data generated and resulting weight-of-evidence conclusions will be reliable.

Choosing the most appropriate sediment sampler for a study will depend on the sediment's characteristics, the volume and efficiency required, and the study's objective (Tables 5.16 through 5.18; Figures 5.61 through 5.63). Numerous sediment samplers are available. Two general categories include core samplers (which can obtain samples that can be analyzed by depth) and surface grab samplers (which only collect surface sediment). ASTM (1995) standard 4823 contains much information concerning core sampling in unconsolidated sediments that is applicable to urban streams. ASTM standard E 1391 also presents additional useful information concerning the sampling of sediment for toxicological testing. The preferred sampling method is to use core samplers whenever possible. However, they collect relatively little sediment and represent only a very small area. In addition, it may be difficult to retain samples in the samplers for retrieval in some types of bottom conditions (especially sandy sediment).

Grab samplers only collect samples from the surface layers of the sediment (10 to 50 cm in depth, at maximum). They also greatly disturb the sediment that is being sampled. Common problems include shallow depth of penetration and presence of a shock wave that results in loss of the fine surface sediments. However, they are much easier to use than corers under a wide variety of conditions. A common grab sampler is the Ponar sampler (Figures 5.64 through 5.67). It comes in a standard size and a "petite" size that weighs substantially less and is more practical for urban streams. The Ponar sampler is useful for sand, silt, and clay sediments and can be used in relatively deep water or shallow waters. It has a flexible cover over a top screen that helps to minimize the loss of fines during sampling. Forestry Suppliers, Inc. (800-543-4203) sells a petite 6" × 6" Wildco Ponar bottom dredge (catalog #77250 for about $450) and a larger 9" × 9" Wildco Ponar bottom dredge (catalog #77249 for about $800). The Peterson grab sampler is similar to the Ponar, but doesn't have a screened top plate. It is heavy and is more suitable for deeper water and harder clay bottoms than the Ponar sampler. Because of its weight, it requires the use of a winch. Cole Parmer (800-323-4340) sells a Peterson dredge sampler (catalog #H-05472-00 for about $1000). An Ekman sampler is also commonly used in small urban streams and ponds, but is limited to sampling soft bottoms. Forestry Suppliers, Inc. sells a light 6" × 6" Wildco–Ekman bottom dredge (catalog #77251 for about $350, including line, messenger, and case). Cole Parmer also sells a larger 9" × 9" Ekman dredge (catalog #H-05470-10 for about $600).

Dredge samplers that quantitatively sample surface sediments have been described (Grizzle and Stegner 1985). The depth profile of the sample may be lost in the removal of the sample from the

Table 5.16 Popular Sediment Samplers: Strengths and Weaknesses

Sampler	Strengths	Weaknesses
Core Samplers		
Hand and gravity corers 0–30 cm depth 0.1–1.5 L volume	Maintains sediment layering of inner core. Fine surficial sediments retained. Replicate samples efficiently obtained. Removable liners. Inert liners may be used. Quantitative sampling allowed.	Small sample volume. Liner removal required for repetitive sampling. Not suitable in large-grain or consolidated sediments. Spillage possible.
Freeze core sampler 0–1 m depth 1 L volume	Maintains sediment layering of core. Fine sediments retained. Replicates samples efficiently obtained. Can be made of inert materials.	Small sample volume. Freezing may disturb sediment. Uses liquid CO_2 or dry ice for collecting sample. Requires several minutes to obtain each sample. May not collect large material. Not suitable for consolidated sediments.
Box corer 0–50 cm depth 1–30 L volume	Maintains sediment layering of large volume of sediment. Surficial fines retained relatively well. Quantitative sampling allowed.	Size and weight require power winch, difficult to handle and transport. Not suitable in consolidated sediments.
Vibratory corers 3–6 m depth 6–13 L volume	Samples deep sediments for historical analyses. Samples consolidated sediments. Minimal disturbance. May be used on small vessels.	Expensive and requires winch. Outer core integrity slightly disrupted.
Grab Samplers		
Ekman or box dredge 0–10 cm depth Up to 3.5 L volume	Relatively large volume may be obtained. May be subsampled through lid. Lid design reduces loss of surficial sediments as compared to many dredges. Usable in moderately compacted sediments of varying grain sizes.	Loss of fines may occur during sampling. Incomplete jaw closure occurs in large-grain sediments or with large debris. Sediment integrity disrupted. Not an inert surface.
Ponar 0–10 cm depth Up to 1 L volume (petite) Up to 7.5 volume (standard)	Commonly used. Large volume obtained. Adequate on most substrates. Weight allows use in deep waters.	Loss of fines and sediment integrity occurs. Incomplete jaw closure occurs occasionally. Not an inert surface.
Van Veen or Young Grab 0–30 cm depth Up to 75 L volume	Useful in deep waters and on most substrates. Young grab coated with inert polymer. Large volume obtained.	Loss of fines and sediment integrity occurs. Incomplete jaw closure possible. Van Veen has metal surface. Young is expensive. Both may require winch.
Peterson 0–30 cm depth Up to 9.5 L volume	Large volume obtained from most substrates in deep waters.	Loss of fines and sediment integrity. Not an inert surface. Incomplete jaw closure may occur. May require winch.
Orange-Peel 0–30 cm depth 10–20 L volume	Large volume obtained from most substrates. Efficient closure.	Loss of fines and sediment integrity. Not an inert surface. Requires winch.
Shipek 0–10 cm depth Up to 3 L volume	Adequate on most substrates.	Small volume. Loss of fines and sediment integrity. Not an inert surface.

Modified from ASTM (American Society for Testing and Materials). *Standard Guide for Collection, Storage, Characterization, and Manipulation of Sediments for Toxicological Testing.* American Society for Testing and Materials, Philadelphia, Standard E 1391. 1991.

Table 5.17 Sediment and Interstitial Water Sampler Selection Guidelines

1. Sediment grain size effects on sampler selection
   - Silt-clay = core, grab, or peeper*
   - Sand = grab or peeper
   - Cobble = peeper
2. Sediment compacted: powered core
3. Sediment vertical gradient must be maintained: core or peepers
4. Sediment volumes
     Large volumes over small vertical gradients: dredge
     Small to moderate volumes: dredge, core, or peeper
5. Optimal samplers, in order of maintaining original sediment characteristics:
   1. *In situ* peeper*
   2. *In situ* suction*
   3. Core
   4. Grab
   5. Dredge
6. Optimal methods of collecting interstitial water (in order of preference, see Table 5.18)
   1. *In situ* peepers
   2. *In situ* suction (airstone or core-port)
   3. Centrifugation @ 10,000 $\times$ g (4°C) (without subsequent filtration)
   4. Centrifugation @ lower speeds
   5. Basal cup
   6. Squeezing or pressurization
   7. Suction or filtration

* For interstitial water collection only.

sampler. Dredge sampling promotes loss of not only fine sediments, but also water-soluble compounds and volatile organic compounds present in the sediment (ASTM 1991a). A comparison of sampler precision for macrobenthic purposes showed the Van Veen sampler to be the least precise; the most precise were the corers and Ekman dredge (Figures 5.68 and 5.69). The Smith–McIntyre and Van Veen samplers are more commonly used in marine studies, due to their weight. Shipek samplers are also used in marine investigations but may lose the top 2 to 3 cm of sediment fines from washout (Mudroch and MacKnight 1991).

Many of the problems associated with dredge samplers are largely overcome with the corers. The best corers for most sediment studies are hand-held polytetrafluoroethylene plastic, high-density polyethylene, or glass corers (liners), or large box corers. Corer samplers can penetrate the sediment by several meters, but that is rarely necessary (or possible) in urban receiving water studies. Their most important advantage is that samples collected by corers can be separated by depth for analyses. However, conventional corer samplers are difficult to use in the highly variable bottom sediment conditions commonly found in urban streams. The freezing core samplers, described later, overcome many of the sample loss and disturbance problems associated with conventional corers.

If used correctly, box corers can maintain the integrity of the sediment surface while collecting a sufficient depth for most toxicity studies. Conventional gravity corers may compress the sediment as evidenced by altered pore water alkalinity gradients, and box coring was superior for studies of *in situ* gradients (Lebel et al. 1982). The box core can be subcored or sectioned at specific depth intervals, as required by the study. Unfortunately, the box corer is large and cumbersome; thus, it is difficult to use and usually requires a lift capacity of 2000 to 3000 kg. Box cores typically require fine-grained sediments of at least a 30 cm depth. Other coring devices that have been used successfully include the percussion corer (Gilbert and Glew 1985), vibratory corers (Imperato 1987; Figure 5.70), and freeze corers (Pitt 1979; Spliethoff and Hemond 1996; Figures 5.71 and 5.72).

When only chemical testing is to be conducted (that is, not toxicity testing), a useful type of corer sampler is the freezing core sampler. Sediments to be used for SOD, BOD, or toxicity

Table 5.18 Optimal Interstitial Water Collection Methods

Device	Sediment Depth (cm)	Volume (cm³)	Advantages	Disadvantages
Peeper	0.2–10	1–500	Most accurate method, reduced artifacts, no lab processing; relatively free of temperature, oxidation, and pressure effects; inexpensive and easy to construct; some selectivity possible on nature of sample via specific membranes, wide range of membrane/mesh pore sizes, and/or internal solutes or substrates.	Deployment easiest by hand. in >0.6-m depth waters; allow hours to days for equilibration, which will vary with site and chamber; methods not standardized and used infrequently; some membranes such as dialysis/cellulose are subject to biofouling; must deoxygenate chamber and materials to prevent oxidation effects; some chambers only allow small sample volumes; care must be used on collection to prevent sample oxidation.
In situ suction	0.2–30	1–250	Reduced artifacts, gradient definition; shallow water (<60 m) air stone method ease; core method deployment may not require diving in deep water, rapid collection, no lab processing; closed system possible which prevents contamination; methods include air stone, syringes, probes, and cores.	Requires custom, nonstandard collection devices; small volumes; limited to softer sediments; core method may require diving for waters; methods used infrequently and by limited numbers of laboratories.
Centrifugation — Sampler dependent			Most accurate of lab processing methods; allows anoxic/cold processing; large volumes; commonly used.	Some chemical loss/alteration; results depend on centrifugation conditions; requires high-speed centrifuge; difficult with sandy sediments.
Suction — Sampler dependent			Use with all sediment types; may process in field; large volumes possible with some sediments; closed system possible.	Alteration of chemical characteristics may occur; increased loss of metals and organics; loss of vertical gradient resolution.
Squeezing — Sampler dependent			Use with all sediment types; may process in field; large volumes possible with some.	Alteration of chemical characteristics may occur; increased loss of metals and organics; loss of vertical gradient resolution sediments.

Note: Incorporation of filtration into any of the collection methods may result in loss of metal and organic compounds.

testing should not be frozen, as the bioavailability of nutrients and toxicants is altered. All of the freezing core samplers rely on CO_2 (either as a liquid or a solid — dry ice). The use of CO_2 must be carefully evaluated and minimized in consideration of its role as a greenhouse gas. Pitt (1979) devised a freezing core sampler to collect profiles in sandy deposits of catchbasins that would also work well in shallow streams. This sampler was a 19-mm-diameter stainless steel tube, with a stainless steel point attached to one end. This was pushed into the sediment. A length of flexible 6 mm copper tubing was then inserted into the free end of the stainless probe (which is above the water depth), extending to the bottom of the stainless probe. The other end of the copper tubing was attached to a high-pressure hose and to a valve on a CO_2 fire extinguisher. The fire extinguisher was modified with a valve in place of the standard squeeze release, and

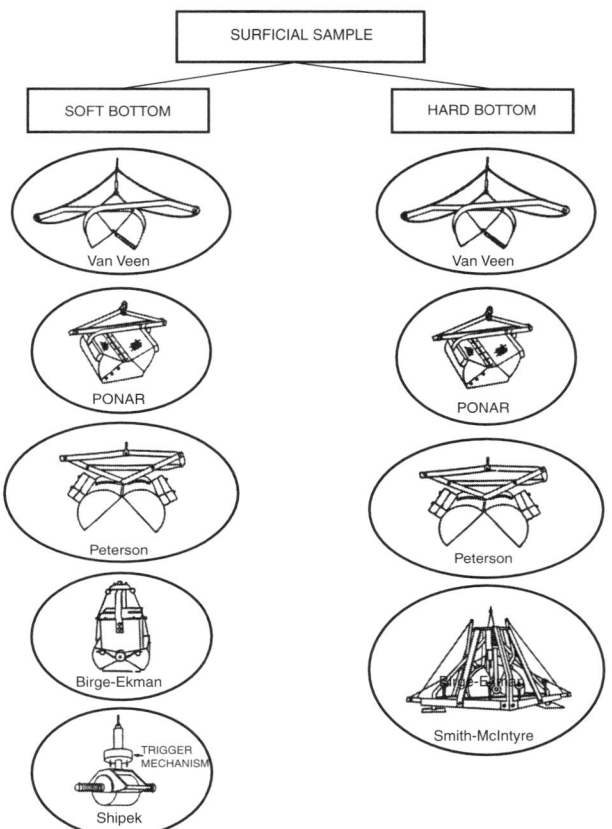

Figure 5.61 Some recommended devices for collecting surficial sediments. (From EPA. *Methods for Collection, Storage and Manipulation of Sediments for Chemical and Toxicological Analyses.* Office of Water. U.S. Environmental Protection Agency. Washington, D.C. In press.)

with an internal "delivery" tube that extended to the bottom of the fire extinguisher. This enabled liquid CO_2 to be delivered to the probe sampler, instead of gaseous CO_2 from the top of the fire extinguisher tank (the fire extinguisher is kept upright during operation). The valve was opened slightly and a continuous flow of CO_2 was delivered to the stainless steel probe (Figure 5.71). Care must be taken to turn off the flow of CO_2 at the fire extinguisher if it appears that a jam has occurred inside the probe (such as from ice forming due to water inside the probe sampler). The vaporization of the liquid CO_2 quickly chills the probe and freezes the sediment sample to the outside of the tube. In operation, the CO_2 is allowed to flow for about 1 min, but this can be changed depending on specific conditions and desired sample thickness. The probe is then removed from the sediment (with the sediment frozen to the outside) after the CO_2 flow is terminated and the copper tube is withdrawn. The probe with frozen sample is then laid on a stainless steel tray and the sample is removed by section and bottled separately, according to desired depth. A flame torch can be used to gently heat the probe uncovered by sample to allow the easier removal of the sample. It may be difficult to separate the sample into precise segments unless the sample is allowed to warm slightly first.

Another version of a freezing core sampler suitable for deeper water use was described by Spliethoff and Hemond (1996). They developed two versions of core samplers using dry ice within a probe that was used to measure the history of heavy metal contamination in an urban lake. One sampler (Figure 5.72) was made of a 96-cm length of 7.6-cm-diameter aluminum tubing. The bottom half of the tube was cut away lengthwise, and a flat aluminum plate was welded to act as a freezing surface. Stabilizing fins were also attached, along with weights to control penetration. PVC was also used to insulate the sampler where sample was not wanted. The sampler nose piece

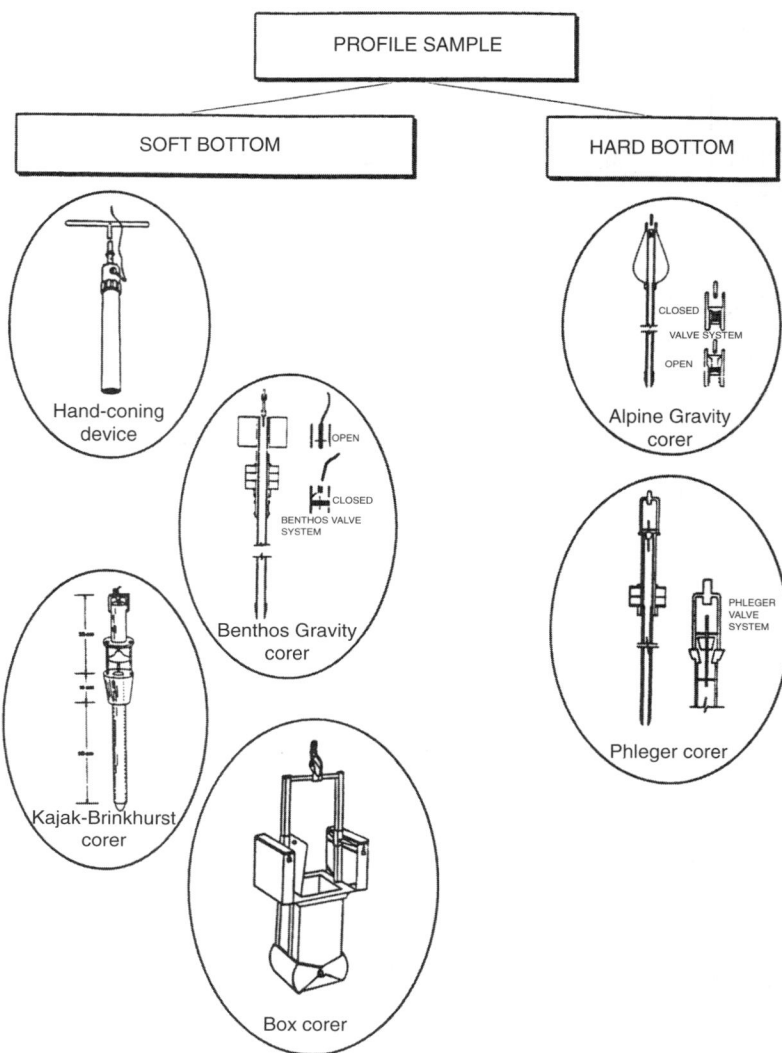

Figure 5.62 Some recommended devices for obtaining sediment profiles. (From EPA. *Methods for Collection, Storage and Manipulation of Sediments for Chemical and Toxicological Analyses*. Office of Water. U.S. Environmental Protection Agency. Washington, D.C. In press.)

Figure 5.63 Gravity and hand corers.

Figure 5.64 Petite Ponar dredge.

Figure 5.65 Petite Ponar sediment dredge being lifted from water after sampling.

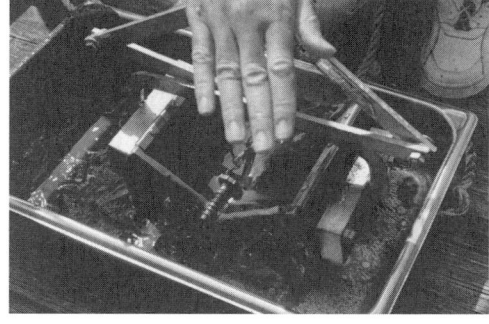

Figure 5.66 Emptying Ponar sample into stainless steel sample pan.

Figure 5.67 Winch with Ponar dredge.

Figure 5.68 Hand-held corer and Ekman dredge.

Figure 5.69 Collecting sediment with an Ekman dredge.

Figure 5.70 Shallow water vibratory core collection.

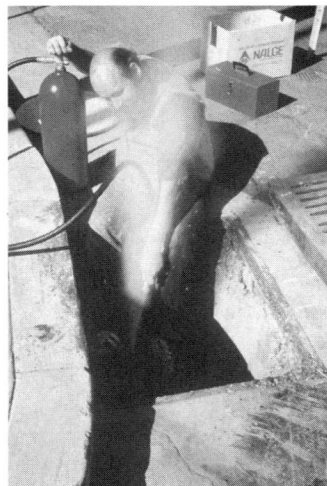

Figure 5.71 Freezing core sampler venting CO_2 used to sample catchbasin sediment in San Jose, CA.

was of solid aluminum. A screw cap was fitted to the other end which had a vent hole drilled in it. Another sampler was also constructed by Spliethoff and Hemond that allowed longer samples to be obtained (also in Figure 5.72). This sampler was made using a 125-cm length of 7.6-cm-square Extren tubing (a fiberglass reinforced resin). One side of the square tubing was machined off and an aluminum plate was attached to act as a freezing surface. A point-shaped lead weight was attached to one end and a cap with gas relief valve was attached to the other end. They used a slurry of dry ice and denatured ethanol to act as a coolant in both samplers. The samplers were dropped from the lake surface to test the penetration depth. The samplers were then retrieved, filled with the coolant mixture, and dropped again. After about 15 min, the CO_2 bubbles reaching the lake surface subsided, and the corers were retrieved. The samplers were then cleaned of unfrozen sediment and filled with warm lake water to help in releasing the frozen sample from the sampler. The frozen samples were sealed in plastic wrap and transported to the lab in dry ice filled coolers where they were separated into segments for analysis.

The above described freezing core samplers result in relatively undisturbed cores for analyses; plus they enable effective sampling in conditions where sample retention using conventional core samplers is difficult (unconsolidated coarse-textured sediment).

Figure 5.72 Freezing core samplers. (From Spliethoff, H.M. and H.F. Hemond. History of toxic metal discharge to surface waters of the Aberjona watershed. *Environ. Sci. Tech.*, 30(1): 121. January 1996. Copyright 1995 American Chemical Society. Reprinted with permission.)

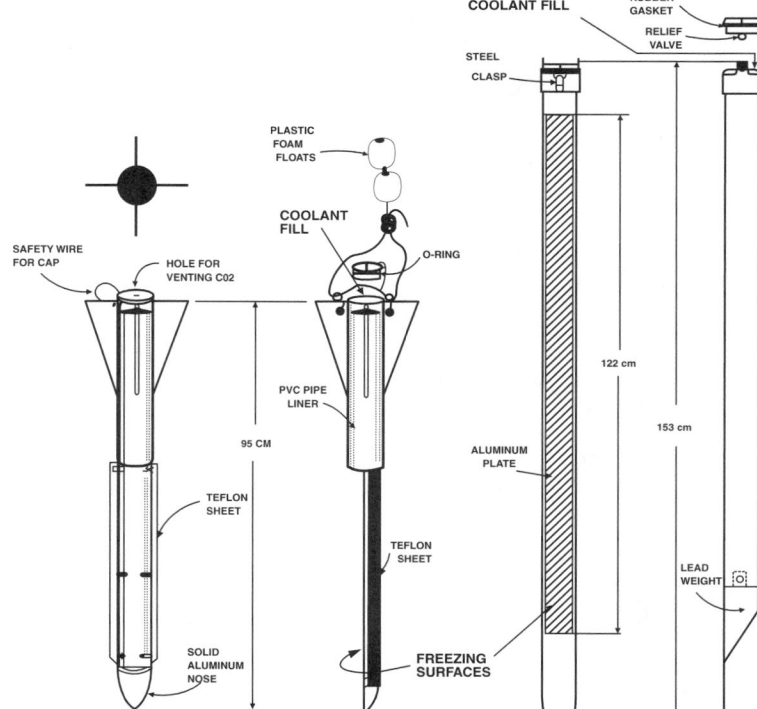

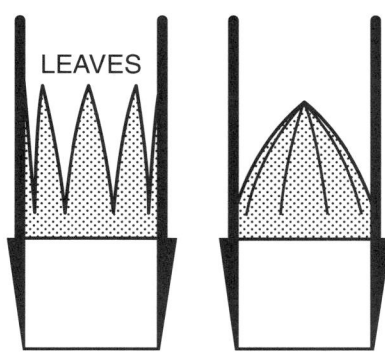

Figure 5.73 Leaf core catcher. (From American Society for Testing and Materials). *ASTM Standards on Environmental Sampling.* ASTM Pub Code No. 03-418095-38. ASTM, Philadelphia. 1995. Copyright ASTM. Reprinted with permission.)

ASTM (1995) in standard D 4823 describes many other types of core samplers. The most common sampler is the open tube sampler with a core catcher. This sampler is commonly used in shallow waters where it is manually pushed into the sediment. When the desired penetration depth is reached, the sampler is carefully withdrawn. A leaf core catcher is commonly used to help retain the sample in the corer (Figure 5.73). The leaves separate and fold against the inside walls of the sampler when the corer penetrates the sediment. The leaves fold closed when the sampler is withdrawn, holding the sample in the corer. Plastic liners are also commonly used inside the sampler, simplifying core removal from the corer. The liners usually have plastic end caps that can be placed on the liner ends, holding the cores inside until analyses. These conventional core samplers are most effective with clayey sediments. Sandy sediments tend to easily wash out of most corers upon retrieval, irrespective of the core catcher used. ASTM (1995) mentions excavating around a core sampler and sliding a flat plate under the bottom of the corer before retrieval in shallow water to capture most of the sample. Forestry Suppliers, Inc. sells the Wildco hand core sediment sampler that is 2" in diameter and 20" long, made of stainless steel with a plastic core liner tube and eggshell catcher (catalog #77258 for about $340). Extra plastic liners are also available (catalog # 77260) for about $12 each. They also sell stainless steel liners and core catchers (catalog #77303 for the stainless steel liner for about $70 each and catalog #77304 for the stainless steel eggshell sample catcher for about $40 each).

Corer samplers also have limitations in some situations (ASTM 1991a). Most corers do not work well in sandy sediments or in extremely soft (high water content) sediments; dredge samplers or diver-collected material remain the only current alternatives. In general, corers collect less sediment than dredge samplers that may provide inadequate quantities for some toxicity studies. Small cores tend to increase bow (pressure) waves (disturbance of surface sediments) and compaction, thus altering the vertical profile. However, these corers provide better confidence limits and spatial information when multiple cores are obtained (EPA 1983b; Elliott and Drake 1981). As shown by Rutledge and Fleeger (1988) and others, care must be taken in subsampling from core samples, since surface sediments might be disrupted even in hand-held core collection. They recommend subsampling *in situ* or homogenizing core sections before subsampling. Slowing the velocity of entry of coring equipment also reduces vertical disturbance. Samples are frequently of a mixed depth, but a 2-cm sample is recommended and the most common depth obtained, although depths up to 40 ft have been used in some dredging studies.

For dredging, remediation, and/or historical pollution studies, it is sometimes necessary to obtain cores of depths up to several meters. This often requires the use of vibracores that are somewhat destructive to sediment integrity but are often the only feasible alternative for deep or hard sediment sampling (Figures 5.74 through 5.76). In most studies of sediment toxicity, it is advantageous to subsample the inner core area (not contacting the sampler) since this area is most likely to have maintained its integrity and depth profile and not be contaminated by the sampler. Subsamples from the depositional layer of concern, for example, the top 1 or 2 cm, should be collected with a nonreactive sampling tool, such as a polytetrafluoroethylene-lined calibration scoop

Figure 5.74 Vibratory core collection.

Figure 5.75 Lowering vibratory corer.

(Long and Buchman 1989). Subsamples are placed in a nonreactive container and mixed until texture and color appear uniform. Due to the large volume of sediment that is often needed for toxicity or bioaccumulation tests and chemical analyses, it might not be possible to use subsampled cores because of sample size limitations. In those situations, the investigator should be aware of the above considerations and their possible effect on test results as they relate to *in situ* conditions.

Once sediment samples are collected, it is important, in most situations, to reduce the possibility of sediment oxidation. The majority of fine-grained sediments that are of concern in toxicity assessments are anaerobic below the top few millimeters (Carlton and Klug 1990), and any introduction of oxygen will likely alter the valence state of many ionic chemicals. This alteration may significantly change the bioavailability and toxicity of the sample. To protect sediments from oxygenation, the use of a glove box or bags with an inert gas supply for subsampling and processing, e.g., preparation of sediments for centrifugation, might be necessary.

While coring is preferred for maintaining a sediment's vertical integrity, care must be taken to reduce the possibility of spillage. Sediment cores should be stoppered immediately upon retrieval to prevent accidental loss of sediment. During all handling procedures, cores should be kept in an upright position as a general precaution against disturbance of the sediment. This is particularly important to prevent mixing of the uppermost part of the sediment column, which usually consists of very fine, soft, and unconsolidated material. The intact core samples (liners) should also be capped or stoppered and taped closed, secured in an upright position (e.g., rack), and labeled with appropriate information regarding sampling site, location, sample number and/or identification, time and date of collection, method of collection, and name or initials of the collector. When using clear plastic liners, the appearance of each sediment core should be recorded prior to any subsampling, along with other descriptive features such as the length of the core, thickness of various sediment units, occurrence of fauna, presence of

Figure 5.76 Emptying vibratory corer.

oil or noticeable odor, and sediment color, texture, and structure (Environment Canada 1994; Mudroch and Azcue 1995; Figure 5.77).

Once samples are collected, some form of subsampling and/or compositing is often performed. Removal of a portion of the collected sediment from the grab sampler (i.e., subsampling) can be performed using a spoon or scoop made of inert, noncontaminating material (e.g., Teflon, titanium, or high-quality stainless steel). It is recommended when subsampling to exclude sediment that is in direct contact with the sides of the grab sampler as a general precaution against any potential contamination from the device. Each subsample may be placed into a separate clean, prelabeled container. As a general rule, each labeled sample container must be tightly sealed and the air excluded. However, if the sample is to be frozen, it is advisable to leave a small amount of headspace in the container to accommodate expansion and avoid breakage.

Figure 5.77 Vertical layers of a sliced core.

Compositing of core samples or subsamples, if necessary, can be done in the field or laboratory, such as by using a drill auger mixer shown in Figure 5.78. The quality of the core sample must be acceptable and only sediment depths with similar stratigraphy should be combined. Although there might be occasions when it is desirable to composite incremental core depths, it is recommended that only horizons of similar stratigraphy be composited. Depending on the study objectives and desired sampling resolution, individual horizons within a single core can be homogenized to create one or more depth composites for that core, or corresponding horizons from two or more cores might be composited. Thorough homogenization of the composite sample, by hand or using a mechanical mixer, is recommended prior to analysis or testing.

The type of sediment characterization needed will depend on the study objectives and the contaminants of concern; however, a minimum set of parameters should be included which are known to influence toxicity and will aid data interpretation. At a minimum, the following physical and chemical characterization of sediment is recommended: total solids (dry weight), total organic carbon (TOC), acid volatile sulfides (AVS) (when metals are of concern), ammonia, and grain size fractionation. The following parameters are also frequently useful in characterization and data interpretation of contaminant effects: pH, ORP (oxidation–reduction potential), temperature, salinity-conductivity, hardness, total volatile solids (ash free weight), nitrogen and phosphorus species, cation exchange activity (CEC), sediment or suspended solids biochemical oxygen demand (BOD), and/or chemical oxygen demand (COD). Many of the characterization methods have been based on analytical techniques for soils, wastewaters, and waters, and the literature should be consulted for further information (EPA 1977; Black 1965; USGS 1969; ASTM 1989; Page et al. 1982).

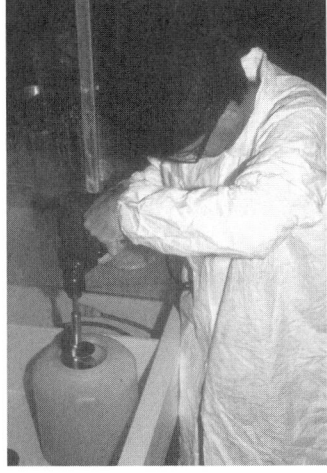

Figure 5.78 Mixing sediment with a drill auger.

Interstitial Water and Hyporheic Zone Sampling

Interstitial water (pore water) is defined as the water occupying space between sediment or soil particles and is often isolated to provide either a matrix for toxicity testing or provide an indication of the concentration and partitioning of contaminants within the sediment matrix. U.S. EPA sediment quality criteria are based on the assumption that the primary route of exposure to benthic organisms is via the interstitial water (Di Toro et al. 1991). However, this route of exposure does not include uptake from ingestion of contaminated sediment particles. In addition, contaminants in interstitial waters can be transported into overlying waters through diffusion, bioturbation, and resuspension processes (Van Rees et al. 1991). The usefulness of interstitial water sampling for determining chemical contamination and/or toxicity will depend on the study objectives and nature of the sediments at the study site. Sediments that are either very large grain-sized (such as gravel or cobble) or hard, compacted clays will likely not have interstitial waters that are significantly contaminated. Therefore, sampling of interstitial waters should be restricted to sediments ranging from sandy to noncompacted clays. Interstitial waters from depositional zones containing smaller-sized sediments (clays) are usually the most contaminated.

Frequently, surface waters and groundwaters intermix via upwelling or downwelling transition zones (TZ). The ecosystem associated with this transition zone is sometimes referred to as the hyporheic zone or hyporheous. It can be a very important zone for many reasons: provides essential habitat and refugia for micro-, meio-, and macrofauna or flora; affects contaminant attenuation, removal, or transport; cycles nutrients and carbon; and provides trophic links between the microbes and invertebrates and their macrofaunal predators (Duncan 1999). To date these zones have largely been ignored in environmental contaminant assessments and conceptual models, even though they are quite common. They provide a challenge in that their assessment requires collaboration of hydrogeologists, hydrologists, ecologists, chemists, and toxicologists.

The biological and physicochemical conditions within the groundwater and surface water are different, and hence may affect the partitioning (e.g., bound or freely dissolved), mobility, and bioavailability of sediment-associated contaminants. For example, changes in pH may affect the binding of metals, whereas the rate and extent of microbial processing of sediment organic matter may affect the partitioning of persistent organic contaminants. Upwelling zones (where groundwater and interstitial water move up toward surface water) are generally anoxic, with low pH. Anaerobic microbial processes dominate and may include reductions, denitrification, ammonification, and methanogenesis. Dissolved organic carbon (DOC) is of low quality and species diversity is often quite low in upwelling zones. However, benthic consumers are attracted to this habitat. Downwelling zones (the downward movement of surface water into the stream bed) are generally higher in oxygen content and pH. Aerobic microbial processes such as oxidation and nitrification are dominant. DOC quality, species diversity, and productivity are high in downwelling habitats. The hydrological interface between upwelling groundwater and downwelling surface water within the stream bed contains large gradients for a variety of physicochemical parameters (e.g., temperature, dissolved oxygen, pH, and pE). Previous studies have shown that organic contaminant and metals concentrations can vary over several orders of magnitude (Benner 1995).

There are several scenarios in which data on groundwater–surface water interactions would be useful in evaluations of the fate and dynamics of sediment contaminants and the *in situ* exposure of biota. Upwelling groundwater can affect benthos and surface water biota if either or both the groundwater and sediments are contaminated. Aqueous phase chemicals (e.g., freely dissolved, colloid-bound) in the upward flowing groundwater and/or the mobilization of sediment-bound contaminants by upwelling groundwater are the potential inputs to the surficial environs under these conditions. Downwelling surface water can affect benthic, hyporheic, and phreatic (groundwater-associated) biota if either or both the surface water and sediments are contaminated. Under such conditions, the potential exists for the transport of sediment contaminants to deep layers within the stream bed and the contamination of groundwater by the downward-flowing contaminant load. We have observed this at sites contaminated by PCBs and chlorinated benzenes (Greenberg and Burton 1999).

Selection of Measurement Methods for Interstitial Water

Isolation of sediment interstitial water can be accomplished by a wide variety of methods, which can be grouped as laboratory or field (*in situ*) based. The common laboratory-based methods can be categorized as (1) centrifugation, (2) pressurization, or (3) suction. Field-based methods include suction and "peepers" (for reviews, see Adams et al. 1991; ASTM 1994; Burton et al. 2001; Environment Canada 1994). Peepers are small chambers with membrane or mesh walls, which are buried in sediments, and surrounding interstitial water then equilibrates within the chamber. Chambers are typically retrieved from 2 to 20 days after deployment.

It is important to work with the analytical and toxicity testing laboratories to determine the least amount of sample needed, because of the difficulty of obtaining large amounts of interstitial water for analyses. As an example, the use of an anodic stripping voltammeter is suitable for direct analyses (undigested) of heavy metals in interstitial water using only about 5 mL of water for several metals (at least copper and lead) simultaneously, instead of about 50 mL typically required. Organic analyses may be conducted with about 250 mL of water, using the modified methods described in Chapter 6, instead of the typically required 1-L sample sizes, but with loss of sensitivity. The use of an automated water analyzer (such as the TrAAcs 2000 analyzer from Bran+Luebbe) can dramatically reduce the water volume needed for conventional nutrient analyses. Ion chromatography also requires only a very small amount of sample for complete cation and anion analyses. Microtox, from Azur Environmental, is also a very useful indicator of toxicity and requires only a very small amount of sample (about 1 mL). Bacteria tests can also be conducted using small sample volumes (using methods from IDEXX, Inc., for example), especially if the bacteria densities are high, as is likely in contaminated urban streams, allowing dilution of the samples.

When relatively large volumes of water are required (such as 20 mL or greater), only grab and core sampling with subsequent centrifugation and sediment squeezing methods are typically used. Other methods such as suction and *in situ* samplers do not easily produce sufficient volumes for most required analyses. However, larger-sized peepers (500 mL volume) have been used for collecting samples for chemical analyses and for exposing test organisms *in situ* (Burton 1992a,b; Sarda and Burton 1995; see also Chapter 6).

Most sediment collection and processing methods have been shown to alter interstitial water chemistry (e.g., Schults et al. 1992; Bufflap and Allen 1995a,b; Sarda and Burton 1995) and, therefore, can potentially alter contaminant bioavailability and toxicity. Some important interstitial water constituents, e.g., dissolved organic carbon, dimethylsulfide, ammonia, major cations, and trace metals can be significantly altered by the collection method (e.g., Martin and McCorkle 1993; Carignan et al. 1994; Bufflap and Allen 1995a,b; Sarda and Burton 1995). Increased sample handling associated with methods such as grab or core sampling and centrifugation, squeezing, or suction may cause significant increases in key constituents, such as ammonia, sulfide, and DOC concentrations, as compared to those collected via *in situ* "peepers" or core-port suction. Other constituents, such as salinity, dissolved inorganic carbon, sulfide, and sulfate, might not be affected by collection, providing oxidation is prevented. If sediments are anoxic, as most depositional sediments are, all steps involved in sample processing should be conducted in inert atmospheres or by limited contact with the atmosphere to prevent oxidation (and subsequent sorption/precipitation) of reduced species. When anoxic sediments are exposed to air, volatile sulfides will be lost which may increase the availability of sulfide-bound metals. In addition, iron and manganese oxyhydroxides are quickly formed which readily complex with trace metals, thus altering metals-related toxicity (e.g., Bray et al. 1973; Troup et al. 1974; Burton 1991). There is no need for maintaining anoxic processing conditions when the study objectives are concerned only with exposures to oxic sediments, or if target contaminants are unaffected by oxidation in short-term toxicity or bioaccumulation testing. For example, often studies of dredged material toxicity do not consider ammonia-related toxicity, and oxidation is actually promoted to remove ammonia from overlying waters of the toxicity test beakers.

Immediate collection and analysis of interstitial water is recommended since chemical changes might occur even when sediments are stored for short periods (e.g., 24 h) at *in situ* temperatures (Sarda and Burton 1995). Coagulation and precipitation of the humic material was noted when interstitial water was stored at 4°C for more than 1 week. Oxidation of reduced arsenic species in interstitial water of stored sediments was unaffected for up to 6 weeks when samples were acidified and kept near 0°C, without deoxygenation. When samples were not acidified, deoxygenation was necessary. Others have recommended interstitial waters be frozen after extraction, prior to toxicity testing, to prevent changes, but others have recommended against freezing samples that will undergo toxicity testing. The optimal collection method will depend upon the purpose of the sample (e.g., acidification for metal analysis and not toxicity testing), characteristics of the sediment, and the contaminants of concern. Sediments that are highly contaminated with strongly nonpolar organics (such as PCBs) are not likely to change in toxicity during storage.

The conditions for isolation of interstitial waters by centrifugation have varied considerably. For toxicity testing, interstitial waters have been isolated over a range of centrifugal forces and temperature ranges (Ankley and Schubauer-Berigan 1994; Schults et al. 1992) with centrifuge bottles of various compositions. When centrifugation followed by filtration has been compared with *in situ* dialysis, higher speed centrifugation followed by filtration with 0.2 membrane filters has produced results that were more similar for metals and organic carbon. Centrifugation at low speeds or use of a larger pore size filtration membrane (e.g., 45 μm mesh) will result in retention of dissolved contaminants, colloidal materials, and aquatic bacteria in the pore water sample. High-speed centrifugation (e.g., $10,000 \times g$) is necessary to remove colloids and dispersible clays (Ankley and Schubauer-Berigan 1994). Typically, toxicity is reduced with high-speed centrifugation or filtration due to the removal of particle-associated contaminants (Ankley and Schubauer-Berigan 1994; Schults et al. 1992; Bufflap and Allen 1995a). While the duration of the centrifugation has been variable, 30 min is relatively common. The temperature for the centrifugation should reflect the ambient temperature of collection to ensure that the equilibrium between particles and interstitial water is not shifted.

Filtration through glass fiber or polycarbonate membranes may cause the loss of some dissolved metals and organics (Schults et al. 1992). If filtration is employed, a nonfiltered sample should also be tested for toxicity and contaminant concentrations. The effects of centrifugation speed, filtration, and oxic conditions on some chemical concentrations in interstitial waters have been well documented (e.g., Ankley and Schubauer-Berigan 1994; Schults et al. 1992; Bufflap and Allen 1995b; Bray et al. 1973). It is recommended that, for routine toxicity testing of interstitial waters, sediments should be centrifuged at $10,000 \times g$ for a 30-min period at 4°C. It is difficult to collect interstitial water from sediments that are predominantly coarse sand. A modified centrifuge bottle has been developed with an internal filter which can recover 75% of the interstitial water as compared to 25 to 30% from squeezing.

Polytetrafluoroethylene (PTF) bottles will collapse at 3000 g but have been used successfully up to 2500 g when filled to 80% of capacity. Isolation of interstitial water in this case should be at the temperature of collection, at a slower speed of $2500 \times g$ for 30 min. This material will contain colloidal material as well as dissolved compounds. At low centrifugation speeds, without filtration, removal of the colloids may not be possible. The influence of dissolved and colloidal organic carbon may be estimated by measuring the organic carbon content. If small volumes of water are required for testing, higher speed centrifugation can be performed with glass tubes (up to $10,000 \times g$). If metal analysis of toxicity is not a concern, then high-speed centrifugation in stainless steel centrifuge tubes is an option. When working with samples contaminated with organics, efforts should be made to reduce sample exposure to light to reduce photo-related degradation or alteration of any potentially toxic compounds. This can be accomplished by using amber bottles and yellow lights.

Isolation of interstitial water by squeezing has been performed with a variety of procedures. In all cases, the interstitial water is passed through a filter that is a part of the apparatus. Filters have different sorptive capacities for different compounds. Numerous studies have shown filters reduce toxicity and contaminant concentrations by retaining contaminant-associated particles and also by contaminant sorption onto the filter matrix (Schults et al. 1992; Bray et al. 1973; Troup et al. 1974; Sasson-Brickson and Burton 1991). The characteristics of filters and the filtering apparatus should be carefully considered. Squeezing has been shown to produce a number of artifacts due to shifts in equilibrium from pressure, temperature, and gradient changes (e.g., Schults et al. 1992; Troup et al. 1974; Mangelsdorf et al. 1969; Fanning and Pilson 1971; Figure 5.79). Squeezing can affect the electrolyte concentration in the interstitial water with a decrease near the end of the squeezing process. It is there-

Figure 5.79 Pore water squeezer — stainless steel with Teflon liner.

fore recommended that moderate pressures be used with electrolyte (conductivity) monitoring during extraction. Significant alterations to interstitial water composition occurred when squeezing was conducted at temperatures different from ambient (e.g., Mangelsdorf et al. 1969). Other sources of alteration of interstitial water when using the squeezing method are contamination from overlying water, internal mixing of interstitial water during extrusion, and solid-solution reactions as interstitial water is expressed through the overlying sediment. As interstitial waters are displaced into upper sediment zones, they come in contact with solids with which they are not in equilibrium. This intermixing causes solid-solution reactions to occur. The chemistry of the sample may be altered due to the fast kinetics (minutes to hours) of these reactions. Most interstitial water species are out of metastable equilibrium with overlying sediments and are rapidly transformed, as observed with ammonia and trace metals. Bollinger et al. found elevated levels of several ions and dissolved organic carbon in squeezed samples as compared to samples collected by peepers. The magnitude of the artifact will depend on the element, sediment characteristics, and redox potential. It is unlikely that reactive species gradients can be established via squeezing of sediment cores.

Many studies have demonstrated the usefulness of *in situ* collection methods (e.g., Barnes 1973; Belzile et al. 1989; Bottomley and Bayly 1984; Buddensiek et al. 1990; Howes et al. 1985; Jahnke 1988; Mayer 1976; Murray and Grundmanis 1980; Sayles et al. 1973; and Whiticar 1982). These methods of interstitial water collection are superior to more traditional methods in that they are less likely to alter the chemistry of the sample. The principal methods of interstitial water collection are through the use of peepers (e.g., Bufflap and Allen 1995a,b; Carignan 1984; Bottomley and Bayly 1984) or *in situ* suction techniques. These methods have the greatest likelihood of maintaining *in situ* conditions and have been used to sample dissolved gases (Sarda and Burton 1995) and volatile organic compounds.

Suction using an aquarium air stone recovered up to 1500 mL from 4 L of sediment suctioned in an anoxic environment (Galli 1997). Hand vacuuming using an aquarium stone has shown to be an effective method of collecting interstitial water (Sarda and Burton 1995). The air stone is attached to a 50-mL syringe via plastic tubing. The stone is inserted in the sediment to the desired depth and then suction applied. Clogging of the air stone is a problem in some sediments; however, it is effective in most tested. The collection system can be purged of oxygen prior to leaving the laboratory. Ammonia concentrations in water obtained by this system were similar to those collected

Figure 5.80 Disassembled small-volume, high-resolution peepers.

Figure 5.81 Small-volume peeper assembly showing 75-mm nylon screening.

with *in situ* peepers (Sarda and Burton 1995). Problems common to suction methods are loss of
equilibration between the interstitial water and the solids, filter clogging, and oxidation. However,
in situ suction or suction via core ports has been shown to accurately define small gradients of
some sediment-associated compounds, including ammonia, the concentrations of which can change
by an order of magnitude over a 1-cm depth interval. However, these small-scale suction methods
may not provide an adequate volume for conducting most standard toxicity test procedures.

Small-volume, high-resolution peepers, made by the University of Alabama at Birmingham,
were designed for chemical and bacteriological analyses of interstitial water (Lalor and Pitt 1998).
These peepers were made from Delrin and are about 10 to 15 cm wide and 45 to 60 cm long, with
one end tapered to a point (Figures 5.80 through 5.83). The main body is made of 20-mm-thick
stock and has numerous deep and wide slots (not cut through), spaced 1 cm apart, that hold about
5 to 10 mL of water each. This common peeper design enables vertical stratification of pore water
quality to be determined. However, because the water volume for each separate chamber is very
small, special laboratory analysis procedures are needed that minimize water volume requirements.
In order to collect larger volumes of water, these peepers are frequently placed in a cluster
arrangement allowing compositing from similar depth slots from adjacent peepers.

The slots should not extend any closer than about 20 mm from the edge, to prevent cracking
of the thinner cover piece (common in peepers made from Plexiglas, for example). A nylon screen
having 75-μm apertures is placed over this thick piece and is then covered with a thinner sheet of

Figure 5.82 Peeper placement near shore in urban
lake.

Figure 5.83 Ten replicate high-resolution peepers (to
obtain larger water composite samples).

Delrin that is 6 mm thick. This cover piece has identically located slots cut through the material and has countersunk holes matching tapped holes in the main body. For use, the cavities in the main body are filled with distilled or deionized water, covered with the nylon screen, and the two Delrin pieces are screwed together using plastic screws, sandwiching the nylon screen (Figures 5.80 and 5.81). The unit is then pushed into the stream or lake sediment, gently pushing down on the unit until resistance prevents further penetration, leaving about five slots above the sediment/water interface (Figure 5.82). The unit is left in place until equilibrium is established, and is then removed (several hours using the large aperture screening). The unit may require up to 2 weeks for equilibrium to become established when using small aperture screenings (such as 0.45 or 2 μm membrane filter material). A recent modification has added a thin stainless steel cover to the peeper that slides over the front slots to protect them while inserting or withdrawing the peepers in sediment. The cover is slid off after the peepers are pushed into the sediment to the appropriate depth. In addition, the water is extracted from the peeper wells after disassembling the units and carefully rolling back the nylon screening, instead of puncturing the screening and inserting a syringe for sample withdrawal. These modifications have significantly reduced the disturbance to the sediments when using the peepers and have reduced contamination of the sample water.

The optimal equilibration time for *in situ* peepers is a function of membrane aperture, sediment type, contaminants of concern, and temperature. There are several artifact problems associated with peepers which use dialysis membranes. Total organic carbon may be elevated in peepers (4 to 8 μm pore size) due to biogenic production; however, colloidal concentrations are lower than centrifuged samples. Cellulose membranes are unsuitable because they decompose too quickly. A variety of polymer materials have been used, some of which may be inappropriate for studies of certain nonpolar compounds.

More recently, larger pore sized mesh has been used (Figures 5.84 through 5.87) which dramatically shortens equilibration time (Fisher 1992; Sarda and Burton 1995), as illustrated in Figure 5.88 during tests at UAB. In this test, 75-μm nylon screening was used on a peeper placed in a bucket of saline water (about 5.5 mS/cm). Every few minutes, the peeper was removed, and a syringe was used to remove water from an individual cell. This was then measured for conductivity.

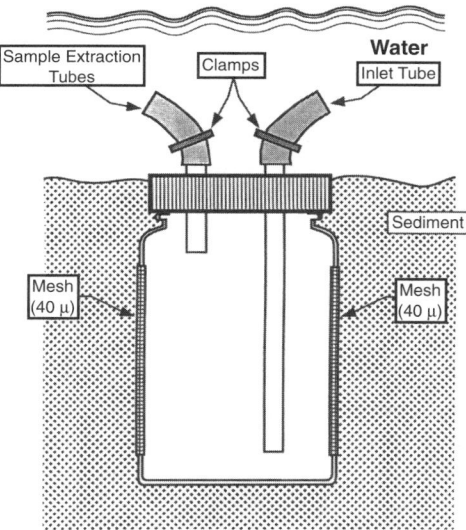

Figure 5.84 Large-volume peeper with large aperture mesh. (From Burton, G. A., Jr., Ed. *Sediment Toxicity Assessment.* Lewis Publishers. Boca Raton, FL. 1992b. With permission.)

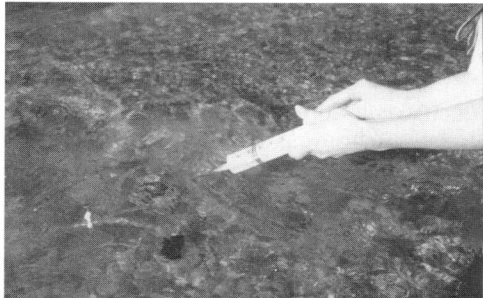

Figure 5.85 Withdrawing interstitial water sample from large-volume peeper.

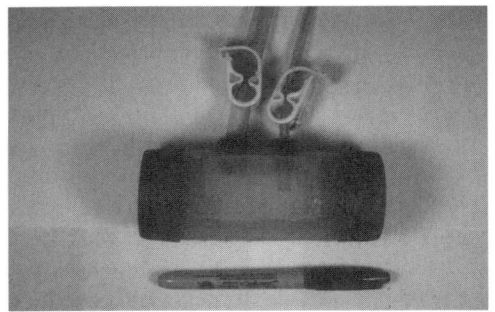

Figure 5.86 Medium-volume peeper with large aperture mesh for water sampling.

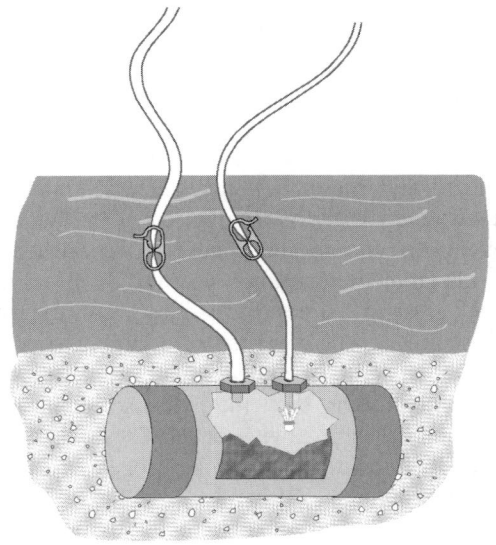

Figure 5.87 Medium-volume peeper buried in sediment.

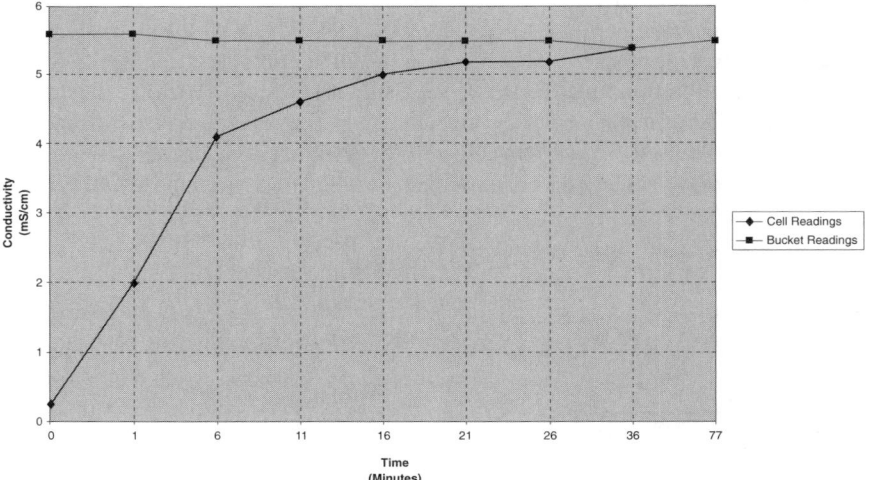

Figure 5.88 Equilibrium plots for 75-μm nylon screening in small-volume peeper.

Effective equilibrium was reached after about 20 min. In comparison, Figure 5.89 is an equilibrium plot for a 0.22-μm polyethersulfone membrane filter used in a diffusion peeper (Easton 2000). This test was conducted in a small laboratory flume with water flowing about 1 ft/s. Saline water was placed in the peeper (about 18 mS/cm), and the flume water was regular tap water (about 200 μS/cm). Samples were withdrawn from the peeper frequently at the beginning of the test, and at longer intervals later, and analyzed for conductivity. In this case, it required about 20 hours to reach equilibrium, although about 90% of the equilibrium was established at 10 hours.

When using sampler peepers and 75-μm membrane material, we commonly leave the peepers in place for about 2 to 24 hours to ensure equilibrium. Solids that pass through the mesh tend to settle to the bottom of the peeper chamber. Long exposure times may be impractical due to security problems and high flows in streams. The samplers need to be taken to the laboratory where the water

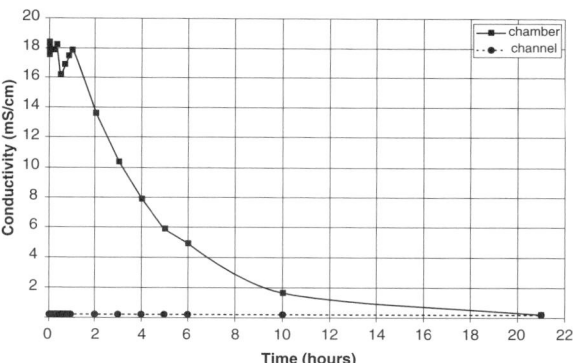

Figure 5.89 Equilibrium plot for 0.22-mm polyethersulfone membrane filter in diffusion peeper (From Easton, J. *The Development of Pathogen Fate and Transport Parameters for Use in Assessing Health Risks Associated with Sewage Contamination.* Ph.D. dissertation, Department of Civil and Environmental Engineering, University of Alabama at Birmingham. 2000. With permission)

is immediately analyzed. It is also possible to remove the samples from the slots in the field (using a syringe and needle), transferring the water into sealed and full bottles (such as small VOC vials). Four or five high-resolution peepers located close together can provide a 20 to 50 mL composite sample of pore water in 1-cm depth increments for chemical analyses (as shown on Figure 5.83).

When ionizable compounds, e.g., metals, are to be collected, it is important to preequilibrate the samplers with an inert atmosphere to avoid introducing oxygen into the sediments, thereby changing the equilibrium. Plastic samplers can contaminate anoxic sediments with diffusable oxygen and should be stored before testing in inert atmospheres (Carignan et al. 1994). In addition, when samples are collected and processed, they should also be kept under an inert atmosphere and processed quickly. Metals sampling of interstitial waters can be accomplished using a polyacrylamide gel probe (Krom et al. 1994) More recently, semipermeable membrane devices (SPMDs) filled with a nonpolar sorbant have been used effectively to show potential for bioaccumulation of nonpolar organic compounds.

Recently, test organisms have been exposed within peeper chambers where larger mesh sizes of 149 μm were used successfully in oxic sediments. Chambers can be buried several centimeters or in surficial sediment depending on the study objectives (Figures 5.90 and 5.91). Equilibration of conductivity was observed within hours of peeper insertion into the sediment (Fisher 1992). Replicate peepers revealed extreme heterogeneity in sediment interstitial water concentrations of ammonia and dissolved oxygen (Frazier et al. 1996; Sarda and Burton 1995; Sherman et al. 1994). Sediments that were high in clay and silt fractions usually were anoxic and did not allow for organism exposure *in situ* (Fisher 1992).

The Birmingham SSO (sanitary sewer overflow) evaluation project is a recent example of the use of peepers with large apertures. *Enterococcus, Escherichia coli,* total coliform bacteria, Micro-

Figure 5.90 Medium-volume peepers *in situ* with sampling tubes exposed.

Figure 5.91 Surficial sediment chambers.

tox toxicity screening, heavy metals (copper and lead), major ions, and nutrients are being analyzed on most of the pore water samples by combining water from three adjacent 10-cm chambers, and by using five replicate peepers located close together. This allows a total of about 150 mL of water for analysis. The careful selection of test methods (and dilution of water for the bacteria analyses) allows a relatively comprehensive evaluation of pore water chemical and bacteriological conditions. Changes in pore water chemical and bacteriological quality for different depths can be used to calculate diffusion coefficients and kinetic rate coefficients.

In situ and real-time chemical measurements of interstitial water are also possible using continuously recording *in situ* water quality sondes. The University of Alabama at Birmingham is currently using YSI 6000 monitoring probes to continuously monitor interstitial water pH, ORP, conductivity, DO, and temperature in urban streams as part of an EPA-sponsored research project investigating SSO impacts. These instruments are capable of unattended operation for several weeks. The probe end of the instrument is wrapped with a nylon screen having 150-μm apertures. Equilibrium should be obtained within a few hours using this large aperture. The instrument can be placed vertically with the probe end buried several hundred mm in the sediment in slow-moving streams for short periods. The instrument is completely buried horizontally for longer periods or for higher flows. The use of a direct readout (hand-held readout from YSI, or a portable computer) is useful in determining equilibrium times during preliminary trials. The available turbidity probe is also used to indicate the effects of placement of the probe by measuring the exchange of water in the probe chamber. A similar unit placed simultaneously in the water column can be used to measure the lag time of any chemical changes (such as conductivity) in response to storm events and to directly determine diffusion coefficients. Of course, this method does not provide accurate vertical placement of the analytical results, but it is expected to be generally representative of near-surface conditions where most of the benthic organism activity occurs. These probes are extremely useful to illustrate the variation of these parameters with time, especially during wet weather events, and to measure the recovery of conditions after events.

Mini-Piezometer Measurements of Pore Water Conditions

Mini-piezometers (Lee and Cherry 1978) are useful tools because they allow for the detection of upwelling groundwater and downwelling surface water on a local scale (i.e., cm to m). Additionally, these simple, inexpensive devices allow for samples of pore water to be withdrawn from desired depths within the stream bed for chemical analysis. Mini-piezometers are comprised of lengths of 1/8" ID plastic tubing that is perforated and screened with 300-μm mesh along the bottom 5 cm (Figures 5.92 through 5.94). A nest is a group of mini-piezometers of different lengths attached to a 1-m dowel rod that will sample at desired levels beneath the sediment surface (e.g., 10, 25, 50, 75, and 100 cm). Once piezometers are installed, they can be left in place indefinitely for repeated sampling and measurements. To detect areas of upwelling and downwelling, transects of nested mini-piezometers are installed in the riffle and pool areas of *in situ* test sites. Hydraulic heads (in cm) are determined by measuring the heights of water columns drawn simultaneously from the inserted mini-piezometer and overlying surface water into a manometer (Winter et al. 1988; Figure 5.94). Relative to surface water, a positive or negative hydraulic head indicates an upwelling or downwelling zone, respectively.

The hydrologic data from mini-piezometer pore water samples and hydraulic head measurements have improved our ability to interpret often complex exposure–effects relationships that result from *in situ* toxicity tests. We have found that contaminant concentrations in samples of sediments and pore water are not always predictive of *in situ* chamber (actual) exposure levels and observed effects in the test species. For example, in an *in situ* study of three sites in a stream system with similar levels of sediment contamination by chlorinated benzenes, one site was downwelling at all mini-piezometer nest locations and two sites had no net hydraulic pressure differences. Total chlorinated benzenes in water samples taken from the piezometer nests ranged

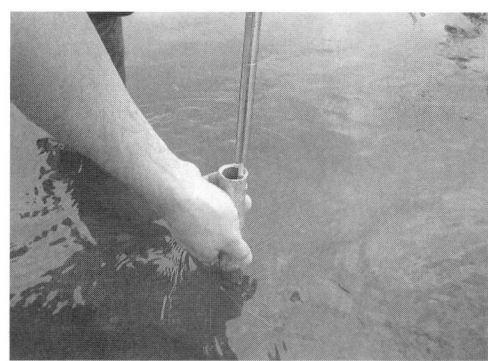

Figure 5.92 Placement of mini-piezometers into support tubing.

Figure 5.93 Placement of mini-piezometer array into sediments via temporary support pipe.

from 100 to 1300 µg/L at all sites. The highest concentrations generally occurred in piezometers installed 30 cm or deeper into the stream bed. Concentrations of total chlorinated benzenes in water samples taken from the chambers used during 4-day *in situ* exposures of *Ceriodaphnia dubia, Hyalella azteca,* and *Chironomus tentans* to surficial sediments were near 100 µg/L at the two no-exchange sites, whereas the level was only 3 µg/L at the downwelling site. Survival of all three test species was significantly higher at the downwelling site (>80%) than at the no-exchange sites (<20%). For *C. dubia* and *H. azteca,* survival between the downwelling and reference sites was not significantly different. It appears the downward flow of surface water through the sediments might have removed bioavailable contaminants in the surficial sediments to deeper zones within the stream bed (Greenberg and Burton 1999). However, this condition places transition zone species and groundwater resources at risk.

Sediment chemists, toxicologists, and risk managers have primarily focused their research efforts and the development of sediment quality guidelines on the effects of contaminants on benthic and water column organisms associated with the surficial sediments (0 to 10 cm depth). Implicit in this approach is that the historical contamination buried beneath the top sediment bed layer is biologically unavailable and hence poses little to no ecological risk. However, deeper sediments (ca. 10 to 100 cm depth), and more specifically sediments within the transition zone, serve important ecosystem functions and therefore may be sensitive to chemical perturbation. Vertical transport of dissolved or colloid-bound contaminants within the sediment interstices can potentially exert deleterious effects in the surficial sediments, surface water, or groundwater, or it can exacerbate preexisting degraded conditions. Therefore, ecosystem integrity can be more effectively evaluated if the scientific and regulatory community adopts a holistic approach to stream health that includes focusing on the transition zone. At the present time, we have begun to incorporate this added hydrologic perspective in our *in situ* sediment toxicity research program through the use of mini-piezometers. Continuing this line of research by developing assessment tools capable of measuring biological effects within the transition zone is the next step.

Figure 5.94 Field manometer connected to mini-piezometer to measure vertical flow through sediments.

Case Example 1. Sediment Sampling for Interstitial Pore Water in an Ice-Covered Lake

A site was sampled in northern Minnesota in January which had depositional sediments (non-consolidated silts and clays) and was ice-covered with water depths of 50 to 60 ft. Site conditions prevented use of peeper sampling and no *in situ* core-port sampling equipment was available. The study design required collection of 30 L of sediment. Based on these restrictions, a Ponar grab sampler was most appropriate for sediment collection.

Replicate Ponar grabs were collected through holes drilled in the ice and were deposited into a 20-L high-density polyethylene bucket and gently stirred to homogenize. Nitrogen gas was bubbled into any overlying water and added to the headspace prior to lid closure. Sediments were placed in ice chests at approximately 4°C and returned to the laboratory for processing.

Interstitial waters were collected using centrifugation. Sediments were distributed to the appropriate type of centrifuge bottles under a nitrogen atmosphere and centrifuged at $10,000 \times g$ at 4°C for 30 min. The supernatant was gently decanted under nitrogen atmosphere. Note: if solids are resuspended with the supernatant, a second centrifugation of the interstitial water should be conducted. The interstitial water from all bottles was combined under nitrogen and then split for chemical analyses and toxicity testing. Chemical samples were preserved and stored as appropriate. Toxicity testing was initiated within 48 hours, at which time the sample temperature was raised from 4°C to the required test temperature and dissolved oxygen checked to ensure adequate levels.

Case Example 2. Shallow Stream with Contaminated Sediments

A shallow stream in Ohio with sediment contamination was studied to develop site-specific sediment quality criteria. Site conditions allowed the placement of peeper samplers. The sediment depth of concern was from 0 to 5 cm. Peepers were constructed from high-density polyethylene bottles with 70- to 140-μm PTF mesh windows on the chamber walls, 1 to 5 cm from the top of the chamber (similar to Figure 5.84). Chambers were filled with sterile deionized water and placed in a nitrogen atmcsphere for 24 to 48 hours prior to site placement. Five replicate (total volume approximately 2.5 L) chambers were placed at the site by removing a plug of sediment the size of the chamber, inserting the chamber and gently packing the sediment around the chamber so that only the lid was exposed. Equilibration time can be reduced and time series sampling of the interstitial water is possible by constructing an outlet tube into the chamber lid (Sarda and Burton 1995). Degassed syringe samplers can then be attached to the outlet port and interstitial water removed without disturbing the peeper unit. Equilibration time with 140-μm mesh windows occurs within several hours. However, it may take days for the sediment gradients to reestablish adjacent to the chamber. Sampling of interstitial waters at the sediment surface (0 to 1 cm depth) is not readily feasible when large samples are required. However, toxicity may be determined on surficial sediment using *in situ* toxicity test chambers which expose organisms either directly to the sediments or via mesh barriers (Burton 1992a,b; see also Chapter 6). Microanalytical sampling of near-surface sediments is possible using narrow plate chamber designs (see reviews in Adams 1991, Burton 1991, and above citations). Samples are returned to the laboratory on ice and then processed by the appropriate chemical and toxicity test methods.

SUMMARY: BASIC SAMPLE COLLECTION METHODS

This chapter presented methods to determine the needed sampling effort, including the number of samples and the number of sampling locations. These procedures can be utilized for many different conditions and situations, but some prior knowledge of the conditions to be monitored is

Table 5.19 General Sampling Guidelines[a]

Location

1. Locate stations at sites representative of least and greatest impact from each pollutant source and for the total system, considering each ecosystem component (e.g., substrate, flow, biota).
2. Sample depositional areas and critical habitats such as riffles and spawning areas.
3. Collect replicate samples at each station which characterize the site spatially.
4. Sample during baseflow and various stormflow conditions.
5. Sample during different seasons.
6. Sample during recovery periods (following storm events) noting different periods of disturbance (i.e., storm recurrence period).
7. Note diurnal, weekly, monthly, and seasonal cycles of various ecosystem components-endpoints (e.g., DO, redox, tissue residues, toxicity, life stage).

Type

8. In areas where effects are uncertain, use a "weight-of-evidence" integrated approach (see Chapter 8). Characterize the inputs and receiving water system both physically (e.g., flow, solids, temperature, habitat) and chemically (e.g., oxygen, hardness, organics, metals). Measure key indigenous biological communities (indices), indicators (e.g., trout), and endpoints (e.g., fish abnormalities). Measure toxicity of effluents, waters, and sediments using sensitive and relevant species representing multiple levels of biological organization (e.g., fish, zooplankton, algae, benthic macroinvertebrates). *In situ* toxicity testing is the preferred approach.

Method

9. Process samples quickly (refrigerate and/or preserve immediately upon collection).
10. Reduce sample manipulation whenever possible (e.g., mixing, sieving, aeration, filtration).
11. Maintain sample integrity when possible (e.g., using core rather than grab [dredge] collection).
12. Characterize key components of all sample replicates.
13. Follow proper QA/QC practices.

[a] All sampling issue decisions must be based on the study objectives and their associated data quality objectives.

needed. A phased sampling approach is therefore recommended, allowing some information to be initially collected and used to make preliminary estimates of the sampling effort. Later sampling phases are then utilized to obtain the total amount of data expected to be needed.

Descriptions of data quality objectives and associated QA/QC requirements are also given. The use of different sample blanks and other quality control samples are described, along with dealing with typical problems associated with detection limits.

The main component of this chapter covers sampling methods, including water, source area, sediment, and pore water sampling options. Numerous examples are given illustrating the use of the many sampling methods and approaches. There are few universal methods that can be used for all sampling activities, and much discretion and professional judgment is needed to select the most appropriate methods for any specific project. However, there are some general guidelines for sampling streams and lakes which should apply to most studies, as listed in Table 5.19. Each of the points listed in this table are also discussed in greater detail elsewhere in this handbook, especially in Chapter 6 and the appendices.

There are a number of factors to consider when selecting a sampling site after preliminary surveys and design elements are completed. The selection factors and their relative importance are often study specific, but some general considerations do exist, as shown in Table 5.20. The factors that influence the representativeness of a sample are numerous and cross many disciplines, as do all ecosystem evaluations. Therefore, it is important to select sampling stations based on professional judgment(s) from an individual(s) with expertise in aquatic ecosystem assessments (hydrology, environmental chemistry, biosurveys, and ecotoxicology), taking into account spatial and temporal variation and the characteristics of base- and stormwater flow; habitat; pollutant loadings, fate and effects; aquatic communities; and sensitive indicator species.

These same selection criteria should then be used to establish reference area sampling, if preexisting reference data are not available. The reference station (upstream), stream or lake, and

Table 5.20 Sampling Size Selection, Sampling Media, and Sampling Frequency Considerations

Consideration	Sample	Influencing Factors
Heterogeneity	Ambient water, sediment effluent, runoff, biotic communities	Flow, mixing, depth, particle size distribution, land use patterns, runoff coefficients, season, life-cycle, behavior, patch dynamics, pollutant partitioning (fate)
Pollutant sources	Upstream-downstream, tributary mouths, sensitive habitats, dilution gradient, beneficial uses, "typical" habitats	Pollutant partitioning (fate), mixing, loading characteristics, toxicity target species and endpoints, habitat complexity
Beneficial uses	"Beneficial" component (e.g., water supply, fishery, swimming) at critical areas	Above factors

watershed should, ideally, have baseline characteristics identical to those of the test system when the pollutant problem (e.g., stormwater) being assessed is removed. However, since no two eco-systems are identical, this reference should be considered as a *general* benchmark from which to determine relative effect.

The next chapter presents much detail and information on evaluating samples and conditions (flow, rainfall, soil, aesthetics, habitat, water, sediment, microorganisms, benthos, zooplankton, fish, and toxicity), heavily supported with case study examples. Chapter 7 discusses statistical evaluations of the data, and Chapter 8 discusses data interpretation.

REFERENCES

Adams, D.D. Sampling sediment pore water, in *CRC Handbook of Techniques for Aquatic Sediments Sampling.* Edited by A. Murdoch and S.D. MacKnight. CRC Press, Boca Raton, FL, pp. 117–202. 1991.

Albert, R. and W. Horwitz. Coping with sampling variability in biota: percentiles and other strategies, in *Principles of Environmental Sampling.* Edited by L.H. Keith. American Chemical Society. 1988.

Alley, W.M. Determination of the decay coefficient in the exponential washoff equation. *International Symposium on Urban Runoff,* University of Kentucky, Lexington, KY, July 1980.

Alley, W.M. Estimation of impervious-area washoff parameters. *Water Resourc. Res.,* 17(4), 1161–1166. 1981.

Ankley, G.T. and M.K Schubauer-Berigan. Comparison of techniques for the isolation of sediment pore water for toxicity testing. *Arch. Environ. Contam. Toxicol.,* 27:507–512. 1994.

ASTM (American Society for Testing and Materials). Water environmental technology, in *ASTM Book of Standards,* Vol. 11.02 American Society for Testing and Materials. Philadelphia. 1989.

ASTM (American Society for Testing and Materials). *Standard Guide for Collection, Storage, Characterization, and Manipulation of Sediments for Toxicological Testing.* American Society for Testing and Materials, Philadelphia, Standard E 1391. 1991.

ASTM (American Society for Testing and Materials). Standard Guide for Collection, Storage, Characterization, and Manipulation of Sediments for Toxicological Testing, in *Water Environmental Technology, ASTM Book of Standards,* Vol. 11.04 American Society for Testing and Materials, Philadelphia, Standard No. E 1391. 1994.

ASTM (American Society for Testing and Materials). *ASTM Standards on Environmental Sampling.* ASTM Pub Code No. 03-418095-38. ASTM, Philadelphia. 1995.

ASTM (American Society of Testing and Materials). *1996 Annual Book of ASTM Standards.* West Consho-hocken, PA. ASTM, Vol. 04.08. 1996.

Bannerman, R., K. Baun, M. Bohn, P.E. Hughes, and D.A. Graczyk. *Evaluation of Urban Nonpoint Source Pollution Management in Milwaukee County, Wisconsin,* Vol. I. PB 84-114164. U.S. Environmental Protection Agency, Water Planning Division. November 1983.

Barkdoll, M.P., D.E. Overton, and R.P. Beton. Some effects of dustfall on urban stormwater quality. *Water Pollut. Control Fed.,* 49(9):1976–84. 1977.

Batley, G.E. and D. Gardner. Sampling and storage of natural waters for trace metal analysis, *Water Res.* 11:745. 1977.

Batley, G.E. Collection, preparation and storage of samples for speciation analysis, in *Trace Element Speciation: Analytical Methods and Problems*. Edited by G.E. Batley. CRC Press, Boca Raton, FL. 1989.

Barnes R.O. An *in situ* interstitial water sampler for use in unconsolidated sediments. *Deep-Sea Res.,* 20:1125–1128. 1973.

Belzile, N., P. Lecomte, and A. Tessier. Testing readsorption of trace elements during partial extractions of bottom sediments. *Environ. Sci. Technol.,* 23:1015. 1989.

Benner, S.G., E.W. Smart, and J.N. Moore. Metal behavior during surface-groundwater interaction, Silver Bow Creek, Montana. *Environ. Sci. Technol.,* 29(7):1789–1795. 1995.

Berg, G., Ed. *Transmission of Viruses by the Water Route*. Interscience Publishers, New York. 1965.

Berthouex, P.M. and L.C. Brown. *Statistics for Environmental Engineers*. Lewis Publishers, Boca Raton, FL. 1994.

Black, C.A., Ed. *Methods of Soil Analysis*. American Society of Agronomy, Agronomy Monograph No. 9, Madison, WI. 1965.

Bottomley, E.Z. and I.L. Bayly A sediment porewater sampler used in root zone studies of the submerged macrophyte, *Myriophyllum spicaum*. *Limnod. Oceanogr.,* 29:671–673. 1984.

Box, G.E.P., W.G. Hunter, and J.S. Hunter. *Statistics for Experimenters*. John Wiley & Sons, New York. 1978.

Bray, J.T., O.P. Bricker, and B.M. Troup. Phosphate in interstitial waters of anoxic sediments: oxidation effects during sampling procedure. *Science,* 180:1362–1364. 1973.

Buddensiek, V., H. Engel, S. Fleischauer-Rossing, S. Olbrich, and K. Wachtleer. Studies on the chemistry of interstitial water taken from defined horizons in the fine sediments of bivalve habitats in several northern German lowland waters: 1. Sampling techniques. *Arch. Hydrobiol.,* 119:55–64. 1990.

Bufflap, S.E. and H.E. Allen. Sediment pore water collection methods for trace metal analysis: a review. *Water Res.,* 29:165–177. 1995a.

Bufflap, S.E. and H.E. Allen. Comparison of pore water sampling techniques for trace metals. *Water Res.,* 29:2051–2054. 1995b.

Burton, G.A., Jr. Assessing freshwater sediment toxicity. *Environ. Toxicol. Chem.,* 10:1585–1627. 1991.

Burton, G.A., Jr. Sediment collection and processing factors affecting realism, in *Sediment Toxicity Assessment*. Edited by G.A. Burton, Jr. Lewis Publishers. Boca Raton, FL. 1992a.

Burton, G.A., Jr., Ed. *Sediment Toxicity Assessment*. Lewis Publishers. Boca Raton, FL. 1992b.

Burton, G.A., Jr., and J. Scott. Sediment toxicity evaluations. *Environ. Sci. Technol.,* 25. 1992.

Burton, G.A., Jr., C.D. Rowland, M.S. Greenberg, D.R. Lavoie, J.F. Nordstrom, and L.M. Eggert. A tiered, weight-of-evidence approach for evaluating aquatic ecosystems. *Aquat. Ecosyst. Health Manage.,* To be published in 2001.

Cairns, J., Jr., and K.L. Dickson. A simple method for the biological assessment of the effects of waste discharges on aquatic bottom-dwelling organisms. *J. Water Pollut. Control Fed.,* 43:755–772. 1971.

Carignan, R. Interstitial Water Sampling by Dialysis: Methodological Notes. *Limnol. Oceangr.,* 29:667–670. 1984.

Carignan, R., S. St. Pierre, R. Gachter. Use of diffusion samplers in oligotrophic lake sediments: effects of free oxygen in sampler material. *Limnol. Oceanogr.,* 39:468–474. 1994.

Carlton, R.G. and M.J. Klug. Spatial and temporal variation in microbial processes in aquatic sediments: implications for the nutrient status of lakes, in *Sediments: Chemistry and Toxicity of In-Place Pollutants*. Edited by R. Baudo, J. Giesy, and H. Muntau. pp. 107–130. Lewis Publishers. Boca Raton, FL. 1990.

Cochran, W.C. *Sampling Techniques*. 2nd edition. John Wiley & Sons, New York. 1963.

COE (U.S. Corps of Engineers), Hydraulic Engineering Center. *Urban Storm Water Runoff: STORM. Generalized Computer Program*. 723-58-L2520, Davis, CA. May 1975.

Cowgill, U.M. Sampling waters, the impact of sample variability on planning and confidence levels, in *Principles of Environmental Sampling*. Edited by L.H. Keith. ACS Professional Reference Book. American Chemical Society. pp. 171–189. 1988.

Cowherd, C.J., C.M. Maxwell, and D.W. Nelson. *Quantification of Dust Entrainment from Paved Roadways*. EPA-450 3-77-027, U.S. Environmental Protection Agency, Research Triangle Park, NC. July 1977.

Day, J. *Selection of Appropriate Analytical Procedures for Volunteer Field Monitoring of Water Quality*. MSCE thesis, Department of Civil and Environmental Engineering, University of Alabama at Birmingham. 1996.

DiToro, D.M., C.S. Zarba, D.J. Hansen, W.J. Berry, R.C. Swartz, C.E. Cowan, S.P. Paviou, H.E. Allen, N.A. Thomas, and P.R. Paquin. Technical basis for establishing sediment quality criteria for nonionic organic chemicals by using equilibrium partitioning. *Environ. Toxicol. Chem.*, 10: 1991.

Donigian, A.S., Jr., and N.H. Crawford. *Modeling Nonpoint Pollution from the Land Surface.* EPA-600/3-76-083, U.S. Environmental Protection Agency, Athens, GA. July 1976.

Duncan, P.B. Groundwater-surface water interactions: no longer ignoring the obvious at Superfund sites. *SETAC News,* 19(5): 20–21. 1999.

Easton, J. *The Development of Pathogen Fate and Transport Parameters for Use in Assessing Health Risks Associated with Sewage Contamination.* Ph.D. dissertation, Department of Civil and Environmental Engineering, University of Alabama, Birmingham. 2000.

EC (Environment Canada). *Guidance Document on Collection and Preparation of Sediments for Physico-chemical Characterization and Biological Testing.* Environmental Protection Series Report, EPS 1/RM/29. Ottawa, Canada. pp. 111–113, December, 1994.

Elliott, J.M. and C.M. Drake. A comparative study of seven grabs for sampling benthic macroinvertebrates of rivers. *Freshwater Biol.*, 11: 99–120. 1981.

EPA. *Ecological Evaluation of Proposed Discharge of Dredged Material into Ocean Waters*, Environmental Effects Laboratory, U.S. Army Engineer Waterways, Experiment Station, U.S. Environmental Protection Agency, Corps of Engineers, Vicksburg, MI. 1977.

EPA. *Handbook for Sampling and Sample Preservation of Water and Wastewater*, Environmental Monitoring and Support Laboratory, U.S. Environmental Protection Agency, Cincinnati, OH, EPA 600/4-82/029. 1982.

EPA. *Results of the Nationwide Urban Runoff Program.* Water Planning Division, U.S. Environmental Protection Agency, PB 84-185552, Washington, D.C. 1983a.

EPA. *Technical Support Manual: Waterbody Surveys and Assessments for Conducting Use Attainability Analyses*, Office of Water Regulations and Standards, U.S. Environmental Protection Agency, Washington, D.C. 1983b.

EPA. *Sampling Guidance Manual for the National Dioxin Manual*, Draft, Office of Water Regulations and Standards, Monitoring and Data Support Division, U.S. Environmental Protection Agency/Corps of Engineers, Washington, D.C. 1984.

EPA. *Sediment Sampling Quality Assurance Users Guide*, Environmental Monitoring and Support Laboratory, U.S. Environmental Protection Agency, Las Vegas, NV. EPA/600/4-85/048. 1985.

EPA. *Handbook for Stream Sampling for Waste Load Allocation Applications*, Office of Research and Development, U.S. Environmental Protection Agency, Washington, D.C. EPA/625/6-86/013. 1986.

EPA. *A Compendium of Superfund Field Operations Methods*, Office of Emergency and Remedial Response, U.S. Environmental Protection Agency, Washington, D.C. EPA 540/P-87/001. 1987.

EPA. *Rapid Bioassessment Protocols for Use in Streams and Rivers: Benthic Macroinvertebrates and Fish*, Office of Water, U.S. Environmental Protection Agency, Washington, D.C. EPA 444/4-89/001. 1989.

EPA. *Technical Support Document for Water Quality-based Toxics Control.* Office of Water, U.S. Environmental Protection Agency, Washington, D.C. 1990a.

EPA. Milwaukee River South declared a priority watershed in Wisconsin. *Nonpoint Source EPA News-Notes*, #9. December 1990b.

EPA. *Macroinvertebrate Field and Laboratory Methods for Evaluating the Biological Integrity of Surface Waters*, Office of Research and Development, U.S. Environmental Protection Agency, Washington, D.C. EPA 600/4-90/030. 1990c.

EPA. *Procedures for Assessing the Toxicity and Bioaccumulation of Sediment-Associated Contaminants with Freshwater Invertebrates*, EPA 600/R-94/024, U.S. Environmental Protection Agency, Duluth, MN. 1994.

EPA. *EPA's Contaminated Sediment Management Strategy*, EPA-823-R-98-001. Washington, D.C. 1998.

EPA. *Methods for Collection, Storage and Manipulation of Sediments for Chemical and Toxicological Analyses.* Office of Water. U.S. Environmental Protection Agency. Washington, D.C. To be published in 2001.

Fanning, K.A. and M.E.Q. Pilson. Interstitial silica and pH in marine sediments: some effects of sampling procedures. *Science,* 173: 1228. 1971.

Finlayson, B. The analysis of stream suspended loads as a geomorphological teaching exercise. *J. Geog. Higher Ed.* 5, 23–25. 1981.

Fisher, R. *Use of* Daphnia magna *for* in Situ *Toxicity Testing*. M.S. thesis, Wright State University, Dayton, OH. 1992.

Ford, P.J. and P.J. Turina. *Characterization of Hazardous Waste Sites — A Methods Manual, Vol. I: Site Investigations*, Office of Advanced Monitoring Systems Division, U.S. Environmental Protection Agency, Las Vegas, NV. EPA 600/4-84/075. 1985.

Frazier, B.E., T.J. Naimo, and M.B. Sandheninrich. Temporal and vertical distribution of total ammonium nitrogen and unionized ammonia nitrogen in sediment pore water from the upper Mississippi. *Environ. Toxicol. Chem.,* 15: 92–99. 1996.

Galli, J. Development and application of the rapid stream assessment technique (RSAT) in the Maryland piedmont. Presented at the *Effects of Watershed Developments and Management on Aquatic Ecosystems* conference. Snowbird, UT, August 4–9, 1996. Edited by L.A. Roesner. ASCE, New York, pp. 295 – 305. 1997.

Gilbert, R.O. *Statistical Methods for Environmental Pollution Monitoring*. Van Nostrand Reinhold, New York. 1987.

Gilbert, R. and J. Glew. A portable percussion coring device for lacustrine and marine sediments. *J. Sed. Petrol.,* 55: 607–608. 1985.

Graczyk, D.J., D.M. Robertson, W.J. Rose, and J.J. Steuer. *Comparison of Water Quality Samples Collected by Siphon Samplers and Automatic Samplers in Wisconsin*. USGS Fact Sheet FS-067-00. U.S. Geological Survey, Middleton, WI. July 2000.

Greenberg and Burton. Evaluation of the role of groundwater upwelling in the toxicity of contaminated sediments. *SETAC Abstracts*, 20th Annual Meeting, 14–18 November, 1999, Philadelphia, PA, p. 115, #552. 1999.

Grey, G. and F. Oliveri. Catch basins — effective floatables control devices. Presented at the *Advances in Urban Wet Weather Pollution Reduction* conference. Cleveland, OH, June 28–July 1, 1998. Water Environment Federation, Alexandria, VA. 1998.

Grizzle, R.E. and W.E. Stegner. A new quantitative grab for sampling benthos. *Hydrobiologia,* 126: 91–95. 1985.

Håkanson, L. and M. Jansson. *Principles of Lake Sedimentology*. Springer-Verlag, New York. 1983.

Horton, R.E. An approach toward a physical interpretation of infiltration capacity. *Trans. Am. Geophys. Union,* 20, 693–711. 1939.

Howes, B.L., J.W.H. Dacey, and S.G. Wakeham. Effects of sampling technique on measurements of porewater constituents in salt marsh sediments. *Limnol. Oceanogr.,* 30: 221–227. 1985.

Huber, W.C. and J.P. Heaney. The USEPA storm water management model, SWMM: a ten year perspective. *Second International Conference on Urban Storm Drainage*, Urbana, IL. June 1981.

HydroQual, Inc. *Floatables Pilot Program Final Report: Evaluation of Non-Structural Methods to Control Combined and Storm Sewer Floatable Materials*. City-Wide Floatables Study, Contract II. Prepared for New York City, Department of Environmental Protection, Bureau of Environmental Engineering, Division of Water Quality Improvement. NYDP2000. December 1995.

Imperato, D. A modification of the vibracoring technique for sandy sediment. *J. Sed. Petrol.,* 57: 788–789. 1987.

Jahnke, R.A. A simple, reliable, and inexpensive pore-water sampler. *Limnol. Oceanogr.,* 33: 483–487. 1988.

Jewell, T.K., D.D. Adrian, and D.W. Hosmer. Analysis of stormwater pollutant washoff estimation techniques. *International Symposium on Urban Storm Runoff*, University of Kentucky, Lexington, KY. July 1980.

Keith, L.H., W. Crummett. J. Deegan, Jr., R.A. Libby, J.K. Taylor, and G. Wentler. Principles of environmental analyses. *Anal. Chem.,* 55, 2210–2218. 1983.

Keith, L.H. *Environmental Sampling and Analysis: A Practical Guide*. Lewis Publishers, Chelsea, MI. 1991.

Kittrell, F. W. *A Practical Guide to Water Quality Studies of Streams*, U.S. Department of Interior, Federal Water Pollution Control Administration, Washington, D.C. CWR-5 1969.

Krom, M.D., P. Davidson, H. Zhang, and W. Davidson. High resolution pore-water sampling with a gel sampler. *Limnol. Oceanogr.,* 39: 1967–1972. 1994.

Lalor, M. and R. Pitt. *Assessment Strategy for Evaluating the Environmental and Health Effects of Sanitary Sewer Overflows from Separate Sewer Systems*. First Year Report. Wet-Weather Flow Management Program, National Risk Management Research Laboratory, U.S. Environmental Protection Agency, Cincinnati, OH. January 1998.

Lebel, J., N. Silverberg, and B. Sundby. Gravity core shortening and pore water chemical gradients. *Deep Sea Res.,* 29: 1365. 1982.

Lee, D.R. and J.A. Cherry. A field exercise on groundwater flow using seepage meters and mini-piezometers. *J. Geol. Educ.,* 27: 6–10. 1978.

Long, E.R. and M.F. Buchman. *An Evaluation of Candidate Measures of Biological Effects for the National Status and Trends Program.* NOAA Tech. Memorandum NOS OMA 45, Seattle, WA. 1989.

Malmquist, Per-Arne. *Atmospheric Fallout and Street Cleaning — Effects on Urban Stream Water and Snow.* Prog. Wat. Tech., 10(5/6): 495–505, 1978. Pergamon Press, London. September 1978.

Mangelsdorf, P.C. and T.R.S. Wilson. Potassium enrichments in interstitial waters of recent marine sediments. *Science,* 165: 171. 1969.

Martin, W.R. and B.C. McCorkle. Dissolved organic carbon concentrations in marine pore waters determined by high temperature oxidation. *Limnol. Oceanogr.,* 38: 1464–1479. 1993.

Mayer, L.M. Chemical water sampling in lakes and sediments with dialysis bags. *Limnol. Oceanogr.,* 21: 909–911. 1976.

McGee, T.J. *Water Supply and Sewerage.* McGraw-Hill, New York. 1991.

McCrone, W.C. Case for polarized light microscopy. *Am. Lab.,* 28(9): 12. June 1996.

Mudroch, A. and S.D. MacKnight. Bottom sediment sampling, in *CRC Handbook of Techniques for Aquatic Sediments Sampling.* Edited by A. Mudroch and S. D. MacKnight. CRC Press, Boca Raton, FL. pp. 29–95. 1991.

Mudroch, A. and J.M. Azcue, *Manual of Aquatic Sediment Sampling.* Lewis Publishers, Boca Raton, FL. 1995.

Murphy, W. *Roadway Particulate Losses.* American Public Works Assoc. Unpublished. 1975.

Murray, J.W. and L.U. Grundmanis. Hydrogen consumption in pelagic marine sediments. *Science,* 209: 1527–1530. 1980.

Nebeker, A.V., M. Cairns, J.H. Gakstatter, K.W. Malueg, G.S. Schuytema, and D.F. Krawczyk. Biological methods for determining toxicity of contaminated freshwater sediments to invertebrates. *Environ. Toxicol. Chem.,* 3: 617–630. 1984.

Page, A.L., R.H. Miller, and D.R. Keeney (Eds). *Methods of Soil Analysis,* Parts 1 and 2, American Society of Agronomy, Madison, WI. 1982.

PEDCo-Environmental, Inc. *Control of Re-entrained Dust from Paved Streets.* EPA-907/9-77-007, U.S. Environmental Protection Agency, Kansas City, MO. 1977.

Pitt, R. *Demonstration of Nonpoint Pollution Abatement through Improved Street Cleaning Practices,* EPA-600/2-79-161, U.S. Environmental Protection Agency, Cincinnati, OH. 270 pp. 1979.

Pitt, R. and G. Shawley. *A Demonstration of Non-Point Source Pollution Management on Castro Valley Creek.* Alameda County Flood Control and Water Conservation District and the U.S. Environmental Protection Agency, Water Planning Division (Nationwide Urban Runoff Program), Washington, D.C. June 1982.

Pitt, R. *Characterizing and Controlling Urban Runoff through Street and Sewerage Cleaning.* U.S. Environmental Protection Agency, Storm and Combined Sewer Program, Risk Reduction Engineering Laboratory. EPA/600/S2-85/038. PB 85-186500. Cincinnati, OH. 467 pp. June 1985.

Pitt, R. and J. McLean. *Toronto Area Watershed Management Strategy Study: Humber River Pilot Watershed Project.* Ontario Ministry of the Environment, Toronto, Ontario. 486 pp. 1986.

Pitt, R. *Small Storm Urban Flow and Particulate Washoff Contributions to Outfall Discharges.* Ph.D. dissertation submitted to the Department of Civil and Environmental Engineering, University of Wisconsin, Madison. 1987.

Pitt, R. and K. Parmer. *Quality Assurance Project Plan: Effects, Sources, and Treatability of Stormwater Toxicants.* Contract No. CR819573. U.S. Environmental Protection Agency, Storm and Combined Sewer Program, Risk Reduction Engineering Laboratory. Cincinnati, OH. February 1995.

Pitt, R., R. Field, M. Lalor, and M. Brown. Urban stormwater toxic pollutants: assessment, sources and treatability. *Water Environ. Res.,* 67(3): 260–275. May/June 1995.

Pitt, R., J. Lantrip, R. Harrison, C. Henry, and D. Hue. *Infiltration through Disturbed Urban Soils and Compost-Amended Soil Effects on Runoff Quality and Quantity.* U.S. Environmental Protection Agency, Water Supply and Water Resources Division, National Risk Management Research Laboratory. EPA 600/R-00/016. Cincinnati, OH. 231 pp. December 1999a.

Pitt, R., B. Robertson, P. Barron, A. Ayyoubi, and S. Clark. *Stormwater Treatment at Critical Areas: The Multi-Chambered Treatment Train (MCTT)*. U.S. Environmental Protection Agency, Wet Weather Flow Management Program, National Risk Management Research Laboratory. EPA/600/R-99/017. Cincinnati, OH. 505 pp. March 1999b.

Pitt, R. and M. Lalor. *Identification and Control of Non-Stormwater Discharges into Separate Storm Drainage Systems. Development of Methodology for a Manual of Practice*. U.S. Environmental Protection Agency, Water Supply and Water Resources Division, National Risk Management Research Laboratory, Cincinnati, OH. 451 pp. To be published in 2001.

Reynoldson, T.B. Interactions between sediment contaminants and benthic organisms. *Hydrobiologia*, 149 (53–66). 1987.

Roa-Espinosa, A. and R. Bannerman. Monitoring BMP effectiveness at industrial sites. *Proceedings of the Engineering Foundation Conference on Stormwater NPDES Related Monitoring Needs*. Aug 7–12 1994. Sponsored by U.S. Environmental Protection Agency; The Engineering Foundation; U.S. Geological Survey; Water Environment Federation; American Institute of Hydrology, ASCE. pp. 467–486. 1994.

Rolfe, G.L. and K.A. Reinhold. *Vol. I: Introduction and Summary. Environmental Contamination by Lead and Other Heavy Metals*. Institute for Environmental Studies, University of Illinois. Champaign-Urbana, IL. July 1977.

Rossum, J.R. Checking the accuracy of water analyses through the use of conductivity. *J. Am. Water Works Assoc.*, 67(4): 204–205. April 1975.

Rubin, A.J. (Ed.). *Aqueous-Environmental Chemistry of Metals*. Ann Arbor Science Publishers, Ann Arbor, MI. 1976.

Rutledge, P.A. and J. Fleeger. Laboratory studies on core sampling with application to subtidal meiobenthos collection. *Limnol. Oceanogr.*, 30: 422–426. 1985.

Sarda, N. and G.A. Burton. Ammonia variation in sediments: spatial, temporal and method-related effects. *Environ. Toxicol. Chem.*, 14: 1499–1506. 1995.

Sartor, J. and G. Boyd. *Water Pollution Aspects of Street Surface Contaminants*. EPA-R2-72-081, U.S. Environmental Protection Agency. November 1972.

Sasson-Brickson, G. and G.A.J. Burton, Jr. *In situ* and laboratory sediment toxicity testing with *Ceriodaphnia dubia. Environ. Toxicol. Chem.*, 10: 201–207. 1991

Sayles, F.L., T.R.S. Wilson, D.N. Hume, and P.C. Mangelsdorf, Jr. *In situ* sampler for marine sedimentary pore waters: evidence for potassium depletion and calcium enrichment. *Science*, 180: 154–156. 1973.

Schults, D.W., S.P. Ferraro, L.M. Smith, F.A. Roberts, and C.K. Poindexter. A comparison of methods for collecting interstitial water for trace organic compounds and metals analyses. *Water Res.* 26: 989–995. 1992.

Schuytema, G.S., A.V. Nebeker, W.L. Griffis, and C.E. Miller. Effects of freezing on toxicity of sediments contaminated with DDT and endrin. *Environ. Toxicol. Chem.*, 8: 883–891. 1989.

Sherman, L.A., L.A. Baker, E.P. Weir, and P.L. Brezonik, Sediment pore water dynamics of Little Rock Lake, WI: geochemical processes and seasonal and spatial variability. *Limnol. Oceanogr.*, 39: 1155–1171. 1994.

Spliethoff, H.M. and H.F. Hemond. History of toxic metal discharge to surface waters of the Aberjona watershed. *Environ. Sci. Tech.*, 30(1): 121. January 1996.

Standard Methods for the Examination of Water and Wastewater. 19th edition. Water Environment Federation, Washington, D.C. 1995.

Stemmer, B.L., G.A. Burton, Jr., and S. Leibfritz-Frederick. Effect of sediment test variables on selenium toxicity to *Daphnia magna. Environ. Toxicol. Chem.*, 9: 381–389. 1990a.

Stemmer, B.L., G.A. Burton, Jr., and G. Sasson-Brickson. Effect of sediment spatial variance and collection method on cladoceran toxicity and indigenous microbial activity determinations. *Environ. Toxicol. Chem.*, 9: 1035–1044. 1990b.

Swartz, R.C., W.A. DeBen, J.K. Jones, J.O. Lamberson, and F.A. Cole. Phoxocephalid amphipod bioassay for hazard assessment. Seventh Symposium, ASTM STP 854. Edited by R.D. Cardwell, R. Purdy, and R.C. Bahner. American Society for Testing and Materials, Philadelphia. pp. 284–307. 1985.

Thomann, R.V. and J.A. Mueller. *Principles of Surface Water Quality Modeling and Control*. Harper & Row., New York. 1987.

Troup, B.N., O.P. Bricker, and J.T. Bray. Oxidation effect on the analysis of iron in the interstitial water of recent anoxic sediments. *Nature*, 249: 237–239. 1974.

U.S. Geological Survey (USGS). Techniques of water-resources investigations of the U.S.G.S. Chp. C1, in *Laboratory Theory and Methods for Sediment Analysis*, Book 5. Edited by H. P. Guy. U.S. Geological Survey, Arlington, VA. p. 58, 1969.

Van Rees, K.C.J., E.A. Sudlicky, P. Suresh, C. Rao, and K.R. Reddy. Evaluation of laboratory techniques for measuring diffusion coefficients in sediments. *Environ. Sci. Technol.*, 25: 1605–1611. 1991.

Whiticar, M.J. Determination of interstitial gases and fluids in sediment collected with an *in situ* sampler. *Anal. Chem.,* 54: 1796–1798. 1982.

Winter, T.C., J.W. LaBaugh, and D.O. Rosenberry. The design and use of a hydraulic potentiomanometer for direct measurement of differences in hydraulic head between groundwater and surface water. *Limnol. Oceanogr.,* 33(5): 1209–1214. 1988.

Ecosystem Component Characterization

"Things don't turn up in this world until somebody turns them up."

James A. Garfield

CONTENTS

Overview 346
    Ecosystem Structure and Integrity, Chaos and Disturbance 346
Flow and Rainfall Monitoring 349
    Flow Requirements for Aquatic Biota 350
    Urban Hydrology...351
    Pollutant Transport 356
    Flow Monitoring Methods 357
    Rainfall Monitoring...377
Soil Evaluations 388
    Case Study to Measure Infiltration Rates in Disturbed Urban Soils 389
    Observations–Infiltration Rates in Disturbed Urban Soils 394
    Water Quality and Quantity Effects of Amending Soils with Compost 397
Aesthetics, Litter, and Safety..398
    Safety Characteristics 398
    Aesthetics, Litter/Floatables, and Other Debris..398
Habitat 400
    Factors Affecting Habitat Quality 403
    Channelization 404
    Substrate 406
    Riparian Habitats 409
    Field Habitat Assessments...410
    Temperature 410
    Turbidity 413
    Dissolved Oxygen 417
Water and Sediment Analytes and Methods...423
    Selection of Analytical Methods...423
    Use of Field Methods for Water Quality Evaluations 425
    Conventional Laboratory Analyses 447
    Use of Tracers to Identify Sources of Inappropriate Discharges to Storm Drainage
    and Receiving Waters 459

        Hydrocarbon Fingerprinting for Investigating Sources of Hydrocarbons...........................483
Microorganisms in Stormwater and Urban Receiving Waters...485
        Determination of Survival Rates for Selected Bacterial and Protozoan Pathogens 487
Benthos Sampling and Evaluation in Urban Streams 491
        Periphyton Sampling 493
        Protozoan Sampling...494
        Macroinvertebrate Sampling 494
Zooplankton Sampling..502
Fish Sampling 502
        Indices of Fish Populations...504
Toxicity and Bioaccumulation 507
        Why Evaluate Toxicity? 507
        Stormwater Toxicity 513
        Pulse Exposures...514
        Measuring Effects of Toxicant Mixtures in Organisms..515
        Standard Testing Protocols: Waters...517
        Standard Testing Protocols: Sediments..527
        In Situ Toxicity Testing 530
        Bioaccumulation 534
        Emerging Tools for Toxicity Testing 536
Summary 546
References 550

OVERVIEW

Ecosystem Structure and Integrity, Chaos and Disturbance

It is impossible to produce meaningful, representative, and reliable data to be used in decisions regarding the status of, or possible impacts to, the environment without first defining the environment, critical receptors, influencing factors, and natural dynamics. This requires the measurement of many aspects of the watershed, as previously described in this book. Simplistic and rapid approaches are fine for initial assessments, but may fall short in providing understanding of the causes of the observed problems. Therefore, later phases of watershed assessment projects generally need to examine more detailed aspects of a study area in order to obtain a better understanding of possible interactions. As an example, the majority of studies dealing with aquatic toxicity have used surrogate species (or a small number of species) and have not attempted to investigate ecosystem interactions *a priori*, such as ecosystem energetics or stress–productivity–predation relationships. For example, surrogate responses have simply been quantified based on sample toxicity, and then effects have been extrapolated to *in situ* conditions. While these exercises might satisfy the study objectives of defining sample toxicity to the test species, they do little to document or define ecosystem disturbance. Ecological processes can be ignored, to a degree, when acute toxicity scenarios are studied, such as in sediments that are severely degraded. However, "significant cases of acute toxic effects have been encountered infrequently" (Chapman 1986), and the more common situations in which effects and zones of contamination are "gray" (Chapman 1986) dictate that natural and anthropogenic effects be separated. This cannot be done accurately without an understanding of ecosystem dynamics such as spatial and temporal variance of chemical, physical, and biological systems and their interactive processes.

Community ecology in lotic and lentic systems has progressed substantially in recent years. "Biotic dynamics and interactions are intimately and inextricably linked to variation in abiotic factors" (Power et al. 1988), and lotic systems are not in equilibrium due to natural disturbances which may occur frequently or infrequently (Resh et al. 1988). Disturbance can be defined as a discrete event that alters community structure and changes the physical environment and resource availability. These disturbances vary in type, frequency, and severity, both among and within ecoregions. The frequency and intensity of disturbances cannot be predicted (Resh et al. 1988). Intermediate levels of disturbance maximizes species richness (Resh et al. 1988). Equilibrium or steady-state conditions will tend to occur if disturbances are infrequent, thus excluding opportunistic species (Minshall 1988). In stream ecology, disturbance is the dominating organizing factor, having a "major impact on productivity, nutrient cycling and spiraling, and decomposition" (Resh et al. 1988). Disturbances such as storm events or the presence of toxicants can eliminate biota (Power et al. 1988). Recovery and succession of these systems between disturbances is typified by recurrent or divergent patterns (Pringle et al. 1988; Resh et al. 1988). Despite this inherent variability, benthic communities have been used effectively to classify community structure and functioning in aquatic ecosystems.

Ecotones are defined as zones of transition between adjacent ecosystems. Disturbance plays a major role in determining the structure and dynamics of ecotones, such as stream bank riparian zones. High relief areas are less stable due to more frequent and diverse disturbances combined with complex topographic effects. Both fluvial and geomorphic processes influence vegetation development along stream and lake embankments (Decamps et al. 1990).

The major role that natural and anthropogenic disturbances have on aquatic ecosystems increases the level of spatial and temporal variance. Spatial and temporal dimensions span 16 orders of magnitude in stream ecology (Minshall 1988; Pringle et al. 1988). Some suggest that spatial heterogeneity enhances the ability of an ecosystem to resist and recover after a disturbance (Fisher 1990). Significant spatial variance in sediments is common (Stemmer et al. 1990). Each level of the system has different dimensions, has different variances associated with it, and is interacting simultaneously with other ecosystem levels and their respective dimensions and variances. This complex reality is difficult, if not impossible, to define accurately but must be considered in all assessments of water quality or ecosystem health.

Orians (1980) stated that one of the greatest challenges in ecology (and ecotoxicology) is bridging the conceptual gap between micro- and macroecology. Aquatic systems can be considered as a mosaic of patches (Pringle et al. 1988). "A patch is a spatial unit that is determined by the organism and problems in question" (Pringle et al. 1988). The heterogeneous environment has highly clumped distributions (patches) of organisms whose spatial and temporal patterns and relationships change seasonally due to factors such as food (resource) patterns (Findlay 1981). These clumped distributions, therefore, pose severe sampling problems. The appropriate sampling scale will depend on the organism size, density, distribution, life cycle, and question being asked (Pringle et al. 1988), which, unfortunately, are often not considered. Aquatic ecosystems are open nonequilibrium systems (Carpenter et al. 1985; Pringle et al. 1988) where patches are in transitory steady state with other patches (Sheldon 1984). Many "ecosystems" are not independent units, and some processes (e.g., nutrient cycling) show no spatial threshold. That is, no one area bounds all processes, showing that ecosystems have both an open nature and are connected in many complex ways. Most aquatic organisms are aggregated at certain spatial scales and are random on other scales. In order to accurately determine total organism numbers and distribution patterns (patches) within and among sites, presampling should be conducted whereby the site is divided into quadrants, sampled, and coefficients of variation (standard deviation divided by the mean) determined. This will detect heterogeneity in density measurements (Westman 1985). Unfortunately, this level of accuracy is often beyond the resource capabilities of typical studies. Different life histories and variable interactions between species may prevent equilibrium (Carpenter et al. 1985).

Ecosystems tend to restore balance (homeostasis or resilience). Diversity does not equate to integrity. Biological integrity may be defined as the ability of species to interact and maintain their structure and function in some self-regulating, homeostatic fashion (Westman 1985). The rate, manner, and extent of recovery following a disturbance is a measure of resilience.

The influence of storm events and watershed characteristics on chemical element dynamics is poorly understood, particularly because some are lumped into operationally defined units such as dissolved or total organic carbon. Significant heterogeneity (62 to 100%) has been observed between adjacent sediment cores in concentrations of organic matter, water, and total phosphorus (Downing and Roth 1988). Some heterogeneity is likely due to small-scale variations in bottom profiles.

In stream benthic communities, hydraulics appear to be more important than substrate in determining distribution (Statzner et al. 1988). As in fish communities, populations will vary in type and number between pool and riffle areas. Most benthic macroinvertebrate testing occurs in riffle areas where continual flow exists and more types of organisms are present. Small-scale sampling is more likely to define benthic invertebrate patches than large-scale sampling, which homogenizes patchiness differences. The replicate number needed to obtain a given precision decreases with increased density and sample size, and the optimal sample size (considering cost and precision) depends on mean density (Morin 1985).

Other important considerations in valid hazard assessments are contaminant interactions and subsequent distribution in the aquatic system via solids. Sediment contaminant data should be evaluated based on grain size correction, which reduces the inert fractions (e.g., hydrates, sulfides, amorphous, and fine-grained organics). The most useful size fraction for contaminant assessments appears to be <63 μm (Håkanson 1984). This size fraction will tend to predominate in deposition areas and will play a major role in the transport, deposition, and resuspension of the fine-grained sediments. Particle diameters of suspended solids vary over two orders of magnitude and settling speeds in waters vary over four orders of magnitude (Gailani et al. 1991). Predicting transport is complicated by the lack of understanding of sizes and settling speeds, floc disaggregation due to shear, processes governing entrainment and deposition, and turbulence description (Gailani et al. 1991).

When resuspension events occur, predicting metal remobilization may be possible in site-specific studies; however, remobilization is dependent on particle residence time in the water column, which varies between sites, storms, and systems. In most systems, however, remobilization of metals from resuspended sediments is likely to be insignificant due to the slow reaction rates (Kersten and Forstner 1987).

Though resuspension effects appear limited if one considers the scavenging effects of solids, laboratory studies of bioturbation effects on contaminant movement and toxicity to planktonic species have shown otherwise. Bioturbation by benthic and epibenthic invertebrates occurs in many ways: by pumping pore water constituents out of the sediment into overlying waters; by injecting water into the sediment; by pumping particulates to the sediment-water interface; by depositing fecal pellets on the sediment surface; and by disrupting horizontal and vertical layering (Petr 1977).

Given the above discussion on the complexities of aquatic ecosystems, it is evident that it is no longer adequate to simply study separate components of the ecosystem, such as planktonic species in water-only systems or chemical dynamics in a water-only or sediment slurry system. This "reductionist" approach is essential for defining processes, but does not provide an accurate picture of the component–ecosystem interactions and, in fact, may produce misleading results. Examples of this disparity are becoming increasingly obvious, particularly in the field of aquatic toxicology, as more "holistic" types of studies are published (Chapman et al. 1992).

Sediments play a major role in ecosystem processes and ecosystem health (Chapman et al. 1992). Generally speaking, the surficial layer (upper few centimeters) is the active portion of the ecosystem, while deeper sediments are passive and more permanently "in-place." These deeper layers are of interest as a historical record of ecosystem activity, but may also be reintroduced into the active portion of the ecosystem via dredging activities and severe storm or hydrogeological events. The usefulness of a sediment monitoring station as an indicator of contaminant presence is a function of the interactions between the change in contaminant net deposition rate, sediment accumulation rate, mixing zone depth and dynamics, sampling method and frequency, the type of laboratory method, and its precision and accuracy (Larsen and Jensen 1989). Sediments and soils typically exhibit more spatial variability from overlying waters but less temporal variation. This reality affects sampling design and statistical analyses.

This chapter describes a wide variety of tools that can be used for assessing the ecosystem and watershed physical characteristics because of the likely need to consider a broad range of assessment procedures. This chapter starts with discussions of rainfall and flow monitoring, as it is difficult to understand pollutant transport, fate, and effects without appreciating the physical movement of the water. The main sections of this chapter pertain to examinations of specific receiving water uses and associated ecosystem components: aesthetics and safety, habitat, water and sediment, microorganisms, benthos, zooplankton, fish, and toxicity and bioaccumulation.

FLOW AND RAINFALL MONITORING

It is essential that there be an accurate description of the system's hydrodynamics when assessing the effects of stormwater runoff on receiving waters. Flow represents the pollutant loading mechanism, and its power and frequency of occurrence can degrade the physical habitat. One of the principal reasons there is a relatively poor understanding of stormwater runoff effects is because of the difficult logistics involved in measuring short-term, high-flow events quickly and accurately. Flow and rainfall monitoring are considered separately from other physical characteristics, which are discussed in the following section on habitat. The hydrology of the stream, reservoir, or lake which receives stormwater runoff is interrelated, directly and indirectly, with many other characteristics, such as substrate composition, temperature, suspended solids, channel morphology, and biological communities. Hydrology, as discussed here, is composed of flow, velocity, power, turbulence, mixing, sedimentation, and resuspension subcomponents. Each of these subcomponents is important to varying degrees depending on the site and study objectives, and each is relatively difficult to quantify accurately during storm events.

As with other major ecosystem components, the storm event hydrodynamics of the receiving water must be evaluated based on references for comparison. References may include an upstream station, present day baseflow conditions, predevelopment conditions, and/or a least disturbed watershed of similar natural characteristics (e.g., soil, topography, drainage area, stream order, stream substrate, biological communities). The assessment should attempt to characterize the hydrology of the system by defining the loading dynamics (i.e., magnitude, duration, frequency) and the receiving system response (e.g., flow, spatial-temporal patterns). The physical characterization of the loading and system response will dictate the chemical sampling from which to determine pollutant (stressor) loading dynamics and optimal stormwater control programs and associated remediation measures.

The rate of stream discharge (flow) (Q) is a function of the channel cross-sectional area (A) and the mean velocity (V), which is usually expressed as cubic feet per second (cfs). So, $Q = AV$. Velocity is a function of runoff quantity, stream width, depth, and gradient, and channel roughness. Roughness is affected by channel sinuosity, substrate size, bottom topography, stream vegetation, debris, and other obstructions. Channelization increases velocity and also tends to make velocity more uniform (EPA 1983). Channelization practices, such as straightening, vegetation and debris removal, berming and leveeing, all increase drainage efficiency. These practices also produce sharper flow hydrographs, with much greater peak flow rates. The resulting higher flow rate and power increases the impact of storm events, including increased scour, erosion, bank cutting, sediment transport, flooding below the channelized section, reduced groundwater levels and stream dewatering, degraded habitat and water quality, promotion of land development, and lowered recreational values. Assessing channelization effects on habitat quality is discussed more fully in the following section on habitat.

Stream staff gauges, which measure stream depth, may be used to indirectly measure flow through the use of a rating curve which shows the relationship between stream depth and flow rate. The rating curve is developed by making velocity measurements in a cross-sectional area of the stream channel where the channel morphology and flow patterns are simple. This is done over a range of flows so that the curve can be constructed. This is discussed in a following subsection.

Stream power is the rate of potential energy expenditure per unit weight of water in a channel and is calculated as:

$$SP = \frac{\Delta Y}{\Delta t} = \frac{\Delta X}{\Delta t} \frac{\Delta Y}{\Delta X} = VS_f$$

where SP = stream power (ft-lbs/lb H_2O/s)
     t = time (s)
     V = velocity (ft/s)
     S_f = stream friction slope (ft/ft) (energy gradient)
     Y = energy grade line elevation above a point, equivalent to potential energy (ft-lb)/lb/H_2O) = water surface elevation and velocity head ($V^2/2g$)
     X = longitudinal distance
     g = gravitational constant

Stream power can be used to estimate the energy available for sediment transport. This energy can be reduced by other habitat factors (e.g., bank and substrate stability, vegetation, or surface erosion).

"Time of passage" has been recommended as a parameter of pollutant movement through a stream more useful than the kinematic wave velocity that is typically used in hydrograph routing calculations (Velz 1970). The distinction is that the kinematic wave (hydrograph crest) moves faster than the waste in the body of water, particularly in large, deep water systems. Time of passage (as seconds or days) is based on the average flow rates that are measured when using current meters. It is determined by dividing the occupied channel volume (from cross-sectional area) (as cubic feet) by the runoff (from drainage area and yield) (as cfs).

Flow Requirements for Aquatic Biota

A popular evaluation tool for evaluating flow effects on aquatic communities was published by Tennant (1976). He found the following in 11 streams of three western states:

- Changes in habitat were similar among streams with similar flow regimes.
- A depth of at least 0.3 m and velocity of at least 0.75 ft/s were required for most fish.
- Thirty percent of the annual flow provided good habitat.
- Sixty percent of the annual flow provided outstanding habitat.

Stream velocity plays a major role in determining the composition of benthic communities (Cummins 1975): invertebrate drift increases as the velocity increases (Minshall and Winger 1968; Walton 1977; Zimmer 1977).

The U.S. Fish and Wildlife Service developed the Instream Flow Incremental Methodology (IFIM) computer program to evaluate changes in aquatic life from alteration of channel morphology, water quality, and hydraulic components. Each species has a range of habitat (including flow) conditions it can tolerate, which can be defined (or is defined) for the species, as can stream conditions. IFIM simulates hydraulic conditions — habitat availability for a species and size class, or usable waters for a particular recreational activity. This is done through use of the Physical Habitat Simulation Model (PHABSIM), which relates changes in flow and channel structure to changes in physical habitat availability.

The basic steps in the IFIM can be summarized by the following:

- Project scoping — Define objectives for the delineation of study area boundaries, determine the species, and define their life history, food types, water quality tolerances, and microhabitat.
- Study reach and site selection — Identify and delineate critical reaches to be sampled, delineate major changes and transition zones and the distribution of the evaluation species.

- Data collection — Transects are selected to adequately characterize the hydraulic and in-stream habitat conditions. Data gathering must be compatible with IFIH computer models.
- Computer simulation — Reduce field data and input into programs described above.
- Interpretation — The output is expressed as the Weighted Usable Area (WUA), a discrete value for each representative and critical study reach, for each life stage and species, and for each flow regime.

For further information on IFIM and PHABSIM, consult *A Guidance to Stream Habitat Analysis Using the Instream Flow Incremental Methodology*. U.S. FWS/OBS-82/26, June 1982.

Urban Hydrology

Basic watershed characteristics need to be known in order to understand stream flow conditions. These include topography (watershed divide plus stream and land slopes), drainage efficiency (stream orders and types of artificial drainage systems), and, to a lesser extent in urban areas, soil characteristics (disturbed or compacted, age since development, type of ground cover, soil texture, etc.). It is important that characteristics throughout the watershed be evaluated when studying streams. Looking only at characteristics adjacent to the stream is very misleading, as urban drainage systems are very efficient transporting systems, capable of carrying water and pollutants to the stream from locations far away. These topics are beyond the scope of this book, but several good books are available that describe urban hydrology and associated drainage design (including McCuen 1989; WEF and ASCE 1992; Debo and Reese 1995; and Wanielista et al. 1997).

Urban hydrology can be used to divide rain into different major categories, each reflecting distinct portions of the long-term rainfall record (Pitt et al. 1999). When monitoring runoff, it is therefore important to include a sampling effort that represents each of these categories. All too often, the small rains are not sampled because of misunderstandings of their significance. It is easy to ignore these small events, considering the problems that occur when trying to program automatic sampling equipment. However, small events are extremely important when conducting a receiving water investigation. As an example, consider the following rainfall and runoff data for Milwaukee, WI, what were obtained during the Nationwide Urban Runoff Program (EPA 1983). Figure 6.1 shows measured rain and runoff distributions for Milwaukee during the 1981 NURP monitored rain year. Rains between 0.05 and 5 in were monitored during this period. Two very large events

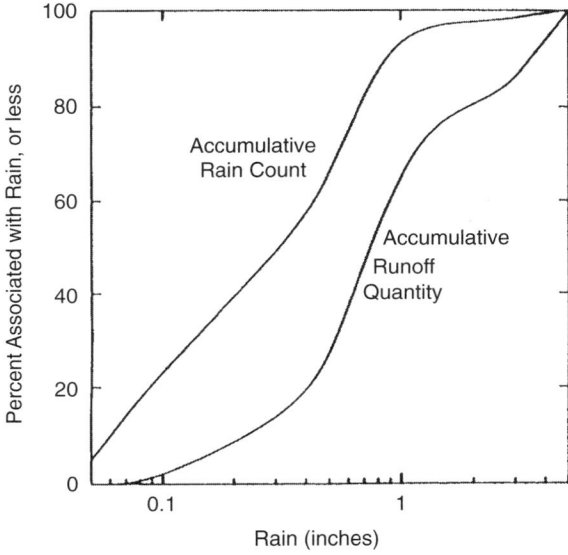

Figure 6.1 Milwaukee rain and runoff distributions.

(greater than 3 in) occurred during this monitoring period, which greatly bias this distribution, compared to typical rain years. The following observations are evident:

- The median rain depth was about 0.3 in.
- 66% of all Milwaukee rains are less than 0.5 in depth.
- For the medium-density residential area, 50% of the runoff was associated with rains less than 0.75 in for Milwaukee.
- Observable runoff occurred with rain as small as 0.05 in in depth.

In addition, a 100-year, 24-hour rain of 5.6 in for Milwaukee could produce about 15% of the typical annual runoff volume, but it only contributes about 0.15% of the average annual runoff volume, when amortized over 100 years. Typical 25-year drainage design storms (4.4 in in Milwaukee) produce about 12.5% of typical annual runoff volume but only about 0.5% of the average runoff volume.

Figure 6.2 shows measured Milwaukee pollutant discharges associated with different rain depths for a monitored medium-density residential area. Suspended solids, COD, lead, and phosphate discharges are seen to closely follow the runoff distribution shown in Figure 6.1. Therefore, the concentrations of most runoff pollutants do not vary significantly for runoff events associated with different rain depths.

The monitored rains at this Milwaukee medium-density residential location can be divided into four categories:

- <0.5 inch. These rains account for most of the events, but little of the runoff volume. They produce much less pollutant mass discharge and probably have fewer receiving water effects than other rains. However, the runoff pollutant concentrations likely exceed regulatory standards for several categories of critical pollutants, especially bacteria and some total recoverable metals. They also cause large numbers of overflow events in uncontrolled combined sewers. These rains are very common, occurring once or twice a week (accounting for about 60% of the total rainfall events and about 45% of the total runoff events that occurred), but they only account for about 20% of the annual runoff and pollutant discharges. Rains less than about 0.05 in did not produce noticeable runoff.
- 0.5 to 1.5 inches. These rains account for the majority of the runoff volume (about 50% of the annual volume for this Milwaukee example) and produce moderate to high flows. They account for about 35% of the annual rain events, and about 20% of the annual runoff events. These rains

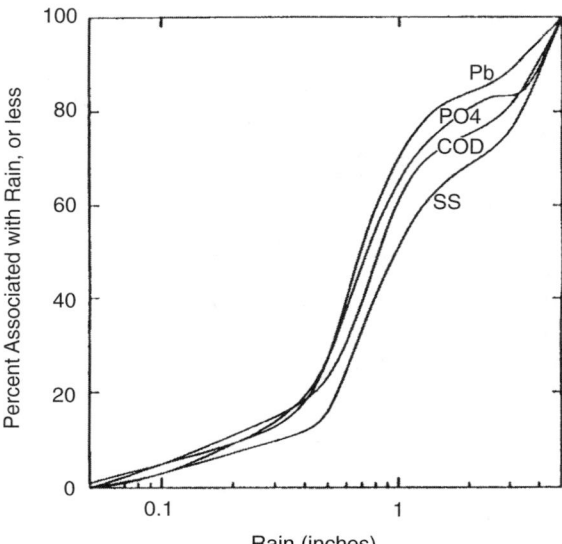

Figure 6.2   Milwaukee pollutant discharge distributions.

occur on the average about every 2 weeks from spring to fall and subject the receiving waters to frequent high pollutant loads and moderate to high flows.

- 1.5 to 3 inches. These rains produce the most damaging flows, from a habitat destruction standpoint, and occur every several months (at least once or twice a year). These recurring high flows, which were historically associated with much less frequent rains, establish the energy gradient of the stream and cause unstable stream banks. Only about 2% of the rains are in this category, and they are responsible for about 10% of the annual runoff and pollutant discharges. Typical storm drainage design events fall in the upper portion of this category.
- >3 inches. The smallest rains in this category are included in design storms used for drainage systems in Milwaukee. These rains occur only rarely (once every several years to once every several decades, or even less frequently) and produce extremely large flows. The monitoring period during the Milwaukee NURP program was unusual in that two of these events occurred. Less than 2% of the rains were in this category (typically <<1%), and they produced about 15% of the annual runoff quantity and pollutant discharges. During a "normal" period, these rains would produce only a very small fraction of the annual average discharges. However, when they do occur, great property and receiving water damage results. Receiving waters can conceivably recover naturally from this damage (mostly associated with habitat destruction, sediment scouring, and the flushing of organisms great distances downstream and out of the system) and return to before-storm conditions within a few years, depending on riparian vegetation growth rates and nearby "reservoir or refugia" areas for aquatic organisms.

The above specific rain values are given for Milwaukee, WI, selected because of the occurrence of two very rare rains during an actual monitoring period. Obviously, the critical values defining the design storm regions would be highly dependent on local rain and development conditions. Computer modeling analyses from 24 urban locations from throughout the United States were conducted by Pitt et al. (1999) to examine these patterns nationwide. These locations represent most of the major river basins and much of the rainfall variations in the country. These simulations were based on 5 to 10 years of rainfall records, usually containing about 500 individual rains each. The rainfall records were from certified NOAA weather stations and were obtained from CD-ROMs distributed by EarthInfo of Boulder, CO. Hourly rainfall depths for the indicated periods were downloaded from the CD-ROMs into an Excel spreadsheet. This file was then read by a utility program included in the Source Loading and Management Model (SLAMM) package (Pitt and Voorhees 1995). This rainfall file utility combined adjacent hourly rainfall values into individual rains, based on user selections (at least 6 hours of no rain was used as the criterion to separate adjacent rain events and all rain depths were used, with the exception of the "trace" values that were <0.01 in). These rain files for each city were then used in SLAMM for typical medium-density and strip commercial developments. SLAMM utilizes unique prediction methods that were especially developed by Pitt (1987) to accurately predict runoff during these small rains. Conventional runoff prediction methods are based on drainage design storms (of several inches in depth) and are not accurate when predicting runoff during small rains.

Table 6.1 summarizes these rain and runoff distributions for these different U.S. locations. Lower and upper runoff distribution breakpoints were identified on all of the individual distributions. The breakpoints separate the distributions into the following three general categories (similar to the regions identified for the Milwaukee rains):

- Less than lower breakpoint: small, but frequent rains. These generally account for 50 to 70% of all rain events (by number), but only produce about 10 to 20% of the runoff volume. The rain depth for this breakpoint ranges from about 0.10 in in the arid Southwest, to about 0.5 in in the wet Southeast. These events are most important because of their frequencies, not because of their mass discharges. They are therefore of great interest where water quality violations associated with urban stormwater occur. This would be most common for bacteria (especially fecal coliforms) and for total recoverable heavy metals, which typically exceed receiving water numeric criteria during practically every rain event in heavily urbanized drainages having separate stormwater drainage systems.

Table 6.1 Rainfall and Runoff Distribution Characteristics for Different Locations from Throughout the U.S. (Pitt, et al. 1999)

	Median Rain Depth, by Count (in)	Corresponding Percentage of Runoff for the Median Rain Depth	Rain Depth Associated with Median Runoff Depth (in)	Lower Breakpoint Rain Depth (in)	Percentage of Rain Events Less Than Lower Breakpoint	Percentage of Runoff Volume Less Than Lower Breakpoint	Upper Breakpoint Rain Depth (in)	Percentage of Rain Events Less Than Upper Breakpoint	Percentage of Runoff Volume Less Than Upper Breakpoint	Percentage of Runoff Volume Between Breakpoints	Percentage of Rain Events Between Breakpoints
Columbia North Pacific											
Boise, ID	0.07	3–5	0.30–0.35	0.10	52	9–11	0.91	99	89–93	80–82	47
Seattle, WA	0.12	4–6	0.62–0.80	0.18	60	8–11	3.4	99	92–96	84–85	39
California											
Los Angeles, CA	0.18	3–5	1.2–1.5	0.29	64	7–10	3.5	99	92–98	85–88	35
Great Basin											
Reno, NV	0.07	3–5	0.35–0.41	0.10	61	8–10	1.7	99	93–95	85	38
Lower Colorado											
Phoenix, AZ	0.10	4–6	0.55–0.68	0.19	64	9–12	2.3	99	94–98	85–87	35
Missouri											
Billings, MT	0.06	2–4	0.55–0.60	0.12	64	8–10	1.6	99	89–93	81–83	35
Denver, CO	0.08	2–4	0.50–0.60	0.19	71	13–17	1.8	99	91–95	78	28
Rapid City, SD	0.06	2–4	0.50–0.55	0.15	69	10–13	1.9	99	92–96	82–83	30
Arkansas-White-Red											
Wichita, KS	0.13	2–5	1.1–1.4	0.31	65	10–13	3.0	99	88–93	78–80	34
Texas Gulf											
Austin, TX	0.14	2–3	1.4–1.8	0.50	72	8–12	6.0	99	88–94	80–82	27
Upper Mississippi											
Minneapolis, MN	0.11	3–5	0.73–1.0	0.22	65	9–13	2.8	99	94–96	83–85	34
Madison, WI	0.12	3–5	0.78–0.98	0.23	65	9–13	3.5	99	97–99	86–88	34
Milwaukee, WI	0.12	2–4	0.9–1.1	0.25	65	9–12	2.5	99	89–95	80–83	34
St. Louis, MO	0.14	4–6	1.0–1.2	0.31	65	10–13	2.8	99	90–95	80–82	34

Great Lakes										
Detroit, MI	0.20	7–11	0.72–0.81	0.20	50	7–11	2.4	99	92–95	85–84	49
Buffalo, NY	0.11	2–4	0.61–0.72	0.12	64	8–12	2.1	99	88–93	80–81	35
Ohio										
Columbus, OH	0.12	3–5	0.80–1.0	0.22	63	8–12	2.2	99	85–91	77–79	36
North Atlantic										
Portland, ME	0.15	2–4	1.1–1.5	0.30	64	8–12	4.5	99	90–96	82–84	35
Newark, NJ	0.28	6–12	1.2–1.5	0.33	54	8–12	3.3	99	89–94	81–82	45
Lower Mississippi										
New Orleans, LA	0.25	3–5	1.7–2.2	0.45	62	7–11	4.0	99	88–93	81–82	37
South Atlantic Gulf										
Atlanta, GA	0.22	3–5	1.2–1.7	0.32	58	5–9	4.0	99	91–95	86	41
Birmingham, AL	0.20	3–5	1.2–1.5	0.40	64	8–13	5.0	99	90–96	82–83	35
Raleigh, NC	0.18	4–6	1.0–1.2	0.26	60	7–11	2.5	99	87–93	80–82	39
Miami, FL	0.13	3–5	1.2–1.6	0.30	67	9–13	4.0	99	87–93	78–80	32

From Pitt, R. et al. *Guidance Manual for Integrated Wet Weather Flow (WWF) Collection and Treatment Systems for Newly Urbanized Areas (New WWF Systems). Second year project report: model integration and use.* Wet Weather Research Program, U.S. Environmental Protection Agency. Cooperative agreement #CX 824933-01-0. February 1999.

- Between the lower and upper breakpoint: moderate rains. These rains generally account for 30 to 50% of all rain events (by number), but produce 75 to 90% of all the runoff volume. The rain depths associated with the upper breakpoint range from about 1 to 2 in in the arid parts of the United States and up to 5 or 6 in in wetter areas. These runoff volume distributions are approximately the same as the pollutant distributions. Therefore, these intermediate rains also account for most of the pollutant mass discharges and many of the actual receiving water problems associated with stormwater discharges.
- Above the upper breakpoint: large but rare rains. These rains include the typical drainage design events and are therefore quite rare. During the period analyzed, many of the sites only had one or two, if any, events above this breakpoint. These rare events do account for about 5 to 10% of the runoff on an annual basis. Obviously, these events must be evaluated to ensure adequate drainage.

The fourth category, evident in the Milwaukee monitoring results and shown in Figures 6.1 and 6.2, was not obvious during these computer analyses. These extremely rare events, which exceed the drainage capacity of most areas, do not significantly affect these long-term probability distributions. During the isolated years when they occur, such as during the monitoring period in Milwaukee, they have significant effects, but when averaged over long periods, their contributions diminish rapidly.

The small rains, generally less than about 0.5 in, are very important in a wet-weather monitoring program. They represent the vast majority of rains that occur in an area, and may represent the majority of runoff events. Water quality violations associated with wet-weather flows are typically common for these events. Similarly, the medium-sized events (from the 0.5-in rains to rains of several inches in depth) contribute the majority of runoff volume and mass pollutant discharges and are therefore likely responsible for most of the biological effects (especially habitat destruction and sediment contamination) in receiving waters. The largest rains (greater than several inches) are the primary focus of drainage design. Therefore, efforts must be made to characterize runoff and receiving water conditions in each of these different categories in order to understand the varying receiving water problems that may be occurring.

Pollutant Transport

The routing of pollutants through a watershed is a complex issue and beyond the scope of this book. One of the most important goals of a monitoring effort is collecting representative samples. In many cases, pollutant routing can affect pollutant concentration distributions. At outfalls, or in receiving waters, stormwater pollutant concentrations are random, with little of the observed variations being explainable by normal parameters (such as time since the event started or rain depth). As noted by Roa-Espinosa and Bannerman (1994), obtaining many discrete subsamples over the event duration likely results in a composite sample that has pollutant concentrations very similar to a flow-weighted composite sample. However, if collecting samples from a relatively small homogeneous area (such as a paved parking area), high concentrations of practically all pollutants are commonly observed near the beginning of the rain.

This "first-flush" phenomenon is most prevalent for rains having relatively constant intensities and for small areas. As a drainage area size increases (or as the surfaces become more complex, such as in a residential area), multiple first-flush waves travel through the drainage system, arriving at a single downstream location at different times. This moderates obvious concentration trends with time during the event. Also, as the rain intensity varies throughout an event, the washoff of pollutants at the sources also varies. Peak washoff occurs during periods of peak rain energy (high rain intensity). Therefore, periods of high concentrations may also occur later in a rain, as high intensities occur. Generally, lighter (more soluble) hydrocarbons and the smallest particles will "always" show a first-flush of high concentrations from small paved areas, while larger particulates and heavier hydrocarbons will wash off more effectively with high rain energies, which may occur randomly during a rain.

Sampling strategies must therefore consider these possible scenarios. The most effective sampling (but most expensive) is flow-weighted composite sampling throughout the entire storm event.

However, compositing many discrete subsamples collected throughout the event is likely to result in similar concentration values. If sampling a small critical source area (such as a gas station or convenience store), it may be useful to obtain an initial sample during the first few minutes of the event, and a composite over the complete event. In all cases, it would be difficult to justify analyzing many individual discrete samples collected throughout an event.

Flow Monitoring Methods

There are a wide variety of methods (Table 6.2) to determine flow in open and closed (e.g., conduit) channels. For additional information, see EPA (1982 and 1987a). Most flow measurements to assess receiving water effects from stormwater are conducted in relatively small streams. Often, channel cross-sectional area is determined and the velocity measured at intervals across the channel using a current meter. In some situations, discharge from a pipe, notched weir, or small dam can be caught in a container of known volume and mean fill-up time used to calculate flow (e.g., liters per second). A variety of flumes and weirs have been used successfully in assessing flow and runoff.

Mechanical current meters are commonly used because they are simple, rugged, accurate at low velocities (0.03 m/s, 0.1 ft/s) and operate at shallow depth (0.1 m). A manufacturer's calibration table converts the meter rotation number into meters or feet per second. Many modern meters are direct reading. The mean velocity at each cross-sectional interval is multiplied by the area of the subsection to calculate volumetric flow for each subsection. These are then summed to obtain the total stream flow.

Salt or fluorescent dyes have been used effectively to estimate velocities and time of passage when other methods are not practical, especially for highly irregular stream shapes or highly turbulent low flows. They depend on determining the amount of dilution that a known concentration of tracer receives as it mixes in the stream. The velocity between two stations is determined by knowing the travel time of the dye, or by comparing the dilute dye concentration to the injected dye concentration. The tracer may be added continuously or as a slug. A common tracer is Rhodamine WT dye which is measured with a fluorometer.

Flow monitoring in streams and other open channels is usually a necessary component of receiving water investigations. Flow estimates need to be made whenever any in-stream measurements are made, or samples collected, for example. In addition, equipment for continuous flow monitoring must be periodically calibrated using manual procedures. The following paragraphs briefly describe several common manual flow monitoring procedures.

Drift Method

The drift method is simply watching and timing debris floating down the stream. This velocity is then multiplied by the estimated or measured stream cross-sectional area to obtain the stream discharge rate. Of course, this method is usually the least accurate flow estimation method. The accuracy can be improved by choosing drift material that floats barely under the stream surface (such as an orange). Do not use material that floats high in the water (such as Styrofoam debris, for example), as it will be strongly influenced by winds. Drift measurements made in the center of a stream will tend to be the highest stream velocities, so the values should be reduced (by roughly 0.6, but highly variable) to better represent average stream flow rates.

Current Meter Method

The most traditional method of measuring flow is using a mechanical current meter. This method requires at least two people (one person should never be working alone near a stream anyway), a current meter, and simple surveying equipment. The stream discharge is measured at a cross section, usually selected along a relatively straight stretch (about 10 stream widths downstream from any major bends). If the stream discharge is being used to calibrate a stage recorder for continuous

Table 6.2 Methods for Flow Measurement and Their Application to Various Types of Problems

Device or Method	Flow Range Measurement	Applicable to Type of Water and Wastewater	Cost	Ease of Installation	Accuracy of Data[a]	Pressure Loss thru the Device	Volumetric Flow Detector	Flow Rate Sensor	Transmitter Available
Formula	Small to large	All	Low	NA	Fair	NA	NA	NA	NA
Bucket and stopwatch	Small	All	Low	Fair	Good	NA	NA	NA	NA
Floating objects	Small to medium	All	Low	NA	Good	NA	NA	NA	NA
Rotating elements current meters	Small to medium	All	Low	NA	Good	NA	Yes	NA	Yes
Dyes	Small to medium	All	Low	NA	Fairly good	NA	NA	NA	NA
Salt dilution	Small to medium	All	Low	NA	Fair	NA	NA	NA	NA
Magnetic flowmeters	Small to large	All	High	Fair	Excellent $1/2$–1%	None	Yes	Yes	Yes
Weirs	Small to large	All	Medium	Difficult	Good to excellent 2–5%	Minimal	Yes	Yes	Yes
Flumes	Small to large	All	High	Difficult	Good to excellent 2–5%	Minimal	Yes	Yes	Yes
Acoustic flowmeters	Small to large	All	High	Fair	Excellent 1%	None	Yes	Yes	Yes

[a] Assume proper installation and maintenance.

Data from Blasso, L. Flow measurement under and conditions, *Instruments Control Syst.*, 48: 45–50. 1975; Thorsen, T. and R. Owen. How to measure industrial wastewater flow. *Chem. Eng.* 82: 95–100. 1975.

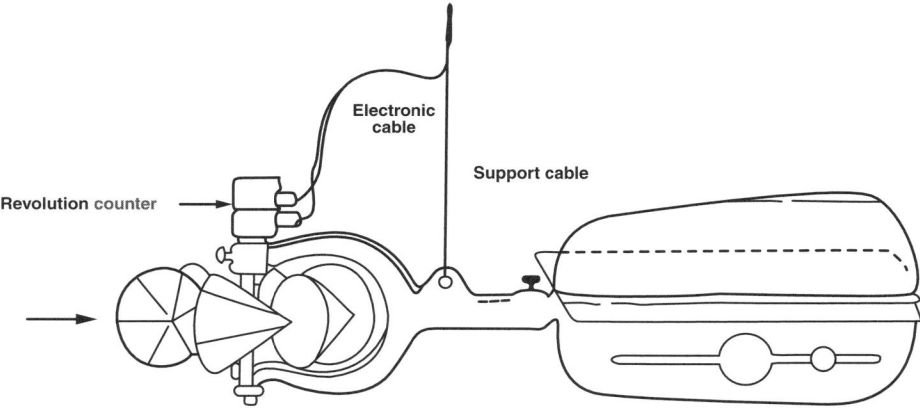

Figure 6.3 Price meter.

flow monitoring, the cross section being measured must not be affected by backwater conditions. If the selected cross section is in the vicinity of sampling and will not be used to calibrate a flow equation but will be used to determine the instantaneous current conditions at the time of sampling, then backwater influences and affects from meanders need to be included in the measurements. Instantaneous flows are determined using current meters to document flows occurring in a sampling period. However, this procedure can also be used to calibrate a state–discharge curve that can be used in conjunction with a conventional continuous stage recording device for long-term studies. Figures 6.3 and 6.4 illustrate common current meters used for stream studies.

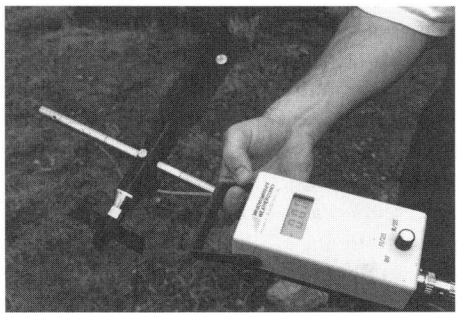

Figure 6.4 Student current meter.

In order to calibrate a flow or discharge model (especially the Manning's equation), the stream is assumed to have normal flow where the water surface is parallel with the stream bottom. This is unusual under real stream conditions, where actual water surface profiles exist. In this case, Manning's equation can still be used, but by substituting the friction slope for the water surface (or stream bed) slope. The friction slope is elevated above the water surface by the velocity head ($v^2/2g$). It is therefore easy to adjust the surveyed water surface slope to the friction slope by adding the velocity heads at the upstream and downstream locations. The calibration procedure usually involves calculating the Manning's roughness factor (n) in the stream stretch. Manning's equation is (in U.S. customary units):

$$V = 1.49(R^{2/3})S_f^{0.5}/n$$

where $V =$ velocity of the open channel flow (ft/s)
        $R =$ hydraulic radius (area/wetter perimeter, ft)
        $S_f =$ friction slope
        $n =$ Manning's roughness coefficient

Biological monitoring is normally conducted during relatively low flow periods, whereas Manning's equation was developed for channel design for large, rare events. Manning's equation is a

Table 6.3   Example Calculation for Flow and Current Measurements

Section Interval (ft)	Midpoint, Distance from Shore (ft)	Depth at Midpoint (ft)	Section Area (ft²)	Velocity at 0.2 Depth (ft/s)	Velocity at 0.8 Depth (ft/s)	Average Velocity (ft/s)	Discharge (ft³/s)
0–1	0.5	0.21	0.2	0	0	0	0
1–3	2	0.74	1.5	0.3	0.4	0.4	0.6
3–5	4	1.42	2.8	1.1	1.6	1.4	3.9
5–7	6	1.70	3.4	1.8	2.0	1.9	6.5
7–9	8	1.93	3.9	1.5	2.5	2.0	7.8
9–11	10	1.94	3.9	1.4	2.5	2.0	7.8
11–13	12	1.79	3.6	2.0	3.0	2.5	9.0
13–15	14	1.54	3.1	1.5	2.2	1.9	5.9
15–17	16	1.19	2.4	0	0	0	0
17–18.5	17.75	0.46	0.7	0	0	0	0
Total			26			1.6	42

conservative design formula (when using the published roughness coefficients). It is not an analysis method and it must be used with care during low flow conditions. During low flows, the roughness coefficient is usually much greater than during high flows, for example, requiring equation calibration at different stream stages.

Current meter flow monitoring requires that the stream be divided into several sections. About 10 sections from 1 to several feet wide are usually adequate, depending on overall stream width. The depth of the stream is measured at each section edge, and the current velocity is measured in a vertical profile in the center of each section. The average velocities in each section are multiplied by the section areas to obtain the discharge rates for each section. These are then summed to obtain the total stream discharge. Table 6.3 is an example calculation for a section on Cahaba Valley Creek, in Shelby County, AL, that is generally used as a field demonstration site for UAB hydrology classes. Figure 6.5 is a cross-sectional diagram of this site, also showing the flow profile distributions. It is interesting to note that the peak water velocity for this stream section is seen to be near the bottom of the stream, close to the middle, but off-set, likely due to the slight meandering of the stream at this location. This is in contrast to the typically assumed velocity profile where the peak velocity is very near the top of the stream (and near the center). Figures 6.6 and 6.7 are photographs of a UAB hydrology class obtaining current measurements at this location.

Stream discharge monitoring is obviously a multiperson job, both from a safety standpoint and in order to take the actual measurements. A safety throw rope should always be ready, and great care should be exercised when working in a fast-moving or deep stream. If a stream has too great a velocity (especially greater than about 2.5 ft/s), or if it is too deep, then current measurements

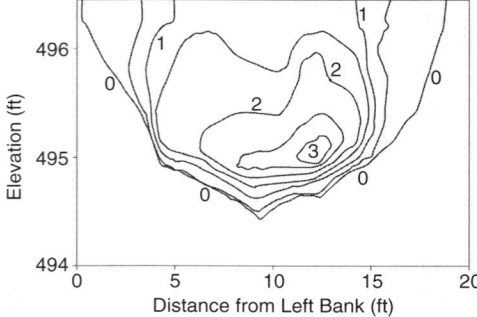

Figure 6.5   Cross section of stream velocities (ft/s) at Cahaba Valley Creek, Shelby County, AL.

Figure 6.6   Obtaining elevation contours at Cahaba Valley Creek, Shelby County, AL.

Figure 6.7 Obtaining current readings across Cahaba Valley Creek, Shelby County, AL.

should be conducted from a bridge, or cable system, and personnel should not be allowed to enter the water. Urban streams are also known for hidden debris and very soft bottoms. As in all work in urban streams, waders are necessary to minimize water contact and to prevent injuries from sharp objects. Riparian plants (such as poison oak and poison ivy) and slippery banks can also present additional hazards near streams. And do not step on any water moccasins.

A suitable current meter is obviously needed for a stream discharge study. Direct-reading digital meters (instead of older audible counter meters, where the operator must count clicks that are related to the water velocity) are now most commonly used. The current meter should be able to measure to 0.1 ft/s, have a threshold velocity of at least 0.2 ft/s, and preferably have an averaging mode in addition to an instantaneous mode. The meter should also be capable of measuring in very shallow water and next to the stream bottom (within a few inches of the stream bottom). The readout should also be selectable between metric and U.S. customary units. The meter must be recalibrated at least every year, preferably in the manufacture's tow tank or in an open channel test facility. Numerous hand-held current meters are available. Forestry Suppliers, Inc. (800-543-4203) has several different mechanical models, as listed below:

Swoffer Model 2100-1514 (#94161)	0.1 to 25 ft/s 1% accuracy	$2300
Handheld Flowmeter (#94303)	0.5 to 25 ft/s ± 0.5 ft/s accuracy	$700
Gurley Model 625 Pigmy (#94993)	0.05 to 3 ft/s audible counter	$1320
Gurley Model 625 Pigmy (#94983)	0.05 to 3 ft/s digital indicator	$2600
Gurley Model D622F Meter (#94982)	0.2 to 32 ft/s digital indicator	$2940

All of these current meters meet the desirable performance criteria, except for the much less expensive flowmeters. Newer portable meters are available that have no moving parts, typically using sonic pulses and Doppler measurements of reflected sound waves from moving particles in the water. These meters are costly (>$3000) and may have a more limited life span than the traditional current-driven meters.

An engineering level, rod, stakes, and tape are also needed to measure the water surface slope between adjacent cross sections when calibrating Manning's equation. Fiberglass tapes are suitable for measuring the stream widths, and rigid (but thin) rules are useful for measuring water depth at the stream sections. When measuring water velocities with a current meter, operators must stand to the side and behind the meter and ensure that no turbulence from their legs (or from others) affects the measurements.

Flow Monitoring Using Tracers

The most precise method of stream current measurements is through the use of tracers. This method is especially important when measuring flows in areas having karst conditions where surface waters frequently lose and/or gain substantial flows to and from underground flows. A single upstream dye injection location and multiple downstream sampling stations through the study area are used in this situation. Tracers are also needed if there is an obviously large fraction of inter-bed flow or if the stream flow is very turbulent. The flow in very shallow streams, especially when the stream is cobble lined, is also very difficult to monitor with current meters, requiring the use of tracers. Another common use of tracers is when measuring the transport and diffusion of a discharge into a receiving water. Hydraulic detention times in small ponds and lakes can also be determined using tracers. Orand and Colon (1993) state that the use of tracers for water discharge measurements is not a new concept. They admit that the use of current meters is usually much simpler and therefore more common. However, current meters are not applicable in many situations, as noted above. As an example, they routinely use dye tracers and a field fluorometer with continuously recording output to measure the discharge of very turbulent mountain streams, which would not be possible with current meters.

Unfortunately, tracers are rarely useful for continuously monitoring flows, but they can be used for instantaneous flow determinations or for calibrating conventional continuous flow monitoring equipment in actual installations.

Brassington (1990) lists the desired traits for a tracer:

- An ideal tracer should be detectable in very small concentrations.
- It should not be naturally occurring.
- If an artificial tracer is being used, it should exhibit conservative behavior.
- It should be safe to use and produce no harmful environmental effects.
- It should be relatively inexpensive and readily available.

Three main classifications of tracers are generally used. Dyes give a specific and distinctive color to the water that can be detected easily. Chemicals, especially naturally occurring salts, can be used effectively if a discharge into a receiving water has a unique water chemistry and the tracer study objective is to determine the behavior of the discharge. Mechanical tracers can also be used to tag the water, much like the drift method described previously.

The most common mechanical tracer is a spore of *Lycopodium*, a club moss (Brassington 1990). The spores can be dyed and used to measure the surface and groundwater interactions in complex systems. Another approach in monitoring complex surface–groundwater interaction is to use bacteriophages to trace groundwater movement, including the role of septic tank discharges on local receiving waters. Paul et al. (1995) injected prepared bacteriophage cultures (ϕHSIC-1 and Salmonella phage PRD1) as viral tracers, along with 1-μm fluorescent spheres and fluorescein dye, into septic tanks and injection wells and identified their presence in local surface waters in Key Largo, FL. They found relatively rapid movement of the viral tracers (from 0.5 to 25 m/h) in the subsurface limestone environment into the surrounding marine waters. These rates were more than 500 times faster than had been previously measured. They concluded that the subsurface flows may not have reflected uniform diffusion through a homogeneous matrix, but were rather "channeling" through the limestone. Another possibility they suggested was that viruses travel like colloids through the subsurface, moving faster than the bulk water flow. They concluded that the bacteriophages were much more efficient than the fluorescent tracers due to their much better detection limits.

The most efficient tracer is a naturally occurring one. Johnson (1984) concluded that using naturally occurring materials (such as salinity, turbidity, temperature, or other suspended or dissolved materials) allows much more data to be collected and is usually relatively inexpensive (compared to using artificial tracers). In order to use a natural tracer, the material must be:

- Conservative
- Highly soluble under a variety of conditions
- Not amenable to sorption or precipitation or degradation
- Linear with mixing
- Present in greatly contrasting concentrations in the two water bodies that are mixing

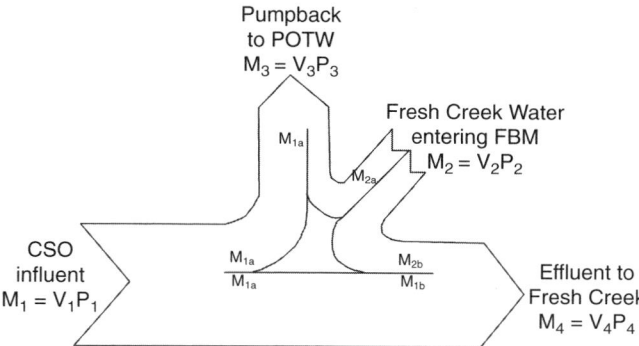

Figure 6.8   Schematic depicting mass balance at Fresh Creek, NY. (From Field, R. R. Pitt, D. Jaeger, and M. Brown. Combined sewer overflow control through in-receiving water storage: an efficiency evaluation. *Water Resources Bulletin, Journal of the American Water Resources Association,* Vol. 30, No. 5. pp. 921–928. October 1994. With permission.)

The tracer must also be easily and cheaply analyzed. In many cases, specific conductivity can be used. Specific conductivity is especially useful when examining freshwater inflows into saline receiving waters. Field et al. (1994) described the use of specific conductivity to measure the effectiveness of a combined sewer overflow (CSO) capture and control device in Brooklyn, NY. In this example, the CSO (which had a specific conductivity of about 1000 µS/cm and a standard deviation of about 250 µS/cm) was contrasted with Fresh Creek water (which had a specific conductivity of about 20,600 µS/cm and a standard deviation of about 2600 µS/cm). Standard conductivity meters were used to trace the CSO water as it displaced the Fresh Creek water in the treatment facility during rains, and to measure the leakage of Fresh Creek water into the treatment facility between rains, as shown in Figure 6.8. The mass (M) of the tracer is equal to the water volume (V) times the concentration (P). It does not matter that there is no adequate conversion for specific conductivity to be expressed as a mass, as specific conductivity concentrations were shown to be linearly related to dilution with the receiving water. A Monte Carlo mixing model was used to calculate the unknowns in this diagram, considering the variabilities of the concentrations in the two water bodies. Stable isotopes have been used successfully as tracers by some researchers with access to sensitive mass spectrophotometers, if the waters being distinguished have a sufficiently different source (Sangal et al. 1996). Ratios of major ions have also been used successfully to identify different waters, especially in groundwater studies.

In most cases, naturally occurring tracers cannot be effectively used because of their non-conservative behavior, insufficient concentration contrasts, or expense. A later section in this chapter discusses the use of natural tracers to identify sources of discharges. Commercially produced fluorescent dyes have been available for many years and have been extensively used for water tracer analyses. Fluorescein (a green fluorescent dye) has been used since the late 1800s, for example, but is not very stable in sunlight. However, it is still commonly used in visual leak detection tests and to visually trace discharge connections (such as determining if floor drains are connected to the sanitary wastewater lines or the storm drain system). Color Figures 6.1 and 6.2* show fluorescein being used to trace sanitary sewage connections to a storm drainage system in Boston.

Rhodamine B was used in the 1950s for water tracing in Chesapeake Bay because it was more stable in sunlight than fluorescein, but it readily adsorbed to sediments, making quantitative measurements difficult (Johnson 1984). Forestry Suppliers, Inc. (800-543-4203) sells liquid and compressed tablets and cakes of Rhodamine B and fluorescein for visual tracer work (but not for use

_____
* Color Plates follow p. 370.

near water intakes). Bottles of 200 tablets of either dye, having a total weight of about 10 oz., or a 3" donut, also weighing 10 oz., of either dye costs about $35.

The most common artificial tracer currently used is Intracid Rhodamine WT dye, a 20% (by weight) stock of dye in water and other solvents having a specific gravity of 1.2. It is available from Crompton and Knowles (Reading, PA, 215-582-8765), at about $400 per 10 L. It is greatly diluted before use in the working stock solution for continuous dye injection studies. Chemical and laboratory suppliers also sell much more dilute mixtures (but at a much greater cost per unit of dye). Forestry Suppliers, Inc., sells a 1-gallon bottle of Rhodamine WT, unspecified dilution, (catalog #92969) for about $100, and bottles of 200 compressed Rhodamine WT tablets (catalog #92991) (weighing 11 oz.) for about $36.

Rhodamine WT was specifically developed in the 1960s for water tracing applications and is much superior for quantitative work compared to the earlier dyes. It is generally easier to detect in much lower concentrations, less toxic, has lower sorption to particles, and exhibits slower decay. Even though it is very expensive by volume, its very low detection limit (about 0.01 ppb of the 20% stock solution) and conservative behavior make it cost-effective.

Rhodamine WT is generally thought to have low toxicity; however, the USGS limits its concentrations at water supply intakes to 10 ppb (Johnson 1984). The biggest toxicity problem associated with Rhodamine WT is apparently associated with reactions with very high concentrations of nitrates. In all cases, it is important to contact local drinking water and state water regulatory agencies when planning a dye tracer study. The largest concern is probably associated with complaints of red water (which should not occur if proper dye concentrations are used).

The Corps of Engineers (Johnson 1984) has published a comprehensive description of the use of water tracing using fluorescent dyes. This report stresses monitoring inflows into reservoirs, with information applicable for a wide range of surface water conditions, including small streams, large rivers, and lakes. Johnson (1984) reports that no significant decay of Rhodamine WT is likely to occur due to chemical or photochemical decay for conditions found in natural waters. However, high chlorine levels (several mg/L, such as are found in many drinking waters) can cause significant decay during long exposure tests (tens of hours). As an example, Johnson reports that chlorine concentrations of 5 mg/L in tests run over 20 hours caused about a 5% decay of fluorescent activity. If operating in urban areas, where the chlorine concentrations may be periodically high or the turbidity variable, it is important to test decay and sorption of the dye. This is best done by using actual receiving water collected at the time of the tracer study as the dilution water when preparing the dye standards. These standards should be compared to standards using proper laboratory dilution water (preferably prepared using ion exchange, and/or reverse osmosis, as laboratory distilled water can contain very high chlorine levels).

Johnson (1984) states that total fluorescent decay of Rhodamine WT is probably about 0.04/day, from both sorption and photochemical decay. Almost all of this loss is likely associated with sorption. The sorption of Rhodamine WT onto particles, according to Orland and Colon (1993), had less than a 7% effect on the measured stream discharges (overestimated) in water having suspended solids concentrations ranging from 200 to 2000 mg/L (particle diameter <200 μm).

Johnson (1984) also reports the effects of pH, temperature, and salinity on the fluorescence of Rhodamine WT. The most serious problem with precise measurements is that the fluorescent intensity decreases with increasing temperature, requiring temperature corrections. This change is a decrease in fluorescence by about 5% for every 2°C increase in temperature. If collecting discrete samples that are brought back to the laboratory for analysis, the samples and the standards can be kept at the same temperature for analysis, eliminating this problem. *In situ* fluorescent measurements require temperature corrections (available as an option in the Turner Designs 10-AU, for example). It is recommended that discrete samples also be periodically collected, along with the continuous field measurements, for temperature-controlled laboratory analysis to confirm the automatic corrections.

The pH of the receiving water affects the sorption of the Rhodamine WT to organic material. Below a pH value of 5.5, the carboxyl acid group of the dye becomes protonated, increasing adsorption. Johnson (1984) reviewed studies that showed that humic sediment solutions of 2.0 and 20 g/L and 100 ppb Rhodamine WT caused 18 and 89% decreases in fluorescence, respectively. The high humic concentrations lowered the pH values of the water and increased the organic content of the water. In similar solutions using a kaolinite clay, the fluorescent losses were only 11 and 23%. These clay concentrations are very high (2000 and 20,000 mg/L) and would be likely to occur only in construction site runoff in urban areas. The very high associated turbidity of these samples would also greatly complicate fluorescent measurements. The samples would likely have to be clarified (by centrifuge or filtering) before measurement (see also below).

The most commonly used fluorescent measurement instrumentation for fluorescent dye studies has been the older and obsolete Turner model 111 fluorometer that is still available in many laboratories, and the newer Turner Designs (408-749-0994) model 10-AU fluorometer (Figures 6.9 and 6.10). Both of these instruments are filter fluorometers and are very sensitive. The Turner Designs 10-AU is a much superior unit for field measurements, as it is designed to operate on 12-volt batteries, has newer and more stable electronics, a wider dynamic range, and has a water-resistant case. It is also suitable for laboratory measurements. The Turner Designs unit also has a flow-though cell, plus built-in temperature correction and data logging options, which are convenient for field use.

The downstream dye concentrations should be measured over a long period and at many locations across the stream to obtain the best flow estimate. In practice, an automatic water sampler is used to obtain samples, or manual grab samples are obtained, at the downstream location for laboratory analyses, or less commonly, a flow-through fluorometer is used to measure the dye concentrations on a real-time basis. If manual sampling is used, subsamples can be obtained from several locations across the stream for compositing. If a flow-through instrument is used, the intake can be moved to various locations across the stream to investigate mixing conditions. In all cases, the downstream location should be well beyond the predicted fully mixed area. Variations in dye concentrations observed are therefore assumed to be associated with flow variations in the stream.

Background fluorescence in the water must be determined before and during the test. During some tests, we have detected residual background fluorescence. In receiving waters affected by sanitary sewage (such as from raw overflows, inappropriate connections, leaks, septic tank influences, and treated effluent), background fluorescent can be very high due to detergents in the water. Almost all of this interference is eliminated by using specific Rhodamine WT filter sets in the fluorometer. The use of the actual water being tested as the injection water diluent during a continuous test reduces background problems, as do the highly selective optics available for Rhodamine WT analyses. However, background water samples need to be collected for analyses before any dye is added to the water. In addition, it is a good idea to collect upstream water

Figure 6.9 Older Turner model 111 fluorometers used in the laboratory.

Figure 6.10 Current model Turner 10-AU fluorometer being calibrated in the laboratory before field deployment.

periodically during the test to check for changing background conditions (especially important when conducting a tracer test in a sanitary sewer where background water quality can change dramatically over a relatively short period of time). If turbidity levels vary greatly during the test, Johnson (1984) recommends that the samples be filtered or centrifuged prior to analysis. Continuous dye analysis in the field does not allow a correction for turbidity (like the built-in temperature correction option available from Turner Designs), but periodic grab samples analyzed in the laboratory after turbidity reduction enables these effects to be determined.

An example of continuous background corrections was described by Dekker et al. (1998) using Rhodamine WT in Detroit to accurately calibrate flow-metering equipment. They found that abrupt changes in suspended solids in the sewage were very common and that this could radically change the fluorescent response. They therefore collected background (upstream) sewage samples every 15 min during the dye tests and prepared calibration curves with this water, changing the response factors for the measurements accordingly. They also monitored the light absorbance at the Rhodamine WT excitation wavelength (550 nm) simultaneously with the dye concentrations to screen out periods of abrupt changes in suspended solids that would affect the calibration curves.

The careful calibration of fluorometers is critical because of their great sensitivity. Calibration solutions from about 0.1 to 500 ppb should be used (these concentrations are in relation to the 20% stock solution). Two sets of calibration solutions need to be prepared. The initial laboratory series is prepared using laboratory-grade clean water, and another set must be prepared using the receiving water. As noted previously, if using distilled water, ensure that the chlorine concentrations are very low. Never use tap water. Deionized water (at 18 meg-ohm resistance) is probably the best. Preparing such low concentration standards requires a great deal of care, especially when withdrawing the stock and making the initial dilution. Needless to say, the largest hazard associated with working with Rhodamine WT is the mess that it can make if splashed or spilled. The stock solution is stratified in the shipping container, requiring stirring, but trying to stir or shake the stock container is a challenge, as it is heavy and minor spills or leaks are a great nuisance.

It is recommended that the amount of dye needed for the test be withdrawn from the stock shipping container, including the minor amount needed for preparing the standards. This will be only a very small amount, usually only a few hundred mLs for a slug dose test, or a few liters if conducting a continuous injection experiment in an urban stream. This aliquot doesn't have to be perfectly representative of the stock solution. The goal is to withdraw the amount needed without spilling any, with minimal mixing. The initial dilution is usually made using 10 mL of the stock diluted in a liter of dilution water, using a volumetric flask. The 10 mL of stock is very dark and viscous, making it almost impossible to measure with a standard pipette. Many people weigh the initial amount, correcting for the 1.2 specific gravity, but unless the aliquot was from a well-mixed stock container, the specific gravity can be quite different. An automatic pipette (capable of handling viscous fluids) is probably better, as volume dilutions are being measured during the test. Serial dilutions are then usually made, making weaker and weaker standards. The strong concentrations foam if violently mixed, making it difficult to fill the volumetric flasks accurately to the calibration marks.

Analytical chemists do not approve of serial dilutions, as errors are easily compounded, but the nature of Rhodamine WT and the great dilutions needed would otherwise require measuring very small quantities of stock. Using a 1-μL pipette and a 1-L volumetric flask would only produce a 1 ppm (1000 ppb) solution, by volume. At least a second (serial) dilution would still have to be made to obtain a 1 ppb concentration, and a third dilution to obtain a fraction of a ppb standard. Inaccuracies associated with serial dilutions are probably less of a problem than trying to pipette such small amounts.

Fluorescent analyses can be conducted in the field or in the laboratory. *In situ* (flow-through) dye analyses (for which the Turner Designs 10-AU is specifically designed) can be much more efficient than collecting water samples and bringing them back to the laboratory for analyses. However, a combination approach is usually best, where periodic samples are collected and brought to the laboratory for temperature controlled analyses for comparison to the *in situ* values. The *in situ* analyses allow immediate evaluation of the sampling program, especially when the dye is being used at proper concentrations, making it nearly invisible to the eye, or if complex hydraulics (such as in an estuary with strong currents) prohibit easy prediction of the flow path. However, using a fluorometer in flow-through mode presents special problems. Johnson (1984) stresses the need to ensure that all water connections are air tight to prevent bubbles from entering the flow path. In addition, the pump should be located above the light cell to decrease bubbles from leaky pump seals. The intake of the water delivery system should also be screened to decrease the chance of sand and other debris from scratching the instrument optics.

The two main types of dye injection include instantaneous or continuous releases. Instantaneous dye releases are much more efficient in the use of dye. The amount of dye quickly added to the water usually results in a visible dye cloud that is easy to follow manually. In addition, no special dye injection equipment is required, as the dye is simply poured quickly into the water body. However, continuous releases of dye, especially in conjunction with *in situ* analyses, is necessary when simply tracking the dye is challenging. Continuous dye releases require substantially more dye and usually more field personnel, but changing conditions can be easily measured (Color Figures 6.3 and 6.4).

Thomann and Mueller (1987) present a USGS method used to estimate the amount of Rhodamine WT dye needed for an instantaneous release experiment. The amount is usually much less than needed for a continuous release experiment. They also present several methods to evaluate the observations and obtain estimates of flow, diffusion coefficients, and recovery of dye.

Continuous release rates of dye are dependent on the desired downstream concentration of dye, the concentration of the dye being released, the injection rate, and the estimated stream discharge. Figure 6.11 shows a basic mass balance for a discharge into a river or stream. This can be easily applied to a dye injection experiment, with the dye being considered as the effluent being discharged into the receiving water.

The mass balance for this situation is:

$$\text{upstream mass} + \text{effluent mass} = \text{downstream mass, or}$$
$$Q_u s_u + Q_e s_e = Qs$$

where
Q_u = upstream flow rate
s_u = upstream concentration
Q_e = effluent discharge (or dye injection) rate
s_e = effluent (or dye injection solution) concentration
Q = resulting downstream discharge rate (equal to $Q_u + Q_e$)
s = resulting downstream concentration

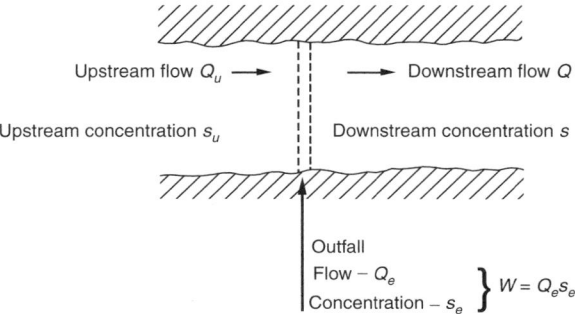

Figure 6.11 Notation for mass balance calculations for dye injection current measurements. (From Thomann, R.V. and J.A. Mueller. *Principles of Surface Water Quality Modeling and Control.* Harper & Row. New York. 1987. With permission.)

Solving for Q, the downstream discharge rate:

$$Q = (Q_u s_u + Q_e s_e)/s$$

If the background concentration (s_u) is zero (as desired in a tracer experiment), this further reduces to:

$$Q = Q_e(s_e/s)$$

where (s_e/s) is the dilution ratio of the dye

Therefore, the stream discharge (Q) is the ratio of the concentration of the dye injection solution (s_e) to the measured downstream dye concentration (s), multiplied by the dye injection rate. As an example, assume the following conditions:

Q_e = 10 cm³/s

s_e = 1.0 (injection dye solution concentration, a given arbitrary concentration of 1.0)

s = 12 ppm$_{vol}$ compared to injection concentration (average dye concentration from numerous samples collected).

The average value for s was determined to be 12 ppm (relative to the injection dye concentrations); therefore, the calculated stream discharge rate is:

$$Q = Q_e(s_e/s) = 10 \text{ cm}^3/\text{s } (1.0/12 \times 10^{-6}) = 830,000 \text{ cm}^3/\text{s}$$

This is equal to 830 L/s, or about 29 ft³/s (cfs). As noted in this example, the absolute concentration of the injection solution does not need to be known, as long as calibration solutions are made using the injection solution and the receiving water.

The injection solution needs to be discharged at a constant rate. This is made much easier by using a special metering pump (as supplied by Turner Designs, for example, or a battery-operated peristaltic pump available from Cole-Parmer). In all cases, someone must be at the injection site for the duration of the experiment to ensure that the discharged dye is well mixed and that constant pumping of the injection solution is occurring. This is achieved by periodically measuring the time needed to fill an appropriate graduated cylinder (retain some of the solution from the filled cylinder for use in later calibration solutions, and dump the remainder of the material from the cylinder into the stream when finished timing). The injection solution samples should be analyzed to detect variations in injection dye concentration during the study period.

Fortunately, as is evident from the above equation, everything is relative to the injection concentration, or the mass of dye used, with tracer work. The stock solution concentrate is never directly used in dye studies because the intense color would make the injection plume visible for a large downstream distance; also, the high 1.2 specific gravity affects the plume buoyancy, and precisely pumping very small dye injection rates is difficult. The stock is therefore greatly diluted (by about 10 to 100 times) to create an injection solution to minimize these problems. When conducting a continuous injection experiment, one measures the ratio in concentrations between the injection dye stream and the resulting receiving water concentration. This initial dilution causes a loss of sensitivity, so more dye is required in a continuous injection experiment. In small urban streams, this loss of efficiency is not too serious. When conducting a large-scale injection experiment, specific gravity adjustments are usually made and close to full-strength dye is injected to minimize costs. In a slug discharge test, much less dye is usually needed, and the full amount of tracer dye is introduced into the water as rapidly as possible (within a few seconds). During instantaneous tests, the strength of the dye solution is not important. It is only necessary to know the mass of the dye used. Therefore, the small amount of dye needed can be effectively diluted in a several-gallon container that can be rapidly poured into the stream to initiate the test.

Experimental conditions needed for various estimated stream discharges can be predetermined by knowing the injection pump rates available and the sensitivity of the fluorometer. A Cole-Parmer Masterflex peristaltic pump can supply a wide range of dye injection rates, depending on the pump rotational speed and the size of tubing. With #13 tubing, the pump can be set to deliver between 0.2 and 0.5 mL/s. Number 16 tubing has a useful range of between 2.0 and 8.0 mL/s, while #18 tubing can be used between 10 and 40 mL/s. A Turner model 111 fluorometer has a range of sensitivity from less than 1 to more than 150 ppb Rhodamine WT, depending on the sensitivity setting. The newer Turner Designs model 10-AU has a much wider dynamic range. The combination of these settings allows measurement of a wide range of flow rates. Table 6.4 illustrates some of the flow rates that can be measured using some of these combinations. The downstream concentrations shown on this table are in relation to the injection concentration, which should be diluted by at least 10 times compared to the 20% stock solution. Therefore, the downstream concentration of 10 ppb shown may actually be closer to 1 ppb of the 20% stock. Intermediate downstream concentrations should be targeted to ensure that variations in stream flow can be accommodated. If a needed injection rate is too low, it may be unstable. The concentration of the dye being injected should then be decreased so a higher pumping rate can be used.

As an example, consider a stream having an estimated discharge rate of 25 cfs and the target downstream concentration is 25 ppb (compared to the injection dye strength which is diluted 10 times from the 20% stock solution; the actual downstream dye concentration is therefore about 2.5 ppb, which would be about mid-scale on the most sensitive setting for a Turner model 111 fluorometer). An injection rate of about 20 mL/s will therefore be required. Therefore, 2 mL of 20% stock will be used per second, or 120 mL of stock per minute of the test, or 7.2 L of stock per hour of the test — a large amount of dye. The injection duration depends on the duration of the steady flow period to be monitored. This should be long in comparison to the flow duration from the injection location to the monitoring location to minimize sampling problems. The sampling location must be located far enough downstream to ensure complete mixing. This length (in feet) can be estimated using the equation presented by Thomann and Mueller (1987):

$$L_m = (2.6 \ UB^2)/H$$

where U = the stream velocity in ft/s
       B = the average stream width in feet
       H = average stream depth in feet

As an example, the discharge rate is estimated to be 25 cfs, the stream velocity is estimated to be about 1 ft/s, the stream width about 25 ft, and the depth about 1 ft. The "complete mixing" length is therefore about 1600 ft. About half of this distance would be needed if the dye injection

Table 6.4   Stream Discharge Rates (cfs) That Can Be Measured for Different Experimental Conditions

Injection Rate (mL/s)	Downstream Conc. = 50 ppb	Downstream Conc. = 25 ppb	Downstream Conc. = 10 ppb	Downstream Conc. = 1 ppb
0.3	0.21	0.42	1.1	11
0.5	0.35	0.71	1.8	18
2	1.4	2.8	7.1	71
10	7.1	14	35	350
20	14	28	71	710
30	21	42	110	1100

point is located at the centerline of the stream. The travel time needed (if injected at midstream) is about 13 min, at least. Therefore, an hour-long injection period would not be unusually long, requiring about 7 L of 20% Rhodamine WT dye, for this example.

The Use of a Multiparameter Probe to Indicate the Presence, Duration, Severity, and Frequency of Wet-Weather Flows

Most receiving water problems are highly dependent on the duration, severity, and frequency of wet-weather events. Habitat effects, for example, are greatly dependent on the frequency of erosive flows that cause bank instability. Sediment scour and deposition is also dependent on the flow energy. Bacteria, turbidity, and other water quality standard violations are much more serious if they occur commonly. Toxicity effects on receiving water organisms are also greatly dependent on the frequency and duration of exposure to excessive concentrations. Knowing when an event occurred, plus knowing the duration and severity of the event, is critical when conducting a long-term exposure experiment using many of the techniques described in this book. Therefore, knowing these basic wet weather event parameters is very important and enables a more complete evaluation of wet-weather problems in receiving waters. The following discussion presents a simple way to automatically monitor these important hydraulic characteristics in a stream without installing a permanent flow monitoring station.

Continuous sondes for water quality monitoring have been available for some time, but current models are vastly improved compared to earlier ones. It is now possible to deploy a water quality sonde for up to several weeks, with little drift and other degradation in performance. This allows the units to be left unattended for extended periods to obtain diurnal variations of constituents (such as DO, temperature, conductivity, turbidity, and water depth) for varying environmental conditions. One application is to examine the duration of degraded receiving water quality conditions following rains.

The following example is based on work by Easton et al. (1998) as part of an investigation studying the effects of SSOs (sanitary sewer overflows) on small urban streams. This study used YSI 6000 UPG water quality sondes to indicate the duration, frequency, and magnitude of wet-weather events in both surface waters and surficial sediments. Short-term, or runoff-induced, pollution effects can be studied in detail using these instruments. Long deployment time and the continuous monitoring capability of the YSI 6000 enables acquisition of data for multiple events, i.e., as many as occur during the time of deployment. The YSI 6000UPG sonde is a multiparameter water quality monitor manufactured by YSI Incorporated, Yellow Springs, OH. The 6000 UPG is capable of performing a subset of the following measurement parameters: dissolved oxygen, conductivity, specific conductance, salinity, total dissolved solids, resistivity, temperature, pH, ORP (oxidation reduction potential), depth, ammonium/ammonia, nitrate, and turbidity. The 6000 UPG can be left unattended in the field for approximately 45 days, depending on the frequency of data logging and parameters being recorded. The instrument is constructed of PVC and stainless steel and is 3.5 in in diameter and 19.5 in in length. It weighs approximately 6.5 lb, with batteries. The sonde is capable of interfacing with an IBM PC-compatible computer for downloading data, or a hand-held unit can be used for direct observations. In addition, a software package, Ecowatch for Windows, is available for sonde setup, data acquisition, and data presentation/analysis. The sondes used in these experiments were configured to acquire the following parameters: dissolved oxygen, conductivity, specific conductance, temperature, pH, ORP, turbidity, and depth.

Five-Mile Creek (which is actually about 50 miles long) is a typical medium-sized Alabama stream, originating in a rural area, then flowing through a suburban, and then a heavily urbanized area. The flow in the creek ranged from approximately 2 to 10 m³/s, depending on recent rainfall conditions. At each test site, one sonde was located on the creek bottom and the second was buried under approximately 6 in of sediment. The buried sondes were protected by placing them inside 75-mm-aperture nylon mesh bags and were used to measure interstitial water characteristics *in situ* and continuously. The sondes were anchored to the bottom by a chain attached to cinder

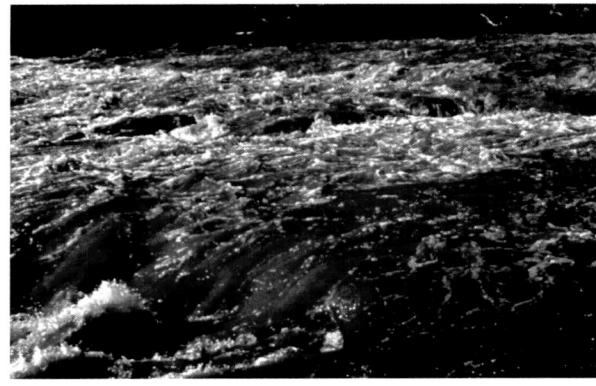

Color Figure 2.1 Extremely turbid high-water flows on Shades Creek, Birmingham, AL.

Color Figure 2.2 Algal bloom in urban stream.

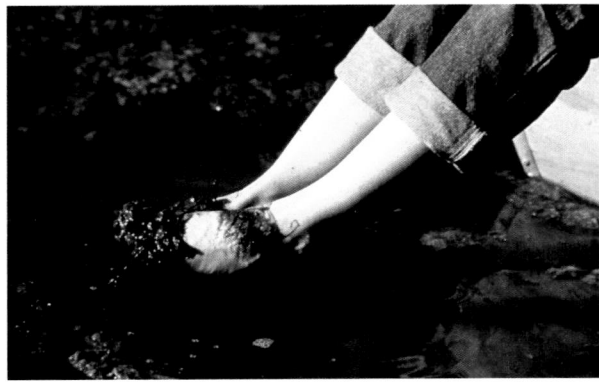

Color Figure 2.3 Highly eutrophic urban lake conditions, Wisconsin. (Photo courtesy of Wisconsin Department of Natural Resources.)

Color Figure 4.1 Mixing of SSO discharge into Five-Mile Creek during high flow conditions.

Color Figure 6.1 Fluorescein dye at outfall from cross-connected sanitary sewer with storm drain.

Color Figure 6.2 Fluorescein dye from outfall toward receiving water during low tide.

Color Figure 6.3 Rhodamine WT solution being injected into stream for current study.

Color Figure 6.4 High levels of rhodamine WT near injection location (should not be visible).

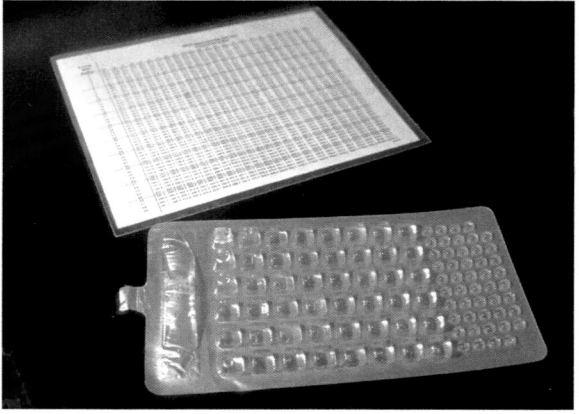

Color Figure 6.5 Multiple-tube response in white light (for total coliforms).

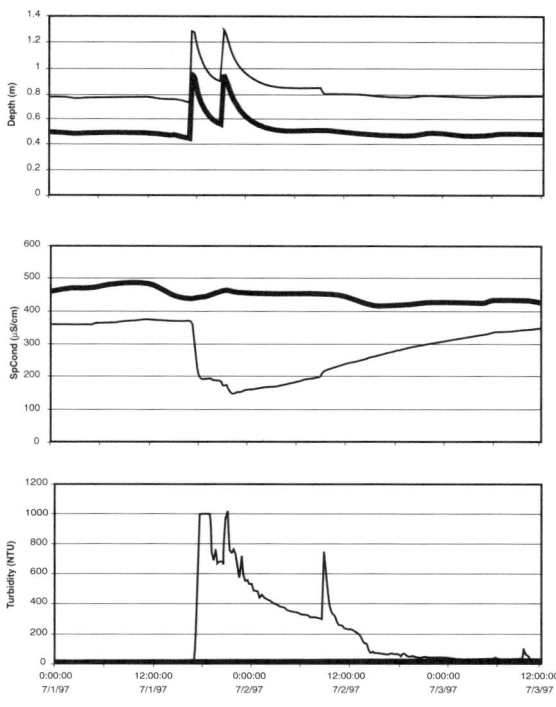

Figure 6.12 Event plots of depth, specific conductance, and turbidity at fine sediment site. (Note: the maximum range of the turbidity probe is 1000 NTU). (From Easton, J.H. et al. The use of a multiparameter water quality monitoring instrument to continuously monitor and evaluate runoff events. Presented at *Annual Water Resources Conference of the AWRA*, Point Clear, AL. 1998. With permission.)

blocks. The cinder blocks were then attached to a tree to prevent the sondes from being washed downstream during major events. One set of sondes was located in an area having coarse sediments (stones of about 1 in in diameter), while the other set was located in an area having finer sediments (sandy grained).

The duration, frequency, and magnitude of runoff events is apparent from an examination of plots constructed from the sonde data (Figures 6.12 and 6.13). These sonde data show a large fluctuation in depth, specific conductance, and turbidity in the water column at both sites on July 1 at 5:00 pm, roughly corresponding to the 0.6 in of rain observed at the Birmingham International Airport several miles away. No site-specific rain information was available, as may be typical for many small-scale studies.

The rise period for all of the parameters was very rapid, and the peaks occurred very early in the runoff event. They then returned to previous levels within 1 to 2 days, depending upon the parameter. The data acquired for water depth are obviously the parameters that best correlate to tracking runoff hydrographs as they pass. There is an obvious change in flood stage (approximately 0.5 m increase in depth), as indicated on these figures. There were two slightly separated, but very similar, runoff hydrographs that passed through the creek; the depth data show two obvious peaks spaced about 3 hours apart. The other two parameters do not distinguish between these two separate, but close events, as is evident in the time taken to return to baseline (Tables 6.5 and 6.6). The turbidity and specific conductance data also substantiate the presence of a runoff event, but with an additional perspective on the duration of the potential effects from elevated turbidity levels and possibly other pollutants. Notice the almost immediate increase in depth and turbidity, and corresponding decrease in specific conductance. These changes are easily explained by a sudden increase in runoff water within the creek. Furthermore, the depth sensors indicate the timing and severity of the runoff event from a hydrologic perspective, while the specific conductance and turbidity sensors indicate the extended duration of probable adverse water quality conditions due to contaminated baseflows entering the stream.

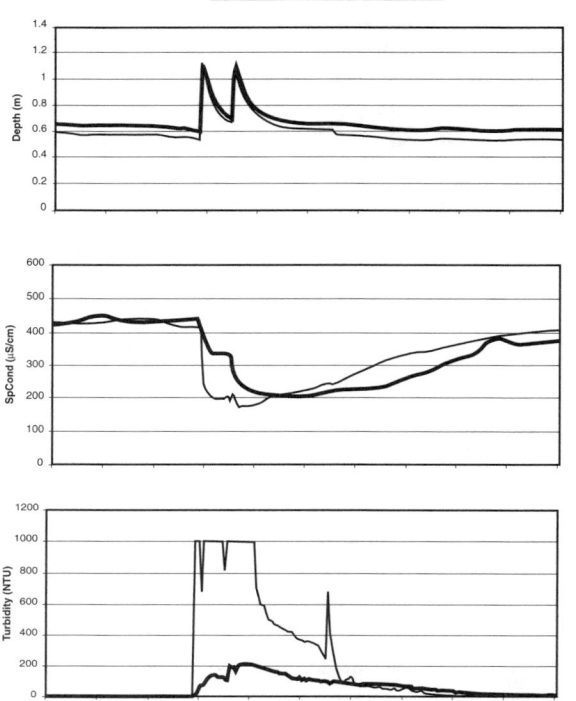

Figure 6.13 Event plots of depth, specific conductance, and turbidity at coarse sediment site. (From Easton, J.H., et al., The use of a multiparameter water quality monitoring instrument to continuously monitor and evaluate runoff events. Presented at *Annual Water Resources Conference of the AWRA,* Point Clear, AL. 1998. With permission.)

The data in Tables 6.5 and 6.6 show the differences in water exchange between the water column and the interstitial water occurring in the two different sediment types (coarse and fine). These experiments show that the interstitial water at the coarse sediment site changes with the water column, although at a slightly reduced magnitude, while the interstitial water at the fine sediment site shows no change. Most urban streams have sediments represented by the fine sediment site (sand sized) or finer. Therefore, very little direct water exchange occurs between the water column and the interstitial water. The interstitial water quality is much more affected by the quality of the deposited sediments (especially decomposable material and toxicants) than by the water column quality directly. This rapid fluctuation of interstitial water in coarse-grained sediments has important implications on evaluations of sediment quality. The benthic micro-, meio-, and macrofaunal exposures in these environments will be more dynamic than typically assumed. Interstitial water sampling and sediment sampling were discussed in Chapter 5.

Table 6.5   Values for Magnitude of Change and Time to Return to Baseline for Specific Conductance, Due to Period of High Flow

Sonde Location	Magnitude of Change (μS/cm)	Time to Return to Baseline (hr)
Water column	210	42
Fine sediment	Not obvious	Not obvious
Water column	260	44
Coarse sediment	230	46

From Easton, J.H., Lalor, M., Pitt, R., and Newman, D.E., The use of a multiparameter water quality monitoring instrument to continuously monitor and evaluate runoff events. Presented at *Annual Water Resources Conference of the AWRA,* Point Clear, AL. 1998. With permission.

Table 6.6 Values for Magnitude of Change and Time to Return to Baseline for Turbidity, Due to Period of High Flow

Sonde Location	Magnitude of Change (NTU)	Time to Return to Baseline (hr)
Water column	>1000	30
Fine sediment	0	0
Water column	>1000	30
Coarse sediment	210	30

From Easton, J.H., Lalor, M., Pitt, R., and Newman, D.E., The use of a multi-parameter water quality monitoring instrument to continuously monitor and evaluate runoff events. Presented at *Annual Water Resources Conference of the AWRA,* Point Clear, AL. 1998. With permission.

The duration of the water column effects from the wet-weather events is seen to be much greater than the duration indicated by the high flows alone (30 to 45 hours vs. 12 hours). This has a major impact on evaluating biological effects of the receiving waters. As an example, rains only occur for about 4.5% of all hours in Birmingham. Periods of extended high flows in Five-Mile Creek may occur for about 15% of the time. However, periods of elevated turbidity (and likely other constituents of concern) may occur for about 40% of the time. This extended time has a significant effect on in-stream beneficial uses and risk assessments from wet-weather toxicants and pathogens.

In-Stream and Outfall Flow Monitoring

Monitoring of flows in storm drainage systems is typically done to supplement stormwater sampling activities. In most cases, flow monitoring equipment available from the same vendor that supplied the automatic water samplers is selected. The flow sensors typically measure depth of flow in the sewerage and apply Manning's equation to calculate the flow rate and discharge. Unfortunately, Manning's equation was developed as a design equation and not as an analysis equation. It was not intended for accurate measurements for shallow flows and does not consider debris that accumulates in sewerage. A better approach is to use a control section in the sewerage and calibrate a stage-discharge relationship. The ultimate solution is to use a special prefabricated manhole that contains a flume. Plasti-Fab (503-692-5460) offers many options of manhole and flume sizes and types for a broad range of sites and conditions. A less expensive alternative (and more suitable for temporary installations) is a manhole flume insert. These are available from Plasti-Fab and from Badger Meter (918-836-8411). These are installed in the discharge sewer line from a manhole, causing a backwater in the manhole that provides an accurate stage-discharge relationship that can be measured. Acoustical flowmeters (measuring water surface distances from a reference location above the water using reflected sound) or bubbler flowmeters (measuring the depth of water above the sensor based on hydrostatic pressure) are usually used to measure the water depth. If the storm sewer line is debris and obstruction free, Manning's equation can be used, but a site-specific stage–discharge relationship must be developed and calibrated over a wide range of depths. Flow calibration is most effectively conducted using Rhodamine WT dye as a tracer, as described previously.

It is critical that the flow monitoring sites be selected to provide accurate flow measurements, along with providing safe and easy access. Sites for flow monitoring must meet numerous criteria in order to obtain accurate results. The most critical criteria require the absence of backwater conditions at the monitoring location and a reasonably straight and homogeneous stream character upstream of the monitoring location for a length of at least 10 times the stream width. Since the stream depth measurements will need to be translated into flow values using a depth–discharge curve, the stream banks and stream bottom need to be reasonably stable at the monitoring locations. The best way to provide the stability and constant stage–discharge relationship at a flow monitoring station is to construct a control section, usually a flume or a weir. If the stream to be monitored is moderate in size and in a natural setting, especially with important in-stream biological resources, constructing a flume or weir is usually not practical.

The electronic components of typical in-stream flowmeters need to be secured near the stream edge, but outside the zone of common flooding. It would be best to secure them within a heavy steel contractor's box permanently mounted onto an oversized concrete slab. A heavy padlock normally provides adequate security. This enclosure can also contain the necessary deep-cycle batteries recommended for power. If an external data logger is needed, it can also be secured within the box. In many instances, a solar panel can be installed to provide a trickle charge to the battery (but the solar panel would be exposed to vandalism, and riparian locations might be heavily shaded). The bubble tube can be easily run inside a steel pipe (2 to 3 in in diameter) buried in the stream bank. The upper end can come through the concrete pad directly into the steel instrument shelter. The lower end must terminate below the lowest expected stream depth, coming up through a moderate-sized concrete pad to protect the pipe and bubbler tube. The bubbler tube end must lie on top of the in-stream concrete pad and needs a heavy, but shallow, wire cage covering. This covering needs to be relatively easy to remove (while submerged) in order to provide intermittent service to the end of the bubbler tube. This installation can be easily upgraded to include an automatic water sampler, with the sampler (and its deep cycle battery) also enclosed in the steel shelter and the sampler tube also running down the pipe. If a water sampler is also to be used, a galvanized steel pipe must not be used because of zinc contamination. A very heavy-duty plastic pipe, sufficiently buried and protected may be suitable, or a much more expensive stainless steel pipe could be used to encase the bubbler and sampler tubes.

Another option for a shelter is to use concrete pipe rings stacked to a sufficient height and a steel plate padlocked to the top. This is a more temporary (and cheaper) alternative that usually works well. The bubbler tube should also be protected, if possible, within a large-diameter heavy plastic pipe. Another alternative is to mount the flowmeter and ancillary components on a road crossing where a stilling well can be run down into the water, usually on the downstream side of a bridge pier. The equipment can be mounted inside a heavy plywood box on top of the stilling well and accessed from the bridge. In this case, the pier may cause water level interferences.

Many flow measurement equipment vendors now offer simultaneous stage and velocity sensors. The velocity sensors directly measure the flow rate of the water, reducing the need for a stage–discharge relationship. The two major types of velocity sensors are the time-of-transit flowmeter and the Doppler flowmeter. Time-of-transit flowmeters use acoustical signals directed diagonally across the water flow path to a receiver. The acoustical signal travel time can be very accurately predicted. Any difference between the predicted and measured travel time is associated with the water motion. Accusonic (508-548-5800) is one vendor of these devices, which have been reliably used in large conduits. A series of three Accusonic sensors is placed in each of three parallel 10 ft × 15 ft CSO outfalls in Brooklyn, NY, as part of the Fresh Creek CSO treatment study (Field et al. 1995). The three sensor and receiver pairs in each outfall are placed in three vertical zones in each outfall, representing three layers of flow that can measure the severe backwater conditions due to daily tides. As an example, the individual sensors can measure tidal flows entering the bottom of the outfall and any floating CSO discharging on top of the saline receiving water.

Rob Washbusch and Dave Owens of the USGS in Madison, WI, recently (1998) tested several different flow monitoring devices simultaneously in a single storm drain pipe for comparison (Figures 6.14 to 6.19). A unique aspect of these tests was the use of continuous dye injection and downstream water sampling that was automatically activated when rainfall started. The samples were then brought to the laboratory for fluorometric determinations and actual flow values. These actual flows were then compared with the flows indicated by the different flow monitoring equipment. The box plots show the observations from 60 events examined over a 6-month period of time. Flow measurement errors of ±25% were not uncommon. They emphasize that these results

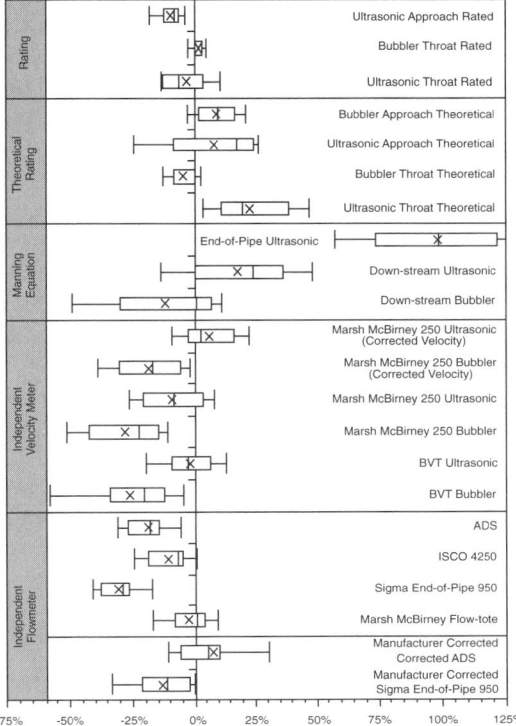

Figure 6.14 Box plots of differences observed when using different flow monitoring methods. (From Waschbusch, R. and D. Owens. *Comparison of Flow Estimation Methodologies in Storm Sewers.* Report prepared by the USGS for the FHWA. Madison, WI. January 1998.)

Figure 6.15 Sigma bubbler flowmeter at USGS test site.

Figure 6.16 ADS acoustic flowmeter at USGS test site.

are for only one site (an industrial area in Madison, WI) and are not likely directly indicative of conditions that might be found elsewhere. They recommend that all runoff flow monitoring equipment be carefully calibrated at the time of installation and periodically rechecked.

Doppler velocity sensors are more commonly used in small storm and sanitary sewer lines. These reflect acoustic signals from particles flowing toward the sensors. The signals reflect off the fastest moving particles, and signal processing then determines the average water velocity. Several vendors sell Doppler units that are constantly improving in accuracy and ease of use. ADS Environmental

Figure 6.17 Acoustic flowmeter at USGS test site.

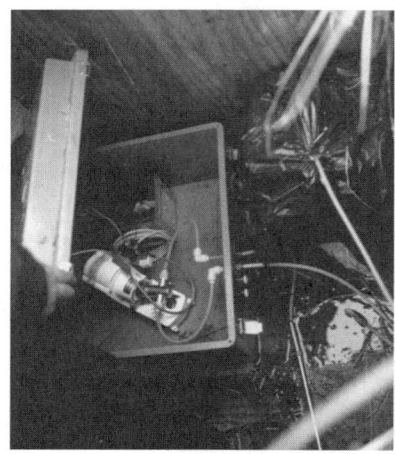

Figure 6.18 Automatic dye injection installation at USGS flowmeter test site.

Figure 6.19 Automatic sampler at outfall at USGS flowmeter test site for collecting real-time dye samples for calibration of flow-meters.

Services, Inc. (800-633-7246) maintains many large-scale flow monitoring networks around the world using its Doppler velocity and ultrasonic level sensors. ISCO (800-228-4373) also sells a Doppler unit that can be used in conjunction with its automatic water samplers. Unidata America (503-697-3570) sells the Starflow ultrasonic/Doppler flowmeter that is very compact and can be used in small open channels and sewer and drainage lines.

Summary of Flow Monitoring Methods

Table 6.7 is a list of some of the advantages and disadvantages of the different flow monitoring/measurement techniques that are most commonly used in urban receiving water studies. The previous discussion presented both manual flow monitoring procedures and methods for flow monitoring that can be used in conjunction with automatic water samplers. In most cases, standard bubble depth sensors supplied by the sampler manufacturer are probably the best choice for an automated station. However, these should be placed in a control section where the stage–discharge curve is specifically known and has been calibrated. Time-of-travel (sonic) current meters can be extremely valuable in situations where stratified flow may occur, but custom interfaces with the sampling equipment may be needed. Basic velocity meters are best used for more casual flow measurements, especially when flow measurements are being taken simultaneously with biological sampling. Dye testing is usually reserved for absolute calibration of flow monitoring setups and to measure in difficult situations, especially during low flow conditions in rocky streams where much of the flow may be actually occurring within gravel deposits, and in streams in karst areas where the interactions between surface and subsurface flows can be dramatic.

Table 6.7 Comparisons of Available Flow Measurement Instruments

Flow Monitoring Instrument Type	Advantages	Disadvantages
Manual Instruments	Simple and rapid results	Instantaneous results, not long-term
Velocity meters	Direct readout of current velocity	Requires multiple measurements across stream to obtain average condition. Can be dangerous during high flows.
Tracers (fluorescent dye)	Considered the standard flow calibration procedure	May be subject to interferences from changing water quality (solids and temperature) or pipe materials. May be difficult to design and to conduct measurements for large systems. Required fluorometer is expensive.
Tracers (naturally occurring salts)	Used for mixing and dilution studies. Inexpensive if using naturally occurring salts in major flow components.	Requires unique and conservative tracer material in mixing components, such as mixing studies for outfalls in marine environment, or industrial discharges.
Automated Instruments	Long-term placement	More expensive and needed for each monitoring location
Bubble sensor depth indicators	Simple and easy to interface with automatic samplers. Most choice and experience from many vendors.	Only measures depth; requires stage-discharge relationship. Should be used in conjunction with a control section (weir or flume) and be verified with frequent velocity meter studies (not commonly done).
Propeller velocity meters	Direct measurement of current velocity.	Foul easily and only indicate velocity at location of propeller.
Time-of-travel (sonic) velocity meters	Direct measurement of velocity. Can be used to measure velocity of specific layer of the water to indicate shear; especially useful in tidal conditions with stratified water moving in different directions.	Relatively expensive and several may be needed to accurately measure flow in different flow strata.
Acoustic velocity meters	Direct measurement of current velocity. Usually measures the peak velocity, and the average velocity for the relatively large sensing zone is calculated as a fraction of the peak velocity.	Current models with supporting software enable relatively easy interpretation of the monitoring results. However, these units generally suffer from a lack of precision and seem to be more subject to error than traditional flow monitoring units.

Rainfall Monitoring

Rainfall data are very important when monitoring receiving water quality and quantity. As an example, the rainfall history in a watershed is needed before interpretation of biological monitoring data can be used to identify possible sources of degraded conditions. The hydrology texts listed previously all contain excellent summaries of rainfall aspects of importance in runoff studies. An especially important reference on rainfall depth measurements and interpretation is the *National Engineering Handbook Series* (Part 630, Chapter 4, Storm Rainfall Depth) published by the USDA (Soil Conservation Service, SCS, now the Natural Resources Conservation Service, NRCS), commonly referred to as NEH-4. This is available from the Consolidated Forms and Distribution Center, 3222 Hubbard Road, Landover, MD 20785. This handbook is supplied in a three-ring binder and sections are periodically updated.

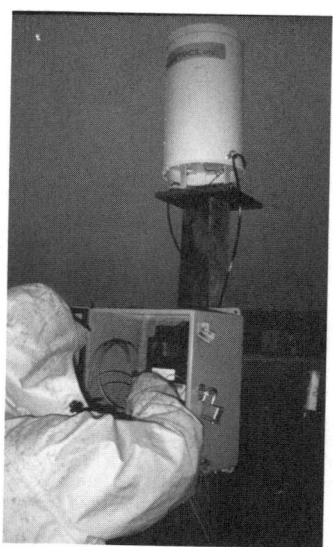

Figure 6.20 Tipping bucket rain gauge with data logger.

Figure 6.21 Close-up of tipping bucket rain gauge mechanism.

Placement and selection of rain gauges are described in these references, along with calculating and interpreting watershed-wide rainfall. This section briefly summarizes several important aspects of rainfall monitoring not usually discussed in available reference texts, especially selecting the proper rain gauge network density and the need for calibration.

Rain gauges suitable for stormwater monitoring are available from many sources. A new small and self-contained weather station is available from Hazco (800-332-0435) that contains sensors for wind speed, wind direction, temperature, relative humidity, dew point, barometric pressure, and rainfall. It has a built-in data logger for up to 6 months of recording and is even available with a modem for connecting to a cellular telephone for telemetry. The cost is about $8500 (catalog #B-W010010M) with a modem and $6600 (catalog #B-W010010) without a modem. Tipping bucket recording rain gauges and data loggers, standard 8" rain gauges, and wind screens are available without the other sensors from several sources, including Quali-metrics, Inc. (800-824-5873) and Global Water (916-638-3429) (Figures 6.20 and 6.21).

The other extreme in rainfall monitoring is the "Clear View" rain gauge from Cole-Parmer (800-323-4340) that is only about $35 (catalog #H-03319-10). This is a nonrecording rain gauge (having a 4" funnel diameter) requiring manual readings of the rain depth. Many other types of "garden store" accumulative rainfall gauges (Figure 6.22) are also available for as little as $5 each, including simple ones that can be made using 3-L plastic soft drink bottles (requiring the collected rain to be poured out and measured). As noted below, relatively few recording rain gauges (for accurate rainfall intensity measurements and start and end rain times) are needed for most urban catchment

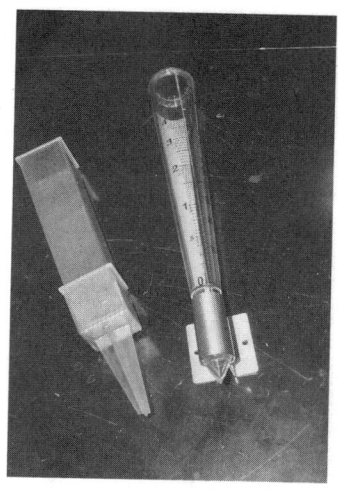

Figure 6.22 Inexpensive "garden/household" rain gauges.

studies. However, numerous nonrecording gauges should be placed throughout the study area to indicate rainfall variations, especially for small rains.

Determining Watershed Averaged Rainfall Depths

Three methods are most commonly used to determine representative watershed-wide rainfall amounts from several point observations. These include the station-average method, the Thiessen polygon method, and the isohyetal method. These methods are briefly described in the following paragraphs.

Station-Average Method

The simplest and easiest method of estimating watershed-wide rainfall amounts is simply to compute the numerical average of all observed values in the watershed. Only those rain gauges physically located in the watershed of interest are usually considered. This method yields good estimates if most of the following conditions are present: the watershed has little topographical relief, a sufficient number of rain gauges are present, the rain gauges are reasonably uniformly distributed throughout the area, and the individual rain depths observed for the different rain gauges do not vary widely from the overall mean. The most important criterion is the need for a large number of rain gauges uniformly distributed throughout the area.

Thiessen Polygon Method

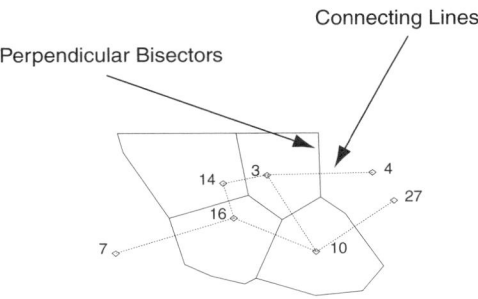

Figure 6.23   Thiessen polygon construction.

The Thiessen method uses a weighted average for the rain gauge network, based on the area assumed to be represented by each rain gauge. Closely spaced rain gauges have smaller weightings than do rain gauges spaced farther apart. The area weightings generally do not consider topography, or other watershed characteristics, although the polygons can be manually adjusted to account for these potential effects, with experience. The area represented by each station is assumed to be the area that is closer to it than to any other station. These areas are determined by drawing connecting lines between all adjacent rain gauges. These connecting lines are then bisected. The perpendicular bisectors then describe a polygon surrounding each rain gauge. Figure 6.23 is a simple illustration of the construction of the polygons surrounding each rain gauge. Figure 6.24 is an example of a Thiessen polygon system for the Toronto, Ontario, metropolitan area which has 35 rain gauges over an area of about 4000 km^2. These polygons were prepared using the SYSTAT computer program.

Results from the Thiessen polygon method are usually assumed to be more accurate than those obtained by the simple station-average method because the Thiessen method accounts for non-uniform distributions of stations. Rain gauge measurements from surrounding areas are also used in the analysis. The polygons also do not change for different rains, unless data are missing from one or more rain gauges. The weightings therefore are relatively constant, making the calculations reasonably simple for multiple rains, after the polygons are initially determined and measured.

Isohyetal Method

This is the most complex method for determining rainfall depths over a watershed and is usually considered the most accurate. It was rarely used before the common availability of computers that

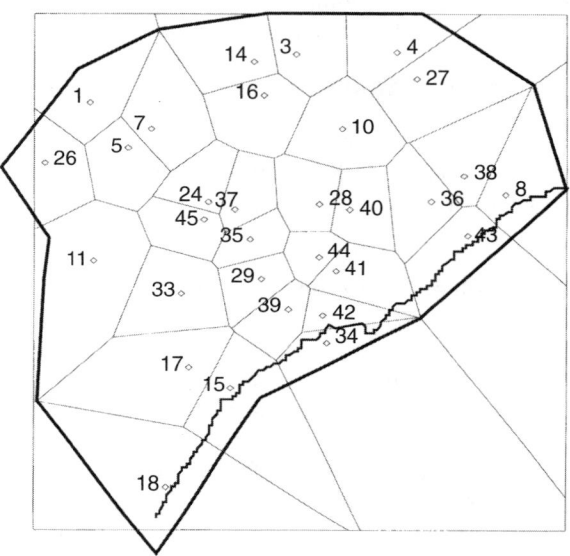

Figure 6.24 Thiessen polygons for Toronto rain gauges.

simplified the necessary calculations. In contrast to the Thiessen polygon method, the isohyetal method requires extensive calculations for each individual rain event. In this method, contours of equal precipitation depth are constructed over the watershed. The construction of the contours can consider the presence of topographic or lake effects. The precipitation averaged over the entire area is computed by multiplying the area enclosed between adjacent isohyetal lines by the average rain depth values of the two adjacent isohyetal lines. Figure 6.25 is an isohyetal map (rain depths in mm) for a single rainfall over the Toronto area, using data from many individual rain gauges. This map was also prepared using SYSTAT.

The Toronto rain gauge network density resulted in small differences between the three averaging methods because of the large number of rain gauges available. The use of the 35 rain gauges was a lot compared to available rain gauge networks in most urban areas. The resulting errors in using the simple averaging method or the Thiessen polygon method, compared to the

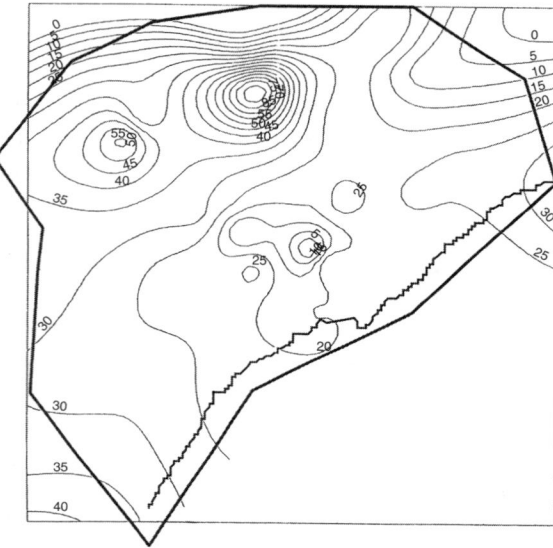

Figure 6.25 Isohyetals prepared for a single Toronto rainfall (mm).

isohyetal method, were all less than 1 mm in rain depth for rains of just a few mm in depth to over 25 mm in depth.

Rain Monitoring Errors

There are several common aspects of rainfall monitoring that can cause measurement errors. Most of these errors result in decreased rainfall values compared to true conditions. These include too few rain gauges for the area, poor placement of the rain gauges, wind effects, splashing of rain out of the gauge during high-intensity rains, tipping rate of tipping bucket rain gauge not keeping up with high-intensity rains, and calibration errors. These problems can usually be identified when reviewing the data. The errors can be corrected during the monitoring period, one hopes; otherwise the rain data might not be usable.

The easiest way to identify questionable rainfall data is to compare the site data with data collected from nearby and independent rain gauge locations. Residual analyses (differences between the site data and surrounding data) may indicate a consistent bias. This may be expected if there is a good reason for the bias (such as topographic differences or nearby large water bodies). The residuals also need to be examined for changes with time. This pattern should also be random, with no obvious trends or abrupt changes. In all cases, a recording rain gauge (especially a tipping bucket rain gauge) must have a standard rain gauge located in close proximity. The total rainfall recorded between observation times of the tipping bucket rain gauge is adjusted based on the standard gauge readings. These adjustment factors should be reasonably consistent. Another way to check rain gauge data is by comparing the watershed rainfall quantity with the stream flow quantity. This relationship should follow a reasonable rainfall–runoff pattern, with no abrupt deviations. Finally, recording rain gauges need to be periodically calibrated against different artificial rain intensities. The measured rainfall causing a tip of the bucket in a tipping bucket rain gauge should remain constant for a wide range of rain intensities. This quantity should also not change abruptly with time.

Needed Rain Gauge Density

One of the most common problems with rainfall monitoring is simply not having enough rain gauges in the watershed. Typical guidance for appropriate rain gauge densities does not consider the likely errors associated with too few gauges located in relatively small urban watersheds. The absolute number of rain gauges is probably more important than the simple rain gauge density. In all cases, multiple rain gauges are needed, even in the smallest study area. The number of rain gauges required depends on local conditions (Curtis 1993). Areas of higher rainfall variability require a greater number of rain gauges to adequately estimate rainfall over a watershed. As an example, mountainous areas will require more gauges than flat lands, and areas subject to convective storms will require more gauges than areas subject to frontal-type storms.

The spatial variability and intended use of the data should be used in determining the needed number of rain gauges. Typical guidance for flat terrain indicates rain gauge spacing of about 25 to 30 km, while this spacing is reduced to 10 to 15 km for mountainous areas. Most monitored urban watershed areas are quite small: almost all are less than 100 km^2, and typically less than 10 ha in area. These small areas seem to justify only a single rain gauge. Wullshleger et al. (1976) made one of the earliest recommendations for the number of rain gauges needed in small urban runoff catchments. They found about one gauge was needed in 0.5- to 1-km^2 watersheds, and about 12 gauges for larger (25-km^2) watersheds. However, multiple rain gauges are needed in all monitored watersheds. This should include a tipping bucket rain gauge and a single standard rain gauge, at least, for the smallest area, if rain intensities are to be monitored. When the study area increases, and if smaller rains are of interest, the number of rain gauges must be increased to compensate for the increased variation in the rain depth throughout the area. These additional rain gauges can be

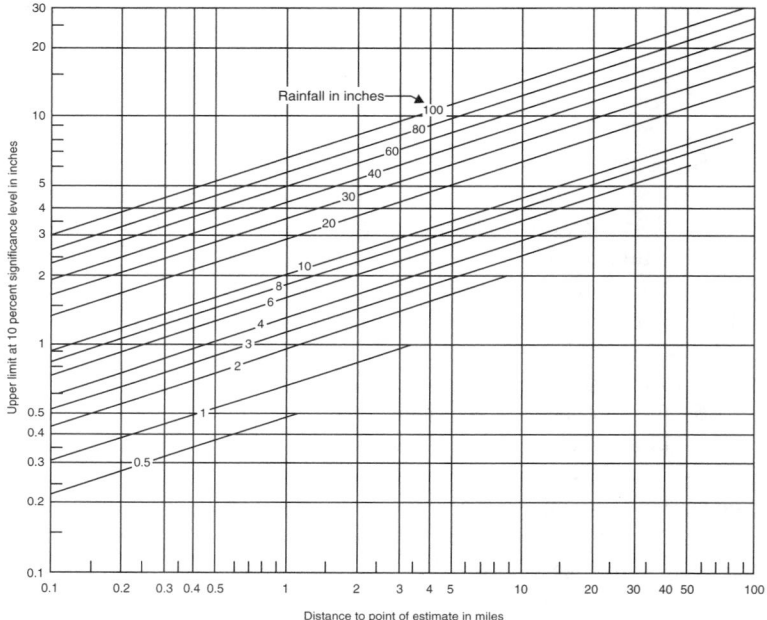

Figure 6.26 Confidence limits based on rain gauge spacing. (From NEH (National Engineering Handbook). Part 630, Chapter 4, *Storm Rainfall Depth* (NEH-4). USDA (Natural Resources Conservation Service), Consolidated Forms and Distribution Center, 3222 Hubbard Road, Landover, MD 20785. Periodically updated.)

additional pairs of tipping bucket and standard rain gauges, or simple accumulative (garden-store type) rain gauges, if intensities are not needed.

The *National Engineering Handbook Series* contains a simple chart, shown here as Figure 6.26, that can be used to estimate the 90% confidence limits of a rainfall located a specific distance from a rain gauge (NEH undated). As an example, if the measured rainfall at a rain gauge is 2 in, the 90% confidence limit in rain depth for a location 0.5 miles away can be estimated as:

- The "plus error" is about 0.8 in, or 2.8 in for the upper limit.
- The "minus error" is assumed to be about one half this amount, or 0.4 in, with a lower limit of 1.6 in.

The NEH also contains a nomograph (Figure 6.27) that can be used to estimate the error in measurement of watershed average rainfall depth, based on the size of the watershed, the number of rain gauges, the annual average precipitation depth, and the storm rainfall depth of concern. The example shown in this figure is for a watershed of 200 acres, having two rain gauges. In the example shown, the annual rainfall is about 33 in, and the rain of interest is 5 in. The average error is estimated to be about ±12%, or ±0.6 in.

Lei and Schilling (1993) studied the rainfall distribution in two urban watersheds located in Essen, Germany. The catchment had an area of 34 km² and was represented by 17 rain gauges. Rainfall data for five summers (1980–1984) were analyzed. They only examined rains that had all stations represented and that had at least 0.5 mm of rain. They compared catchment-wide averaged rain depth using subsets of the complete rain gauge network against the data from all 17 rain gauges as a reference. Figure 6.28 shows the basin-wide runoff volume errors that would result if only one rain gauge was used in rainfall–runoff modeling. It shows that relative errors of computed runoff volume decreased with increasing rain depth. Rains greater than about 8 mm had about ±20% errors in modeled runoff volume with a single rain gauge over the 34 km²

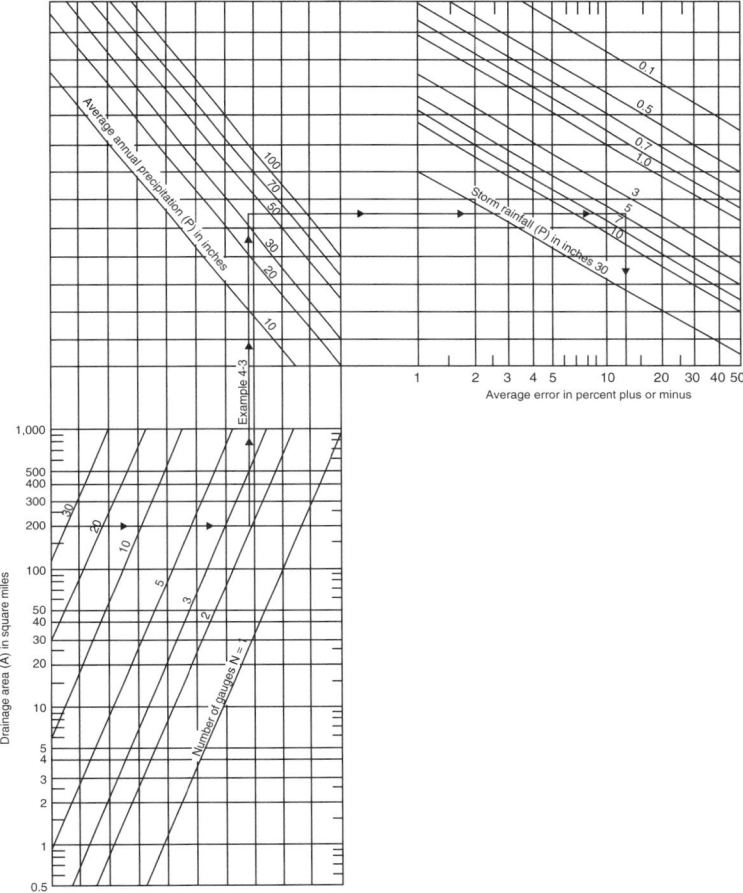

Figure 6.27 Errors in watershed rain depth. (From NEH (National Engineering Handbook). Part 630, Chapter 4, *Storm Rainfall Depth* (NEH-4). USDA (Natural Resources Conservation Service), Consolidated Forms and Distribution Center, 3222 Hubbard Road, Landover, MD 20785. Periodically updated.)

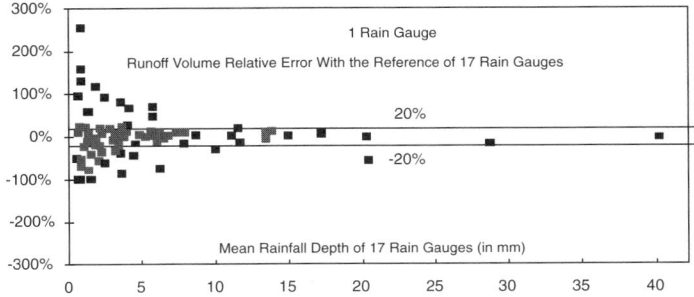

Figure 6.28 Relative runoff volume errors while using one rain gauge in Essen, Germany. (From Lei, J.L. and W. Schilling. Requirements of spatial rain data resolution in urban rainfall runoff simulation, in *Proceedings of the 6th International Conference on Urban Storm Drainage,* Niagara Falls, Ontario, Sept. 12–17, 1993. pp. 447–452. IAHR/IAWQ. Seaport Publishing, Victoria, B.C. 1993. With permission.)

drainage area. However, smaller rains could have rain depth errors of up to 250% with only a single rain gauge.

Ciaponi et al. (1993) studied rainfall variability in the 11.4-ha Cascina Scala experimental urban catchment watershed in Pavia, Italy, for a 3-year period. Two rain gauges separated by 310 m were used in this study. During this period, 233 storm events were selected for analysis, all greater than 1 mm in depth. The following list shows the percentage differences between the rain depths measured at the two monitoring locations for three rain depth categories:

- For 1 mm < h < 5 mm (135 storms), the average error was 31%.
- For 5 mm < h < 20 mm (75 storms), the average error was 10%.
- For h > 20 mm (23 storms), the average error was 8%.

These results show that the rainfall monitoring variations over even a very small watershed and with two closely spaced rain gauges can be quite large for small rain depths (<5 mm), with the differences decreasing for larger rains.

The National Weather Service guideline (Curtis 1993) used to determine the minimum number of gauges required in a local flood warning system is:

$$N = A^{0.33}$$

where A is the basin area in square miles. As an example, a 10-mi^2 watershed would require at least two rain gauges, while a 100-mi^2 watershed would require at least five.

Figure 6.29 shows the expected coefficients of variation for different rain gauge numbers and watershed sizes (Curtis 1993). For a fast-responding watershed, a coefficient of variation (the standard deviation divided by the mean) goal of 0.10 would require about six rain gauges for a 50-mi^2 watershed, while a 500 mi^2 watershed would require about 13 rain gauges for the same COV of observed rain depths in the watershed. Average and slow-responding watersheds would require slightly fewer rain gauges for the same watershed areas.

Rodda (1976) presented recommendations (Tables 6.8 and 6.9) for the minimum number of rain gauges required for small and moderate-sized watersheds and for larger watersheds. Table 6.8 shows the number of rain gauges needed for observations of daily rain depth totals and for monthly rain depth totals.

According to Chow (1964), one rain gauge per 625 mi^2 is the minimum for general climatological purposes, while for hydrologic purposes, each study basin should have at least one rain gauge per 100 mi^2. However, one rain gauge per 1 mi^2 was recommended for the analysis of thunderstorms.

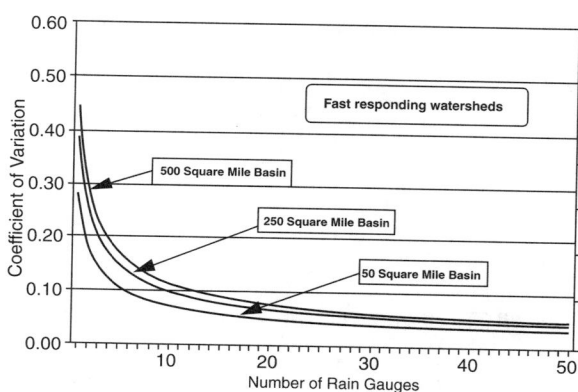

Figure 6.29 Areal rainfall accuracies for fast-responding watersheds.

Table 6.8 Recommended Minimum Numbers of Rain Gauges Needed in Small and Medium-Sized Watersheds

Area (mi²)	Daily	Monthly	Total
1	1	2	3
2	2	4	6
8	3	7	10
16	4	11	15
31	5	15	20
47	6	19	25
63	8	22	30

Data from Rodda, D.W.C. Water data collection and use. *Water Pollution Control,* Maidstone, England. Vol. 75, No. 1, pp. 115–123. 1976. With permission.

Table 6.9 Recommended Minimum Number of Rain Gauges Needed for Large Watersheds

Area (mi²)	Number of Rain Gauges
10	1
100	2
500	3
1000	4
2000	5
3000	6

Data from Rodda, D.W.C. Water data collection and use. *Water Pollution Control,* Maidstone, England. Vol. 75, No. 1, pp. 115–123. 1976. With permission.

Pitt and McLean (1986) investigated rainfall distributions in the Toronto area as part of the Humber River pilot watershed study. Rainfall data were available for 35 rain gauges over an area of about 4000 km². This high number of gauges allowed sensitivity calculations to be made to determine the appropriate number of rain gauges that may be needed. Numerous random subsets of these rain gauge data were used to analyze potential errors associated with using fewer gauges for 46 different rains greater than 1 mm in depth. Figure 6.30 shows the likely errors for different numbers of rain gauges over this area. The largest rains (>20 mm) had the smallest rainfall variations over the area and therefore had the smallest errors for a specific number of rain gauges. The smallest rains (<5 mm in depth) had much greater errors because their variations were much larger throughout the area. This plot shows that the errors would be very large (several hundred percent in error) for all rains with only one rain gauge for the complete area. The errors somewhat leveled off after about 12 rain gauges were used. However, the rain depth errors for the largest rain category would remain greater than 10% even for 25 rain gauges, and the smallest rains may still have about 50% errors associated with this large number of gauges.

The small catchment monitoring effort by Pitt and McLean (1986) in Toronto illustrated the need to include multiple rain gauges even in very small areas. The two urban watersheds monitored were 39 and 154 ha in area and were located about 3 km apart. Rainfall was monitored at one of the areas only, and the rainfall at the airport several kilometers away was used for comparison. Part way through the monitoring program, a large deviation was noted between the local and airport monitored rain depths. The local rain gauge was then recalibrated, with a 40% increase in the

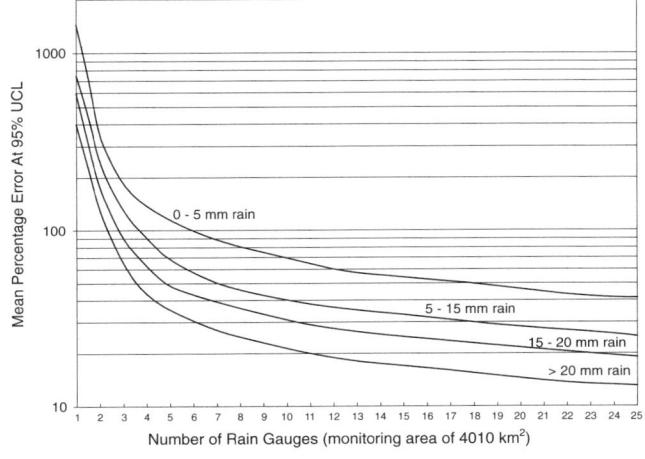

Figure 6.30 Calculated errors for using various rain gauge densities for different sized storms in Toronto.

volume needed for a single bucket tip compared to the initial calibration value. This of course had a significant effect on the rainfall quantity monitored, and much time was spent in identifying why and when the rain gauge had changed so much since its initial calibration. After much analysis using surrounding rainfall data and investigating the history of the specific rain gauge, it was determined that the rain gauge used had a historical problem with its bearings and several repairs had been made in an attempt to correct it. Unfortunately, the gauge calibration was found to be highly variable, and all the locally monitored data were therefore questionable and not used. Thankfully, the Toronto rain gauge network had six other rain gauges surrounding the two study areas within a few km. These data were extensively evaluated, including examining the storm tracks across the city during all monitored rains, to derive suitable rain depth and intensity values for the storms of interest. This analysis required much time, but was possible because of the additional rain gauge data. This problem could have been prevented with the use of a standard rain gauge located next to the tipping bucket rain gauge (as required in professional rain monitoring installations) for more frequent checks on the calibration factor. Nonrecording rain gauges could also have been located in several locations in the small test watersheds to indicate variations throughout the drainage. Both of these options would have cost a fraction of the amount associated with the additional detailed rainfall analysis required during this project and would have alerted the field personnel to the rainfall monitoring problem much sooner.

Proper Placement of Rain Gauges

Precipitation measurements are greatly influenced by wind. Careful placement and shielding of rain gauges are both necessary to reduce wind-induced errors. The upward movement of air over a rain gauge reduces the amount of precipitation captured in a rain gauge. Proper placement is needed to minimize wind-induced turbulence (and to minimize rain shadow effects) from nearby obstructions.

Linsley et al. (1982) concluded that reliable measurements of wind-induced errors are difficult because of problems involved in determining the actual amounts of precipitation reaching the ground. They reported that wind-induced errors during rainfall monitoring exceed about 10% for winds greater than about 8 mph, for both shielded and unshielded rain gauges. This error increases to about 20% during 20 mph winds. Shielded rain gauges perform slightly better, with a wind-induced error about 3% less than for an unshielded rain gauge during 10 mph winds, and about 5% less during 20 mph winds. The effects of winds on snowfall is much greater, with shielded gauges having about half the magnitude of errors as unshielded gauges when monitoring snowfalls. Snowfall errors (all underreported) for unshielded gauges may be about 50% for 10 mph winds and increase to about 70% for 20 mph winds. Various types of wind shields have been used, but the Alter shield (loose-hanging vanes in a circle around the rain gauge) has been adopted as a standard in the United States. Its open and flexible construction provides less opportunity than solid shields for snow buildup, and the flexible design allows wind movement to help keep the shield free of accumulated snow and ice.

Rain gauge exposure and placement are very important to reduce rainfall measurement errors. The higher the rain gauge is located above the ground, the greater the wind error. It is therefore best to locate the rain gauge on level ground, definitely avoiding roof installations and steep hillsides. Linsley et al. (1982) and Shaw (1983) both recommend a partially sheltered site. Brassington (1990) stated that the rain gauge should be located at a distance that is at least twice the height of surrounding obstructions: the vertical angle from the rain gauge to the top of the surrounding trees and buildings should be no greater than 30°. Also, Shaw (1983) recommended that a turf wall be used in overexposed locations where natural shelter is rare. A surrounding small grassed embankment decreases wind turbulence around the rain gauge which can inhibit raindrops from falling into an unprotected gauge. The turf wall should form a circle having an inside diameter of about 3 m, and be built up to the top of the rain gauge. The inside wall should have vertical walls, while the outside should have a slope of about 1 to 4. The inner area must be drained to the outside to prevent flooding.

Rain gauges must also be placed level. If a rain gauge is inclined 10° from the vertical, it will catch 1.5% less than it should due to a decreased open area exposed to the rain. In addition, if a rain gauge is inclined slightly toward the wind, it will catch more rain than the true amount.

Proper Calibration of Rain Gauges

The standard U.S. Weather Bureau rain gauge is a nonrecording, but accumulating rain gauge that has an 8-in-diameter funnel opening. The opening directs the water into a measuring tube that has 1/10 the cross-sectional area of the gauge opening. The depth of accumulated rain in the measuring tube is therefore 10 times the depth of rain that fell since the gauge was last checked. This gauge is usually used to measure the 24-hour total rain depths, usually read at 8:00 am each day. This standard gauge should be located adjacent to any recording rain gauge to check the total amount of rain that has fallen during the observation period.

A tipping bucket rain gauge is the most common type that measures rainfall intensity. This gauge has an internal tipping mechanism that fills with water from the funnel connected to the standard 8-in-diameter opening (see Figure 6.21). The tipping mechanism is balanced to dump its contents after a specific amount of water has accumulated (usually 0.01 in). Upon dumping, another small bucket rises to collect the next increment of rainfall. Each tipping motion is recorded on an event recorder, along with its time. Rainfall intensity is therefore related to the number of tips per time period.

Tipping bucket rain gauges must be periodically calibrated by measuring the number of tips associated with a specific amount of water slowly introduced into the gauge. The calibration water must be introduced at a rate comparable to that of the rainfall of interest. Several rainfall rates should be checked over the range of interest. This calibration should be conducted in the field, with the gauge installed, at least every 6 months. As noted previously, tipping bucket rain gauges are most accurate for small to moderate rain intensities. Significant rain can be missed during the time that the tipping action is moving and before the other bucket is in place. Heavy rains also tend to hold the buckets in intermediate positions for long periods, preventing the rain from accumulating in the buckets. The use of a standard accumulating rain gauge adjacent to any recording rain gauge is therefore highly recommended.

Table 6.10 shows the water delivery rate to a tipping bucket rain gauge needed for calibration for different equivalent rainfall intensities, assuming a standard 8-in opening. The rates needed to calibrate a tipping bucket rain gauge for the smallest rainfall intensities shown on this table are very low and would require special low flow pumps. As an example, a Masterflex® portable pump can pump from 0.06 to 1100 mL/min, depending on pump head, tubing size, and pump speed (available from Forestry Suppliers, catalog #76899, model 7570-10 variable speed pump with rechargeable battery, and #76888 pump head with #16 tubing, for 0.80 to 320 mL/min, at a total cost of about $900). This pump can therefore be used for all the rainfall intensity calibrations listed in Table 6.10. Of course, other available peristaltic pumps can also be used for this calibration.

Table 6.10 Water Delivery Rates for Recording Rain Gauge Calibration (standard 8-in opening)

Rainfall Intensity (mm/hour)	Rainfall Intensity (in/hour)	Water Delivery Rate for Calibration (mL/min)
2	0.078	1.1
5	0.20	2.7
10	0.39	5.4
25	0.98	14
50	2.0	27
100	3.9	54
200	7.9	110

When the rainfall intensity becomes great, the tipping bucket mechanism cannot keep up, resulting in a decreased amount of rain recorded. As an example, Ciaponi et al. (1993) used a peristaltic pump to calibrate two gauges in an urban test watershed in Pavia, Italy. The calibrations showed that the rain gauges could accurately measure rainfall intensities at 44 mm/hour (the lowest rate calibrated with the pump) with errors less than 1%. However, at rain intensities of about 250 mm/hour, the errors were about 10%, and at 400 mm/hour, the errors increased to about 15%. The measured rain intensities were all less than the actual intensities due to missing rain during the tipping time of the individual buckets. Of course, very few rains would be expected to have prolonged large intensities that would cause errors greater than about 10%. However, short-duration, very high rain intensities are much more likely, and accurate rates in these high-intensity ranges may be needed. Therefore, care must be taken when calibrating rain gauges to use appropriate water delivery rates that correspond to a wide range of expected rainfall intensities.

Summary of Rainfall Monitoring Methods

Table 6.11 lists the main advantages and disadvantages of the different basic types of rainfall monitoring methods. In all cases, a tipping bucket rain gauge is needed in an urban study area, with a standard gauge located nearby for proper calibration. In addition, at least several rain gauges (need not be recording, but that would obviously be most helpful) must be placed throughout the study area. For large areas, many gauge installations are needed. In areas of snowfall, special modifications are also required. Proper placement and shielding of the rain gauges are also needed but frequently overlooked. Radar rainfall information can be valuable, but only as a supplement to standard rain gauges in a study area. Proper use of radar rainfall data generally requires an expert and specialized software, and it is only useful relatively close to the radar installation.

SOIL EVALUATIONS

Knowing local soil properties is critical for many aspects of watershed evaluations. Soil properties are extremely important for less-developed areas, because they control many of the hydrologic and sediment aspects of the stormwater. As a watershed becomes developed, however, soil characteristics may become less important than other aspects (especially the nature and extent of the paved and roofed surfaces). Nonetheless, it is important to acknowledge that soils become dramat-

Table 6.11 Advantages and Disadvantages of Different Rainfall Monitoring Methods

Rainfall Monitoring Method	Advantages	Disadvantages
Tipping bucket rain gauges	Most commonly used and available gauge. Obtains high resolution rainfall intensity data. Relatively inexpensive for current versions of recording models.	Must be frequently calibrated and located adjacent to a standard rain gauge (not usually done). Usually insufficient numbers of recording gauges in most local networks.
Standard rain gauges	Standard rain gauge and most accurate. Can be heated and used for monitoring snowfall.	Does not obtain rain intensity information. Must be manually read at least once a day.
"Garden store" rain gauges	Inexpensive and can be placed throughout a study area. Best use to supplement standard and tipping bucket rain gauges.	Does not obtain rain intensity information. Must be manually read.
Radar rainfall measurements (such as NEXRAD)	High resolution data over a large area. Real-time measurements.	Most indicative of severe weather conditions. Can be very inaccurate and requires substantial calibration from standard rain gauges. Only suitable for areas relatively close to a radar installation.

Figure 6.31 Double-ring infiltrometer measurements in disturbed urban soils in Oconomowac, WI.

Figure 6.32 Infiltration test apparatus at University of Essen, Germany.

ically altered with typical urban development and to understand how these changes affect local stormwater. The following paragraphs describe the unusual soil conditions found during some studies of urban soils and the methodologies that were used.

Local USDA Soil Conservation Service (SCS) (now NRCS, Natural Resources Conservation Service) offices have a wealth of information pertaining to soils in all areas of the nation. The county soil surveys should be carefully reviewed for important information. However, urbanization typically alters many "natural," or mapped, soil characteristics beyond recognition through removing vegetation and topsoil, large-scale cut-and-fill operations, compaction, and artificial landscaping. Unfortunately, these changes usually all adversely affect the soils' abilities to infiltrate runoff and to retain soil during storms. It may therefore be important to directly measure some of these critical soil characteristics in watersheds undergoing study. This section briefly describes the experimental design and numerous test procedures and some results for a recent EPA-sponsored research project (Pitt et al. 1999a) that investigated adverse soil changes with urbanization (mostly compaction) and possible mitigation methods (amending soil with compost).

Numerous methods have been used to measure infiltration in urban areas. Figure 6.31 is a double-ring infiltration apparatus used to measure infiltration through disturbed urban soils in Oconomowac, WI. Figures 6.32 and 6.33 are photographs of an infiltration test apparatus developed by Dr. Wolfgang Geiger at the University of Essen, Germany, and Figures 6.34 through 6.35 are photographs of the Pac Forest soil infiltration test site developed by Dr. Rob Harrison of the Ecosystem Science and Conservation Division, College of Forest Resources at the University of Washington, Seattle.

Case Study to Measure Infiltration Rates in Disturbed Urban Soils

The soil characteristics of most interest for a receiving water investigation include the soil texture, the soil erosion factors (NRCS K and T factors), and the soil infiltration rates. Because soils in urban areas are greatly disturbed during construction activities, the information contained in the county soil surveys will not be directly applicable, requiring site investigations. Soil infiltration may be related to the time since the soil was disturbed by construction or grading operations (turf age). In most new developments, compact soils are expected to be dominant, with reduced infiltration compared to preconstruction conditions. In older areas, the soil may have recovered

Figure 6.33 Adjustments being made to rain drop tubes at Essen infiltration test apparatus.

Figure 6.34 Soil infiltration test plot at University of Washington, Seattle.

some of its infiltration capacity due to root structure development and from soil insects and other digging animals.

The following discussion presents a case study that was conducted by Pitt et al. (1999) that investigated infiltration rates in disturbed urban soils. These types of data can be used to more accurately predict watershed hydrology and associated receiving water problems, compared to using published information for natural soil conditions. The results presented in the following example

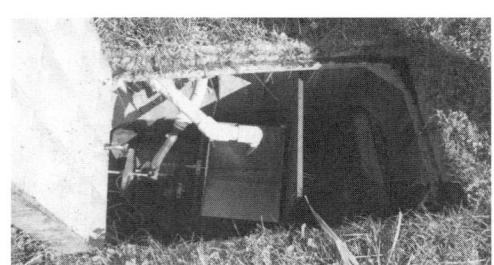

Figure 6.35 Tipping bucket flow measurement device for measuring groundwater flows at UW test plot.

Figure 6.36 Weather station at UW soil infiltration test plot.

show how site measurements can be significantly different from published and traditional data. This case study is presented as an example of how this type of study can be conducted to obtain this critical, site-specific information.

Experimental Design

A series of 153 double-ring infiltrometer tests were conducted in disturbed urban soils in the Birmingham and Mobile, AL, areas. The tests were organized in a complete 2^3 factorial design (Box et al. 1978) to examine the effects of soil moisture, soil texture, and soil compactness on water infiltration through historically disturbed urban soils. Moisture and soil texture conditions were determined by standard laboratory soil analyses. Compaction was measured in the field using a cone penetrometer (Dickey-John Corp. 1987) and confirmed by the site history. Moisture levels were increased using long-duration surface irrigation before most of the saturated soil tests. From 12 to 27 replicate tests were conducted in each of the eight experimental categories in order to measure the variations within each category for comparison to the variation between the categories.

Table 6.12 shows the analytical measurement methods used for measuring the infiltration rates, and supporting measurements, during the tests of infiltration at disturbed urban sites. Table 6.13 defines the different levels for the experimental factors used during these tests.

Infiltration Rate Measurements

The infiltration test procedure included several measurements. Before a test was performed, the compaction of the soil was measured with the DICKEY-john Soil Compaction Tester and a sample was obtained to analyze moisture content. TURF-TEC Infiltrometers (1989) were used to measure the soil infiltration rates. These small devices have an inner ring about 64 mm (2.5 in) in diameter and an outer ring about 110 mm (4.25 in) in diameter. The water depth in the inner compartment starts at 125 mm (5 in) at the beginning of the test, and the device is pushed 50 mm (2 in) into the ground. The rings are secured in a frame with a float in the inner chamber and a pointer next to a stopwatch. These units are smaller than standard double-ring infiltrometers, but their ease of use allowed many tests to be conducted under a wide variety of conditions. The use of three infiltrometers placed close together also enabled better site variability to be determined than if larger, standard-

Table 6.12 QA Objectives for Detection Limits, Precision, and Accuracy for Critical Infiltration Rate Measurements in Disturbed Urban Soils

Measurement	Method[a]	Reporting Units	MDL	Precision
Double-ring infiltration rate measurements	ASTM D3385-94	in/hr	0.05	10%
Soil texture	ASTM D 422-63, D 2488-93, and 421	plots	na	10%
Soil moisture (analytical balance)	ASTM D 2974-87	Percentage of moisture in soil (mg)	5% (0.1 mg)	10% (1%)
Soil compaction	History of site activities and cone penetrometer	psi	5	10%
Soil age	Age of development	years	na	na

[a] ASTM 1994 and Dickey-John Corp. 1987.

Table 6.13 Experimental Test Levels during Infiltration Rate Tests

	Moisture	Disturbance[a]	Soil Texture[b]
Enhanced infiltration	Dry (<20% moisture)	Uncompacted (<300 psi)	Sandy (per ASTM D 2487)
Decreased infiltration	Wet (>20% moisture)	Compact (>300 psi)	Clayey (per ASTM D 2487)

[a] Dickey-John Corp. 1987.
[b] ASTM 1994.

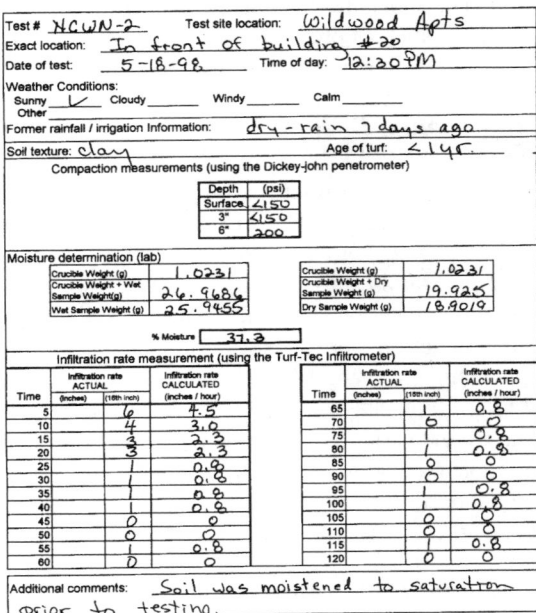

Figure 6.37   Field observation sheet. (From Pitt et al., 1999.)

sized units were used. These small units are available from Forestry Suppliers, Inc., while the standard-sized units are available from Gilson, or other soil engineering equipment suppliers.

Three infiltrometers were inserted into the turf within a meter of each other to indicate the infiltration rate variability of soils in close proximity. Both the inner and outer compartments were filled with clean water by first filling the inner compartment and allowing it to overflow into the outer compartment. As soon as the measuring pointer reached the beginning of the scale, the timer was started. Readings were taken every 5 min for 2 hours. The instantaneous infiltration rates were calculated by noting the decline in the water level in the inner compartment over the 5-min period.

Tests were recorded on a field observation sheet as shown in Figure 6.37. Each document contained information such as relative site information, testing date and time, compaction data, moisture data, and water level drops over time, with the corresponding calculated infiltration rate for the 5-min intervals.

All measurements were taken in soils in the field (leaving the surface sod in place), with no manipulation besides possibly increasing the moisture content before "wet" soil tests are conducted (if needed).

Soil Moisture Measurements

Moisture values relating to dry or wet conditions are highly dependent on soil texture and are mostly determined by the length of antecedent dry period before the test. Soil moisture was determined in the laboratory using the ASTM D 2974-87 (1994) method (basically weighing a soil before and after oven drying). For typical sandy and clayey soil conditions at the candidate test areas, the dry soils had moisture contents ranging from 5 to 20% (averaging 13%) water, while wet soils had moisture contents ranging from 20 to 40% (averaging 27%) water.

The moisture condition at each test site was an important test factor. The weather occurring during the testing enabled most site locations to produce a paired set of dry and wet tests. The dry tests were taken during periods of little rain, which typically extended for as long as 2 weeks with no rain and with sunny, hot days. The saturated tests were conducted after thorough artificial soaking of the ground, or after prolonged rain. The soil moisture was measured in the field using a portable moisture meter (for some tests) and in the laboratory using standard soil moisture methods (for all tests). The moisture content, as defined by Das (1994), is the ratio of the weight of water to the weight of solids in a given volume of soil. This was obtained using ASTM method D 2974-87 (1994), by weighing the soil sample with its natural moisture content and recording the mass. The sample was then oven-dried and its dry weight recorded.

Soil Texture Measurements

At each site location, a soil sample was obtained for a texture classification. The texture of the samples was determined by ASTM standard sieve analyses (1994) to verify the soil conditions estimated in the field and for comparison to the NRCS soil maps. The sieve analysis used was the ASTM D 422-63 *Standard Test Method for Particle Size Analysis of Soils* for the particles larger than the No. 200 sieve, along with ASTM D 2488-93 *Standard Practice for Description and Identification of Soils (Visual — Manual Procedure)*. The sample was prepared based on ASTM 421 *Practice for Dry Preparation of Soil Samples for Particle Size Analysis and Determination of Soil Constants.*

The texture analyses required a representative dry sample of the soil to be tested. After the material was dried and weighed, it was crumbled to allow a precise sieve analysis. The sample was then treated with a dispersing agent (sodium hexametaphosphate) and water at the specified quantities. The mixture was then washed over a No. 200 sieve to remove all soil particles smaller than the 0.075 mm openings. The sample was then dried again and a dry weight obtained. At that point, the remaining sample was placed in a sieve stack containing No. 4, No. 8, No. 16, No. 30, No. 50, No. 100, No. 200 sieves, and the pan. The sieves were then placed in a mechanical shaker and allowed to separate onto their respective sieve sizes. The cumulative weight retained on each sieve was then recorded.

The designation for the sand or clay categories follows the *Unified Soil Classification System*, ASTM D 2487. Sandy soils required that more than half of the material be larger than the No. 200 sieve, and more than half of that fraction be smaller than the No. 4 sieve. Similarly, for clayey soils, more than half of the material is required to be smaller than the No. 200 sieve. Figure 6.38 is the standard soil texture triangle defining the different soil texture categories.

Soil Compaction Measurements

The extent of compaction at each site was also measured before testing using a cone penetrometer. The compaction of the test areas was obtained by pushing a DICKEY-john Soil Compaction Tester (available from Forestry Suppliers, Inc.) into the ground and recording the readings from

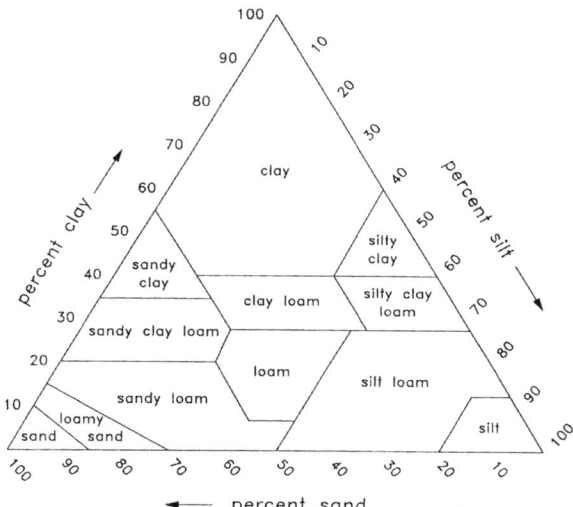

Figure 6.38 Standard soil triangle.

the gauge. For these tests, compact soils were defined as a reading of greater than 300 psi at a depth of 3 in, while uncompacted soils had readings of less than 300 psi.

Compaction was confirmed based on historical use of the test site location, as moisture levels affected the cone penetrometer readings. Soils, especially clay soils, are obviously more spongy and soft when wet compared to when they are extremely dry. Therefore, the penetrometer measurements were not made for saturated conditions, and the degree of soil compaction was also determined based on the history of the specific site (especially the presence of parked vehicles, unpaved lanes, well-used walkways, etc.). Other factors that were beyond the control of the experiments, but also affect infiltration rates, include bioturbation by ants, gophers and other small burrowing animals, worms, and plant roots.

Bulk Density

Bulk density was estimated using a coring device of known volume (bulk density soil sampler). The core was removed, oven dried, and weighed. Bulk density was calculated as the oven-dry weight divided by the core volume. Particle density was determined by using a gravimetric displacement. A known weight of soil or soil/compost mixture was placed in a volumetric flask containing water. The volume of displacement was measured and particle density was calculated by dividing the oven-dry weight by displaced volume.

Gravimetric water-holding capacity was determined using a soil column extraction method that approximates field capacity by drawing air downward through a soil column. Soil or soil/compost mixture was placed into 50 mL syringe tubes and tapped down (not compressed directly) to achieve the same bulk density as the field bulk density measured with coring devices. The column was saturated by drawing 50 mL of water through the soil column, then brought to approximate field capacity by drawing 50 mL of air through the soil or soil/compost column.

Volumetric water-holding capacity was calculated by multiplying gravimetric field capacity by the bulk density. Total porosity was calculated by using the following function:

$$\text{total porosity} = 1 - \left(\frac{\text{bulk density}}{\text{particle density}} \right) \times 100\% \tag{6.1}$$

Particle size distribution was determined both by sieve analysis and sedimentation analysis for particles less than 0.5 mm in size. Due to the light nature of the organic matter amendment, particle size analysis was sometimes difficult, and possibly slightly inaccurate. Soil structure was determined using the feel method and comparing soil and soil/compost mixture samples to known structures.

Subsurface Flow Measurements

Subsurface flows and surface runoff during rains were measured and sampled using special tipping bucket flow monitors collecting the samples from the tubing shown in Figure 6.39 (Harrison et al. 1997). The flow amounts and rates were measured by tipping-bucket-type devices attached to an electronic recorder, as shown in Figure 6.40 (a close-up of the tipping bucket flowmeters shown previously in Figure 6.35), taken at the University of Washington installation. Each tip of the bucket was calibrated for each site and checked on a regular basis to give rates of surface and subsurface runoff from all plots.

Observations — Infiltration Rates in Disturbed Urban Soils

The initial exploratory analyses of the data showed that sand was most affected by compaction, with little change due to moisture levels. However, the clay sites were affected by a strong interaction

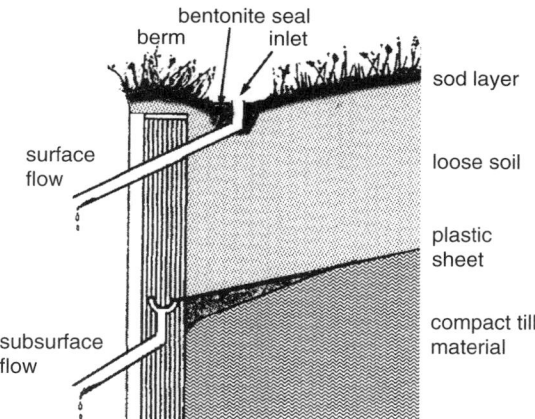

Figure 6.39 Drawing of surface and subsurface flow collectors for use in field sites. (From Harrison, R.B., M.A. Grey, C.L. Henry, and D. Xue. *Field Test of Compost Amendment to Reduce Nutrient Runoff.* Prepared for the City of Redmond. College of Forestry Resources, University of Washington, Seattle. May 1997.)

Figure 6.40 Picture of the tipping bucket installation for monitoring surface runoff and subsurface flows at the University of Washington.

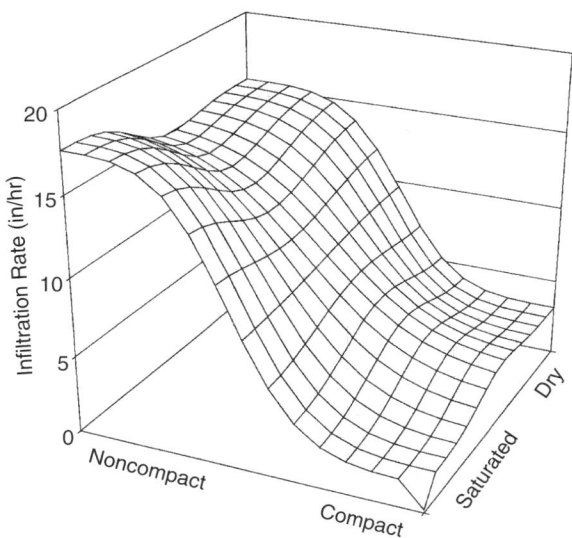

Figure 6.41 Three-dimensional plot of infiltration rates for sandy soil conditions. (From Pitt et al. 1999.)

of compaction and moisture (see Figures 6.41 and 6.42). The variations in the observed infiltration rates in each category were relatively large, but four soil conditions were found to be distinct, as shown in Table 6.14.

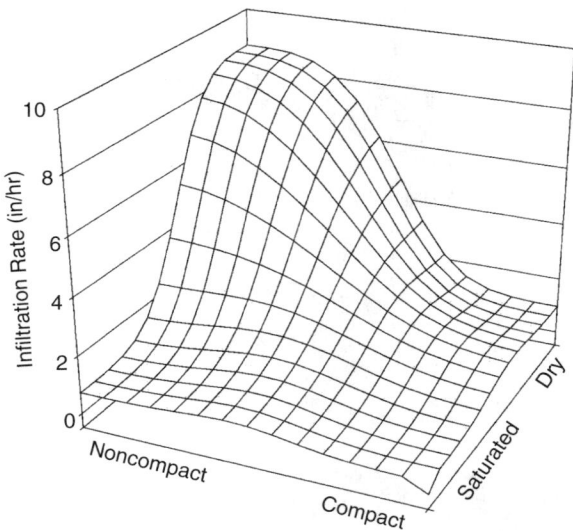

Figure 6.42 Three-dimensional plot of infiltration rates for clayey soil conditions. (From Pitt et al. 1999.)

The data from each individual test were fitted to the Horton (1939) equation (Table 6.15), but the resulting equation coefficients were relatively imprecise, and it may not matter much which infiltration model is used, as long as the uncertainty is considered in the evaluation. Therefore, when modeling runoff from urban soils, it may be best to assume relatively constant infiltration rates throughout an event, and to utilize Monte Carlo procedures to describe the observed random variations about the predicted mean value, possibly using time-averaged infiltration rates and COV values.

Importance of Field Tests of Soil Infiltration Characteristics

Very large errors in soil infiltration rates can easily be made if published soil maps and most available models are used for typically disturbed urban soils, because these tools ignore compaction.

Table 6.14 Infiltration Rates for Distinct Groupings of Soil Texture, Moisture, and Compaction Conditions

Group	Number of Tests	Average Infiltration Rate (in/hr)	COV
Noncompacted sandy soils	36	16.3	0.4
Compact sandy soils	39	2.5	0.2
Noncompacted and dry clayey soils	18	8.8	1.0
All other clayey soils (compacted and dry, plus all saturated conditions)	60	0.7	1.5

From Pitt et al. 1999.

Table 6.15 Observed Horton Equation Parameter Values for Sandy and Clayey Soils

	f_o (in/hr)		f_c (in/hr)		k (l/min)	
	Mean	Range	Mean	Range	Mean	Range
Observed noncompacted sandy soils	39	4.2 to 146	15	0.4 to 25	9.6	1.0 to 33
Observed compact sandy soils	15	0.1 to 86	1.8	0.1 to 9.5	11	1.8 to 37
Observed dry noncompacted clayey soils	18	2.5 to 58	6.6	0.1 to 24	8.8	−6.2 to 19
Observed for all other clayey soils (compacted and dry, plus all saturated conditions)	3.4	0 to 48	0.4	−0.6 to 6.7	5.6	0 to 46

From Pitt et al. 1999.

Knowledge of compaction (which can be mapped using a cone penetrometer, or estimated based on expected activity on grassed areas) can be used to more accurately predict stormwater runoff quantity.

It is therefore recommended that certain site-specific soil measurements be made in the watershed being studied. These tests should at least include actual soil texture near the surface and the shallow root zone area. Soil compaction greatly affects runoff rates and amounts and should be measured during moderately dry to moist conditions. Care should be taken when using a cone penetrometer during excessively dry or wet soil conditions. The simple double-ring infiltrometer tests, such as described for the Alabama tests, can be easily used to examine the effects of disturbing soils during development and use.

Water Quality and Quantity Effects of Amending Soils with Compost

Surface runoff decreased by five to 10 times after amending the soil with compost (4 in of compost tilled 8 in into the soil), compared to unamended sites. However, the concentrations of many pollutants increased in the surface runoff, especially associated with leaching of nutrients from the compost. The surface runoff from the compost-amended soil sites had greater concentrations of almost all constituents, compared to the surface runoff from the soil-only test sites. The only exceptions were some cations (Al, Fe, Mn, Zn, Si) and toxicity, which were all lower in the surface runoff from the compost-amended soil test sites. The concentration increases in the surface runoff and subsurface flows from the compost-amended soil test site were quite large, typically in the range of five to 10 times greater. Subsurface flow concentration increases for the compost-amended soil test sites were also common and about as large. The only exceptions were for Fe, Zn, and toxicity. Toxicity tests indicated reduced toxicity with filtration at both the soil-only and at the compost-amended test sites, likely due to the sorption or ion exchange properties of the compost.

Compost-amended soils caused increases in concentrations of many constituents in the surface runoff. However, the compost amendments also significantly decreased the amount of surface runoff leaving the test plots. Table 6.16 summarizes these expected changes in surface runoff and subsur-

Table 6.16 Changes in Pollutant Discharges from Surface Runoff and Subsurface Flows at New Compost-Amended Sites, Compared to Soil-Only Sites

Constituent	Surface Runoff Discharges, Amended-Soil Compared to Unamended Soil	Subsurface Flow Discharges, Amended-Soil Compared to Unamended Soil
Runoff volume	0.09	0.29
Phosphate	0.62	3.0
Total phosphorus	0.50	1.5
Ammonium nitrogen	0.56	4.4
Nitrate nitrogen	0.28	1.5
Total nitrogen	0.31	1.5
Chloride	0.25	0.67
Sulfate	0.20	0.73
Calcium	0.14	0.61
Potassium	0.50	2.2
Magnesium	0.13	0.58
Manganese	0.042	0.57
Sodium	0.077	0.40
Sulfur	0.21	1.0
Silica	0.014	0.37
Aluminum	0.006	0.40
Copper	0.33	1.2
Iron	0.023	0.27
Zinc	0.061	0.18

From Pitt et al. 1999.

face flow mass pollutant discharges associated with compost-amended soils. All of the surface runoff mass discharges were reduced from 2 to 50% of the unamended discharges. However, many of the subsurface flow mass discharges increased, especially for ammonia (340% increase), phosphate (200% increase), plus total phosphorus, nitrates, and total nitrogen (all with 50% increases). Most of the other constituent mass discharges in the subsurface flows decreased.

Importance of Measuring Chemical Properties of Soils When Making Soil Modifications

The use of soil amendments, or otherwise modifying soil structure and chemical characteristics, is becoming an increasingly popular stormwater control practice. However, little information is available to reasonably quantify benefits and problems associated with these changes. An examination of appropriate soil chemical characteristics, along with surface and subsurface runoff quantity and quality, was done during these Seattle tests. It is recommended that researchers considering soil modifications as a stormwater management option conduct similar local tests, including at least the detail contained in this case study, in order to understand the effects these soil changes may have on runoff quality and quantity. During the Seattle tests, the compost was found to have significant sorption and ion-exchange capacity that was responsible for pollutant reductions in the infiltrating water. However, the compost also leached large amounts of nutrients to the surface and subsurface waters. Related tests with older test plots in the Seattle area found much less pronounced degradation of surface and subsurface flows with aging of the compost amendments. In addition, it is likely that the use of a smaller fraction of compost would have resulted in fewer negative problems, while providing most of the benefits. Again, local studies using locally available compost and soils would be needed to examine this emerging stormwater management option more thoroughly.

AESTHETICS, LITTER, AND SAFETY

Safety Characteristics

Chapter 3 discussed safety-related problems associated with urban receiving waters. This is a critical beneficial use and should therefore be considered in evaluations of receiving water use impairment studies. The important safety-related information should be collected as part of the habitat survey process, as the recognized safety hazards are also indicative of poor habitat conditions for aquatic life. These include rapidly changing flows and common high flows, steep or cut banks, muddy and slippery banks, and fine-grained/mucky stream sediments. The presence of trash and other hazardous debris should also be noted as part of stream habitat surveys. Most of these problems are related to high flows from developed areas and erosion from developing areas. Watershed surveys may therefore also be important in identifying these specific sources and the necessary preventive measures to reduce safety hazards associated with urban stormwater.

Aesthetics, Litter/Floatables, and Other Debris

Aesthetics and these other elements were also described in Chapter 3 as important basic receiving water uses. Again, they should be considered in any urban receiving water evaluation investigation. Stream habitat surveys typically collect information relating to general aesthetics, including litter and other debris. An example of a beach litter survey was reported by Williams and Simmons (1997) who conducted surveys at 50 sites in South Wales and 20 sites in Devon, U.K., over a 1-year period. The surveys were conducted in both winter and summer. At each site, three transects were made, each 5 m wide, perpendicular to the beach and covering all litter strand lines. The number and types of litter were recorded in each transect. Supplemental surveys were also carried out along 1-km lengths of beach specifically for containers (material, size, color, original contents, age, and geographical origin). Plastic fragments, bags, and plastic sheeting were the most numerous litter items found. Investigators determined that little of this material accumulates along U.K. riverbanks, leaving more for deposition along marine beaches.

HydroQual (1995) reported New York City's major efforts in characterizing litter loadings and in measuring the effectiveness of litter control devices. New York City has a Scorecard Litter Rating (SLR) Program with regular inspections of sidewalks and streets. The SLR has a numeric rating of 1.0 for streets with no litter to 3.0 for streets with a continuous line of litter. An acceptably clean rating is 1.5, with 70 to 75% of all New York City streets meeting this criterion since 1986. An extensive field monitoring program to quantify litter loadings was conducted in the summer of 1993, simultaneous with SLR inspections. This monitoring program quantified the amounts and characteristics of floatable litter. Ninety blockfaces (each 80 ft in length) were selected throughout the city for monitoring. The cleanest rating was between 1.10 and 1.19, while the dirtiest was between 2.00 and 2.09. Five to six visits were made that summer to each test area, resulting in almost 7 miles of street being directly sampled. Litter samples were collected Monday through Friday, with about half collected in the morning and half in the afternoon. At each test area, the streets and the sidewalks were individually swept with push brooms and the litter collected. The litter was then brought to a central laboratory where it was weighed and separated into 13 floatable categories (listed in Table 6.17) and nonfloatables. The material in each category was counted, weighed, and the accumulative surface areas (after laying out on a table) were measured.

The sampling procedure involved a two-person crew, one cleaning the street and the other cleaning the sidewalk. Each person used a push broom, a long-handled sweep pan, and a wheeled garbage bin lined with a plastic sample bag. The loose litter was collected and deposited into the appropriate bin labeled for the street or the sidewalk. Natural materials (sticks, leaves, etc.), gravel, sand, bricks, animal droppings and remains, and items pinned under parked vehicles were ignored. Hazardous items (syringes, glass shards, etc.) were retrieved with tongs and placed inside hard plastic containers for safe handling. Bulky items (large appliances, tires, etc.) were noted on the field sheets and not collected. Containers having liquids were drained (unless they were tightly capped or contained petroleum) before collecting. The collection bags were carefully tagged. Several bags were sometimes needed for any one sampling site. The sample bags were brought to the laboratory for analysis.

Table 6.17 Discharged Litter Material Categories Captured during New York City Tests

	Category	Examples
1	Sensitives	Syringes, crack vials, baby diapers
2	Paper-coated/waxed	Milk cartons, drink cups, candy wrappers
3	Paper-cigarette	Cigarette butts, cigarettes
4	Paper-other products	Newspaper, cardboard, napkins
5	Plastic	Spoons, straws, sandwich bags
6	Polystyrene	Cups, packing material, some soda bottle labels
7	Metal/foil	Soft drink cans, gum wrappers
8	Rubber	Pieces from automobiles, pieces from toys
9	Glass	Bottles, light bulbs
10	Wood	Popsicle sticks, coffee stirrers
11	Cloth/fabric	Clothing, seat covers, flags
12	Misc. floatables	Citrus peels. Pieces of foam
13	Non-floatables	Opened food cans, broken bottles, bolts

Data from Grey, G. and F. Oliveri. Catch basins — effective floatables control devices. Presented at the *Advances in Urban Wet Weather Pollution Reduction* conference. Cleveland, OH, June 28–July 1, 1998. Water Environment Federation. Alexandria, VA. 1998.

Three laboratory technicians would then weigh an unopened bag before pouring the contents onto a sorting table. A water-filled test bucket was also available to determine if an item was floatable or nonfloatable, if in doubt. After sorting, counting, and measuring areas (using a grid on the table), the material was placed into 20-gallon bins where the sorted material was weighed and the total volume measured. Periodically, individual categories were further subdivided into 47 subcategories to attempt to "fingerprint" the types of material found on streets and sidewalks to compare to the similarly monitored material being collected during the floatable capture activities in the receiving waters.

HABITAT

Habitat can be defined as the total physical and chemical environment where organisms live. Some of these environmental components of habitat are very dynamic, such as flow, and may change by an order of magnitude within minutes, while others change on a seasonal basis (e.g., riparian vegetation) or annual (e.g., channel morphology). As noted in the preceding discussion on flow, watershed development may dramatically alter the temporal dynamics of many of these habitat components, in addition to changing spatial relationships (e.g., patch dynamics) and general habitat character. These habitat alterations play a major role, if not *the* major role, in determining the type, size, and diversity of species, populations, and communities that will reside in the affected water system (Figure 6.43).

The other major determinants (stressors) of ecosystem quality are the pollutant types and loading dynamics that are present. Habitat and pollution stress are often interwoven, interacting components which are difficult to separate. Fortunately, it usually is not necessary to accurately determine the nature, type, and/or degree (quantity) of each individual stressor. It is often necessary, however, to determine to what degree runoff effects are due to development (anthropogenic activities) or to particular sources (e.g., construction site) as compared to natural, predevelopment, or least-impacted conditions. This necessitates use of qualitative or quantitative measures of habitat conditions in a test and reference site, as discussed in Chapter 3 and Appendix A.

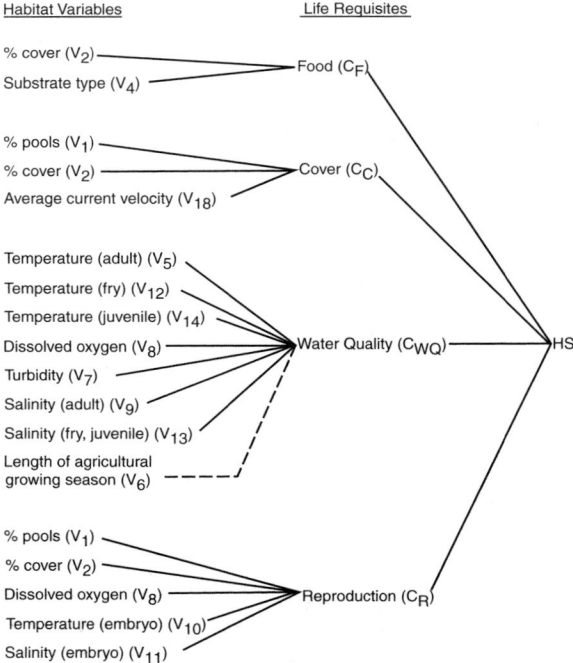

Figure 6.43 The relationship between habitat and biological condition. (From EPA. *Ecological Assessment of Hazardous Waste Sites.* Environmental Research Laboratory, U.S. Environmental Protection Agency, Corvallis, OR. EPA 600/3-89/013. 1989a; EPA. *Rapid Bioassessment Protocols for Use in Streams and Rivers: Benthic Macroinvertebrates and Fish.* Office of Water, U.S. Environmental Protection Agency, Washington, D.C., EPA 444/4-89/001. 1989c.)

Figure 6.44 Pool and riffle area in Milwaukee. (Courtesy of Wisconsin Department of Natural Resources.)

Figure 6.45 Long riffle in Milwaukee.

Figures 6.44 through 6.48 illustrate various relatively natural habitat types found in urban areas. These various types, plus the heavily modified urban streams that are also common (see Figures 3.7 through 3.11), all require investigation and specialized sampling techniques, because all are expected to be significantly different biologically. Habitat plays an important role in the natural ecosystem, and these natural differences must be evaluated when trying to understand the specific effects associated with urbanization. The following discussion will show the usefulness of characterizing physical habitat in evaluations of stormwater runoff effect, while later sections of this chapter will address specific biological sampling methods that should be used for the different habitat types.

For some studies, quantification of habitat effects is useful and necessary to meet the Data Quality Objectives (DQOs). These methods do, of course, require more resources (time, equipment, expertise, and/or expense) than qualitative assessments. Quantitative approaches include the Habitat Suitability Indices (HSI) (Figure 6.49) (Terrell 1984), Habitat Quality Index (Binns and Eiserman 1979), and the Physical Habitat Simulation Model (PHABSIM) (Hilgart 1982).

Figure 6.46 Long pool in Milwaukee.

Figure 6.47 Dry creek in Austin, TX.

Figure 6.48 Rocky substrate in Milwaukee area stream. (Courtesy of Wisconsin Department of Natural Resources.)

The HSI were developed on a species-specific basis and may be useful when particularly sensitive or economically important species are of concern. The HSI models provide information on species habitat requirements and are effective tools for beneficial use attainability analyses. These models are based on two assumptions: an HSI value has a positive relationship to potential animal numbers and a positive relationship between habitat quality and some measure of carrying capacity (EPA 1983b). HSI values range from 0 to 1, with 1 equating to optimal conditions. When comparing before and after impact data, "habitat units" may be used.

Habitat area × habitat quality (HSI) = Habitat units

Since these methods are models, they contain subjective data and such components as determining which habitat variables to include; using incomplete data sets; using data from different species of different life stages; determining independence or codependence of variables; determining when, where, and how variables should be measured; and converting assumed relationships into an aggregate suitability index (Terrell et al. 1982). The subjectivity level has been reduced, however, through extensive peer-review by the USFWS (EPA 1983b).

Most runoff effect assessments can be successful, however, without quantifying habitat effects. Rather, structured qualitative assessments exist which have been used successfully in a wide variety of ecoregions across the United States (EPA 1989c). The Qualitative Habitat Evaluation Index (QHEI) (OEPA) and the Habitat Quality Assessment Procedure (EPA 1989c) of the Ohio EPA and EPA are similar and effective at measuring six to nine interrelated metrics, including substrate, stream canopy, channel morphology, riparian and bank condition, pool and riffle quality, and gradient characteristics. All of these parameters have been shown to be related to fish and benthic macroinvertebrate community composition (OEPA 1989; EPA 1989a,c).

A key component in effectively evaluating habitat effects is the availability of baseline (predevelopment) non-(least)-impacted, reference condition information. These data are seldom available for predevelopment periods at the test site. Usually, the reference site must be in a nearby watershed that has the desired, unimpacted conditions. This approach falls within the "ecoregion" concept, which has been recommended by the EPA and successfully used by Ohio and Arkansas in their surface water quality programs (EPA 1989a).

Ecoregions are defined based on regional patterns in land-surface topology, soil and vegetation types, and land use (Omernik 1987). The biotic communities within each ecoregion are expected to be relatively similar due to habitat similarities. Studies in Ohio, Arkansas, and Oregon have suggested that fish community patterns coincide with ecoregions (Hughes et al. 1986, 1987; Larsen et al. 1986; Rohm et al. 1987; Whittier et al. 1988; Omernik 1987). Benthic macroinvertebrates show smaller habitat distribution patterns than fish (Omernik 1987) and may be influenced more by stream size, hydrologic regime, and riparian vegetation (EPA 1989a).

The QHEI, used by the State of Ohio (OEPA 1989), has shown good relationships between macrohabitat quality and fish community composition, and has been an effective tool both for implementing a biological criteria program and for assessing use impairment. Table 6.18 shows the metrics that are used in the assessment with their associated scoring ranges. When fish communities were evaluated using the Index of Biotic Integrity (IBI) (Karr 1981) and scores were less than 20, impacts were usually from a "toxicant(s)" source(s), showing greatly reduced abundance, biomass, species diversity, or other community components. However, when habitat was severely

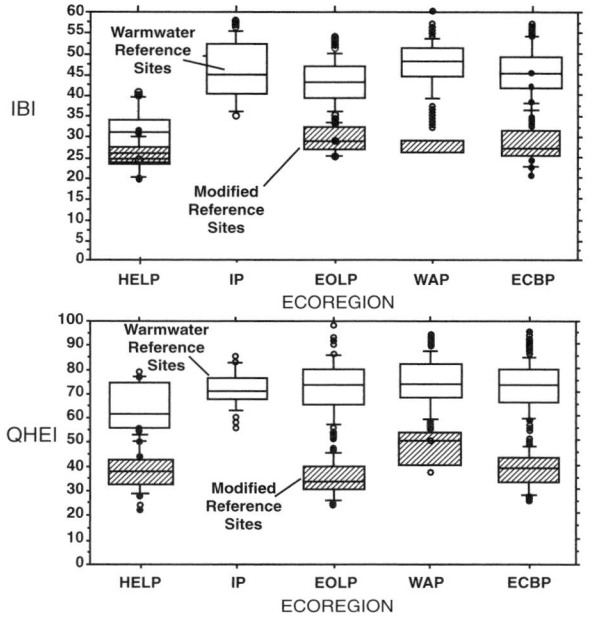

Figure 6.49 Box and whisker plots (medians, 25th and 75th percentiles, maximum value, minimum value, and outliers > two interquartile ranges from the median) for warm-water (open boxes) and modified (shaded boxes) reference sites for the IBI (top panel) and QHEI (bottom panel). (From Ohio Environmental Protection Agency. *The Qualitative Habitat Evaluation Index (QHEI): Rationale, Methods, and Application.* Ecological Assessment Section, Ohio Environmental Protection Agency, Columbus, OH. 1989.)

modified, the fish community usually responded by a shift in community function, such as from insectivore to omnivore species dominance. IBI scores rarely dropped below 20 in these situations when toxicants were absent. By utilizing individual IBI metrics and another index, the modified Index of Well-Being (mIWB), community response due to habitat or toxic impacts can be further separated (OEPA 1989).

By doing extensive surveys of habitat and aquatic communities in each ecoregion, reference site conditions can be quantified, with associated variances (for example, see Figure 6.49). Reference conditions can be tailored to meet different criteria. For example, in many states there has been extensive channel modification during the previous century.

These areas may be unable to ever recover to premodification conditions, particularly if low gradients exist (<5 ft/mi), or maintenance activities (e.g., dredging) recur. For areas where there is no evidence of or expected recovery over extended periods (i.e., decades), a channel modified reference station may be appropriate (Table 6.19, Figure 6.50) (OEPA 1989). These "irretrievable anthropogenic modifications" do not allow waters to be degraded, but rather attempt to manage historically modified streams in a realistic manner.

Factors Affecting Habitat Quality

The degree to which any habitat characteristic controls the "use" or quality of the aquatic ecosystem will vary with the site and ecoregion. There are, however, some general relationships that have been observed in a wide variety of stream systems. Small streams are more likely to be affected by riparian conditions and modifications than larger streams. Removal of riparian vegetation in headwater streams may increase water temperature 6 to 9°C and disrupt allochthonous inputs (Karr and Schlosser 1977). Another factor affecting biotic community indices is the presence of refuge areas and nearby unaffected "sources" of species (Palliam 1988; Levin 1989). If an upstream reach or tributary is unimpacted, species from this "source area or refuge" may drift or migrate into the impacted area and both assist in recovery and complicate the assessment process. Refuge areas in urban streams tend to be quite small and more limited to a protective function (e.g., debris piles) rather than a source of unaffected organisms. There are enough site-specific habitat variables to prevent the use of habitat alone as an absolute site-specific predictor of fish community quality (OEPA 1989).

**Table 6.18 Metrics and Scoring Ranges for the
Qualitative Habitat Evaluation Index**

Metric	Score
Substrate	**20 pts**
1) Type	0–21
2) Quality	–5–3
In-stream Cover	**20 pts**
1) Type	0–10
2) Amount	1–11
Channel Quality	**20 pts**
1) Sinuosity	1–4
2) Development	1–7
3) Channelization	1–6
4) Stability	1–3
Riparian/Erosion	**10 pts**
1) Width	0–4
2) Floodplain quality	0–3
3) Bank erosion	1–3
Pool Riffle	**20 pts**
1) Max depth	0–6
2) —	—
3) Current available	–2–4
4) Pool morphology	0–2
5) Riffle/run depth	0–4
6) Riffle substrate stab.	0–2
7) Riffle embeddedness	–1–2
Drainage Area	Not included
Gradient	0–15 pts
Total score	0–100 pts.

From OEPA (Ohio Environmental Protection
Agency). *The Qualitative Habitat Evaluation Index
(QHEI): Rationale, Methods, and Application.* Eco-
logical Assessment Section, Ohio Environmental
Protection Agency, Columbus, OH. 1989.

Surveys of five different ecoregions in Ohio by three fish collection methods found some significant relationships between habitat components (metrics) and fish community quality (Table 6.20). Three metrics were frequently related to the IBI, namely, pool, channel, and substrate quality (OEPA 1989).

Channelization

The process of channelizing a stream alters flow (Figure 6.51), channel morphology, and stream bank and adjacent riparian zone characteristics. When these projects cover small areas, such as for road or bridge construction, adverse impacts may be limited to the short term and affect only tens to hundreds of meters. The long-reach channelization projects, however, may cause severe ecosystem quality degradation. The most significant ecosystem alterations are usually the loss of the run–riffle–pool sequence, refuge areas, substrate composition characteristics change (e.g., particle

Table 6.19 Habitat Characteristics of Modified Warm-Water Streams and Warm-Water Streams in Ohio

Modified Warm-Water Streams	Warm-Water Streams
1. Recent channelization[1] or recovering[2]	1. No channelization or recovered
2. Silt/muck substrates[1] or heavy to moderate silt covering other substrates[2]	2. Boulder, cobble, or gravel
3. Sand substrates — Boat[2], Hardpan origin[2]	3. Silt free
4. Fair–poor development[2]	4. Good–excellent development
5. Low–No sinuosity[2], Headwater[1]	5. Moderate–high sinuosity
6. Only 1–2 cover types[2], cover sparse to none[1]	6. Cover extensive to moderate
7. Intermittent or interstitial — with poor pools[2]	7. Fast current, eddies
8. Lack of fast current[2]	8. Low–normal substrate embeddedness
9. Max. depth <40- Wading[1], -Headwater[2]	9. Max. depth > 40
10. High embeddedness of substrates[2]	10. Low/no embeddedness

Note: Superscripts for MWH streams refer to the influence of a particular characteristic in determining the use (1 = high influence, 2 = moderate influence).

From OEPA (Ohio Environmental Protection Agency). *The Qualitative Habitat Evaluation Index (QHEI): Rationale, Methods, and Application.* Ecological Assessment Section, Ohio Environmental Protection Agency, Columbus, OH. 1989.

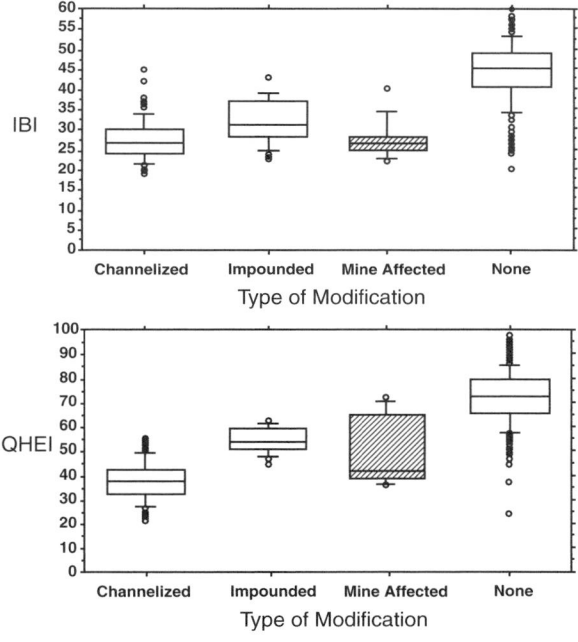

Figure 6.50 Box and whisker plots (medians, 25th and 75th percentiles, maximum value, minimum value, and outliers > two interquartile ranges from the median) from modified reference sites with channel modifications, impoundments, and mine affects (crosshatched) and warmwater reference sites for the IBI (top panel) and QHEI (bottom panel). (From Ohio Environmental Protection Agency. *The Qualitative Habitat Evaluation Index (QHEI): Rationale, Methods, and Application.* Ecological Assessment Section, Ohio Environmental Protection Agency, Columbus, OH. 1989.)

Figure 6.51 Stream flow-altering channel conditions.

size reduction, increased embeddedness), and increased temperature, and an altered productivity and trophic level regime (EPA 1983b). By straightening a stream channel, length, habitat diversity (e.g., edge habitat), and quantity are all reduced. Since fish and benthic invertebrates are habitat selective, they are directly affected by these alterations. Numerous studies have documented stream modification effects on ecosystem, structure, function, and quality (see OEPA 1989; EPA 1983b, 1977).

Table 6.20 Relative Ranking by the Magnitude of Significant (P < 0.05) Correlation Coefficients (r) between the QHEI and IBI for Ohio Ecoregions and Fish Sampling Methods

Ecoregion[a]	N[b]	Metric Ranking
		Boat Methods
HELP	28	Substrate > Channel > Riffle
IP	7	*No significant correlations*
EOLP	22	Channel > Riffle > Substrate > Pool > Gradient > Riparian > Cover
WAP	26	Substrate > Gradient > Channel > Cover > Riparian > Riffle > Pool
ECBP	56	Pool > Channel > Gradient > Substrate > Riffle > Cover
		Wading Methods
HELP	16	*No significant correlations*
IP	20	Gradient
EOLP	28	Gradient > Riffle > Channel > Pool > Substrate
WAP	47	Substrate > Cover > Channel > Gradient
ECBP	73	Cover > Channel > Pool > Gradient > Substrate > Riffle > Riparian
		Headwater Methods
HELP	8	*No significant correlations*
IP	13	Pool
EOLP	35	Channel > Cover > Substrate > Pool
WAP	31	Substrate > Channel > Cover
ECBP	52	Channel > Cover > Pool > Substrate > Riffle > Riparian > Gradient

[a] Ecoregion classifications = HELP, Huron/Erie Lake Plain; IP, Interior Plateau; EOLP, Erie/Ontario Lake Plain; WAP, Western Allegheny Plateau; and ECBP, Eastern Corn Belt Plains
[b] Number of sample data sets.

From OEPA (Ohio Environmental Protection Agency). *The Qualitative Habitat Evaluation Index (QHEI): Rationale, Methods, and Application.* Ecological Assessment Section, Ohio Environmental Protection Agency, Columbus, OH. 1989.

Substrate

The substrate composition is a direct function of watershed and channel characteristics and to a large extent controls the composition of benthic macroinvertebrates, meio- and microfauna, periphyton, and fish communities (e.g., EPA 1983b). Algal (phytoplankton) and zooplankton communities are indirectly affected by nutrient availability, which changes as the rate of cycling changes in different sediment environments. Microbial communities are influenced structurally and functionally by sediment quality (see Benthos section).

Though substrates consist of any inorganic or organic material that is utilized as a growth surface or is solid in nature, most substrate classifications are based on inorganic particle sizes (Table 6.21). Generally, mean particle sizes decrease (get finer) farther downstream due to reduced bottom shear stress and stream power. Current velocities of >50 cm/s on steep gradients typically result in substrate that is gravel size or larger. Velocities of 20 to 50 cm/s result in substrate that is sandy, while <20 cm/s velocities result in substrate dominated by silt and clay-sized particles. Channelization impacts are often greater in headwater streams that have high gradients and where coarse substrates are necessary to provide protection from strong currents (EPA 1983b). Few to no impacts have been observed in low gradient, high order, large streams where particle sizes are smaller and food sources for sensitive species are fewer.

Large-grained (e.g., gravel) sediments typically have macrobenthic communities indicative of higher quality water. These substrates have a greater amount of living space, provide protection, trap more organic material, and are well oxygenated. High flows (storm events) tend to wash out

Table 6.21 Substrate Particle Size Classification for Sieve Analysis

Name	Particle Size (mm)	(μm)	U.S. Standard Sieve Number
Boulder	>256 (10 in)		
Rubble	64 to 256		
Coarse gravel	32 to 64		
Medium gravel	8 to 32		
Fine gravel	2 to 8		10
Coarse sand	0.5 to 2	500–2000	35
Medium sand	0.25 to 0.5	250–500	60
Fine sand	0.125 to 0.25	125–250	120
Very fine sand	0.0625 to 0.125	62–125	230
Silt	0.0039 to 0.0625	4–62	
Clay	<0.0039	<4	

Modified from Wentworth, 1922; see Cummins, 1962 (EPA 1990c).

organic matter and thereby decrease food availability. Other important substrates include cobble, macrophytes, roots, and organic debris (sticks to leaves), which are used by numerous groups of organisms (e.g., periphyton, protozoa, filamentous algae, fungi, bacteria, and invertebrates) for attachment and as a food source.

Siltation is a significant stressor for many desirable species. Silt and clay have been shown to decrease habitat diversity by filling interstitial spaces (embeddedness), standing crop, density, taxa richness, reproductive success, and productivity, and to increase pollution-tolerant species (EPA 1983b).

In unchannelized, nonsandy streams, there is often an alternating pool–riffle structure. Riffles are stationary, comprised of gravel, cobble, and boulders, which may move. The increased flow, habitat space, and food in riffle areas support greater benthic macroinvertebrate populations than pool areas. For many fish, a 1:1 ratio of pool to riffle run areas is optimal for survival and reproduction.

The importance and heterogeneity of substrates in stormwater assessments necessitates the collection of multiple samples at each site and characterization of both organic and inorganic constituents. Useful characterization parameters are listed in Table 6.22.

Table 6.22 Substrate Characterization

Parameter	Method		Ref.
Particle size distribution	Sieve: wet sieve	Sample	Welch 1948
	Sedimentation: Pipette		Allen 1975
	Hydrometer		APHA 1985
	Particle size: Coulter counter		ASTM 1991
	Laser		ASTM D854-83
Dry weight	60-105MC 24 h		APHA 1985
			ASTM D4318-84
Volatile solids (ash-free)	500MC 1 h		ASTM 1987
Total organic carbon			APHA 1985
Acid volatile sulfides	Spectrophotometric or gravimetric		EPA 1990
Synthetic organics	Variety		EPA SW-846 8010-8310 3510-3550
Metals	Variety		EPA SW-846 7040-7951 3010-3060
Total organic halides			
Cation exchange capacity			
Total nitrogen			APHA 1985
Ammonia			APHA 1985
Total phosphorus			APHA 1985
Extractable phosphorus			

Scour of Bottom Sediments

A critical component of habitat quality is substrate stability. Frequent scouring or sedimentation is obviously detrimental to benthic organisms. Classical methods to monitor scour have been to use standard surveying procedures and carefully measure stream cross sections and slope. This should be supplemented with scour pins and scour chains. Scour pins are long rods (with a scale) driven deeply into stream banks, with a bright end exposed. Frequent visits are needed to note the length of pin exposed at any time. With receding banks, the pins will become more and more exposed, and the bank loss rate can be calculated. They can be reset when too much of the pin is exposed. Scour pins should be used at several locations at any cross section.

Scour pins cannot be effectively used in the stream to measure scour and sedimentation separately. The use of scour chains, as described by May et al. (1999) in the following comments for work on salmon streams in the Pacific Northwest, works well in many stream locations. Nawa and Frissell (1993) monitored stream bed stability using bead-type scour monitors installed in salmonid spawning riffles in selected reaches. Figure 6.52 illustrates these devices. They found that larger scour and/or fill events normally resulted from larger storms and the resultant higher flows, as would be expected. Cooper (1996) found that increased urbanization leads to increased stream power and stream bed instability and that basin urbanization in Puget Sound lowland streams was found to have the potential to cause locally excessive scour and fill. Urban streams in the Puget Sound lowland area having gradients greater than 2% and lacking in large woody debris (LWD) were found to be more susceptible to scour than undeveloped streams.

May et al. (1999) used a stream stability classification similar to Booth (1996): stream segments with >75% of the reach classified as stable were given a score of 4; stream segments having between 50 and 75% stable banks were scored as a 3; those with 25 to 50% stable banks were scored as a 2; and those having less than 25% stable banks were scored as a 1. The presence of artificial stream bank protection (such as rip-rap) was considered a sign of bank instability and scored as a 1. Researchers found that only two undeveloped reference stream segments (watershed areas having total imperviousness area < 5%) had a stability rating less than 3. In basins that had from 5 to 10% imperviousness, the stream bank ratings were generally 3 or 4. However, in basins having between 10 and 30% impervious area, there was a fairly even mixture of stream bank conditions, from stable and natural to highly eroded or artificially "protected." For basins having total imperviousness areas of 30%, there were no segments having stream bank stability ratings of 4 and very few with ratings of 3 (only found in segments with intact and wide riparian corridors). Artificial stream bank protection was a common feature of all highly urbanized streams (those that had total imperviousness areas > 45%). May et al. (1999) also found that stream bank stability was influenced by the condition of the riparian vegetation surrounding the stream, with the stream bank stability rating being strongly related to the width of the riparian buffer and inversely related to the number of breaks in the riparian corridor.

Sediment Transport

Sediment may be composed of organic or inorganic material ranging in size from colloidal humus (<1 µm) to boulders. Total sediments are the sum of suspended, bedload, and consolidated

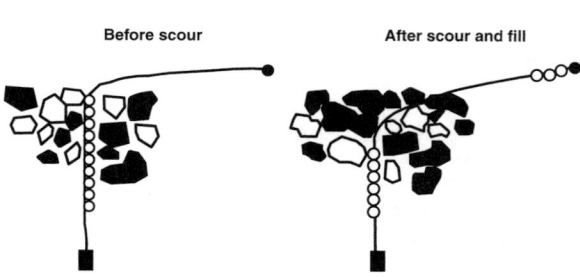

Figure 6.52 Sliding-bead type scour monitors. (From May, C., R.R. Horner, J.R. Karr, B.W. Mar, and E.B. Welch. *The Cumulative Effects of Urbanization on Small Streams in the Puget Sound Lowland Ecoregion.* University of Washington, Seattle. 1999. With permission.)

sediments, each of which may have deleterious effects on ecosystem quality. Sediment erosion, watershed yields, or loading can be estimated by various simple sampling methods. The total sediment yield includes both suspended and bedload sediment. This provides a good indicator of land-use changes. Bedload sediments are more of a concern in high flow waters, as they can scour, abrade, and smother benthic biota. Sediments may also release or adsorb nutrients and toxicants. The partitioning coefficients and controlling conditions are not well understood. Sedimentation and resuspension are affected by biological and physical processes. The physical processes include bioturbation and fluid flow (laminar or turbulent). Particle movement and settling will depend on particle size, shape, and density, cohesion-flocculation properties, temperature, solids concentration, and water velocity and turbulence. Organic settleable solids can accumulate at velocities of 0.6 ft/s or less. The sediment particle size distribution is directly related to the system's hydraulics. The most significant changes in particle size distributions occur when flow dynamics change in the stream or receiving water body, e.g., river mouth, riffle–pool boundary, or river bend. By knowing the watershed and substrate particle size characteristics and channel velocity, areas of sediment accumulation or scour may be predicted. Combining this information with time of passage data, sludge deposit areas were located (Velz 1970). As time since deposition increases, solids will tend to compact and higher velocities (e.g., 1.5 fps) will be required to induce scour.

Typical automatic water samplers are limited in their ability to sample particles in the water, as discussed in Chapter 5. If particles larger than several hundred μm need to be included in the sampling program, then manual depth-integrated (Helley–Smith) or bedload samples also need to be used, as described in Chapter 5. The Helley–Smith sampler (Helley and Smith 1971) effectively collects water and bed sediment at the same flow velocity that occurs at the stream bottom. Samples must be collected at several intervals across the channel bottom and integrated for total transport. With a depth-integrated sampler, water passes into the vented sampler at the same flow velocity as the stream, so as it is lowered it collects in proportion to the total discharge. Suspended sediment is then measured by filtering and weighing (Guy and Normal 1970; Guy 1969; Kunkle and Comer 1971).

Bedload sediment moves along the streambed by traction and saltation mechanisms (slide, roll, bounce, or hop) (Davis 1983) and may comprise approximately 10% of the total sediment load. It is more difficult to measure than suspended sediment. Bedload in streams varies greatly with stormwater discharge conditions and by season. These measurements must therefore be repeated frequently. Bedload trapping samplers can be used to measure the material moving along stream bottoms over a period of time. There are several designs for these samplers. A simple sampler is made by burying cans (bottom intact, top removed) in sediment (top flush with sediment surface). The cans are filled with large, uniformly sized marbles to provide an effective trap and prevent scour of the finer material that filters down between the much larger marbles. More exotic samplers are scoop shaped and face upstream to allow moving sediment to enter the trap and accumulate in a deeper sump.

Riparian Habitats

The importance of lake, streamside, or wetland (riparian) ecosystems in determining water quality is well known. The relationship or correlation is essentially a holistic system. However, no one riparian component or parameter can be used to predict water quality (EPA 1983b). Obviously, the effect of the riparian zone is much greater in small stream systems (i.e., high riparian area:stream area ratio). These unique ecosystems are often described as ecotones, a gradient of changing habitat between terrestrial and aquatic systems, which supports greater diversity and abundance of terrestrial species than adjacent areas.

The three principal stressors that result when riparian zones are removed are: (1) elevated temperatures from lack of shading; (2) increased siltation from the ecoregion with associated nutrients (salts, metals, pesticides, and other synthetic organics); and (3) more dynamic changes in flow-runoff. Solids, nutrients, and toxicant loadings may increase orders of magnitude when riparian zones are removed (EPA 1991c; Lowrance et al. 1983). Another less noticeable yet important ecosystem perturbation that might occur when riparian zones are removed from small watersheds or small streams is the loss of allochthonous inputs of organic matter. The principal energy process in these systems is detritus processing and is accomplished by several biotic groups. Benthic macroinvertebrates, called "shredders," produce fine particulate organic matter which is used by "collectors." The organic matter processing is assisted by fungi and bacteria. When coarse particulate organic matter inputs are reduced, light and temperature are increased. The ecosystem changes to one of herbivorous grazers which feed on the periphytic algal populations (Cummins et al. 1973, 1974, 1975; Marzolf 1978; Vannote et al. 1980). The other interactive effects are discussed in previous sections.

Accurate assessments of riparian zone measures and their contribution to water quality are difficult and require extensive sampling and expertise. Some of the many variables factors of importance are listed in Table 6.23.

Field Habitat Assessments

When conducting qualitative assessments, the procedures outlined by the Ohio EPA or in the EPA Rapid Bioassessment Protocols (OEPA 1989; EPA 1989c, 1999) should be used. The methods are very similar and are presented with field data sheets in Appendix A.

Recommended Stream Bed/Sediment Monitoring

Unstable stream sediments may be one of the most common causes of degradation of biological uses in urban streams. It is therefore important that indicators of unstable stream beds be included in any habitat evaluation effort. As discussed above, the use of scour pins can be used to indicate unstable stream banks, while sliding bead scour monitors can be used to indicate sediment deposition and scour. These techniques can be used to supplement conventional cross section surveying at established stream stations. Pins and chain monitors can be much more rapidly examined for many intermediate locations between survey stations, enabling a better overall understanding of the magnitude and location of unstable stream bed conditions. If bedload samples are desired, or if bedload movement needs to be quantified, special bedload samplers (traps) should be used, because automatic water quality samplers cannot adequately collect the larger material that comprises bedload.

Temperature

Elevated temperatures of urban streams caused by heated stormwater has caused much concern. Much-needed research is currently being conducted by Steve Greb, of the WI Department of Natural Resources, Madison, WI. Figures 6.53 through 6.57 show some of the temperature measurement equipment he is using to investigate surface temperatures and sheetflow temperature increases from many different urban surfaces.

Fish are cold-blooded poikilothermic organisms and are sensitive to water temperature changes. Gradual changes can induce metamorphosis, migration, and spawning behavior. As with many stressors, effects are greater during the sensitive early life stages. Fish may survive in suboptimal temperatures which may favor competitors, predators, parasites, and disease, and alter food sources. Metabolic activity is increased at warmer temperatures, which increases feeding until threshold levels are reached, and it also affects toxicokinetics.

Temperature profiles in streams and rivers are generally more homogeneous than deeper, less turbulent reservoirs and lakes. As previously discussed, stormwater runoff from developed land (commercial, residential, or agricultural) is often significantly warmer than from vegetated non-developed areas. In a small receiving water system, this may quickly raise water temperatures.

Table 6.23 Riparian Zone Components That May Affect Water Quality

Geomorphology (erosion, runoff rates and variations, sediment loads)
  Slope
  Topography
  Parent material
Soils (sediment loads, nutrient inputs, runoff rates)
  Particle size distribution
  Porosity
  Field saturation
  Organic component
  Profile (presence or absence of mottling)
  Cation exchange capacity
  Redox (Eh)
  pH
Hydrology (water budget, flooding potential, nutrient loads)
  Groundwater
    a. Elevation
    b. Chemical quality
    c. Rate of movement
  Climatic factors
    a. Total annual rainfall and temporal distribution
      – Chemical quality
    b. Temperature
    c. Humidity
    d. Light
Vegetative and Faunal Characteristics
  Floristics ("community health," disturbance levels)
    a. Presence/absence
    b. Nativity
  Vegetation (nutrient loads, "community health," disturbance levels)
    a. Production
    b. Biomass
    c. Decomposition
    d. Litter (Detritus) dynamics
      – Size
      – Transportability
      – Quantity
    e. Plant size classes
      – Grasses, herbs (forbs), shrubs, trees
    f. Canopy density and cover
      – Light intensity
    g. Cover values
  Fauna (community disturbance, community health)
    a. Production
    b. Biomass
    c. Mortality
  Community Structure
    a. Diversity
    b. Evenness
  Physiological Processes
    a. Transpirational water loss (community health)
    b. Photosynthetic rates (community health)
  Stream bank characteristics
    a. Stream sinuosity
    b. Stream bank stability (sediment loads, habitat availability)

EPA. *Technical Support Manual: Waterbody Surveys and Assessments for Conducting Use Attainability Analyses.* Office of Water Regulations and Standards, U.S. Environmental Protection Agency, Washington, D.C. 1983b.

Figure 6.53 Rain temperature monitoring by WI DNR in Madison, WI.

Figure 6.54 Roof temperature data loggers being used by WI DNR.

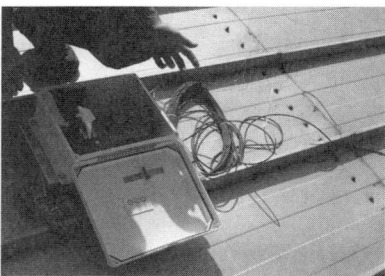

Figure 6.55 Rooftop temperature data logging used by WI DNR.

Figure 6.56 Pavement temperature monitoring by WI DNR.

This change may not exceed the temperature threshold of the species but could exceed its acclimation ability. Many urban channels also have had their natural cover removed, causing further temperature increases.

A sizable database exists on temperature effects on fish. In areas where temperature patterns change, fish populations can be expected to change. Table 6.24 shows preferred temperatures for some fish species.

Temperature also affects physical stratification (water density) in reservoirs and lakes, and thus mixing, partitioning, and the fate of feeder stream loadings. Productivity and organic matter cycling are dramatically affected by temperature both through changes in metabolic rates and changes in species (planktonic and benthic microorganisms and algae) composition. These factors combined with the physical effect of temperature on dissolved oxygen concentrations will affect macrofaunal distribution, community composition, BOD rates (waste assimilation capacity), and metal-nutrient partitioning, and thus bioavailability as soluble or insoluble species. Dissolved oxygen levels should not drop below 5 mg/L during spawning seasons (EPA 1991c) in most areas of the United States where desirable habitats exist.

Figure 6.57 Pavement temperature data loggers used by WI DNR.

Temperature is an easy parameter to define in stormwater assessments. It should accompany the collection of all samples at all sites. Background data are frequently available from nearby areas, but the land use similarity to current test conditions should be known due to its significant effect on temperature. The diurnal and seasonal patterns should be defined at reference and test sites, during baseflow, stormflow, and post-storm event conditions.

Turbidity

In many developing urban areas, urban receiving waters are typically characterized by high turbidity levels caused by high erosion rates from ongoing construction activities. Large discharges of sediment in urban runoff are mostly associated with poorly controlled construction sites, where 30 to 300 tons of sediment per acre per year of exposure may be lost. These high rates can be 20 to 2000 times the unit area sediment discharge rates associated with other land uses. Unfortunately, much of this sediment reaches urban receiving waters, where massive impacts on the aquatic environment can result. With complete development, sediment discharges from urban stormwater are significantly reduced. Unfortunately, high rates of sediment loss can also be associated with later phases of urbanization, where receiving water channel banks widen to accommodate the increased runoff volume and frequency of highly erosive flow rates. The associated increased levels of turbidity can interfere with algal productivity and with aquatic life. Increased turbidity is also typically associated with increases in settleable solids that can smother the natural bottom material and benthic organisms. These changes in the bottom characteristics of streams and lakes can produce dramatic interferences with spawning and rearing of fish.

Schueler (1997a) listed the impacts that can be associated with suspended sediment:

- "Abrades and damages fish gills, increasing risk of infection and disease
- Scouring of periphyton from streams (plants attached to rocks)
- Loss of sensitive or threatened fish species when turbidity exceeds 25 NTU
- Shifts in fish communities toward more sediment-tolerant species
- Decline in sunfish, bass, chub, and catfish when monthly turbidity exceeds 100 NTU
- Reduces sight distance for trout, with reduction in feeding efficiency
- Reduces light penetration that causes reduction in plankton and aquatic plant growth
- Reduces filtration efficiency of zooplankton in lakes and estuaries
- Adversely impacts aquatic insects, which are the base of the food chain
- Slightly increases stream temperature in summer
- Suspended sediments are major carriers of nutrients and metals
- Turbidity increases the probability of boating, swimming, and diving accidents
- Increased water treatment to meet drinking water standards
- Increased wear and tear on hydroelectric and water intake equipment
- Reduces anglers chances of catching fish
- Diminishes direct and indirect recreational experience of receiving waters"

Bolstad and Swank (1997) examined the in-stream water quality at five sampling stations in Cowetta Creek in western North Carolina over a 3-year period. The watershed is 4350 ha and is relatively undeveloped (forested) in the area above the most upstream sampling station and becomes

Table 6.24 Preferred Temperature of Some Fish Species

Common Name (Species)	Life Stage[a]	Acclimation Temperature, °C	Preferred Temperature, °C
Alewife (*Alosa pseudoharengus*)	J	18	20
	J	21	22
	A	24	23
	A	31	23
Threadfin shad (*Dorosoma petenense*)	A		>19
Sockeye salmon (*Oncorhynchus nerka*)	J		12–14
	A		10–15
Pink salmon (*O. gorbuscha*)	J		12–14
Chum salmon (*O. keta*)	J		12–14
Chinook salmon (*O. tshawytscha*)	J		12–14
Coho salmon (*O. kisutch*)	J		12–14
	A		13
Cisco (*Coregonus artedii*)	A		13
Lake whitefish (*C. clupeaformis*)	A		13
Cutthroat trout (*Salmo clarki*)	A		9–12
Rainbow trout (*S. gairdneri*)	J	—	14
	J	18	18
	J	24	22
	A		13
Atlantic salmon (*S. salar*)	A		14–16
Brown trout (*S. trutta*)	A		12–18
Brook trout (*Salvelinus fontinalis*)	J	6	12
	J	24	19
	A		14–18
Lake trout (*Salvelinus namaycush*)	J		8–15
Rainbow smelt (*Osmerus mordax*)	A		6–14
Grass pickerel (*Esox americanus vermiculatus*)	J, A		24–26
Muskellunge (*Esox masquinongy*)	J		26
Common carp (*Cyprinus carpio*)	J	10	17
	J	15	25
	J	20	27
	J	25	31
	J	35	32
	A		33–35
Emerald shiner (*Notropis atherinoides*)	J		25
White sucker (*Catostomus commersoni*)	A		19–21
Buffalo (*Ictiobus* sp.)	A		31–34
Brown bullhead (*Ictalurus nebulosus*)	J	18	21
	J	23	27
	J	26	31
	A		29–31
Channel catfish (*Ictalurus punctatus*)	J	22–29	35
	A		30–32
White perch (*Morone americana*)	J	6	10
	J	15	20
	J	20	25
	J	26–30	31–32
White bass (*M. chrysops*)	A		28–30
Striped bass (*M. saxatilis*)	J	5	12
	J	14	22
	J	21	26
	J	28	28
Rock bass (*Ambloplites rupestris*)	A	26–30	
Green sunfish (*Lepomis cyanellus*)	J	6	16
	J	12	21
	J	18	25

Table 6.24 Preferred Temperature of Some Fish Species *(Continued)*

Common Name (Species)	Life Stage[a]	Acclimation Temperature, °C	Preferred Temperature, °C
	J	24	30
	J	30	31
Pumpkinseed (*L. gibbosus*)	J	8	10
	J	19	21
	J	24	31
	J	26	33
	A		31–31
Bluegill (*L. machrochirus*)	J	6	19
	J	12	24
	J	18	29
	J	24	31
	J	30	32
Smallmouth bass (*Micropterus dolomieui*)	J	15	20
	J	18	23
	J	24	30
	J	30	31
Spotted bass (*M. punctulatus*)	J	6	17
	J	12	20
	J	18	27
	J	24	30
	J	30	32
Largemouth bass (*M. salmoides*)	J		26–32
White crappie (*Pomoxis annularis*)	J	5	10
	J	24	26
	J	27	28
	A		28–29
Black crappie (*P. nigromaculatus*)	J		27–29
	A		24–31
Yellow perch (*Perca flavescens*)	J, A		19–24
Sauger (*Stizostedion canadense*)	A		18–28
Walleye (*S. vitreum*)	J, A		20–25
Freshwater drum (*Aplodinotus grunniens*)	A		29–31

[a] J = juvenile, A = adult.

EPA. *Technical Support Manual: Waterbody Surveys and Assessments for Conducting Use Attainability Analyses.* Office of Water Regulations and Standards, U.S. Environmental Protection Agency, Washington, D.C. 1983b.

more urbanized at the downstream sampling station. Baseflow water quality was good, while most constituents increased in concentration during wet weather. Water quality was compared to building density for the different monitoring stations. Stormwater pollutant-related concentrations of turbidity increased as building densities increased. Baseflow concentrations also typically increased with density, but at a much lower rate. In addition, the highest concentrations observed during individual events corresponded to the highest flow rates.

There has been conflicting evidence on the role of elevated turbidity levels on eutrophication processes and resulting highly fluctuating DO levels. Because of the high sediment loads, urban lakes are quite different compared to most impoundments. Burkholder et al. (1998) described a series of enclosure experiments they conducted in Durant Reservoir, near Raleigh, NC. Secchi disk transparency ranged from 0.5 to 1.3 m during the summer of 1990 when these experiments were conducted. The algal communities are P-limited until late summer, when N becomes the primary limiting nutrient. The phytoplankton biomass significantly increases during the summer growing season. Several 2-m-diameter enclosures were constructed, isolating sediment to water surface columns of water. The experimental design allowed investigating the effects of different levels of sediment and nutrients on algal productivity. They found that the effects (reduction of light and coflocculation of clay and phosphate) of low (about 5 mg/L) and moderately high clay (about 15 mg/L) loadings added every 7 to 14 days did not significantly reduce the algal productivity simulation caused by high phosphate loadings. However, higher clay loadings (about 25 mg/L added every 2 days) did produce depressed effects of phosphorus enrichment on the test lake. They concluded that dynamically turbid systems, such as is represented in southeastern urban lakes, have complex interacting mechanisms between discharged clay and nutrients that make simple predictions of the effects of eutrophication much more difficult than in the more commonly studied clear lakes. In general, increased turbidity will either have no effect, or will have a mitigating effect, on the cultural eutrophication process.

Sediment is typically listed as one of the most important pollutants causing receiving water problems in the nation's waters, and turbidity is therefore an important indicator of water quality. Turbidity, along with associated water column transparency, are two of the most commonly monitored water quality parameters in receiving water studies. Transparency is easily measured using Secchi disks by minimally trained volunteers (Figures 6.58 and 6.59). This has resulted in long-term transparency data being available for many urban lakes. Unfortunately, Secchi disk readings are instantaneous measurements and are usually obtained only during dry weather, with little high-resolution transparency information available. Measurements of water turbidity, however, can be readily obtained from both manual and automatic water sampling efforts, plus from continuous long-term monitoring sondes. Both laboratory and field nephelometers are available for measuring water turbidity (Figures 6.60 and 6.61).

A discussion earlier in this chapter presented the results of a small study conducted along Five-Mile Creek in Jefferson County, AL, where a YSI 6000 sonde, having continuous turbidity monitoring capabilities, was used to indicate the frequency, duration, and severity of wet-weather flow events. Increases in turbidity, along with attendant decreases in specific conductivity, were a much more accurate indicator of the durations of wet-weather flow impacts than flow rate and stream stage. Turbidity immediately increased from base levels (about 10 NTU) to more than 1000 NTU (the upper limit of the instrument) with the initial increases in stream stage. Elevated turbidity levels (greater than 100 NTU) persisted long after the flow subsided. The actual duration of the detrimental effects of the wet-weather flow was two to three times longer than the duration of the elevated flows in the streams. In addition, interstitial water turbidity levels also substantially and rapidly increased (to levels of about 200 NTU) in areas having coarse sediment. The interstitial water turbidity levels remained elevated for a much longer period than the water column turbidity

Figure 6.58   Secchi disk being lowered into lake for transparency measurement.

Figure 6.59   Underwater Secchi disk showing slow disappearance of contrasting disk sectors.

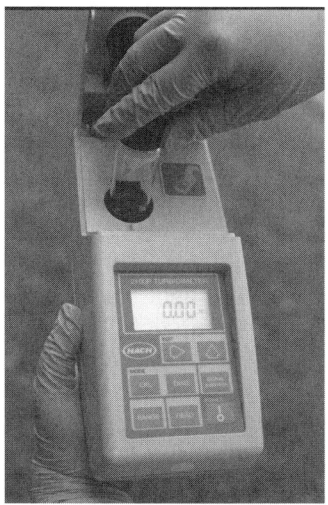

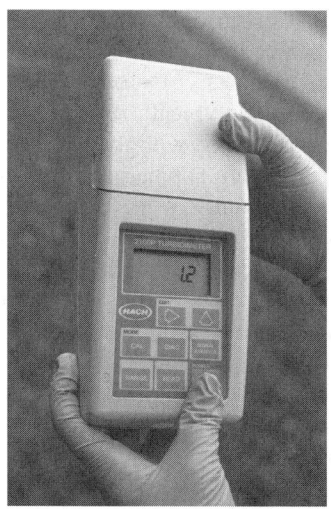

Figure 6.60 HACH 2100P field turbidimeter.   **Figure 6.61** HACH turbidity reading.

levels in the creek. There were no indicated interstitial water quality changes in areas having fine-grained (sandy) sediment. Therefore, turbidity can have much more prolonged effects on in-stream (and possibly in-sediment) conditions than is typically assumed, based solely on water flow measurements. The use of continuous turbidity measurements to supplement biological observations in wet-weather receiving water studies is therefore highly desirable.

Dissolved Oxygen

The adverse effects of low dissolved oxygen on aquatic life are well known, and reliable modeling techniques exist that predict DO levels in waters which receive wastewaters (EPA 1986). However, oxygen demand dynamics associated with stormwater events are not well understood. Peak oxygen demand may occur days after storm events, and miles downstream due to BOD and sediment oxygen demand (SOD) loading and transport.

The measurement of SOD is often overlooked in stream surveys and methods are not standardized, but it may be a critical measurement. Research reported by Werblow (2000) has shown that SOD may be a very large sink of DO in Tualatin Basin in Oregon, for example. In systems or reaches where small particle sizes (i.e., silts and clays) dominate and where organic matter and nutrient inputs may be elevated, SOD may be an important stressor. Station selection for SOD measurements should be based on deposition zones and sources of loadings. SOD may be measured in the laboratory or *in situ* (Edberg and Von Hofsten 1973; O'Connor and DiToro 1970; Bowie et al. 1985; Whittemore 1986; Davis et al. 1987). Given the importance of maintaining sediment integrity (Burton 1991; ASTM 1991; Stemmer et al. 1990; Sasson and Burton 1991) in contamination assessments, *in situ* measures are preferred. The precision of SOD measurements is largely a function of the level of operator experience.

The range of diurnal variation must be defined during baseflow and post-event conditions. By sampling three to four times daily over 2 or 3 days, this range may be established (EPA 1986). If DO variations are extreme, then sampling and modeling requirements will be more complex.

Photosynthesis/Respiration (P/R) Rate Analyses

Photosynthesis/respiration measurements are needed to measure local eutrophication problems and to evaluate the potential effects of discharges on receiving waters. Many receiving water quality models also require photosynthesis and respiration rates in order to calculate dissolved oxygen conditions. Accurate values are important, and "default" values can be very misleading. Therefore, local measurements are strongly recommended. Traditional P/R analyses require the use of light and dark bottles (typically BOD bottles, one set clear, the other set wrapped in aluminum foil). The bottles are filled with the test water, an initial DO is measured, the dark bottles are wrapped, and the bottles are submerged in the waterbody of concern. Every few hours, a set of light and dark bottles (usually at least three of each) is removed and the bottle DO is measured. This is repeated during the day, typically from late morning until midafternoon, obtaining from three to five sets of observations. The DO values are plotted and the trends are measured. Thomann and Mueller (1987) describe the test and data evaluation procedures. The light bottles undergo both photosynthesis and respiration, while the dark bottles only undergo respiration. The P/R rates vary greatly depending on the local conditions. As an example, tests can be conducted in urban streams in full sun, in the shade, in shallow water, and in deep water. Weather conditions (cloud cover, obviously, and temperature) all affect the P/R rates. These variations can all be very important and should be considered when modeling oxygen conditions in urban streams.

Case Study to Measure in Situ P/R Rates

The following is a discussion of a more efficient method of measuring P/R rates *in situ*, using a plastic bag test chamber and a continuous water quality monitoring probe, as demonstrated by Easton et al. (1998) as part of an EPA-sponsored project investigating SSO discharge effects in Birmingham, AL. The advantage of this method is that a tremendous amount of data can be collected in a very efficient manner. The only personnel time required is that needed to calibrate the instruments, set up the chambers, retrieve the chambers, and evaluate the data. The probes can be programmed to obtain DO (along with other parameters of interest) every 5 to 15 min for an extended period (up to several weeks). This allows the effects of changing weather (cloud cover, temperature shifts, rains, etc.) on the P/R rates to be directly measured. In addition, numerous replicates of the rates can be easily obtained when the probes are left out for an extended period. These are all significant advantages over conventional light and dark bottle P/R tests. The following case study demonstrates the type of information that can be obtained using this technique, along with the appropriate data analysis procedures.

This study used YSI 6000 UPG sondes. The important aspect of this sonde that allows these tests to be conducted is the rapid-pulse DO probe that consumes very little dissolved oxygen. Measured DO changes in the test chambers are therefore associated with the oxidation of wastes and are not significantly affected by measurement artifacts (including drift). In addition, the long-term monitoring capability of the unit enables many diurnal cycles to be measured efficiently. Also, the other measurements (especially pH, ORP, and conductivity) are very useful in indicating associated water quality changes in the test chamber and offer additional insight into the local P/R process. During this study, the YSI 6000 sondes were used to evaluate *in situ* P/R rates of different mixtures of raw sewage and fresh water. The sondes were calibrated for the following experimental parameters: depth, specific conductance, dissolved oxygen, turbidity, pH, oxidation-reduction potential (ORP), and temperature. The sondes were also programmed to acquire data in unattended mode for 2 weeks at 15-min intervals.

There are several biological processes that were apparent from monitoring the water quality. During the daylight hours, photosynthetic organisms, such as algae, use energy derived from the sun to produce ATP (adenosine triphosphate) and NADPH (reduced nicotine adenine dinucleotide phosphate) — reactions that generate oxygen. Then, the energy (ATP) and reducing power (NADPH) are used to fix carbon dioxide (CO_2) into carbohydrate (Filip and Alberts 1994). Simultaneously, photosynthetic organisms and any other aerobic organism, such as fish and certain types of microorganisms, use oxygen to break down carbohydrates for energy. This process occurs during the daylight and nighttime hours. Therefore, there is a constant drain on levels of dissolved oxygen in the water that must be replenished by photosynthesis and/or exchange with the atmosphere. The net effect of these processes is that the dissolved oxygen level in the water rises during the daylight and falls at night. In addition, the pH of typical receiving waters is governed by the carbonic acid/bicarbonate/carbonate buffering system. Increases in the dissolved CO_2 concentration causes corresponding decreases in pH, and vice versa. Therefore, the pH increases during the daytime hours because CO_2 is being fixed by photosynthetic organisms and is thereby removed from the water. Then, at night, pH drops because atmospheric CO_2, and CO_2 being produced by respiration, increase the concentration of CO_2 in the water.

Figure 6.62 *In situ* P/R tests being conducted.

The raw sewage was obtained at a local sewage treatment plant. The site for the tests was a small private lake that rarely, if ever, received sanitary sewage. Four different mixtures of sewage and fresh lake water (0/100%, 33/67%, 67/33%, and 100/0% sewage/fresh water) were prepared in their respective test chambers (20-L clear plastic bags containing 15 L of the test water mixture). The sondes were sealed in the bag with as little air trapped inside as possible. The test chambers and sondes were placed on the lake bottom in approximately 1 to 2 ft of water near the shore with full sun during daylight hours (Figure 6.62).

The temperature results showed increasing temperatures with time, consistent with typical spring conditions. The range on day 1 was 20 to 23°C; while the range on day 10 was 23 to 25°C. A diurnal variation of about 3 to 4°C was also observed — again, typical for the day/night solar cycle. It is important to note that the last 2 days were overcast with scattered heavy rains and variable winds, and therefore the diurnal variation was less than it was on days with full sun. The temperature data also show that the results for each of the four probes were quite consistent, except that the 33% sewage chamber did not reach as high a daily peak as the others. It is possible that differences in the temperatures may have been due to differences in the color of the water/algae mixture. The large amount of green biomass observed in the 33% sewage chamber may have acted to moderate the extreme temperature levels found in the other chambers that did not have such a large algal growth.

The pH results were also as predicted. There was a diurnal variation, at least in the test chambers that had photosynthesis occurring: 33% sewage (daily pH change $\cong$ 1 to 2, after day 7) and 0% sewage (daily pH change $\cong$ 0.25). This is due to the change in CO_2 concentrations from photosynthesis (removal of dissolved CO_2) and respiration (addition of dissolved CO_2). An increase in dissolved CO_2 causes the pH to decrease from the formation of more carbonic acid, while removal of dissolved CO_2 increases pH.

The results for oxidation-reduction potential (ORP) were also as expected. The test chambers with high oxygen demand, and corresponding reducing environment (67 and 100% sewage), dropped rapidly to less than −400 mV within the first few hours of the experiment, and stayed there. The ORP in the 33% sewage chamber was similar to the 67 and 100% for the first 5 days, but then began to climb, reaching positive ORP values by day 6. This result is well correlated with the DO data, showing that after an initial acclimation period, the algae and other microorganisms began to respire and photosynthesize. The 0% sewage chamber showed a definite diurnal trend and stayed above 300 mV, two factors that correlate well with the diurnal DO cycle resulting from P/R.

The dissolved oxygen data were used to calculate P/R rates for the microorganisms in the test chambers (Figure 6.63). The 0% sewage test chamber contained a 5-day biochemical oxygen demand (BOD_5) of approximately 2.5 mg/L. Therefore, there was a general downward trend in

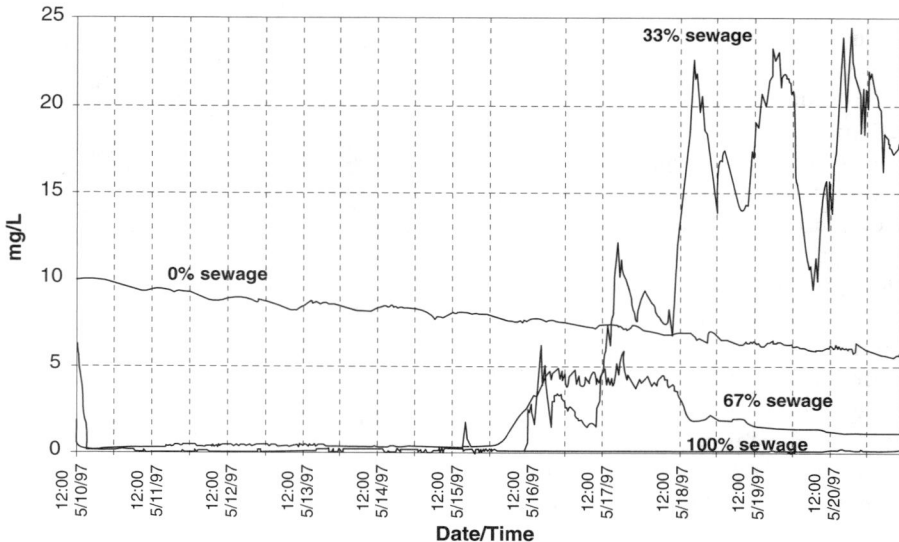

Figure 6.63 Dissolved oxygen data for all four probes over 10-day experiment. (From Easton, J.H., Lalor, M., Pitt, R., and Newman, D.E., The use of a multi-parameter water quality monitoring instrument to continuously monitor and evaluate runoff events. Presented at *Annual Water Resources Conference of the AWRA*, Point Clear, AL. 1988. With permission.)

dissolved oxygen levels over the 10-day period of the experiment, typical for a eutrophic lake. The water body where this study was conducted rarely, if ever, received sanitary sewage. In this case, an acclimation period was expected. However, if the water body had received regular discharges of sewage, the long acclimation period would most likely not be observed. The 33% sewage chamber had initial anoxic conditions, but after acclimating for approximately 5 days, there was a pronounced diurnal P/R variation. Indeed, the DO levels in this chamber were supersaturated during the daylight hours, as photosynthesis rates were very high. When this chamber was pulled at the end of the experiment, there was a large amount of green biomass, indicating the presence of photosynthesizing organisms. The 67% sewage test chamber stayed at anoxic DO levels, as expected. However there was an increase in DO on the 5th and 6th days. Possibly, the organisms in this chamber began photosynthesizing after acclimating to the sewage, but the oxygen demand of the waste quickly drove the DO levels to anoxic levels shortly thereafter. The 100% test chamber stayed anoxic throughout the experiment, as anticipated.

The rates of P/R were analyzed using the following methods. First, after analyzing the data for the entire length of the experiment, it was determined that the data from only the last 5 days would be used to calculate rates of P/R; these days occurred after the acclimation period. An analysis of the dissolved oxygen data given in Figure 6.63 showed that the rates of P/R would be impossible to determine from the 67 and 100% sewage samples because the DO levels were essentially zero. Therefore, the methods were applied only to the 0 and 33% sewage samples. In the future, further experiments should be done to look at sewage dilutions between the 0 and 33% levels. Most examples of raw sewage discharges into receiving waters (such as for SSOs) likely only comprise a few percent of the receiving flow. Plots were then created of the 0 and 33% sewage results for this 5-day period, as shown in Figures 6.64 and 6.65, for detailed analysis.

These plots were inspected visually, and lines were drawn on positive slope portions and negative slope portions of the graphs. The positive slopes (occurring during daylight hours) represented periods of photosynthesis minus respiration (p_{net}), while the negative slopes (occurring during nighttime hours) represented periods of respiration (R). The R rate was then subtracted from the p_{net} rate to obtain an hourly photosynthesis rate. The mean values were:

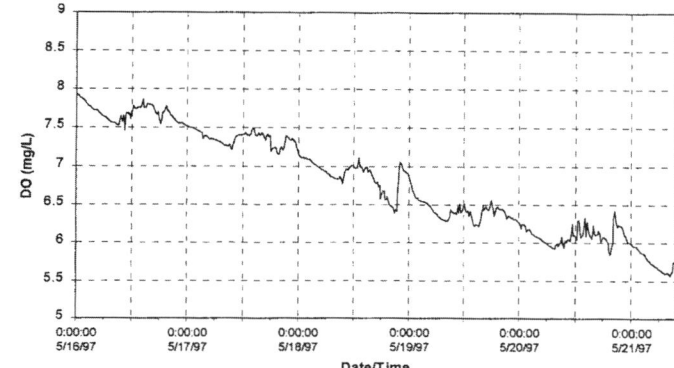

Figure 6.64 Dissolved oxygen data for 0% sewage. (From Easton, J.H., Lalor, M., Pitt, R., and Newman, D.E., The use of a multi-parameter water quality monitoring instrument to continuously monitor and evaluate runoff events. Presented at *Annual Water Resources Conference of the AWRA,* Point Clear, AL. 1998. With permission.)

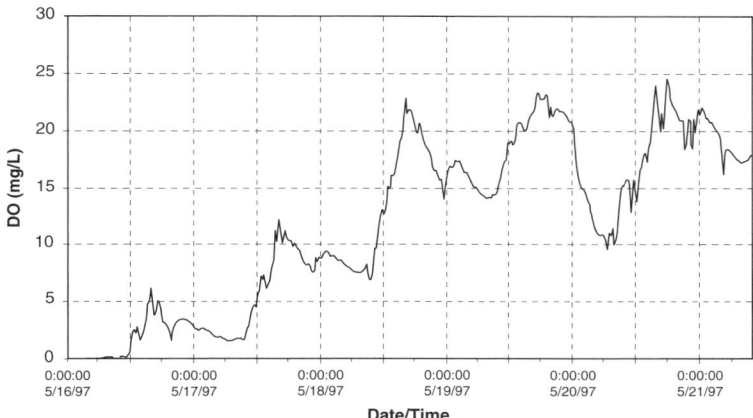

Figure 6.65 Dissolved oxygen data for 33% sewage. (From Easton, J.H., Lalor, M., Pitt, R., and Newman, D.E., The use of a multi-parameter water quality monitoring instrument to continuously monitor and evaluate runoff events. Presented at *Annual Water Resources Conference of the AWRA,* Point Clear, AL. 1998. With permission.)

0% sewage: p_{net} = 0.04 mg/L·hr, R = –0.05 mg/L·hr, P = 0.09 mg/L·hr, and

33% sewage: p_{net} = 1.16 mg/L·hr, R = –0.47 mg/L·hr, P = 1.63 mg/L·hr

The next step in determining the photosynthesis rate was to apply the daily average DO model (Thomann and Mueller 1987). The respiration rate is assumed constant throughout the day. The hourly rates determined previously were multiplied by 24 hours to give a respiration rate in units of mg/L·day. The photosynthetic oxygen production is assumed sinusoidally distributed over the photoperiod from 6:00 am to 7:00 pm for these conditions. These results are given in Table 6.25.

The photosynthesis rates for the 33% sewage were extremely high and variable, ranging from 12 to 30 mg/L·day; and the rates for the 0% sewage (100% lake water) were typical, approximately 1 to 2 mg/L·day. Typical local surface water photosynthesis values of approximately 1 to 4 mg/L·day have been obtained from previous experiments with light and dark bottles in local natural waters (Lake Purdy and the Cahaba River during other student projects at UAB).

Table 6.25 Calculated Values for the Estimated Daily Averaged Photosynthetic Oxygen Production Rate (p_a)

Date	0% Sewage				33% Sewage			
	p_{net} (mg/L-day)	Respir (mg/L-day)	p' (mg/L-day)	p_a (mg/L-day)	p_{net} (mg/L-day)	Respir (mg/L-day)	p' (mg/L-day)	p_a (mg/L-day)
5/16/97	1.19	0.91	2.10	1.06	19.37	5.25	24.62	12.47
5/17/97	0.94	0.85	1.79	0.90	28.62	6.30	34.92	17.68
5/18/97	1.19	1.70	2.89	1.46	47.57	12.59	60.17	30.47
5/19/97	0.85	1.00	1.85	0.94	18.32	22.39	40.70	20.61
5/20/97	0.91	1.60	2.51	1.27	25.19	10.40	35.59	18.02
Mean	1.01	1.21	2.23	1.13	27.81	11.39	39.20	19.85
Std. dev.	0.16	0.41	0.47	0.24	11.83	6.84	13.09	6.63
COV	0.16	0.34	0.21	0.21	0.43	0.60	0.33	0.33

From Easton, J.H., Lalor, M., Pitt, R., and Newman, D.E., The use of a multi-parameter water quality monitoring instrument to continuously monitor and evaluate runoff events. Presented at *Annual Water Resources Conference of the AWRA*, Point Clear, AL. 1998. With permission.

Recommendations for P/R Investigations

Site-specific photosynthesis and respiration measurements are needed whenever an in-depth DO investigation (especially to support TMDL analyses) is required. DO has traditionally been one of the most significant indicators of poor receiving water conditions, and many regulatory agencies heavily rely on DO predictions. However, wet-weather flow effects on DO are typically unclear, especially considering the relatively slow effect stormwater has on BOD. Nutrient discharges associated with wet-weather flows can also dramatically affect P/R conditions in a receiving water. Actual measurements of these rates for all of the wastewaters affecting a receiving water can lead to much more accurate in-stream DO predictions.

The *in situ* sonde method for measuring P/R described above is an improved procedure for studying P/R compared to conventional methods (light/dark bottle testing). The data collected are far more useful because they are continuous and collected over multiple day/night cycles. This enables daily variations to be quantified and to account for weather changes. The high-resolution data also enable the identification of periods of questionable data associated with the acclimation period at the beginning of the test period.

WATER AND SEDIMENT ANALYTES AND METHODS

Selection of Analytical Methods

Environmental researchers need to be concerned with many attributes of numerous analytical methods when selecting the most appropriate methods to use for analyses of their samples. The main factors that affect the selection of an analytical method include: cost, reliability (the "data quality objectives," or DQO, discussed earlier in Chapter 5, which includes sensitivity, selectivity, repeatability), and safety. Another factor to be considered is whether the analyses should/can be conducted in the field or in the laboratory. These items can be subdivided into many categories including:

- Capital cost, costs of consumables, training costs, method development costs, age before obsolescence, age when needed repair parts or maintenance supplies are no longer available, replacement costs, other support costs (data management, building and laboratory requirements, waste disposal, etc.)
- Sensitivity, interferences, selectivity, repeatability, quality control, and quality assurance reporting, etc.
- Sample collection, preservation, and transportation requirements, etc.
- Long-term chemical exposure hazards, waste disposal hazards, chemical storage requirements, etc.

Most of these issues are not well documented in the literature for environmental sample analyses. Aspects of analytical reliability have received the most attention in the literature, but most of the other aspects noted above have not been adequately discussed for the many analytical alternatives available, especially for field analytical methods. It is therefore difficult for a water quality analyst to decide which methods to select, or even if a choice exists.

The selection of the appropriate procedure depends on the use of the data and how false negatives or false positives would affect water use decisions or regulatory questions. The QA objectives for the method detection limit (MDL) and precision (RPD) for the compounds of interest have been shown to be a function of the anticipated median concentrations in the samples (Pitt and Lalor 1998). The MDL objectives should generally be about 0.25, or less, of the median value for sample sets having typical concentration variations (COV values ranging from 0.5 to 1.25), based on many Monte Carlo evaluations to examine the rates of false negatives and false positives. The precision goal is estimated to be in the range of 10 to 100% (Relative Percent Difference of duplicate analyses), depending on

Table 6.26 Summary of Quantitative QA Objectives (MDL and RPD) Required for an Example Stormwater Characterization Project

Constituent	Units	Example COV Category[a]	Example Median Conc.	Calculated MDL Requirement	Calculated RPD Requirement
pH	pH units	Very low	7.5	Must be readable to within 0.3 unit	<0.3 unit
Specific conductance	μmhos/cm	Low	100	80	<10%
Hardness	mg/L as $CaCO_3$	Low	50	40	<10%
Color	HACH units	Low	30	24	<10%
Turbidity	NTU	Low	5	4	<10%
COD	mg/L	Medium	50	12	<30%
Suspended solids	mg/L	Medium	50	12	<30%
Particle size	size distribution	Medium	30 μm	7 μm	<30%
Alkalinity	mg/L as $CaCO_3$	Low	35	30	<10%
Chloride	mg/L	Low	2	1.5	<10%
Nitrates	mg/L	Low	5	4	<10%
Sulfate	mg/L	Low	20	16	<10%
Calcium	mg/L	Low	20	16	<10%
Magnesium	mg/L	Low	2	1.5	<10%
Sodium	mg/L	Low	2	1.5	<10%
Potassium	mg/L	Low	2	1.5	<10%
Microtox toxicity screening	I20 or EC50	Medium	I20 of 25%	I20 of 6%	<30%
Chromium	μg/L	Medium	40	9	<30%
Copper	μg/L	Medium	25	6	<30%
Lead	μg/L	Medium	30	7	<30%
Nickel	μg/L	Medium	30	7	<30%
Zinc	μg/L	Medium	50	12	<30%
1,3-Dichlorobenzene	μg/L	Medium	10	2	<30%
Benzo(a)anthracene	μg/L	Medium	30	8	<30%
Bis(2-ethylhexyl)phthalate	μg/L	Medium	20	5	<30%
Butyl benzyl phthalate	μg/L	Medium	15	3	<30%
Fluoranthene	μg/L	Medium	15	3	<30%
Pentachlorophenol	μg/L	Medium	10	2	<30%
Phenanthrene	μg/L	Medium	10	2	<30%
Pyrene	μg/L	Medium	20	5	<30%
Lindane	μg/L	Medium	1	0.2	<30%
Chlordane	μg/L	Medium	1	0.2	<30%

[a] COV value:

COV value	Multiplier for MDL:	RDL Objective:
<0.5 (low)	0.8	<10%
0.5 to 1.25 (medium)	0.23	<30%
>1.25 (high)	0.12	<50%

From Pitt and Lalor 1998.

the sample variability. Table 6.26 lists the typical median stormwater runoff constituent concentrations and the associated calculated MDL and RPD goals, for a typical stormwater monitoring project.

In some cases, field test kits, or especially continuous *in situ* monitors, may be preferred over conventional laboratory methods. Table 6.27 lists some of the benefits and problems associated with each general approach. The advantages of field analytical methods can be very important, but their limitations must be recognized and considered.

The environmental researcher also must be concerned with sampling costs (discussed in Chapter 5), in addition to analytical costs. Most environmental research efforts are not adequately supported to provide the necessary numbers of samples needed for statistically reliable results to support typical (lofty) project goals. Expensive recommendations are therefore commonly made based on too small an analytical investment. The number of samples needed to simply characterize a water quality constituent can be estimated based on the expected variability of the constituent and on the

Table 6.27 Comparisons of Field and Laboratory Analytical Methods

Field Analytical Methods		Conventional Laboratory Methods	
Advantages	Disadvantages	Advantages	Disadvantages
Minimal change in sample character because no transport and storage. Opportunity to collect replacement sample if questionable results, or if sample is damaged. Results generally available soon after sample collection. Continuous *in situ* monitors result in large numbers of observations with fine resolution.	Difficult to control environmental variables affecting analytical measurements and working conditions. Individual samples usually analyzed separately with more time required per sample. Additional time needed to set up equipment and standardize procedure for each location. Analytical hazardous waste (and sharps) management may be a problem. Many field analytical reagent sets are sensitive to storage conditions that may be difficult to meet. Documentation can be incomplete and hazards not described. Generally poor limits of detection and limited working range. Some of the most sensitive tests are very complex with analytical errors common.	Good control of laboratory working conditions and use of in-place hazardous waste management. Can analyze several samples in one batch. More precise equipment generally used for analyses, and less time to set up for analyses. Easier to conduct and meet QA/QC requirements. Usually much lower limits of detection.	Need to preserve samples and conduct analyses in prescribed period of time. Results may not be available for an extended time after sample collection. Minimal opportunity to re-sample due to errors. Generally more expensive and sample numbers are therefore limited. Sample storage space-consuming and requires logging system for sample tracking.

allowable error of the result. As an example, 40 samples are needed to estimate the average concentration with an allowable error of 25%, if the coefficient of variation of the constituent measurements is about 0.8. If only 10 samples are evaluated, the error increases to a possibly unusable 100%. Analyses of toxicants of great interest in many research activities currently can cost many hundreds of dollars per sample for a short list of organic and heavy metal compounds. A simple effort to adequately characterize the conditions at a single location can therefore cost more than $25,000, as shown in Chapter 4. Clearly, there is a great need to be able to afford to collect and analyze a sufficient number of samples. The following discussion therefore presents several methods of collecting the needed data, including continuous *in situ* monitors, simple field test kits, and conventional laboratory analyses.

Use of Field Methods for Water Quality Evaluations

There are many problems with current environmental sampling and analysis programs that can be met by conducting water quality evaluations in the field, especially if continuous, *in situ* procedures are used. Foremost among these problems is the need to collect many samples in order to obtain the desired accuracy of the characteristics of interest. Other concerns involve inadvertent changes that may affect the sample characteristics between sample collection and analysis. The high cost of analyzing trace levels of organic and metallic toxicants using conventional laboratory procedures is also restrictive, but field methods for these analytes are very expensive, complex to use, or not very sensitive. The following discussion covers *in situ* monitoring and the use of field test kits.

In Situ *Physicochemical Monitoring*

One way to collect adequate data is to use simple field analytical methods, preferably continuously recording *in situ* analyses. These methods allow a great amount of data to be collected without sample collection, transportation, or laboratory problems. However, new problems arise, specifically related to long-term reliability and costs of the instrumentation. Many of these instruments are currently available, but they are restricted to only a few of the common constituents (usually temperature, conductivity, dissolved oxygen, and pH, plus turbidity on a few units) and can cost from $3000 to $7000. The newest and most reliable units can be placed in a water body and left unattended for several weeks to months before requiring service. They can continuously record these constituents over this time with very high resolution, enabling a much greater understanding of the dynamics of the pollutant behavior in the water body. Unfortunately, the constituents currently capable of being continuously and automatically monitored do not include many of the most interesting. Some ion-selective electrode (ISE) probes are being offered as options on some of the continuous *in situ* probes. Unfortunately, their reliability is not well established, but they may be very useful for shorter-term and specialized projects.

In situ monitors give continuous and relatively rapid results, in contrast to typical field test kits, which require time and patience to evaluate the chemical parameters of interest. Unfortunately, these are all relatively costly instruments. However, their capabilities cannot be matched using other procedures. These instruments can be separated into two general categories. *In situ* probes, having real-time display capabilities, but with limited data logging capabilities, are designed for real-time monitoring. The other category includes continuously recording probe units that are designed for long-term unattended operation, but are commonly also available with direct read-out displays for real-time use. Examples of both types have been available for more than 20 years.

In Situ *Direct-Reading Probes*

The simplest direct-reading probes that perform their analyses *in situ*, with no sample preparation, include the classical series of field instruments from YSI, such as their DO probe and SCT (salinity, conductivity, and temperature) probe. These are very robust instruments that have been in use at many institutions for decades. The original models of the DO probes did require practice to replace the membranes, and they required relatively frequent (but simple) recalibration. Newer YSI models, especially these utilizing the rapid pulse current probe, exhibit much slower drift and are designed for long-term unattended operation.

Other direct-reading instrumentation includes pH and ORP instruments. These generally are more sensitive to storage conditions and require frequent maintenance or calibrations. Some of the newer dry pH electrodes are very robust and much more reliable and easier to use. Ion selective electrodes (ISE) are sometimes included in this category and various equipment vendors offer them as options for their direct-reading *in situ* probes. It is suggested that careful and frequent evaluations be made of any electrode-equipped equipment (especially pH and ISE) to ensure that the instrument is operating properly and that the probe has not dried out or been damaged by oils or detergents.

Some direct-reading *in situ* probes are available that have the capability to measure several parameters. Most of these are designed for long-term unattended operation, but somewhat less expensive versions are also available that have minimal data logging capabilities and are designed for real-time measurements. The Horiba U-10, for example, was evaluated by Day (1996). It costs about $2500 from Hazco (800-332-0435, catalog # B-H020001) and can simultaneously measure conductivity, temperature, salinity, dissolved oxygen, turbidity, and pH. Hazco also rents the Horiba U-10. It is especially useful for real-time profiling of shallow lakes and small urban streams. Relatively few probes offer turbidity, which is helpful when examining light penetration and algal activity. Solomat and YSI also have hand-held instruments having capabilities similar to those of the Horiba U-10.

Other instrumentation is also available that can monitor hydrocarbon conditions in water on a real-time basis. The Turner 10-AU field fluorometer with "oil in water" optics is extremely sensitive

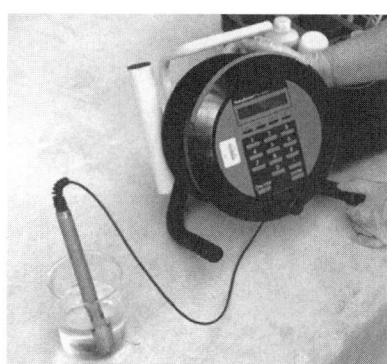

Figure 6.66 Petrosense being calibrated.

Figure 6.67 Petrosense used to measure hydrocarbons in manhole water.

and is used with no sample preparation. It can be used in a flow-through mode to map hydrocarbon concentrations in real time. It can also be used as a stand-alone instrument for long-term unattended operation, if properly housed. This instrument costs from $8000 to $16,000, depending on housing, data logging, and filter options, and is therefore not likely to be readily available. This instrument is also used for fluorescent tracer analyses (using Rhodamine WT) for primary calibration of water flow equipment. It can also be used for limited real-time chlorophyll a analyses, when using appropriate optics and filters.

The Petrosense hydrocarbon probe from FCI Environmental is also available for real-time hydrocarbon analyses (Figures 6.66 and 6.67). This instrument costs about $7000, has a slower response time (about 5 min), and is not nearly as sensitive (about 100 μg/L, as xylene) as the fluorometer. It can also be used in real time to monitor "total" hydrocarbons in water, with no sample preparation. It quantifies hydrocarbons by measuring changes in optical properties caused by hydrocarbon adsorption onto an exposed fiber optic.

Continuously Recording and Long-Term In Situ Measurements of Water Quality Parameters

Several classical instruments have long been available to measure various water quality parameters with unattended instruments for relatively long periods of time. Hydrolab and YSI have long offered equipment that can monitor dissolved oxygen, pH, temperature, and conductivity unattended. The early instruments were plagued with stability problems and were usually most suited for unattended operation over a period of only about a day. This was still a major breakthrough, as it enabled diurnal fluctuations of these important parameters to be obtained accurately and relatively conveniently.

Currently available equipment, in contrast, has been demonstrated to be capable of unattended operation for longer than a month. These are relatively expensive instruments that can cost up to $7000 each, depending on options selected. Examples of equipment currently available include the 803 probe series from Solomat, which can have up to eight sensors installed. These may include pH, ORP, DO, temperature, conductivity, depth, ammonium, nitrite, and other ions by ISE. Several meters and data loggers are available for hand-held real-time measurements, or for long-term unattended operation. YSI also offers several *in situ* probe instruments. The original YSI unit available many years ago (Figure 6.68) was a breakthrough unit that enabled overnight DO measurements. The current 6000 series sonde is much improved (Figure 6.69). It is self-contained, measuring and logging up to nine separate

428

STORMWATER EFFECTS HANDBOOK

Figure 6.68 Older YSI DO meter for continuous monitoring. (Courtesy of Wisconsin Department of Natural Resources.)

Figure 6.69 YSI 6000 sonde detail showing several probes.

parameters simultaneously, including DO, conductivity, temperature, pH, depth, ORP, nitrate, ammonium, and turbidity. The rapid pulse DO and self-wiping turbidity sensors enable very long unattended operations (up to 45 days), with minimal fouling or drift. Hazco (800-332-0435) sells the YSI 6000 basic sonde (catalog # B-6001) for about $7000. The unit without the depth sensor is about $500 less. The performance specifications for the more common sensors, provided by the manufacturer, are given in Table 6.28. Appendix E contains detailed instructions for calibrating and setting up this sonde.

These unattended instruments are capable of collecting high-resolution data (typically with observations every 5 to 15 min) over long periods. This is extremely useful in receiving water studies affected by stormwater. Even though few dissolved oxygen problems have ever been associated with stormwater (in contrast to CSOs), these probes are unexcelled in documenting the

Table 6.28 YSI6000 Specifications

Parameter	Sensor Type	Range	Accuracy	Resolution
Dissolved oxygen % saturation	Rapid pulse — Clark-type, polarographic	0 to 200% air saturation	±2% air saturation	0.1% air saturation
Conductivity[a]	4 electrode cell with autoranging	0 to 100 mS/cm	±0.5% of reading + 0.001 mS/cm	0.01 mS/cm
Temperature	Thermistor	−5 to 45°C	±0.15°C	0.01°C
pH	Glass combination electrode	2 to 14 units	±0.2 units	0.01 units
ORP	Platinum ring	−999 to 999 mV	±20 mV	mV
Turbidity	Optical, 90° scatter, mechanical cleaning	0 to 1000 NTU	±5%	0.1 NTU
Depth — Medium	Stainless steel strain gauge	0 to 61 m	±0.12 m	0.001 m
Depth — Shallow	Stainless steel strain gauge	0 to 9.1 m	±0.06 m	0.001 m

[a] Report outputs of specific conductance (conductivity corrected to 25°C).

exposure periods and gross variations in receiving water conditions over many separate storm events. These data are very important when used in conjunction with *in situ* toxicity test chambers that are exposed for relatively long periods of time. In addition, the YSI self-contained probes with rapid-pulse DO sensors (the probes consume very little power and oxygen themselves) can be used in light and dark chambers to conveniently obtain necessary data pertaining to sediment and water photosynthesis and respiration, as previously described.

Field Test Kits

Field test kits cover a wide range of instrumentation and methods. They range from very simple visual comparator tests (which use colored paper, colored solutions in small vials, or color wheels to match against the color developed with the test) to miniaturizations of standard laboratory tests (using small spectrophotometers or other specialized instruments). Appendix E contains listings and photographs of selected field procedures. Appendix E also contains a summary of the tested performance of several representative field test kits, highlighting their performance (limits of detection, repeatability, and recovery), hazards associated with their use, complications and time requirements, approximate costs, and other notes (Day 1996).

The least expensive test kits use small droppers or spoons to measure reagents into a reaction tube where the color is developed. More sophisticated tests use small filter colorimeters to more precisely measure the color developed during the test. HACH also offers continuous wavelength field spectrophotometers that are capable of measuring a wide variety of chemical parameters using a single instrument (Figure 6.70). La Motte has a filter colorimeter that contains several filter sets, also enabling many different chemical analyses to be conducted with the one instrument. HACH also has a field titration kit that is also very flexible, providing additional capabilities not available with spectrophotometric methods. These multiparameter instruments are usually superior to the simple dedicated test kits because of the increased sensitivity and precision that is achievable with the better equipment. They, of course, cost more. If only one or two parameters are to be monitored in the field, then it might be hard to justify the added cost of the more flexible instruments. However, if the best quality data are needed, the cost may be justified, especially if more than a few parameters are to be measured.

Also included in the category of field test kits are very sophisticated methods that are laboratory instrumentation and procedures that have been miniaturized and simplified. Some of these tests even meet the EPA reporting requirements for NPDES permit compliance. However, some of the field procedures skip certain sample cleanup or digestion steps that would be impractical to conduct in the field and are therefore not suitable for compliance monitoring. It is important to check with the field equipment suppliers and the reviewing regulatory agency to verify the current status of a field method for various reporting purposes. Many of the spectrophotometer and titration methods fall into this category of simplified laboratory methods. Several new instruments are also available that permit sensitive and precise heavy metal (especially copper and lead) analyses in the field.

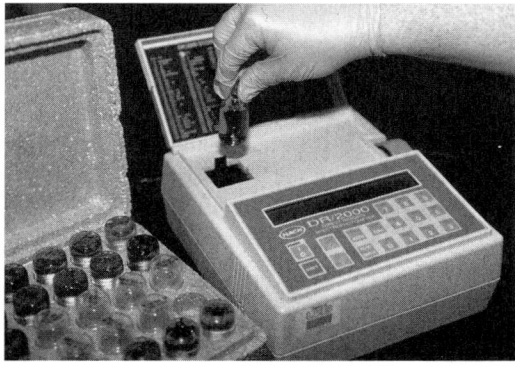

Figure 6.70   HACH DR/2000 field spectrophotometer.

However, these instruments are expensive (equipment costs of $2000 to $4000 and per sample costs of $5 to $15). They are also not sensitive to particulate-bound metals (which may be an advantage, depending on study objectives).

The biggest difficulty with almost all of these field test kits is that they can require a substantial amount of time to evaluate the water sample, especially when only one sample at a time is being analyzed. Continuous and *in situ* monitors eliminate field analytical time. Some of the simple *in situ* monitors are included in this test kit discussion (such as conductivity meters, pH meters, and DO meters), while the more complex continuously recording units were discussed previously. Even though these field test kits enable personnel to evaluate samples at the point of collection, that may not be desirable. Lalor (1993) and Pitt et al. (1994) found that test kit performance was greatly enhanced by bringing the collected samples to a temporary "laboratory" for analyses. This greatly increased sample analytical through-put, as many of the test kits enabled multiple samples to be analyzed at one time. This is especially critical if sampling locations are widely spaced and the alternative is to analyze many parameters at each location before moving to the next sampling location. It may take more than an hour to conduct a relatively few chemical tests at each location, including setting up equipment and restandardizing procedures. However, if many samples are being collected in a small area, the equipment can be left in one place and simultaneous sample analyses would be possible in the field. Indoor facilities should be sought, because protection from weather, available electricity, good lighting, and water enhance analytical performance. Make sure that adequate ventilation is available, however, wherever the tests are conducted. Many of the field test kits are not well labeled, especially concerning hazardous materials in the kit that require special protection and disposal practices.

Safety issues, along with test kit performance, have been examined (Pitt et al. 1994; Day 1996). The test kit evaluations were based on "fatal flaws" of the alternative equipment available for each parameter category. In the series of tests conducted by Day (1996), 50 test kits were subjected to preliminary evaluations with half further subjected to more detailed tests. His results are summarized in the following discussions. Safety hazards, cost, poor detection limits, matrix interferences, limited concentration ranges, poor response factors, and complexity of the test kits were all reasons for rejection. The most suitable test kits in each category were then identified, after rejecting those kits that were much more expensive than alternatives in each category. The comparison of field screening equipment is a somewhat objective process. Some parameters of interest are easily quantified; other features that should be evaluated require more objective evaluation techniques. Therefore, these evaluations were made using both subjective and objective information. The evaluation of the kits was based on five major tests:

1. Subjective evaluations of the health and safety features (kit reagent contents, design features to minimize operator exposure to hazardous reagents, disposal problems, and warnings)
2. Performance using samples spiked with known pollutant additions in "clean" and "dirty" water
3. Comparisons with standard laboratory procedures using parallel analyses of typical samples
4. Repeatability and precision using replicate analyses
5. Complexity of each method

The first tests for each method used spiked samples. The reported ranges for each kit were used to define a gross range of all methods for each parameter. The gross range was bounded by the lowest reported detection limit and the highest upper limit reported by the manufacturers. Two series of samples were prepared, one using reverse osmosis (RO) water and another using a composite of parking lot runoff water. The number of samples prepared varied by parameter depending on the magnitude of the gross range. RO and runoff water blanks were also prepared and tested for each parameter. RO water served as a control for identifying optimal test kit performance (assuming low ionic strength effects did not adversely affect the test). The parking lot runoff water was used to detect any significant matrix interferences. The runoff water was collected from a UAB parking lot.

The spiked standards were evaluated by all methods for each parameter. Due to the large number of methods that were evaluated, no replicate analyses were initially made. In most cases, these kit methods are used as field screening methods to detect potential problem pollutants in relatively high concentrations. During these analyses, data were collected on "useful" range, capital costs, expendable costs, analysis time, health and safety considerations, and "usability." These parameters are described below:

- Useful range: the range of concentrations that the instrument can measure with some certainty. The lower limit is defined by the detection limit. The upper limit is defined by the highest concentration the method can measure without dilution of the sample. The upper limit values were determined as the lowest spike concentration producing an "over range" error, or the lowest concentration that obviously deviated from the linear range of spike concentration to instrument response. If neither problem was identified, the manufacturer's reported upper limit was reported.
- Capital costs: the initial costs associated with purchasing the capital equipment required to use the method. Prices were obtained from the manufacturers during April 1996.
- Expendable costs: the costs associated with buying replacement reagents for the method. The value reported is per sample. The costs do not include general glassware, tissues, gloves, and other generic equipment required for many of the tests. The prices were obtained from the manufacturers during April 1996. The costs reported are based on list price of the smallest quantity of reagent available, and, therefore, the costs do not reflect bulk discounts which might be available.
- Analysis time: the approximate time to analyze one sample at a time with the method. In some cases, additional time must be allotted to prepare the method for measurement. For example, all analyses assume any needed instrument has been properly calibrated before analysis begins. In some cases, multiple tests can be performed simultaneously.
- Health and safety considerations: the health and safety considerations are a broad scope of factors that represent potential hazards to the user or the environment. The factors considered in this analysis include the hazardous nature of the reagents used, the packaging of the reagents, required disposal of reagent and sample wastes and waste glass, and the potential exposures or any feature of the kit requiring special attention.
- Usability: this ubiquitous term is a subjective evaluation of the expertise required to perform an acceptable analysis. Under this heading, an attempt was made to describe any feature of the kit that may not represent a hazard, but could affect the quality of the test. Examples of factors affecting usability include the number of steps, complexity of the procedure, additional equipment to make the procedure easier, limited shelf life of the kit, or any special skill required to complete the analysis.

From three to seven spiked samples were analyzed using each method for both RO and runoff water sample matrices. For each matrix, a plot of instrument response to spike concentration was made. This was used to estimate the range of linear response of the instrument. Spike responses showing a significant departure from a linear response indicate the range of the method. A regression analysis was performed on the data providing further information about the method. Ideally, the slope generated from these regression analyses (response factor) should be 1. A slope significantly different from 1 indicates a bias in the method. Also, the slope of the regression in the RO water matrix should be the same as the slope of the regression in the runoff water matrix. The difference in the slopes between matrices indicates the magnitude of matrix interference associated with the method. The value of the standard error of the regression was used to estimate the detection limit of the method, using the following equation (McCormick and Roach 1987):

$$\text{D.L.} = y_0 + s_y z_a$$

where D.L. = detection limit of the method
y_0 = the intercept of the regression equation
s_y = standard error of the regression
z_a = the area under the normal curve associated with a one-tail probability for a given confidence level (these analyses used the 95% confidence level, with a = 0.05)

Concentrations exceeding the detection limit only indicate the presence of the parameter. The equation may be modified to calculate the limit of quantification. Reported concentrations exceeding the limit of quantification may be used to quantify the results. The modified equation is:

$$LOQ = y_0 + 2s_y z_a$$

Therefore, the LOQ is approximately twice the D.L., if the intercept of the regression line is very small (as it should be). For example, if the D.L. is calculated to be 0.5 mg/L and the LOQ is calculated to be 1.0 mg/L, the following statements are true.

1. A response of 0.25 mg/L does not positively indicate the presence of the pollutant with the desired confidence.
2. A response of 0.75 mg/L does indicate the presence of the pollutant with the desired confidence, but the measured concentration does not have the desired level of confidence.
3. A response of 1.25 mg/L does indicate the presence of the pollutant, and its measured concentration is within the desired level of confidence.

The residuals of the regressions were also examined to identify any evidence of bias. A plot of residual vs. predicted spike concentration should produce a random band of points with an average value representing the concentration of the parameter of interest in the blank sample. Narrow error bands indicate a more precise method. A plot of residuals vs. the order of analysis indicates if a bias is time dependent. For example, the calibration of a pH meter will drift over time. A plot of residuals vs. the order of measurement will show a linear trend if the meter is not regularly recalibrated.

From these analyses, the most suitable set of equipment was identified for further study. These were selected based on the measured detection limits, safety considerations, and shortest analysis time. This subset of methods was then evaluated by parallel analysis for 25 runoff water samples. The test kit results were compared to the results obtained using standard laboratory procedures. This set of analyses was also analyzed by a regression technique to identify the correlation between field measurements and laboratory analyses.

The precision of the selected methods was also evaluated by testing five replicates of a composite polluted water sample. The average, standard deviation, and relative standard deviation (RSD, also known as the coefficient of variation or COV, the ratio of the standard deviation to the mean) for the methods were determined for each test kit.

Assembling an appropriate set of field test kits is obviously dependent on the specific uses of the data. In most cases, several colorimetric analyses will be included in the monitoring program, and the purchase of a good field spectrophotometer or filter colorimeter will be easily justified. The two major choices include the HACH DR/2100 field spectrophotometer (which costs about $1500), or the La Motte Smart Colorimeter (which costs about $800). The use of specific filter colorimeters (which cost from $250 to $400) may only be suited to very simple programs. The use of most manual color comparator tests will limit the utility of the data, but may still be justifiable.

A more important problem, besides cost, is probably associated with the time and expertise needed to conduct the analyses. Many of the analyses can be conducted together (especially those with extensive color development times, such as the immunoassays and the bacteria tests, plus the ammonia, copper, detergents, lead, and potassium tests). However, there will be a limit, as some of the tests are very complex (especially the immunoassays and the LeadTrak, which also require extensive expertise to obtain good results).

Appendix E contains summarizes of the information from the field test kit evaluations conducted by Day (1996), and includes information for the following constituents:

Ammonia
BTEX and PAHs
Chloride

Copper
Detergents
Fluoride
Lead
Nitrate
Potassium
Zinc

Most of the field test kits evaluated performed very well, with significant response factors and recoveries close to 1.0 (slopes of the regression lines when comparing known concentrations with test responses). In addition, the response factors were very close for spiked sample analyses in both RO and runoff sample water, indicating few matrix interference problems. The precision of the tests was generally excellent, with almost all replicate analyses having COV values of less than 20%, and many were much less than 10%. The exceptions were for tests that had very poor detection limits compared to the concentrations in the samples being tested. However, the detection limits of almost all of the analytical methods were much worse than reported by the manufacturers. The limits of quantification are all about twice as large as the detection limits shown in Appendix E. In some cases, this resulted in a very narrow workable range for the method before dilution is needed.

The following comments pertain to several groups of parameters of special interest when using field test kits. These comments stress the need to carefully select and evaluate field test kits used in monitoring programs, especially since there have been few independent evaluations of their capabilities and limitations. Many of the procedures (including some that were relatively inexpensive) were found to be surprisingly good in our tests. In all cases, careful tests, such as performed by Day (1996), should be conducted using samples and conditions representing specific characteristics of the field monitoring program.

Bacteria

Bacteria analysis is an important parameter for many monitoring programs. Unfortunately, conventional laboratory tests are time-consuming (typically requiring at least 24 hours under very controlled temperature conditions). IDEXX supplies a simple procedure for monitoring enterococcus, *E. coli*, and total coliforms for general field work (described later) that can be adapted for field work (Figures 6.71 through 6.74, including Color Figure 6.5). Millipore probably has the most complete selection of field equipment and supplies to conduct bacteria analyses in the field. HACH also supplies suitable field equipment for many types of bacteria tests. However, these tests also require the same standard incubation times as the time-consuming laboratory tests. There are a few procedures that can indicate the presence of very large populations of bacteria in water samples in relatively short periods of time. Most of these require UV light analyses and controlled incubation temperatures. An interesting alternative is the KoolKount Assayer from Industrial Municipal Equipment, Inc. This is a visual colorimetric test that costs about $4 per test. It is unique in that it only requires from 30 min to 13 hours for a determination at "room temperature" incubation. Very high bacteria populations will be evident in a short period of time. This is not a selective test, but sensitive to a mixed microbial population. The test was developed to analyze gross bacterial contamination of cooling waters, but may also be useful in receiving water studies. There are now DNA-based procedures being developed (see a later section on emerging technologies) that offer promise for much more rapid, inexpensive, and easy analysis of bacteria.

In all cases, the user must be aware of the inherent problems in interpreting bacteria data, especially if one is using bacteria as an indicator of sewage contamination. As an example, fecal coliform bacteria are in very high populations is many waters, including stormwater that is not contaminated by sanitary sewage. The use of the fecal streptococci to fecal coliform ratio to indicate sources of contamination is also inherently inaccurate, unless the source of contamination is very

Figure 6.71 Pouring sample into IDEXX analytical tray.

Figure 6.72 Placing analytical tray into heat sealing unit.

recent. O'Shea and Field (1992) reviewed many of these issues for stormwater. A better indication of potential sanitary sewage contamination in surface waters is the use of a small battery of chemical tracer analyses (detergents, fluoride, ammonia, and potassium), as developed and tested by Lalor (1993) and Pitt et al. (1994) and described later.

Organic Compounds

The analysis of organic compounds using field test kits is also of great interest because of the high costs of conventional laboratory analyses and the importance of these compounds. The organic compounds of most interest in studies of receiving waters affected by stormwater include BTEX, PAHs, and pesticides.

The two BTEX test kits evaluated by Day (1996) include the Dexsil PetroFlag (Figures 6.75 and 6.76) and the Dtech immunoassay test kit for BTEX (Figure 6.77). The PetroFlag is a simple solvent extraction test for sediment analyses. It requires a $700 reader that is only used for this test. Each test costs about $10 and requires about 10 min. It has poor detection limits and is not very

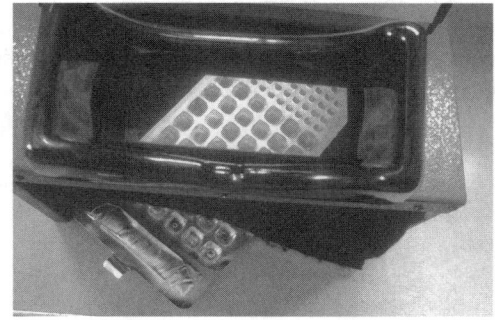

Figure 6.73 Placing prepared tray into incubator.

Figure 6.74 IDEXX analysis for *E. coli* in ultraviolet light.

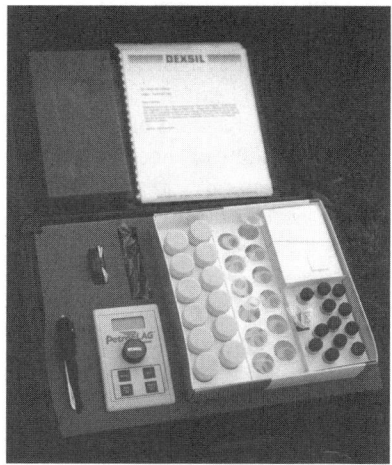

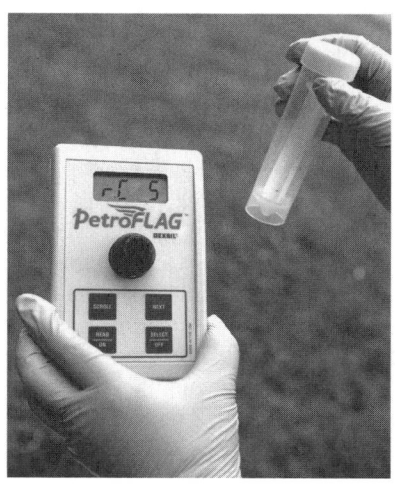

Figure 6.75 PetroFlag kit for field analyses of BTEX. **Figure 6.76** Sample being read using PetroFlag.

selective. An immunoassay test may be the only selective and sensitive option currently available. The Dtech (EM Science) BTEX Test Kit is an example of an immunoassay kit. It has an extremely low detection limit and reasonable selectivity that can be used for both water and sediment BTEX analyses. However, it is very complex and requires up to an hour. An initial cost of $500 for the Dtech reader can be used for both soil and water analyses and for both BTEX and PAH analyses for more precise results. The per-sample cost is about $25 for water samples and about $50 for sediment samples (requiring an additional soil extraction kit). The Dtech reagents have a relatively short shelf life (as little as a few weeks if not refrigerated, to several months if refrigerated).

The only selective option for PAH analyses is probably an immunoassay procedure. One example is the EM Science Dtech PAH Test Kit. Unfortunately, this test is also quite complex, requires more training than most other field test kits, and costs from $25 to $50 per sample. The Dtech reagent also expires in about 1 to 2 months and needs refrigeration.

Strategic Diagnostics, Inc. (www.sdix.com) also offers a number of test tube, magnetic particle immunoassay kits sold under the name RaPID Assay®. Kits are available for the detection of BTEX/TPH in environmental samples ($605/100 samples). Quantitative results can be obtained for BTEX in soil (assay range 0.9 to 30 ppm) or water (0.02 to 3.0 ppm), and if the operator knows the fuel source, total petroleum hydrocarbons (TPH) can be analyzed. The analytical range of this test kit is comparable to EPA GC method 8015 for TPH. Two immunoassay kits for PAHs are available. The PAHs RaPID Assay tests for 16 common PAHs ($1275/100 samples) and is comparable to EPA SW-846 Method #4035 and GC method 8270 or HPLC method 8310, with assay ranges in soil and water of 0.2 to 5.0 ppm and 0.93 to 66.5 ppb, respectively. Results are normalized to phenanthrene. The Carcinogenic RaPID Assay offers increased sensitivity to the seven most carcinogenic PAHs and is normalized to benzo[a]pyrene ($1395/100 samples). As of March 2000, 29 RaPID Assays for commonly used pesticides were available for prices ranging from $435 to $545. These

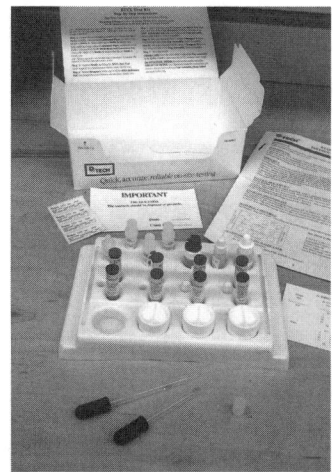

Figure 6.77 Dtech Immunoassay test kit for BTEX.

kits include alachlor, aldicarb, atrazine and five major metabolites, benomy/carbendazim, captan, carbofuran, chlorothalonil, chlorpyrifos, cyanazine, 2,4-D, endothall, fluridone, methomyl, metolachlor, metribuzin, organophosphates/carbamates, paraquat, picloram, procymidone, silvex, simazine, spinosad, TNT, TCP, and trichlopyr. For RaPID Assay kits, an optional soil extraction kit (12 samples), based on a 2-min methanol extraction procedure is available for $120. Kits are often sold for 30 or 100 samples, and results are usually obtained within 60 min. It is not always necessary to purchase the reader, the RPA-I RaPID Analyzer, as many tests can be quantitated on a spectrophotometer.

Two additional instruments were also recently examined at UAB for "total" hydrocarbon analyses. A Turner 10-AU field fluorometer with oil in water optics (see Figure 6.12) is extremely sensitive and is used with no sample preparation. It can be used in a flow-through mode to map hydrocarbon concentrations in real time. It is not very selective for different hydrocarbons. This instrument (which is also used for flow measurements using Rhodamine as a tracer) costs from $8000 to $16,000, depending on options, and is therefore not likely to be readily available to most people conducting field monitoring programs. A Petrosense PHA-100 probe from FCI Environmental, Inc. (see Figures 6.66 and 6.67) was also recently evaluated for real-time hydrocarbon analyses. This instrument costs about $7000 and has a slower response time (about 5 min), and it is not nearly as sensitive (about 100 µg/L, as xylene) as the fluorometer. It can also be used in real time, with no sample preparation.

EnviroLogix, Inc. (www.envirologix.com) offers antibody-based, enzyme-linked immunosorbent assay (ELISA) 96-well plate kits for pesticide detection. Pesticides include acetanilides (alachlor), aldicarb, benomyl/carbendazim, chlorpyrifos, cholinesterases (organophosphates and carbamates), cyanazine, cyclodienes (chlordane), DDT, fenarimol, fluometuron, imidacloprid, iprodione, isoproturon, metalaxyl, methoprene acid, organophosphates (cyclodienes and DDT), paraquat, parathion, synthetic pyrethroids, triazines (atrazine), and 3,5,6-trichloropyridonol. Accessories including soil extraction kits and a miniphotometer are available. The pesticides above cost $396/96-well-plate kit. Broad screening kits for cholinesterase inhibitors and organochlorine pesticides are available for $240 and $340, respectively.

Heavy Metals

Heavy metals are also of great interest in receiving water studies because they are possibly the most important toxic pollutants present. However, most of the metals in stormwater are associated with particulates (Pitt et al. 1995), with the exception of zinc, while all of the field test kits examined are only sensitive to "soluble" forms of the metals.

The HACH Bicinchonate Copper Method using AccuVac ampoules is the most suitable field method available (at a reasonable price) for measuring copper that was evaluated. This test uses the HACH DR/2000 spectrophotometer (at $1495) (or a less expensive dedicated filter spectrophotometer at $400), and the unit test cost is $0.56. It uses AccuVac ampoules that are very easy to use and makes the test very repeatable. However, the glass ampoules do produce glass wastes. The method detects the presence of a copper bicinchonate complex in the sample solution. An AccuVac ampoule is immersed in approximately 50 mL of sample and the tip is broken, which draws a known volume of sample into the ampoule. After a 2-min reaction time, the ampoule is scanned to determine the copper complex concentration, compared to a blank sample. Other metal ions present in large concentrations may also compete with copper for bicinchonate ligands. This interference will most likely produce a reported concentration larger than the true value if the metal complex absorbs in the same range as the copper complex. This method only indicates the presence of ionized copper. Any metallic or chelated copper will not be detected.

The HACH LeadTrak system is by far the most sensitive low-cost lead field test kit available (Figure 6.78). It is capable of detecting lead concentrations as low as 1 µg/L. Unfortunately, it is also quite complex and requires extensive experience. The test also takes about 45 min to conduct, which may be reduced to about 15 min with experience and if conducting several analyses at one

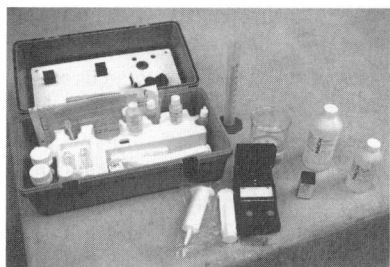

Figure 6.78   HACH LeadTrak field test kit.

time. The initial test kit costs about $400 (including a dedicated filter spectrophotometer) and the per-sample cost is about $5. The LeadTrak system determines lead concentrations through colorimetric determination of a lead complex extracted from the sample. The test procedure is quite complicated, requires a large amount of space compared to the other tests, and uses hazardous chemicals. However, it does produce good results.

The test requires 100 mL of sample, which is treated with an acid preservative (a nitric acid solution buffered with potassium nitrate). The solution is then treated with a solution of trishydroxymethylaminomethane, potassium nitrate, succinic acid, and imidazole. The prepared sample is then filtered through a solid-phase extractor (basically a syringe with a fabric plug). The lead in solution is held by the filter in the extractor. The lead is then removed from the plug with the eluant solution, another nitric acid solution. The eluant is allowed to pass over the plug until it stops flowing. The remaining eluant is forced through with the syringe plunger. This produces approximately 30 mL of extracted solution containing the lead from the sample. The extract is neutralized with a solution of trishydroxyaminomethane, tartaric acid, and sodium hydroxide. One powder pillow, containing potassium chloride and meso-tetra(-4-N-methylpryidyl)-porphine tetratosylate is added to the elutant. Two 10-mL portions are taken. A decolorizing solution is added to one portion; this portion is now the blank. Both sample portions are then analyzed using a spectrophotometer.

The La Motte Zinc test was the only acceptable zinc method investigated. This test uses a dilute solution containing cyanide, whereas the alternative tests use full-strength granular cyanide. The tests cost about $0.60 each and require about 5 min.

UAB recently evaluated two very sensitive electrochemical heavy metal field methods. The Palintest SA-1000 Scanning Analyzer is an anodic stripping voltammeter that uses preprepared electrode cards that come precalibrated (Figure 6.79). The instrument costs $2000 (available from AZUR Environmental), and each analysis for copper and lead costs about $5.50. The test is extremely sensitive (lead

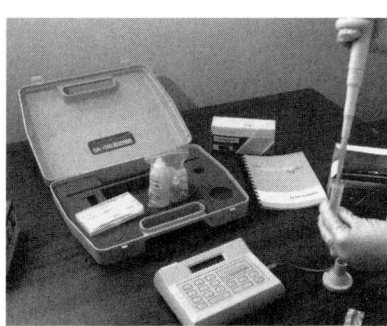

Figure 6.79   Palintest SA-1000 Scanning Analyzer.   **Figure 6.80**   Metalyzer 3000 metals analyzer.

to about 5 µg/L and copper to about 75 µg/L) and relatively rapid (3 min). Another field method recently evaluated is the Metalyzer 3000 from Environmental Technologies Group, Inc. (Figure 6.80). This is a potentiometric voltammeter that is also capable of very sensitive simultaneous analyses of copper and lead. This instrument (which includes a built-in data logger) costs about $4200 and each analysis for copper and lead costs about $15. Since neither of these instruments detects particulate-bound heavy metals, their best use may be in evaluating rainwater, most groundwaters, and finished drinking water, where particulate metal forms are not significant. Most surface waters and wastewaters have large fractions of the metals bound to particulates, and any metal analysis procedure that does not include sample digestion will likely severely underreport the total metal content. However, if one is interested only in "dissolved" metal conditions, these procedures may be quite suitable.

Emerging Analytical Methods for Heavy Metals — An important pollutant category that is not represented with any real-time instrumentation is heavy metals. Samples require digestion in order to release all of the particulate-bound heavy metals for analysis. In addition, most metals are not amenable to real-time analyses. Some colorimetric procedures, such as the diethyldithiocarbamate copper method (as available from La Motte) or the bicinchonate copper method (as available from HACH), could be conducted on a real-time basis with an automated chemical mixing and analysis procedure. Recent research at the University of Alabama at Birmingham (in conjunction with the General Physics Institute of the Russian Academy of Sciences and Alabama Laser) sponsored by the National Science Foundation developed and demonstrated a laser-based instrument that may be capable of continuous heavy metal analyses in water (Mirov et al. 1999) (Figure 6.81). This instrument is extremely sensitive, as it is based on atomic fluorescence. The use of lasers enables the specific wavelengths most critical for analysis to be precisely used in the instrument. In addition, automated digestion of the samples may also be possible.

An initial demonstration of the extreme sensitivity of the laser atomic fluorescence (LAF) instrument procedure for copper used selective excitation of the $^2P_{3/2}$ level of Cu and the strong absorption transition $^2S_{1/2}$ ($3d^{10}4s$) $\rightarrow$ $^2P_{3/2}$ ($3d^{10}4p$) at 324.754 nm. The fluorescence signal was detected at the emission transition $^2P_{3/2} \rightarrow {}^2D_{5/2}$ ($3d^94s^2$) at 510.554 nm. The average power of the excitation beam was 10 mW at 324.754 nm (third harmonic of alexandrite laser pumped LiF:F_2^{+**} laser). The repetition rate of the laser was 20 Hz, and the pulse duration was about 50 ns. Spectral resolution of the spectrometer during the experiments was about 0.1 nm. The spectrum accumulation time was set to 5 s (slightly less than the atomization time set by the graphite furnace controller), which allowed for signal collection during approximately 100 laser excitation pulses. Typical examples of the observed fluorescence spectra for water solutions with different concentrations of Cu are shown in Figure 6.82. All three spectra were measured under the same experimental conditions. The graphite furnace was heated to 2800 to 3000°C before and between measurements, in order to clean the graphite tube from possible residuals. The spectral peak at 511.46 nm is due to some scattered light of the third harmonic (255.73 nm) of alexandrite laser in the second diffraction order. This peak is reasonably constant during the experiment, so it was used as an amplitude reference signal. As shown in this figure, it was possible to detect extremely low levels of Cu, even in the RO and distilled water samples (<<10 µg/L).

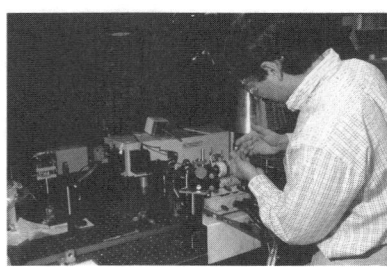

Figure 6.81   Laser fluorescence analysis of heavy metals being adapted for field analyses by UAB Physics Department.

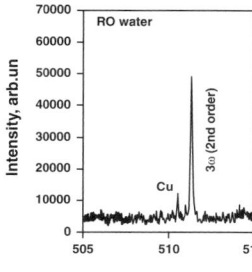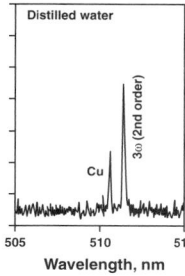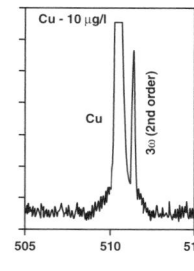

Figure 6.82 Fluorescence of Cu atoms under 324.75 nm laser excitation. (From Mirov, S.B., R.E. Pitt, A. Dergachev, W. Lee, D.V. Martyshkin, O.D. Mirov, J.J. Randolph, L.J. DeLucas, C.G. Brouillette, TT. Basiev, Y.V. Orlovskii, and O.K. Alimiv. A novel laser breakdown spectrophotometer for environmental monitoring. In: *Air Monitoring and Detection of Chemical and Biological Agents,* J. Leonelli and M.L. Althouse, Eds. Society of Photo-Optical Instrumentation Engineers (SPIE). Vol. 3855. pp. 34–41. September 1999. With permission.)

Solids

Analysis of the amount of solids in water samples in the field is another highly desired objective. Unfortunately, that is not practical. However, dissolved solids can be estimated using a simple conductivity meter, while suspended solids may be qualitatively estimated using a field nephelometer. Secchi disk transparency has also been used historically as an indication of suspended solids (especially related to algal activity). An excellent field nephelometer is available from HACH (for $800), while turbidity "probes" (miniaturized nephelometers) are now available on several *in situ* continuously recording multiwater quality probes (the Horiba HU-10 for $2800 and the YSI 6000 series, for about $7000). Numerous pocket conductivity meters are available that have "TDS" scales. These should be avoided in lieu of standard conductivity meters, as site-specific correlations between conductivity and TDS are usually required.

The Horiba Twin is a very small conductivity meter that has done very well in evaluation tests (Figure 6.83). It costs about $250, but the sensor should be replaced every 6 months at a cost of $60. This meter automatically compensates for temperature effects and is suited for very small sample volumes (3 to 4 drops). The meter includes a standard calibration solution. The procedure is to calibrate the meter using the provided standard solution and to select the conductivity mode. The user may partially immerse the probe in the sample or cover the probe with a few drops of sample.

pH

pH is usually considered an easy parameter to measure in the field. Unfortunately, the use of most "pocket" pH meters results in very inaccurate results, as the inexpensive probes included with these meters are not very reliable or robust, especially with storage. Recently available "dry" pH probes offer some hope for better field pH measurements. However, the most common FET transistor-based probes are delicate and can be irrecoverably damaged with abrasion or through contamination with oils and detergents. The Sentron field pH meter (at $600) is very sturdy, stores dry, and can be easily cleaned with a brush. Although the Horiba Twin pH meter is more likely to break, having a thin glass cover, it has worked well and is much less expensive (about $300) (Figures 6.84 and 6.85). Most field pH evaluations

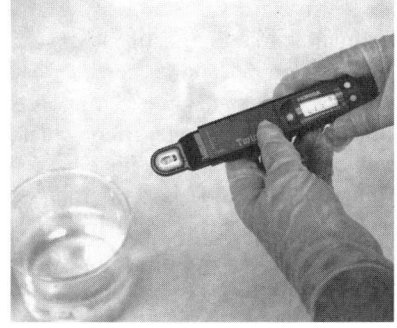

Figure 6.83 Horiba Twin conductivity meter.

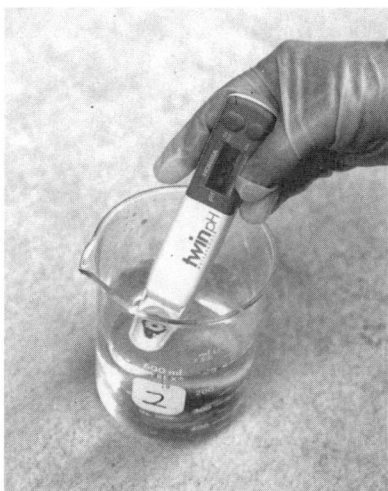

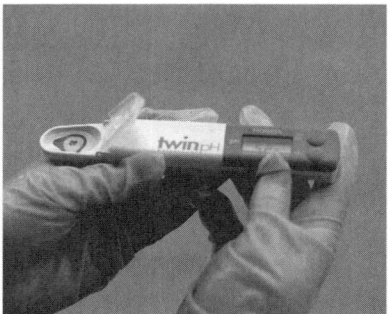

Figure 6.84   Horiba Twin pH meter dipped in sample.   **Figure 6.85**   Horiba Twin pH meter reading sample.

can probably be conducted using standard pH paper, as long as laboratory pH tests are also conducted. Fisher Scientific Alkacid Test Strips, for example, are very simple to use and inexpensive (<$1 per test), but the pH value is only readable to within ±1 pH unit (0.3 would be preferable). However, this sensitivity may be acceptable for many situations.

Dissolved Oxygen

DO is a parameter that is most commonly determined in the field. The YSI line of instruments is probably the best known and most commonly used among DO meters. The newer rapid-pulsed current DO probes from YSI are much superior to the older Clark membranes, especially if long-term monitoring is needed (such as with the *in situ* continuously recording probes). Many companies supply DO probes that work well, but with varying numbers of problems associated with storage, membrane replacement, and calibration. Winkler titration is not commonly used in the field, but HACH's digital field titrator even makes that feasible. The titration procedures work best with BOD analyses, including field titrations of BOD bottles used for *in situ* photosynthesis/respiration tests.

Detergents

The CHEMetrics Detergents (Anionic Surfactants) test kit was the only practical test for detergents investigated (Figure 6.86). The tests cost about $2.38 each and require about 10 min. The test uses a chloroform extraction, but it is very well designed to minimize exposure to the operator and uses a very small amount of chemical (Figures 6.87 through 6.90). The CHEMetrics procedure uses a visual comparator to determine the concentration of the detergents in the sample (Figures 6.91 and 6.92). A small volume of sample (5 mL) is required. An ampoule containing methylene blue and chloroform is mixed with the sample. Anionic detergents complex with the methylene blue and are extracted into the chloroform layer. Cationic detergents and sulfides interfere with the reaction and lead to decreased readings. The method is very quick and easy. However, the it uses chloroform, a known carcinogen. Users must conduct the test in well-ventilated areas. Furthermore, the waste must be disposed of properly. The kit is well designed to minimize the use and exposure of the chloroform. The reagent packs do have a limited shelf life, however. One method that can be used to detect the presence of detergents in outfalls tests for optical brighteners. This method was originally developed by researchers in Massachusetts for detecting inappropriate discharges at outfalls, especially from septic tanks. This method is described at www.thecompass.org/8TB/pages/SamplingContents.html. Untreated cotton

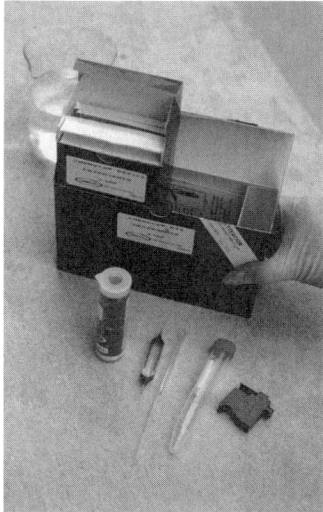

Figure 6.86 CHEMetrics detergent test kit.

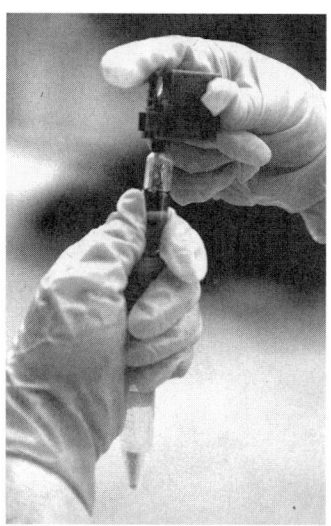

Figure 6.87 Extraction step 1 for use of the CHEMetrics detergent test kit.

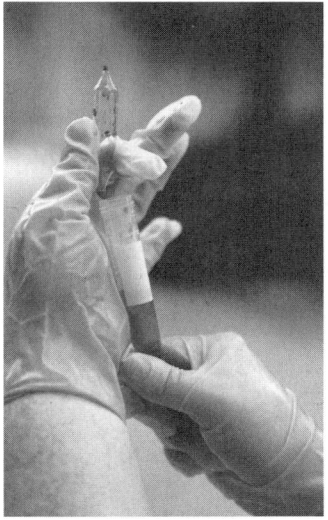

Figure 6.88 Extraction step 2 for use of the CHEMetrics detergent test kit.

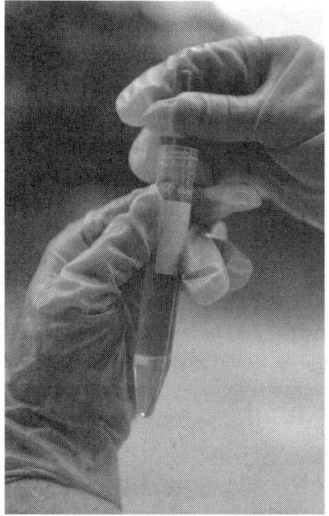

Figure 6.89 Extraction step 3 for use of the CHEMetrics detergent test kit.

pads are secured at the test locations where they are left exposed for several days, recovered, and examined under a UV lamp. Optical brighteners adsorb to the cotton, if present in the flowing water. This method is not quantitative but should indicate gross contamination associated with wash waters, septage, and sewage.

Fluoride

The HACH Fluoride SPADNS Reagent test using AccuVac Ampoules is another AccuVac test that shares the DR/2000 (Figure 6.93). The tests cost about $1.17 each and require about 5 min. The test does produce a small amount of glass waste, and the SPADNS reagent is hazardous, requiring special disposal considerations.

Figure 6.90 Extraction step 4 for use of the CHEM-etrics detergent test kit.

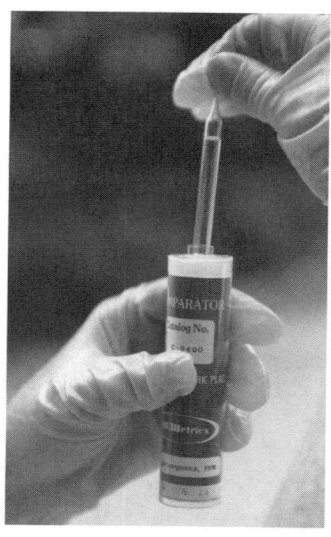

Figure 6.91 Inserting sample into color comparator.

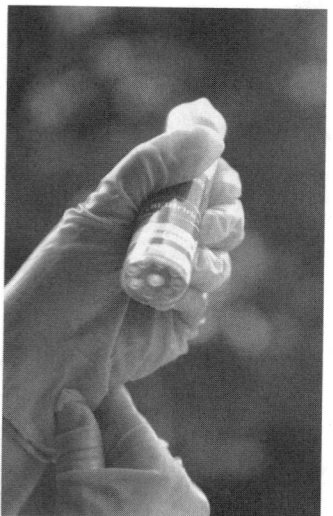

Figure 6.92 Reading the detergent concentration with the color comparator.

Figure 6.93 HACH AccuVac fluoride method.

Potassium

The La Motte Potassium Reagent Set was tested in the HACH DR/2000 spectrophotometer. This is an example of a hybrid test that was tested successfully by combining the very good La Motte reagents with the very good (and needed for many other tests) HACH DR/2000. The cost per test is about $0.29, and the test should take about 15 min. Potassium (K^+) can be used as an indicator of sewage contamination in water bodies, especially by examining the ratio of ammonia to potassium concentrations (Lalor 1993; Pitt et al. 1994).

The HACH and La Motte kits both determine potassium concentrations using tetraphenylborate salts. These procedures add large doses of sodium tetraphenylborate to the sample. The potassium in the sample reacts with the sodium tetraphenylborate to form insoluble potassium tetraphenylborate.

The insoluble potassium tetraphenylborate increases the turbidity of the sample solution. The presence of magnesium (Mg^{2+}), ammonium (NH_4^+) and calcium (Ca^{2+}) ions can interfere with the reaction by competing in the reaction with tetraphenylborate. These salts will result in a reported potassium concentration larger than is actually present in the sample. Both methods measure this increase in turbidity, using a spectrophotometer. To compensate for not using a nephelometer to measure this turbidity, both procedures include very specific timing requirements. The reaction and settling times must be followed exactly in order to obtain repeatable results.

Nutrients

The most common nutrient tests are for ammonia nitrogen, nitrate nitrogen, and phosphorus. The HACH Nitrate, MR test also shares the DR/2100 spectrophotometer and uses AccuVacs. The test is therefore very simple and quick, but produces glass debris and a hazardous reagent waste. The test costs about $0.56 per test and takes about 7 min.

The HACH Ammonia method using salicylate without distillation is a colorimetric determination of ammonia using salicylate. This method requires a DR/2100 spectrophotometer (usable for several other parameters) and a per sample cost of $2.88. It is also available as a self-contained test kit with a colorimeter for about $400.

Numerous simple field test kits are available for phosphorus. HACH, for example, has eight separate spectrographic tests and 11 colorimetric tests available for different forms and concentration ranges for phosphorus. Reactive phosphorus (orthophosphate) is probably of greatest interest for most simple environmental monitoring activities. The HACH AccuVac ascorbic acid method with the DR/2100 spectrophotometer is probably the simplest test procedure available. The tests cost about $0.56 each, after purchase of the spectrophotometer.

Two ion selective electrode (ISE) probes were also evaluated for fluoride analyses, with disappointing results. Probe problems were mostly associated with the lack of stability of the probe, especially with storage, and time-consuming standardization. Similar problems were found with ISE probes for ammonia, detergents, and potassium. ISE probes may work well in controlled laboratory settings, especially with proper care of the probes, but their use in the field is probably restricted to trained electrochemists who know how to take proper care of the probes and who know how to calibrate them more efficiently. Exceptions were the Horiba Cardy ISE probes for potassium (Figure 6.94) and nitrates that have worked very well in the field, although they are not very sensitive.

Figure 6.94 Horiba Cardy ion selection electrode for potassium.

Selection of Appropriate Field Test Kits

The most appropriate field test kit for a specific use can be selected based on the criteria presented earlier in this section, and in Chapter 5, and summarized in Appendix E. In most cases, the limits of detection are the most critical criteria. It is quite possible that the simplest field test kits may be useful for some studies, as most were found to be generally free from interferences (Day 1996). However, during tests using actual stormwater samples and spiked waters, their sensitivity was found to be generally poor, even less sensitive than typically advertised. This will likely lead to false negatives if actual limits of detection are not determined through sensitivity tests using local waters. The needed limits of detection must be known before analytical methods are selected, using methods presented earlier in Chapter 5.

The field test kits highlighted in the above discussion were selected based on our (Pitt et al. 1993; Day 1996) laboratory and field comparison tests and have been found to generally best meet our needs during investigations of stormwater problems, although other field test kits are also likely suitable.

If field test kits or *in situ* methods are suitable and available to meet the project objectives, other criteria must also be considered, especially the amount of time required for analyses, complexity and training needs, hazardous wastes and sharps produced, and cost. As indicated in Appendix E, some analyses are virtually instantaneous, while other tests may require almost an hour. Obviously, if multiple samples can be evaluated at the same time, the longer times required for some of the tests may not be as critical. A more serious concern is the use and production of hazardous reagents and wastes, and glass sharps. Unless personnel are especially well trained and have suitable facilities, these field test kits need to be avoided. The complex tests, such as the immunoassay kits for organics, may also require specialized training, as indicated in Appendix E, also eliminating their use except for the most patient and skilled analysts. If the field test kits are suitable for the needed monitoring activity, conventional laboratory procedures, discussed in the following section, are available.

The following example illustrates how this information can be used to select the most appropriate field testing methods, or to rely on conventional laboratory analyses. Table 4.37 was a simple matrix showing which parameters would be of greatest concern when evaluating receiving waters having different beneficial uses. In this example, biological life and integrity uses are of concern. Table 6.29 lists the water quality parameters of most interest for this use, expected concentrations of most concern (from Appendix G, a discussion of water quality criteria) and their associated assumed variation, and corresponding needed limits of detection. Obviously, the listed parameters shown on this table are only a portion of the needed field study for this assessment, as habitat destruction, high/low flow durations, inappropriate discharges, benthic macroinvertebrate and fish sampling, sediment investigations, and bioaccumulation of toxicants should also be considered (as listed in Table 4.37).

The only other primary water quality constituents noted on Table 4.37 of great interest for receiving water assessments include the microorganisms. These currently cannot be analyzed in the field, although portable sample preparation and field incubators are available from HACH and others. Because of the long incubation periods required (typically 18 to 24 hours for preliminary results), these methods are not really considered field methods here. Therefore, the analyses that might be conducted using field test kits that meet basic sensitivity requirements include:

Conventional Constituents:
- Hardness (using field titration equipment)
- Alkalinity (using field titration equipment)
- Turbidity (possible using moderately expensive field nephelometer, or expensive *in situ* recording probes)
- pH (easily conducted using electrodes, or expensive *in situ* recording probes)
- Conductivity (easily conducted using electrodes, or expensive *in situ* recording probes)
- DO (easily conducted using electrodes, or expensive *in situ* recording probes)
- Temperature (easily conducted using electrodes, thermometers, or expensive *in situ* recording probes)

Nutrients:
- Ammonia (several simple field test kits available)
- Nitrates (several simple field test kits available)
- Phosphates (several simple field test kits available)

Toxicants:
- Lead (but difficult, time-consuming, or expensive)
- Toxicity screening (expensive instrument)
- Pesticides, PAHs, PCBs, etc., by immunoassays (but difficult, time-consuming, and expensive)

Of these, DO (field probe preferred to titration in most cases), temperature (mandatory), and pH (within a few hours) may need to be conducted in the field to meet QA/QC requirements, while conductivity is very easy to measure in the field (and therefore commonly done). The decision to measure the other constituents listed above in the field should be based on other considerations,

Table 6.29 Water Quality Measurements of Interest and Expected Analytical Requirements for Hypothetical Receiving Water Investigation Assessing Aquatic Life Use Impairments

Water Quality Parameter	Example Water Quality Objectives Associated with Aquatic Life Beneficial Uses (short-term exposures)	Expected Coefficient of Variation (COV) Category[a]	Estimated Needed MDL[a]	Suitable Field Measurement Methods Providing Estimated Needed MDL (mostly from Table E-2, from Day 1996, also from text).
Zinc	<120 µg/L (CMC^2)	Medium	28 µg/L	No available field method could approach this desired MDL. The lowest MDL found was about 140 µg/L for Zn. Most of the field test methods also require toxic (cyanide) reagents.
Copper	<13 µg/L (CMC)	Medium	3 µg/L	No available field method could approach this desired MDL. The lowest MDL found was about 100 µg/L for Cu.
Lead	<65 µg/L (CMC)	Medium	15 µg/L	The HACH LeadTrak system has a MDL of about 5 µg/L, although it is a time-consuming test and relatively expensive. The Metalyzer 3000 and Palintest SA-1000 both have lead MDLs of about 5 µg/L and would therefore be suitable, but are expensive instruments.
Microtox screening test	n/a: indicative of toxicants that may be present (such as pesticides), desire low value; l20 of <25%.	Medium	l20 of 6%	Deltatox (expensive instrument, but field portable).
Hardness	Narrative (want moderate to hard water conditions to reduce effect of some toxicants), would like to detect hardness to at least 50 mg/L.	Low	40 mg/L	HACH Digital Titrator and CHEMetrics EDTA titration methods would both likely be suitable field methods.
Alkalinity	n/a (would like moderate to high levels of alkalinity to reduce effects of some toxicants), would like to detect alkalinity to at least 25 mg/L.	Low	20 mg/L	Field titration methods available, but not evaluated.
Ammonia	<3.8 mg/L (2.5 × chronic at 30°C)	Low	3 mg/L	All 4 field test kits investigated have limits of detection lower than this estimated needed MDL. However, one requires refrigeration, and others contain mercury in waste.
Nitrates	n/a (rarely toxic to aquatic life in natural streams, but indicative of potential eutrophication problems in nitrogen limited streams), would like to detect NO$_3$ to at least 1 mg/L.	Low	0.8 mg/L	The La Motte and CHEMetrics nitrate tests, and likely the HACH low range nitrate test, can meet this MDL objective. Sharps and cadmium containing wastes are common with these methods.
Phosphates	Narrative, <25 µg/L to prevent eutrophication.	Low	20 µg/L	Numerous phosphate field test kits are available, although not reviewed by Day (1996). It is expected that there are several that can meet these performance objectives.

Table 6.29 Water Quality Measurements of Interest and Expected Analytical Requirements for Hypothetical Receiving Water Investigation Assessing Aquatic Life Use Impairments *(Continued)*

Water Quality Parameter	Example Water Quality Objectives Associated with Aquatic Life Beneficial Uses (short-term exposures)	Expected Coefficient of Variation (COV) Category[a]	Estimated Needed MDL[a]	Suitable Field Measurement Methods Providing Estimated Needed MDL (mostly from Table E-2, from Day 1996, also from text).
Suspended solids	Narrative: <100 mg/L settleable fraction to prevent smothering of stream bed.	Large	12 mg/L	No field instruments known for measuring suspended solids (requires drying ovens and analytical balance), but can be predicted/tracked using turbidity.
COD	n/a (indication of organic matter), would like to be <5 mg/L.	Medium	1 mg/L	No field instruments known for measuring COD (requires digestion).
pH	Between 6.5 and 9 desired (harmless to fish in this range).	Very low	Readable to 0.3 pH units	All of the pH electrode methods investigated should meet this readability objective, but the pH paper methods are not likely suitable.[c]
Conductivity	n/a (variation should be minimal), would like to determine conductivity at 100 µS/cm.	Low	80 µS/cm	All three conductivity probes investigated had limits of detection about equal to this objective and would be suitable.[c]
Turbidity	Narrative: <50 NTU increase above background conditions.	Large	6 NTU	The HACH portable nephelometer, or the Horiba HU-10 and YSI in-situ probes can measure turbidity in the field, although these are all moderate to very expensive options.[c]
DO	>5.0 mg/L	Low	Readable to 0.25 mg/L	Most modern field DO meters could be used to meet these objectives.[c]
Temperature	Narrative (variation from natural conditions should be minimal).	Low	Readable to 0.5°C	Most modern field DO meters also have temperature readouts and would be suitable, alternatively, simple pocket thermometers could be used.[c]

[a] If the COV is low (<0.5), the multiplier for the MDL is 0.8 × the desired median value of the observations, in this case taken to be the water quality objectives. If the COV is medium (0.5 to 1.25), the multiplier is 0.23, and if the COV is large (>1.25), the multiplier is 0.12; see Table 6.26 and corresponding discussion.
[b] CMC: criterion maximum concentration (exposure period of 1 hr)
[c] The combination probes (such as the YSI 6000) should be considered as they can monitor several needed constituents: pH, conductivity, turbidity, DO, and temperature).

mainly safety, cost, time, and difficulty. In many cases, it is not practical to conduct field measure-ments at the time of sample collection due to the time needed to set up equipment, standardize the procedures, and conduct the individual constituent analyses at each sampling location. However, it might be very reasonable to use these field methods in a temporary field laboratory when conducting sampling in remote areas. In this case, samples collected over a short period of time (such as during the day) can be analyzed together, minimizing the time requirements. In addition, the use of continuous recording *in situ* probes should be seriously considered for turbidity, conductivity, pH, DO, and temperature, in addition to possibly ORP and stream stage (depth). Although expensive (can be rented for short periods), these probes have been extremely useful when monitoring these key constituents over several weeks that include both wet and dry periods. The high resolution data (measurements typically are taken and logged every 15 min) dramatically illustrate the variabilities of these constituents over short periods of time (as discussed in the narratives for some of the water quality criteria) and help to understand the duration of exposure to wet-weather-related discharges.

The earlier Table 4.37 also included additional water quality measurements that were not listed as primary constituents and therefore not discussed above. Many of these should also be periodically evaluated as part of an assessment project, but few are amenable to safe, inexpensive, and rapid field measurements. These other constituents (such as the PAHs and pesticides, other metals, and microorganisms), plus those listed above that are not generally suited (or selected) for field mea-surements, must be analyzed using conventional laboratory methods. Of course, a good QA/QC plan would also require that samples being analyzed in the field be periodically split and analyzed using conventional laboratory methods for comparison. In some cases, it may be appropriate to use some of the more difficult field test kits (such as the immunoassay tests) due to the lack of conventional laboratory facilities, or for faster turn-around time.

Conventional Laboratory Analyses

Table 6.30 lists standard analytical methods that may be used for stormwater analyses. Several methods need to be modified to effectively analyze stormwater samples, especially if only small sample volumes are available (such as from pore water from stream sediments, from bench-scale treatability tests, or to reduce sample shipping costs). Modifications to the standard methods are described in Appendix E and are necessary because of the large particulate fractions of the organic toxicants which interfered with conventional extraction methods. Reducing the sample volumes (especially for the organic analyses) also significantly reduces the volumes of hazardous laboratory wastes. Appendix E also contains information pertaining to heavy metal analysis options and laboratory safety. This table should not be considered as a complete listing of laboratory methods for stormwater analyses, but is an example of some analyses and the associated standard methods.

Quality control and quality assurance activities (see Chapter 5 and Appendix E) require a substantial effort in most analytical laboratories. EPA analytical guidelines published in the *Federal Register* for the various tests specify the types and magnitude of QA/QC analyses. These analyses supplement the standardization efforts as they are used to measure the efficiency of the sample preparation and analysis procedures. Blanks are used to identify possible contamination problems, while matrix spikes added to the samples prior to any preparation steps indicate the efficiency of the complete analytical process. Spikes added to the samples prior to analyses are also used to identify interferences, mainly associated with other compounds in the sample. In heavy metal analyses, for example, it is not uncommon to increase the sample analysis effort by an extra 50% for standards and QA/QC samples in production work. Method development activities require an even greater additional analytical effort.

Appendix E contains descriptions of the modifications to the standard methods for the organic toxicants noted in the above table that are needed for effective measurements of stormwater characteristics. These modifications are needed to obtain necessary levels of recovery of the organics that are bound to particulates in the stormwater. The following discussions present summaries of special aspects of laboratory tests of possible interest in receiving water investigations, especially

Table 6.30 Typical List of Standard and Modified Methods for Wet-Weather Flow Analyses

Parameter	Method
Physical Analyses	
Color, spectrophotometric	EPA 110.3
Conductance, specific conductance	EPA 120.1
Particle size analysis by Coulter Counter Multi Sizer IIe	Coulter method
pH, Electrometric	EPA 150.1
Residue, filterable, gravimetric, dried at 180°C	EPA 160.1
Residue, nonfilterable, gravimetric, dried at 103–105°C	EPA 160.2
Residue, total, gravimetric, dried at 103–105°C	EPA 160.3
Residue, volatile, gravimetric, ignition at 550°C	EPA 160.4
Turbidity, nephelometric	EPA 180.1
Inorganic Analyses	
Hardness, total (mg/L as $CaCO_3$), Titrimetric EDTA	EPA 130.2
Aluminum, arsenic, cadmium, chromium, copper, iron, lead, nickel, and zinc	EPA 200.9
Chloride, fluoride, nitrate, nitrite, phosphate, and sulfate	EPA 300.0
Ammonium, calcium, lithium, magnesium, potassium, and sodium	EPA 300.0 modified
Alkalinity, titrimetric (pH 4.5)	EPA 310.1
Organic Analyses	
Chemical oxygen demand, colorimetric	EPA 410.4
Aldrin, Chlordane-alpha, Chlordane-gamma, 4,4'-DDD, 4,4'-DDE, 4,4'-DDT, Dieldrin, Endosulfan I, Endosulfan II, Endosulfan sulfate, Endrin, Endrin aldehyde, Endrin ketone, HCH-alpha, HCH-beta, HCH-gamma (Lindane), Heptachlor, Heptachlor epoxide, and Methoxychlor	EPA 608 modified
Acenaphthene, Acenaphthylene, Anthracene, Azobenzene, Benzo(a)anthracene, Benzo(b)fluoranthene, Benzo(g,h,i)perylene, Benzo(k)fluoranthene, Benzo(a)pyrene, 4-Bromophenyl-phenylether, Bis-(2-chloroethyl)ether, Bis-(2-chloroethoxy)methane, Bis-(2-ethylhexyl)phthalate, Butylbenzyl phthalate, Carbazole, 4-Chloro-3-methylphenol, 2-Chloronaphthalene, 2-Chlorophenol, 4-Chlorophenyl-phenylether, Chrysene, Coprostanol, Dibenzo(a,h)anthracene, 1,2-Dichlorobenzene, 1,3-Dichlorobenzene, 1,4-Dichlorobenzene, 2,4-Dichlorophenol, Diethyl phthalate, 2,4-Dimethylphenol, Dimethyl phthalate, Di-n-butyl phthalate, 2,4-Dinitrophenol, 2,4-Dinitrotoluene, 2,6-Dinitrotoluene, Di-n-octyl phthalate, Fluoranthene, Fluorene, Hexachlorobenzene, Hexachlorobutadiene, Hexachlorocyclopentadiene, Hexachloroethane, Indeno(1,2,3-cd)pyrene, Isophorone, 2-Methylnaphthalene, 2-Methylphenol, 4-Methylphenol, Naphthalene, Nitrobenzene, 2-Nitrophenol, 4-Nitrophenol, N-Nitroso-di-n-propylamine, N-Nitroso-diphenylamine, Pentachlorophenol, Phenanthrene, Phenol, Pyrene, 1,2,4-Trichlorobenzene, 2,4,5-Trichlorophenol, and 2,4,6-Trichlorophenol	EPA 625 modified
Toxicity Analyses	
Microtox 100% toxicity screening analysis (using reagent salt for osmotic adjustments)	Azur Environmental method

methods suitable for large numbers of samples, particle size analyses, and laboratory tests to identify associations of metal compounds that determine their effects on receiving water uses.

Automated Methods Suitable for Large Numbers of Samples

There are a number of laboratory instruments suitable for rapidly analyzing large numbers of samples for common constituents. Two instruments that have been especially helpful in the Environmental Engineering Laboratories at the University of Alabama at Birmingham have been a Dionex Ion Chromatograph (we use an older DX-100 with an autosampler) and a Bran + Luebbe TRAACS 2000 Continuous-Flow Analyzer (we use a basic 2-channel unit, with XYZ autosampler and syringe diluter). These instruments are relatively expensive and are most suitable for rapidly analyzing many samples for a few constituents at one time. The sample volume requirements are very small (less than 10 mL) and expendable analytical cost per analysis is also very small (typically less than $0.10).

Figure 6.95 TRAACS 2000 instrument showing older linear auto sampler and main module.

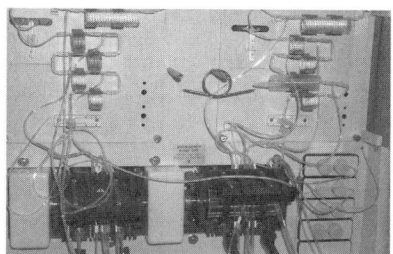

Figure 6.96 TRAACS manifold.

Unfortunately, required sample cleanup for the ion chromatograph adds several dollars per sample, and required filtration of surface water samples for the TRAACS also adds several dollars per sample. However, if many samples are to be analyzed in a short time, especially when working with small sample volumes, these instruments are very cost effective. However, necessary operator training and skill is much more than required for most conventional manual analyses.

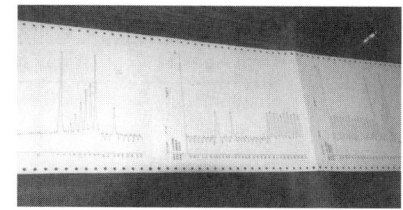

Figure 6.97 Quality control output for TRAACS.

Bran + Luebbe TRAACS 2000 Continuous-Flow Analyzer — This is a new instrument in our laboratory, and we are still learning its capabilities (and requirements). We are using the TRAACS mostly for dissolved nutrient analyses (phosphate, ammonia, nitrate, and nitrite), plus hardness and alkalinity (Figures 6.95 through 6.97). The instrument is capable of analyzing many other analytes, requiring a several-hour period to switch reagents and tubing, and for other initial setup activities. A block digester is available that would also enable total forms of the nutrients (specifically total Kjeldahl nitrogen and total phosphorus) to be rapidly analyzed. The instrument draws samples and needed reagents through specialized manifolds where mixing takes place before colorimetric determinations. The basic two-channel instrument enables two different analytes to be determined simultaneously. It is possible to add more channels, allowing additional simultaneous analyses. There are some restrictions on the analytes that can be simultaneously analyzed on the parallel channels, however. The XYZ autosampler, containing samples, plus standard solutions and blanks, allows several hundred samples to be evaluated at one setup. The syringe diluter automatically adjusts sample strength if a sample goes over-range.

Most analyses require only a few minutes per sample and very small amounts of standard reagents. The cost per analysis is therefore very low, but the setup time and other maintenance requirements make the instrument most suitable when a relatively large number of samples are to be analyzed at one time.

Dionex DX-100 Ion Chromatograph — We have much more experience with the DX-100, having used it for several years in support of many of our recent stormwater research projects. It can be used for the determination of the following common inorganic ions in drinking water, surface water, mixed domestic and industrial wastewaters, groundwater, reagent waters, solids (after extraction), and leachates (when no acetic acid is used):

- Anions: fluoride, chloride, nitrate-N, nitrite-N, ortho-phosphate-P, and sulfate
- Cations: lithium, sodium, potassium, ammonium, magnesium, and calcium

A small volume of sample (0.5 mL) is introduced into the ion chromatograph using the autosampler. The ions of interest are separated and measured, using a system comprised of a guard

Table 6.31 Anion Chromatographic Conditions and Detection Limits in Water

Analyte	Peak No.	Retention Time (min)	MDL (mg/L)
Fluoride	1	1.2	0.027
Chloride	2	1.7	0.080
Nitrite-N	3	2.0	0.111
Nitrate-N	4	3.2	0.040
o-Phosphate-P	5	5.4	0.084
Sulfate	6	7.0	0.083

Standard conditions: pump rate 2.0 mL/min, sample loop 25 μL.

Table 6.32 Cation Chromatographic Conditions and Detection Limits in Water

Analyte	Peak No.	Retention Time (min)	MDL (mg/L)
Lithium	1	1.3	0.0138
Sodium	2	2.0	0.454
Ammonium	3	3.2	0.123
Potassium	4	4.8	0.081
Magnesium	5	5.7	0.055
Calcium	6	7.9	0.318

Standard conditions: pump rate 1.0 mL/min, sample loop 25 μL.

column, analytical column, suppressor device, and conductivity detector. The difference between the methods for determining anion and cation concentrations are the separator columns, guard columns, and sample preparation procedures.

Tables 6.31 and 6.32 give the single laboratory method detection limit (MDL) for each ion (based on analyses at UAB).

These detection limits can be easily improved by changing sample loop lengths (and therefore the sample volume introduced into the IC), but resolution may suffer (and the ability to separate some ions) with increased volumes, and the upper limits also decrease correspondingly when the detection limits are improved.

Substances with retention times that are similar to and/or overlap those of the ion of interest can interfere with the analysis. Any ion that is not retained by the column or only slightly retained will elute in the area of fluoride or lithium and interfere. Known co-elution is caused by carbonate and small organic ions. At concentrations of fluoride and lithium above 1.5 mg/L, this interference is likely not significant. However, quality control is required to show whether this interference occurred. Do not attempt to quantify unretained peaks, such as low-molecular-weight organic acids (formate, acetate, propionate, etc.) which are conductive and co-elute with or near fluoride. These will bias the fluoride measurement in some drinking water and most wastewaters. The acetate anion elutes early during the chromatographic run, and the retention times of the anions seem to differ when large amounts of acetate are present. Therefore, do not use this method for leachates of solid samples when acetic acid is used. Residual chlorine dioxide present in the sample will result in the formation of additional chlorite. If chlorine dioxide is suspected in the sample, purge the sample with an inert gas (argon or nitrogen) for about 5 min or until no chlorine dioxide remains. The water dip or negative peak that elutes near, and can interfere with, the fluoride peak can usually be eliminated by the addition of the equivalent of 1 mL of concentrated eluent to 100 mL of each standard and sample.

Large amounts of an ion can interfere with the peak resolution of an adjacent ion. Sample dilution and/or fortification can be used to solve most interference problems associated with retention times. However, this method is not recommended for samples containing snowmelt runoff where chloride is used as a deicer. Samples that contain particles larger than 0.45 μm and reagent solutions that contain particles larger than 0.20 μm require filtration to prevent damage to instrument columns and flow systems.

Particle Size Measurements

Knowing the settling velocity characteristics associated with stormwater particulates is necessary when designing numerous stormwater control devices. In addition, particle size can be critical when understanding the effects and sources of stormwater sediments. There is a wide range of methods for determining particle size based on different principles and assumptions. No one method is ideal for all applications. For a review of sediment grain size methods, see ASTM Standard E 1391-94 (1994).

Particle size is directly related to settling velocity (using Stokes law, for example, and using appropriate shape factors, specific gravity, and viscosity values) and is usually used in the design of detention facilities. Particle size can also be much more rapidly measured in the laboratory than settling velocities. Settling tests for stormwater particulates need to be conducted over a period of about 3 days in order to quantify the smallest particles that are of interest in stormwater. If designing rapid treatment systems (such as grit chambers or vortex separators for CSO treatment), then much more rapid settling tests can be conducted. Probably the earliest description of conventional particle settling tests for stormwater samples was made by Whipple and Hunter (1981).

Randall et al. (1982), in settleability tests of urban runoff, found that nonfilterable residue (suspended solids) behaves liked a mixture of discrete and flocculant particles. The discrete particles settled rapidly, while the flocculent particles were very slow to settle. Therefore, simple particle size information may not be sufficient when flocculant particles are also present. Particle size analyses should include identification of the particle by microscopic examination to predict the extent of potential flocculation.

Figure 6.98 shows approximate stormwater particle size distributions derived from several upper Midwest and Ontario analyses, from all of the U.S. EPA's National Urban Runoff Program (NURP) data (Driscoll 1986), and for several eastern sites that reflect various residue concentrations (Grizzard and Randall 1986). Pitt and McLean (1986) microscopically measured the particles in selected stormwater samples collected during the Humber River Pilot Watershed Study in Toronto. The upper Midwest data sources were two NURP projects: Terstriep et al. (1982), in Champaign/Urbana, IL, and Akeley (1980) in Washtenaw County, MI. Figure 6.99 also shows the particle size distributions of stormwater particulates from the Monroe St. detention pond study in Madison, WI.

The Monroe St. project was a joint effort of the Wisconsin Department of Natural Resources and the U.S. Geological Survey. It obtained a number of stormwater particle size distributions for 46 storms having a wide range of characteristics. Bedload samplers were also used to obtain measurements representing the larger particles that are commonly not sampled by most researchers. The observed median particle sizes ranged from about 2 to 26 μm, with a median size of about 9 μm.

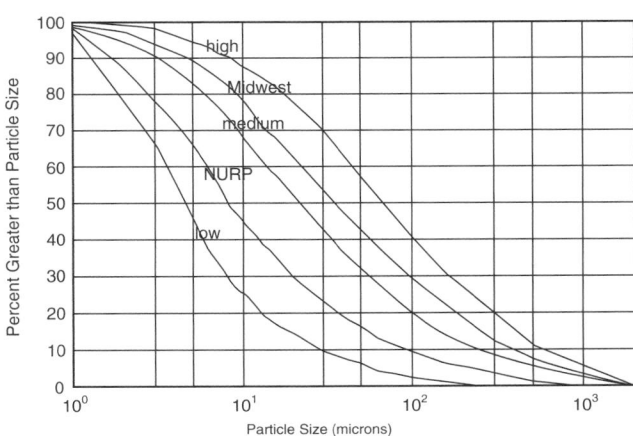

Figure 6.98 Particle size distributions for various stormwater sample groups.

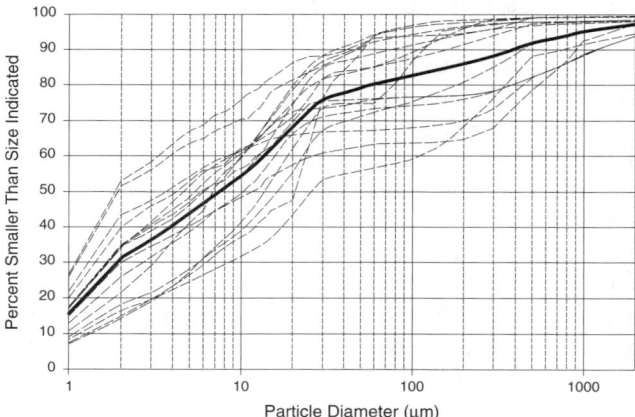

Figure 6.99 Monroe St. particle size distributions.

These distributions included bedload material that was also sampled and analyzed during these tests. This distribution is generally comparable to the "all NURP" particle size distribution presented previously. The 50th and 90th percentile particle size values are as follows for the different data groups:

	90%	50%
Monroe St.	0.8	9.1 μm
All NURP	1	8
Midwest	3.2	34
Low solids concentration	1.4	4.4
Medium solids concentration	3.1	21
High solids concentration	8	66

For many urban runoff conditions, the median stormwater particle size is estimated to be about 30 μm (which can be much smaller than the median particle size of some source area particulates). Very few particles larger than 1000 μm are found in stormwater, but particles smaller than 10 μm are expected to make up more than 20% of the stormwater total residue weight.

Specific conditions (such as source area type, rain conditions, and upstream controls) have been shown to have dramatic effects on particle size distributions. Randall et al. (1982) monitored particle size distributions in runoff from a shopping mall that was cleaned daily (by street cleaning). Their data (only collected during the rising limb of the hydrographs) showed that about 80% of the particles were smaller than 25 μm, in contrast to about 40% which were smaller than 25 μm during the outfall studies. They also only found about 2% of the runoff particles in sizes greater than 65 μm, while the outfall studies found about 35% of the particles in sizes greater than 65 μm.

Limited data are available concerning the particle size distribution of erosion runoff from construction sites. Hittman (1976) reported erosion runoff having about 70% of the particles (by weight) in the clay fraction (less than 4 μm), while the exposed soil being eroded had only about 15 to 25% of the particles (by weight) in the clay fraction. When the available data are examined, it is apparent that many factors affect runoff particle sizes. Rain characteristics, soil type, and on-site erosion controls are all important.

The particle size distributions of stormwater at different locations in an urban area greatly affect the ability of different source area and inlet controls to reduce the discharge of stormwater pollutants. A series of U.S. Environmental Protection Agency-funded research projects has examined the sources and treatability of urban stormwater pollutants (Pitt et al. 1995). This research included particle size analyses of 121 stormwater inlet samples from three states (southern New Jersey; Birmingham, AL; and at several cities in Wisconsin) that were not affected by stormwater controls. Particle sizes were measured using a Coulter® Multisizer™ IIe and verified with microscopic, sieve, and settling column tests. In all cases, the New Jersey samples had the smallest particle sizes associated with specific occurrence frequencies (even though they were collected using manual "dipper" samplers and not automatic samplers, which might miss the largest particles), followed by Wisconsin, and then Birmingham, AL, which had the largest particles (which were collected using automatic samplers). The New Jersey samples were obtained from gutter flows in a residential neighborhood that was xeriscaped; the Wisconsin samples were obtained from several source areas, including parking areas and gutter flows mostly from residential, but from some commercial areas, while the Birmingham samples were collected from a long-term parking area on the UAB campus.

The median particle sizes ranged from 0.6 to 38 μm and averaged 14 μm. The 90th percentile sizes ranged from 0.5 to 11 μm and averaged 3 μm. These particle sizes are all substantially smaller than have been typically assumed for stormwater. Stormwater particle size distributions typically do not include bedload components because automatic sampler intakes are usually located above the bottom of the pipe where the bedload occurs. During the Monroe St. (Madison, WI) detention pond monitoring, the USGS and WI DNR installed special bedload samplers that trapped the bedload material for analysis. This additional bedload comprised about 10% of the annual total solids loading. This is not a large fraction of the solids, but it represents the largest particle sizes flowing in the stormwater, and it can be easily trapped in most detention ponds or catchbasins. The bedload component in Madison was most significant during the early spring rains when much of the traction control sand that could be removed by rains was being washed from the streets.

The settling velocities of discrete particles are shown in Figure 6.100, based on Stokes' and Newton's settling relationships. Probably more than 90% of all stormwater particulates are in the 1 to 100 μm range, corresponding to laminar flow conditions. This figure also shows the effects of different specific gravities on the settling rates. In most cases, stormwater particulates have specific gravities in the range of 1.5 to 2.5. This corresponds to a relatively narrow range of settling rates for a specific particle size.

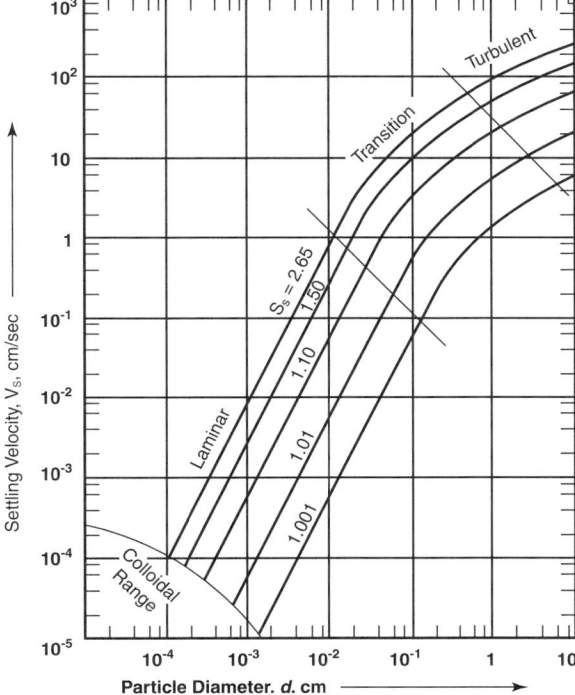

Figure 6.100 Settling rates of stormwater particulates.

Methods to Measure Stormwater Particle Sizes and Settling Velocities

Particle size is much easier to measure than settling rates. Automated (but expensive) particle sizing equipment is recommended because it enables very fast and accurate measurements, especially if supplemented with periodic settling column tests to determine deviations from standard settling theory. The following paragraphs briefly describe some of the particle settling options that have been used successfully for stormwater analyses. The most critical aspect of these analyses is obtaining an accurate sample, representing all particle sizes of interest. Automatic water samplers are suitable for obtaining samples having particles up to several hundred μm in size, but they cannot adequately sample particles much larger than about 1 mm in size. These large sizes are rare in stormwater, but they should be included in analyses in order to make suitable conclusions based on the data analyses. Automatic samplers can be supplemented using bedload samplers, as described in Chapter 5. However, the bedload samplers normally have to collect samples over an extended period of time to obtain sufficient samples for analysis. Manual sampling is usually easiest for representative sediment sampling, but is representative of only very short periods of time. Effective manual sampling must represent the complete water column, including bedload. This is easiest to accomplish if a "dipper" or "bucket" is used to collect flowing stormwater as it drops from an outfall or into an inlet.

Figure 6.101 Sieve analysis for stormwater particle determinations.

Sieve Analyses — This is probably the best procedure for laboratories that do not have access to expensive automated equipment, but have typical solids analysis balances, drying ovens, etc. (Figure 6.101). The basic procedure is as follows, using a 1- to 2-L well-mixed stormwater sample and a set of small sieves (usually about eight sieves, from 25 to 2000 μm, each having about one half the sieve opening as the next largest sieve):

1. Remove 100 mL of the sample for standard TDS and suspended solids analysis. The TDS sample is obtained by filtering the 100 mL through a 0.45-μm glass fiber filter. The filtrate, after passing through the filter, is placed in a dried and preweighed crucible for evaporation and final weighing. The filter is placed in a clean and preweighed small aluminum foil dish for drying and final weighing for the suspended solids analysis. Another 100-mL sample is placed directly in a preweighed crucible for evaporation and final weighing for a total solids analysis (and to check for errors associated with the separate TDS and SS analyses).
2. The remaining complete sample is then poured through the largest-sized sieve (the 2000 μm), and collected in another beaker. The sieved water captured in the second beaker is then sampled for total solids. After another 100-mL sample is removed for analysis, the remaining water is poured through the next smallest sieve, and another sample for total solids is removed. This process is repeated until water has been poured through all of the sieves and appropriate samples have been obtained for total solids analysis for each fraction.
3. All of the total solids samples are then oven dried, placed in a desiccator for cooling, and then weighed. The total solids content of each size fraction is then calculated, using the amount of water sample evaporated. The TDS content of the sample is subtracted from each total solids value, resulting in the suspended solids concentration for each particle size. An accumulative particle size distribution can then be prepared for the sample.

Unfortunately, this straightforward procedure requires a lot of time per sample and is limited as to the smallest particle size that can be measured, because the smallest sieve size available is about 25 μm. There is therefore a large gap between this particle size and the 0.45-μm "TDS" size, and much of the sample may be in this size range. It is possible to obtain higher resolution data in this range by using a series of Teflon or nylon filters (mounted on a vacuum filtering setup, as for the TDS filtration) in this size range. These are relatively expensive filters.

If the filtered water is to be analyzed for other pollutants (usually heavy metals, COD, and nutrients are the primary constituents of concern for particle size analysis of stormwater), stainless steel sieves and plastic or Teflon membrane filters should be used on a plastic filter stand. Standard glass fiber filters used for suspended solids analyses and glass filter stands cause zinc contamination from the glass, and standard brass sieves cause contamination of many heavy metals. In all cases, blank water should be subjected to the sampling processing apparatus and tested for contamination potential.

Automated Particle Size Analyses — This is the fastest, easiest, but most expensive (in terms of equipment) procedure for determining particle sizes in stormwater. There are many instruments capable of automated particle size analyses, but most are designed for high concentration suspensions and slurries that are not suitable for stormwater analyses, unless extraordinary sample preparation significantly concentrates the sample. The most common methods used for stormwater samples are laser-based diffraction instruments and the "electrical sensing zone method" (the Coulter Multisizer, Figure 6.102). The following briefly describes the features of the Coulter method used in the UAB Environmental Engineering Laboratory. This method is intended to characterize particles and agglomerated state particles in water. This technique uses the Electrical Sensing Zone Method, which has been utilized and verified for many decades in the medical and health services industries.

Figure 6.102 Coulter Multisizer for stormwater particle size analyses.

The Coulter Multisizer method determines the number and size of particles suspended in a conductive liquid (a saline solution containing several mL of the sample) by monitoring the electrical current between two electrodes immersed in the conductive liquid (Isoton) on either side of a small aperture. The continuously stirred liquid containing the sample is forced to flow through the aperture by a pump in the unit. As a particle passes through the aperture, it changes the impedance between the electrodes and produces an electrical pulse of short duration having a magnitude proportional to the particle volume. The series of pulses is electronically scaled, counted, and accumulated in a number of size-related channels which, when their contents are displayed on an integral visual display, produces a size distribution curve.

This method provides accurate particle size distribution curves within a 30:1 dynamic range by diameter from any one aperture. Size distributions from 0.4 to 1200 μm can be evaluated, depending on the orifice tube aperture size. Aperture sizes larger than 200 μm require a modification of sample viscosity using Karo corn syrup to prevent the particles from settling during the test. Each aperture allows the measurement of particles in the nominal diameter range of 2 to 60% of the aperture diameter.

When more than one particle passes through the aperture at the same time, it is called *coincidence*. Coincidence is detected by the Multisizer II by the unique properties of coincident signals. The instrument reports the level of coincidence as the measurement is being made. Coincidence levels of 5 to 10% are normal and acceptable. The Multisizer II reports coincidence level, raw count, and coincidence corrected count as part of the size distribution report. If coincidence levels are too high, the sample must be diluted. If there is no coincidence, the sample is not concentrated enough and a larger aliquot of sample must be pipetted into the Isoton solution.

We have found it most accurate to prefilter the sample before analyses with our Coulter Counter. We separate the sample into three size fractions: <0.45 μm, >120 μm, and 0.45 to 120 μm, with the intermediate size fraction further analyzed on the Coulter Counter using both the 50- and 200-µm aperture tubes. This results in four particle size distributions for each sample. These are manually combined (based on particle mass values for each size increment) using a spreadsheet. The most size data (highest resolution) is obtained from the intermediate sample fraction, which represents the majority of the particles (by mass) found in normal stormwater samples. This multistep approach is needed to ensure that the sample portions outside the normal working range of the Coulter Counter are included in the final size distribution. The sample is prepared as follows (about 300 to 500 mL of sample are needed for this analysis):

1. Remove 100 mL of the sample for standard TDS and SS analysis. The TDS sample is obtained by filtering 100 mL of sample through a 0.45-µm glass fiber filter (precleaned and preweighed). The filtrate (the sample after passing through the filter) is placed in a dried and preweighed crucible for evaporation, and final weighing (for the TDS determination). The filter is placed in a clean and preweighed small aluminum foil dish for drying and final weighing for the suspended solids analyses.
2. Another 100-mL sample is placed directly in a preweighed crucible for evaporation and final weighing for a total solids analysis (to check for errors associated with the separate TDS and SS analyses.)
3. The remaining sample (several hundred mL) is then poured through a moderate-sized sieve (with about 120-µm openings), and collected in another beaker. The sieved water captured in the second beaker is then sampled for total solids by removing another 100 mL sample for evaporation in a clean and preweighed crucible.
4. Finally, a sample is removed from the sieved water for the Coulter Counter analysis. This sieved sample should contain only particles up to about 1 to 120 µm, the range for the 50- and 200-µm aperture tubes that we commonly use.

The total solids fractions representing the three main sample portions are therefore known. The mid-fraction is further divided into very small increments using the data from the Coulter Counter tests. The final distribution of particle sizes is therefore well known over the entire range of particulates in the collected sample.

Settling Column Tests for Settling Velocity — Small-scale settling columns (using 50-cm-diameter Teflon columns about 0.7 m long) can be used to directly measure settling rate distributions of particles using basic engineering test procedures described in most wastewater and water treatment texts. Type one (discrete) settling is the predominant settling process for discrete stormwater particulates, and a simple settling column can be used with only a single sample port near the bottom of the tube. If Type two (hindered) settling is expected (due to high concentrations of flocculant particulates near the settling zone), then multiple sampling ports are needed along the settling column. For a simple settling apparatus, extended settling periods are needed to obtain information for the small particles that may be of most interest in stormwater. The test typically lasts about 3 days, with frequent samples (for total solids analyses) taken near the beginning of the test, tapering off as the test progresses. This is therefore a time-consuming (and expensive) test, but it should be conducted in conjunction with more frequent simpler particle size tests to confirm the relationship between size and settling velocity.

Much simpler hydrometer analyses of stormwater may not be effective because these procedures are intended for solutions having very high concentrations of particulates (Figure 6.103); however, they are useful for quantifying the clay size component of sediments (ASTM 1994). Some of the most polluted construction site runoff water (having suspended solids concentrations of several tens of thousands of mg/L) can be used with this method. Other settling rate monitoring methods, such as the Andreason pipette, are also rarely useful for the same reason (Figure 6.104). These are normally soil texture procedures where high concentrations of the soil particles can be mixed with water for the tests.

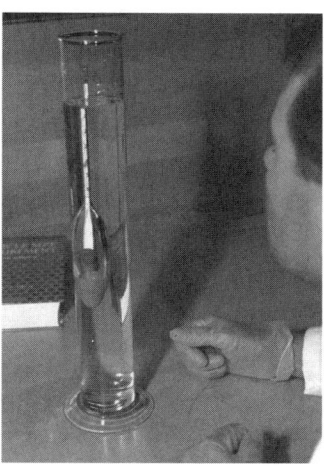

Figure 6.103 Hydrometer analyses for particle size determinations.

Figure 6.104 Andreason pipette for particle size determinations.

Visual Observations of Particle Characteristics — Microscopic observations of stormwater and receiving water particulates can yield much information. Standard laboratory microscopes, especially if equipped with a CCD camera and connected to a computer having particle analysis software (such as SigmaScan Pro by SPSS Software) can be used to measure particle sizes, particle morphology, and even origin (Figure 6.105). *The Particle Atlas,* both in print and the software version, from McCrone Assoc., Chicago, has a wealth of information to enable identification of particles. Most particles in stormwater are of erosion (mineral) origin that have become contaminated. As shown in Figure 6.106 (a typical microscopic view of stormwater particulates), relatively few particulates are from plants, and some are obviously from asphalt degradation and automobile exhaust. This photograph covers an area of about 600 by 800 µm, so the largest particles noted are about 100 µm in length. The polarized light images show asphalt particles dark, while minerals are generally much lighter.

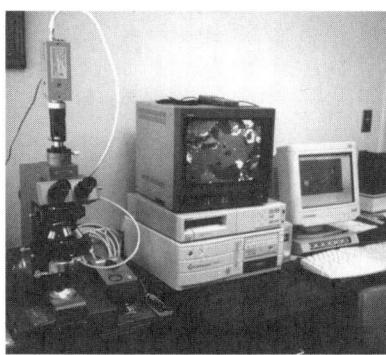

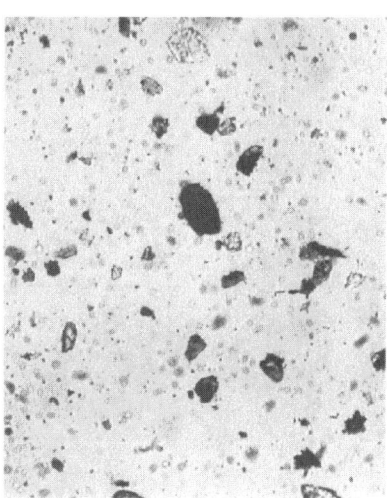

Figure 6.105 Microscope, video camera, and computer analyses of stormwater particles.

Figure 6.106 Typical microscopic view of particles in stormwater.

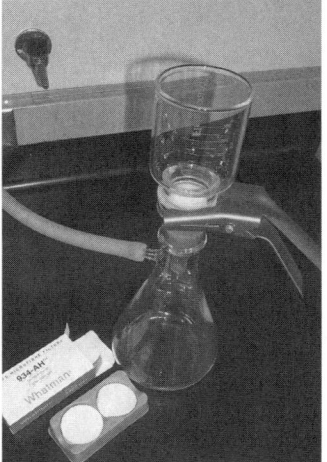

Figure 6.107 Standard filtration setup used with different membrane filters.

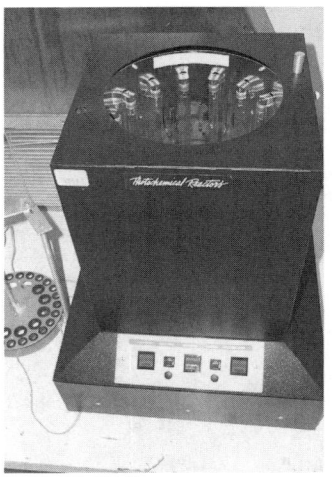

Figure 6.108 UV light digestion for controlled photo-oxidation.

Special Analytical Methods and Sample Preparation Procedures for Identifying Specific Forms of Metals

Sequential extraction has been used to separate the metals in a sample into various forms, such as separating the fraction bound to organic material from the fraction bound to mineral particulates, and to identify the fraction of the metals that may accumulate in aquatic organisms (Florence and Batley 1980). Figures 6.107 through 6.109 show various equipment used in the UAB environmental labs for treating samples for sequential analyses.

Several types of sequential extraction procedures were summarized by Bott (1995) to identify the form of heavy metals that may exist in a water sample (Figures 6.110 from Figura and McDuffie 1980; 6.111 from Florence and Batley 1980; and 6.112 from Nurnberg 1985). These procedures are useful to supplement the Toxicity Identification Evaluation (TIE) scheme noted later if metals are found to be the causative agent for stormwater toxicity (highly likely). The TIE scheme resulted

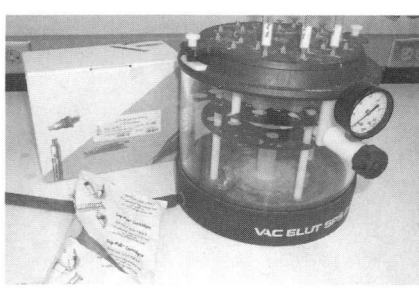

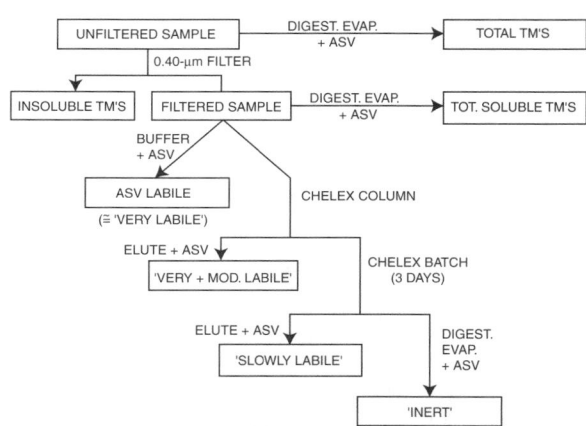

Figure 6.109 Solid phase extraction manifold for resin exchange experiments.

Figure 6.110 Figura and McDuffie scheme for chemical speciation analyses of trace metals. (From Figura, P. and B. McDuffie. *Anal. Chem.*, 52, 1433, 1980. Copyright 1980. American Chemical Society. Reprinted with permission.)

in sample components having specific toxicities. The most toxic sample components can then be subjected to further analyses to measure the toxicant concentrations. Organic analyses using GC/MSD or HPLC technology are very sensitive and can identify specific organic compounds present in the water. Unfortunately, the heavy metal analysis methods are only capable of measuring the total and filterable forms of each metal. However, heavy metals have greatly varying toxicities depending on their form. These sequential extraction procedures can result in a better understanding of the forms of the metals present in the sample and can identify the likely toxic forms present. These schemes typically separate the metals into functional categories, depending on the sample handling. As an example, Figure 6.111 shows a 0.45-μm filtration step to separate particulate from "soluble" forms. The soluble forms are further subjected to acetate buffer digestion (at pH 4.9) to identify labile forms of the metals, then to Chelex-100 extraction columns to identify forms that are sorbed onto inorganics or organics, and finally to UV digestion to identify the organic bound fraction. Anodic stripping voltammetric (ASV) methods are available to further identify the oxidation state of many of the metals of interest and can result in much more information than if graphite furnace atomic adsorption spectrophotometry is used for the metal analyses with these schemes. The sequential extraction procedures have been widely reported for studies of nutrient and metal availability in agricultural soils and for studies of sediments and dredged materials (for example see Tessier and Turner 1999).

Use of Tracers to Identify Sources of Inappropriate Discharges to Storm Drainage and Receiving Waters

Sources of Inappropriate Discharges into Storm Drainage

The need to identify inappropriate sources of discharges to storm drainage is critical for all stormwater management activities and is required by the EPA's stormwater discharge permits. Prior research (as summarized in EPA 1983a; Lalor 1993; Pitt et al. 1993) has shown that dry-weather flows from storm drainage may contribute a larger annual discharge mass for many pollutants than stormwater. These dry-weather sources may include direct connections to the storm drainage and sources that enter the drainage mainly through infiltration. Direct connections refer to physical

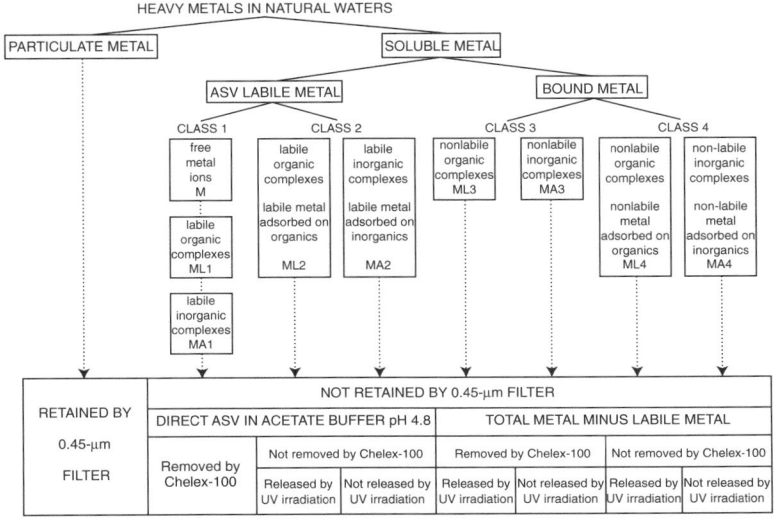

Figure 6.111 Florence and Batley scheme for chemical speciation analyses of trace metals. (From Bott, A. Voltammetric determination of trace concentrations of metals in the environment. *Current Separations,* 14(1): 24–30. July 1995. With permission.)

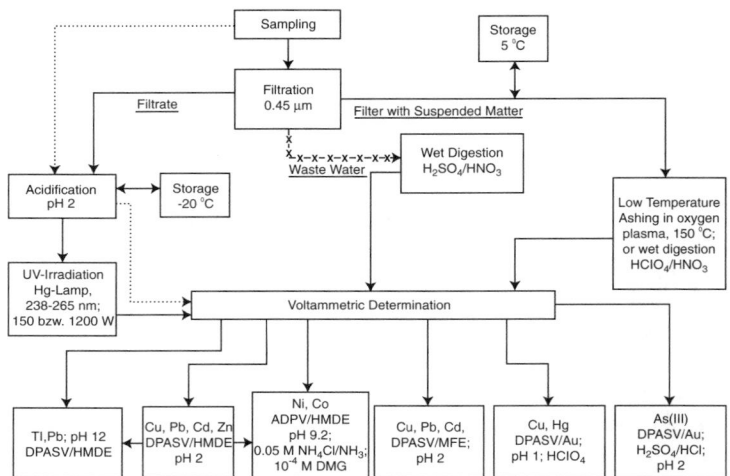

Figure 6.112 Nurnberg scheme for chemical speciation analyses of trace metals. Note: the last row of boxes contains notations describing the typical anodic stripping voltammetry (ASV) methods for each group of heavy metals. (From Nurnberg, H.W. Applications and potentialities of voltammetry in environmental chemistry of ecotoxic metals. In *Electrochemistry in Research and Development.* Edited by R. Kalvoda and R. Parsons. Plenum Publishing. pp. 121. New York. 1985. With permission.)

connections of sanitary, commercial, or industrial piping carrying untreated or partially treated wastewaters to a separate storm drainage system. These connections are usually unauthorized. They may be intentional, or may be accidental due to mistaken identification of sanitary sewer lines. They represent the most common source of entries to storm drains by industry. Direct connections can result in continuous or intermittent dry-weather entries of contaminants into the storm drain. Some common situations are:

- Sanitary sewers that tie into a storm drain.
- Foundation drains or residential sump-pump discharges that are frequently connected to storm drains. While this practice may be quite appropriate in many cases, it can be a source of contamination when the local groundwater is contaminated, as for example by septic tank failures.
- Commercial laundries and car wash establishments that may route process wastewaters to storm drains rather than sanitary sewers.

Continuous dry-weather flows may be caused by groundwater infiltration into storm drains when the storm sewers are located below the local groundwater table. These continuous discharges generally are not a pollution threat to surface waters, since most groundwaters which infiltrate into storm sewers are not contaminated, but these flows will have variable flow rates due to fluctuations in the level of the water table and percolation from rainfall events. Underground potable water main breaks are a potential clean source of releases to storm drains. While such occurrences are not a direct pollution source, they should obviously be corrected. However, when groundwater pollution does occur, such as from leaky underground storage tanks, storm drains may become a method of conveyance for these contaminants to the surface waters. Infiltration into storm drains most commonly occurs through leaking pipe joints and poor connections to catchbasins, but can also be due to other causes, such as damaged pipes and subsidence. Storm drains, as well as natural drainage channels, can therefore intercept and convey subsurface groundwater and percolating waters. Groundwater may be contaminated, either in localized areas or on a relatively widespread basis. In cases where infiltration into the storm drains occurs, it can be a source of excessive contaminant levels in the storm drains. Potential sources of groundwater contamination include, but are not limited to:

- Failing or nearby septic tank systems
- Exfiltration from sanitary sewers in poor repair
- Leaking underground storage tanks and pipes
- Landfill seepage
- Hazardous waste disposal sites
- Naturally occurring toxicants and pollutants due to surrounding geological or natural environment

Leaks from underground storage tanks and pipes are a common source of soil and groundwater pollution and may lead to continuously contaminated dry-weather entries. These situations are usually found in commercial operations, such as gasoline service stations, or industries involving the piped transfer of process liquids over long distances and the storage of large quantities of fuel, e.g., petroleum refineries. Pipes that are plugged or collapsed, as well as leaking storage tanks, may cause pollution when they release contaminants underground which can infiltrate through the soil into stormwater pipes.

The most common potential nonstormwater entries, which have been identified by a review of documented case studies for commercial and residential areas by Lalor (1993) and Pitt et al. (1993) included:

- Sanitary wastewater sources:
  - Raw sanitary wastewater from improper sewerage connections, exfiltration, or leakage
  - Effluent from improperly operating, designed, or nearby septic tanks
- Automobile maintenance and operation sources:
  - Car wash wastewaters
  - Radiator flushing wastewater
  - Engine degreasing wastes
  - Improper oil disposal
  - Leaky underground storage tanks
- Relatively clean sources:
  - Lawn runoff from over-watering
  - Direct spraying of impervious surfaces
  - Infiltrating groundwater
  - Water routed from preexisting springs or streams
  - Infiltrating potable water from leaking water mains
- Other sources:
  - Laundry wastewaters
  - Noncontact cooling water
  - Metal plating baths
  - Dewatering of construction sites
  - Washing of concrete ready-mix trucks
  - Sump pump discharges
  - Improper disposal of household toxic substances
  - Spills from roadway and other accidents

From the above list, sanitary wastewater is the most significant source of bacteria, while automobile maintenance and plating baths are the most significant sources of toxicants. Waste discharges associated with the improper disposal of oil and household toxicants tend to be intermittent and low volume. These wastes may therefore not reach the stormwater outfalls unless carried by higher flows from another source, or by stormwater during rains.

Human Health Problems Caused by Inappropriate Discharges

There are several mechanisms through which exposure to stormwater can cause potential human health problems. These include exposure to stormwater contaminants at swimming areas affected by stormwater discharges, drinking water supplies contaminated by stormwater discharges, and the consumption of fish and shellfish that have been contaminated by stormwater pollutants. In receiving waters having only stormwater discharges, it is well known that inappropriate sanitary and other wastewaters are also discharging through the storm drainage system. The most serious problems appear to be associated with the presence of potential pathogens in problematic numbers. Contact recreation in pathogen-contaminated waters has been studied at many locations. The sources of the pathogens are typically assumed to be sanitary sewage effluent or periodic industrial discharges from certain food preparation industries (especially meat packing and fish and shellfish processing). However, several studies have investigated pathogen problems associated with stormwater discharges. It has generally been assumed that the source of the pathogens in the stormwater is inappropriate sanitary connections. However, stormwater unaffected by these inappropriate sources still contains high counts of pathogens that are also found in surface runoff samples from many urban surfaces. Needless to say, sewage contamination of urban streams is an important issue that needs attention during an urban water assessment investigation. Obviously, inappropriate discharges must be identified and corrected as part of any effort to clean up urban streams. If these sources are assumed to be nonexistent in an area and are therefore not considered in the stormwater management activities, incorrect and inefficient management decisions are likely, with disappointing improvements in the receiving waters.

A number of issues emerged from the individual projects of the U.S. EPA's NURP (EPA 1983a). One of these issues involved illicit connections to storm drainage systems and was summarized as follows in the Final Report of the NURP executive summary: "A number of the NURP projects identified what appeared to be illicit connections of sanitary discharges to stormwater sewer systems, resulting in high bacterial counts and dangers to public health. The costs and complications of locating and eliminating such connections may pose a substantial problem in urban areas, but the opportunities for dramatic improvement in the quality of urban stormwater discharges certainly exist where this can be accomplished. Although not emphasized in the NURP effort, other than to assure that the selected monitoring sites were free from sanitary sewage contamination, this BMP (best management practice) is clearly a desirable one to pursue." The illicit discharges noted during NURP were especially surprising, because the monitored watersheds were carefully selected to minimize factors other than stormwater. Presumably, illicit discharge problems in typical watersheds would be much worse. Illicit entries into urban storm sewerage were identified by flow from storm sewer outfalls following substantial dry periods. Such flow could be the result of direct "illicit connections" as mentioned in the NURP final report, or could result from indirect connections (such as contributions from leaky sanitary sewerage through infiltration to the separate storm drainage). Many of these dry-weather flows are continuous and would therefore also occur during rain-induced runoff periods. Pollutant contributions from the dry-weather flows in some storm drains have been shown to be high enough to significantly degrade water quality because of their substantial contributions to the annual mass pollutant loadings to receiving waters.

In many cases, sanitary sewage was an important component (although not necessarily the only component) of the dry-weather discharges from storm drainage systems that have been investigated. From a human health perspective (associated with pathogens), it may not require much raw or poorly treated sewage to cause a receiving water problem. However, at low discharge rates, the DO receiving water levels may be minimally affected. The effects these discharges have on the receiving waters is therefore highly dependent on many site-specific factors, including frequency and quantity of sewage discharges and the creek flows. In many urban areas, the receiving waters are small creeks in completely developed watersheds. These creeks are the most at risk from these discharges as dry baseflows may be predominantly dry-weather flows from the drainage systems. In Tokyo (Fujita 1998), for example, numerous instances were found where correcting inappropriate sanitary sewage discharges resulted in the urban streams losing all of their flow. In cities adjacent to large receiving waters, these discharges likely have little impact (such as DO impacts from Nashville CSO discharges on the Cumberland River; Cardozo et al. 1994). The presence of pathogens from raw or poorly treated sewage in urban streams, however, obviously presents a potentially serious public health threat. Even if the receiving waters are not designated as water contact recreation, children often play in small city streams.

Assessment Strategies for Identifying Inappropriate Discharges to Storm Drainage

The following is a summary of the strategy developed by Lalor (1993) and Pitt et al. (1993) for the EPA to support the outfall screening activities required by the National Pollutant Discharge Elimination System (NPDES) Stormwater Permit Program to identify and correct inappropriate discharges to storm drainage systems. Those documents should be consulted for more detailed information. The methods summarized here require the use of multiple indicators used in combination. The evaluation procedures outlined range from the most basic, requiring minimal information, to more complex, requiring additional analyses.

The detection and identification of flow components require the quantification of specific characteristics of the observed combined flow. Lalor (1993) developed a simple test suite that tested very reliably in field verification trials. This method requires the analysis of detergents, fluoride, ammonia, and potassium, plus noting obvious indicators. The characteristics of most interest should be relatively unique for each potential flow source. This will enable the presence of each flow source to be indicated, based on the presence (or absence) of these unique characteristics. The selected characteristics are termed *tracers*, because they have been selected to enable the identification of the sources of these waters. These methods can be used in many areas, although the selection of the specific tracers might vary if the likely source flows are different. This section also discusses other methods used to indicate sources of contaminants, such as fingerprinting hydrocarbon residuals and newly available analytical methods that are very specific to individual sources.

Investigations designed to determine the contribution of urban stormwater runoff to receiving water quality problems have led to a continuing interest in inappropriate connections to storm drainage systems. Urban stormwater runoff is traditionally defined as that portion of precipitation which drains from city surfaces and flows via natural or man-made drainage systems into receiving waters. In fact, urban stormwater runoff also includes waters from many other sources which find their way into storm drainage systems. Sources of some of this water can be identified and accounted for by examining current NPDES permit records for permitted industrial wastewaters that can be legally discharged to the storm drainage system. However, most of the water comes from other sources, including illicit and/or inappropriate entries to the storm drainage system. These entries can account for a significant amount of the pollutants discharged from storm sewerage systems (Pitt and McLean 1986).

In response to the early studies that indicated the importance of stormwater discharge effects on receiving waters, provisions of the Clean Water Act (1987) now require NPDES permits for stormwater discharges. Permits for municipal separate storm sewers include a requirement to effectively prohibit problematic nonstormwater discharges, thereby placing emphasis on the elimination of inappropriate connections to urban storm drains. Section 122.26 (d)(1)(iv)(D) of the rule specifically requires an initial screening program to provide means for detecting high levels of pollutants in dry-weather flows, which should serve as indicators of illicit connections to the storm sewers. To facilitate the application of this rule, the EPA's Office of Research and Development's Storm and Combined Sewer Pollution Control Program and the Environmental Engineering & Technology Demonstration Branch, along with the Office of Water's Nonpoint Source Branch, supported research for the investigation of inappropriate entries to storm drainage systems (Pitt et al. 1993). This research was designed to provide information and guidance to local agencies by (1) identifying and describing the most common potential sources of nonstormwater pollutant entries into storm drainage systems; and (2) developing an investigative methodology that would allow a user to determine whether significant nonstormwater entries are present in a storm drain, and then to identify the type of source, as an aid to determining the location of the source. An important premise for the development of this methodology was that the initial field screening effort would require minimal effort and expense, but would have little chance of missing a seriously contaminated outfall. This screening program would then be followed by a more in-depth analysis to more accurately determine the significance and source of the nonstormwater pollutant discharges.

The approach presented in this research was based on the identification and quantification of clean baseflow and the contaminated components during dry weather. If the relative amounts of potential components are known, then the importance of the dry-weather discharge can be determined. As an example, if a baseflow is mostly uncontaminated groundwater, but contains 5% raw sanitary wastewater, it is likely an important source of pathogenic bacteria. Typical raw sanitary wastewater parameters (such as BOD_5 or suspended solids) would be in low concentrations and the sanitary wastewater source would be difficult to detect. Fecal coliform bacteria measurements would not help much because they originate from many possible sources, in addition to sanitary wastewater. Expensive unique microorganism or biochemical measurements would probably be needed to detect the presence of the wastewater directly. However, a tracer may be identified that can be used to identify relatively low concentrations of important source flows in storm drain dry-weather baseflows.

The ideal tracer should have the following characteristics:

- Significant difference in concentrations between possible pollutant sources
- Small variations in concentrations within each likely pollutant source category
- A conservative behavior (i.e., no significant concentration change due to physical, chemical, or biological processes)
- Ease of measurement with adequate detection limits, good sensitivity, and repeatability

In order to identify tracers meeting the above criteria, literature characterizing potential inappropriate entries into storm drainage systems was examined. Several case studies which identified procedures used by individual municipalities or regional agencies were also examined. Though most of the investigations resorted to expensive and time-consuming smoke or dye testing to locate individual illicit pollutant entries, a few provided information regarding test parameters or tracers. These screening tests were proven useful in identifying drainage systems with problems before the smoke and dye tests were used. The case studies also revealed the types of illicit pollutant entries most commonly found in storm drainage systems.

Selection of Parameters for Identifying Inappropriate Discharge Sources

Table 6.33 is an assessment of the usefulness of candidate field survey parameters in identifying different potential nonstormwater flow sources. Natural and domestic waters should be uncontaminated (except in the presence of contaminated groundwaters entering the drainage system, for example). Sanitary sewage, septage, and industrial waters can produce toxic or pathogenic conditions. The other source flows (wash and rinse waters and irrigation return flows) may cause nuisance conditions or degrade the ecosystem. The parameters marked with a plus sign can probably be used to identify the specific source flows by their presence. Negative signs indicate that the potential source flow probably does not contain the listed parameter in adverse or obvious amounts, and may help confirm the presence of the source by its absence. Parameters with both positive and negative signs for a specific source category would probably not be very helpful due to expected wide variations.

Fecal Coliform Bacteria as Indicators of Inappropriate Discharges of Sanitary Sewage

Several investigations have relied on fecal coliform measurements as indicators of sanitary sewage contamination of stormwater. However, the use of fecal coliforms has been shown to be

Table 6.33   Candidate Field Survey Parameters and Associated Non-Stormwater Flow Sources

Parameter	Natural Water	Potable Water	Sanitary Sewage	Septage Water	Indus. Water	Wash Water	Rinse Water	Irrig. Water
Fluoride	−	+	+	+	+/−	+	+	+
Hardness change	−	+/−	+	+	+/−	+	+	−
Surfactants	−	−	+	+	−	+	+	−
Fluorescence	−	−	+	+	−	+	+	−
Potassium	−	−	+	+	−	−	−	−
Ammonia	−	−	+	+	−	−	−	+/−
Odor	−	−	+	+	+	+/−	−	−
Color	−	−	−	−	+	−	−	−
Clarity	−	−	+	+	+	+	+/−	−
Floatables	−	−	+	−	+	+/−	+/−	−
Deposits and stains	−	−	+	−	+	+/−	+/−	−
Vegetation change	−	−	+	+	+	+/−	−	+
Structural damage	−	−	−	−	+	−	−	−
Conductivity	−	−	+	+	+	+/−	+	+
Temperature change	−	−	+/−	−	+	+/−	+/−	−
pH	−	−	−	−	+	−	−	−

Note: − implies relatively low concentration; + implies relatively high concentration; +/− implies variable conditions.

From Pitt et al. 1993.

an inadequate indicator of sewage except in gross contamination situations (see also Chapter 3). Low fecal coliform levels may also cause false negative findings, as was indicated during the Inner Grays Harbor study where a storm drain outfall with a confirmed domestic sewage connection was not found to have elevated fecal coliform levels (Pelletier and Determan 1988). High fecal coliform bacteria populations were observed at storm sewer outfalls at all times in both industrial and residential/commercial areas during a study in Toronto (Pitt and McLean 1986). During the warm-weather storm sampling period, surface sheetflows were shown to be responsible for most of the observations of bacteria at the outfalls. However, during cold weather, very few detectable surface snowmelt sheetflow or snow pack fecal coliform observations were obtained, while the outfall observations were still quite high. High fecal coliform bacteria populations were also observed during dry-weather flow conditions at the storm sewer outfalls during both warm and cold weather. Leaking, or cross-connected, sanitary sewerage was therefore suspected at both study areas. Contaminated sump-pump water (from poorly operating septic tank systems in medium-density residential areas) in the Milwaukee area has been noted as a potentially significant source of bacteria to storm drainage systems (R. Bannerman, WI DNR, personal communication).

The presence of bacteria in stormwater runoff, dry-weather flows, and in urban receiving waters has caused much concern, as described in Chapter 3. However, there are many potential sources of fecal coliforms in urban areas, besides sanitary sewage. Research projects conducted in Toronto, Ontario (Pitt and McLean 1986), and in Madison, WI (R. Bannerman, WI DNR, personal communication) have investigated the abundance of common indicator bacteria, potential pathogenic bacteria, and bacterial types that may indicate the source of bacterial contamination. The monitoring efforts included sampling from residential, industrial, and commercial areas. As in many previous studies, fecal coliforms were commonly found to exceed water quality standards by large amounts during the Toronto investigations. Fecal coliform populations were very large at all land uses investigated during warm weather (typical median outfall values were 10,000 to 30,000 organisms per 100 mL). Dry-weather baseflow fecal coliform populations were found to be statistically similar to the stormwater runoff populations. The cold-weather fecal coliform populations were much lower (300 to 10,000 per 100 mL), but still exceeded the water quality standards.

Samples were obtained from many potential sources, in addition to the outfall, during the Toronto study (Pitt and McLean 1986). Source area fecal coliform populations were very similar for different land uses for the same types of areas, but different source areas within the watersheds varied significantly. Generally, roof runoff had the lowest fecal coliform populations, while roads and roadside ditches had the largest populations.

The types and concentrations of different bacteria biotypes vary for different animal sources. Quresh and Dutka (1979) found that pathogenic bacteria biotypes are present in urban runoff and are probably from several different sources. The sources (nonhuman vs. human) of bacteria in urban runoff are difficult to determine. Geldreich and Kenner (1969) caution against using the ratio of fecal coliform to fecal streptococci as an indicator, unless the waste stream is known to be "fresh." Unfortunately, urban runoff bacteria may have been exposed to the environment for some time before rain washed it into the runoff waters. Delays may also be associated with some dry-weather bacteria sources. This aging process can modify the fecal coliform to fecal streptococci ratio to make the bacteria appear to be of human origin. In fact, samples collected in runoff source areas usually have the lowest FC/FS ratio in a catchment, followed by urban runoff, and finally the receiving water (Pitt 1983). This transition probably indicates an aging process and not a change in bacteria source.

Debbie Sargeant of the Washington State Department of Ecology has prepared a summary of different methods for fecal contamination source identification. Her report is available at www.ecy.wa.gov/biblio/99345.html. She concluded that there is no easy, low-cost method for differentiating between human and nonhuman sources of bacterial contamination. Genetic fingerprinting and newly emerging PCR methods, plus combinations of indicators, are some of the recommendations made in this report to further investigate bacterial sources.

Therefore, bacteria are usually poor indicators of the presence of sanitary sewage contamination. Past use of fecal strep to fecal coliform ratios to indicate human vs. nonhuman bacteria sources in mixed and old wastewaters (such as most nonpoint waters) has not been successful and should be used with extreme caution. There may be some value in investigating specific bacteria types, such as fecal strep biotypes, but much care should be taken in the analysis and interpretation of the results. A more likely indicator of human wastes may be the use of certain molecular markers, specifically the linear alkylbenzenes and fecal sterols, such as coprostanol and epicoprostanol (Eaganhouse et al. 1988), although these may also be discharged by other carnivores (especially dogs) in a drainage ditch. Recent discussions of specific tracers for indicating sanitary sewage contamination is presented later in this discussion. The following discussion presents a more generally useful approach for identifying inappropriate discharges to storm drainage, relying on easily evaluated chemical tracers and visual observations.

Tracer Characteristics of Local Source Flows

Table 6.34 is a summary of tracer parameter measurements for Birmingham, AL by Pitt et al. (1993). This table is a summary of the "library" that describes the tracer conditions for each potential source category. The important information shown on this table includes the median and coefficient of variation (COV) values for each tracer parameter for each source category. The COV is the ratio of the standard deviation to the mean. A low COV value indicates a much smaller spread of data compared to a data set having a large COV value. It is apparent that some of the generalized relationships shown in Table 6.33 did not exist during the demonstration project. This emphasizes the need for obtaining local data describing likely source flows.

The fluorescence values shown in Table 6.34 are direct measurements from a fluorometer having general-purpose filters and lamps and at the least sensitive setting (number 1 aperture). The toxicity screening test results are expressed as the toxicity response noted after 25 minutes of exposure using an Azur Environmental Microtox unit which measures toxicity using the light output from phosfluorescent algae. The I_{25} values are the percentage light output decreases observed after 25 minutes of exposure to the sample, compared to a reference. Fresh potable water has a relatively high toxicity response because of the chlorine levels present. Dechlorinated, potable water has much smaller toxicity responses.

Table 6.34   Tracer Concentrations Found in Birmingham, AL, Waters (Mean, Standard Deviation, and Coefficient of Variation, COV)

	Spring Water	Treated Potable Water	Laundry Wastewater	Sanitary Wastewater	Septic Tank Effluent	Car Wash Water	Radiator Flush Water
Fluorescence	6.8	4.6	1020	250	430	1200	22,000
(% scale)	2.9	0.35	125	50	100	130	950
	0.43	0.08	0.12	0.20	0.23	0.11	0.04
Potassium (mg/L)	0.73	1.6	3.5	6.0	20	43	2800
	0.070	0.059	0.38	1.4	9.5	16	375
	0.10	0.04	0.11	0.23	0.47	0.37	0.13
Ammonia (mg/L)	0.009	0.028	0.82	10	90	0.24	0.03
	0.016	0.006	0.12	3.3	40	0.066	0.01
	1.7	0.23	0.14	0.34	0.44	0.28	0.3
Ammonia/potassium	0.011	0.018	0.24	1.7	5.2	0.006	0.011
(ratio)	0.022	0.006	0.050	0.52	3.7	0.005	0.011
	2.0	0.35	0.21	0.31	0.71	0.86	1.0
Fluoride (mg/L)	0.031	0.97	33	0.77	0.99	12	150
	0.027	0.014	13	0.17	0.33	2.4	24
	0.87	0.02	0.38	0.23	0.33	0.20	0.16
Toxicity	<5	47	99.9	43	99.9	99.9	99.9
(% light decrease	n/a	20	<1	26	<1	<1	<1
after 25 min, I_{25})	n/a	0.44	n/a	0.59	n/a	n/a	n/a
Surfactants	<0.5	<0.5	27	1.5	3.1	49	15
(mg/L as MBAS)	n/a	n/a	6.7	1.2	4.8	5.1	1.6
	n/a	n/a	0.25	0.82	1.5	0.11	0.11
Hardness (mg/L)	240	49	14	140	235	160	50
	7.8	1.4	8.0	15	150	9.2	1.5
	0.03	0.03	0.57	0.11	0.64	0.06	0.03
pH (pH units)	7.0	6.9	9.1	7.1	6.8	6.7	7.0
	0.05	0.29	0.35	0.13	0.34	0.22	0.39
	0.01	0.04	0.04	0.02	0.05	0.03	0.06
Color (color units)	<1	<1	47	38	59	220	3000
	n/a	n/a	12	21	25	78	44
	n/a	n/a	0.27	0.55	0.41	0.35	0.02
Chlorine (mg/L)	0.003	0.88	0.40	0.014	0.013	0.070	0.03
	0.005	0.60	0.10	0.020	0.013	0.080	0.016
	1.6	0.68	0.26	1.4	1.0	1.1	0.52
Specific conductivity	300	110	560	420	430	485	3300
(μS/cm)	12	1.1	120	55	311	29	700
	0.04	0.01	0.21	0.13	0.72	0.06	0.22
Number of samples	10	10	10	36	9	10	10

From Pitt et al. 1993.

Appropriate tracers are characterized by having significantly different concentrations in flow categories that need to be distinguished. In addition, effective tracers also need low COV values within each flow category. Table 6.33 shows the expected changes in concentrations per category, and Table 6.34 indicates how these expectations compared with the results of an extensive local sampling effort. The study indicated that the COV values were quite low for each category, with the exception of chlorine, which had much greater COV values. Chlorine is therefore not recommended as a quantitative tracer to estimate the flow components. Similar data should be collected in each community where these procedures are to be used. Recommended field observations include color, odor, clarity, presence of floatables and deposits, and rate of flow, in addition to chemical measurements for fluoride, potassium, ammonia, and detergents (or fluorescence).

Collection of Samples and Field Analyses

All outfalls should be evaluated, not just those larger than a certain size. Lalor (1994) found that the smallest outfalls were typically the most contaminated because they were likely to be associated with creek-side small automotive businesses that improperly disposed of their wastes through small pipes. Figure 6.113 illustrates the simple sample collection methods used. The creeks are walked and all outfalls observed are evaluated. Generally, three-person crews are used, two walking the creek with waders, sampling equipment, and notebooks, while the third person drives the car to the next downstream meeting location (typically about ¹/₂ mile). It requires several (typically at least three) trips along a stream to find all the outfalls. Multiple sampling visits are also needed throughout the year to verify changing discharge conditions. Outfalls may be dry during some visits, but flowing during others.

We have found it to be much more convenient and efficient to collect samples in the field and return them to the laboratory where groups of samples can be evaluated together. Some simple field analyses are appropriate. Figure 6.114 shows a portable gas analyzer that can indicate explosive conditions, lack of oxygen, and the presence of H_2S. This is important from a safety standpoint in areas having little ventilation, and the H_2S can also be used to indicate sewage problems. Most of the field test kits examined during this research (and as summarized earlier in this chapter) would take much too long to conduct correctly and safely in the field.

Simple Data Evaluation Methods to Indicate Sources of Contamination

Negative Indicators Implying Contamination

Indicators of contamination (negative indicators) are clearly apparent visual or physical parameters indicating obvious problems and are readily observable at the outfall during the field screening activities. Relying only on these indicators can lead to an unacceptably high rate of false negatives and false positives and must therefore be supplemented with additional confirmatory methods. However, these indicators are easy to measure, are useful for indicating gross contamination, are easy to describe to nontechnical decision makers, and are therefore highly recommended as an important part of a field screening effort.

These observations are very important during the field survey because they are the simplest method of identifying grossly contaminated dry-weather outfall flows. The direct examination of outfall characteristics for unusual conditions of flow, odor, color, turbidity, floatables, deposits/stains, vegetation conditions, and damage to drainage structures is therefore an important part of these investigations. Table 6.35 presents a summary of these indicators, along with narratives of the descriptors to be selected in the field.

Figure 6.113 Collecting outfall samples for inappropriate discharge evaluations.

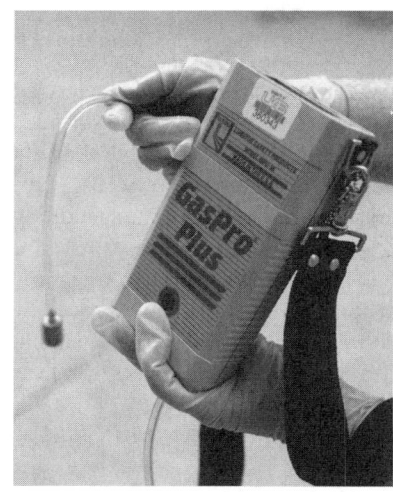

Figure 6.114 Portable gas analyzer for H_2S and explosive conditions.

Table 6.35   Interpretations of Physical Observation Parameters and Possible Associated Flow Sources

Odor — Most strong odors, especially gasoline, oils, and solvents, are likely associated with high responses on the toxicity screening test. Typical obvious odors include gasoline, oil, sanitary wastewater, industrial chemicals, decomposing organic wastes, etc.

sewage: smell associated with stale sanitary wastewater, especially in pools near outfall or septic system drainage.

sulfur ("rotten eggs"): industries that discharge sulfide compounds or organics (meat packers, canneries, dairies, etc.).

oil and gas: petroleum refineries or many facilities associated with vehicle maintenance or petroleum product storage.

rancid-sour: food preparation facilities (restaurants, hotels, etc.).

Color — Important indicator of inappropriate industrial sources. Industrial dry-weather discharges may be of any color, but dark colors, such as brown, gray, or black, are most common.

yellow: chemical plants, textile and tanning plants.

brown: meat packers, printing plants, metal works, stone and concrete, fertilizers, and petroleum refining facilities.

green: chemical plants, textile facilities.

red: meat packers or iron oxide from groundwater seeps, e.g., acid mine drainage.

gray: dairies, sewage.

Turbidity — Often affected by the degree of gross contamination. Dry-weather industrial flows with moderate turbidity can be cloudy, while highly turbid flows can be opaque. High turbidity is often a characteristic of undiluted dry-weather industrial discharges or soil erosion.

cloudy: sanitary wastewater, concrete or stone operations, fertilizer facilities, automotive dealers.

opaque: food processors, lumber mills, metal operations, pigment plants.

Floatable Matter — A contaminated flow may contain floating solids or liquids directly related to industrial, sanitary wastewater pollution, or agricultural feed lots. Floatables of industrial origin may include animal fats, spoiled food, oils, solvents, sawdust, foams, packing materials, or fuel.

oil sheen: petroleum refineries or storage facilities and vehicle service facilities.

sewage: sanitary wastewater.

Deposits and Stains — Refers to any type of coating near the outfall and are usually of a dark color. Deposits and stains often will contain fragments of floatable substances. These situations are illustrated by the grayish-black deposits that contain fragments of animal flesh and hair which often are produced by leather tanneries, or the white crystalline powder which commonly coats outfalls due to nitrogenous fertilizer wastes.

sediment: construction site or agricultural soil erosion.

oily: petroleum refineries or storage facilities, vehicle service facilities, and large parking lot runoff.

Vegetation — Vegetation surrounding an outfall may show the effects of industrial pollutants. Decaying organic materials coming from various food product wastes would cause an increase in plant life, while the discharge of chemical dyes and inorganic pigments from textile mills could noticeably decrease vegetation. It is important not to confuse the adverse effects of high stormwater flows on vegetation with highly toxic dry-weather intermittent flows.

excessive growth: food product facilities, sewage, or agricultural operations.

inhibited growth: high stormwater flows, beverage facilities, printing plants, metal product facilities, drug manufacturing, petroleum facilities, vehicle service facilities and automobile dealers, pesticide spraying.

Damage to Outfall Structures — Another readily visible indication of industrial contamination. Cracking, deterioration, and spalling of concrete or peeling of surface paint, occurring at an outfall are usually caused by severely contaminated discharges, usually of industrial origin. These contaminants are usually very acidic or basic in nature. Primary metal industries have a strong potential for causing outfall structural damage because their batch dumps are highly acidic. Poor construction, hydraulic scour, and old age may also adversely affect the condition of the outfall structure.

concrete cracking: industrial flows.

concrete spalling: industrial flows.

peeling paint: industrial flows.

metal corrosion: industrial flows.

From Pitt et al. 1993.

This method does not allow quantifiable estimates of the flow components, and it will very likely result in many incorrect negative determinations (missing outfalls that have important levels of contamination). These simple characteristics are most useful for identifying gross contamination. Only the most significant outfalls and drainage areas would therefore be recognized from this method. The other methods, requiring chemical determinations, can be used to quantify the flow contributions and to identify the less obviously contaminated outfalls.

Indications of intermittent flows (especially stains or damage to the structure of the outfall) could indicate serious illegal toxic pollutant entries into the storm drainage system that will be very difficult to detect and correct. Highly irregular dry-weather outfall flow rates or chemical characteristics could indicate industrial or commercial inappropriate entries into the storm drain system.

Correlation tests were conducted to identify relationships between outfalls that were known to have severe contamination problems and the negative indicators (Lalor 1994). Pearson correlation tests indicated that high turbidity (lack of clarity) and odors appeared to be the most useful physical indicators of contamination when contamination was defined by toxicity and the presence of detergents. Lack of clarity best indicated the presence of detergents, with an 80% correlation. As noted later, the detergent test was the single most useful of the chemical tests for distinguishing between contaminated and uncontaminated flows. The Pearson correlation tests also showed that noticeable odor was the best indicator of toxicity, with a 77% correlation. There is no theoretical connection between the physical indicators and these problems. High turbidity was noted in 74% of the contaminated source flow samples. This represented a 26% false negative rate (indication of no contamination when contamination actually exists), if one relied on turbidity alone as an indicator of contamination. High turbidity was noted in only 5% of the uncontaminated source flow samples. This represents the rate of false positives (indication of contamination when none actually exists) when relying on turbidity alone. Noticeable odor was indicated in 67% of flow samples from contaminated sources, but in none of the flow samples from uncontaminated sources. This translates to 37% false negatives, but no false positives. Obvious odors identified included gasoline, oil, sanitary wastewater, industrial chemicals or detergents, decomposing organic wastes, etc. A 65% correlation was also found to exist between color and Microtox toxicity. Color is an important indicator of inappropriate industrial sources, but it was also associated with some of the residential and commercial flow sources. Color was noted in 100% of the flow samples from contaminated sources, but it was also noted in 40% of the flow samples from uncontaminated sources. This represents 60% false positives, but no false negatives. Finally, a 63% correlation between the presence of sediments (assessed as settleable solids in the collection bottles of these source samples) and Microtox toxicity was also found. Sediments were noted in 34% of the samples from contaminated sources and in none of the samples from uncontaminated sources.

False negatives are more of a concern than a reasonable number of false positives when working with a screening methodology. Screening methodologies are used to direct further, more detailed investigations. False positives would be discarded after further investigation. However, a false negative during a screening investigation results in the dismissal of a problem outfall for at least the near future. Missed contributors to stream contamination may result in unsatisfactory in-stream results following the application of costly corrective measures elsewhere.

The method of using physical characteristics to indicate contamination in outfall flows does not allow quantifiable estimates of the flow components and, if used alone, will likely result in many incorrect determinations, especially false negatives. However, these simple characteristics are most useful for identifying gross contamination: only the most significantly contaminated outfalls and drainage areas would therefore be recognized using this method.

Detergents as Indicators of Contamination

Results from Mann–Whitney U tests (at the $\alpha = 0.05$ confidence level) indicated that samples from any of the dry-weather flow sources could be correctly classified as clean or contaminated based only on the measured value of any one of the following parameters: detergents, color, or conductivity (Lalor 1994). Color and high conductivity were present in samples from clean sources as well as contaminated sources, but their levels of occurrence were significantly different between the two groups. If samples from only one source were expected to make up outfall flows, the level of color or conductivity could be used to distinguish contaminated outfalls from clean outfalls. However, since multisource flows occur, measured levels of color or conductivity could fall within acceptable levels because of dilution, even though a contaminating source was contributing to the flow. Detergents (anionic surfactants), on the other hand, can be used to distinguish between clean and contaminated outfalls simply by their presence or absence, using a detection limit of 0.06 mg/L. All samples analyzed from contaminated sources contained detergents in excess of this amount (with the exception of three septage samples collected from homes discharging only toilet flushing water). No clean source samples were found to contain detergents. Contaminated sources would be detected in mixtures with uncontaminated waters if they made up at least 10% of the mixture.

The HACH detergents test was used during these analyses and was found to work very well. Unfortunately, this test uses a large amount of benzene for sample extractions and so great care is needed with the analysis and waste disposal. Only the most highly trained analysts, understanding the dangers of using benzene, should be allowed to use this test. An alternative method examined by CHEMetrics uses relatively small amounts of chloroform (well contained) for sample extractions and is therefore much safer, although care is also needed during the test and in disposal of waste. However, this method has a poorer detection limit (about 0.15 mg/L) than the HACH method, leading to less sensitivity (and possible false negatives).

Because of the hazardous problems associated with using these simple detergent (anionic surfactant) tests, we have investigated numerous alternative, but related, tests. Standard tests for boron are relatively simple, safe, and sensitive. Historically, boron was an important component in laundry detergents and tests were conducted to see if this analysis would be a suitable substitute for the detergent tests. Unfortunately, boron appears to have been replaced in most U.S. detergents, as numerous tests of commercial laundry products found little boron. In addition, boron tests of sewage mixtures and from numerous mixed wastewaters from throughout the country also indicated little boron. Fluorescence of test waters, using an extremely sensitive, but expensive, fluorometer (Turner 10-AU), was also evaluated, but with mixed results. The analyses of sewage and detergents found highly variable fluorescence values because of the highly variable amounts of fabric whiteners found in detergents. However, it is possible to use fluorescence as a good presence/absence test, like the initial detergent evaluations. The previous discussion of optical brighteners (as a field test kit) indicated the potential usefulness of this method, but more work is needed to determine its sensitivity. As indicated later, more sophisticated tests for detergent components (LAS and perfumes, especially) have been successfully used as sewage tracers in many waters, but these analyses require expensive and time-consuming HPLC analyses.

Simple Checklist for Major Flow Component Identification

Table 6.36 is a simplification of the analysis strategy to separate the major nonstormwater discharge sources for areas having no industrial activity. The first indicator is the presence or absence of flow. If no dry-weather flow exists at an outfall, then indications of intermittent flows must be investigated. Specifically, stains, deposits, odors, unusual stream-side vegetation conditions, and outfall structural damage can all indicate intermittent nonstormwater flows. However, multiple visits to outfalls over long time periods are needed to confirm that only stormwater flows occur. The following paragraphs summarize the rationale used to distinguish between treated potable water and sanitary wastewater, the two most common dry-weather flow sources in storm drainage systems in residential and commercial areas.

Table 6.36   Simplified Checklist to Identify Residential Area Non-Stormwater Flow Sources

1. Flow? If yes, go to 2; if no, go to 3.
2. Fluorides (or different hardness)? If yes, probably treated water (may be contaminated), go to 4; if no, then untreated natural water (probably uncontaminated), or untreated industrial water (may be toxic).
3. Check for intermittent dry-weather flow signs (may be contaminated). If yes, recheck outfall at later date; if no, then not likely a significant non-stormwater source.
4. Surfactants (or fluorescence)? If yes, may be sanitary wastewater, laundry water, or other wash water (may be pathogenic, or nuisance), go to 5; if no, then may be domestic water line leak, irrigation runoff, or rinse water (probably not a contaminated non-stormwater source, but may be a nuisance).
5. Elevated potassium (or ammonia)? If yes, then likely sanitary wastewater source (pathogenic); if no, likely wash water (probably not a contaminated non-stormwater source, but may be a nuisance).

From Pitt et al. 1993.

Treated Potable Water — A number of tracer parameters may be useful for distinguishing treated potable water from natural waters:

- Major ions or other chemical/physical characteristics of the flow components can vary substantially, depending on whether the water supply sources are groundwater or surface water, and whether the sources are treated or not. Specific conductance may also serve as an indicator of the major water source.
- Fluoride can often be used to separate treated potable water from untreated water sources. This latter group may include local springs, groundwater, regional surface flows, or nonpotable industrial waters. If the treated water has no fluoride added, or if the natural water has fluoride concentrations close to potable water fluoride concentrations, then fluoride may not be an appropriate indicator. Water from treated water supplies (that test positive for fluorides or other suitable tracers) can be relatively uncontaminated (domestic water line leakage or irrigation runoff), or it may be heavily contaminated. If the drainage area has industries that have their own water supplies (quite rare for most urban drainage areas), further investigations are needed to check for industrial nonstormwater discharges. Toxicity screening methods would be very useful in areas known to have commercial or industrial activity, or to check for intermittent residential area discharges of toxicants. Fluoride can be very high in some commercial wash waters and industrial wastewaters.
- Hardness can also be used as an indicator if the potable water source and the baseflow are from different water sources. An example would be if the baseflow is from hard groundwater and the potable water is from softer surface supplies.
- If the concentration of chlorine is high, then a major leak of disinfected potable water is probably close to the outfall. Because of the rapid loss of chlorine in water (especially if some organic contamination is present), it is not a good parameter for quantifying the amount of treated potable water observed at the outfall.

Water from potable water supplies (that test positive for fluorides, or other suitable tracers) can be relatively uncontaminated (domestic water line leakage or irrigation runoff) or heavily contaminated (sanitary wastewater).

Sanitary Wastewaters — In areas containing no industrial or commercial sources, sanitary wastewater is probably the most important dry-weather source of storm drain flows. In addition, septic systems often do not operate properly and might be a significant source of contamination in rural areas. The following parameters can be used for quantifying the sanitary wastewater components of the treated domestic water portion:

- Surfactant (detergent) analyses may be useful in determining the presence of sanitary wastewaters, as noted previously. However, surfactants present in water originating from potable water sources could indicate sanitary wastewaters, laundry wastewaters, car washing wastewater, or any other waters containing surfactants. If surfactants are not present, then the potable water could be relatively uncontaminated (domestic water line leaks or irrigation runoff).

- The presence of fabric whiteners (as measured by fluorescence) can also be used in distinguishing laundry and sanitary wastewaters.
- Sanitary wastewaters often exhibit predictable trends during the day in flow and quality. In order to maximize the ability to detect direct sanitary wastewater connections into the storm drainage system, it would be best to survey the outfalls during periods of highest sanitary wastewater flows (mid to late morning hours).
- The ratio of surfactants to ammonia or potassium concentrations may be an effective indicator of the presence of sanitary wastewaters or septic tank effluents. If the surfactant concentrations are high, but the ammonia and potassium concentrations are low, the contaminated source may be laundry wastewaters. Conversely, if ammonia, potassium, and surfactant concentrations are all high, sanitary wastewater is the likely source. Some researchers have reported low surfactants in septic tank effluents. Therefore, if surfactants are low but potassium and ammonia are both high, septic tank effluent may be present. However, research in the Birmingham, AL, area found high surfactant concentrations in septic tank effluent, further stressing the need to obtain local characterization data for potential contaminating sources.
- Obviously, odor and other physical appearances such as turbidity, coarse and floating "tell-tale" solids, foaming, color, and temperature would also be very useful in distinguishing sanitary wastewater from wash water or laundry wastewater sources, as noted previously. However, these indicators may not be very obvious for small levels of sanitary wastewater contamination.

Flowchart for Most Significant Flow Component Identification

A further refinement of the above checklist is the flowchart shown on Figure 6.115. This flow chart describes an analysis strategy that may be used to identify the major component of dry-weather flow samples in residential and commercial areas. This method does not attempt to distinguish among all potential sources of dry-weather flows identified earlier, but rather the following four major groups of flow are identified: (1) tap waters (including domestic tap water, irrigation water, and rinse water), (2) natural waters (spring water and shallow groundwater), (3) sanitary wastewaters (sanitary sewage and septic tank discharge), and (4) wash waters (commercial laundry waters, commercial car wash waters, radiator flushing wastes, and plating bath wastewaters). The use of this method would not only allow outfall flows to be categorized as contaminated or uncontaminated, but would also allow outfalls carrying sanitary wastewaters to be identified. These outfalls could then receive the highest priority for further investigation leading to source control. This flowchart was designed for use in residential and/or commercial areas only. Investigations in industrial or industrial/commercial land use areas must be approached in an entirely different manner.

In residential and/or commercial areas, all outfalls should be located and examined. The first indicator is the presence or absence of dry-weather flow. If no dry-weather flow exists at an outfall, indications of intermittent flows must be investigated. Specifically, stains, deposits, odors, unusual stream-side vegetation conditions, and damage to outfall structures can all indicate intermittent nonstormwater flows. However, frequent visits to outfalls over long time periods, or the use of other monitoring techniques, may be needed to confirm that only stormwater flows occur. If intermittent flow is not indicated, the outfall probably does not have a contaminated nonstormwater source. The other points on the flowchart serve to indicate if a major contaminating source is present, or if the water is uncontaminated. Component contributions cannot be quantified using this method, and only the "most contaminated" type of source present will be identified.

If dry-weather flow exists at an outfall, then the flow should be sampled and tested for detergents. If detergents are not present, the flow is probably from a noncontaminated nonstormwater source. The lower limit of detection for detergent should be about 0.06 mg/L.

If detergents are not present, fluoride levels can be used to distinguish between flows with treated water sources and flows with natural sources in communities where water supplies are fluoridated and natural fluoride levels are low. In the absence of detergents, high fluoride levels

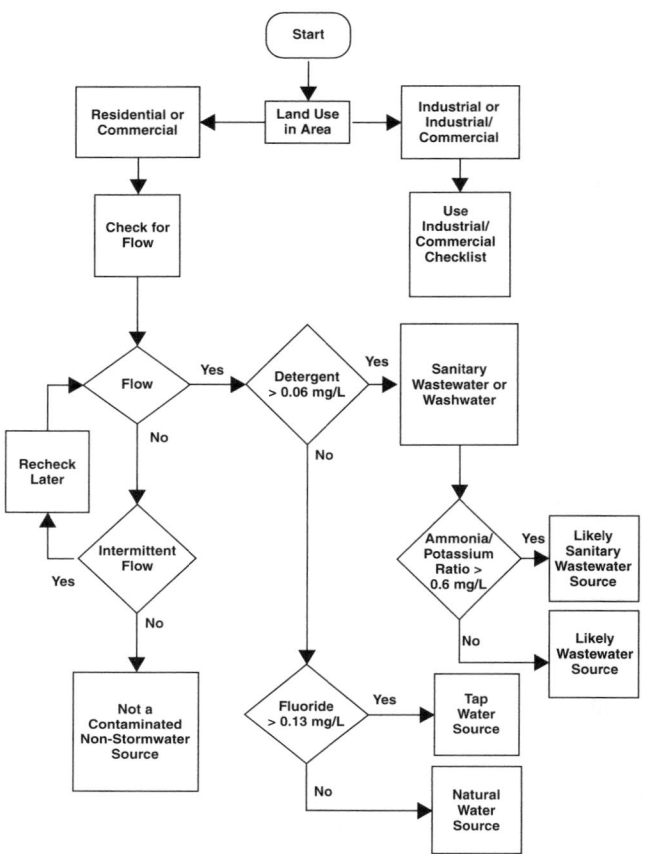

Figure 6.115 Simple flowchart method to identify significant contaminating sources. (From Lalor, M. *An Assessment of Non-Stormwater Discharges to Storm Drainage Systems in Residential and Commercial Land Use Areas.* Ph.D. dissertation. Department of Civil and Environmental Engineering. Vanderbilt University. 1994. With permission.)

would indicate a potable water line leak, irrigation water, or wash/rinse water. Low fluoride levels would indicate waters originating from springs or shallow groundwater. Based on the flow source samples tested in this research (Table 6.34), fluoride levels above 0.13 mg/L would most likely indicate that a tap water source was contributing to the dry-weather flow in the Birmingham, AL, study area.

If detergents are present, the flow is probably from a contaminated nonstormwater source, as indicated on Table 6.34. The ratio of ammonia to potassium can be used to indicate whether or not the source is sanitary wastewater. Ammonia/potassium ratios greater than 0.60 would indicate likely sanitary wastewater contamination. Ammonia/potassium ratios were above 0.9 for all septage and sewage samples collected in Birmingham (values ranged from 0.97 to 15.37, averaging 2.55). Ammonia/potassium ratios for all other samples containing detergents were below 0.7, ranging from 0.00 to 0.65, averaging 0.11. One radiator waste sample had an ammonia/potassium ratio of 0.65.

Noncontaminated samples collected in Birmingham had ammonia/potassium ratios ranging from 0.00 to 0.41, with a mean value of 0.06 and a median value of 0.03. Using the mean values for noncontaminated samples (0.06) and sanitary wastewaters (2.55), flows comprised of mixtures containing at least 25% sanitary wastes with the remainder of the flow from uncontaminated sources would likely be identified as sanitary wastewaters using this method. Flows containing a smaller percentage of contributions from sanitary wastewaters might be identified as having a wash water source, but would not be identified as uncontaminated.

General Matrix Algebra Methods to Indicate Sources of Contamination through Fingerprinting

Other approaches can also be used to calculate the source components of mixed outfall flows. One approach is the use of matrix algebra to simultaneously solve a series of chemical mass balance equations. This method can be used to predict the most likely flow source, or sources, making up an outfall sample. It is possible to estimate the outfall source flow components using a set of simultaneous equations. The number of unknowns should equal the number of equations available, resulting in a square matrix. If there are seven likely source categories, then there should be seven tracer parameters used. If there are only four possible sources, then only the four most efficient tracer parameters should be used. Only tracers that are linearly related to mixture components can be used. As an example, pH cannot be used in these equations, because it is not additive.

Further site-specific statistical analyses may be needed to rank the usefulness of the tracers for distinguishing different flow sources. As an example, chlorine is generally not useful for these analyses because the concentration variability within many source categories is high (it is also not a conservative parameter). Chlorine may still be a useful parameter, but only to identify possible large potable water line leaks. Another parameter having problems for most situations is pH. The variation of pH between sources is very low (they are all very similar). pH may still be useful to identify industrial wastewater problems, but it cannot be used to quantify flow components. Toxicity is another parameter used during this research that was found not to be linearly additive.

This method estimates flow contributions from various sources using a "receptor model," based on a set of chemical mass balance equations. Such models, which assess the contributions from various sources based on observations at sampling sites (the receptors), have been applied to the investigation of air pollutant sources for many years (Lee et al. 1993; Cooper and Watson 1980). The characteristic "signatures" of the different types of sources, as identified in the library of source flow data, allow the development of a set of mass balance equations. These equations describe the measured concentrations in an outfall's flow as a linear combination of the contributions from the different potential sources. A major requirement for this method is the physical and chemical characterization of waters collected directly from potential sources of dry-weather flows (the "library"). This allows concentration patterns (fingerprints) for the parameters of interest to be established for each type of source. Theoretically, if these patterns are different for each source, the observed concentrations at the outfall would be a linear combination of the concentration patterns from the different component sources, each weighted by a source strength term (m_n). This source strength term would indicate the fraction of outfall flow originating from each likely source. By measuring a number of parameters equal to, or greater than, the number of potential source types, the source strength term could be obtained by solving a set of chemical mass balance equations of the type:

$$C_p = \sum_n m_n x_{pn}$$

where C_p is the concentration of parameter p in the outfall flow and x_{pn} is the concentration of parameter p in source type n.

As an example of this method, consider eight possible flow sources and eight parameters, as presented in Table 6.37. The number of parameters evaluated for each outfall must equal the number of probable dry-weather flow sources in the drainage area. Mathematical methods are available which provide for the solution of over-specified sets of equations (more equations than unknowns), but these are not addressed here.

The selection of parameters for measurement should reflect evaluated parameter usefulness. Evaluation of the Mann–Whitney U Test results (Lalor 1994) suggested the following groupings of parameters, ranked by their usefulness for distinguishing between all the types of flow sources sampled in Birmingham, AL:

- First set (most useful): potassium and hardness
- Second set: fluorescence, conductivity, fluoride, ammonia, detergents, and color
- Third set (least useful): chlorine

If parameter variations within the sources are not accounted for, the equations would take the form presented in Table 6.38. Here, the x terms, representing parameter concentrations within the specified source, have been replaced with the mean concentrations noted in the source library. After measured values are substituted into the equations for parameter concentrations in the outfall flow (C_p), this set of simultaneous equations can be solved using matrix algebra. The use of mean concentration values in the equation set was evaluated by entering the potential dry-weather flow source samples from Birmingham as unknowns (as if they were outfall samples) and solving for fractions of flow (the m terms in Table 6.38). This exercise resulted in four false negatives (6%) and 27 false positives (73%). The results of these simple preliminary tests indicated that there was too much variation of parameter concentrations within the various source types to allow them to be adequately characterized by simple use of the mean concentrations alone. Another method, recognizing variations in source flow characteristics in a Monte Carlo model, is presented by Lalor (1994). Both of these methods listed the likely multiple contaminating sources and estimated their relative contributions. Unfortunately, confirmation testing indicated inaccurate results much of the time, implying the greater usefulness of the simpler methods described previously. However, these matrix algebra methods may be very useful in other situations or locations and should be investigated as part of a local screening project to identify inappropriate discharges to storm drainage.

There are numerous other statistical analysis methods suitable for identifying sources of flows. Salau et al. (1997) present a review of several advanced statistical methods also derived from air pollution source identification research (see Chapter 7 for illustrations from his paper). Principal component analysis and hierarchical cluster analysis are shown as tools that can identify common sources of contamination by examining a set of well-selected tracer compounds (in northwest Mediterranean marine sediments in their example). These are used to develop the alternating least squares approach, similar to Lalor's (1994) use of these same techniques to identify the best parameters for the simultaneous equation solutions described above.

Table 6.37   Set of Chemical Mass Balance Equations

	Source 1	Source 2	Source 3	Source 4	Source 5	Source 6	Source 7	Source 8 Outfall
Parameter 1:	(m1)(x11) + (m2)(x12) + (m3)(x13) + (m4)(x14) + (m5)(x15) + (m6)(x16) +(m7)(x17) + (m8)(x18) = C1
Parameter 2:	(m1)(x21) + (m2)(x22) + (m3)(x23) + (m4)(x24) + (m5)(x25) + (m6)(x26) +(m7)(x27) + (m8)(x28) = C2
Parameter 3:	(m1)(x31) + (m2)(x32) + (m3)(x33) + (m4)(x34) + (m5)(x35) + (m6)(x36) +(m7)(x37) + (m8)(x38) = C3
Parameter 4:	(m1)(x41) + (m2)(x42) + (m3)(x43) + (m4)(x44) + (m5)(x45) + (m6)(x46) +(m7)(x47) + (m8)(x48) = C4
Parameter 5:	(m1)(x51) + (m2)(x52) + (m3)(x53) + (m4)(x54) + (m5)(x55) + (m6)(x56) +(m7)(x57) + (m8)(x58) = C5
Parameter 6:	(m1)(x61) + (m2)(x62) + (m3)(x63) + (m4)(x64) + (m5)(x65) + (m6)(x66) +(m7)(x67) + (m8)(x68) = C6
Parameter 7:	(m1)(x71) + (m2)(x72) + (m3)(x73) + (m4)(x74) + (m5)(x75) + (m6)(x76) +(m7)(x77) + (m8)(x78) = C7
Parameter 8:	(m1)(x81) + (m2)(x82) + (m3)(x83) + (m4)(x84) + (m5)(x85) + (m6)(x86) +(m7)(x87) + (m8)(x88) = C8

Equations of the form $C_p = \sum_n m_n x_{pn}$

where: C_p = the concentration of parameter p in the outfall flow

m_n = the fraction of flow from source type n

x_{pn} = the mean concentration of parameter p in source type n

From Lalor, M. *An Assessment of Non-Stormwater Discharges to Storm Drainage Systems in Residential and Commercial Land Use Areas.* Ph.D. dissertation. Department of Civil and Environmental Engineering. Vanderbilt University. 1994. With permission.

Once sources are identified, it is important to confirm their source and to ensure that corrective action is undertaken. Figures 6.116 and 6.117 show TV surveying being conducted in Boston to confirm the likely source of inappropriate discharges. Normally, the TV camera is remotely operated and pulled through small pipes (Figure 6.116). However, in the coastal area and in large pipes, crews were required to conduct the surveys manually (Figure 6.117).

Emerging Tools for Identifying Sources of Discharges

Coprostanol and Other Fecal Sterol Compounds Utilized as Tracers of Contamination by Sanitary Sewage

A more likely indicator of human wastes than fecal coliforms and other "indicator" bacteria may be the use of certain molecular markers, specifically the fecal sterols, such as coprostanol and epicoprostanol (Eaganhouse et al. 1988). However, these compounds are also discharged by other carnivores (especially dogs) in a drainage. A number of research projects have used these compounds to investigate the presence of sanitary sewage contamination. The most successful application may be associated with sediment analyses instead of water analyses. As an example, water analyses of coprostanol are difficult due to the typically very low concentrations found, although the concentrations in many sediments are quite high and much easier to quantify. Unfortunately, the long persistence of these compounds in the environment easily confuses recent contamination with historical or intermittent contamination.

Particulates and sediments collected from coastal areas in Spain and Cuba receiving municipal sewage loads were analyzed by Grimalt et al. (1990) to determine the utility of coprostanol as a chemical marker of sewage contamination. Coprostanol cannot by itself be attributed to fecal matter inputs. However, relative contributions of steroid components can be useful indicators. When the relative concentrations of coprostanol and coprostanone are higher than their 5α epimers, or more realistically, other sterol components of background or natural occurrence, they can provide useful information.

Sediment cores from Santa Monica Basin, CA, and effluent from two local municipal wastewater discharges were analyzed by Venkatesan and Kaplan (1990) for coprostanol to determine the degree of sewage addition to sediment. Coprostanols were distributed throughout the basin sediments in association with fine particles. Some stations contained elevated levels, either due to their proximity to outfalls or because of preferential advection of fine-grained sediments. A noted decline of coprostanols relative to total sterols from outfalls seaward indicated dilution of sewage by biogenic sterols.

Other chemical compounds have been utilized for sewage tracer work. Saturated hydrocarbons with 16 to 18 carbons, and saturated hydrocarbons with 16 to 21 carbons, in addition to coprostanol, were chosen as markers for sewage in water, particulate, and sediment samples near the Cocoa, FL, domestic wastewater treatment plant (Holm et al. 1990). The concentration of the markers was highest at points close to the outfall pipe and diminished with distance. However, the concentration of C16 to C21 compounds was high at a site 800 m from the outfall, indicating that these compounds were unsuitable markers for locating areas exposed to the sewage plume. The concentrations for the other markers were very low at this station.

The range of concentrations of coprostanol found in sediments and mussels of Venice, Italy, were reported by Sherwin et al. (1993). Raw sewage is still discharged directly into the Venice lagoon. Coprostanol concentrations were determined in sediment and mussel samples from the lagoon using gas chromatography/mass spectroscopy. Samples were collected in interior canals and compared to open-bay concentrations. Sediment concentrations ranged from 0.2 to 41.0 µg/g (dry weight). Interior canal sediment samples averaged 16 µg/g compared to 2 µg/g found in open-bay sediment samples. Total coprostanol concentrations in mussels ranged from 80 to 620 ng/g (wet weight). No mussels were found in the four most polluted interior canal sites.

Table 6.38 Chemical Mass Balance Equations with Parameter Means

Parameters	Spring Water 1	Ground-Water 2	Tap Water 3	Irrigation Water 4	Sanitary Sewage 5	Septic Tank 6	Car Wash 7	Laundry Water 8	Unknown Sample
Potassium	(m1)(0.73)	+ (m2)(1.19)	+ (m3)(1.55)	+ (m4)(6.08)	+ (m5)(5.97)	+ (m6)(18.82)	+ (m7)(42.69)	+ (m8)(3.48)	= (1)(C$_p$)
Hardness	(m1)(240)	+ (m2)(27)	+ (m3)(49)	+ (m4)(40)	+ (m5)(143)	+ (m6)(57)	+ (m7)(157)	+ (m8)(36)	= (1)(C$_p$)
Fluorescence	(m1)(6.8)	+ (m2)(29.9)	+ (m3)(4.6)	+ (m4)(214)	+ (m5)(251)	+ (m6)(382)	+ (m7)(1190)	+ (m8)(1024)	= (1)(C$_p$)
Conductivity	(m1)(301)	+ (m2)(51)	+ (m3)(112)	+ (m4)(105)	+ (m5)(420)	+ (m6)(502)	+ (m7)(485)	+ (m8)(563)	= (1)(C$_p$)
Fluoride	(m1)(0.03)	+ (m2)(0.06)	+ (m3)(0.97)	+ (m4)(0.90)	+ (m5)(0.76)	+ (m6)(0.93)	+ (m7)(12.3)	+ (m8)(32.82)	= (1)(C$_p$)
Ammonia	(m1)(0.01)	+ (m2)(0.24)	+ (m3)(0.03)	+ (m4)(0.37)	+ (m5)(9.92)	+ (m6)(87.21)	+ (m7)(0.24)	+ (m8)(0.82)	= (1)(C$_p$)
Detergents	(m1)(0.00)	+ (m2)(0.00)	+ (m3)(0.00)	+ (m4)(0.00)	+ (m5)(1.50)	+ (m6)(3.27)	+ (m7)(49.00)	+ (m8)(26.90)	= (1)(C$_p$)
Color	(m1)(0.0)	+ (m2)(8.0)	+ (m3)(0.0)	+ (m4)(10.0)	+ (m5)(37.9)	+ (m6)(70.6)	+ (m7)(221.5)	+ (m8)(46.7)	= (1)(C$_p$)

Equations of the form $C_p = \sum_n m_n x_{pn}$

where: C_p = the concentration of parameter p in the outfall flow

    m_n = the fraction of flow from source type n

    x_{pn} = the mean concentration of parameter p in source type n

From Lalor, M. *An Assessment of Non-Stormwater Discharges to Storm Drainage Systems in Residential and Commercial Land Use Areas*. Ph.D. dissertation. Department of Civil and Environmental Engineering. Vanderbilt University. 1994. With permission.

Figure 6.116 Remotely operated TV camera surveys of storm sewers in Boston to locate inappropriate discharges.

Figure 6.117 Manual surveys conducted in Boston in large tidally influenced storm drains.

Nichols et al. (1996) also examined coprostanol in stormwater and the sea-surface microlayer to distinguish human vs. nonhuman sources of contamination. Other steroid compounds in sewage effluent were investigated by Routledge et al. (1998) and Desbrow et al. (1998), who both examined estrogenic chemicals. The most commonly found were 17β-estradiol and estrone, which were detected at concentrations in the tens of nanograms per liter range. These were identified as estrogenic through a toxicity identification and evaluation approach, where sequential separations and analyses identified the sample fractions causing estrogenic activity using a yeast-based estrogen screen. GC/MS was then used to identify the specific compounds.

Estimating Potential Sanitary Sewage Discharges into Storm Drainage and Receiving Waters Using Detergent Tracer Compounds

As described above, detergent measurements (using methylene blue active substance, MBAS, test methods) were the most successful individual tracers to indicate contaminated water in storm sewerage dry-weather flows. Unfortunately, the MBAS method uses hazardous chloroform for an extraction step. Different detergent components, especially linear alkylbenzene sulfonates (LAS) and linear alkylbenzenes (LAB), have also been tried to indicate sewage dispersal patterns in receiving waters. Boron, a major historical ingredient of laundry chemicals, can also potentially be used. Boron has the great advantage of being relatively easy to analyze using portable field test kits, while LAS requires chromatographic equipment. LAS can be measured using HPLC with fluorescent detection, after solid-phase extraction, to very low levels. Fujita et al. (1998) developed an efficient enzyme-linked immunosorbent assay (ELISA) for detecting LAS at levels from 20 to 500 µg/L.

LAS from synthetic surfactants (Terzic and Ahel 1993) which degrade rapidly, as well as nonionic detergents (Zoller et al. 1991) which do not degrade rapidly, have been utilized as sanitary sewage markers. LAS was quickly dispersed from wastewater outfalls except in areas where wind was calm. In these areas, LAS concentrations increased in fresh water but were unaffected in saline water. After time, the lower alkyl groups were mostly found, possibly as a result of degradation or settling of longer alkyl chain compounds with sediments. Chung et al. (1995) also describe the distribution and fate of LAS in an urban stream in Korea. They examined different LAS compounds having carbon ratios of C12 and C13 compared to C10 and C11, plus ratios of phosphates to MBAS and the internal to external isomer ratio (I/E) as part of their research. González-Mazo et al. (1998)

examined LAS in the Bay of Cádiz off the southwest coast of Spain. They found that LAS degrades rapidly. Fujita et al. (1998) found that complete biodegradation of LAS requires several days and is also strongly sorbed to particulates. In areas close to shore and near the untreated wastewater discharges, there was significant vertical stratification of LAS: the top 3 to 5 mm of water had LAS concentrations about 100 times greater than those found at 0.5 m.

Zeng and Vista (1997) and Zeng et al. (1997) describe a study off San Diego where LAB was measured, along with polycyclic aromatic hydrocarbons (PAHs) and aliphatic hydrocarbons (AHs) to indicate the relative pollutant contributions of wastewater from sanitary sewage, nonpoint sources, and hydrocarbon combustion sources. They developed and tested several indicator ratios (alkyl homologue distributions and parent compound distributions) and examined the ratios of various PAHs (such as phenanthrene to anthracene, methylphenanthrene to phenanthrene, fluoranthene to pyrene, and benzo(a)anthracene to chrysene) as tools for distinguishing these sources. They concluded that LABs are useful tracers of domestic waste inputs to the environment due to their limited sources. They also describe the use of the internal to external isomer ratio (I/E) to indicate the amount of biodegradation that may have occurred to the LABs. They observed concentrations of total LABs in sewage effluent of about 3 μg/L, although previous researchers have seen concentrations of about 150 μg/L in sewage effluent from the same area.

The fluorescent properties of detergents have also been used as tracers by investigating the fluorescent whitening agents (FWAs), as described by Poiger et al. (1996) and Kramer et al. (1996). HPLC with fluorescence detection was used in these studies to quantify very low concentrations of FWAs. The two most frequently used FWAs in household detergents (DSBP and DAS 1) were found at 7 to 21 μg/L in primary sewage effluent and at 3 to 9 μg/L in secondary effluent. Raw sewage contains about 10 to 20 μg/L FWAs. The removal mechanisms in sewage treatment processes is by adsorption to activated sludge. The type of FWAs varies from laundry applications to textile finishing and paper production, making it possible to identify sewage sources. The FWAs were found in river water at 0.04 to 0.6 μg/L. The FWAs are not easily biodegradable, but they are readily photodegraded. Photodegradation rates have been reported to be about 7% for DSBP and 71% for DAS 1 in river water exposed to natural sunlight, after 1-hour exposure. Subsequent photodegradation is quite slow.

Other Compounds Found in Sanitary Sewage That May Be Used for Identifying Contamination by Sewage

Halling-Sørensen et al. (1998) detected numerous pharmaceutical substances in sewage effluents and in receiving waters. Their work addressed human health concerns of these low-level compounds that can enter downstream drinking water supplies. However, the information might also be used to help identify sewage contamination. Most of the research has focused on clofibric acid, a chemical used in cholesterol-lowering drugs. It has been found in concentrations ranging from 10 to 165 ng/L in a Berlin drinking water sample. Other drugs commonly found include aspirin, caffeine, and ibuprofen. Current FDA guidance mandates that the maximum concentration of a substance or its active metabolites at the point of entry into the aquatic environment be less than 1 μg/L (Hun 1998).

Caffeine has been used as an indicator of sewage contamination by several investigators (Shuman and Strand 1996). The King County, WA, Water Quality Assessment Project is examining the impacts of CSOs on the Duwamish River and Elliott Bay. They are using both caffeine (representing dissolved CSO constituents) and coprostanol (representing particulate-bound CSO constituents), in conjunction with heavy metals and conventional analyses, to help determine the contribution of CSOs to the river. The caffeine is unique to sewage, while coprostanol is from both humans and carnivorous animals and is therefore also in stormwater. They sampled upstream of all CSOs, but with some stormwater influences, 100 m upstream of the primary CSO discharge (but downstream of other CSOs), within the primary CSO discharge line, and 100 m downriver of the CSO discharge location. The relationship between caffeine and coprostanol was fairly consistent for the four sites (coprostanol was about 0.5 to 1.5 µg/L higher than caffeine). Similar patterns were found between the three metals, chromium was always the lowest and zinc was the highest. King County is also using clean transported mussels placed in the Duwamish River to measure the bioconcentration potential of metal and organic toxicants and the effects of the CSOs on mussel growth rates (after 6-week exposure periods). Paired reference locations are available near the areas of deployment, but outside the areas of immediate CSO influence. *U.S. Water News* (1998) also described a study in Boston Harbor that found caffeine at levels of about 7 µg/L in the harbor water. The caffeine content of regular coffee is about 700 mg/L, in contrast.

DNA Profiling to Measure Impacts on receiving water Organisms and to Identify Sources of Microorganisms in Stormwater

This rapidly emerging technique seems to have great promise in addressing a number of nonpoint source water pollution issues. Kratch (1997) summarized several investigations on cataloging the DNA of *E. coli* to identify their source in water. The procedure, developed at the Virginia Polytechnic Institute and State University, has been used in Chesapeake Bay. In one example, it was possible to identify a large wild animal population as the source of fecal coliform contamination of a shellfish bed, instead of suspected failing septic tanks. DNA patterns in fecal coliforms vary among animals and birds, and it is relatively easy to distinguish between human and nonhuman sources of the bacteria. However, some wild animals have DNA patterns that are not easily distinguishable. Some researchers question the value of *E. coli* DNA fingerprinting, believing that there is little direct relationship between *E. coli* and human pathogens. However, this method should be useful to identify the presence of sewage contamination in stormwater or in a receiving water.

One application of the technique, as described by Krane et al. (1999) of Wright State University, used randomly amplified polymorphic DNA polymerase chain reaction (RAPD-PCR) generated profiles of naturally occurring crayfish. They found that changes in the underlying genetic diversity of these populations were significantly correlated with the extent to which they have been exposed to anthropogenic stressors. They concluded that this rapid and relatively simple technique can be used to develop a sensitive means of directly assessing the impact of stressors on ecosystems. These Wright State University researchers have also used the RAPD-PCR techniques on populations of snails, pill bugs, violets, spiders, earthworms, herring, and some benthic macroinvertebrates, finding relatively few obstacles in its use for different organisms. As noted above, other researchers have used DNA profiling techniques to identify sources of *E. coli* bacteria found in coastal waterways. It is possible that these techniques can be expanded to enable rapid detection of many different types of pathogens in receiving waters, and the most likely sources of these pathogens.

Stable Isotope Methods for Identifying Sources of Water

Stable isotopes had been recommended as an efficient method to identify illicit connections to storm sewerage. A demonstration was conducted in Detroit as part of the Rouge River project to identify sources of dry-weather flows in storm sewerage (Sangal et al. 1996). Naturally occurring stable isotopes of oxygen and hydrogen can be used to identify waters originating from different geographical sources (especially along a north–south gradient). Ma and Spalding (1996) discuss this approach by using stable isotopes to investigate recharge of groundwaters by surface waters. During water vapor transport from equatorial source regions to higher latitudes, depletion of heavy isotopes occurs with rain. Deviation from a standard relationship between deuterium and ^{18}O for a specific area indicates that the water has undergone additional evaporation. The ratio is also affected by seasonal changes. As discussed by Ma and Spalding (1996), the Platte River water is normally derived in part from snowmelt from the Rocky Mountains, while the groundwater in parts of Nebraska is mainly contributed from the Gulf air stream. The origins of these waters are sufficiently different and allow good measurements of the recharge rate of the surface water to the groundwater. In Detroit, Sangal et al. (1996) used differences in origin between the domestic water supply, local surface waters, and the local groundwater to identify potential sanitary sewage contributions to the separate storm sewerage. Rieley et al. (1997) used stable isotopes of carbon in marine organisms to distinguish the primary source of carbon being consumed (sewage sludge vs. natural carbon sources) in two deep sea sewage sludge disposal areas.

Stable isotope analyses would not be able to distinguish between sanitary sewage, industrial discharges, wash waters, and domestic water, as they all have the same origin. Nor would it be possible to distinguish sewage from local groundwaters if the domestic water supply was from the same local aquifer. This method works best for situations where the water supply is from a distant source and where separation of waters into separate flow components is not needed. It may be an excellent tool to study the effects of deep well injection of stormwater on deep aquifers having distant recharge sources (such as in the Phoenix area). Few laboratories can analyze for these stable isotopes, requiring shipping the samples and a long wait for the analytical results. Sangal et al. (1996) used Geochron Laboratories, in Cambridge, MA.

Dating of sediments using ^{137}Cs was described by Davis et al. (1997). Arsenic-contaminated sediments in the Hylebos Waterway in Tacoma, WA, could have originated from numerous sources, including a pesticide manufacturing facility, a rock-wool plant, steel slags, powdered metal plant, shipbuilding facilities, marinas and arsenic-based boat paints, and the Tacoma Smelter. Dating the sediments, combined with knowing the history of potential discharges and conducting optical and electron microscopic studies of the sediments, was found to be a powerful tool to differentiate the metal sources to the sediments.

Comparison of Parameters That Can Be Used for Identifying Inappropriate Discharges to Storm Drainage

In almost all cases, a suite of analyses is most suitable for effective identification of inappropriate discharges. An example was reported by Standley et al. (2000), where fecal steroids (including coprostanol), caffeine, consumer product fragrance materials, and petroleum and combustion by-products were used to identify wastewater treatment plant effluent, agricultural and feedlot runoff, urban runoff, and wildlife sources. They studied numerous individual sources of these wastes from throughout the United States. A research-grade mass spectrophotometer was used for the majority of the analyses in order to achieve the needed sensitivities, although much variability was found when using the methods in actual receiving waters affected by wastewater effluent. This sophisticated suite of analyses did yield much useful information, but the analyses are difficult to conduct and costly and may be suitable for special situations, but not for routine survey work.

Another series of tests examined several of these potential emerging tracer parameters, in conjunction with the previously identified parameters, during a project characterizing stormwater that had collected in telecommunication manholes, funded by Tecordia (previously Bellcore), AT&T, and eight regional telephone companies throughout the country (Pitt and Clark 1999). Numerous conventional constituents, plus major ions, and toxicants were measured, along with candidate tracers to indicate sewage contamination of this water. Boron, caffeine, coprostanol, *E. coli*, enterococci, fluorescence (using specific wavelengths for detergents), and a simpler test for detergents were evaluated, along with the use of fluoride, ammonia, potassium, and obvious odors and color. About 700 water samples were evaluated for all of these parameters, with the exception of bacteria and boron (about 250 samples), and only infrequent samples were analyzed for fluorescence. Coprostanol was found in about 25% of the water samples (and in about 75% of the 350 sediment samples analyzed). Caffeine was found in very few samples, while elevated *E. coli* and enterococci (using IDEXX tests) were observed in about 10% of the samples. Strong sewage odors in water and sediment samples were also detected in about 10% of the samples. Detergents and fluoride (at >0.3 mg/L) were found in about 40% of the samples and are expected to have been contaminated by industrial activities (lubricants and cleansers) and not sewerage. Overall, about 10% of the samples were therefore expected to have been contaminated with sanitary sewage, about the same rate previously estimated for stormwater systems.

Additional related laboratory tests, funded by the University of New Orleans and the EPA (Barbé et al. 2000), were conducted using many sewage and laundry detergent samples, and it was found that the boron test was a poor indicator of sewage, possibly due to changes in formulations in modern laundry detergents. Laboratory tests did find that fluorescence was an excellent indicator of sewage, especially when using specialized "detergent whitener" filter sets, but this was not very repeatable. Researchers also examined several UV absorbance wavelengths as sewage indicators and found excellent correlations with 228 nm, a wavelength having very little background absorbance in local spring waters, but with a strong response factor with increasing strengths of sewage.

Table 6.39 summarizes the different measurement parameters discussed above. We recommend that our originally developed and tested protocol (including measurement of obvious indicators, detergents, fluoride, ammonia, and potassium) still be used as the most efficient routine indicator of sewage contamination of stormwater drainage systems, with the possible addition of specific *E. coli* and enterococci measurements and UV absorbance at 228 nm. The numerous exotic tests requiring specialized instrumentation and expertise do not appear to warrant their expense and long analytical turn-around times, except in specialized research situations, or when special confirmation is economically justified (such as when examining sewer replacement or major repair options).

Hydrocarbon Fingerprinting for Investigating Sources of Hydrocarbons

Fingerprinting to identify the likely source of hydrocarbon contamination is a unique process that recognizes degradation of the material by examining a wide variety of parameters, usually by sophisticated chromatography methods. There are numerous experts who have developed and refined the necessary techniques. The following is a list of some of these expert groups, from recommendations from the Internet environmental engineering list serve group, enveng-L:

- Friedman & Bruya, Seattle, WA
- Arthur D. Little, Inc., Cambridge, MA
- GW/S Environmental Consulting, Tulsa, OK
- Public & Environmental Affairs, Indiana University, Bloomington, IN
- Graduate School of Oceanography, University of Rhode Island, Narragansett, RI
- Louisiana State University, Baton Rouge, LA
- Geological and Environmental Research Group, Texas A&M University, College Station, TX
- Trillium, Inc., Coatesville, PA
- McLaren/Hart, Inc., Albany, NY
- Phoenix Laboratories, Chicago, IL
- Golder Assoc., Mississauga, Ontario, Canada
- Daniel B. Stephens & Assoc., Albuquerque, NM
- Global Geochemistry Corp., Canoga Park, CA
- Fluor Daniel GTI, Kent, WA
- Battelle, Inc., Duxbury, MA

In addition, the University of Wisconsin, Madison, Department of Engineering Professional Development (608-262-1299) periodically offers extension classes specifically on hydrocarbon pattern recognition and dating, led by experts in the field. The IBC Group (Southborough, MA, 508-481-6400) also offers an executive forum on environmental forensics, also led by many of the above experts, that addresses many issues pertaining to the legal implications of hydrocarbon tracing.

Stout et al. (1998) prepared an overview of environmental forensics, describing how systematic investigation of a contaminated site or an event can make it possible to determine the true origin and nature of complex chemical conditions. Chemical fingerprinting, generally using high-resolution gas

Table 6.39 Comparison of Measurement Parameters Used for Identifying Inappropriate Discharges into Storm Drainage

Parameter Group	Comments	Recommendation
Fecal coliform bacteria and/or use of fecal coliform to fecal streptococci ratio	Commonly used to indicate presence of sanitary sewage.	Not very useful as many other sources of fecal coliforms are present, and ratio not accurate for old or mixed wastes.
Physical observations (odor, color, turbidity, floatables, deposits, stains, vegetation changes, damage to outfalls)	Commonly used to indicate presence of sanitary and industrial wastewater.	Recommended due to easy public understanding and easy to evaluate, but only indicative of gross contamination, with excessive false negatives (and some false positives). Use in conjunction with chemical tracers for greater sensitivity and accuracy.
Detergents presence (anionic surfactant extractions)	Used to indicate presence of wash waters and sanitary sewage.	Recommended, but care needed during hazardous analyses (only for well-trained personnel). Accurate indicator of contamination during field tests.
Fluoride, ammonia and potassium measurements	Used to identify and distinguish between wash waters and sanitary sewage.	Recommended, especially in conjunction with detergent analyses. Accurate indicator of major contamination sources and their relative contributions.
TV surveys and source investigations	Used to identify specific locations of inappropriate discharges, especially in industrial areas.	Recommended after outfall surveys indicate contamination in drainage system.
Coprostanol and other fecal sterol compounds	Used to indicate presence of sanitary sewage.	Possibly useful. Expensive analysis with GC/MSD. Not specific to human wastes or recent contamination. Most useful when analyzing particulate fractions of wastewaters or sediments.
Specific detergent compounds (LAS, fabric whiteners, and perfumes)	Used to indicate presence of sanitary sewage.	Possibly useful. Expensive analyses with HPLC. A good and sensitive confirmatory method.
Fluorescence	Used to indicate presence of sanitary sewage and wash waters.	Likely useful, but expensive instrumentation. Rapid and easy analysis. Very sensitive.
Boron	Used to indicate presence of sanitary sewage and wash waters.	Not very useful. Easy and inexpensive analysis, but recent laundry formulations in U.S. have minimal boron components.
Pharmaceuticals (colfibric acid, aspirin, ibuprofen, steroids, illegal drugs, etc.)	Used to indicate presence of sanitary sewage.	Possibly useful. Expensive analyses with HPLC. A good and sensitive confirmatory method.
Caffeine	Used to indicate presence of sanitary sewage.	Not very useful. Expensive analyses with GC/MSD. Numerous false negatives, as typical analytical methods not suitably sensitive.
DNA profiling of microorganisms	Used to identify sources of microorganisms	Likely useful, but currently requires extensive background information on likely sources in drainage. Could be very useful if method can be simplified, but with less specific results.
UV absorbance at 228 nm	Used to identify presence of sanitary sewage.	Possibly useful, if UV spectrophotometer available. Simple and direct analyses. Sensitive to varying levels of sanitary sewage, but may not be useful with dilute solutions. Further testing needed to investigate sensitivity in field trials.
Stable isotopes of oxygen	Used to identify major sources of water.	May be useful in area having distant domestic water sources and distant groundwater recharge areas. Expensive and time consuming procedure. Cannot distinguish between wastewaters if all have common source.
E. coli and enterococci bacteria	More specific indicators of sanitary sewage than coliform tests.	Recommended in conjunction with chemical tests. Relatively inexpensive and easy analyses, especially if using the simple IDEXX methods.

chromatography coupled with mass spectroscopy, is usually supplemented with site information on soils and groundwater conditions. The presentation of masses of data is usually highly visually oriented to make complex patterns and associations easier to comprehend. In addition to GC/MS, stable isotope analyses may be conducted to identify origins of very similar materials. Historical records also need to be reviewed to understand the changes that a site has undergone over the years ("corporate archaeology"). Sanborn Fire Insurance Maps (Geography and Map Division, Library of Congress) are commonly used to identify site activities during the second half of the 19th century, for example. This type of approach can be used to identify sources of contaminated sediments in urban streams, especially in areas having historical industrial activities.

Other techniques can be used to date deposits and to indicate the extent of the weathering of petroleum (Whittaker and Pollard 1997). The weathered state of spilled (or discharged) hydrocarbons can be determined using biomarkers (pristane, phthane, hopanes, and steranes) which are quite resistant to weathering processes (biotransformations and evaporation). These are therefore relatively conservative materials and can be compared to less stable oil components to indicate the extent of weathering that has occurred, and hence the approximate time since the petroleum was deposited. Other biomarkers can also be used as unique fingerprints to identify the likely source of the oil. Hurst et al. (1996) also describe how lead isotopes ($^{206}Pb/^{207}Pb$ ratio) can be used to age spilled gasoline, based on changes in gasoline additives with time.

MICROORGANISMS IN STORMWATER AND URBAN RECEIVING WATERS

As discussed in Chapter 3, microorganisms frequently interfere with beneficial uses in urban receiving waters. The use of conventional indicator organisms may be helpful, but investigations of specific pathogens is also becoming possible with new analytical technologies. The following discussion contains some background on the development of water quality standards for indicator organisms, describes some new analytical procedures, and presents an approach that measures organism die-off *in situ*, which is important for assessing the public health risk associated with water contact in urban receiving waters.

Pathogens in stormwater and urban receiving waters are a significant concern potentially affecting human health. The use of indicator bacteria is controversial for stormwater, as is the assumed time of typical exposure of swimmers to contaminated receiving waters. However, recent epidemiological studies have shown significant health effects associated with stormwater-contaminated marine swimming areas. Protozoan pathogens, especially those associated with likely sewage-contaminated stormwater, is also a public health concern.

Human health standards for body contact recreation (and for fish and water consumption) are based on indicator organism monitoring. Monitoring for the actual pathogens, with few exceptions, requires an extended laboratory effort, is very costly, and not very accurate. Therefore, the use of indicator organisms has become established. Dufour (1984a) presents an excellent overview of the history of indicator bacterial standards and water contact recreation.

Total coliforms were initially used as indicators for monitoring outdoor bathing waters, based on a classification scheme presented by W.J. Scott in 1934. Total coliform bacteria, refers to a number of bacteria including *Escherichia*, *Klebsiella*, *Citrobacter*, and *Enterobacter*. They are able to grow at 35°C and ferment lactose. They are all Gram-negative asporogenous rods and have been associated with feces of warm-blooded animals. They are also present in soil.

The fecal coliform test is not specific for any one coliform type, or groups of types, but instead has an excellent positive correlation for coliform bacteria derived from the intestinal tract of warm-blooded animals (Geldreich et al. 1968). The fecal coliform test measures *Escherichia coli* as well as all other coliforms that can ferment lactose at 44.5°C and are found in warm-blooded fecal discharges. Geldreich (1976) found that the fecal coliform test represents over 96% of the coliforms derived from human feces and from 93 to 98% of those discharged in feces from other warm-blooded animals, including livestock, poultry, cats, dogs, and rodents. In many urban runoff studies, all of the fecal coliforms were *E. coli* (Quresh and Dutka 1979). *E. coli*, a member of the fecal coliform group, has been used as a better indicator of fresh fecal contamination, compared to fecal coliforms. Table 6.40 indicates the species and subspecies of the *Streptococcus* and *Enterococcus* groups of bacteria that are used as indicators of fecal contamination.

Table 6.40   Streptococcus Species Used as Indicators of Fecal Contamination

Indicator Organism	Enterococcus Group	Streptococcus Group
Group D antigen		
Streptococcus faecalis	X	X
S. faecealis subsp. *liquifaciens*	X	X
S. faecalis subsp. *zymogenes*	X	X
S. faecium	X	X
S. bovis		X
S. equinus		X
Group Q antigen		
S. avium		X

Fecal streptococci bacteria are all of the intestinal streptococci bacteria from warm-blooded animal feces (Geldreich and Kenner 1969). The types and concentrations of different bacteria biotypes vary for different animal sources. Fecal streptococci bacteria are indicators of fecal contamination. The enterococci group is a subgroup that is considered a better indication of human fecal contamination. *S. bovis* and *S. equinus* are considered related to feces from nonhuman warm-blooded animals (such as from meat processing facilities, dairy wastes, and feedlot and other agricultural runoff), indicating that enterococcus may be a better indication of human feces contamination. However, *S. faecalis* subsp. *liquifaciens* is also associated with vegetation sources, insects, and some soils.

The EPA's evaluation of the bacteriological data indicated that using the fecal coliform indicator group at the maximum geometric mean of 200 organisms per 100 mL, as recommended in *Quality Criteria for Water* would cause an estimated eight illnesses per 1000 swimmers at freshwater beaches. Additional criteria, using *E. coli* and enterococci bacteria analyses, were developed using these currently accepted illness rates. See Appendix G for specific details of these criteria. These bacteria are assumed to be more specifically related to poorly treated human sewage than the fecal coliform bacteria indicator. It should be noted that these indicators only relate to gastrointestinal illness, and not other problems associated with waters contaminated with bacterial or viral pathogens. Common swimming beach problems associated with contamination by stormwater include skin and ear infections caused by *Pseudomonas aeruginosa* and *Shigella*.

Viruses may also be important pathogens in urban runoff. Very small amounts of a virus are capable of producing infections or diseases, especially when compared to the large numbers of bacterial organisms required for infection (Berg 1965). The quantity of enteroviruses which must be ingested to produce infections is usually not known (Olivieri et al. 1977b). Viruses are usually detected at low levels in urban receiving waters and storm runoff. Researchers have stated that even though the minimum infective doses may be small, the information available indicates that stormwater virus threats to human health are small. Because of the low levels of virus necessary for infection, dilution of viruses does not significantly reduce their hazard.

States et al. (1997) examined *Cryptosporidium* and *Giardia* in river water serving as Pittsburgh's water supply. They collected monthly samples from the Allegheny and Youghiogheny Rivers for 2 years. They also sampled a small stream flowing through a dairy farm, treated sanitary sewage effluent, and CSOs. The CSO samples had much greater numbers of the protozoa than any of the other samples. No raw sewage samples were obtained, but they were assumed to be very high because of the high CSO sample values. The effluent from the sewage treatment plant was the next highest, at less than half the CSO values. The dairy farm stream was not significantly different from either of the two large rivers. The water treatment process appeared to effectively remove *Giardia*, but some *Cryptosporidium* was found in the filtered water. Settling the river water seemed to remove some of the protozoa, but the removal would not be adequate by itself. States et al. (1997) also reviewed *Giardia* and *Cryptosporidium* monitoring data. Raw drinking water supplies were shown to have highly variable levels of these protozoa, typically up to several hundred *Giardia* cysts and *Cryptosporidium* oocysts per 100 L, and were found in 5 to 50% of the samples evaluated. Conventional water treatment appeared to remove about 90% of the protozoa.

A microorganism monitoring program for stormwater-impacted urban receiving waters could therefore be very complex and expensive if all the above organisms were to be evaluated. The bacteria (especially total coliforms, fecal coliforms, *E. coli*, enterococci, and hopefully *Pseudomonas aeruginosa*) should probably all be adequately covered in a monitoring program. Total coliforms are of most interest in marine environments based on epidemiological studies conducted in Santa Monica Bay (see case study in Chapter 4). In most cases, total coliform data could be misleading because of its ubiquitous nature (see Chapter 8). Protozoa, and especially viruses, require highly specialized analytical skills and are therefore not likely to be routinely investigated. However, protozoa are being more commonly monitored, especially with new federal regulations to protect drinking water supplies.

Sampling for microorganism evaluations is more challenging than for most constituents, requiring sterile sample containers and tools, plus rapid shipment of the samples to the laboratory and immediate initiation of analyses. Bacteriological analyses are becoming much more simplified with special procedures and methods developed by HACH, Millipore, and IDEXX Corp., for example. Available methods require little more than mixing a freeze-dried "reagent" with a measured amount of sample, pouring the mixture into special incubation trays and sealing them, and finally placing them into incubators for the designated time (usually 18 to 48 hours).

The IDEXX method for *E. coli*, Colilert-18 (see Figures 6.71 through 6.74), is used by many state agencies for EPA reporting purposes. It is used for the simultaneous detection, specific identification, and confirmation of total coliforms and *E. coli* in water. It is based on IDEXX's patented Defined Substrate Technology® (DST™). It is a most probable number (MPN) method. Colilert-18 utilizes nutrient indicators that produce color and/or fluorescence when metabolized by total coliforms and *E. coli*. When the Colilert-18 reagent is added to a sample and incubated, it can detect these bacteria at 1 cfu in 100 mL within 18 hours with as many as 2 million heterotrophic bacteria per 100 mL present. The required apparatus includes the Quanti-tray sealer, an incubator, a 6-watt 365-nm UV light, and a fluorescence comparator. This procedure requires 100 mL of sample, which should be analyzed ASAP after sampling. Marine water samples must be diluted at least tenfold with sterile fresh water to reduce the salinity. Quality control includes testing with cultures of *E. coli*, *Klebsiella pneumoniae*, and *Pseudomonas aeruginosa*.

The Enterolert procedure, also from IDEXX, is very similar to the Colilert method outlined above. Enterolert is used for the detection of enterococci such as *Enterococcus faecium* or *E. faecalis* in fresh and marine water. When the Enterolert reagent is added to a sample and incubated, bacteria down to 1 CFU in a 100 mL sample can be detected within 24 hours. This method also has a quality control procedure that should be conducted on each lot of Enterolert, using test cultures of *E. faecium*, *Serratia marcescens* (Gram-negative), and *Aerococcus viridans* (Gram-positive).

Determination of Survival Rates for Selected Bacterial and Protozoan Pathogens

The following discussion was prepared by John Easton while he was a Ph.D. student at the University of Alabama at Birmingham and describes some of the experiments he has conducted concerning the survival of wet-weather flow bacteria and pathogens after being discharged to urban receiving waters (Easton 2000). This section is not intended to be a comprehensive review of survival of microorganisms in the environment, but is intended to illustrate how actual site-specific survival rates can be determined, especially for unusual conditions (affected by water temperature, turbidity, natural predation, local sources and receptors, etc.). This information is necessary for human health assessments when predicting resulting downstream pathogen conditions. Much of the literature on microorganism survival is based on laboratory investigations that might not be applicable to actual field conditions. The simple tests described in this section allow more accurate in-stream predictions.

Pathogenic organisms found in sewage can adversely impact public health when the sewage is discharged to waters that humans come in contact with when wading, swimming, fishing, drinking, etc. UAB is conducting research to develop a risk assessment methodology for evaluating varying degrees of risk related to human contact with pathogenic microorganisms found in sewage-contaminated waters, especially those caused by separate sanitary sewer overflows (SSOs). One component of this research is to study the fate and transport of these microorganisms in the environment. The survivability, or die-off, rates for these organisms are critical to understanding their fate in the environment, e.g., from an SSO discharge through a receiving water.

Microorganisms have varying degrees of stability within the environment. Their numbers are dependent upon population dynamics, which is controlled by several criteria (McKinney 1992): (1) competition for food (limited food sources limit microbial numbers), (2) predator–prey relationships (some organisms consume others for food sources), (3) nature of organic matter (carbohydrates, organic acids, and proteins all stimulate different organisms), and (4) environmental conditions (oxygen concentration, nutrient levels, temperature, pH, etc.). Since there are a multitude of factors that contribute to microorganism survivability, the use of an *in situ* method to characterize the rates of growth and death is necessary to account for variable environmental conditions.

Several experiments were conducted to evaluate the rate of die-off, or decay, for the study microorganisms. These *in situ* experiments were conducted in specially designed chambers (Figure 6.118). These were designed to allow passage of water and nutrients between the inside of the chamber and the outside environment (Five-Mile Creek in Jefferson County, AL), while sequestering the microorganisms inside to allow enumeration at various times during the experiment.

These experiments included exposures over a 21-day period. A polyethersulfone (Supor®, Gelman Sciences) membrane filter, which is not susceptible to biological degradation, was used. This membrane material was clamped onto either end of a piece of acrylic tubing in a design devised by researchers at UAB (Figure 6.119). The membrane pore size is 0.22 μm, allowing exchange of ions with the surrounding water while sequestering the microorganisms inside the test chamber.

Multiple chambers containing sewage samples were placed in the creek and removed after 0, 1, 3, 7, 10, 14, and 21 days. For each time point, three separate chambers were removed and composited for analysis. Once the samples were composited, they were blended (Waring blender for 2 min) to minimize agglomeration of the microorganisms.

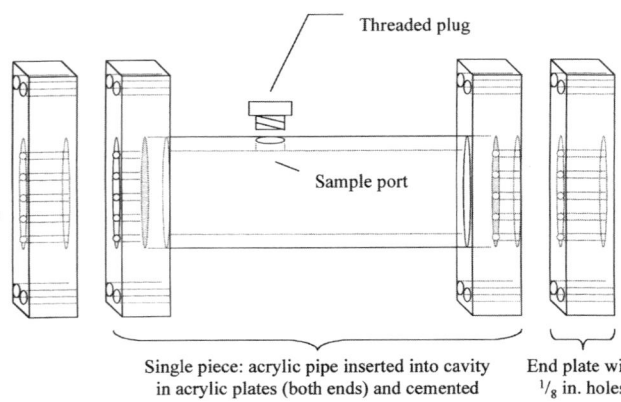

Threaded plug

Sample port

Single piece: acrylic pipe inserted into cavity in acrylic plates (both ends) and cemented

End plate with $1/_8$ in. holes

Figure 6.118 Acrylic components of *in situ* chamber. (From Easton, J. *The Development of Pathogen Fate and Transport Parameters for Use in Assessing Health Risks Associated with Sewage Contamination.* Ph.D. dissertation, the Department of Civil and Environmental Engineering, University of Alabama at Birmingham. 2000. With permission.)

The experiments conducted to evaluate degradation of *G. lamblia* were conducted *in situ*. The sewage matrix was spiked with approximately 10,000 cysts per liter to enable detection after significant die-off. These cysts were formalinized in order not to risk releasing a potentially infectious pathogen into the environment. Since these organisms are in cyst form, i.e., relatively inert, it was hypothesized that the mechanism of die-off would be predation by other organisms and formalinized organisms would be a suitable surrogate for "live" ones.

The results of these experiments show that the microorganisms die off at a constant, rapid rate (assumed in most receiving models) only for an initial short period. As time progressed, the die-off rate slows. Figure 6.120 is a plot of the levels of *Giardia* cysts vs. time. The method used to enumerate these organisms (EPA method 1623) requires a presumptive test followed by a confirmed test. The presumptive test consists of identifying objects of the correct size and shape which are stained by a *Giardia*-specific antibody bound to a fluorescent probe. Next, the organisms are confirmed by identification of internal structures stained by the nuclear stain DAPI (4′,6-diamindino-2-phenylindole). Unfortunately, problems were encountered with the confirmation test in these

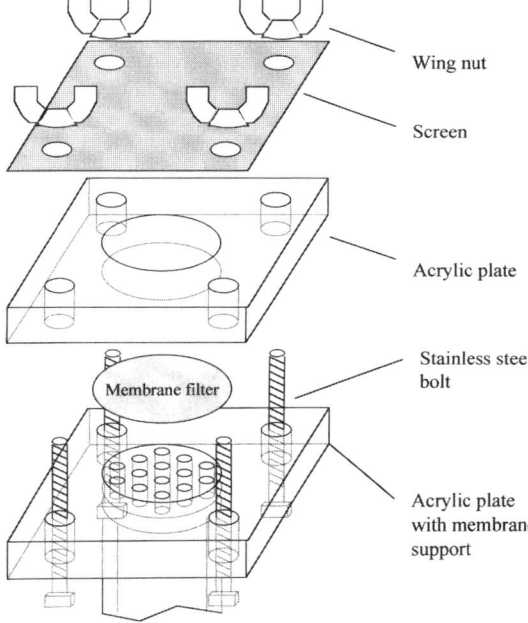

Figure 6.119 End-plate of *in situ* chamber showing the location of membrane filter. (From Easton, J.H. et al. The use of a multi-parameter water quality monitoring instrument to continuously monitor and evaluate runoff events. Presented at *Annual Water Resources Conference of the AWRA,* Point Clear, AL. 1998. With permission.)

Figure 6.120 Degradation plot of *Giardia* cysts. (From Easton, J.H. et al. The use of a multi-parameter water quality monitoring instrument to continuously monitor and evaluate runoff events. Presented at *Annual Water Resources Conference of the AWRA,* Point Clear, AL. 1998. With permission.)

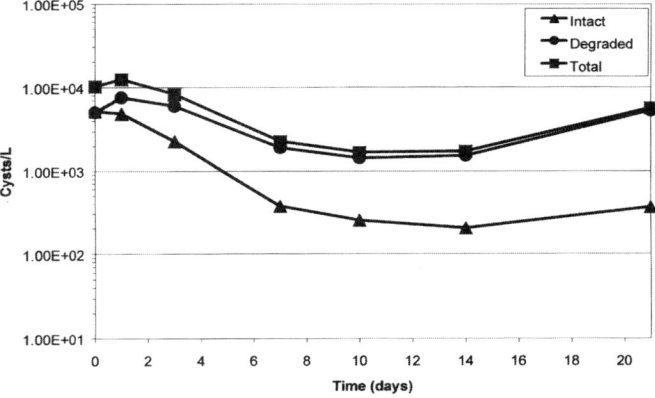

experiments (the DAPI stain of the background was too intense to enable identification of internal structures). However, using the presumptive stain, which binds to the cyst cell wall, it was possible to detect differences in these presumptive *Giardia* cysts. Some cysts were intact (i.e., the stain covered the cell wall continuously), and some cysts were present but degraded (i.e., the staining of the cell wall was less intense and not continuous). The levels of the former, "intact cysts," are plotted along with the levels of the latter, "degraded cysts," in Figure 6.120.

Since the microorganisms' rate of die-off seems to decrease over time, the regression model was applied stepwise, starting with the first three data points and adding one additional point until the entire 21-day, or 7-point, data set was used. In general, the die-off rates decreased, and T_x values correspondingly increased as data over longer time periods are included in the regression analyses. The T_{90} values (time needed for 90% die-off) for the indicator bacteria, total coliforms and *E. coli*, are in accordance with conventional wisdom. Many studies have shown T_{90} values for these organisms to be in the range of several hours to a few days (Droste and Gupgupoglu 1982; Geldreich et al. 1968; Geldreich and Kenner 1969). The initial, rapid die-off occurred, generally, within the first 7 days of the experiment. Table 6.41 gives a first-order die-off constant, k (days^{-1}), and its associated 95% confidence interval, for each of the microorganisms. In addition, the results of the Mann–Kendall Test (a nonparametric test for trend) are given. All of the die-off constants (slope of the regression line) are statistically significant except for enterococci.

Table 6.41   Die-off Rates Determined Using Day 0 to Day 7 Data

Organism	Die-off Rate (day^{-1})	95% CI	Mann–Kendall Trend[a]
Total coliforms	−0.310	± 0.152	p = 0.042
E. coli	−0.331	± 0.049	p = 0.042
Enterococci	−0.078	± 0.189	p = 0.375[b]
Giardia	−0.171	± 0.074	p = 0.042

[a] p < 0.05 indicates significant downward trend.
[b] Not significant, no trend (die-off).

From Easton, J. *The Development of Pathogen Fate and Transport Parameters for Use in Assessing Health Risks Associated with Sewage Contamination.* Ph.D. dissertation, the Dept. of Civil and Environmental Engineering, University of Alabama at Birmingham. 2000. With permission.

The data generated by this study suggest that if one were using die-off constants from indicator bacteria studies, then one may tend to underpredict the length of time or distance downstream in which adverse health effects due to pathogens in sewage are present. In addition, these data indicate that assumptions regarding the constancy of die-off rates may be invalid. There seems to be a modulation of the rate of die-off with increased time, as all of the test organisms showed a pattern of leveling off toward some equilibrium level with increasing time.

The *Enterococcus* results are quite different from the others, with no rapid initial die-off, as generally reported in the literature (Facklam and Sahm 1995). This persistence is due to the enterococci being Gram-positive and is therefore a better indicator of virus survival. For these reasons, the EPA has selected enterococci as an indicator organism in their new guidance documents.

The *Giardia* results were not as expected. The descriptions of this organism found in the literature seem to predict that *Giardia* will persist much longer than observed in these tests. This study seems to show that *Giardia*, and perhaps other protozoan pathogens, exhibits die-off characteristics similar to the bacteria included in this study. However, these cysts were treated with formalin and therefore may have been less resistant to degradation in the environment.

There are many stormwater microorganisms of interest when conducting a receiving water study. However, besides characterizing microorganism conditions, it is also necessary to understand population dynamics when predicting fate and exposures. This section briefly described some of the currently used analytical methodologies for measuring microorganism counts, along with an example *in situ* die-off experiment.

BENTHOS SAMPLING AND EVALUATION IN URBAN STREAMS

Ecosystem degradation via water, sediment, and habitat alteration affects food resources, reproduction, growth, and survival of aquatic biota, thereby altering the structure and functioning of the system. Structural indicators include the number and kinds of individuals, species, population, and communities as measured by a variety of metrics. The structural alterations may impact ecosystem functions such as productivity, respiration, organic matter degradation, nutrient cycling, and energy flow, which, unfortunately, are often difficult to quantify and are resource-demanding. A useful way to measure functional changes is an indirect method whereby organisms are placed into trophic categories (e.g., predators/consumers, producers, omnivores, detritivores), which allows production and consumption dynamics to be measured. This concept has been described by Cummins (1974, 1975) and Vannote et al. (1980) in stream ecosystems as a predictable and continuous gradient of interrelated physical, structural, and functional characteristics (Table 6.42). When conditions deviate from those in reference streams of a similar stream order for that ecoregion, then impacts may be occurring.

Bottom-dwelling organisms comprise all the major trophic levels including decomposers, photosynthetic organisms (algae and macrophytes), and herbivorous and carnivorous animals. These communities live on or in the sediment or other solid surfaces (e.g., roots, decaying wood, rocks) for significant parts of their life cycle. The fauna and flora studied in environmental quality assessments have ranged from small to large, using bacteria, phytoplankton, macrophytes, protozoa, worms, crustaceans, molluscs, insects, and fish (Burton 1991). Fish will be discussed in a following subsection. The major component of benthic fauna is often the bacteria, segmented worms (e.g., oligochaetes), microcrustacea (e.g., ostracods), macrocrustacea (e.g., isopods, decapods, amphipods), and immature insects (e.g., chironomids, plecoptera, trichoptera, and ephemeroptera). Of these major groups, the immature insects have received the greatest amount of study. Consequently, there is a large database concerning life history information and relative pollution sensitivity. The major aquatic insect groups are Odonata (dragonflies), Ephemeroptera (mayflies), Plecoptera (stoneflies), Trichoptera (caddisflies), Coleoptera (beetles), and Diptera (flies, midges, mosquitoes). Each group varies in its pollution sensitivity. Each goes through multiple life stages and molts, often emerging from the water as adults. Life cycles range from a few weeks to 2 years. See Pennak (1989) and Merritt and Cummins (1984) for more information on life histories. The sedentary (nontransitory) nature of most benthic species makes them ideal chronic, long-term pollution indicators, as compared to migratory fish or other species, such as zooplankton.

The micro-, meiofauna and flora may play a major role in the aquatic ecosystem's functioning, such as photosynthetic production by periphyton, and organic matter and nutrient processing–cycling by a variety of microbial populations and communities. These groups have temporal spatial dynamics and microhabitat requirements that are much different from the macrofauna and flora, and in many respects are more difficult to study (Burton 1991). For holistic, integrative ecosystem assessments of stormwater impacts, it is necessary to define effects on the benthic microbial communities, which will require additional expertise and resources. Further information is available from Burton et al. (2000) and the annual review issue of *Water Environment Research*. Most studies, however, whose objectives are to assess stormwater effects on receiving waters, will focus on the macroinvertebrate component of the benthos. This is not because they are more important than the meio- or microbenthos, but rather because they are more effectively used in pollution assessments. The following discussions highlight some of the important benthic groups and the characteristics one should consider in their sampling and evaluation.

Table 6.42 General Characteristics of Running Water Ecosystems According to Size of Stream

Stream Size	Primary Energy Source	Production (trophic) State	Light and Temperature Regime	Trophic Status of Dominant Insects	Fish
*Small headwater streams (stream order 1–3)	Coarse particular organic matter (CPOM) from the terrestrial environmental	Heterotrophic	Heavily shaded	Shredders	Invertivores
	Little primary production	P/R < 1	Stable temperatures	Collectors	
*Medium-sized streams (4–6)	Fine particulate organic matter (FPOM), mostly	Autotrophic	Little shading	Collectors	Invertivores
	Considerable primary production	P/R < 1	High daily temperature variation	Scrapers (grazers)	Piscivores
Large rivers (7–12)	FPOM from upstream	Heterotrophic	Little shading	Planktonic	Planktivores
		P/R < 1	Stable temperatures	Collectors	

* Streams are typically subdivided into three size classes based on the stream order classification system of Kuehne (1962).

Modified from Cummins, K.W. Ecology of running waters: theory and practice, in *Proc. Sandusky River Basin Symposium.* Edited by D.B. Baker. Heidelburg College, Tiffin, OH. 1975.

Periphyton Sampling

Periphyton is a general descriptor which can encompass epipelic (sediment surface), epilithic (stone surface), and epiphytic (plant surface) algae and other benthic meio-, microorganisms. Most periphyton studies have focused on the diatom group, which frequently dominates. Green algae, blue-green algae (cyanobacteria), and flagellates are also dominant species in some sediments, with diatoms favoring calcareous sediments (Wetzel 1975). The animal communities which may be present include protozoa, rotifers, nematodes, and bryozoans. A major controlling factor is light. In turbid, eutrophic, shaded, or deep waters, the low light levels may restrict photosynthetic activity (Wetzel 1975). Some epipelic algae appear to have a diurnal migration pattern through the top few centimeters of sediment in response to light availability. Their photosynthetic activity causes a diurnal change in oxygen concentrations with the upper few millimeters of sediment (Carlton and Klug 1990), which may affect metal bioavailability. They serve as an important transformation link for nitrogen, assimilating pore water ammonia and excreting organic nitrogen to overlying waters, and may be the primary productivity source (Wetzel 1975). These communities have less temporal fluctuation in a lake than planktonic algae and may have one to two biomass peaks per year (Wetzel 1975). Some algae present on sediment surfaces may have settled from the water column and can resuspend to overlying waters.

The algal community is not only extremely important in aquatic ecosystems, but has several attributes as a monitoring tool. Algae have short life cycles. Therefore, they indicate recent-to-present water quality conditions. They are directly affected by physical and chemical conditions since they are primary producers. Sampling of indigenous algae is nondestructive, easy, and inexpensive, and traditional assessment methods exist. Finally, they represent a unique level of biological organization and are sensitive to contaminants which may not be detected with nonalgal surrogates.

Periphyton is difficult to study in a quantitative manner when collecting from natural substrates, as small particle size-surface area differences between samples or sites can have significant effects. Often-used taxonomic references for algae and diatoms include Smith (1950), Prescott (1962, 1970), and Patrick and Reimer (1966). The use of artificial substrates for periphyton and other benthic communities removes the substrate variable. Natural substrates may be sampled using the methods of Stevenson and Lowe (1986) or Hamala et al. (1981). A commonly used artificial sampler (diatometer) consists of multiple glass slides suspended from a floating holding frame (APHA 1985; Figure 6.121; also see Figure 4.11 illustrating the use of a diatometer in Coyote Creek, San Jose, CA). Not all species will colonize the glass slides, but the advantages of efficient and precise evaluations outweigh this

Figure 6.121 Diatometer for artificial substrate periphyton sampling.

disadvantage in most cases. Valid station comparisons are only possible when the key variables affecting periphyton communities are similar; these include flow, turbidity, temperature, dissolved oxygen, alkalinity, hardness, conductivity, nutrients (APHA 1985), and photosynthetically active radiation (LiCor 1979).

The periphyton community can be evaluated for stormwater effects using several endpoints. When using a diatometer, slides should be left *in situ* for 6 to 14 days, then placed in formalin upon collection. Evaluation endpoints may include: number, richness, relative abundance, diversity, chlorophyll *a*, and other community or productivity indices (APHA 1985; Crossey and LaPoint

1988). Stevenson and Lowe (1986) recommend counting 200 cells for dominant, 500 for uncommon, and 1000 cells for rare species, or an additional 100 cells for each new species encountered (EPA 1989a,b,c, 1999).

Periphyton community analyses may be of a structural and functional nature. Structural measures include diversity indices, taxa richness, indicator species, and biomass (Rodgers et al. 1979; Wetzel 1979; Palmer 1977; Patrick 1973). Functional measures which have been used are primary productivity (e.g., chlorophyll *a*), or respiration (Rodgers 1979). Integrating the structural and functional characteristic provides the best means of evaluating ecosystem health, as demonstrated in the macroinvertebrate and fish approaches below.

Protozoan Sampling

Protozoans, like algae, exist in the planktonic and benthic communities. Because their biomass is relatively low compared to that of other aquatic communities, their contribution as a food source to higher trophic levels is probably limited; however, their function as predators or decomposers may fill important ecosystem niches and assist in maintaining–stabilizing decomposition and cycling processes. When protozoan cropping of bacteria is removed, the sediments can function as a carbon sink and microbial community structure–function relationships could alter, affecting nutrient availability to higher trophic levels (Griffiths 1983; Porter et al. 1987).

Several studies have shown the effective use of artificial (polyurethane) substrates in water and sediment pollution studies (Pontasch et al. 1989; Henebry and Ross 1989). This approach allows the foam substrates to colonize at reference sites for several days. Then they are exposed to toxicants either in the laboratory or *in situ* to test sample waters and compared to reference responses. The test endpoints of this multispecies assay include decolonization, protozoan abundance, taxa number, phototroph and heterotroph abundance, respiration, and island-epicenter colonization rates. Both stimulatory and inhibitory results are observed, and careful interpretation is required (Henebry and Ross 1989).

Macroinvertebrate Sampling

This group is operationally defined as those invertebrates retained on sieve mesh sizes greater than 0.2 mm (Hynes, 1970); however, the larger size of 0.5 or 0.95 mm (U.S. Standard No. 30) is used routinely (EPA 1989c). More representative benthos samples may be collected using smaller mesh sizes, such as 0.25 mm (U.S. Standard No. 60), which collect early life stages, chironomids, and nadid and tubificid oligochaetes (EPA 1990b). The major freshwater taxonomic groups may be separated into the trophic levels — functional feeding group descriptors of herbivores, omnivores, carnivores; or deposit and detritus feeders, collectors, shredders, grazers; or scrapers, parasites, scavengers, and predators (EPA 1990b). In most studies of high-to-medium-quality waters, species level identification will be necessary, with tolerant species only dominating in polluted systems. Each taxonomic group may contain a variety of functional feeding groups (Table 6.43). Some common pollution indicators are shown in Figure 6.122.

The benthic macroinvertebrate community has been used for many years to qualitatively and, more recently, to quantitatively assess water quality and pollution effects. There are advantages and disadvantages in using macrobenthos in water quality assessments (Table 6.44). However, except in cases of extreme and obvious pollution, they should always be a component of a stormwater effect assessment.

There is a wealth of reference information available to assist in the use of macroinvertebrates as monitoring tools, including Armitage (1978), Benke et al. (1984), Brinkhurst (1974), Cairns (1979), Cummins et al. (1984), Cummins and Wilzbach (1985), Edmondson and Winberg (1971), Goodnight and Whitley (1960), Hart and Fuller (1974), Hellawell (1978, 1986), Hilsenhoff (1977), Howmiller and Scott (1977), Hynes (1960, 1970), Holme and McIntyre (1971), Hulings and Gray (1971),

Table 6.43 Trophic Mechanisms and Food Types of Aquatic Insects

General Category Based on Feeding Mechanism	General Particle Size Range of Food (μm)	Subdivision Based on Feeding Mechanisms	Subdivision Based on Dominant Food	Aquatic Insect Taxa Containing Predominant Examples
Shredders	>103	Chewers and miners	Herbivores: living vascular plant tissue	Trichoptera (*Phryganeidae, Leptoceridae*) Lepidoptera Coleoptera (*Chrysomelidae*) Diptera (*Tipulidae, Chironomidae*)
		Chewer and miners	Detritivores (large particle detritivores): decomposing vascular plant tissue	Plecoptera (*Filipalpia*) Trichoptera (*Limnephilidae, Lepidostomatidae*) Diptera (*Tipulidae, Chironomidae*)
Collectors	<103	Filter or suspension feeders	Herbivores-detritivores: living algal cells, decomposing vascular plant tissue	Ephemeroptera (*Siphlonuridae*) Trichoptera (*Philopotamidae, Psychomyidae, Hydropsychidae, Brachycentridae*) Lepidoptera Diptera (*Simuliidae, Chironomidae, Culicidae*)
		Sediment or deposit (surface) feeders	Detritivores (fine particle detritivores): decomposing organic particulate matter	Ephemeroptera (*Caenidae, Ephemenidae, Leptophlebiidae, Baetidae, Ephemerellidae, Heptageniidae*) Hemiptera (*Gerridae*) Coleoptera (*Hydrophilidae*) Diptera (*Chironomidae, Ceratopogonidae*)
Scrapers	<103	Mineral scrapers	Herbivores: algae and associated microflora attached to living and nonliving substrates	Ephemeroptera (*Heptageniidae, Baetidae, Ephemerellidae*) Triochoptera (*Glossosomatidae, Helicopsychidae, Molannidae, Odontoceridae, Goreridae*) Lepidoptera Coleoptera (*Elmidae, Psephenidae*) Diptera (*Chironomidae, Tabanidae*)
		Organic scrapers	Herbivores: algae and associated attached microflora	Ephemeroptera (*Caenidae, Leptophlebiidae, Heptageniidae, Baetidae*) Hemiptera (*Corixidae*) Trichoptera (*Leptoceidae*) Diptera (*Chironomidae*)
Predators	>103	Swallowers	Carnivores: whole animals (or parts)	Odonata Plecoptera (*Setipalpia*) Megaloptera Trichoptera (*Rhyacophilidae, Polycentropidae, Hydropsychidae*) Coleoptera (*Dytiscidae, Gyrinnidae*) Diptera (*Chironomidae*)
		Piercers	Carnivores: cell and tissue fluids	Hemiptera (*Belastomatidae, Nepidae, Notonectidae, Naucoridae*) Diptera (*Chironomidae*)

Lenat (1983), Lind (1985), Merritt and Cummins (1984), Mason (1981), Metcalfe (1989), Milbrink (1983), Meyer (1990), Neuswanger et al. (1982), Pennak (1989), Posey (1990), Resh (1979), Resh and Roseberg (1984), Resh and Unzicker (1975), Reynoldson et al. (1989), Ward and Stanford (1979), Warren (1971), Waters (1977), Welch (1948), Welch (1980), Winner et al. (1975), EPA (1989a,c, 1990a,c, 1999), and OEPA (1989). Previous discussions highlighted the importance of attempting to control habitat (e.g., substrate), flow dynamics, and seasonal variables when monitoring — particularly the benthic macroinvertebrate community. Other, obviously critical issues, include the sampling procedure's precision and accuracy, taxonomic identification, and data evaluation.

Substrates can be sampled with nets, grab (dredge), core, and vegetation collection devices (Table 6.45). The Hess and Surber samples are often used to sample stream riffle habitats, whereby the substrates within a confined 0.1 m² area are vigorously disrupted and scrubbed down to a depth of approximately 10 cm. A flow velocity of at least 0.5 m/s is required for effective use of these net samplers. See also ASTM (1987) for additional information.

Sampling is frequently of a qualitative to semiquantitative nature that is relatively easy to conduct. The objective here is to determine differences between sites. Semiquantitative methods incorporate a level-of-effort constant or use quantitative methods in a nonrandom manner (EPA 1990b). Quantitative methods sample unit areas or volumes of habitat in a random manner. The approach chosen should depend on the data quality objectives (DQOs).

Semi- and quantitative sampling may use grab samplers (see Chapter 5 and Table 6-45), stream net samplers (Figure 6.123 and Table 6.46), and artificial substrates (Figures 6.124 through 6.127 and Table 6.47).

In large streams, deep waters, and areas of slow current velocities, it is necessary to use core or dredge samplers, which are also used for sediment sampling as discussed previously (Chapter 5). See also ASTM (1987, 1991), Lind (1979), APHA (1985), Downing (1984), and Wetzel and Likens (1991), for additional sampler information. The Ekman and Ponar grab samplers are commonly used in relatively soft sediments of clay to gravel size, with relatively good efficiency (Elliott and Drake 1981). The hand and gravity corers are preferred in soft sediments because pressure waves and loss of surficial fines are reduced, variance can be determined horizontally and vertically, sieving volume is reduced, precision is increased, and sediment structure-integrity is maintained to

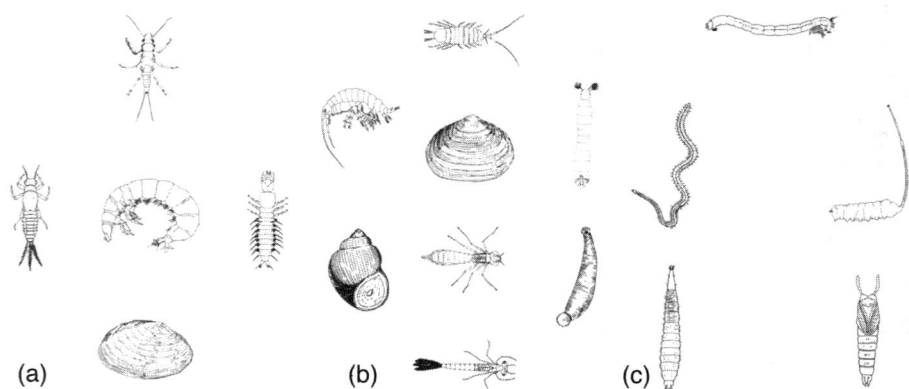

(a)   (b)   (c)

Figure 6.122 Representatives of stream bed animals commonly associated with various degrees of organic pollution. (a) The clean water (sensitive) group (from left): stonefly nymph, mayfly naiad, caddisfly larvae, hellgrammite, unionid clam. (b) The intermediately tolerant group (from left): scud, sowbug, blackfly larvae, fingernail clam, snail, dragonfly nymph, leech, damselfly nymph. (c) The very tolerant group (from left): bloodworm or midge larvae, sludgeworm, rattailed maggot, sewage fly larvae, sewage fly pupae. (From *The Practice of Water Pollution Biology*. U.S. Department of the Interior. Washington, D.C. 1969.)

Table 6.44   Advantages and Disadvantages of Using Macroinvertebrates and Fish in Evaluation of the Biotic Integrity of Freshwater Aquatic Communities

Advantages	Disadvantages
Macroinvertebrates	
Fish, highly valued by humans, are dependent on benthic invertebrates as a food source.	They require taxonomic expertise for identification, which is also time-consuming.
Many species are sensitive to pollution	Background life-history information is lacking for some species and groups.
Bottom fauna often have a complex life cycle of a year or more, and, therefore, represent long-term exposure periods to water and sediment conditions.	Results are difficult to translate into values meaningful to the general public.
Many have an attached or sessile mode of life and are not subject to rapid migrations, therefore serve as resident monitors of test site quality.	May not detect short-term or recent chronic pollution events.
	Not as sensitive a pollution indicator in large rivers, bays, lakes, and marine systems.
	Natural levels of spatial and temporal variation may make detection of significant effects difficult.
Fish	
Life history information is extensive for most species.	Sampling fish communities is selective in nature.
Fish communities generally include a range of species that represent a variety of trophic levels (omnivores, herbivores, insectivores, planktivores, piscivores) and utilize foods of both aquatic and terrestrial origin. Their position at the top of the aquatic food web also helps provide an integrated view of the watershed environment.	Fish are highly mobile. This can cause sampling difficulties and also creates situations of preference and avoidance. Fish also undergo movements on diurnal and seasonal time scales. This increases spatial and temporal variability, which makes detection of significant effects or trends difficult.
Fish are highly valued by the public.	There is a high requirement for manpower and equipment for field sampling.
Fish are relatively easy to identify. Most samples can be sorted and identified in the field, and then released.	
Both lethality and stress effects (depressed growth, lesions, abnormalities, and reproductive success) can be evaluated. Careful examination of recruitment and growth dynamics among ages of fish can help pinpoint periods of unusual stress.	

Modified from Cairns, J., Jr. and K.L. Dickson. A simple method for the biological assessment of the effects of waste discharges on aquatic bottom-dwelling organisms. *J. Water Pollut. Control Fed.*, 43: 755–772. 1971; Karr, J.R. and D.R. Dudley. Ecological perspective on water quality goals. Ecological perspective on water quality goals. *Environ. Manage.*, 5: 55–68. 1981. With permission.

Table 6.45   Sampling Methods for Macroinvertebrates

Method	Habitat	Substrate Type	Effort Required[a] Persons	Time (hr)	Ref.
Hess, Surber	Stream riffle (<0.5 m deep)	Sand, gravel, cobble	1	0.50	ASTM (1987)
Ponar grab	Rivers, lakes, estuaries	Mud, silt, sand, fine gravels	2	0.50	ASTM (1987)
Ekman grab	Stream pools, shallow lakes	Mud, silt, sand	1	0.25	ASTM (1987)
Corers	Rivers, lakes	Mud, silts	1–2	0.25	Downing (1984)
Sweep net	Littoral	Vegetation	1	0.25	Downing (1984)
Macan McCauley Minto Wilding	Littoral	Vegetation	1	0.50	Downing (1984)
Standardized substrates	All	All	1	0.25–1.0[b]	APHA (1985)

[a] Effort includes time spent in field to collect, sieve, and isolate one sample. Laboratory time required to remove and identify organisms ranges from 1 to 5 per sample, depending on expertise level, and taxonomic resolution sought.

[b] Two- to six-week colonization time ended before sample is removed.

Modified from EPA. *Ecological Assessment of Hazardous Waste Sites.* Environmental Research Laboratory. U.S. Environmental Protection Agency, Corvallis, OR. EPA 600/3-89/013. 1989a.

a much higher degree than in dredge samples. The principal disadvantages, however, are their ineffective sampling of coarse, large-grained sediments and the small volumes that are collected.

The efficiency of benthic collection samplers has been compared, and, in general, the grab samplers are less efficient than the corers (ASTM 1991a). The Ekman dredge is the most commonly used sampler for benthic investigations (Downing 1984). The Ekman is limited to less compacted, fine-grained sediments, as are the corer samplers. However, these are usually the sediments of greatest concern in toxicity assessments. The most commonly used corer is the Kajak–Brinkhurst, or hand corer. In more resistant sediments, the Petersen, Ponar, Van Veen, and Smith–McIntyre dredges are used most often (Downing 1984). Based on studies of benthic macroinvertebrate populations, the sediment corers are the most accurate samplers, followed by the Ekman dredge, in most cases (Downing 1984). For consolidated sediments, the Ponar dredge was identified as the most accurate, while the Petersen was the least effective (Downing 1984).

Quantitative benthic macroinvertebrate sampling of small streams may be improved by also using small to large emergence traps. These samplers trap the dominant stream insects as they leave the water as flying adults. In this way, effects from habitat heterogeneity are reduced, time-consuming "bug" picking from substrate samples is avoided, and most adult stages can be identified to the species level. See also Wetzel and Likens (1991), Illies (1971), Hall et al. (1980), and Peckarsky (1984).

Semiquantitative methods also include the traveling kick method (Horning and Pollard 1978) and the Rapid Bioassessment Protocols II and III (kicknets) (EPA 1990b). Readers should note that the EPA Rapid Bioassessment Protocol manual has been revised (EPA 1999) and no longer differentiates Protocols I through III (EPA 1989c). As with other sediment-associated components, quantitative evaluations are complicated by often high degrees of variability. By using multimetric (indice) assessment endpoints, the impact of population variability can be reduced (EPA 1990b). Nevertheless, it is essential that replicate sampling of each habitat niche be conducted at each site, allowing measures of precision. Precision may also be increased by collecting larger samples, thus the influence of reducing small patches. Three to five replicates are a minimum requirement. Use of a transect to select replicate sites may result in different habitats being selected.

A number of artificial substrate samplers have been used to assess benthic macroinvertebrate conditions, i.e., flow, depth, light, and temperature. These samplers remove the substrate variable and provide known sampling areas and exposure times. Unfortunately, there are some disadvantages which may be significant, including some taxa may not utilize the substrate; proportional relationships may be altered; substrates are colonized primarily by upstream "drift" organisms, and effects from contact with possibly contaminated bed sediments is reduced or eliminated; they require 4- to 8-week exposures and two sampling trips; and they may be lost due to high flow or vandalism (see Figures 6.123 to 6.127). As with the periphyton samplers, care must be taken to ensure uniformity.

For most studies, semiquantitative approaches using the EPA's Rapid Bioassessment Protocols II or III (RBP) and the Ohio EPA Hester–Dendy samplers (Figure 6.127) are preferred, with habitat evaluations. The RBP II method samples 1 m² riffle areas, and 100 organisms are randomly picked and identified to the family level (EPA 1989c). The Ohio EPA method uses 10 metrics (nine based on Hester–Dendy results and one based on dip net sampling) to compute an Invertebrate Community Index in wadeable streams (Ohio 1989). In streams where rocks are the dominant habitat, it may be useful to use a basket sampler (Figure 6.124) containing approximately 30 rocks of equal size or a particle size distribution similar to the test or reference site. This approach is used by the State of Maine and by other investigators (e.g., Clements et al. 1996). It is the most realistic artificial substrate method. When high quantities of biomass are needed, such as for tissue residue analyses, the grill-basket sampler containing 3M polyethylene mesh is useful (Stauffer et al. 1974). All of the artificial substrates are set out in triplicate and secured to concrete blocks in shallow waters for 4 to 8 weeks. The metrics vary in their ability to detect organic material or toxicant-related impacts. They overlap in ranges of sensitivity and thereby reinforce final conclusions regarding the condition of the system's biological communities (EPA 1989c). The RBP II methods, organism pollution tolerance levels, and indices calculated in the RBP and Ohio EPA methods are described in detail in Appendix B. Note that in many states, special collection permits are required to collect macroinvertebrates.

Table 6.46 Comparison of Stream-Net Samplers

Type	Habitats and Substrates Sampled	Effectiveness of Device	Advantages	Limitations
Surber sampler	Shallow, flowing streams, less than 32 cm in depth with good current; rubble substrate, mud, sand, gravel	Relatively quantitative when used by experienced biologist; performance depends on current and substrate	Encloses area sampled; easily transported or constructed; samples a unit area	Difficult to set in some substrate types, that is, large rubble; cannot be used efficiently in still slow-moving streams
Portable invertebrate box sampler, Hess stream bottom sampler, and stream-bed fauna sampler	Same as Surber	Same as Surber	Same as above except completely enclosed with stable platform; can be used in weed beds	Same as Surber
Drift nets	Flowing rivers and streams; all substrate types	Relatively quantitative and effective in collecting all taxa which drift in the water column; performance depends on current velocity and sampling period	Low sampling error; less time, money, effort; collects macroinvertebrates from all substrates, usually collects more taxa	Unknown where organisms come from; terrestrial species may make up a large part of sample in summer and periods of wind and rain; does not collect nondrifting organisms

From EPA. *Macroinvertebrate Field and Laboratory Methods for Evaluating the Biological Integrity of Surface Waters.* Office of Research and Development, U.S. Environmental Protection Agency, Washington, D.C., EPA 600/4-90/030. 1990.

Table 6.47   Comparison of Substrate Samplers

Type of Substrate	Advantages	Limitations
Artificial		
General characteristics	Reduce habitat substrate variability influence Eliminate subjectivity in collection process Patchiness reduced Skill level required is less Long exposure periods (6–8 weeks) Discriminate between sediment and water toxicity	Habitats may be different, thus promotes growth of different species, not representative of site. Two trips needed Long exposure periods (6–8 wks) Sediment substrate effects, including toxicity, reduced Sampler loss through vandalism or sedimentation
Modified	Reduces compounding effects of substrate differences, multiplate sampler	Long exposure time, difficult to anchor, easily vandalized
Fullner	Wider variety of organisms	Same as modified Hester–Dendy
Basket Type	Comparable date, limited extra material for quick lab processing. Large amount of biomass.	No measure of pollution on strata, only community formed in sampling period, long exposure time, difficult to anchor, easily vandalized
Periphyton	Floats on surface, easily anchored, glass slides exposed just below surface	May be damaged by craft or flows, easily vandalized
Natural		
Any bottom or sunken material	Indicate effects of pollution, gives indication of long-term pollution	May be difficult to quantitate; possible lack of growth, not knowing previous location or duration of exposure

Modified from EPA. *Handbook for Sampling and Sample Preservation of Water and Wastewater.* Environmental Monitoring and Support Lab, U.S. Environmental Protection Agency, Cincinnati, OH, EPA 600/4-82/029. 1982; EPA. *Ecological Assessment of Hazardous Waste Sites.* Environmental Research Laboratory, U.S. Environmental Protection Agency, Corvallis, OR. EPA 600/3-89/013. 1989a.

Figure 6.123 Stream net sampler.

Figure 6.124 Artificial substrates (polyethylene mesh) in BBQ baskets secured to cinder blocks.

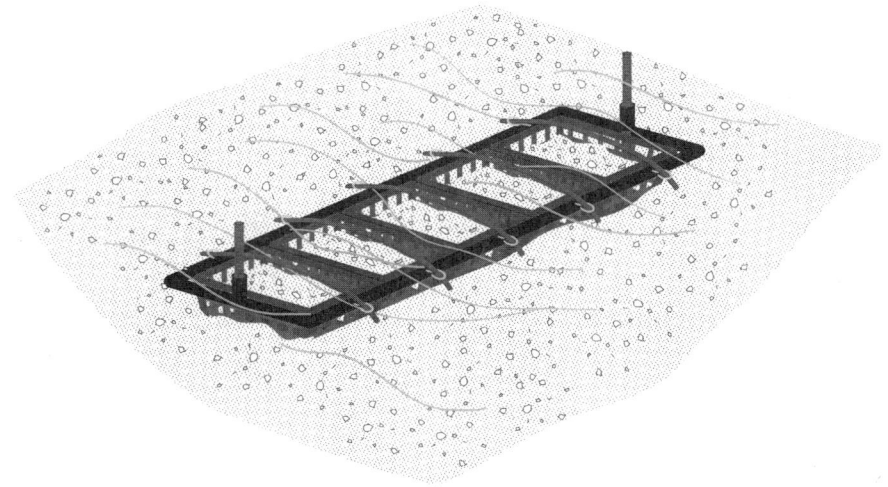

Figure 6.125 Colonization trays buried to stream sediment surface and secured with iron rods. Baskets are filled with cleaned substrates representative of the reference or test site.

Figure 6.126 Periphyton sampler, two styrofoam floats with eight glass microscope slides in rack.

Figure 6.127 Periphyton sampler in place, plus Hester–Dendy samplers.

Table 6.48   Comparison of Net Sampling Devices

Devices	Application	Advantages	Disadvantages
Wisconsin net	Zooplankton	Efficient shape concentrates samples	Qualitative
Clarke-Bumpus	Zooplankton	Quantitative	No point sampling, difficult to measure depth of sample accurately

From EPA. *Handbook for Sampling and Sample Preservation of Water and Wastewater*. Environmental Monitoring and Support Lab, U.S. Environmental Protection Agency, Cincinnati, OH, EPA 600/4-82/029. 1982.

ZOOPLANKTON SAMPLING

The zooplankton community plays a major role in the food web and aquatic ecosystem dynamics. Its use as an indicator of pollution in lotic systems has been limited. Studies are more common in lentic systems; however, they are complicated by a high degree of spatial and temporal variability, and less knowledge of pollution tolerances as compared to the benthos. The cladocerans, *Daphnia magna*, *Daphnia pulex,* and *Ceriodaphnia dubia*, have been useful as sensitive toxicity surrogate species. If an intensive lake–reservoir ecosystem effect study is to be conducted, they should be included. Commonly used sampling nets are listed in Table 6.48 and Figure 6.128.

Figure 6.128   Net sampler for plankton.

FISH SAMPLING

The fish community is perhaps the most important component of the ecosystem as viewed by public opinion, commercial interests, and regulatory requirements. In reality, however, it is no more important than any other major level of biological organization and is not as useful as other groups when evaluating stormwater effects. Fish, by nature, are in general a more transitory species than other aquatic organisms and, therefore, produce more variable results in biosurveys. Because they are mobile, they are often able to avoid polluted waters. This avoidance behavior makes evaluations of site-specific sources and problems more difficult. Sampling methods vary in their degree of efficiency and selectivity and compound data variance problems (EPA 1989c). They do, however, possess many advantages in the assessment process:

- Beneficial uses of stream segments characterized in terms of fisheries
- Many endangered species exist
- Effective collection methods exist
- Effective quality assessments are possible using community structure and functional metrics to form an index of integrity
- Used as regulatory and monitoring tools for decades; an extensive database exists on life history, distribution, and effects
- Indicators of long-term exposures and watershed conditions
- Comprise multiple trophic levels
- Drive ecosystem dynamics in the "top-down" approach theory and may integrate effects from lower trophic levels ("bottom-up" approach)
- Contaminant source to humans
- Useful for sublethal, chronic pollutant exposure effect studies

Many fish communities contain multiple trophic levels, such as invertivores, planktivores, herbivores, omnivores, and piscivores (Table 6.49; Karr et al. 1983). Trophic guild information is useful for evaluating system alterations at a functional and structural level. The omnivore component typically increases as water quality declines. Streams with fewer than 20% omnivores are often of good quality, and poor if greater than 45% are true omnivores (Karr 1981). There is also often a strong inverse correlation between the abundance of insectivorous cyprinids (minnows) and water quality (more abundant minnow populations indicate worse water quality). Another generality of feeding type and water quality is the presence/absence of top carnivores, which are at the top of the aquatic food chain and thereby integrate lower trophic level effects. They are most likely to show biomagnified toxicants in their tissue, but might not necessarily show effects from those toxicants. The validity of these generalizations has been well documented in the agricultural Midwest. However, there are exceptions nationwide. Tissue residues are good indicators of exposure for some nonpolar organics and methyl mercury; however, many metals and organics that can be metabolized cannot be assessed well with tissue information.

Sampling of fish communities is relatively labor intensive, often requiring special equipment and expertise. But, given the importance of fish in ecosystem structure and functioning, sport and commercial fishing, and public perceptions, they should be monitored.

Generally, the preferred sampling season is mid to late summer, when stream and river flows are moderate to low, and less variable than during other seasons (EPA 1990b). Although some fish species are capable of extensive migration, fish populations and individual fish may remain in the same area during summer (Funk 1957; Gerking 1959; Cairns and Kaesler 1971). However, large river, lake, and harbor habitats promote greater migration ranges. Ross et al. (1985) and Matthews (1986) found that stream fish assemblages were stable and persistent for 10 years, recovering rapidly from droughts and floods, indicating that large population fluctuations are unlikely to occur in response to purely natural environmental phenomena. However, comparison of data collected during different seasons is discouraged, as are data collected during or immediately after major flow changes (EPA 1989a).

Although various collection methods are routinely used to sample fish; electrofishing (Figures 6.129 through 6.131), seines (Figure 6.132), and rotenone (a poison) are the most commonly used methods in freshwater habitats (Tables 6.50 and 6.51). Each method has advantages and disadvantages (Nielsen and Johnson 1983; Hendricks et al. 1980). However, electrofishing is recommended for most fish field surveys because of its greater applicability and efficiency, and the good recoverability of stunned fish that are returned to the water (EPA 1989a,c).

Table 6.49   Trophic Guilds Used by Schlosser (1981, 1982a, 1982b) to Categorize Fish Species

Herbivore–Detritivores (HD)	HD species feed almost entirely on diatoms or detritus.
Omnivores (OMN)	OMN species consume plant and animal material. They differ from GI species in that, subjectively, greater than 25% of their diet is composed of plant or detritus material.
Generalized insectivores (GI)	GI species feed on a range of animal and plant material including terrestrial and aquatic insects, algae, and small fish. Subjectively, less than 25% of their diet is plant material.
Surface and water column insectivores (SWI)	WSI species feed on water column drift or terrestrial insects at the water surface.
Benthic insectivores (BI)	BI species feed predominantly on immature forms of benthic insects.
Insectivore–Piscivores (IP)	IP species feed on aquatic invertebrates and small fish. Their diets range from predominantly fish to predominantly invertebrates.

Figure 6.129 Electrofishing with backpack unit in main stream reach (notice nearby seine to capture stunned fish).

Figure 6.130 Electrofishing with backpack unit in near-shore areas.

Figure 6.131 Boat electrofishing unit. (Courtesy of Wisconsin Department of Natural Resources.)

Figure 6.132 Fish seining.

Metrics	Biological Condition		
	Non-Impaired		Severely Impaired
Species	├──────────────────────────┤		
Darters	├──────────────┤		
Sunfishes	├──────────────────┤		
Suckers	├────────────────────┤		
Intolerants	├──────┤		
% Green Sunfish			├──────┤
% Omnivores		├───────────────┤	
% Insectivorous Cyprinids		├────────────┤	
% Piscivores	├──────────┤		
Number		├────────────┤	
% Hybrids			├──────┤
% Diseased			├──────┤

Figure 6.133 Range of sensitivities of Rapid Bioassessment Protocol V fish metrics in assessing biological condition. (Modified from EPA. *Ecological Assessment of Hazardous Waste Sites.* Environmental Research Laboratory. U.S. Environmental Protection Agency, Corvallis, OR. EPA 600/3-89/013. 1989a.)

Indices of Fish Populations

Perhaps the most popular index is the IBI. A slightly modified version is used in the EPA Rapid Bioassessment Protocols for fish. The IBI is weighted on the basis of individual species' tolerances to water and habitat quality. The IBI is comprised of 12 metrics, as follows:

A. Species richness and composition
   1. Species number
   2. Darter species number
   3. Sucker species number
   4. Sunfish species number
   5. Intolerant species number
   6. Green sunfish proportion
B. Abundance and condition
   1. Individual numbers
   2. Hybrid proportion
   3. Proportion with disease anomalies
C. Trophic composition
   1. Omnivore proportion
   2. Insectivorous cyprinid proportion
   3. Piscivore proportions

Table 6.50 Fish Sampling Methods

Methods	Advantages	Disadvantages
Electrofishing	Greater standardization of catch per unit of effort Less time and manpower than some sampling methods Less selective than seining (although it is selective toward size and species) Adverse effects on fish are minimized Appropriate in a variety of habitats	Sampling efficiency is affected by turbidity and conductivity. Initial cost of equipment Although less elective than seining, electrofishing is size and species selective. Effects of electrofishing increase with body size. Species specific behavioral and anatomical differences also determine vulnerability to electroshocking A hazardous operation that can injure field personnel if proper safety procedures are ignored
Reformed seining	Relatively inexpensive Lightweight and are easily transported and stored Repair and maintenance are minimal and can be accomplished on-site Restricted water quality parameters Effects on the fish population are minimal because fish are collected alive and are generally unharmed	Previous experience and skill, knowledge of fish habitats and behavior, and sampling effort are probably more important in seining than in the use of any other gear Sample effort and results are more variable than sampling with electrofishing or rotenoning Generally restricted to slower water with smooth bottoms, and is most effective in small streams or pools with little cover Standardization of unit of effort to ensure data comparability is difficult
Rotenoning	Effective use independent of habitat complexity Greater standardization of unit of effort than seining Provides more complete censusing of the fish population than seining or electrofishing	Kills all fish and possibly nontarget species, should only be used if other methods are not appropriate and if the data are essential Prohibited in many states Application and detoxification can be time and manpower intensive Effective use affected by temperature, light, dissolved oxygen, alkalinity, and turbidity High environmental impact; concentration miscalculations can produce substantial fish kills downstream of the study site

From EPA. *Ecological Assessment of Hazardous Waste Sites.* Environmental Research Laboratory, U.S. Environmental Protection Agency, Corvallis, OR. EPA 600/3-89/013. 1989a.

Table 6.51 Sampling Methods for Fish[a]

Method	Habitat	Persons	Time (hr)
Electrofishing	Small streams	2	0.25–1
	Large streams, rivers, lakes	2	0.25–1
Seining	Small streams or impoundments	2–3	0.50–1
Hoop net	Streams or rivers	2–3	2[b]
Gill, trammel nets	Lakes[d]	2–3	2–4[c]
Fyke net	Lakes[d]	2–3	2[c]

[a] Taken from Lagler (1978); Hendricks et al. (1980); Hubert (1983); Nielsen and Johnson (1985).

[b] Time for obtaining fish sample; time for stationary netting techniques includes time spent setting and receiving nets. It does not include time required to process sample (weighing, measuring, or taxonomic identification), which can range from 1 to 4 hours depending on taxonomic resolution and number of fish obtained.

[c] Time for hoop, gill, trammel, and fyke nets does not include 24 hours or period for which net is left in water to obtain sample.

[d] Gill, trammel, and fyke nets can also be used in some cases in flowing water if properly anchored; however, debris usually makes these applications troublesome.

From EPA. *Protocols for Short-Term Toxicity Screening of Hazardous Waste Sites.* Environmental Research Laboratory, U.S. Environmental Protection Agency, Corvallis, OR. EPA 600.3-88/029. 1989b.

Each metric is scored as 1 (worst), 3 (moderate), or 5 (best) as compared to the reference site or other data (see Fausch et al. 1984) showing regional norms (Table 6.52). Therefore, the index may range from 12 to 60 after all metric scores are totaled. Regional modifications have been developed by Hughes and Gammon (1987), Leonard and Orth (1986), Steedman (1988), and Wade and Stalaup (1987). The IBI is shown generally in Figure 6.134 and described in detail in Appendix C.

The Index of Well-Being (IWB), developed by Gammon (1976), was also developed in the midwestern United States to evaluate environmental stress effects on riverine fish. It is simpler than the IBI, using four measures: numbers of individuals, biomass, and the Shannon diversity index based on numbers and weight. Unfortunately, in some systems, high numbers and biomass of pollution-tolerant species produce a high index value, yet quality is reduced. To deal with this problem the Ohio EPA (1989) and Gammon (1989) developed a modified IWB which eliminates highly tolerant species, exotic species, or hybrids from the numbers and biomass components of the IWB, but retained in the Shannon index calculations. This modification has proven to be an effective assessment tool, which is consistent and sensitive to a wide range of environmental stresses. These equations are listed below:

Index of Well-Being:

$$IWB = 0.5 \ln N + 0.5 \ln B + H \text{ (no.)} + H \text{ (wt.)}$$

where   N   = relative number of all species
        B   = relative weight of all species
        H (no.) = Shannon index based on relative numbers
        H (wt.) = Shannon index based on relative weight

Shannon Diversity Index:

$$\overline{H} = -\sum \left(\frac{n_i}{N} \right) \ln \left(\frac{n_i}{N} \right)$$

where n_i = relative numbers or weight of the ith species
       N = total number or weight of the sample

The IBI and mIWB require that indigenous fish species be classified in terms of environmental tolerance (to both natural and anthropogenic stressors). Tolerance levels (Appendix C) vary with each species, between ecoregions, seasonally, at different life stages, and they depend on the presence of other stressors, organism health, and the type of stressor. This group of critical variables makes any "tolerance" classification crude and tenuous. Nonetheless, the use of these classifications has been effective in evaluating ecosystem impairment. For many systems, shifts in dominant species and trophic classification away from sensitive, nonomnivores (e.g., trout, walleye) to tolerant omnivores (e.g., carps), clearly and easily show impairment exists. In other areas, where impairment is just beginning, as in a stream reach downstream of acute effects ("gray" zone), and where ecosystem recovery is beginning, the species tolerance levels will be uncertain.

TOXICITY AND BIOACCUMULATION

Why Evaluate Toxicity?

Toxicity and bioaccumulation evaluations are important and often essential components of storm-water impact assessments. They produce information that cannot be accurately determined or extrapolated from other assessment components. Toxicity tests have strengths and weaknesses that must be recognized (Table 6.53). If there is a clear understanding of the test responses and associated assumptions, and if proper QA/QC is followed, toxicity testing will allow for sensitive, meaningful, and efficient assessments of ecosystem quality and will identify stressor magnitude frequency, and duration. The science of aquatic toxicology has progressed rapidly in recent years and is now an integral component of many EPA regulatory programs. Toxicity testing may evaluate effects and address a wide variety of study objectives, using any of several general and specific monitoring approaches (Table 6.54). This variety of approaches allows for a high number of different component combinations, with each possibly providing unique information and having different assumptions associated with them. Many different approaches and organisms have been used for toxicity testing, and these will be discussed in the following section. Figures 6.135 through 6.138 show several test setups used in the Environmental Health Sciences laboratories at Wright State University, while Figures 6.139 and 6.140 are two of the Azur Environmental procedures, using phosphorescent phytoplankton, used in the environmental engineering labs at the University of Alabama at Birmingham.

Odum (1992) stated that stress is usually first detected in sensitive species at the population level. Natural population and community responses are not measured directly with whole effluent toxicity (WET) tests (La Point et al. 1996, 2000). The traditional surrogates (*P. promelas* and *C. dubia*) may not be as sensitive as indigenous species (Cherry et al. 1991). Indirect effects of toxicity on species, population, and community interactions can be important (Clements et al. 1989; Clements and Kiffney, 1996; Day et al. 1995; Fairchild et al. 1992; Giesey et al. 1979; Gonzalez and Frost 1994; Hulbert 1975; La Point et al. 2000; Schindler 1987; Wipfli and Merritt 1994), and may not be detected by WET testing. A huge ecological database exists showing the importance of species interactions in structuring communities (e.g., Dayton 1971; Power et al. 1988; Pratt et al. 1981).

It is less likely that strong relationships will exist between WET test responses and indigenous communities at sites where there are other pollutant sources, where effluent toxicity is low to moderate, or where dilution is high. Based on fish and benthic invertebrate responses, several studies suggest that WET tests are not always predictive of receiving water impacts (Clements and Kiffney 1994; Cook et al. 1999; Dickson et al. 1992, 1996; Niederlehner et al. 1985; Ohio EPA 1987); however, many studies have shown WET tests to be predictive of aquatic impacts (e.g., Birge et al. 1989; Diamond et al. 1997; Dickson et al. 1992, 1996; Eagleson et al. 1990; Schimmel and Thursby 1996; Waller et al. 1996). These differences should not be surprising however, as it is likely a result of WET test organisms and field populations experiencing different exposures (Burton et al. 2000; EPA 1991e). In an effluent-dominated system, the in-stream exposure is very similar to a WET test. A less degraded watershed, or one that is not dominated by point sources, may have sensitive indigenous populations that are exposed to "toxic" effluents at nontoxic concentrations. Conversely, if sensitive species have already been lost from a watershed, a toxic effluent may be inhibiting their return. In highly degraded sites, virtually any traditional assessment tool (acute toxicity testing, chemical concentrations, indigenous communities) will show effects and strong correlations with other tools. The WET tests were not developed to evaluate all natural and anthropogenic stressors nor to show all biological responses (such as mutagenicity, carcinogenicity, teratogenicity, endocrine disruption, or other important subcellular responses). In addition, highly nonpolar compounds may elicit an effect in short-term exposures. These issues dictate that additional assessment tools be utilized in order to protect aquatic ecosystems (Waller et al. 1996).

Table 6.52 Regional Variations of IBI Metrics

Variations in IBI Metrics	Midwest	New England	Ontario	Central Appalachia	Colorado Front Range	Western Oregon	Sacramento San Joaquin
1. Total Number of Species	X	X		X	X		X
# native fish species			X			X	X
# salmonid age classes						X	X
2. Number of Darter Species							
# sculpin species						X	
# benthic insectivore species		X					
# darter and sculpin species			X				
# salmonid yearlings (individuals)						X	X
% round-bodied suckers	X						
# sculpins (individuals)	X						X
3. Number of Sunfish Species	X				X		
# cyprinid species						X	
# water column species		X					
# sunfish and trout species			X				
# salmonid species							X
# headwater species	X						
4. Number of Sucker Species	X	X				X	
# adult trout species						X	X
# minnow species	X						
# sucker and catfish species			X		X		
5. Number of Intolerant Species	X				X	X	X
# sensitive species	X	X					
# amphibian species							X
Presence of brook trout			X				X
6. % Green Sunfish							
% common carp							
% white sucker		X			X	X	
% tolerant species	X						
% creek chub				X			
% dace species			X				

Metric						
7. % Omnivores	X		X	X	X	X
% yearling salmonids	X			X	X	X
8. % insectivorous Cyprinids	X				X	
% insectivores		X	X	X	X	
% specialized insectivores				X	X	X
# juvenile trout						
% insectivorous species	X					
9. % Top Carnivores	X	X	X	X	X	
% catchable salmonids					X	X
% catchable wild trout						X
% pioneering species	X					
Density catchable wild trout						X
10. Number of Individuals	X	X	X	X	X	X
Density of individuals		X				
11. % Hybrids	X	X				
% introduced species				X	X	X
# simple lithophils	X					
% simple lithophilic species	X					
% native species						X
% native wild individuals						X
12. % Diseased Individuals	X	X	X	X	X	
13. Total Fish Biomass		X				

Note: X = metric used in region. Many of these variations are applicable elsewhere.

From EPA, *Rapid Bioassessment Protocols for Use in Streams and Rivers: Benthic Macroinvertebrates and Fish*. Office of Water, U.S. Environmental Protection Agency, Washington, D.C. EPA 444/4-89/001. 1989c.

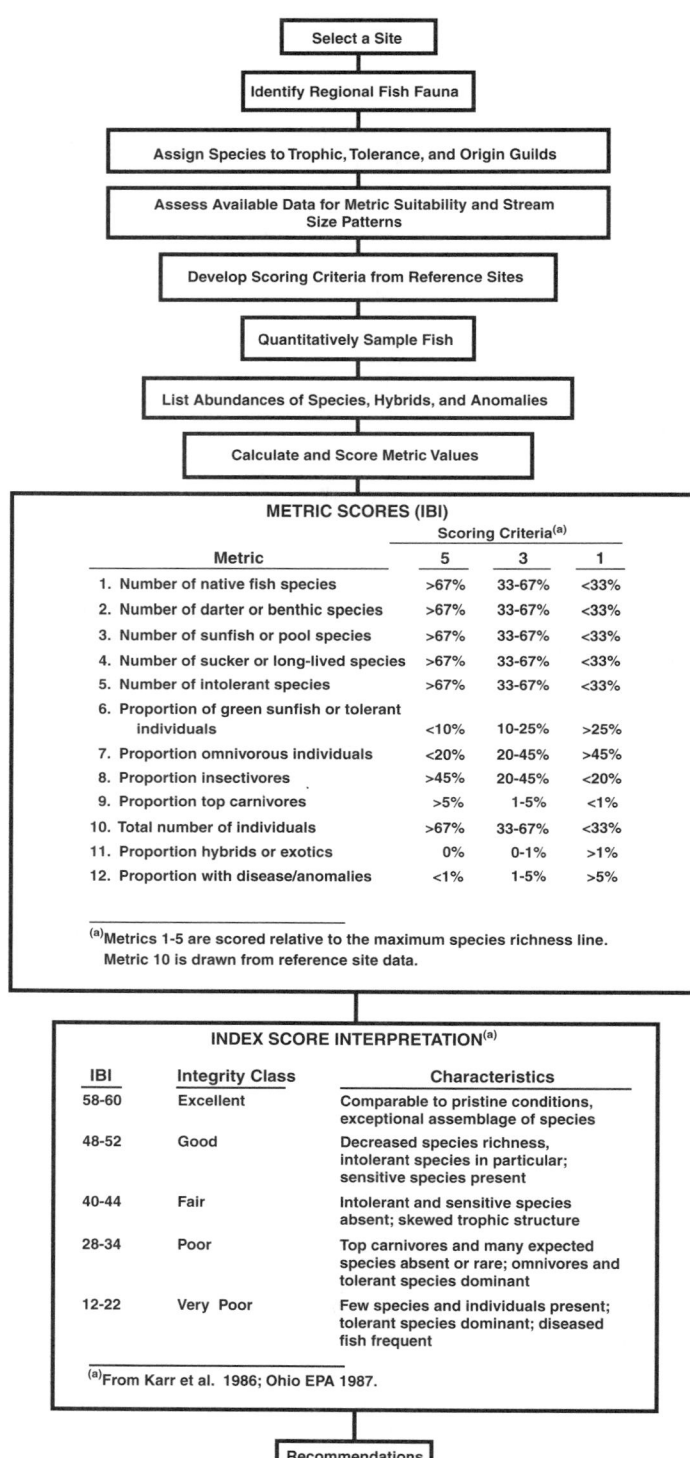

Figure 6.134 Flowchart of bioassessment approach advocated for Rapid Bioassessment Protocol V. (From EPA. *Rapid Bioassessment Protocols for Use in Streams and Rivers: Benthic Macroinvertebrates and Fish.* Office of Water, U.S. Environmental Protection Agency, Washington, D.C. EPA 444/4-89/001. 1989c.)

Table 6.53   Strengths and Weaknesses of Toxicity Tests in Stormwater Assessments

Strengths	Weaknesses
Toxicity can be quantified and linked to the presence of specific or multiple contaminants, sources, or affected media (i.e., soil, water, sediment, vegetation, aquatic biota); an important assessment component needed to establish causality.	Measure of potential toxic effects on resident biota at the test site; however, cannot always be directly translated into an expected magnitude of effects on populations in the field.
Response is an integrated index of bioavailable contamination, whereas chemical analyses measure only total concentrations of specific compounds.	Results are dependent on specific techniques, e.g., test species, collection method, water or sediment quality, test duration, etc.
More sensitive than biosurvey methods.	If surrogate species used, there is a question of their response relationship to indigenous species.
Sensitive in all types of aquatic ecosystems.	Single species test responses may not relate to community structure and ecosystem function impacts.
Results are specific to the location at which the sample was collected; thus they can be used to develop maps of the extent and distribution of bioavailable contamination and toxic conditions.	May not detect long-term toxicity, bioaccumulation, sublethal effects, or persistent, hydrophobic contaminants.
Temporal toxicity dynamics of stormwater events can be quantified and correlated with flow and other physicochemical characteristics.	Laboratory exposure conditions in toxicity tests are not directly comparable to field exposures; additional confounding variables and other stresses are important in the field.
Indigenous species may be tested in the laboratory or *in situ.*	
Approach effectively used by the EPA and many states to regulate point source pollution.	
Multiple species, multiple trophic levels, and multiple levels of biological organization (e.g., plant, bacteria to fish) may be evaluated.	
Results are easily interpreted and amenable to QA/QC; within- and among-laboratory precision estimates are already available for several tests.	
May be tested *in situ,* thus reducing laboratory-sample handling related artifacts.	
Acute toxicity tests are relatively quick, easy, and inexpensive to conduct; results from acute tests are used as a guide in the design of chronic toxicity tests.	
Chronic and short-term chronic toxicity tests are generally more sensitive than are acute tests, and can be used to define "no effect" levels; in addition, chronic tests provide a better index of field population responses and more closely mimic actual exposures in the field.	
In situ and laboratory exposures may be used to assess bioaccumulation.	
May reveal recent short-term toxicity events that are not detected in biosurveys.	
Have a long regulatory use in the NPDES program	

Table 6.54   Problem Definition: Toxicity Test Approaches

Assessment Component	Monitoring Approach
Test media	Effluent (e.g., point source discharges of wastewater or runoff)
	Ambient water
	Sediment
	Interstitial water
	Extractable fraction (e.g., elutriate)
	Soil
	Sludge
	Sample fractionation (e.g., the EPA's Toxicity Identification Evaluation procedures)
Test organism	Surrogate
	Indigenous to ecoregion
	Resident
	Single species
	Multiples of single species
	Communities or populations
	Multitrophic and/or multiple levels of biological organization
Effect level	Acute (lethality endpoint)
	Short-term chronic (e.g., growth or reproduction during partial life cycle
	Chronic (sublethal endpoint during full life cycle)
	Biomarker (sublethal endpoint in short-term exposure)
	Concentration response defined (e.g., LC50, NOEL[a]) vs. exposure to undiluted (100%) sample
Test environment	Laboratory:
	Static, static-renewal, recirculating, or flow-through
	Water only
	Water (reconstituted or site water)[b] and sediment (suspended[c] or bedded[d])
	In situ:
	Effluent mixing zone
	Ambient water only
	Sediment only
	Water and sediment
	Artificial substrate
Measured endpoints	Functional
	Population-community structure
	Organism
	Cellular or molecular

[a] Sample concentration with 50% lethality, no observable effect level.
[b] Allows separation of water and sediment toxicity.
[c] Suspended solids concentration physically maintained or fluctuates.
[d] Mixed, sieved, or intact core.

Figure 6.135 Fathead minnow rearing tanks at Environmental Health Sciences laboratories at Wright State University.

Figure 6.136 Adult fathead minnow rearing tank at Wright State.

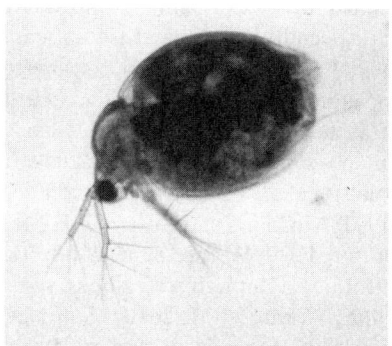

Figure 6.137 *Ceriodaphnia dubia* used for toxicity tests at Wright State.

Figure 6.138 Sediment toxicity tests at Wright State.

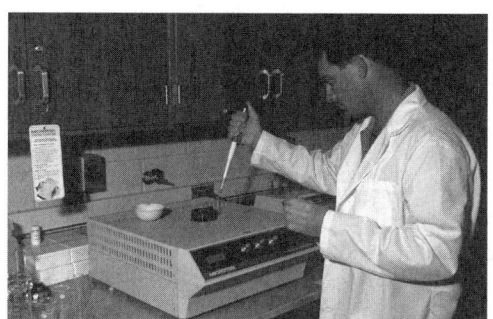

Figure 6.139 Microtox screening toxicity test at environmental engineering labs at UAB.

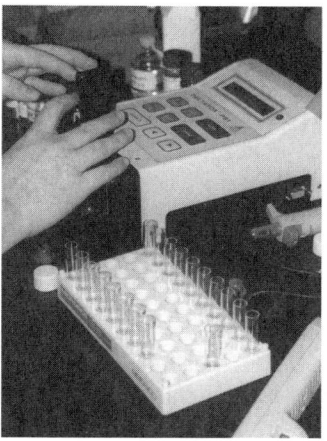

Figure 6.140 Deltatox screening toxicity test at environmental engineering labs at UAB.

Stormwater Toxicity

The water quality of stormwater, or of ambient waters immediately following high flow events, has been shown to be degraded in many studies with chemical concentrations which may exceed toxicity thresholds (e.g., Horner et al. 1994; Makepeace et al. 1995; Morrison et al. 1993; Waller et al. 1995a). Stormwater toxicants are primarily associated with particulate fractions and are typically assumed to be "unavailable." Toxicity tests with sediment removed have found reduced levels of toxicity in stormwater, compared to stormwater that has not undergone sediment removal (Crunkilton et al. 1996), as described in Chapter 3.

Also confusing is that typically short and intermittent runoff events cannot be easily compared to the criteria or standards developed and tested for traditional "long" duration point source discharges. Chemical analyses, without biological analyses, typically underestimate the severity of the problems because the water column quality varies rapidly, while the major problems were associated with sediment quality and effects on macroinvertebrates (Lenat and Eagleson 1981; Lenat et al. 1981).

Standardized toxicity tests have been used for many years in the United States to evaluate effluents in the National Pollutant Discharge Elimination System (NPDES) (EPA 1991e). These whole-effluent toxicity (WET) tests have been shown to be useful for evaluating stormwaters. The use of toxicity tests on stormwater and receiving waters, especially *in situ* and side-stream tests that also reflect changing conditions for extended periods, has added greatly to our knowledge of toxicant problems associated with stormwater. While some stormwaters may not be toxic, there is a large body of evidence that suggests many are. Laboratory testing of runoff samples has shown acute and chronic toxicity to a variety of species (Connor 1995; Cooke et al. 1995; Dickerson et al. 1996; Hatch and Burton 1999; Ireland et al. 1996; Katznelson et al. 1995; Kuivila and Foe 1995; McCahon and Pascoe 1990, 1991; McCahon et al., 1990, 1991; Medeiros and Coler 1982; Medeiros et al. 1984; Mote Marine Laboratory 1984; Tucker and Burton 1999; Werner et al. 2000; Vlaming et al. 2000; Bailey et al. 2000). Pesticide pulses have been followed through watersheds, remaining toxic for days to weeks from runoff (Kuivila and Foe 1995; Werner et al. 2000). Samples from urban streams in southern California showed 85% exceeded diazinon criteria and 80% exceeded chlorpyrifos criteria. Of these samples, 76.6% produced 100% *C. dubia* mortality within 72 hours of exposure. Toxicity Identification Evaluations (TIE) confirmed the toxicity was due to the pesticides. Diazinon has been implicated as the primary toxicant in runoff causing acute toxicity to *C. dubia*, *P. promelas*, and *in situ Corbicula fluminea* assays (Kuivila and Foe 1995; Connor 1995; Waller et al. 1995a,b; Cooke et al. 1995). Organophosphate (chlorpyrifos, diazinon, malathion) and carbamate (carbofuran, carbaryl) pesticides in a delta draining urban and agricultural areas were the primary toxicants causing acute toxicity in 9.8 to 19.6% of water samples sampled between 1993 and 1995 (Werner et al. 2000). *C. dubia* reproduction and growth of *C. fluminea in situ* closely paralleled the health of the indigenous communities (Dickson et al. 1992; Waller et al. 1995b). A simulation of farm waste effluent (increased ammonia and reduced dissolved oxygen) found amphipod precopula disruption to be the most sensitive indicator of stress (McCahon et al. 1991). Mortality occurred only when D.O. fell to 1 to 2 mg/L and feeding rates recovered after exposure to ammonia (5 to 7 mg/L) ended. Elevations of major ion concentrations were toxic to *C. dubia* and *P. promelas* in some irrigation drainage waters (Dickerson et al. 1996).

Toxicity may also be reduced in runoff. When turbidity increased during high flow, photoinduced toxicity of PAHs was reduced *in situ*, as compared to baseflow conditions (Ireland et al. 1996). A recent study of the chronic toxicity of fenoxycarb to *Daphnia magna* showed a realistic single-pulse exposure resulted in an MATC of 26 µg/L, as compared to 0.0016 µg/L from a standard, constant-exposure study (Hosmer et al. 1998).

WET tests have also been used to evaluate the toxicity of effluents from stormwater runoff treatment systems. An evaluation of an urban runoff treatment marsh found strong relationships between *C. dubia* time-to-death, conductivity, and storm size, and time from storm flow initiation (Katznelson et al. 1995). Airport runoff containing glycol-based deicer/anti-icer mixtures was toxic to *P. promelas* and *D. magna* during high-use winter months; however, during summer months runoff toxicity only coincided with fuel spills (Fisher et al. 1995). Anti-icer was more toxic to *P. promelas*, *D. magna*, *D. pulex,* and *C. dubia* than deicer. Additives were more toxic than glycols (Hartwell et al. 1995). Stormwater detention ponds reduced *P. promelas* and Microtox toxicity 50 to 90% when particles greater than 5 µm were removed (Crunkilton et al. 1996; Pitt et al. 1999a).

Pulse Exposures

Some have suggested that relatively short periods of exposure to the toxicant concentrations in stormwater are not sufficient to produce the receiving water effects that are evident in urban receiving waters, especially considering the relatively large portion of the toxicants that are associated with particulates (Lee and Jones-Lee 1995a,b). Lee and Jones-Lee (1995b) suggest that the biological problems evident in urban receiving waters are mostly associated with illegal discharges and that the sediment-bound toxicants are of little risk. Mancini and Plummer (1986) have long been advocates of numeric water quality standards for stormwater that reflect the partitioning of the toxicants and the short periods of exposure during rains. Unfortunately, this approach attempts to isolate individual runoff events and does not consider the cumulative adverse effects caused by the frequent exposures of receiving water organisms to stormwater (Davies 1986, 1991, 1995; Herricks 1995; Herricks et al. 1996).

A growing preponderance of data, however, is showing that toxicity is commonly observed during stormwater runoff and that short-term, pulse exposures can be more toxic than long-term, continuous exposures (e.g., Brent and Herricks 1998; Crunkilton et al. 1996; Curtis et al. 1985). Short pulse exposures in stormwater produced lethality several days to weeks later (Abel 1980; Bascombe 1988; Bascombe et al. 1989; Brent and Herricks 1998; Ellis et al. 1992). Some of this apparent response delay may be a result of uptake and accumulation kinetics (Bascombe et al. 1989, 1990; Borgmann and Norwood 1995; Borgmann et al. 1993). Recent investigations have identified acute toxicity problems and the importance of an adequate post-exposure observation period in side-stream studies with *P. promelas* in urban streams (Crunkilton et al. 1996), and in laboratory spiking studies (Cd, Zn, phenol) with *Ceriodaphnia dubia, Pimephales promelas,* and *Hyalella azteca* (Brent and Herricks 1998; Van Der Hoeven and Gerritsen 1997). Other laboratory studies have also shown acute and chronic toxicity of short-term exposures using fish and amphipods exposed to chloroamines, metals, and pesticides (Abel 1980; Abel and Gardner 1986; Holdway et al. 1994; Jarvinen et al. 1988a,b; McCahon and Pascoe 1991; Meyer et al. 1995; Parsons and Surgeoneer 1991a,b; Pascoe and Shazili 1986). In general, it appears that exposure to higher concentrations of toxicants for brief periods is more important that exposure to lower concentrations for longer periods (Brent and Herricks 1998; McCahon and Pascoe 1990; Meyer et al. 1995). However, increased amphipod depuration or metallothionein induction in the presence of Zn allowed greater tolerance (Borgmann and Norwood 1995; Brent and Herricks 1998).

Not all pulsed exposures are more toxic. If there is adequate time for organism recovery between pulsed exposures to toxicants, the effects of the pulsed exposure of some toxicants are diminished (Brent and Herricks 1998; Kallander et al. 1997; Mancini 1983; Wang and Hanson 1985). This difference may be attributed to the mechanism of toxicity. For example, organophosphates are relatively irreversible inhibitors of acetylcholinesterase (AChE), while carbamate inhibition may be reversible (Kuhr and Dorough 1976; Matsumura 1985). So little difference is observed between continual exposures and pulsed exposures (Kallander et al. 1997). Trout were observed to acclimate to ammonia if pulsed exposures were below their toxicity threshold (Thurston et al. 1981). Fenoxycarb was four orders of magnitude less toxic in a single pulsed exposure to *Daphnia magna* compared to a standard WET exposure (Hosmer et al. 1998). Complicating predictions of effects are synergistic interactions that occur between some contaminants such as pesticides and metals (Forget et al. 1999) and between herbicides and insecticides (Pape-Lindstrom and Lydy 1997). Organisms recovered to varying degrees given adequate time in clean water following pulsed exposures to phenol, permethrin, fenitothion, and carbamates (Brent and Herricks 1998; Green et al. 1988; Kallander et al. 1997; Kuhr and Dorough 1976; Parsons and Surgeoner 1991a,b).

Measuring Effects of Toxicant Mixtures in Organisms

Toxicant exposure is dependent on toxicant, organism, and habitat characteristics, such as toxicant partitioning (fugacity), the organisms' direct contact with substrates, and their feeding mechanisms. The toxicant target site and effect within the organism will be toxicant, species, and life stage dependent. The mixed function oxygenase (MFO) system and metallothionein production are well-known metabolic processes which often detoxify compounds, converting them to excretable metabolites (Rand and Petrocelli 1985). These metabolic systems vary dramatically among aquatic species, so it is difficult to predict aquatic toxicity to multiple species without actual testing each species. All of the above uncertainties associated with toxicant differences and interactions, exposure pathways, and organism responses support the use of multiple species in stormwater assessments.

There are mixtures of chemicals in stormwaters. Since chemical water quality criteria and standards only consider effects from one chemical, the question arises as to what effects may result to organisms when they are exposed to a mixture of potentially toxic chemicals. Mixture effects have been studied for decades. Sprague and Ramsay (1965) proposed a toxic unit (TU) that defined the strength of a toxicant. One toxic unit is equal to the incipient LC50 (the level of a toxicant that is lethal to 50% of the individuals exposed for a period of time where acute lethal effects have ceased). The strength of a toxicant, or the TU, is calculated as actual toxicant concentration in solution divided by the LC50. If the calculated sum of toxic units in a mixture of chemicals is one or larger, the mixture is said to be lethal.

The EPA (1991e) assumes that chemical toxicants act in an additive fashion, as opposed to being antagonistic (less toxicity than predicted) or synergistic (greater toxicity than predicted). A great deal of experimentation has been completed in this area, and some general principles have emerged. Overall, it appears that joint toxicity often occurs among chemicals with a similar mode of action. Within similar modes of action, the concentration-addition model (often called the TU concept) often describes the interaction

$$\text{TU mixture} = \sum_{i=1}^{n} TU_i$$

Additivity or near additivity has been demonstrated for many groups of chemicals, such as narcotics, organophosphate pesticides, pyrethroid pesticides, polynuclear aromatic hydrocarbons, major ions, and metals (Sprague 1968; Sprague and Ramsay 1965; Broderius and Kahl 1985; Carder and Hoagland 1998; Deneer et al. 1988a; Hermens and Leewangh 1982; Hermens et al. 1984a,b,c; Konemann 1981; Muska and Weber 1977).

In contrast to mixtures of chemicals with similar modes of action, chemicals with dissimilar modes of action (e.g., zinc and diazinon) show antagonistic, little, or no interaction, such that the toxicity of a binary mixture shows toxicity equal to or less than that of the most toxic component (Howell 1985; Herbes and Beauchamp 1977; Schultz and Allison 1979; Deneer et al. 1988b; Spehar and Fiandt 1986; Alabaster and Loyd 1982).

Extreme interactions of chemical mixtures, such as synergy (TU mixture $\gg \Sigma TU_i$) have also been frequently reported (Sprague and Ramsay 1965; Spehar and Fiandt 1986; Sharma et al. 1999; Christensen 1984; Vasseur et al. 1988; Marking 1977; Christen 1999; Marking and Dawson 1975; Anderson and Weber 1977; Doudoroff 1952; Wink 1990; Pape-Linstrom and Lydy 1997; Forget et al. 1999). One mechanism for synergism is where one chemical has a potentiating effect on the physiological pathway that is the target of a second toxicant. The classic example is piperonyl butoxide and pyrethroid pesticides; piperonyl butoxide blocks the detoxification pathway for pyrethroids, thereby greatly exacerbating their toxicity. In fact, this interaction is used intentionally in pyrethroid pesticide formulations.

While laboratory experiments have demonstrated approaches for mixture assessment, the test of the approach lies in its effectiveness when applied to mixtures occurring in the field, and experience suggests that the approach of assuming addition *within* modes of action and independence *between* different modes of action is adequate in many cases. For example, in studies of over 80 municipal and industrial effluents, toxicity identification studies showed no instances where observed toxicity was greater than would be predicted by this approach (D.R. Mount and J.R. Hockett, unpublished data).

The finding that mixture models are necessary to account for the potency of PAHs and dioxin-like compounds in the field provides excellent insights into the circumstances necessary for the expression of interactive toxicity in the environment. In addition to sharing a common mode of action (narcosis for PAHs; Ah-receptor agonism for dioxins/furans/PCBs), the sources for these contaminants and their environmental fate are such that they occur in mixture compositions where multiple components contribute meaningfully to the toxicity. The absence of the latter attribute greatly simplifies the assessment of many mixtures. In cases where one component of the mixture dominates, ignoring toxic interactions within the mixture adds little uncertainty to the overall assessment. Metals provide an excellent example. In practice, many metal mixtures are dominated by a particular metal. Hence, assessing the potency of the mixture on the basis of its most potent component is often effective. In the case of PAHs, however, multiple individual PAHs contribute substantially to toxicity, and the additive toxicity must be taken into account to adequately assess the mixture.

Unfortunately, the many studies cited above suggest that toxicity resulting from stressor mixtures cannot be accurately predicted simply based on additivity or chemical type. A number of studies have shown that interactions of chemical mixtures can change from antagonistic to synergistic based on the life stage of the organisms, concentrations or levels of the contaminants, or length of exposures (Sprague and Ramsay 1965; Eaton 1973; Spehar and Fiandt 1986; Marking and Dawson 1975; Munawar et al. 1987; Sharma et al. 1999; Cairns et al. 1978). This suggests that site-specific *in situ* assessments of toxicity and biological communities, as discussed later in this section, are necessary for establishing the effects of stormwater runoff.

Standard Testing Protocols: Waters

As with any of the preceding assessment methods and approaches, it is usually important that standard methods and proper QA/QC practices be followed. This helps ensure the production of valid data that are comparable to other similar study results, are reproducible, and may be usable in enforcement actions. For many of the toxicity test applications, standard methods exist, either as EPA, state, APHA, or ASTM methods. However, the absence of a standard method, such as for *in situ* or multispecies assays, does not preclude their use. These "nonstandard" assays should be based on methods published in peer-reviewed scientific periodicals that have been demonstrated as valid and useful. Since this science is relatively young, the standardization process is also young and ongoing. Standard test species have been shown to represent the sensitive range of ecosystems analyzed (EPA 1991e). In addition, resident species testing is more difficult and subject to variability than standardized testing, and many important quality assurance–quality control requirements (e.g., same life stage, sensitive life stage, reference toxicant testing, interlaboratory variation, acclimation) cannot be met (EPA 1991e).

The preferred assessment design is to have toxicity tests as a screening and definitive tool, using acute and short-term chronic toxicity measures from multiple levels of biological organization. This approach has been the foundation of chemical-specific water quality criteria development and modification. Most toxicity test requirements in NPDES permits require the use of the fathead minnow (*Pimephales promelas*) and cladoceran *Ceriodaphnia dubia* (Figures 6.141 through 6.143). However, the EPA recommends that three species be tested in whole-effluent toxicity (WET) calculations including a fish, an invertebrate, and an algae (EPA 1991e). EPA guidance on hazardous waste site evaluations suggested the fish, *Daphnia*, and green algal (*Selenastrum capricornutum*) assays (Figure 6.144), along with terrestrial testing of seed germination and root elongation,

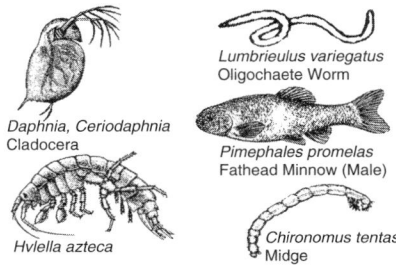
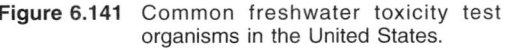

Figure 6.141 Common freshwater toxicity test organisms in the United States.

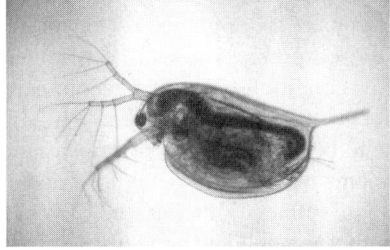

Figure 6.142 *Ceriodaphnia dubia*, the water flea.

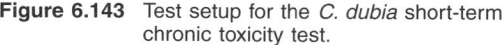

Figure 6.143 Test setup for the *C. dubia* short-term chronic toxicity test.

Figure 6.144 Culturing *Selenastrum capricornutum*.

earthworm survival, and soil respiration (Table 6.55; EPA 1989a,b; Porcella 1983). The ASTM and EPA now have standardized methods for sediment toxicity and bioaccumulation evaluations using benthic macroinvertebrates (EPA 2000c; ASTM 2000). They recommend a multispecies approach that is essential, as no one organism can serve as a surrogate for all species. An analysis of species sensitivity ranges observed in the National Water Quality Criteria documents found that when four or more species were tested, the LC50 of all was within one order of magnitude for 71 of the 73 pollutants tested (EPA 1991e). No one species was consistently the most sensitive (EPA 1991e).

A wide variety of useful and sensitive assays exists for toxicity evaluations of waters (Table 6.56) and sediments (Table 6.57). The optimal assay(s) is dependent on several issues, which will vary with the geographic area, study objectives, and pollutant problem (Table 6.58). For typical stormwater assessments, a tiered assessment approach is warranted, where the initial runoff is tested using a toxicity screening technique using the water flea (*D. magna*, *D. pulex*, or *C. dubia*) in 24- to 48-hour exposures. Additionally, if depositional (clay-silt) sediments exist downstream of stormwater outfalls, they should be evaluated for toxicity using EPA 10-day whole-sediment methods. If no toxicity is detected, however, the community indices of the benthic macroinvertebrate or fish communities indicate impairment, additional toxicity testing should be conducted, such as short-term chronic toxicity (EPA 7-day assays) and/or *in situ* toxicity exposures (described below and Appendix D). If toxicity problems are identified in the stormwater samples from the screening tests, definitive testing is conducted that may consist of acute to chronic laboratory, on-site, and/or *in situ* exposures; testing whole sediment, ambient water, or effluent; testing additional species such as bacteria (e.g., Microtox), photosynthetic organisms (e.g., duckweed, green algae), and fish (e.g., fathead minnow); and/or TIE evaluations to identify specific toxicants.

Defining stormwater toxicity at both a spatial and temporal scale may require large numbers of samples, which would surpass the resource capabilities of most projects if attempting to run conventional EPA-approved surrogate species (e.g., *P. promelas* and *C. dubia*). Stressor variability, as discussed previously, will be substantial through the course of a storm event and the return to baseflow conditions. The EPA recommends that for sampling of effluents and for annual monitoring of effluents using grab sampling, a minimum of four to six samples be collected in 1 day, once per month, to better define short-term variation. Sewage treatment plant effluents typically have shown coefficients of variation (COV) for acute toxicity of 20 to 42% and 0 to 88% for chronic toxicity. Among oil refinery effluents, the COVs ranged from 19 to 54% for acute and 30 to 60% for chronic data. Other manufacturing facility effluents had acute toxicity COVs of 20 to 100% (EPA 1991e). It may be useful to split definitive samples and run Microtox in tandem with the macrofaunal assays. If a consistent relationship is observed, i.e., few false positive or false negatives using

Table 6.55 Toxicity Evaluation Categories for Hazardous Waste Sites

Assay	Activity Measured	Sample Type[a]	MAD[b]	Units	Response Levels for LC50 or EC50 Concentrations[c]		
					High	Moderate	Low or Not Detectable
Freshwater fish	96-hr LC50 (lethality)	S	1	g/L	<0.01	0.01–0.1	0.1–1
		L	100	%	<20	20–75	75–100
Freshwater invertebrate	46-hr EC50 (immobilization)	S	1	g/L	<0.01	0.01–0.1	0.1–1
		L	100	%	<20	20.75	75–100
Freshwater algae	96-hr EC50 (growth inhibition)	S	1	g/L	<0.01	0.1–01	0.1–1
		L	100	%	<20	20–72	75–100
Seed germination and root elongation	115-hr EC50 (inhibited root elongation)	L	100	%	<20	20–75	75–100
Earthworm test	336-hr LC50	S	500	g/kg	<50	50–500	500
Soil respiration test	336-hr EC50	S	500	g/kg	<50	50–500	500
		L	100	%	<50	20–75	75–100

[a] S = solid, L = aqueous liquid, includes water samples and elutriate or leachate. Nonaqueous liquids are evaluated on an individual basis because of variations in samples, such as vehicle, percent organic vehicle, and percent solids.

[b] MAD = Maximum applicable dose.

[c] LC50 = Calculated concentration expected to kill 50% of population within the specified time interval. EC50 = Calculated concentration expected to produce effect in 50% of population within the specified time interval.

From Porcella, D.B. *Protocol for Bioassessment of Hazardous Waste Sites*, prepared for Environmental Research Laboratory, U.S. Environmental Protection Agency, Corvallis, OR, EPA 600/2-83/054, NTIS Publ. No. PB83-241737. 1983.

Table 6.56   Useful Species and Life Stages for Aqueous Phase Testing

Species		Life Stage
Fish		
Cold Water		
Brook trout	*Salvelinus fontinalis*	30 to 90 days
Coho salmon	*Oncorhynchus kisutch*	30 to 90 days
Rainbow trout	*Salmo gairdneri*	30 to 90 days
Warm Water		
Bluegill	*Lepomis macrochirus*	1 to 90 days
Channel catfish	*Ictalurus punctatus*	1 to 90 days
Fathead minnow	*Pimephales promelas*	Embryo to 90 days
Benthic Invertebrates		
Cold Water		
Stoneflies	*Pteronarcys* spp.	Larvae
Crayfish	*Pacifastacus leniusculus*	Juveniles
Mayflies	*Baetis* spp. or *Ephemerella* spp.	Nymphs
Warm Water		
Amphipods	*Hyalella azteca*	Juveniles (<.250 mm)
	Gammarus lacustris, G. fasciatus,	Juveniles
	or *G. pseudolimnaeus*	Juveniles
Cladocera	*Daphnia magna* or *D. pulex,*	1 to 24 hours
	Ceriodaphnia spp.	1 to 24 hours
Crayfish	*Orconectes* spp., *Cambarus* spp.,	Juveniles
	Procambarus spp.	Juveniles
Mayflies	*Hexagenia limbata* or *H. bilineata*	Nymphs
Midges	*Chironomus tentans* or *C. riparius*	Larvae (1st or 2nd instar)
Algae		
Green algae	*Selenastrum capricornutum*	Log-phase growth
Bacteria		
Microtox	*Photobacterium phosphoreum*	Log-phase growth (freeze-dried culture)

Modified from EPA. *A Compendium of Superfund Field Operations Methods.* Office of Emergency and Remedial Response, U.S. Environmental Protection Agency, Washington, D.C., EPA 540/P-87/001. 1987.

Microtox, then the assumption may be made that Microtox responses are related (noting statistical confidence) to the other surrogate responses. This will allow for the analysis of many more samples, because Microtox requires a few hours rather than days to complete, and many samples can conveniently be evaluated at one time.

When conducting ecotoxicity evaluations, it is important that one understand what effects sample collection, processing manipulation, and exposure design have on the observed toxicity response. Is this response similar to what is occurring in the field or is it simply an artifact of the method? A thorough discussion of this critical issue is beyond the scope of this book. See ASTM (1991) and Burton (1991) for additional information. For sediment testing, these effects are particularly significant, as sample integrity is easily disrupted, altering bioavailability and partitioning of toxicants. Sediment test phases include whole sediments, interstitial water, elutriate, or other extractable phases. Each has associated strengths and weaknesses (Table 6.57) (Burton 1991).

Table 6.57 Sediment Phases Used in Toxicity Tests

Phase	Strengths	Weaknesses	Routine Uses
Extractable phase (XP) (solutes vary)	• Use with all sediment types • Sequentially extract different degrees of bioavailable fractions • Greater variety of available assay endpoints • Determine dose response	• Ecosystem realism: Bioavailability unknown, chemical alteration	• Rapid screen • Unique endpoints component of test battery
Elutriate phase (EP) (water extractable)	• Use with all sediment types • Readily available fraction • Mimics anoxic toxic environmental process • Large variety of available assay endpoints • Methods relatively standardized • Determine dose response	• Ecosystem realism: Only one oxidizing condition used; only one solid: water ratio; exposure for extended period of one-phase condition that never occurs in situ or never occurs in equilibrium in situ • Extract conditions vary with investigator • Filtration affects response, sometimes used	• Rapid screen • Endpoints not possible with WS • Dredging evaluations
Interstitial water (IW)	• Direct route of uptake for some species • Semi-direct exposure phase for some species • Large variety of available assay endpoints • Methods of exposure relatively standardized • Determine dose response • Sediment quality criteria	• Cannot collect IW from some sediments • Limited volumes can be collected efficiently • Optimal collection method unknown, constituents altered by all methods • Exposure phase altered chemically and physically when isolated from WS • Flux between overlying water and sediment unknown • Relationship to and between some organisms uncertain: burrowers, epibenthic, water column species; filter feeders, selective filtering, life cycle vs. pore water exposure	• Rapid screen • Endpoints not possible with WS • Initial surveys • Sediment criteria
Whole sediment (WS)	• Use with all sediment • Relative realism high • Determine dose response • Holistic (whole) versus reductionist toxicity approach (water, IW, EP, and XP) • Use site or reconstituted water to isolate WS toxicity	• Some physical/chemical/microbiological alteration from field collection • Dose–response methods tentative • Testing more difficult with some species and some sediments • Few standard methods • Indigenous biota may be present in sample	• Rapid screen • Chronic studies • Initial surveys • Sediment criteria
In situ[a] (NS)	• Real measure integrating all key components, eliminating extraneous influences • Sediment quality criteria may be determined • Resuspension/suspended solids effects assessed.	• Fee methods and endpoints • Not as rapid as some assay systems • Mesocosms variable • Predation by indigenous biota	• Resuspension effects • Intensive system monitoring • Sediment criteria

[a] Organism exposed in situ in natural systems, pond/stream mesocosms, or lake limnocorrals.

Reprinted with permission from Burton, G.A., Jr. Assessing freshwater sediment toxicity. *Environ. Toxicol. Chem.*, 10: 1585–1267, 1991. © SETAC, Pensacola, FL, U.S.A.

Table 6.58   Optimal Toxicity Assay Considerations

1. Verification components
   Ecosystem relevance
   Species sensitivity patterns
   Appropriate test phase
   Short or long exposure period
   Definitive response dynamics

2. Resource components
   Organism availability
   Laboratory availability
   Expertise required
   Expense and time required

3. Standardization components
   Approved standard methods
   Reference database
   Interlaboratory validation
   Quality assurance and quality control criteria

Reprinted with permission from Burton, G.A., Jr. Assessing freshwater sediment toxicity. *Environ. Toxicol. Chem.*, 10: 1585–1627, 1991. © SETAC, Pensacola, FL, U.S.A.

Case Study: Example Use of Microtox to Identify Sources and Controllability of Stormwater Toxicants

A series of projects were sponsored by the EPA to investigate sources and treatability of toxicants in stormwater (Pitt et al. 1995, 1999). The first project phase investigated typical toxicant concentrations in stormwater, the origins of these toxicants, and storm and land-use factors that influenced these toxicant concentrations. The second project phase investigated the control of stormwater toxicants using a variety of conventional bench-scale treatment processes. The Microtox 100% sample toxicity screening test by Azur Environmental (was Microbics, Inc.) was selected for this research because of its unique capabilities: it is a rapid procedure (requiring about 1 hour) and only requires small (<40 mL) sample quantities. The Microtox toxicity test uses marine bioluminescence bacteria and monitors the light output for different sample concentrations. About 1 million bacteria organisms are used per sample, resulting in highly repeatable results. The more toxic samples produce greater stress on the bacteria test organisms, which results in a greater light attenuation compared to the control sample. It must be stressed that the Microtox toxicity screening test was not used to indicate the absolute toxicities of the samples nor to predict the toxic effects of the stormwater runoff on receiving waters during this research. It was used as a control parameter to indicate relative toxicities of different source flows and to measure relative benefits of different control options. The precision and bias of the Microtox test were easy to measure and control during these tests, which also strongly favored its use for our purposes. The following paragraphs describe the results of these tests and indicate the types of information that can be obtained using a toxicity screening procedure, such as the Microtox test.

Phase 1 — Sources of Stormwater Toxicants

The first project phase included the collection and analysis of 87 urban stormwater runoff samples from a variety of source areas under different rain conditions. All of the samples were analyzed in filtered (0.45-μm filter) and nonfiltered forms to enable partitioning of the toxicants into particulate and filterable forms. The samples were all obtained from the Birmingham, AL, area. Samples were obtained from shallow flows originating from homogeneous sources. These data were used to evaluate the effects of different land uses and source areas, plus the effects of rain characteristics, on sample toxicant concentrations and toxicity. Organic pollutants were analyzed using two gas chromatographs, one with a mass selective detector (GC/MSD) and another with an electron capture detector (GC/ECD). The pesticides were analyzed according to EPA method 505, while the base neutral compounds were analyzed according to EPA method 625 (but using only 100-mL samples). The pesticides were analyzed on a Perkin Elmer Sigma 300 GC/ECD using a J&W DB-1 capillary column (30 m by 0.32 mm ID with a 1-μm film thickness). The base neutrals were analyzed on a Hewlett Packard 5890 GC with a 5970 MSD using a Supelco DB-5 capillary column (30 m by 0.25 mm ID with a 0.2-μm film thickness).

Metallic toxicants were analyzed using a graphite furnace-equipped atomic absorption spectrophotometer (GFAA). EPA methods 202.2 (Al), 213.2 (Cd), 218.2 (Cr), 220.2 (Cu), 239.2 (Pb), 249.2 (Ni), and 289.2 (Zn) were followed in these analyses. A Perkin Elmer 3030B atomic absorption spectrophotometer was used after nitric acid digestion of the samples. Previous research (Pitt and McLean 1986; EPA 1983a) indicated that low detection limits were necessary in order to measure the filtered sample concentrations of the metals, which would not be achieved by use of a standard flame atomic absorption spectrophotometer. Low detection limits would enable partitioning of the metals between the solid and liquid phases to be investigated, an important factor in assessing the fates of the metals in receiving waters and in treatment processes.

Comparison of Microtox with Other Toxicity Tests — The Microtox procedure was compared with about 20 different laboratory bioassay tests using 20 stormwater and CSO samples. Conventional bioassay tests were conducted using freshwater organisms at the EPA's Duluth, MN, laboratory and using marine organisms at the EPA's Narragansett Bay, RI, laboratory. In addition, other toxicity tests were also conducted at the Environmental Health Sciences Laboratory at Wright State University, Dayton, OH. The comparison tests were all short-term tests. However, some of the tests were indicative of chronic toxicity (life cycle tests and the marine organism sexual reproduction tests, for example), whereas the others are classically considered as indicative of acute toxicity (Microtox and the fathead minnow tests, for example). The following list shows the major tests that were conducted by each participating laboratory:

- University of Alabama at Birmingham, Environmental Engineering Laboratory
    Microtox bacterial luminescence tests (10-, 20-, and 35-min exposures) using the marine
        Photobacterium phosphoreum
- Wright State University, Biological Sciences Department
    Macrofaunal toxicity tests:
        Daphnia magna (water flea) survival
        Lemma minor (duckweed) growth
        Selenastrum capricornutum (green alga) growth
    Microbial activity tests (bacterial respiration):
        Indigenous microbial electron transport activity
        Indigenous microbial inhibition of β-galactosidase activity
        Alkaline phosphatase for indigenous microbial activity
        Inhibition of β-galactosidase for indigenous microbial activity
        Bacterial surrogate assay using *O*-nitrophenol-β-D-galactopyranside activity and *Escherichia coli*
- EPA Environmental Research Laboratory, Duluth, MN
    Ceriodaphnia dubia (water flea) 48-hour survival
    Pimephales promelas (fathead minnow) 96-hour survival
- EPA Environmental Research Laboratory, Narraganset Bay, RI
    Champia parvula (marine red alga) sexual reproduction (formation of cystocarps after 5 to 7 days exposure)
    Arbacua punctulata (sea urchin) fertilization by sperm cells

Therefore, the tests represented a range of organisms that included fish, invertebrates, plants, and microorganisms.

Table 6.59 summarizes the results of the toxicity tests. The *C. dubia. P. promelas,* and *C. parvula* tests experienced problems with the control samples, and those results are therefore uncertain. The *A. pustulata* tests on the stormwater samples also had a potential problem with the control samples. The CSO test results (excluding the fathead minnow tests) indicated that from 50 to 100% of the samples were toxic, with most tests identifying the same few samples as the most toxic. The toxicity tests for the stormwater samples indicated that 0 to 40% of the samples were toxic. The Microtox screening procedure gave rankings similar to those of the other toxicity tests.

All of the Birmingham samples represented separate stormwater. However, as part of the Microtox evaluation, several CSO samples from New York City were also tested to compare the different toxicity tests.

Table 6.59   Fraction of Samples Rated as Toxic

Sample Series	Combined Sewer Overflows, %	Stormwater, %
Microtox marine bacteria	100	20
C. dubia	60	0[a]
P. promelas	0[a]	0[a]
C. parvula	100	0[a]
A. punctulata	100	0[a]
D. magna	63	40
L. minor	50[a]	0

[a] Results uncertain, see text.

Source Area Sampling Results — Thirteen organic compounds, out of more than 35 targeted compounds analyzed, were detected in over 10% of all samples. The greatest detection frequencies were for 1,3-dichlorobenzene and fluoranthene, which were each detected in 23% of the samples. The organics most frequently found in these source area samples (i.e., PAHs, especially fluoranthene and pyrene) were similar to the organics most frequently detected at outfalls in prior studies (EPA 1983a). Roof runoff, parking area, and vehicle service area samples had the greatest detection frequencies for the organic toxicants. Vehicle service areas and urban creeks had several of the observed maximum organic compound concentrations. Most of the organics were associated with the nonfiltered sample portions, indicating an association with the particulate sample fractions. The compound 1,3-dichlorobenzene was an exception, having a significant dissolved fraction.

In contrast to the organics, the heavy metals analyzed were detected in almost all samples, including the filtered sample portions. The nonfiltered samples generally had much higher concentrations, with the exception of zinc, which was associated mostly with the dissolved sample portion (i.e., not associated with the suspended solids). Roof runoff generally had the highest concentrations of zinc, probably from galvanized roof drainage components, as previously reported by Bannerman et al. (1983). Parking and storage areas had the highest nickel concentrations, while vehicle service areas and street runoff had the highest concentrations of cadmium and lead.

Replicate samples were collected from several source areas at three land uses during four different storm events to statistically examine toxicity and pollutant concentration differences due to storm and site conditions. These data indicated that variations in Microtox toxicities and organic toxicant concentrations may be better explained by rain characteristics than by differences in sampling locations. As an example, high concentrations of many of the PAHs were more likely associated with long antecedent dry periods and large rains, than by any other storm or sampling location parameter.

Phase 2 — Laboratory-Scale Toxicant Reduction Tests

The Phase 2 tests examined toxicant treatability for a variety of conventional bench-scale treatment processes. The data from Phase 1 identified the critical source areas (storage/parking and vehicle service areas, which generally had the highest toxicant concentrations) for study during the second research phase.

The objective of the second research phase was to obtain relative measurements of sample toxicity improvements for different stages of each bench-scale treatment method to indicate the relative effectiveness of different treatment efforts and processes. To meet this objective and considering resource restraints on cost and time, the Microtox screening toxicity test was chosen to indicate relative changes in toxicity.

The selected source area runoff samples all had elevated toxicant concentrations compared to other urban source areas, allowing a wide range of laboratory partitioning and treatability analyses to be conducted. The treatability tests conducted were:

1. Settling column (37 mm × 0.8 m Teflon column)
2. Flotation (series of eight glass, narrow-neck, 100-mL volumetric flasks)
3. Screening and filtering (series of 11 stainless steel sieves, from 20 to 106 μm, and a 0.45-μm membrane filter).
4. Photodegradation (2-L glass beaker with a 60-watt, broad-band, incandescent light placed 25 cm above the water, stirred with a magnetic stirrer with water temperature and evaporation rate also monitored)
5. Aeration (the same beaker arrangement as above, without the light, but with filtered compressed air keeping the test solution supersaturated and well mixed)
6. Photodegradation and aeration combined (the same beaker arrangement as above, with compressed air, light, and stirrer)
7. Undisturbed control sample (a sealed and covered glass jar at room temperature)

Each test (except for filtration, which was an "instantaneous" test) was conducted over a duration of 3 days. Plots of the toxicity reductions observed during each treatment procedure examined, including the control measurements, were prepared. The plots were grouped according to source area sampling location and the treatment type. Figures 6.145 through 6.147 are plots of toxicity reductions associated with filtering selected samples through different sized screens. Significant and important toxicity reductions are associated with screening using the smaller apertures.

The highest toxicant reductions were obtained by settling for at least 24 hours (providing at least 50% reductions for all but two samples), screening through at least a 40-μm screen (20 to 70% reductions), and aeration and/or photodegradation for at least 24 hours (up to 80% reductions). Increased settling, aeration or photodegradation times, and screening through finer meshes, all reduced sample toxicities further. The flotation tests produced floating sample layers that generally increased in toxicity with time and lower sample layers that generally decreased in toxicity with time, as expected; however, the benefits were quite small (less than 30% reduction).

These tests indicate the wide-ranging behavior of these related samples for the different treatment tests. Some samples responded poorly to some tests, while other samples responded well to all of the treatment tests. Any practical application of these treatment unit processes would therefore require a treatment train approach, subjecting critical source area runoff to a combination of processes in order to obtain relatively consistent overall toxicant removal benefits.

Phase 3 – Pilot-Scale Demonstration of the Multichambered Treatment Train (MCTT)

The last research phase included a pilot-scale test of the most promising treatment processes suitable for small critical source areas. This device consists of a series of chambers, including an initial grit and aeration chamber, an intermediate tube settler with oil sorbents, and a final mixed sand/peat filter. Extensive testing of PAHs, phthalate esters, phenols, pesticides, metals, toxicity screening, chemical oxygen demand, pH, conductivity, turbidity, hardness, sodium adsorption ratio, major ions, particle sizes, solids, and nutrients was performed on filtered and unfiltered samples during 12 rains at the inlets and outlets of each component of the treatment train. The results from this pilot-scale test were confirmed by full-scale installations in Wisconsin constructed and monitored by the WI DNR. The MCTT units have been shown to be extremely effective, with >90%

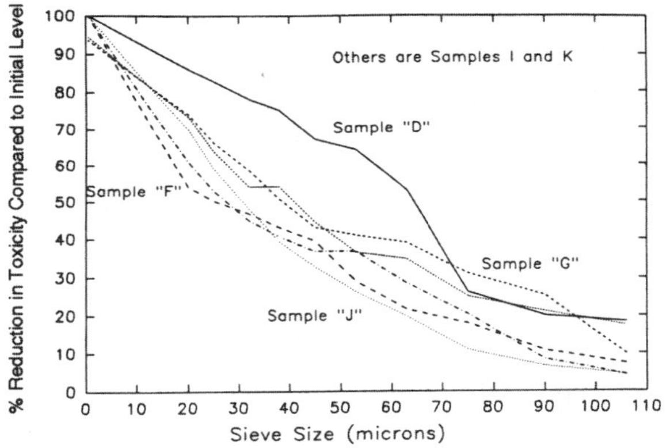

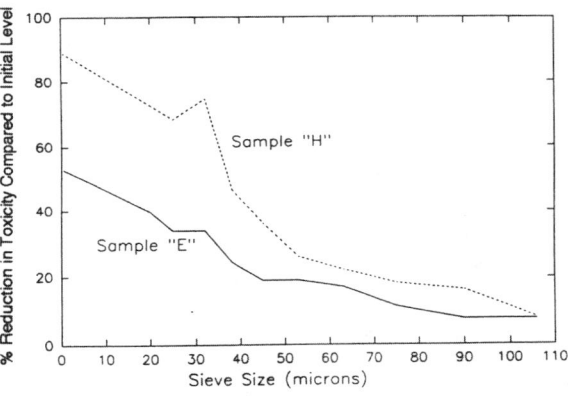

Figure 6.145 Toxicity reductions during sieving of industrial loading and parking area runoff samples. (From Pitt, R., B. Robertson, P. Barron, A. Ayyoubi, and S. Clark. *Stormwater Treatment at Critical Areas: The Multi-Chambered Treatment Train (MCTT).* U.S. Environmental Protection Agency, Wet Weather Flow Management Program, National Risk Management Research Laboratory. EPA/600/R-99/017. Cincinnati, OH. 505 pp. March 1999b.)

Figure 6.146 Toxicity reductions during sieving of automobile service facility runoff samples. (From Pitt, R., B. Robertson, P. Barron, A. Ayyoubi, and S. Clark. *Stormwater Treatment at Critical Areas: The Multi-Chambered Treatment Train (MCTT).* U.S. Environmental Protection Agency, Wet Weather Flow Management Program, National Risk Management Research Laboratory. EPA/600/R-99/017. Cincinnati, OH. 505 pp. March 1999b.)

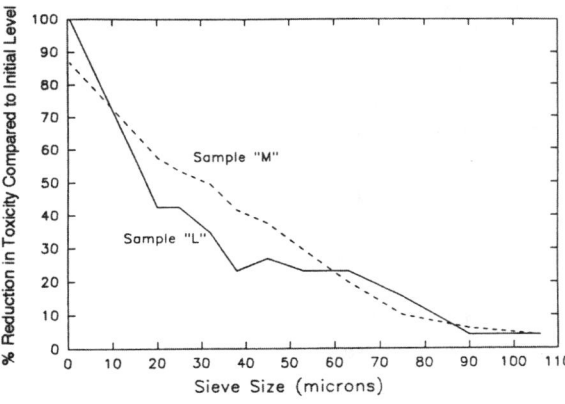

Figure 6.147 Toxicity reductions during sieving of automobile salvage yard runoff samples. (From Pitt, R., B. Robertson, P. Barron, A. Ayyoubi, and S. Clark. *Stormwater Treatment at Critical Areas: The Multi-Chambered Treatment Train (MCTT).* U.S. Environmental Protection Agency, Wet Weather Flow Management Program, National Risk Management Research Laboratory. EPA/600/R-99/017. Cincinnati, OH. 505 pp. March 1999b.)

removal of heavy metals and most organic toxicants. Caltrans (California Department of Transportation) is currently constructing and monitoring three MCTT units for treatment of runoff from a maintenance area and from parking lots in Los Angeles.

This research showed the usefulness of a toxicity screening test in evaluating sources of stormwater toxicants and in developing and testing control technologies. It would have been prohibitively expensive to base this research solely on chemical analyses of specific metallic and organic toxicants, although toxicants were specifically monitored as part of the demonstration projects to show applicability of results.

Standard Testing Protocols: Sediments

The release of the EPA Contaminated Sediment Management Strategy and Sediment Quality Inventory compiled the limited sediment data (only 4% of monitored sites had toxicity data) and documented that adverse effects are probable from sediments at 26% (>5000) of sites surveyed (EPA 1998). A recent random survey of sediments in North Carolina's estuaries found from 19 to 36% had contaminant levels known to cause toxicity and 13% had few to no living organisms (Pelly 1999). These areas are dominated by agricultural watershed inputs. The paucity of sediment toxicity information and the focus of past sediment surveys on industrialized waterways raises the question of whether the extent of sediment contamination is actually much greater than envisioned. Since chemicals, nutrients, and pathogens readily sorb to sediments, sediment contamination is likely in depositional areas of urban and agricultural watersheds (Burton 1992a,b; Burton et al. 1987). Contaminated sediments have been shown to severely impact aquatic ecosystems (e.g., Burton 1992a,b; EPA 1998) and are the source of fish contamination and advisories in many parts of the nation (EPA 1998). For this reason, it is essential that their contribution to use impairment be determined.

By the mid-1990s, standardized methods for whole-sediment toxicity testing occurred within the EPA, American Society for Testing and Materials (ASTM), and Environment Canada. These tests measured acute (short-term ≤ 10 days) toxicity in benthic macroinvertebrates such as the amphipods *Hyalella azteca, Rhepoxynius abronius, Ampelisca abdita, Eohaustorius estuarius,* and *Leptocheirus plumulosus,* and the midges *Chironomus tentans* and *Chironomus riparius* (EPA 2000c). The primary response measured was mortality, but in the case of the midge, growth was included and reburial was an additional endpoint for the *Rhepoynmius abronius.* These whole-sediment tests have been useful at testing sediment contamination (Figure 6.148) and provide information on chemical bioavailability. A large number of other species have been used for determining the toxicity of sediments, ranging from bacteria to fish and amphibians (Burton 1991). Comparisons of their sensitivities have shown a wide range of responses to different types of sediment contamination, with an equally wide range of discriminatory power (ability to detect differences between samples) (Burton et al. 1996a). This reality suggests that more than one or two species may be necessary to determine with certainty whether or not sediment contamination is ecologically significant (EPA 1994c).

Unfortunately, most of the test methods are focused on acute and not chronic toxicity. The measures of acute toxicity are often not adequate to detect the impacts on benthic communities. For instance, the 10-day test with *Rhepoxynius abronius* was not sensitive enough to describe the loss of amphipods from the Lauritizen Channel in San Francisco Bay (Swartz et al. 1994). In reality, chronic toxicity is the more pervasive problem, and it is the chronic responses, such as changes in reproduction, that lead to population level responses. Late in 1999, the EPA released its first standardized methods for determining chronic toxicity, specifically focused on growth and reproduction in *Hyalella azteca* and *Chironomus tentans* (as described in Benoit et al. 1997; Ingersoll et al. 1998). While these methods greatly aid our ability to determine if sediments are chronically toxic, their long duration and increased costs may impede their widespread adoption by state agencies.

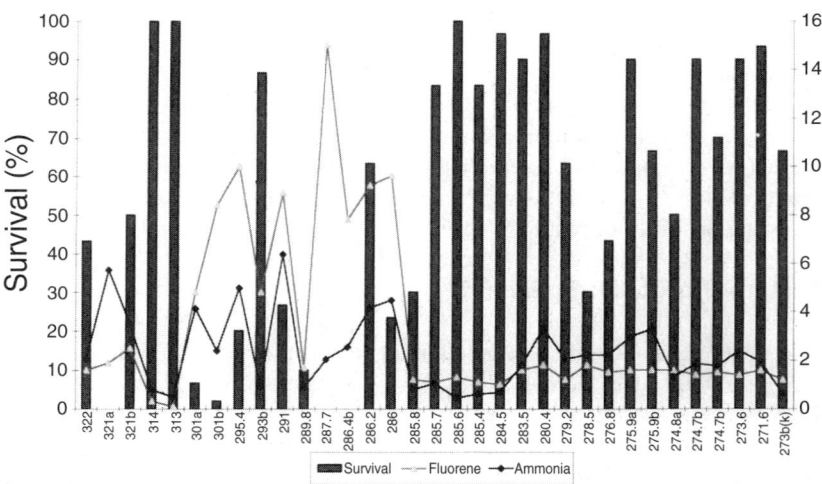

Figure 6.148 *Chironomus tentans* response in sediments from the DuPage River below Chicago in 10-day EPA exposures. Significant relationships with ammonia and fluorene sediment concentrations.

Beyond the standard tests, there have been a large number of tests with a wide range of marine benthos that may lead to better, or at least more effective, measures of chronic toxic response. For example, tests with marine amphipods have already been described in the literature to optimize the conditions for a 28-day test to examine growth and reproduction with *Leptochierius plumolosus* (Gray et al. 1998). Additional tests make use of organisms with shorter life spans, such as marine copepods, and can sort out differential response to different life stages (Green et al. 1996). These copepods are also useful in more community structure-based assessments, such as in the use of microcosms (Chandler et al. 1997). These meiobenthos may well be useful for developing standardized chronic tests since life cycle tests can be completed in 15 to 25 days and the organisms have been found to be sensitive to sediment-associated toxicants under laboratory and field conditions (Coull and Chandler 1992). Tests with organisms having shorter life spans and methods that include mixed assemblages in microcosms linked with single species tests provide insight into the functioning of communities. These new methods will help bridge the gap between our field observations and the cause–effect links that can be established in the laboratory.

There are several reasons the "water column" species used in WET tests are useful for assessments of sediments. Aquatic organisms rarely exclusively inhabit one media during their life cycle. Many "pelagic" organisms may graze on surficial sediments and even encounter pore waters. For example, the often-used "water column" surrogate, the fathead minnow (*Pimephales promelas*) is an omnivore, ingesting a mixture of detritus and invertebrates (Lemke and Bowan 1998) and frequently feeding on sediment surfaces. The zooplankton, *Daphnia magna,* grazes on surficial sediments in whole-sediment toxicity assays. The responses of WET tests have been highly predictive of indigenous benthic community responses at many sites (Dickson et al. 1996; Eagleson et al. 1990). Many vertebrate and invertebrate species have some link to sediments and have been shown to be adversely affected by sediment contamination through toxicity and effects of bioaccumulation (e.g., Baumann and Harshbarger 1995; Benson and Di Giulio 1992; Burgess and Scott 1992; Burton 1989, 1991, 1992a,b, 1999; Burton and Scott 1992; Burton and Stemmer 1988; Burton et al. 1989, 1996a,b,c; Chapman et al. 1992; Lamberson et al. 1992; Lester and McIntosh 1994; Ludwig et al. 1993; Mac and Schmitt 1992; Maruya and Lee 1998).

For most stormwater effect evaluations, sediment toxicity determinations should focus on sampling surficial sediments (approximately to 2 cm) during low flow conditions and use whole-sediment exposures. During high flow conditions, suspended-sediment assays can be conducted in the laboratory or *in situ*. The EPA and ASTM has developed standard guides for whole-sediment toxicity and bioaccumulation testing using invertebrates (ASTM 2000; EPA 2000c). Specific species guidance exists for *H. azteca*, *C. tentans*, and *C. riparius* (Figures 6.149 through 6.151). ASTM methods are also available for *Daphnia* and *Ceriodaphnia* spp. and resuspension testing. For additional test method references, see Burton (1991). Appendix D includes summaries of toxicity test methods for aqueous samples, using fish, cladocerans, algae, benthic invertebrates, and Microtox, which may be modified for sediment testing (Burton 1991). Testing suspended-sediment toxicity in the laboratory presents a logistical challenge. It is difficult to maintain a constant suspended solids concentration yet keep flow velocity and mixing turbulence reduced so as not to overly stress the test species, such as *Daphnia* sp. or *P. promelas* larvae. Relatively simple recirculation systems have been described by Hall (1986), Schuytema et al. (1984), and Schrap and Opperhuizen (1990). A preferred method of testing suspended solids is either with on-site mobile laboratories (using a flow-through pump system) or with *in situ* exposure chambers (Sasson and Burton 1991; Ireland et al. 1996; Burton and Moore 1999).

Standardized test methods have been developed for chronic toxicity testing of freshwater sediments. The EPA and ASTM have nearly identical methods (EPA 2000c; ASTM 2000). These methods are for *H. azteca* and *C. tentans* and extend for 42 to 60 days.

Hyalella azteca are routinely used to assess the toxicity of chemicals in sediments (e.g., Burton et al. 1989, 1996c; Burton 1991). Test duration and endpoints recommended in previously developed standard methods for sediment testing with *H. azteca* include 10-day survival and 10- to 28-d survival and growth. Short-term exposures, which only measure effects on survival, can be used to identify high levels of contamination, but may not be able to identify moderately contaminated sediments.

This method can be used to evaluate potential effects of contaminated sediment on survival, growth, and reproduction of *H. azteca* in a 42-day test. The sediment exposure starts with 6- to 8-day-old amphipods. On Day 28, amphipods are isolated from the sediment and placed in water-only chambers where reproduction is measured on Day 35 and 42. Typically, amphipods are first in amplexus at about Day 21 to 28 with release of the first brood between Day 28 to 42. Endpoints measured include survival (Day 28, 35, and 42), growth (dry weight measured on Day 28 and 42), and reproduction (number of young/female produced from Day 28 to 42). The EPA and ASTM state that a subset of endpoints may be measured with minor method modifications.

Reproduction in amphipods is measured by exposing them in sediment until a few days before the release of the first brood. The amphipods are then sieved from the sediment and held in water to determine the number of young produced. This test design allows a quantitative measure of reproduction. One limitation to this design is that amphipods might recover from effects of sediment exposure during this holding period in clean water; however, amphipods are exposed to sediment during critical developmental stages before release of the first brood in clean water.

Figure 6.149 EPA whole sediment, overlying water renewal design.

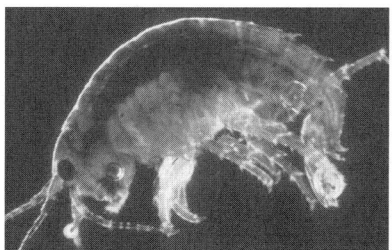

Figure 6.150 The amphipod *Hyalella azteca*, also known as the scud.

The midge *Chironomus tentans* has been used extensively in the short-term assessment of chemicals in sediments (e.g., Burton 1991; Burton et al. 1996c), and standard methods have been developed for testing with this midge using 10-day exposures (EPA 2000c). *Chironomus tentans* is a good candidate for long-term toxicity testing because it normally completes its life cycle in a relatively short period of time (25 to 30 days at 23°C), and a variety of developmental (growth, survivorship) and reproductive (fecundity) endpoints can be monitored. In addition, emergent adults can be readily collected, so it is possible to

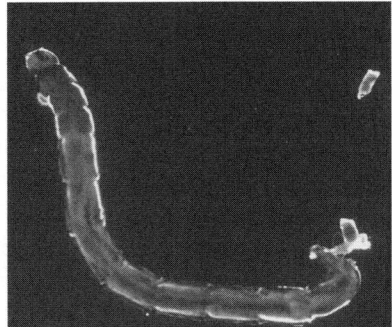

Figure 6.151 The midge *Chironomus tentans.*

transfer organisms from the sediment test system to clean, overlying water for direct quantification of reproductive success. In Europe and Canada, the chronic midge method ends after emergence.

The long-term sediment toxicity test with the midge, *Chironomus tentans*, is a life-cycle test in which the effects of sediment exposure on survival, growth, and emergence are measured. In addition, reproduction endpoints may be assessed. Survival is determined at 20 days and at the end of the test (about 50 to 65 days). Growth is determined at 20 days, which corresponds to the 10-day endpoint in the 10-day *C. tentans* growth test started with 10-day-old larvae. From Day 23 to the end of the test, emergence is monitored daily. Each treatment of the life-cycle test is ended separately when no additional emergence has been recorded for 7 consecutive days (the 7-day criterion). When no emergence is recorded from a treatment, ending of that treatment should be based on the control sediment using this 7-day criterion. EPA and ASTM state that minor modifications to the basic methods and a subset of endpoints may be used.

In Situ Toxicity Testing

An effective and accurate way to determine stormwater effects is through *in situ* toxicity testing. This may be done by placement of either artificial substrates (e.g., Hester–Dendy [OEPA 1989], rock- or mesh-filled baskets [EPA 1990b], foam [Henebry and Ross 1989], glass slides [APHA 1985]), side-stream chambers, or placing chambers-cages containing the test species into the stream or lake. The substrates or chambers must be secured to the stream bottom and be able to withstand high flow conditions. Some form of protective barrier might be necessary which might complicate flow-related effects on colonization.

In situ assessments of toxicity using confined organisms, while not new, have not been used traditionally in contaminant assessments (Burton et al. 1996b). A limited number of *in situ* exposures have been conducted to assess water or effluent toxicity. These assays have utilized adult fish, phytoplankton, amphipods, oligochaetes, and protozoans. Recent studies have shown the usefulness of *in situ* toxicity testing (Burton et al. 1996b; Chappie and Burton 1997; Crane et al. 1995; Monson et al. 1995; Sasson-Brickson and Burton 1991; Ireland et al. 1996; Bascombe et al. 1990; Ellis et al. 1995; Maltby et al. 1995; Sarda and Burton 1995; Schulz 1996; Nichols et al. 1999; Pereira et al. 1999; Maltby et al. 2000; Schulz and Liess 1999; Sibley et al. 1999). Determining the significance of sediment-associated contaminants requires an assessment of overlying water toxicity as organisms are exposed to both. This water-column exposure includes low and high flow conditions, in which water quality can vary markedly (Figure 6.152). Laboratory testing of wet-weather runoff samples has shown acute and chronic toxicity to a variety of species (e.g., Portele et al. 1982; Medeiros and Coler 1982; Medeiros et al. 1984; Ireland et al. 1996; Tucker and Burton 1999; Bailey et al. 2000). However, it is difficult to extrapolate results of these constant exposures with actual time-scale events

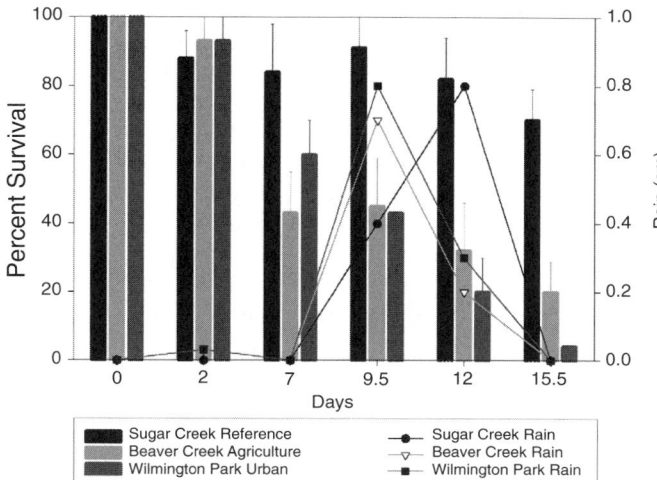

Figure 6.152 Decreased survival in urban and agricultural waters following a rain event.

in the field (Burton et al. 1996b; Tucker and Burton 1999; Burton and Moore 1999). Other *in situ* studies which have been used successfully in runoff studies include exposure of fish eggs (Pitt and Bissonnette 1984), artificial substrates for benthic invertebrate colonization and protozoa (e.g., Sayre et al. 1986), and use of transplants (Cherry 1996; Malley 1996).

There are several advantages to *in situ* testing. This approach removes sampling and laboratory-related errors from the assessment process, negating laboratory-to-field extrapolation uncertainties. Field conditions which may affect organism response and toxicity (and which are difficult to simulate in the laboratory) include sunlight, diurnal effects of temperature and oxygen, suspended solids, stressor(s) magnitude, frequency and duration, sediment integrity, spatial and temporal variation of physicochemical constituents, resident meio–microfaunal interactions, and other unknowns. Significant differences have been observed between laboratory and field testing. For example, acute toxicity to *C. dubia* in 48-hour exposures (Figure 6.153) was increased and overlying water reduced in the laboratory as compared to simultaneous *in situ* exposures (Figures 6.154 and 6.155) (Sasson-Brickson and Burton 1991). Ellis et al. (1992) observed acute and chronic toxicity to the amphipod, *Gammarus* sp., following storm event exposures in an urban stream. Death occurred up to 3 weeks following the storm and was related to elevated zinc concentrations in high-flow waters. Effects could also be correlated with *Gammarus* tissue levels of Zn. Kocan and Landolt (1990) exposed herring embryos both in the laboratory and *in situ* by placing 20 to 25 eggs on five glass slides, covering the slide holder with mesh and placing *in situ*. This system was not tested in fresh waters or in flowing waters.

Artifacts associated with sampling and manipulation (e.g., sieving and mixing of sediments) of the test samples are reduced in *in situ* assays. Such manipulations may disrupt sediment vertical contaminant gradients, thereby altering the contaminant exposure regime that organisms face in

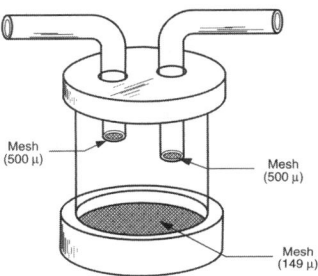

Figure 6.153 Sediment exposure chamber for invertebrates. (Reprinted with permission from Sasson-Brickson, G. and Burton, G.A., Jr. *In situ* and laboratory toxicity testing with *Ceriodaphnia dubia*. *Environ. Toxicol. Chem.*, 10: 201–207, 1991. © SETAC, Pensacola, FL, U.S.A.

the field (Sasson-Brickson and Burton 1991). *In situ* collection of interstitial water by deploying "peeper" devices has shown chemistry differences when compared to traditional collection methods using grab or core sampling (e.g., Adams 1991; Sarda and Burton 1995) and also when used for organism exposures (Fisher 1992; Figure 6.156).

In situ toxicity tests are more realistic than laboratory tests at integrating stressors (both measured and unmeasured), and can be used to study a variety of effects, such as photoinduced toxicity of PAHs (interactions with sunlight, solids, and contaminants), stormwater runoff (interactions of contaminants, suspended and dissolved solids, flow, and food), sediment-associated contaminants and physicochemical stressors, point source effluents, and contaminant gradients (Sasson-Brickson and Burton 1991; Ireland et al. 1996; Jones et al. 1995; Absil et al. 1996; Postma et al. 1994; Dickson et al. 1992; Roper and Hickey 1995; Hickey et al. 1995). Worms, bivalves, and fish have all been used *in situ* in bioaccumulation studies (e.g., Monson et al. 1995; Warren et al. 1995) with a need for linking critical body burdens to biological responses (Borgman 1996). Multiple stressors in the field usually occur in nonlinear, nonorthogonal combinations, challenging

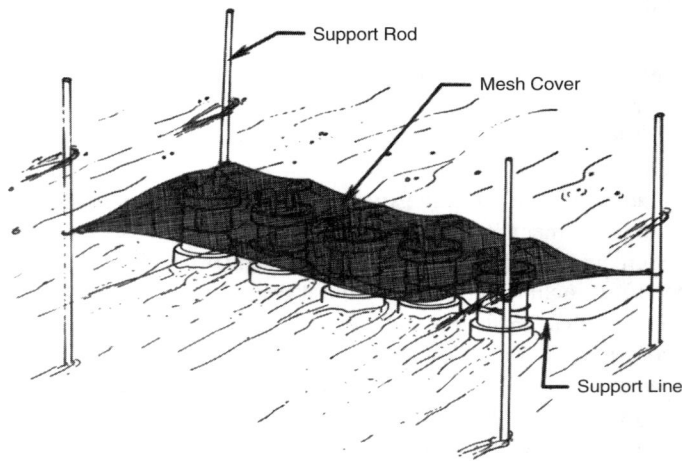

Figure 6.154 Sediment exposure chamber units secured in stream bed. (Reprinted with permission from Sasson-Brickson, G. and Burton, G.A., Jr. *In situ* and laboratory toxicity testing with *Ceriodaphnia dubia. Environ. Toxicol. Chem.,* 10: 201–207, 1991. © SETAC, Pensacola, FL, U.S.A.)

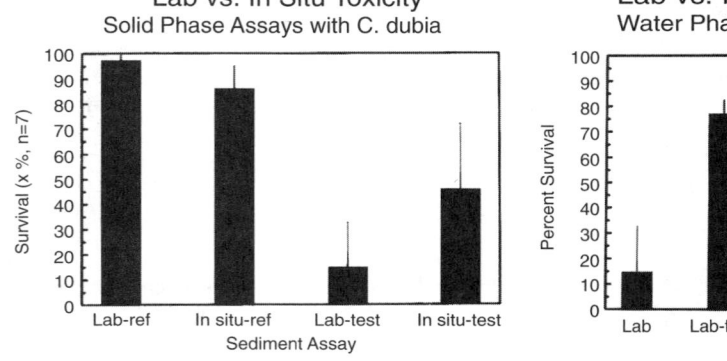
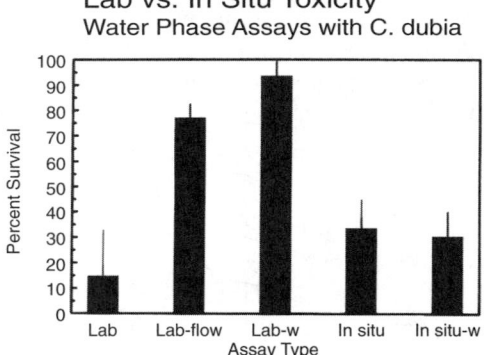

Figure 6.155 *Ceriodaphnia dubia* survival in laboratory (static and flow-through) whole sediment and site water (W) exposures; as compared to *in situ* exposures (whole sediment and overlying water, and overlying water (W) only).

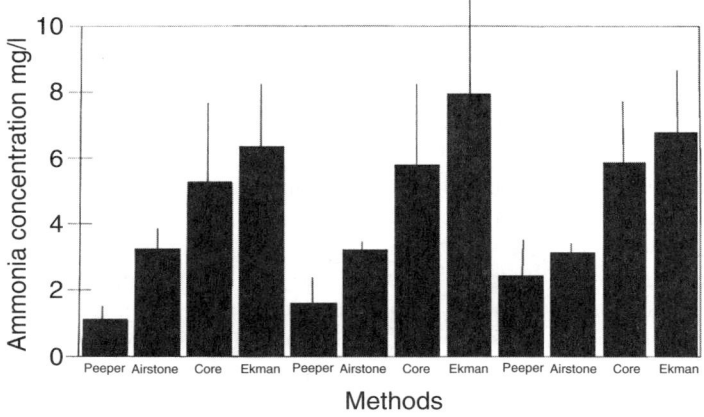

Figure 6.156 Differences in ammonia concentrations associated with various water collection methods.

biological systems in ways that are difficult at best to reproduce in the laboratory. So methods for teasing out the relative contributions of each stressor are often best conducted using a combination of *in situ* and laboratory-based experiments.

The integration of time-varying stressors (such as those related to wet-weather flow, pesticide runoff, or tidal inundation) is best conducted with field-deployed tests allowing continual exposure, as opposed to the grab sampling, static-type exposures of the laboratory. The first-flush of stormwater or pesticide runoff produces acute to sublethal responses to organisms exposed *in situ* (e.g., Herricks et al. 1994; Maltby et al. 1995; Crane et al. 1995; Waller et al. 1995b). Bivalve gape monitoring appears to be useful as an early warning indicator of effluent or stormwater toxicity (Waller et al. 1995a).

In situ methodologies can be extended to examine toxicological responses at the community level, for which they are much more cost effective than mesocosm studies (i.e., the laboratory analog). Typically, these experiments have been carried out by placing dosed sediments into the field (Berge 1990; Watzin et al. 1994) or by carrying out contaminant dosing *in situ* (Pridmore et al. 1991; Morrisey et al. 1996).

At the same time, the limitations of *in situ* toxicity tests should be recognized. Laboratory tests control variability of nontreatment factors much better than their *in situ* analogs. Deployment of caged organisms introduces the possibility of acclimation and transportation stress. If this is not monitored, data interpretation could be flawed. The *in situ* tests incorporate spatial and temporal variation, so the appropriate sampling design and analytical methods must be adapted to ensure there is adequate sensitivity and discriminatory power. The ease and practicality of *in situ* testing is site dependent. Deployment in intertidal or shallow water systems is easier than in deeper waters. Shallow subtidal deployment has the advantages of its inaccessibility to the public and reduced disturbance of sediment, especially in the case of very soft muds where trampling of intertidal sites can be a major problem. However, subtidal studies may be impacted by fishing trawls (e.g., Morrisey et al. 1996). In some areas, destruction of cages by vandals is problematic.

Primary considerations in the design and analysis of *in situ* testing approaches are the availability of food and potential starvation associated with exposures. The bioaccumulation and toxicity of contaminants is strongly influenced by food or feeding (Absil et al. 1996; Postma et al. 1994). Laboratory feeding often cannot duplicate either the quality or quantity of food present in the field. Stimulatory or inhibitory effects in these situations will likely be most marked for filter- or deposit-feeding organisms (Roper and Hickey 1995; Hickey et al. 1995).

Stressor exposures may be altered due to caging effects. Primary among these would be reduced flow, altered suspended solids or food, and interactions with predators, communities, or the food web. Depending on the organisms and the flow dynamics, cage design restricts flow to varying degrees associated with flow-through screens (Nowell and Jumars 1984). It is essential in stormwater evaluations to reduce flow velocity to protect cages and organisms. This, however, increases the uncertainty concerning flow-related interactions in the receiving water (Vogel 1994). Predator–prey effects, suspended solids concentration, and settling within the cage may be increased or reduced depending on the mesh size. Artifacts associated with *in situ* experiments are further discussed by DeWitt et al. (1996).

Other important issues with *in situ* toxicity testing are the controls and reference sites. Selection of the appropriate controls and references is partially dictated by the questions being addressed in the study. In order to ascertain where stressors exist, site controls may be needed as well as reference sites. *A priori* impressions of what constitutes a "good" reference site may be incorrect. Multiple reference sites may be desirable to adequately interpret the impact data and accommodate unexpected loss of *in situ* devices. Artificial (formulated) sediments are also useful tools for investigating effects of food and bioavailability controls in conjunction with *in situ* deployments.

In situ testing provides unique information that may not be provided by laboratory testing or community surveys. The laboratory environment is superior for mechanistic and single-stressor effect delineation. However, complex exposure dynamics and stressor interactions are difficult or impossible to reproduce in the laboratory and may best be studied *in situ*. Significant advancements in understanding ecotoxicological processes and in conducting site assessments will come from the creative use of laboratory and *in situ* testing, and community survey approaches. When properly used in an integrated weight-of-evidence approach, *in situ* testing should help reduce the uncertainties associated with evaluating contaminant and natural stressor effects in complex ecosystems.

Bioaccumulation

Why Evaluate Bioaccumulation?

Aquatic organisms are exposed to chemicals through their contact with water and sediment and ingestion of food. Many inorganic and organic chemicals have been found to accumulate in organisms. These chemicals may accumulate to levels that cause chronic toxicity or even death. One of the most common sources of tissue contamination is sediment-associated contaminants. This contamination has been linked via food web transfer to impacts on upper trophic levels. Such transfer occurs with mercury and some organochlorines, such as PCBs and DDT, that are not well biotransformed and are hydrophobic; however, with other chemicals, these connections are more difficult to establish. Some organics such as PAHs are metabolized by many organisms, so detection in tissues may indicate recent exposures. Metals are difficult to evaluate in tissues since many are essential and can be regulated by organisms. Bioconcentration factors cannot be used with metals (with the exception of methyl mercury) because they can be high or low depending on the organism, their surrounding media, the metals, and their adaptation — most of which are not clearly defined in a study. From modeling exercises, food web transfer of persistent contaminants is important for maintaining the chemical concentrations observed in upper tropic levels, and the benthic component is essential in accounting for the observed concentrations (Thomann et al. 1992; Morrison et al. 1996; see Chapter 8). Further, trophic transfer of sediment-associated contaminants has been documented in both freshwater systems (e.g., Lester and McIntosh 1994) and marine systems (e.g., Maruya and Lee 1998). This food web transfer does not have to be limited to the aquatic environment and connections have been made to terrestrial species, particularly birds (Froese et al. 1998). In Saginaw Bay, Lake Huron, tree swallows were found to accumulate PCBs from sediments. In some areas of the Great Lakes and in the Hudson River, NY, system reproductive damage has been observed for this species directly linked to PCBs (Bishop et al. 1999; McCarty and Secord 1999).

Determining Bioaccumulation

A useful way to establish a link between beneficial use impairment and contamination is by showing that exposure to sediment or stormwater runoff contaminants results in tissue residue and adverse effects in organisms. Because many factors appear to alter the bioavailability of contaminants in sediments and stormwaters, approaches to establish links between the body-residue concentrations and effects in aquatic organisms provide the insight to better link the toxic response directly to contaminants. The concept is based on the understanding that it is the dose at the receptor that is responsible for the toxic response and that the receptor concentration is proportional to the contaminant concentration in the organism. This leads to development of a database of the concentrations of contaminants responsible for toxic responses in organisms (McCarty and Mackay 1993). Data have been amassed over the course of the past several years that allow the direct comparison of some residue levels with acute and chronic effects (McCarty and Mackay 1993; Jarvinen and Ankley 1999; www.wes.army.mil/el/ered). However, the database is very limited at this time, and there is still need to establish a weight-of-evidence approach for developing the link between the observed response and the presence of contaminants in sediments. Currently, there is only one standardized EPA test for sediment bioaccumulation. It is a 28-day test with the oligochaete *Lumbriculus variegatus* (Figure 6.157).

Bioaccumulation has often been assessed with *in situ* studies to determine site-specific effects. These studies have primarily used caged mussels (marine) or fish (EPA 1987; Mac et al. 1990). In one approach, adult fish (*P. promelas*) are placed in mesh cages (10 fish per compartment, 4 compartments per cage) and exposed for 10 days *in situ*. This may also be done with benthic invertebrates (e.g., mussels, amphipods, and oligochaetes (e.g., *Lumbriculus variegatus*), providing there is adequate biomass for chemical analyses. Caution should be exercised when formulating conclusions from

Figure 6.157 *Lumbriculus variegatus* 28-day bioaccumulation test.

these studies because the organisms are not exposed for extended periods, they may not be able to ingest foods and surficial sediments due to their mesh-cage barrier, and they may be stressed due to caging. These factors alter toxicokinetics. These weaknesses can be addressed by also collecting resident target species (Table 6.60) and analyzing tissues (EPA 2000a,b). Target species should be large adults that are upper trophic level (top predator) and/or bottom feeders, and they should be collected prior to winter yet well after spawning. Nonmigratory species are preferred, and their commercial or sport fishing importance should be considered. Samples should be processed as described in Appendix D. The decision of whether to analyze whole fish or select target organs (e.g., gills, liver, kidneys) depends on the study objective and concerns over food chain or human health effects.

Residue information should be interpreted with caution (as discussed above with metals); however, guidance exists for calculating fish consumption advisories (EPA 2000a). There is little information available on what constitutes a significant tissue concentration, and correlations with adverse effects are usually lacking. Many contaminants are present for days or less (e.g., synthetic pyrethroids), rapidly metabolized (e.g., synthetic pyrethroids, organophosphates), biotransformed (e.g., polycyclic aromatic hydrocarbons), or only present in the environment seasonally (e.g., herbicides, insecticides). The U.S. Food and Drug Administration and the U.S. Fish and Wildlife Service have some effect-level information for a few common contaminants. For further information, see EPA (1982, 2000a), Carlton and Klug (1990), and Mac and Schmidt (1992).

Table 6.60   Target Fish Species for Use in Tissue Analysis

I. Target Species (East of Appalachian Mountains)

***Brook trout (*Salvelinus fontinalis*)
***Small mouth bass (*Micropterus dolomieui*)
***Large mouth bass (*Micropterus salmoides*)
***Channel catfish (*Ictalurus punctatus*)
**Brown trout (*Salmo trutta*)
**Rainbow trout (*Salmo gairdnerii*)

**Bluegill (*Lepomis macrochirus*)
**Pumpkinseed (*Lepomis gibbosus*)
**Black crappie (*Pomoxis nigromaculatus*)
**Striped bass (*Morone saxatilis*)
*Carp (*Cyprinus carpio*)

II. Target Species (West of Appalachian Mountains and East of Rocky Mountains)

***Rainbow trout (*Salmo gairdnerii*)
***Brook trout (*Salvelinus fontinalis*)
***Small mouth bass (*Micropterus dolomieui*)
***Large mouth bass (*Micropterus salmoides*)
***Channel catfish (*Ictalurus punctatus*)
**Striped bass (*Morone saxatilis*)

**Yellow perch (*Perca flavescens*)
**Walleye (*Stizostedion vitreum*)
**Bluegill (*Lepomis macrochirus*)
**Brown trout (*Salmo trutta*)
*Carp (*Cyprinus carpio*)

III. Target Species (West of and including Rocky Mountains)

***Rainbow trout (*Salmo gairdnerii*)
***Brook trout (*Salvelinus fontinalis*)
***Small mouth bass (*Micropterus dolomieui*)
***Large mouth bass (*Micropterus salmoides*)
***Channel catfish (*Ictalurus punctatus*)

**Bluegill (*Lepomis macrochirus*)
**Striped bass (*Morone saxatilis*)
*Cutthroat trout (*Salmo clarki*)
*Brown trout (*Salmo trutta*)
*Carp (*Cyprinus carpio*)

*** Preferred target species.
** Good target species.
*   Acceptable target species.

From Freed, J. et al. *Sampling Protocol for Analysis of Toxic Pollutants in Ambient Water, Bed Sediments, and Fish, Interim Final Report.* Office of Water Planning and Standards, U.S. Environmental Protection Agency, Washington, D.C. 1980.

Emerging Tools for Toxicity Testing

Semipermeable Membrane Devices (SPMDs)

While no standard methods exist, SPMDs have been reported widely in recent years as an excellent passive, *in situ* sampling device for organic contaminants in water and in air (Huckins et al. 1999; Axelman et al. 1999; Peterson et al. 1995; Sabaliunas et al. 1998; Petty et al. 1998; Woolgar and Jones 1999; Zabik et al. 1992; Prest et al. 1995, 1992). Granmo et al. (2000) recently conducted tests in marine waters in Sweden using SPMDs for comparison with bioaccumulation of organochlorine compounds (chlorobenzenes, chlorophenols, and PCBs) in feral and caged mussels and the concentrations found in sediment and the associated water column. Short-term exposures (30-day) of SPMDs and caged mussels were used to find whether the high pollutant concentrations found in the sediments were associated with recent or older industrial discharges. Feral mussels were also analyzed after longer exposure periods. They found that the test approach using the combination of SPMDs and mussels allowed the detection of short-term changes of discharges of these organochlorine compounds, especially considering that the SPMDs were found to be more effective at concentrating some of the target compounds.

The devices are generally polymeric (such as low-density polyethylene) tube bags containing a neutral lipid (such as triolein, iso-octane, 2,2,4-trimethylpentane). These bags are placed in receiving waters for a period of days and then recovered and the contents analyzed using gas chromatography/mass spectrometry, or high-pressure liquid chromatography for target compounds. The concentrations accumulated in the bags have been found to be relatively similar to what is accumulated in resident fish and shellfish. However, the concentrations may be higher or lower by several-fold and vary in their relationship to each other. This method has the advantage of being easy to deploy and retrieve, and can sample compounds found at a specific site that are in the water column during a specified period of time (unlike fish, which migrate to different areas). In addition, biological organisms are not sacrificed for the analyses. Extended exposures may result in biofouling of the bag and care must be taken to ensure adequate field blanks are used to assess that no water-related contamination has occurred.

DNA Fingerprinting

Another novel assessment tool to measure stress is genetic markers (as discussed above). Randomly amplified polymorphic DNA (RAPD) markers have proven promising for determining differences in genetic variability in populations (Williams et al. 1990, 1991). Other studies (Sternberg et al. 1996; Ellis et al. 1997) showed highly significant differences in DNA profiles between benthic invertebrates from stressed and nonstressed sites. This inexpensive and quick assay shows the number and size of distinctive DNA profiles of genomic DNA from each organism. Because RAPD-PCR primers are not designed to amplify specific target sequences, the amplified loci are anonymous, scattered throughout the genome, and are not associated with stressor adaptation (neutral markers) (Williams et al. 1990). RAPD-PCR products are often highly polymorphic within naturally occurring populations and have proven to be excellent indicators of genetic diversity (Clark and Lanigan 1993).

Biological Toxicity Fractionations

After toxicity is identified in receiving waters, researchers commonly attempt to identify the toxicants responsible for the observed effects through toxicity identification evaluation (TIE) studies. Numerous TIE protocols have been used. Figure 6.158, from Lopes et al. (1995), is one example that was used in association with a stormwater toxicity study conducted in Phoenix, AZ. Acute toxicity of stormwater was found to occur, especially to fathead minnows, and was likely to degrade the quality of the receiving water (the Salt River).

This test protocol involved first conducting toxicity tests to identify stormwater that was toxic (>20% mortality after 24 hours). The toxic stormwater was then subjected to different extractions to selectively remove various pollutants from the stormwater, after which additional toxicity tests were conducted. The first extractions were with activated carbon to remove oil and grease. The water was then split by filtering through 0.45- and 0.7-μm filters and further treated to remove metals (by chelation extraction) and organics (by solid-phase extraction). This procedure enabled the pollutant phase causing the toxicity to be identified: particulate bound pollutants, filterable metals, or filterable organics.

The EPA TIE protocols consist of three levels of confirmation (EPA 1991a). These methods were designed for analyzing wastewater effluents; however, they have been used for stormwaters, sediment pore waters, and whole sediments (e.g., EPA 1991d; Bailey et al. 2000; Werner et al. 2000; Vlaming et al. 2000; USGS 1999; Burgess et al. 1997; Ho et al. 1999; Kosian et al. 1999; Boucher and Watzin 1999). Usually, only Phase I is conducted due to the time and expense required. The TIE Phase I is a physical and chemical fractionation process that separates chemical groups by their properties. The principal groups of contaminants include pH-sensitive and volatile compounds (such as ammonia), metals, and nonpolar organics. This consists of exposing *Ceriodaphnia dubia* neonates and/or *P. promelas* larvae to water fractions for 48-hour periods. If toxicity is removed in any fraction, subsequent chemical analyses can be used to confirm the removal of compounds which may be contributing to toxicity. These methods are relatively complex and should be conducted by an experienced laboratory.

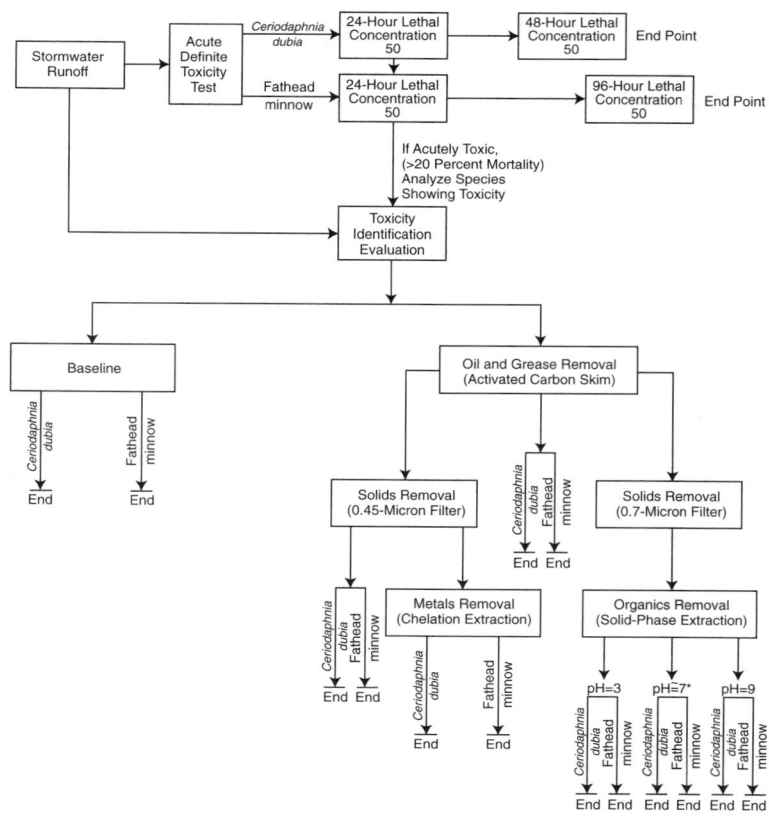

Figure 6.158 Toxicity identification evaluation (TIE) protocol. (From Lopes, T.J. et al. *Statistical Summary of Selected Physical, Chemical, and Microbial Characteristics, and Estimates of Constituent Loads in Urban Stormwater, Maricopa County, Arizona.* U.S. Geological Survey Water-Resources Investigations Report 94-4240. Tucson, AZ. 1995.)

Examples of Identifying Stressors

Diazinon was shown to be the primary toxicant in stormwater samples using *C. dubia* (Ohio EPA 1987). Anderson et al. (1991) compared numerous stormwater outfalls in the lower San Francisco Bay, CA. They found that nonpolar compounds in the most toxic stormwater (from a small, heavily industrialized drainage area) were the most important components of the toxicity, with lesser effects associated with suspended solids, metal chelates, and cationic metals. In another toxic stormwater study (from large parking areas surrounding an airport and industry), toxicity was most strongly influenced by cationic metals. Diazinon and chlorpyrifos in urban stormwater showed additive toxicity to *C. dubia* in a TIE (Bailey et al. 1997). TIE evaluations in the Sacramento–San Joaquin River basins confirmed that several organophosphate and carbamate pesticides were responsible for acute toxicity to *C. dubia* in water samples (Werner et al. 2000; Vlaming et al. 2000; Bailey et al. 2000). A TIE of pore water from a stormwater detention pond using *C. dubia* 48-hour exposures showed ammonia to be the primary toxicant, with some effects from metals (Zn, Fe, and Cu). The high level of ammonia may have obscured the metal toxicity (Wenholz and Crunkilton 1995).

Jirik et al. (1998) also used selected Phase 1 TIE studies to identify the toxicants most responsible for stormwater toxicity in the Santa Monica Bay area. Sea urchin fertilization tests

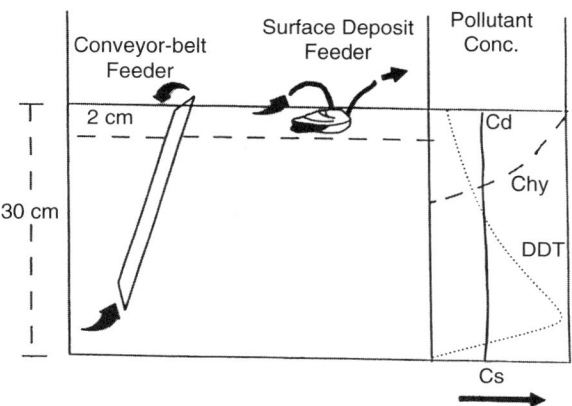

Figure 6.159 Different organism exposures should be matched with sampling of correct sediment depths. (From Lee, H., II. A clam's eye view of the bioavailability of sediment-associated pollutants, in *Organic Substances and Sediments in Water*, Vol. 3, *Biological*. Edited by R.A. Baker. Lewis Publishers, Boca Raton, FL. 1991. With permission.)

indicated EC50 values of stormwater of about 12 to 20%. Santa Monica Bay receiving waters were also found to be toxic, with the level of toxicity generally corresponding to the amount of stormwater in the receiving water. EDTA addition removed virtually all of the toxicity, implying that divalent metals were the likely toxicant component. Spiking studies showed that zinc, and sometimes copper, were the most likely metallic toxicants. Further studies, using EDTA vs. sodium thiosulfate for toxicity removal, also strongly implicated zinc as the likely cause of toxicity.

In situ tests also provide an excellent means for identifying the source and nature of the stressor by simply altering the exposure via chamber design and placement. It is essential to relate organism responses (e.g., mortality) with their correct, realistic exposure, such as overlying water, surficial sediment, or deeper sediments and pore waters (Figure 6.159). Useful *in situ* approaches to separating media effects and characterizing contaminant sources, pathways, and effects include characterization of benthic communities, *in situ* toxicity testing, and groundwater/surface water interactions (Greenberg et al. 2000; Figure 6.160). In a simplistic TIE approach, stressors can be partitioned out: overlying water, bulk sediment, interstitial water, light, suspended solids, flow velocity, and predator effects (Burton and Moore 1999) (see also Chapter 5). Strategic placement of chambers at reference and potentially impacted sites can identify both natural and anthropogenic stressors. Placement along known or suspected contamination gradients can provide an exposure–response relationship when combined with physicochemical measurements. For example, utilization of naturally occurring gradients (e.g., within and beyond a mixing zone) may facilitate an exposure–response characterization and regression analysis rather than a paired comparison (e.g., ANOVA) (Liber et al. 1992).

Useful *in situ* chambers for assessing stormwaters and surficial sediments are shown in Figures 6.161 through 6.164. Chambers are also buried in surficial sediments to assess sediment and groundwater associated contamination (Figures 6.165, 6.166, and 6.168) where chamber mesh

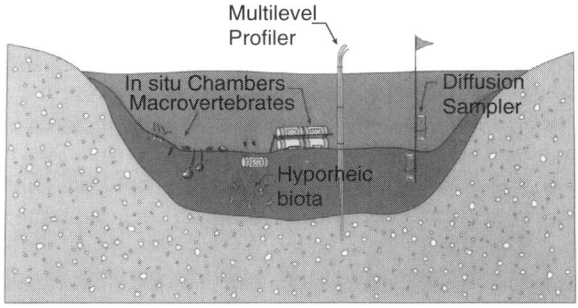

Figure 6.160 Integrated assessments of surface waters, sediments, and groundwater/pore waters.

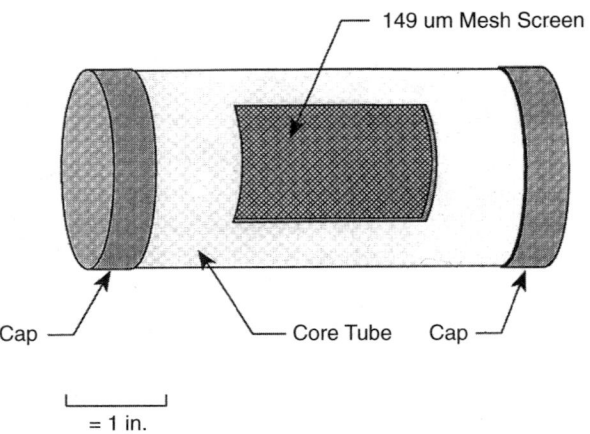

Figure 6.161 *In situ* chambers optimized for surface water exposures on top of chambers optimized for surficial sediment exposures.

Figure 6.162 *In situ* chamber design components.

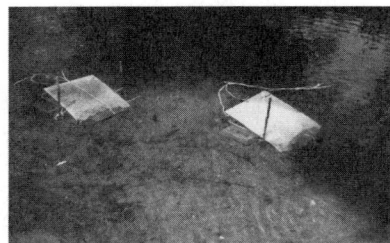

Figure 6.163 *In situ* chambers with water or sediment exposures with high-flow deflector.

Figure 6.164 *In situ* chambers with water or sediment exposures with high-flow deflector.

windows contact surficial sediments (bottom tray) or overlying water (top tray). Test organisms are placed within the chambers during low flow (Figure 6.169). Following organism addition, high flow guards (aluminum sheet metal) are attached to stakes to protect the chambers (Figures 6.163 and 6.164).

Assessments of PAH-contaminated sediments have demonstrated why both laboratory and field toxicity exposures were essential to adequately identify key stressors and characterize exposure dynamics (Ireland et al. 1996; Sasson-Brickson and Burton 1991; Stemmer et al. 1990). Sediment-associated toxicity increased in the laboratory exposure of *P. promelas, C. dubia, D. magna,* and *H. azteca* as compared to *in situ* exposures, whereas toxicity decreased in overlying waters (Figure 6.156). Photoinduced toxicity from PAH and UV interactions and sampling-induced artifacts accounted for these laboratory-to-field differences. Toxicity was also reduced significantly in the presence of UV light when the organic fraction of the stormwater was removed. Photoinduced toxicity occurred frequently during low flow conditions, but was reduced during high turbidity associated with high flow conditions. Toxicity was also higher in overlying waters near the contaminated sediment surface as opposed to waters several centimeters above the sediment–water interface.

An elevation in temperature of Des Plaines River water accentuated the toxicity of the water and of sediments, using both water column and benthic species (Brooker and Burton 1998; Burton and Rowland 1999; Lavoie and Burton 1998). Responses were replicated in laboratory, *in situ*, and

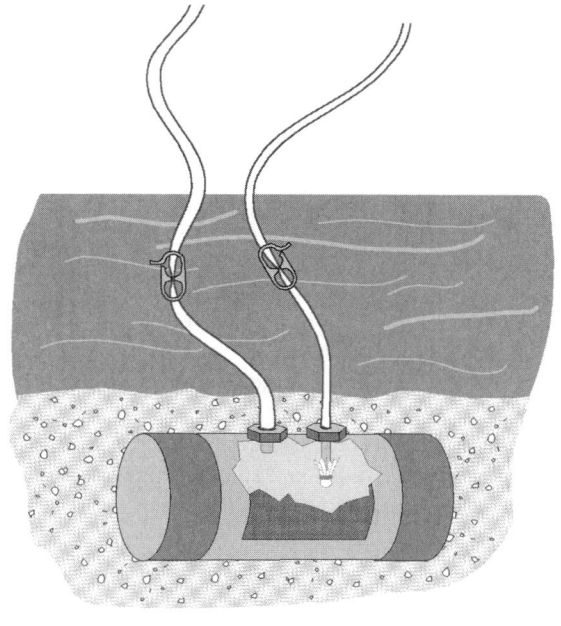

Figure 6.165 *In situ* chamber used as a "peeper" (buried for pore water exposure) or sediment–water interface (half-buried) exposure.

Figure 6.166 *In situ* sediment–water interface chambers buried.

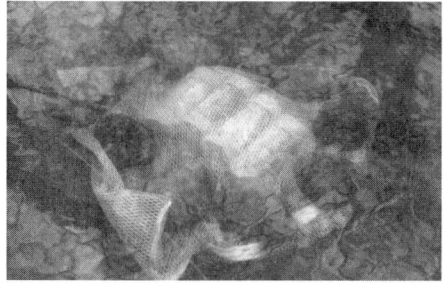

Figure 6.167 *In situ* chambers optimized for surface water and photoinduced toxicity effects from PAHs and UV light.

Figure 6.168 Chambers for conducting sediment bioaccumulation studies.

Figure 6.169 Loading *in situ* chambers that are peepers for sediment–water interface exposures.

artificial, side-stream exposures. The laboratory exposures helped define exact threshold temperatures, critical exposure times, and interactions with ammonia (Figure 6.170). Field exposures, on
the other hand, better defined real-world exposures and interactions with other stressors, such as
suspended solids and fluctuating temperatures. Conclusions based on laboratory exposures would
have underestimated stream effects.

An urban site receiving large loadings of residential, commercial, and industrial stormwater
runoff was assessed using an integrated low and high flow assessment (Moore and Burton 1999).
The effects of turbidity and flow were shown by reducing the mesh size in the *in situ* chambers
(Figures 6.171 and 6.172). A survey of sediment quality during baseflow conditions found one
depositional area where sediments were acutely toxic and contained elevated levels of contaminants. An *in situ* toxicity assessment found that low flow water was not toxic, but high flows were
toxic, and suspended solids and flow contributed significantly to overall stress. However, indigenous communities appeared to be affected more strongly by contaminated sediments than high
flow conditions.

Newer TIE methods include whole-sediment manipulations, exposure to UV (Kosian et al.
1998), or *in situ* exposures with various stressor partitioning methods and substrates (Burton et al.
1998; Greenberg et al. 1998; Moore and Burton 1999), and may reduce the likelihood of artifacts.

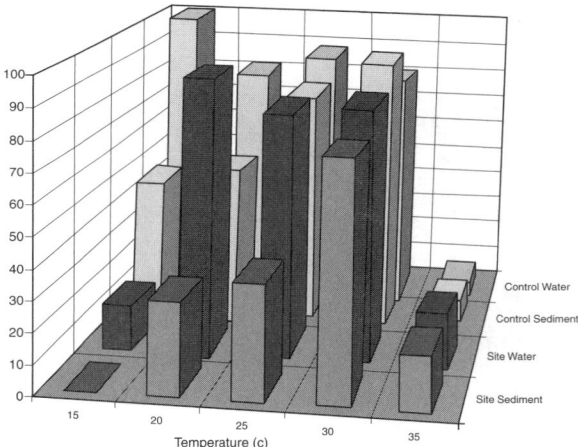

Figure 6.170 Temperature threshold determination in the presence of contaminated site water and
sediment vs. control waters and
sediment. Survival (%) of
Hyalella azteca.

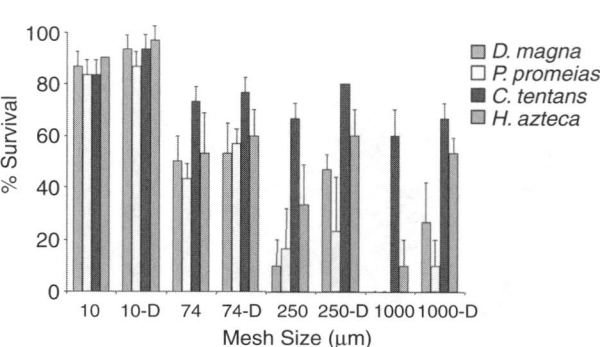

Figure 6.171 Relationship between toxicity
and suspended solids/flow.
In situ exposures in chambers with smaller mesh sizes
decreased solids and flow
and increased survival.

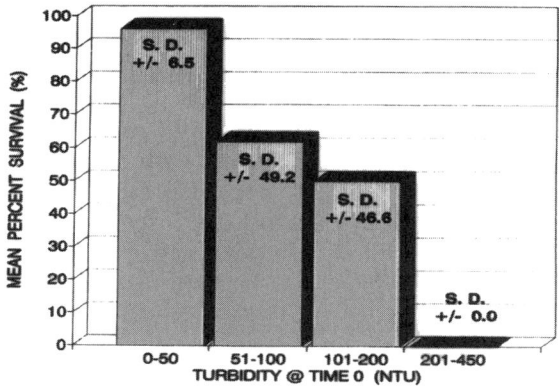

Figure 6.172 Relationship between turbidity and *Daphnia magna* toxicity in peeper exposures.

Toxicant Sampling and In-Stream Modeling Considerations

When sampling for, or predicting the fate of, toxicants, it is helpful to consider whether the likely contaminants tend to sorb to particulates, such as suspended solids or bedded sediments, or whether they tend to remain dissolved. Though metals will sorb to sediments in most waterways, if the water pH is acidic or if suspended colloids and solids concentrations are low, metals may remain in the water column. Dissolved metals do not necessarily equate with toxicity, as they may be complexed (e.g., carbonates, hydroxides) in less toxic forms. Many organics can be transported in dissolved forms at low suspended solid concentrations (EPA 1986). Adsorption can be predicted by knowing the octanol-water coefficient (K_{ow}), the organic carbon content of the suspended sediment, and then calculating the partition coefficient (K_p) (EPA 1986), as shown in Figure 6.173 (Novotny and Olem 1994). The K_p, however, is a site-specific value which varies at the site spatially and temporally during storm events and should thus be used with caution.

Sediment resuspension (scour) is an important mechanism affecting water column concentrations of many problematic constituents that tend to accumulate in stream sediments (especially pathogens, toxicants, and nutrients). Scouring of sediments can also be an important factor influencing water

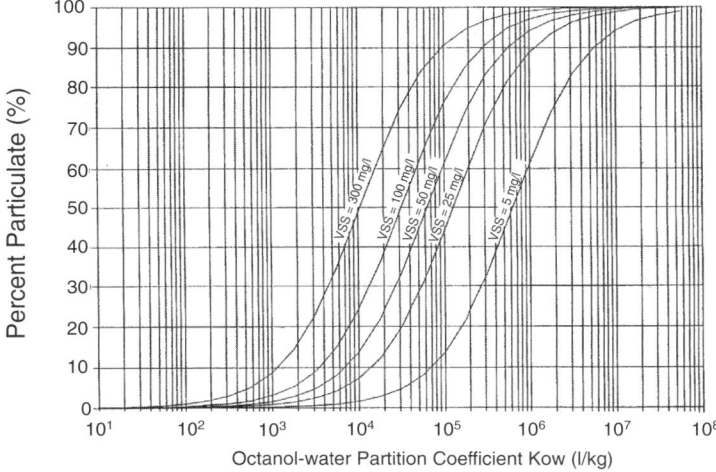

Figure 6.173 Relationship of dissolved and total concentrations of organic priority pollutant related to the octanol partioning number and volatile suspended solids content of runoff. (From Novotny, V. and H. Olem. *Water Quality: Prevention, Identification, and Management of Diffuse Pollution.* Van Nostrand Reinhold, New York. Copyright 1994. Used with permission of John Wiley & Sons.)

turbidity in some cases. Methods for measuring sediment scour were discussed previously in this chapter in the general habitat discussion. In that case, the significant role that scour has on habitat was stressed. The measurement methods described there (used in conjunction with sediment quality information) can also be used to measure the resuspension of contaminated sediments. Modeling of sediment resuspension can only be crudely predicted because site-specific details are rarely available in sufficient detail and local scour "hot spots" (small areas where the flowing water has excessive shear stress) are extremely difficult to predict. However, scour around bridge piers has been investigated for several thousand years, and there are methods to reduce sediment losses in those situations. In most cases, it is only possible to grossly predict average sediment resuspension based on average stream bed conditions. Therefore, careful scour measurements should be conducted to indicate likely sediment resuspension rates for different flows for specific streams.

Many organic toxicants move in and through an ecosystem being controlled primarily by one fate process. Volatilization controls the fate of compounds such as trichloroethylene, toluene, xylene, acetone, and benzene. Adsorption dominates the fate of polychlorinated biphenyls, dioxins, and furans. For many common contaminants, such as the metals, metalloids, polycyclic aromatic hydrocarbons, and nutrients, multiple processes (e.g., biodegradation, methylation, photolysis, hydrolysis) dominate at different stages in different microenvironments.

A number of stream models exist for predicting pollutant fates, ranging from simple to complex, which may in limited cases be useful tools for stormwater effect studies. A summary of screening approach data requirements for metals and organics are listed in Tables 6.61 and 6.62, respectively.

Contaminants may move from their source through the receiving system, (e.g., stream, lake, wetland), in a conservative or nonconservative manner depending on the fate processes that dominate in that system and are characteristics of that particular toxicant. Generalized toxicant concentration profiles shown in Figure 6.174a reflect stream dilution and toxicant decay. This profile is not representative of reactive (nonconservative) constituents, such as highly volatile compounds, nutrients, species, or dissolved oxygen concentrations. Effects from these stressors must always be considered when toxicant fate and effects are being assessed. As shown in Figure 6.174b, during high flow conditions, contaminated sediment scour may increase concentrations in some stream segments before dilution and first-order decay profiles return. By constructing suspended solids profiles at low and high flow conditions, both sources and erosion- and scour-related stressors (e.g., sorbed toxicants and nutrients, oxygen demand, solids-related filter/gill clogging, or suffocation) can be better defined (see Figure 6.175).

Table 6.61   Summary of Data Requirements for Screening Approach for Metals in Rivers

Data	Calculation Methodology Where Data Are Used*	Remarks
	Hydraulic Data	
1. Rivers:		
River flow rate, Q	D, R, S, L	An accurate estimation of flow rate is very important because of dilution considerations. Measure or obtain from USGS gauge.
Cross-sectional area, A	D, R, S	
Water depth, h	D, R, S, L	The average water depth is cross-sectional area divided by surface width.
Reach lengths, x	R, S	
Stream velocity, U	R, S	The required velocity is distance divided by travel time. It can be approximated by Q/A only when A is representative of the reach being studied.
2. Lakes:		
Hydraulic residence time, T	L	Hydraulic residence times of lakes can vary seasonally as the flow rates through the lakes change.
Mean depth, H	L	Lake residence times and depths are used to predict settling of absorbed metals in lakes.

Table 6.61   Summary of Data Requirements for Screening Approach for Metals in Rivers   *(Continued)*

Data	Calculation Methodology Where Data Are Used*	Remarks
Source Data		
1. Background		
Metal concentrations, C_T	D, R, S, L	Background concentrations should generally not be set to zero without justification.
Boundary flow rates, Q_U	D, R, S, L	
Boundary suspended solids, S_U	D, R, S, L	One important reason for determining suspended solids concentrations is to determine the dissolved concentration, C, of metals, based on C_T, S, and K_p. However, if C is known along with C_T and S, this information can be used to find K_p.
Silt, clay fraction of suspended solids	L	
Locations	D, R, S, L	
2. Point Sources		
Locations	D, R, S, L	
Flow rate, Q_w	D, R, S, L	
Metal concentration, C_{Tw}	D, R, S, L	
Suspended solids, S_w	D, R, S, L	
Bed Data		
Depth of contamination		For the screening analysis, the depth of contamination is most useful during a period of prolonged scour when metal is being input into the water column from the bed.
Porosity of sediments, n		
Density of solids in sediments (e.g., 2.7 for sand), ρ_s		
Metal concentration in bed during prolonged scour period, C_{T2}		
Derived Parameters		
Partition coefficient, Kp	All	The partition coefficient is a very important parameter. Site-specific determination is preferable.
Settling velocity, ws	S, L	This parameter is derived based on suspended solids vs. distance profile.
Resuspension velocity, Wrs	R	This parameter is derived based on suspended solids vs. distance profile.
Equilibrium Modeling		
Water quality characterization of river:	E	Equilibrium modeling is required only if predominant metal species and estimated solubility controls are needed.
pH		
Suspended solids		
Conductivity		
Temperature		
Hardness		Water quality criteria for many metals are keyed to hardness, and allowable concentrations increase with increasing hardness.
Total organic carbon		
Other major cations and anions		

* D = dilution (includes total dissolved and adsorbed phase concentration predictions); R = dilution and resuspension; S = dilution and settling; L = lake; E = equilibrium modeling.

From EPA. *Handbook — Stream Sampling for Waste Load Allocation Applications.* U.S. Environmental Protection Agency. Office of Research and Development. Washington, D.C. EPA/625/6-86/013. 1986.

Table 6.62 Summary of Data Requirements for Screening Approach for Toxic Organics in Rivers

Data	Methodology Where Data Are Used	Remarks
River Hydraulic Data		
Flow rate, Q	D, DA, DAK	An accurate estimate of flow rate is very important because of dilution, which for many organics is the most important process that influences their fate. Measure or obtain from USGS gauge.
Cross-sectional area, A	D, DA, DAK	
Water depth, h	DAK	Water depth can influence rate processes such as volatilization and photolysis.
Reach lengths, x	DAK	
Stream velocity, U	DAK	U = Q/A should be used only where A is representative of the reach being analyzed. Otherwise dye tracers, measured from centroid to centroid of the dispersing dye, are a better method of finding velocity (indirectly as distance divided by travel time).
Source Data		
1. Background		
Toxicant concentrations	D, DA, DAK	Concentrations of organic toxicants may be negligible in areas not influenced by man.
Boundary flow rates	D, DA, DAK	
Boundary suspended solids	DA, DAK	Suspended solids are used to help determine the dissolved and adsorbed phase concentrations.
2. Point Source		
Locations	D, DA, DAK	
Flow rates, Q_w	D, DA, DAK	
Total toxicant concentration, C_T	D, DA, DAK	
Suspended solids, S_w	DA, DAK	
Partition Coefficient and Rate Constant Data		Difficult to calculate accurately.

From EPA. *Handbook — Stream Sampling for Waste Load Allocation Applications.* U.S. Environmental Protection Agency. Office of Research and Development. Washington, D.C. EPA/625/6-86/013. 1986.

Dye studies (as discussed earlier) are recommended in waste load allocation (WLA), or total maximum daily load (TMDL) studies to study point source mixing, movement of conservative pollutants, and to construct ambient toxicity profiles (EPA 1986; Figure 6.176). Multiple samples on a transect are necessary immediately downstream of sources or in wide streams (Figure 6.177). Samples of effluent from point sources (e.g., sewer overflow, culverts, tributaries [months], and stormwater) should be collected prior to dye studies, and both acute and chronic toxicity should be measured using EPA-recommended species (i.e., *Pimephales promelas*, *Ceriodaphnia dubia*), key surrogates (e.g., *Hyalella azteca*, *Selenastrum capricornutum*), and/or important resident species (e.g., trout). The dilution required to reach the no-observable-effects level (NOEL) in the toxicity tests should be the final sample points for constructing the dye isopleth (Figure 6.178; EPA 1986). These data may then be used to guide station location selection for ambient toxicity sample collection. In this manner, toxicity decay or persistence can be defined for various flow conditions.

SUMMARY

As indicated in many discussions in this book, multiple approaches are needed to effectively evaluate receiving water impacts in urban areas. This chapter presents details in collecting information pertaining to different ecosystem components and specific beneficial use impairments, including rainfall and flow monitoring; soil characteristics; aesthetics, litter, and safety; habitat

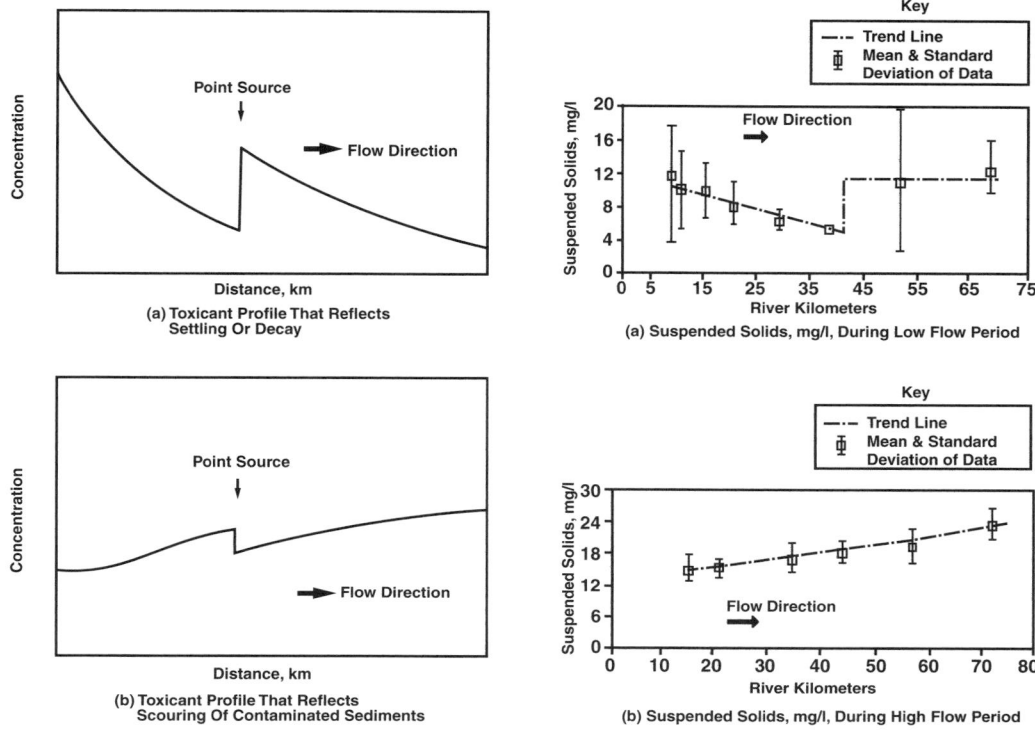

Figure 6.174 Typical concentration profiles of toxicants in rivers. (From EPA. *Handbook — Stream Sampling for Waste Load Allocation Applications.* U.S. Environmental Protection Agency. Office of Research and Development. Washington, D.C. EPA/625/6-96/013. 1986.)

Figure 6.175 Typical suspended solids concentrations during (a) low flow and (b) high flow periods. (From EPA. *Handbook — Stream Sampling for Waste Load Allocation Applications.* U.S. Environmental Protection Agency. Office of Research and Development. Washington, D.C. EPA/625/6-96/013. 1986.)

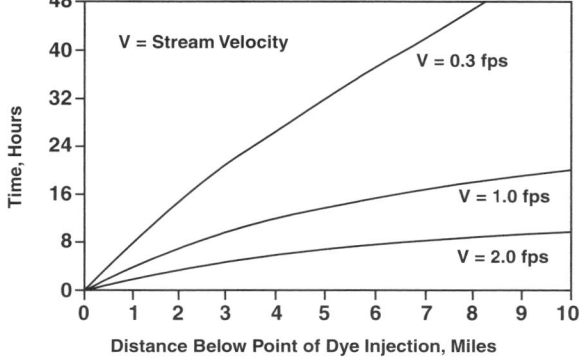

Figure 6.176 Time required for a continuous release of dye to reach steady-state concentrations at selected locations below the point of discharge. Note: the curves are based on a solution to the advection-dispersion equation which is used to predict when dye concentrations are 95% of steady-state levels. (From EPA. *Handbook — Stream Sampling for Waste Load Allocation Applications.* U.S. Environmental Protection Agency. Office of Research and Development. Washington, D.C. EPA/625/6-96/013. 1986.)

conditions; water and sediment chemical analyses; microorganism evaluations; benthos, zooplankton, and fish collecting; and tests for toxicity and bioaccumulation. This information supplements the information provided in Chapter 5 concerning collecting samples and selecting an experimental design. Chapter 7 briefly presents some statistical analyses tools, while Chapter 8 presents data interpretation for the complete study.

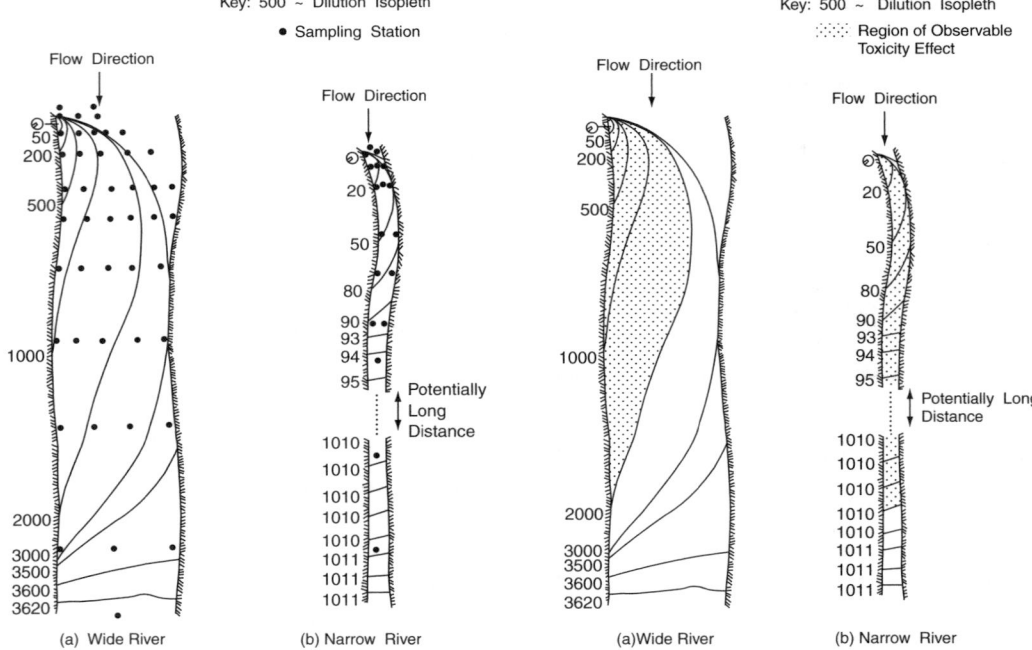

Figure 6.177 Example of sampling locations in wide and narrow rivers. (From EPA. *Handbook — Stream Sampling for Waste Load Allocation Applications.* U.S. Environmental Protection Agency. Office of Research and Development. Washington, D.C. EPA/625/6-96/013. 1986.)

Figure 6.178 Regions of observable toxicity in wide and narrow rivers. (From EPA. *Handbook — Stream Sampling for Waste Load Allocation Applications.* U.S. Environmental Protection Agency. Office of Research and Development. Washington, D.C. EPA/625/6-96/013. 1986.)

It is essential that there be an accurate description of the system's hydrodynamics when assessing the effects of stormwater runoff on receiving waters. Flow represents the pollutant loading mechanism, and its power and frequency of occurrence can degrade the physical habitat. Instantaneous flow can be measured using traditional current meters, while long-term flow monitoring is usually conducted using stage recorders. Tracer methods are also useful, especially where the flows are quite shallow and the stream channel very rough. Tracers can also be used to effectively indicate diffusion and transport of pollutant discharges into small streams. Flow is also of primary consideration in supporting aquatic life, as minimum depths and velocities are needed for their survival. With urbanization, flow changes can be dramatic, with excessive flows occurring during wet periods and significantly reduced flows occurring during dry months.

The role that different rains have on wet weather-related receiving water effects is also important to understand through evaluation of local data. As an example, small rains (less than about 0.5 in in the upper Midwest) are important because they are associated with the majority of runoff events and they frequently exceed heavy metal and bacteria objectives, although they only account for a small fraction of the annual pollutant discharges. Intermediate-sized rains (from about 0.5 to 1.5 in in the upper Midwest) account for the majority of the pollutant discharges and subject the receiving waters to frequent high pollutant loads and moderate-to-high flow rates. Larger rains (from about 1.5 to 3 in in the upper Midwest) produce relatively small amounts of the annual pollutant discharges, but produce the most damaging flows from a habitat destruction standpoint. The largest rains are critical from a drainage aspect and must be controlled to provide safe conditions for inhabitants of the watershed. These rains must be controlled in the primary drainage systems, while excessive flows that exceed the capacities of these systems must be safely controlled in secondary drainages (such as temporary flooding of some roads, parking areas, vacant fields, etc.). Therefore, the type of receiving water problem being addressed is likely associated with a specific set of rain conditions, typically much smaller than the rains used in the design of storm drainage.

Soils can play a significant role in watershed and receiving water assessments. Most of the particulates being transported in stormwater originate as local soil, and their texture can have dramatic effects on stream turbidity levels and the amounts of erosion from nonpaved areas. In addition, soils in urban areas undergo significant modifications and are generally greatly compacted compared to natural soil profiles. The compacted soils provide much less infiltration for the rain water, increasing the runoff flow rates. Soil surveys can describe the soil types, textures, depths, chemical quality, and amounts of compaction, which are all useful measures. Soil modifications to enhance infiltration, to capture pollutants during percolation above the groundwater, and improve the fertility of the soil to enhance plant growth with minimal fertilization can therefore be important stormwater control practices.

Aesthetics, litter, and safety are all critical receiving water attributes that need to be quantified to indicate if basic beneficial uses (such as noncontact recreation) are being met. Many municipalities currently suffer large litter accumulations along public streams that significantly detract from their use and respect. Habitat problems are probably some of the most important impairments to aquatic life beneficial uses. Unfortunately, "standards" for habitat goals are not likely to become possible, requiring local investigations to compare receiving waters to local reference conditions. The role that highly fluctuating flows have on habitat is beginning to be understood. The amount of large woody debris, and other channel-forming materials, can be directly measured in streams, along with the rate of channel enlargement. Stormwater controls can possibly be designed to overcome habitat problems if the role of the causative impairment factors in local waters is better understood.

Water quality measurements also need to be made in a comprehensive receiving water assessment. Historically, most studies overly relied on expensive water quality measurements, with little supportive information. Currently, many areas are almost totally eliminating water quality analyses in stream assessments and only examining several basic stream biological conditions. As noted in this book, it is important that a balanced set of parameters be included in an effective program, requiring a basic set of traditional, plus specialized water quality measurements. The specific water quality parameters to be monitored should be selected based on the beneficial uses of the stream, along with additional indicator parameters that can identify the presence of inappropriate discharges and other unusual conditions. This chapter describes different field monitoring options, along with modifications that may be needed for conventional laboratory methods to be most effective for stormwater samples. Needed detection limits, along with safety and complexity, are presented as the most important factors that determine the most appropriate analytical methods that should be used for the selected parameters.

Microorganism measurements are needed in most receiving water assessments, especially in areas having water-contact recreation and consumption of aquatic life beneficial uses. Newly available microorganism measurement methods and changes in guidance on target organisms require a reexamination of traditional approaches in the assessment of these important parameters in receiving waters.

Benthos sampling is one of the most important measurements in receiving water assessments (along with habitat evaluations). Much guidance is now available on obtaining and evaluating appropriate samples. Fish sampling, although more complex to conduct and evaluate, is an important assessment tool, especially when relating to beneficial uses that are easier for the interested public to understand. Currently accepted methods for benthos and fish sampling are described in detail in this chapter and in related appendices.

Toxicity and bioaccumulation measurements can be important tools, especially when trying to identify cause-and-effect relationships between different stressors and receiving water impacts. Recently developed *in situ* toxicity test methods are especially useful tools because they subject the test organisms to natural conditions, such as fluctuations in receiving water conditions, and to the toxicity effects of in-place sediments. Traditional and newly developed methods for toxicity testing is presented in this chapter.

Chapter 6 presents a wide range of tools for characterizing many different components of ecosystems. Case studies also illustrate these procedures and show how they can be effectively utilized. Summaries of the advantages and disadvantages of the different methods are also frequently presented. Several appendices also present supportive information for the techniques given in this chapter.

REFERENCES

Abel, P.D. Toxicity of y-hexachlorocyclohexane (lindane) to *Gammarus pulex*: mortality in relation to concentration and duration of exposure. *Freshwater Biol.*, 10: 251–259. 1980.

Abel, P.D. and S.M. Gardner. Comparisons of median survival times and medial lethal exposure times for *Gammarus pulex* exposed to cadmium, permethrin, and cyanide. *Water Res.*, 20: 579–582. 1986.

Absil, M.C.P., M. Berntssen, and L.J.A. Gerringa. The influence of sediment, food and organic ligands on the uptake of copper by sediment-dwelling bivalves. *Aquat. Toxicol.*, 34: 13–29. 1996.

Adams, D.D. Sampling sediment pore water, in *CRC Handbook of Techniques for Aquatic Sediments Sampling.* Edited by A. Mudroch and S.D. MacKnight. Lewis Publishers. Boca Raton, FL. pp. 171–202. 1991.

Akeley, R.P. Retention basins for control of urban stormwater quality, in *Proceedings: National Conference on Urban Erosion and Sediment Control: Institutions and Technology.* EPA-905/9-80-002, Chicago. January 1980.

Alabaster, J.S. and R. Lloyd (Eds.). *Water Quality Criteria for Fish, 2nd edition.* Butterworth Scientific, London. 1982.

Allen, T. *Particle Size Measurement.* John Wiley & Sons. New York. 1975.

Anderson, J.W., M. Stevenson, R.P. Markel, and M. Singer. *Results of a "tie" for samples of urban runoff entering San Francisco Bay.* Oceans [New York] Ocean Technologies and Opportunities in the Pacific for the 90s. Proceedings of Oceans '91, Oct. 1–3, 1991, IEEE Oceanic Engineering Soc. Piscataway, NJ. 1991.

Anderson, P.D. and L.J. Weber. The toxicity of aquatic populations of mixtures containing certain heavy metals. *Proc. Int. Conf. on Heavy Metals in the Environment.* Toronto, Ontario, Canada. 2: 933–953. 1977.

APHA. *Standard Methods for the Examination of Water and Wastewater.* Washington, D.C. 1985.

Armitage, P. D. Downstream changes in the composition, numbers and biomass of bottom fauna in the Tees below Cow Green Reservoir and in an unregulated tributary Maize Beck, in the first five years after impoundment. *Hydrobiologia,* 58: 145–156. 1978.

ASTM (American Society for Testing and Materials). Annual book of ASTM standards, *Water Environ. Technol.*, Vol. 11.04, ASTM, Philadelphia, PA. 1987.

ASTM (American Society for Testing and Materials). *Standard Guide for Collection, Storage, Characterization, and Manipulation of Sediments for Toxicological Testing.* American Society for Testing and Materials, Philadelphia, Standard No. E 1391. 1991.

ASTM (American Society of Testing and Materials). *1996 Annual Book of ASTM Standards.* West Conshohocken, PA. ASTM, Vol. 04.08, 1994.

ASTM. *Test Method for Measuring the Toxicity of Sediment-Associated Contaminants with Freshwater Invertebrates.* ASTM Standard E 1706-00. American Society for Testing and Materials, Philadelphia, PA. 2000.

Axelman, J., K. Naes, C. Naf, and D. Broman. Accumulation of polycyclic aromatic hydrocarbons in semipermeable membrane devices and caged mussels (*Mytilus edulis* L.) in relation to water column phase distribution. *Environ. Toxicol. Chem.*, 18: 2454–2461. 1999

Bailey, H.C., J.L. Miller, M.J. Miller, L.C. Wiborg, L. Deanovic, and T. Shed. Joint acute toxicity of diazinon and chlorpyrifos to *Ceriodaphnia dubia. Environ. Toxicol. Chem.*, 16: 2304–2308. 1997.

Bailey, H.C., L. Deanovic, E. Reyes, T. Kimball, K. Larson, K. Cortright, V. Connor, and D.E. Hinton. Diazinon and chlorpyrifos in urban waterways in northern California, USA. *Environ. Toxicol. Chem.*, 19: 82–87. 2000

Bannerman, R., K. Baun, M. Bohn, P.E. Hughes, and D.A. Graczyk. *Evaluation of Urban Nonpoint Source Pollution Management in Milwaukee County, Wisconsin*, Vol. I. PB 84-114164. U.S. Environmental Protection Agency, Water Planning Division. November 1983.

Barbé, D.E., R.E. Pitt, M.M. Lalor, J.P. Harper, and C.M. Nix. *Including New Technologies into the Investigation of Inappropriate Pollutant Entries into Storm Drainage Systems — A User's Guide*. Urban Waste Management & Research Center, University of New Orleans, and the U.S. Environmental Protection Agency. January 2000.

Bascombe, A.D. *Urban Pollution, Research Report 11: Biological Monitoring of Benthic Invertebrates for the Assessment of Heavy Metal Pollution in Urban Rivers*. Middlesex Polytechnic Urban Pollution Research Centre, England. 1988.

Bascombe, A.D., M.A. House, and J.B. Ellis. The utility of chemical and biological monitoring techniques for the assessment of urban pollution. *River Basin Management*, H. Laikari. Pergamon Press, Oxford, U.K. pp. 59–69. 1989.

Bascombe, A.D., J.B. Ellis, D.M. Revitt, and R.B.E. Shutes. Development of ecotoxicological criteria in urban catchments. *Water Sci. Technol.*, 22(10/1): 173–179. 1990.

Baumann P.C. and J.C. Harshbarger. Decline in liver neoplasms in wild brown bullhead catfish after coking plant closes and environmental PAHs plummet. *Environ. Health Perspect.*, 103: 168–170. 1995.

Benke, A.C., D.M. Gillespie, and T.C. Van Arsdall. Invertebrate productivity in a subtropical Blackwater River: the importance of habitat and life history. *Ecol. Monogr.*, 545: 25–63. 1984.

Benoit, D.A., P.K. Sibley, J.L. Juenemann, and G.T. Ankley. *Chironomus tentans* life-cycle test: design and evaluation for use in assessing toxicity of contaminated sediments. *Environ. Toxicol. Chem.*, 16: 1165–1176. 1997.

Benson, W.H. and R.T.D. Guilio. Biomarkers in hazard assessments of contaminated sediments, in *Sediment Toxicity Assessment*. Edited by G. A. Burton, Jr. Lewis Publishers, Boca Raton, FL. pp. 241–266. 1992.

Berg, G., Ed. *Transmission of Viruses by the Water Route*. Interscience Publishers, New York. 1965.

Berge, J.A. Macrofaunal recolonization of subtidal sediments. Experimental studies on defaunated sediment contaminated with crude oil in two Norwegian fjords with unequal eutrophication status. I. Community responses. *Mar. Ecol. Prog. Ser.,* 66: 103–115. 1990.

Birge, W.J., J.A. Black, T.M. Short, and A.G. Westerman. A comparative ecological and toxicological investigation of a secondary wastewater treatment plant effluent and its receiving stream. *Environ. Toxicol. Chem.,* 8: 437–449. 1989.

Bishop, C.A., N.A. Mahony, S. Trudeau, and K.E. Pettit. Reproductive success and biochemical effects of tree swallows (*Tachycineta bicolor*) exposed to chlorinated hydrocarbon contaminants in wetlands of the Great Lakes and St. Lawrence River Basin, USA and Canada. *Environ. Toxicol. Chem.,* 18: 263–271. 1999.

Blasso, L. Flow measurement under arid conditions, *Instruments Control Syst.,* 48: 45–50. 1975.

Bolstad, P.V. and W.T. Swank. Cumulative impacts of landuse on water quality in a southern Appalachian watershed. *J. Am. Water Resourc. Assoc.,* 33: 519–533. 1997.

Borgmann, U. A recommendation for greater emphasis on body-concentration based toxicity assessment for sediments. *SETAC News,* 16: 13. 1996.

Borgmann, U. and W.P. Norwood. Spatial and temporal variability in toxicity of Hamilton Harbor sediments: Evaluation of the *Hyalella azteca* 4-week chronic toxicity test. *J. Great Lakes Res.*, 19: 72–82. 1993.

Borgmann, U. and W.P. Norwood. Kinetics of excess (above background) copper and zinc in *Hyalella azteca* and their relationship to chronic toxicity. *Can. J. Fish. Aquat. Sci.,* 52: 864–874. 1995.

Bott, A. Voltammetric determination of trace concentrations of metals in the environment. *Curr. Separations*, 14(1): 24–30. 1995.

Boucher, A.M. and M.C. Watzin. Toxicity identification evaluation of metal-contaminated sediments using an artificial pore water containing dissolved organic carbons. *Environ. Toxicol. Chem.,* 18: 509–518. 1999.

Bowie, G., W. Mills, D. Porcella, C. Campbell, J. Pagenkopf, G. Rupp, C. Chamberlin, K. Johnson, and S. Gherini. *Rates, Constants, and Kinetics Formations in Surface Water Quality Modeling, Edition 2.* U.S. Environmental Protection Agency, Athens, GA. 1985.

Box, G.E.P., W.G. Hunter, and J.S. Hunter. *Statistics for Experimenters*. John Wiley & Sons, New York. 1978.

Brent, R.N. and E.E. Herricks. Postexposure effects of brief cadmium, zinc, and phenol exposures on freshwater organisms. *Environ. Toxicol. Chem.,* 17: 2091–2099. 1998

Brinkhurst, R. O. *The Benthos of Lakes*. St. Martin's Press, New York. 190 pp. 1974.

Broderius, S. and M. Kahl. Acute toxicity of organic chemical mixtures to the fathead minnow. *Aquat. Toxicol.*, 6: 307–322. 1985.

Brooker, J.A. and G.A. Burton, Jr. *In Situ Exposures of Asiatic Clams (Corbicula fluminea) and Mayflies (Hexagenia limbata) to Assess the Effects of Point and Nonpoint Source Pollution.* Abstract: Annual Meeting, Society of Environmental Toxicology and Chemistry, Pensacola, FL. 1998

Burgess, R.M. and K.J. Scott. The significance of in-place contaminated marine sediments on the water column: processes and effects, in *Sediment Toxicity Assessment.* Edited by G. A. Burton, Jr. Lewis Publishers, Boca Raton, FL. pp. 129–166. 1992.

Burgess, R.M., J.B. Charles, A. Kuhn, K.T. Ho, L.E. Patton, and D.G. McGovern. Development of a cation-exchange methodology for marine toxicity identification evaluation applications. *Environ. Toxicol. Chem.*, 16: 1203–1211. 1997.

Burkholder, J.M., L.M. Larsen, J.H.B. Glasgow, K.M. Mason, P. Gama, and J.E. Parsons. Influence of sediment and phosphorus loading on phytoplankton communities in an urban piedmont reservoir. *J. Lake Reservoir Manage.*, 14(1): 110–121. 1998.

Burton, G.A., Jr. Evaluation of seven sediment toxicity tests and their relationships to stream parameters. *Toxicity Assess.*, 4: 149–159. 1989.

Burton, G.A., Jr. Assessing freshwater sediment toxicity. *Environ. Toxicol. Chem.*, 10: 1585–1627. 1991a.

Burton, G.A. Jr. Assessing freshwater sediment toxicity. *Environ. Toxicol. Chem.*, 10: 1585–1627. 1991b.

Burton, G.A., Jr. Realistic assessments of ecotoxicity using traditional and novel approaches. *Aquat. Ecosyst. Health Manage.*, 2: 1–8, 1999.

Burton, G.A., Jr. Quality assurance issues in assessing receiving waters, in *Proc. of the Conf. on Effects of Urban Runoff on Receiving Systems.* Edited by J. Saxena, Engineering Foundation Publ., New York. 1992a.

Burton, G.A., Jr., Ed. *Sediment Toxicity Assessment.* Lewis Publishers, Boca Raton, FL. 1992b.

Burton, G.A., Jr. Sediment collection and processing factors affecting realism, in *Sediment Toxicity Assessment,* Edited by G.A. Burton, Jr., Lewis Publishers, Boca Raton, FL. 1992c.

Burton, G.A., Jr., M.K. Nelson, and G. Ingersoll. Freshwater benthic toxicity tests, in *Sediment Toxicity Assessment.* Edited by G.A. Burton, Jr. Lewis Publishers, Boca Raton, FL. pp. 213–240. 1992.

Burton, G.A., Jr. and J. Scott. Sediment toxicity evaluations. *Environ. Sci. Technol.*, Vol. 25. 1992.

Burton, G.A., Jr. and L. Moore. *An Assessment of Stormwater Runoff Effects in Wolf Creek, Dayton, OH. Final Report.* City of Dayton, OH. 1999.

Burton, G.A. Jr. and C. Rowland. *Assessment of in Situ Stressors and Sediment Toxicity in the Lower Housatonic River.* Final Report to R.F. Weston, Manchester, NH. 1999.

Burton, G.A., Jr., D. Gunnison, and G.R. Lanza. Survival of pathogenic bacteria in various freshwater sediments. *Appl. Environ. Microbiol.*, 53: 633–638. 1987.

Burton, G.A., C.G. Ingersoll, L.C. Burnett, M. Henry, M.L. Hinman, S.J. Klaine, P.F. Landrum, P. Ross, and M. Tuchman. A comparison of sediment toxicity test methods at three Great Lakes areas of concern. *J. Great Lakes Res.*, 22: 495–511, 1996a.

Burton, G A., Jr., C. Hickey, T. DeWitt, D. Morrison, D. Roper, and M. Nipper. *In situ* toxicity testing: teasing out the environmental stressors. *SETAC News*, 16(5): 20–22. 1996b.

Burton, G.A., Jr., T.J. Norberg-King, C.G. Ingersoll, G.T. Ankley, P.V. Winger, J. Kubitz, J.M. Lazorchak, M.E. Smith, I.E. Greer, F.J. Dwyer, D.J. Call, K.E. Day, P. Kennedy, and M. Stinson. Interlaboratory study of precision: *Hyalella azteca* and *Chironomus tentans* freshwater sediment toxicity assay. *Environ. Toxicol. Chem.*, 15: 1335–1343, 1996c.

Burton, G.A., Jr., R. Pitt, and S. Clark. The role of whole effluent toxicity test methods in assessing stormwater and sediment contamination. *CRC Crit. Rev. Environ. Sci. Tech.* 30: 413–447.

Burton, G.A., Jr., C. Rowland, K. Kroeger, M. Greenberg, D. Lavoie, and J. Brooker. *Determining the Effect of Ammonia at Complex Sites: Laboratory and in Situ Approaches.* Abstr. Ann. Meet. Soc. Environ. Toxicol. Chem., Pensacola, FL. 1998.

Burton, G.A., B.L. Stemmer, K.L. Winks, P.E. Ross, and L.C. Burnett. A multitrophic level evaluation of sediment toxicity in Waukegan and Indiana Harbors. *Environ. Toxicol. Chem.*, 8: 1057–1066. 1989.

Cairns, J., Jr. and K.L. Dickson. A simple method for the biological assessment of the effects of waste discharges on aquatic bottom-dwelling organisms. *J. Water Pollut. Control Fed.*, 43: 755–772. 1971.

Cairns, J., Jr. and R.L. Kaesler. Cluster analysis of fish in a portion of the upper Potomac River. *Trans. Am. Fish. Soc.*, 100: 750–756. 1971.

Cairns, J., Jr., A.L. Buikema Jr., A.G. Heath, and B.C. Parker. Effects of temperature on aquatic organism sensitivity to selected chemicals. *Va. Water Resourc. Res. Ctr. Bull.*, 106: 9–70. 1978.

Cairns, J., Jr. A strategy for use of protozoans in the evaluation of hazardous substances, in *Biological Indicators of Water Quality.* John Wiley & Sons, New York. 1979.

Carder, J.P. and K.D. Hoagland. Combined effects of alachlor and atrazine on benthic algal communities in artificial stream. *Environ. Toxicol. Chem.*, 17: 1415–1420. 1998.

Carlton, R.G. and M.J. Klug. Spatial and temporal variation in microbial processes in aquatic sediments: implications for the nutrient status of lakes, in *Sediments: Chemistry and Toxicity of In-Place Pollutants.* Edited by R. Baudo, J. Giesy, and H. Muntau. Lewis Publishers, Boca Raton, FL. pp. 107–130, 1990.

Carpenter, S.R., J.F. Kitchel, and J.R. Hodgson. Cascading trophic interactions and lake productivity. *Bioscience,* 35: 634–639. 1985.

Chandler, G.T., B.C. Coull, N.V. Schizas, and T.L. Donelan. A culture-based assessment of the effects of chlorpyrifos on multiple meiobenthic copepods using microcosms of intact estuarine sediments. *Environ. Toxicol. Chem.*, 16: 2339–2346. 1997.

Chapman, P.M., E.A. Power, and G.A. Burton, Jr. Integrative assessments in aquatic ecosystems, in *Sediment Toxicity Assessment*, Edited by G. A. Burton, Jr., Lewis Publishers, Boca Raton, FL. 1992.

Chapman, PM. Sediment quality criteria from the sediment quality triad: an example. *Environ. Toxicol. Chem.,* 5: 967–964. 1986.

Chappie D.J. and G.A. Burton, Jr. Optimization of *in situ* bioassays with *Hyalella azteca* and *Chironomus tentans. Environ. Toxicol. Chem.*, 16: 559–564. 1997.

Cherry, D.S. State of the art of *in situ* testing (transplant experiments) in hazard evaluation. *SETAC News*, 16: 24–25. 1996.

Cherry, D.S., J.L. Farris, and R.J. Neves. Laboratory and field ecotoxicological studies at the Clinch River Plant, Virginia. Columbus, OH, Biology Department, Virginia Tech, Blacksburg VA. Final report to American Electric Power Company, Columbus, OH. 1991.

Chow, V.T. *Handbook of Applied Hydrology: A Compendium of Water Resource Technology.* McGraw-Hill, New York. 1964.

Christen, K. Synergistic effects of chemical mixtures and degradation byproducts not reflected in water quality standards, USGS finds. *Environ. Sci. Technol.*, 33: 230A–231A. 1999.

Christensen, E.R. Dose-response functions in aquatic toxicity testing and the Weibull model. *Water Res.*, 18: 213–221. 1984.

Chung, K.H., K.S. Ro, and S.U. Hong. Synthetic detergent in an urban stream in Korea. *Proceedings of the 65th Water Environment Federation Technical Exhibition and Conference,* Miami Beach, FL. *Volume 4, Surface Water Quality and Ecology.* pp. 429–434. Alexandria, VA. 1995.

Ciaponi, C., U. Moisello, and S. Papiri. Rainfall measurements and spatial variability in a small urban catchment, in *Proceedings of the 6th International Conference on Urban Storm Drainage*, Niagara Falls, Ontario, Sept. 12–17, 1993. pp. 158–163. IAHR/IAWQ. Seaport Publishing, Victoria, B.C., 1993.

Clark, A.G. and C.M.S. Lanigan. Prospects for estimating nucleotide divergence with RAPDs. *Mol. Biol. Evol.*, 10: 1096–1111. 1993.

Clements, W.H., D.S. Cherry, and J. Cairns, Jr. The influence of copper exposure on predator-prey interactions in aquatic insect communities. *Freshwater Biol.*, 21: 483–488. 1989.

Clements, W.H. and P.M. Kiffney. Integrated laboratory and field approach for assessing impacts of heavy metals at the Arkansas River, Colorado. *Environ. Toxicol. Chem.*, 13: 397–404. 1994.

Clements, W.H. and P.M. Kiffney. Validation of whole effluent toxicity tests: integrated studies using field assessments, microcosms, and mesocosms, in *Whole Effluent Toxicity Testing: An Evaluation of Methods and Prediction of Receiving System Impacts,* Edited by D.R. Grothe, K.L. Dickson, and D.K. Reed-Judkins. Soc. Environ. Toxicol. Chem. Press, Pensacola, FL. pp. 229–244. 1996.

Coakley, J.P. et al. Specific organic components as tracers of contaminated fine sediment dispersion in Lake Ontario near Toronto. *Hydrobiologia,* 235–236, 85. July 1992.

Connor, V. Pesticide toxicity in stormwater runoff. Technical memorandum. Sacramento, CA, California Regional Water Quality Control Board. Central Valley Region. Sacramento, CA. 1995.

Cook, R.B., G.W. Sutter II, and E.R. Swain. Ecological risk assessment in a large river reservoir. 1. Introduction and background. *Environ. Toxicol. Chem.*, 18: 581–588. 1999.

Cooke, T.D., D. Drury, R. Katznelson, C. Lee, P. Mangarella, and K. Whitman. Stormwater NPDES monitoring in Santa Clara Valley. ASCE Engineering Research Foundation, Crested Butte, CO. 1995.

Cooper, J.A. and J.G., Watson, Jr. Receptor oriented methods of air particulate source apportionment. *J. Air Pollut. Control Assoc.*, 30(10): 1116–1125. Oct. 1980.

Cooper, C. *Hydrologic Effects of Urbanization on Puget Sound Lowland Streams.* Masters thesis. University of Washington, Seattle, WA. 1996.

Coull, B.C. and G.T. Chandler. Pollution and meiofauna: field, laboratory and mesocosm studies. *Oceanogr. Mar. Biol. Annu. Rev.,* 30: 191–271. 1992.

Crane, M., P. Delaney, C. Mainstone, and S. Clarke. Measurement by *in situ* bioassay of water quality in an agricultural catchment. *Water Res.,* 29: 2441–2448. 1995.

Crossey, M.H. and T.W. LaPoint. A comparison of periphyton community structural and functional responses to heavy metals, *Hydrobiologia,* 162: 109–121. 1988.

Crunkilton, R., J. Kleist, J. Ramcheck, B. DeVita, and D. Villeneuve. Assessment of the response of aquatic organisms to long-term *in situ* exposures to urban runoff, in *Effects of Watershed Development & Management on Aquatic Ecosystems,* Engineering Foundation Conference, Snowbird, UT. ASCE, New York. August 1996.

Cummins, K.A. An evaluation of some techniques for the collection and analysis of benthic samples with special emphasis on lotic waters. *Am. Midl. Nat.,* 67: 477–504. 1962.

Cummins, K.W. Ecology of running waters: theory and practice, in *Proc. Sandusky River Basin Symposium.* Edited by D. B. Baker. Heidelburg College, Tiffin, OH. 1975.

Cummins, K.W. Trophic relations of aquatic insects. *Annu. Rev. Entomol.,* 18: 183–206. 1973.

Cummins, K.W. Structure and function of stream ecosystems. *BioScience,* 24: 631–641. 1974.

Cummins, K.W., G.W. Minshall, J.R. Sedell, C.E. Cushing, and R.C. Petersen. Stream ecosystem theory. *Verh. Internatl. Verein. Limnol.,* 22: 1818–1827. 1984.

Cummins, K.W., and M.A. Wilzbach. *Field Procedures for Analysis of Functional Feeding Groups of Stream Macroinvertebrates,* Contribution 1611, Appalachian Environmental Laboratory, University of Maryland, Frostburg, MD. 1985.

Curtis, D.C. Designing rain gage networks for automated flood warning systems, presented at *Flood Plain Management Association Conference,* Solvang, CA. Mar. 30–Apr. 2, 1993.

Curtis, L.R., W.K. Seim, and G.A. Chapman. Toxicity of fenvalerate to developing steelhead trout following continuous or intermittent exposure. *J. Toxicol. Environ. Health,* 15: 445–457. 1985.

Das, B.M. *Principals of Geotechnical Engineering.* PWS Publishing Co., Boston. 1994.

Davies, P.H. Toxicology and chemistry of metals in urban runoff, in *Urban Runoff Quality: Impact and Quality Enhancement Technology.* Engineering Foundation Conference, Henniker, NH. ASCE, New York. 1986.

Davies, P.H. Synergistic effects of contaminants in urban runoff, in *Effects of Urban Runoff on Receiving Systems: An Interdisciplinary Analysis of Impact, Monitoring, and Management.* Engineering Foundation Conference. Mt. Crested Butte, CO. ASCE, New York. 1991.

Davies, P.H. Factors in controlling nonpoint source impacts, in *Stormwater Runoff and Receiving Systems: Impact, Monitoring, and Assessment.* Edited by E.E. Herricks, CRC/Lewis Publishers, Boca Raton, FL. pp. 53–64. 1995.

Davis, A., P. DeCurnou, and L.E. Eary. Discriminating between sources of arsenic in the sediments of a tidal waterway, Tacoma, Washington. *Environ. Sci. Technol.,* 31(7): 1985. 1997.

Davis, R.A., Jr. *Depositional Systems: A Genetic Approach to Sedimentary Geology.* Prentice-Hall, Englewood Cliffs, NJ. 1983.

Davis, W.S., L.A. Fay, and C.E. Herdendorf. Overview of USEPA/Clear Lake Erie sediment oxygen demand investigations during 1979. *J. Great Lakes Res.,* 13: 731–737. 1987.

Day, J. *Selection of Appropriate Analytical Procedures for Volunteer Field Monitoring of Water Quality.* MSCE thesis, Department of Civil and Environmental Engineering, University of Alabama at Birmingham. 1996.

Day, K.E., B.J. Dutka, K.K. Kwan, N. Batista, T.B. Reynoldson, and J.L. Metcalfe-Smith. Correlations between solid-phase microbial screening assays, whole-sediment toxicity tests with macroinvertebrates and *in situ* benthic community structure. *J. Great Lakes Res.,* 21: 192–206. 1995.

Dayton, P.K. Competition, disturbance, and community organization: the provision and subsequent utilization of space in a rocky intertidal community. *Ecol. Monogr.,* 41: 351–389. 1971.

Debo, T.N. and A.J. Reese. *Municipal Storm Water Management*. Lewis Publishers, Boca Raton, FL. 1995.

Decamps, H., F. Fournier, F. Naiman, R.J. Naiman, and R.C. Petersen. Wetland management and restoration. *Ambio*, 19(3): 175–176. May 1990.

Dekker, T., M. TenBroek, and R. Karsan. Dye dilution testing in the greater Detroit regional sewer system — laboratory investigation and field screening techniques. Presented at the *Advances in Urban Wet Weather Pollution Reduction* conference. Cleveland, OH. June 28–July 1, 1998. pp. 367–376. Water Environment Federation. Alexandria, VA. 1998.

Deneer, J.W., W. Seinen, and J.L.M. Hermens. Growth of *Daphnia magna* exposed to mixtures of chemicals with diverse modes of action. *Ecotoxicol. Environ. Saf.*, 15: 72–77. 1988a.

Deneer, J.W., T.L. Sinnige, W. Seinen, and J.L.M. Hermens. 1988. The joint acute toxicity to *Daphnia magna* of industrial organic chemicals at low concentrations. *Aquat. Toxicol.*, 12: 33–38. 1988b.

Desbrow, C., E.J. Routledge, G.C. Brighty, J.P. Sumpter, and M. Waldock. Identification of estrogenic chemicals in STW effluent. 1. Chemical fractionation and in vitro biological screening. *Environ. Sci. Technol.*, 32(11): 1549. 1998.

Dewitt, T., D.J. Morrisey, D. Roper, and M. Nipper. Fact or artefact: the need for appropriate controls in ecotoxicological field experiments. *SETAC News*, 1996.

Diamond, J.M., C. Gerardi, E. Leppo, and T. Miorelli. Using a water effect ratio approach to establish effects of an effluent influenced stream on copper toxicity to the fathead minnow. *Environ. Toxicol. Chem.*, 16: 1480–1487. 1997.

Dickerson, K.K., W.A. Hubert, and H.L. Bergman. Toxicity assessment of water from lakes and wetlands receiving irrigation drain water. *Environ. Toxicol. Chem.*, 15: 1097–1101. 1996.

DICKEY-john Corporation. *Installation Instructions Soil Compaction Tester.* Auburn, IL. 1987.

Dickson, K.L., W.T. Waller, J.H. Kennedy, and L.P. Ammann. Assessing the relationship between ambient toxicity and instream biological response. *Environ. Toxicol. Chem.*, 11: 1307–1322. 1992.

Dickson, K.L., W.T. Waller, J.H. Kennedy, L.P. Ammann, R. Guinn, and T.J. Norberg-King. Relationships between effluent toxicity, ambient toxicity, and receiving system impacts: Trinity River dechlorination case study, in *Whole Effluent Toxicity Testing: An Evaluation of Methods and Prediction of Receiving System Impacts*. Edited by D.R. Grothe, K.L. Dickson, and D.K. Reed-Judkins. Soc. Environ. Toxicol. Chem. Press, Pensacola, FL. pp. 287–305. 1996.

Doudoroff, P. and M. Katz. Critical review of literature on the toxicity of industrial wastes and their components to fish. II. The metals, as salts. *Sewage Ind. Waste*, 25: 802–839.

Downing, J.A. Sampling the benthos of standing waters, a manual on methods for the assessment of secondary productivity in freshwaters, in *IBP Handbook 12, 2nd Ed.,* Edited by J.A. Downing and F.H. Rigler, Blackwell Scientific Publications, Boston, MA. pp. 87–130. 1984.

Downing, J.A. and L.C. Roth. Spatial patchiness in the lacustrine sedimentary environment. *Limnol Oceanogr.*, 33: 447–458. 1988.

Driscoll, E.D. Detention and retention controls for urban stormwater. Engineering Foundation Conference: *Urban Runoff Quality — Impact and Quality Enhancement Technology*. Henniker, NH. Edited by B. Urbonas and L.A. Roesner. pp. 145–163. Published by the American Society of Civil Engineers, New York, June 1986.

Droste, R.L. and Gupgupoglu, A.I. *Indicator Bacteria Die-off in the Rideau River*. University of Ottawa, Ottawa, Ontario 1982.

Dufour, A.P. Bacterial indicators of recreational water quality. *Can, J, Publ, Health*, 75: 49–56. January/February 1984.

Eaganhouse, R.P., D.P. Olaguer, B.R. Gould, and C.S. Phinney. Use of molecular markers for the detection of municipal sewage sludge at sea. *Mar. Environ. Res.*, 25(1): 1–22. 1988.

Eagleson, K.W., D.L. Lenat, L.W. Ausley, and F.B. Winborne. Comparison of measured instream biological responses with responses predicted using the *Ceriodaphnia dubia* chronic toxicity test. *Environ. Toxicol. Chem.*, 9: 1019–1028. 1990.

Easton, J.H., M. Lalor, R. Pitt, and D.E. Newman. The use of a multi-parameter water quality monitoring instrument to continuously monitor and evaluate runoff events. Presented at *Annual Water Resources Conference of the AWRA*, Point Clear, AL. 1998.

Easton, J. *The Development of Pathogen Fate and Transport Parameters for Use in Assessing Health Risks Associated with Sewage Contamination*. Ph.D. dissertation, Department of Civil and Environmental Engineering, University of Alabama at Birmingham. 2000.

Eaton, J.G. Chronic toxicity of a copper, cadmium and zinc mixture to the fathead minnow (*Pimephales promelas* Rafinesque). *Water Res.*, 7: 1723–1736. 1973.

Eckner, K.F. Comparison of membrane filtration and multiple-tube fermentation by the Colilert and Enterolert methods for detection of waterborne coliform bacteria, *Escherichia coli*, and enterococci used in drinking and bathing water quality monitoring in southern Sweden. *Appl. Environ. Microbiol.*, 64(8): 3079–3083, 1998.

Edberg, N. and Hofstan, B.V. Oxygen uptake of bottom sediment studied *in-situ* and in the laboratory. *Water Res.*, 7: 1285. 1973.

Edberg, S.C., M.J. Allen, D.B. Smith, and N.J. Kriz. Enumeration of total coliforms and *Escherichia coli* from source water by the defined substrate technology. *Appl. Environ. Microbiol.*, 56(2): 366–369, 1990.

Edmondson, W.T., and G.G. Winberg, Eds. *A Manual on Methods for the Assessment of Secondary Productivity in Fresh Water.* International Biological Programme Handbook 17. Blackwell Scientific Publications, Oxford. 358 pp. 1971.

Eganhouse, R.P., D.P. Olaguer, B.R. Gould, and C.S. Phinney. Use of molecular markers for the detection of municipal sewage sludge at sea. *Mar. Environ. Res.*, 25(1): 1–22, 1988.

Elliott, J.M. and C.M. Drake. A comparative study of seven grabs for sampling benthic macroinvertebrates of rivers. *Freshwater Biol.*, 11: 99–120. 1981.

Ellis, J., R. Shutes, and D. Revitt. Ecotoxicological approaches and criteria for the assessment of urban runoff impacts on receiving waters, in *Proc. of the Effects of Urban Runoff on Receiving Systems: An Interdisciplinary Analysis of Impact, Monitoring and Management.* Edited by E. Herricks, J. Jones, and B. Urbonas, Engineering Foundation, New York. 1992.

Ellis, G.S., J.N. Huckins, C.E. Rostad, C.J. Schmitt, J.D. Petty, and P. MacCarthy. Evaluation of lipid-containing semipermeable membrane devices for monitoring organochlorine contaminants in the upper Mississippi River. *Environ. Toxicol. Chem.*, 14: 1875–1884. 1995.

Ellis, D.H., G.A. Burton, Jr., and D. Krane. RAPD-PCR is a sensitive measure of changes in genetic diversity induced by pollution stress. *Abstr. Annu. Meet. Soc. Environ. Toxicol. Chem.*, San Francisco, CA. 1997.

EPA (U.S. Environmental Protection Agency). *Suspended and Dissolved Solids Effects on Freshwater Biota: A Review,* Environmental Research Laboratory, U.S. Environmental Protection Agency, Corvallis, OR, EPA 600/3–77/042. 1977.

EPA. *Handbook for Sampling and Sample Preservation of Water and Wastewater.* Environmental Monitoring and Support Lab, U.S. Environmental Protection Agency, Cincinnati, OH, EPA 600/4-82/029. 1982.

EPA. *Results of the Nationwide Urban Runoff Program.* Water Planning Division, PB 84-185552, Washington, D.C., December 1983a.

EPA. *Technical Support Manual: Waterbody Surveys and Assessments for Conducting Use Attainability Analyses.* Office of Water Regulations and Standards, U.S. Environmental Protection Agency, Washington, D.C. 1983b.

EPA. *Methods for Chemical Analysis of Water and Wastes.* EPA-600/4-79-020, U.S. Environmental Protection Agency, Cincinnati, Ohio. 1983c.

EPA. *Handbook — Stream Sampling for Waste Load Allocation Applications.* U.S. Environmental Protection Agency. Office of Research and Development. Washington, D.C. EPA/625/6-86/013. 1986.

EPA. *A Compendium of Superfund Field Operations Methods.* Office of Emergency and Remedial Response, U.S. Environmental Protection Agency, Washington, D.C. EPA 540/P-87/001. 1987.

EPA. *Ecological Assessment of Hazardous Waste Sites.* Environmental Research Laboratory, U.S. Environmental Protection Agency, Corvallis, OR. EPA 600/3-89/013. 1989a.

EPA. *Protocols for Short-Term Toxicity Screening of Hazardous Waste Sites.* Environmental Research Laboratory, U.S. Environmental Protection Agency, Corvallis, OR. EPA 600/3-88/029. 1989b.

EPA. *Rapid Bioassessment Protocols for Use in Streams and Rivers: Benthic Macroinvertebrates and Fish.* Office of Water, U.S. Environmental Protection Agency, Washington, D.C., EPA 444/4-89/001. 1989c.

EPA. *Upper Great Lakes Connecting Channels Study. Volume II, Final Report.* Great Lakes National Program Office, Chicago, IL. 1989d.

EPA. *Biological Criteria: National Program Guidance for Surface Waters.* EPA-440-5-90-004. U.S. Environmental Protection Agency. Office of Water Regulations and Standards. Washington, D.C. 1990a.

EPA. *Macroinvertebrate Field and Laboratory Methods for Evaluating the Biological Integrity of Surface Waters.* Office of Research and Development, U.S. Environmental Protection Agency, Washington, D.C., EPA 600/4-90/030. 1990b.

EPA. *Methods for Aquatic Toxicity Identification Evaluations. Phase 1, Toxicity Characterization Procedures 2nd edition.* EPA/600/6-91/003. Office of Research and Development., Washington, D.C. 1991a.

EPA. *Modeling of Nonpoint Source Water Quality in Urban and Non-urban Areas.* Office of Research and Development, U.S. Environmental Protection Agency, Washington, D.C. EPA 600/3-91/039. 1991b.

EPA. *Proposed Guidance Specifying Management Measures for Sources of Nonpoint Pollution in Coastal Waters.* Office of Water, U.S. Environmental Protection Agency, Washington, D.C. 1991c.

EPA. *Sediment Toxicity Identification Evaluations: Phase 1 (Characterization), Phase 2 (Identification), and Phase 3 (Confirmation). Modification of Effluent Procedures.* Environmental Research Laboratory, Duluth, MN. Tech. Report No. 08-91. EPA 600/6-91-007. 1991d.

EPA. *Technical Support Document for Water Quality-based Toxics Control.* U.S. Environmental Protection Agency. Office of Water. EPA/5052-90-001. 1991e.

EPA. Wisconsin has had a priority watershed program in operation for more than a decade now; eleven new watersheds named this spring. *Nonpoint Source EPA News-Notes,* #12. April–May 1991f.

EPA. *Methods for Measuring the Toxicity and Bioaccumulation of Sediment-associated Contaminants with Estuarine and Marine Amphipods.* Office of Research and Development. EPA/600/R-94/025. Washington, D.C. 1994a.

EPA. *Assessment and Remediation of Contaminated Sediments (ARCS) Program: Assessment Guidance Document.* EPA 905-B94-002. U.S. Environmental Protection Agency, Great Lakes National Program Office, Chicago, IL. 1994b.

EPA. *Short-term Methods for Estimating the Chronic Toxicity of Effluents and Receiving Water to Freshwater Organisms.* Research and Development. Cincinnati, OH. EPA/600/4-91/002. 1995.

EPA. *The Incidence and Severity of Sediment Contamination in Surface Waters of the United States.* U.S. Environmental Protection Agency. Office of Science and Technology. EPA 823-R-97-006. Washington, D.C. 1997a.

EPA. *Volunteer Stream Monitoring: A Methods Manual.* EPA 841-B-97-003. U.S. Environmental Protection Agency, Office of Water, Washington, D.C. 1997b.

EPA. *EPA's Contaminated Sediment Management Strategy.* U.S. Environmental Protection Agency. Office of Water. EPA-823/R-98-001. 1998.

EPA. *Rapid Bioassessment Protocols for Use in Wadeable Streams and Rivers: Periphyton, Benthic Macroinvertebrates, and Fish, 2nd Edition.* U.S. Environmental Protection Agency. Office of Water. EPA 841-B-99-002. Download at: http: //www.epa.gov/owow/monitoring/rbp/download.html. Washington, D.C. 1999.

EPA. *Bioaccumulation Testing and Interpretation for the Purpose of Sediment Quality Assessment. Status and Needs.* EPA-823-R-00-001. Office of Water and Office of Solid Waste. Washington, D.C. 2000a.

EPA. *Appendix to Bioaccumulation Testing and Interpretation for the Purpose of Sediment Quality Assessment. Status and Needs.* Chemical Specific Summary Tables. EPA-823-R-00-002. Office of Water and Office of Solid Waste. Washington, D.C. 2000b.

EPA. *Methods for Measuring the Toxicity and Bioaccumulation of Sediment-associated Contaminants with Freshwater Invertebrates. 2nd Edition.* U.S. Environmental Protection Agency Office of Research and Development and Office of Water. EPA/600/R-99/064. 2000c

Facklam, R.R. and Salim, D.F. Enterococcus, in *Manual of Clinical Microbiology, 6th ed.* Edited by P.R. Murray. American Society for Microbiology, Washington, D.C. pp. 308–316. 1995.

Fairchild, J.F., T.W. La Point, J.L. Zajicek, M.K. Nelson, F.J. Dwyer, and P.A. Lovely. Population-community- and ecosystem-level responses of aquatic mesocosms to pulsed doses of a pyrethroid insecticide. *Environ. Toxicol. Chem.,* 11: 115–129. 1992.

Fausch, D.D., J.R. Karr, and P.R. Yant. Regional application of an index of biotic integrity based on stream fish communities. *Trans. Am. Fish. Soc.,* 113: 39–55. 1984.

Field, R., R. Pitt, D. Jaeger, and M. Brown. Combined sewer overflow control through in-receiving water storage: an efficiency evaluation. *Water Resourc. Bull. J. Am. Water Res. Assoc.,* 30(5): 921–928. October 1994.

Field, R., R. Pitt, M. Brown, and T. O'Conner. Combined sewer overflow control using storage in seawater. *Water Res.,* 29(6): 1505–1514. 1995.

Figura, P. and B. McDuffie. *Anal. Chem.,* 52: 1433. 1980.

Fillip, Z. and J.J. Alberts. Microbial utilization resulting in early diagenesis of salt-marsh humic acids. *Sci. Total Environ.*, 144(1–3): 121. April 29, 1994.

Findlay, S.E. Small-scale spatial distribution of meiofauna on a mud- and sandflat. *Estuarine Coastal Shelf Sci.*, 12: 471–484. 1981

Fisher, S.G. Recovery processes in lotic ecosystems: limits of successional theory. *Environ. Manage.*, 14: 725–736. 1990.

Fisher, R. Use of *Daphnia magna* for *in Situ* Toxicity Testing. M.S. thesis, Wright State University, Dayton, OH. 1992.

Fisher, D.J., M.H. Knott, S.D. Turley, B.S. Turley, L.T. Yonkos, and G.P. Ziegler. The acute whole effluent toxicity of storm water from an international airport. *Environ. Toxicol. Chem.*, 14: 1103–1111. 1995.

Florence, T.M. and G.E. Batley. *CRC Crit. Rev. Anal. Chem.* 9: 219. 1980.

Forget, J., J.-F. Pavillon, B. Beliaeff, and G. Bocquene. Joint action of pollutant combinations (pesticides and metals) on survival (LC50 value) and acetylcholinesterase activity of *Tigriopus brevicornia* (copepoda, harpacticoida). *Environ. Toxicol. Chem.*, 18: 912–918. 1999

Freed, J., P. Abell, D. Dixon, and R. Huddleston, Jr. *Sampling Protocol for Analysis of Toxic Pollutants in Ambient Water, Bed Sediments, and Fish, Interim Final Report.* Office of Water Planning and Standards, U.S. Environmental Protection Agency, Washington, D.C. 1980.

Froese K.L., D.A. Verbrugge, G.T. Ankley, G.J. Niemi, C.P. Larsen, and J.P. Giesy, Bioaccumulation of polychlorinated biphenyls form sediment to aquatic insecta and tree swallow eggs and nestlings in Saginaw Bay, Michigan, USA. *Environ. Toxicol. Chem.* 17: 484–492. 1998.

Fujita, M., M. Ike, Y. Goda, S. Fujimoto, Y. Toyoda, and K.-I. Miyagawa. An enzyme-linked immunosorbent assay for detection of linear alkylbenzene sulfonate: development and field studies. *Environ. Sci. Technol.*, 32(8): 1143–1146. 1998.

Funk, J.L. Movement of stream fishes in Missouri. *Trans. Am. Fish. Soc.*, 85: 39–57. 1957.

Gailani, J., C.K. Ziegler, and W. Lick. Transport of suspended solids in the lower Fox River. *J. Great Lakes Res.*, 17: 479–494. 1991.

Gammon, J.R. *The Fish Populations of the Middle 340 km of the Wabash River,* Technical Report 86, Water Resources Research Center, Purdue University, West Lafayette, IN. 1976.

Gammon, J.R. Personal communication, Department of Biological Sciences, DePauw University, Greencastle, IN. 1989.

Geldreich, E.E., L.C. Best, B.A. Kenner, and D.J. Van Donsel. The bacteriological aspects of stormwater pollution. *J. Water Pollut. Control Fed.*, 40(11): 1861–1872, 1968.

Geldreich, E.E. and B.A. Kenner. Concepts of fecal streptococci in stream pollution. *J. Water Pollut. Control Fed.*, 41(8): R336–R352. 1969.

Geldreich, E.E. Fecal coliform and fecal streptococcus density relationships in waste discharges and receiving waters. *Crit. Rev. Environ. Control*, 6(4): 349. Oct. 1976.

Gerking, S.D. The restricted movement of fish populations, *Biol. Rev.*, 34: 221–242. 1959.

Giesy, J.P., H.J. Kania, J.W. Bowling, R.L. Knight, S. Mashburn, and S. Clarkin. *Fate and Biological Effects of Cadmium Introduced into Channel Microcosms.* U.S.EPA, Athens, GA. 1979.

Gilbert, R.O., *Statistical Methods for Environmental Pollution Monitoring.* Van Nostrand Reinhold, New York. 1987.

González-Mazo, E., J.M. Forja, and A. Gómez-Parra. Fate and distribution of linear alkylbenzene sulfonates in the littoral environment. *Environ. Sci. Technol.*, 32(11): 1636–1641. 1998.

Gonzalez, M.J. and T.M. Frost. Comparisons of laboratory bioassays and a whole-lake experiment: rotifer responses to experimental acidification. *Ecol. Appl.*, 4: 69–80. 1994.

Goodnight, C.J. and L.S. Whitley. Oligochaetes as indicators of pollution. *Proc. 15th Industrial Waste Conference*, Purdue University, Lafayette, IN, pp. 139–142. 1960.

Granmo, Å., R. Ekelund, M. Berggren, E. Brorström-Lundrén, and P.-A. Berqvist. Temporal trend of organochlorine marine pollution indicated by concentrations in mussels, semipermeable membrane devices, and sediment. *Environ. Sci. Technol.*, 34(16): 3323–3329. Aug. 15, 2000.

Gray, B.R., V.L. Emery Jr., D.L. Brandon, R.B. Wright, B. M. Duke, J.D. Farrar, and D.W. Moore. Selection of optimal measures of growth and reproduction for the sublethal *Leptocheirus plumulosus* sediment bioassay. *Environ. Toxicol. Chem.*, 17: 2288–2297. 1998.

Green, D.W.J., K.A. Williams, D.R.L. Hughes, G.A.R. Shaik, and D. Pascoe. Toxicity of phenol to *Asellus aquaticus* (L.) — effects of temperature and episodic exposure. *Water Res.*, 22: 225–231. 1988.

Green, M.B. and J.R. Martin. Constructed reed beds clean up storm overflows on small wastewater treatment works. *Water Environ. Res.*, 68(6): 1054–1060. Sept./Oct. 1996.

Greenberg, M., C. Rowland, G.A. Burton, Jr., C. Hickey, W. Stubblefield, W. Clements, and P. Landrum. *Isolating Individual Stressor Effects at Sites with Contaminated Sediments and Waters*. Abstract: Annual Meeting, Society of Environmental Toxicology and Chemistry, Pensacola, FL. 1998.

Grey, G. and F. Oliveri. Catch basins — effective floatables control devices. Presented at the *Advances in Urban Wet Weather Pollution Reduction* conference. Cleveland, OH, June 28–July 1, 1998. Water Environment Federation. Alexandria, VA. 1998.

Griffiths, R.P. The importance of measuring microbial enzymatic functions while assessing and predicting long-term anthropogenic perturbations. *Mar. Pollut. Bull.*, 14: 162–165. 1983.

Grimalt, J.O. et al. Assessment of fecal sterols and ketones as indicators of urban sewage inputs to coastal waters. *Environ. Sci. Technol.*, 24(3): 357. March 1990.

Grizzard, T.J., C.W. Randall, B.L. Weand, and K.L. Ellis. Effectiveness of extended detention ponds. *Engineering Foundation Conference: Urban Runoff Quality — Impact and Quality Enhancement Technology*, Henniker, NH. Edited by B. Urbonas and L.A. Roesner. Published by the American Society of Civil Engineers, New York, June 1986.

Guy, H. P. Laboratory theory and methods for sediment analysis, in *Water Resources Inventory*, Book 5, Chap. C-1. U.S. Geological Survey, Washington, D.C. 1969.

Guy, H.P. and Normal, V.W. Field methods for the measurement of fluvial sediment, in *Techniques for Water Resources Investigations*, Book 3, Chap. C-2. U.S. Geological Survey, Washington, D.C. 1970.

Håkanson, L. and M. Jansson. *Principles of Lake Sedimentology*. Springer-Verlag, New York. 1983.

Håkanson, L. Sediment sampling in different aquatic environments: statistical aspects. *Water Resour. Res.*, 20: 41–46. 1984.

Hall, R. J., S. B. Fiance, and G. R. Hendrey. Experimental acidification of a stream in the Hubbard Brook Experimental Forest, New Hampshire, *Ecology*, 61: 976–989. 1980.

Hall, W.S. Effects of suspended solids on the bioavailability of chlordane to *Daphnia magna. Arch. Environ. Contam. Toxicol.*, 15: 529–534. 1986.

Halling-Sørensen, B., S. Nors Nielsen, P.F. Lanzky, F. Ingerslev, H.C. Holten Lützhoft, and S.E. Jørgensen. Occurrence, fate and effects of pharmaceutical substances in the environment — a review. *Chemosphere*, 36(2): 357–393. 1998.

Hamala, J., S. Duncan, and D. Blinn. A portable pump sampler for lotic periphyton. *Hydrobiologia*, 80: 189–191. 1981.

Hart, C.W., Jr., and S.L.H. Fuller. *Pollution Ecology of Freshwater Invertebrates*. Academic Press, New York. 389 pp. 1974.

Harrison, R.B., M.A. Grey, C.L. Henry, and D. Xue. *Field Test of Compost Amendment to Reduce Nutrient Runoff*. Prepared for the City of Redmond. College of Forestry Resources, University of Washington, Seattle. May 1997.

Hartwell, S.I., D.M. Jordahl, J.E. Evans, and E.B. May. Toxicity of aircraft de-icer and anti-icer solutions to aquatic organisms. *Environ. Toxicol. Chem.*, 14: 1375–1386. 1995.

Hatch, A.C. and G.A. Burton. Sediment toxicity and stormwater runoff in a contaminated receiving system: consideration of different bioassays in the laboratory and field. *Chemosphere*, 39: 1001–1017. 1999.

Hellawell, J.M. *Biological Surveillance of Rivers*. Water Research Center, Stevenage, England. 332 pp. 1978.

Hellawell, J.M. *Biological Indicators of Freshwater Pollution and Environmental Management*. Elsevier Applied Science Publishers, New York. 546 pp. 1986.

Helley, E.J. and W. Smith. *Development and Calibration of a Pressure-Difference Bedload Sampler,* Open File Report, U.S. Geological Survey, Menlo Park, CA. 1971.

Hendricks, M.L., C.H. Hocutt, and J.R. Stauffer, Jr. Monitoring of fish in lotic habitats, in *Biological Monitoring of Fish*. Edited by C.H. Hocutt and J.R. Stauffer, Jr. D.C. Heath Company, Lexington, MA. pp. 205–233. 1980.

Henebry, M.S. and P.E. Ross. Use of protozoan communities to assess the ecotoxicological hazard of contaminated sediments. *Toxicity Assess.*, 4: 209–227. 1989.

Herbes, S.E. and J. Beauchamp. Toxic interaction of mixtures of two coal conversion effluent components (resorcinol and 6-methylquinoline) to *Daphnia magna. Bull. Environ. Contam. Toxicol.*, 17: 25–32. 1977

Hermens, J. and P. Leewangh. Joint toxicity of mixtures of 8 and 24 chemicals to the guppy (*Poecilia reticulata*). *Ecotoxicol. Environ. Saf.*, 6: 302–310. 1982

Hermens, J., H. Canton, P. Janssen, and R. De Jong. Quantitative structure-activity relationships and toxicity studies of mixtures of chemicals with anaesthetic potency: acute lethal and sublethal toxicity to *Daphnia magna. Aquat. Toxicol.*, 5: 143–154. 1984a.

Hermens, J., H. Canton, N. Steyger, and R. Wegman. Joint effects of a mixture of 14 chemicals on mortality and inhibition of reproduction of *Daphnia magna. Aquat. Toxicol.*, 5: 315–322. 1984b.

Hermens, J., P. Leeuwangh, and A. Musch. Quantitative structure-activity relationships and mixture toxicity studies of chloro- and alkylanilines at an acute lethal toxicity level to the guppy (*Poecilia reticulata*). *Ecotoxicol. Environ. Saf.,* 8: 388–394. 1984c.

Herricks, E.E., I. Milne, and I. Johnson. *Selecting Biological Test Systems to Assess Time Scale Toxicity.* Water Environment Research Foundation Project 92-BAR-1. Interim Report. Alexandria, VA. 1994.

Herricks, E.E., Ed. *Stormwater Runoff and Receiving Systems: Impact, Monitoring, and Assessment.* Engineering Foundation and ASCE. CRC/Lewis. Boca Raton, FL. 1995.

Herricks, E.E, I. Milne, and I. Johnson. A protocol for wet weather discharge toxicity assessment. Volume 4, pp. 13–24. *WEFTEC'96: Proceedings of the 69th Annual Conference & Exposition.* Dallas, TX. 1996a.

Herricks, E.E., R. Brent, I. Milne, and I. Johnson. Assessing the response of aquatic organisms to short-term exposures to urban runoff, in *Effects of Watershed Development & Management on Aquatic Ecosystems,* Engineering Foundation Conference, Snowbird, UT. ASCE, New York. August 1996b.

Hickey, C.W., D.S. Roper, and S. Buckland. Metal concentrations of resident and transplanted freshwater mussels *Hyridella menziesi* (Unionacea: Hyriidae) and sediments in the Waikato River, New Zealand. *Sci. Total Environ.* 175: 163–177. 1995.

Hilgert, P. *Evaluation of Instream Flow Methodologies for Fisheries in Nebraska.* Nebraska Game and Park Commission Technical Bulletin No. 10, Lincoln, NE. 1982.

Hilsenhoff, W.L. *Use of Arthropods to Evaluate Water Quality of Streams.* Tech. Bull. 100, Wisconsin Department of Natural Resources, 15 pp. 1977.

Hittman Assoc. *Methods to Control Fine-Grained Sediments Resulting from Construction Activity.* U.S. Environmental Protection Agency, Pb-279 092, Washington, D.C. December 1976.

Ho, K.T., R.A. McKinney, A. Kuhn, M.C. Pelletier, and R.M. Burgess. Identification of acute toxicants in New Bedford Harbor sediments. *Environ. Toxicol. Chem.,* 16: 551–558. 1997.

Holdway, D.A., M.J. Barry, D.C. Logan, D. Robertson, V. Yong, and J.T. Ahokas. Toxicity of pulse-exposed fenvalerate and esfenvalerate to larval Australian crimson-spotted rainbow fish (*Melanotaenia fluviatilis*). *Aquatic Toxicol.,* 28: 169–187. 1994.

Holm, S.E. and J.G. Windsor. Exposure assessment of sewage treatment plant effluent by a selected chemical marker method. *Arch. Environ. Contam. Tox.,* 19(5): 674. Sept.–Oct. 1990.

Holme, NA., and A.D. McIntyre, Eds. *Methods for the Study of Marine Benthos*, International Biological Programme Handbook 16. Blackwell Scientific Publications, Oxford. 346 pp. 1971.

Horner, R.R., J.J. Skupien, E.H. Livingston, and H.E. Shaver. *Fundamentals of Urban Runoff Management: Technical and Institutional Issues.* Terrene Institute, Washington, D.C. 1994.

Hornig, C.E. and J.E. Pollard. *Macroinvertebrate Sampling Techniques for Streams in Semi-arid Regions. Comparison of the Surber Method and a Unit-Effort Traveling Kick Method.* U.S. Environmental Protection Agency, Las Vegas, NV, EPA 600/4-78/040. 1978.

Horton, R.E. An approach toward a physical interpretation of infiltration capacity. *Trans. Am. Geophy. Union.* 20: 693–711. 1939.

Hosmer, A.J., L.W. Warren, and T.J. Ward. Chronic toxicity of pulse-dosed fenoxycarb to *Daphnia magna* exposed to environmentally realistic concentrations. *Environ. Toxicol. Chem.,* 17: 1860–1866. 1998.

Howell, R. Effect of zinc on cadmium toxicity to the amphipod *Gammarus pulex. Hydrobiologia*, 123: 245–249. 1985

Howmiller, R.P. and M.A. Scott. An environmental index based on relative abundance of oligochaete species. *J. Water Pollut. Control Fed.,* 49: 809–815. 1977.

Hubbard, T.P. and T.E. Sample. Source tracing of toxicants in storm drains, in *Design of Urban Runoff Quality Controls.* Proceedings of an Engineering Foundation Conference, pp. 436. Trout Lake, Potosi, MO, July 1988. ASCE, New York. 1988.

Hubert, W.A. Passive capture techniques, in *Fisheries Techniques.* Edited by L.A. Nielsen, and D.L. Johnson. American Fisheries Society, Bethesda, MD. pp. 95–122. 1983.

Huckins, J.H., J.D. Petty, C.E. Orazio, J.A. Lebo, R.C. Clark, V.L. Gibson, W.R. Gala, and K.R. Echols. Determination of uptake kinetics (sampling rates) by lipid-containing semipermeable membrane devices (SPMDs) for polycyclic aromatic hydrocarbons (PAHs) in water. *Environ. Sci. Technol.*, 33: 3918–3923. 1999.

Hughes, R.M., D.P. Larsen, and J.M. Omernik. Regional reference sites: a method for assessing stream potentials. *Environ. Manage.,* 10: 629–635. 1986.

Hughes, R.M., and J.R. Gammon. Longitudinal changes in fish assemblages and water quality in the Willamette River, Oregon. *Trans. Am. Fish. Soc.,* 116: 196–209. 1987.

Hughes, R.M., E. Rexstad, and C.E. Bond. The relationship of aquatic ecoregions, river basins, and physiographic provinces to the ichthyogeographic regions of Oregon. *Copeia,* 197: 423–432. 1987.

Hulbert, S.H. Secondary effects of pesticides on aquatic systems. *Residue Rev.,* 58: 81–148. 1975.

Hulings, N.C. and J.S. Gray. *A Manual for the Study of Meiofauna.* Smithsonian Contr. Zool. No. 78, Smithsonian Institution Press, Washington, D.C. 84 pp. 1971.

Hun, T. NewsWatch: water quality, studies indicate drugs in water may come from effluent discharges. *Water Environ. Technol.,* pp. 17–22. July 1998.

Hurst, R.W., T.E. Davis, and B.D. Chinn. The lead fingerprints of gasoline contamination. *Environ. Sci. Technol.,* 30(7): 304A–306A. 1996.

HydroQual, Inc. *Floatables Pilot Program Final Report: Evaluation of Non-Structural Methods to Control Combined and Storm Sewer Floatable Materials.* City-Wide Floatables Study, Contract II. Prepared for New York City, Department of Environmental Protection, Bureau of Environmental Engineering, Division of Water Quality Improvement. NYDP2000. December 1995.

Hynes, H. B. N. *The Biology of Polluted Waters.* Liverpool University, Liverpool. 202 pp. 1960.

Hynes, H. B. N. *The Ecology of Running Waters.* University of Toronto, Toronto, Ontario. 555 pp. 1970.

Illies, J. Emergenz 1969 in Breitenbach, Schlitzer Produktions Biologische Studien. *Arch. Hydrobiol.,* 69: 14–59. 1971.

Ingersoll, C.G., E.L. Brunson, F.J. Dwyer, D.K. Hardesty, and N.E. Kemble. Use of sublethal endpoints in sediment tests with the amphipod, *Hyalella azteca. Environ. Toxicol. Chem.,* 17: 1508–1523. 1998.

Ireland, D.S., G.A. Burton, Jr., and G.G. Hess. *In-situ* toxicity evaluations of turbidity and photoinduction of polycyclic aromatic hydrocarbons. *Environ. Toxicol. Chem.,* 15(4): 574–581. April 1996.

Ireland, D.S., J. G.A. Burton et al. *In-situ* toxicity evaluations of turbidity and photoinduction of polycyclic aromatic hydrocarbons. *Environ. Toxicol. Chem.,* 15(4): 574–581. 1996.

Jarvinen, A.W., D.K. Tanner, and E.R. Kline. Toxicity of chlorpyrifos, endrin, or fenvalerate to fathead minnows following episodic or continuous exposure. *Ecotoxicol. Environ. Saf.,* 15: 78–95. 1988a.

Jarvinen, A.W., D.K. Tanner, E.R. Kline, and M.L. Knuth. Acute and chronic toxicity of triphenyltin hydroxide to fathead minnows (*Pimephales promelas*) following brief or continuous exposure. *Environ. Pollut.,* 52: 289–301. 1988b.

Jarvinen, A.W. and G.T. Ankley. *Linkage of Effects to Tissue Residues: Development of a Comprehensive Database for Aquatic Organisms Exposed to Inorganic and Organic Chemicals.* SETAC Press, Pensacola, FL. p. 358. 1999.

Jenkins, D. and L.L. Russel. Heavy metals contribution of household washing products to municipal wastewater. *Water Environ. Res.,* 66: 805. September/October 1994.

Jirik, A.W., S.M. Bay et al. Application of TIEs in studies of urban stormwater impacts on marine organisms. *7th Symposium on Toxicology and Risk Assessment: Ultraviolet Radiation and the Environment.* St. Louis, MO. ASTM Special Technical Publication, Philadelphia, PA. 1998.

Johnson, M.C. *Fluorometric Techniques for Tracing Reservoir Inflows.* Instruction Report E-84-1. U.S. Army Engineer Waterways Experiment Station. Corps of Engineers. Vicksburg, MS. 1984.

Jones, J. Beware the sediment scare. *Civ. Eng.,* pp. 56–57. July 1995.

Kallander, D.B., S.W. Fisher, and M.J. Lydy. Recovery following pulsed exposure to organophosphorus and carbamate insecticides in the midge, *Chironomus riparius. Arch. Environ. Contam. Toxicol.,* 33: 29–33. 1997.

Karr, J.R. and I.J. Schlosser. *Impact of Nearstream Vegetation and Stream Morphology on Water Quality and Stream Biota.* U.S. Environmental Protection Agency, EPA 600/3–77/097. 1977.

Karr, J.R. Assessment of biotic integrity using fish communities. *Fisheries,* 6: 21. 1981.

Karr, J.R. and D.R. Dudley. Ecological perspective on water quality goals. Ecological perspective on water quality goals. *Environ. Manage.,* 5: 55–68. 1981.

Karr, J.R. et al. *Habitat Preservation for Midwest Stream Fishes: Principles and Guidelines.* U.S. Environmental Protection Agency, Corvallis, OR, EPA 600/3-83/006. 1983.

Karr, J.R., K.D. Fausch, P.L. Angermeier, P.R. Yant, and I.J. Schlosser. *Assessing Biological Integrity in Running Waters: A Method and Its Rationale.* Special Publication 5, Illinois Natural History Survey. 1986.

Katznelson, R., W.T. Jewell, and S.L. Anderson. Spatial and temporal variations in toxicity in an urban-runoff treatment marsh. *Environ. Toxicol. Chem.,* 14: 471–482. 1995.

Kersten, M. and U. Forstner. Cadmium associations in freshwater and marine sediments, in *Cadmium in the Aquatic Environment.* Edited by J.O. Nriagu and J.B. Sprague. John Wiley & Sons, New York. 1987.

Kocan, R.M. and M.L. Landolt. Use of herring embryos for *in vitro* monitoring of marine pollution, in *In Situ Evaluation of Biological Hazards of Environmental Pollutants.* Edited by S.S. Sandhu, W.R. Lower, F.J. de Serres, W.A. Suk, and R.R. Tire. Plenum Press, New York. pp. 49–60. 1990.

Konemann, H. Fish toxicity tests with mixtures of more than two chemicals: a proposal for a quantitative approach and experimental results. *Toxicology,* 19: 229–238. 1981.

Kosian, P.A., E.A. Makynen, P.D. Monson, D.R. Mount, A. Spacie, O.V. Mekenyan, and G.T. Ankley. Application of toxicity-based fractionation techniques and structure-activity relationship models for the identification of phototoxic polycyclic aromatic hydrocarbons in sediment pore water. *Environ. Toxicol. Chem.,* 17: 1021–1033. 1998.

Kosian, P.A., E.A. Makynen, P.D. Monson, D.R. Mount, A. Spacie, O.G. Mekenyan, and G.T. Ankley. Application of toxicity-based fractionation techniques and structure-activity relationship models for the identification of phototoxic polycyclic aromatic hydrocarbons in sediment pore water. *Environ. Toxicol. Chem.,* 17: 1021–1033. 1999.

Kramer, J.B., S. Canonica, J. Hoigé, and J. Kaschig. Degradation of fluorescent whitening agents in sunlit natural waters. *Environ. Sci. Technol.,* 30(7): 2227–2234. 1996.

Krane, D.E., D.C. Sternberg, and G.A. Burton. Randomly amplified polymorphic DNA profile-based measures of genetic diversity in crayfish correlated with environmental impacts. *Environ. Toxicol. Chem.,* 18(3): 504–508. March 1999.

Kratch, K. NewsWatch, Water Quality: Cataloging DNA of *E. coli* sources pinpoints contamination causes. *Water Environ. Technol.,* 9(8): 24–26. August 1997.

Kuhr, R.J. and H.W. Dorough. *Carbamate Insecticides: Chemistry, Biochemistry, and Toxicology.* CRC Press, Cleveland, OH. 1976.

Kuivila, K.M. and C.G. Foe. Concentrations, transport and biological effects of dormant spray pesticides in the San Francisco estuary, California. *Environ. Toxicol. Chem.,* 14: 1141–1150. 1995.

Kunkle, S.H. and G.H. Commer. Estimating suspended sediment concentrations in streams by turbidity measurements. *J. Soil Water Conserv.,* 26: 18–20. 1971.

La Point, T.W., M.T. Barbour, D.L. Borton, D.S. Cherry, W.H. Clements, J.M. Diamond, D.R. Grothe, M.A. Lewis, D.K. Reed-Judkins, and G.W. Saalfeld. Field assessments: discussion synopsis. Whole effluent toxicity testing: an evaluation of methods and prediction of receiving system impact, in *Whole Effluent Toxicity Testing.* Edited by D. Grothe, K. Dickson, and D. Reed-Judkins. Soc. Environ. Toxicol. Chem. Press, Pensacola, FL. pp. 191–227. 1996.

La Point, T.W., W.H. Clements, and W.T. Waller. Field assessments in conjunction with WET testing. *Environ. Toxicol. Chem.,* 19: 14–24. 2000.

Lagler, K.F. Capture, sampling, and examination of fishes, in *Fish Production in Fresh Waters,* IBP Handbook No. 3. Edited by T. Bagenal. Blackwell Scientific Publishers, London, England. pp. 7–47. 1978.

Lalor, M. *An Assessment of Non-Stormwater Discharges to Storm Drainage Systems in Residential and Commercial Land Use Areas.* Ph.D. dissertation. Department of Civil and Environmental Engineering. Vanderbilt University. 1994.

Lalor, M. and R. Pitt. *Assessment Strategy for Evaluating the Environmental and Health Effects of Sanitary Sewer Overflows from Separate Sewer Systems.* First Year Report. Wet-Weather Flow Management Program, National Risk Management Research Laboratory, U.S. Environmental Protection Agency, Cincinnati, OH. January 1998.

Lamberson, J.O., T.H. DeWitt, and R.C. Swartz. Assessment of sediment toxicity to marine benthos, in *Sediment Toxicity Assessment.* Edited by G. A. Burton, Jr. Lewis Publishers, Boca Raton, FL. pp. 183–212. 1992.

Larsen, B. and A. Jensen. Evaluation of the sensitivity of sediment stations in pollution monitoring. *Mar. Pollut. Bull.,* 20: 556–560. 1989.

Larsen, D.P., J.M. Omernik, R.M. Hughes, C.M. Rohm, T.R. Whittier, A.J. Kinney, A.L. Gallant, and D.R. Dudley. The correspondence between spatial patterns in fish assemblages in Ohio streams and aquatic ecoregions. *Environ. Manage.,* 10: 815–828. 1986.

Lavoie, D.R. and G.A. Burton, Jr. The effects of temperature on the sensitivity of two novel test species, *Hydra attenuata* and *Lophopodella carteri: in situ* and laboratory studies. *Abstract: Annual Meeting, Society of Environmental Toxicology and Chemistry,* Pensacola, FL. 1998.

Lee, H.S., R.A. Wadden, and P.A. Scheff. Measurement and evaluation of acid air pollutants in Chicago using an annular denuder system. *Atmos. Environ. Part A Gen. Top.,* 27A(4): 543–553. March 1993.

Lee, G.F. and A. Jones-Lee. Issues in managing urban stormwater runoff quality. *Water Eng. Manage.,* 142(5): 51–53. 1995a.

Lee, G.F. and A. Jones-Lee. Deficiencies in stormwater quality monitoring, in *Stormwater NPDES Related Monitoring Needs,* Engineering Foundation Conference, Mt. Crested Butte, CO. ASCE, New York. pp. 651–662. 1995b.

Lee, H., II. A clam's eye view of the bioavailability of sediment-associated pollutants, in *Organic Substances and Sediments in Water,* Vol. 3, *Biological.* Edited by R.A. Baker. Lewis Publishers, Boca Raton, FL. 1991.

Lei, J.L. and W. Schilling. Requirements of spatial rain data resolution in urban rainfall runoff simulation, in *Proceedings of the 6th International Conference on Urban Storm Drainage,* Niagara Falls, Ontario, Sept. 12–17, 1993. pp. 447–452. IAHR/IAWQ. Seaport Publishing, Victoria, B.C. 1993.

Lemke, M.J. and S.H. Bowan. The nutritional value of organic detrital aggregate in the diet of fathead minnows. *Freshwater Biol.,* 39: 447–453. 1998.

Lenat, D.R., D.L. Penrose, and K.W. Eagleson. Variable effects of sediment addition on stream benthos. *Hydrobiologia,* 79, 187–194. 1981.

Lenat, D. and K. Eagleson. *Ecological Effects of Urban Runoff on North Carolina Streams. North Carolina Division of Environmental Management, Biological Series #104.* North Carolina Dept. of Natural Resources and Community Development, Raleigh, NC. 1981.

Lenat, D.R. Chironomid taxa richness: natural variation and use in pollution assessment. *Freshwater Invertebr. Biol.,* 2: 192–198. 1983.

Leonard, P.M. and D.J. Orth. Application and testing of an index of biotic integrity in small, cool-water streams. *Trans. Am. Fish. Soc.,* 115: 401–414. 1986.

Lester D.C. and A. McIntosh. Accumulation of polychlorinated biphenyl congeners from Lake Champlain sediments by *Mysis relicta. Environ. Toxicol. Chem.,* 13: 1825–1841. 1994.

Levin, R. Sources and sinks complicate ecology. *Science,* 243: 477–478. 1989.

Liber, K., N.K. Kaushik, K.R. Solomon, and J.H. Carey. Experimental designs for aquatic mesocosm studies: a comparison of the ANOVA and regression design for assessing the impact of tetrachlorophenol on zooplankton populations in limnocorals. *Environ. Toxicol. Chem.,* 11: 61–77. 1992.

Li-Cor. *Radiation Measurement.* Publ. RMR2-1084, Li-Cor, Inc., Lincoln, NE. 1979.

Lind, O.T. *Handbook of Common Methods in Limnology.* C. V. Mosby Company, St. Louis, MO. 154 pp. 1979.

Lind, O.T. *Handbook of Common Methods in Limnology. 2nd edition.* Kendall Hunt Publ. Co., Dubuque, IA. 1985.

Linsley, R.K. Rainfall-runoff models — an overview, in *Rainfall-Runoff Relationships.* Edited by V.P. Singh. Water Resources Publications, Highlands Ranch, CO. 1982.

Lopes, T.J., K.D. Fossum, J.V. Phillips, and J.E. Monical. *Statistical Summary of Selected Physical, Chemical, and Microbial Characteristics, and Estimates of Constituent Loads in Urban Stormwater, Maricopa County, Arizona.* U.S. Geological Survey Water-Resources Investigations Report 94-4240. Tucson, AZ. 1995.

Lowrance, R., R. Todd, and L. Asmussen. Waterborne nutrient budgets for the riparian zone of an agricultural watershed. *Agric. Ecosyst. Environ.,* 20: 371–384. 1983.

Ludwig, J.P., H.J. Auman, H. Kurita, M.E. Ludwig, L.M. Campbell, J.P. Giesy, D. Tillitt, P. Jones, N. Yamashita, S. Tanabe, and R. Tatsukawa. Caspian tern reproduction in the Saginaw bay ecosystem following a 100-year flood event. *J. Great Lakes Res.,* 96–108. 1993.

Ma, L. and R.F. Spalding. Stable isotopes characterization of the impacts of artificial ground water recharge. *Water Resources Bulletin,* 32(6): 1273–1282. December 1996.

Mac, M., G. Noguchi, R. Hesselberg, C. Edsall, J. Shoesmith, and J. Bowker. A bioaccumulation bioassay for freshwater sediments. *Environ. Toxicol. Chem.,* 9: 1405–1414. 1990.

Mac, M.J. and C.J. Schmitt. Sediment bioaccumulation testing with fish, in *Sediment Toxicity Assessment.* Edited by G.A. Burton, Jr. Lewis Publishers, Boca Raton, FL. pp. 295–308. 1992.

Makepeace, D.K., D.W. Smith, and S.J. Stanley. Urban stormwater quality: summary of contaminant data. *Crit. Rev. Environ. Sci. Tech.,* 25: 93–139. 1995.

Malley, D.F. Transplantation of unioinid mussels: a powerful biomonitoring technique when used judiciously. *SETAC News,* 16: 23–24. 1996

Maltby, L., D.M. Forrow, A.B.A. Boxall, P. Calow, and C.I. Betton. The effects of motorway runoff on freshwater ecosystems: 1. field study. *Environ. Toxicol. Chem.,* 14: 1079–1092. 1995.

Maltby, L., S.A. Clayton, H. Yu, N. McLoughlin, R.M. Wood, and D. Yin. Using single-species toxicity tests, community-level responses, and toxicity identification evaluations to investigate effluent impacts. *Environ. Toxicol. Chem.,* 19: 151–157. 2000.

Mancini, J.L. A method for calculating effects on aquatic organisms of time varying concentrations. *Water Res.,* 17: 1255–1362. 1983.

Mancini, J. and A. Plummer. Urban runoff and water quality criteria, in *Urban Runoff Quality — Impact and Quality Enhancement Technology.* Edited by B. Urbonas and L.A. Roesner. Engineering Foundation Conference, Henniker, NH. ASCE, New York. pp. 133–149. June 1986.

Marking, L.L and V.K. Dawson. *Method for Assessment of Toxicity or Efficacy of Mixtures of Chemicals.* Investigations in Fish Control Report No. 67. U.S. Fish and Wildlife Service. 1975.

Marking, L.L. Method for assessing additive toxicity of chemical mixtures, in *Aquatic Toxicology and Hazard Evaluation.* Edited by F.L. Mayer and J.L. Hamelink. ASTM STP 634. American Society for Testing and Materials. Philadelphia. pp. 99–108. 1977

Maruya K.A. and R.F. Lee. Biota-sediment accumulation and trophic transfer factors for extremely hydrophobic polychlorinated biphenyls. *Environ. Toxicol. Chem.,* 17: 2463–2469. 1998.

Marzolf, G.R. *The Potential Effects of Clearing and Snagging on Stream Ecosystems.* U.S. Fish and Wildlife Service, Washington, D.C., FWS OB5-78-14. 1978.

Mason, C. *Biology of Freshwater Pollution.* Longmans, London. 1981.

Matsumura, F. *Toxicology of Insecticides.* Plenum, New York. 1985.

Matthews, W.J. Fish faunal structure in an Ozark stream: stability, persistence, and a catastrophic flood. *Copeia,* 1986: 388–397. 1986.

Matusik, J.E. et al. Gas chromatographic/mass spectrometric confirmation of coprostanol in *Mercenaria mercenaria* (bivalvia) taken from sewage-polluted water. *Assoc. Official Anal. Chem. J.,* 71(5): 994. Sept.–Oct. 1988.

May, C., R.R. Horner, J.R. Karr, B.W. Mar, and E.B. Welch. *The Cumulative Effects of Urbanization on Small Streams in the Puget Sound Lowland Ecoregion.* University of Washington, Seattle. 1999.

McCahon, C.P. and D. Pascoe. Episodic pollution: causes, toxicological effects and ecological significance. *Funct. Ecol.,* 4: 375–383. 1990.

McCahon, C.P., S.F. Barton, and D. Pascoe. The toxicity of phenol to the freshwater crustacean *Asellus aquaticus* (L.) during episodic exposure-relationship between sub-lethal responses and body phenol concentrations. *Arch. Environ. Contam. Toxicol.,* 19: 926–929. 1990.

McCahon, C.P. and D. Pascoe. Brief exposure of first and fourth instar *Chironomus riparius* larvae to equivalent assumed doses of cadmium: effects on adult emergence. *Water Air Soil Pollut.,* 60: 395–403. 1991.

McCarty L.S. and D. Mackay. Enhancing ecotoxicological modeling and assessment. *Environ. Sci. Technol.,* 27: 1719–1728. 1993.

McCarty, J.P. and A.L. Secord. Reproductive ecology of tree swallows (*Tachycineta bicolor*) with high levels of polychlorinated biphenyl contamination. *Environ. Toxicol. Chem.,* 18: 1433–1439. 1999.

McCormick, D. and A. Roach. *Measurement, Statistics and Computation.* John Wiley & Sons, Chicester, U.K. 1987.

McCuen, R.H. *Hydrologic Analysis and Design.* Prentice-Hall, Englewood Cliffs, NJ. 1989.

McKinney, R.E. *Microbiology for Sanitary Engineers.* McGraw-Hill Book Company, Inc., New York. 1962.

McMahon, T.E., and J.W. Terrell. *Habitat Suitability Index Models: Channel Catfish.* U.S. Fish and Wildlife Service, Ft. Collins, CO, FSW OB5-82/1012. 1982.

Medeiros, C. and R.A. Coler. *A Laboratory/Field Investigation into the Biological Effects of Urban Runoff.* Water Resources Research Center, University of Massachusetts, Amherst, MA. July 1982.

Medeiros, C., R.A. Coler, and E.J. Calabrese. A laboratory assessment of the toxicity of urban runoff on the fathead minnow (*Pimephales promelas*). *Journal of Environmental Science Health*, A19(7): 847–861. 1984.

Merritt, R.W. and K.W. Cummins. *An Introduction to the Aquatic Insects of North America, 2nd Edition.* Kendall/Hunt Publishing Company, Dubuque, IA. 1984.

Metcalfe, J.L. Biological water quality assessment of running waters based on macroinvertebrate communities: history and present status in Europe. *Environ. Pollut.*, 60: 101–139. 1989.

Meyer, J.L. A blackwater perspective on riverine ecosystems. *BioScience*, 40: 643–651. 1990.

Meyer, J.S., D.D. Gulley, M.S. Goodrich, D.C. Szmania, and A.S. Brooks. Modeling toxicity due to intermittent exposure of rainbow trout and common shiners to monochloramine. *Environ. Toxicol. Chem.*, 14: 165–175. 1995.

Milbrink, G. An improved environmental index based on the relative abundance of oligochaete species. *Hydrobiologia*, 102: 89–97. 1983.

Minshall, G.W. and P.V. Winger. The effect of reduction in stream flow on invertebrate drift. *Ecology*, 49: 580. 1968.

Minshall, G.W. Stream ecosystem theory: a global perspective. *J. N. Amer. Benthol. Soc.*, 7: 263–288. 1988.

Mirov, S.B., R.E. Pitt, A. Dergachev, W. Lee, D.V. Martyshkin, O.D. Mirov, J.J. Randolph, L.J. DeLucas, C.G. Brouillette, TT. Basiev, Y.V. Orlovskii, and O.K. Alimiv. A novel laser breakdown spectrophotometer for environmental monitoring. In: *Air Monitoring and Detection of Chemical and Biological Agents.* Edited by J. Leonelli and M.L. Althouse. Society of Photo-Optical Instrumentation Engineers (SPIE). Vol. 3855, pp. 34–41. September 1999.

Monson, P.D., G.T. Ankley, and P.A. Kosian. Phototoxic response of *Lumbriculus variegatus* to sediments contaminated by polycyclic aromatic hydrocarbons. *Environ. Toxicol. Chem.*, 14: 891–894. 1995.

Moore, L. and G.A. Burton, Jr. *Assessment of stormwater runoff effects on Wolf Creek, Dayton, OH.* Dayton, OH, City of Dayton. 1999.

Morin, A. Variability of density estimates and the optimization of sampling programs for stream benthos. *Canadian J. Fisheries and Aquatic Sci.*, 42: 1530–1534. 1985.

Morrisey, D.J., A.J. Underwood, and L. Howitt. Effects of copper on the faunas of marine soft-sediments: an experimental field study. *Mar. Biol.*, 125: 199–213. 1996.

Morrison, G.M., C. Wei, and M. Engdahl. Variations of environmental parameters and ecological response in an urban river. *Water Sci. Tech.*, 27: 191–194. 1993.

Mote Marine Laboratory. *Biological and Chemical Studies on the Impact of Stormwater Runoff upon the Biological Community of the Hillsborough River, Tampa, Florida*, Stormwater Management Division, Dept. of Public Works, Tampa, FL. 1984.

Munawar, M., I.F. Munawar, and C.I. Mayfield. Differential sensitivity of natural phytoplankton size assemblages to metal mixture toxicity. *Arch. Hydrobiol. Beih.*, 25: 123–139. 1987.

Muska, C.F. and L.J. Weber. An approach for studying the effects of mixtures of environmental toxicants on whole organisms performances, in *Recent Advances in Fish Toxicology, a Symposium.* Edited by R.A. Tubb. EPA-600/3-77-085. Corvallis, OR. pp. 71–87. 1977.

Nawa, R.K. and C.A. Frissell. Measuring scour and fill of gravel streambeds with scour chains and sliding-bead monitors. *North American Journal of Fisheries Management*, 13: 634–639. 1993.

NEH (*National Engineering Handbook*). Part 630, Chapter 4, Storm Rainfall Depth (NEH-4). USDA (Natural Resources Conservation Service), Consolidated Forms and Distribution Center, 3222 Hubbard Road, Landover, MD 20785. Periodically updated.

Neuswanger, D.J., W.W. Taylor, and J.B. Reynolds. Comparison of macroinvertebrate herptobenthos and haptobenthos in side channel and slough in the upper Mississippi River. *Freshwat. Invertebr. Biol.*, 1: 13–24. 1982.

Nichols, K.M., S.R. Miles-Richardson, E.M. Snyder, and J.P. Giesy. Effects of exposure to municipal wastewater *in situ* on the reproductive physiology of the fathead minnow (*Pimephales promelas*). *Environ. Toxicol. Chem.*, 18: 2001–2012. 1999.

Nichols, P.D., R. Leeming. M.S. Rayner, and V. Latham. Use of capillary gas-chromatography for measuring fecal-derived sterols application to stormwater, the sea-surface microlayer, beach greases, regional studies, and distinguishing algal blooms and human and nonhuman sources of sewage pollution. *Journal of Chromatography*, 733(1–2): 497–509. May 10, 1996.

Niederlehner, B.R., J.R. Partt, A.L. Buikema, Jr., and J. Cairns, Jr. Laboratory tests evaluating the effects of cadmium on freshwater protozoan communities. *Environ. Toxicol. Chem.,* 4: 155–165. 1985.

Nielsen, L.A. and D.L. Johnson, Eds. *Fisheries Techniques.* American Fisheries Society, Bethesda, MD. 1983.

Novotny, V. and H. Olem. *Water Quality: Prevention, Identification, and Management of Diffuse Pollution.* Van Nostrand Reinhold, New York. 1994.

Nowell, A.R.M. and P.A. Jumars. Flow environments of aquatic benthos. *Ann. Rev. Ecol. Syst.,* 15: 303–328. 1984.

Nurnberg, H.W. Applications and potentialities of voltammetry in environmental chemistry of ecotoxic metals. In *Electrochemistry in Research and Development.* Edited by R. Kalvoda and R. Parsons. Plenum Publishing. pp. 121. New York. 1985.

O'Connor, D.J. and D.M. DiToro. Photosynthesis and oxygen balance in streams. *J. Sanitary Engineering Division, ASCE.* 96(SA2): 547–571. 1970.

Odum, E. Great ideas in ecology for the 1990s. *Bioscience,* 42: 542–545. 1992.

OEPA (Ohio Environmental Protection Agency). *The Qualitative Habitat Evaluation Index (QHEI): Rationale, Methods, and Application.* Ecological Assessment Section, Ohio Environmental Protection Agency, Columbus, OH. 1989.

Olivieri, V.P., C.W. Kruse, and K. Kawata. *Microorganisms in Urban Stormwater.* USEPA Rept. No. EPA-600/2-77-087. July 1977b.

Omernik, J.M. Ecoregions of the conterminous United States. *Ann. Assoc. Amer. Geograph.,* 77: 118–125. 1987.

Orians, G.H. Micro and macro in ecological theory. *BioScience,* 30: 79. 1980.

O'Shea, M. and R. Field. Detection and disinfection of pathogens in storm-generated flows. *Can. J. Microbiol.,* 48(4): 267–276. April 1992.

Orand, A. and M. Colon. Use of fluorometric techniques for streamflow measurements in turbulent mountain streams. *Revue des Sciences de L'Eau,* 6(2): 195–209. 1993.

Pallium, H.R. Sources, sinks, and population regulation. *Amer. Naturalist,* 132: 652–661. 1988.

Palmer, C.M. *Algae and Water Pollution.* U.S. Environmental Protection Agency, Cincinnati, OH, EPA 600/9-77/036. 1977.

Pape-Lindstrom, P.A. and M.J. Lydy. Synergistic toxicity of atrazine and organophosphate insecticides contravenes the response addition mixture model. *Environ. Toxicol. Chem.,* 16: 2415–2420. 1997.

Parsons, J. and G. Surgeoner. Acute toxicities of permethrin, fenitrothion, carbaryl and carbofuran to mosquito larvae during single- or multiple-pulse exposures. *Environ. Toxicol. Chem.,* 10: 1229–1233. 1991a.

Parsons, J.T. and G.A. Surgeoner. Effect of exposure time on the acute toxicities of permethrin, fenitrothion, carbaryl, and carbofuran to mosquito larvae. *Environ. Toxicol. Chem.,* 10: 1219–1227. 1991b.

Pascoe, D. and N.A.M. Shazili. Episodic pollution — a comparison of brief and continuous exposure of rainbow trout to cadmium. *Ecotoxicol. Environ. Safety,* 12: 189–198. 1986.

Patrick, R. and C.W. Reimer. *The Diatoms of the United States,* Vol. 1. Monographs of the Academy of Natural Sciences, Philadelphia, PA. 1966.

Patrick, R. Use of algae, especially diatoms in the assessment of water quality, in *Biological Methods for the Assessment of Water Quality.* Edited by J. Cairns, Jr. and K. L. Dickson. Special Technical Publication 528. American Society for Testing and Materials, Philadelphia, PA. 1973.

Paul, J.H., J.B. Rose, J. Brown, E.A. Shinn, S. Miller, and S.R. Farrah. Viral tracer studies indicate contamination of marine waters by sewage disposal practices in Key Largo, Florida. *Appl. Environ. Microbiol.,* 61(6): 2230–2234. June 1995.

Peckarsky, B.L. Sampling the stream benthos, in *A Manual on Methods for the Assessment of Secondary Productivity in Fresh Waters.* Edited by J. Downing and F. Rigler. IBP Handbook 17. Blackwell Scientific Publishers, Oxford, England. pp. 131–160. 1984.

Pelletier, G.J. and T.A. Determan. *Urban Storm Drain Inventory, Inner Gray Harbor.* Prepared for Washington State Department of Ecology, Water Quality Investigations Section, Olympia, WA. 1988.

Pelley J. North Carolina considers controls to protect contaminated waters. *Environ. Sci. Technol.,* 33: 10A. 1999.

Pennak, R.W. *Freshwater Invertebrates of the United States: Protozoa to Mollusca.* John Wiley& Sons, Inc., New York. 628 pp. 1989.

Pereira, A.M.M., A.M.V.D.M. Soares, F. Goncalves, and R. Ribeiro. Test chambers and test procedures for *in situ* toxicity testing with zooplankton. *Environ. Toxicol. Chem.,* 18: 1956–1964. 1999.

Peterson, S.M., S.C. Apte, G.E. Batley and G. Coade. Passive sampler for chlorinated pesticides in estuarine waters. *Chemical Speciation and Bioavailability,* 7: 83–88. 1995

Petr, T. Bioturbation and exchange in the mud-water interface, in *Interface Between Sediment and Freshwater.* Edited by H.L. Golterman. Junk, The Hague, The Netherlands. 1977

Petty, J.D., B.C. Poulton, C.S. Charbonneau, J.N. Huckins, S.B. Jones, J.T. Cameron, and H.R. Prest. Determination of bioavailable contaminants in the lower Missouri River following the flood of 1993. *Environ. Sci. Technol.,* 32: 837–842. 1998

Pitt, R. *Small Storm Urban Flow and Particulate Washoff Contributions to Outfall Discharges.* Ph.D. dissertation. Department of Civil and Environmental Engineering, University of Wisconsin, Madison. 1987.

Pitt, R. and P. Bissonnette. *Bellevue Urban Runoff Program, Summary Report.* PB84 237213, Water Planning Division, U.S. Environmental Protection Agency and the Storm and Surface Water Utility, Bellevue, WA. 1984.

Pitt, R. and S. Clark. *Communication Manhole Water Study: Characteristics of Water Found in Communications Manholes.* Final Draft. Office of Water, U.S. Environmental Protection Agency. Washington, D.C. July 1999.

Pitt, R. and J. McLean. *Humber River Pilot Watershed Project.* Ontario Ministry of the Environment, Toronto, Canada. 483 pp. June 1986.

Pitt, R. and J. Voorhees. Source loading and management model (SLAMM). *Seminar Publication: National Conference on Urban Runoff Management: Enhancing Urban Watershed Management at the Local, County, and State Levels.* March 30–April 2, 1993. Center for Environmental Research Information, U.S. Environmental Protection Agency. EPA/625/R-95/003. Cincinnati. OH. pp. 225–243. April 1995.

Pitt, R., M. Lalor, R. Field, D.D. Adrian, and D. Barbe. *A User's Guide for the Assessment of Non-Stormwater Discharges into Separate Storm Drainage Systems.* Jointly published by the Center of Environmental Research Information, US EPA, and the Urban Waste Management & Research Center (UWM&RC). EPA/600/R-92/238. PB93-131472. Cincinnati, OH. January 1993.

Pitt, R., S. Clark, and K. Parmer. *Protection of Groundwater from Intentional and Nonintentional Stormwater Infiltration.* U.S. Environmental Protection Agency, EPA/600/SR-94/051. PB94-165354AS, Storm and Combined Sewer Program, Cincinnati, OH. 187 pp. May 1994.

Pitt, R., R. Field, M. Lalor, and M. Brown. Urban stormwater toxic pollutants: assessment, sources and treatability. *Water Environ. Res.,* 67(3): 260–275. May/June 1995.

Pitt, R., S. Clark, K. Parmer, and R. Field. *Groundwater Contamination from Stormwater.* Ann Arbor Press, Chelsea, MI. 219 pages. 1996.

Pitt, R., S. Nix, J. Voorhees, S.R. Durrans, and S. Burian. *Guidance Manual for Integrated Wet Weather Flow (WWF) Collection and Treatment Systems for Newly Urbanized Areas (New WWF Systems). Second year project report: model integration and use.* Wet Weather Research Program, U.S. Environmental Protection Agency. Cooperative agreement #CX 824933-01-0. February 1999a.

Pitt, R., B. Robertson, P. Barron, A. Ayyoubi, and S. Clark. Stormwater Treatment at Critical Areas: The Multi-Chambered Treatment Train (MCTT). U.S. Environmental Protection Agency, Wet Weather Flow Management Program, National Risk Management Research Laboratory. EPA/600/R-99/017. Cincinnati, OH. 505 pp. March 1999b.

Pitt, R., J. Lantrip, R. Harrison, C. Henry, and D. Hue. *Infiltration through Disturbed Urban Soils and Compost-Amended Soil Effects on Runoff Quality and Quantity.* U.S. Environmental Protection Agency, Water Supply and Water Resources Division, National Risk Management Research Laboratory. Cincinnati, OH. 338 pp. 2000.

Poiger, T., J.A. Field, T.M. Field, and W. Giger. Occurrence of fluorescent whitening agents in sewage and river determined by solid-phase extraction and high-performance liquid chromatography. *Environ. Sci. Technol.,* 30(7): 2220–2226. 1996.

Pontasch, K.W., B.R. Niederlehner, and J. Cairns, Jr. Comparisons of single-species microcosm and field responses to a complex effluent. *Environ. Toxicol. Chem.,* 8: 521–532. 1989.

Porcella, D.B. *Protocol for Bioassessment of Hazardous Waste Sites,* prepared for Environmental Research Laboratory, U.S. Environmental Protection Agency, Corvallis, OR, EPA 600/2-83/054, NTIS Publ. No. PB83-241737. 1983.

Portele, G.J., B.W. Mar, R.R. Horner, and E.W. Welch. *Effects of Seattle Area Highway Runoff on Aquatic Biota.* Washington State Dept. of Transportation. WA-RD-39.11. PB8-3-170761. Olympia, WA. January 1982.

Porter, K.G., H. Paerl, R. Hodson, M. Pace, J. Priscu, B. Riemann, D. Scavia, and J. Stockner. Microbial interactions in lake food webs, in *Complex Interactions in Lake Communities.* Edited by S.R. Carpenter. Springer-Verlag, New York. pp. 209–227. 1987.

Posey, M.H. Functional approaches to soft-substrate communities: how useful are they? *Rev. Aquat. Sci.,* 2: 343–356. 1990.

Postma, J.F., M.C. Buckertdejong, N. Staats, and C. Davids. Chronic toxicity of cadmium to *Chironomus riparius* (Diptera, chironomidae) at different food levels. *Arch. Environ. Contam. Toxicol.,* 26: 143–148. 1994.

Power, M.E., R.J. Stout, C.E. Cushing, P.P. Harper, F.R. Hauer, W.J. Matthews, P.B. Moyle, B. Statzner, and I.R. Wars de Badgen. Biotic and abiotic controls in river and stream communities. *J. North Am. Benthol. Soc.,* 7: 456–479. 1988.

Practice of Water Pollution Biology, The. U.S. Department of the Interior. Washington, D.C. 1969.

Pratt, J.M., R.A. Coler, and P.J. Godfrey. Ecological effects of urban stormwater runoff on benthic macroin-vertebrates inhabiting the Green River, Massachusetts. *Hydrobiologia,* 83: 29–42. 1981.

Prescott, G.W. *Algae of the Western Great Lakes Area.* William C. Brown, Dubuque, IA. 1962.

Prescott, G.W. *Algae of the Western Great Lakes Area, 2nd edition.* William C. Brown Publ., Dubuque, IA. 1970.

Prest, H.F., W.M. Jarman, S.A. Burns, T. Weimuller, M. Martin, and J.N. Huckins. Passive water sampling via semipermeable membrane devises (SPMDs) in concert with bivalves in the Sacramento/San Joaquin River Delta. *Chemosphere,* 25: 1811–1823. 1992.

Prest, H.F., B.R. Richardson, LA. Jacobson, J. Vedder, and M. Martin. Monitoring organochlorines with semi-permeable membrane devices (SPMDs) and mussels (*Mytilus edulis*) in Corio Bay, Victoria, Australia. *Mar Pollut. Bull.,* 30: 543–554. 1995.

Pridmore, R.D., S.F. Thrush, R.J. Wilcock, T.J. Smith, J.E. Hewitt, and V.J. Cummings. Effect of the orga-nochloride pesticide technical chlordane on the population structure of suspension and deposit feeding bivalves. *Mar. Ecol. Prog. Ser.,* 76: 261–271. 1991.

Pringel, C.M., R.J. Naiman, G. Brelschko, J.R. Karr, M.W. Oswood, J.R. Webster, R.L. Welcomme, and M.J. Winterbourn. Patch dynamics in lotic systems: the stream as a mosaic. *J. North Am. Benthol. Soc.,* 7: 503–524. 1988.

Qureshi, A.A. and B.J. Dutka. Microbiological studies on the quality of urban stormwater runoff in southern Ontario, Canada. *Water Res.,* 13: 977–985. 1979.

Rand, G.M. and S.R. Petrocelli (Eds.). *Fundamentals of Aquatic Toxicology.* McGraw-Hill, New York. 1985

Randall, C.W. Stormwater detention ponds for water quality control. *Proceedings of the Conference on Stormwater Detention Facilities, Planning, Design, Operation, and Maintenance,* Henniker, NH, Edited by W. DeGroot. Published by the American Society of Civil Engineers, New York, August 1982.

Resh, V.H. and J.D. Unzicker. Water quality monitoring and aquatic organisms: the importance of species identification. *J. Water Pollut. Control Fed.,* 47: 9–19. 1975.

Resh, V.H. Sampling variability and life history features: basic considerations in the design of aquatic insect studies. *J. Fish. Res. Bd. Can.,* 36: 290–311. 1979.

Resh, V.H. and D.M. Rosenberg. *The Ecology of Aquatic Insects.* Praeger Publishers, New York. 625 pp. 1984.

Resh, V.H. Variability, accuracy, and taxonomic costs of rapid assessment approaches in benthic biomonitoring. Presented at the 36th Annual North American Benthological Society Meeting at Tuscaloosa, AL. 1988.

Reynoldson, T.B., D.W. Schloesser, and B.A. Manny. Development of a benthic invertebrate objective for mesotrophic Great Lakes waters. *J. Great Lakes Res.,* 15: 669–686. 1989.

Rieley, G., C.L. VanDiver, and G. Eglinton. Fatty acids as sensitive tracers of sewage sludge carbon in a deep-sea ecosystem. *Environ. Sci. Technol.,* 31(4): 1018. 1997.

Roa-Espinosa, A. and R. Bannerman. Monitoring BMP effectiveness at industrial sites. *Proceedings of the Engineering Foundation Conference on Stormwater NPDES Related Monitoring Needs.* Aug 7–12 1994. Sponsored by U.S. Environmental Protection Agency; The Engineering Foundation; U.S. Geological Survey; Water Environment Federation; American Institute of Hydrology, ASCE. pp. 467–486. 1994.

Rodda, D.W.C. Water data collection and use. *Water Pollut. Control* (Maidstone, England). 75(1): 115–123. 1976.

Rodgers, J.H., Jr., K.L. Dickson, and J. Cairns, Jr. A review and analysis of some methods used to measure functional aspects of periphyton, in *Methods and Measurements of Periphyton Communities: A Review.* Special Technical Publication 690. Edited by R.L. Weitzel. American Society for Testing and Materials, Philadelphia, PA. 1979.

Roper, D.S. and C.W. Hickey. Effects of food and silt on filtration, respiration and condition of the freshwater mussel *Hyridella menziesi* (Unionacea: Hyriidae): implications for bioaccumulation. *Hydrobiologia,* 312: 17–25. 1995.

Ross, S.T., W.J. Matthews, and A.E. Echelle. Persistence of stream fish assemblages, effects of environmental change. *Am. Nat.,* 126: 24–40. 1985.

Routledge, E.J., D. Sheahan, C. Desbrow, C.C. Brighty, M. Waldock, and J.P. Sumpter. Identification of estrogenic chemicals in STW effluent. 2. In vivo responses in trout and roach. *Environ. Sci. Technol.,* 32(11): 1559. 1998.

Sabaliunas, D.J., Lazutka, I. Sabaliuniene, and A. Sodergren. Use of semipermeable membrane devices for studying effects of organic pollutants: comparison of pesticide uptake by semipermeable membrane devices and mussels. *Environ. Toxicol. Chem.,* 17: 1815–1824. 1998.

Salau, J.S., R. Tauler, J.M. Bayona, and I. Tolosa. Input characterization of sedimentary organic contaminants and molecular markers in the Northwestern Mediterranean Sea by exploratory data analysis. *Environ. Sci. Technol.,* 31(12): 3482. 1997.

Sangal, S., P.K. Aggarwal, and D. Tuomari. *Identification of Illicit Connections in Storm Sewers: An Innovative Approach Using Stable Isotopes.* Rouge River Studies. Detroit, MI. 1996.

Sarda, N. and G.A. Burton. Ammonia variation in sediments: spatial, temporal and method-related effects. *Environ. Toxicol. Chem.,* 14: 1499–1506. 1995.

Sasson-Brickson, G. and G.A. Burton Jr. In situ and laboratory sediment toxicity testing with *Ceriodaphnia dubia. Environ. Toxicol. Chem.,* 10: 201–207. 1991.

Sayre, P.G., D.M. Spoon, and D.G. Loveland. Use of *Heliophrya* sp., a sesile suctorian protozoan, as a biomonitor of urban runoff, in *Aquatic Toxicology and Environmental Fate: Ninth Volume.* ASTM Special Technical Publication 921. 1986.

Schimmel, S.C. and G.B. Thursby. Predicting receiving system impacts from effluent toxicity: a marine perspective, in *Whole Effluent Toxicity Testing: An Evaluation of Methods and Prediction of Receiving System Impacts.* Edited by D.R. Grothe, K.L. Dickson, and D.K. Reed-Judkins. Soc. Environ. Toxicol. Chem. Press, Pensacola, FL. pp. 322–330. 1996.

Schindler, D.W. Detecting ecosystem responses to anthropogenic stress. *Can. J. Fish Aquat. Sci.,* (Suppl.) 1: 6–25. 1987.

Schlosser, I.J. *Effects of Perturbations by Agricultural Land Use on Structure and Function of Stream Ecosystems.* Ph.D. dissertation, University of Illinois, Urbana, IL. 1981.

Schlosser, I.J. Trophic structure, reproductive success, and growth rate of fishes in a natural and modified headwater stream. *Can. J. Fish. Aquat. Sci.,* 36: 968. 1982a.

Schlosser, I.J. Fish community structure and function along two habitat gradients in a headwater stream. *Ecol. Monogr.,* 52: 395. 1982b.

Schrap, S.M. and A. Opperhuizen. Relationship between bioavailability and hydrophobicity: reduction of the uptake of organic chemicals by fish due to the sorption on particles. *Environ. Toxicol. Chem.,* 9: 715–724. 1990.

Schueler, T. (Ed.). Comparison of forest, urban and agricultural streams in North Carolina. *Watershed Protect. Tech.,* 2(4): 503–506. June 1997.

Schultz, T.W. and T.C. Allison. Toxicity and toxic interaction of aniline and pyridine. *Bull. Environ. Contam. Toxicol.,* 23: 814–819. 1979.

Schulz, R. Use of an *in situ* bioassay to detect insecticide effects related to agricultural runoff. *Abstr. Int. Symp. Environ. Chem. Toxicol.,* Sydney, Australia. 1996.

Schulz, R. and M. Liess. Validity and ecological relevance of an active *in situ* bioassay using *Gammarus pulex* and *Limnephilus lunatus. Environ. Toxicol. Chem.,* 18: 2243–2250. 1999.

Schuytema, G.S., P.O. Nelson, K.W. Malueg, A.V. Nebeker, D.F. Krawczyk, A.K. Ratcliff, and J.H. Gakstatter. Toxicity of cadmium in water and sediment slurries to *Daphnia magna. Environ. Toxicol. Chem.,* 3: 293–308. 1984

Sharma, D. C. and C. F. Forster. Removal of hexavalent chromium using sphagnum moss peat. *Water Res.,* 27(7): 1201–1208. 1993.

Shaw, E.M. *Hydrology in Practice.* Van Nostrand Reinhold, Wokingham, U.K. 1983.

Sheldon, A.L. Colonization dynamics of aquatic insects, in *The Ecology of Aquatic Insects.* Edited by V.H. Resh and D.M. Rosenber. Praeger, New York. pp. 401–429. 1984.

Sherwin, M.R. et al. Coprostanol (5β-cholestan-3β-ol) in lagoonal sediments and mussels of Venice, Italy. *Mar. Pollut. Bull.,* 26(9): 501. September 1993.

Shuman, R. and J. Strand. King County water quality assessment: CSO discharges, biological impacts being assessed. *Wet Weatherx.* Water Environment Research Foundation. Fairfax, VA. 1(3): 10–14. Fall 1996.

Sibley, P.K., D.A. Benoit, M.D. Balcer, G.L. Phipps, C.W. West, R.A. Hoke, and G.T. Ankley. *In situ* bioassay chamber for assessment of sediment toxicity and bioaccumulation using benthic invertebrates. *Environ. Toxicol Chem.,* 18: 2325–2336. 1999.

Simon, A. *Practical Hydraulics.* John Wiley & Sons, New York. 1976.

Skalski, C. *Laboratory and in Situ Sediment Toxicity Evaluations with Early Life Stages of Pimephales promelas.* M.S. thesis, Wright State University, Dayton, OH. 1991.

Smith, G.M. *The Fresh Water Algae of the United States, 2nd edition.* McGraw-Hill, New York. 1950.

Spehar, R.L. and J.T. Fiandt. Acute and chronic effects of water quality criteria-based metal mixtures on three aquatic species. *Environ. Toxicol. Chem.,* 5: 917–931. 1986.

Sprague, J.B. and B.A. Ramsay. Lethal levels of mixed copper-zinc solutions for juvenile salmon. *J. Fish. Res. Bd. Can.,* 22: 425–432. 1965.

Sprague, J.B. Promising anti-pollutant: chelating agent NTA protects fish from copper and zinc. *Nature* (London), 220: 1345–1346. 1968

Standley, L., L.A. Kaplan, and D. Smith. Molecular tracers of organic matter sources to surface water resources. *Environ. Sci. Technol.,* 34(15): 3124–3130. Aug. 1, 2000.

States, S., K. Stadterman, L. Ammon, P. Vogel, J. Baldizar, D. Wright, L. Conley, and J. Sykora. Protozoa in river water: sources, occurrence, and treatment. *J. Am. Water Works Assoc.,* 89(9): 74–83. Sept. 1997.

Statzner, B., J.A. Gore, and V.H. Resh. Hydraulic stream ecology: observed patterns and potential applications. *J. North Am. Benthol. Soc.,* 7: 307–360. 1988.

Stauffer, J.R., H.A. Beiles, J.W. Cox, K.L. Dickson, and D.E. Simonet. Colonization of macrobenthic communities on artificial substrates. *Rev. Biol.,* 10: 49–61. 1974–1976.

Steedman, R.J. Modification and assessment of an index of biotic integrity to quantify stream quality in southern Ontario. *Can. J. Fish. Aquat. Sci.,* 45: 492–501. 1988.

Stemmer, B.L., G.A. Burton, Jr., and S. Leibfritz-Frederick. Effect of sediment test variables on selenium toxicity to *Daphnia magna. Environ. Toxicol. Chem.,* 9: 381–389. 1990.

Sternberg, D.C., G.A. Burton, Jr., D.E. Krane, and K. Grasman. Randomly amplified polymorphic DNA markers in determinations of genetic variation in populations affected by stressors. *Abstr. Annu. Meeting Soc. Environ. Toxicol. Chem.,* Washington, D.C., P0757, p. 259. 1996.

Stevenson, R.J. and R.L. Lowe. Sampling and interpretation of algal patterns for water quality assessments, in *Rationale for Sampling and Interpretation of Ecological Data.* ASTM STP 894. Edited by B.G. Isom. American Society for Testing and Materials, Philadelphia, PA. pp. 118–149. 1986.

Stout, S.A., A.D. Uhler, T.G. Naymik, and K.J. McCarthy. Environmental forensics: unraveling site liability. *Environ. Sci. Technol.,* pp. 260a–264a. June 1, 1998.

Swartz, R.C., F.A. Cole, J.O. Lamberson, S.P. Ferraro, D.W. Schults, W.A. DeBen, H. Lee II, and R.J. Ozretich. Sediment toxicity, contamination and amphipod abundance at a DDT- and Dieldrin-contaminated site in San Francisco Bay. *Environ. Toxicol. Chem.,* 13: 949–962. 1994.

Tennant, P.L. Instream flow regimens for fish, wildlife, recreation, and related environmental resources, in *Proc. of the Symposium and Speciality Conference on Instream Flow Needs, Vol. II.* Edited by J. Osborn, and C. Allman. American Fisheries Society, Bethesda, MD. 1976.

Terrell, J.W. et al. Habitat suitability index models: appendix A. Guidelines for riverine and lacustrine applications of fish his. 1982. *Proc. Workshop on Fish Habitat Suitability Models,* Edited by J.W. Terrell. Western Energy Land Use Team, U.S. Department of the Interior, Biological Report 85(6) 1984.

Terstriep, M.L., G.M. Bender, and D.C. Noel. *Nationwide Urban Runoff Project, Champaign, Illinois: Evaluation of the Effectiveness of Municipal Street Sweeping in the Control of Urban Storm Runoff Pollution.* Contract No. 1-5-39600, U.S. Environmental Protection Agency, Illinois Environmental Protection Agency, and the State Water Survey Division, University of Illinois, December 1982.

Terzic, S. and M. Ahel. Determination of alkylbenzene sulphonates in the Krka River estuary. *Bull. Environ. Contam. Toxicol.,* 50(2): 241. February 1993.

Thomann R.V., J.P. Connolly, and T.F. Parkerton. An equilibrium model of organic chemical accumulation in aquatic food webs with sediment interaction. *Environ. Toxicol. Chem.,* 11: 615–629. 1992.

Thomann, R.V. and J.A. Mueller. *Principles of Surface Water Quality Modeling and Control.* Harper and Row. New York. 1987.

Thorsen, T. and R. Owen. How to measure industrial wastewater flow. *Chem. Eng.* 82: 95–100. 1975.

Thurston, R.V., C. Chakoumakos, and R.C. Russo. Effect of fluctuating exposures on the acute toxicity of ammonia to rainbow trout (*Salmo gairdneri*) and cutthroat trout (*S. clarki*). *Water Res.,* 15: 911–917. 1981.

Tucker K.A. and Burton G.A., Jr. Assessment of nonpoint source runoff in a stream using *in situ* and laboratory approaches. *Environ. Toxicol. Chem.,* 18: 2797–2803, 1999.

Turf Tec International. *Turf Tec Instructions.* Oakland Park, FL. 1989.

US Water News. Scientists say Boston Harbor water full of caffeine. p. 3. July 1998.

USFWS. *A Guidance to Stream Habitat Analysis Using the Instream Flow Incremental Methodology.* U.S. FWS/OBS-82/26, 1982.

USGS. *The Quality of Our Nation's Waters: Nutrient and Pesticides.* U.S. Geological Survey Circular 1225. Reston, VA. 1999

Vannote, R.L. et al. The river continuum concept. *Can. J. Fish. Aquat. Sci.,* 37: 130. 1980.

Vasseur, P., D. Dive, Z. Sokar, and H. Bonnemain. Interactions between copper and some carbamates used in phytosanitary treatments. *Chemosphere,* 17: 767–782. 1988.

Velz, C.J. *Applied Stream Sanitation.* John Wiley & Sons, New York. 1970.

Venkatesan, M.I. and I.R. Kaplan. Sedimentary coprostanol as an index of sewage addition in Santa Monica Basin, Southern California. *Environ. Sci. Technol.,* 24(2): 208. February 1990.

Vlaming, V.D., V. Connor, C. DiGiorgiio, H.C. Bailey, L.A. Deanovic, and D.E. Hinton. Application of whole effluent toxicity test procedures to ambient water quality assessment. *Environ. Toxicol. Chem.,* 19: 42–62. 2000.

Vogel, S. *Life in Moving Fluids: The Physical Biology of Flow, 2nd edition.* Princeton University Press, Princeton, NJ. 1994.

Wade, D.C., and S.B. Stalcup. *Assessment of the Sport Fishery Potential for the Bear Creek Floatway: Biological Integrity of Representative Sites, 1986.* TVA/ONRED/AWR-87/30, Tennessee Valley Authority, Muscle Shoals, AL. 1987.

Waller, W.T., L.P. Ammann, W.J. Birge, K.L. Dickson, P.B. Dorn, N.E. LeBlanc, D.I. Mount, B.R. Parkhurst, H.R. Preston, S.C. Schimmel, A. Spacie, and G.B. Thursby. Predicting instream effects from WET tests: discussion synopsis, in *Whole Effluent Toxicity Testing: An Evaluation of Methods and Prediction of Receiving System Impacts.* Edited by D.R. Grothe, K.L. Dickson, and D.K. Reed-Judkins. Soc. Environ. Toxicol. Chem. Press, Pensacola, FL. pp. 271–286. 1996.

Waller, W.T., M.F. Acevedo, H.J. Allen, J.H. Kennedy, K.L. Dickson, L.P. Ammann, and E.L. Morgan. The use of remotely sensed bioelectric action potentials to evaluate episodic toxicity events and ambient toxicity. Water for Texas Conference, *Proceedings of the 24th Water for Texas Conference: Research Leads the Way.* Texas Water Development Board, Texas Water Resources Institute, Texas Water Conservation Association, Austin, TX. 726 pp. 1995a.

Waller, W.T., M.F. Acevedo, E.L. Morgan, K.L. Dickson, J.H. Kennedy, L.P. Ammann, H.J. Allen, and P.R. Keating. Biological and chemical testing of stormwater, in *Stormwater NPDES Related Monitoring Needs. Proceedings of an Engineering Foundation Conference.* Water Resources Planning and Management Division/ASCE. August 7–12, 1994. Mount Crested Butte, CO. 1995b.

Walton, O.E., Jr. The effects of density, sediment size, and velocity on drift of *Acroneuria abnormis* (Plecoptera). *OIKOS,* 28: 291. 1977.

Wang, M.P. and S.A. Hanson. The acute toxicity of chlorine on freshwater organisms: time-concentration relationships of constant and intermittent exposures. *Aquatic Toxicology and Hazard Assessment: Eighth Symposium.* STP 891, American Society for Testing and Materials, Philadelphia, PA. 1985.

Wanielista, M., R. Kersten, and R. Eaglin. *Hydrology: Water Quantity and Quality Control.* John Wiley & Sons, New York. 1997.

Ward, J.V. and J.A. Stanford, Eds. *The Ecology of Regulated Streams.* Plenum Publishers, New York. 398 pp. 1979.

Warren, C.E. *Biology and Water Pollution Control.* W.B. Saunders Publisher, Philadelphia, PA. 434 pp. 1971.

Warren, L.W., S.J. Klaine, and M.T. Finley. Development of a field bioassay with juvenile mussels. *J. North Am. Benthol. Soc.,* 14: 341–346. 1995.

Waschbusch, R. and D. Owens. *Comparison of Flow Estimation Methodologies in Storm Sewers*. Report prepared by the USGS for the FHWA. Madison, WI. January 1998.

Waters, T.F. Secondary production in inland waters. *Adv. Ecol. Res.,* 10: 1–164. 1977.

Watzin, M.C., P.F. Roscigno, and W.D. Burke. Community-level field method for testing the toxicity of contaminated sediments in estuaries. *Environ. Toxicol. Chem.,* 13: 1187–1193. 1994.

WEF and ASCE. *Design and Construction of Urban Stormwater Management Systems*. ASCE Manuals and Reports of Engineering Practice No. 77 and WEF Manual of Practice FD-20. ASCE, New York, and WEF, Alexandria, VA. 1992.

Welch, E.B. *Ecological Effects of Waste Water*. Cambridge University Press, Cambridge, England. 1980.

Welch, P.S. *Limnological Methods*. McGraw-Hill, New York, 381 pp. 1948.

Wenholz, M. and R. Crunkilton. Use of toxicity identification evaluation procedures in the assessment of sediment pore water toxicity from an urban stormwater retention pond in Madison, Wisconsin. *Bull. Environ. Contam. Toxicol.,* 54: 676–682. 1995.

Werblow, S. A closer look: detailed river study spurs Oregon to reassess total maximum daily loads in the Tualatin River. *Water Environ. Technol.,* pp. 40–44. Feb. 2000.

Werner, I., L.A. Deanovic, V. Connor, V.D. Vlaming, H.C. Bailey, and D.E. Hinton. Insecticide-caused toxicity to *Ceriodaphnia dubia* (Cladocera) in the Sacramento-San Joaquin River delta, California, USA. *Environ. Toxicol. Chem.,* 19: 215–227. 2000

Westman, W.E. *Ecology, Impact Assessment and Environmental Planning*. John Wiley & Sons, New York. 1985.

Wetzel, R.G. *Limnology*. W.B. Saunders Publishing, Philadelphia, PA. 1975.

Wetzel, R.G. and E. Likens. *Limnology Analyses, 2nd edition*. Springer-Verlag, New York. 1991.

Wetzel, R. The role of littoral zone and detritus in lake metabolism, in Symposium on Lake Metabolism and Lake Management. Edited by G.E. Likens, W. Rodhe, and C. Serrugy. Ergebnisse Lijmnol., *Arch. Hydrobiol.,* 13: 145–161. *Limnology.* 1979

Whipple, W. and J.V. Hunter. Settleability of urban runoff pollution. *J. WPCF,* 53(12): 1726, 1981.

Whittaker, M. and J.T. Pollard. A performance assessment of source correlation and weathering indices for petroleum hydrocarbons in the environment. *Environ. Toxicol. Chem.,* 16(6): 1149–1158. 1997.

Whittemore, R.C. Implementation of in-situ and laboratory SOD measurements, in *Water Quality Modeling, in Sediment Oxygen Demand: Processes, Modeling, and Measurements*. Edited by K. Hatcher. Institute of Natural Resources, University of Georgia, Athens. 1986.

Whittier, T.R., R.M. Hughes, and D.P. Larsen. Correspondence between ecoregions and spatial patterns in stream ecosystems in Oregon. *Can. J. Fish. Aquat. Sci.,* 45: 1264–1278. 1988.

Williams, A.T. and S.L. Simmons. Estuarine litter at the river/beach interface in the Bristol Channel, United Kingdom. *J. Coast. Res.,* 13(4): 1159–1165. Fall 1997.

Williams, J.G.K, A.R. Kubelik, K.J. Livak, J.A. Rafalski, and S.V. Tingey. DNA polymorphisms amplified by arbitrary primers are useful as genetic markers. *Nucleic Acids Res.,* 18: 6531–6535. 1990.

Williams, J.G.K., M.K. Kubelik, J.A. Rafalski, and S.V. Tingey. Genetic data analysis with RAPD markers, in *More Gene Manipulations in Fungi*. Edited by Bennett, J.W. and L.L. Lasure. Academic Press, San Diego, CA. pp. 433–439. 1991.

Wink, K.L. *Effects of Metal Mixtures on Pimephales Promelas Larval Growth in Water and Sediment Exposures*. M.S. thesis. Wright State University, Dayton, OH. 1990

Winner, R.W., J.S. Van Dyke, N. Caris, and M.P. Farrell. Response of the macroinvertebrate fauna to a copper gradient in an experimentally polluted stream. *Verh. Int. Verein. Limnol.,* 19: 2121–2127. 1975.

Wipfli, M.S. and R.W. Merritt. Disturbance to a stream food web by a bacterial larvicide specific to black flies: feeding responses of predatory macroinvertebrates. *Freshwater Biol.,* 32: 91–103. 1994.

Woolgar, P.J. and K.C. Jones. Studies on the dissolution of polycyclic aromatic hydrocarbons from contaminated materials using a novel dialysis tubing experimental method. *Environ. Sci. Technol.,* 33: 2118–2126. 1999

Wullshleger, R.E., A.E. Zanoni, and C.A. Hansen. *Methodology for the Study of Urban Storm Generated Pollution and Control*. EPA-600/2-76-145. Municipal Environmental Research Laboratory, U.S. Environmental Protection Agency. Cincinnati, OH. Aug. 1976.

Zabik, J.M., L.S. Aston, and J.N. Seiber. Rapid characterization of pesticide residues in contaminated soils by passive sampling devices. *Environ. Toxicol. Chem.,* 11: 765–770. 1992.

Zeng, E. and C.L. Vista. Organic pollutants in the coastal environment off San Diego, CA. 1. Source identification and assessment by compositional indices of polycyclic aromatic hydrocarbons. *Environ. Toxicol. Chem.,* 16(2): 179–188. 1997.

Zeng, E., A.H. Khan, and K. Tran. Organic pollutants in the coastal environment off San Diego, California. 2. Using linear alkylbenzenes to trace sewage-derived organic materials. *Environ. Toxicol. Chem.,* 16(2): 196–201. 1997.

Zimmer, D.W. Observations of invertebrate drift in the Skunk River, Iowa. *Proc. Iowa Acad. Sci.,* 82: 175. 1977.

Zoller, U. et al. Nonionic detergents as tracers of groundwater pollution caused by municipal sewage. *Chemistry for the Protection of the Environment* (Plenum). p. 225. 1991.

## CHAPTER 7

Statistical Analyses of Receiving Water Data

"Get your facts first, then you can distort them as much as you please."

Mark Twain

CONTENTS

Selection of Appropriate Statistical Analysis Tools and Procedures 575
    Computer Software and Recommended Statistical References to Assist in Data Analysis ..576
    Selection of Statistical Procedures..580
Comments on Selected Statistical Analyses Frequently Applied to Receiving Water Data 582
    Determination of Outliers 582
    Exploratory Data Analyses...583
    Comparing Multiple Sets of Data with Group Comparison Tests 588
    Data Associations 591
    Regression Analyses...596
    Analysis of Trends in Receiving Water Investigations..601
    Specific Methods Commonly Used for Evaluation of Biological Data.............................603
Summary of Statistical Elements of Concern when Conducting a Receiving Water Investigation ..605
References 606

SELECTION OF APPROPRIATE STATISTICAL
ANALYSIS TOOLS AND PROCEDURES

The appropriate selection of statistical analyses must be an integral aspect of the experimental design activities for an effective data collection effort. Chapter 5 examined various sampling strategies and presented methods that can be used to estimate the sampling effort. This chapter reviews some typically used statistical analysis procedures that have been very effective in receiving water studies.

Statistical software packages have become an indispensable tool for research, but their selection and use can be frustrating. There are numerous comprehensive statistical software programs available that contain both conventional and specialized tools of interest to environmental researchers. The number of choices is almost overwhelming and covers a wide range of cost (from freeware to several thousand dollars). The selection process can therefore become difficult without some guidance. It is highly recommended that the selection of a software program be made based on consultation with a colleague who has experience with these tools, especially if that colleague can be relied on for assistance later.

When computerized statistical packages were first made available to typical users, it was very easy to rely too much on the wealth of available options, to produce copious piles of irrelevant printouts, and then to become dismayed at the prospect of sorting through the material to find what was important. This was exacerbated if there was also no appropriate experimental design developed at the onset of the data collection effort, or if one needed to rely solely on existing data that were not collected for the objectives at hand. However, examining existing data is an important initial step in experimental design activities, as described in Chapter 5. "Data mining" to identify trends and relationships hidden in large amounts of retrospective data that can be exploited has now become a common household term, even showing up in Dilbert® comic strips. Obviously, collecting haphazard data and relying on powerful computer programs to ferret out conclusions is not a very efficient experimental design for ongoing research. The need for carefully stated project objectives, an appropriate experimental design, and an understanding of the likely statistical tools that will be used for data analysis are all important initial steps in any research activity.

In addition to having access to appropriate software tools, it is imperative that the researcher have some knowledge of applied statistics. Most professionals involved in environmental research have been required to take some type of introductory statistical course for their degree. Few, however, have likely been exposed to the broad range of options that should be examined to select the few procedures that may be most efficient for the specific project objectives. Luckily, well-written articles are available in many technical journals that do an excellent job of describing the statistical tests that were used. In addition, statistical consultants are available through most university statistics and biostatistics departments, in addition to private statistical consultants and experienced colleagues, who are readily available to consult on environmental research projects examining receiving water impacts. Obviously, self-study by the researcher is also necessary, as the person involved in the specific study must take an active role and be ultimately responsible for the experimental design and data evaluation. This chapter therefore lists several applied statistics texts that the authors have found to be extremely valuable and that are well written and at a level understandable to those who are not experts in the statistics field.

Computer Software and Recommended Statistical References to Assist in Data Analysis

The following sections present brief information for useful print and software resources that the receiving water impact investigator may find useful.

Statistical Reference Books

The following books comprise a basic library in applied statistics and have proven very useful in environmental research. These are especially helpful in that they contain many example applications in the environmental sciences and engineering. The use of these books, along with consultation with statistical experts as needed, will enable more efficient experimental designs and data analyses.

- Exploratory data analysis

    Exploratory Data Analysis. John W. Tukey. Addison-Wesley Publishing Co. 1977. This is a basic book with many simple ways to examine data to find patterns and relationships.

    The Visual Display of Quantitative Information. Edward R. Tufte. Graphics Press, Box 430, Cheshire, CT 06410. 1983. This is a beautiful book with many examples of how to and how not to present graphical information. Tufte has two other books that are sequels: *Envisioning Information,* 1990, and *Visual Explanations: Images and Quantities, Evidence and Narrative*, 1997.

    Visualizing Data. William S. Cleveland. Hobart Press, P.O. Box 1473, Summit, NJ 07902, 1993, and *The Elements of Graphing Data*, 1994, are both continuations of the concept of beautiful and informative books on elements of style for elegant graphical presentations of data.

- Experimental design (and some basic methods)

  Statistics for Experimenters. George E. P. Box, William G. Hunter, and J. Stuart Hunter. John Wiley & Sons, 1978. This book contains detailed descriptions of basic statistical methods for comparing experimental conditions and model building.

  Statistical Methods for Environmental Pollution Monitoring. Richard O. Gilbert. Van Nostrand Company, 1987. This book contains a good summary of sampling designs and methods to identify trends, unusual conditions, etc.

- General statistics

  Statistics for Environmental Engineers. Paul Mac Berthouex and Linfield C. Brown. Lewis. 1994. This excellent book reviews the shortcomings and benefits of many common statistical procedures, enabling much more thoughtful evaluation of environmental data.

  Biostatistical Analysis. Jerrold H. Zar. Prentice-Hall. 1996. A highly recommended basic statistics text for the environmental sciences, especially with its many biological science examples.

  Primer on Biostatistics. Stanton A. Glantz. McGraw-Hill. 1992. This is one of the easiest to read and understand introductory texts on applied statistics available.

- Specialized statistical methods

  Nonparametrics: Statistical Methods Based on Ranks. E.L. Lehman and H.J.M. D'Abrera. Holden-Day and McGraw-Hill. 1975. This is a good discussion, with many examples of nonparametric methods for the analysis and planning of comparative studies.

  Applied Regression Analysis. Norman Draper and Harry Smith. John Wiley & Sons. 1981. Thorough treatment of one the most commonly used (and misused) statistical tools.

Statistical Software Programs

There are several tiers of software available for statistical analyses, although the distinctions in their capabilities are becoming blurred. Freeware and shareware (or otherwise inexpensive) software packages have traditionally been developed and made available by private individuals, or are "obsolete" versions of enhanced commercial products. The individually developed packages were typically created to solve a specific problem, or for cost-effective or straightforward use in classrooms. Many of these products are very good, but documentation is likely minimal.

Modern spreadsheet programs also contain many built-in statistical routines (at least regression analyses and simple comparison tests) and graphing options. Spreadsheets are now ubiquitous on all microcomputers, are familiar to users, and the available statistical capabilities should therefore be considered before purchasing additional software. Spreadsheets are extremely helpful for preliminary analyses and for concurrent data evaluation as the data observations are being collected and organized in the spreadsheet (especially critical for laboratory QA/QC control plots as described in Chapter 5). Relatively inexpensive spreadsheet add-ons are also available for decision analysis and Monte Carlo sampling routines, plus some contain rather complete sets of sophisticated statistical routines and graphing templates. Spreadsheets can also be programmed by the user with macros for "customized" statistical routines.

There are also many very elegant and easy-to-use commercial software packages that contain almost all that one would likely need. There is a wide range in price for these products, and some offer specialized capabilities. For comprehensive research, it is common for several different software products to be used for specific data evaluation objectives. Reviews of statistical software packages are commonly available in technical journals and should be consulted. In addition, much information is also available on the Internet. One outstanding example is the "Statistics on the Web" Internet page, created by Clay Helberg, which presents links to many statistical resources. The URL for this page is http://www.execpc.com/~helberg/statistics.html.

The following is a list of groupings of links available on Helberg's Web page:

- Professional organizations
- Institutes and consulting groups
- Educational resources (web courses and online textbooks)
- Publications and publishers
- Statistics book list
- Software-oriented pages
- Mailing lists and discussion groups
- Other lists of links
- Statisticians and other statistical people

Of special interest is his list of software-oriented pages where short reviews and descriptions are given for numerous freeware, shareware, and commercial statistical software. In addition, links are given to the sources of this software, enabling the user to download freeware and shareware packages, and in many cases, trial versions of the commercial packages. Another comprehensive listing of freeware, shareware, and commercial statistical software (for Windows, UNIX, DOS, and Macintosh computers) is available from *St@tServ* — Statistical Software, whose URL is http://www.statserv.com/.

The following programs are briefly mentioned here because of the authors' experience with them, and to indicate some of their major features. There are obviously many other suitable programs, including highly specialized programs emphasizing specific methods.

SYSTAT

SYSTAT (now available through SPSS, Inc., Chicago, IL) has been available to users of small computers for many years and is available for both Windows and Macintosh. It was one of the first comprehensive software packages that was competitive with the early mainframe statistical software packages. Not only did it offer a cost-effective alternative to other programs, but it was also noticeably easier to use and contained several areas of strength not readily available to many (especially nonlinear and multivariate analyses). Many of the examples in this book were prepared using various versions of SYSTAT. Recent versions of SYSTAT include many tools including cluster analysis, correlations and distance measures, factor analysis, multidimensional scaling, regression, analysis of variance, multivariate models, nonlinear models, nonparametric statistics, time series, and basic statistics. Numerous graphical options are also available, often integrated with the statistical methods. Three-dimensional graphing and multiminiatures are especially valuable. SYSTAT graphing is usually easy to use, such as when repeating many basic graphs for numerous parameters (e.g., automatically preparing probability plots for all constituents measured). However, fine-tuning the graphs was not straightforward in the earlier versions of SYSTAT. All data are entered (or imported) in spreadsheet-like tables, making large-scale analyses using very large data matrices easy. Numerous data transformations are also available. A very large number of options are usually available for each statistical tool, but basic setups are typically suitable for most analyses. The comprehensive documentation contains a great deal of information and some guidance, but the user should be reasonably knowledgeable about the techniques selected (as in all computer-based statistical programs).

SigmaStat and SigmaPlot

These programs were originally developed and distributed by Jandel Scientific, but are now owned and updated by SPSS, Inc., Chicago, IL. Probably the most important feature of these programs is their ease of use, especially the built-in guidance and evaluation of data pertaining to the selection of the most appropriate statistical procedure. In addition, it is easy to produce final publication-quality graphs. Exploratory data analysis is especially well covered by these programs.

Although not as comprehensive as some of the other available statistical programs (such as SYSTAT), SigmaStat and SigmaPlot offer complementary strengths. Recent versions include tests for comparing two or more groups (parametric and nonparametric tools), comparing reseated measures of the same individuals, comparing frequencies, rates, and proportions, prediction and correlation, computing power and sample size, and nonlinear regressions. Numerous data transformations are also available, and the data are also imported and managed in a spreadsheet-like table. SigmaPlot is a standalone program that can also be integrated with SigmaStat, offering a powerful data analysis package. The numerous graphing display options make for a very powerful and flexible tool, but also make it somewhat more difficult to prepare routine plots. However, they also offer several graphing templates for exploratory data analysis based on Tufte's excellent book (*The Visual Display of Quantitative Information*, Graphics Press, Cheshire, CT. 1983).

SAS

This statistical program is available from SAS Institute and covers a wide variety of statistical procedures. Although it is generally considered to emphasize business applications (such as database marketing, customer relationship management, clinical trials, quality improvement, fraud detection, etc.), it is also commonly used by researchers from all technical areas. This is one of the best supported statistical software packages, backed by many independent reference books covering SAS programming language for custom applications to specific statistical topics. It does require some training for most users. Because of its long history in academic computer centers, most users will be able to find assistance from experienced SAS users on most university campuses.

SPSS

SPSS, from SPSS, Inc., Chicago, IL, has also long been popular in the academic world and enjoys wide support. It also has numerous independently written books available for reference. The basic module contains the data management utilities, numerous basic statistical tools and graphs, and demographic analyses. Several add-on options are also available: advanced models for complex relationships, regression models, tables, and trends for forecasting. Again, a user should have little trouble finding assistance with this comprehensive program from university-based statistical consultants.

Statistica

Statistica, from StatSoft, of Tulsa, OK, is quickly gaining favor among environmental scientists for its ease of use, comprehensive set of tools, and intensive graphically based options. There are several levels of the program — student versions, a "quick" version, and the "full" version and associated optional add-on packages, including one for designs of experiments and another for quality control charts. Statistica's background was in basic statistical software written specifically for and by social science researchers using microcomputers. The ease of use and ease of interpretation objectives of the early programs are still very much evident in the current Windows- and Macintosh-based versions.

A great public service from StatSoft is the *Downloadable Electronic Statistics Textbook* (Stat-Soft, Inc. Tulsa, OK, 1999). This text can be used on-line through their web page, or downloaded to a local hard drive. According to StatSoft, this textbook offers training in the understanding and application of statistics. The material was developed at the StatSoft R&D department based on many years of teaching undergraduate and graduate statistics courses and covers a wide variety of applications, including laboratory research (biomedical, agricultural, etc.), business statistics and forecasting, social science statistics and survey research, data mining, engineering, and quality control applications. The *Electronic Textbook* begins with an overview of relevant elementary concepts and continues with more in-depth explorations of specific areas of statistics representing classes of analytical techniques. A glossary of statistical terms and a list of references for further study are also included. The text requires about 30 min to download at 28.8 Kbps from http://www.statsoftinc.com/textbook/stathome.html. The textbook is lengthy and covers many subjects, and it is very well written for novice statistical software users. Surprisingly, however, it has few numeric examples, although it contains numerous graphical outputs from Statistica.

Selection of Statistical Procedures

Most of the objectives of receiving water studies can be examined through the use of relatively few statistical evaluation tools. The following briefly outlines some simple experimental objectives and a selected number of statistical tests (and their data requirements) that can be used for data evaluation.

Statistical Power

Type 1 and Type 2 errors (along with statistical confidence and power) are discussed in Chapter 5 in the experimental design section. Errors in decision making are usually divided into Type 1 (α: alpha) and Type 2 (β: beta) errors:

α (alpha) (Type 1 error) — a false positive, or assuming something is true when it is actually false. An example would be concluding that a tested water was adversely contaminated, when it was actually clean. The most common value of α is 0.05 (accepting a 5% risk of having a Type 1 error). Confidence is $1 - \alpha$, or the confidence of not having a false positive.

β (beta) (Type 2 error) — a false negative, or assuming something is false when it is actually true. An example would be concluding that a tested water was clean when it was actually contaminated. If the sample was an effluent, it would therefore be an illegal discharge with the possible imposition of severe penalties from a regulatory agency. In most statistical tests, β is usually ignored (if ignored, β is 0.5). If it is considered, a typical value is 0.2, implying accepting a 20% risk of having a Type 2 error. Power is $1 - \beta$, or the certainty of not having a false negative.

When evaluating data using a statistical test, power is the sensitivity of the test for rejecting the hypothesis. For an ANOVA test, it is the probability that the test will detect a difference among the groups if a difference really exists.

Comparison Tests

Probably the most common situation is to compare data collected from different locations, or seasons. Comparison of test with reference sites, of influent with effluent, of upstream with downstream locations, for different seasons of sample collection, of different methods of sample collection can all be made with comparison tests. If only two groups are to be compared (above/below; in/out; test/reference), then the two group tests can be used effectively, such as the simple Student's t-test or nonparametric equivalent. If the data are collected in "pairs," such as concurrent influent and effluent samples, or concurrent above and below samples, then the more powerful and preferred paired tests can be used. If the samples cannot be collected to represent similar conditions (such as large physical separation in sampling location, or different time frames), then the independent tests must be used.

If multiple groupings are used, such as from numerous locations along a stream, but with several observations from each location; or from one location, but for each season, then a one-way ANOVA is needed. If one has seasonal data from each of the several stream locations for multiple seasons, a two-way ANOVA can be used to investigate the effects of location, season, and the interaction of location and season together. Three-way ANOVA tests can be used to investigate another dimension of the data (such as contrasting sampling methods or weather for the different seasons at each of the sampling locations), but that would obviously require substantially more data to represent each condition. (See the discussion on stratified random sampling in Chapter 5, for example.)

There are various data characteristics that influence which specific statistical test can be used for comparison tests. The parametric tests require the data to be normally distributed and that the different data groupings have the same variance, or standard deviation (checked with probability plots and appropriate test statistics for normality, such as the Kolmogorov–Smirnov one-sample test, the chi-square goodness of fit test, or the Lilliefors test). If the data do not meet the requirements for the parametric tests, the data may be transformed to better meet the test conditions (such as taking the $\log_{10}$ of each observation and conducting the test on the transformed values). The nonparametric tests are less restrictive, but are not free of certain requirements. Even though the parametric tests have more statistical power than the associated nonparametric tests, they lose any advantage if inappropriately applied. If uncertain, nonparametric tests should be used.

A few example statistical tests (as available in SigmaStat, SPSS, Inc.) are indicated below for different comparison test situations:

- Two groups
    Paired observations
        Parametric tests (data require normality and equal variance)
            – Paired Student's t-test (more power than nonparametric tests)
        Nonparametric tests
            – Sign test (no data distribution requirements, some missing data accommodated)
            – Friedman's test (can accommodate a moderate number of "nondetectable" values, but no missing values are allowed
            – Wilcoxon signed rank test (more power than sign test, but requires symmetrical data distributions)
    Independent observations
        Parametric tests (data require normality and equal variance)
            – Independent Student's t-test (more power than nonparametric tests)
        Nonparametric tests
            – Mann–Whitney rank sum test (probability distributions of the two data sets must be the same and have the same variances, but do not have to be symmetrical; a moderate number of "nondetectable" values can be accommodated)
- Many groups (use multiple comparison tests, such as the Bonferroni t-test, to identify which groups are different from the others if the group test results are significant).
    Parametric tests (data require normality and equal variance)
        One-way ANOVA for single factor, but for >2 "locations" (if 2 locations, use Student's t-test)
        Two-way ANOVA for two factors simultaneously at multiple "locations"
        Three-way ANOVA for three factors simultaneously at multiple "locations"
        One factor repeated measures ANOVA (same as paired t-test, except that there can be multiple treatments on the same group)
        Two factor repeated measures ANOVA (can be multiple treatments on two groups)
    Nonparametric test
        Kurskal–Wallis ANOVA on ranks (use when samples are from non-normal populations or the samples do not have equal variances)
        Friedman repeated measures ANOVA on ranks (use when paired observations are available in many groups)
    Nominal observations of frequencies (used when counts are recorded in contingency tables)
        Chi-square (χ^2) test (use if more than two groups or categories, or if the number of observations per cell in a 2×2 table are > 5)
        Fisher Exact test (use when the expected number of observations is <5 in any cell of a 2×2 table)
        McNamar's test (use for a "paired" contingency table, such as when the same individual or site is examined both before and after treatment)

Data Associations and Model Building

These activities are an important component of the "weight-of-evidence" approach used to identify likely cause-and-effect relationships. The following list illustrates some of the statistical tools (as available in SigmaStat and/or SYSTAT, SPSS, Inc.) that can be used for evaluating data associations and subsequent model building:

- Data Associations
    - Simple
        - Pearson Correlation (residuals, the distances of the data points from the regression line, must be normally distributed. Calculates correlation coefficients between all possible data variables. Must be supplemented with scatterplots, or scatterplot matrix, to illustrate these correlations. Also identifies redundant independent variables for simplifying models)
        - Spearman Rank Order Correlation (a nonparametric equivalent to the Pearson test)
    - Complex (typically only available in advanced software packages)
        - Hierarchical Cluster Analyses (graphical presentation of simple and complex interrelationships. Data should be standardized to reduce scaling influence. Supplements simple correlation analyses)
        - Principal Component Analyses (identifies groupings of parameters by factors so that variables within each factor are more highly correlated with variables in that factor than with variables in other factors. Useful to identify similar sites or parameters).
- Model building/equation fitting (these are parametric tests and the data must satisfy various assumptions regarding behavior of the residuals)
    - Linear equation fitting (statistically-based models)
        - Simple linear regression ($y = b_0 + b_1 x$, with a single independent variable, the slope term, and an intercept. It is possible to simplify even further if the intercept term is not significant)
        - Multiple linear regression ($y = b_0 + b_1 x_1 + b_2 x_2 + b_3 x_3 + \dots + b_k x_k$, having k independent variables. The equation is a multidimensional plane describing the data)
        - Stepwise regression (a method generally used with multiple linear regression to assist in identifying the significant terms to use in the model)
        - Polynomial regression ($y = b_0 + b_1 x^1 + b_2 x^2 + b_3 x^3 + \dots + b_k x^k$, having one independent variable describing a curve through the data)
    - Nonlinear equation fitting (generally developed from theoretical considerations)
        - Nonlinear regression (a nonlinear equation in the form: $y = b^x$, where x is the independent variable. Solved by iteration to minimize the residual sum of squares)
- Data trends
    - Graphical methods (simple plots of concentrations vs. time of data collection)
    - Regression methods (perform a least-squares linear regression on the above data plot and examine ANOVA for the regression to determine if the slope term is significant. Can be misleading due to cyclic data, correlated data, and data that are not normally distributed)
    - Mann–Kendall test (a nonparametric test that can handle missing data and trends at multiple stations. Short-term cycles and other data relationships affect this test and must be corrected)
    - Sen's estimator of slope (a nonparametric test based on ranks closely related to the Mann–Kendall test. It is not sensitive to extreme values and can tolerate missing data).
    - Seasonal Kendall test (preferred over regression methods if the data are skewed, serially correlated, or cyclic. Can be used for data sets having missing values, tied values, censored values, or single or multiple data observations in each time period. Data correlations and dependence also affect this test and must be considered in the analysis)

COMMENTS ON SELECTED STATISTICAL ANALYSES FREQUENTLY APPLIED TO RECEIVING WATER DATA

Determination of Outliers

Outliers in data collection can be recognized in the tails of the probability distributions. Observations that do not perfectly fit the probability distributions in the tails are commonly considered outliers. They can be either very low or very high values. These values always attract considerable attention because they don't fit the mathematical probability distributions exactly and are usually assumed to be flawed and are then discarded. Certainly, these values (like any other suspect values) require additional evaluation to confirm that simple correctable errors (transcription, math, etc.) are not responsible. If no errors are found, then these values should be included in the data analyses because they represent rare conditions that may be very informative.

Analytical results less than the practical quantification limit (PQL) or the method detection limit (MDL) need to be flagged, but the result (if greater than the instrument detection limit, or IDL) should still be used in most of the statistical calculations. In some cases, the statistical test procedures can handle some undetected values with minimal modifications. In most cases, however, commonly used statistical procedures behave badly with undetected values. In these cases, results less than the IDL should be treated according to Berthouex and Brown (1994). Generally, the statistical procedures should be used twice, once with the less than detection values (LDV) equal to zero, and again with the LDV equal to the IDL. This procedure will determine if a significant difference in conclusions would occur with handling the data in a specific manner. In all cases of substituting a single value for LDV, the variability is artificially reduced, which can significantly affect comparison tests. It may therefore be best to use the actual instrument reported value for many statistical tests, even if it is below the IDL or MDL. This value may be considered a random value, but it is probably closer to the true value than a zero or other arbitrary value, plus it retains some aspects of the variability of the data sets. Of course, these values should not be "reported" in the project report, or to a regulatory agency, as they obviously do not meet the project QA/QC requirements.

Similarly, unusually high values need to be examined critically to identify possible errors. In most cases, the sample should be reanalyzed. This is a good reason to retain any "left over" sample until satisfied with the results. Of course, long-stored samples may not be very representative of actual conditions for many constituents, so care will have to be taken when using these reanalyzed values if they exceeded the recommended storage periods. It is difficult to reject wet-weather constituent observations solely because they are unusually high, as wet weather flows can easily have wide-ranging constituent observations. High values should not automatically be considered as outliers and therefore worthy of rejection, but as rare and unusual observations that may shed some light on the problem.

Exploratory Data Analyses

Exploratory data analysis (EDA) is an important tool to quickly review available data before a specific data collection effort is initiated. It is also an important first step in summarizing collected data to supplement the specific data analyses associated with the selected experimental designs. A summary of the data's variation is most important and can be presented using several simple graphical tools. *The Visual Display of Quantitative Information* (Tufte 1983) is a book with many examples of how to and how not to present graphical information. *Envisioning Information*, also by Tufte (1990), supplements his earlier book. Another important reference for basic analyses is *Exploratory Data Analysis* (Tukey 1977), which is the classic on this subject and presents many simple ways to examine data to find patterns and relationships. Cleveland (1993 and 1994) has also published two books related to exploratory data analyses: *Visualizing Data* and *The Elements of Graphing Data*. The basic plots described below can obviously be supplemented by many others presented in these books. Besides plotting the data, exploratory data analyses should always include corresponding statistical test results, if available.

Basic Data Plots

There are several basic data plots that need to be prepared as data are being collected and when all of the data are available. These plots are basically for QA/QC purposes and to demonstrate basic data behavior. These basic plots include: time series plots (data observations as a function of time), control plots (generally the same as time series plots, but using control samples and with standard deviation bands, as described in Chapter 5), probability plots (described below), scatterplots (described below), and residual plots (needed for model building activity, especially for regression analyses, also described below).

Probability Plots

The most basic exploratory data analysis method is to prepare a probability plot of the available data. The plots indicate the possible range of the values expected, their likely probability distribution type, and the data variation. It is difficult to recommend another method that results in so much information using the available data. Histograms, for example, cannot accurately indicate the probability distribution type very accurately, but they more clearly indicate multimodal distributions. The observations are ranked in ascending order and probability values are calculated for each observation using the following formula:

$$P = (i - 0.5)/n$$

where   i = the rank position
        n = the total number of observations.

If 11 observations are available, the 6th ranked value would have a probability of 0.50 (50%), using the above formula. The values and corresponding probability positions are plotted on special normal-probability paper. This paper has a y-axis whose values are spread out for the extreme small and large probability values. When plotted on this paper, the values form a straight line if they are normally distributed (Gaussian). If the points do not form an acceptably straight line, they can then be plotted on log-normal probability paper (or the data observations can be log transformed and plotted on normal probability paper). If they form a straight line on the log-normal plot, then the data are log-normally distributed. Other data transformations are also possible for plotting on normal-probability paper, but these two (normal and log-normal) are usually sufficient for most receiving water analyses.

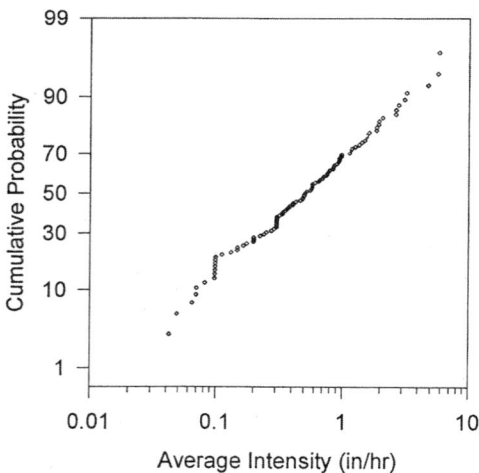

Figure 7.1 Log-normal probability distribution for 1976 Birmingham, AL, average rain intensities. (From Pitt, R. and S.R. Durrans. *Drainage of Water from Pavement Structures*. Alabama Department of Transportation. Research Project 930-275. Montgomery, AL. Sept. 1995.)

Figure 7.1 is an example of a probability plot of average rain intensity for Birmingham, AL, for 1976 (Pitt and Durrans 1995). This is a log-normal probability plot, as the rain intensity values are plotted on a log scale. These intensities plot along a reasonably straight line, indicating that they are generally log-normally distributed. Figure 7.2 shows three types of results that can be observed when plotting pollutant reduction observations on probability plots, using data collected at the Monroe St. wet detention pond in Madison, WI, by the USGS and the WI DNR. Figure 7.2a for suspended solids (particulate residue) shows that SS are effectively removed over a wide range of influent concentrations, ranging from 20 to over 1000 mg/L. A simple calculation of percentage reduction would not show this consistent removal over the wide range. In contrast, Figure 7.2b for total dissolved solids (filtered residue) shows poor removals of TDS for all concentration conditions, as expected for this wet detention pond. The average percentage

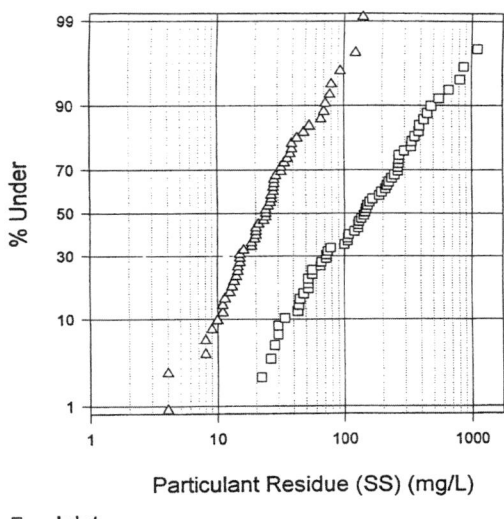

Particulant Residue (SS) (mg/L)

□ Inlet
△ Outlet

a

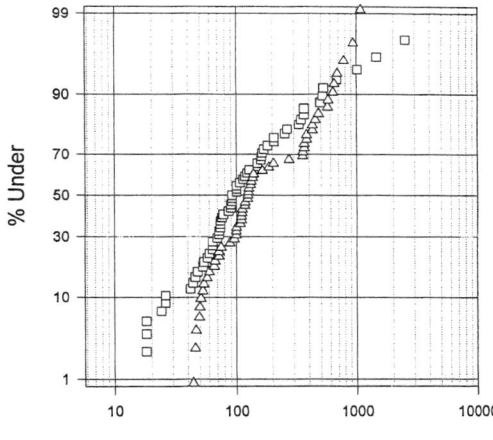

Filtered Residue (TDS) (mg/L)

□ Inlet
△ Outlet

b

removal for TDS would be close to zero and no additional surprises are indicated on this plot. Figure 7.2c, however, shows a wealth of information that would not be available from simple numerical statistical summaries. In this plot, filtered COD is seen to be poorly removed for low concentrations (less than about 20 mg/L, but the removal increases substantially for higher concentrations. In addition, the COV of the effluent concentrations is much smaller than for the influent concentrations. Although not indicated on these plots, the rank order of concentrations was similar for both influent and effluent distributions for all three pollutants.

Generally, water quality observations do not form a straight line on normal probability paper, but do (at least from about the 10 to 90 percentile points) on log-normal probability paper. This indicates that the samples generally have a log-normal distribution, and many parametric statistical tests can probably be used, but only after the data are log-transformed. These plots indi-

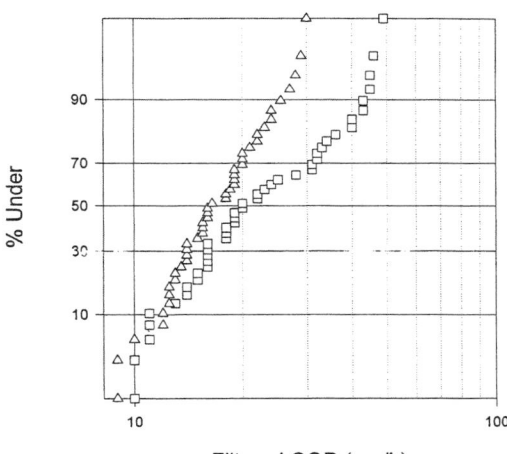

□ Inlet
△ Outlet

Filtered COD (mg/L)

c

Figure 7.2 Influent and effluent observations for suspended solids, dissolved solids, and filtered COD at the Monroe St., Madison, WI, stormwater detention pond.

cate the central tendency (median) of the data, along with their possible distribution type and variance (the steeper the plot, the smaller the COV, and the flatter the slope of the plot, the larger the COV for the data). Multiple data sets can also be plotted on the same plot (such as for different sites, different seasons, different habitats, etc.) to indicate obvious similarities (or differences) in the data sets. Most statistical methods used to compare different data sets require that the sets have the same variances, and many require normal distributions. Similar variances would be indicated by generally parallel plots of the data on the probability paper, while normal distributions would be reflected by straight line plots on normal probability paper.

Probability plots should be supplemented with standard statistical tests that determine if the data are normally distributed. These tests (at least some are usually available in most software packages) include the Kolmogorov–Smirnov one-sample test, the chi-square goodness of fit test, and the Lilliefors variation of the Kolmogorov–Smirnov test. They basically are paired tests comparing data points from the best-fitted normal curve to the observed data. The statistical tests may be visualized by comparing the best-fitted normal curve data and the observed data plotted on normal probability paper. If the observed data cross the fitted curve data numerous times, it is much more likely to be normally distributed than if it only crossed the fitted curve a few times.

Digidot Plot

Berthouex and Brown (1994) point out that since the best way to display data is with a plot, it makes little sense to present the data in a table. They highly recommend a digidot plot, developed by Hunter (1988) based on Tukey (1977), as a basic presentation of characterization data. This plot indicates the basic distribution of the data, shows changes with time, and presents the actual values, all in one plot. A data table is therefore not needed in addition to the digidot plot. A stem and leaf plot of the data is presented as the y-axis, and the data are presented in a time series (in the order of collection) along the x-axis. Figure 7.3 is an example of a digidot plot, as presented by Berthouex and Brown (1994). The stem and leaf plot is constructed by placing the last digit of the value on the y-axis between the appropriate tic marks. In this example, the value 47 is represented with a 7 placed in the division between 45 and 50. Similarly, 33 is represented with a 3 placed in the division between 30 and 35. Values from 30 to 34 are placed between the 30 and 35 tic marks, while values from 35 to 39 are placed between the 35 and 40 tic marks. Simultaneously, the values are plotted in a time series in the order of collection. This plot can therefore be constructed in real time as the data are collected, and obvious trends can be noted with time. This plot also presents the actual numerical data that can also be used in later statistical analyses.

Scatterplots

According to Berthouex and Brown (1994), the majority of the graphs used in science are scatterplots. They stated that these plots should be made before any other analyses of the data are performed. Scatterplots are typically made by plotting the primary variable (such as a water quality constituent) against a factor that may influence its value (such as time, season, flow, another constituent like suspended solids, etc.). Figure 7.4 is a scatterplot showing COD values plotted against rain depth to investigate the possibility of a "first-flush," where higher concentrations are assumed to be associated with small runoff events (Pitt 1985). In this example, the smallest rains appear to have the highest COD concentrations associated with them, but the distribution of values is very wide. This may simply be a result of the much greater number of events having small rains and an increased likelihood of events having unusual observations when more observations are

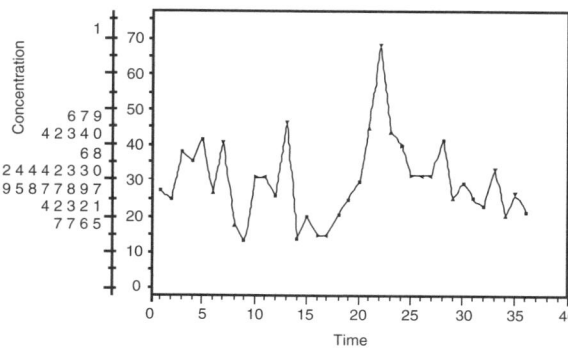

Figure 7.3   Digidot plot. (From Berthouex, P.M. and L.C. Brown. *Statistics for Environmental Engineers.* Lewis Publishers, Boca Raton, FL. 1994. With permission.)

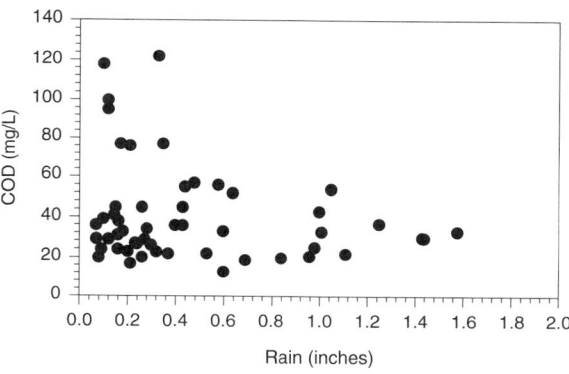

Figure 7.4 Scatterplot for Bellevue, WA, COD stormwater concentrations, by rain depth. (From Pitt, R. *Characterizing and Controlling Urban Runoff through Street and Sewerage Cleaning.* U.S. Environmental Protection Agency, Storm and Combined Sewer Program, Risk Reduction Engineering Laboratory. EPA/600/S2-85/038. PB 85-186500. Cincinnati, OH. 467 pp. June 1985.)

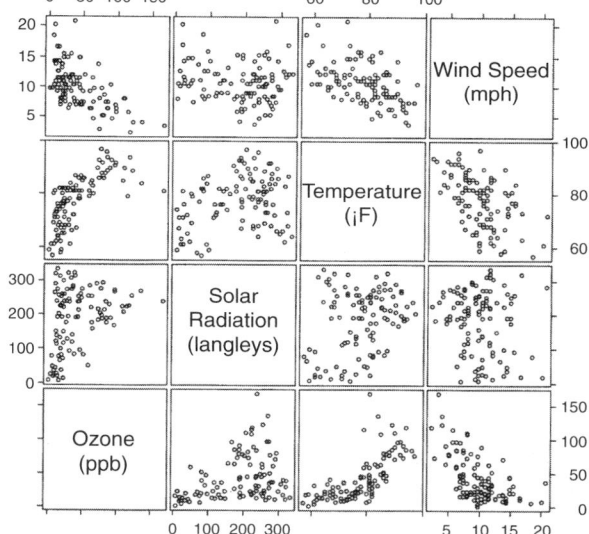

Figure 7.5 Grouped scatterplot for ozone, solar radiation, temperature, and wind speed. (From Cleveland, W.S. *The Elements of Graphing Data.* Hobart Press, Summit, NJ. 1994. With permission.)

made. When many data are observed for many sites, generally smaller rains do seem to be associated with the highest concentrations observed, but it is not a consistent pattern.

Grouped scatterplots (miniatures) of all possible combinations of constituents can be organized as in a correlation matrix (Figure 7.5; Cleveland 1994). This arrangement allows obvious relationships to be seen easily, and even indicates if the relationships are straight-lined or curvilinear. In this example, the highest ozone values occur on days having the highest temperatures, and the lowest ozone concentrations occur on days having brisk winds and low temperatures, for example.

Grouped Box and Whisker Plots

Another primary exploratory data analysis tool, especially when differences between sample groups are of interest, is the use of grouped box and whisker plots. Examples of their use include examining different sampling locations (such as above and below a discharge), influent and effluent of a treatment process, different seasons, etc. These plots indicate the range and major percentile locations of the data, as shown on Figure 7.6 (Pitt 1985). In this example, seasonal groupings of stormwater quality observations for COD (chemical oxygen demand) from Bellevue, WA, were plotted to indicate obvious differences in the values. If the 75 and 25 percentile lines of the boxes are higher or lower than the medians of other box and whisker plots, then the data groupings are likely significantly different (at least at the 95% level). When large numbers of data sets are plotted using box and whisker plots, the relative overlapping (or separation) of the plots can be used to

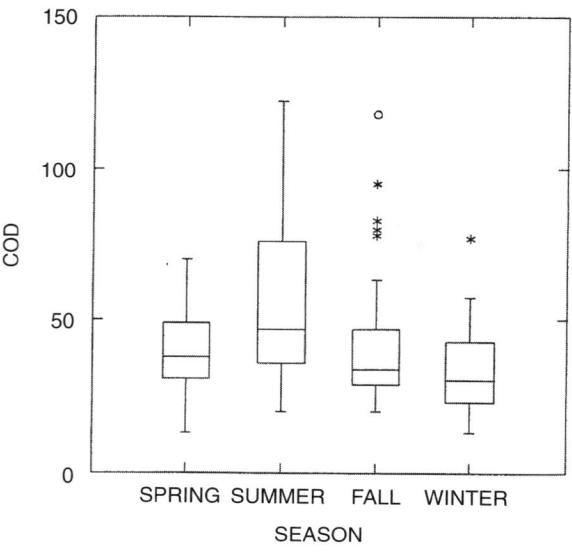

Figure 7.6 Grouped box and whisker plot for Bellevue, WA, COD stormwater concentrations, by season. (From Pitt, R. *Characterizing and Controlling Urban Runoff through Street and Sewerage Cleaning*. U.S. Environmental Protection Agency, Storm and Combined Sewer Program, Risk Reduction Engineering Laboratory. EPA/600/S2-85/038. PB 85-186500. Cincinnati, OH. 467 pp. June 1985.)

identify possible groupings of the separate sets. In this case, the winter has lower concentrations than the summer.

To supplement the visual presentation with the grouped box and whisker plots, a one-way ANOVA test (or the Kurskal–Wallis ANOVA on ranks test) should be conducted to determine if there are any statistically significant differences between the different boxes on the plot. ANOVA doesn't specifically identify which sets of data are different from others, however. A multiple comparison procedure (such as the Bonferroni *t*-test) can be used to identify significant differences between all cells if the ANOVA finds that a significant difference exists. Both of these tests (ANOVA and Bonferroni *t*-test) are parametric tests and require that the data be normally distributed. It may therefore be necessary to perform a log-transformation on the raw data. These tests will identify differences in sample groupings, but similarities (to combine data) are probably also important to know.

Comparing Multiple Sets of Data with Group Comparison Tests

Making comparisons of data sets is a fundamental objective of many receiving water investigations. Different habitats and seasons can produce significant affects on the observations. The presence of influencing factors, such as pollutant discharges or control practices, also affects the data observations. Berthouex and Brown (1994) and Gilbert (1987) present excellent summaries of the most common statistical tests that are used for these comparisons in environmental investigations. The significance of the test results (the α value, the confidence factor, along with the β value, the power factor, both discussed in Chapter 5) will indicate the level of confidence and power that the two sets of observations are the same. In most cases, an α level of less than 0.05 has been traditionally used to signify significant differences between two sets of observations, although this is an arbitrary criterion. In most cases, β is ignored (resulting in a default value of $1 - \beta$ of 0.5), although some use a $1 - \beta$ value of 0.8. An α value of 0.05 implies that the interpretation will be in error an average of 1 in 20 times. In some cases, this may be too conservative, while in others (such as where health and welfare implications are involved), it may be too liberal. The selection of the critical α value should be decided beforehand, while the calculated values for α should always be presented in the data evaluation (not simply stating that the results were significant or not significant at the 0.05 level, as is common). Even if the α level is significant, the magnitude of the difference, such as the pollutant reduction, may not be very important. The importance of the level of pollutant reductions should also be graphically presented using grouped box plots indicating the range and variations of the concentrations at each of the sampling locations, as described previously.

Comparison tests are divided into simple tests between two groups (such as Student's t-test) and tests that examine larger numbers of groups and interactions (such as analysis of variance tests, or ANOVA).

Simple Comparison Tests with Two Groups

The main types of simple comparison tests are separated into independent and paired tests. These can be further separated into tests that require specific probability distribution characteristics (parametric tests) and tests that do not have as many restrictions based on probability distribution characteristics of the data (nonparametric data). If the parametric test requirements can be met, they should be used because they have more statistical power. However, if information concerning the probability distributions is not available, or if the distributions do not behave correctly, then the somewhat less powerful nonparametric tests should be used. Similarly, if the data gathering activity can allow for paired observations, they should be used preferentially over independent tests.

In many cases, observations cannot be related to each other, such as a series of observations at two locations during all of the rains during a season. Unless the sites are very close together, the rains are likely to vary considerably at the two locations, disallowing a paired analysis. However, if data can be collected simultaneously, such as at influent and effluent locations for a (rapid) treatment process, paired tests can be used to control all factors that may influence the outcome, resulting in a more efficient statistical analysis. Paired experimental designs ensure that uncontrolled factors basically influence both sets of data observations equally (Berthouex and Brown 1994).

The parametric tests used for comparisons are the Student's t-tests (both independent and paired t-tests). All statistical analysis software and most spreadsheet programs contain both of these basic tests. These tests require that the variances of the sample sets be the same and constant over the range of the values. These tests also require that the probability distributions be Gaussian (normal). Transformations can be used to modify the data sets to these conditions. Log-transformations can be used to produce Gaussian distributions of most water quality data. Square root transformations are also commonly used to make the variance constant over the data range, especially for biological observations (Sokal and Rohlf 1969). In all cases, it is necessary to confirm these requirements before the standard t-tests are used.

Nonparametrics: Statistical Methods Based on Ranks by Lehman and D'Abrera (1975) is a comprehensive general reference on nonparametric statistical analyses. Gilbert (1987) presents an excellent review of nonparametric alternatives to the Student's t-tests, especially for environmental investigations from which the following discussion is summarized. Even though the nonparametric tests remove many of the restrictions associated with the t-tests, the t-tests should be used if justifiable. Unfortunately, seldom are the Student's t-test requirements easily met with environmental data, and the slight loss of power associated with using the nonparametric tests is much more acceptable than misusing the Student's t-tests. Besides having few data distribution restrictions, many of the nonparametric tests can also accommodate a few missing data, or observations below the detection limits. The following paragraphs briefly describe the features of the nonparametric tests used to compare data sets.

Nonparametric Tests for Paired Data Observations

The sign test is the basic nonparametric test for paired data. It is simple to compute and has no requirements pertaining to data distributions. A few "not detected" observations can also be accommodated. Two sets of data are compared and the differences are used to assign a positive sign if the value in data set #1 is greater than the corresponding value in data set #2, or a negative sign is assigned if opposite. The number of positive signs are added and a statistical table (such as in Lehman and D'Abrera 1975) is used to determine if the number of positive signs found is unusual for the number of data pairs examined.

The Mann–Whitney signed rank test has more power than the sign test, but it requires that the data distributions be symmetrical (but with no specific distribution type). Without transformations, this requirement may be difficult to justify for water quality data. This test requires that the differences between the data pairs in the two data sets be calculated and ranked before checking with a special statistical table (as in Lehman and D'Abrera 1975). In the simplest case for monitoring the effectiveness of treatment alternatives, comparisons can be made of inlet and outlet conditions to determine the level of pollutant removal and the statistical significance of the concentration differences. StatXact-Turbo (CYTEL, Cambridge, MA) is a microcomputer program that computes exact nonparametric levels of significance, without resorting to normal approximations. This is especially important for the relatively small data sets that will typically be evaluated during most environmental research activities.

Friedman's test is an extension of the sign test for several related data groups. There are no data distribution requirements and the test can accommodate a moderate number of "nondetectable" values, but no missing values are allowed.

Nonparametric Tests for Independent Data Observations

As for the *t*-tests, paired test experimental designs are superior to independent designs for nonparametric tests because of their ability to cancel out confusing properties. However, paired experiments are not always possible, requiring the use of independent tests. The Wilcoxon rank sum test is the basic nonparametric test for independent observations. The test statistic is also easy to compute and compare to the appropriate statistical table (as in Lehman and D'Abrera 1975). The Wilcoxon rank sum test requires that the probability distributions of the two data sets be the same (and therefore have the same variances). There are no other restrictions on the data distributions (they do not have to be symmetrical, for example). A moderate number of "nondetectable" values can be accommodated by treating them as ties.

The Kruskal–Wallis test is an extension of the Mann–Whitney rank sum test and allows evaluations of several independent data sets, instead of just two. Again, the distributions of the data sets must all be the same, but they can have any shape. A moderate number of ties and nondetectable values can also be accommodated.

Comparisons of Many Groups

If there are more than two groups of data to be compared (such as in-stream concentrations at several locations along a river, each with multiple observations), one of the analysis of variance, or ANOVA, tests should be used. The commonly available one-way, two-way, and three-way ANOVA tests are parametric tests and require that the data in each grouping be normally distributed and that the variances be the same in each group. This can be visually examined by preparing a probability plot for the data in each group displayed on the same chart. The probability plots would need to be parallel and straight. Obviously, log transformations of the data can be used if assumptions are met when the data is plotted using log-normal probability axes. In Figure 7.2a, the influent and effluent probability plots for suspended solids at the Monroe St. wet detention pond site in Madison, WI, the probability plots are reasonably parallel and straight when plotted as log-normal plots. However, Figure 7.2c, a similar plot for dissolved COD, indicates that the plots are not parallel. Of course, these figures contain only two groupings of data (influent and effluent), and one of the previous two-group tests would be more efficient for this data.

If data from multiple stations along a river were collected during different seasons, it would be possible to use the two-way ANOVA test to examine the effects of different seasons and different locations, along with the interaction of these parameters. Three-way ANOVA tests can be used to evaluate the results of similar field sampling data (different locations, different seasons) and another factor, such as natural vs. artificial substrate samplers for benthic macroinvertebrates (or seining vs. electroshocking for fish sampling). These tests would then indicate if the results from these different sampling procedures varied significantly by season, or sampling location. These analyses are more flexible than the factorial tests described earlier in Chapter 5, as the factorial tests are most commonly only used for two levels (such as winter vs. summer; pools vs. riffles; and artificial substrate vs. natural substrate samplers). Factorial tests are more complicated when intermediate, or more than 2 levels, are being considered. However, the ANOVA tests are parametric tests and require multiple observations in each group, while the factorial tests are not and can be used with single observations per group (although that may not be a good idea considering the expected high variability in most environmental sampling).

A nonparametric test, usually included in statistical programs for comparing many groups, is the Kruskal–Wallis ANOVA on ranks test. This is only a one-way ANOVA test and is only suitable for comparing data from different sampling sites alone, for example. This would be a good test to supplement grouped box and whisker plots.

Grouped comparison tests indicate only that at least one of the groups is significantly different from at least one other, they do not indicate which ones. For that reason, some statistical programs also conduct multiple comparison tests. SigmaStat, for example, offers: the Tukey test, Student–Newman–Keuls test, Bonferroni *t*-test, Fisher's LDS, Dunner's test, and Duncan's multiple range test. These tests basically conduct comparisons of each group against each other group and identify which are different.

Data Associations

Identifying patterns and associations in data may be considered a part of exploratory data analyses, but many of the tools (especially cluster, principal component, and factor analyses) may require specialized procedures having multiple data handling options that are not available in all statistical software packages, while some (such as correlation matrices discussed here) are commonly available.

Identifying data associations, and possible subsequent model building, is another area of interest to many investigators examining receiving water conditions. This is a critical component of the "weight-of-evidence" approach for identifying possible cause and effect relationships. The following are possible steps for investigating data associations:

1. Reexamine the hypothesis of cause and effect (an original component of the experimental design previously conducted and the basis for the selected sampling activities).
2. Prepare preliminary examinations of the data, as described previously (most significantly, prepare scatterplots and grouped box/whisker plots).
3. Conduct comparison tests to identify significant groupings of data. As an example, if seasonal factors are significant, then cause and effect may vary for different times of the year.
4. Conduct correlation matrix analyses to identify simple relationships between parameters. Again, if significant groupings were identified, the data should be separated into these groupings for separate analyses, in addition to an overall analysis.
5. Further examine complex interrelationships between parameters by possibly using combinations of hierarchical cluster analyses, principal component analyses (PCA), and factor analyses.
6. Compare the apparent relationships observed with the hypothesized relationships and with information from the literature. Potential theoretical relationships should be emphasized.
7. Develop initial models containing the significant factors affecting the parameter outcomes. Simple apparent relationships between dependent and independent parameters should lead to reasonably simple models, while complex relationships will likely require further work and more complex models.

The following sections briefly describe these tools and present some interesting examples of their use.

Correlation Matrices

Knowledge of the correlations between data elements is very important in many environmental data analysis efforts. They are especially important when model building, such as with regression analysis. When constructing a model, it is important to include the important factors in the model, but the factors should be independent. Correlation analyses can assist by identifying the basic structure of the model.

Table 7.1 (Pitt 1987) is a standard correlation matrix that shows the relationships between measured rain and measured runoff parameters. This is a common Pearson correlation matrix, constructed using the microcomputer program SYSTAT (SPSS, Inc., Chicago, IL). It measures the strength of association between the variables. The Pearson correlation coefficients vary from –1 to +1. A coefficient of 0 indicates that neither of the two variables can be predicted from the other using a linear equation, while values of –1 or +1 indicate that perfect predictions can be made of one variable by only using the other variable. This example shows several very high correlations between pairs of parameters (>0.9). The paired parameters having high correlations are the same for both sites, possibly indicating the same basic processes for rainfall-runoff. High correlations are seen between total runoff depth (RUNTOT) and rain depth (RAINTOT) and between runoff duration (RUNDUR) and rain duration (RAINDUR).

It is very important not to confuse correlation with causation. Box et al. (1978) presents a historical example of a plot (Figure 7.7) of the population of Oldenburg, Germany, against the number of storks observed in each year. In this example, few would conclude that the high correlation between the increased number of storks observed and the simultaneous increase in population is a cause and effect relationship. The two variables observed are most likely related to another factor (such as time in this example, as both sets of populations increased over the years from 1930 to 1936). However, many investigators make similar improper assumptions of cause and effect from their observations, especially if high correlations are found. It is extremely important that theoretical knowledge of the system being modeled be considered. If this knowledge is meager, then specific tests to directly investigate cause and effect relationships must be conducted.

Hierarchical Cluster Analyses

Another method to examine correlations between measured parameters is by using hierarchical cluster analyses. Figure 7.8 (Pitt 1987) is a tree diagram (dendogram) produced by SYSTAT using the same data as presented in the correlation matrix. A tree diagram illustrates simple and complex correlations between parameters. Parameters with short branches linking them are more closely correlated than parameters linked by longer branches. In addition, the branches can encompass more than just two parameters. The length of the short branches linking only two parameters is indirectly comparable to the correlation coefficients (short branches signify correlation coefficients close to 1). The main advantage of a cluster analyses is the ability to identify complex correlations that cannot be observed using a simple correlation matrix. In this example, the rain total — runoff total and runoff duration — rain duration high correlation coefficients found previously are also seen to have simple relationships. In contrast, predicting peak runoff rates (PEAKDIS) requires more complex information. Therefore, the model used to predict peak runoff would have to be more complex, requiring additional information than required to merely predict total runoff.

Principal Component Analyses (PCA) and Factor Analyses

Another important tool to identify relationships and natural groupings of samples or locations is with principal component analyses (PCA). Normally, data are autoscaled before PCA in order to remove the artificially large influence of constituents having large values compared to constituents having small values. PCA is a sophisticated procedure where information is sorted to determine

Table 7.1 Pearson Correlation Matrix

	RAINTOT	RAINDUR	AVEINT	PEAKINT	DRYPER	RUNTOT	RUNDUR	AVEDIS	PEAKDIS	LAG
					Emery (Industrial)					
RAINTOT	1.000									
RAINDUR	0.533	1.000								
AVEINT	0.138	-0.387	1.000							
PEAKINT	0.512	-0.039	0.675	1.000						
DRYPER	0.169	0.273	-0.096	-0.132	1.000					
RUNTOT	**0.906**	0.562	0.007	0.405	0.075	1.000				
RUNDUR	0.501	**0.965**	-0.348	0.035	0.184	0.556	1.000			
AVEDIS	0.709	-0.013	0.480	0.654	-0.095	0.680	-0.026	1.000		
PEAKDIS	0.729	0.129	0.372	**0.748**	0.041	0.699	0.150	**0.849**	1.000	
LAG	0.135	0.220	-0.217	-0.217	0.052	0.205	0.134	0.098	0.107	1.000
					Thistledowns (Residential/Commercial)					
RAINTOT	1.000									
RAINDUR	0.553	1.000								
AVEINT	0.321	-0.295	1.000							
PEAKINT	0.564	-0.104	**0.827**	1.000						
DRYPER	0.281	0.308	-0.190	-0.122	1.000					
RUNTOT	**0.903**	0.448	0.187	0.551	0.283	1.000				
RUNDUR	0.508	**0.989**	-0.322	-0.148	0.337	0.402	1.000			
AVEDIS	0.398	-0.178	0.593	0.817	-0.037	0.585	-0.227	1.000		
PEAKDIS	0.600	-0.051	0.659	**0.917**	0.009	0.702	-0.106	**0.946**	1.00	
LAG	-0.192	-0.037	-0.114	-0.202	-0.122	-0.184	-0.094	-0.138	-0.173	1.000

From Pitt, R. *Small Storm Urban Flow and Particulate Washoff Contributions to Outfall Discharges*. Ph.D. dissertation submitted to the Department of Civil and Environmental Engineering, University of Wisconsin, Madison. 1987.

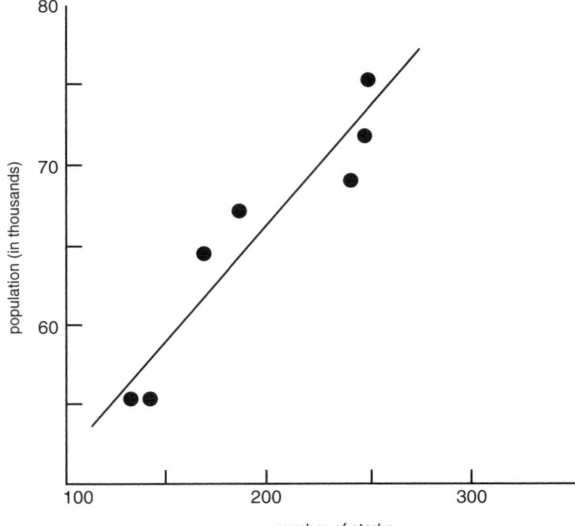

Figure 7.7   Possible cause and effect confu-
sion from correlation tests. (From
Box, G.E.P., W.G. Hunter, and
J.S. Hunter. *Statistics for Experi-
menters*. John Wiley & Sons.
New York. Copyright © 1978.
John Wiley & Sons. This material
is used by permission of John
Wiley & Sons, Inc.)

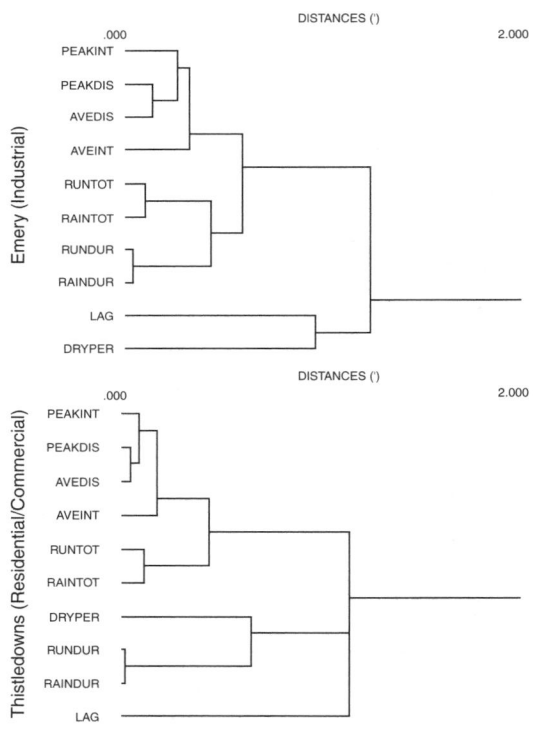

(1)   Distance metric is 1-Pearson correlation
coefficient (normalized) and the linkage method is
nearest neighbor

Figure 7.8   Tree diagram from cluster analyses of Toronto rainfall and runoff parameters. (From Pitt, R. *Small
Storm Urban Flow and Particulate Washoff Contributions to Outfall Discharges*. Ph.D. dissertation
submitted to the Department of Civil and Environmental Engineering, University of Wisconsin,
Madison. 1987. With permission.)

	%	cum		%	cum
PC1	75.4	75.4	PC3	5.2	89.4
PC2	8.8	84.2	PC4	3.8	93.2

%, percent of variance; cum, cumulative variance.

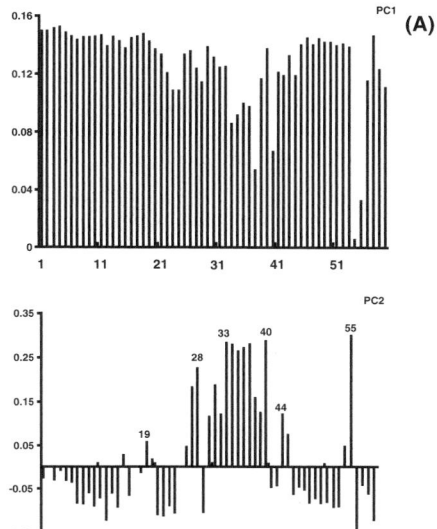

Figure 7.9A Loadings of principal components. (Reprinted with permission from Salau, J.S., R. Tauler, J.M. Bayona, and I. Tolosa. Input characterization of sedimentary organic contaminants and molecular markers in the northwestern Mediterranean Sea by exploratory data analysis. *Environ. Sci. Technol.*, Vol. 31, No. 12, pp. 3482. 1997. Copyright 1997. American Chemical Society.)

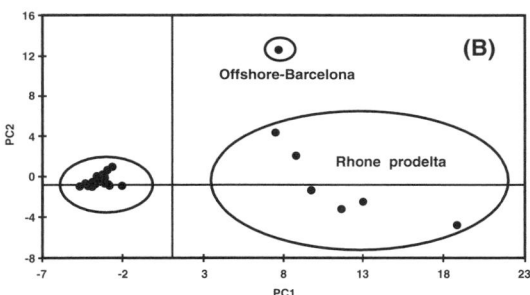

Figure 7.9B Score plots of principal components. (Reprinted with permission from Salau, J.S., R. Tauler, J.M. Bayona, and I. Tolosa. Input characterization of sedimentary organic contaminants and molecular markers in the northwestern Mediterranean Sea by exploratory data analysis. *Environ. Sci. Technol.*, Vol. 31, No. 12, pp. 3482. 1997. Copyright 1997. American Chemical Society.)

the components (usually constituents) needed to explain the variance of the data. Typically, very large numbers of constituents are available for PCA analyses and a relatively small number of sample groups are to be identified. Salau et al. (1997) used PCA (and then cluster analyses) to identify characteristics of sediment off Spain. Figure 7.9A shows the first two component loadings (collectively comprising most of the information) for 59 constituents. The first principal component (PC1) is seen to be a near reversed image of the second principal component (PC2) (if a constituent is very important in one PC, it should be much less important in the other). Figure 7.9B shows a scatterplot of PC1 vs. PC2 values for different sample locations, showing how there are three main groups of samples, which generally correspond to two sampling areas, plus a third group. The third

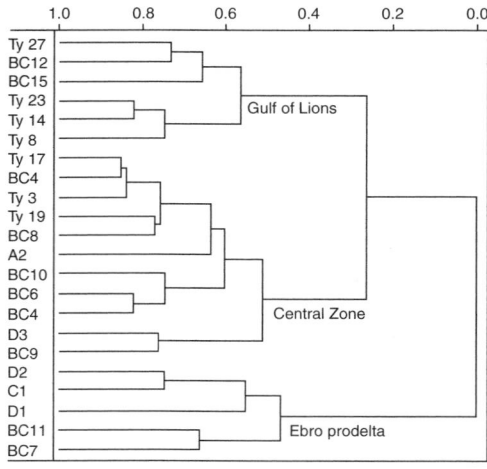

Figure 7.10 Dendogram of data, without two major groupings. (Reprinted with permission from Salau, J.S., R. Tauler, J.M. Bayona, and I. Tolosa. Input characterization of sedimentary organic contaminants and molecular markers in the northwestern Mediterranean Sea by exploratory data analysis. *Environ. Sci. Technol.*, Vol. 31, No. 12, pp. 3482. 1997. Copyright 1997. American Chemical Society.)

group was then further analyzed using cluster analysis to examine more complex groupings and sampling subareas, as shown in the dendogram of Figure 7.10.

Regression Analyses

Requirements for the Use of Regression Analyses

Regression analyses are a very popular, but commonly misused, statistical analysis tool. All statistical packages and most spreadsheets contain regression analysis routines. An excellent reference for regression analysis is *Applied Regression Analysis* by Draper and Smith (1981), while Berthouex and Brown (1994) have extensive discussions concerning misapplications and suggestions for proper use of regression analysis.

Regression analyses are best used to fit data to a theoretically derived equation that has some physical meaning. Theoretically derived equations often result in a nonlinear equation that cannot be evaluated using standard regression procedures, and many of the statistical programs available do not have any, or have only limited, nonlinear regression capabilities. Nonlinear regression analyses require assumptions and analyses steps similar to the more conventional regression analyses. Statistically based models (such as are common with stepwise regression or multiparameter polynomial regression equations) are very important and useful for many applications, but they are more limited in their transferability to other conditions and do not result in as useful understandings of the processes.

Regression models are most commonly misused when used to establish cause and effect, as illustrated in Figure 7.7, which showed an excellent correlation between stork and human populations. As described in Chapter 8, a weight-of-evidence approach (independent evaluations with a preponderance of supporting data) is typically needed to establish confidence in a proposed cause and effect relationship. Regression analyses are important components of most weight-of-evidence approaches, but they should not be overly relied upon. Besides these basic problems in objectives for conducting the test and in interpretation of regression analyses, many apply regression analyses improperly.

The following steps should be followed when conducting a regression (curve-fitting or model building) analysis:

1. Formulate the objectives of the curve-fitting exercise (a subset of the experimental design previously conducted).
2. Prepare preliminary examinations of the data, as described previously. (Most significantly, prepare scatterplots and probability plots of the data, plus correlation evaluations to examine independence between multiple parameters that may be included in the models.)

3. Identify alternative models from the literature that have been successfully applied for similar problems (part of the previously conducted experimental design activities in order to identify which parameters to measure, or to modify or control).
4. Evaluate the data to ensure that regression is applicable and make suitable data transformations.
5. Apply regression procedures to the selected alternative models.
6. Evaluate the regression results by examining the coefficient of determination (R^2) and the results of the analysis of variance of the model (standard error analyses and probability values for individual equation parameters and overall model).
7. Conduct an analysis of the residuals (as described below).
8. Evaluate the results and select the most appropriate model(s).
9. If not satisfied, it may be necessary to examine alternative models, especially those based on data patterns (through cluster analyses and principal component analyses) and to reexamine and modify the theoretical basis of existing models. Statistically based models can be developed using step-wise regression routines.

The following discussion presents the necessary assumptions and proper verification steps needed when using regression analyses. Draper and Smith (1981) list the following requirements for proper use of regression analyses:

- The residuals are independent
- The residuals have zero mean
- The residuals have a constant variance (σ^2)
- The residuals have a normal distribution (required for making F-tests)

Residuals are the unexplained variation of a model and are calculated as the differences between what is actually observed and what is predicted by the model (equation). Examination of the residuals should confirm if the fitted model is correct. The easiest method to confirm residual behavior is through graphical analyses, as described below. The examination of residuals applies to any model situation, not just regression models.

The Need for Graphical Analyses of Residuals

In all cases, graphical analysis of model residuals is necessary to confirm most of these requirements and to verify the use of the model. Berthouex and Brown (1994) list the following required residual graphical analyses for a regression model:

- Check for normality of the residuals (preferably by constructing a probability plot on normal probability paper and having the residuals form a straight line, or at least use an overall plot, as in Figure 7.11a).
- Plot the residuals against the predicted values (Figure 7.11b).
- Plot the residuals against the predictor variables (similar to Figure 7.11b).
- Plot the residuals against time in the order the measurements were made (Figure 7.11c).

Examples of these plots are shown in Figure 7.11 (Draper and Smith 1981) and in Figure 7.12 (Pitt 1987). The residuals need to be random and have the same variance for all these plots, as indicated in Figure 7.12a. If the residuals spread out (as in Figure 7.12b), then data transformations or a weighted least-squares analysis may be needed. If a trend is evident (as in Figure 7.12c), then a linear term should have been added to the model. If the residuals are curved (as in Figure 7.12d), then a higher level model (if a polynomial) may be needed.

Figure 7.13 shows a fitted regression model relating runoff volume to rain depth for 60 observations (Pitt 1987). Figure 7.13a shows the predicted and the observed runoff volumes, while Figure 7.13b is a probability plot of the model residuals. All of the 60 residuals fit the normal distribution, except for one low value and three high values. Figures 7.13c and 7.13d are plots of

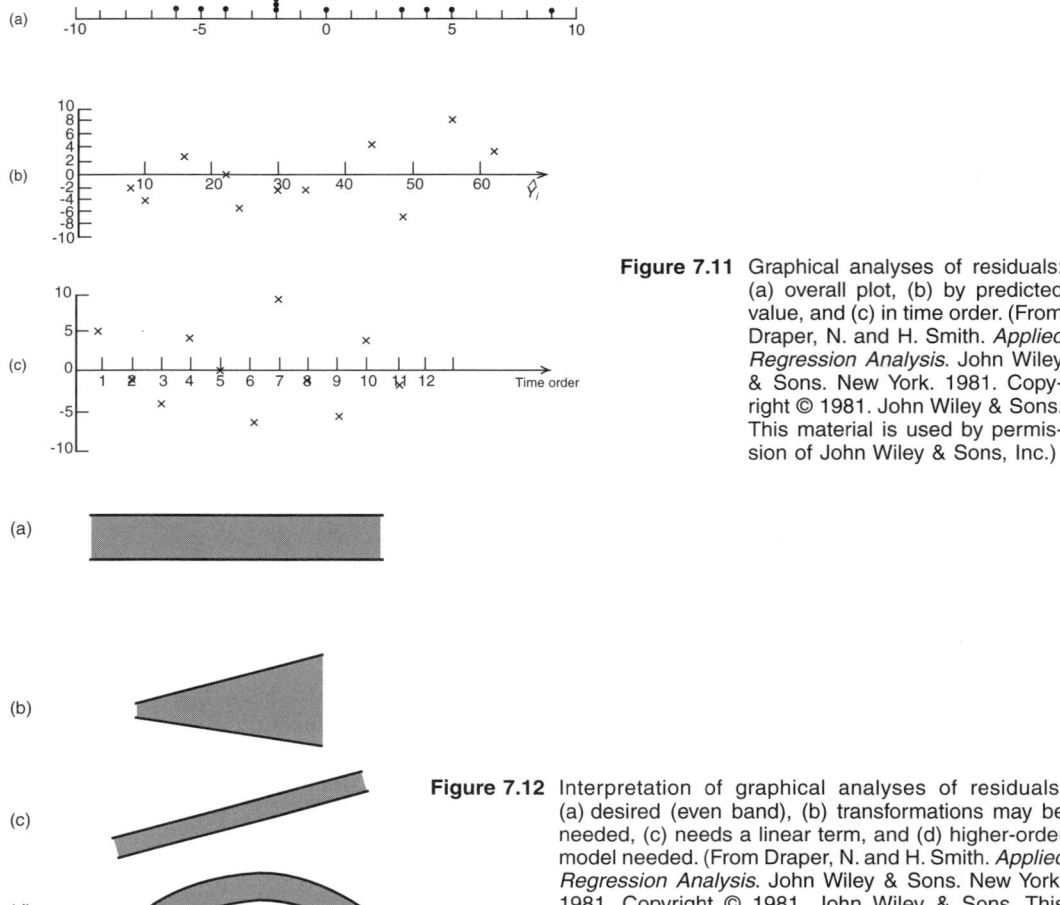

Figure 7.11 Graphical analyses of residuals: (a) overall plot, (b) by predicted value, and (c) in time order. (From Draper, N. and H. Smith. *Applied Regression Analysis*. John Wiley & Sons. New York. 1981. Copyright © 1981. John Wiley & Sons. This material is used by permission of John Wiley & Sons, Inc.)

Figure 7.12 Interpretation of graphical analyses of residuals: (a) desired (even band), (b) transformations may be needed, (c) needs a linear term, and (d) higher-order model needed. (From Draper, N. and H. Smith. *Applied Regression Analysis*. John Wiley & Sons. New York. 1981. Copyright © 1981. John Wiley & Sons. This material is used by permission of John Wiley & Sons, Inc.)

the residuals with time and against the predicted runoff volume. All observations, except for 5, fall within one standard deviation of the mean residual (zero) (as expected, since ±1 standard deviation contains about ⅔ of the data). The trends appear to be random, although there are many more observations associated with the smaller runoff volumes.

Simple lag plots should also be constructed to identify serial correlations of the residuals. Figure 7.14 (Draper and Smith 1981) shows two lag-1 serial correlation plots. To make lag-1 plots, the residuals are plotted against the preceding residual value. A lag-2 plot is prepared in a similar manner, by plotting a value against a preceding value skipping one. Different lag plots are normally prepared, although the lag-1 plot is usually the most informative. However, if daily samples are collected, sometimes lag-7 plots can be interesting by indicating some repeatable feature (such as associated with an industrial wastewater discharge), or if monthly samples are taken, lag-12 plots indicate seasonal changes. If these patterns are evident, then the model should be expanded to consider these possibly significant effects. If the resulting plot has a negative slope (as in Figure 7.14a), then the residuals are negatively serially correlated. If the resulting plot has a positive slope (as in Figure 7.14b), then the residuals are positively correlated. Both of these behaviors are undesirable for residuals because they indicate that the measurements are not independent. Serial correlation plots should be supplemented with a statistical procedure, such as the Durbin–Watson test for independence.

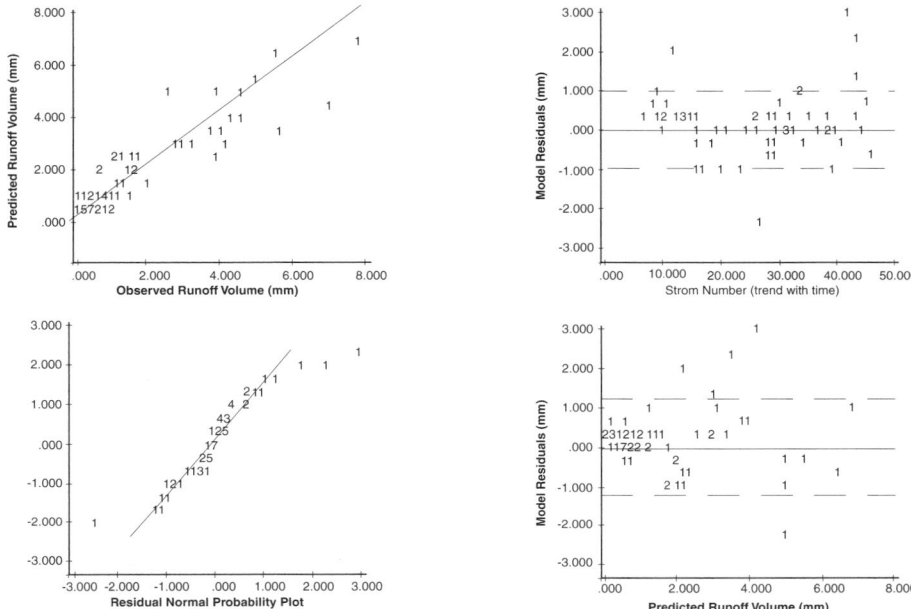

Figure 7.13 Example residual analysis for simple rainfall-runoff model. (From Pitt, R. *Small Storm Urban Flow and Particulate Washoff Contributions to Outfall Discharges.* Ph.D. dissertation submitted to the Department of Civil and Environmental Engineering, University of Wisconsin, Madison. 1987. With permission.)

Problems with Interpreting Regression Analysis Results

Berthouex and Brown (1994) present a fascinating discussion on the coefficient of determination (R^2) commonly used to "verify" a regression model. The following is a brief summary of that discussion. The coefficient of determination is the proportion of the total variability in the dependent variables that the regression equation accounts for. An R^2 of 1.0 indicates that the equation accounts for all of the variability of the dependent variables. Unfortunately, a high R^2 value, even if the model is statistically significant, doesn't guarantee that the model has any predictive value. Figure 7.15 shows plots of four data sets (from Anscombe 1973) having identical predicted regression equations with significant coefficients, the same R^2 values (0.67), and the same standard error values. However, the plots show that the relationships are vastly different from each other, stressing the need to always prepare basic scatterplots of the data and to perform residual analyses for the fitted equation (as described earlier).

Berthouex and Brown (1994) also show that having a low R^2 doesn't mean that the regression model is useless. The significance of the regression coefficients (presented in an ANOVA test of the regression equation) is highly dependent on the number of data observations. Highly significant equation coefficients are possible with a concurrent very low R^2 value if the number of data observations is large. The opposite is also true: a high R^2 value can occur with insignificant equation coefficients if only a few data observations are available. This leads to their comment that practical significance and statistical significance are not equivalent: a modest and unimportant true relationship may be established as statistically significant if a large number of observations are available. Conversely, a strong and important relationship may not be shown to be significant if only a few data are available. They therefore stress that great care needs to be exercised if a regression equation

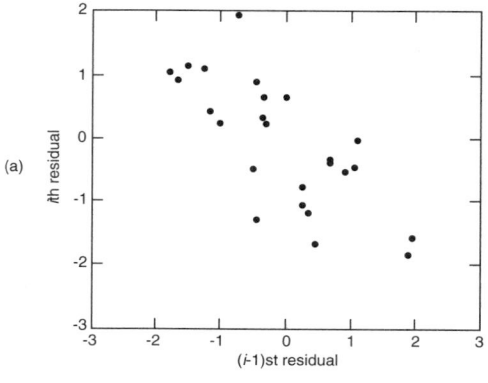

(a)

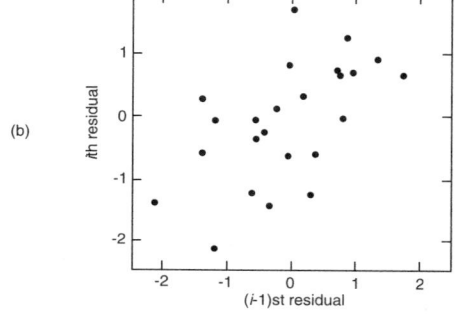

(b)

Figure 7.14 Example serial correlation analyses of residuals: (a) negative serial correlation, and (b) positive serial correlation. (From Draper, N. and H. Smith. *Applied Regression Analysis*. John Wiley & Sons. New York. 1981. Copyright © 1981. John Wiley & Sons. This material is used by permission of John Wiley & Sons, Inc.)

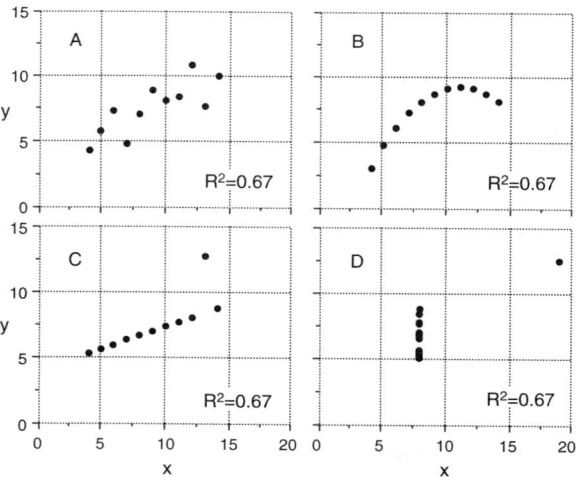

Figure 7.15 Problems when relying on coefficient of determination (R^2) to verify model. (From Anscombe, F.J. Graphs in statistical analysis. *Am. Stat.*, 27: 17–21. 1973. Reprinted with permission from *The American Statistician*. Copyright 1997 by the American Statistical Association. All rights reserved.)

is to be used for predictions because it is not possible to determine how accurate predictions will be based on the value of R^2. They strongly suggest that the model (such as a regression equation) be evaluated by: (1) examining the data and resultant model residuals graphically (as described previously), and (2) by using the standard error of the estimate (as in an ANOVA evaluation) as a more useful measure of the prediction capability of the model instead of relying only on R^2. The standard error of the estimate is computed from the variance of the predicted values using the model, so it is a more accurate indicator of the ability of the model to predict dependent variables.

Analysis of Trends in Receiving Water Investigations

The statistical identification of trends is very demanding. Several publications have excellent descriptions of statistical trend analyses for water quality data (as summarized by Pitt 1995). In addition to containing detailed descriptions and examples of experimental design methods to determine a required sampling effort, Gilbert (1987) devotes a large portion of his book to detecting trends in environmental data and includes the code for a comprehensive computer program for trend analysis. Reckhow and Stow (1990) present a comprehensive assessment of the effectiveness of different water quality monitoring programs in detecting water quality trends using EPA STORET data for several rivers and lakes in North Carolina. They found that most of the data (monthly phosphorus, nitrogen, and specific conductance values were examined) exhibited seasonal trends and inverse relations with flow. In many cases, large numbers of samples would be needed to detect changes of 25% or less (typical for stormwater retrofitting activities).

Spooner and Line (1993) present recommendations for monitoring requirements in order to detect trends in receiving water quality associated with nonpoint source pollution control programs, based on many years' experience with the Rural Clean Water Program. These recommendations, even though derived from rural experience, should also be applicable to urban receiving water trend analyses. The following is a general list of their recommended data needs for associating water quality trends with land use/treatment trends:

- Appropriate and sufficient control practices must be implemented. A high level of participation/control implementation is needed in the watershed to result in substantial and more easily observed water quality improvement. Controls need to be used in areas of greatest benefit (critical source areas, or in drainages below major sources), and most of the area must be treated.
- Control practice and land use monitoring is needed to separate and quantify the effects of changes in water quality due to the implemented controls by reducing the statistical confusion from other major factors. Monitor changes in land use and other activity on a frequent basis to observe temporal changes in the watershed. Seasonal variations in runoff quality can be great, along with seasonal variations in pollutant sources (monitor during all flow phases, such as during dry weather, wet weather, cold weather, warm weather, for example). Collect monitoring data and implement controls on a watershed basis.
- Monitor the pollutants affecting the beneficial uses of the receiving waters. Conduct the trend analyses for pollutants of concern, not just for easy, convenient parameters.
- Monitor for multiple years (at least 2 to 3 years for both pre- and post-control implementation) to account for year-to-year variability. Utilize a good experimental design, with preferable use of parallel watersheds (one must be a control and the other undergoing treatment).

Preliminary Evaluations before Trend Analyses Are Used

Gilbert (1987) illustrates several sequences of water quality data that can confuse trend analyses. It is obviously easiest to detect a trend when the trend is large and the random variation is very small. Cyclic data (such as seasonal changes) are often perceived as trends when no trends exist (Type 1 error), or they can mask trends that do exist (Type 2 error) (Reckhow and Stow 1990; Reckhow 1992). Three data characteristics need to be addressed before the data can be analyzed for trends because of confusing factors. These include:

- Measure data correlations, as most statistical tests require uncorrelated data. If data are taken close together (in time or in location), they are likely partially correlated. As an example, it is likely that a high value is closely surrounded by other relatively high values. Close data can therefore influence each other and do not provide unique information. This is especially important when determining confidence limits of predicted values or when determining the amount of data needed for a trend analyses (Reckhow and Stow 1990). Test statistics developed by Sen can use dependent data, but they may require several hundred data observations to be valid (Gilbert 1987).

- Remove any seasonal (or daily) effects or select a data analysis procedure that is unaffected by data cycles. The nonparametric Sen test can be used when no cycles are present or if cyclic effects are removed, while the seasonal Kendall test is not affected by cyclic data (Gilbert 1987).
- Identify any other likely predictable effects on concentrations and remove their influence. Normally occurring large variations in water quality data easily mask commonly occurring subtle trends. Typical relations between water quality and flow rate (for flowing water) can be detected by fitting a regression equation to a concentration vs. flow plot. The residuals from subtracting the regression from the data are then tested for trends using the seasonal Kendall test (Gilbert 1987).

Reckhow (1992) presents a chart listing specific steps that need to be taken to address the above problems. These steps are as follows:

1. Check the data for deterministic patterns of variability (such as concentration vs. flow by using graphical and statistical methods). If deterministic patterns exist, subtract the modeled pattern from the original data, leaving the residuals for subsequent seasonality analyses.
2. Examine the remaining residuals (or data, if no deterministic patterns exist) for seasonal (can be a short period, such as daily) variations. Again use graphical and statistical methods. If "seasonality" exists, subtract the modeled seasonality from the data (residuals from #1 above), leaving the remaining residuals for subsequent trend analyses.
3. Conduct the trend analysis on the residuals from #2 above, using the standard seasonal Kendall test. If a trend exists, subtract the trend, leaving the remaining residuals for subsequent autocorrelation analyses.
4. Test the remaining residuals from #3 above (or the raw data, if no deterministic or cyclic patterns or trends were found) for autocorrelation. If the autocorrelation is significant, reevaluate the trends using an autocorrelated-corrected version of the seasonal Kendall (or regular Kendall) test. If no autocorrelation was found, use the standard seasonal Kendall test if seasonality was identified, or the standard Kendall test if no seasonality was identified. The final residual variation is then used (after correcting for autocorrelation) in calculating the required number of samples needed to detect trends for similar situations.

Statistical Methods Available for Detecting Trends

Graphical Methods

Several sophisticated graphical methods are available for trend analyses that use special smoothing routines to reduce short-term variations so the long-term trends can be seen (Gilbert 1987). In all cases, simple plots of concentrations vs. time of data collection should be made. This will enable obvious data gaps, potential short-term variations, and distinct long-term trends to be possibly seen.

Regression Methods

A time-honored approach in trend analysis is to perform a least-squares linear regression on the quality vs. time plot and to conduct a t-test to determine if the true slope is not different from zero (Gilbert 1987). However, Gilbert (1987) points out that the t-test can be misleading due to cyclic data, correlated data, and data that are not normally distributed.

Mann–Kendall Test

This test is useful when missing data occur (due to gaps in monitoring, such as if waters freeze during the winters, equipment fails, or when data are reported as below the limit of detection). Besides missing data, this test can also consider multiple data observations per time period. This test also examines trends at multiple stations (such as surface waters and deep waters, etc.) and enables comparisons of any trends between the stations. This method also is not sensitive to the data distribution type. This test can be considered a nonparametric test for zero slope of water quality vs. time of sample collection (Gilbert 1987). Short-term (such as seasonal changes) cycles and other data relationships (such as flow vs. concentration) affect this test and must be corrected. If data are highly correlated, then this test can be applied to median values in each discrete time grouping.

Sen's Nonparametric Estimator of Slope

Being a nonparametric test based on ranks, this method is not sensitive to extreme values (or gross data errors) when calculating slope (Gilbert 1987). This test can also be used when missing data occur in the set of observations. It is closely related to the Mann–Kendall test.

Seasonal Kendall Test

This method is preferred to most regression methods if the data are skewed, serially correlated, or cyclic (Gilbert 1987). This test can be used for data sets having missing values, tied values, censored values (less than detection limits), or single or multiple data observations in each time period. The testing of homogeneity of trend direction enables one to determine if the slopes at different locations are the same when seasonality is present. Data correlations (such as flow vs. concentration) and dependence also affect this test and must be considered in the analysis.

The code for the computer program contained in Gilbert (1987) computes Sen's estimator of slope for each station–season combination, along with the seasonal Kendall test, Sen's aligned test for trends, the seasonal Kendall slope estimator for each station, the equivalent slope estimator for each season, and confidence limits on the slope.

Chapter 4 contains a case study of receiving water improvements with time for a Swedish urban lake after the implementation of watershed controls. The above steps were used to identify and measure nutrient and transparency improvements after stormwater control to remove phosphates was installed.

Specific Methods Commonly Used for Evaluation of Biological Data

Many of the above examples reflect water quality data analyses. However, in many areas of science, specialized tests are often used to great advantage based on specific conditions that are commonly encountered. Biological data analysis is certainly one field where some of these specialized tests are worth noting. The following discussion specifically considers toxicity data and some of the unique statistical approaches that are useful.

Typically, there are a few differences between analyzing laboratory and field (*in situ*) toxicity data. Regardless of where an evaluation takes place, the focus of any toxicity test design is to determine if environmental stressors are affecting a biological system and to what degree they are doing so. Once a test design is chosen, relevant chemical (e.g., pH, conductivity, ammonia, and turbidity, etc.) and physical (e.g., temperature, flow rate, stage, rainfall, etc.) data should always be collected throughout testing. For *in situ* biomonitoring, physical and chemical characteristics should generally be monitored each day the exposure takes place. It is recommended that initial (i.e., Day 0 or exposure commencement) and final (i.e., the final day of exposure or end of the bioassay) measurements be made at a minimum. The same approach should be made for any laboratory testing. The field conditions at the time of environmental sample (e.g., sediments, effluents, or receiving waters) collection must be monitored. Once any effluent or receiving water bioassay commences in the laboratory, daily physical and chemical measurements should be compiled. Following an exposure, for either laboratory or *in situ* experiments, routine descriptive statistics are always calculated. At a minimum, means (e.g., survival, reproduction, or growth), standard deviations, and coefficients of variation should be calculated from resulting test data. In many situations, these descriptive statistics are sufficient for making an assessment of environmental impact, especially for a short-term, one-time-only exposure. However, in most cases further statistical analysis is needed to better explain the status of a biological community. These supplemental data and descriptive statistics are usually very useful in supporting statistical analysis or conclusions.

For most laboratory-derived toxicity data, it is recommended in the USEPA chronic (1993) and acute (1995) freshwater laboratory test methods that either hypothesis testing or point estimate approaches be used for analyzing resulting endpoints (e.g., survival, growth, and reproduction). Hypothesis testing is most frequently used to determine whether one or more biological responses resulting from exposure to a particular treatment differs as compared to the control response. These statistical tests can be done when effluents, receiving waters, or sediments are tested in the laboratory, and when field sites are evaluated *in situ*. Intuitively, the control response for any exposure should be representative of the condition being evaluated. Some hypothesis testing procedures require that the experiment yield a dose response or be conducted using a dilution series (e.g., effluent and receiving water tests). For experiments with a dilution series, hypothesis tests are used to yield specific effect levels, or concentrations at which either no effect or the first detection of an effect in the testing population appears. Therefore, the effect levels are either a No Observed Effect Concentration (NOEC) or a Lowest Observed Effect Concentration (LOEC). Prior to assigning NOEC and LOEC values, an analysis of variance (ANOVA) must be conducted on test data. An ANOVA allows the investigator to determine whether treatments differ from one another statistically; however, it does not identify which group(s) are different, only that there is at least one group that is statistically different from at least one other group. If statistical significance is detected after an ANOVA, the NOEC and LOEC values can be identified using a Student's *t*-test, or an equivalent nonparametric test. The NOEC is the highest concentration not significantly different from the control and the LOEC is the first concentration that is significantly different from the control. If the data are parametric (i.e., normally distributed and homogeneous) and test replicate numbers are equal, Dunnett's test is the appropriate choice. If test replicates are unequal, a *t*-test with Bonferroni adjustment is appropriate. Nonparametric data with an equal number of replicates require Steel's Many-One Rank test, and Wilcoxon Rank Sum test if they are not.

In situ toxicity tests may represent a natural, more "realistic," exposure period but never provide the luxury of the controlled laboratory bioassay. Dose–response restrictions are rarely possible during *in situ* evaluations, and toxicity (i.e., contaminant concentrations) at field sites usually varies greatly in no particular order. Currently, no EPA guidance exists for statistical analysis of *in situ* toxicity data, but hypothesis testing can be implemented quite easily. For most *in situ* biomonitoring studies, a weight-of-evidence approach utilizing a suite of established statistical tools and scientific judgment is the general process. In many cases, it is very useful to use ANOVAs in conjunction with various *post hoc* tests for a simple and useful means to detect significant differences between sample treatments. The *post hoc* multiple comparison tests are then required to differentiate those treatments. Opinion varies widely on which *post hoc* tests are best in certain situations. However, Tukey's honest significant difference (HSD) test or Duncan's multiple range test is sufficient in most cases for defining where significance lies in the data. Both Tukey's and Duncan's compare all treatments (i.e., control and contaminated treatments) against one another and can allow one to show all significant difference for all the data. Dunnett's can also be used again as a useful *post hoc* test to detect significant differences between all the treatments and only the control. Furthermore, it is sometimes recommended that to better meet the ANOVA assumptions of homogeneity of variance and normality, transforming binomial data (i.e., survival data) is sometimes needed. Typically, the square root, log, and arcsin-square root transformation are utilized most.

Almost all point estimate data analysis is conducted on data from laboratory effluent, receiving water, or reference toxicant testing. Data used to calculate point estimates are required to have a continuous, dose-response relationship, usually a function of a dilution series. Traditionally, they allow the investigator to describe the relationship between two variables (e.g., a sample concentration and biological response), in order to relate any adverse effects of known or suspected toxicants to a concentration or dose. Point estimation results are recorded as a lethal concentration (LC) for acute toxicity tests, and effective or inhibition concentration (EC or IC) for chronic tests. An LC is usually expressed as the concentration at which there is 50% mortality in the testing population (i.e., an LC50 value). The EC and IC values are generally expressed as the concentration at which there is a 25% effect in a response, such as growth or reproduction (i.e., EC25 and IC25 values). Probit analysis is the only parametric, point estimate model where it is assumed the data are binomial (e.g., dead or alive, deformed or not) and normally distributed. For probit analysis, it is also required that there be at least two partial responses (i.e., no "all or nothing" responses). Probit effect levels are also reported as LC50 or EC50 values. A chi-square test (χ^2) for heterogeneity can be used to determine whether or not data will fit the probit model. The Spearman–Karber model is the preferred nonparametric model and yields an LC50 or EC50 value. However, no mathematical relationship for the concentration response is assumed for Spearman–Karber. A symmetrical distribution around the mean, including no response in the lowest concentration and 100% response in the highest concentration, is required for the untrimmed model, but the trimmed model is employed when the zero and/or 100% response is not met.

When a response variable or endpoint is dependent upon another variable(s), linear regression analysis may be useful. For example, for an *in situ* biomonitoring study where turbidity caused from suspended sediment is suspected of degrading water quality following storm events, numerous measurements must be taken to adequately assess impacts. After representative field sites are chosen, multiple measures of turbidity, flow, and particulate-associated contaminants throughout the exposure period would be needed. Trends can be detected by correlating the response of surrogate organisms (e.g., *Pimephales promelas*, *Hyalella azteca*, or *Chironomus tentans*) and physical or chemical measurements to strengthen a judgment of water quality and biological health of the waterway. A linear regression may be drawn between an endpoint and a single predictor variable (e.g., pH, temperature, or concentration of contaminant) in order to identify which independent variable is most closely related to the response. Multiple regression can be used to assess how an endpoint is related to multiple factors in a complex system. Linear regressions can be derived using many different functions (e.g., simple linear, exponential, hyperbolic). Least-squares estimates are used to determine the equation for the best fit line through the data, and this procedure is followed by computing the sum of squares (measures of the amount of variation in the response variable) and an ANOVA table. The ANOVA table partitions the variability of the responses and thus distinguishes what can be explained by regression and what remains unexplained (i.e., error). A large F value resulting from an ANOVA suggests that there is a significant linear relationship between the response (endpoint) and the predictor variable. However, a significant F value is not an indication that the regression equation used is the "best fit" model. Calculation of the Pearson's correlation (r), the coefficient of determination (R^2), and the coefficient of multiple correlation (in the case of multiple regression) indicate the fitness or strength of the regression. The SAS package offers a MAXR procedure for determining the best regression equation for a response variable and many predictor variables by optimizing R^2 while maintaining parsimony (i.e., yielding an equation with the fewest predictor variables). Further evaluation of the adequacy of the regression relationship is always needed through hypothesis testing (t-tests) of the equation constants (e.g., slope and intercept values), determination of confidence intervals for the response variables, and inspection of the plot of residuals. It should be noted that the above regression approach assumes only a single, or simple, interaction between expected causes and the observed effect. As described earlier, several tests that consider complex interactions (such as hierarchical cluster analyses or principal component analyses and factor analyses) may be necessary supplements to this traditional approach.

SUMMARY OF STATISTICAL ELEMENTS OF CONCERN WHEN CONDUCTING A RECEIVING WATER INVESTIGATION

This chapter briefly presented a number of tools available to the environmental researcher. These have been selected as having special utility when conducting experiments that are not easily controlled. The experimental design methods presented in Chapter 5 included simple and robust experimental designs and stressed an adequate sampling effort to help ensure successful data analyses. Various exploratory data analyses procedures have been briefly presented in this chapter, along with several cautionary examples of common problems encountered when using popular statistical methods. In almost all cases, the researcher will need to rely on the methods as presented in the references, as this discussion has been mostly descriptive. The applied statistical reference books included in the reference list comprise a fundamental library to which the environmental researcher should have access.

Exploratory data analysis is a very useful tool for preliminary evaluations of historical data needed to help design data-gathering experiments, and, it should also be used as the first step in evaluating newly collected data. The comparison of data from multiple situations (upstream and downstream of an outfall, summer vs. winter observations, etc.) is a very common experimental objective. Similarly, the use of regression analyses is also a very common statistical tool for receiving water investigations. Trend investigations of water quality or biological conditions with time are also commonly conducted. The experimental design determines the location and conditions of the sampling for these statistical objectives, but several errors are commonly made when conducting the statistical evaluations of the collected data.

In all cases, statistical analyses should not be considered a last-minute thought. Even in the best of conditions, with carefully controlled experiments and simple project objectives, it is mandatory that a general outline of the proposed statistical analysis procedures be developed before the initial experimental design is developed. It is only possible to collect adequate and sufficient data if a comprehensive objective is available and if the most appropriate statistical methods are identified. Of course, it is likely that additional analyses, or even substitutions, will be used during the final data analysis activities, and some of these modifications may require the collection of additional data that was not anticipated at the beginning of the project.

A general strategy in data analysis should include several phases and layers of analyses. Graphical presentations of the data (using exploratory data analyses) should be conducted initially. Simple-to-complex relationships between variables may be more easily identified through visual data presentations for most people, compared to relying only on descriptive statistical summaries. Of course, graphical presentations should be supplemented with statistical test data to quantify the significance of any patterns observed.

This chapter outlined several basic approaches to data analysis divided into major categories (multiple data sets, data associations, regression analyses, and trends) that are generally of the most interest in receiving water assessments. There is a great number of statistical references, software products, and consultants available to assist the data analyst. Several are presented in this chapter for additional information.

REFERENCES

Anscombe, F.J. Graphs in statistical analysis. *Am. Stat.*, 27: 17–21. 1973.

Berthouex, P.M. and L.C. Brown. *Statistics for Environmental Engineers.* Lewis Publishers, Boca Raton, FL. 1994.

Box, G.E.P., W.G. Hunter, and J.S. Hunter. *Statistics for Experimenters.* John Wiley & Sons. New York. 1978.

Cleveland, W.S. *Visualizing Data.* Hobart Press, Summit, NJ. 1993.

Cleveland, W.S. *The Elements of Graphing Data.* Hobart Press, Summit, NJ. 1994.

Draper, N. and H. Smith. *Applied Regression Analysis.* John Wiley & Sons. New York. 1981.

Ellersieck, M.R. and T.W. La Point. Statistical analysis, in *Fundamentals of Aquatic Toxicology: Effects, Environmental Fate and Risk Assessment, 2nd edition.* Edited by G.M. Rand. pp. 307–344. 1995.

EPA. *Methods for Measuring the Acute Toxicity of Effluents and Receiving Water to Freshwater and Marine Organisms.* Office of Research and Development, Cincinnati, OH. EPA/600/4-90/027F. 1993.

EPA. *Short-term Methods for Estimating the Chronic Toxicity of Effluents and Receiving Water to Freshwater Organisms.* Research and Development. Cincinnati, OH. EPA/600/4-91/002. 1995.

Gilbert, R. O., *Statistical Methods for Environmental Pollution Monitoring*. Van Nostrand Reinhold, New York. 1987.

Glantz, S.A. *Primer on Biostatistics*. McGraw-Hill, New York. 1992.

Grothe, D.R., K.L. Dickson, and D.K. Reed-Judkins. Whole effluent toxicity testing: an evaluation of methods and prediction of receiving system impacts. *Proceedings of the Pellston Workshop on Whole Effluent Toxicity*, 16–21 September 1995. SETAC Press. 346 pp. 1996.

Hunter, J.S. The digidot plot. *Am. Stat.*, 42: 54. 1988.

Lehman, E.L. and H.J.M. D'Abrera. *Nonparametrics: Statistical Methods Based on Ranks*. Holden-Day and McGraw-Hill, New York. 1975.

Newman, M. C. *Quantitative Methods in Aquatic Ecotoxicology*. CRC Press, Boca Raton, FL. 426 pp. 1995.

Pitt, R. *Characterizing and Controlling Urban Runoff through Street and Sewerage Cleaning*. U.S. Environmental Protection Agency, Storm and Combined Sewer Program, Risk Reduction Engineering Laboratory. EPA/600/S2-85/038. PB 85-186500. Cincinnati, OH. 467 pp. June 1985.

Pitt, R. *Small Storm Urban Flow and Particulate Washoff Contributions to Outfall Discharges*. Ph.D. dissertation submitted to the Department of Civil and Environmental Engineering, University of Wisconsin, Madison. 1987.

Pitt, R. Water quality trends from stormwater controls, in *Stormwater NPDES Related Monitoring Needs*. Edited by H.C. Torno. Engineering Foundation and ASCE, New York. pp. 413–434. 1995.

Pitt, R. and S.R. Durrans. *Drainage of Water from Pavement Structures*. Alabama Department of Transportation. Research Project 930-275. Montgomery, AL. Sept. 1995.

Reckhow, K.H. and C. Stow. Monitoring design and data analysis for trend detection. *Lake Reservoir Manage.*, 6(1): pp. 49–60, 1990.

Reckhow, K.H., K. Kepford, and W. Warren-Hicks. *Methods for the Analysis of Lake Water Quality Trends*, School of the Environment, Duke University. Prepared for the U.S. Environmental Protection Agency. October 1992.

Salau, J.S., R. Tauler, J.M. Bayona, and I. Tolosa. Input characterization of sedimentary organic contaminants and molecular markers in the northwestern Mediterranean Sea by exploratory data analysis. *Environ. Sci. Technol.*, 31(12): 3482. 1997.

Sokal, R.R and F.J. Rohlf. *Biometry: The Principles and Practice of Statistics in Biological Research*. W.H. Freeman and Co., New York. 1969.

Spooner, J. and D.E. Line. Effective monitoring strategies for demonstrating water quality changes from nonpoint source controls on a watershed scale. *Water Sci. Technol.*, 28(3–5): 143–148, 1993.

Tufte, E. R. *The Visual Display of Quantitative Information*. Graphics Press, Cheshire, CT. 1983.

Tufte, E. R. *Envisioning Information*. Graphics Press, Cheshire, CT. 1990.

Tufte, E. R. *Visual Explanations*. Graphics Press, Cheshire, CT. 1997.

Tukey, J. W. *Exploratory Data Analysis*. Addison-Wesley, Reading, MA. 1977.

Zar, J.H. *Biostatistical Analysis, 3rd edition*. Prentice-Hall, Englewood Cliffs, NJ. 662 pp. 1996.

## CHAPTER 8

Data Interpretation

"If you get all the facts, your judgment can be right; if you don't get all the facts, it can't be right."

Bernard M. Baruch

CONTENTS

Is There a Problem?..609
Evaluating Biological Stream Impairments Using the Weight-of-Evidence Approach...............611
    The Process...611
    Benchmarks 612
    Ranking and Confirmatory Studies...615
    Comments Pertaining to Habitat Problems and Increases in Stream Flow 617
Evaluating Human Health Impairments Using a Risk Assessment Approach 619
    Deterministic Approach...619
    Probabilistic Approach 619
    Example Risk Assessment for Human Exposure to Stormwater Pathogens 620
Identifying and Prioritizing Critical Stormwater Sources...626
    Sources of Urban Stormwater Contaminants 626
    Case Study: Wisconsin Nonpoint Source Program in Urban Areas 628
    Use of SLAMM to Identify Pollutant Sources and to Evaluate Control Programs 629
Summary 636
References 637

IS THERE A PROBLEM?

Unit 1 (Chapters 1 through 3) described problems associated with stormwater runoff. Unit 2 (Chapters 4 though 8) described the development of appropriate experimental designs that included selecting the components of the assessment process and determining an appropriate level of effort, plus specific sampling and monitoring activities to assess receiving water impacts. Unit 3 (the appendices) includes additional guidance on conducting specific field activities. There are numerous case study examples throughout these chapters showing how the recommended approach has functioned during previous successful projects. In this concluding chapter, these important issues are highlighted for the data interpretation process. Now that an assessment has been conducted, how does one determine whether or not the receiving waters are impaired and, if so, what is the source, or sources, of the impairment?

As indicated in Chapters 2 and 3, there is a variety of receiving water problems that may be associated with stormwater. The specific problems in any area are dependent on many site conditions and objectives. There are many documented cases, previously described, where stormwater has caused detrimental impairments on receiving water uses and goals. Probably the most common problem is associated with stormwater conveyance (flood prevention) caused by increased amounts of pavement in the drainage area. The increased flows, however, are also responsible for many habitat problems related to the increased stream power and associated unstable stream environment. Other common receiving water problems in urban waters are associated with noncontact recreation (linear parks, aesthetics, boating, etc.). The seemingly simple task of preventing floatable debris from being discharged can be very difficult to accomplish. Much of this book has addressed environmental health issues associated with biological uses (warm-water fishery, biological integrity, etc.). In addition to the habitat destruction problems associated with increased flows and increased stream power, contaminated sediment may be a significant causative agent affecting biological uses. Poor water quality obviously can also significantly affect most of the above uses, in addition to interfering with water contact recreation (swimming) and water supply uses. It is unlikely that these human health uses would be appropriate in any waterway located in a heavily urbanized watershed.

The study design is dependent on the expected problems likely to be encountered (see also Chapter 4). Without having that information at the beginning of a study, the initial list of parameters to be monitored has to be based on best judgment. The parameters to be monitored can be grouped into general categories, depending on expected beneficial use impairments, as follows:

- Flooding and drainage: debris and obstructions affecting conveyance are parameters of concern.
- Biological life/integrity: habitat destruction, high/low flows, taxonomic composition of existing aquatic life, inappropriate discharges, polluted sediment (texture, SOD, and toxicants), and wet weather quality (toxicity, bioaccumulation, toxicants, nutrients, DO) are key parameters.
- Noncontact recreation: odors, trash, high/low flows, aesthetics, and public access are the key parameters.
- Swimming and other contact recreation: pathogens, and above listed noncontact parameters, are key.
- Water supply: water quality standards (especially pathogens and toxicants) are key parameters.
- Shellfish harvesting and other consumptive fishing: pathogens, toxicants, and those listed under biological life/integrity, are key parameters.

Obviously, there are definite problems in receiving waters that will dictate many components of the sampling program and measures against which the data are to be compared. These problems may be minor if the watershed is relatively undeveloped, but they can be extreme for fully developed urban or agricultural areas. In addition, local use objectives also dramatically affect the definition of a "problem." In all cases, however, basic receiving water objectives should include safe drainage, noncontact recreation (acceptable aesthetics), and basic biological life objectives. It is unlikely that contact recreation or biological integrity, with the stream being able to support a full mixture of native organisms, would be reasonable receiving goals in a fully developed urban or agricultural watershed.

The information and guidance provided in this book should enable a researcher to investigate local conditions to identify local use impairments and to identify the most likely causes of these problems. Depending on the magnitude of the effort expended and the clarity of the problems in the local area, it may also be possible to quantify the magnitude of stream use improvements with different levels of reduction of the causative agent. Once the causes and sources of the problems are identified, choices pertaining to improvement, or prevention measures in other areas, can be examined.

The following sections outline the concept of "weight-of-evidence" as a tool to assemble a large amount of data to help in obtaining needed information pertaining to environmental health. An example risk assessment is also provided to show how risks associated with exposures to humans can be examined.

EVALUATING BIOLOGICAL STREAM IMPAIRMENTS
USING THE WEIGHT-OF-EVIDENCE APPROACH

The Process

The term "weight-of-evidence" (WOE) has been used frequently during the past several years in the environmental assessment arena. However, there is no clear definition or approach accepted, and approaches have varied from those that are crude and qualitative to very complex and quantitative. As discussed in Chapter 4, no one assessment approach is adequate for drawing conclusions on the quality of a waterway because of the associated uncertainties and weaknesses of each approach. Therefore, there is now widespread acceptance that multiple approaches (lines of evidence) are essential in order to reach reliable conclusions of whether a problem exists. Using the WOE approach, however, does not ensure that accurate conclusions will be obtained. It is critical that a well-designed assessment design be used (see Chapter 4) and that the key ecosystem components (biological, chemical, and physical) be characterized correctly, noting their associated uncertainties. The following discussion presents useful approaches for WOE evaluations.

One of the first WOE approaches to gain widespread attention was the "sediment quality triad" (Chapman et al. 1987). In this approach, sediment toxicity, indigenous biota, and sediment chemistry were characterized at each test site and normalized as a percentage of the reference (background) site condition. Results were presented graphically in an X-Y-Z axis type format. Comparing test site conditions to reference sites has long been used, but the primary contribution of the triad was to promote the notion that components must be assessed together. In stormwater assessments, the triad approach should be expanded to include the physical conditions (i.e., habitat), water and sediment conditions, and the associated temporal dynamics of each assessment component (Table 8.1).

Unfortunately, it is difficult to make quantitative evaluations of significant differences from this original "triad" approach. The comparisons between sites are particularly difficult at intermediate levels of contamination or if significant variability exists in the monitoring data. This "weight-of-evidence" approach can be evaluated using both parametric and nonparametric procedures to address the following study objectives: which stations are significantly different (impacted) relative to other stations?; how do the stations relate to each other?; and which parameters (monitoring components)

Table 8.1 Summary of Key Weight-of-Evidence Components for Assessing Stormwater Effects on Receiving Waters

Component	Media	Priority	Flow Level	Difficulty[a]
Benthic community	Sediment	High	Low	Low
Fish community	Water	Medium	Low	Medium
Toxicity				
Lab-based	Sediment	Medium	Low	Low–Med.
	Water	High	Low and High	Low–Med.
In situ-based	Sediment	High	Low	Low–Med.
	Water	High	Low and High	Low–Med.
Bioaccumulation				
Benthic species	Organism	Medium	Low	Med.–High
Fish species	Organism	Medium	Low	Med.–High
In situ passive	Water	Low	Low and High	Med.–High
Chemistry (metals,	Sediment	High	Low	Med.–High
organics, conventional	Water	High	Low and High	Med.–High
physicochemistry)				
Physical				
Flow	Water	High	Low and High	Low
Habitat	Whole stream	High	Low	Low

[a] Difficulty rating considers both level of effort and cost to measure by typical approaches described in Appendices.

are significantly different (impacted) relative to other stations? Initial exploratory data analyses should be used to identify relationships between variables, identify and rank important variables, and identify weighting factors or redundant variables (i.e., responses mimic each other). These analyses may include correlation analyses, scatterplots, and other ordination tests. Results can also be ranked, whereby endpoint measures are averaged at each station and stations are then ranked by performance. Sample average ranks can be compared to a critical value to determine if significant differences exist between stations. Ordination procedures can be used to determine distances among stations and endpoints (e.g., multidimensional scaling). Scatterplots will show similarity of ranked groups and the magnitude of relationships among measured endpoints.

The Massachusetts Department of Environmental Protection formalized the WOE approach (Menzie et al. 1996) for relating measurement endpoints to assessment endpoints in ecological risk assessments. They identified three major components:

1. Weight assigned to each measurement endpoint: measurement endpoints (e.g., mortality, growth) may vary in the degree they relate to the assessment endpoints, or their quality, and may therefore be assigned differing levels of weight (i.e., importance).
2. Magnitude of response in the measurement endpoint: a greater weight is assigned to strong responses.
3. Concurrence among measurement endpoints: there tends to be greater confidence in findings that agree with other lines of evidence. However, disagreement between components does not negate their validity or importance. For example, aquatic species have varying levels of sensitivity to different chemicals, or sampling may induce artifacts. Concordance of findings is more likely when very high levels of contamination are present, causing acute toxicity, as opposed to lower chronic toxicity exposures.

Numerical weighting values (e.g., 1 to 3 or 1 to 5) are assigned to elements of the process via professional judgment. This weighting of relative importance of the various tools has been done using Delphi techniques where a group of environmental professionals is surveyed. For example, each measurement endpoint (such as species population number) could be rated as high, medium, or low for three attributes (strength of relationship to an assessment endpoint, such as fish catch, data quality, and study design). These three attribute ratings may then be summed to get an overall measurement weight (of 1 to 3). The reliability of this best professional judgment approach is obviously related to the quality and comprehensive expertise of the survey group. After weighting values are assigned, measured responses are multiplied by their respective weights and summarized. The evidence showing the relationship between exposure to a stressor and a biological response (e.g., an assessment endpoint) is then assessed for risk. *This leads the assessor to the most critical point of the assessment where the question is asked: what is the relationship between exposure to the stressor of concern (e.g., suspended solids, zinc, pesticides, stormwater) and adverse biological effects?* The WOE process will help answer this question. While the WOE is the preferred approach, it is not without its shortcomings. Aside from not being a simple standardized protocol, the WOE is also not strictly quantitative, requiring best professional judgment. Statistically significant differences and relationships cannot readily be determined for the overall, integrated process. Certainty and accuracy are ensured via greater weight that is obtained through sound, comprehensive, integrated assessments. More importantly, as the WOE process is used in an area, it becomes "calibrated" through experience and observation and can become fine-tuned to better represent actual changes that may be occurring.

Benchmarks

In the process of interpreting exposure and effect relationships, there are a number of tools that can be used, ranging from "benchmarks," or deterministic approaches, to probabilistic methods. These are discussed in the following sections. Benchmarks refer to concentrations or levels of physical and chemical parameters above which adverse biological effects may occur. These are often derived from large scientific databases linking biological responses with exposures to

Table 8.2   Categories of Biological Impairment Benchmarks

Regional or National Water/Sediment Quality Criteria or Standards
State, Provincial, or Regional Water Quality Standards
Biological Criteria
Threshold (Toxicity) Effect Levels for Water, Sediment, or Tissues
Hazard Quotients (Threshold Level or Site Concentration vs. No Effect Level, Reference or Background Site)
Percentile Distributions
Statistical Significance of Test vs. Control or Reference

compounds. Examples of commonly used benchmarks are listed in Table 8.2. Specific benchmarks/criteria for water and sediment criteria and biota are also discussed in Appendices B, C, and G. For each of these benchmark categories, there exists chemical specific benchmarks calculated by a variety of methods. These methods vary in the amount of biological effect (toxicity) information they include, ranging from only acute toxicity information on one species, to acute and chronic toxicity on many species. In addition, the toxicity information generated in these benchmarks ranges from a site-specific nature to being applicable to large geographical areas (such as north America). As with any assessment tool, each has associated uncertainties that should be recognized and considered by the assessor. The optimal approach is to use multiple benchmarks to better ascertain whether impairment exists.

The most important issue to remember when using benchmarks is that they are simply "benchmarks" to use in the chemical-physical data interpretation process. They do not unequivocally determine whether adverse effects are occurring. Often, these benchmarks do not include site-specific biological effects data. In addition, the biological effect benchmarks may not be applicable to the conditions at your study site. For example, a suite of stressors may exist at your sites that interact to produce antagonistic or synergistic effects or conditions may alter the bioavailability of the chemical of concern. However, the use of multiple benchmarks that have been derived from large, scientifically valid, databases will assist in the weighting and data interpretation process.

The optimal method of establishing a relationship between biological effects and a site-specific parameter(s) is to thoroughly characterize exposure and effects. Benchmarks, unfortunately, only suggest that effects may be occurring if they are exceeded. If exceeded, they should at least be treated as "red flags," emphasizing areas where additional investigation is warranted. They do not address spatial and temporal variability or site-specific interactions. This requires carefully designed biological and toxicity studies during low and high flow conditions (Chapters 4 and 6). In the absence of site-specific effects data, use of probability modeling is preferred if adequate site data exist for determining spatial and temporal exposure-effects interactions (see Ecological Risk Assessment section below).

Perhaps the best recognized and accepted benchmarks are the U.S. EPA's National Ambient Water Quality Criteria (see Appendix G), which many states have adopted as their ambient water quality standards. The results of stormwater quality analyses have commonly been compared to water quality criteria in order to identify potentially toxic waters, and likely problematic pollutants. This has led to numerous problems with the interpretation of the data, especially concerning the "availability" of the toxicants to receiving water organisms and the exposure durations in receiving waters. The quality of stormwater, or of ambient waters immediately following high flow events, has been shown to be degraded in many studies with chemical concentrations that may exceed toxicity thresholds (e.g., Horner et al. 1994; Makepeace et al. 1995; Morrison et al. 1993; Waller et al. 1995). Stormwater toxicants are primarily associated with particulate fractions and are typically assumed to be "unavailable." Typically short and intermittent runoff events can also not be easily compared to the "long" duration criteria or standards. Chemical analyses, without biological analyses, would have underestimated the severity of the problems because the water column quality varied rapidly, while the major problems were associated with sediment quality and effects on macroinvertebrates (Lenat and Eagleson 1981; Lenat et al. 1981).

The contradictions noted between in-stream biological effects and water quality criteria should not be surprising, given the assumptions used by the EPA:

1. Single acute and chronic average exposure period that does not account for pulse or repeated exposures for short time periods
2. Single bioavailability normalization factors (such as hardness)
3. Laboratory-derived toxicity values for surrogate species are protective of indigenous species
4. Effects derived from single chemical exposures in clean solutions where the toxicant is in the dissolved form
5. Chemical exposures in the field based on limited grab sample analyses

To address magnitude and duration issues, the EPA developed the "Criterion Maximum Concentration" concept, with an exposure period assumption of 1 hour, and the "Criterion Continuous Concentration," with an average period assumption of 4 days. Yet, these assumptions do not accurately describe most wet weather runoff exposures. Tests with pentachloroethane (Erickson et al. 1989, 1991) showed that with intermittent exposures, higher pulse concentrations were needed to affect growth, and when averaged over the entire test, effects were elicited at concentrations lower than when under constant exposure. The simplest toxicity model (with first-order, single-compartment toxicokinetics and a fixed lethal threshold) could not completely describe the data. Erickson et al. (1989) concluded that kinetic models which predict mortality were reasonable; however, chronic toxicity effects were much more complicated, and no adequate models existed. Hickie et al. (1995) describe a one-compartment, first-order kinetics, pulse exposure model for residue-based toxicity of pentachlorophenol to *P. promelas*. Pulse exposures were of 2 min to 24 hours with durations of 2 to 24 hours, repeated 2 to 15 times. A comparison of three models (Cxt, Mancini, Breck 3 dimensional range repair) showed reasonable prediction of fish toxicity following 1 to 4 monochloramine pulses (2-h pulse, 22-h recovery). However, predictive capability decreased with greater than 4 pulses (Meyer et al. 1995). Beck et al. (1991) examined the transient nature of receiving water effects associated with stormwater, stressing the weaknesses associated with more typical steady-state approaches. They felt that there were still major misconceptions associated with modeling these effects.

Despite these limitations, water quality criteria and standards have been used effectively to identify potential stormwater problems and direct further assessment studies (see Chapter 6). Nonetheless, it is apparent that the use of water quality criteria to identify potential receiving water problems should be done with care. In many cases, the most direct comparison is made for concentrations of the soluble forms of the pollutants only and to use the short-term acute exposure criteria. This seems to be the most conservative approach, and if any measured pollutant exceeds this critical value, a problem pollutant is easily flagged. However, this approach is fraught with false negatives, as many chronic problems may still exist that are not recognized. As an example, numerous in-stream receiving water investigations (described in Chapter 3) have identified severe problems (indicated by lack of sensitive species) where the measured water quality met the criteria. Because the toxicants are strongly associated with particulates, secondary sediment contamination occurs that may be more important than water column conditions for aquatic life effects. In addition, habitat degradation caused by urbanization and agricultural activity (including highly fluctuating flows) are also likely responsible for many of the recognized receiving water problems. Finally, the irregular, but frequent, exposures of pollutant concentrations lower than the criteria may cause a greater problem than relatively constant, but higher, concentrations (see also Chapters 4 and 6). Therefore, direct comparisons of water quality criteria with monitored in-stream concentrations should be carefully conducted and used as adjuncts to direct in-stream biological use observations, plus evaluations of habitat and sediment quality. Human health criteria (such as pathogens for water-contact recreation and toxicants for drinking water supplies of fish/shellfish consumption) are more applicable to wet weather conditions and can be more directly used to flag potential problem pollutants.

Table 8.3 NURP Reported Median and 90th Percentile Event Mean Concentrations (EMC) (mg/L, unless otherwise noted) for Urban Runoff

Constituent	Median Urban Site EMC	Event to Event Variability in EMC (COV)	90th Percentile Urban Site EMC
Suspended solids	100	1–2	300
BOD$_5$	9	0.5–1.0	15
COD	65	0.5–1.0	140
Total P	0.33	0.5–1.0	0.70
Soluble P	0.12	0.5–1.0	0.21
TKN	1.5	0.5–1.0	3.3
NO$_2$ + NO$_3$ (as N)	0.68	0.5–1.0	1.8
Total copper (µg/L)	34	0.5–1.0	93
Total lead (µg/L)	144	0.5–1.0	350
Total zinc (µg/L)	160	0.5–1.0	500

From EPA (U.S. Environmental Protection Agency). *Final Report for the Nationwide Urban Runoff Program.* Water Planning Division, Washington, D.C. December 1983.

Appendix G presents a summary of the human health and aquatic health criteria for pollutants that commonly occur in urban runoff and receiving waters. Most of the criteria are expressed with a recommended exceedance frequency of 3 years. This is the EPA's best scientific judgment of the average amount of time it will take an unstressed system to recover from a detrimental event in which exposure to the pollutant exceeds the criterion. A stressed system, for example, one in which several outfalls occur in a limited area, would be expected to require more time for recovery. Obviously, if criteria are exceeded for most rain events (such as can occur for bacteria and total recoverable heavy metals), then 3 years are not available for recovery before the next runoff event.

The discussions on the effects of the pollutants on aquatic life and human health presented in Appendix G are summarized from the U.S. EPA's *Quality Criteria for Water, 1986* (EPA 1986). The criteria were also reviewed using the EPA's web page (http:/www.epa.gov) on the Internet for more recent changes. Some minor changes have been made since 1986 (chloride standards, for example, in 1988). Numeric criteria for heavy metals have been proposed as part of the states' Compliance for Toxic Pollutants (for the nine states subject to EPA's 1992 National Toxics Rule) as an interim rule. In most cases, only the short-term criteria are applicable for wet-weather receiving water conditions. Most runoff events last only a few hours; very few last for several days. However, degraded in-stream conditions can occur for several times the duration of the rain event itself. In addition, frequent exposures to concentrations less than the critical short-term criteria may result in significant problems that would not be predicted based on these criteria alone.

In some instances, acceptable stormwater concentration guidelines may be based on typical data as obtained during the Nationwide Urban Runoff Program (NURP) (EPA 1983). These data were almost solely represented by medium-density residential area runoff, with some data from other areas (such as shopping centers and light industrial areas). Useful benchmarks include the event mean concentrations, or EMC, (average of all observed concentrations) and the 90th percentile values of common parameters as measured during NURP (Table 8.3). The 90th percentile values are sometimes used as an upper limit for acceptable concentrations.

Ranking and Confirmatory Studies

If an adequate stormwater runoff study design is implemented (Chapter 4) and the weight-of-evidence process followed with the ensuing monitoring data, then sound decisions can be made. In reality, few comprehensive stormwater assessments look at all possible stressors and species of concern while characterizing the spatial and temporal dynamics of the system. Most environmental assessments are resource limited, requiring a tiered approach, where potential problem sites are identified, ranked, and then decisions made as to what further assessments are necessary. This is

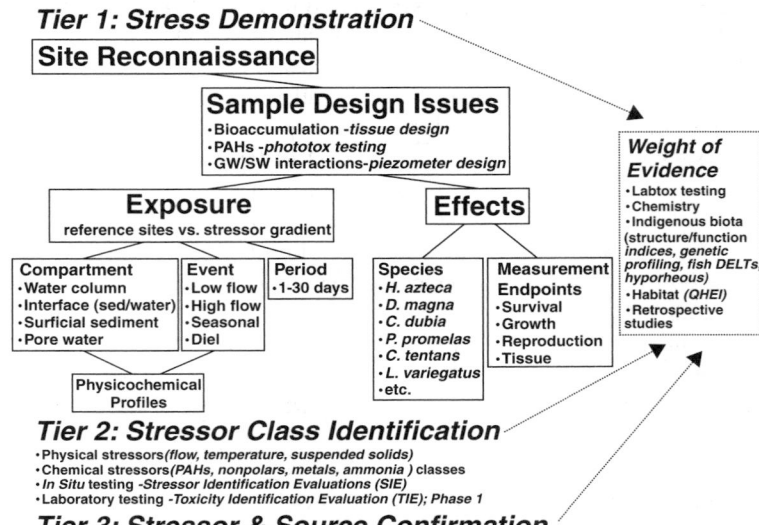

Figure 8.1 Example of a tiered weight-of-evidence approach used by Wright State University to evaluate stormwater runoff and aquatic ecosystem contamination.

particularly true for stormwater studies, where historical funding mechanisms do not exist and where the watershed-receiving water relationships are complex. It is important to be realistic in the expectations of initial screening studies. The goals should be to simply rank problem sites through the WOE approach (see above WOE discussion). Then, follow-up confirmatory (Tier 2) studies can focus on fewer sites, allowing for more quantitative characterization of the temporal dynamics and resulting effects of runoff events (Figure 8.1) (see also Chapter 4 example outline of a comprehensive runoff effect study).

For example, an approach used to identify stressors in aquatic ecosystems used by Wright State University is shown in Figure 8.1. During initial site reconnaissance, a determination is made as to whether three common sample design issues need to be incorporated: (1) Do pollutants (such as PCBs) that readily bioaccumulate likely occur? (2) Do PAHs likely occur? and/or, (3) Are there likely groundwater–surface water transition zones occurring in the area of contamination? If any of these three issues are present, then the typical Tier 1 sampling design may be modified to include: tissue residue or bioaccumulation testing, phototoxicity evaluations, and piezometer measures of groundwater movement (with concurrent chemical and toxicity testing of those compartments). The typical Tier 1 design will involve toxicity testing of two to four species which are exposed to three to four compartments (overlying water, sediment-water interface, surficial sediment, pore water) during low flow. At high flows, these same species are exposed to overlying waters. During their exposures, basic water quality measures are made, such as DO, conductivity, alkalinity, hardness, pH, temperature, turbidity (or TSS), and ammonia. If toxic effects are noted following these exposures, then Tier 2 testing may commence to better identify the type of stressor. Tier 2 testing may then require more in-depth chemical analyses and try to separate out stressors such as ammonia, metals, and nonpolar organics. Finally, in Tier 3, the focus can be to determine the significance of the dominant stressor(s) via a WOE approach.

The WOE process lends itself easily to ranking sites — particularly using broad categories such as high, medium, and low priority. For instance, this may separate sites that have acute toxicity and few pollution-sensitive benthic organisms from those with possible chronic toxicity and marginal benthic communities. The decision maker may then choose to immediately pursue installation of stormwater controls at the worst site, while conducting confirmatory studies at the marginal site to establish the extent and/or cause of the problem. Confirmatory studies are frequently necessary to establish the: (1) dominant stressor(s); (2) exposure pathways/dynamics; (3) receptor organisms; (4) food web interactions; (5) environmental risk (human and ecological significance of effects); and (6) stressor sources. Confirmatory or Tier 2 studies are designed to answer very focused questions and use many of the same tools described for the more routine stormwater assessments. However, as the questions may be more focused, more specific and novel assessment techniques may be employed, such as DNA fingerprinting (RAPD PCR), toxicity identification evaluations (TIEs), or SPMDs (all described in Chapter 6). The environmental quality of many of our agricultural and urban waterways will also be less than pristine where anthropogenic influences are minimal. Therefore, the issue of whether significant ecological impairment exists will be more of a challenge in these human-dominated watersheds. The point of comparison for determinations of impairment should be an appropriate ecoregion reference or criteria, where manageable stressors have been removed (such as high temperature, erosion, pesticides, lack of riparian zone).

Comments Pertaining to Habitat Problems and Increases in Stream Flow

Habitat changes due to urbanization and agricultural activities are likely the cause for much of the degradation noted in biological conditions in streams. Appendix A outlines habitat evaluation schemes, while Chapter 6 also included descriptions on characterizing habitat. Understanding the effect that habitat has on stream biological uses is very important if these changes are to be minimized. This understanding needs to come from detailed local investigations, as our ability to predict habitat changes associated with stormwater discharges is rather poor. With site studies, some researchers have been able to recommend local guidelines to minimize habitat degradation.

MacRae (1997) found that stream bed and bank erosion is controlled by the frequency and duration of the mid-depth flows (generally occurring more often than once a year), not the bank-full condition (approximated by the 2-year event). During monitoring near Toronto, he found that the duration of the geomorphically significant predevelopment mid-bankfull flows increased by a factor of 4.2, after 34% of the basin had been urbanized, compared to before-development flow conditions. The channel had responded by increasing in cross-sectional area by as much as three times in some areas, and was still expanding. He also reported other studies that found channel cross-sectional areas began to enlarge after about 20 to 25% of the watershed was developed, corresponding to about a 5% impervious cover in the watershed. When the watersheds are completely developed, the channel enlargements were about five to seven times the original cross-sectional areas. Changes from stable stream bed conditions to unstable conditions appear to occur with basin imperviousness of about 10%, similar to the value reported previously for serious biological degradation. MacRae concluded that an effective criterion to protect stream stability (a major component of habitat protection) must address mid-bankfull events, especially by requiring similar durations and frequencies of stream power at these depths, compared to satisfactory reference conditions.

Much research on habitat changes and rehabilitation attempts in urban streams has occurred in the Seattle area of western Washington over the past 20 years. Sovern and Washington (1997) described the in-stream processes associated with urbanization in this area. The important factors that affect the direction and magnitude of changes in a steam's physical characteristics due to urbanization include:

- The depths and widths of the dominant discharge channel will increase directly proportional to the water discharge. The width is also directly proportional to the sediment discharge. The channel width divided by the depth (the channel shape) is also directly related to sediment discharge.
- The channel gradient is inversely proportional to the water discharge rate and is directly proportional to the sediment discharge rate and the sediment grain size.
- The sinuosity of the stream is directly proportional to the stream's valley gradient and is inversely proportional to the sediment discharge.
- Bedload transport is directly related to the stream power and the concentration of fine material and inversely proportional to the fall diameter of the bed material.

In their natural state, small streams in forested watersheds in western Washington have small low-flow channels (the aquatic habitat channel) with little meandering (Sovern and Washington 1997). The stream banks are nearly vertical because of clayey bank soils and heavy root structures, and the streams have numerous debris jams from fallen timber. The widths are also narrow, generally from 3 to 6 feet wide. Stable forested watersheds also support about 250 aquatic plant and animal species along the stream corridor. Pool/riffle habitat is dominant along streams having gradients less than about 2% slope, while pool/drop habitat is dominant along streams having gradients from 4 to 10%. The pools form behind large organic debris (LOD) or rocks. The salmon and trout in western Washington have evolved to take advantage of these stream characteristics. Sovern and Washington (1997) point out that less athletic fish species (such as chum and pink salmon) cannot utilize the steeper gradient, upper reaches of the streams. However, coho, steelhead, and cutthroat can use these upper areas.

Urbanization radically affects many of these natural stream characteristics. Pitt and Bissonnette (1984) reported that coho and cutthroat were affected by the increased nutrients and elevated temperatures of the urbanized streams in Bellevue, as studied by the University of Washington as part of the EPA's NURP project (EPA 1983). These conditions were probably responsible for accelerated growth of the fry, which were observed to migrate to Puget Sound and the Pacific Ocean sooner than their counterparts in the control forested watershed, which was also studied. However, the degradation of sediments, from decreased particle sizes, adversely affected their spawning areas in streams that had become urbanized.

Sovern and Washington (1997) reported that in western Washington, frequent high flow rates can be 10 to 100 times the predevelopment flows in urbanized areas, but that the low flows in the urban streams are commonly lower than the predevelopment low flows. They have concluded that the effects of urbanization on western Washington streams are dramatic, in most cases permanently changing the stream hydrologic balance by: increasing the annual water volume in the stream, increasing the volume and rate of storm flows, decreasing the low flows during dry periods, and increasing the sediment and pollutant discharges from the watershed. With urbanization, the streams increase in cross-sectional area to accommodate these increased flows, and headwater downcutting occurs to decrease the channel gradient. The gradients of stable urban streams are often only about 1 to 2%, compared to 2 to 10% gradients in natural areas. These changes in width and the downcutting result in very different and changing stream conditions. The common pool/drop habitats are generally replaced by pool/riffle habitats, and the stream bed material is comprised of much finer material, for example. Along urban streams, fewer than 50 aquatic plant and animal species are usually found. Researchers have concluded that once urbanization begins, the effects on stream shape are not completely reversible. Developing and maintaining quality aquatic life habitat is possible under urban conditions, but it requires human intervention and it will not be the same as for forested watersheds.

Other Seattle area researchers have specifically examined the role that large woody debris (LWD) has in stabilizing the habitat in urban streams. Booth and Jackson (1997) found that LWD performs key functions in undisturbed streams that drain lowland forested watersheds in western Washington. These important functions include dissipation of the flow energy, channel bank and bed stabilization, sediment trapping, and pool formation. Urbanization typically results in the almost complete removal of this material. They point out that logs and other debris have long been removed from channels in urban areas for many reasons, especially because of their potential for blocking culverts or forming jams at bridges. Also, they may increase bank scour, and many residents favor "neat" stream bank areas (a lack of woody debris in and near the water and even with mowed grass to the water's edge).

It is clear that stream hydraulics, sediment transport, and riparian vegetation dramatically affect habitat in streams. Water quality evaluations, by themselves, obviously do not consider these important factors. Evaluations of habitat conditions and effects of changing habitat on biological uses obviously require combinations of stream studies, modeling, and comparison with local reference streams. The ability to predict habitat changes associated with urbanization, and the general success of habitat restoration efforts, is currently very poor. However, it is clear that detailed local investigations are critical and that habitat changes are likely one of the most important detrimental effects associated with urbanization.

EVALUATING HUMAN HEALTH IMPAIRMENTS USING A RISK ASSESSMENT APPROACH

The risk assessment paradigm is now well established in North America. The approach is basically the same for human health and ecological risk assessments (EPA 1989, 1998). The risk assessment paradigm is comprised of the following components: problem formulation, exposure and effects characterization followed by risk characterization, then the final risk management decisions. Risk assessment is a broad term which encompasses both risk characterization and risk management. The distinction between these two terms is an important one. The National Research Council's 1983 report on risk assessment in the federal government distinguished between risk assessment and risk management.

> Broader uses of the term [risk assessment] than ours also embrace analysis of perceived risks, comparisons of risks associated with different regulatory strategies, and occasionally analysis of the economic and social implications of regulatory decision functions that we assign to risk management. (EPA 1995)

The U.S. EPA has made the additional distinction of separating risk assessment from risk characterization. Risk characterization is the last step in risk assessment, is the starting point for risk managers, and is the foundation for regulatory decision making. The risk characterization identifies and highlights the noteworthy risk conclusions and related uncertainties (EPA 1995). The process described above is similar, but we have used different terminology. If the stormwater assessor is more comfortable using the EPA risk assessment approach, it can incorporate the guidelines of this handbook. The EPA guidance for conducting ecological risk assessments (ERAs) is quite general and does not provide specific methodologies and processes (EPA 1998). A number of good references (e.g., Suter 1993) exist that describe risk assessment approaches and considerations which are beyond the scope of this handbook. The two principal approaches for assessing adverse effects (hazard) in risk assessments are briefly described below.

Deterministic Approach

The simplest approach is the benchmark approach. This method (described above) basically ignores temporal exposure issues and focuses on point-in-time evaluations where threshold effect levels are compared to site contamination levels to ascertain risk. Many of the commonly used benchmarks (Appendix G) can be found in databases such the EPA's Ambient Water Quality Database, state water quality standards, ECOTOX, and the U.S. Department of Energy's Oak Ridge National Laboratory web site (http://www.hsrd.ornl.gov/ecorisk/ecorisk.html). This approach uses the quotient method for screening-level risk assessments. For compounds that bioaccumulate, it is easy to rearrange exposure equations involving uptake to back calculate ecotoxicity criteria for sediment, surface water, or soil (e.g., Pastorok et al. 1996).

Probabilistic Approach

A potentially more accurate and powerful assessment approach uses probability estimates to link likelihood of exposure with effects. This approach has been used frequently at Superfund sites, looking at exposure pathway analysis and risk modeling to assess chemical risks to humans, and aquatic and terrestrial wildlife. Since food is a primary source of toxicants, food web models are important tools to describe potential ecosystem effects (Pastorok et al. 1996). The more advanced wildlife exposure models now contain three attributes: habitat spatial structure, food web complexity, and receptor behavior and physiology ranging from Tier 1 (steady-state, worst-case conservative) to Tier III (dynamic, stochastic). For assessments of aquatic stressor impacts, probabilistic assessments of pesticide effects have been conducted using the following steps (Solomon et al. 1996; World Wildlife Fund 1992):

1. Characterize sensitivity effects (select appropriate measurement endpoints and rank effect, e.g., EC50 or no observed effect levels, vs. concentration)
2. Characterize exposure (plot distribution of chemical concentrations vs. site vs. frequency of occurrence)
3. Risk characterization: compare exposure distribution with overlap of the sensitivity distribution, while considering uncertainty, confounding stressors, variables, and ecological relevance of the assessment

Example Risk Assessment for Human Exposure to Stormwater Pathogens

The following discussion, summarized from Meyland et al. (1998), describes waterborne pathogens in separate sewer overflow (SSO) discharges as an example of the risk assessment process applied to a wet-weather problem. SSOs are generally sanitary sewage discharges that occur at "relief" locations, resulting in untreated wastewater being discharged directly into receiving waters.

Hazard Identification

The first step in a risk assessment, hazard identification, can be examined by gathering information regarding waterborne disease outbreaks. The agent that causes disease could be chemical, physical, or biological. However, in this case we will focus on biological causes, or infectious agents, i.e., pathogenic microorganisms. Table 8.4 shows the agents that have caused waterborne disease outbreaks in the United States, from 1971 to 1990. Notice that the vast majority of known agents are microorganisms. Table 8.5 shows additional data compiled from waterborne disease outbreaks. This table shows the agent associated with the disease.

The Centers for Disease Control (CDC) keep detailed records regarding notifiable, or reportable, diseases. There are legal requirements for reporting cases of these diseases. This list of notifiable diseases includes cryptosporidiosis. The fact that this disease is notifiable means that it is recognized as being extremely hazardous. As of mid-April 1998, there were 520 cases of cryptosporidiosis (not notifiable in all 50 states) (CDC 1998).

Dose–Response

The concept of dose–response, the second step in a risk assessment, is critical. Briefly, dose–response describes a relationship between a given level of contaminant and the biological response induced. This relationship is usually incremental; i.e., increase in the dose causes an increase in the response. In this particular case, the dose is the number of pathogenic microorganisms that the human subject is exposed to (through ingestion, swimming, wading, etc.), and the response is the level of infection. Generally, there is a minimum infective dose threshold that must be reached in order to infect a given individual. Once an individual has been infected, there are increasing degrees of infection severity. A *subclinical* infection describes the case where the pathogen produces a detectable immune response or organisms may be found growing in the human host, but the subject exhibits no clinical signs or symptoms, e.g., diarrhea, vomiting, etc. A *clinical* infection

Table 8.4 Causative Agents of Waterborne Disease Outbreaks, 1971 to 1990

	Outbreaks		Illness	
	Number of Cases	Percentage of Total	Number of Cases	Percentage of Total
Gastroenteritis (unknown cause)	293	49.66	67,367	47.60
Giardiasis	110	18.64	26,531	18.75
Chemical poisoning	55	9.32	3877	2.74
Shigellosis	40	6.78	8806	6.22
Viral gastroenteritis	27	4.58	12,699	8.97
Hepatitis A	25	4.24	762	0.54
Salmonellosis	12	2.03	1370	0.97
Camplylobacterosis	12	2.03	5233	3.70
Typhoid fever	5	0.85	282	0.20
Yersiniosis	2	0.34	103	0.07
Cryptosporidiosis	2	0.34	13,117	9.27
Chronic gastroenteritis	1	0.17	72	0.05
Toxigenic *E. coli*	2	0.34	1243	0.88
Cholera	1	0.17	17	0.01
Dermatitis	1	0.17	31	0.02
Amebiasis	1	0.17	4	0.00
Cyanobacteria-like bodies	1	0.17	21	0.01
Total	590	100	141,535	100

Data from Committee on Groundwater Recharge, NRC (National Research Council), National Academy of Science. *Ground Water Recharge Using Waters of Impaired Quality.* National Academy Press, Washington, D.C. 284 pp. 1994.

Table 8.5 Waterborne Disease Outbreaks Due to Microorganisms[a]

Disease	Agent	Outbreaks[b] (%)	Cases[c] (%)
Bacteria			
Typhoid fever	*Salmonella typhi*	10	0.1
Shigellosis	*Shigella* spp.	9	2.6
Salmonellosis	*Salmonella paratyphi* and other *Salmonella* species	3	3.5
Gastroenteritis	*Escherichia coli*	0.3	0.7
	Campylobacter spp.	0.3	0.7
Viruses			
Infectious hepatitis	Hepatitis A virus	11	0.5
Diarrhea	Norwalk virus	1.5	0.6
Protozoa			
Giardiasis	*Giardia lamblia*	7	3.8
Cryptosporidiosis[d]	*Cryptosporidium parvum*	0.2	71
Unknown etiology			
Gastroenteritis		57	16.7

[a] Compiled from data provided by the Centers for Disease Control, Atlanta, GA.
[b] Of more than 650 outbreaks in recent decades.
[c] Of 520,000 cases over the same period.
[d] A single outbreak of cryptosporidiosis in 1993 caused illness in 370,000 individuals from Milwaukee, WI. This is the largest single recorded outbreak of a waterborne disease in history.

Data from Madigan, M.T., J.M. Martinko, and J. Parker. *Brock Biology of Microorganisms, 8th ed.* Prentice-Hall, Upper Saddle River, NJ. 1997.

refers to the condition in which there are clinical signs and symptoms present. In layman's terms, one would refer to a person with a clinical infection as being "ill." The most severe response to infection would be death, i.e., a *fatality*. Therefore, one usually refers to the MID50, that is, the minimum infective dose that will cause subclinical infection in 50% of people exposed to that number of pathogens. The minimal infective dose (MID) varies widely with the type of pathogen, as shown in Table 8.6 (Bitton 1994). Of those infected, a percentage will show clinical signs; this

Table 8.6   Minimal Infective Doses for Some Pathogens
             and Parasites

Organism	Minimal infective Dose
Salmonella spp.	10^4 to 10^7
Shigella spp.	10^1 to 10^2
Escherichia coli	10^6 to 10^8
Vibrio cholerae	10^3
Giardia lamblia	10^1 to 10^2 cysts
Cryptosporidium	10^1 cysts
Entamoeba coli	10^1 cysts
Ascaris	1–10 eggs
Hepatitis A virus	1–10 PFU

Data from Bitton, G. *Wastewater Microbiology.* John Wiley & Sons,
Inc. New York. 1994.

Table 8.7   Values Used to Calculate Risks of Infection, Illness, and Mortality from Selected Enteric
             Microorganisms

	Probability of Infection from Exposure to One Organism (per million)	Ratio of Clinical Illness to Infection (%)	Mortality Rate (%)	Secondary Spread (%)
Campylobacter	7000			
Salmonella typhi	380			
Shigella	1000			
Vibrio cholerae	7			
Coxsackieviruses		5–96	0.12–0.94	76
Echoviruses	17,000	50	0.27–0.29	40
Hepatitis A virus		75	0.6	78
Norwalk virus			0.0001	30
Poliovirus 1	14,900	0.1–1	0.9	90
Poliovirus 3	31,000			
Rotavirus	310,000	28–60	0.01–0.12	
Giardia lamblia	19,800			

Data from Committee on Groundwater Recharge, NRC (National Research Council), National Academy of
Science. *Ground Water Recharge Using Waters of Impaired Quality.* ISBN 0-309-05142-8. National Academy
Press, Washington, D.C. 284 pp. 1994.

is referred to as the ratio of clinical illness to infection. In addition, a percentage of those infections
will result in fatalities; this is referred to as the case fatality rate. Table 8.7 shows example values
for these various levels of response to infection.

Notice that higher probabilities, rates, or percentages correspond to pathogens with higher virulence. For example, if 1 million people are exposed to one rotavirus each, then 310,000 may be
infected. In contrast, if 1 million people are exposed to one *Vibrio cholerae* bacterium each, only
seven may be infected. In general, viral pathogens are much more virulent than bacterial pathogens.

Table 8.8 shows another example of data that can be obtained from published studies. These
data show, for instance, that once infected by *Salmonella* bacteria, approximately 41% will
exhibit clinical infection. In addition, *Cryptosporidium* infection results in a 71% clinical
infection frequency.

Another study (DuPont et al. 1995) published results pertaining to infection rates from the oral
introduction of *Cryptosporidium* oocysts into healthy volunteers. Various doses of oocysts, from
30 to 1 million, were given to volunteers in gelatin capsules, and these subjects were followed up
to record the incidence of infection. Table 8.9 gives these results. A linear regression analysis of
the data yielded a correlation coefficient of 0.983 and an infectious dose of 50 of 132 oocysts. This
is an excellent example of the dose–response relationship, as increasing doses of oocysts caused
increasing rates of infection.

Table 8.8 Ratio of Clinical to Subclinical Infections and Case Fatality Rates
for Enteric Microorganisms

Microorganism	Frequency of Clinical Illness (%)	Case:Fatality Rate (%)
Viruses		
Hepatitis (adults)	75	0.6
Rotavirus	25–60	0.01
Astrovirus (adults)	12.5	0.12
Coxsackie A16	50	0.59–0.94
Coxsackie B	5–96	
Bacteria		
Salmonella	41	0.1
Shigella	46	0.2
Protozoan parasites		
Giardia	50–67	
Cryptosporidium	71	

Data from Gerba, C.P., J.B. Rose, C.N. Hass, and K.D. Crabtree. Waterborne rotavirus:
a risk assessment. *Water Research*, Vol. 30, No 12, pp. 2929–2940. Dec. 1996.

Table 8.9 Rate of Infection, Enteric Symptoms, and Clinical Cryptosporidiosis,
According to the Intended Dose of Oocysts

Intended Dose of Oocysts	No. of Subjects	Infection	Enteric Symptoms	Cryptosporidiosis
		Number (percent)		
30	5	1 (20)	0	0
100	8	3 (37.5)	3 (37.5)	3 (37.5)
300	3	2 (66.7)	0	0
500	6	5 (83.3)	3 (50)	2 (33.3)
>1000	7	7 (100)	5 (71.4)	2 (28.6)
Total	29	18	11	7

Data from DuPont, H.L., C.L. Chappell, C.R. Sterling, P.C. Okhuysen, J.B. Rose, and
W. Jakubowski. The infectivity of *Cryptosporidium parvum* in healthy volunteers. *N. End.
J. Med.*, Vol. 332, No. 13, pp. 855–859. 1995.

Exposure Assessment

The exposure assessment is the third step in a risk assessment. Several factors contribute to
whether or not contact with a particular pathogen may cause disease. Among these factors are
virulence, mode of transmission, portal of entry, and host susceptibility. Virulence is defined as a
particular organism's ability to cause disease in humans and is related to the dose of infectious
agent necessary for host infection and causing disease (Bitton 1994). The mode of transmission is
the particular method in which the organism is transported from the reservoir of pathogens (such
as a contaminated outfall) to the host, i.e., person-to-person, waterborne, or foodborne.

The research conducted by Meyland et al. (1998) concentrated on the waterborne transmission
route of SSOs to exposed individuals, but exposure assessment can also be evaluated based on
portal of entry. The portal of entry is dictated by the mechanism of contact; examples or entry
portals are access through the gastrointestinal tract, respiratory tract, or skin. Host susceptibility is
dependent upon resistance to infectious agents, which consists of the roles of the immune system
and nonspecific factors (Bitton 1994). Immunity can be both natural (genetic) and acquired from
previous contact with the pathogen.

There are many documented examples of waterborne transmission of pathogenic microorganisms. Recently, in the United States, there has been widespread concern about *Cryptosporidium* contamination of water supplies. This is an example of waterborne transmission via a
contaminated drinking water supply. Table 8.10 summarizes the available information regarding
Cryptosporidium outbreaks in the United States. Outbreaks caused by this organism are a

Table 8.10 *Cryptosporidium* Outbreaks: Affected Populations and Characteristics of the Raw Water Supply

County, State (City)	Date	Estimated No. of People Affected (Confirmed Cases)[a]	Raw Water Source	Suspected Sources of Contamination
Bexar County, TX (Braun Station)	May–Jul 1984	2,000 (47)	Well	Raw sewage[b]
Bernalillo County, NM (Albuquerque)	Jul–Oct 1986	(78)	Surface water	Surface runoff from livestock grazing areas
Carroll County, GA (Carrollton)	Jan–Feb 1987	13,000	River	Raw sewage and runoff from cattle grazing areas
Berks County, PA	Aug 1991	551	Well	Septic tank effluent, nearby creek
Jackson County, OR (Talent and Medford)	Jan–Jun 1992	15,000	Spring/river	Surface water, treated wastewater,[b] or runoff from agricultural areas
Milwaukee County, WI (Milwaukee)	Jan–Apr 1993	403,000	Lake	Cattle wastes, slaughterhouse wastes, and sewage carried by tributary rivers
Yakima County, WA	Apr 1993	7 (3)	Well	Infiltration of runoff from cattle, sheep, or elk grazing areas
Cook County, MN (Grand Marais)	Aug 1993	27 (5)	Lake	Backflow of sewage or septic tank effluent into distribution, raw water inlet lines, or both
Clark County, NV (Las Vegas)	Jan–Apr 1994	(78)[c]	Lake	Treated wastewater, sewage from boats
Walla Walla County, WA	Aug–Oct 1994	86 (15)	Well	Treated wastewater[b]
Alachua County, FL	Jul 1995	(72)	N/A	Backflow of contaminated water

[a] Estimates are based on epidemiologic studies; confirmed cases correspond to patients whose stool samples tested positive for *Cryptosporidium*.
[b] Strong evidence to support effect of wastewater.
[c] 103 laboratory-confirmed cases were associated with the outbreak; 78 of these were documented during the epidemiologic study period.
Data from Solo-Gabriele, H. and S. Neumeister. US outbreaks of cryptosporidiosis. *Journal American Water Works Association*, Vol. 88, No. 9, pp. 76–86. Sept. 1996.

significant health threat (more than 400,000 people were infected during the 1993 Milwaukee outbreak). Moreover, notice that the suspected source of contamination is likely to be sewage. In fact, municipal wastewater was implicated as the source in roughly half of the outbreaks (Solo-Gabriele and Neumeister 1996). The remaining outbreaks were likely caused by contaminated agricultural runoff.

Another important mode of transmission is water-contact recreation. This type of transmission is usually associated with swimming beach exposures. Many studies have shown an association between illness and swimming near stormwater, SSO, or CSO (combined sewer overflow) outfalls. In general, most of these studies found an increased risk of illness resulting from swimming in waters that contained fecal contamination indicators or pathogenic microorganisms. The SMBRP (Santa Monica Bay Restoration Project) Study is unique in that it found a distance-dependent association between contamination sources and health effects. In this study, there was a higher rate of enteric illness in swimmers who swam within 400 ft of a stormwater outfall than in those who swam more than 400 ft away. In many urban receiving waters, children frequently play in and near potentially contaminated small streams and creeks, well away from designated swimming beaches. In most cases, this exposure route is not considered, because this is not a designated use of the water. The most important pathogens contained in stormwater are likely from sewage contamination.

Table 8.11   Sensitive Populations in the United States

Population	Individuals	Year
Pregnancies	5,657,900	1989
Neonates	4,002,000	1989
Elderly people (over 65)	29,400,000	1989
Residences in nursing homes or related care facilities	1,553,000	1986
Cancer patients (non-hospitalized)	2,411,000	1986
Transplant organ patients (1981–1989)	110,270	1981–1989
AIDS patients	142,000	1981–1990

Data from Gerba, C.P., J.B. Rose, C.N. Hass, and K.D. Crabtree. Waterborne rotavirus: a risk assessment. *Water Research*, Vol. 30, No 12, pp. 2929–2940. Dec. 1996.

Pitt et al. (1993) conducted a study of inappropriate pollutant entries into storm drainage systems in which many illegal sanitary sewer connections to storm drain systems were commonly found.

Another possible exposure route is through the consumption of contaminated fish or shellfish. One example of this type of outbreak occurred in Louisiana in 1993 (Kohn 1995). This outbreak was caused by contaminated oysters that were consumed raw. The agent implicated in this outbreak was Norwalk virus, which causes gastroenteritis. Seventy of the 84 people (83% infection rate) who ate the raw oysters became ill. The epidemiologic investigation found that this outbreak was probably caused by overboard sewage disposal by harvesters near the oyster bed.

An additional consideration that one must account for when assessing the adverse health effects of contact with pathogenic microorganisms is that certain individuals within the population are at higher risk for serious infections. Individuals who are at higher risk are the very young, elderly people, pregnant individuals, and immunocompromised (organ transplants, cancer, and AIDS) patients (Gerba et al. 1996). This collective group represents almost 20% of the current U.S. population (Table 8.11). In addition, elderly and immunocompromised people are an increasing segment of the population.

Calculation of Risk

The fourth step in a risk assessment is the calculation of risk for a specific situation. As indicated in the first three steps, it is possible to identify which components pose a risk for a specific activity, to estimate the doses necessary to cause different responses, and to conduct the exposure assessment. For an urban receiving water study, it is possible to consider an important, but commonly overlooked, scenario: exposure to pathogens by children who are wading in an urban stream. The list of potential pathogens may be lengthy, depending on the likelihood of sanitary sewage contamination. Even with separate stormwater, numerous pathogens are still commonly present (such as *Pseudomonas aeruginosa* and *Shigella*). Sanitary sewage contamination lengthens the list considerably, as shown on the above lists. The dose–response curves can be examined for each likely pathogen and route of exposure (such as water contact for *P. aeruginosa*, or ingestion for *Giardia* and *Cryptosporidium*). The difficulty is estimating the magnitude of discharge for each pathogen and its fate after the discharge to the likely point of contact. Exposure duration, or ingestion quantity, of the water also needs to be estimated, along with likely time between discharge and exposure. This is especially important for wading exposures, for example, because water-contact recreation is unlikely during a runoff event but can certainly occur soon after a rain has ended. However, exposure to pathogens when wading is likely to be made more severe through stirring up contaminated sediments. Bacteria have been shown to survive for long periods in stream sediments in areas of deposition (pools, backwaters, or behind small dams), many of which attract children because of the deeper water (Burton et al. 1987; Burton 1982). Therefore, it would be necessary to predict areas where the particulates, with which pathogens are associated, are likely to accumulate.

The local monitoring program must therefore consider characterization studies of the pathogens in the discharge, along with the hydraulic characteristics of the receiving water that would affect transport and fate of the pathogens. This information is used in a receiving water model to predict the conditions (concentrations in the water) at the time and location of exposure. Appendix H summarizes representative receiving water models that may be applicable for certain fate and route pathways. HSPF, one of the more complex urban stream models, contains many options, but requires a great deal of monitoring data for calibration and verification. Other, simpler models, such as WASP5, may be suitable for this task. The predicted conditions at the likely sites of exposure can be used, along with assumptions pertaining to exposure duration. This information is then compared to the dose–response information to estimate the likelihood of contacting disease.

IDENTIFYING AND PRIORITIZING CRITICAL STORMWATER SOURCES

An important goal of a receiving water study is to learn enough about local problems to be able to reduce them in the future, and to prevent new problems from occurring. The tools and techniques presented in this book enable local watershed managers to obtain a good understanding of their local problems and their likely causes.

After receiving water problems (beneficial use impairments) are described, the likely causes for these problems must be identified. Many of the problems are directly associated with measured parameters in excessive quantities and are therefore inherently obvious (such as high bacteria concentrations interfering with water-contact recreation and fish consumption; high flows and debris blockages affecting drainage; trash or odors affecting noncontact recreation; and high toxicant concentrations affecting water supplies). The most difficult task is identifying the possible causes for declining aquatic life conditions, which can be associated with numerous causes, including habitat destruction, high/low flows, inappropriate discharges, polluted sediment, and water quality. The weight-of-evidence approach, described above, has therefore become a useful framework to understand possible cause-and-effect relationships for biological resources.

Once the critical pollutants (or flow conditions) are identified, it is more straightforward to identify the likely sources in the watershed contributing these problem constituents. Classical approaches include watershed modeling to develop mass balances for targeted pollutants. The following discussion describes this approach, along with a case study describing how the Wisconsin Nonpoint Source Program has integrated field monitoring, data analysis, watershed modeling, and the development of stormwater controls to reduce future receiving water problems. It is also important to recognize additional potential sources of contaminants in a watershed that are not easily addressed by most watershed models. The most important include snowmelt and inappropriate sources. These two sources can be much greater contributors of some critical pollutants than warm-weather runoff alone. Field investigations to locate and quantify inappropriate sources (sanitary sewage and discharges from small industries and commercial establishments are the most important) were described in Chapter 6 and must be conducted along with conventional watershed modeling activities. Snowmelt contributions should also be quantified for comparison with the more traditional sources.

Sources of Urban Stormwater Contaminants

Urban runoff is comprised of many separate source area flow components that are combined within the drainage area and at the outfall before entering the receiving water. It may be adequate to consider the combined outfall conditions alone when evaluating the long-term, areawide effects of many separate outfall discharges to a receiving water. However, if better predictions of outfall characteristics (or the effects of source area controls) are needed, the separate source area components must be characterized. The discharge at the outfall is made up of a mixture of contributions from different source areas. The "mix" depends on the characteristics of the drainage area and the specific rain event. The effectiveness of source area controls is therefore highly site and storm specific.

Various urban source areas contribute different quantities of runoff and pollutants, depending on their specific characteristics. Impervious source areas may contribute most of the runoff during small rain events. Examples of these source areas include paved parking lots, streets, driveways, roofs, and sidewalks. Pervious source areas become important contributors for larger rain events. These pervious source areas include gardens, lawns, bare ground, unpaved parking areas and driveways, and undeveloped areas. The relative importance of each individual source is a function of their areas, their pollutant washoff potentials, and the rain characteristics.

The washoff of debris and soil during a rain is dependent on the energy of the rain and the properties of the material. Pollutants are also removed from source areas by winds, litter pickup, or other cleanup activities. The runoff and pollutants from the source areas flow directly into the drainage system, onto impervious areas that are directly connected to the drainage system, or onto pervious areas that will attenuate some of the flows and pollutants before they discharge to the drainage system.

Sources of pollutants on paved areas include on-site particulate storage that cannot be removed by the usual processes e.g., rain, wind, street cleaning, etc. Atmospheric deposition, deposition from activities on these paved surfaces (auto traffic, material storage, etc.), and the erosion of material from upland areas that discharge flows directly onto these areas are the major sources of pollutants to the paved areas. Pervious areas contribute pollutants mainly through erosion processes where the rain energy dislodges soil from between plants. The runoff from these source areas enters the storm drainage system where sedimentation in catchbasins or in the sewerage may affect their ultimate discharge to the outfall. In-stream physical, biological, and chemical processes affect the pollutants after they are discharged to the ultimate receiving water.

It is important to know when the different source areas become "active" (when runoff initiates from the area, carrying pollutants to the drainage system). If pervious source areas are not contributing runoff or pollutants, the prediction of urban runoff quality is much simplified. The mechanisms of washoff, and delivery yields of runoff and pollutants from paved areas, are much better known than from pervious urban areas (Novotny and Chesters 1981). In many cases, pervious areas are not active runoff contributors except during rain events greater than at least 5 or 10 mm. For smaller rain depths, almost all of the runoff and pollutants originate from impervious surfaces (Pitt 1987). However, in many urban areas, pervious areas may contribute the majority of the runoff, and some pollutants, when rain depths are greater than about 20 mm. The actual importance of the different source areas is highly dependent on the specific land use and rainfall patterns. Obviously, in areas having relatively low-density development, especially where moderate- and large-sized rains occur frequently (such as in the Southeast), pervious areas typically dominate outfall discharges. In contrast, in areas having significant paved areas, especially where most rains are relatively small (such as in the arid West), the impervious areas dominate outfall discharges. The effectiveness of different source controls is therefore quite different for different land uses and climatic patterns.

If the number of events exceeding a water quality objective is important, then the small rain events are of most concern. Stormwater runoff typically exceeds some water quality standards for practically every rain event (especially for bacteria and some heavy metals). In the upper Midwest, the median rain depth is about 6 mm, while in the Southeast, the median rain depth is about twice this depth. In most urban areas, no runoff is observed until the rain depth exceeds 2 or 3 mm. For these small rain depths and for most urban land uses, directly connected paved areas usually contribute most of the runoff and pollutants. However, if annual mass discharges are more important, e.g., for long-term effects, then the moderate rains are more important. Rains from about 10 to 50 mm produce most of the annual runoff volume in many areas of the United States. Runoff from both impervious and pervious areas can be very important for these rains. The largest rains (greater than about 100 mm) are relatively rare and do not contribute significant amounts of runoff pollutants during normal years, but are very important for drainage design. The specific source areas that are most important (and controllable) for these different conditions vary widely.

Case Study: Wisconsin Nonpoint Source Program in Urban Areas

The urban stormwater evaluation methodology used by the Nonpoint Source Program at the Wisconsin Department of Natural Resources was developed to supplement the extensive ongoing rural aspects of the statewide watershed planning program (Pitt 1986). This comprehensive urban methodology includes:

- Evaluation of receiving water goals, based on established water use objectives and through meetings of citizen groups and technical experts
- Identification of current problems and sources of problem pollutants and flows through monitoring and modeling
- Identification and evaluation of suitable source area and outfall treatment options, including the development of model ordinances for construction site erosion control and stormwater management, plus the development of design manuals for constructing controls
- Demonstration projects evaluating alternative controls
- Receiving water evaluations to confirm or to modify recommendations

An important element of this methodology was to extensively modify SLAMM, the Source Loading and Management Model (Pitt and Voorhees 1995; www.winslamm.com). This model was developed to identify sources of pollutants in urban areas and to evaluate many alternative stormwater control programs. This methodology is generally used in the development of the watershed plans and to determine suitable cost-sharing aspects of the Wisconsin Nonpoint Source Program. The Milwaukee area was the first urban watershed planning effort to use this methodology. Milwaukee has a great deal of stormwater quality information that was used in this planning and implementation effort.

Stormwater quality management in the Milwaukee area was initiated as part of the Wisconsin Priority Watershed Program. This program was developed in 1978 to help combat both urban and rural nonpoint sources of pollution (EPA 1990a, 1991). This program is one of the oldest in the nation funding nonpoint pollution abatement. An important element of this program is retrofitting control practices in both rural and urban areas. The program was initially heavily involved in rural areas, with technical assistance from the NRCS. A unique aspect of the program is that it is implemented on a watershed, and not on a political jurisdiction basis. Of the state's 330 watersheds, 130 (mostly located in the southern part of Wisconsin) will likely require comprehensive management activities to control nonpoint pollutants. A 25-year plan was developed in 1982 which would require the start-up of about eight or nine new watershed abatement efforts per year. The watershed plans are prepared by the state with cooperation and reviews by local government agencies. They contain detailed analyses of the water resources objectives (existing and desired beneficial uses, including the problems and threats to these uses), the critical sources of problem pollutants, and the control practices that can be applied within each watershed. The plans also include implementation schedules and budgets to meet the pollution reduction objectives.

Each plan requires 1 year to prepare, including the necessary fieldwork. Various field inventory activities are needed to prepare the plans, including aquatic biology and habitat surveys to identify existing and potential fishery uses, stream bank surveys to identify the nature and magnitude of stream bank erosion problems and to help design needed controls, field and barnyard surveys to supply information needed to estimate and rank their pollution potentials and to design farm control practices, and urban surveys needed to evaluate urban runoff pollution potential and its control.

Urban planning was initiated in 1983 in the Milwaukee and Madison areas, with other urban areas of the state following. The urban practices eligible for cost sharing identified in these plans have included stream bank protection, detention basins, and infiltration devices for existing urbanized areas. Construction site erosion controls are also usually required as a condition for a grant agreement in an urban area, but they are not eligible for state cost sharing. About $3 to $5 million per year will be used by the nonpoint source program over a 20-year period in controlling urban runoff. An outcome of the Milwaukee River South Watershed plan included goals for reducing urban stormwater discharges (D'Antuono 1998). These goals were 50% reductions for suspended solids and heavy metals, and 50 to 70% reductions for phosphorus.

Detailed studies on toxicant sources, effects, and controls have also been conducted in Milwaukee, including a study conducted in the heavily urbanized Lincoln Creek (having a 19-mi^2 watershed and being 9 mi long). A seven-tiered indicator program, incorporating many physical, chemical, and biological tests, was simultaneously conducted which identified long-term toxicity problems, likely associated with resuspended contaminated sediments having high levels of organic compounds (Claytor 1996). It was found that discharges of these fine sediments could be significantly reduced through the use of well-designed and maintained wet detention basins. The in-stream toxicity monitoring methods developed and used during the Lincoln Creek study can be used by other municipalities to answer the following basic questions:

- Are toxic conditions present?
- What is causing the toxicity?
- How much is too much urbanization?
- Can stormwater controls reduce these problems?

The benefits of stormwater controls have also been evaluated in Milwaukee, especially grass swales, wet detention ponds, and underground devices for critical source areas. The Southeastern Wisconsin Regional Planning Commission also prepared a comprehensive report documenting costs associated with construction site erosion and stormwater control.

Use of SLAMM to Identify Pollutant Sources and to Evaluate Control Programs

A logical approach to stormwater management requires knowledge of the problems to be solved, the sources of the problem pollutants, and the effectiveness of stormwater management practices that can control the problem pollutants at their sources and at outfalls. The Source Loading and Management Model (SLAMM) is designed to provide information on these last two aspects of this approach. The first versions of SLAMM were developed by Pitt in the mid-1970s to help evaluate the results from early EPA stormwater projects (Pitt and Voorhees 1995). Further information on SLAMM, especially its integration with GIS systems, is included in Appendix H and at www.winslamm.com.

The development of SLAMM began in the mid-1970s, primarily as a data reduction tool for use in early street cleaning and pollutant source identification projects sponsored by the EPA's Storm and Combined Sewer Pollution Control Program (Pitt 1979, 1984; Pitt and Bozeman 1982). Additional information contained in SLAMM was obtained during the EPA's Nationwide Urban Runoff Program (NURP) (EPA 1983), especially the early Alameda County, CA (Pitt and Shawley 1982), and the Bellevue, WA (Pitt and Bissonnette 1984) projects. The completion of the model was made possible by the remainder of the NURP projects and additional field studies and programming support sponsored by the Ontario Ministry of the Environment (Pitt and McLean 1986), the Wisconsin Department of Natural Resources (Pitt 1986), and Region V of the U.S. Environmental Protection Agency. Early users of SLAMM included the Ontario Ministry of the Environment's Toronto Area Watershed Management Strategy (TAWMS) study (Pitt and McLean 1986) and the Wisconsin Department of Natural Resources' Priority Watershed Program (Pitt 1986). SLAMM can now be effectively used as a tool to enable watershed planners to obtain a better understanding of the effectiveness of different control practice programs.

Some of the major users of SLAMM have been associated with the Nonpoint Source Pollution Control Program of the Wisconsin Department of Natural Resources, where SLAMM has been used for a number of years to support their extensive urban stormwater planning and cost-sharing program (Thum et al. 1990; Kim et al. 1993a,b; Ventura and Kim 1993; Bachhuber 1996; Bannerman et al. 1996; Haubner and Joeres 1996; Legg et al. 1996). Many of these applications have included the integrated use of SLAMM with GIS models, as illustrated in Appendix H.

SLAMM was developed primarily as a planning-level tool to generate information needed for planning-level decisions, while not generating or requiring superfluous information. Its primary capabilities include predicting flow and pollutant discharges that reflect a broad variety of development conditions and the use of many combinations of common urban runoff control practices. Control practices evaluated by SLAMM include detention ponds, infiltration devices, porous pavements, grass swales, catchbasin cleaning, and street cleaning. These controls can be evaluated in many combinations and at many source areas as well as the outfall location. SLAMM also predicts the relative contributions of different source areas (roofs, streets, parking areas, landscaped areas, undeveloped areas, etc.) for each land use investigated. As an aid in designing urban drainage systems, SLAMM also calculates correct NRCS curve numbers that reflect specific development and control characteristics. These curve numbers can then be used in conjunction with available urban drainage procedures to reflect the water quantity reduction benefits of stormwater quality controls.

SLAMM is normally used to predict source area contributions and outfall discharges. However, it has been used in conjunction with a receiving water model (HSPF) to examine the ultimate receiving water effects of urban runoff (Ontario 1986), and has been recently been modified to be integrated with SWMM (Pitt et al. 1999) to more accurately consider the joint benefits of source area controls on drainage design. The following example illustrates how SLAMM is used to identify the most important source areas and how to select the most appropriate control programs.

The areas of the different surfaces in each land use are very important for SLAMM evaluations. Figure 8.2 is an example showing the areas of different surfaces for a medium-density residential area in Milwaukee. As shown in this example, streets make up between 10 and 20% of the total area, while landscaped areas can make up about half of the drainage area. The variation of these different surfaces can be very large within a designated area. The analysis of many candidate areas might therefore be necessary to understand how effective or how consistent the model results might be for a general land use classification.

One of the first problems in evaluating an urban area for stormwater controls is the need to understand where the pollutants of concern are originating under different rain conditions.

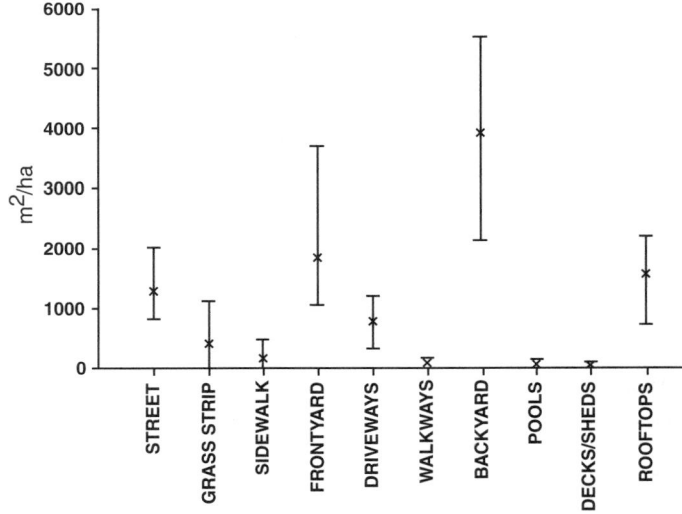

Figure 8.2   Source areas — Milwaukee medium-density residential areas (without alleys). (From Pitt, R. *Small Storm Flow and Particulate Washoff Contributions to Outfall Discharges*. Ph.D. dissertation, Department of Civil and Environmental Engineering, the University of Wisconsin, Madison, November 1987. With permission.)

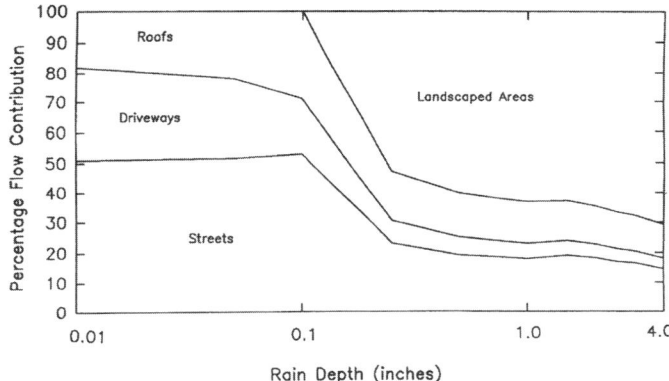

Figure 8.3   Flow sources for example medium-density residential area having clayey soils. (From Pitt, R. and J. Voorhees. Source loading and management model (SLAMM). *Seminar Publication: National Conference on Urban Runoff Management: Enhancing Urban Watershed Management at the Local, County, and State Levels.* March 30–April 2, 1993. Center for Environmental Research Information, U.S. Environmental Protection Agency. EPA/625/R-95/003. Cincinnati. OH. pp. 225–243. April 1995.)

Figure 8.3 is an example of a typical medium-density residential area, showing the percentage of runoff originating from different major sources, as a function of rain depth. For storms of up to about 0.1 in in depth, street surfaces contribute about one half of the total runoff to the outfall. This contribution decreased to about 20% for storms greater than about 0.25 in in depth. This decrease in the significance of streets as a source of water is associated with an increase of water contributions from landscaped areas (which make up more than 75% of the area and have clayey soils). Similarly, the significance of runoff from driveways and roofs also starts off relatively high and then decreases with increasing storm depth. Obviously, this is just an example plot, and the source contributions would vary greatly for different land uses/development conditions, rainfall patterns, and the use of different source area controls.

A major use of SLAMM is to better understand the role of different sources of pollutants and the suitability of controls that can be applied at the sources and at outfalls. As an example, to control suspended solids, street cleaning (or any other method to reduce the washoff of particulates from streets) may be very effective for the smallest storms, but would have very little benefit for storms greater than about 0.25 in in depth. However, erosion control from landscaped surfaces may be effective over a wider range of storms. The following list shows the different control programs that were investigated in this hypothetical medium-density residential area:

- Base level (as built in 1961–1980 with no additional controls)
- Catchbasin cleaning
- Street cleaning
- Grass swales
- Roof disconnections
- Wet detention pond
- Catchbasin and street cleaning combined
- Roof disconnections and grass swales combined
- All of the controls combined

This residential area, which was based upon actual Birmingham, AL, field observations for homes built between 1961 and 1980, has no controls, including no street cleaning or catchbasin cleaning. The use of catchbasin cleaning in the area, in addition to street cleaning, was evaluated. Grass swale use was also evaluated, but swales are an unlikely retrofit option and would only be appropriate for newly developing areas. However, it is possible to disconnect some of the roof drainages and divert the roof runoff away from the drainage system and onto grass surfaces for infiltration in existing developments. In addition, wet detention ponds can be retrofitted in different areas and at outfalls. Besides those controls examined individually, catch basin and street cleaning controls combined were also evaluated, in addition to the combination of disconnecting some of the rooftops and the use of grass swales. Finally, all of the controls were also examined together.

The following list shows a general description of this hypothetical area:

- All curb and gutter drainage (in fair condition)
- 70% of roofs drain to landscaped areas
- 50% of driveways drain to lawns
- 90% of streets are intermediate texture (remaining are rough)
- No street cleaning
- No catchbasins

About one half of the driveways currently drain to landscaped areas, while the other half drain directly to the pavement or the drainage system. Almost all of the streets are of intermediate texture, and about 10% are rough textured. As noted earlier, there currently is no street cleaning or catchbasin cleaning.

The level of catchbasin use that was investigated for this site included 950 ft^3 of total sump volume per 100 acres (typical for this land use), with a cost of about $50 per catchbasin cleaning. Typically, catchbasins in this area could be cleaned about twice a year, for a total annual cost of about $85 per acre of the watershed. Street cleaning could also be used, with a monthly cleaning effort for about $30 per year per watershed acre. Grass swale drainage was also investigated. Assuming that swales could be used throughout the area, there could be 350 ft of swales per acre (typical for this land use), and the swales could be 3.5 ft wide. Because of the clayey soil conditions, an average infiltration rate of about 0.5 in per hour was used in this analysis, based on many different double ring infiltrometer tests of typical soil conditions. Swales cost much less than conventional curb and gutter systems, but have an increased maintenance frequency. Again, the use of grass swales is appropriate for new development, but not for retrofitting in this area.

Roof disconnections could also be utilized as a control measure by directing all roof drains to landscaped areas. The objective would be to direct *all* the roof drains to landscaped areas. Since 70% of the roofs already drain to the landscaped areas, only 30% would be further disconnected, at a cost of about $125 per household. The estimated total annual cost for roof disconnections would be about $10 per watershed acre. An outfall wet detention pond suitable for 100 acres of this medium-density residential area would have a wet pond surface of 0.5% of drainage area to provide about 90% suspended solids control. It would need 3 ft of dead storage and live storage equal to runoff from 1.25 in of rain. The total annual cost for wet detention ponds was estimated to be about $130 per watershed acre.

Table 8.12 summarizes the SLAMM results for runoff volume, suspended solids, filterable phosphate, and total lead for 100 acres of this medium-density residential area. The only control practices evaluated that would reduce runoff volume are the grass swales and roof disconnections. All of the other control practices evaluated do not infiltrate stormwater. Table 8.12 also shows the total annual average volumetric runoff coefficient (Rv) for these different options. The base level of control has an annual flow-weighted Rv of about 0.3, while the use of swales would reduce the Rv to about 0.1. Only a small reduction of Rv (less than 10%) would be associated with complete roof disconnections, compared to the existing situation, because of the large number of roof disconnections that already exist. The suspended solids analyses show that catchbasin cleaning alone could result in about 14% suspended solids reductions. Street cleaning would have very little benefit, while the use of grass swales would reduce the suspended solids discharges by about 60%. Grass swales would have minimal effect on the reduction of suspended solids concentrations at the outfall, but provide about 60% reductions in annual pollutant mass discharges (they are primarily an infiltration device, having very little filtering benefit). Wet detention ponds would remove about 90% of the mass and concentrations of suspended solids. Similar observations can be made for filterable phosphates and total lead.

Table 8.12 SLAMM Predicted Runoff and Pollutant Discharge Conditions for Example[a]

Birmingham 1976 rains: (112 rains, 55 in. total 0.01–3.84 in. each)	Runoff Volume			Suspended Solids		Filterable Phosphate		Total Lead	
	Annual ft³/acre	Flow-wtg. Rv	CN Range	Flow-wtg. mg/L	Annual lbs/acre	Flow-wtg. µg/L	Annual lbs/acre	Flow-wtg. µg/L	Annual lbs/acre
Base (no controls)	59800	0.3	77–100	385	1430	157	0.58	543	2.0
Catchbasin cleaning	59800	0.3	77–100	331	1230	157	0.58	468	1.7
reduction (lbs or ft³)	0				200		0		0.29
reduction (%)	0			14	14	0	0	14	14
cost ($/lb or $/ft³) ($85/acre/yr)	N/A				0.43		N/A		293
Street cleaning	59800	0.3	77–100	385	1430	157	0.58	543	2.0
reduction (lbs or ft³)	0				0		0		0.01
reduction (%)	0			0	0	0	0	0	0.49
cost ($/lb or $/ft³) ($30/acre/yr)	N/A				N/A		N/A		3000
Grass swales	23300	0.12	63–100	380	554	151	0.22	513	0.75
reduction (lbs or ft³)	36500				876		0.36		1.28
reduction (%)	61			1	61	1	62	6	63
cost ($/lb or $/ft³) ($minimal/acre/yr)	minimal				minimal		minimal		minimal
Roof disconnections	56000	0.28	76–100	410	1430	156	0.55	443	1.6
reduction (lbs or ft³)	3800				0		0.03		0.48
reduction (%)	6			–6	0	1	5	18	24
cost ($/lb ir $/ft³) (10/acre/yr)	0				N/A		333		21
Wet detention pond	59800	0.3	77–100	49	185	157	0.58	69	0.26
reduction (lbs or ft³)	0				1250		0		1.8
reduction (%)	0			87	87	0	0	87	87
cost ($/lb or $/ft³) ($130/acre/yr)	N/A				0.10		N/A		73
CB and street cleaning	59800	0.3	77–100	331	1230	157	0.58	468	1.7
reduction (lbs. or $/ft³)	0				200		0		0.29
reduction (%)	0			14	14	0	0	14	14
cost ($/lb or $/ft³) ($115/acre/yr)	N/A				0.58		N/A		397
Roof dis. and swales	20900	0.1	63–100	403	526	139	0.18	352	0.46
reduction (lbs or ft³)	38900				904		0.4069		1.6
reduction (%)	65			–5	63	11	25	35	77
cost ($/lb or $/ft³) ($10/acre/yr)	0.00026				0.01				6.4

Table 8.12 SLAMM Predicted Runoff and Pollutant Discharge Conditions for Example[a] (continued)

Birmingham 1976 rains: (112 rains, 55 in. total 0.01–3.84 in. each)	Runoff Volume			Suspended Solids		Filterable Phosphate		Total Lead	
	Annual ft³/acre	Flow–wtg. Rv	CN Range	Flow–wtg. mg/L	Annual lbs/acre	Flow–wtg. µg/L	Annual lbs/acre	Flow–wtg. µg/L	Annual lbs/acre
All above controls	20900	0.1	63–100	42	55	139	0.18	36	0.05
reduction (lbs or ft³)	38900			89	1375	11	0.40	93	1.98
reduction (%)	65				96		69		97
cost ($/lb or $/ft³) ($255/acre/yr)	0.0066				0.19		638		129

[a] Medium-density residential area, developed in 1961–1980, with clayey soils (curbs and gutters); new development controls (not retrofit).

From Pitt, R. and J. Voorhees. Source loading and management model (SLAMM). *Seminar Publication: National Conference on Urban Runoff Management: Enhancing Urban Watershed Management at the Local, County, and State Levels.* March 30–April 2, 1993. Center for Environmental Research Information, U.S. Environmental Protection Agency. EPA/625/R-95/003. Cincinnati. OH. pp. 225–243. April 1995.

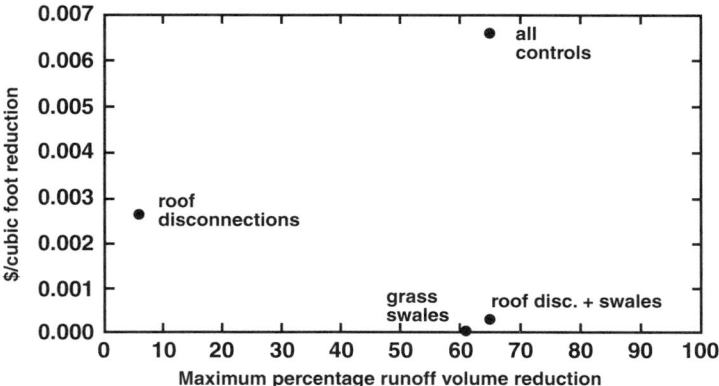

Figure 8.4 Cost-effectiveness data for runoff volume reduction benefits. (From Pitt, R. and J. Voorhees. Source loading and management model (SLAMM). *Seminar Publication: National Conference on Urban Runoff Management: Enhancing Urban Watershed Management at the Local, County, and State Levels.* March 30–April 2, 1993. Center for Environmental Research Information, U.S. Environmental Protection Agency. EPA/625/R-95/003. Cincinnati. OH. pp. 225–243. April 1995.)

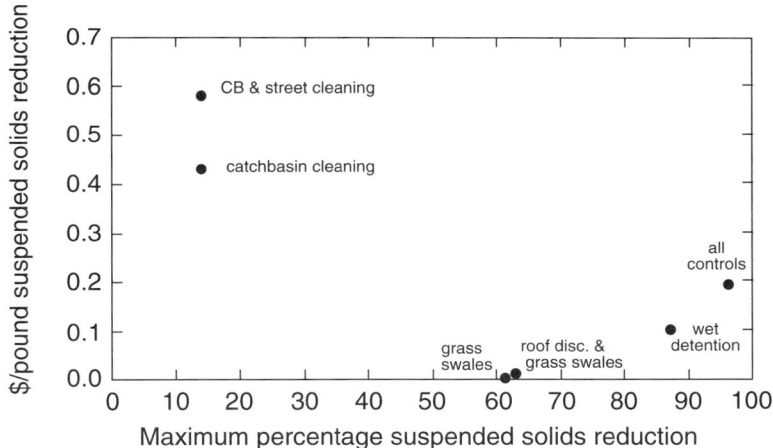

Figure 8.5 Cost-effectiveness data for suspended solids reduction benefits. (From Pitt, R. and J. Voorhees. Source loading and management model (SLAMM). *Seminar Publication: National Conference on Urban Runoff Management: Enhancing Urban Watershed Management at the Local, County, and State Levels.* March 30–April 2, 1993. Center for Environmental Research Information, U.S. Environmental Protection Agency. EPA/625/R-95/003. Cincinnati. OH. pp. 225–243. April 1995.)

Figures 8.4 through 8.7 show the maximum percentage reductions in runoff volume and pollutants, along with associated unit removal costs. As an example, Figure 8.4 shows that roof disconnections would have a very small potential maximum benefit for runoff volume reduction and at a very high unit cost compared to the other practices. The use of grass swales could have about a 60% reduction at minimal cost. The use of roof disconnections plus swales would slightly increase the maximum benefit to about 65%, at a small unit cost. Obviously, the use of roof disconnections alone, or all control practices combined, is very inefficient for this example. For suspended solids control, catchbasin cleaning and street cleaning would have minimal benefit at high cost, while the use of grass swales would produce a substantial benefit at very small cost. However, if additional control is necessary, the use of wet detention ponds may be necessary at a higher cost. If close to 95% reduction of suspended solids was required, then all of the controls investigated could be used together, but at substantial cost.

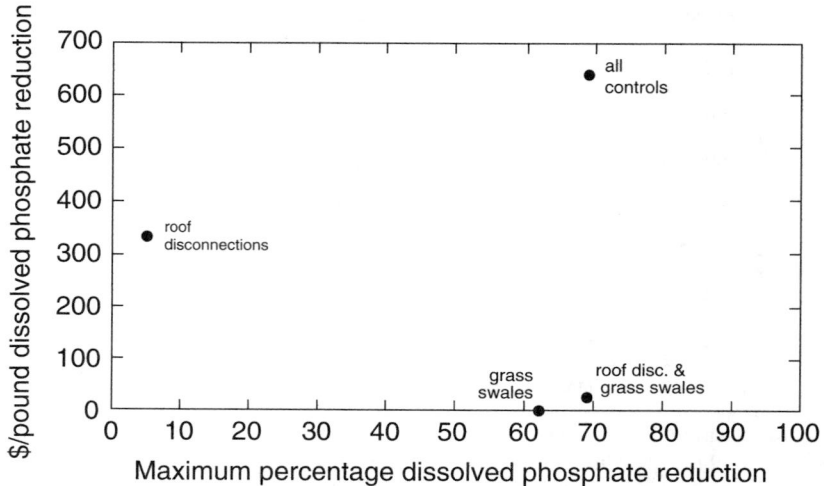

Figure 8.6 Cost-effectiveness data for dissolved phosphate reduction benefits. (From Pitt, R. and J. Voorhees. Source loading and management model (SLAMM). *Seminar Publication: National Conference on Urban Runoff Management: Enhancing Urban Watershed Management at the Local, County, and State Levels.* March 30–April 2, 1993. Center for Environmental Research Information, U.S. Environmental Protection Agency. EPA/625/R-95/003. Cincinnati. OH. pp. 225–243. April 1995.)

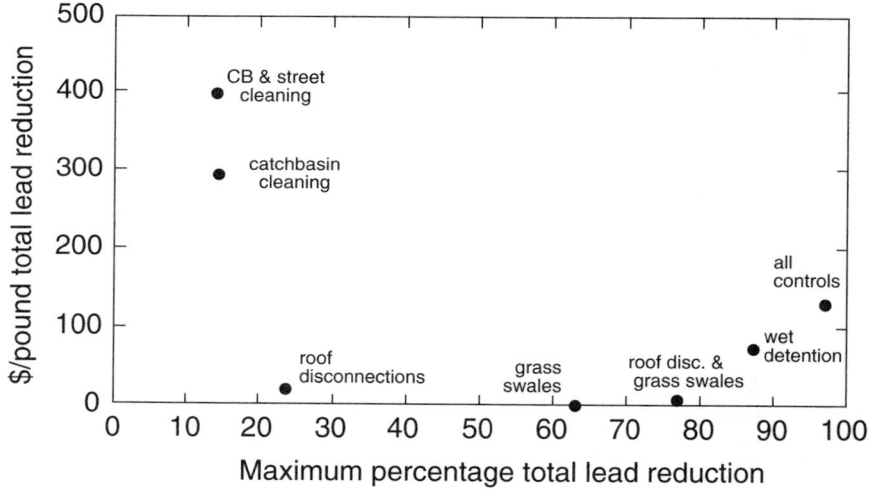

Figure 8.7 Cost-effectiveness data for total lead reduction benefits. (From Pitt, R. and J. Voorhees. Source loading and management model (SLAMM). *Seminar Publication: National Conference on Urban Runoff Management: Enhancing Urban Watershed Management at the Local, County, and State Levels.* March 30–April 2, 1993. Center for Environmental Research Information, U.S. Environmental Protection Agency. EPA/625/R-95/003. Cincinnati. OH. pp. 225–243. April 1995.)

SUMMARY

This chapter has provided some, but limited, insight into how an investigator of urban receiving waters can interpret collected data and develop appropriate conclusions. This is obviously not an easy task. The main chapters of the book include many case studies and a large number of references to illustrate how prior investigators have accomplished this difficult task. The investigator should consult selected references that are similar in location, scope, and/or objectives. Some major theses of this book are summarized below.

It is critical that the investigator have a good idea of what is to be accomplished and develop a suitable experimental design with an appropriate tiered approach. Shortcomings of many investigations can be traced to a lack of initial thought and suitable study hypotheses. In addition, while it is critical to retain flexibility and increase attention given to newly uncovered interesting phenomena, it is important not to keep changing direction based on preliminary conclusions. Of course, the reality of limited resources also precludes continued increases in project scope. Most receiving water investigations are probably only initial investigations, with little prior specific data for the location being studied, and it is natural for many new questions to develop during the studies. A tiered, weight-of-evidence approach enables the most significant objectives to be adequately addressed, with resources available for more detailed investigations to clarify issues.

It is difficult to examine the collected data and clearly identify some beneficial use impairments, especially considering the dynamic and seasonal nature of watersheds and receiving waters. It is easy to miss important short-term events and to misjudge the significance of other seemingly obvious events. Careful and complete sampling, especially if conducted using a stratified random sampling strategy, can help reduce these problems. Well-calibrated and verified models are also important because they allow a long-term perspective of the discretely collected data.

Cause-and-effect relationships tying together stressors and biological impairments are especially difficult to identify and quantify, requiring specialized tests and the weight-of-evidence approach. The use of adequate and appropriate reference sites is very important for biological evaluations, as comparisons to water quality criteria are uncertain for stormwater-related problems, habitat guidance is in its infancy, and contaminated stream sediment guidelines are unclear. The use of available criteria is needed, obviously, but criteria exceedances should be considered red flags to focus site-specific investigations. Human risks should be much more closely related to water quality criteria for water contact, drinking water supplies, and fish consumption, but there is still a potential for error when predicting actual exposure associated with stormwater sources.

The implications of receiving water investigations can be extremely important, especially if remedial action is warranted or if problems are not to be worsened in the future. It is therefore necessary that the whole watershed and associated urban infrastructure be considered. It would be very unusual to find an urban or agricultural receiving water in a completely developed watershed that has significant acceptable uses in the absence of dramatic stormwater controls. In most cases, these streams are managed, it at all, solely for flood control and drainage, with no acknowledgment of other reasonably acceptable uses of noncontact recreation and biological life. Unfortunately, many urban and agricultural streams attract children who play around and in them. Obviously, water-contact recreation and fish consumption are not appropriate uses of most urban and agricultural streams, but these uses do occur, often by members of the community most at risk. These concerns and challenges can be effectively addressed by linking assessments of stormwater effects with progressive watershed management.

REFERENCES

Bachhuber, J.A. A decision making approach for stormwater management measures: a case example in the City of Waukesha, Wisconsin. *North American Water and Environment Congress*, American Society of Civil Engineers, Reston, VA. C-184-1. 1996.

Bannerman, R.T., A.D. Legg, and S.R. Greb. *Quality of Wisconsin Stormwater, 1989–94*. U.S. Geological Survey. Open-file report 96-458. Madison, WI. 26 pp. 1996.

Beck, S., S. Wang, and W. Lee. Two-dimensional transport of resuspended, dredged sediments. *Proceedings of the International Symposium on Environmental Hydraulics Proceedings of the International Symposium on Environmental Hydraulics*. Dec. 16–18 1991. Vol 1. pp 633. A.A. Balkema, Rotterdam, The Netherlands. 1991.

Bitton, G. *Wastewater Microbiology*. John Wiley & Sons, New York. 1994.

Booth, D.B. and C.R. Jackson. Urbanization of aquatic systems: degradation thresholds, stormwater detection, and the limits of mitigation. *J. Am. Water Resour. Assoc.*, 33(5): 1077–1090. October 1997.

Burton, G.A., Jr. *Microbiological Water Quality of Impoundments: A Literature Review*. Miscellaneous Paper E-82-6. Environmental and Water Quality Operational Studies. U.S. Army Engineer Waterways Experiment Station, Vicksburg, MS. 1982.

Burton, G.A., Jr., D. Gunnison, and G.R. Lanza. Survival of pathogenic bacteria in various freshwater sediments. *Appl. Environ. Microbiol.*, 53: 633–638. 1987.

CDC (Centers for Disease Control and Prevention). *CDC Surveillance Summaries*. 1998.

Chapman P.M., R.C. Barrick, J.M. Neff, and R.C. Swartz. Four independent approaches to developing sediment quality criteria yield similar values for model contaminants. *Environ. Toxicol. Chem.*, 6: 723–725. 1987.

Claytor, R. Multiple indicators used to evaluate degrading conditions in Milwaukee County. *Watershed Prot. Tech.*, 2(2): 348–351. Spring 1996.

Committee on Groundwater Recharge, NRC (National Research Council), National Academy of Science. *Ground Water Recharge Using Waters of Impaired Quality*. National Academy Press, Washington, D.C. 284 pp. 1994.

D'Antuono, J.R. Storm water permitting in the Milwaukee River basin, in *Watershed Management: Moving from Theory to Implementation*. May 3–6, 1998. pp. 655–662. Water Environment Federation. 1998.

DuPont, H.L., C.L. Chappell, C.R. Sterling, P.C. Okhuysen, J.B. Rose, and W. Jakubowski. The infectivity of *Cryptosporidium parvum* in healthy volunteers. *N. Eng. J. Med.*, 332(13): 855–859. 1995.

EPA (U.S. Environmental Protection Agency). *Final Report for the Nationwide Urban Runoff Program*. Water Planning Division, Washington, D.C. December 1983.

EPA (U.S. Environmental Protection Agency). *Quality Criteria for Water*. EPA 440/5-86-001. Washington, D.C. U.S. Environmental Protection Agency. May 1986.

EPA. *Risk Assessment for Superfund Volume 1 Human Health Evaluation Manual (Part A)*. EPA/540/1-89/002. U.S. Environmental Protection Agency. Office of Emergency and Remedial Response. Washington, D.C. 1989.

EPA. Wisconsin legislature establishes a nonpoint pollution committee. *Nonpoint Source EPA News-Notes*, #8. October 1990a.

EPA. Milwaukee River South declared a priority watershed in Wisconsin. *Nonpoint Source EPA News-Notes*, #9. December 1990b.

EPA. Wisconsin has had a priority watershed program in operation for more than a decade now: eleven new watersheds named this spring. *Nonpoint Source EPA News-Notes*, #12. April-May 1991.

EPA. *Short-term Methods for Estimating the Chronic Toxicity of Effluents and Receiving Water to Freshwater Organisms*. Research and Development. Cincinnati, OH. EPA/600/4-91/002. 1995.

EPA. *Guidelines for Ecological Risk Assessment*. U.S. Environmental Protection Agency Risk Assessment Forum. Washington, D.C. EPA/630/R-95-002F. 1998.

Erickson, R., C. Kleiner, J. Fiandt, and T. Highland. *Report on the Feasibility of Predicting the Effects of Fluctuating Concentrations on Aquatic Organisms and Possible Application to Water Quality Criteria*. U.S. EPA, Environmental Research Laboratory, Deliverable 7170, Duluth, MN. 1989

Erickson, R., C. Kleiner, J. Fiandt, and T. Highland. *Report on the Use of Toxicity Models to Reduce Uncertainty in Aquatic Hazard Assessments: Effects of Exposure Conditions on Pentachloroethane Toxicity to Fathead Minnows*. U.S. EPA, Environmental Research Laboratory, Deliverable A8215, Duluth, MN. 1991.

Gerba, C.P., J.B. Rose, C.N. Hass, and K.D. Crabtree. Waterborne rotavirus: a risk assessment. *Water Res.*, 30(12): 2929–2940. Dec. 1996.

Haubner, S.M. and E.F. Joeres. Using a GIS for estimating input parameters in urban stormwater quality modeling. *Water Res. Bull.*, 32(6): 1341–1351. December 1996.

Hickie, B.E., L.S. McCarty, and D.G. Dixon. A residue-based toxicokinetic model for pulse-exposure toxicity in aquatic systems. *Environ. Toxicol. Chem.*, 14: 2187–2197. 1995.

Horner, R.R., J.J. Skupien, E.H. Livingston, and H.E. Shaver. *Fundamentals of Urban Runoff Management: Technical and Institutional Issues*. Terrene Institute, Washington, D.C. 1994.

Kim, K., P.G. Thum, and J. Prey. Urban non-point source pollution assessment using a geographical information system. *J. Environ. Manage.*, 39(39): 157–170. 1993.

Kim, K. and S. Ventura. "Large-scale modeling of urban nonpoint source pollution using a geographical information system." *Photogrammetric Eng. Remote Sensing*. 59(10): 1539–1544. October 1993.

Kohn, M.A., T.A. Farley, T. Ando, M. Curtis, S.A. Wilson, Q. Jin, S.S. Monroe, R.C. Baron, L.M. McFarland, and R.I. Glass. An outbreak of Norwalk virus gastroenteritis associated with eating raw oysters. *JAMA*, 272(6): 466–471. 1995.

Legg, A.D., R.T. Bannerman, and J. Panuska. *Variation in the Relation of Rainfall to Runoff from Residential Lawns in Madison, Wisconsin, July and August 1995*. U.S. Geological Survey. Water-Resources Investigations Report 96-4194. Madison, WI. 11 pp. 1996.

Lenat, D.R., D.L. Penrose, and K.W. Eagleson. Variable effects of sediment addition on stream benthos. *Hydrobiologia*, 79, 187–194. 1981.

Lenat, D. and K. Eagleson. *Ecological Effects of Urban Runoff on North Carolina Streams*. North Carolina Division of Environmental Management, Biological Series #104. North Carolina Dept. of Natural Resources and Community Development, Raleigh, NC. 1981.

MacRae, C.R. Experience from morphological research on Canadian streams: is control of the two-year frequency runoff event the best basis for stream channel protection? Presented at the *Effects of Watershed Developments and Management on Aquatic Ecosystems* conference. Snowbird, UT, August 4–9, 1996. Edited by L.A. Roesner. ASCE, New York. 1997.

Madigan, M.T., J.M. Martinko, and J. Parker. *Brock Biology of Microorganisms, 8th ed.* Prentice-Hall, Upper Saddle River, NJ. 1997.

Makepeace, D.K., D.W. Smith, and S.J. Stanley. Urban stormwater quality: summary of contaminant data. *Crit. Rev. Environ. Sci. Technol.*, 25: 93–139. 1995.

Menzie, C., M.H. Henning, J. Cura, K. Finkelstein, J. Gentile, J. Maughan, D. Mitchell, S. Petron, B. Potocki, S. Svirsky, and P. Tyler. Report of the Massachusetts weight-of-evidence workgroup: a weight-of-evidence approach for evaluating ecological risks. *Hum. Ecol. Risk Assess.*, 2: 277–304. 1996.

Meyer, J.S., D.D. Gulley, M.S. Goodrich, D.C. Szmania, and A.S. Brooks. Modeling toxicity due to intermittent exposure of rainbow trout and common shiners to monochloramine. *Environ. Toxicol. Chem.*, 14: 165–175. 1995.

Meyland, S., M. Lalor, and R. Pitt. Environmental and public health impacts on sanitary sewer overflows. *34th Annual American Water Resources Association Conference: Applications of Water Use Information.* Point Clear, AL. November 1998.

Morrison, G.M., C. Wei, and M. Engdahl. Variations of environmental parameters and ecological response in an urban river. *Water Sci. Technol.*, 27: 191–194. 1993.

Novotny, V. and G. Chesters. *Handbook of Nonpoint Pollution Sources and Management*. Van Nostrand Reinhold, New York. 1981.

NRC (National Research Council). *Risk Assessment in the Federal Government: Managing the Process*. National Academy Press, Washington, D.C. 1983.

Ontario Ministry of the Environment. *Humber River Water Quality Management Plan*. Toronto Area Watershed Management Strategy. Toronto, Ontario, 1986.

Pastorok, R.A., M.K. Butcher, and R.D. Nielsen. Modeling wildlife exposure to toxic chemicals: trends and recent advances. *Hum. Ecol. Risk Assess.*, 2: 444–480. 1996.

Pitt, R. *Demonstration of Nonpoint Pollution Abatement through Improved Street Cleaning Practices*. EPA-600/2-79-161, U.S. Environmental Protection Agency, Cincinnati, OH. August 1979.

Pitt, R. and M. Bozeman. *Sources of Urban Runoff Pollution and Its Effects on an Urban Creek*. EPA-600/S2-82-090, U.S. Environmental Protection Agency, Cincinnati, OH. 1982.

Pitt, R. and G. Shawley. *A Demonstration of Non-Point Source Pollution Management on Castro Valley Creek*. Alameda County Flood Control and Water Conservation District (Hayward, CA) for the Nationwide Urban Runoff Program, U.S. Environmental Protection Agency, Water Planning Division, Washington, D.C. June 1982.

Pitt, R. and P. Bissonnette. *Bellevue Urban Runoff Program, Summary Report*. Storm and Surface Water Utility, Bellevue, WA, November 1984.

Pitt, R. *Characterization, Sources, and Control of Urban Runoff by Street and Sewerage Cleaning*. Contract No. R-80597012, U.S. Environmental Protection Agency, Office of Research and Development, Cincinnati, OH, 1984.

Pitt, R. Runoff controls in Wisconsin's priority watersheds. *Conference on Urban Runoff Quality — Impact and Quality Enhancement Technology*. Henniker, NH. Edited by B. Urbonas and L.A. Roesner. Proceedings published by the American Society of Civil Engineering, New York. June 1986.

Pitt, R. and J. McLean. *Toronto Area Watershed Management Strategy Study — Humber River Pilot Watershed Project*. Ontario Ministry of the Environment, Toronto, Ontario. June 1986.

Pitt, R. *Small Storm Flow and Particulate Washoff Contributions to Outfall Discharges*. Ph.D. dissertation, Department of Civil and Environmental Engineering, University of Wisconsin, Madison, November 1987.

Pitt, R.E., R.I. Field, M.M. Lalor, D.D. Adrian, and D. Barbe. *Investigation of Inappropriate Pollutant Entries into Storm Drainage Systems: A User's Guide*. Rep. No. EPA/600/R-92/238, NTIS Rep. No. PB93-131472/AS, U.S. EPA, Storm and Combined Sewer Pollution Control Program (Edison, NJ). Risk Reduction Engineering Lab., Cincinnati, OH. 1993

Pitt, R. and J. Voorhees. Source loading and management model (SLAMM). *Seminar Publication: National Conference on Urban Runoff Management: Enhancing Urban Watershed Management at the Local, County, and State Levels*. March 30–April 2, 1993. Center for Environmental Research Information, U.S. Environmental Protection Agency. EPA/625/R-95/003. Cincinnati. OH. pp. 225–243. April 1995.

Pitt, R., M. Lilburn, S. Nix, S.R. Durrans, S. Burian, J. Voorhees, and J. Martinson. *Guidance Manual for Integrated Wet Weather Flow (WWF) Collection and Treatment Systems for Newly Urbanized Areas (New WWF Systems)*. U.S. Environmental Protection Agency. 612 pp. 1999.

Solo-Gabriele, H. and S. Neumeister. US outbreaks of cryptosporidiosis. *J. Am. Water Works Assoc.*, 88(9): 76–86. Sept. 1996.

Solomon, K.R., D.B. Baker, P. Richards, K.R. Dixon, S.J. Klaine, T.W. La Point, R.J. Kendall, C.P. Weisskopf, J.M. Giddings, J. P. Giesy, L.W. Hall, Jr., and W.M. Williams. Ecological risk assessment of atrazine in North American surface waters. *Environ. Toxicol. Chem.*, 15: 31–76. 1996.

Sovern, D.T. and P.M. Washington. Effects of urban growth on stream habitat. *Proceedings of the Engineering Foundation Conference: Effects of Watershed Development and Management on Aquatic Ecosystems*. Aug 4–9 1996, Snowbird, UT. Sponsored by ASCE, New York. pp. 163–177. 1997.

Suter, G. *Ecological Risk Assessment*. CRC Press, Boca Raton, FL. 1993.

Thum, P.G., S.R. Pickett, B.J. Niemann, Jr., and S.J. Ventura. LIS/GIS: integrating nonpoint pollution assessment with land development planning. *Wisconsin Land Information Newsletter*, University of Wisconsin, Madison. Vol., No. 2, pp. 1–11. 1990.

Ventura, S.J. and K. Kim. Modeling urban nonpoint source pollution with a geographical information system. *Water Resour. Bull.*, 29(2): 189–198. April 1993.

Waller, W.T., M.F. Acevedo, E.L. Morgan, K.L. Dickson, J.H. Kennedy, L.P. Ammann, H.J. Allen, and P.R. Keating. Biological and chemical testing of stormwater, in *Stormwater NPDES Related Monitoring Needs. Proceedings of an Engineering Foundation Conference, Water Resources Planning and Management Division/ASCE*. August 7–12, 1994. Mount Crested Butte, CO. 1995.

Toolbox of Assessment Methods

Habitat Characterization

CONTENTS

The Qualitative Habitat Evaluation Index (QHEI) (OEPA 1989) 643

    Geographical Information 643

    Riffle and Run Habitats...645

    Pool and Glide Habitats 646

    Computing the Total QHEI Score...650

    Stream Map 652

The USEPA Habitat Assessment for the Rapid Bioassessment Protocols (EPA 1999)..............652

    Water Quality...656

    Physical Characterization 658

    Procedure for Performing the Habitat Assessment...662

    Quality Assurance Procedures...662

References 662

While more advanced habitat quantification methods are available (see Chapters 5 and 6), rapid methods have proven to be quite useful. Two of the more popular and similar approaches for assessing the habitat of streams and rivers are described below. These methods are not particularly useful for large rivers, lakes, or coastal areas. Habitat effects may be qualitatively evaluated using methods of the Ohio EPA (OEPA 1989) or the USEPA Rapid Bioassessment Protocol (EPA 1999), both of which are described in this appendix. The Qualitative Habitat Evaluation Index (QHEI) was issued by the OEPA for fish sampling, but it may be also used in any type of stream survey.

THE QUALITATIVE HABITAT EVALUATION INDEX (QHEI) (OEPA 1989)

A general evaluation of macrohabitat is made while sampling each location using the Ohio EPA Site Description Sheet — Fish (Figures A.1 and A.2). This form is used to tabulate data and information for calculating the Qualitative Habitat Evaluation Index (QHEI). The following guidance should be used when completing the site evaluation form.

Geographical Information

    1. *Stream, River Mile (RM), Date*
       The official stream name may be found in the *State Gazetteer* of streams or on USGS 7.5 minute topographic maps.

Ohio EPA Site Description Sheet -   **QHEI SCORE:** [　]

Stream_____ FM_____ Date_____ River Code_____
Location_____ Crew_____

1) SUBSTRATE (Check *Only* Two Substrate TYPE BOXES. Check all types present): **SUBSTRATE SCORE:** [　]

TYPE   Pool Ruffle   Pool Ruffle   **SUBSTRATE QUALITY**
[][]-Blder /Slabs(10)____ ____   [][]-Gravel(7) ____ ____   **Substrate Origin**(Check all)   **Silt Cover** (Check One)
[][] -Boulder (9)   [][]-Sand(6) ____ ____   []-Limestone(1) []-Riprap (0)   []-Silt Heavy(-2)
[][]-Cobble(8)   ____ ____   [][]-Bedrock(5) ____ ____   []-Tells(1)   []-Hardpan(0)   []-Silt Moderate(-1)
[][]-Hardpan(4) ____ ____   [][]-Detritus(3) ____ ____   []-Sandstone(0)   []-Silt Normal (0) []-Silt Free(1)
[][]-Muck(2)   ____ ____   [][]-Artific(0) ____ ____   []-Shale(-1)   **Extent of Embeddness** (Check One)
Total Number of Substrate Types: []>4 (2), []<=4 (0)   []-Coal Fines(-2)   []-Extensive (-2) []Moderate(-1)
NOTE: (Ignore Sludge that originates from point-sources; score based on natural substances) []-Low(0)   []-None(1)
Comments:_____

2) INSTREAM COVER   **COVER SCORE:** [　]

TYPE (Check *All* That Apply)   **Amount:** (Check *ONLY* One or check 2 and AVERAGE)
[]-Undercut Banks(1)   []-Deep Pools(2)   []-Oxbows(1)   []-Extensive > 75% (11)
[]-Overhanging Vegetation(1) []-Rootwads(1)   []-Aquatic Macrophytes(1) []-Moderate 25-75% (7)
[]-Shallows (In Slow Water)(1) []-Boulders(1) []-Logs or Woody Debris(1) []-Sparse 5-25% (3)
                                                         []-Nearly Absent < 5% (1)
Comments:_____

3) CHANNEL MORPHOLOGY: (Check *ONLY* One PER Category OR check 2 and AVERAGE) **CHANNEL:** [　]

SINUOSITY	DEVELOPMENT	CHANNELIZATION	STABILITY	MODIFICATIONS/OTHER	
[]-High (4)	[]-Excellent (7)	[]-None (6)	[]-High (3)	[]-Snagging	[]-Impound
[]-Moderate(3)	[]-Good (5)	[]-Recovered (4)	[]-Moderate(2)	[]-Relocation	[]-Islands
[]-Low (2)	[]-Fair (3)	[]-Recovering (3)	[]-Low (1)	[]-Canopy Removal	[]-Leveed
[]-None (1)	[]-Poor (1)	[]-Recent or No Recovery(1)		[]-Dredging	[]-Bank Shaping
				[]-One Side Channel Modifications	

Comments:_____

4) RIPARIAN ZONE AND BANK EROSION: (Check *ONE* box per bank or check 2 and AVERAGE per bank) **RIPARIAN:** [　]
River Right Looking Downstream

RIPARIAN WIDTH	EROSION/RUNOFF - FLOOD PLAIN QUALITY	BANK EROSION	
L R (Per Bank)	L R (Most Predominant Per Bank)	L R (Per Bank)	
[][]-Wide>50m (4)	[][]-Forest, Swamp (3)	[][]-Urban or Industrial(0)	[][]-None or Little (3)
[][]-Moderate10-50m (3)	[][]-Open Pasture/Rowcrop (0)	[][]-Shrub or Old Field(2)	[][]-Moderate (2)
[][]-Narrow 1-5m (2)	[][]-Resid. Park, New Field (1)	[][]-Conserv. Tillage (1)	[][]-Heavy or Severe(1)
[][]-Very Narrow1-5m(1)	[][]-Fenced Pasture (1)	[][]-Mining/construction(0)	
[][]-None (0)			

Comments:_____

5) POOL/GLIDE AND RIFFLE/RUN QUALITY   **POOL:** [　]

MAX DEPTH (Check 1)	MORPHOLOGY (Check 1)	POOL/RUN/RIFFLE CURRENT VELOCITY (Check *All* That Apply)	
[]- >1m (6)			
[]- 0.7-1m (4)	[]-Pool Width>Riffle Width(2)	[]-Torrential (-1)	[]-Eddies (1)
[]- 0.4-0.7m (2)	[]-Pool Width=Riffle Width(1)	[]-Fast (1)	[]-Interstitial (-1)
[]- <0.4m (1)	[]-Pool Width<Riffle Width(0)	[]-Moderate (1)	[]-Intermittent (-2)
[]- <0.2m (Pool=0)		[]-Slow (1)	[]-No Pool (0)

Comments:_____

RIFFLE: [　]

RIFFLE/RUN DEPTH	RIFFLE/RUN SUBSTRATE	RIFFLE/RUN EMBEDDEDNESS	
[]-Generally>10cm,Max>50(4)	[]-Stable (e.g. Cobble, Boulder) (2)	[]-Extensive (-1)	
[]-Generally>10cm,Max<50(3)	[]-Mod. Stable (e.g. Pea Gravel)(1)	[]-Moderate (0)	
[]-Generally5-10cm (1)	[]-Unstable (Gravel, Sand) (0)	[]-Low (1)	
[]Generally <5cm (Riffle=0)		[]-None (2)	[]-No Ruffle (0)

Comments:_____
6) Gradient (feet/mile):_____   %Pool:_____   %Ruffle:_____   %Run:_____   **GRADIENT:** [　]

Figure A.1 Front side of the Ohio EPA Site Description Sheet for the Qualitative Habitat Evaluation Index (QHEI). (From Ohio Environmental Protection Agency. *The Qualitative Habitat Evaluation Index (QHEI): Rationale, Methods, and Application.* Ecological Assessment Section, Ohio Environmental Protection Agency, Columbus, OH. 1989.)

2. *Specific Location*
   A brief description of the sampling location should include proximity to a local landmark such as a bridge, road, discharge outfall, railroad crossing, park, tributary, dam, etc.
3. *Field Sampling Crew*
   The field crew involved with the sampling is noted on the sheet, with the person who filled out the sheet listed first.

Is Reach Representative of Stream? (Y/N) _____   If Not: _____

_____
_____
_____
_____

	GEAR	DISTANCE	WATER QUALITY	WATER STAGE		
FIRST PASS	_____	_____	_____	_____	_____	_____
SECOND PASS	_____	_____	_____	_____		
THIRD PASS	_____	_____	_____	_____		

                                                                Subjective   Aesthetic
                                                                  Rating   Rating
                                                                  (1-10)   (1-10)

CANOPY (% OPEN) _____   GRADIENT: O-LOW O-MODERATE O-HIGH   PHOTOS:
                _____

STREAM MEASUREMENTS: Average Width: _____   Average Depth: _____   Max. Depth: _____
   LENGTH   WIDTH   DEPTHS >   POOL:Gld;Rif;Run

_____
_____
_____
_____
_____
_____
_____
_____
_____

                              CROSS-SECTIONS OF STREAM

_____   _____   _____

_____   DRAWING OF STREAM   _____

Figure A.2   Reverse side of the Ohio EPA Site Description Sheet for evaluating the geographical and physical characteristics of fish sampling locations. This is used to record additional information about the sampling site and adjacent area. (From Ohio Environmental Protection Agency). *The Qualitative Habitat Evaluation Index (QHEI): Rationale, Methods, and Application.* Ecological Assessment Section, Ohio Environmental Protection Agency, Columbus, OH. 1989.)

4. *Habitat Characteristics: QHEI Metrics*

The Qualitative Habitat Evaluation Index (QHEI) is a physical habitat index designed to provide an empirical, quantified evaluation of the general lotic *macrohabitat* characteristics that are important to fish communities. The QHEI is composed of six principal metrics each of which is described below. The maximum possible QHEI site score is 100. Each of the metrics is scored individually and then summed to provide the total QHEI site score. This is completed at least once for each sampling site during each year of sampling. An exception to this convention is when substantial changes to the macrohabitat have occurred between sampling passes. Standardized definitions for pool, run, and riffle habitats, for which a variety of existing definitions and perceptions exist, are essential for using the QHEI accurately. It is recommended that this reference also be consulted prior to scoring individual sites.

Riffle and Run Habitats

Riffle — areas of the stream with fast current and shallow depth; the water surface is visibly broken.

Run — areas of the stream that have a rapid, nonturbulent flow; runs are deeper than riffles, with a faster current velocity than pools, and are generally located downstream from riffles where the stream narrows; the stream bed is often flat beneath a run and the water surface is not visibly broken.

Pool and Glide Habitats

Pool — an area of the stream with slow current velocity and a depth greater than riffle and run areas; the stream bed is often concave and stream width frequently is the greatest; the water surface slope is nearly zero. If a pool or glide has a maximum depth of less than 20 cm, it is deemed to have lost its functionality and the metric is scored a 0.

Glide — an area common to most modified stream channels that does not have distinguishable pool, run, and riffle habitats; the current and flow are similar to that of a canal; the water surface gradient is nearly zero.

The following is a description of each of the six QHEI metrics and the individual metric components. Guidelines on how to score each are presented. Generally, metrics are scored by checking boxes. In certain cases, the biologist completing the QHEI sheet may interpret a habitat characteristic as being intermediate between the possible choices; in cases where this is allowed (denoted by the term "Double-Checking"), two boxes may be checked and their scores averaged.

Metric 1: Substrate

This metric includes two components, *substrate type* and *substrate quality*.

Substrate Type — Check the two most common substrate types in the stream reach. If one substrate type predominates (greater than approximately 75 to 80%) of the bottom area *or* is clearly the most *functionally* predominant substrate), then this substrate type should be checked twice. **DO NOT CHECK MORE THAN TWO BOXES.** *Note the category for artificial substrates.* Spaces are provided to note the presence (by check marks, or estimates of %, if time allows) of *all* substrate types present in pools and riffles that each comprise at least 5% of the site (i.e., they occur in sufficient quantity to support species that may commonly be associated with the habitat type). This section must be filled out completely to permit future analyses of this metric. If there are more than four substrate types in the zone that are present in greater than approximately 5% of the sampling area, check the appropriate box.

Substrate Quality — Substrate *origin* refers to the "parent" material that the stream substrate is derived from. Check **ONE** box under the substrate origin column *unless* the parent material is from multiple sources (*e.g.*, limestone and tills). **Embeddedness** is the degree to which cobble, gravel, and boulder substrates are surrounded, impacted in, or covered by fine materials (sand and silt). Substrates should be considered embedded if >50% of surface of the substrates is embedded in fine material. Embedded substrates cannot be easily dislodged. This also includes substrates that are concreted or "armor-plated." Naturally sandy streams are not considered embedded; however, a sand-predominated stream that is the result of anthropogenic activities that have buried the natural coarse substrates is considered embedded. Boxes are checked for *extensiveness* (area of sampling zone) of the embedded substrates as follows: **Extensive**: >75% of site area, **Moderate**: 50 to 75%, **Sparse**: 25 to 50%, **Low**: <25%.

Silt Cover — the extent to which substrates are covered by a silt layer (i.e., more than 1 inch thick). **Silt Heavy** means that nearly all of the stream bottom is layered with a deep covering of silt. **Moderate** includes extensive coverings of silts, but with some areas of cleaner substrate (e.g., riffles). **Normal** silt cover includes areas where silt is deposited in small amounts along the stream margin *or* is present as a "dusting" that appears to have little functional significance. If substrates are exceptionally clean, the **Silt Free** box should be checked.

Substrate types are defined as:

a. *Bedrock* — solid rock forming a continuous surface.
b. *Boulder* — rounded stones over 250 mm in diameter (10 in) or large "slabs" more than 256 mm in length (*boulder slabs*).
c. *Cobble* — stones from 64 to 256 mm (2½ to 10 in) in diameter.
d. *Gravel* — mixture of rounded coarse material from 2 to 64 mm (0.8 to 2½ in) in diameter.
e. *Sand* — materials 0.06 to 2.0 mm in diameter, gritty texture when rubbed between fingers.
f. *Silt* — 0.004 to 0.06 mm in diameter; generally this is fine material which feels "greasy" when rubbed between fingers.
g. *Hardpan* — particles less than 0.004 mm in diameter, usually clay, which form a dense, gummy surface that is difficult to penetrate.
h. *Marl* — calcium carbonate; usually grayish-white; often contains fragments of mollusc shells.
i. *Detritus* — dead, unconsolidated organic material covering the bottom, which could include sticks, wood, and other partially or undecayed coarse plant material.
j. *Muck* — black, fine, flocculent, completely decomposed organic matter (*does not include* sewage sludge).
k. *Artificial* — substrates such as rock baskets, gabions, bricks, trash, concrete, etc., placed in the stream for reasons *OTHER* than habitat mitigation.

Sludge is defined as thick layers of organic matter, that is decidedly of human or animal origin. **NOTE: SLUDGE THAT ORIGINATES FROM POINT SOURCES IS NOT INCLUDED; THE SUBSTRATE SCORE IS BASED ON THE UNDERLYING MATERIAL.**

Substrate Metric Score — Although the theoretical maximum metric score is > 20, the maximum score allowed for the QHEI is limited to 20 points.

Metric 2: In-Stream Cover

This metric consists of *in-stream cover type* and *in-stream cover amount*. All of the cover types that are present in amounts greater than approximately 5% of the sampling area (i.e., they occur in sufficient quantity to support species that may commonly be associated with the habitat type) should be checked. Cover should not be counted when it is in areas of the stream with insufficient depth (usually <20 cm) to make it useful. For example, a logjam in 5 cm of water contributes very little if any cover and may be dry at low flow. Other cover types with limited utility in shallow water include *undercut banks and overhanging vegetation, boulders, and rootwads.* Under *amount*, one or two boxes may be checked. *Extensive* cover is that which is present throughout the sampling area, generally greater than about 75% of the stream reach. Cover is *moderate* when it occurs over 25 to 75% of the sampling area. Cover is *sparse* when it is present in less than 25% of the stream margins (sparse cover usually exists in one or more isolated patches). Cover is *nearly absent* when no large patch of any type of cover exists anywhere in the sampling area. This situation is usually found in recently channelized streams or other highly modified reaches (e.g., ship channels). If cover is thought to be intermediate in amount between two categories, *check two boxes and **average** their scores.* Cover types include: (1) undercut banks, (2) overhanging vegetation, (3) shallows (in slow water), (4) logs or woody debris, (5) deep pools (>70 cm), (6) oxbows, (7) boulders, (8) aquatic macrophytes, and (9) rootwads (tree roots that extend into stream). Do not check undercut banks AND rootwads unless undercut banks exist *along with* rootwads as a major component.

Cover Metric Score — Although the theoretical maximum score is >20, the maximum score assigned for the QHEI for the in-stream cover metric is limited to 20 points.

Metric 3: Channel Morphology

This metric emphasizes the quality of the stream channel that relates to the creation and stability of macrohabitat. It includes channel sinuosity (i.e., the degree to which the stream meanders), channel development, channelization, and channel stability. One box under each should be checked unless conditions are considered to be intermediate between two categories; in these cases *check two boxes and **average** their scores.*

- a. *Sinuosity* — **No** sinuosity is a straight channel. **Low** sinuosity is a channel with only one or two poorly defined outside bends in a sampling reach, or perhaps slight meandering within modified banks. **Moderate** sinuosity is more than two outside bends, with at least one bend well defined. **High** sinuosity is more than two or three well-defined outside bends with deep areas outside and shallow areas inside. Sinuosity may be more conceptually described by the ratio of the stream distance between these same two points, taken from a topographic map. Check *only* one box.
- b. *Development* — This refers to the development of riffle/pool complexes. **Poor** means *riffles* are absent, or if present, shallow with sand and fine gravel substrates; *pools*, if present, are shallow. Glide habitats, if predominant, receive a **Poor** rating. **Fair** means riffles are poorly developed or absent; however, pools are more developed with greater variation in depth. **Good** means better defined riffles present with larger substrates (gravel, rubble, or boulder); pools vary in depth and there is a distinct transition between pools and riffles. **Excellent** means development is similar to the Good category except the following characteristics must be present: pools must have a maximum depth of >1 m and deep riffles and runs (>0.5 m) must also be present. In streams sampled with wading methods, a sequence of riffles, runs, and pools must occur more than once in a sampling zone. Check *one* box.
- c. *Channelization* — This refers to anthropogenic channel modifications. **Recovered** refers to streams that have been channelized in the past, but which have recovered most of their natural channel characteristics. **Recovering** refers to channelized streams which are still in the process of regaining their former, natural characteristics; however, these habitats are still degraded. This category also applies to those streams, especially in the Huron/Erie Lake Plain ecoregion (NW Ohio), that were channelized long ago and have a riparian border of mature trees, but still have **Poor** channel characteristics. **Recent** or **No Recovery** refers to streams that were recently channelized or those that show no significant recovery of habitats (e.g., drainage ditches, grass lined or rock riprap banks, etc.). The specific type of habitat modification is also checked in the two columns, but not scored.
- d. *Stability* — This refers to channel stability. Artificially stable (concrete) stream channels receive a **High** score. Even though they are generally a negative influence on fish, the negative effects are related to features other than their stability. Channels with **Low** stability are usually characterized by fine substrates in riffles that often change location, have unstable and severely eroding banks, and a high bedload that slowly creeps downstream. Channels with **Moderate** stability are those that appear to maintain stable riffle/pool and channel characteristics, but which exhibit some symptoms of instability, e.g., high bedload, eroding or false banks, or show the effects of wide fluctuations in water level. Channels with **High** stability have stable banks and substrates, and little or no erosion and bedload.
- e. *Modifications/Other* — Check the appropriate box if impounded, islands present, or leveed (these are not included in the QHEI scoring) as well as the appropriate source of habitat modifications.

The maximum QHEI metric score for Channel Morphology is **20 points**.

Metric 4: Riparian Zone and Bank Erosion

This metric emphasizes the quality of the riparian buffer zone and quality of the floodplain vegetation. This includes riparian zone width, floodplain quality, and extent of bank erosion. Each of the three components requires scoring the left *and* right banks (looking downstream). The *average* of the left and right banks is taken to derive the component value. One box per bank should be checked unless conditions are considered to be intermediate between two categories; in these cases *check two boxes and **average** their scores.*

    a. *Width of Floodplain Vegetation* — This is the width of the riparian (stream side) vegetation. Width estimates are only done for forest, shrub, swamp, and old field vegetation. Old field refers to a fairly mature successional field that has stable, woody plant growth; this generally does not include weedy urban or industrial lots that often still have high runoff potential. Two boxes, one each for the left and right bank (looking downstream), should be checked and then averaged.

    b. *Floodplain Quality* — The two most predominant floodplain quality types should be checked, one each for the left and right banks (includes urban, residential, etc.), and then averaged. By floodplain we mean the areas *immediately outside* the riparian zone *or greater than 100 ft from the stream,* whichever is wider on each side of the stream. These are areas adjacent to the stream that can have direct runoff and erosional effects during normal wet weather. We do not limit it to the riparian zone, and it is much less encompassing than the stream basin.

    c. *Bank Erosion* — The following Streambank Soil Alteration Ratings should be used; check one box for each side of the stream and average the scores. *False banks* mean banks that are no longer adjacent to the normal flow of the channel but have been moved back into the floodplain, most commonly as a result of livestock trampling.

        1. **None** — stream banks are stable and not being altered by water flows or animals (e.g., livestock) — Score **3**.
        2. **Little** — stream banks are stable, but are being lightly altered along the transect line; less than 25% of the stream bank is receiving any kind of stress, and if stress is being received it is very light; less than 25% of the stream bank is false, broken down, or eroding — Score **3**.
        3. **Moderate** — stream banks are receiving moderate alteration along the transect line; at least 50% of the stream bank is in a natural stable condition; less than 50% of the stream bank is false, broken down, or eroding; false banks are rated as altered — Score **2**.
        4. **Heavy** — stream banks have received major alterations along the transect line; less than 50% of the stream bank is in a stable condition; over 50% of the stream bank is false, broken down, or eroding — Score **1**.
        5. **Severe** — stream banks along the transect line are severely altered; less than 25% of the stream bank is in a stable condition; over 75% of the stream bank is false, broken down, or eroding — Score **1**.

The maximum score for Riparian Zone and Erosion metric is **10 points**.

Metric 5: Pool/Glide and Riffle-Run Quality

This metric emphasizes the quality of the pool, glide, and/or riffle-run habitats. This includes pool depth, overall diversity of current velocities (in pools *and* riffles), pool morphology, riffle-run depth, riffle-run substrate, and riffle-run substrate quality.

    A. POOL/GLIDE QUALITY
        1. *Maximum depth of pool or glide* — check one box only (Score **0** to **6**). Pools or glides with maximum depths of less than 20 cm are considered to have lost their function and the total *metric* is scored a **0**. No other characteristics need be scored in this case.
        2. *Current Types* — check each current type that is present in the stream (including riffles and runs; score — **2** to **4**), definitions are:
        Torrential — extremely turbulent and fast flow with large standing waves; water surface is very broken with no definable, connected surface; usually limited to gorges and dam spillway tail-waters.
        Fast — mostly nonturbulent flow with small standing waves in riffle-run areas; water surface may be partially broken, but there is a visibly connected surface.
        Moderate — nonturbulent flow that is detectable and visible (i.e., floating objects are readily transported downstream); water surface is visibly connected.

Slow — water flow is perceptible, but very sluggish.

Eddies — small areas of circular current motion usually formed in pools immediately downstream from riffle-run areas.

Interstitial — water flow that is perceptible only in the interstitial spaces between substrate particles in riffle-run areas.

Intermittent — no flow is evident anywhere leaving standing pools that are separated by dry areas.

3. *Morphology* — Check *Wide* if pools are wider than riffles, *Equal* if pools and riffles are the same width, and *Narrow* if the riffles are wider than the pools (Score **0** to **2**). If the morphology varies throughout the site, *average* the types. If the entire stream area (including areas outside of the sampling zone) is pool or riffle, then check riffle = pool.

Although the theoretical maximum score is >12, the maximum score assigned for the QHEI for the Pool Quality metric is limited to **12 points**.

B. RIFFLE-RUN QUALITY

(score **0** for this metric if no riffles are present)

1. *Riffle/Run Depth* — Select one box that most closely describes the depth characteristics of the riffle (Score **0** to **4**). If the riffle is generally less than 5 cm in depth, riffles are considered to have lost their function and the entire riffle metric is scored a 0.

2. *Riffle/Run Substrate Stability* — Select one box from each that best describes the substrate type and stability of the riffle habitats (Score **0** to **2**).

3. *Riffle/Run Embeddedness* — **Embeddedness** is the degree that cobble, gravel, and boulder substrates are surrounded or covered by fine material (sand, silt). We consider substrates embedded if >50% of the surface of the substrates is embedded in fine material, as these substrates cannot be easily dislodged. This also includes substrates that are concreted. Boxes are checked for *extensiveness* (riffle area of sampling zone) with embedded substrates: **Extensive**: >75% of stream area, **Moderate**: 50 to 75%, **Sparse**: 25 to 50%, **Low**: <25%.

The maximum score assigned for the QHEI for the Riffle/Run Quality metric is **8 points**.

Metric 6: Map Gradient

Local or map gradient is calculated from USGS 7.5 minute topographic maps by measuring the elevation drop through the sampling area. This is done by measuring the stream length between the first contour line upstream and the first contour line downstream of the sampling site and dividing the distance by the contour interval. If the contour lines are closely "packed," a minimum distance of at least 1 mile should be used. Some judgment may need to be exercised in certain anomalous areas (e.g., in the vicinity of waterfalls, impounded areas, etc.), and this can be compared to an in-field, visual estimate, which is recorded on the back of the habitat sheet.

Scoring for ranges of stream gradient takes into account the varying influence of gradient with stream size (measured as drainage area in square miles or stream width). Gradient classifications (Table A.1) were modified from Trautman (1981), and scores were assigned, by stream size category, after examining scatterplots of IBI vs. natural log of gradient in ft/mile. Scores are listed in Table A.1.

The maximum QHEI metric score for Gradient is **10 points**.

Computing the Total QHEI Score

To compute the total **QHEI** score, add the components of each metric to obtain the metric scores and then sum the metric scores to obtain the total **QHEI** score. The **QHEI** metric scores cannot exceed the Metric Maximum Score indicated below:

Additional Information

Additional information is recorded on the reverse side of the Site Description Sheet (Figure A.2) and is described as follows:

Table A.1 Classification of Stream Gradients for Ohio, Corrected for Stream Size. Scores Were Derived from Plots of IBI vs. the Natural Log of Gradient for Each Stream Size Category

Stream Width (m)	Drainage Area (mi²)	Very Low	Low	Low–Moderate	Moderate	Moderate–High	High	Very High[a]
0.3–4.7	0–9.2	0–1.0	1.1–5.0	5.1–10.0	10.1–15.0	15.1–20	20.1–30	30.1–40
		2	4	6	8	10	10	8
4.8–9.2	9.2–41.6	0–1.0	1.1–3.0	3.1–6.0	6.1–12.0	12.1–18.0	18.0–30	30.1–40
		2	4	6	10	10	8	6
9.2–13.8	41.6–103.7	0–1.0	1.1–2.5	2.6–5.0	5.1–7.5	7.6–12.0	12.1–20	20.1–30
		2	4	6	8	10	8	6
3.9–30.6	103.7–622.9 15.1–25		0–1.0	1.1–2.0	2.1–4.0	4.1–6.0	6.1–10.0	10.1–15
		4	6	8	10	10	8	6
>30.6	> 622.9	—	0–0.5	0.6–1.0	1.1–2.5	2.6–4.0	4.1–9.0	> 9.0
			6	8	10	10	10	8

[a] Any site with a gradient greater than the upper bound of the "very high" gradient classification is assigned a score of 4.

1. *Additional Comments/Pollution Impacts* — Different types of pollution sources (e.g., wastewater treatment plant, feedlot, industrial discharge, nonpoint source inputs) are noted with their proximity (in 0.1-mile increments) to the sampling site; any evidence of litter, either in-stream or on the stream bank, is also noted.

2. *Sampling Gear/Distance Sampled* — The type of fish sampling gear used during each pass is specified, and any variation in sampling procedures is noted (e.g., sampler type A specifies sampling along one shoreline of 0.5 km, but due to local restriction, sampling may be performed on both shorelines to accumulate 0.5 km); the total sampling distance in kilometers for each sampling site for each pass is recorded.

3. *Water Clarity* — The following descriptions can be used as a guide:
   a. Clear — bottom is clearly visible (if shallow enough), and the water contains no apparent color or staining.
   b. Stained — usually a brownish (or other) color to the water; the bottom may be visible in shallow areas.
   c. Turbid — bottom seldom visible at more than a few inches; caused by suspended sediment particles. The apparent source of stained (e.g., tannic acid, leaf decay, etc.) and turbid (e.g., runoff [clay/silt], algae/diatoms, sewage, etc.) water may be specified under additional comments.

4. *Water Stage* — This is the general water level of the stream during each pass; suggested descriptors are: a) flood, b) high, c) elevated, d) normal, e) low, and f) interstitial. (Note: sampling should not be conducted during flood or high flows.)

5. *Canopy* — This is the percentage of the sampling site that is not covered or shaded by woody bank vegetation. In wide streams and rivers, this determination should be made along both sides of the river or stream (i.e., the percent of the sampling path that is open).

6. *Gradient* — Check the box that best describes the gradient at the site. This will be used to check the accuracy of gradients taken from topographic maps.

7. *Field Crew* — The names of all individuals involved with the sampling/site description at each site are included.

8. *Photographs* — The number of each photograph taken is recorded; the subject of the photograph is briefly described.

9. *Stream Measurements (optional)* — When measuring the individual sampling sites, length, width, and average and maximum depth information should be recorded; each measurement should be recorded as either a riffle, run, or pool or glide by placing an X in the correct box to the right of where measurements are recorded (Figure A.2); see the introduction for definitions of riffles, runs, etc.

   The number of width measurements is left to the discretion of the field crew leader. Short riffles may require only one or two width measurements, while long pools will probably require more, depending on the degree of variation that exists in the stream's width. Depth measurements should be made in association with individual width measurements. Depths should be taken at the stream margins and various points across the stream. Up to nine depth measurements may be taken, depending on the variability in the stream bottom. Maximum depth is the deepest spot in the stream section sampled. One purpose of this information is to calculate pool volume.

10. *Stream Diagram — Cross sections*: Two or three cross sections of the stream are drawn to provide information on features of the stream bank, stream bottom, stream channel, and floodplain.

   Channel — The cross section containing the stream that is distinct from the surrounding area due to breaks in the general slope of the land, lack of terrestrial vegetation, and changes in the composition of the substrate materials. The channel is made up of stream banks and stream bottoms.

   Banks — The portion of the channel cross section that tends to restrict lateral movement of water. The banks often have a slope steeper than 45° and exhibit a distinct break in slope from the stream bottom. Also, an obvious change in substrate materials is a reliable delineation of the bank.

   Stream Bottom — The portion of the channel cross section not classified as bank. The bottom is usually composed of stream sediments or water transported debris and may be covered by rooted or clinging aquatic vegetation. In some geologic formations, the stream bottom may consist of bedrock rather than sediments.

   Floodplain — The area adjacent to the channel that is seasonally submerged under water. Usually the floodplain is a low area covered by various types of riparian vegetation.

Stream Map

The entire sampling zone is sketched in the area provided. Important physical features are noted on the map with standard symbols used where possible. The sampling path taken is described, along with any other pertinent information.

THE USEPA HABITAT ASSESSMENT FOR THE RAPID BIOASSESSMENT PROTOCOLS (EPA 1999)

Rosgen (1985, 1994, 1996) presented a stream and river classification system that is founded on the premise that dynamically stable stream channels have a morphology that provides appropriate distribution of flow energy during storm events. Further, he identifies eight major variables that affect the stability of channel morphology, but are not mutually independent: channel width, channel depth, flow velocity, discharge, channel slope, roughness of channel materials, sediment load, and sediment particle size distribution. When streams have one of these characteristics altered, some of their capability to dissipate energy properly is lost (Leopold et al. 1964; Rosgen 1985) and will result in accelerated rates of channel erosion. Some of the habitat structural components that function to dissipate flow energy are sinuosity, roughness of bed and bank materials, presence of point bars (slope is an important characteristic), vegetative conditions of stream banks and the riparian zone, condition of the floodplain (accessibility from bank, overflow, and size are important characteristics).

Measurement of these parameters or characteristics serves to stratify and place streams into distinct classifications. However, none of these habitat classification techniques attempts to differentiate the quality of the habitat and the ability of the habitat to support the optimal biological condition of the region. Much of our understanding of habitat relationships in streams has emerged from comparative studies that describe statistical relationships between habitat variables and abundance of biota (Hawkins et al. 1993). A rapid and qualitative habitat assessment approach has been developed to describe the overall quality of the physical habitat (Ball 1982; Ohio EPA 1987; Plafkin et al. 1989; Barbour and Stribling 1991, 1994; Rankin 1991, 1995). For a more detailed guidance, please refer to the original document (USEPA 1999, www.epa.gov/owow/monitoring/rbp/).

The habitat assessment matrix developed for the Rapid Bioassessment Protocols (RBPs) in Plafkin et al. (1989) were originally based on the Stream Classification Guidelines for Wisconsin developed by Ball (1982) and "Methods of Evaluating Stream, Riparian, and Biotic Conditions" developed by Platts et al. (1983). Barbour and Stribling (1991, 1994) modified the habitat assessment approach originally developed for the RBPs to include additional assessment parameters for high-gradient streams and a more appropriate parameter set for low-gradient streams. All parameters are evaluated and rated on a numerical scale of 0 to 20 (highest) for each sampling reach. The ratings are then totaled and compared to a reference condition to provide a final habitat ranking. Scores increase as habitat quality increases. To ensure consistency in the evaluation procedure, descriptions of the physical parameters and relative criteria are included in the rating form (Figures A.3 through A.8).

A biologist who is well versed in the ecology and zoogeography of the region can generally recognize optimal habitat structure as it relates to the biological community. The ability to accurately assess the quality of the physical habitat structure using a visual-based approach depends on several factors: the parameters selected to represent the various features of habitat structure need to be relevant and clearly defined; a continuum of conditions for each parameter must exist that can be characterized from the optimum for the region or stream type under study to the poorest situation reflecting substantial alteration due to anthropogenic activities; the judgment criteria for the attributes of each parameter should minimize subjectivity through either quantitative measurements or specific categorical choices, in which the investigators are experienced or adequately trained, for stream assessments in the region under study (Hannaford et al. 1997); adequate documentation and ongoing training must be maintained to evaluate and correct errors resulting in outliers and aberrant assessments.

Habitat evaluations are first made on in-stream habitat, followed by channel morphology, bank structural features, and riparian vegetation. Generally, a single, comprehensive assessment is made that incorporates features of the entire sampling reach as well as selected features of the catchment. Additional assessments may be made on neighboring reaches to provide a broader evaluation of habitat quality for the stream ecosystem. The actual habitat assessment process involves rating the 10 parameters as optimal, suboptimal, marginal, or poor, based on the criteria included on the Habitat Assessment Field Data Sheets. Some state programs, such as Florida Department of Environmental Protection (DEP) (1996) and Mid-Atlantic Coastal Streams Workgroup (MACS) (1996) have adapted this approach using somewhat fewer and different parameters.

Reference conditions are used to scale the assessment to the "best attainable" situation. This approach is critical to the assessment because stream characteristics will vary dramatically across different regions (Barbour and Stribling 1991). The ratio between the score for the test station and the score for the reference condition provides a percent comparability measure for each station. The station of interest is then classified on the basis of its similarity to expected conditions (reference condition), and its apparent potential to support an acceptable level of biological health. Use of a percent comparability evaluation allows for regional and stream-size differences which affect flow or velocity, substrate, and channel morphology. Some regions are characterized by streams having a low channel gradient, such as coastal plains or prairie regions.

Other habitat assessment approaches or a more rigorously quantitative approach to measuring the habitat parameters may be used (see Klemm and Lazorchak 1994; Kaufmann and Robison 1994; Meador et al. 1993). However, holistic and rapid assessment of a wide variety of habitat attributes along with other types of data is critical if physical measurements are to be used to best advantage in interpreting biological data.

A generic habitat assessment approach based on visual observation can be separated into two basic approaches — one designed for high-gradient streams, and one designed for low-gradient streams. High-gradient or riffle/run prevalent streams are those in moderate- to high-gradient landscapes. Natural high-gradient streams have substrates primarily composed of coarse sediment particles

PHYSICAL CHARACTERIZATION/WATER QUALITY FIELD DATA SHEET
(FRONT)

STREAM NAME	LOCATION
STATION #_____ RIVERMILE_____	STREAM CLASS
LAT _____ LONG _____	RIVER BASIN
STORET #	AGENCY
INVESTIGATORS	

FORM COMPLETED BY	DATE _____ TIME _____ AM PM	REASON FOR SURVEY

WEATHER CONDITIONS	Now		Past 24 hours	Has there been a heavy rain in the last 7 days? ❑ Yes ❑ No
	❑	storm (heavy rain)	❑	
	❑	rain (steady rain)	❑	Air Temperature_____ ° C
	❑	showers (intermittent)	❑	
	____%❑	%cloud cover	❑ ____ %	Other_____
	❑	clear/sunny	❑	

SITE LOCATION/MAP	Draw a map of the site and indicate the areas sampled (or attach a photograph)

STREAM CHARACTERIZATION	**Stream Subsystem** ❑ Perennial ❑ Intermittent ❑ Tidal	**Stream Type** ❑ Coldwater ❑ Warmwater
	Stream Origin ❑ Glacial ❑ Non-glacial montane ❑ Swamp and bog ❑ Spring-fed ❑ Mixture of origins ❑ Other_____	**Catchment Area**_____km²

Figure A.3   For use in Rapid Bioassessment Protocols. (From EPA. *Rapid Bioassessment Protocols for Use in Wadeable Streams and Rivers: Periphyton, Benthic Macroinvertebrates and Fish.* Office of Water, U.S. Environmental Protection Agency, Washington, D.C. EPA 841/B-99/002. 1999.)

(i.e., gravel or larger) or frequent coarse particulate aggregations along stream reaches. Low-gradient or glide/pool prevalent streams are those in low- to moderate-gradient landscapes. Natural low-gradient streams have substrates of fine sediment or infrequent aggregations of more coarse (gravel or larger) sediment particles along stream reaches. The entire sampling reach is evaluated for each parameter. A brief set of decision criteria is given for each parameter corresponding to each of the four categories, reflecting a continuum of conditions on the field sheet (optimal, suboptimal, marginal, and poor).

PHYSICAL CHARACTERIZATION/WATER QUALITY FIELD DATA SHEET
(BACK)

WATERSHED FEATURES	**Predominant Surrounding Landuse** ❏ Forest ❏ Commercial ❏ Field/Pasture ❏ Industrial ❏ Agricultural ❏ Other _____ ❏ Residential	**Local Watershed NPS Pollution** ❏ No evidence ❏ Some potential sources ❏ Obvious sources **Local Watershed Erosion** ❏ None ❏ Moderate ❏ Heavy
RIPARIAN VEGETATION (18 meter buffer)	**Indicate the dominant type and record the dominant species present** ❏ Trees ❏ Shrubs ❏ Grasses ❏ Herbaceous dominant species present _____	
INSTREAM FEATURES	**Estimated Reach Length** _____ m **Estimated Stream Width** _____ m **Sampling Reach Area** _____ m² **Area in km² (m²x1000)** _____ km² **Estimated Stream Depth** _____ m **Surface Velocity** _____ m/sec **(at thalweg)**	**Canopy Cover** ❏ Partly open ❏ Partly shaded ❏ Shaded **High Water Mark** _____ m **Proportion of Reach Represented by Stream Morphology Types** ❏ Riffle_____% ❏ Run_____% ❏ Pool_____% **Channelized** ❏ Yes ❏ No **Dam Present** ❏ Yes ❏ No
LARGE WOODY DEBRIS	**LWD** _____ m² **Density of LWD** _____ m²/km² (LWD/ reach area)	
AQUATIC VEGETATION	**Indicate the dominant type and record the dominant species present** ❏ Rooted emergent ❏ Rooted submergent ❏ Rooted floating ❏ Free floating ❏ Floating Algae ❏ Attached Algae dominant species present _____ **Portion of the reach with aquatic vegetation _____%**	
WATER QUALITY	**Temperature** _____ ° C **Specific Conductance** _____ **Dissolved Oxygen** _____ **pH** _____ **Turbidity** _____ **WQ Instrument Used** _____	**Water Odors** ❏ Normal/None ❏ Sewage ❏ Petroleum ❏ Chemical ❏ Fishy ❏ Other_____ **Water Surface Oils** ❏ Slick ❏ Sheen ❏ Globs ❏ Flecks ❏ None ❏ Other_____ **Turbidity (if not measured)** ❏ Clear ❏ Slightly turbid ❏ Turbid ❏ Opaque ❏ Stained ❏ Other_____
SEDIMENT/ SUBSTRATE	**Odors** ❏ Normal ❏ Sewage ❏ Petroleum ❏ Chemical ❏ Anaerobic ❏ None ❏ Other_____ **Oils** ❏ Absent ❏ Slight ❏ Moderate ❏ Profuse	**Deposits** ❏ Sludge ❏ Sawdust ❏ Paper fiber ❏ Sand ❏ Relict shells ❏ Other_____ **Looking at stones which are not deeply embedded, are the undersides black in color?** ❏ Yes ❏ No

INORGANIC SUBSTRATE COMPONENTS (should add up to 100%)			ORGANIC SUBSTRATE COMPONENTS (does not necessarily add up to 100%)		
Substrate Type	Diameter	% Composition in Sampling Reach	Substrate Type	Characteristic	% Composition in Sampling Area
Bedrock			Detritus	sticks, wood, coarse plant materials (CPOM)	
Boulder	> 256 mm (10")				
Cobble	64-256 mm (2.5"-10")		Muck-Mud	black, very fine organic (FPOM) .	
Gravel	2-64 mm (0.1"-2.5")				
Sand	0.06-2mm (gritty)		Marl	grey, shell fragments	
Silt	0.004-0.06 mm				
Clay	< 0.004 mm (slick)				

Figure A.4 For use in Rapid Bioassessment Protocols. (From EPA. *Rapid Bioassessment Protocols for Use in Wadeable Streams and Rivers: Periphyton, Benthic Macroinvertebrates and Fish.* Office of Water, U.S. Environmental Protection Agency, Washington, D.C. EPA 841/B-99/002. 1999.)

Use of a percent comparability evaluation allows for regional and stream-size differences that affect flow or velocity, substrate, and channel morphology. Some regions are characterized by streams having a low channel gradient. Such streams are typically shallower, have a greater pool/riffle or run/bend ratio, and have a less stable substrate than streams with a steep channel gradient. Although some low-gradient streams do not provide the diversity of habitat or fauna afforded by steeper-

HABITAT ASSESSMENT FIELD DATA SHEET—HIGH GRADIENT STREAMS (FRONT)

STREAM NAME		LOCATION	
STATION #_____ RIVERMILE_____		STREAM CLASS	
LAT _____ LONG _____		RIVER BASIN	
STORET #		AGENCY	
INVESTIGATORS			
FORM COMPLETED BY		DATE _____ TIME _____ AM PM	REASON FOR SURVEY

	Habitat Parameter	Condition Category			
		Optimal	Suboptimal	Marginal	Poor
Parameters to be evaluated in sampling reach	1. Epifaunal Substrate/ Available Cover	Greater than 70% of substrate favorable for epifaunal colonization and fish cover; mix of snags, submerged logs, undercut banks, cobble or other stable habitat and at stage to allow full colonization potential (i.e., logs/snags that are not new fall and not transient).	40-70% mix of stable habitat; well-suited for full colonization potential; adequate habitat for maintenance of populations; presence of additional substrate in the form of newfall, but not yet prepared for colonization (may rate at high end of scale).	20-40% mix of stable habitat; habitat availability less than desirable; substrate frequently disturbed or removed.	Less than 20% stable habitat; lack of habitat is obvious; substrate unstable or lacking.
	SCORE	20 19 18 17 16	15 14 13 12 11	10 9 8 7 6	5 4 3 2 1 0
	2. Embeddedness	Gravel, cobble, and boulder particles are 0-25% surrounded by fine sediment. Layering of cobble provides diversity of niche space.	Gravel, cobble, and boulder particles are 25-50% surrounded by fine sediment.	Gravel, cobble, and boulder particles are 50-75% surrounded by fine sediment.	Gravel, cobble, and boulder particles are more than 75% surrounded by fine sediment.
	SCORE	20 19 18 17 16	15 14 13 12 11	10 9 8 7 6	5 4 3 2 1 0
	3. Velocity/Depth Regime	All four velocity/depth regimes present (slow-deep, slow-shallow, fast-deep, fast-shallow). (Slow is < 0.3 m/s, deep is > 0.5 m.)	Only 3 of the 4 regimes present (if fast-shallow is missing, score lower than if missing other regimes).	Only 2 of the 4 habitat regimes present (if fast-shallow or slow-shallow are missing, score low).	Dominated by 1 velocity/ depth regime (usually slow-deep).
	SCORE	20 19 18 17 16	15 14 13 12 11	10 9 8 7 6	5 4 3 2 1 0
	4. Sediment Deposition	Little or no enlargement of islands or point bars and less than 5% of the bottom affected by sediment deposition.	Some new increase in bar formation, mostly from gravel, sand or fine sediment; 5-30% of the bottom affected; slight deposition in pools.	Moderate deposition of new gravel, sand or fine sediment on old and new bars; 30-50% of the bottom affected; sediment deposits at obstructions, constrictions, and bends; moderate deposition of pools prevalent.	Heavy deposits of fine material, increased bar development; more than 50% of the bottom changing frequently; pools almost absent due to substantial sediment deposition.
	SCORE	20 19 18 17 16	15 14 13 12 11	10 9 8 7 6	5 4 3 2 1 0
	5. Channel Flow Status	Water reaches base of both lower banks, and minimal amount of channel substrate is exposed.	Water fills >75% of the available channel; or <25% of channel substrate is exposed.	Water fills 25-75% of the available channel, and/or riffle substrates are mostly exposed.	Very little water in channel and mostly present as standing pools.
	SCORE	20 19 18 17 16	15 14 13 12 11	10 9 8 7 6	5 4 3 2 1 0

Figure A.5 For use with Rapid Bioassessment Protocols. (From EPA. *Rapid Bioassessment Protocols for Use in Wadeable Streams and Rivers: Periphyton, Benthic Macroinvertebrates and Fish.* Office of Water, U.S. Environmental Protection Agency, Washington, D.C. EPA 841/B-99/002. 1999.)

gradient streams, they are characteristic of certain regions. Using the approach presented here, these streams may be evaluated relative to other low-gradient streams (USEPA 1989).

Assessment Category	Percent of Comparability
Comparable to reference	≥90%
Supporting	75–88%
Partially supporting	60–73%
Nonsupporting	≤58%

Water Quality

Information requested in this section is standard to many aquatic studies and allows for some comparison between sites. Additionally, conditions that may significantly affect aquatic biota are

HABITAT ASSESSMENT FIELD DATA SHEET—HIGH GRADIENT STREAMS (BACK)

Habitat Parameter	Condition Category			
	Optimal	Suboptimal	Marginal	Poor
6. Channel Alteration	Channelization or dredging absent or minimal; stream with normal pattern.	Some channelization present, usually in areas of bridge abutments; evidence of past channelization, i.e., dredging, (greater than past 20 yr) may be present, but recent channelization is not present.	Channelization may be extensive; embankments or shoring structures present on both banks; and 40 to 80% of stream reach channelized and disrupted.	Banks shored with gabion or cement; over 80% of the stream reach channelized and disrupted. Instream habitat greatly altered or removed entirely.
SCORE	20 19 18 17 16	15 14 13 12 11	10 9 8 7 6	5 4 3 2 1 0
7. Frequency of Riffles (or bends)	Occurrence of riffles relatively frequent; ratio of distance between riffles divided by width of the stream <7:1 (generally 5 to 7); variety of habitat is key. In streams where riffles are continuous, placement of boulders or other large, natural obstruction is important.	Occurrence of riffles infrequent; distance between riffles divided by the width of the stream is between 7 to 15.	Occasional riffle or bend; bottom contours provide some habitat; distance between riffles divided by the width of the stream is between 15 to 25.	Generally all flat water or shallow riffles; poor habitat; distance between riffles divided by the width of the stream is a ratio of >25.
SCORE	20 19 18 17 16	15 14 13 12 11	10 9 8 7 6	5 4 3 2 1 0
8. Bank Stability (score each bank) Note: determine left or right side by facing downstream.	Banks stable; evidence of erosion or bank failure absent or minimal; little potential for future problems. <5% of bank affected.	Moderately stable; infrequent, small areas of erosion mostly healed over. 5-30% of bank in reach has areas of erosion.	Moderately unstable; 30-60% of bank in reach has areas of erosion; high erosion potential during floods.	Unstable; many eroded areas; "raw" areas frequent along straight sections and bends; obvious bank sloughing; 60-100% of bank has erosional scars.
SCORE ___ (LB)	Left Bank 10 9	8 7 6	5 4 3	2 1 0
SCORE ___ (RB)	Right Bank 10 9	8 7 6	5 4 3	2 1 0
9. Vegetative Protection (score each bank)	More than 90% of the streambank surfaces and immediate riparian zone covered by native vegetation, including trees, understory shrubs, or nonwoody macrophytes; vegetative disruption through grazing or mowing minimal or not evident; almost all plants allowed to grow naturally.	70-90% of the streambank surfaces covered by native vegetation, but one class of plants is not well-represented; disruption evident but not affecting full plant growth potential to any great extent; more than one-half of the potential plant stubble height remaining.	50-70% of the streambank surfaces covered by vegetation; disruption obvious; patches of bare soil or closely cropped vegetation common; less than one-half of the potential plant stubble height remaining.	Less than 50% of the streambank surfaces covered by vegetation; disruption of streambank vegetation is very high; vegetation has been removed to 5 centimeters or less in average stubble height.
SCORE ___ (LB)	Left Bank 10 9	8 7 6	5 4 3	2 1 0
SCORE ___ (RB)	Right Bank 10 9	8 7 6	5 4 3	2 1 0
10. Riparian Vegetative Zone Width (score each bank riparian zone)	Width of riparian zone >18 meters; human activities (i.e., parking lots, roadbeds, clear-cuts, lawns, or crops) have not impacted zone.	Width of riparian zone 12-18 meters; human activities have impacted zone only minimally.	Width of riparian zone 6-12 meters; human activities have impacted zone a great deal.	Width of riparian zone <6 meters: little or no riparian vegetation due to human activities.
SCORE ___ (LB)	Left Bank 10 9	8 7 6	5 4 3	2 1 0
SCORE ___ (RB)	Right Bank 10 9	8 7 6	5 4 3	2 1 0

Parameters to be evaluated broader than sampling reach

Total Score _____

Figure A.6 For use with Rapid Bioassessment Protocols. (From EPA. *Rapid Bioassessment Protocols for Use in Wadeable Streams and Rivers: Periphyton, Benthic Macroinvertebrates and Fish.* Office of Water, U.S. Environmental Protection Agency, Washington, D.C. EPA 841/B-99/002. 1999.)

documented. Documentation of recent and current weather conditions is important because of the potential impact that weather may have on water quality. To complete this phase of the bioassessment, a photograph may be helpful in identifying station location and documenting habitat conditions. Any observations or data not requested but deemed important by the field observer should be recorded. This section is identical for all protocols, and the specific data requested are described below:

Temperature (C), Dissolved Oxygen, pH, Conductivity — Measure and record values for each of the water quality parameters indicated, using the appropriate calibrated water quality instrument(s). Note the type of instrument and unit number used.

HABITAT ASSESSMENT FIELD DATA SHEET—LOW GRADIENT STREAMS (FRONT)

STREAM NAME	LOCATION	
STATION #_____ RIVERMILE_____	STREAM CLASS	
LAT _____ LONG _____	RIVER BASIN	
STORET #	AGENCY	
INVESTIGATORS		
FORM COMPLETED BY	DATE _____ TIME _____ AM PM	REASON FOR SURVEY

	Habitat Parameter	Condition Category			
		Optimal	Suboptimal	Marginal	Poor
Parameters to be evaluated in sampling reach	1. Epifaunal Substrate/ Available Cover	Greater than 50% of substrate favorable for epifaunal colonization and fish cover; mix of snags, submerged logs, undercut banks, cobble or other stable habitat and at stage to allow full colonization potential (i.e., logs/snags that are not new fall and not transient).	30-50% mix of stable habitat; well-suited for full colonization potential; adequate habitat for maintenance of populations; presence of additional substrate in the form of newfall, but not yet prepared for colonization (may rate at high end of scale).	10-30% mix of stable habitat; habitat availability less than desirable; substrate frequently disturbed or removed.	Less than 10% stable habitat; lack of habitat is obvious; substrate unstable or lacking.
	SCORE	20 19 18 17 16	15 14 13 12 11	10 9 8 7 6	5 4 3 2 1 0
	2. Pool Substrate Characterization	Mixture of substrate materials, with gravel and firm sand prevalent; root mats and submerged vegetation common.	Mixture of soft sand, mud, or clay; mud may be dominant; some root mats and submerged vegetation present.	All mud or clay or sand bottom; little or no root mat; no submerged vegetation.	Hard-pan clay or bedrock; no root mat or vegetation.
	SCORE	20 19 18 17 16	15 14 13 12 11	10 9 8 7 6	5 4 3 2 1 0
	3. Pool Variability	Even mix of large-shallow, large-deep, small-shallow, small-deep pools present.	Majority of pools large-deep; very few shallow.	Shallow pools much more prevalent than deep pools.	Majority of pools small-shallow or pools absent.
	SCORE	20 19 18 17 16	15 14 13 12 11	10 9 8 7 6	5 4 3 2 1 0
	4. Sediment Deposition	Little or no enlargement of islands or point bars and less than <20% of the bottom affected by sediment deposition.	Some new increase in bar formation, mostly from gravel, sand or fine sediment; 20-50% of the bottom affected; slight deposition in pools.	Moderate deposition of new gravel, sand or fine sediment on old and new bars; 50-80% of the bottom affected; sediment deposits at obstructions, constrictions, and bends; moderate deposition of pools prevalent.	Heavy deposits of fine material, increased bar development; more than 80% of the bottom changing frequently; pools almost absent due to substantial sediment deposition.
	SCORE	20 19 18 17 16	15 14 13 12 11	10 9 8 7 6	5 4 3 2 1 0
	5. Channel Flow Status	Water reaches base of both lower banks, and minimal amount of channel substrate is exposed.	Water fills >75% of the available channel; or <25% of channel substrate is exposed.	Water fills 25-75% of the available channel, and/or riffle substrates are mostly exposed.	Very little water in channel and mostly present as standing pools.
	SCORE	20 19 18 17 16	15 14 13 12 11	10 9 8 7 6	5 4 3 2 1 0

Figure A.7 For use with Rapid Bioassessment Protocols. (From EPA. *Rapid Bioassessment Protocols for Use in Wadeable Streams and Rivers: Periphyton, Benthic Macroinvertebrates and Fish.* Office of Water, U.S. Environmental Protection Agency, Washington, D.C. EPA 841/B-99/002. 1999.)

Stream Type — Note the appropriate stream designation according to state water quality standards.

Water Odors — Note those odors described (or include any other odors not listed) that are associated with the water in the sampling area.

Water Surface Oils — Note the term that best describes the relative amount of any oils present on the water surface.

Turbidity — Note the term which, based upon visual observation, best describes the amount of material suspended in the water column.

Physical Characterization

Physical characterization parameters include estimations of general land use and physical stream characteristics such as width, depth, flow, and substrate. The evaluation begins with the riparian zone (stream bank and drainage area) and proceeds in-stream to sediment/substrate descriptions. Such information will provide insight as to what organisms may be present or are expected to be

HABITAT ASSESSMENT FIELD DATA SHEET—LOW GRADIENT STREAMS (BACK)

Parameters to be evaluated broader than sampling reach

Habitat Parameter	Condition Category			
	Optimal	Suboptimal	Marginal	Poor
6. Channel Alteration	Channelization or dredging absent or minimal; stream with normal pattern.	Some channelization present, usually in areas of bridge abutments; evidence of past channelization, i.e., dredging, (greater than past 20 yr) may be present, but recent channelization is not present.	Channelization may be extensive; embankments or shoring structures present on both banks; and 40 to 80% of stream reach channelized and disrupted.	Banks shored with gabion or cement; over 80% of the stream reach channelized and disrupted. Instream habitat greatly altered or removed entirely.
SCORE	20 19 18 17 16	15 14 13 12 11	10 9 8 7 6	5 4 3 2 1 0
7. Channel Sinuosity	The bends in the stream increase the stream length 3 to 4 times longer than if it was in a straight line. (Note - channel braiding is considered normal in coastal plains and other low-lying areas. This parameter is not easily rated in these areas.)	The bends in the stream increase the stream length 1 to 2 times longer than if it was in a straight line.	The bends in the stream increase the stream length 1 to 2 times longer than if it was in a straight line.	Channel straight; waterway has been channelized for a long distance.
SCORE	20 19 18 17 16	15 14 13 12 11	10 9 8 7 6	5 4 3 2 1 0
8. Bank Stability (score each bank)	Banks stable; evidence of erosion or bank failure absent or minimal; little potential for future problems. <5% of bank affected.	Moderately stable; infrequent, small areas of erosion mostly healed over. 5-30% of bank in reach has areas of erosion.	Moderately unstable; 30-60% of bank in reach has areas of erosion; high erosion potential during floods.	Unstable; many eroded areas; "raw" areas frequent along straight sections and bends; obvious bank sloughing; 60-100% of bank has erosional scars.
SCORE ___ (LB)	Left Bank 10 9	8 7 6	5 4 3	2 1 0
SCORE ___ (RB)	Right Bank 10 9	8 7 6	5 4 3	2 1 0
9. Vegetative Protection (score each bank) Note: determine left or right side by facing downstream.	More than 90% of the streambank surfaces and immediate riparian zone covered by native vegetation, including trees, understory shrubs, or nonwoody macrophytes; vegetative disruption through grazing or mowing minimal or not evident; almost all plants allowed to grow naturally.	70-90% of the streambank surfaces covered by native vegetation, but one class of plants is not well-represented; disruption evident but not affecting full plant growth potential to any great extent; more than one-half of the potential plant stubble height remaining.	50-70% of the streambank surfaces covered by vegetation; disruption obvious; patches of bare soil or closely cropped vegetation common; less than one-half of the potential plant stubble height remaining.	Less than 50% of the streambank surfaces covered by vegetation; disruption of streambank vegetation is very high; vegetation has been removed to 5 centimeters or less in average stubble height.
SCORE ___ (LB)	Left Bank 10 9	8 7 6	5 4 3	2 1 0
SCORE ___ (RB)	Right Bank 10 9	8 7 6	5 4 3	2 1 0
10. Riparian Vegetative Zone Width (score each bank riparian zone)	Width of riparian zone >18 meters; human activities (i.e., parking lots, roadbeds, clear-cuts, lawns, or crops) have not impacted zone.	Width of riparian zone 12-18 meters; human activities have impacted zone only minimally.	Width of riparian zone 6-12 meters; human activities have impacted zone a great deal.	Width of riparian zone <6 meters: little or no riparian vegetation due to human activities.
SCORE ___ (LB)	Left Bank 10 9	8 7 6	5 4 3	2 1 0
SCORE ___ (RB)	Right Bank 10 9	8 7 6	5 4 3	2 1 0

Total Score _____

Figure A.8   For use with Rapid Bioassessment Protocols. (From EPA. *Rapid Bioassessment Protocols for Use in Wadeable Streams and Rivers: Periphyton, Benthic Macroinvertebrates and Fish.* Office of Water, U.S. Environmental Protection Agency, Washington, D.C. EPA 841/B-99/002. 1999.)

present, and to presence of stream impacts. The information requested in the Physical Characterization section of the Field Data Sheet is briefly discussed below:

Predominant Surrounding Land Use — Observe the prevalent land-use type in the vicinity (noting any other land uses in the area which, although not predominant, may potentially affect water quality).

Local Watershed Erosion — The existing or potential detachment of soil within the local watershed (the portion of the watershed that drains directly into the stream) and its movement into a stream are noted. Erosion can be rated through visual observation of watershed and stream characteristics. (Note any turbidity observed during water quality assessment below.)

Local Watershed Nonpoint Source Pollution — This item refers to problems and potential problems other than siltation. Nonpoint source pollution is defined as diffuse agricultural and urban runoff. Other compromising factors in a watershed that may affect water quality are feedlots, wetlands, septic systems, dams and impoundments, and/or mine seepage.

Estimated Stream Width (m) — Estimate the distance from shore to shore at a transect representative of the stream width in the area.

Estimated Stream Depth (m) — Riffle, run, and pool. Estimate the vertical distance from water surface to stream bottom at a representative depth at each of the three habitat types.

High Water Mark (m) — Estimate the vertical distance from the stream bank to the peak overflow level, as indicated by debris hanging in bank or floodplain vegetation, and deposition of silt or soil. In instances where bank overflow is rare, a high water mark may not be evident.

Velocity — Record an estimate of stream velocity in a representative run area.

Dam Present — Indicate the presence or absence of a dam upstream or downstream of the sampling station. If a dam is present, include specific information relating to alteration of flow.

Channelized — Indicate whether or not the area around the sampling station is channelized.

Canopy Cover — Note the general proportion of open to shaded area which best describes the amount of cover at the sampling station.

Sediment Odors — Disturb sediment and note any odors described (or include any other odors not listed) which are associated with sediment in the area of the sampling station.

Sediment Oils — Note the term which best describes the relative amount of sediment oils observed in the sampling area.

Sediment Deposits — Note those deposits described (or include any other deposits not listed) which are present in the sampling area. Also indicate whether the undersides of rocks not deeply embedded are black (which generally indicates low dissolved oxygen or anaerobic conditions).

Inorganic Substrate Components — Visually estimate the relative proportion of each of the several substrate/particle types listed that are present in the sampling area.

Organic Substrate Components — Indicate relative abundance of each of the three substrate types listed.

Listed below is a general explanation of some major habitat parameters to be evaluated.

Substrate and In-Stream Cover

The in-stream habitat characteristics directly pertinent to the support of aquatic communities consist of substrate type and stability, availability of refugia, and migration/passage potential.

Bottom Substrate — This refers to the availability of habitat for support of aquatic organisms. A variety of substrate materials and habitat types is desirable. The presence of rock and gravel in flowing streams is generally considered the most desirable habitat. However, other forms of habitat may provide the niches required for community support. For example, logs, tree roots, submerged or emergent vegetation, undercut banks, etc., will provide excellent habitat for a variety of organisms, particularly fish. Bottom substrate is evaluated and rated by observation.

Embeddedness — The degree to which boulders, rubble, or gravel are surrounded by fine sediment indicates suitability of the stream substrate as habitat for benthic macroinvertebrates and for fish spawning and egg incubation. Embeddedness is evaluated by visual observation of the degree to which larger particles are surrounded by sediment. In some western areas of the United States, embeddedness is regarded as the stability of cobble substrate by measuring the depth of burial of large particles (cobble, boulders).

Stream Discharge and/or Stream Velocity — Stream discharge relates to the ability of a stream to provide and maintain a stable aquatic environment. Stream discharge (and water quality) is most critical to the support of aquatic communities when the representative low flow is ≤0.15 cms (5 cfs). In these small streams, discharge should be estimated in a straight stretch of run area where banks are parallel and bottom contour is relatively flat. Even where a few stations may have discharges in excess of 0.15 cms, discharge may still be the predominating constraint. Therefore, the evaluation is based on discharge rate rather than velocity.

In larger streams and rivers (>0.15 cms), velocity, in conjunction with depth, has a more direct influence than the discharge rate on the structure of benthic communities (Osborne and Hendricks 1983) and fish communities (Oswood and Barber 1982). The quality of the aquatic habitat can therefore be evaluated in terms of a velocity and depth relationship. As patterned after Oswood and Barber (1982), four general categories of velocity and depth are optimal for benthic and fish communities: (1) slow (<0.3 m/s), shallow (<0.5 m); (2) slow (<0.3 m/l), deep (>0.5 m); (3) fast (>0.3 m/s), deep (0.5 m); and (4) fast (>0.3 m/s), shallow (<0.5 m). Habitat quality is reduced in the absence of one or more of these four categories.

Channel Morphology

Channel morphology is determined by the flow regime of the stream, local geology, land surface form, soil, and human activities (Platts et al. 1983). The sediment movement along the channel, as influenced by the tractive forces of flowing water and the sinuosity of the channel, also affects habitat conditions.

Channel Alteration — The character of sediment deposits from upstream is an indication of the severity of watershed and bank erosion and stability of the stream system. The growth, or appearance, of sediment bars tends to increase with continued watershed disturbance. Channel alteration also results in deposition, which may occur on the inside of bends, below channel constrictions, and where stream gradient flattens out. Channelization (e.g., straightening, construction of concrete embankments) decreases stream sinuosity, thereby increasing stream velocity and the potential for scouring.

Bottom Scouring and Deposition — These parameters relate to the destruction of in-stream habitat resulting from the problems described above. Characteristics to observe are scoured substrate and degree of siltation in pools and riffles. Scouring results from high-velocity flows. The potential for scouring is increased by channelization. Deposition and scouring result from the transport of sediment or other particulates and may be an indication of large-scale watershed erosion. Deposition and scouring are rated by estimating the percentage of an evaluated reach that is scoured or silted (i.e., 50-ft silted in a 100-ft stream length equals 50%).

Pool/Riffle or Run/Bend Ratio — These parameters assume that a stream with riffles or bends provides more diverse habitat than a straight (run) or uniform depth stream. Bends are included because low-gradient streams may not have riffle areas, but excellent habitat can be provided by the cutting action of water at bends. The ratio is calculated by dividing the average distance between riffles or bends by the average stream width. If a stream contains riffles and bends, the dominant feature with the best habitat should be used.

Riparian and Bank Structure

Well-vegetated banks are usually stable regardless of bank undercutting; undercutting actually provides excellent cover for fish (Platts et al. 1983). The ability of vegetation and other materials on the stream banks to prevent or inhibit erosion is an important determinant of the stability of the stream channel and in-stream habitat for indigenous organisms. Because riparian and bank structure indirectly affect the in-stream habitat features, they are weighted less than the primary or secondary parameters.

The upper bank is the land area from the break in the general slope of the surrounding land to the normal high water line. The upper bank is normally vegetated and covered by water only during extreme high water conditions. Land forms vary from wide, flat floodplains to narrow, steep slopes. The lower bank is the intermittently submerged portion of the stream cross section from the normal high water line to the lower water line. The lower channel defines the stream width.

Bank Stability — Bank stability is rated by observing existing or potential detachment of soil from the upper and lower stream bank and its potential movement into the stream. Steeper banks are generally more subject to erosion and failure, and may not support stable vegetation. Streams with poor banks will often have poor in-stream habitat. Adjustments should be made in areas with clay banks where steep, raw areas may not be as susceptible to erosion as other soil types.

Bank Vegetative Stability — Bank soil is generally held in place by plant root systems. Erosional protection may also be provided by boulder, cobble, or gravel material. An estimate of the density of bank vegetation (or proportion of boulder, cobble, or gravel material) covering the bank provides an indication of bank stability and potential in-stream sedimentation.

Streamside Cover — Streamside cover vegetation is evaluated in terms of stream-shading and escape cover or refuge for fish. A rating is obtained by visually determining the dominant vegetation type covering the exposed stream bottom, bank, and top of bank. Platts (1974) found that streamside cover consisting primarily of shrub had a higher fish standing crop than similar-size streams having tree or grass streamside cover. Riparian vegetation dominated by shrubs and trees provides the CPOM source in allochthonous systems.

Procedure for Performing the Habitat Assessment

1. Select the reach to be assessed. The habitat assessment is performed on the same 100-m reach (or other reach designation [e.g., 40 × stream wetted width]) from which the biological sampling is conducted. Some parameters require an observation of a broader section of the catchment than just the sampling reach.
2. Complete the station identification section of each field data sheet and habitat assessment form (Figures A.2 through A.7).
3. It is best for the investigators to obtain a close look at the habitat features to make an adequate assessment. If the physical and water quality characterization and habitat assessment are done before the biological sampling, care must be taken to avoid disturbing the sampling habitat.
4. Complete the Physical Characterization and Water Quality Field Data Sheet. Sketch a map of the sampling reach on the back of the form.
5. Complete the Habitat Assessment Field Data Sheet, in a team of two or more biologists, if possible, to come to a consensus on determination of quality. Those parameters to be evaluated on a scale greater than a sampling reach require traversing the stream corridor to the extent deemed necessary to assess the habitat feature. As a general rule-of-thumb, use two lengths of the sampling reach to assess these parameters.

Quality Assurance Procedures

1. Each biologist is to be trained in the visual-based habitat assessment technique for the applicable region or state.
2. The judgment criteria for each habitat parameter are calibrated for the stream classes under study. Some text modifications may be needed on a regional basis.
3. Periodic checks of assessment results are completed using pictures of the sampling reach and discussions among the biologists in the agency.

REFERENCES

Ball, J. *Stream Classification Guidelines for Wisconsin*. Wisconsin Department of Natural Resources Technical Bulletin, Wisconsin Department of Natural Resources, Madison, WI. 1982.

Barbour, M.T. and J.B. Stribling. Use of habitat assessment in evaluating the biological integrity of stream communities, in *Biological Criteria: Research and Regulation*. Edited by G. Gibson. Proceedings of a Symposium, Office of Water, U.S. Environmental Protection Agency, Washington, D.C. EPA-440-5-91-005. 1991.

Barbour, M.T. and J.B. Stribling. A technique for assessing stream habitat structure, in *Conference Proceedings, Riparian Ecosystems in the Humid U.S.: Functions, Values and Management*. National Assoc. of Conservation Districts, Washington, D.C., Atlanta, GA., pp. 156–178. 1994.

EPA. *Methods for Measuring the Toxicity and Bioaccumulation of Sediment-Associated Contaminants with Freshwater Invertebrates. 2nd edition*. EPA series number pending, Duluth, MN. 1999.

EPA. *Rapid Bioassessment Protocols for Use in Wadeable Streams and Rivers: Periphyton, Benthic Macroinvertebrates and Fish.* Office of Water, U.S. Environmental Protection Agency, Washington, D.C. EPA 841/B-99/002. 1999.

Hannaford, M.J., M.T. Barbour, and V.H. Resh. Training reduces observer variability in visual-based assessments of stream habitat. *J. North Am. Benthol. Soc.*, 16: 853–860. 1997.

Hawkins, C.P., J.L. Kersher, P.A. Bisson, M.D. Bryant, L.M. Decker, S.V. Gregory, D.A. McCullough, C.K. Overton, G.H. Reeves, R.J. Steedman, and M.K. Young. A hierarchical approach to classifying stream habitat features. *Fisheries*, 18: 3–12. 1993.

Kaufmann, P.R. and E.G. Robison. Physical habitat assessment, in *Environmental Monitoring and Assessment Program*. Edited by D.J. Klemm and J.M. Lazorchak. 1994 Pilot Field Operations Manual for Stream. Environmental Monitoring Systems Laboratory, Office of Research and Development, U.S. Environmental Protection Agency, Cincinnati, OH. EPA/620/R-94/004. pp. 6-1 to 6-38. 1994.

Klemm, D.J. and J.M. Lazorchak. *Environmental Monitoring and Assessment Program*. 1994. Pilot Field Operations Manual for Stream. Environmental Monitoring Systems Laboratory, Office of Research and Development, U.S. Environmental Protection Agency, Cincinnati, OH. EPA/620/R-94/004. 1994.

Leopold, L.B., M.G. Wolman, and J.P. Miller. *Fluvial Processes in Geomorphology*. W.H. Freeman, San Francisco, CA. 1964.

MACS (Mid-Atlantic Coastal Streams Workgroup). *Standard Operating Procedures and Technical Basis: Macroinvertebrate Collection and Habitat Assessment for Low-Gradient Nontidal Streams*. Delaware Dept. Natural Resources and Environ. Conservation, Dover, DE. 1996.

Meador, M.R., C.R. Hupp, T.F. Cuffney, and M.E. Gurtz. *Methods for Characterizing Stream Habitat as Part of the National Water-Quality Assessment Program*. U.S. Geological Survey Open-File Report, Raleigh, NC. USGS/OFR 93-408. 1993.

OEPA (Ohio Environmental Protection Agency). *Biological Criteria for the Protection of Aquatic Life: Volume II. User's Manual for Biological Assessment of Ohio Surface Waters*. Ohio Environmental Protection Agency, Columbus, OH. 1987.

OEPA (Ohio Environmental Protection Agency). *The Qualitative Habitat Evaluation Index (QHEI): Rationale, Methods, and Application*. Ecological Assessment Section, Ohio Environmental Protection Agency, Columbus, OH. 1989.

Osborne, L.L. and E.E. Hendricks. *Streamflow and Velocity as Determinants of Aquatic Insect Distribution and Benthic Community Structure in Illinois*. Water Resources Center, University of Illinois, Report No. UILU-WRC-83-183, U.S. Department of the Interior, Bureau of Reclamation. 1983.

Oswood, M.E. and W.E. Barber. Assessment of fish habitat in streams: goals, constraints, and a new technique. *Fisheries*, 7: 8–11 (1982).

Plafkin, J.L., M.T. Barbour, K.D. Porter, S.K. Gross, and R.M. Hughes. *Rapid Bioassessment Protocols for Use in Streams and Rivers: Benthic Macroinvertebrates and Fish*. EPA-440-4-89-001. U.S. Environmental Protection Agency. Office of Water Regulations and Standards. Washington, D.C. 1989.

Platts, W.S. *Geomorphic and Aquatic Conditions Influencing Salmonids and Stream Classification with Application to Ecosystem Management*. U.S. Department of Agriculture SEAM program, Billings, MT. 1974.

Platts, W.S., W.F. Megahan, and G.W. Minshall. *Methods for Evaluating Stream, Riparian and Biotic Conditions*. U.S. Department of Agriculture General Technical Report INT-138, Ogden, UT. 1983.

Rankin, E.T. The use of the qualitative habitat evaluation index for use attainability studies in streams and rivers in Ohio, in *Biological Criteria: Research and Regulation*. Edited by G. Gibson. Office of Water, U.S. Environmental Protection Agency, Washington, D.C. EPA 440/5-91-005. 1991.

Rankin, E.T. Habitat indices in water resource quality assessments, in *Biological Assessment and Criteria: Tools for Water Resource Planning and Decision Making*. Edited by W.S. Davis, and T.P. Simon. Lewis Publ., Boca Raton, FL., pp. 181–208. 1995.

Rosgen, D. A stream classification system, in *Riparian Ecosystems and their Management*. First North American Riparian Conference. Rocky Mountain Forest and Range Experiment Station, RM-120. pp. 91–95. 1985.

Rosgen, D. A classification of natural rivers. *Catena*, 22: 169–199. 1994.

Rosgen, D. *Applied River Morphology*. Wildland Hydrology Books, Pagosa Spring, CO. 1996.

Benthic Community Assessment

CONTENTS

Rapid Bioassessment Protocol: Benthic Macroinvertebrates (EPA 1989, 1999) 665
    Sample Collection 666
    Sample Sorting and Identification..667
    Data Analysis Techniques 669
    Guidance for Data Summary Sheets for Benthic RBP 680
The Ohio EPA Invertebrate Community Index Approach (OEPA 1989)....................................681
    Field Methods — Quantitative Sampling 681
    Laboratory Methods — Quantitative Sampling...682
    Macroinvertebrate Data Analysis 683
A Partial Listing of Agencies That Have Developed Tolerance Classifications and/or
    Biotic Indices...687
References 690

RAPID BIOASSESSMENT PROTOCOL:
BENTHIC MACROINVERTEBRATES (EPA 1989, 1999)

As with the habitat assessments, there are more advanced and complex methods for characterizing benthic communities than what is presented below. However, the Rapid Bioassessment Protocols (RBP) outlined by the U.S. EPA (EPA 1989, 1999) have been proven to be efficient and effective in small streams and rivers. The EPA is currently developing guidance for benthic characterization in lakes, large rivers, and coastal areas. States such as Ohio, Maine, and North Carolina use approaches that are also very useful, and similar in many ways. The following are direct excerpts from EPA (1989, 1999; www.epa.gov/owow/monitoring/rbp) and Ohio EPA (1989) guidance manuals. For more extensive information, the reader should refer directly to those manuals. In addition to the references given in the following text, other useful information for identifying benthic macroinvertebrates is found in Barbour et al. 1999; Beck 1977; Harris and Lawrence 1978; Hubbard and Peters 1978; Surdick and Gaufin 1978; USDA 1985.

Rapid Bioassessment Protocol (RBP) utilizes the systematic field collection and analysis of major benthic taxa. The data are compiled into various metrics. The optimal metrics will vary across (and even within) ecoregions, so a qualified benthic ecologist should be used to select the most appropriate metrics. The protocol can be used to prioritize sites for more intensive evaluation (i.e., replicate sampling, ambient toxicity testing, chemical characterization). The EPA 1989 guidance described three levels of RBPs, each with more accurate taxonomic resolution. This approach also recommended sampling a single habitat type. The 1999 guidance describes methods for multi-habitat assessments, which are more appropriate in low-gradient streams and rivers where there is little cobble and riffle area. The description below focuses on single habitat characterization.

Sample Collection

The collection procedure provides representative samples of the macroinvertebrate fauna from comparable habitat types at all stations constituting a site evaluation, and is supplemented with separate coarse particulate organic matter (CPOM) samples (e.g., leaves, decaying vegetation). This RBP single habitat approach focuses on the riffle/run habitat because it is the most productive habitat available in stream systems and includes many pollution-sensitive taxa of the scraper and filtering collector functional feeding groups. The CPOM sample provides a measure of effects (particularly toxicity effects) on a third trophic component of the benthic community, the shredders.

In sampling situations where a riffle/run habitat with a rock substrate is not available, any submerged fixed structure will provide a substrate for the scraper and filtering collector functional groups emphasized here. This allows for the same approach to be used in non-wadable streams and large rivers and wadable streams and rivers with unstable substrates.

Riffle/Run Sample

Riffle areas with relatively fast currents and cobble and gravel substrates generally provide the most diverse community. Riffles should be sampled using a kick net to collect from an approximately 1-m^2 area. A minimum of two 1-m^2 riffle samples should be collected at each station: one from an area of fast current and one from an area of slower current. The samples are composited for processing. In streams lacking riffles, run areas with cobble or gravel substrate are also appropriate for kick net sampling.

Where riffle/run communities with a rock substrate are not available, other submerged fixed structures (e.g., submerged boulders, logs, bridge abutments, pier pilings) should be sampled by hand picking. These structures provide suitable habitat for the scrapers and filtering collectors and will allow use of the RBP in a wider range of regions and stream orders.

CPOM Sample

In addition to the riffle/run sample collected for evaluation of the scraper and filtering collector functional feeding groups, a CPOM sample should also be collected to provide data on the abundance of shredders at the site. Large particulate shredders are important in forested areas of stream ecosystems ranging from stream orders 1 through 4 (Minshall et al. 1985). The absence of shredders of large particulate material is characteristic of unstable, poorly retentive headwater streams in disturbed watersheds or in dry areas where leaf material processing is accomplished by terrestrial detritivores (Minshall et al. 1985). McArthur et al. (1988) reported that very few shredders were found in summer leaf packs in South Carolina because processing was so rapid.

The CPOM sample is processed separately from the riffle/run sample and used only for characterizing the functional feeding group representation. Sampling the CPOM component requires a composite collection of various plant parts such as leaves, needles, twigs, bark, or their fragments. Potential sample sources include leaf packs, shore zones, and other depositional areas where CPOM may accumulate. Only the upper surface of litter accumulation in depositional areas should be sampled to ensure that it is from the aerobic zone. For the shredder community analysis, several handfuls of material should be adequate. A variety of CPOM forms should be collected if available. CPOM collected may be washed in a dip net or a sieve bucket.

Shredder abundance is maximum when the CPOM is partially decomposed (Cummins et al. 1989). Care must be taken to *avoid* collecting recent or fully decomposed leaf litter to optimize collection of the shredder community. For this CPOM collection technique, seasonality may have an important influence on shredder abundance data. For instance, fast-processing litter (e.g., basswood, alder, maples, birch) would have the highest shredder representation in the winter (Cummins et al. 1989). The slow-processing litter (e.g., oaks, rhododendrons, beech, conifers) would have the highest shredder representation in the summer.

Sample Sorting and Identification

Riffle/Run Sample

Sorting and enumeration in the field to obtain a 100 (or higher) -count organism subsample is recommended for the riffle/run sample. After processing in the field, the organisms and sample residue should be preserved for archiving. Thus, a reanalysis (for quality control) or more thorough processing (e.g., larger counts, more detailed taxonomy) would be possible. The subsampling method described in this protocol is based on Hilsenhoff's Improved Biotic Index (Hilsenhoff 1987) and is similar to that used by the New York Department of Environmental Conservation (Bode 1988). This subsampling technique provides for a consistent unit of effort and a representative estimate of the benthic fauna (modified from Hilsenhoff 1987):

1. Thoroughly rinse sample in a (500-μm) screen or the sampling net to remove fine sediments. Any large organic material (whole leaves, twigs, algal or macrophyte mats) should be rinsed, visually inspected, and discarded.
2. Place sample contents in a large, flat pan with a light-colored (preferably white) bottom. The bottom of the pan should be marked with a numbered grid pattern, each block in the grid measuring 5×5 cm. (Sorting using a gridded pan is only feasible if the organism movement in the sample can be slowed by the addition of club soda or tobacco to the sample. If the organisms are not anesthetized, 100 organisms should be removed from the pan as randomly as possible.) A 30×45 cm pan is generally adequate, although pan size ultimately depends on sample size. Larger pans allow debris to be spread more thinly, but they are unwieldy. Samples too large to be effectively sorted in a single pan may be thoroughly mixed in a container with some water, and half of the homogenized sample placed in each of two gridded pans. Each half of the sample must be composed of the same kinds and quantity of debris, and an equal number of grids must be sorted from each pan to ensure a representative subsample.
3. Add just enough water to allow complete dispersion of the sample within the pan; excessive water will allow sample material to shift within the grid during sorting. Distribute sample material evenly within the grid.
4. Use a random numbers table to select a number corresponding to a square within the gridded pan. Remove all organisms from within that square and proceed with the process of selecting squares and removing organisms until the total number sorted from the sample is within 10% of 100. Any organism that is lying over a line separating two squares is considered to be in the square containing its head. In those instances where it is not possible to determine the location of the head (worms for instance), the organism is considered to be in the square containing the largest portion of its body. Any square sorted must be sorted in its entirety, even after the 100 count has been reached. In order to lessen sampling bias, the investigator should attempt to pick smaller, cryptic organisms as well as the larger, more obvious ones.

An alternative method of subsampling live samples in the field is to simply sort 100 organisms in a random manner. Narcotization to slow the organisms is less important with this subsampling technique. To lessen sampling bias, the investigator should pick smaller, cryptic organisms, as well as the larger, more obvious organisms.

BENTHIC MACROINVERTEBRATE FIELD DATA SHEET

STREAM NAME	LOCATION	
STATION #_____ RIVERMILE_____	STREAM CLASS	
LAT _____ LONG _____	RIVER BASIN	
STORET #	AGENCY	
INVESTIGATORS		LOT NUMBER
FORM COMPLETED BY	DATE _____ TIME _____ AM PM	REASON FOR SURVEY

HABITAT TYPES	Indicate the percentage of each habitat type present ❑ Cobble____% ❑ Snags____% ❑ Vegetated Banks____% ❑ Sand____% ❑ Submerged Macrophytes____% ❑ Other ( )____%
SAMPLE COLLECTION	Gear used ❑ D-frame ❑ kick-net ❑ Other _____ How were the samples collected? ❑ wading ❑ from bank ❑ from boat Indicate the number of jabs/kicks taken in each habitat type. ❑ Cobble_____ ❑ Snags_____ ❑ Vegetated Banks_____ ❑ Sand_____ ❑ Submerged Macrophytes_____ ❑ Other ( )_____
GENERAL COMMENTS	

QUALITATIVE LISTING OF AQUATIC BIOTA
Indicate estimated abundance: 0 = Absent/Not Observed, 1 = Rare, 2 = Common, 3= Abundant, 4 = Dominant

Periphyton	0	1	2	3	4	Slimes	0	1	2	3	4
Filamentous Algae	0	1	2	3	4	Macroinvertebrates	0	1	2	3	4
Macrophytes	0	1	2	3	4	Fish	0	1	2	3	4

FIELD OBSERVATIONS OF MACROBENTHOS
Indicate estimated abundance: 0 = Absent/Not Observed, 1 = Rare (1-3 organisms), 2 = Common (3-9 organisms), 3= Abundant (>10 organisms), 4 = Dominant (>50 organisms)

Porifera	0	1	2	3	4	Anisoptera	0	1	2	3	4	Chironomidae	0	1	2	3	4
Hydrozoa	0	1	2	3	4	Zygoptera	0	1	2	3	4	Ephemeroptera	0	1	2	3	4
Platyhelminthes	0	1	2	3	4	Hemiptera	0	1	2	3	4	Trichoptera	0	1	2	3	4
Turbellaria	0	1	2	3	4	Coleoptera	0	1	2	3	4	Other	0	1	2	3	4
Hirudinea	0	1	2	3	4	Lepidoptera	0	1	2	3	4						
Oligochaeta	0	1	2	3	4	Sialidae	0	1	2	3	4						
Isopoda	0	1	2	3	4	Corydalidae	0	1	2	3	4						
Amphipoda	0	1	2	3	4	Tipulidae	0	1	2	3	4						
Decapoda	0	1	2	3	4	Empididae	0	1	2	3	4						
Gastropoda	0	1	2	3	4	Simuliidae	0	1	2	3	4						
Bivalvia	0	1	2	3	4	Tabinidae	0	1	2	3	4						
						Culcidae	0	1	2	3	4						

Figure B.1 Benthic macroinvertebrate field data sheet. (From EPA. *Rapid Bioassessment Protocols for Use in Wadeable Streams and Rivers: Periphyton, Benthic Macroinvertebrates and Fish.* Office of Water, U.S. Environmental Protection Agency, Washington, D.C. EPA 841/B-99/002. 1999.)

All organisms in the subsample should be classified according to functional feeding group. Field classification is important because many families comprise genera and species representing a variety of functional groups. Knowing the family-level identification of the organisms will generally be insufficient for categorization by functional feeding group. Functional feeding group classification can be done in the field, on the basis of morphological and behavioral features, using Cummins and Wilzbach (1985). Care should be taken in noting early instars, which may constitute different functional feeding groups from the later instars. Recommended forms for recording benthic data are presented in Figures B.1 through B.4 (EPA 1999).

The scraper and filtering collector functional groups are the most important indicators in the riffle/run community. Numbers of individuals representing each of these two groups are recorded on the Benthic Macroinvertebrate Field Data Sheet (Figure B.1) (EPA 1999). The Benthic

BENTHIC MACROINVERTEBRATE SAMPLE LOG-IN SHEET

Date Collected	Collected By	Number of Containers	Preservation	Station #	Stream Name and Location	Date Received by Lab	Lot Number	Date of Completion		
								sorting	mounting	identification

Serial Code Example: B0754001(1)

B = Benthos (F = Fish; P = Periphyton) ■ 0754 = project number ■ 001 = sample number ■ (1) = lot number (e.g., winter 1996 =1; summer 1996 = 2)

Figure B.2 Benthic macroinvertebrate sample log-in sheet. (From EPA. *Rapid Bioassessment Protocols for Use in Wadeable Streams and Rivers: Periphyton, Benthic Macroinvertebrates and Fish*. Office of Water, U.S. Environmental Protection Agency, Washington, D.C. EPA 841/B-99/002. 1999.)

Macroinvertebrate Sample Log-In Sheet (Figure B.2) (EPA 1999) is used to record all collections and is an important part of the QA/QC and sample tracking activities.

All organisms in the subsample should be identified to family or order, enumerated, and recorded, along with any observations on abundance of other aquatic biota, on this data sheet. A summary of all benthic data to be used in the final analysis will be recorded on the Benthic Macroinvertebrate Laboratory Bench Sheet (Figures B.3 and B.4) (EPA 1999) upon return to the laboratory. The use of family-level identification in this protocol is based on Hilsenhoff's Family Biotic Index, which uses higher taxonomic levels of identification (Hilsenhoff 1988).

CPOM Sample

Organisms collected in the supplemental CPOM sample are classified as shredders or non-shredders. Taxonomic identification is not necessary for this component. The composited CPOM sample may be field sorted in a small pan with a light-colored bottom or in the net or sieve through which it was rinsed. (If a large number of benthic macroinvertebrates have been collected, a representative subsampling of 20 to 60 organisms may be removed for functional feeding group classification.) Numbers of individuals representing the shredder functional group, as well as total number of macroinvertebrates collected in this sample, should be recorded for later analysis. The shredder/nonshredder metric may be deemed optional in rivers or in some regions where shredder abundance is naturally low. However, the potential utility of such a metric for assessing toxicant effects warrants serious consideration in this bioassessment approach.

Data Analysis Techniques

Biological impairment of the benthic community may be indicated by the absence of generally pollution-sensitive macroinvertebrate taxa such as Ephemeroptera, Plecoptera, and Trichoptera (EPT); excess dominance by any particular taxon, especially pollutant-tolerant forms such as some

BENTHIC MACROINVERTEBRATE LABORATORY BENCH SHEET (FRONT)

page _____ of _____

STREAM NAME		LOCATION
STATION #_____ RIVERMILE_____		STREAM CLASS
LAT _____ LONG _____		RIVER BASIN
STORET #		AGENCY
COLLECTED BY DATE_____		LOT #
TAXONOMIST DATE_____		SUBSAMPLE TARGET ❑ 100 ❑ 200 ❑ 300 ❑ Other ____

Enter Family and/or Genus and Species name on blank line.

Organisms	No.	LS	TI	TCR	Organisms	No.	LS	TI	TCR
Oligochaeta					Megaloptera				
Hirudinea					Coleoptera				
Isopoda									
Amphipoda					Diptera				
Decapoda									
Ephemeroptera									
					Gastropoda				
					Pelecypoda				
Plecoptera									
					Other				
Trichoptera									
Hemiptera									

Taxonomic certainty rating (TCR) 1-5:1=most certain, 5=least certain. If rating is 3-5, give reason (e.g., missing gills). LS= life stage: I = immature; P = pupa; A = adult TI = Taxonomists initials

Total No. Organisms _____   Total No. Taxa _____

Figure B.3 Benthic macroinvertebrate laboratory bench sheet (front). (From EPA. *Rapid Bioassessment Protocols for Use in Wadeable Streams and Rivers: Periphyton, Benthic Macroinvertebrates and Fish*. Office of Water, U.S. Environmental Protection Agency, Washington, D.C. EPA 841/B-99/002. 1999.)

BENTHIC MACROINVERTEBRATE LABORATORY BENCH SHEET (BACK)

SUBSAMPLING/SORTING INFORMATION	Number of grids picked: _____

SUBSAMPLING/SORTING INFORMATION

Sorter _____

Date _____

Number of grids picked: _____

Time expenditure _____   No. of organisms _____

Indicate the presence of large or obviously abundant organisms:

QC: ❑ YES   ❑ NO   QC Checker _____

$$\frac{\text{\# organisms originally sorted}}{}\ \div\ \left(\text{\# organisms recovered by checker} + \text{\# organisms originally sorted} \right) = \text{\% sorting efficiency}$$

≥ 90%, sample passes _____

< 90%, sample fails, action taken _____

TAXONOMY

ID _____

Date _____

Explain TCR ratings of 3-5:

Other Comments (e.g. condition of specimens):

QC: ❑ YES   ❑ NO   QC Checker _____

Organism recognition   ❑ pass   ❑ fail
Verification complete   ❑ YES   ❑ NO

General Comments (use this space to add additional comments):

_____

_____

_____

_____

_____

_____

_____

Figure B.4 Benthic macroinvertebrate laboratory bench sheet (back). (From EPA. *Rapid Bioassessment Protocols for Use in Wadeable Streams and Rivers: Periphyton, Benthic Macroinvertebrates and Fish.* Office of Water, U.S. Environmental Protection Agency, Washington, D.C. EPA 841/B-99/002. 1999.)

Chironomidae and Oligochaeta taxa; low overall taxa richness; or appreciable shifts in community composition relative to the reference condition. Impairment may also be indicated by an overabundance of fungal slimes or filamentous algae, or an absence of expected populations of fish. All of these indicators can be evaluated using the sampling data generated. A number of useful metrics exist (Tables B.2 and B.3), while Figure B.5 (EPA 1999) is a preliminary assessment score sheet.

On the basis of observations made in the assessment of habitat, water quality, physical characteristics, and the qualitative biosurvey, the investigator concludes whether impairment is detected. If impairment is detected, an estimation of the probable cause and source should be made. The aquatic biota that indicated an impairment, are noted along with observed indications of potential

PRELIMINARY ASSESSMENT SCORE SHEET
(PASS)

page ____ of ____

STREAM NAME		LOCATION
STATION # _____ RIVERMILE _____		STREAM CLASS
LAT _____ LONG _____		RIVER BASIN
STORET #		AGENCY
COLLECTED BY _____ DATE _____		LOT # _____ NUMBER OF SWEEPS _____
HABITATS: ☐ COBBLE ☐ SHOREZONE ☐ SNAGS ☐ VEGETATION		

Enter Family and/or Genus and Species name on blank line.

Organisms	No.	LS	TI	TCR	Organisms	No.	LS	TI	TCR
Oligochaeta					Megaloptera				
Hirudinea					Coleoptera				
Isopoda									
Amphipoda					Diptera				
Decapoda									
Ephemeroptera									
					Gastropoda				
					Pelecypoda				
Plecoptera									
					Other				
Trichoptera									
					Taxonomic certainty rating (TCR) 1-5: 1=most certain, 5=least certain. If rating is 3-5, give reason (e.g., missing gills). LS= life stage: I = immature; P = pupa; A = adult TI = Taxonomists initials				
Hemiptera									

	Site Value	Target Threshold	If 2 or more metrics are ≥ target threshold, site is
Total No. Taxa			**HEALTHY**
EPT Taxa			If less than 2 metrics are within target range, site is
Tolerance Index			**SUSPECTED IMPAIRED**

Figure B.5 Preliminary assessment score sheet. (From EPA. *Rapid Bioassessment Protocols for Use in Wadeable Streams and Rivers: Periphyton, Benthic Macroinvertebrates and Fish.* Office of Water, U.S. Environmental Protection Agency, Washington, D.C. EPA 841/B-99/002. 1999.)

problem sources. The downstream extent of impact is estimated and multiplied by appropriate stream width to provide an estimate of the areal extent of the problem.

The data analysis scheme used in this RBP integrates several community, population, and functional parameters into a single evaluation of biotic integrity. Each parameter, or metric, measures a different component of community structure and has a different range of sensitivity to pollution stress (Figure B.6). This integrated approach provides more assurance of a valid assessment because a variety of parameters are evaluated. Deficiency of any one metric in a particular situation should not invalidate the entire approach.

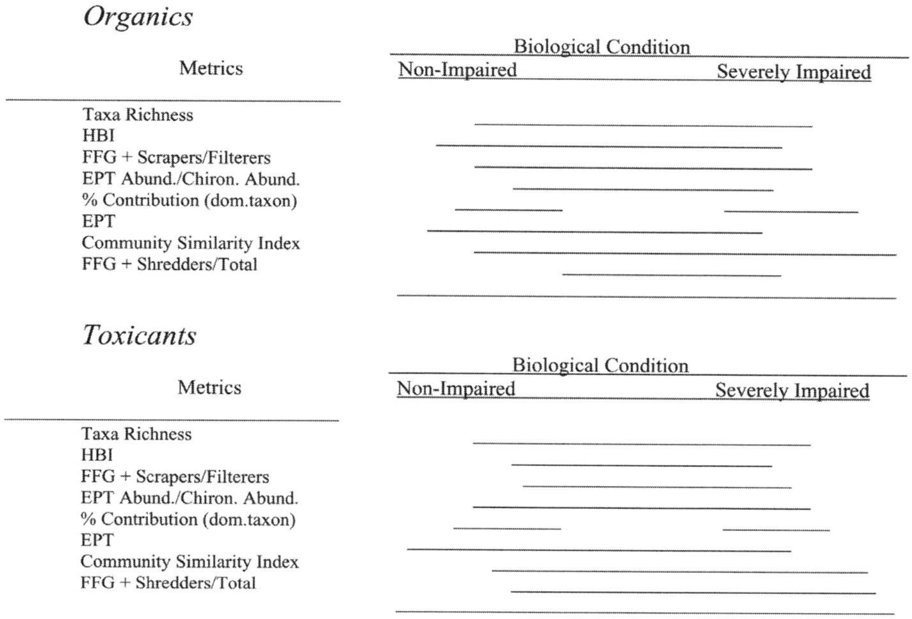

Organics

Metrics	Biological Condition	
	Non-Impaired	Severely Impaired
Taxa Richness		
HBI		
FFG + Scrapers/Filterers		
EPT Abund./Chiron. Abund.		
% Contribution (dom.taxon)		
EPT		
Community Similarity Index		
FFG + Shredders/Total		

Toxicants

Metrics	Biological Condition	
	Non-Impaired	Severely Impaired
Taxa Richness		
HBI		
FFG + Scrapers/Filterers		
EPT Abund./Chiron. Abund.		
% Contribution (dom.taxon)		
EPT		
Community Similarity Index		
FFG + Shredders/Total		

Figure B.6   Range of sensitivities of Rapid Bioassessment Protocol II and III benthic metrics in assessing biological condition in response to organics and toxicants.

The integrated data analysis (Figure B.7) is performed as follows. Using the raw benthic data, a numerical value is calculated for each metric. Calculated values are then compared to values derived from either a reference site within the same region, a reference database applicable to the region, or a suitable control station on the same stream. Each metric is then assigned a score according to the comparability (percent similarity) of calculated and reference values. Scores for the eight metrics are then totaled and compared to the total metric score for the reference station. The percent comparison between the total scores provides a final evaluation of biological condition. The criteria to be used for scoring the eight metrics *may need to be adjusted for use in particular regions.*

Inherent variability in each metric was considered in establishing percent comparability criteria (Figure B.6). The metrics based on taxa richness, FBI, and EPT Indices have low variability (Resh 1988). This variability is accounted for in the criteria for characterization of biological condition, based on existing data. For metrics based on standard taxa richness and FBI and EPT Indices, differences of 10 to 20% relative to the reference condition would be considered nominal, and the station being assessed would receive the maximum metric score. Because increasing FBI values denote worsening biological condition, percent difference for this metric is calculated by dividing the reference value by the value for the station of comparison.

Metrics that utilize ratios fluctuate more widely, however, and comparing percent differences between ratios (ratios of ratios) will compound the variability. Scoring increments are therefore set at broad intervals of 25% or greater. For metrics based on functional feeding group ratios, Cummins (1987, personal communication) contends that differences as great as 50% from the reference may be acceptable, but differences in the range of 50 to 100% are not only important, but discriminate degrees of impact more clearly.

The contribution of the dominant taxon to total abundance is a simple estimator of evenness. Scoring criteria are based on theoretical considerations rather than direct comparison with a reference.

The Community Loss Index (a representative similarity index) already incorporates a comparison with a reference. Therefore, actual index values are used in scoring.

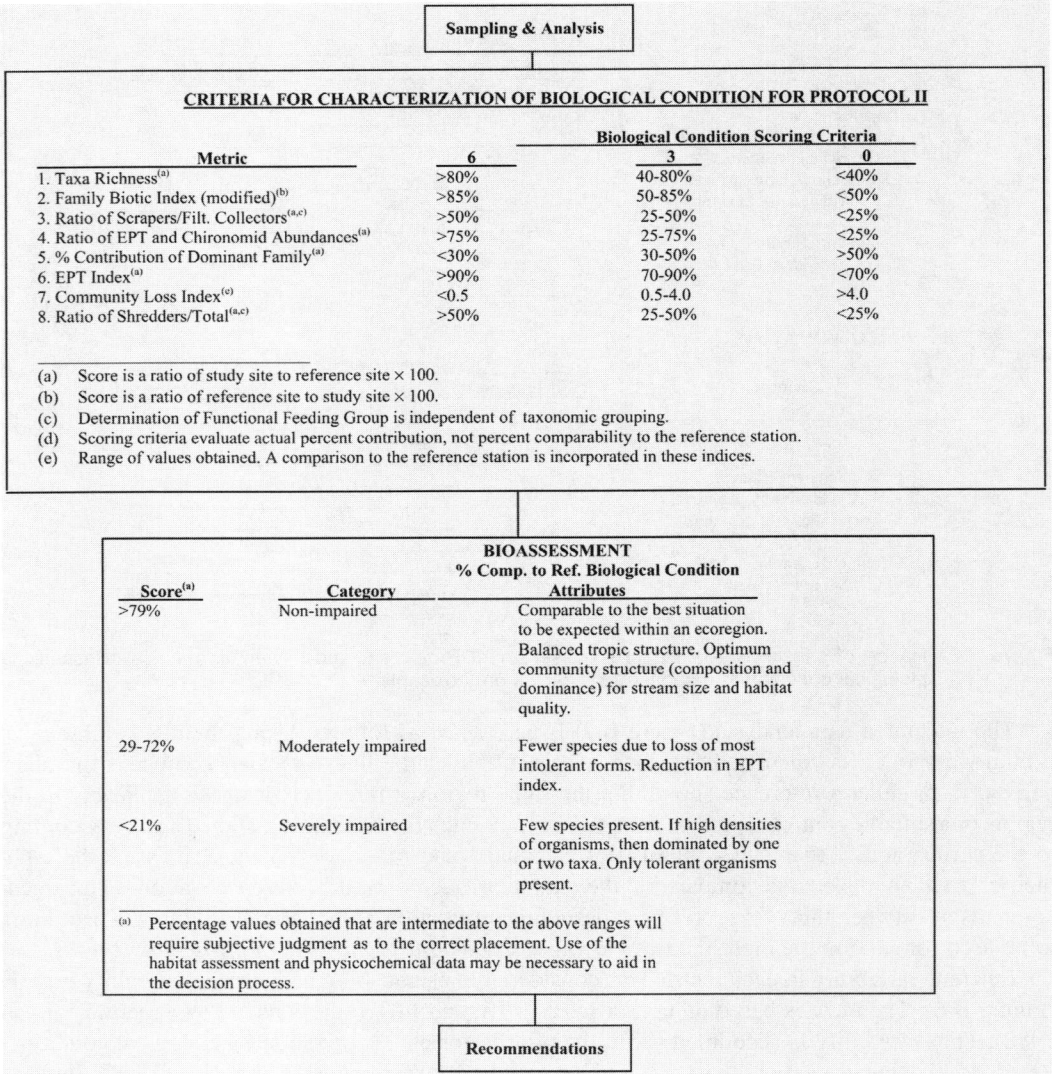

Figure B.7 Flowchart of bioassessment approach advocated for a Rapid Bioassessment Protocol.

The metrics used to evaluate the benthic data and their significance are explained below and in Tables B.1 and B.2.

Riffle/Run Sample

Metric 1. Taxa Richness

Reflects health of the community through a measurement of the variety of taxa (total number of families) present. Generally increases with increasing water quality, habitat diversity, and habitat suitability. Sampling of highly similar habitats will reduce the variability in this metric attributable to factors such as current speed and substrate type. Some pristine headwater streams may be naturally unproductive, supporting only a very limited number of taxa. In these situations, organic enrichment may result in an increased number of taxa (including EPT taxa).

Table B.1 Definitions of Best Candidate Benthic Metrics and Predicted Direction of Metric Response to Increasing Perturbation

Category	Metric	Definition	Predicted Response to Increasing Perturbation
Richness measures	Total No. taxa	Measures the overall variety of the macroinvertebrate assemblage	Decrease
	No. EPT taxa	Number of taxa in the insect orders Ephemeroptera (mayflies), Plecoptera (stoneflies), and Trichoptera (caddisflies)	Decrease
	No. Ephemeroptera taxa	Number of mayfly taxa (usually genus or species level)	Decrease
	No. Plecoptera taxa	Number of stonefly taxa (usually genus of species level)	Decrease
	No. Trichoptera taxa	Number of caddisfly taxa (usually genus or species level)	Decrease
Composition measures	% EPT	Percent of the composite of mayfly, stonefly, and caddisfly larvae	Decrease
	% Ephemeroptera	Percent of mayfly nymphs	Decrease
Tolerance/intolerance measures	No. intolerant taxa	Taxa richness of those organisms considered to be sensitive to perturbation	Decrease
	% tolerant organisms	Percent of macrobenthos considered to be tolerant of various types of perturbation	Increase
	% dominant taxon	Measures the dominance of the single most abundant taxon. Can be calculated as dominant 2, 3, 4, or 5 taxa.	Increase
Feeding measures	% filterers	Percent of the macrobenthos that filter FPOM from either the water column or sediment	Variable
	% grazers and scrapers	Percent of the macrobenthos that scrape or graze upon periphyton	Decrease
Habit measures	No. clinger taxa	Number of taxa of insects	Decrease
	% clingers	Percent of insects having fixed retreats or adaptations for attachment to surfaces in flowing water.	Decrease

Data from DeShon 1995; Barbour et al. 1996b; Fore et al. 1996; Smith and Voshell 1997.

Metric 2. Modified Family Biotic Index

Tolerance values range from 0 to 10 for families and increase as water quality decreases. The index was developed by Hilsenhoff (1988) to summarize the various tolerances of the benthic arthropod community with a single value. The Modified Family Biotic Index was developed to detect organic pollution and is based on the original species-level index (Hilsenhoff 1982). Tolerance values for each family were developed by weighting species according to their relative abundance in the State of Wisconsin.

The family-level index has been modified for this document to include organisms other than just arthropods using the genus and species-level biotic index developed by the State of New York (Bode 1988). The formula for calculating the Family Biotic Index is:

$$\text{HBI} = \frac{\sum x_i t_j}{n}$$

where x_i = number of individuals within a taxon

t_j = tolerance value of a taxon

n = total number of organisms in the sample

Table B.2 Definitions of Additional Potential Benthic Metrics and Predicted Direction of Metric Response to Increasing Perturbation

Category	Metric	Definition	Predicted Response to Increasing Perturbation	References
Richness measures	No. *Pteronarcys* species	The presence or absence of a long-lived stonefly genus (2–3 year life cycle)	Decrease	Fore et al. 1996
	No. Diptera taxa	Number of "true" fly taxa, which includes midges	Decrease	DeShon 1995
	No. Chironomidae taxa	Number of taxa of chironomid (midge) larvae	Decrease	Hayslip 1993; Barbour et al. 1996b
Composition measures	% Plecoptera	Percent of stonefly nymphs	Decrease	Barbour et al. 1994
	% Trichoptera	Percent of caddisfly larvae	Decrease	DeShon 1995
	% Diptera	Percent of all "true" fly larvae	Increase	Barbour et al. 1996b
	% Chironomidae	Percent of midge larvae	Increase	Barbour et al. 1994
	% Tribe Tanytarsini	Percent of Tanytarisinid midges to total fauna	Decrease	DeShon 1995
	% Other Diptera and noninsects	Composite of those organisms generally considered to be tolerant to a wide range of environmental conditions	Increase	DeShon 1995
	% *Corbicula*	Percent of Asiatic clam in the benthic assemblage	Increase	Kerans and Karr 1994
	% Oligochaeta	Percent of aquatic worms	Variable	Kerans and Karr 1994
Tolerance/intolerance measures	No. intol. snail and mussel species	Number of species of molluscs generally thought to be pollution intolerant	Decrease	Kerans and Karr 1994
	% sediment tolerant organisms	Percent of infaunal macrobenthos tolerant of perturbation	Increase	Fore et al. 1996
	Hilsenhoff Biotic Index	Uses tolerance values to weight abundance in an estimate of overall pollution; originally designed to evaluate organic pollution	Increase	Barbour et al. 1992; Hayslip 1993; Kerans and Karr 1994
	Florida Index	Weighted sum of intolerant taxa, which are classed as 1 (least tolerant) or 2 (intolerant); Florida Index = 2 × Class 1 taxa + Class 2 taxa	Decrease	Barbour et al. 1996b
	% Hydropsychidae to Trichoptera	Relative abundance of pollution tolerant caddisflies (metric could also be regarded as a composition measure)	Increase	Barbour et al. 1992; Hayslip 1993
Feeding measures	% omnivores and scavengers	Percent of generalists in feeding strategies	Increase	Kerans and Karr 1994
	% ind. gatherers and filterers	Percent of collector feeders of CPOM and FPOM	Variable	
	% gatherers	Percent of the macrobenthos that "gather"	Variable	Barbour et al. 1996b
	% predators	Percent of the predator functional feeding group; can be made restrictive to exclude omnivores	Variable	Kerans and Karr 1994
	% shredders	Percent of the macrobenthos that "shreds" leaf litter	Decrease	Barbour et al. 1992; Hayslip 1993
Life cycle measures	% multivoltine	Percent of organisms having short (several per year) life cycle	Increase	Barbour et al. 1994
	% univoltine	Percent of organisms relatively long-lived (life cycles of 1 or more years)	Decrease	Barbour et al. 1994

Hilsenhoff's family-level tolerance values may require modification for some regions. Alternative tolerance classifications and biotic indices have been developed by some state agencies. Additional biotic indices are listed in EPA (1983).

Although the FBI may be applicable for toxic pollutants, it has only been evaluated for organic pollutants. The State of Wisconsin is conducting a study to evaluate the ability of Hilsenhoff's index to detect nonorganic effects.

Metric 3. Ratio of Scraper and Filtering Collector Functional Feeding Groups

The scraper and filtering collector metric reflects the riffle/run community foodbase. When compared to a reference site, shifts in the dominance of a particular feeding type indicate a community responding to an overabundance of a particular food source. The predominant feeding strategy reflects the type of impact detected. Assignment of individuals to functional feeding groups is independent of taxonomy, with some families representing several functional groups.

A description of the functional feeding group concept can be found in Cummins (1973) and Merritt and Cummins (1984). Functional feeding group designations for most aquatic insect families may be found in Merritt and Cummins (1984). Most aquatic insects can also be classified to functional feeding group in the field, on the basis of morphological and behavioral features, using Cummins and Wilzbach (1985).

The relative abundance of scrapers and filtering collectors in the riffle/run habitat is an indication of the periphyton community composition, availability of suspended fine particulate organic material (FPOM), and availability of attachment sites for filtering. Scrapers increase with increased diatom abundance and decrease as filamentous algae and aquatic mosses (which scrapers cannot efficiently harvest) increase. However, filamentous algae and aquatic mosses provide good attachment sites for filtering collectors, and the organic enrichment often responsible for overabundance of filamentous algae can also provide FPOM that is utilized by the filterers.

Filtering collectors are also sensitive to toxicants bound to fine particles and should be the first group to decrease when exposed to steady sources of such bound toxicants. This situation is often associated with point-source discharges where certain toxicants adsorb readily to dissolved organic matter (DOM), forming FPOM during flocculation. Toxicants thus become available to filterers via FPOM. The scraper to filtering collector ratio may not be a good indicator of organic enrichment if adsorbing toxicants are present. In these instances the FBI and EPT Index may provide additional insight. Qualitative field observations on periphyton abundance may also be helpful in interpreting results.

Metric 4. Ratio of EPT and Chironomidae Abundances

The EPT and Chironomidae abundance ratio uses relative abundance of these indicator groups (Ephemeroptera, Plecoptera, Trichoptera, and Chironomidae) as a measure of community balance. Good biotic condition is reflected in communities with an even distribution among all four major groups and with substantial representation in the sensitive groups Ephemeroptera, Plecoptera, and Trichoptera. Skewed populations having a disproportionate number of the Chironomidae relative to the more sensitive insect groups may indicate environmental stress (Ferrington 1987; Shackleford 1988). Certain species of some genera such as *Cricotopus* are highly tolerant (Lenat 1983; Mount et al. 1984), and as opportunists may become numerically dominant in habitats exposed to metal discharges where EPT taxa are not abundant, thereby providing a good indicator of toxicant stress (Winner et al. 1980). Clements et al. (1988) found that mayflies were more sensitive than chironomids to exposure levels of 15 to 32:g/L of copper. Chironomids tend to become increasingly dominant in terms of percent taxonomic composition and relative abundance along a gradient of increasing enrichment or heavy metals concentration (Ferrington 1987).

An alternative to the ratio of EPT and Chironomidae abundance metric is the Indicator Assemblage Index (IAI) developed by Shackleford (1988). The IAI integrates the relative abundances of the EPT taxonomic groups and the relative abundances of chironomids and annelids upstream and downstream of a pollutant source to evaluate impairment. The IAI may be a valuable metric in areas where the annelid community may fluctuate substantially in response to pollutant stress.

Metric 5. Percent Contribution of Dominant Family

The percent contribution of the dominant family to the total number of organisms uses abundance of the numerically dominant taxon relative to the rest of the population as an indication of community balance at the family level. A community dominated by relatively few families would indicate environmental stress. This metric may be redundant if the Pinkham and Pearson Similarity Index is used as a community similarity index for metric number 7.

Metric 6. EPT Index

The EPT Index generally increases with increasing water quality. The EPT Index value is the total number of distinct taxa within the groups Ephemeroptera, Plecoptera, and Trichoptera. The EPT Index value summarizes the taxa richness within the insect groups that are generally considered pollution sensitive. This was developed for species-level identifications; however, the concept is valid for use at family-level identifications.

Headwater streams which are naturally unproductive may experience an increase in taxa (including EPT taxa) in response to organic enrichment.

Metric 7. Community Similarity Indices

Community Similarity Indices are used in situations where a reference community exists, either through sampling or through prediction for a region. Data sources or ecological data files may be available to predict a reference community to be used for comparison. The combined information provided through a regional analysis and EPA's ERAPT ecological database (Dawson and Hellenthal 1986) may be useful for this analysis. These indices are designed to be used with either species level identifications or higher taxonomic levels. Three of the many community similarity indices available are discussed below:

- Community Loss Index. Measures the loss of benthic taxa between a reference station and the station of comparison. The Community Loss Index was developed by Courtemanch and Davies (1987) and is an index of compositional dissimilarity, with values increasing as the degree of dissimilarity with the reference station increases. Values range from 0 to "infinity." Based on preliminary data analysis, this index provides greater discrimination than either of the following two community similarity indices.
- Jaccard Coefficient of Community Similarity. Measures the degree of similarity in taxonomic composition between two stations in terms of taxon presence or absence. The Jaccard Coefficient discriminates between highly similar collections. Coefficient values, ranging from 0 to 1.0, increase as the degree of similarity with the reference station increases. See Jaccard (1912), Boesch (1977), and EPA (1983) for more detail. The formulae for the Community Loss Index and the Jaccard Coefficient are

$$\text{Community Loss} = \frac{d - a}{e}$$

$$\text{Jaccard Coefficient} = \frac{a}{a + b + c}$$

where

    a = number of taxa common to both samples
    b = number of taxa present in Sample B but not A
    c = number of taxa present in Sample A but not B
    d = total number of taxa present in Sample A
    e = total number of taxa present in Sample B
  Sample A = reference station (or mean of reference database)
  Sample B = station of comparison

- Pinkham and Pearson Community Similarity Index Incorporates abundance and compositional information and can be calculated with either percentages or numbers. A weighting factor can be added that assigns more significance to dominant taxa. See Pinkham and Pearson (1976) and EPA (1983) for more detail. The formula is

$$S.I._{ab} = \sum \frac{min(x_{ia}, x_{ib})}{max(x_{ia}, x_{ib})} \left[\frac{\frac{x_{ia}}{x_a} \cdot \frac{x_{ib}}{x_b}}{2} \right]$$

where x_{ia}, x_{ib} = number of individuals in the ith taxon in Sample A or B

Other community similarity indices include Spearman's Rank Correlation (Snedecor and Cochran 1967), Morisita's Index (Morisita 1959), Biotic Condition Index (Winget and Mangum 1979), and Bray-Curtis Index (Bray and Curtis 1957; Whittaker 1952). Calculation of a chi-square "goodness of fit" (Cochran 1952) may also be appropriate.

CPOM Sample

Metric 8. Ratio of Shredder Functional Feeding Group and Total Number of Individuals Collected

Also based on the Functional Feeding Group concept, the abundance of the shredder functional group relative to the abundance of all other functional groups allows evaluation of potential impairment as indicated by the CPOM-based shredder community. Shredders are sensitive to riparian zone impacts and are particularly good indicators of toxic effects when the toxicants involved are readily adsorbed to the CPOM and either affect microbial communities colonizing the CPOM or the shredders directly (Cummins 1987, personal communication).

The degree of toxicant effects on shredders vs. filterers depends on the nature of the toxicants and the organic particle adsorption efficiency. Generally, as the size of the particle decreases, the adsorption efficiency increases as a function of the increased surface to volume ratio (Hargrove 1972). Because waterborne toxicants are readily adsorbed to FPOM, toxicants of a terrestrial source (e.g., pesticides, herbicides) accumulate on CPOM prior to leaf fall, thus having a substantial effect on shredders (Swift et al. 1988a,b). The focus on this approach is on a comparison to the reference community which should have a reasonable representation of shredders as dictated by seasonality, region, and climate. This allows for an examination of shredder or collector "relative" abundance as indicators of toxicity.

The data collected in the 100-organism riffle/run subsample and the CPOM sample are summarized according to the information required for each metric and entered on the Data Summary Sheet.

Each metric result is given a score based on percent comparability to a reference station. Scores are totaled and compared to the total metric score for the reference station. The percent comparison between the total scores provides a final evaluation of biological condition. Values obtained may sometimes be intermediate to established ranges and require some judgment as to assessment of biological condition. In these instances, habitat assessment, physical characterization, and water quality data may aid in the evaluation process.

Guidance for Data Summary Sheets for Benthic RBP

Station Number: Indicate station number for each data set recorded.

Station Location: Record brief description of sampling site relative to established landmarks (i.e., roads, bridges).

Taxa Richness: Record total number of families (or higher taxa) collected in the 100-organism riffle subsample.

FBI (modified): Record the Family Biotic Index value (Hilsenhoff 1988) calculated for the 100-organism riffle subsample using the formula presented in RBP II. Tolerance classification values can be entered into the computer database to simplify calculation.

Functional Feeding Group: Functional feeding group classifications may be entered into the computer database to simplify calculations.

    Riffle Community: Scrapers/filtering collectors: enter the value obtained by dividing the number of individuals in the riffle subsample representing the scraper functional group, by the number representing the filtering collector functional group.

    CPOM Community: Shredders/total: enter the value obtained by dividing the number of individuals in the CPOM sample (or subsample) representing the shredder functional group, by the total number of organisms in the sample (or subsample).

EPT/Chironomidae: Enter the value obtained by dividing the number of individuals in the 100-organism riffle subsample in the family Chironomidae, by the total number of individuals in the orders Ephemeroptera, Plecoptera, and Trichoptera.

Percent Contribution (Dominant Family): Record the value obtained by dividing the number of individuals in the family that is most abundant in the 100-organism riffle subsample, by the total number of individuals in the sample.

EPT Index: Record the total number of taxa in the 100-organism riffle subsample representing the orders Ephemeroptera, Plecoptera, and Trichoptera.

Community Similarity Index: Enter the value calculated for the appropriate community similarity index, using data from the 100-organism riffle subsample.

Values obtained for each metric should be assigned a score based on percent comparability to the control or reference station data. Scores are summed for both the impaired and reference station. The percent comparison between the total scores provides the final evaluation of biological condition.

Family-Level Tolerance Classification

The original RBP II (EPA 1989) is based on family-level identifications. The adequate assessment of biological condition for RBP II requires the use of a tolerance classification for differentiating among responses of the benthic community to pollutants. Hilsenhoff's Family Biotic Index (FBI) is used as a basis for the family-level tolerance classification.

The biotic index (BI) of organic pollution is adapted (Hilsenhoff 1987) for rapid evaluation by providing tolerance values for families (Tables B.3 and B.4) to allow a family-level biotic index (FBI) to be calculated in the field. The FBI is an average of tolerance values of all arthropod families in a sample. It is not intended as a replacement for the BI and can be effectively used in the field only by biologists who are familiar enough with arthropods to be able to identify families without using keys.

Using the same method and more than 2000 stream samples from throughout Wisconsin that were used to revise tolerance values for species and genera (Hilsenhoff 1987) family-level tolerance values were established by comparing occurrence of each family with the average BI of streams in which they occurred in the greatest numbers. Thus, family-level tolerance values tend to be a weighted average of tolerance values of species and genera within each family based on their relative abundance in Wisconsin.

Table B.3 Tolerance Values for Families of Stream Arthropods in the Western Great Lakes Region

Plecoptera	Capniidae 1, Chloroperlidae 1, Leuctridae 0, Nemouridae 2, Perlidae 1, Perlodidae 2, Pteronarcyidae 0, Taeniopterygidae 2
Ephemeroptera	Baetidae 4, Baetiscidae 3, Caenidae 7, Ephemerellidae 1, Ephemeridae 4, Heptageniidae 4, Leptophlebiidae 2, Metretopodidae 2, Oligoneuriidae 2, Polymitarcyidae 2, Potomanthidae 4, Siphlonuridae 7, Tricorythidae 4
Odonata	Aeshnidae 3, Calopterygidae 5, Coenagrionidae 9, Cordulegastridae 3, Corduliidae 5, Gomphidae 1, Lestidae 9, Libellulidae 9, Macromiidae 3
Trichoptera	Brachycentridae 1, Glossosomatidae 0, Helicopsychidae 3, Hydropsychidae 4, Hydroptilidae 4, Lepidostomatidae 1, Leptoceridae 4, Limnephilidae 4, Molannidae 6, Odontoceridae 0, Philopotamidae 3, Phryganeidae 4, Polycentropodidae 6, Psychomyiidae 2, Rhyacophilidae 0, Sericostomatidae 3
Megaloptera	Corydalidae 0, Sialidae 4
Lepidoptera	Pyralidae 5
Coleoptera	Dryopidae 5, Elmidae 4, Psephenidae 4
Diptera	Athericidae 2, Blephariceridae 0, Ceratopogonidae 6, Blood-red Chironomidae (Chironomini) 8, other (including pink) Chironomidae 6, Dolochopodidae 4, Empididae 6, Ephyridae 6, Psychodidae 10, Simuliidae 6, Muscidae 6, Syrphidae 10, Tabanidae 6, Tipulidae 3
Amphipoda	Gammaridae 4, Talitridae 8
Isopoda	Asellidae 8

Data from Hilsenhoff, W.L. Rapid field assessment of organic pollution with a family-level biotic index. *J. North Am. Benthol. Soc.*, 7: 65–68. 1988; EPA. *Rapid Bioassessment Protocols for Use in Streams and Rivers: Benthic Macroinvertebrates and Fish*. Office of Water, U.S. Environmental Protection Agency, Washington, D.C. EPA 444/4-89/001. 1989.

THE OHIO EPA INVERTEBRATE COMMUNITY INDEX APPROACH (OEPA 1989)

Field Methods — Quantitative Sampling

The primary sampling equipment used for the collection of benthic macroinvertebrates is the modified Hester–Dendy multiple-plate artificial substrate sampler. The sampler is constructed of 1/8-in tempered hardboard cut into 3-in square (or circular) plates and 1-in square spaces. A total of eight plates and 12 spacers are used for each sampler. The plates and spacers are placed on a 1/4-in stainless steel eyebolt so that there are three single spaces, three double spaces, and one triple space between the plates. The total surface area of the sampler, excluding the eyebolt, is 145.6 in².

Samplers placed in streams are tied to a concrete construction block, which anchors them in place and prevents the multiple-plates from coming into contact with the natural substrates. In water deeper than 4 ft, a float (1 quart cubitainer) is attached to the samplers to keep them within 4 ft of the surface. Whenever possible, the samplers are placed in runs rather than pools or riffles and an attempt is made to establish stations in as similar an ecological situation as possible. All samplers are exposed for a 6-week period. A set of samplers consists of three multiple-plate samplers (about 3 ft² of surface area) at National Ambient Water Quality Monitoring Network (NAWQMN) stations and five multiple-plate samplers at all other sampling locations. All NAWQMN stations and most routine monitoring stations are sampled from June 15 to September 30.

Retrieval of the sampler is accomplished by cutting them from the block and placing them in 1-quart, wide-mouth plastic containers while still submersed. Care is taken to avoid disturbing the samplers and thereby dislodging any organisms. Enough formalin is added to each container to equal an approximate 10% solution.

Qualitative samples of macroinvertebrates inhabiting the natural substrates are also collected at the time of sampler retrieval. In shallow water, samples are taken in a stream segment covering all available habitats near where the samplers were placed. Samples are collected using triangular ring frame 30-mesh dip nets and hand picking with forceps. Grab samplers (i.e., Ekman, Peterson, or Ponar) can also be used in deep water. The qualitative sampling continues until, by gross examination, no new taxa are being taken. A station description sheet is filled out by the collector at the time

Table B.4 Tolerance Values for Some Macroinvertebrates Not Included in Hilsenhoff (1982, 1987)

Acariformes	4
Decapoda	6
Gastropoda	
Amnicola	8
Bithynia	8
Ferrissia	6
Gyraulus	8
Helisoma	6
Lymnaea	6
Physa	8
Sphaeriidae	8
Oligochaeta	
Chaetogaster	6
Dero	10
Nais barbata	8
Nais behningi	6
Nais bretscheri	6
Nais communis	8
Nais elinguis	10
Nais pardalis	8
Nais simplex	6
Nais variabilis	10
Pristina	8
Stylaria	8
Tubificidae	
Aulodrilus	8
Limnodrilus	10
Hirudinea	
Helobdella	10
Turbellaria	4

From Bode, R.W. *Quality Assurance Workplan for Biological Stream Monitoring in New York State.* New York State Department of Environmental Conservation, Albany. 1988; EPA. *Rapid Bioassessment Protocols for Use in Streams and Rivers: Benthic Macroinvertebrates and Fish.* Office of Water, U.S. Environmental Protection Agency, Washington, D.C. EPA 444/4-89/001. 1989.

of sampler retrieval. The substrate is described using the categories for substrate characterization indicated in the U.S. EPA biological field manual (Weber 1973).

In situations where quantitative biological samples are collected from the natural substrates using a Surber square foot sampler (30-mesh netting), the collector stands on the downstream side of the sampler and works the substrate using a hand cultivator with 2-in tines. Large rocks are gently scrubbed with a brush. The material collected is placed in sealed containers, preserved in 10% formalin, and transported to the laboratory. Three to five Surber samples are taken at each site.

In situations where Ekman, Peterson, or Ponar grab samples are used for quantitative purposes, three to five samples are collected and then treated in essentially the same manner as the Surber samples. The material collected with the grab is washed through a bucket with a 30-mesh screen bottom, placed in sealed containers, preserved in 10% formalin, and returned to the laboratory.

Laboratory Methods — Quantitative Sampling

Samples are coded and sample numbers are immediately entered into a log book upon arrival at the laboratory. Samples are given a log number derived from the date, e.g., 871108-10, where 87 represents the year, 11 represents the month, and 08 the day. The number following this six-digit date, i.e., the number 10 in the previous example, indicates that this was the 10th sample logged that day. Other information in the log book includes the name(s) of field personnel who collected the sample, date, stream or lake name, basin name, entity (where applicable), general location, sample type, sampling method(s) used, the person who conducted the analyses, and any other comments considered pertinent to the collection and analysis of the sample.

Macroinvertebrate Counts and Identifications

Composite samples consisting of five multiple-plate samplers are used in station evaluations for routine monitoring. However, replicate samples (three multiple-plate samplers) are reported to the EPA for NAWQMN stations. Replicate sets of five multiple-plate samplers can be used if deemed necessary in cases where sampling is for litigation purposes. In all cases, the multiple-plate samplers are disassembled in a bucket of water and cleaned of organisms and debris. The organism/debris mixture is then passed through U.S. Standard Testing Sieves number 30 (0.589-mm openings) and number 40 (0.425-mm openings). The material retained in each sieve is preserved in properly labeled and coded jars containing 70% alcohol.

The following procedures are used during the course of analyzing an artificial substrate, Surber, or grab sample:

1. Sorting the sample is done in a white enamel pan followed by scanning under the dissecting microscope (10× magnification). Subsamples are produced using the following guidelines:
   a. A Folsom sample splitter is used for all subsampling. In an effort to determine the accuracy of the Folsom sample splitter, a sample composed of 200 individuals of five frequently collected organisms was prepared and repeatedly split. Statistical analysis of the data yielded a chi-square value of 2.56, df = 4, indicating that the subsamples were not significant at the 95% probability level.
   b. After an entire sample has been sorted, subsampling within families containing unmanageable numbers is acceptable.
   c. Very large samples may be subsampled prior to sorting, but only after examination in a white enamel pan to remove obvious rare taxa, e.g., hellgramites, non-hydropsychid caddisflies.
   d. A minimum of 250 organisms are identified, with at least 50 to 100 midges, 70 caddisflies, 70 mayflies.
2. Dipterans of the family Chironomidae are prepared for identification by clearing the larvae in hot 10% KOH for 30 min and then mounting in water on microscope slides. Permanent slides for the voucher collection are mounted in Euparol mounting medium.
3. Material retained in the #40 screen is counted and identified or counted and extrapolated when identification is impossible or impractical. (Artificial substrate sample only.)
4. Organisms determined to be dead before the time of collection are discarded.
5. When only one sex or life stage can be identified, it is assumed that the other sex or stage is the same species.
6. Sections of bryozoan colonies are removed from the plates and saved for identification. Only colonies, not individuals, are counted. (Artificial substrate sample only.)
7. Early instars that cannot be identified are extrapolated where possible.
8. Species-level identifications are made where possible and practical. Generic or higher level classifications are made if specimens are damaged beyond identification, in those cases where taxonomy is incomplete or laborious and time-consuming, or where the specimen is an unidentifiable early instar.
9. Organisms are listed in tables following the laboratory table format.
10. Two end fragments of an oligochaete are counted as one individual. Fragments without ends are not counted.
11. Any taxonomic key in the laboratory may be used as an aid in the identification of an organism. Also indicated is the level of taxonomy attainable with the keys listed.

Macroinvertebrate Data Analysis

Invertebrate Community Index

The principal measure of overall macroinvertebrate community condition used by the Biological Field Evaluations Group is the Invertebrate Community Index (ICI), a measurement derived in-house from information collected over many years. The ICI is a modification of the Index of Biotic Integrity (IBI) for fish developed by Karr (1981). The ICI consists of 10 structural community metrics, each with four scoring categories of 6, 4, 2, and 0 points (Table B.5). The point system evaluates a sample against a database of 247 relatively undisturbed reference sites throughout Ohio. Six points will be scored if a given metric has a value comparable to those of exceptional stream communities, 4 points for those metric values characteristic of more typical good communities, 2 points for metric values slightly deviating from the expected range of good values, and 0 points for metric values strongly deviating from the expected range of good values. The summation of the individual metric scores (determined by the relevant attributes of an invertebrate sample with some consideration given to stream drainage area) results in the ICI value. Metrics 1 through 9 are all generated from the artificial substrate sample data, while Metric 10 is based solely on the

Table B.5 Invertebrate Community Index (ICI) Metrics and Scoring Criteria Based on Macroinvertebrate Community Data from 247 Reference Sites throughout Ohio

Metric		Scoring Criteria			
		0	2	4	6
1.	Total number of taxa	Scoring of each metric varies			
2.	Total number of mayfly taxa	with drainage area; see			
3.	Total number of caddisfly taxa	Ohio EPA (1987)			
4.	Total number of dipteran taxa				
5.	Percent mayflies				
6.	Percent caddisflies				
7.	Percent tribe tanytarsini midges				
8.	Percent other dipterans and non-insects				
9.	Percent tolerant organisms				
10.	Total number of qualitative Ephemeroptera, Plecoptera, and Trichoptera (EPT) taxa				

qualitative sample data from natural substrates. More discussion of the derivation of the ICI including descriptions of each metric and the data plots and other information used to score each metric can be found in Ohio EPA (1987).

Community Similarity Index

A coefficient of similarity between two stations can be calculated using Van Horn's (1950) equation modified from the general formula described by Gleason (1920):

$$c = \frac{2w}{a+b}$$

The variables in this expression can be based either on the number of taxa present or absent at each station or on actual numerical data collected at each site. If the presence/absence method is being used:

a = the number of taxa collected at one station
b = the number of taxa collected at the other station
w = the number of taxa common to both stations

When actual numerical data are being used, each taxon is assigned a prominence value calculated by multiplying the density of the taxon by the square root of its frequency of occurrence (Beals 1961; Burlington 1962). In this case:

a = the sum of the prominence values of all of the taxa at one station
b = the sum of the prominence values of all of the taxa at the other station
c = the sum of the prominence values of all of the taxa of one station which it has in common with the other station. The lower of the two resulting values of w is used in the equation.

Rank Correlation Coefficient

A rank correlation coefficient between measured biological, chemical, or other physical data can be calculated using the formula defined by Spearman (1904):

$$r_s = 1 - \frac{6\sum_{i-1}^{n} D_i^2}{n(n^2-1)}$$

where n = the number of paired observations $(x_i y_i)$ and D_i = the rank of x_i minus the rank of y_i.

Table B.6 Benthic Macroinvertebrate Equipment and Supplies

Item	Unit	Source[a]
Boat, flat bottom, 14–16 ft, snatch-block meter, wheel and trailer, 18 hp outboard motor. Life jackets, other accessories	1	(7,15)
Boat crane kit and winch	1	(3,15)
Boat, inflatable with oar set	1	(1,15)
Cable fastening tools		(4,15)
Cable clamps, 1/8"	25	
Nicro-press clamps, 1/8"	100	
Nicro-press tool, 1/8"	1	
Wire cutter, Felco	1	
Wire thimbles, 1/8"	25	
Cable, 1/8", galvanized steel	1000 ft	(3,15)
Large capacity metal wash tub	1	
Sample wash bucket (sieve)	1	(8,14)
Core sampler, hand held	1	(3,8,14)
Box corer	1	(14)
K-B corer	1	(8)
Wide-barrel gravity corer	1	(14)
Phleger corer	1	(8,14)
Ballchek single or multiple corer	1	(8,14)
Ewing portable piston corer	1	(14)
Hardboard multiplate sampler	10	(3,8)
Ceramic multiplate sampler	10	(14)
Trawl net	1	(8)
Dredge	1	(3,8,14)
Rectangular box sediment sampler	1	(14)
Drift net, stream	6	(8,14)
Triple-net drift sampler	2	(14)
Stream bottom sampler, Surber type	2	(3,8,14)
Portable invertebrate box sampler	2	(13)
Stream-bed fauna sampler, Hess type	2	(14)
Hess stream bottom sampler	2	(8)
Grab sampler, Ponar	1	(3,8,14)
Wildco box corer	1	(8)
Grab sampler, Ekman	1	(3,8,14)
Grab sampler, Petersen	1	(3,8,14)
Grab sampler, Smith-McIntyre	1	(14)
Grab sampler, Van Veen	1	(14)
Grab sampler, Orange Peel	1	(14)
Grab sediment sampler, Shipek	1	(8)
Basket, bar B-Q, tumbler (#740-0035)	12	(9,11)
Sieves, US Standard No. 30	2	(5)
Flowmeter, mechanical	1	(3)
Mounting media, CMCP-9/9AF with stain	4 oz	No longer available
Mounting medium, CMCP-9	4 oz	(6)
Mounting medium, CMCP-10	4 oz	(6)
Fuchsin basic, C.I. dye	25 g	(6)
Mounting medium, Aquamount	4 oz	(12)
Refrigerated circulator	1	(5)
Water pump, epoxy-coated	2	(1)
Holding tank, constant temp	1	(10)
Balance, top-loading	1	(5)
Counter, 12-unit, 2 × 6	1	(3)
Counter, hand tally	2	(3)
Waders, with suspenders	1 pr	(1,15)
Boots, hip	1 pr	(1,15)
Raincoat	1	(3,15)

Table B.6   Benthic Macroinvertebrate Equipment and Supplies (continued)

Item	Unit	Source[a]
Magni-focuser, 2×	1	(5)
Microscope, field	1	(3)
Magnifier, illuminated + base	1	(3)
Magnifier, pocket, 5×, 10×, and 15×	1	(3)
Microscope, compound, with phase and bright-field, trinocular, 10× and 15× eyepieces, 4×, 10×, 20×, 45× and 100× objectives	1	(5)
Microscope, stereoscopic, with stand	1	(2)
Microscope slide dispenser	1	(1)
Microscope slides and cover slips, 12 and 15 mm circles	10 gross	(1)
Photographic system, photostar	1	(5)
Camera, photomicrographic, with 50 mm lens	1	(1,15)
Stirrer, magnetic	1	(5)
Aquarium, 10 gal., with cover, air pump and filter	1	(1,15)
Aquatic dip net, Model 412D	2	(3)
Jars, screw cap, specimen	5 dz	(1)
Bottles, wide mouth, 32 oz	1 case	(1)
Specimen jars, wide mouth, 4 oz	48	(1)
Specimen jars, wide mouth, 6 oz	48	(1)
Vials, specimen, 1 oz	10 gross	(1)
Petri dish, ruled grid	4	(1)
Petri dish, compartmented	1 case	(1)
Watch glasses	10	(1)
Vacuum oven	1	(5)
Sounding lead and calibrated line	1	(3)
Forceps, watchman's, stainless	1 pr	(1)
Forceps, microdissection	2 pr	(1)
Dissecting set, basic	1	(1)
Water test kit, limnology	1	(1)
Thermometer, digital	1	(1)
Wash bottle, wide mouth, 500 mL	4	(1)
Wash bottle, polyethylene, 4 oz	2	(1)
Dropper bottle, polystop, 30 mL	2	(2)
Desiccator, polypropylene	1	(1)
Clipboard with cover	2	(3,15)
Calculator, scientific	1	(3,15)
Marker, permanent, black	2	(3,15)
Pen set, slim pack, Koh-i-noor	1	(3,15)
Heavy paper tags with string	1000	(1,15)
Ice chest, insulated, 48 qt	2	(3,15)
Blue ice, soft pack	10	(3,15)
Plastic bags	100	(3,15)
Formalin, 10%	4 L	(2)
Ethyl alcohol	20 L	(2)
Trays, polypropylene, sorting	6	(5)

Note: Listed above are equipment and supplies needed for the collection and analysis of macroinvertebrate samples. The data quality objectives and sampling and analysis methods should determine the type of equipment and supplies needed. The source numbers refer to the companies that are listed at the end of the table. Mention of these sources or products does not constitute endorsement by the U.S. Environmental Protection Agency.

Table B.6   Benthic Macroinvertebrate Equipment and Supplies (continued)

[a] Sources of equipment and supplies:

1. Carolina Biological Supply Co.
   2700 York Road
   Burlington, NC 27215
2. Fisher Scientific
   50 Fadem Road
   Springfield, NJ 07081
3. Forestry Suppliers, Inc.
   205 West Rankin Street
   Jackson, MS 39284-8397
4. Industrial Rope Supply
   5250 River Road
   Cincinnati, OH 45233
5. Curtin Matheson Scientific, Inc.
   9999 Veterans Memorial Drive
   Houston, TX 77038-2499
6. Polyscience
   400 Valley Road
   Warrington, PA 18976
7. MonArk Boat Company
   Monticello, AK 71655
8. Wildlife Supply Company
   301 Case Street
   Saginaw, MI 48602

9. Tenaco
   2007 NE 27th Avenue
   Gainesville, FL 32609
10. Frigid Units, Inc.
    3214 Sylvania Avenue
    Toledo, OH 43613
11. W.C. Bradley Enterprises, Inc.
    P.O. Box 1240
    Columbus, GA 31993
12. Gallard-Schlesinger Chemical Mfg. Corp.
    584 Mineola Avenue
    Carle Place, NY 11514
13. Ellis-Rutter Associates
    P.O. Box 401
    Punta Gorda, FL 33950
14. Kahl Scientific Instrument Corp.
    P.O. Box 1166
    El Cajon, CA 92022-1166
15. Locally

From EPA. *Biological Criteria: Guide to Technical Literature*. U.S. Environmental Protection Agency, Washington, D.C. EPA-440-5-91-004. 1991.

Coefficient of Variation

In cases where replicate analyses are conducted (e.g., litigation purposes of NAWQMN stations), a coefficient of variation (CV or COV) between replicates is determined following the procedures outlined by Li (1964) using the formula:

$$CV = \frac{s}{x} \bullet 100\%$$

where   s = the sample standard deviation
        x = the sample mean.

A PARTIAL LISTING OF AGENCIES THAT HAVE DEVELOPED TOLERANCE CLASSIFICATIONS AND/OR BIOTIC INDICES

Florida Department of Environmental Regulation
Illinois EPA
New York Department of Environmental Conservation
North Carolina Department of Environmental Management
Ohio EPA
U.S. Department of Agriculture, Forest Service, Intermountain Region
U.S. EPA Region V
Vermont Department of Environmental Conservation

Table B.7 Phylogenetic Order for Macroinvertebrate Listing Including Level of Taxonomy Generally Used

Porifera:	Species	Plecoptera	
Coelenterata:	Genus	Pteronarcyidae:	Genus
Platyhelminthes:	Class	Peltoperfidae:	Genus
Nematomorpha:	Genus	Taeniopterygidae:	Genus
Bryozoa:	Species	Nemounidae:	Species
Entoprocta:	Species	Leuctridae:	Genus
Annelida		Capniidae:	Genus
Oligochaeta:	Class	Perfidae:	Species
Hirudinea:	Species	Perlodidae:	Species
Arthropoda		Chloroperfidae:	Genus
Crustacea		Hemiptera	
Isopoda:	Genus	Belostomatidae:	Genus
Amphipoda:	Genus/Species	Nepidae:	Genus
Decapoda:	Species	Pleidae:	Genus
Arachnoidea		Naucoridae:	Genus
Hydracarina:	Class	Corixidae:	Genus
Insecta		Notonectidae:	Genus
Ephemeroptera		Megaloptera	
Siphlonuridae:	Genus	Sialidae:	Genus
Baetidae:	Genus	Corydalidae:	Species
Oligoneuriidae:	Genus	Neuroptera:	Genus
Heptageniidae:	Genus/Species	Trichoptera	
Leptophlebiidae	Genus	Philopotamidae:	Genus/Species
Ephemerelidae:	Species	Psychomyiidae:	Species
Tricorythidae:	Genus	Polycentropodidae:	Genus
Caenidae:	Genus	Hydropsychidae:	Genus/Species
Baetiscidae:	Species	Rhyacophilidae:	Genus/Species
Potamanthidae:	Genus	Glossosomatidae:	Genus
Ephemeridae:	Genus	Hydroptidae:	Genus/Species
Polymitarchidae:	Species	Phryganeidae:	Genus
Odonata		Brachycentridae:	Genus
Zygoptera		Limnophilidae:	Genus
Calopterygidae:	Genus	Lepidostomatidae:	Genus
Lestidae:	Species	Beraeidae:	Genus
Coenagrionidae:	Family/Genus	Sericostomatidae:	Genus
Anisoptera		Odontocaridae:	Genus
Aeshnidae:	Species	Molannidae:	Genus
Gomphidae:	Species	Helicopsychidae:	Species
Cordulegastridae:	Species	Calamoceratidae:	Genus
Macromiidae:	Species	Leptocaridae:	Genus/Species
Corduliidae:	Species	Lepidoptera:	Genus
Libellulidae:	Species		

Table B.8 Level of Macroinvertebrate Taxonomy Attainable Using Keys

Coleoptera	
Gynnidae:	Genus
Haliplidae:	Genus
Dytiscidae:	Genus
Noteridae:	Genus
Hydrophilidae:	Genus
Hydraenidae:	Genus
Psepheriidae:	Species
Dryopidae:	Genus
Scirtidae:	Family
Elmidae:	Genus/Species
Limnichidae:	Genus
Heteroceridae:	Family
Ptilodactylidae:	Family
Chrysomelidae:	Family
Curculionidae:	Family
Lampyridae:	Family
Diptera	
Tipulidae:	Genus
Psychodidae:	Genus
Ptychopteridae:	Genus
Dixidae:	Genus
Chaoboridae:	Genus
Culicidae:	Genus
Thaumaleidae:	Genus
Simuliidae:	Genus
Certopogonidae:	Family/Genus/Species
Chironomidae	
Tanypodinae:	Genus/Species
Diamesinae:	Genus/Species
Prodiamesinae:	Genus/Species
Orthocladinae:	Genus/Species
Chironominae	
Chironomini:	Genus/Species
Pseudochironomini:	Genus/Species
Tanytarsini:	Genus/Species
Tabanidae:	Genus/Species
Athericidae:	Species
Stratiomyidae:	Genus
Empididae:	Family
Dolichopodidae:	Family
Syrphidae:	Family/Genus
Sciomyzidae:	Family/Genus
Ephydridae:	Family/Genus
Muscidae:	Species
Mollusca	
Gastropoda:	Family/Genus/Species
Pelecypoda:	Family/Genus/Species

REFERENCES

Barbour, M.T., J. Gerritsen, B.D. Snyder, and J.B. Stribling. *Rapid Bioassessment Protocols for Use in Streams and Wadeable Rivers: Periphyton, Benthic Macroinvertebrates and Fish, 2nd Edition.* EPA 841-B-99-002. U.S. Environmental Protection Agency, Office of Water, Washington, D.C. 1999.

Barbour, M.T., J.L. Plafkin, B.P. Bradley, C.G. Graves, and R.W. Wisseman. Evaluation of EPA's rapid bioassessment benthic metrics: metric redundancy and variability among reference stream sites. *Environ. Toxicol. Chem.*, 11: 437–449. 1992.

Barbour, M.T., M.L. Bowman, and J.S. White. *Evaluation of the Biological Condition of Streams in the Middle Rockies — Central Ecoregion.* Prepared for Wyoming Dept. Environ. Quality. Casper, WY. 1994.

Barbour, M.T., J. Gerritsen, G.E. Griffith, R. Frydenborg, E. McCarron, J.S. White, and M.L. Bastian. A framework for biological criteria for Florida streams using benthic macroinvertebrates. *J. North Am. Benthol. Soc.*, 15: 185–211. 1996.

Beck, W.M., Jr. *Environmental Requirements and Pollution Tolerance of Common Freshwater Chironomidae.* Environmental Monitoring and Support Laboratory, Report No. EPA-600/4-77-024. U.S. EPA, Cincinnati. 1977.

Bode, R.W. *Quality Assurance Workplan for Biological Stream Monitoring in New York State.* New York State Department of Environmental Conservation, Albany. 1988.

Boesch, D.F. *Application of Numerical Classification in Ecological Investigation of Water Pollution*, U.S. Environmental Protection Agency, Corvallis, OR. EPA 600/3-77/033. 1977.

Bray, J.R. and J.T. Curtis. An ordination of the upland forest communities of southern Wisconsin. *Ecol. Monogr.*, 27: 325. 1957.

Burlington, R.F. Quantitative biological assessment of pollution, *J. Water Pollut. Control Fed.*, 34: 179–183. 1962.

Clements, W.H., D.S. Cherry, and J. Cairns, Jr. Structural alterations in aquatic insect communities exposed to copper in laboratory streams. *Environ. Toxicol. Chem.*, 7: 715–722. 1988.

Cochran, W.C. *Sampling Techniques. 2nd edition.* John Wiley & Sons, New York. 1963.

Courtemanch, D.L., and S.P. Davies. A coefficient of community loss to assess detrimental change in aquatic communities, *Water Res.*, 21:217–222. 1987.

Cummins, K.W. Trophic relations of aquatic insects. *Annu. Rev. Entomol.*, 18: 183–206. 1973.

Cummins, K.W. and M.A. Wilzbach. Field procedures for analysis of functional feeding groups of stream macroinvertebrates. Contribution 1611, Appalachian Environmental Laboratory, University of Maryland, Frostburg, MD. 1985.

Cummins, K.W., M.A. Wilzbach, D.M. Gates, J.B. Perry, and W.B. Taliaferro. Shredders and riparian vegetation. *BioScience*, 39: 24–30. 1989.

Dawson, C.L. and R.A. Hellenthal. *A Computerized System for the Evaluation of Aquatic Habitats Based on Environmental Requirements and Pollution Tolerance Association of Resident Organisms.* Report No. EPA-600/S3-86/019. 1986.

DeShon, J.E. Development and application of the invertebrate community index (ICI), in *Biological Assessment and Criteria: Tools for Water and Resource Planning and Decision Making.* Edited by W.S. Davis and T.P. Simon. Lewis Publ. Boca Raton, FL, pp. 217–243. 1995.

EPA. *Technical Support Manual: Waterbody Surveys and Assessments for Conducting Use Attainability Analyses.* Office of Water Regulations and Standards, U.S. Environmental Protection Agency, Washington, D.C. 1983.

EPA. *Rapid Bioassessment Protocols for Use in Streams and Rivers: Benthic Macroinvertebrates and Fish.* Office of Water, U.S. Environmental Protection Agency, Washington, D.C. EPA 444/4-89/001. 1989.

EPA. *Biological Criteria: Guide to Technical Literature.* U.S. Environmental Protection Agency, Washington, D.C. EPA-440-5-91-004. 1991.

EPA. *Rapid Bioassessment Protocols for Use in Wadeable Streams and Rivers: Periphyton, Benthic Macro-invertebrates and Fish.* Office of Water, U.S. Environmental Protection Agency, Washington, D.C. EPA 841/B-99/002. 1999.

Ferrington, L.C. *Collection and Identification of Floating Exuviae of Chironomidae for Use in Studies of Surface Water Quality.* U.S. Environmental Protection Agency. Kansas City, KS. SOP No. FW 130A. 1987.

Fore, L.S., J.R. Karr, and R.W. Wisseman. Assessing invertebrate responses to human activities: evaluating alternative approaches. *J. North Am. Benthol. Soc.*, 15: 212–231. 1996.

Gleason, H.A. Some applications of the quadrat method. *Torrey Bot. Club Bull.*, 47: 21–33. 1920.

Hargrove, B.T. Aerobic decomposition of sediment and detritus as a function of particle surface area and organic content. *Limnol. Oceanogr.*, 17: 583–596. 1972.

Harris, T.L. and T.M. Lawrence. *Environmental Requirements and Pollution Tolerance of Trichoptera*. Report No. EPA-600/4-78-063. U.S. EPA, Washington, D.C. 1978.

Hayslip, G.A. *EPA Region 10 In-Stream Biological Monitoring Handbook (For Wadable Streams in the Pacific Northwest)*. U.S. Environmental Protection Agency–Region 10, Environmental Services Division. Seattle, WA. EPA-910-9-92-013. 1993.

Hilsenhoff, W.L. *Using a Biotic Index to Evaluate Water Quality in Streams*. Technical Bulletin No. 132. Department of Natural Resources, Madison, WI. 1982.

Hilsenhoff, W.L. An improved biotic index of organic stream pollution. *Great Lakes Entomol.*, 20: 31–39. 1987.

Hilsenhoff, W.L. Rapid field assessment of organic pollution with a family-level biotic index. *J. North Am. Benthol. Soc.*, 7: 65–68. 1988.

Hubbard, M.D. and W.L. Peters. *Environmental Requirements and Pollution Tolerance of Ephemeroptera*. Report No. EPA-600/4-78-061. U.S. EPA, Washington, D.C. 1978.

Jaccard, P. The distribution of flora in an alpine zone. *New Phytol.*, 11: 37. 1912.

Karr, J.R. Assessment of biotic integrity using fish communities. *Fisheries*, 6: 21. 1981.

Kerans, B.L. and J.R. Karr. A benthic index of biotic integrity (B-IBI) for rivers of the Tennessee Valley. *Ecol. Appl.*, 4: 768–785. 1994.

Lenat, D.R. Benthic macroinvertebrates of Cane Creek, North Carolina, and comparisons with other southeastern streams. *Brimleyana*, 9: 53–68. 1983.

McArthur, J.V., J.R. Barnes, B.J. Hansen, and L.G. Leff. Seasonal dynamics of leaf litter breakdown in a Utah alpine stream. *J. North Am. Benthol. Soc.*, 7: 44–50. 1988.

Merritt, R.W. and K.W. Cummins. *An Introduction to the Aquatic Insects of North America, 2nd edition*. Kendall/Hunt Publishing Company, Dubuque, IA. 1984.

Minshall, G., K. Cummins, R. Petersen, C. Cushing, D. Bruns, J. Sedell, and R. Vannote. Development in stream ecosystem theory. *Can. J. Fish. Aquat. Sci.*, 42: 1045–1055. 1985.

Moriseta, M. Measuring of interspecific association and similarity between communities. *Memoirs Faculty Sci.*, Kyoshu University, Series I. Biol. 3–65. 1959.

Mount, D.I., N.A. Thomas, T.J. Norberg, M.T. Barbour, T.H. Roush, and W.F. Brandes. *Effluent and Ambient Toxicity Testing and Instream Community Response on the Ottawa River, Lima, Ohio*. U.S. Environmental Protection Agency. Duluth, MN. EPA 600/3-84/080. 1984.

OEPA (Ohio Environmental Protection Agency). *Biological Criteria for the Protection of Aquatic Life: Volume II. User's Manual for Biological Assessment of Ohio Surface Waters*. Ohio Environmental Protection Agency, Columbus, OH. 1987.

Resh, V.H. *Variability, Occurrence, and Taxonomic Costs of Rapid Assessment Approaches in Benthic Biomonitoring*. Presented at the 36th Annual North American Benthological Society Meeting at Tuscaloosa, AL. 1988.

Shackleford, B. *Rapid Bioassessment of Lotic Macroinvertebrate Communities: Biocriteria Development*. Arkansas Department of Pollution Control and Ecology, Little Rock. 1988.

Smith, E.P. and J.R. Voshell, Jr. *Studies of Benthic Macroinvertebrates and Fish in Streams within EPA Region 3 for Development of Biological Indicators of Ecological Condition*. Virginia Polytechnic Institute and State University, Blacksburg, VA. 1997.

Snedecor, G.W. and W.G. Cochran. *Statistical Methods*. Iowa State University Press, Ames, IA. 1967.

Spearmon, C. The proof and measurement of association between two things. *Am. J. Psychol.*, 15: 72–101. 1904.

Surdick, R.F. and A.R. Gaufin. *Environmental Requirements and Pollution Tolerance of Plecoptera*. Report No. EPA-600/4-78-062. U.S. EPA, Cincinnati. 1978.

Swift, M.C., K.W. Cummins, and R.A. Smucker. Effects of dimilin on stream leaf-litter processing rates. *Verh. Int. Verein. Limnol.*, 23: 1255–1260. 1988a.

Swift, M.C., R.A. Smucker, and K.W. Cummins. Effects of dimilin on freshwater litter decomposition. *Environ. Toxicol. Chem.*, 7: 161–166. 1988b.

USDA (U.S. Department of Agriculture). *Fisheries Survey Handbook, Aquatic Ecosystem Inventory, Chapter 5 Aquatic Macroinvertebrate Surveys*. Document No. R-4 FSH 2609.23. U.S. Department of Agriculture, Forest Service, Ogden, UT. 1985.

Weber, C.I. *Biological Field and Laboratory Methods for Measuring the Quality of Surface Waters and Effluents*. Report No. EPA-670/4-73-001. U.S. EPA, Cincinnati. 1973.

Whittaker, R.H. A study of summer foliage insect communities in the Great Smokey Mountains. *Ecol. Monogr.* 22:6. 1952.

Winget, R.M. and F.A. Mangum. *Biotic Condition Index: Integrated Biological, Physical, and Chemical Stream Parameters for Management*. U.S. Department of Agriculture, Forest Service, Ogden, UT. 1979.

Fish Community Assessment

CONTENTS

Rapid Bioassessment Protocol V — Fish (EPA 1989, 1999)..693
    Sample Processing..694
    Data Analysis Techniques 694
References 707

RAPID BIOASSESSMENT PROTOCOL V — FISH (EPA 1989, 1999)

The following are excerpts from U.S. EPA (1989, 1999; www.epa.gov/owow/monitoring/rbp) guidance manuals. For more extensive information, the reader should refer directly to those manuals.

Rapid Bioassessment Protocol V (RBP V) is a rigorous approach similar to species-level identi-fication with the macroinvertebrate RBP in accuracy and effort, but focuses on fish. Electrofishing, the most common technique used by agencies that monitor fish communities, and the most widely applicable approach for stream habitats, is the sampling technique recommended for use with RBP V.

The fish community biosurvey data are designed to be representative of the fish community at all station habitats, similar to the "representative qualitative sample" proposed by Hocutt (1981). The sampling station should be representative of the reach, incorporating at least one (preferably two) riffle(s), run(s), and pool(s) if these habitats are typical of the stream in question. Sampling of most species is most effective near shore and cover (macrophytes, boulders, snags, brush). Sampling procedures effective for large rivers are described in Gammon (1980), Hughes and Gammon (1987), and Ohio EPA (1987).

Typical sampling station lengths range from 100 to 200 m for small streams and 500 to 1000 m in rivers, but are best determined by pilot studies. The size of the reference station should be sufficient to produce 100 to 1000 individuals and 80% of the species expected from a 50% increase in sampling distance. Sample collection is usually done during the day, but night sampling can be more effective if the water is especially clear and there is little cover (Reynolds 1983). Use of block nets set (with as little wading as possible) at both ends of the reach increases sampling efficiency for large, mobile species sampled in small streams.

The RBP V fish community assessment requires that all fish species (not just gamefish) be collected. Small fish that require special gear for their effective collection may be excluded. Exclusion of young-of-the-year fish during collection has only a minor effect on IBI scores (Angermeier and Karr 1986), but lowers sampling costs and reduces the need for laboratory identification. Karr et al. (1986) recommended exclusion of fish less than 20 mm in length. However, this may prevent detection of species-specific effects, or early life stage effects from recent pollution incidents. This recommendation should be considered on a regional basis and is also applicable to large fish requiring special gear for collection (e.g., sturgeon). The intent of the sample (as with the entire protocol) is to obtain a representative estimate of the species present, and their abundances, for a reasonable amount of effort. However, if threatened or endangered species are present, special attention should be given to documenting their presence and numbers.

Sampling effort among stations is standardized as much as possible. Regardless of the gear used, the collection method, site length (or area), and work hours expended must be comparable to allow comparison of fish community status among sites. Major habitat types (riffle, run, and pool) sampled at each site and the proportion of each habitat sampled should also be comparable. Generally 1 to 2 hours of actual sampling time are required, but this varies considerably with the gear used and the size and complexity of the site.

Atypical conditions, such as high flow, excessive turbidity or turbulence, heavy rain, drifting leaves, or other unusual conditions that affect sampling efficiency, are best avoided. Glare, a frequent problem, is reduced by wearing polarized glasses during sample collection.

Sample Processing

A field collection data sheet (Figure C.1) is completed for each sample. Sampling duration and area or distance sampled are recorded in order to determine level of effort. Species may be separated into adults and juveniles by size and coloration; then total numbers and weights and the incidence of external anomalies are recorded for each group. Reference specimens of each species from each site are preserved in 10% formaldehyde, the jar labeled, and the collection placed with the state ichthyological museum to confirm identifications and to constitute a biological record. This is especially important for uncommon species, for species requiring laboratory identification, and for documenting new distribution records. If retained in a live well, most fish can be identified, counted, and weighed in the field by trained personnel and returned to the stream alive. In warmwater sites, where handling mortality is highly probable, each fish is identified and counted, but for abundant species, subsampling (weigh, measure, observe for abnormalities, and return) may be considered. When subsampling is employed, the subsample is extrapolated to obtain a final value. Subsampling for weight is a simple, straightforward procedure, but failure to examine all fish to determine frequency of anomalies (which may occur in about 1% of all specimens) can bias results. The trade-off between handling mortality and data bias must be considered on a case-by-case basis. If a site is to be sampled repeatedly over several months (i.e., monitoring), the effect of sampling mortality might outweigh data bias. Holding fish in live-boxes in shaded, circulating water will substantially reduce handling mortality. More information on field methods is presented in Karr et al. (1986) and Ohio EPA (1987).

Data Analysis Techniques

Based on observations made in the assessment of habitat, water quality, physical characteristics, and the fish biosurvey, the investigator concludes whether impairment is detected. If impairment is detected, the probable cause and source are estimated and recorded on an Impairment Assessment Sheet (Figure C.2).

Data can be analyzed using the Index of Biotic Integrity (IBI) (or individual IBI metrics), the Index of Well Being (IWB) (Gammon 1980), or modified IWB (OEPA 1989; Gammon 1989), and multivariate statistical techniques to determine community similarities. Detrended correspondence analysis (DCA) is a useful multivariate analysis technique for revealing regional community patterns and patterns among multiple sites. It also demonstrates assemblages with compositions differing from others in the region or reach. See Gauch (1982) and Hill (1979) for descriptions of, and software for, DCA. Data analyses and reporting, including parts of the IBI, can be computer generated. Computerization reduces the time needed to produce a report and increases staff capability to examine

FISH FIELD COLLECTION DATA SHEET

Page _____ of _____

Drainage _____   Date_____

Sampling Duration (min)_____

Sampling Distance (m)_____   Sampling Area (m2)_____   Crew_____

Habitat Complexity/Quality (excellent   good   fair   poor   very poor)

Weather_____   Flow (flood   bankfull   moderate   low)

Gear Used_____   Gear/Crew Performance_____

Comments_____

Fish (preserved)   Number of Individuals_____   Number of Anomalies_____

Genus/Species	Adults		Juveniles		Anomalies(*)	
	No.	Wt.	No.	Wt.	No.	Wt.

(*) Discoloration, deformities, eroded fins, excessive mucus, excessive external parasites, fungus, poor condition, reddening, tumors, and ulcers

Figure C.1   Fish field collection data sheet for use with Rapid Bioassessment Protocol V. (From EPA. *Rapid Bioassessment Protocols for Use in Streams and Rivers: Benthic Macroinvertebrates and Fish*. Office of Water, U.S. Environmental Protection Agency, Washington, D.C. EPA 444/4-89/001. 1989.)

data patterns and implications. Illinois EPA has developed software to assist professional aquatic biologists in calculating IBI values in Illinois streams (Bickers et al. 1988). (Use of this software outside Illinois without modification is not recommended.) However, hand calculation in the initial use of the IBI promotes understanding of the approach and provides insight into local inconsistencies. Metrics should be optimized for specific ecoregions. See EPA (1999) for a range of alternative IBI metrics.

Each metric is scored against criteria based on expectations developed from appropriate regional reference sites. Metric values approximating, deviating slightly from, or deviating greatly from values occurring at the reference sites are scored as 5, 3, or 1, respectively. The scores of the 12 metrics are added for each station to give an IBI of 60 (excellent) to 12 (very poor). Trophic and tolerance classifications of many species are listed below. Additional classifications can be derived from information in state and regional fish texts or by objectively assessing a large statewide database. Use of the IBI in the southeastern and southwestern United States and its widespread use by water resource agencies may result in further modifications. Past modifications have occurred (Miller et al. 1988) without changing the IBI's basic theoretical foundations.

IMPAIRMENT ASSESSMENT SHEET

1. Detection of impairment: Impairment detected   No impairment
                                        (Complete Item 2-6)   detected
                                                                         (Stop here)

2. Biological impairment indicator:
    Fish   Other aquatic communities

    _____ sensitive species reduced/absent   _____ Macroinvertebrates

    _____ dominance of tolerant species   _____ Periphyton

    _____ skewed trophic structure   _____ Macrophytes

    _____ abundance reduced/unusually high

    _____ biomass reduced/unusually high

    _____ hybrid or exotic abundance unusually high

    _____ poor size class representation

    _____ high incidence of anomalies

3. Brief description of problem: _____
    Year and date of previous surveys: __

    Survey data available in: _____

4. Cause (indicate potential cause):   organic enrichment   toxicants   flow
                                        sediment   temperature   poor habitat

                                        other _____

5. Estimated areal extent of problem (m2) and length of stream reach affected (m)
    where applicable: _____

6. Suspected source(s) of problem
    _____ point source   _____ mine

    _____ urban runoff   _____ dam or diversion

    _____ agricultural runoff   _____ channelization or snagging

    _____ silvicultural runoff   _____ natural

    _____ livestock   _____ other

    _____ landfill   _____ unknown

    Comments: _____

    _____

Figure C.2   Impairment assessment sheet for use with Fish Rapid Bioassessment Protocol V. (From EPA. *Rapid Bioassessment Protocols for Use in Streams and Rivers: Benthic Macroinvertebrates and Fish.* Office of Water, U.S. Environmental Protection Agency. Washington, D.C. EPA 444/4-89/001. 1989.)

The steps in calculating the IBI are explained below:

1. Assign species to trophic guilds; identify and assign species tolerances. Where published data are lacking, assignments are made based on knowledge of closely related species and morphology.
2. Develop scoring criteria for each IBI metric. Maximum species richness (or density) lines are developed from a reference database.
3. Conduct field study and identify fish; note anomalies, eroded fins, poor condition, excessive mucus, fungus, external parasites, reddening, lesions, and tumors. Complete field data sheets.
4. Enumerate and tabulate number of fish species and relative abundances.
5. Summarize site information for each IBI metric.

6. Rate each IBI score to one of the five integrity classes.
7. Translate total IBI score to one of the five integrity classes.
8. Interpret data in the context of the habitat assessment. Individual metric analysis may be necessary to ascertain specific trends.

Species Richness and Diversity

These metrics assess the species richness component of diversity and the health of the major taxonomic groups and habitat guilds of fishes. Two of the metrics assess community composition in terms of tolerant or intolerant species. Scoring for the first five of these metrics and their substitutes requires development of species–water body size relationships for different zoogeographic regions. Development of this relationship requires data sufficient to plot the number of species collected from regional reference sites of various stream sizes against a measure of stream size (watershed area, stream order) of those sites. A line is then drawn with slope fit by eye to include 95% of the points. Finally the area under the line is trisected into areas that are scored as 5, 3, or 1. A detailed description of these methods can be found in Fausch et al. (1984), Ohio EPA (1987), and Karr et al. (1986).

Metric 1. Total number of fish species — Substitute metrics: total number of native fish species, and salmonid age classes.

This number decreases with increased degradation; hybrids and introduced species are not included. In cold-water streams supporting few fish species, the age classes of the species found represent the suitability of the system for spawning and rearing. The number of species is strongly affected by stream size at small stream sites, but not at large river sites (Karr et al. 1986; Ohio EPA 1987). Thus, scoring depends on developing species–waterbody size relationships.

Metric 2. Number and identity of darter species — Substitute metrics: number and identity of sculpin species, benthic insectivore species, salmonid yearlings (individuals); number of sculpins (individuals); percent round-bodied suckers, sculpin and darter species.

These species are sensitive to degradation resulting from siltation and benthic oxygen depletion because they feed or reproduce in benthic habitats (Kuehne and Barbour 1983; Ohio EPA 1987). Many smaller species live within the rubble interstices, are weak swimmers, and spend their entire lives in an area of 100 to 400 m² (Hill and Grossman 1987; Matthews 1986). Darters are appropriate in most Mississippi basin streams; sculpins and yearling trout occupy the same niche in western streams. Benthic insectivores and sculpins or darters are used in small Atlantic slope streams that have few sculpins or darters, and round-bodied suckers are suitable in large midwestern rivers. Scoring requires development or species–water body size relationships.

Metric 3. Number and identity of sunfish species — Substitutes: number and identity of cyprinid species, water column species, salmonid species, headwater species, and sunfish and trout species.

These pool species decrease with increased degradation of pools and in-stream cover (Gammon et al. 1981; Angermeier 1983; Platts et al. 1983). Most of these fishes feed on drifting and surface invertebrates and are active swimmers. The sunfishes and salmonids are important sport species. The sunfish metric works for most Mississippi basin streams, but where sunfish are absent or rare, other groups are used. Cyprinid species are used in cool-water western streams; water column species occupy the same niche in northeastern streams; salmonids are suitable in cold-water streams; headwater species serve for midwestern headwater streams; and trout and sunfish species are used in southern Ontario streams. Karr et al. (1986) and Ohio EPA (1987) found the number of sunfish species to be dependent on stream size in small streams, but Ohio EPA (1987) found no relationship between stream size and sunfish species in medium to large streams, nor between stream size and headwater species in small streams. Scoring of this metric requires development of species–water body size relationships.

Metric 4. Number and identity of sucker species — Substitutes: number of adult trout species, number of minnow species, and number of suckers and catfish.

These species are sensitive to physical and chemical habitat degradation and commonly comprise most of the fish biomass in streams. All but the minnows are long-lived species and provide a multiyear integration of physicochemical conditions. Suckers are common in medium and large streams; minnows dominate small streams in the Mississippi basin; and trout occupy the same niche in cold-water streams. The richness of these species is a function of stream size in small and medium-sized streams but not in large rivers. Scoring of this metric requires development of species–water body size relationships.

Metric 5. Number and identity of intolerant species — Substitutes: number and identity of sensitive species (5), amphibian species, and presence of brook trout.

This metric distinguishes high- and moderate-quality sites using species that are intolerant of various chemical and physical perturbations. Intolerant species are typically the first to disappear following a disturbance. Species classified as intolerant or sensitive should only represent the 5 to 10% most susceptible species; otherwise this becomes a less discriminating metric. Candidate species are determined by examining regional ichthyological books for species that were once widespread but have become restricted to only the highest quality streams. Ohio EPA (1987) uses number of sensitive species (which includes highly intolerant and moderately intolerant species) for headwater sites because highly intolerant species are generally not expected in such habitats. Moyle (1976) suggested using amphibians in northern California streams because of their sensitivity to silvicultural impacts. This also may be a promising metric in Appalachian streams which may naturally support few fish species. Steedman (1988) found that the presence of brook trout had the greatest correlation with IBI score in Ontario streams. The number of sensitive and intolerant species increases with stream size in small and medium-sized streams but is unaffected by size of large rivers. Scoring this metric requires development of species–water body size relationships.

Metric 6. Proportion of individuals as green sunfish — Substitutes: proportion of individuals as common carp, white sucker, tolerant species, creek chub, and dace.

This metric is the reverse of Metric 5. It distinguishes low- from moderate-quality waters. These species show increased distribution or abundance despite the historical degradation of surface waters, and they shift from incidental to dominant in disturbed sites. Green sunfish are appropriate in small midwestern streams; creek chubs were suggested for central Appalachian streams; common carp were suitable for a cool-water Oregon river; white suckers were selected in the Northeast and Colorado where green sunfish are rare to absent; and dace (*Rhinichthys* species) were used in southern Ontario. To avoid weighting the metric on a single species, Karr et al. (1986) and Ohio EPA (1987) suggest using a small number of highly tolerant species. Scoring of this metric may require development of expectations based on water body size.

Trophic Composition Metrics

These three metrics assess the quality of the energy base and trophic dynamics of the community. Traditional process studies, such as community production and respiration, are time-consuming, and the results are equivocal; distinctly different situations can yield similar results. The trophic composition metrics offer a means to evaluate the shift toward more generalized foraging that typically occurs with increased degradation of the physicochemical habitat.

Metric 7. Proportion of individuals as omnivores — Substitutes: proportion of individuals as yearlings.

The percent of omnivores in the community increases as the physical and chemical habitat deteriorates. Omnivores are defined as species that consistently feed on substantial proportions of plant and animal material. Ohio EPA (1987) excludes sensitive filter-feeding species such as paddle-fish and lamprey ammocoetes and opportunistic feeders like channel catfish. Where omnivorous species are nonexistent, such as in trout streams, the proportion of the community composed of yearlings, which initially feed omnivorously, may be substituted.

Metric 8. Proportion of individuals as insectivorous cyprinids — Substitutes: proportion of individuals as insectivores, specialized insectivores, and insectivorous species; and number of juvenile trout.

Insectivores or invertivores are the dominant trophic guild of most North American surface waters. As the invertebrate food source decreases in abundance and diversity due to physicochemical habitat deterioration, there is a shift from insectivorous to omnivorous fish species. Generalized insectivores and opportunistic species, such as blacknose dace and creek chub, were excluded from this metric by the Ohio EPA (1987). This metric evaluates the midrange of biotic integrity.

Metric 9. Proportion of individuals as top carnivores — Substitutes: proportion of individuals as catchable salmonids, catchable wild trout, and pioneering species.

The top carnivore metric discriminates between systems with high and moderate integrity. Top carnivores are species that, as adults, feed predominantly on fish, other vertebrates, or crayfish. Occasional piscivores, such as creek chub and channel catfish, are not included. In trout streams, where true piscivores are uncommon, the percent of large salmonids is substituted for percent piscivores. These species often represent popular sport fish such as bass, pike, walleye, and trout. Pioneering species are used by Ohio EPA (1987) in headwater streams typically lacking piscivores.

Fish Abundance and Condition Metrics

The last three metrics (plus the final optional matrix) indirectly evaluate population recruitment mortality, condition, and abundance. Typically, these parameters vary continuously and are time-consuming to estimate accurately. Instead of such direct estimates, the final results of the population parameters are evaluated. Indirect estimation is less variable and much more rapidly determined.

Metric 10. Number of individuals in sample — Substitutes: density of individuals.

This metric evaluates population abundance and varies with region and stream size for small streams. It is expressed as catch per unit effort, either by area, distance, or time sampled. Generally sites with lower integrity support fewer individuals, but in some nutrient-poor regions, enrichment increases the number of individuals. Steedman (1988) addressed this situation by scoring catch per minute of sampling greater than 25 as a three, and less than 4 as a one. Unusually low numbers generally indicate toxicity, making this metric most useful at the low end of the biological integrity scale. Hughes and Gammon (1987) suggest that in larger streams, where sizes of fish may vary in orders of magnitude, total fish biomass may be an appropriate substitute or additional metric.

Metric 11. Proportion of individuals as hybrids — Substitutes: proportion of individuals as introduced species, simple lithophils, and number of simple lithophilic species.

This metric is an estimate of reproductive isolation or the suitability of the habitat for reproduction. Generally, as environmental degradation increases, the percent of hybrids and introduced species also increases, but the proportion of simple lithophils decreases. However, minnow hybrids are found in some high-quality streams; hybrids are often absent from highly impacted sites; and hybridization is rare and difficult for many to detect. Thus, Ohio EPA (1987) substitutes simple lithophils for hybrids. Simple lithophils spawn where their eggs can develop in the interstices of sand, gravel, and cobble substrates without parental care. Hughes and Gammon (1987) and Miller et al. (1988) proposed using percent introduced individuals. This metric is a direct measure of the loss of species segregation between midwestern and western fishes that existed before the introduction of midwestern species to western rivers.

Metric 12. Proportion of individuals with disease, tumors, fin damage, and skeletal anomalies — this metric depicts the health and condition of individual fish. These conditions occur infrequently or are absent from minimally impacted reference sites but occur frequently below major pollutant sources. They are excellent measures of the subacute effects of chemical pollution and the aesthetic value of game and nongame fish.

Metric 13. Total fish biomass (optional) — Hughes and Gammon (1987) suggest that in larger areas, where sizes of fish may vary in orders of magnitude, this additional metric may be appropriate.

Because the IBI is an adaptable index, the choice of metrics and scoring criteria is best developed on a regional basis through use of available publications (Karr et al. 1986; Ohio EPA 1987; Miller et al. 1988). Several steps are common to all regions. The fish species must be listed and assigned to trophic and tolerance guilds. Scoring criteria are developed through use of high-quality historical data and data from minimally impacted regional reference sites. This has been done for much of the country, but continued refinements are expected as more fish community ecology data become available. Once scoring criteria have been established, a fish sample is evaluated by listing the species and their abundances, calculating values for each metric, and comparing these values with the scoring criteria. Individual metric scores are added to calculate the total IBI score (Figure C.3).

Station No.   _____

Site   _____

	Scoring Criteria[b]				
Metrics[a]	5 %	3 %	1 %	Metric Value	Metric Source
1. Number of Native Fish Species	>67	33-67	<33		
2. Number of Darter or Benthic Species	>67	33-67	<33		
3. Number of Sunfish or Pool Species	>67	33-67	<33		
4. Number of Sucker or Long-Lived Species	>67	33-67	<33		
5. Number of Intolerant Species	>67	33-67	<33		
6. % Green Sunfish or Tolerant Individuals	<10	10-25	>25		
7. % Omnivores	<20	20-45	>45		
8. % Insectivores or Invertivores	>45	20-45	<20		
9. % Top Carnivores	>5	1 - 5	<33		
10. Total Number of Individuals	>67	33-67	<33		
11. % Hybrids or Exotics	0	0 - 1	>1		
12. % Anomalies	<1	1 - 5	>5		

Scorer   _____   IBI Score   _____

Comments: _____
_____
_____

_____

[a] Karr's original metrics or commonly used substitutes. See text for other possibilities.

[b] Karr's original scoring criteria or commonly used substitutes. These may require refinement in other ecoregions.

Figure C.3 Data summary sheet for Rapid Bioassessment Protocol V. (From EPA. *Rapid Bioassessment Protocols for Use in Streams and Rivers: Benthic Macroinvertebrates and Fish*. Office of Water, U.S. Environmental Protection Agency. Washington, D.C. EPA 444/4-89/001. 1989.)

Hughes and Gammon (1987) and Miller et al. (1988) suggest that scores lying at the extremes of scoring criteria can be modified by a plus or minus; a combination of three pluses or three minuses results in a two-point increase or decrease in IBI. Ohio EPA (1987) scores proportional metrics as 1 when the number of species and individuals in samples are fewer than 6 and 75, respectively, when their expectations are of higher numbers.

Table C.1 Tolerance, Trophic Guilds, and Origins of Selected Fish Species

	Trophic Level	Tolerance	Origin
Willamette Species			
Salmonidae			
Chinook salmon	piscivore	intolerant	native
Cutthroat trout	insectivore	intolerant	native
Mountain whitefish	insectivore	intolerant	native
Rainbow trout	insectivore	intolerant	native
Cyprinidae			
Chiselmouth	herbivore	intermediate	native
Common carp	omnivore	tolerant	exotic
Goldfish	omnivore	tolerant	exotic
Leopard dace	insectivore	intermediate	native
Longnose dace	insectivore	intermediate	native
Northern squawfish	piscivore	tolerant	native
Peamouth	insectivore	intermediate	native
Redside shiner	insectivore	intermediate	native
Speckled dace	insectivore	intermediate	native
Catostomidae			
Largescale sucker	omnivore	tolerant	native
Mountain sucker	herbivore	intermediate	native
Ictaluridae			
Brown bullhead	insectivore	tolerant	exotic
Yellow bullhead	insectivore	tolerant	exotic
Percopsidae			
Sand roller	insectivore	intermediate	native
Gasterosteidae			
Threespine stickleback	insectivore	intermediate	native
Centrarchidae			
Bluegill	insectivore	tolerant	exotic
Largemouth bass	piscivore	tolerant	exotic
Smallmouth bass	piscivore	intermediate	exotic
White crappie	insectivore	tolerant	exotic
Percidae			
Yellow perch	insectivore	intermediate	exotic
Cottidae			
Paiute sculpin	insectivore	intolerant	native
Prickly sculpin	insectivore	intermediate	native
Reticulate sculpin	insectivore	tolerant	native
Torrent sculpin	insectivore	intolerant	native
Midwest Species			
Petromyzontidae			
Silver lamprey	piscivore	intermediate	native
Northern brook lamprey	filterer	intolerant	native
Mountain brook lamprey	filterer	intolerant	native
Ohio lamprey	piscivore	intolerant	native
Least brook lamprey	filterer	intermediate	native
Sea lamprey	piscivore	intermediate	exotic
Polyodontidae			
Paddlefish	filterer	intolerant	native

Table C.1   Tolerance, Trophic Guilds, and Origins of Selected Fish Species (continued)

	Trophic Level	Tolerance	Origin
Acipenseridae			
Lake sturgeon	invertivore	intermediate	native
Shovelnose sturgeon	insectivore	intermediate	native
Lepisosteidae			
Alligator gar	piscivore	intermediate	native
Shortnose gar	piscivore	intermediate	native
Spotted gar	piscivore	intermediate	native
Longnose gar	piscivore	intermediate	native
Amiidae			
Bowfin	piscivore	intermediate	native
Hiodontidae			
Goldeye	insectivore	intolerant	native
Mooneye	insectivore	intolerant	native
Clupeidae			
Skipjack herring	piscivore	intermediate	native
Alewife	invertivore	intermediate	exotic
Gizzard shad	omnivore	intermediate	native
Threadfin shad	omnivore	intermediate	native
Salmonidae			
Brown trout	insectivore	intermediate	exotic
Rainbow trout	insectivore	intermediate	exotic
Brook trout	insectivore	intermediate	native
Lake trout	piscivore	intermediate	native
Coho salmon	piscivore	intermediate	exotic
Chinook salmon	piscivore	intermediate	exotic
Lake herring	piscivore	intermediate	native
Lake whitefish	piscivore	intermediate	native
Osmeridae			
Rainbow smelt	invertivore	intermediate	exotic
Umbridae			
Central mudminnow	insectivore	tolerant	native
Esocidae			
Grass pickerel	piscivore	intermediate	native
Chain pickerel	piscivore	intermediate	native
Northern pike	piscivore	intermediate	native
Muskellunge	piscivore	intermediate	native
Cyprinidae			
Common carp	omnivore	tolerant	exotic
Goldfish	omnivore	tolerant	exotic
Golden shiner	omnivore	tolerant	native
Horneyhead chub	insectivore	intolerant	native
River chub	insectivore	intolerant	native
Silver chub	insectivore	intermediate	native
Bigeye chub	insectivore	intolerant	native
Streamline chub	insectivore	intolerant	native
Gravel chub	insectivore	intermediate	native
Speckled chub	insectivore	intolerant	native
Blacknose dace	generalist	tolerant	native
Longnose dace	insectivore	intolerant	native
Creek chub	generalist	tolerant	native
Tonguetied minnow	insectivore	intolerant	native
Suckermouth minnow	insectivore	intermediate	native
Southern redbelly dace	herbivore	intermediate	native
Redside dace	insectivore	intolerant	native
Pugnose minnow	insectivore	intolerant	native
Emerald shiner	insectivore	intermediate	native
Silver shiner	insectivore	intolerant	native

Table C.1 Tolerance, Trophic Guilds, and Origins of Selected Fish Species (continued)

	Trophic Level	Tolerance	Origin
Cyprinidae			
Rosyface shiner	insectivore	intolerant	native
Redfin shiner	insectivore	intermediate	native
Rosefin shiner	insectivore	intermediate	native
Striped shiner	insectivore	intermediate	native
Common shiner	insectivore	intermediate	native
River shiner	insectivore	intermediate	native
Spottail shiner	insectivore	intermediate	native
Blackchin shiner	insectivore	intolerant	native
Bigeye shiner	insectivore	intolerant	native
Steelcolor shiner	insectivore	intermediate	native
Spotfin shiner	insectivore	intermediate	native
Bigmouth shiner	insectivore	intermediate	native
Sand shiner	insectivore	intermediate	native
Mimic shiner	insectivore	intolerant	native
Ghost shiner	insectivore	intermediate	native
Blacknose shiner	insectivore	intolerant	native
Pugnose shiner	insectivore	intolerant	native
Silverjaw minnow	insectivore	intermediate	native
Mississippi silvery minnow	herbivore	intermediate	native
Bullhead minnow	omnivore	intermediate	native
Bluntnose minnow	omnivore	tolerant	native
Fathead minnow	omnivore	tolerant	native
Central stoneroller	herbivore	intermediate	native
Popeye shiner	insectivore	intolerant	native
Grass carp	herbivore	intermediate	exotic
Red shiner	omnivore	intermediate	native
Brassy minnow	omnivore	intermediate	native
Central silvery minnow	herbivore	intolerant	native
Catostomidae			
Blue sucker	insectivore	intolerant	native
Bigmouth buffalo	insectivore	intermediate	native
Black buffalo	insectivore	intermediate	native
Smallmouth buffalo	insectivore	intermediate	native
Quillback	omnivore	intermediate	native
River carpsucker	omnivore	intermediate	native
Highfin carpsucker	omnivore	intermediate	native
Silver redhorse	insectivore	intermediate	native
Black redhorse	insectivore	intolerant	native
Golden redhorse	insectivore	intermediate	native
Shorthead redhorse	insectivore	intermediate	native
Greater redhorse	insectivore	intolerant	native
River redhorse	insectivore	intolerant	native
Harelip sucker	invertivore	intolerant	native
Northern hog sucker	insectivore	intolerant	native
White sucker	omnivore	tolerant	native
Longnose sucker	insectivore	intermediate	native
Spotted sucker	insectivore	intermediate	native
Lake chubsucker	insectivore	intermediate	native
Creek chubsucker	insectivore	intermediate	native
Ictaluridae			
Blue catfish	piscivore	intermediate	native
Channel catfish	generalist	intermediate	native
White catfish	insectivore	intermediate	native
Yellow bullhead	insectivore	intolerant	native
Brown bullhead	insectivore	intolerant	native
Black bullhead	insectivore	tolerant	native

Table C.1   Tolerance, Trophic Guilds, and Origins of Selected Fish Species (continued)

	Trophic Level	Tolerance	Origin
Ictaluridae			
Flathead catfish	piscivore	intermediate	native
Stonecat	insectivore	intolerant	native
Mountain madtom	insectivore	intolerant	native
Slender madtom	insectivore	intolerant	native
Freckled madtom	insectivore	intermediate	native
Northern madtom	insectivore	intolerant	native
Scioto madtom	insectivore	intolerant	native
Brindled madtom	insectivore	intolerant	native
Tadpole madtom	insectivore	intermediate	native
Anguillidae			
American eel	piscivore	intolerant	native
Cyprinodontidae			
Western banded killifish	insectivore	intolerant	native
Eastern banded killifish	insectivore	tolerant	native
Blackstripe topminnow	insectivore	intermediate	native
Poeciliidae			
Mosquitofish	insectivore	intermediate	exotic
Gadidae			
Burbot	piscivore	intermediate	native
Percopsidae			
Trout-perch	insectivore	intermediate	native
Aphredoderidae			
Pirate perch	insectivore	intermediate	native
Atherinidae			
Brook silverside	insectivore	intermediate	native
Percichthyidae			
White bass	insectivore	intermediate	native
Striped bass	insectivore	intermediate	exotic
White perch	insectivore	intermediate	exotic
Yellow bass	insectivore	intermediate	native
Centrarchidae			
White crappie	invertivore	intermediate	native
Black crappie	invertivore	intermediate	native
Rock bass	piscivore	intermediate	native
Smallmouth bass	piscivore	intermediate	native
Spotted bass	piscivore	intermediate	native
Largemouth bass	piscivore	intermediate	native
Warmouth	invertivore	intermediate	native
Green sunfish	invertivore	tolerant	native
Bluegill	insectivore	intermediate	native
Orangespotted sunfish	insectivore	intermediate	native
Longear sunfish	insectivore	intolerant	native
Redear sunfish	insectivore	intermediate	native
Pumpkinseed	insectivore	intermediate	native
Percidae			
Sauger	piscivore	intermediate	native
Walleye	piscivore	intermediate	native
Yellow perch	piscivore	intermediate	native
Dusky darter	insectivore	intermediate	native
Blackside darter	insectivore	intermediate	native
Longhead darter	insectivore	intolerant	native
Slenderhead darter	insectivore	intolerant	native
River darter	insectivore	intermediate	native
Channel darter	insectivore	intolerant	native
Gilt darter	insectivore	intolerant	native
Logperch	insectivore	intermediate	native

Table C.1   Tolerance, Trophic Guilds, and Origins of Selected Fish
Species (continued)

	Trophic Level	Tolerance	Origin
Percidae			
Crystal darter	insectivore	intolerant	native
Eastern sand darter	insectivore	intolerant	native
Western sand darter	insectivore	intolerant	native
Johnny darter	insectivore	intermediate	native
Greenside darter	insectivore	intermediate	native
Banded darter	insectivore	intolerant	native
Variegate darter	insectivore	intolerant	native
Spotted darter	insectivore	intolerant	native
Bluebreast darter	insectivore	intolerant	native
Tippecanoe darter	insectivore	intolerant	native
Iowa darter	insectivore	intermediate	native
Rainbow darter	insectivore	intermediate	native
Orangethroat darter	insectivore	intermediate	native
Fantail darter	insectivore	intermediate	native
Least darter	insectivore	intermediate	native
Slough darter	insectivore	intermediate	native
Sciaenidae			
Freshwater drum	invertivore	intermediate	native
Cottidae			
Spoonhead sculpin	insectivore	intermediate	native
Mottled sculpin	insectivore	intermediate	native
Slimy sculpin	insectivore	intermediate	native
Deepwater sculpin	insectivore	intermediate	native
Gasterosteidae			
Brook stickleback	insectivore	intermediate	native

From EPA. *Rapid Bioassessment Protocols for Use in Streams and Rivers:
Benthic Macroinvertebrates and Fish*. Office of Water, U.S. Environmental
Protection Agency. Washington, D.C. EPA 444/4-89/001. 1989.

Table C.2   National List of Intolerant Fish Species[a]

Common Name	Latin Name
Cisco	*Coregonus artedii*
Arctic cisco	*Coregonus autumnalis*
Lake whitefish	*Coregonus clupeaformis*
Bloater	*Coregonus hoyi*
Kiyi	*Coregonus kiyi*
Bering cisco	*Coregonus laurettae*
Broad whitefish	*Coregonus nasus*
Humpback whitefish	*Coregonus pidschian*
Hortnose cisco	*Coregonus reighardi*
Least cisco	*Coregonus sardinella*
Shortjaw cisco	*Coregonus zenithicus*
Pink salmon	*Oncorhynchus gorbuscha*
Chum salmon	*Oncorhynchus keta*
Coho salmon	*Oncorhynchus kisutch*
Sockeye salmon	*Oncorhynchus nerka*
Chinook salmon	*Oncorhynchus tshawytscha*
Pygmy whitefish	*Prosopium coulteri*
Round whitefish	*Prosopium cylindraceum*
Mountain whitefish	*Prosopium williamsoni*
Golden trout	*Salmo aguabonita*
Arizona trout	*Salmo apache*
Cutthroat trout	*Salmo clarki*

Table C.2   National List of Intolerant Fish Species[a] (continued)

Common Name	Latin Name
Rainbow trout	*Salmo gairdneri/O. mykiss*
Atlantic salmon	*Salmo salar*
Brown trout	*Salmo trutta*
Arctic char	*Salvelinus alpinus*
Bull trout	*Salvelinus confluentus*
Brook trout	*Salvelinus fontinalis*
Dolly varden	*Salvelinus malma*
Lake trout	*Salvelinus namaycush*
Inconnu	*Stenodus leucichthys*
Arctic grayling	*Thymallus arcticus*
Largescale stoneroller	*Campostoma oligolepis*
Redside dace	*Clinostomus elongatus*
Cutlips minnow	*Exoglossum maxillingua*
Bigeye chub	*Hybobsis amblops*
River chub	*Nocomis micropogon*
Pallid shiner	*Notropis amnis*
Pugnose shiner	*Notropis anogenus*
Rosefin shiner	*Notropis ardens*
Bigeye shiner	*Notropis boops*
Pugnose minnow	*Notropis emiliae*
Whitetail shiner	*Notropis galacturus*
Blackchin shiner	*Notropis heterodon*
Blacknose shiner	*Notropis heterloepis*
Spottail shiner	*Notropis hudsonius*
Sailfin shiner	*Notropis hypselopterus*
Tennessee shiner	*Notropis leuciodus*
Yellowfin shiner	*Notropis lutipinnis*
Ozark minnow	*Notropis nubilus*
Ozark shiner	*Notropis ozarcanus*
Silver shiner	*Notropis photogenis*
Duskystripe shiner	*Notropis pilsbryi*
Rosyface shiner	*Notropis rubellus*
Safron shiner	*Notropis rubricroceus*
Flagfin shiner	*Notropis signipinnis*
Telescope shiner	*Notropis telescopus*
Topeka shiner	*Notropis topeka*
Mimic shiner	*Notropis volucellus*
Steelcolor shiner	*Notropis whipplei*
Coosa shiner	*Notropis zaenocephalus*
Bleeding shiner	*Notropis zonatus*
Bandfin shiner	*Notropis zonistius*
Blackside dace	*Phoxinus cumberlandensis*
Northern redbelly dace	*Phoxinus eos*
Southern redbelly dace	*Phoxinus erythrogaster*
Blacknose dace	*Rhinichthys atratulus*
Pearl dace	*Semotilus margarita*
Alabama hog sucker	*Hypentelium etowanum*
Northern hog sucker	*Hypentelium nigricans*
Roanoke hog sucker	*Hypentelium roanokense*
Spotted sucker	*Minytrema melanops*
Silver redhorse	*Moxostoma anisurum*
River redhorse	*Moxostoma carinatum*
Black jumprock	*Moxostoma cervinum*
Gray redhorse	*Moxostoma congestum*
Black redhorse	*Moxostoma duquesnei*
Rustyside sucker	*Moxostoma hamiltoni*
Greater jumprock	*Moxostoma lachneri*
Blacktail redhorse	*Moxostoma poecilurum*

Table C.2   National List of Intolerant Fish Species[a] (continued)

Common Name	Latin Name
Torrent sucker	*Moxostoma rhothoecum*
Striped jumprock	*Moxostoma rupiscartes*
Greater redhorse	*Moxostoma valenciennesi*
Ozark madtom	*Noturus albater*
Elegant madtom	*Noturus elegans*
Mountain madtom	*Noturus eleutherus*
Slender madtom	*Noturus exilis*
Stonecat	*Noturus flavus*
Black madtom	*Noturus funebris*
Least madtom	*Noturus hildebrandi*
Margined madtom	*Noturus insignis*
Speckled madtom	*Noturus leptacanthus*
Brindled madtom	*Noturus miurus*
Frecklebelly madtom	*Noturus minitus*
Brown madtom	*Noturus phaeus*
Roanoke bass	*Ambloplites cavifrons*
Ozark rockbass	*Ambloplites constellatus*
Rock bass	*Ambloplites rupestris*
Longear sunfish	*Lepomis megalotis*
Darters[a]	*Ammocrypta* sp.
Darters[a]	*Etheostoma* sp.
Darters[a]	*Percina* sp.
Sculpins[a]	*Cottus* sp.
O'opu alamoo (goby)	*Lentipes concolor*
O'opu nopili (goby)	*Sicydium stimpsoni*
O'opu nakea (goby)	*Awaous stamineus*
Johnny darter	*Etheostoma nigrum*
Bluntnose darter	*E. chlorosomum*
Slough darter	*E. gracile*
Cypress darter	*E. proeliare*
Orangethroat darter	*E. spectabile*
Swamp darter	*E. fusiforme*
River darter	*Percina shumardi*

[a] Reader note that there are inconsistencies between some tolerance rankings with Table C.1 (UEPA 1989).

[b] The United States has 150 species of darters and sculpins, the great majority of which are intolerant species. Possible exceptions include:

From EPA. *Methods for Chemical Analysis of Water and Wastes.* EPA-600/4-79-020, U.S. Environmental Protection Agency, Cincinnati, Ohio. 1983b.

REFERENCES

Angermeier, P.L. *The Importance of Cover and Other Habitat Features to the Distribution and Abundance of Illinois Stream Fishes.* Ph.D. dissertation, University of Illinois, Urbana, IL. 1983.

Angermeier, P.L. and J.R. Karr. Applying an index of biotic integrity based on stream fish communities: considerations in sampling and interpretation. *North Am. J. Fish. Manage.*, 6: 418–429. 1986.

Bickers, C.A., M.H. Kelly, J.M. Levesque, and R.L. Hite. *User's Guide to IBI-AIBI-Version 2.01 (A BASIC Program for Computing the Index of Biotic Integrity with the IBM-PC).* State of Illinois, Environmental Protection Agency, Marion, IL. 1988.

EPA. *Rapid Bioassessment Protocols for Use in Streams and Rivers: Benthic Macroinvertebrates and Fish.* Office of Water, U.S. Environmental Protection Agency. Washington, D.C. EPA 444/4-89/001. 1989.

EPA. *Rapid Bioassessment Protocols for Use in Wadeable Streams and Rivers: Periphyton, Benthic Macroinvertebrates and Fish.* Office of Water, U.S. Environmental Protection Agency. Washington, D.C. EPA 841/B-99/002. 1999.

Fausch, D.D., J.R. Karr, and P.R. Yant. Regional application of an index of biotic integrity based on stream fish communities. *Trans. Am. Fish. Soc.*, 113: 39–55. 1984.

Gammon, J.R. The use of community parameters derived from electrofishing catches of river fish as indicators of environmental quality, in *Seminar on Water Quality Management Tradeoffs*. Report No. 905/9-80/009. U.S. Environmental Protection Agency, Washington, D.C. 1980.

Gammon, J.R., A. Spacie, J.L. Hamelink, and R.L. Kaesler. Role of electrofishing in assessing environmental quality of the Wabash River, in *Ecological Assessments of Effluent Impacts on Communities of Indigenous Aquatic Organisms*. Edited by J.M. Bates, and C.I. Weber. American Society for Testing and Materials. Philadelphia, PA. STP 730, pp. 307–324. 1981.

Gammon, J.R. personal communication, Department of Biological Sciences, DePauw University, Greencastle, IN. 1989.

Gauch, H., Jr. *Multivariate Analysis in Community Ecology*. Cambridge University Press, New York. 1982.

Hill, M.O. *DECORANA: A FORTRAN Program for Detrended Correspondence Analysis and Reciprocal Averaging*. Cornell University, Ithaca, NY. 1979.

Hill, J. and G.D. Grossman. Home range estimates for three North American stream fishes. *Copeia*, 1987:376–380. 1987.

Hocutt C.H. Fish as indicators of biological integrity. *Fisheries*, 6:28–31. 1981.

Hughes, R.M. and J.R. Gammon. Longitudinal changes in fish assemblages and water quality in the Willamette River, Oregon. *Trans. Am. Fish. Soc.*, 116:196–209. 1987.

Karr, J.R., K.D. Fausch, P.L. Angermeier, P.R. Yant, and I.J. Schlosser. *Assessing Biological Integrity in Running Waters: A Method and Its Rationale*. Special Publication 5, Illinois Natural History Survey. 1986.

Kuehne, R.A. and R.W. Barbour. *The American Darters*. University Press of Kentucky, Lexington, KY. 1983.

Matthews, W.J. Fish faunal structure in an Ozark stream: stability, persistence, and a catastrophic flood. *Copeia*, 1986:388–397. 1986.

Miller, D.L., P.M. Leonard, R.M. Hughes, J.R. Karr, P.B. Moyle, L.H. Schrader, B.A. Thompson, R.A. Daniels, K.D. Fausch, G.A. Fitzhugh, J.R. Gammon, D.B. Halliwell, P.L. Angermeier, and D.J. Orth. Regional applications of an index of biotic integrity for use in water resource management. *Fisheries*, 5:12–20. 1988.

Moyle, P.B. *Inland Fishes of California*. University of California Press, Berkeley, CA. 1976.

OEPA (Ohio Environmental Protection Agency). *Biological Criteria for the Protection of Aquatic Life: Volume II. User's Manual for Biological Assessment of Ohio Surface Waters*. Ohio Environmental Protection Agency, Columbus, OH. 1987.

OEPA (Ohio Environmental Protection Agency). *The Qualitative Habitat Evaluation Index (QHEI): Rationale, Methods, and Application*. Ecological Assessment Section, Ohio Environmental Protection Agency, Columbus, OH. 1989.

Platts, W.S., W.F. Megahan, and G.W. Minshall. *Methods for Evaluating Stream, Riparian and Biotic Conditions*. U.S. Department of Agriculture General Technical Report INT-138, Ogden, UT. 1983.

Reynolds, J.B. Electrofishing, in *Fisheries Techniques*. Edited by L.A. Nielsen and D.L. Johnson. American Fisheries Society, Bethesda, MD. 1983.

Steedman, R.J. Modification and assessment of an index of biotic integrity to quantify stream quality in southern Ontario. *Can. J. Fish. Aquat. Sci.*, 45: 492–501. 1988.

Toxicity and Bioaccumulation Testing

CONTENTS

General Toxicity Testing Methods 710
Methods for Conducting Long-Term Sediment Toxicity Tests with *Hyalella azteca* 710
    Placement of Sediment into Test Chambers 710
    Acclimation 713
    Placing Organisms in Test Chambers 713
    Feeding 713
    Monitoring a Test 713
    Ending a Test 714
    Interpretation of Results 716
Methods for Conducting Long-Term Sediment Toxicity Tests with *Chironomus tentans* 718
    Collection of Egg Cases 719
    Hatching of Eggs 719
    Placing Organisms in Test Chambers 720
    Feeding 720
    Dissolved Oxygen 720
    Monitoring Survival and Growth 721
    Monitoring Emergence 722
    Ending a Test 722
    Interpretation of Results 723
In Situ Testing Using Confined Organisms 724
Toxicity Identification Evaluations 729
Toxicity — Microtox Screening Test 730
    Scope and Application 730
    Summary of Method 730
    Sample Handling and Preservation 730
    Interferences 730
    Apparatus 731
    Reagents 731
    Procedure 731
    Calculations 732
    Precision and Accuracy 733
    Health and Safety Information 733
References 733

GENERAL TOXICITY TESTING METHODS

There are a large number of toxicity and bioaccumulation test methods that can be used in laboratory or field (*in situ*) settings. The strengths and weaknesses of the two settings were discussed in Chapter 6. The toxicity test methods most commonly used in North America are those required by the EPA and state environmental protection agencies, such as *Pimephales promelas* and *Ceriodaphnia dubia* for wastewater effluent testing. While these tests have been used successfully to evaluate stormwaters, there are also other options that may be acceptable to the regulatory authorities, since they have been found useful in the scientific peer-reviewed literature. In addition, there are many standardized test methods approved by Environment Canada (Table D.1) and ASTM (Table D.2) that are often quite similar to U.S. EPA procedures. Only a few examples are listed below to help familiarize the user with the procedures and associated quality assurance issues. The project manager should verify that the appropriate test methods are being used to meet any regulatory requirements. These tests should only be conducted by laboratories with documented experience in aquatic toxicology. Given the potential for sampling and method-related artifacts (Chapters 5 and 6), it is important that the project manager ensure that proper study design, sample collection, and testing protocols are adhered to. The categories of assessment tools that are useful in receiving water assessments are shown in Table D.3. The methods recommended for screening are listed on Tables D.4 through D.14.

METHODS FOR CONDUCTING LONG-TERM SEDIMENT TOXICITY TESTS WITH *HYALELLA AZTECA*

The EPA recently finalized methods for long-term chronic toxicity testing of sediments (EPA 2000). These methods have not been widely used but have been found to be more sensitive to sediment contaminants than the 10-day assays. In addition, they were found to have acceptable levels of variability based on interlaboratory variance studies. Since these assays require 42 days and longer to run, they are somewhat costly to perform. Conditions for evaluating sublethal endpoints in a sediment toxicity test with *H. azteca* are summarized in Table D.15. A general activity schedule is outlined in Table D.16.

The 42-day sediment toxicity test with *H. azteca* is conducted at 23°C with a 16L:8D photoperiod at an illuminance of about 500 to 1000 lux. Test chambers are 300-mL high-form lipless beakers containing 100 mL of sediment and 175 mL of overlying water. Amphipods in each test chamber are fed 1.0 mL of YCT daily. Each test chamber receives two volume additions/day of overlying water.

A total of 12 replicates, each containing ten 7- to 8-d-old amphipods are tested for each sample. For the total of 12 replicates, the assignment of beakers is as follows: 12 replicates are set up on Day –1, of which 4 replicates are used for 28-day growth and survival endpoints and 8 replicates for measurement of survival and reproduction on Day 35, and survival, reproduction, or growth on Day 42.

Placement of Sediment into Test Chambers

The day before the sediment test is started (Day –1), each sediment is thoroughly homogenized and added to the test chambers. Sediment is visually inspected to judge the degree of homogeneity.

Each test chamber will contain the same amount of sediment, determined by volume. Overlying water is added to the chambers on Day –1 in a manner that minimizes suspension of sediment. Renewal of overlying water is started on Day –1. A test begins when the organisms are added to the test chambers (Day 0). Hardness, alkalinity, and ammonia concentrations in the water above the sediment within a treatment should not vary by more than 50% during the test.

Table D.1 Status Report — Environment Canada Biological Test Method Development Program[a] (Revised December 1999)

Test Method / Supporting Guidance Documents	Status	Publication Date	Report Number
Universal Test Methods			
1. Acute Lethality Test Using Rainbow Trout	Published	July 1990	EPS 1/RM/9
2. Acute Lethality Test Using Threespine Stickleback	Published	July 1990	EPS 1/RM/10
3. Acute Lethality Test Using *Daphnia* spp.	Published	July 1990	EPS 1/RM/11
4. Test of Reproduction and Survival Using the Cladoceran *Ceriodaphnia dubia*	Published	February 1992	EPS 1/RM/21
5. Test of Larval Growth and Survival Using Fathead Minnows	Published	February 1992	EPS 1/RM/22
6. Toxicity Test Using Luminescent Bacteria (*Photobacterium phosphoreum*)	Published	October 1992	EPS 1/RM/24
7. Growth Inhibition Test Using the Freshwater Alga (*Selenastrum capricornutum*)	Published	November 1992	EPS 1/RM/25
8. Acute Test for Sediment Toxicity Using Marine or Estuarine Amphipods	Published	December 1992	EPS 1/RM/26
9. Fertilization Assay with Echinoids (Sea Urchins and Sand Dollars)	Published	December 1992	EPS 1/RM/27
10. Early Life-Stage Toxicity Tests Using Salmonid Fish (Rainbow Trout) – Second Edition	Published	July 1998	EPS 1/RM/28
11. Survival and Growth in Sediment Using Freshwater Midge Larvae *Chironomus tentans* or *riparius*	Published	December 1997	EPS 1/RM/32
12. Survival and Growth in Sediment Using the Freshwater Amphipod *Hyalella azteca*	Published	December 1997	EPS 1/RM/33
13. Test for Measuring the Inhibition of Growth Using the Freshwater Macrophyte *Lemna minor*	Published	March 1999	EPS 1/RM/37
14. Survival and Growth in Sediment Using Estuarine or Marine Polychaete Worms	Final draft in preparation	Early 2001	—
Reference Methods			
1. Reference Method for Determining Acute Lethality of Effluents to Rainbow Trout	Published	July 1990	EPS 1/RM/13
2. Reference Method for Determining Acute Lethality of Effluents to *Daphnia magna*	Published	July 1990	EPS 1/RM/14
3. Reference Method for Determining Acute Lethality of Sediment to Estuarine or Marine Amphipods	Published	1999	EPS 1/RM/35
Supporting Guidance Documents			
1. Control of Toxicity Test Precision Using Reference Toxicants	Published	August 1990	EPS 1/RM/12
2. Collection and Preparation of Sediment for Physicochemical Characterization and Biological Testing	Published	December 1994	EPS 1/RM/29
3. Measurement of Toxicity Test Precision Using Control Sediments Spiked with a Reference Toxicant	Published	September 1995	EPS 1/RM/30
4. Application and Interpretation of Single-Species Test Data in Environmental Toxicology	Final version in preparation	Spring 2000	EPS 1/RM/34
5. Statistics for the Determination of Toxicity Test Endpoints	Second draft in preparation	Early 2001	—

[a] Documents available in French and English, copies of published documents can be obtained from EPS Publication Section, ETAD, Environment Canada, Fax: (819)953-7253) Tel: (819)953-5921.

Table D.2 ASTM Standards on Toxicity Testing

Std. No.	

Aquatic Toxicity Testing — Water

General
E 1850-97 — Guide for Selection of Resident Species as Test Organisms for Aquatic and Sediment Toxicity Tests
E 1203-98 — Practice for Using Brine Shrimp Nauplii as Food for Test Animals in Aquatic Toxicity
E 1733-95 — Guide for Use of Light in Laboratory Testing

Phytoplankton
D 3978-80 — Practice for Algal Growth Potential Testing with *Selenastrum capricornutum*
E1218-97a — Guide for Conducting Static 96-hour Toxicity Testing with Microalgae
E1913-97 — Guide for Conducting Toxicity Tests with Bioluminescent Dinoflagellates

Plant
E 1841-96 — Guide for Conducting Renewal Phytotoxicity Tests with Freshwater Emergent Macrophytes
E 1498-92 (1998) — Guide for Conducting Sexual Reproduction Tests with Seaweeds
E 1415-91 (1998) — Guide for Conducting Static Toxicity Tests with *Lemna gibba* G3
E 1913-97 — Guide for Conducting Static, Axenic, 14-Day Phytotoxicity Tests in Test Tubes with the Submerged Aquatic Macrophyte, *Myriophyllum sibiricum* Komarov

Invertebrates
E 1440-91 (1998) — Guide for Acute Toxicity Test with the Rotifer Brachionus
E 1562-94 — Guide for Conducting Acute, Chronic, and Life Cycle Aquatic Toxicity Tests with Polychaetous Annelids
E 724-98 — Guide for Conducting Static Acute Toxicity Tests Starting with Embryos of Four Species of Saltwater Bivalve Molluscs
E 1193-97 — Guide for Conducting *Daphnia magna* Life Cycle Toxicity Tests
E 1191-97 — Guide for Conducting Life-Cycle Toxicity Tests with Saltwater Mysids
E 1463-92 (1998) — Guide for Conducting Static and Flow-Through Acute Toxicity Tests with Mysids from the West Coast of the United States
E 1295-89 (1995) — Guide for Conducting Three-Brood, Renewal Toxicity Tests with *Ceriodaphnia dubia*
E 1563-98 — Guide for Conducting Static Acute Toxicity Tests with Echinoid Embryos

Vertebrate
E 729-96 — Guide for Conducting Acute Toxicity Tests on Test Materials with Fishes, Macroinvertebrates, and Amphibians
E 1192-97 — Guide for Conducting Acute Toxicity Tests on Aqueous Ambient Samples and Effluents with Fishes, Macroinvertebrates, and Amphibians
E 1241-98 — Guide for Conducting Early Life-Stage Toxicity Tests with Fishes
E 1439-98 — Guide for Conducting the Frog Embryo Teratogenesis Assay-Xenopus (Fetax)

General
E 1022-94 — Practice for Conducting Bioconcentration Tests with Fishes and Saltwater Bivalve Molluscs
E 1242-97 — Practice for Using Octanol-Water Partition Coefficient to Estimate Median Lethal Concentrations for Fish Due to Narcosis

Microcosm
E 1366-96 — Practice for Standardized Aquatic Microcosm; Freshwater

Behavior
E 1604-94 — Guide for Behavioral Testing in Aquatic Toxicology
E 1711-95 — Guide for Measurement of Behavior during Fish Toxicity Tests
E 1768-95 — Guide for Ventilatory Behavioral Toxicology Testing of Freshwater Fishes

Aquatic Toxicity Testing — Sediment

General
E 1391-94 — Guide for Collection, Storage, Characterization, and Manipulation of Sediments for Toxicological Testing
E 1525-94a — Guide for Designing Biological Tests with Sediments

Marine Sediment Toxicity Tests
E 1611-94 — Guide for Conducting Sediment Toxicity Tests with Marine and Estuarine Polychaetous Annelids
E 1367-99 — Guide for Conducting 10-day Static Sediment Toxicity Tests with Marine and Setuarine Amphipods
E 1688-97a — Guide for Determination of the Bioaccumulation of Sediment-Associated Contaminates by Benthic Invertebrates

Freshwater Sediment Toxicity Tests
E 1706-95b — Test Methods for Measuring the Toxicity of Sediment-Associated Contaminants with Fresh Water Invertebrates

Table D.3   Toxicity and Bioaccumulation Testing Categories

Site	Type Assay	Media	Organisms (Examples)
Laboratory	Acute/Screening Toxicity Short-term Chronic	Low Flow High Flow Outfalls	*P. promelas, C. dubia, Daphnia magna*
	Acute or Chronic	Sediments	*Hyalella azteca, Chironomus tentans, Chironomus riparius*
	Bioaccumulation	Sediments	*Lumbriculus variegatus*
Field	Acute to Chronic Toxicity	Low Flow High Flow Mixing Zones	*P. promelas, C. dubia, D. magna, H. azteca, Gammarus, C. tentans*, or *C. riparius*, bivalves
		Sediment	*H. azteca, Gammarus, C. tentans* or *C. riparius, P. promelas, D. magna*, Bivalves
	Bioaccumulation	Low Flow High Flow Mixing Zones Sediment	*Lumbriculus variegatus*, bivalves, fish
	Bioaccumulation Surrogate	Low Flow High Flow Mixing Zones Interstitial water?	Semipermeable membrane devices

Acclimation

Test organisms are cultured and tested at the same temperature. Test organisms are cultured in the same water that is used in testing, as recommended by EPA (EPA 1994); therefore, no acclimation will be necessary.

Placing Organisms in Test Chambers

Amphipods are introduced into the overlying water below the air–water interface. Weight is measured on a subset of 20 amphipods used to start the test.

Feeding

For each beaker, 1.0 mL of YCT is added daily from Day 0 to Day 42. The amount of food added to the test chambers is kept to a minimum to avoid microbial growth and water fouling. If excess food collects on the sediment, a fungal or bacterial growth may develop on the sediment surface, in which case feeding is suspended for 1 or more days. A drop in dissolved oxygen below 2.5 mg/L during a test may indicate that the food added is not being consumed. Feeding is suspended for the amount of time necessary to increase the dissolved oxygen concentration. If feeding is suspended in one treatment, it should be suspended in all treatments. Detailed records of feeding rates and the appearance of the sediment surface are made daily.

Monitoring a Test

All chambers are checked daily and observations made to assess test organism behavior such as sediment avoidance. However, monitoring effects on burrowing activity of test organisms may be difficult because the test organisms are often not visible during the exposure. The operation of the exposure system is monitored daily.

Measurement of Overlying Water Quality Characteristics

Conductivity, hardness, alkalinity, and ammonia is measured in all treatments at the beginning and at the end of the sediment exposure portion of the test. Water quality characteristics are also

Table D.4 Recommended Toxicity Test Conditions and Test Acceptability Criteria for *Ceriodaphnia dubia*
Screening and Definitive Acute Tests

Test Conditions	Recommended
1. Test type:	Static non-renewal, static renewal or flow through
2. Test duration:	24, 48, or 96 h
3. Temperature:	20°C ± 1°C; or 25°C ± 1°C
4. Light quality:	Ambient laboratory illumination
5. Light intensity:	10–20 µE/m²/s (50–100 ft-c) (ambient laboratory levels)
6. Photoperiod:	16 h light, 8 h darkness
7. Test chamber size:	30 mL (minimum)
8. Test solution volume:	25 mL (minimum) – For whole sediment tests use 5 mL sediment, 20 mL water
9. Renewal of test solutions:	Minimum, after 48 h
10. Age of test organisms:	Less than 24 h old
11. No. organisms per test chamber:	Minimum, 5 for effluent and receiving water tests
12. No. replicate chambers per concentration:	Minimum, 4 for effluent and receiving water tests
13. No. organisms per concentration:	Minimum, 20 for effluent and receiving water tests
14. Feeding regime:	Feed YCT and *Selenastrum* while holding prior to the test; newly-released young should have food available a minimum of 2 h prior to use in a test; add 0.1 mL each of YCT and *Selenastrum* 2 h prior to test solution renewal at 48 h
15. Test chamber cleaning:	Cleaning not required
16. Test solution aeration:	None
17. Dilution water:	Moderately hard synthetic water is prepared using MILLIPORE MILLI-Q® or equivalent deionized water and reagent grade chemicals or 20% DMW, receiving water, groundwater, or synthetic water, modified to reflect receiving water hardness
18. Test concentrations:	Effluents: minimum of five effluent concentrations and a control Receiving waters: 100% receiving water and a control
19. Dilution series:	Effluents: ≥ 0.5 dilution series Receiving Waters: None, or ≥ 0.5 dilution series
20. Endpoint:	Effluents: Mortality (LC50 or NOAEC) Receiving Waters: Mortality (significant difference from control)
21. Sampling and sample holding requirements:	Effluents and Receiving Waters: Grab or composite samples are used within 36 h of completion of the sampling period
22. Sample volume required:	1 L
23. Test acceptability criterion:	90% or greater survival in controls

measured at the beginning and end of the reproductive phase (Day 29 to Day 42). Conductivity will be measured weekly and DO and pH three times/week

Dissolved oxygen is measured a minimum of three times/week and should be at a minimum of 2.5 mg/L. Aeration is used to maintain dissolved oxygen in the overlying water above 2.5 mg/L.

Temperature is measured at least daily in at least one test chamber from each treatment. The daily mean test temperature must be within 1°C of 23°C. The instantaneous temperature must always be within 3°C of 23°C.

Ending a Test

Endpoints monitored include 28-d survival and growth of amphipods and 35-day and 42-day survival, growth, and reproduction (number of young/female) of amphipods. Growth or reproduction of amphipods may be a more sensitive toxicity endpoint compared to survival.

On Day 28, four of the replicate beakers/sediment are sieved with a #40 mesh sieve (425-µm mesh; U.S. standard size sieve) to remove surviving amphipods for growth determinations. Growth

Table D.5 Recommended Toxicity Test Conditions and Test Acceptability Criteria for *Daphnia pulex* and *D. magna* Screening and Definitive Acute Tests

Test Conditions	Recommended
1. Test type:	Static non-renewal, static renewal or flow through
2. Test duration:	24, 48 or 96 h
3. Temperature:	20°C ± 1°C; or 25°C ± 1°C
4. Light quality:	Ambient laboratory illumination
5. Light intensity:	10–20 µE/m²/s (50–100 ft-c) (ambient laboratory levels)
6. Photoperiod:	16 h light, 8 h darkness
7. Test chamber size:	30 mL (minimum)
8. Test solution volume:	25 mL (minimum) — for whole sediment tests, use 10 mL, sediment, 40 mL water
9. Renewal of test solutions:	Minimum, after 48 h
10. Age of test organisms:	Less than 24 h old
11. No. organisms per test chamber:	Minimum, 5 for effluent and receiving water tests
12. No. replicate chambers per concentration:	Minimum, 4 for effluent and receiving water tests
13. No. organisms per concentration:	Minimum, 20 for effluent and receiving water tests
14. Feeding regime:	Feed YCT and *Selenastrum* while holding prior to the test; newly released young should have food available a minimum of 2 h prior to use in a test; add 0.1 mL each of YCT and *Selenastrum* 2 h prior to test solution renewal at 48 h
15. Test chamber cleaning:	Cleaning not required
16. Test solution aeration:	None
17. Dilution water:	Moderately hard synthetic water prepared using MILLIPORE MILLI-Q or equivalent deionized water and reagent grade chemicals, or 20% DMW, receiving water, groundwater, or synthetic water, modified to reflect receiving water hardness
18. Test concentrations:	Effluents: minimum of five effluent concentrations and a control Receiving Waters: 100% receiving water and a control
19. Dilution series:	Effluents: ≥ 0.5 dilution series Receiving Waters: None, or ≥ 0.5 dilution series
20. Endpoint:	Effluents: Mortality (LC50 or NOAEC) Receiving Waters: Mortality (significant difference from control)
21. Sampling and sample holding requirements:	Effluents and Receiving Waters: Grab or composite samples are used within 36 h of completion of the sampling period
22. Sample volume required:	1 L
23. Test acceptability criterion:	90% or greater survival in controls

of amphipods are reported as weight. Dry weight of amphipods in each replicate are determined on Days 28 and 42. Dry weight of amphipods are determined by: (1) transferring rinsed amphipods to a preweighed aluminum pan; (2) drying these samples for 24 hours at 60°C; and (3) weighing the pan and dried amphipods on a balance to the nearest 0.01 mg. Average dry weight of individual amphipods in each replicate is calculated from these data.

On Day 28, the remaining eight beakers/sediment are sieved and the surviving amphipods in each sediment beaker are placed in 300-mL water-only beakers containing 150 to 275 mL of overlying water and a 5 × 5 cm piece of Nitex screen or 3M fiber mat. Each water-only beaker receives 1.0 mL of YCT stock solution and about two volume additions of water daily.

Reproduction of amphipods is measured on Day 35 and Day 42 in the water-only beakers by removing and counting the adults and young in each beaker. On Day 35, the adults are then returned to the same water-only beakers. Adult amphipods surviving on Day 42 are preserved in sugar formalin. The number of adult females is determined by simply counting the adult males (mature male amphipods will have an enlarged second gnathopod) and assuming all other adults are females.

Table D.6 Recommended Toxicity Test Conditions and Test Acceptability Criteria for the Fathead Minnow (*Pimephales promelas*) Screening and Definitive Acute Tests

	Test Conditions	Recommended
1.	Test type:	Static non-renewal, static renewal or flow through
2.	Test duration:	24, 48 or 96 h
3.	Temperature:	20°C ± 1°C or 25°C ± 1°C
4.	Light quality:	Ambient laboratory illumination
5.	Light intensity:	10–20 μE/m²/s (50–100 ft-c) (ambient laboratory levels)
6.	Photoperiod:	16 h light, 8 h darkness
7.	Test chamber size:	250 mL (minimum)
8.	Test solution volume:	200 mL (minimum) — for whole sediment tests, use 150 mL sediment, 600 mL water
9.	Renewal of test solutions:	Minimum, after 48 h
10.	Age of test organisms:	1–14 days: 24 h range in age
11.	No. organisms per test chamber:	Minimum, 10 for effluent test
12.	No. replicate chambers per concentration:	Minimum, 2 for effluent tests Minimum, 4 for receiving water tests
13.	No. organisms per concentration:	Minimum, 20 for effluents tests Minimum, 40 for receiving waters tests
14.	Feeding regime:	*Artemia nauplii* are made available while holding prior to the test; add 0.2 mL *Artemia nauplii* concentrate 2 h prior to test solution renewal at 48 h
15.	Test chamber cleaning:	Cleaning not required
16.	Test solution aeration:	None, unless DO concentration falls below 40% saturation; rate should not exceed 100 bubbles/min
17.	Dilution water:	Moderately hard synthetic water prepared using MILLIPORE MILLI -Q or equivalent deionized water and reagent grade chemicals or 20% DMW, receiving water, groundwater, or synthetic water, modified to reflect receiving water and hardness
18.	Test concentrations:	Effluents: minimum of five effluent concentrations and a control Receiving Waters: 100% receiving water and a control
19.	Dilution series:	Effluents: ≥ 0.5 dilution series Receiving Waters: None, or ≥ 0.5 dilution series
20.	Endpoint:	Effluents: Mortality (LC50 or NOAEC) Receiving Waters: Mortality (significant difference from control)
21.	Sampling and sample holding requirements:	Effluents and Receiving Waters: Grab or composite samples are used within 36 h of completion of the sampling period
22.	Sample volume required:	2 L for effluents and receiving waters
23.	Test acceptability criterion:	90% or greater survival in controls

The number of females is used to determine number of young/female/beaker from Day 28 to Day 42. Growth will also be measured for these adult amphipods.

Interpretation of Results

Endpoints measured in the 42-day *H. azteca* test include survival (Days 28, 35, and 42), growth (Days 28 and 42), and reproduction (number of young/female produced from Days 28 to 42). Reproduction is often more variable than growth. Some investigators have shown growth provides unique information that can help discriminate toxic effects of exposure to contaminants in sediment, while others have not seen differences from survival information.

On rare occasions, test organism responses in control sediments may exhibit responses which are less than reference or test sediments. This may be due to the poor nutritional content of the control sediment or other unknown physicochemical factors. Currently, there are no standard control sediments which can be strongly recommended for chronic toxicity testing due to a lack of testing

Table D.7 Recommended Toxicity Test Conditions and Test Acceptability Criteria for the Rainbow Trout (*Oncorhynchus mykiss*) and Brook Trout (*Salvelinus fontinalis*) Screening and Definitive Acute Tests

Test Conditions	Recommended
1. Test type:	Static non-renewal, static-renewal or flow-through
2. Test duration:	24, 48 or 96 h
3. Temperature:	12 ± 2°C
4. Light quality:	Ambient laboratory illumination
5. Light intensity:	10–20 µE/m²/s (50–100 ft-c) (ambient laboratory levels)
6. Photoperiod:	16 h light, 8 h darkness. Light intensity should be raised gradually over a 15 min period at the beginning of the photoperiod, and lowered gradually at the end of the photoperiod, using a dimmer switch or other suitable control device.
7. Test chamber size:	5 L (minimum) (test chamber should be covered to prevent fish from jumping out)
8. Test solution volume:	4 L (minimum) — for whole sediment tests, use 80 mL sediment, and 320 mL water
9. Renewal of test solutions:	Minimum, after 48 h
10. Age of test organisms:	Rainbow Trout: 5–30 days, "24 h (after yolk sac absorption to 30 days) Brook Trout: 30–60 days
11. No. organisms per test chamber:	Minimum, 10 for effluent and receiving water tests
12. No. replicate chambers per concentration:	Minimum, 2 for effluent tests Minimum, 4 for receiving water tests
13. No. organisms per concentration:	Minimum, 20 for effluent tests Minimum 40 for receiving water tests
14. Feeding regime:	Feeding not required
15. Test chamber cleaning:	Cleaning not required
16. Test solution aeration:	None, unless DO concentration falls below 6.0 mg/L; rate should not exceed 100 bubbles/min
17. Dilution water:	Moderately hard synthetic water is prepared using MILLIPORE MILLI-Q or equivalent deionized water and reagent grade chemicals or 20% DMW, receiving water, groundwater, or synthetic water, modified to reflect receiving water hardness
18. Test concentrations:	Effluents: minimum of five effluent concentrations and a control Receiving Waters: 100% receiving water and a control
19. Dilution series:	Effluents: ≥ 0.5 dilution series Receiving Waters: None, or ≥ 0.5 dilution series
20. Endpoint:	Effluents: Mortality (LC50 or NOAEC) Receiving Waters: Mortality (significant difference from control)
21. Sampling and sample holding requirements:	Effluents and Receiving Waters: Grab or composite samples are used within 36 h of completion of the sampling period
22. Sample volume required:	20 L for effluents 40 L for receiving waters
23. Test acceptability criterion:	90% or greater survival in controls

and research. Should poor responses be observed in a control sediment, a secondary control or reference sediment may be substituted for comparisons of significance. This will not invalidate the test, but simply adds some degree of uncertainty in the determination of ecological significance.

Recently, the U.S. EPA conducted interlaboratory variance testing with the 42-day *H. azteca* assay. In these tests, the draft standard methods were used. The minimum detectable differences for amphipod survival at 28 and 42 days ranged from 8 to 12% in moderately contaminated sediments. Minimum detectable differences for reproductive endpoints were higher, as expected, ranging from 19 to 25%.

Table D.8 Recommended Toxicity Test Conditions and Test Acceptability Criteria for the *Ceriodaphnia dubia* Survival and Reproduction Test

Test Conditions	Recommended
1. Test type:	Static renewal
2. Test duration:	Until 60% of control females have three broods (maximum test duration 8 days)
3. Temperature:	25 ± 1°C
4. Light quality:	Ambient laboratory illumination
5. Light intensity:	10–20 μE/m²/s (50–100 ft-c) (ambient laboratory levels)
6. Photoperiod:	16 h light, 8 h darkness.
7. Test chamber size:	30 mL (minimum)
8. Test solution volume:	15 mL (minimum) — for whole sediment assays, use 5 mL sediments and 20 mL water
9. Renewal of test solutions:	Daily
10. Age of test organisms:	Less than 24 h and all released within an 8 h period
11. No. neonates per test chamber:	1
12. No. replicate test chambers per concentration:	10
13. No. neonates per concentration:	10
14. Feeding regime:	Feed 0.1 mL each of YCT and 0.1 mL of algal suspension per test chamber daily
15. Test solution aeration:	None
16. Dilution water:	Uncontaminated source of receiving water or other natural water, synthetic water prepared using MILLIPORE MILLI-Q or equivalent deionized water and reagent grade chemicals or DMW
17. Test concentrations:	Effluents: Minimum of five effluent concentrations and a control Receiving water: 100% receiving water or minimum of five concentrations and a control
18. Dilution factor:	Effluents ≥ 0.5 Receiving waters: None or ≥ 0.5
19. Endpoints:	Survival and reproduction
20. Sampling and sample holding requirements:	For on-site tests, samples collected daily and used within 24 h of the time they are removed from the sampling device. For off-site tests, a minimum of three samples collected on days one, three, and five with a maximum holding time of 36 h before first use
21. Sample volume required:	1 L
22. Test acceptability criteria:	80% or greater survival and an average of 15 or more young per surviving female in the control solutions; 60% of surviving control organisms must produce three broods

METHODS FOR CONDUCTING LONG-TERM SEDIMENT TOXICITY TESTS WITH *CHIRONOMUS TENTANS*

Conditions for conducting a long-term sediment toxicity test with *C. tentans* are summarized in Table D.17. A general activity schedule is outlined in Table D.18.

The long-term sediment toxicity test with *C. tentans* is conducted at 23°C with a 16L:8D photoperiod at an illuminance of about 500 to 1000 lux. Test chambers, sediment addition, water renewal, and water quality monitoring are as described above for *H. azteca*.

A total of 16 replicates, each containing 12, <24-hour-old larvae are tested for each sample. For the total of 16 replicates, the assignment of beakers is as follows: initially, 12 replicates are set up on Day –1, of which 4 replicates are used for 20-day growth and survival endpoints and 8 replicates for determination of emergence and reproduction. It is typical for males to begin emerging 4 to 7 days before females. Midges in each test chamber are fed 1.5 mL of a 4-g/L Tetrafin™ suspension daily. Endpoints monitored include 20-day survival and ash-free dry weight, emergence; and time to death (adults). Reproduction and egg hatchability are not assessed.

Table D.9 Recommended Toxicity Test Conditions and Test Acceptability Criteria for the Fathead Minnow (*Pimephales promelas*) Larval Survival and Growth Test

Test Conditions	Recommended
1. Test type:	Static renewal
2. Test duration:	7 days
3. Temperature:	25 ± 1°C
4. Light quality:	Ambient laboratory illumination
5. Light intensity:	10–20 µE/m^2/s (50–100 ft-c) (ambient laboratory levels)
6. Photoperiod:	16 h light, 8 h darkness.
7. Test chamber size:	500 mL (minimum)
8. Test solution volume:	250 mL (minimum) — for whole sediment tests use 50 mL sediment and 200 mL water
9. Renewal of test solutions:	Daily
10. Age of test organisms:	Newly hatched larvae less than 24 h old. If shipped, not more than 48 h old, 24 h range in age
11. No. larvae per test chamber:	15 (minimum of 10)
12. No. replicate test chambers per concentration:	4 (minimum of 3)
13. No. larvae per concentration:	60 (minimum of 30)
14. Source of food:	Newly hatched *Artemia nauplii* (less than 24 h old)
15. Feeding regime:	Feed 0.1 mL newly hatched (less than 24 h old) brine shrimp nauplii three times daily at 4 h intervals or, as a minimum 0.15 mL twice daily, 6 h between feedings (at the beginning of the work day following renewal). Sufficient larvae are added to provide an excess. Larvae are not fed during the final 12 h of the test
16. Test chamber cleaning:	Siphon daily, immediately before test solution renewal
17. Test solution aeration:	None, unless DO concentration falls below 4.0 mg/L. Rate should not exceed 100 bubbles/min.
18. Dilution water:	Uncontaminated source of receiving water or other natural water, synthetic water prepared using MILLIPORE MILLI-Q or equivalent deionized water and reagent grade chemicals or DMW
19. Test concentrations:	Effluents: Minimum of five effluent concentration and a control Receiving water: 100% receiving water or minimum of five concentrations and a control
20. Dilution factor:	Effluents: ≥ 0.5 Receiving waters: none or ≥ 0.5
21. Endpoints:	Survival and growth (weight)
22. Sampling and sample handling requirements:	For on-site tests, samples are collected daily, and used within 24 h of the time they are removed from the sampling device. For off-site tests, a minimum of three samples are collected on days one, three, and five with a maximum holding time of 36 h before first use
23. Sample volume required:	2.5 L/day
24. Test acceptability criteria:	80% or greater survival in controls: average dry weight per surviving organism in control chambers equals or exceeds 0.25 mg

Collection of Egg Cases

Egg cases are obtained from adult midges held in a sex ratio of 1:3 male:female. Adults are collected 4 days before starting a test. The day after collection of adults, 6 to 8 of the larger "C"-shaped egg cases are transferred to a petri dish with culture water and incubated (at 23°C). Hatching typically begins around 48 hours and larvae typically leave the egg case 24 hours after the first hatch.

Hatching of Eggs

Hatching of eggs should be complete by about 72 hours. Hatched larvae remain with the egg case for about 24 hours and appear to use the gelatinous component of the egg case as an initial

Table D.10 Toxicity Test Conditions and Test Acceptability Criteria for the Fathead Minnow (*Pimephales promelas*) Embryo-Larval Survival and Teratogenicity Test

Test Conditions	Recommended
1. Test type:	Static renewal
2. Test duration:	7 days
3. Temperature:	25 ± 1°C
4. Light quality:	Ambient laboratory illumination
5. Light intensity:	10–20 μE/m²/s (50–100 ft-c) (ambient laboratory levels)
6. Photoperiod:	16 h light, 8 h darkness.
7. Test chamber size:	150–500 mL
8. Test solution volume:	70 mL (minimum) — for whole sediment tests, use 50 mL sediment and 200 mL water
9. Renewal of test solutions:	Daily
10. Age of test organisms:	Less than 36 h old embryos (maximum 48 h if shipped)
11. No. embryos per test chamber:	15 (minimum of 10)
12. No. replicate test chambers per concentration:	4 (minimum of 3)
13. No. embryos per concentration:	60 (minimum of 30)
14. Feeding regime:	Feeding not required
15. Test solution aeration:	None, unless DO concentration falls below 4 mg/L
16. Dilution water:	Uncontaminated source of receiving water or other natural water, synthetic water prepared using MILLIPORE MILLI-Q or equivalent deionized water and reagent grade chemicals or DMW
17. Test concentrations:	Effluents: Minimum of five effluent concentration and a control Receiving water: 100% receiving water or minimum of five concentrations and a control
18. Dilution factor:	Effluents: ≥ 0.5 Receiving waters: none or ≥ 0.5
19. Endpoint:	Combined mortality (dead and deformed organisms)
20. Sampling and sample handling requirements:	For on-site tests, samples are collected daily, and used within 24 h of the time they are removed from the sampling device. For off-site tests, a minimum of three samples are collected on days one, three, and five with a maximum holding time of 36 h before first use
21. Sample volume required:	1.5 to 2.5 L/day depending on the volume of test solutions used
22. Test acceptability criteria:	80% or greater survival in controls

source of food. After the first 24-hour period with larvae hatched, egg cases are transferred from the incubation petri dish to another dish with clean test water. The action of transferring the egg case stimulates the remaining larvae to leave the egg case within a few hours. These are the larvae that are used to start the test.

Placing Organisms in Test Chambers

To start the test, larvae are collected with a Pasteur pipette from the bottom of the incubation dish with the aid of a dissecting microscope. Test organisms are pipetted directly into overlying water. Larvae are transferred to exposure chambers within 4 hours of emerging from the egg case.

Feeding

Each beaker received a daily addition of 1.5 mL of Tetrafin (4 mg/mL dry solids). Feeding is curtailed under circumstances described in the amphipod methods.

Dissolved Oxygen

Routine chemistries on Day 0 should be taken before organisms are placed in the test beakers. Excursions of DO as low as 1.5 mg/L did not seem to have an effect on midge survival and

Table D.11 Toxicity Test Conditions and Test Acceptability Criteria for the Algal (*Selenastrum capricornutum*) Growth Test

	Test Conditions	Recommended
1.	Test type:	Static non-renewal
2.	Test duration:	48–96 h
3.	Temperature:	25 ± 1°C
4.	Light quality:	"Cool white" fluorescent lighting
5.	Light intensity:	86 ± 8.6 µE/m²/s (400 ± 40 ft-c or 4306 lux)
6.	Photoperiod:	Continuous illumination
7.	Test chamber size:	125 or 250 mL
8.	Test solution volume:	50 or 100 mL
9.	Renewal of test solutions:	None
10.	Age of test organisms:	4 to 7 days
11.	Initial cell density in test chamber:	10,000 cells/mL
12.	No. replicate chambers per sample:	4 (minimum or 3)
13.	Shaking rate:	100 rmp continuous, or twice daily by hand
14.	Test solution aeration:	None
15.	Dilution water:	Algal stock culture medium, enriched uncontaminated source of receiving or other natural water, synthetic water prepared using MILLIPORE MILLI-Q or equivalent deionized water and reagent grade chemicals, or DMW without EDTA or enriched surface water
16.	Test concentrations:	Effluents: Minimum of five effluent concentrations and a control Receiving water: 100% receiving water or minimum of five concentrations and a control
17.	Test dilution factor:	Effluents: $\geq$ 0.5 Receiving waters: None or $\geq$ 0.5
18.	Endpoint:	Growth (cell counts, chlorophyll fluorescence, absorbance, biomass)
19.	Sample requirements:	For on-site tests, one sample collected at test initiation, and used within 24 h of the time being removed from the sampling device. For off-site tests, holding time must not exceed 36 h
20.	Sample volume required:	1 or 2 L depending on test volume
21.	Test acceptability criteria:	1×10^6 cells/mL with EDTA or 2×10^5 cells/mL without EDTA in the controls: Variability of controls should not exceed 20%

development (P.K. Sibley, University of Guelph, Guelph, Ontario, personal communication). Based on these findings, periodic depressions of DO below 2.5 mg/L (but not below 1.5 mg/L) are not likely to adversely affect test results, and thus should not be a reason to discard test data. Nonetheless, tests should be managed toward a goal of DO > 2.5 mg/L to ensure satisfactory performance. If the DO level of the water falls below 2.5 mg/L for any one treatment, aeration is conducted in all replicates for the duration of the test.

Monitoring Survival and Growth

At 20 days, four of the initial 12 replicates are selected for use in growth and survival measurements. Using a #40 sieve (425-µm mesh) to remove larvae from sediment, *C. tentans* is collected. Surviving larvae are kept separated by replicate for weight measurements; if pupae are recovered, these organisms are included in survival data but not included in the growth data.

The AFDW of midges is determined for the growth endpoint. All living larvae per replicate are combined and dried to a constant weight (e.g., 60°C for 24 hours). All weigh boats are ashed before use to eliminate weighing errors due to the pan oxidizing during ashing. The sample is brought to room temperature in a desiccator and weighed to the nearest 0.01 mg to obtain mean weights per surviving organism per replicate. The dried larvae in the pan are then ashed at 550°C for 2 hours. The pan with the ashed larvae is then reweighed and the tissue mass of the larvae is determined as the difference between the weight of the dried larvae plus pan and the weight of the ashed larvae plus pan.

Table D.12 Toxicity Test Conditions and Test Acceptability Criteria for the Amphipod (*Hyalella azteca*) Survival Test

	Test Conditions	Recommended
1.	Test type:	Whole sediment toxicity test with renewal of overlying water
2.	Test duration:	10 d
3.	Temperature:	23 ± 1°C
4.	Light quality:	Wide-spectrum fluorescent lights
5.	Illuminance:	About 100 to 1000 lux
6.	Photoperiod:	16 h light, 8 h dark
7.	Test chamber size:	300 mL high-form lipless beaker
8.	Sediment volume:	100 mL
9.	Overlying water volume:	175 mL
10.	Renewal of overlying water:	2 volumes additions/d; continuous or intermittent (e.g., 1 volume addition every 12 h)
11.	Age of test organisms:	7 to 14 d old at the start of the test (1 to 2 d range in age)
12.	No. organisms per test chamber:	10
13.	No. replicate chambers per treatment:	Depends on the objective of the test. Eight replicates are recommended for routine testing
14.	Feeding regime:	YCT food, fed 1.0 mL daily (1800 mg/L stock) to each test chamber
15.	Test solution aeration:	None, unless DO in overlying water falls below 2.5 mg/L
16.	Overlying water:	Culture water, well water, surface water, site water, or reconstituted water
17.	Test chamber cleaning:	If screens become clogged during a test, gently brush the outside of the screen
18.	Overlying water quality:	Hardness, alkalinity, conductivity, pH, and ammonia at the beginning and end of a test; temperature and dissolved oxygen daily
19.	Endpoint:	Survival and growth
20.	Test acceptability criterion:	Minimum mean control survival of 80% and measurable growth of test organisms in the control sediment

Monitoring Emergence

Emergence traps are placed on the reproductive replicates on Day 20 (emergence traps for the auxiliary beakers are added at the corresponding 20-day time interval for those replicates. At 23°C, emergence in control sediments typically begins on or about Day 23 and continues for about 2 weeks. However, in contaminated sediments, the emergence period may be extended by weeks.

Two categories are recorded for emergence: complete emergence and partial emergence. Complete emergence occurs when an organism has shed the pupal exuviae completely and escapes the surface tension of the water. If complete emergence has occurred but the adult has not escaped the surface tension of the water, the adult will die within 24 hours. Therefore, 24 hours will elapse before this death is recorded. Partial emergence occurs when an adult has only partially shed the pupal exuvia. These adults will also die, an event which can be recorded after 24 hours.

Between Day 23 and the end of the test, emergence of males and females, pupal and adult mortality, and time to death for adults is recorded daily for the reproductive replicates.

Ending a Test

The point at which the life cycle test is ended depends upon the sediments being evaluated. In clean sediments, the test typically requires 40 to 50 days from initial setup to completion if all possible measurement endpoints are evaluated. However, test duration will increase in the presence of environmental stressors that act to reduce growth and delay emergence. Where a strong gradient of sediment contamination exists, emergence patterns between treatments will likely become asynchronous, in which case each treatment needs to be ended separately. For this reason, emergence is used as a guide to decide when to end a test. Testing will be terminated with completion of emergence.

Table D.13 Toxicity Test Conditions and Test Acceptability Criteria for the Midge (*Chironomus tentans*) Survival Test

	Test Conditions	Recommended
1.	Test type:	Whole sediment toxicity test with renewal of overlying water
2.	Test duration:	10 d
3.	Temperature:	23 ± 1°C
4.	Light quality:	Wide-spectrum fluorescent lights
5.	Illuminance:	About 100 to 1000 lux
6.	Photoperiod:	16 h light, 8 h dark
7.	Test chamber size:	300 mL high-form lipless beaker
8.	Sediment volume:	100 mL
9.	Overlying water volume:	175 mL
10.	Renewal of overlying water:	2 volumes additions/d; continuous or intermittent (e.g., 1 volume addition every 12 h)
11.	Age of test organisms:	Second to third instar larvae (about 10 d old larvae; all organisms must be third instar or younger with at least 50% of the organisms at third instar)
12.	No. organisms per test chamber:	10
13.	No. replicate chambers per treatment:	Depends on the objective of the test. Eight replicates are recommended for routine testing
14.	Feeding regime:	Tetrafin goldfish food, fed 1.5 ml daily to each test chamber (1.5 mL contains 6.0 mg of dry solids)
15.	Test solution aeration:	None, unless DO in overlying water falls below 2.5 mg/L
16.	Overlying water:	Culture water, well water, surface water, site water, or reconstituted water
17.	Test chamber cleaning:	If screens become clogged during a test, gently brush the outside of the screen
18.	Overlying water quality:	Hardness, alkalinity, conductivity, pH, and ammonia at the beginning and end of a test; temperature and dissolved oxygen daily
19.	Endpoint:	Survival and growth (ash-free dry weight, AFDW)
20.	Test acceptability criterion:	Minimum mean control survival *must* be *70%*, with *minimum* mean *weight/surviving* control organisms of *0.48* mg *AFDW*

For treatments in which emergence has occurred, the treatment (not the entire test) is ended when no further emergence is recorded over a period of 7 days (the 7-day criterion). At this time, all beakers of the treatment are sieved through a #40 mesh screen (425 μm) to recover remaining larvae, pupae, or pupal casts. When no emergence is recorded in a treatment at any time during the test, that treatment can be ended once emergence in the control sediment has ended using the 7-day criterion.

Interpretation of Results

Endpoints measured in the *C. tentans* test include survival, growth, and emergence. On rare occasions, test organisms in control sediments may exhibit responses which are less than reference or test sediments. This may be due to the poor nutritional content of the control sediment or other unknown physicochemical factors. Currently, there are no standard control sediments that can be strongly recommended for chronic toxicity testing due to a lack of testing and research. Should poor responses be observed in a control sediment, a secondary control or reference sediment may be substituted for comparisons of significance. This will not invalidate the test, but simply adds a degree of uncertainty to the determination of ecological significance.

Recently, the U.S. EPA conducted interlaboratory variance testing with the chronic *C. tentans* assay. In these tests, the draft standard methods were used. The minimum detectable differences have not been calculated at this time, but will be available in the near future to provide a point of comparison for the test assays. It is expected that the minimum detectable difference for 28-day survival and emergence endpoints will be in the 15 to 30% range.

Table D.14 Toxicity Test Conditions and Test Acceptability Criteria for the Oligochaete (*Lumbriculus variegatus*) Survival Test

Test Conditions	Recommended
1. Test type:	Whole sediment toxicity test with renewal of overlying water
2. Test duration:	23 d
3. Temperature:	23 ± 1°C
4. Light quality:	Wide-spectrum fluorescent lights
5. Illuminance:	About 100 to 1000 lux
6. Photoperiod:	16 h light, 8 h dark
7. Test chamber size:	4 to 6 L aquaria with stainless steel screens or glass stand pipes
8. Sediment volume:	1 L or more depending on TOC
9. Overlying water volume:	1 L or more depending on TOC
10. Renewal of overlying water:	2 volumes additions/d; continuous or intermittent (e.g. 1 volume addition every 12 h)
11. Age of test organisms:	Adults
12. No. organisms per test chamber:	Ratio of total organic carbon in sediment to organism dry weight should be no less than 50:1. Minimum of 1 g/ replicate, preferably 5 g/replicate
13. No. replicate chambers per treatment:	Depends on the objective of the test. Five replicates are recommended for routine testing
14. Feeding regime:	None
15. Test solution aeration:	None, unless DO in overlying water falls below 2.5 mg/L
16. Overlying water:	Culture water, well water, surface water, site water, or reconstituted water
17. Test chamber cleaning:	If screens become clogged during a test, gently brush the outside of the screen
18. Overlying water quality:	Hardness, alkalinity, conductivity, pH, and ammonia at the beginning and end of a test; temperature and dissolved oxygen daily
19. Endpoint:	Bioaccumulation
20. Test acceptability criterion:	Performance-based criteria specifications

Four test species will be evaluated *in situ* in exposure chambers. The exposure chambers are constructed on plastic core tubes of ~3-in diameter and 4-in length. Two windows are cut on opposite sides of the chamber and covered with nylon mesh. The mesh size varies with the experimental treatment, ranging from 10- to 1000-μm openings. For high flow testing, only water column chambers will be exposed. One duplicate set of chambers will have reduced mesh size openings to allow determinations of flow and suspended solids effects. Chambers are placed in the stream, either in the overlying water or partially buried in the sediment, with exposures varying with the treatment. Organisms are slowly acclimated to site water temperatures and then added to each test chamber (10 organisms/chamber). The age of the organisms, handling, and culturing follow U.S. EPA toxicity test methods for short-term chronic toxicity testing. For bioaccumulation testing, additional organisms are placed to provide enough tissue mass. For the oligochaete assay, 5 g of tissue are used in each chamber. Chambers are placed in the stream in replicates of four and secured with netting and steel stakes. At Days 2 and 10, chambers will be retrieved and organisms enumerated within 2 hours of collection. Test endpoints are shown in Table D.20.

The effects of water quality during high flow events will be measured at all test sites. This will involve exposures using chambers with small and large mesh sizes to vary the organism exposure to suspended solids. Exposures will be for 48 hours and include *D. magna*, *H. azteca*, and *C. tentans*. Testing will only be conducted when organisms can be exposed to a significant first flush event.

IN SITU TESTING USING CONFINED ORGANISMS

There are many reasons for evaluating toxicity and bioaccumulation *in situ*, such as those shown in Table D.19 and discussed in Section 6. Numerous assessments of stormwater quality have found

Table D.15 Test Conditions for Conducting a 42-day Sediment Toxicity Test with *Hyalella azteca*

	Parameter	Conditions
1.	Test type:	Whole-sediment toxicity test with renewal of overlying water
2.	Temperature:	$23 \pm 1°C$
3.	Light quality:	Wide-spectrum fluorescent lights
4.	Illuminance:	About 500 to 1000 lux
5.	Photoperiod:	16L:8D
6.	Test chamber:	300-mL high-form lipless beaker
7.	Sediment volume:	100 mL
8.	Overlying water volume:	175 mL in the sediment exposure from Day 0 to Day 28 (175 to 275 mL in the water-only exposure from Day 28 to Day 42)
9.	Renewal of overlying water:	2 volume additions/d; continuous or intermittent (e.g., one volume addition every 12 h)
10.	Age of organisms:	7- to 8-d old at the start of the test
11.	Number of organisms/chamber:	10
12.	Number of replicate chambers/treatment:	12 (4 for 28-d survival and growth and 8 for 35- and 42-d survival, growth, and reproduction). Reproduction is more variable than growth or survival; hence, more replicates might be needed to establish statistical differences among treatments
13.	Feeding:	YCT food, fed 1.0 mL (1800 mg/L stock) daily to each test chamber
14.	Aeration:	None, unless dissolved oxygen in overlying water drops below 2.5 mg/L
15.	Overlying water:	Culture water, well water, surface water, or site water. Use of reconstituted water is not recommended
16.	Test chamber cleaning:	If screens become clogged during a test; gently brush the *outside* of the screen
17.	Overlying water quality:	Hardness, alkalinity, conductivity, and ammonia at the beginning and end of a sediment exposure (Day 0 and 28). Temperature daily. Conductivity weekly. Dissolved oxygen (DO) and pH three times/ week. Concentrations of DO should be measured more often if DO drops more than 1 mg/L since the previous measurement.
18.	Test duration:	42 d
19.	Endpoints:	28-d survival and growth; 35- and 42-d survival, growth, reproduction, and number of adult males and females on Day 42
20.	Test acceptability:	Minimum mean control survival of 80% on Day 28

the following study design example useful. The typical assessment will be an upstream–downstream evaluation of an outfall with an additional reference site. The assessment must include both low and high flow periods to separate the role of stormwater and nonpoint source runoff from low flow conditions that may include point sources and groundwater upwelling inputs. For *in situ* toxicity and/or bioaccumulation tasks, a variety of exposure periods can be used, depending on several issues, such as species resilience, meteorological conditions, concern over acute vs. chronic effects, and available resources (longer assessments are more expensive). A great challenge in any stormwater assessment is detecting chronic toxicity effects. The literature has documented (see Chapter 6) that delayed effects may occur days to weeks after pulse exposures to pesticides or metals. This is obviously difficult to determine in routine receiving water assessments. However, given the reality that chronic toxicity may be occurring, it is important to try and assess effects for as long a period as possible. Some test species, such as the cladocerans *C. dubia* and *D. magna* and early life stages of the fathead minnow *P. promelas*, do not survive well within typical *in situ* chambers for more than 4 days. The benthic macroinvertebrates, such as the amphipods *H. azteca* and *Gammarus,* midge *C. tentans*, and bivalves, can be exposed for periods of over a week (Brooker 2000). Fish may also be exposed for longer periods, but often require routine feeding. When determining bioaccumulation potential, the oligochaete worm *L. variegatus* is recommended. It accumulates nonpolar organic chemicals relatively quickly, so exposures as short as 4 days are acceptable.

Table D.16 General Activity Schedule for Conducting a 42-d Sediment Toxicity Test with *Hyalella azteca*

Day	Activity
	Pre-Test
−7	Remove adults and isolate <24-h-old amphipods (if procedures outlined in Section 12.3.4 are followed).
−8	Separate known-age amphipods from the cultures and place in holding chambers. Begin preparing food for the test. The <24-h amphipods are fed 10 mL of YCT (1800 mg/L stock solution) and 10 mL of *Selenastrum capricornutum* (about 3.0 x 107 cells/mL) on the first day of isolation and 5 mL of both YCT and *S. capricornutum* on the 3rd and 5th d after isolation.
−6 to −2	Feed and observe isolated amphipods, monitor water quality (e.g., temperature and dissolved oxygen).
−1	Feed and observe isolated amphipods, monitor water quality. Add sediment into each test chamber, place chambers into exposure system, and start renewing overlying water.
	Sediment Test
0	Measure total water quality (pH, temperature, dissolved oxygen, hardness, alkalinity, conductivity, ammonia). Transfer ten 7- to 8-day-old amphipods into each test chamber. Release organisms under the surface of the water. Add 1.0 mL of YCT (1800 mg/L stock) into each test chamber. Archive 80 amphipods for dry weight determination. Observe behavior of test organisms.
1 to 27	Add 1.0 mL of YCT to each test beaker. Measure temperature daily, conductivity weekly, and dissolved oxygen (DO) and pH three times/week. Observe behavior of test organisms.
28	Measure temperature, dissolved oxygen, pH, hardness, alkalinity, conductivity and ammonia. End the sediment-exposure portion of the test by collecting the amphipods with a #40 mesh sieve (425-μm mesh; U.S. standard size sieve). Use four replicates for growth measurements: count survivors and preserve organisms in sugar formalin for growth measurements. Eight replicates for reproduction measurements: Place survivors in individual replicate water-only beakers and add 1.0 mL of YCT to each test beaker/d and 2 volume additions/d of overlying water.
	Reproduction Phase
29 to 35	Feed daily. Measure temperature daily, conductivity weekly, DO and pH three times a week. Measure hardness and alkalinity weekly. Observe behavior of test organisms.
35	Record the number of surviving adults and remove offspring. Return adults to their original individual beakers and add food.
36 to 41	Feed daily. Measure temperature daily, conductivity weekly, DO and pH three times a week. Measure hardness and alkalinity weekly. Observe behavior of test organisms.
41	Same as Day 1. Measure total water quality (pH, temperature, dissolved oxygen, hardness, alkalinity, conductivity, ammonia).
42	Record the number of surviving adults and offspring. Surviving adult amphipods on Day 42 are preserved in sugar formalin solution. The number of adult males in each beaker is determined from this archived sample. This information is used to calculate the number of young produced per female per replicate from Day 28 to Day 42.

A routine assessment of *in situ* toxicity and bioaccumulation requires that organisms be deployed during low flow conditions; once when the entire exposure period is at baseflow and a second time that captures a high flow event. The organisms at baseflow should be exposed for a period of time greater than or equal to the period of the high flow exposure period (usually 2 to 4 days). Another useful design is to deploy a large number of replicates on Day 0 and then subsample every 2 days for an extended period (such as 14 days). Between one and four species can be evaluated simultaneously, depending on available resources. Often two species are used in each test chamber (as described below). The *in situ* chambers are constructed of clear core sampling tubes (cellulose acetate butyrate) cut to a length of approximately 15 cm. Polyethylene closures cap each end. Two rectangular windows (~85% of the core surface area) are usually covered with 80 μm Nitex® mesh and silicon glued opposite each other. The mesh size varies with the experimental treatment, ranging from 10 to 1000 μm openings. For high flow testing, only water column chambers need be exposed. Duplicate sets of chambers having small vs. large mesh size openings (e.g., 10 vs. 250 μm) allow determinations of flow and suspended solids effects. The source of toxicity/bioaccumulation can also be measured as originating from sediments or overlying water by varying the chamber posi-

Table D.17 Test Conditions for Conducting a Long-Term Sediment Toxicity Test with *Chironomus tentans*

	Parameter	Conditions
1.	Test type:	Whole-sediment toxicity test with renewal of overlying water
2.	Temperature:	23 ± 1°C
3.	Light quality:	Wide-spectrum fluorescent lights
4.	Illuminance:	About 500 to 1000 lux
5.	Photoperiod:	16L:8D
6.	Test chamber:	300-mL high-form lipless beaker
7.	Sediment volume:	100 mL
8.	Overlying water volume:	175 mL
9.	Renewal of overlying water:	2 volume additions/d; continuous or intermittent (e.g., one volume addition every 12 h)
10.	Age of organisms:	<24-hour-old larvae
11.	Number of organisms/chamber:	10
12.	Number of replicate chambers/treatment:	16
13.	Feeding:	Tetrafin goldfish food, fed 1.5 mL daily to each test chamber (1.5 mL contains 6.0 mg of dry solids); starting Day −1
14.	Aeration:	None, unless dissolved oxygen in overlying water drops below 2.5 mg/L
15.	Overlying water:	Culture water, well water, surface water, site water, or reconstituted water
16.	Test chamber cleaning:	If screens become clogged during a test; gently brush the *outside* of the screen
17.	Overlying water quality:	Hardness, alkalinity, conductivity, and ammonia at the beginning and end of a test. Temperature daily (ideally continuously). Dissolved oxygen (DO) and pH three times/week. Concentrations of DO should be measured more often if DO has declined by more than 1 mg/L since previous measurement.
18.	Test duration:	About 40 to 50 d; each treatment is ended separately when no additional emergence has been recorded for seven consecutive days. When no emergence is recorded from a treatment, termination of that treatment should be based on the control sediment using this 7-d criterion.
19.	Endpoints:	20-d survival and AFDW; female and male emergence, adult mortality
20.	Test acceptability:	Minimum average size of *C. tentans* in the control sediment at 20 d must be at least 0.6 mg/surviving organism as dry weights or 0.48 mg/surviving organism as AFDW. Emergence should be ≥ 50%. Time to death after emergence is <6.5 d for males and <5.1 d for females.

tioning and design. Prior to chamber deployment, 10 of each organism (*H. azteca, C. tentans*, and *D. magna*) were gently added to 50-mL test tubes of culture water for ease of transport to field locations (one test tube contained one species only). Transportation of organisms to field sites by this method has proven to minimize handling and travel-related stressors (Chappie and Burton 1997). Upon acclimation, *in situ* chambers capped on one end were immersed into the river, allowing water to fill the chamber by infiltration through the mesh, and test organisms were slowly delivered from the test tubes into the open end and the chambers then capped. Before placement into *in situ* baskets, chambers were held below the water surface and purged of all internal air. Chambers exposed to the sediment interface are secured under wire baskets (see Figure 6.161) and placed with the mesh windows against the sediment. Quadruplicate chambers exposed to overlying waters are secured on top of the wire baskets. The baskets were weighted down with bricks and anchored to the stream bed with rebar. Organisms are acclimated to site water temperatures slowly (1 to 2 degree/hour) and then added to each test chamber (10 organisms/chamber). For example, *C. tentans* and *H. azteca* were placed together in replicate chambers for a total of 20 organisms per chamber. Ground-up laboratory paper toweling is provided as a substrate to reduce stress on these benthic species. Test water for laboratory controls should be the organism culture water. These controls

Table D.18 General Activity Schedule for Conducting a Long-Term Sediment Toxicity Test with *Chironomus tentans*

Day	Activity
−4	Start reproduction flask with cultured adults (1:3 male:female ratio). For example for 15 to 25 egg cases, 10 males and 30 females are typically collected. Egg cases typically range from 600 to 1500 egg/case.
−3	Collect egg cases (a minimum of 6 to 8) and incubate at 23°C.
−2	Check egg cases for viability and development.
−1	1. Check egg cases for hatch and development.
	2. Add 100 mL of homogenized test sediment to each replicate beaker and place in corresponding treatment holding tank. After sediment has settled for at least 1 h, add 1.5 mL Tetrafin slurry (4g/L solution) to each beaker. Overlying water renewal begins at this time.
0	1. Transfer all egg cases to a crystallizing dish containing control water. Discard larvae that have already left the egg cases in the incubation dishes. Add 1.5 mL food to each test beaker with sediment before the larvae are added. Add 12 larvae to each replicate beaker (beakers are chosen by random block assignment). Let beakers sit (outside the test system) for 1 h following addition of the larvae. After this period, gently immerse all beakers into their respective treatment holding tanks.
	2. Measure temperature, pH, hardness, alkalinity, dissolved oxygen, conductivity and ammonia at start of test.
1–End	On a daily basis, add 1.5 mL food to each beaker. Measure temperature daily. Measure the pH and dissolved oxygen three times a week during the test. If the DO has declined more than 1 mg/L since previous reading, increase frequency of DO measurements and aerate if DO continues to be less than 2.5 mg/L. Measure hardness, alkalinity, conductivity, ammonia weekly.
6	For auxiliary male production, start reproduction flask with culture adults (e.g., 10 males and 30 females; 1:3 male to female ratio).
7–10	Set up schedule for auxiliary male beakers (4 replicates/treatment) same as that described above for Day −3 to Day 0.
19	In preparation for weight determinations, ash weigh-pans at 550°C for 2 h. Note that the weigh boats should be ashed before use to eliminate weighing errors due to the pan oxidizing during ashing of samples.
20	Randomly select four replicates from each treatment and sieve the sediment to recover larvae for growth and survival determinations. Pool all living larvae per replicate and dry the sample to a constant weight (e.g., 60°C for 24 h). Install emergence traps on each reproductive replicate beakers.
21	The sample with dried larvae is brought to room temperature in a desiccator and weighed to the nearest 0.01 mg. The dried larvae in the pan are then ashed at 550°C for 2 h. The pan with the ashed larvae is then reweighed and the tissue mass of the larvae determined as the difference between the weight of the dried larvae plus pan and the weight of the ashed larvae plus pan.
23–End	On a daily basis, record emergence of males and females, pupal, and adult mortality, and time to death for previously collected adults.
33–End	Transfer males emerging from the auxiliary male replicates to individual inverted petri dishes. The auxiliary males are used for mating with females from corresponding treatments from which most of the males had already emerged or in which no males emerged.
40–End	After 7 d of no recorded emergence in a given treatment, end the treatment by sieving the sediment to recover larvae, pupae, or pupal exuviae. When no emergence occurs in a test treatment, that treatment can be ended once emergence in the control sediment has ended using the 7-d criterion.

are typically maintained in a hotel room during field assessments. The age of the organisms, handling, and culturing follow U.S. EPA toxicity test methods for short-term chronic toxicity testing. For bioaccumulation testing, additional organisms are placed to provide enough tissue mass. For the oligochaete assay, 1 to 5 g of tissue (equal to approximately 1:10 animal wet wt:sediment organic carbon) is used in each chamber, depending on analytical requirements. After exposures of 1 to 30 days depending on species and objectives, chambers were gently lifted out of the river and placed into coolers of site water and returned to the field laboratory for enumeration. Upon arrival at the lab, chambers were checked for damage, the outsides rinsed, then individually emptied into crystallizing dishes and the survivors of each species enumerated and logged. Typical measurement endpoints are shown in Table D.20.

Table D.19 *In Situ* **Stressor and Sediment Toxicity Tasks and Outcomes**

Task	Rationale and Outcome
1. Sediment toxicity: *H. azteca*, *C. tentans*	Laboratory measure of sediment chronic toxicity. Trigger for comprehensive sediment toxicity survey. Determine the potential for adverse effects on benthic organisms.
2. *In situ* toxicity and uptake: *D. magna*, *H. azteca*, *C. tentans*, *L. variegatus*	Realistic field exposures to water, suspended solids, and sediments. Determine low and high flow responses. Relate to storm flow and food web modeling. Assess the potential for, and source of, adverse effects on the ecosystem.
3. *In situ* partitioning of exposure and *D. magna*, *H. azteca*	In field exposures, determine and rank primary stressors: flow and stressors: turbidity, photoinduced toxicity, ammonia, metals, non-polar organics, overlying water, pore water. Relate to transport and food web modeling. Assess the contribution and source of various stressors that produce adverse effects.
4. *In situ* assessment of bioaccumulation and transport potential: SPMDs and peepers.	In field exposures, determine presence and potential for uptake of nonpolar organics through time with SPMDs in surficial waters and pore waters. Assess the presence and transport of contaminants through time with peepers. Target side channel seepage to support transport and food web modeling.

Table D.20 *In situ* **Toxicity and Bioaccumulation Measurement Endpoints**

Test Organism	Endpoints
Daphnia magna	Survival (2 d)
Hyalella azteca	Survival (2, 7 d) Tissue concentration (7 d)
Chironomus tentans	Survival (2, 7 d), growth (7 d), tissue concentration (7 d)
Lumbriculus variegatus	Tissue concentration (7 d)

The effects of water quality during high flow events should be measured at all test sites. Physicochemical water quality parameters are measured as often as is practical. Preferably, continuous measures of flow and general water quality parameters are made using a data sonde-type instrument. At a minimum, however, measures are made at test initiation, then again at test termination at each field site for each of the following: temperature (°C), dissolved oxygen (mg/L), pH, hardness (mg/L CaCO$_3$), alkalinity (mg/L CaCO$_3$), turbidity (NTU), conductivity (µmhos), and flow. Samples for other potentially useful parameters, such as ammonia, pathogen indicators, BOD, and nutrients, are also collected.

Organisms sampled for tissue analyses are allowed to depurate in culture for several hours. Following that time, organisms are counted, weighed, and frozen. Tissue analyses should be conducted by a laboratory capable of low detection limits with small quantities of tissues.

TOXICITY IDENTIFICATION EVALUATIONS

The toxicity identification evaluation (TIE) is a process by which effluent or pore water samples are fractionated into various classes of contaminants and then tested for toxicity. This allows one to characterize which class of contaminants is primarily responsible for toxicity (EPA 1991a,b). These groups of contaminants include: pH-sensitive and volatile compounds (such as ammonia), metals, oxidant/reductants, and nonpolar organics. Toxicity is determined by exposing *C. dubia* for 24 hours to the various treatment fractionations and then measuring survival. A TIE was conducted following modified draft EPA guidelines for TIEs of sediments (EPA 1991b). Pore water aliquots were used for initial toxicity tests (within 24 hours of sample receipt), baseline ambient pore water, pH adjusted with aeration, pH adjusted with filtration, pH adjusted with C18 filtration, sodium thiosulfate addition, and EDTA addition fractions. If toxicity is removed in any fraction, subsequent chemical analyses will be conducted to confirm the removal of compounds which may be contributing to pore water toxicity. These manipulations and data interpretation can be quite involved and should only be conducted by a laboratory with documented experience.

TOXICITY — MICROTOX SCREENING TEST

Scope and Application

This test measures the reduction of light output at a specific time during the run by bacteria exposed to a water sample. This light output is compared to that of a control sample to calculate relative toxicity. The Microtox Screening Procedure has a range of relative toxicities between 0 and 100% (light output reduction, as compared to the control).

Summary of Method

The Microtox Screening Procedure uses a bioluminescent marine bacteria, *Photobacterium phosphoreum*, to measure the toxicity of a sample relative to a control sample at three times during the 25-min run. At each of the three reading times, the light output of each sample and each control is measured on a chart recorder and recorded as the height of the peak light output on a scale of 0 to 100.

P. phosphoreum emit light as a by-product of respiration. If a sample contains one or more components that interfere with respiration, then the bacteria's light output is reduced proportional to the amount of interference with respiration, or toxicity. The light output reduction is proportional to the toxicity of the sample. The relative toxicity of a sample to the control can then be calculated. These relative toxicities can be compared to toxicity test results using standard reagents specified by this procedure.

For samples that are calculated to be more than 50% toxic, an EC50 concentration is calculated. The EC50 concentration is the fraction of sample, using the Microtox diluent as the dilution solution, that causes a light output from the sample that is 50% of the light output of the control. It is also called the 50% effective concentration.

Sample Handling and Preservation

Glass sample containers must be clean and free of soap residues, and stoppers and lids must not be made of cork. Detergents, cork, and other materials may add chemicals to the sample and may add to its toxicity.

Tap water and distilled water are fatal to the bacteria due to high levels of chlorine. Sample storage containers must be rinsed with deionized or ultra-pure water prior to use, with ultra-pure water being preferable.

Samples should be analyzed soon after arrival at the laboratory. Until they are analyzed, samples should be stored at 4°C. Stored samples may be kept up to 1 week in the refrigerator. Freshwater samples should not be salted until the samples are ready to be analyzed, as salt–metal complexes seem to readily form, reducing the toxicity of the sample. Salted samples can only be stored for approximately 15 to 30 min.

Interferences

Samples having pH values outside the range of 6.3 to 7.8 may be toxic to the bacteria. Normally, the pH of the sample is not adjusted because pH may be the parameter causing toxicity in a natural environment. Color and turbidity will interfere with, and probably reduce, the amount of emitted light leaving the cuvette and reaching the photomultiplier. Organic matter may provide a second food source for the bacteria and may result in a sample whose relative toxicity is calculated to be less than zero.

Apparatus

- Microtox 2055 Analyzer
- 500 μL pipettor (with disposable tips)
- 10 μL pipettor (with disposable tips)
- Glass cuvettes (disposable)

Reagents

- Microtox bacterial reagent
- Microtox reconstitution solution
- Microtox diluent
- Microtox osmotic adjusting solution
- Reagent grade sodium chloride

Procedure

Sampling, Sample Preparation

Note: The older Microtox 2055 instrument has space in its incubator for 15 cuvettes. We label these positions with letters for each of the three rows (A, B, and C) and label the five columns with numbers (1 to 5), giving each position a letter and number, such as A1 for the first position and C5 for the last position. For a normal run, three of the cuvettes (A1, B1, and C1) are reserved for the control solution. One of the remaining 12 cuvettes is reserved for the standard solution whose concentration is approximately the predetermined $ZnSO_4 \cdot 7H_2O$ EC50 concentration. The remaining 11 cuvettes contain the samples to be tested using this screening procedure.

1. Rinse clean 40-mL sample vials, vial caps, and Teflon septa with ultra-pure water.
2. Mix the sample by inverting the container several times.
3. Pour 10 mL of sample into the vial.
4. Add 0.2 g NaCl (reagent grade) to the vial.
5. Mix the sample and salt by inverting the vial until the salt is completely dissolved.

Preparation of Apparatus

1. Discard the cuvettes remaining in the incubator and pre-cool slots from any prior run (used cuvettes are normally left in the incubator to reduce condensation problems).
2. Put new cuvettes into the 15 slots in the incubator and one in the pre-cool slot.
3. Pipette 1.0 mL of diluent into the cuvettes in positions A1, B1, and C1.
4. Pipette 1.0 mL of reconstitution solution into a cuvette in the pre-cool position.
5. Pipette 1.0 mL of each sample (already adjusted for salinity, as specified above) into separate cuvettes in positions A2 through A5, B2 through B5, or C2 through C5.
6. Set the timer for 5 min to allow for temperature stabilization of the reconstitution solution.
7. Get a vial of the Microtox reagent bacteria out of the freezer. (Must be stored in a freezer at no warmer than −20°C.)
8. Tap the reagent vial on the countertop gently several times to break up the contents.
9. After the 5 min temperature stabilization period has expired, open the vial.
10. Quickly, pour the reconstitution solution in the pre-cool slot into the reagent vial.
11. Swirl the contents to mix (all solid reagent should go into solution).
12. Pour the reagent solution back into the pre-cool cuvette.
13. Mix the reagent solution approximately 20 times with a 500 μL pipette.
14. Set the timer for 15 min.

Analysis of Samples

1. Pipette 10 μL of reagent solution into each cuvette in the following order: A1, B1, C1, A2 through A5, B2 through B5, and C2 through C5. Do not immerse the pipette tip in the solutions.
2. Gently mix each cuvette's contents 20 times with a 500 μL pipette. Mix the cuvettes in the same order in which reagent solution was added. Use a single pipette tip for the three controls, but a new tip for each sample and the standard.
3. Push in the "HV" and "HV Check" buttons on the front of the Microtox analyzer. The panel on the front should read between –700 and –800.
4. Push in the "HV Check" button (so it toggles back out) and push in the "Sensitivity X10" and "Run" buttons.
5. Turn on the strip chart recorder.
6. Zero the chart recorder using the knob located on the right side of the machine.
7. Make sure the speed setting is for 1 in/min.
8. Make sure the pen is touching the recorder paper by putting the pen arm down.
9. Place the cuvette in A1 into the turret and close the turret to get a reading on A1.
10. After the reading is obtained, remove the cuvette from the turret.
11. Read the cuvettes in B1 and C1 also to determine which of the three has the largest reading. Place that cuvette back in the turret and close.
12. Adjust the chart reading to between 90 and 100 using the Scan knob on the front of the analyzer. If display reads "1" (not "001"), change the sensitivity setting to "Sensitivity X1."
13. Open the turret and check the zero point again on the chart recorder. Adjust as necessary.
14. Close the turret.
15. Set the timer for 5 min.
16. When the timer rings, read the samples in the following order: A1, B1, C1, A1 through A5, B1 through B5, C1 through C5, A1, B1, and C1.
17. Place the control cuvette (A1, B1, or C1) which has the highest reading in the turret and close.
18. Set the timer for 10 min.
19. When the timer rings, read the samples in the following order: A1, B1, C1, A1 through A5, B1 through B5, C1 through C5, A1, B1 and C1.
20. Place the control cuvette (A1, B1, or C1) which has the highest reading in the turret and close.
21. Set the timer for 10 min.
22. When the timer rings, read the samples in the following order: A1, B1, C1, A1 through A5, B1 through B5, C1 through C5, A1, B1 and C1.
23. Shut off the chart recorder and cap the pen.
24. Return the C1 cuvette to the incubator and close the turret.
25. Push in the "HV" and "Turret" buttons on the front of the analyzer (toggle them off).
26. If, at the end of the test, the light output of any sample is less than half of the light output of the controls, the EC50 concentration of that sample must be found. This is done by rerunning the Microtox test using three to four dilutions of that sample (including one at 100% strength). The previously prepared (salted) sample cannot be used either to create the dilutions or as the 100% strength sample.

Calculations

At each of the three times that a sample is read, each of the three control samples is read three times. The results of these nine analyses are averaged and their standard deviation and coefficient of variation calculated. If the coefficient of variation for the control samples at any time in the run is greater than 0.05 (5%), the run is rejected.

Relative toxicity is calculated as follows:

$$\%\text{Reduction [at time t]} = \frac{\text{Control} - \text{Sample}}{\text{Control}} \times 100\%$$

where Control = average peak height of the control samples at t
      Sample = peak height of sample at t

Precision and Accuracy

The Microtox Analyzer is calibrated using solutions of either zinc sulfate or phenol. A standard solution of approximately 10 mg/L zinc sulfate or of approximately 50 mg/L phenol is made. Four dilutions of the standard solution, with three replicates of each dilution, are used in place of the 12 samples in the normal Microtox screening procedure. The four dilutions should bracket the expected EC50 concentration of the standard solution. However, instead of using sodium chloride to adjust the ionic strength of the sample, the Microtox osmotic adjusting solution (MOAS) should be used. The amount of MOAS used should be 10% of the volume of the standard.

During each run, one of the 12 sample positions is occupied by the standard solution at the EC50 concentration. If the relative toxicity of the standard sample is outside the range of 45 to 55%, the run is rejected and repeated with freshly made standard solution. If the EC50 on the repeat again falls outside the range of 45 to 55%, the calibration is repeated. If the calibrated EC50 is significantly higher than the previous calibrations on that box of reagent, then a new box of reagent is opened and the calibration screening procedure is performed on one of the reagents in that box.

Extensive work has been done to establish the precision and accuracy of this procedure. Please refer to A. Ayyoubi, *Physical Treatment of Urban Stormwater Runoff Toxicants*, pp. 11–23.

Health and Safety Information

Refer to the MSDSs for information regarding the use of the reagents in this procedure.

None of the reagents and materials has OSHA PEL(s), AGGIH TLV(s), or other limits. Oral rat LD50 data have not been established for any of the reagents supplied by Microtox.

Sodium chloride, which is one of the reagents and is a component of most of the reagents supplied by Microtox, has an LD50 of 3000 mg/kg. The sodium chloride, either as a reagent or as a component of the other reagents, may cause eye irritation, and ingestion of large quantities may cause vomiting, diarrhea, and dehydration.

No special storage requirements are needed beyond keeping the freeze-dried bacteria culture in a freezer. Reagents are not considered to be a fire or explosion hazard (water may be used to extinguish a fire), and have no hazardous decomposition products. The reagents are stable under ordinary conditions of use and storage. Spilled reagent, whether reacted or not, may be cleaned up by adsorption with paper towels, and excess fluid may be flushed down a regular sewer drain.

REFERENCES

ASTM. *Standard Test Methods for Measuring the Toxicity of Sediment Associated Contaminants with Freshwater Invertebrates.* E1706-95b, revision in press. American Society for Testing and Materials. Philadelphia, PA. 1999.

Ayyoubi, A. *Physical Treatment of Urban Stormwater Runoff Toxicants,* Master's thesis. University of Alabama at Birmingham, Birmingham, AL. 1993.

Brooker, J. *Use of the Asian Clam and Mayflies as in Situ Toxicity Indicators.* M.S. thesis. Wright State University, Dayton, OH. 2000.

Burton, G.A., B.L. Stemmer, K.L. Winks, P.E. Ross, and L.C. Burnett, A multitrophic level evaluation of sediment toxicity in Waukegan and Indiana harbors. *Environ. Toxicol. Chem.,* 8: 1057–1066. 1989.

Burton, G.A., Jr. Assessing freshwater sediment toxicity. *Environ. Toxicol. Chem.,* 10: 1585–1627. 1991.

Burton G.A., C.G. Ingersoll, L.C. Burnett, M. Henry, M.L. Hinman, S.J. Klaine, P.F. Landrum, P. Ross, and M. Tuchman, A comparison of sediment toxicity test methods at three Great Lakes areas of concern. *J. Great Lakes Res.,* 22: 495–511, 1996a.

Burton G.A., T.J. Norberg-King, C.G. Ingersoll, G.T. Ankley, P.V. Winger, J. Kubitz, J.M. Lazorchak, M.E. Smith, I.E. Greer, F.J. Dwyer, D.J. Call, K.E. Day, P. Kennedy, and M. Stinson. Interlaboratory study of precision: *Hyalella azteca* and *Chironomus tentans* freshwater sediment toxicity assay. *Environ. Toxicol. Chem.*, 15: 1335–1343, 1996b.

Burton, G.A., Jr., C. Hickey, T. DeWitt, D. Morrison, D. Roper, and M. Nipper. *In situ* toxicity testing: teasing out the environmental stressors. *SETAC News*, 16(5): 20–22. 1996c.

Burton, G.A., Jr. *Quality Assurance Project Plan for the U.S. Environmental Protection Agency's Freshwater Sediment Toxicity Methods Evaluation*. EPA Cooperative Agreement No. CR-824161. U.S. EPA Office of Science and Technology. Washington, D.C. 1997.

Chappie, D.J. and G.A. Burton, Jr. Optimization of *in situ* bioassays with *Hyalella azteca* and *Chironomus tentans*. *Environ. Toxicol. Chem.*, 16: 559–564. 1997.

EPA. *Methods for Aquatic Toxicity Identification Evaluations. Phase 1, Toxicity Characterization Procedures, 2nd edition*. EPA/600/6-91/003. Office of Research and Development, Washington, D.C. 1991a.

EPA. *Sediment Toxicity Identification Evaluations: Phase 1 (Characterization, Phase 2 (Identification), and Phase 3 (Confirmation). Modification of Effluent Procedures*. EPA 600/6-91-007. Environmental Research Laboratory, Duluth, MN. Tech Report No. 08-91. 1991b.

EPA. *Procedures for Assessing the Toxicity and Bioaccumulation of Sediment-Associated Contaminants with Freshwater Invertebrates*. EPA 600/R-94/024, U.S. Environmental Protection Agency. Duluth, MN. 1994.

EPA. *Methods for Measuring the Toxicity and Bioaccumulation of Sediment-Associated Contaminants with Freshwater Invertebrates*. Duluth, MN, Draft. 1998.

EPA. *Methods for Measuring the Toxicity and Bioaccumulation of Sediment-Associated Contaminants with Freshwater Invertebrates. 2nd edition*. EPA series number pending, Duluth, MN, 1999.

EPA. *Methods for Measuring the Toxicity and Bioaccumulation of Sediment-Associated Contaminants with Freshwater Invertebrates*. Office of Research and Development and Office of Water. U.S. Environmental Protection Agency. EPA/600/R-99/064. Washington, D.C. 2000.

Huckins, J.N., G.K. Manuweera, J.D. Petty, D. Mackay, and J.A. Lebo. Lipid-containing semipermeable membrane devices for monitoring organic contaminants in water. *Environ. Sci. Technol.*, 27: 2489–2496, 1993.

Manahan, S.E. *Environmental Chemistry, 6th edition*. CRC Press, Boca Raton, FL. 1994.

Microbics Corporation. *How to Run a Standard Microtox Test*. Carlsbad, CA. 1988.

Microbics Corporation. *Microtox 100% Screening Procedure*. Carlsbad, CA. 1990.

Stumm, W. and J.J. Morgan. *Aquatic Chemistry — An Introduction Emphasizing Chemical Equilibria in Natural Waters*. John Wiley & Sons, New York. 1981.

Laboratory Safety, Waste Disposal, and Chemical Analyses Methods

CONTENTS

Introduction 736
Fundamentals of Laboratory Safety 737
    Procurement of Chemicals 737
    Distribution of Chemicals 737
    Laboratory Chemical Storage 737
    Storage Cabinets..738
Basic Rules and Procedures for Working with Chemicals 738
    Laboratory Protocol...738
    Personal Safety Practices 738
    Housekeeping 739
    Personal Protection — Protective Eyewear 739
    Personal Protection — Protective Gloves 739
    Personal Protection — Other Protective Clothing..741
    Avoidance of Routine Exposure 741
    Fume Hoods 741
    Choice of Chemicals 742
    Equipment and Glassware 742
    Labels and Signs 742
    Unattended Operations 743
    Electrical Safety...743
Use and Storage of Chemicals in the Laboratory 743
    Procurement of Chemicals 743
    Working with Allergens...743
    Working with Embryotoxins 744
    Working with Chemicals of Moderate or High Acute Toxicity or High Chronic Toxicity ...744
    Chemical Storage..747
    Transportation 748
Procedures for Specific Classes of Hazardous Materials...748
    Flammable Solvents 749
    Oxidizers..750
    Corrosives 752

        Reactives 754
        Compressed Gas Cylinders 757
Emergency Procedures 758
Chemical Waste Disposal Program 760
        Chemical Waste Containers 760
        Waste Minimization 760
        Disposal of Chemicals down the Sink or Sanitary Sewer System 761
        Chemical Substitution 761
        Neutralization and Deactivation 761
        Elimination of Nonhazardous Waste from Hazardous Waste 761
        Waste Disposal 762
Material Safety Data Sheets (MSDS) 763
        Product Name and Identification 764
        Hazardous Ingredients/Identity Information 764
        Physical/Chemical Characteristics 764
        Fire and Explosion Hazard Data 765
        Reactivity Data 765
        Health Hazard Data 765
        Specific HACH MSDS Information 766
Summary of Field Test Kits 767
Special Comments Pertaining to Heavy Metal Analyses 774
Stormwater Sample Extractions for EPA Methods 608 and 625 779
Calibration and Deployment Setup Procedure for YSI 6000upg Water Quality
        Monitoring Sonde 782
References 785

INTRODUCTION

The laboratory safety discussion included in this appendix is summarized from the *Laboratory Safety and Standard Operating Procedures* manual prepared for use in the Water Quality Laboratories of the Department of Civil and Environmental Engineering at the University of Alabama at Birmingham. It was prepared by Shirley Clark and Robert Pitt to ensure safe laboratory practices during our research. The manual and the excerpted information in this appendix include information concerning safe laboratory practices, the use of personal protective equipment, emergency procedures, use and storage of chemicals, and the proper method of waste disposal. This manual also covers hazard communication and incident response. This information is intended to help those in the laboratory to minimize hazards to themselves and their colleagues.

In view of the wide variety of chemical products handled in laboratories, it should not be assumed that the precautions and requirements stated here are all-inclusive. This information should be updated as needed with supplementary information to better protect the health and safety of anyone working in or visiting the laboratories.

Also included in this appendix is a summary of analytical test kits that have been reviewed as to their ability to be used in the field by a variety of users. These kits were reviewed during projects funded by the EPA (Pitt et al. 1993) and by the telecommunications industry (Day 1996; Pitt and Clark 1999). In addition, comments pertaining to needed stormwater extraction methods for organic analyses are also presented, along with information pertaining to the various methods available for analyzing heavy metals. The appendix concludes with a detailed description of calibration and setup procedures for the YSI 6000 water quality sonde that is frequently referenced in the text.

FUNDAMENTALS OF LABORATORY SAFETY

Procurement of Chemicals

Before a chemical is received, information on proper handling, storage, and disposal must be known to those involved. Refer to the appropriate MSDS for further information. No container may be accepted into a laboratory without an adequate identifying label. This label cannot be removed, defaced, or damaged in any way. All substances should be received in a central location. The date of receipt should be noted on all chemicals. Receipt of all chemicals must be noted in the chemical inventory, as well as the laboratory in which the chemical shall be located.

Distribution of Chemicals

When chemicals are hand-carried between laboratories, place the chemical in an outside (secondary) container or bucket. These secondary containers provide protection to the bottle and help keep it from breaking. They also help minimize spillage if the bottle does break. It is recommended that transport of chemicals inside a building be done using a cart where feasible.

Laboratory Chemical Storage

   a. Read the label carefully before storing a chemical. All chemicals must be stored according to the Chemical Storage Segregation Scheme. Note that this is a simplified scheme and that in some instances, chemicals in the same category may be incompatible.
   b. Store all chemicals by their hazard class. Only within segregation groups can chemicals be stored in alphabetical order. If a chemical exhibits more than one hazard, segregate by using the characteristic that exhibits the *primary hazard*.
   c. Do not store chemicals near heat sources such as ovens or steam pipes. Also, do not store chemicals in direct sunlight.
   d. Date chemicals when received and first opened. This will ensure that the oldest chemicals are used first, which will decrease the amount of chemicals for disposal. If a particular chemical can become unsafe while in storage, an expiration date should also be included. Keep in mind that expiration dates set by the manufacturer do not necessarily imply that the chemical is safe to use up to that date.
   e. Do not use lab benches as permanent storage for chemicals. In these locations, the chemicals can easily be knocked over, incompatible chemicals can be stored alongside one another, and the chemicals are unprotected in the event of a fire. Each chemical must have a proper designated storage location and be returned to it after use.
   f. Inspect chemicals and their containers for any signs of deterioration and for the integrity of the label.
   g. Do not store any chemicals in glass containers on the floor.
   h. Do not use fume hoods as a permanent storage location for chemicals, with the exception of particularly odorous chemicals that may require ventilation. The more containers, boxes, equipment, and other items that are stored in a fume hood, the greater likelihood of having chemical vapors drawn back into the room.
   i. Promptly dispose of any old, outdated, or unused chemicals.
   j. Chemicals that require refrigeration must be sealed with tight-fitting caps and kept in lab-safe refrigerators. Lab-safe refrigerators/freezers must be used for cold storage of flammables.
   k. Do not store chemicals above eye level. If the container breaks, the contents can easily fall on the face and body.
   l. Do not store excessive amounts of chemicals in the lab.

Storage Cabinets

Flammable Material Storage Cabinets

Flammables not in active use must be stored in safe containers inside fire-resistant storage cabinets specifically designed to hold them. Flammable material storage cabinets must be specified for all labs that use flammable chemicals. The cabinets must meet NFPA 30 and OSHA 1910.106 standards. Flammable material storage cabinets are designed to protect the contents of the cabinet from the heat and flames of external fire rather than to confine burning liquids within. They can perform their protective function only if used and maintained properly. Cabinets are generally designed with double-walled construction and doors that are 2 in above the base (the cabinet is liquid-proof up to that point).

Acid Storage Cabinets

Acids should be kept in acid storage cabinets specifically designed to hold them. Such cabinets have the same construction features as flammable materials storage cabinets but are coated with epoxy enamel to guard against chemical attack, and use polyethylene trays to collect small spills and provide additional protection from corrosion for the shelves. Periodically check shelves and support for corrosion. Nitric acid should always be stored by itself or in a separate cabinet compartment.

BASIC RULES AND PROCEDURES FOR WORKING WITH CHEMICALS

Laboratory Protocol

Everyone in the lab is responsible for his or her own safety and for the safety of others. Before starting any work in the lab, make it a point to become familiar with the procedures and equipment that are to be used. Work only with chemical products when you know their flammability, reactivity, toxicity, safe handling, storage, and emergency procedures. If you do not understand or are unclear about something, ASK!

Personal Safety Practices

1. Lab coats and safety glasses are required of all persons in laboratories where chemicals are used. This includes visitors, as well as all laboratory personnel. Safety glasses can be found in a case just inside the door to each laboratory. Safety equipment must be donned before a person crosses the tape line separating the entryway to the lab from the working area. Personal protective equipment is only required in the areas designated.
2. Never wear shorts, short skirts, sandals, or open-toed or perforated shoes in the lab.
3. Minimize skin contact. Disposable gloves are available in all labs. Their use is recommended, especially when handling dangerous chemicals or samples whose properties are unknown. This is especially important since we often work with stormwater samples that may be contaminated by raw sewage. Wash exposed skin before leaving the laboratory.
4. Keep the work area clean and uncluttered.
5. Do not smell or taste chemicals.
6. No horseplay in laboratories. Do not engage in behavior that may distract another worker.
7. Always make sure that the exits from the laboratory are free of obstruction.
8. Do not allow children or pets in the lab.
9. Never pipette anything by mouth.
10. Be aware of dangling jewelry, loose clothing, or long hair that might get caught in the equipment.

11. Store food and drinks in refrigerators that are designated for that use only. Food and drinks shall not be carried into the work areas in the lab. Do not consume food or drinks using glassware or utensils that are used for laboratory procedures.
12. Never work alone in the lab if it is avoidable. If you must work alone, make someone aware of your location and have him or her call or check on you periodically. If you must work alone, do not use large containers of any dangerous chemical (such as acids or solvents).
13. Wash your hands frequently throughout the day and before leaving the lab for the day.
14. Do not wear contact lenses in the lab because chemicals or particulates may get caught behind them and cause severe damage to the eye.

Housekeeping

1. Work areas must be kept clean and free of unnecessary chemicals. Clean your work area throughout the day and before you leave at the end of the day.
2. If necessary, clean equipment after use to avoid the possibility of harming the next person who uses it or of contaminating his/her samples.
3. Keep all aisles and walkways in the lab clear to provide a safe walking surface and an unobstructed exit.
4. Do not block access to emergency equipment and utility controls.

Personal Protection — Protective Eyewear

1. Goggles provide the best all-around protection against chemical splashes, vapors, dusts, and mists.
2. Goggles that have indirect vents or are not vented provide the most protection, but an anti-fog agent might be needed.
3. Standard safety glasses provide protection against impact.
4. If using a laser or strong UV light sources (such as photodegradation equipment), wear safety glasses or goggles that provide protection against the specific wavelengths involved.
5. Prescription glasses are generally not appropriate in a laboratory setting. If you wear prescription glasses, either get and wear a pair of prescription safety glasses from your optician or wear the "over-the-glasses" safety glasses when working in the laboratory.
6. Contact lenses should not be worn in a laboratory because they can trap contaminants behind them and reduce or eliminate the effectiveness of flushing with water from an eyewash. Contact lenses may also increase the amount of chemicals trapped on the surface of the eye and decrease removal of the chemical by tearing. If it is necessary to wear contact lenses in a lab, wear protective goggles at all times.

Personal Protection — Protective Gloves

1. Chemicals can permeate any glove. The vapor form of the liquid chemical will break through to the skin side of the glove in most cases within a matter of minutes. The rate at which this occurs depends on the composition of the glove, the chemicals present and their concentration, and the exposure time. While for most chemicals this vapor exposure will not be particularly harmful, for some of the more toxic chemicals, it can be. In addition, once chemicals reach the skin, the glove then acts as a barrier which aids in the penetration of the chemicals through the skin. Effectively, a process called "occlusion" can occur, by which the chemical penetrates the skin more easily when trapped between the glove and the skin than if the skin were exposed without a glove. Consult glove and chemical compatibility charts (such as Table E.1) to ensure that you are using the most appropriate glove. Be sure to check the most up-to-date recommendations from the glove vendors.
2. If direct chemical contact occurs, replace gloves regularly throughout the day. Wash hands regularly and remove gloves before answering the telephone or opening doors. Make sure that hands are clean before using gloves. If chemicals have contaminated the skin prior to the glove being put on, the glove will then speed up the process of skin penetration.
3. Check gloves for cracks, tears, and holes. If the gloves are not in good condition, replace them.

Table E.1　Chemical Resistance of Glove Materials
(E = Excellent, G = Good, F = Fair, P = Poor)

Chemical	Natural Rubber	Neoprene	Nitrile	Vinyl
Acetaldehyde	G	G	E	G
Acetic acid	E	E	E	E
Acetone	G	G	G	F
Acrylonitrile	P	G	N/A	F
Ammonium hydroxide	G	E	E	E
Aniline	F	G	E	G
Benzaldehyde	F	F	E	G
Benzene*	P	F	G	F
Benzyl chloride*	F	P	G	P
Bromine	G	G	N/A	G
Butane	P	E	N/A	P
Butyraldehyde	P	G	N/A	G
Calcium hypochlorite	P	G	G	G
Carbon disulfide	P	P	G	F
Carbon tetrachloride*	P	F	G	F
Chlorine	G	G	N/A	G
Chloroacetone	F	E	N/A	P
Chloroform	P	F	G	P
Chromic acid	P	F	F	E
Cyclohexane	F	E	N/A	P
Dibenzyl ether	F	G	N/A	P
Dibutyl phthalate	F	G	N/A	P
Diethanolamine	F	E	N/A	E
Diethyl ether	F	G	E	P
Dimethyl sulfoxide**	N/A	N/A	N/A	N/A
Ethyl acetate	F	G	G	F
Ethylene dichloride*	P	F	G	P
Ethylene glycol	G	G	E	E
Ethylene trichloride*	P	P	N/A	P
Fluorine	G	G	N/A	G
Formaldehyde	G	E	E	E
Formic acid	G	E	E	E
Glycerol	G	G	E	E
Hexane	P	E	N/A	P
Hydrobromic acid (40%)	G	E	N/A	E
Hydrochloric acid	G	G	G	E
Hydrofluoric acid (30%)	G	G	G	E
Hydrogen peroxide	G	G	G	E
Iodine	G	G	N/A	G
Methylamine	G	G	E	E
Methyl cellosolve	E	E	N/A	P
Methyl chloride*	P	E	N/A	P
Methyl ethyl ketone	F	G	G	P
Methylene chloride*	F	F	G	F
Monoethaloamine	F	E	N/A	E
Morpholine	F	E	N/A	E
Naphthalene*	G	G	E	G
Nitric acid	P	P	P	G
Perchloric acid	F	G	F	E
Phosphoric acid	G	E	N/A	E
Potassium hydroxide	G	G	G	E
Propylene dichloride*	P	F	N/A	P
Sodium hydroxide	G	G	G	E
Sodium hypochlorite	G	P	F	G
Sulfuric acid	G	G	F	G
Toluene*	P	F	G	F
Trichloroethylene*	P	F	G	F

Table E.1   Chemical Resistance of Glove Materials (continued)
(E=Excellent, G=Good, F=Fair, P=Poor)

Chemical	Natural Rubber	Neoprene	Nitrile	Vinyl
Tricresyl phosphate	P	F	N/A	F
Triethanolamine	F	E	E	E
Trinitrotoluene	P	E	N/A	P

* Aromatic/halogenated hydrocarbons attack all types of glove. Should glove swelling occur, change to fresh gloves.
** No data available regarding resistance to DMSO by natural rubber, neoprene, nitrile, or vinyl; use butyl rubber gloves.

4. Butyl, neoprene, and nitrile gloves are resistant to most chemicals, e.g., alcohols, aldehydes, ketones, most inorganic acids, and most caustics.
5. Disposable latex and vinyl gloves protect against some chemicals, most aqueous solutions, and microorganisms, and reduce the risk of product contamination. DO NOT WEAR LATEX GLOVES IF YOU SHOW SIGNS OF A LATEX ALLERGY.
6. Leather and some knit-gloves will protect against cuts, abrasions, and scratches, but not against chemicals.
7. Temperature-resistant gloves protect against cryogenic liquids, flames, and high temperatures.
8. If the above guidelines are followed and gloves are changed frequently, particularly when liquid comes in contact with the glove, then any of the thin rubber gloves available on the market should serve general laboratory purposes.

Personal Protection — Other Protective Clothing

1. The primary purpose of a lab coat is to protect against splashes and spills. A lab coat should be nonflammable, where necessary, and easily removed.
2. Rubber-coated aprons can be worn to protect against chemical splashes and may be worn over a lab coat for additional protection.
3. Face shields can protect the face, eyes, and throat against impact, dust, particulates, and chemical splashes. However, always wear protective eyewear underneath a face shield. Always wear a face shield when handling large quantities of hazardous chemicals, such as when preparing an acid bath.
4. Shoes that fully cover the feet should always be worn in a lab. If work is going to be performed that includes moving large and heavy objects, steel-toed shoes must be worn.

Avoidance of Routine Exposure

Develop and encourage safe habits. Avoid unnecessary exposure to chemicals by any route. Do not smell or taste chemicals. Vent apparatus that may discharge toxic chemicals (e.g., vacuum pumps, microwaves) into local exhaust devices. Inspect gloves before use. Do not allow release of toxic substances in cold rooms or warm rooms, since these have contained recirculated atmospheres.

Fume Hoods

1. Use the fume hood for all procedures that might result in the release of hazardous chemical vapors or dust. Confirm that the hood is working by holding a Kimwipe® (or other lightweight paper) up to the opening of the hood. The paper should be pulled inward. Leave the hood "on" when it is not in active use if toxic substances are stored inside or if it is uncertain whether adequate general laboratory ventilation will be maintained when it is "off."
2. Equipment and other materials should be placed at least 6 in behind the sash. This will reduce the exposure of personnel to chemical vapors that may escape into the lab due to air turbulence.
3. When the hood is not in use, pull the sash all the way down.

4. While personnel are working in the hood, pull the sash down as far as is practical. The sash is protection against fires, explosions, chemical splashes, and projectiles. Never put the sash above the line marked as the maximum allowable height for safe use.

5. Do not keep loose papers, paper towels, or tissues in the hood. These material can be drawn into the blower and adversely affect the performance of the hood.

6. Do not use a fume hood as a storage cabinet for chemicals. Excessive storage of chemicals and other items will disrupt the designed airflow in the hood. In particular, do not store chemicals against the baffle at the back of the hood because this will interfere with the laminar air flow.

7. Do not place objects directly in front of a fume hood.

8. Minimize the amount of foot traffic immediately in front of a hood. Walking past hoods causes turbulence that can draw contaminants out of the hood and into the room.

Choice of Chemicals

Use only those chemicals for which the quality of the available ventilation system is appropriate. Do not begin any experiment that requires a fume hood if the hood is not working. If the hood is not working, call Maintenance immediately.

Equipment and Glassware

1. Inspect all glassware before use. Repair or discard any broken, cracked, or chipped glassware.
2. Transport all glass chemical containers in rubber or polyethylene bottle carriers.
3. Inspect laboratory apparatus before use. Use only equipment that is free from cracks, chips, or other defects.
4. If possible, place a pan under a reaction vessel or other container to contain the liquid if the glassware breaks.
5. Do not allow burners or any other ignition source nearby when working with flammable liquids.
6. Properly support and secure laboratory apparatus before use.
7. Either work in the fume hood or ensure that the apparatus is venting to the fume hood if there is a possibility of hazardous vapors being evolved.
8. Always work in a fume hood if there is a possibility of an implosion or explosion.
9. If possible, vent vacuum pump exhaust into a fume hood.
10. When using a vacuum pump, place a trap between the pump and the apparatus.
11. Lubricate pump regularly if possible. Check belt condition and do not operate in a fume hood cabinet that is used for storage of flammables.

Labels and Signs

All hazardous chemicals are required by law to be labeled by the manufacturer. The chemical hygiene officer must ensure that each existing container and any incoming containers are properly labeled. The label must provide the following information:

- The identity of the chemical
- Any warnings
- The manufacturer's name and address

Temporary or transfer containers intended for immediate use by the person who transferred the chemical need not be labeled. However, if the chemical is left unattended (such as premade standards), the container must be labeled. Temporary labels must include:

- The identity of the chemical
- Any warnings
- The target organs affected, if applicable

Signs are intended to warn employees of chemical and physical dangers, such as designated areas where carcinogens or highly toxic chemicals are used or stored. All high hazard areas or hazardous chemical storage should be posted with the proper signs.

Unattended Operations

If an experiment/operation is left unattended, place an appropriate sign on the door and provide for containment of toxic substances in the event of equipment or utility service.

Electrical Safety

1. Examine all electrical cords periodically for signs of wear and damage. If damaged electrical cords are discovered, unplug the equipment and repair (or send the equipment out for repair).
2. Properly ground all electrical equipment.
3. If sparks are noticed while plugging in or unplugging equipment or if the cord feels hot, do not use the equipment until it has been serviced.
4. Do not run electrical cords along the floor where they will be a tripping hazard and subject to wear. If a cord must be run along the floor, protect it with a cord cover.
5. Do not run electrical cords along the floor where liquid spills may be a problem (such as around sinks).
6. Do not run electrical cords above the ceiling if possible. The cord should be visible at all times to ensure that it is in good condition.
7. Do not plug too many items into a single outlet. Multistrip plugs can be used only if they are protected with a circuit breaker and if they are not overused.
8. Do not use extension cords for permanent wiring.

USE AND STORAGE OF CHEMICALS IN THE LABORATORY

Procurement of Chemicals

Material Safety Data Sheets (MSDS) must accompany all initial incoming shipments of all chemicals. MSDSs must be readily available to all personnel in the labs where the chemicals are stored and where they are used. MSDSs shall be kept in three-ring binders near the door so that personnel can familiarize themselves with new chemicals before getting them out and using them.

Before ordering a new chemical, laboratory personnel should obtain information on proper handling, storage, and disposal methods for that chemical.

Consumer products used as they would be at home (such as dishwashing detergent) do not require an MSDS.

Sources of MSDSs include:

- Chemical supplier
- Chemical manufacturer
- Internet resources, such as the UAB Department of Occupational Health and Safety webpage http://www.healthsafe.uab.edu

Working with Allergens

A wide variety of substances can elicit skin and lung hypersensitivity. Examples include common substances such as diazomethane, chromium, nickel, bichromates, formaldehyde, isocyanates, and certain phenols. Because of this variety and the varying responses of individuals, suitable gloves should be used whenever there is a potential for contact with chemicals that may cause skin irritation.

Working with Embryotoxins

Embryotoxins are substances that cause adverse effects on a developing fetus. These effects may include embryolethality, malformations, retarded growth, and postnatal function deficits.

A few substances have been demonstrated to be embryotoxic in humans. These include:

Acrylic acid
Aniline
Benzene
Cadmium
Carbon sulfide
N,N-dimethylacetamide
Dimethylformamide
Dimethyl sulfoxide
Diphenylamine
Estradiol
Formaldehyde
Formamide
Hexachlorobenzene
Iodoacetic acid
Lead compounds
Mercury compounds
Nitrobenzene
Nitrous oxide
Phenol
Thalidomide
Toluene
Vinyl chloride
Xylene
Polychlorinated and polybrominated biphenyls

Embryotoxins requiring special controls should be stored in an adequately ventilated area. The container should be labeled in a clear manner such as the following: EMBRYOTOXIN: READ SPECIFIC PROCEDURES FOR USE. If the storage container is breakable, it should be kept in an impermeable, unbreakable secondary container having sufficient capacity to retain the material, should the primary container fail.

Working with Chemicals of Moderate or High Acute Toxicity or High Chronic Toxicity

Before beginning a laboratory operation, each worker is strongly advised to consult the standard compilations that list toxic properties of known substances and learn what is known about the substance to be used. The precautions and procedures described in this section should be followed if any of the substances to be used in significant quantities is known to be moderately or highly toxic. If any of the substances being used is known to be highly toxic, it is desirable to have two people present in the area at all times.

These procedures should be followed if the toxicological properties of any of the substances being used or prepared are UNKNOWN. If any of the substances to be used or prepared are known to have high chronic toxicity (e.g., compounds of heavy metals and other potent carcinogens), then the precautions and procedures described in this section should be supplemented with additional precautions to aid in containing and ultimately destroying the substances having high chronic toxicity.

If you are considering pregnancy, handle these substances only in a hood with a confirmed satisfactory performance, using appropriate protective apparel to prevent skin contact. If you are pregnant, notify your supervisor and consult your physician before working with these materials.

In addition to the safety protocols discussed earlier, the following three steps must be followed when working with one or more of these substances:

1. Label containers of substances having high chronic toxicity as follows: **WARNING! HIGH ACUTE OR CHRONIC TOXICITY OR CANCER SUSPECT AGENT.**
2. Protect the hands and forearms by wearing either gloves and a laboratory coat or suitable long gloves to avoid contact of the toxic material with the skin.
3. Procedures involving volatile toxic substances and those involving solid or liquid toxic substances that may result in the generation of aerosols should be conducted in a fume hood or other suitable containment device.
4. After working with toxic materials, wash the hands and arms immediately. Never eat, drink, chew gum, apply cosmetics, take medicine, or store foods in areas where toxic substances are being used.

These standard precautions will provide laboratory workers with good protection from most toxic substances. In addition, records that include amounts of material used and names of workers involved should be kept as part of the laboratory notebook record of the experiment. For strong carcinogens, an accurate record of such substances being stored and the amounts used, dates of use, and names of users must be maintained.

To minimize hazards from accidental breakage of apparatus or spills of toxic substances in the hood, containers of such substances should be stored in pans or trays made of polyethylene or other chemical-resistant material, and the apparatus should be mounted above trays of the same material. Alternatively, the working surface of the hood can be fitted with a removable liner of adsorbent, plastic-backed paper. Such procedures will make clean up of accidental spills easier. Areas where toxic substances are being used and stored must have restricted access, and warning signs should be posted if a special toxicity hazard exists. If the substance is suspected of having a high chronic toxicity, the storage area must be maintained under negative pressure with respect to its surroundings.

In general, the waste materials and solvents containing toxic substances should be stored in closed, impervious containers so that personnel handling the containers will not be exposed to their contents.

The laboratory worker must be prepared for potential accidents or spills involving toxic substances. If a toxic substance contacts the skin, the area should be washed with water. If there is a major spill outside the hood, the room or appropriate area should be evacuated and necessary measures should be taken to prevent exposures to other workers. Spills must be cleaned by personnel wearing suitable personal protective equipment.

Some examples of potent carcinogens (substances known to have high chronic toxicity), along with their corresponding chemical class, are:

Alkylating Agents:
  α-Halo ethers
    Bis(chloromethyl)ether and chloromethyl ether
    Methyl chloromethyl ether
  Aziridines
    Ethylene imine
    2-Methylaziridine
  Diazo, azo, and azoxy compounds
    4-Dimethylaminobenzene
  Electrophilic alkenes and alkynes
    Acrylonitrile
    Acrolein
    Ethyl acrylate
  Epoxides
    Ethylene oxide
    Diepoxybutane
    Epichlorohydrin

Propylene oxide
Styrene oxide
Acylating Agents:
β-Propiolactone
Dimethylcarbamoyl chloride
β-Butyrolactone
Organohalogen compounds:
1,2-Dibromo-3-chloropropane
Vinyl chloride
Chloroform
Methyl iodide
2,4,6-Trichlorophenol
Bis(2-chloroethyl)sulfide
Carbon tetrachloride
Hexachlorobenzene
1,4-Dichlorobenzene
Natural products:
Adriamycin
Bleomycin
Progesterone
Aflatoxins
Reserpine
Safrole
Inorganic compounds:
Cisplatin
Aromatic amines:
4-Aminobiphenyl
Aniline
o-Anisidine
Benzidine and derivatives
1,1-Bis(p-chlorophenyl)-2,2,2-trichloroethane (DDT)
o-Toluidine
Other Extremely Hazardous Chemicals:
Arsenic, organic arsenic, and derivatives
Arsine and gaseous derivatives
Asbestos
Azathioprine
Bromodeoxyuridine
1,4-Butanediol dimethylsulfonate (Myleran)
N-Butyl-N-(4-hydroxybutyl)nitrosamine (OH-BBN)
Chlorambucil
Chloropicrin in gas mixtures
Cyanogen
Cyanogen chloride
Cyclophosphamide
Diborane
Diisopropylfluorophosphate
9,10-Dimethyl-1,2-benzanthracene (DMBA)
Erionite
Germane
Hexaethyltetraphosphate
Hydrogen cyanide
Hydrogen selenide
Melphalan
N-Methyl-N-benzylnitrosamine
N-Methyl-N-nitrosourea

Mustard gas
2-Naphthylamine
Nitric oxide
Nitrogen dioxide
Nitrogen tetroxide
Parathion
Phosgene
Phosphine
2,3,7,8-Tetrachlorodibenzo-*p*-dioxin
Thorium dioxide

Some examples of compounds normally classified as strong carcinogens include the following:

2-Acetylaminofluorene
Benzo[*a*]pyrene
7,12-Dimethylbenz[*a*]anthracene
Dimethylcarbamoyl chloride
Hexamethylphosphoramide
3-Methylcholanthrene
2-Nitronaphthalene
Propane sultone
Various *N*-nitrosamides

The above substances (in both lists) must be used and stored in areas with restricted access. Special warning signs must be posted in these areas. Containers should be stored in chemical-resistant trays, and work must be performed within or above these trays. Cover surfaces where these substances are used with absorbent, plastic-backed paper. Performance-certified hood or other containment devices must be used when generation of toxic vapor, gases, dusts, or aerosols might occur.

Chemical Storage

The chemical storage area should be posted with an appropriate sign. Chemicals must be stored in appropriate containers and correctly labeled. Chemical compatibility must be determined to reduce the likelihood of hazardous reactions. The following steps should be followed when assessing chemical compatibility:

1. Identify the chemical
2. Determine the hazard class of the chemical: toxic, flammable, reactive, corrosive, oxidizer, low hazard.
3. Segregate the chemicals according to the above classifications. If there is a potential for hazardous interactions within a specific class, further separation is warranted. Label the area for each class of chemical.
4. General rules for compatibility:
   a. Highly toxic or carcinogenic chemicals should be ordered and stored in the smallest practical amount.
   b. Flammable or combustible liquids must be stored in approved containers, flammable material storage cabinets, or in properly designed under-hood storage areas. No more than 10 gallons of flammable liquids may be stored outside an approved flammable material storage cabinet. No more than 60 gallons of flammable liquids may be stored in a laboratory.
   c. Water-reactive chemicals should be located in a cool, dry area away from potential sources of water.
   d. Corrosives should be separated into acid and base subclasses. Large containers of corrosives should be stored on the lowest shelf or in special cabinets. Acids and bases should be separated from active metals and substances that can generate toxic gases upon contact. **NITRIC ACID MUST BE STORED SEPARATELY.**
   e. Oxidizers must be separated from combustible and flammable chemicals as well as reducing agents.

Compressed gas cylinders must be stored in well-ventilated areas where the temperature does not exceed 125°F. Cylinders must be stored in an upright position. Cylinders not in use should have the valve protection caps in place. Cylinders must be chained down to a fixed structure using the appropriate brackets and chains.

Never mix chemicals unless such mixing is part of a documented and approved procedure.

Transportation

1. All chemicals should be labeled before being transported.
2. When chemicals are hand-carried, they should be placed in an outside container or acid-carrying bucket to protect against breakage and spillage.
3. When chemicals are transported by wheeled cart, the cart should be stable under the load and have wheels large enough to negotiate uneven surfaces (such as expansion joints and floor drain depressions) without tipping or stopping suddenly. Incompatible chemicals should never be transported on the same cart.
4. Laboratory moves and transfers of large amounts of chemicals should be coordinated through the Hazardous Materials Facility.
5. Secondary containment should always be used to contain substances if there is a break in the primary container.

The following are conditions for chemical transport in elevators:

Chemicals should be labeled and carried in secure, break-resistant containers with tight-fitting caps. The packing systems supplied by manufacturers are excellent at preventing breakage during transport and may be reused for this purpose. The individual transporting the hazardous chemicals should operate the elevator alone, whenever possible.

The safe transport of small quantities of flammable liquids should include provisions that include the use of rugged, pressure-resistant, nonventing containers, storage during transport in a well-ventilated vehicle, and elimination of potential ignition sources.

If there is a spill or accident, contact the University Chemical Safety Director and state your name, telephone number, location of incident, name and quantity of material involved, and the extent of injuries, if any. Take all necessary emergency measures, such as removing contaminated clothing, washing any chemicals from the skin with soap and water, and seeking prompt medical attention. If it is necessary for the individual transporting the chemicals to leave the scene of an accident or spill, he/she should delegate someone to remain at the scene until emergency personnel arrive. The responsible party should return as soon as possible.

Cylinders that contain compressed gases are primarily shipping containers and should not be subjected to rough handling or abuse. Such misuse can seriously weaken the cylinder and render it unfit for further use or transform it into a missile with sufficient energy to propel it through masonry walls. To protect the valve during transport, the cover cap should be left screwed on hand-tight until the cylinder is in place and ready for actual use. The preferred transport method, even for short distances, is by suitable hand truck with the cylinder strapped into place. Only one cylinder should be handled at a time. After a cylinder has been relocated, straps, chains, or a suitable stand to keep it from falling must restrain it.

PROCEDURES FOR SPECIFIC CLASSES OF HAZARDOUS MATERIALS

This section will address the rules and procedures for handling chemicals that fall into one or more of five fundamental classes of laboratory chemicals: flammables, corrosives, oxidizers, reactives, and compressed gases.

Flammable Solvents

Flammable liquids are the most common chemicals found in a laboratory. The primary hazard associated with flammable liquids is their ability to readily ignite and burn. One should note that it is the vapor of a flammable liquid, not the liquid itself, which ignites and causes a fire.

The rate at which a liquid vaporizes is a function of its *vapor pressure*. In general, liquids with a high vapor pressure evaporate at a higher rate compared to liquids of lower vapor pressure. It should be noted that vapor pressure increases rapidly as the temperature rises, as does the evaporation rate. A reduced-pressure environment also accelerates the rate of evaporation.

The *flash point* of a liquid is the lowest temperature at which a liquid gives off a vapor at a rate sufficient to form an air–vapor mixture that will ignite, but will not sustain ignition. Many common flammable solvents have flash points significantly lower than room temperature.

The limits of *flammability* or *explosivity* define the range of fuel–air mixtures that will sustain combustion. The lower limit of this range is called the *lower explosive limit* (*LEL*), and the higher limit of this range is called the *upper explosive limit* (*UEL*). Materials with very broad flammability ranges are particularly treacherous due to the fact that virtually any fuel–air combination may form an explosive atmosphere.

The *vapor density* of a flammable material is the density of the corresponding vapor relative to air under specific temperature and pressure conditions. Flammable vapors with densities greater than one (and thus "heavier" than air) are potentially lethal because they will accumulate at floor level and flow with remarkable ease, in much the same manner that a liquid would. The obvious threat is that these mobile vapors may eventually reach an ignition source, such as an electrical outlet or a lit Bunsen burner.

Examples of Flammable Liquids

Acetone
Ethyl ether
Toluene
Methyl formate

Use and Storage of Flammables

1. Flammable liquids that are not in active use must be stored in safe containers inside fire-resistant storage cabinets designed for flammables, or inside storage rooms.
2. Minimize the amount of flammable liquids stored in the lab.
3. Use flammables only in areas free of ignition sources.
4. Never heat flammables with an open flame. Instead, use steam baths, water baths, oil baths, hot air baths, sand baths, or heating mantles.
5. Never store flammable chemicals in a standard household refrigerator. There are several ignition sources located inside a standard refrigerator that can set off a fire or violent explosion. Flammables can only be stored cold in a lab safe or explosion-proof refrigerator. Another alternative is to use an ice bath to chill the chemicals. Remember, there is no safety benefit in storing a flammable chemical in a refrigerator if the flash point of that chemical is below the temperature of that refrigerator.
6. The transfer of material to or from a metal container is generally accompanied by an accumulation of static charge on the container. This fact must be kept in mind when transferring flammable liquids, since the discharge of this static charge could generate a spark, thereby igniting the liquid. To make these transfers safer, flammable liquid dispensing and receiving containers must be bonded together before pouring. Large containers such as drums must also be grounded when used as dispensing or receiving vessels. All grounding and bonding connections must be metal to metal.

Health Effects Associated with Flammables

In general, the vapors of many flammables are irritating to mucous membranes of the respiratory system and eyes, and in high concentrations are narcotic. The following symptoms are typical for the respective routes of entry:

Acute Health Effects:
  Inhalation — headache, fatigue, dizziness, drowsiness, narcosis (stupor and unresponsiveness)
  Ingestion — slight gastrointestinal irritation, dizziness, fatigue
  Skin Contact — dry, cracked, and chapped skin
  Eye Contact — stinging, watery eyes, inflammation of the eyelids
Chronic Health Effects:
  The chronic health effects will vary depending on the specific chemical, the duration of the exposure, and the extent of the exposure. However, damage to the lungs, liver, kidneys, heart, and/or central nervous system may occur. Cancer and reproductive effects are also possible.
Flammable Groups Exhibiting These Health Effects:
  Hydrocarbons — aliphatic hydrocarbons are narcotic but their systemic toxicity is relatively low. Aromatic hydrocarbons are all potential narcotic agents, and overexposure to the vapors can lead to loss of muscular coordination, collapse, and unconsciousness. Benzene is toxic to bone marrow and can cause leukemia.
  Alcohols — vapors are only moderately narcotic.
  Ethers — exhibit strong narcotic properties but for the most part are only moderately toxic.
  Esters — vapors may result in irritation to the eyes, nose, and upper respiratory tract.
  Ketones — systemic toxicity is generally not high.

First-Aid Procedures for Exposures to Flammable Materials

Inhalation Exposure — remove person from contaminated area if it is safe to do so. Get medical attention and do not leave person unattended.
Ingestion Exposures — remove the person, if possible, from source of contamination. Get medical attention.
Dermal Exposures — remove person from source of contamination. Remove clothing, jewelry, and shoes from the affected areas. Flush the affected areas with water for at least 15 min and obtain medical attention.
Eye Contact — remove person from source of contamination. Flush the eyes with water for at least 15 min. Obtain medical attention.

Personal Protective Equipment

Always use a fume hood while working with flammable liquids. Nitrile and neoprene gloves are effective against most flammables. Wear a nonflammable lab coat to provide a barrier to your skin, and goggles if splashing is likely to occur.

Oxidizers

Oxidizers or oxidizing agents present fire and explosion hazards on contact with combustible materials. Depending on the class, an oxidizing material may increase the burning rate of combustibles with which it comes in contact; cause the spontaneous ignition of combustibles with which it comes in contact; or undergo an explosive reaction when exposed to heat, shock, or friction. Oxidizers are generally corrosive.

Examples of Common Oxidizers

Peroxides
Nitrites
Nitrates
Chlorates
Perchlorates
Chlorites
Hypochlorites
Dichromates

Use and Storage of Oxidizers

1. In general, store oxidizers away from flammables, organic compounds, and combustible materials.
2. Strong oxidizing agents like chromic acid should be stored in glass or some other inert container, preferably unbreakable. Corks and rubber stoppers should not be used.
3. Reaction vessels containing appreciable amounts of oxidizing materials should never be heated in oil baths, but rather on a heating mantle or sand bath.

Use and Storage of Perchloric Acid

1. Perchloric acid is an oxidizing agent of particular concern. The oxidizing power of perchloric acid increases as concentration and temperature increase. Cold, 70% perchloric acid is a strong, nonoxidizing corrosive. A 72% perchloric acid solution at elevated temperatures is a strong oxidizing agent. An 85% perchloric acid solution is a strong oxidizer at room temperature.
2. Do not attempt to heat perchloric acid if you do not have access to a properly functioning perchloric acid fume hood. Perchloric acid can only be heated in a hood specially equipped with a wash down system to remove any perchloric acid residue. The hood should be washed down after each use and it is preferred to dedicate the hood to perchloric acid use only.
3. Whenever possible, substitute a less hazardous chemical for perchloric acid.
4. Perchloric acid can be stored in a perchloric acid fume hood. Keep only the minimum amount necessary for your work. Another acceptable storage site for perchloric acid is on a metal shelf or in a metal cabinet away from organic or flammable materials. A bottle of perchloric acid should also be stored in a glass secondary container to contain leakage.
5. Do not allow perchloric acid to come in contact with any strong dehydrating agents such as sulfuric acid. The dehydration of perchloric acid is a severe fire and explosion hazard.
6. Do not order or use anhydrous perchloric acid. It is unstable at room temperature and can decompose spontaneously with a severe explosion. Anhydrous perchloric acid will explode upon contact with wood.

Health Effects Associated with Oxidizers

Oxidizers are covered here primarily due to their potential to add to the severity of a fire or to initiate a fire. But there are some generalizations that can be made regarding the health hazards of an oxidizing material. In general, oxidizers are corrosive and many are highly toxic.

Acute Health Effects

Some oxidizers, such as nitric and sulfuric acid vapors, chlorine, and hydrogen peroxide, act as irritant gases. All irritant gases can cause inflammation in the surface layer of tissues when in direct contact. They can also cause irritation of the upper airways, conjunctiva, and throat.

Some oxidizers, such as fluorine, can cause severe burns of the skin and mucous membranes. Chlorine trifluoride is extremely toxic and can cause severe burns to tissue.

Nitrogen trioxide is very damaging to tissue, especially the respiratory tract. The symptoms from an exposure to nitrogen trioxide may be delayed for hours, but fatal pulmonary edema may result.

Osmium tetroxide, another oxidant commonly employed in the laboratory, is also dangerous due to its high degree of acute toxicity. It is a severe irritant of both the eyes and the respiratory tract. Inhalation can cause headache, coughing, dizziness, lung damage, difficulty breathing, and may be fatal.

Chronic Health Effects

Nitrobenzene and chromium compounds can cause hematological and neurological changes. Compounds of chromium and manganese can cause liver and kidney disease. Chromium (VI) compounds have been associated with lung cancer.

First Aid for Oxidizers

In general, if a person has inhaled, ingested, or come into direct contact with these materials, the person must be removed from the source of contamination as quickly as possible when it is safe to do so. Medical help must be summoned. In the case of an exposure directly to the skin or eyes, it is imperative that the exposed person be taken to an emergency shower or eyewash immediately. Flush the affected areas for a minimum of 15 minutes and then get medical attention.

Personal Protective Equipment

1. In many cases, the glove of choice will be neoprene, polyvinyl chloride (PVC), or nitrile. Be sure to consult a glove compatibility chart to ensure that the glove material is appropriate for the particular chemical you are working with.
2. Goggles must be worn if the potential for splashing exists or if exposure to vapor or gas is likely.
3. Always use these materials in a chemical fume hood as most pose a hazard via inhalation.

Corrosives

General Characteristics

1. Corrosives are most commonly acids or alkalis, but many other materials can be severely damaging to living tissue.
2. Corrosives can cause visible destruction or irreversible alterations at the site of contact. Inhalation of the vapor or mist can cause severe bronchial irritation. Corrosives are particularly damaging to the skin and eyes.
3. Certain substances considered noncorrosive in their natural dry state are corrosive when wet, such as when in contact with moist skin or mucous membranes. Examples of these materials are lithium chloride, halogen fluorides, and allyl iodide.
4. Sulfuric acid is a very strong dehydrating agent and nitric acid is a strong oxidizing agent. Dehydrating agents can cause severe burns to the eyes due to their affinity for water.

Examples of Corrosives

Sulfuric acid
Chromic acid
Stannic chloride
Ammonium bifluoride
Bromine
Ammonium hydroxide

Use and Storage of Corrosives

1. Always store acids separately from bases. Also, store acids in acid storage cabinets away from flammables since many acids are also strong oxidizers.
2. Do not work with corrosives unless an emergency shower and continuous flow eyewash are available.
3. Add acid to water, but never water to acid. This is to prevent splashing from the acid due to the generation of excessive heat as the two substances mix.
4. Never store corrosives above eye level. Store on a low shelf or cabinet.
5. It is a good practice to store corrosives in a tray or bucket to contain any leakage.
6. When possible, purchase corrosives in containers that are coated with a protective plastic film that will minimize the danger to personnel if the container is dropped.
7. Store corrosives in a wood cabinet or one that has a corrosion-resistant lining. Corrosives stored in an ordinary metal cabinet will quickly damage it. If the supports that hold up the shelves become corroded, the result could be serious. Acids should be stored in acid storage cabinets specially designed to hold them, and nitric acid should be stored in a separate cabinet or compartment.

Use and Storage of Hydrofluoric Acid

1. Hydrofluoric acid is extremely hazardous. Hydrofluoric acid can cause severe burns, and inhalation of anhydrous hydrogen fluoride can be fatal.
2. Initial skin contact with hydrofluoric acid may not produce any symptoms.
3. Only persons fully trained in the hazards of hydrofluoric acid should use it.
4. Always use hydrofluoric acid in a properly functioning fume hood. Be sure to wear personal protective clothing.
5. If you suspect that you have come in direct contact with hydrofluoric acid: wash the area with water for at least 15 minutes, remove clothing, and then promptly seek medical attention. If hydrogen fluoride vapors are inhaled, move the person immediately to an uncontaminated atmosphere (if safe to do so), keep the person warm, and seek prompt medical attention.
6. NEVER STORE HYDROFLUORIC ACID IN A GLASS CONTAINER BECAUSE IT IS INCOMPATIBLE WITH GLASS.
7. Store hydrofluoric acid separately in an acid storage cabinet and keep only the amount necessary in the lab.
8. Creams for treatment of hydrofluoric acid exposure are commercially available and should be kept on site.

Health Effects Associated with Corrosives

All corrosives are severely damaging to living tissues and also attack other materials, such as metal.

Skin contact with alkali metal hydroxides, e.g., sodium hydroxide and potassium hydroxide, is more dangerous than with strong acids. Contact with alkali metal hydroxides normally causes deeper tissue damage because there is less pain than with an acid exposure. The exposed person may not wash it off thoroughly enough or seek prompt medical attention.

All hydrogen halides are acids that are serious respiratory irritants and also cause severe burns. Hydrofluoric acid is particularly dangerous. At low concentrations, hydrofluoric acids do not immediately show any signs or symptoms upon contact with skin. It may take several hours for the hydrofluoric acid to penetrate the skin before you would notice a burning sensation. However, by this time permanent damage, such as second and third degree burns with scarring, can result.

Acute Health Effects

Inhalation — irritation of mucous membranes, difficulty in breathing, fits of coughing, pulmonary edema
Ingestion — irritation and burning sensation of lips, mouth, and throat; pain in swallowing; swelling of the throat; painful abdominal cramps; vomiting; shock; risk of perforation of stomach
Skin Contact — burning, redness and swelling, painful blisters, profound damage to tissues; and with alkalis, a slippery, soapy feeling
Eye Contact — stinging, watery eyes, swelling of eyelids, intense pain, ulceration of eyes, loss of eyes or eyesight

Chronic Health Effects

Symptoms associated with a chronic exposure vary greatly depending on the chemical. The chronic effect of hydrochloric acid is damage to the teeth; the chronic effects of hydrofluoric acid are decreased bone density, fluorosis, and anemia; the chronic effects of sodium hydroxide are unknown.

First Aid for Corrosives

Inhalation — remove person from source of contamination if safe to do so. Get medical attention. Keep person warm and quiet and do not leave unattended.
Ingestion — remove person from source of contamination if safe to do so. Get medical attention and inform emergency responders of the name of the chemical swallowed.
Skin Contact — remove person from source of contamination if safe to do so and take immediately to an emergency shower or source of water. Remove clothing, shoes, socks, and jewelry from affected areas as quickly as possible, cutting them off if necessary. Be careful to not get any chemical on your skin or to inhale the vapors. Flush the affected area with water for a minimum of 15 minutes. Get medical attention.
Eye Contact — remove person from source of contamination if safe to do so and take immediately to an eyewash or source of water. Rinse the eyes for a minimum of 15 minutes. Have the person look up and down and from side to side. Get medical attention. Do not let the person rub the eyes or keep them tightly shut.

Personal Protective Equipment

Always wear proper gloves when working with acids. Neoprene and nitrile gloves are effective against most acids and bases. Polyvinyl chloride (PVC) is also effective for most acids. A rubber-coated apron and goggles should also be worn. If splashing is likely to occur, wear a face shield over the gloves. Always use corrosives in a chemical fume hood.

Reactives

General Characteristics

Polymerization Reactions

Polymerization is a chemical reaction in which two or more molecules of a substance combine to form repeating structural units of the original molecule. This can result in an extremely high or uncontrolled release of heat. An example of a chemical that can undergo a polymerization reaction is styrene.

Water-Reactive Molecules

When water-reactive materials come in contact with water, one or more of the following can occur:

- Liberation of heat, which may cause ignition of the chemical itself if it is flammable, or ignition of flammables that are stored nearby
- Release of a flammable, toxic, or strong oxidizing gas; release of metal oxide fumes
- Formation of corrosive acids

Water-reactive chemicals can be particularly hazardous to firefighting personnel responding to a fire in a lab, because water is the most commonly used fire-extinguishing medium. Examples of water-reactive materials:

Alkali metals: lithium, sodium, potassium
Magnesium
Silanes
Alkylaluminums
Zinc
Aluminum

Pyrophoric material can ignite spontaneously in the presence of air. Examples of pyrophoric materials:

Diethylzinc
Triethylaluminum
Many organometallic compounds

Peroxide-Forming Materials

Peroxides are very unstable and some chemicals that can form them are commonly used in laboratories. This makes peroxide-forming materials some of the most hazardous substances found in a lab. Peroxide-forming materials are chemicals that react with air, moisture, or impurities to form peroxides. The tendency to form peroxides by most of these materials is greatly increased by evaporation or distillation. Organic peroxides are extremely sensitive to shock, sparks, heat, friction, impact, and light. Many peroxides formed from materials used in laboratories are more shock sensitive than TNT. Just the friction from unscrewing the cap of a container of ether that has peroxides in it can provide enough energy to cause a severe explosion.

Examples of peroxide-forming materials:

Diisopropyl ether
Sodium amide
Dioxane
Tetrahydrofuran
Butadiene
Acrylonitrile
Divinylacetylene
Potassium amide
Diethyl ether
Vinyl ethers
Vinylpyridine
Styrene

Other Shock-Sensitive Materials

These materials are explosive and sensitive to heat and shock. Examples of shock-sensitive materials:

Chemicals containing nitro groups
Fulminates
Hydrogen peroxide (30+%)
Ammonium perchlorate
Benzoyl peroxide (when dry)
Compounds containing the functional groups: acetylide, azide, diazo, halamine, nitroso, and ozonide

Use and Storage of Reactives

1. A good way to reduce the potential risks is to minimize the amount of material used in the experiment. Use only the amount of material necessary to achieve the desired results.
2. Always substitute a less hazardous chemical for a highly reactive chemical whenever possible. If it is necessary to use a highly reactive chemical, order only the amount that is necessary for the work.
3. Store water-reactive materials in an isolated part of the lab. A cabinet far removed from any water sources, such as sinks, emergency showers, and chillers, is an appropriate location. Clearly label the cabinet "**Water-Reactive Chemicals — No Water**."
4. Store pyrophorics in an isolated part of the lab and in clearly marked cabinets. Be sure to routinely check the integrity of the container and dispose of materials in corroded or damaged containers.
5. Do not open the chemical container if peroxide formation is suspected. The act of opening the container could be sufficient to cause a severe explosion. Visually inspect liquid peroxide-forming materials for crystals or unusual viscosity before opening. Pay special attention to the area around the cap. Peroxides usually form upon evaporation, so they will most likely be formed on the threads under the cap.
6. Date all peroxide-forming materials with the date received and the expected shelf life. Chemicals such as diisopropyl ether, divinyl acetylene, sodium amide, and vinylidene chloride should be discarded after 3 months. Chemicals such as dioxane, diethyl ether, and tetrahydrofuran should be discarded after 1 year.
7. Store all peroxide-forming chemicals away from heat, sunlight, and sources of ignition. Sunlight accelerates the formation of peroxides.
8. Secure the lids and caps on these containers to discourage the evaporation and concentration of these chemicals.
9. Never store peroxide-forming chemicals in glass containers with screw cap lids or glass stoppers. Friction and grinding must be avoided. Also, never store these chemicals in a clear glass bottle where they would be exposed to light.
10. Contamination of an ether by peroxides or hydroperoxides can be detected simply by mixing the ether with 10% (w/w) aqueous potassium iodide solution — a yellow color change due to oxidation of iodide to iodine confirms the presence of peroxides. Small amounts of peroxides can be removed from contaminated ethers via distillation from lithium aluminum hydride ($LiAlH_4$), which both reduces the peroxide and removes contaminating water and alcohols. However, if you suspect that peroxides may be present, it is wise to dispose of the material. If you notice crystal formation in the container or around the cap, do not attempt to open or move the container.
11. Never distill an ether unless it is known to be free of peroxides.
12. Store shock-sensitive materials separately from other chemicals and in a clearly labeled cabinet.
13. Never allow picric acid to dry out, as it is extremely explosive. Always store picric acid in a wetted state.

Health Hazards Associated with Reactives

Reactive chemicals are grouped as a category primarily because of the safety hazards associated with their use and storage and not because of similar acute or chronic health effects. For health hazard information on specific reactive materials, consult the MSDS or the manufacturer. However, there are some hazards common to the use of reactive materials. Injuries can occur due to heat or flames, inhalation of fumes, vapors and reaction products, and flying debris.

First Aid for Reactives

If someone is seriously injured, the most important step is to contact emergency responders as quickly as possible. Explain the situation and describe the location clearly and accurately.

If someone is bleeding severely, apply a sterile dressing, clean cloth, or handkerchief to the wound. Then put protective gloves on and place the palm of your hand directly over the wound and apply pressure and keep the person calm. Continue to apply pressure until help arrives.

If a person's clothes are on fire, he or she should drop immediately to the floor and roll. If a fire blanket is available, put it over the individual. An emergency shower, if one is immediately available, can also be used to douse the flames.

If a person goes into shock, have the individual lie down on his/her back, if safe to do so, and raise the feet about 1 ft above the floor.

Personal Protective Equipment

Wear appropriate personal protective clothing while working with highly reactive materials. This might include impact-resistant safety glasses or goggles, a face shield, gloves, a lab coat (to minimize injuries from flying glass or an explosive flash), and a shield. Conduct work within a chemical fume hood as much as possible and pull down the sash as far as is practical. When the experiment does not require you to reach into the fume hood, keep the sash closed.

Barriers can offer protection of personnel against explosion and should be used. Many safety catalogs offer commercial shields that are commonly polycarbonate and are weighted at the bottom for stability. It may be necessary to secure the shields firmly to the work surface.

Compressed Gas Cylinders

Cylinders of compressed gas can pose a chemical as well as a physical hazard. If the valve were to break off a cylinder, the amount of force present could propel the cylinder through a block wall. For example, a small cylinder of compressed breathing air used by SCUBA divers has the explosive force of 1.5 lb of TNT.

Use and Storage of Compressed Gas Cylinders

1. Whenever possible, use flammable and reactive gases in a fume hood or other well-ventilated enclosure. Certain categories of toxic gases must always be stored and used in well-ventilated enclosures.
2. Always use the appropriate regulator on a cylinder. If a regulator will not fit a cylinder's valve, do not attempt to adapt or modify it to fit a cylinder it was not designed for. Regulators are designed to fit only specific cylinders to avoid improper use.
3. Inspect regulators, pressure-relief valves, cylinder connections, and hose lines frequently for damage.
4. Never use a cylinder that cannot be positively identified. Color-coding is not a reliable way to identify cylinders since the color can vary from supplier to supplier.
5. Do not use oil or grease on any cylinder component of an oxidizing gas because a fire or explosion can result.
6. Never transfer gases from one cylinder to another. The gas may be incompatible with the residual gas remaining in the cylinder or may be incompatible with the cylinder material.
7. Never completely empty cylinders during lab operations; rather, leave approximately 25 PSI of pressure. This will prevent any residual gas in the cylinder from becoming contaminated.

8. Place all cylinders so the main valve is accessible.

9. Close the main cylinder valve whenever the cylinder is not in use.

10. Remove regulators from unused cylinder and always put the safety cap in place to protect the valve.

11. Always secure cylinder, whether empty or full, to prevent it from falling over and damaging the valve (or falling on your foot). Secure cylinders by chaining or strapping them to a wall, lab bench, or other fixed support.

12. Oxygen should be stored in an area that is at least 20 feet away from any flammable or combustible materials or separated from them by a noncombustible barrier at least 5 ft high and having a fire-resistant rating of at least $1/2$ hour.

13. To transport a cylinder, put on the safety cap and strap the cylinder to a hand truck in an upright position. Never roll a cylinder.

14. Always clearly mark empty cylinders and store them separately (using chalk to write "MT" on a cylinder in big letters is satisfactory for noting an empty cylinder).

15. Open cylinder valves slowly.

16. Only compatible gases should be stored together in a gas cylinder cabinet.

17. Flammable gases must be stored in properly labeled, secured areas away from possible ignition sources and kept separate from oxidizing gases.

18. Do not store compressed gas cylinders in areas where the temperature can exceed 125°F.

EMERGENCY PROCEDURES

All accidents, hazardous materials spills, or other dangerous incidents should be reported. A list of telephone numbers must be posted on the door to each laboratory (and must be kept up to date). Telephone numbers shall also be posted beside every telephone in the laboratories. The list of telephone numbers must include 24-hour numbers for the following personnel:

Laboratory Supervisor
Principal Investigator(s)
Emergency Medical Services
Police Department
Maintenance
Chemical Response Unit

Callers should explain any emergency situation clearly, calmly, and in detail.

Primary Emergency Procedures for Fires, Spills, and Accidents

1. In the event of a fire, pull the nearest fire alarm. If you are in the laboratory and a fire alarm sounds, quickly secure your work (cap bottles, etc.) so that it is not dangerous to a passer-by, lock the laboratory, and evacuate the building per the fire evacuation instructions. If the emergency is not in the laboratory where you are located, the last person to leave should turn off the lights.

2. If you are unable to control or extinguish a fire, follow the building evacuation procedure.

3. Attend to any person who may have been contaminated and/or injured if it is safe to reach them.

4. Use safety showers and eye washes as appropriate. In the case of eye contact, promptly flush eyes with water for a minimum of 15 minutes and seek immediate medical attention. For ingestion cases, contact the Poison Control Center at 1-800-POISON1. In the case of skin contact, promptly flush the affected area with water and remove any contaminated clothing or jewelry. If symptoms persist after washing, seek medical attention.

5. Notify persons in the immediate area about the spill, evacuating all nonessential personnel from the spill area and adjoining areas that may be impacted by vapors or a potential fire.

6. If the spilled material is flammable, turn off all potential ignition sources. Avoid breathing vapors of the spilled materials. Be aware that some materials either have no odor or create olfactory fatigue, so that you stop smelling the odor very quickly.

7. Leave on or establish exhaust ventilation if it is safe to do so. Close doors to slow the spread of odors.
8. Notify the appropriate authorities (Laboratory Supervisor, Principal Investigator, Chemical Health and Safety) about the spill and the required documentation.
9. IF THERE IS AN IMMEDIATE THREAT TO LIFE OR HEALTH, call Emergency Services at 911.

Building Evacuation Procedures

1. Building evacuation may be necessary if there is a chemical release, fire, explosion, natural disaster, or medical emergency.
2. Be aware of the marked exits from your area and building.
3. To activate the building alarm system, pull the handle on one of the red boxes located in the hallway.
4. Call the appropriate authorities.
5. Walk quickly to the nearest marked exit and ask others to do the same.
6. Outside, proceed to a clear reassembly area that is at least 150 ft from the affected building and that does not interfere with the work of emergency personnel.
7. DO NOT RETURN TO THE BUILDING UNTIL YOU ARE TOLD THAT IT IS SAFE TO DO SO.

Minor Spills

1. Trained personnel should use the spill control kit appropriate to the material spilled to clean up the spill.
2. If the spill is minor and of known limited danger, clean it up immediately. Determine the appropriate cleaning method by referring to the material's MSDS. During cleanup, wear the appropriate protective gear.
3. Cover liquid spills with compatible adsorbent material such as spill pillows or a kitty litter/ vermiculite mix, if it is compatible. If appropriate materials are available, corrosives should be neutralized prior to adsorption. Clean spills from the outer area first, cleaning toward the center.
4. Place the spilled material into an appropriate impervious container and seal. Schedule its disposal.
5. If appropriate, wash the affected surface with soap and water. Mop up the residues and place them in an appropriate container for disposal.
6. If the spilled material is not water soluble, a solvent such as xylene may be necessary to clean the surface(s). Check the solubility of the spilled material in various solvents and use the least toxic effective solvent available. Wear appropriate personal protective equipment.
7. Notify the Laboratory Supervisor about the need to replace the used items from the spill control kit.

Mercury Spills

Mercury is commonly used in many technical procedures. When contained properly, it is of little threat to our health. Immediate attention to mercury spills is important because spilled mercury can accumulate over time, resulting in exposure to mercury vapor.

When a spill occurs, use the following procedure:

1. Restrict the area. Allow no one to enter the room except for trained personnel to help with containment of the spill.
2. Contact the Chemical Safety Director.
3. Broken thermometers that contain small amounts of mercury may be safely collected by trained laboratory personnel in a container that can be sealed. Always wear disposable gloves when cleaning up mercury and dispose of all mercury and mercury contaminated waste through the chemical waste program. Anyone handling mercury or cleaning up mercury spills should wash hands thoroughly using soap and water when finished. Report all mercury spills to the Chemical Safety Director.

CHEMICAL WASTE DISPOSAL PROGRAM

Chemical Waste Containers

Containers used for the accumulation of hazardous waste must be in good condition, free of leaks and compatible with the waste being stored in them. A waste accumulation container should be opened only when it is necessary to add waste, and should otherwise be capped. Hazardous waste must not be placed in unwashed containers that previously held incompatible materials.

If a hazardous waste container is not in good condition (i.e., it leaks), either transfer the waste from the bad container into a good container, pack the container in a larger and nonleaking container, or manage the waste in some other way that prevents the potential for a release of contamination.

A storage container holding a hazardous waste that is incompatible with any waste or other materials stored nearby in other containers must be separated from the other materials or protected from them by means of a wall, partition, or other secondary containment device.

Guidelines for Waste Containers

- Must be marked with the words "waste" or "spent" and its contents indicated. NO container should be marked with the words "hazardous" or "nonhazardous." Paint over or remove old labels from waste containers.
- Must be kept at or near (immediate vicinity) the site of generation and under control of the generator.
- Must be compatible with the contents (i.e., acid should not be stored in metal cans).
- Must be closed at all times except when actively receiving waste.
- Must be properly identified before disposal.
- Must be safe to transport with nonleaking screw-on caps.
- Must be filled to a safe level (not beyond the bottom of the neck of the container or a 2-in headspace for a 55-gallon drum).

NOTE: Do not use RED BAGS or SHARPS CONTAINERS (Biohazard) for hazardous waste collection.

Labeling Containers

Before chemicals can be disposed of, a waste tag is required. It should be filled out by the waste generator and attached to each container. The information on the tag is used to categorize and treat the waste. A manifest is also required. Fill out all paperwork legibly, accurately, and completely.

Waste Minimization

Avoid purchasing and using large quantities when it is not necessary. Implement microscale techniques whenever possible.

Flammable Organic Solvents

Collection for Reuse

Many flammable organics can be reused for fuel unless they are extremely toxic or give off toxic products of combustion. *Do not* combine any other chemicals with the flammable organic solvents listed below. *Halogenated solvents* (solvents containing chlorine, fluorine, or bromine), acutely toxic flammables, acids, bases, heavy metals, oxidizers, and pesticides should be collected in separate containers. The following is a list of the most frequently encountered compounds that are suitable for heat recovery:

Acetone	Methyl alcohol
2-Butanol	Methyl cellosolve
Butyl alcohol	Pentane
Cyclohexane	Petroleum ether
Diethyl ether	2-Propanol
Ethyl acetate	Sec-butyl alcohol
Ethyl alcohol	Tert-butyl alcohol
Heptane	Tetrahydrofuran
Hexane	Xylene

Disposal of Chemicals down the Sink or Sanitary Sewer System

Very few chemical wastes produced in laboratories are acceptable for disposal down the sink or sanitary sewer system. The local Sewer Use/Pretreatment Ordinance establishes uniform requirements for all users of the wastewater treatment system. Many chemicals can interfere with the proper function of the treatment facility and can render them unable to comply with state and federal regulations under the Clean Water Act of 1977.

Generators of laboratory waste are advised to exercise caution with respect to sink disposal of chemical wastes. In general, small-scale research activities (100 mL or less) of certain types of water-soluble, nontoxic, and nonflammable chemicals may be poured if they have been approved by the Chemical Safety Director. It is recommended that such materials be disposed of through the Department of Occupational Health and Safety, even in small quantities.

Chemical Substitution

Whenever possible, it is desirable to substitute nonhazardous, biodegradable chemicals for hazardous chemicals. Use of these chemicals will reduce the volume of hazardous waste generated. Examples of acceptable substitutes include:

1. Citric acid-based cleaning solutions for xylene-, benzene-, and toluene-containing cleaning solutions.
2. Nonhalogenated solvents in parts washers or other solvent processes.
3. Detergent and enzymatic cleaners can be substituted for sulfuric acid/potassium dichromate (chromerge) cleaning solutions and ethanol/potassium hydroxide cleaning solutions.

Neutralization and Deactivation

Certain hazardous chemical wastes can be rendered nonhazardous by specific neutralization or deactivation laboratory procedures. Contact the Chemical Safety Officer to see if the waste you generate is suitable for neutralization.

Elimination of Nonhazardous Waste from Hazardous Waste

The following items *are not* considered to be hazardous. They should be collected in disposable containers or plastic bags, clearly labeled as nonhazardous waste, and put in the wastebasket. All compounds identified by the two letter code "NH" are nonhazardous and should not be disposed of via the chemical waste program unless they are components of a mixture with hazardous materials or are suitable for chemical recycling.

Nonhazardous Waste

Organic Chemicals

Acetates: calcium (Ca), sodium (Na), ammonium (NH₄), and potassium (K)
Amino acids and their salts
Citric acid and salts of sodium (Na), potassium (K), magnesium (Mg), calcium (Ca), and ammonium (NH₄)
Lactic acid and salts of sodium (Na), potassium (K), magnesium (Mg), calcium (Ca), and ammonium (NH₄)
Sugars: glucose, lactose, fructose, sucrose, maltose

Inorganic Chemicals

Bicarbonates: sodium (Na), potassium (K)
Borates: sodium (Na), potassium (K), magnesium (Mg), calcium (Ca)
Bromides: sodium (Na), potassium (K)
Carbonates: sodium (Na), potassium (K), magnesium (Mg), calcium (Ca)
Chlorides: sodium (Na), potassium (K), magnesium (Mg), calcium (Ca)
Fluorides: calcium (Ca)
Iodides: sodium (Na), potassium (K)
Oxides: boron (B), magnesium (Mg), calcium (Ca), aluminum (Al), silicon (Si), iron (Fe)
Phosphates: sodium (Na), potassium (K), magnesium (Mg), calcium (Ca), ammonium (NH₄)
Silicates: sodium (Na), potassium (K), magnesium (Mg), and calcium (Ca)
Sulfates: sodium (Na), potassium (K), magnesium (Mg), calcium (Ca), ammonium (NH₄)

Laboratory Materials

Chromatographic adsorbents
Filter paper without hazardous chemical residue
Non-contaminated glassware
Rubber gloves

Waste Disposal

All laboratories are required to comply with federal and state regulations regarding the packing, labeling, and transport of hazardous materials. Before contacting the Hazardous Materials Facility for waste removal, the following procedures must be completed. *Improperly packed or labeled waste cannot be removed.*

Step One: Packing the Waste

Containers

Collect each chemical waste in a separate screw-top container. *Do not mix* wastes. Use the smallest container size to match the amount of chemical waste generated. The container the chemical was originally shipped in is an ideal waste collection container, if it is an appropriate size. All waste containers must be tightly capped. *Each container must be labeled as to chemical content.* For mixtures, give approximate percentages of each chemical compound. *Milk jugs are not acceptable for chemical storage.* If using a container that originally contained another chemical, completely remove the original label prior to relabeling. Completely fill chemical waste collection containers.

Shock-Sensitive and Water-Reactive Compounds and Lecture Bottles

Shock-sensitive and water-reactive compounds and lecture bottles require special handling. These materials should always be packed separately from other chemicals.

Packing Filled Containers in Boxes

Chemicals that have the potential to react with each other should *not* be packed in the same box. Determine the packing hazard class for each chemical waste. When determining the class for a mixture of chemicals, *reactivity* has priority over *toxicity*. If you have difficulty determining the packing class of a mixture, call the Hazardous Materials Manager.

Segregate the wastes according to the hazard class and pack them into cardboard boxes. *Do not pack different classes in the same box*. Place dividers and shock absorbing materials (newspapers, vermiculite) between the containers.

Step Two: Completing the Manifest

The label for the chemical waste is called a packing *manifest*. A manifest must be completed and attached to *each* box. Laboratory personnel should complete the manifest following the directions below:

1. Laboratory Information: Fill in the generator's name (i.e., principal investigator, lab director), telephone number, department, building, room number, and the date.
2. Waste Information: The contents of each container must be identified on the manifest. *Nonspecified chemical waste* items are extremely difficult for hazardous materials personnel to handle. Good laboratory record keeping and labeling of all chemicals and chemical wastes prevents unknown waste items. Any chemical material that is potentially recyclable should *not* be contaminated with other chemicals for disposal. Where appropriate, note on the manifest if material is unopened.
3. The generator should check the information on the manifest, sign his or her name, and attach it to the corresponding box.

Step Three: Chemical Waste Removal

Attach one copy to the box and retain a copy for laboratory records. Specify where the waste is to be picked up. If your waste is not picked up in a reasonable period of time, call to inquire. Any incomplete or improperly completed manifest will be returned to the generator with an explanation for its return.

MATERIAL SAFETY DATA SHEETS (MSDS)

Since Material Safety Data Sheets (MSDS) are centrally related to the safe handling of hazardous substances, it is imperative that laboratory workers have easy access to them. There are three basic means of obtaining an MSDS:

Chemical manufacturer
Chemical supplier
Internet, such as through the UAB Department of Occupational Health and Safety webpage at:
    http://www.healthsafe.uab.edu

In general, the preferred source for the MSDS is the chemical manufacturer, primarily because these files are actively updated to accurately reflect all that is known about the hazardous material in question.

MSDSs are the cornerstone of chemical hazard communication. They provide most of the information you should know to work with chemicals safely. The following sections describe the information normally contained in an MSDS:

Product Name and Identification

Name of the chemical as it appears on the label
Manufacturer's name and address
Emergency telephone numbers for obtaining further information about a chemical in the event of an
  emergency
Chemical name or synonym
C.A.S. # — the Chemical Abstract Service Registry number, which identifies the chemical
Date of preparation of the MSDS

Hazardous Ingredients/Identity Information

Hazardous Ingredients

Substances which, in sufficient concentration, can produce physical or acute or chronic health hazards to persons exposed to the product. Physical hazards include fire, explosion, corrosion, and projectiles. Health hazards include any health effect, even irritation or development of allergies.

Threshold Limit Value (TLV)

A TLV is the highest airborne concentration of a substance to which nearly all adults can be repeatedly exposed, day after day, without experiencing adverse effects. These are usually based on an 8-hour time-weighted average.

Permissible Exposure Limit (PEL)

The PEL is an exposure limit established by OSHA.

Short-Term Exposure Limit (STEL)

The STEL is a 15-min time-weighted average exposure which should not be exceeded at any time during a workday. A STEL exposure should not occur more than four times per day, and there should be at least 60 min between exposures.

Lethal Dose 50 (LD50)

Lethal single dose (usually oral) in mg/kg (milligrams of chemical per kilogram of animal body weight) of a chemical that results in the death of 50% of a test animal population.

Lethal Concentration 50 (LC50)

Concentration dose expressed in ppm for gases or micrograms per liter of air for dusts or mists that results in the death of 50% of a test animal population administered in one exposure.

Physical/Chemical Characteristics

Boiling point, vapor pressure, vapor density, specific gravity, melting point, appearance, and odor are given in this section and all provide useful information about the chemical. Boiling point and vapor pressure provide a good indication of the volatility of the material. Vapor density indicates whether vapors will sink, rise, or disperse throughout the area. The farther the values are from 1 (the value assigned to atmospheric air), the faster the vapors will sink or rise.

Fire and Explosion Hazard Data

Flashpoint — refers to the lowest temperature at which a liquid gives off enough vapor to form an ignitable mixture with air.

Flammable or Explosive Limits — the range of concentrations over which a flammable vapor mixed with air will flash or explode if an ignition source is present.

Extinguishing Media — the fire-fighting substance that is suitable for use on the substance which is burning.

Unusual Fire and Explosive Hazards — hazards that might occur as the result of overheating or burning of the specific material.

Reactivity Data

Stability — indicates whether the material is stable or unstable under normal conditions of storage, handling, and use.

Incompatibility — lists any materials that would, upon contact with the chemical, cause the release of large amounts of energy, flammable gas or vapor, or toxic vapor or gas.

Hazardous Decomposition Products — any materials that may be produced in dangerous amounts if the specific material is exposed to burning, oxidation, heating, or allowed to react with other chemicals.

Hazardous Polymerization — a reaction with an extremely high or uncontrolled release of energy, caused by the material reacting with itself.

Health Hazard Data

Routes of Entry

Inhalation — breathing in of a gas, vapor, fume, mist, or dust.

Skin Absorption — a possible significant contribution to overall chemical exposure by way of absorption through the skin, mucous membranes, and eyes by direct or airborne contact.

Ingestion — the taking up of the substance through the mouth.

Injection — having the material penetrate the skin through a cut or by mechanical means.

Health Hazards (Acute and Chronic)

Acute — an adverse effect with symptoms developing rapidly

Chronic — an adverse effect that can be the same as an acute effect, except that the symptoms develop slowly over a long period of time or with recurrent exposures.

Carcinogen

A substance that is determined to be cancer producing or potentially cancer producing.

Signs and Symptoms of Overexposure

The most common symptoms or sensations a person could expect to experience from overexposure to a specific material. It is important to remember that only some symptoms will occur with exposures in most people.

Emergency and First-Aid Procedures

Instructions for treatment of a victim of acute inhalation, ingestion, and skin or eye contact with a specific hazardous substance. The victim should be examined by a physician as soon as possible.

Specific HACH MSDS Information

This information is presented here because of the large number of specialized HACH Co. reagents and procedures used in environmental laboratories. HACH MSDSs describe the hazards of their chemical products. Each of their MSDSs has 10 sections.

Header Information

Typically provides the vendor name, company address and telephone number, emergency telephone numbers, vendor's catalog number, date of the MSDS, and version of the MSDS.

Product Information

Product name
Chemical Abstract Services (CAS) number
Chemical name
Chemical formula, where appropriate
Chemical family to which the material belongs

Ingredients (lists all components)

PCT: Percent by weight of each component in product (unless trade secret)
CAS NO: Chemical Abstract Services (CAS) registry number for component
SARA: If component is listed in SARA 313 and more is used than amount listed, must notify EPA.
TLV: Threshold Limit Value. Maximum airborne concentration for 8-hour exposure that is recommended by the American Conference for Governmental Industrial Hygienists (ACGIH).
PEL: Permissible Exposure Limit. Maximum airborne concentration for 8-hour exposure that is regulated by the Occupational Health and Safety Administration (OSHA).
HAZARD: Physical and health hazards of component explained.

Physical Data

Physical state, color, odor, solubility, boiling point, melting point, specific gravity, pH, vapor density, evaporation rate, corrosivity, stability, and storage precautions.

Fire, Explosion Hazard, and Reactivity Data

Flashpoint: Temperature at which liquid will give off enough vapor to ignite. Used to define flammability and ignitability
Lower Flammable Limit (LFL or LEL): Lowest concentration that will produce flash or fire when ignition source is present

Upper Flammable Limit (UFL or UEL): Vapor concentration in air above which the vapor concentration is too great to burn

NFPA Codes: The National Fire Protection Association (NFPA) has a system to rate the degree of hazard presented by a chemical. Codes usually found in colored diamond and range from 0 (minimal hazard) to 4 (extreme hazard). They are grouped into the following hazards: health (blue), flammability (red), reactivity (yellow), and special hazards (white).

Health Hazard Data

Describes how a chemical can enter body (ingestion, inhalation, skin contact), its acute and chronic effects, and lists if a component is a carcinogen, mutagen, or teratogen.

Precautionary Measures

Special storage instructions
Handling instructions
Conditions to avoid
Protective equipment needed

First Aid

Spill and disposal procedures.

Transportation Data

Shipping name, hazard class, and ID number of the product.

References

Supporting references are also included in the HACH MSDS sheets.

SUMMARY OF FIELD TEST KITS

Field test kits can be important analytical tools during receiving water investigations. Chapter 6, among others, described how they can be used to obtain rapid and cost-effective data. However, the careful selection of the test kits to be used is critical. It is important to consider several factors, specifically the sensitivity of the procedure, safety hazards associated with the method, the cost (both capital and expendables) to conduct the analyses, and the time and expertise needed to conduct the test. Table E.2 summarizes these attributes, including results of conducting sensitivity tests using ultra-clean water and stormwater (Pitt and Clark 1999). The useful range is the minimum detection limit found during our tests to the upper limit that does not require dilution. The precision is the coefficient of variation based on replicate analyses, and the recovery is the slope of the regression line comparing analyses of spiked samples using these procedures and standard methods. The recovery tests were conducted using both ultra-clean water prepared using reverse osmosis (RO) and stormwater to identify any matrix interference problems. Any problems noted during the tests are also indicated, especially safety concerns, unusual amounts of expertise needed, and storage requirements.

These tests represent several classes of analytical procedures. The following sets of photos illustrate some of the simpler test kit methods. Figure E.1 illustrates the basic colorimetric procedure with a color wheel to analyze basic water color using a HACH test kit, while Figures E.2 and E.3 show simple color indicator paper strips for alkalinity. Vacuum vials are also used in several test

Table E.2 Summary of All Field Test Kits Evaluated

Method	Manufacturer and Kit Name	Capital Cost	Expendable Cost (per sample)	Time Reqd. (min)	Useful Range	Precision (COV)	Recovery (RO/runoff)	Problems with Test (safety hazards, expertise required, etc.)
Ammonia								
Colorimetric determination of ammonia using Nessler's reaction	CHEMetrics *Ammonia 1 DCR Photometer*	$435 for kit	$0.63	5	0.03–2.5 mg/L	0.15	0.85/1.27	6-month shelf life, with refrigeration; sharps and mercury in waste
Colorimetric determination of ammonia using salicylate	HACH *Nitrogen, Ammonia: Salicylate Method without Distillation*	$1495 for DR/2000	$2.88	20	0.10–0.7	0.17	1.15/1.10	
Colorimetric determination of ammonia using Nessler's reaction	La Motte *Ammonia Nitrogen, High Range*	$895 for Smart Color.	$0.33	10	0.38–3	na	1.22/1.21	Waste contains a mercury compound; high detection limit (0.4 mg/L)
Colorimetric determination of ammonia using salicylate	La Motte *Ammonia Nitrogen, Low Range*	$895 for Smart Color.	$0.76	20	0.17–1.5	na	1.04/0.96	
Bacteria								
Colorimetric	IDEXX *Colilert*	$0.00	$4.00	24 hr	na	na	na	24-hour test period required
Colorimetric	Industrial Municipal Equipment, Inc. *IME Test KoolKount Assayer*			30 min to 13 hr	na	na	na	Not a selective test, but sensitive to a mixed microbial population
BTEX								
Immunoassay	Dtech (EM Science) *BTEX Test Kit*	$500	$25	30–60	na	na	na	Reagents expire in 1 to 2 months and require refrigeration; requires 30–60 min to conduct test; requires extensive expertise; $25 per test
	PetroSense	$6900		5	na	na	na	Expensive instrument ($6900)

Method	Product	Kit cost	Cost/test	No.	Range		Precision	Comments
Chlorides								
Silver nitrate titration	HACH *silver nitrate titration*	$94 for digital titrator	$0.66	not evaluated	na	na	na	Unclear titration endpoint, no useful data obtainable; recommended that conductivity analyses be used as a better indicator of chlorides in a sample
Conductivity								
Electronic probe	YSI *Model 33 SCT*	$600 for kit	$0.00	1	98–? µS/cm	na	0.90/0.93	
Electronic probe	Horiba *Twin*	$250 for kit	$0.00	1	75–50,000 µS/cm	0.04	1.08/1.02	Replace sensor every 6 months for $60
Electronic probe	Horiba *U-10 (Cond., temp., DO, turb., pH)*	$2800 for kit	$0.00	1	87–? µS/cm	na	0.95/0.96	Expensive instrument, but multiparameter
Copper								
Colorimeter	CHEMetrics *Copper 1 DCR Photometer Kit*	$435 for kit	$0.63	15	0.3–3.5 mg/L	na	0.64/0.52	Sharps and poor recovery; not very repeatable
Colorimeter	La Motte *Copper (Diethyldithio-carbamate)*	$895 for Smart Color.	$0.41	10	0.1–3.5	na	1.11/0.93	
Colorimeter	La Motte *Copper (Bicinchoninic Acid)*	$895 for Smart Color.	$0.23	20	0.6–3.5	na	0.94/0.93	Extra time required to dissolve reagent; not very repeatable
Colorimeter	HACH *Bicinchonate Copper Method Using AccuVac Ampoules*	$1495 for DR/2000	$0.28	20	0.5–5.0	na	0.97/0.96	Sharps
Detergents								
Colorimetric	CHEMetrics *Detergents (Anionic Surfactants)*	$60 for 1st 30 tests and standards	$2.38	10	0.15–3 mg/L	na	1.66/1.82	Sharps; chloroform extraction (very small volume and well contained)
Colorimetric	HACH *Surfactants, Anionic, Crystal Violet Method*	$1495 for DR/2000	$1.10	30	na	na	na	Large amounts of benzene required; require laboratory hood; waste disposal problem

Table E.2 Summary of All Field Test Kits Evaluated (continued)

Method	Manufacturer and Kit Name	Capital Cost	Expendable Cost (per sample)	Time Reqd. (min)	Useful Range	Precision (COV)	Recovery (RO/runoff)	Problems with Test (safety hazards, expertise required, etc.)
Fluoride								
Ion selective electrode	Cole-Parmer *Fluoride Tester*	$600 for electrode, meter and calib. kit	$0.25	5–10	0.1–20 mg/L	0.22	0.97/0.96	Requires frequent and time-consuming calibration; too fragile for field use
Spectrophotometric determination of bleaching by fluoride	HACH *Fluoride SPADNS Reagent*	$1495 for DR/2000	$0.37	10	0.3–2	na	1.10/1.07	Should use automatic pipettes, hard to use in field; SPADNS Reagent is hazardous
Spectrophotometric determination of bleaching by fluoride	HACH *Fluoride SPADNS Reagent Using AccuVac Ampoules*	$1495 for DR/2000	$1.17	5	0.1–2	0.05	0.97/0.94	Sharps; SPADNS Reagent is hazardous
Hardness								
EDTA titration	CHEMetrics *Hardness, Total 20–200 ppm*	$0.00	$2.25	5–10	na	0.01	na	Sharps
EDTA titration	HACH *Total Hardness Using Digital Titrator*	$94 for digital titrator	Varies with sample strength	Varies with sample strength	na	na	na	
Lead								
Solid phase extraction, colorimeter	HACH *LeadTrak System*	$395 for DR/100 kit or $1495 for DR/2000	$4.61	45	0.005–0.15	na	0.84/0.87	Requires extensive expertise; complex kit; time-consuming (45 min), but only kit with useful sensitivity
Sulfide staining	Innovative Synthesis Corporation *The Lead Detective*	$3.00		5	na	na	na	Poor sensitivity
	HybriVet Systems *Lead Check Swabs*	$3.00		5	na	na	na	Poor sensitivity
	Carolina Environment Company *KnowLead*	$3.00		5	na	na	na	Poor sensitivity

Test strips	EM Science *Lead*	$500 for Reflecto-Quant Meter	$1.11	na	10	na	na	Not sensitive enough
					Nitrate*			
Colorimeter	La Motte *Nitrate*	$895 for Smart Color.	$1.22	0.8–3 mg/L	20	na	0.81/1.06	
ISE	Horiba *CARDY*	$235 for kit	$60/ sensor (per 6 months)	4.9–?	na	0.97	0.90/0.70	Designed for high concentrations; poor recoveries and precision at lower concentrations
Test strips	EM Science *Nitrate Quant Test Strips*	$500 for Reflecto-Quant Meter	$0.49	1.7–500	2	na	1.00/1.61	Reagents must be refrigerated; more scatter than most other tests
Spectrophotometric	HACH *Nitrate, LR*	$1495 for DR/2000		na	na	na	na	Sharps; too sensitive of a test
Spectrophotometric	HACH *Nitrate, MR*	$1495 for DR/2000	$0.56	2.8–16	7	na	0.93/1.06	Sharps
Colorimeter	CHEMetrics *Nitrate (Nitrogen)*	$48 for 1st 30 tests and standards	$0.73	0.5–22	30	na	1.06/1.02	Sharps

* Nitrite and nitrate tests have a Cd-based reagent that is hazardous.

					PAH			
Immunoassay	EM Science *Dtech PAH Test Kit*	$500	$25	na	30–60	na	na	Reagents expire in 1 to 2 months and require refrigeration; requires 30–60 min to conduct test; requires extensive expertise; $25 per test
					pH			
Electrode	Cole-Parmer *pH Wand*	$155 for kit	$92/ electro.	0–14	5	0.01	na	Daily calibration; fragile meter
Electrode	Horiba *Twin pH*	$235 for kit	$70 for sensor. $25 for stand.	0–12	1	<0.01	na	Daily calibration
Electrode	Sentron *pH Probe*	$595 for meter and electrode	None	0–14	1	<0.01	na	Expensive, but rugged instrument ($595)

Table E.2　Summary of All Field Test Kits Evaluated (continued)

Method	Manufacturer and Kit Name	Capital Cost	Expendable Cost (per sample)	Time Reqd. (min)	Useful Range	Precision (COV)	Recovery (RO/runoff)	Problems with Test (safety hazards, expertise required, etc.)
Test paper	EM Science *ReflectoQuant pH*	$500 for Reflecto-Quant Meter	$0.89	2	4–9	0.08	na	Optics of expensive instrument ($500) are difficult to keep clean
Spectrophotometric	La Motte *pH*	$895 for Smart Color.	$0.22	5	5–9.5	na	na	
Test paper	Fisher Scientific *Alkacid Test Strips*	$0.00		1	0–12	0.07	na	Only readable to within ±1 pH unit, poor comparison to pH meters for actual samples
Potassium								
Spectrophotometric	HACH *Potassium Tetraphenylborate*	$1495 for DR/2000	$3	30	0.5–7 mg/L	na	0.81/0.90	
ISE	Horiba *CARDY*	$235 for kit	$60/ sensor (per 6 months)	5	2.0–?	0.04	0.53/0.46	Method designed for much higher concentrations; more scatter than other tests
Colorimeter	La Motte *Potassium*	$895 for Smart Color.	$0.29	15	3.3–10	na	1.35/1.05	
Spectrophotometric	*La Motte Potassium Reagent Set*	$895 for Smart Color.	$0.29	15	1.3–7	0.06	?/0.90	
Zinc								
Spectrophotometric	La Motte *Zinc*	$895 for Smart Color.	$0.59	5	0.14–3 mg/L	na	0.88/0.85	Dilute indicator expires in a month; uses dilute cyanide
Spectrophotometric	HACH *Zinc, Zincon Method*	$1495 for DR/2000	$0.37	10	na	na	na	Uses granular cyanide and is unacceptable for field use
Test strips	EM Science *ReflectoQuant Zinc*	$500 for Reflecto-Quant Meter	$0.56	5	na	na	na	Reflectoquant requires frequent cleaning and test has high detection limit

From Day, J. *Selection of Appropriate Analytical Procedures for Volunteer Field Monitoring of Water Quality.* MSCE thesis, Department of Civil and Environmental Engineering, University of Alabama at Birmingham. 1996. With permission.

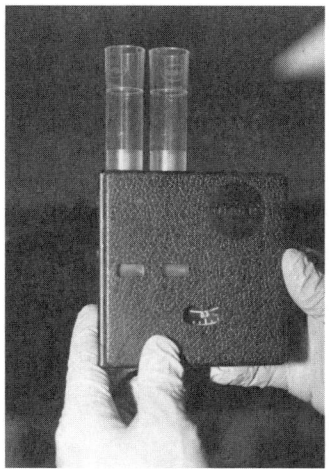

Figure E.1 HACH color test kit.

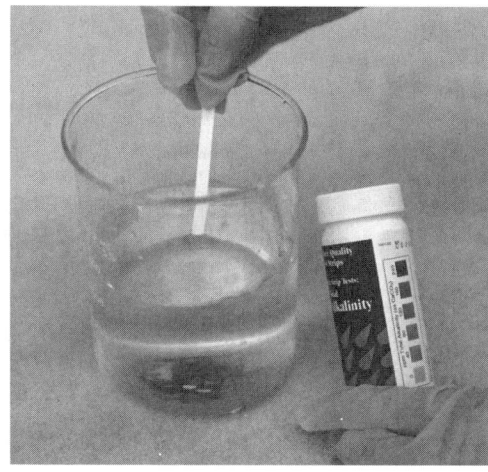

Figure E.2 Quantistrip method for alkalinity.

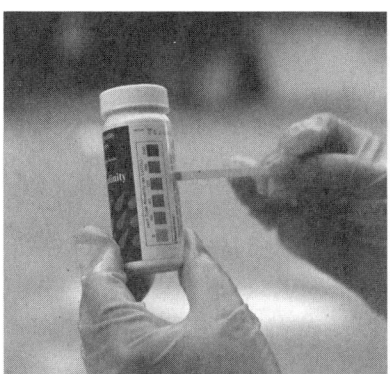

Figure E.3 Comparing Quantistrip against color standards.

Figure E.4 CHEMetrics copper test kit.

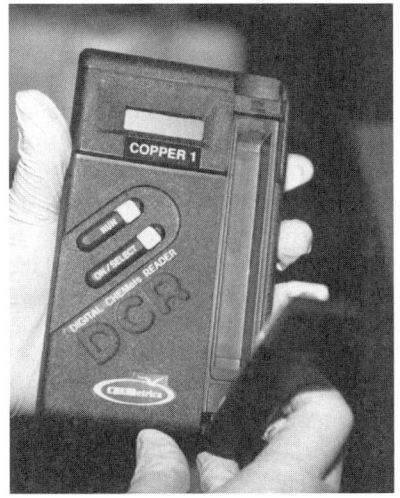

Figure E.5 CHEMetrics color reader.

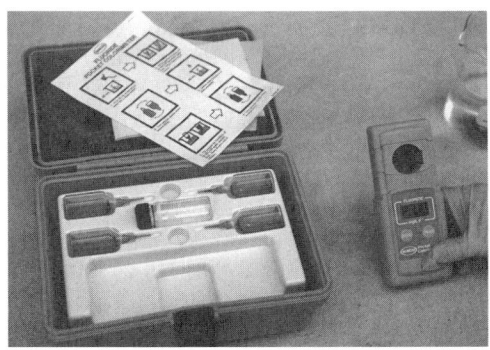

Figure E.6 HACH AccuVac kit for fluoride.

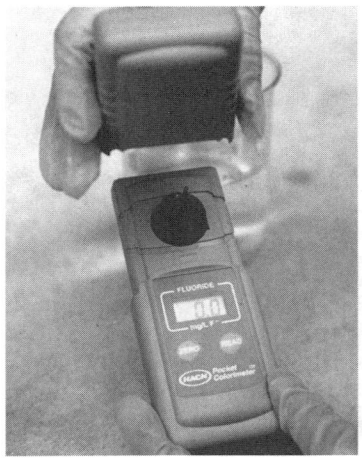
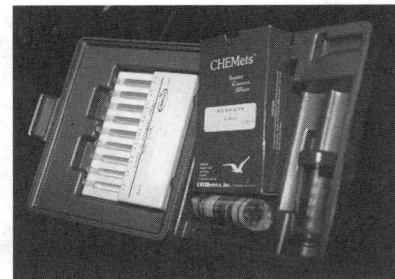
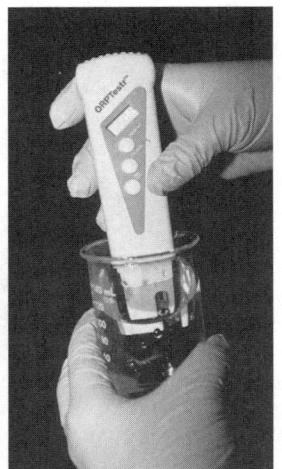

Figure E.7 Reading AccuVac absor-   **Figure E.8** CHEMetrics nitrate test kit.   **Figure E.9** Cole Parmer
bance.   ORP probe.

kits to automatically draw a sample into an evacuated ampoule that contains a specific amount of reagent. Figures E.4 through E.8 are different examples of these types of kits. Figure E.9 is an example of a simple probe used to directly measure ORP of a water sample (a necessary field analysis because of changes occurring after sample collection and transport to the laboratory). Many of other types of test kits are more complex and require several steps for the analyses. Some of the most complex procedures may require as many as 10 steps and more than 30 min for analyses.

While many of the simple methods are quite useful for field monitoring, the more complex (and expensive) procedures must be more carefully weighed against traditional (and more accurate) laboratory methods. In general, we found that the field test kits were more accurate than we had originally expected. However, the sensitivities of many of the field test kits were much poorer than expected, making them much less useful. In addition, numerous safety hazards can exist with these kits, sharps and hazardous reagents and wastes being the most serious.

SPECIAL COMMENTS PERTAINING TO HEAVY METAL ANALYSES

The above discussion on field test kits points out the obvious shortcomings of trying to obtain meaningful heavy metal data using simple procedures. There are a number of methods available for heavy metals, with the traditional methods restricted to the laboratory. The following discussion summarizes these available methods, especially their sensitivities.

Table E.3 lists the metals and associated methods included in the 1995 version of *Standard Methods for the Examination of Water and Wastewater*. Other listings of environmental analytical methods are published by ASTM (American Society of Testing Materials) and by the U.S. Environmental Protection Agency (in the *Code of Federal Regulations*, especially 40 CFR, 136 "Guidelines Establishing Test Procedures for the Analysis of Pollutants"). Methods listed in these references are generally taken as approved for many purposes. Table E.3 lists about 40 different metals and 12 different basic analytical methods. Most all of the metals can be analyzed using atomic absorption spectrometry (AAS) and inductively coupled plasma emission spectrometry (ICP). In addition, many of the metals have specific chemical tests that use spectrophotometric or titration methods. For most stormwater investigations, only a relatively few of these metals are routinely evaluated, including arsenic, cadmium, chromium, copper, lead, mercury, nickel, selenium, and zinc.

Table E.3 Metal Methods Included in the 1995, 19th Edition of *Standard Methods for the Examination of Water and Wastewater*

	Color	AAS	Flame	C-V AAS	ET AAS	Hydride	ICP	ASV	Other
Aluminum	×	×					×		
Antimony		×					×		
Arsenic	×					×	×		
Barium		×					×		
Beryllium	×	×					×		
Bismuth		×							
Cadmium	×	×					×	×	
Calcium	×	×					×		
Cesium		×							
Chromium	×	×					×		IC
Cobalt		×					×		
Copper	×	×					×		
Gold		×							
Iridium		×							
Iron	×	×					×		
Lead	×	×					×	×	
Lithium		×	×				×		
Magnesium		×					×		grav
Manganese	×	×					×		
Mercury	×			×					
Molybdenum		×					×		
Nickel		×					×		
Osmium		×							
Palladium		×							
Platinum		×							
Potassium		×	×				×		ISE
Rhenium		×							
Rhodium		×							
Ruthenium		×							
Selenium	×				×	×	×		fluro
Silver	×	×					×		
Sodium		×	×				×		
Strontium		×	×				×		
Thallium		×					×		
Thorium		×							
Tin		×							
Titanium		×							
Vanadium	×	×					×		
Zinc	×	×					×		

Note: Color: Specific chemical colorimetric methods; AAS: Atomic absorption spectrometry; Flame: Flame emission photometry; ASV: Anodic stripping voltammetry; C-V AAS: Cold-vapor AAS; ET AAS: Electrothermal AAS; ICP: Inductively coupled plasma emission spectrometry; Hydride: Hydride generation AAS; Other: IC (ion chromatography), grav (gravimetric), ISE (ion selective electrode), and fluro (fluorometric)

Table E.4 compares the optimal metal concentration ranges for AAS and ICP, the most commonly used instrumentation (*Standard Methods* 1995). Instrument detection limits are about 15 times less than the lower values shown on this table, which represent the lower limits of quantification. The lower limits of the flame AAS optimal concentration ranges are generally about the same as for the plasma AES, while the electrothermal AAS lower limits are 10 to 1000 times lower. However, the plasma AES instrument has a much greater dynamic range than either AAS instrument. The plasma AES also has fewer interferences and can analyze many elements simultaneously. Because of these differences, many laboratories use a plasma AES for general

Table E.4   Optimal Concentration Ranges of Metals in Samples

	Flame AAS (mg/L)	Electrothermal AAS (mg/L)	Inductively Coupled Plasma AES (mg/L)
Aluminum	5–100	0.02–0.2	0.6–100
Antimony	1–40	0.02–0.3	0.45–100
Arsenic		0.005–0.1	0.75–100
Barium	1–20	0.01–0.2	0.030–50
Beryllium	0.05–2	0.001–0.03	0.005–10
Bismuth	1–5		
Cadmium	0.05–2	0.0005–0.01	0.06–50
Calcium	0.2–20		0.15–100
Cesium	0.5–15		
Chromium	0.2–10	0.005–0.1	0.1–50
Cobalt	0.5–10	0.005–0.1	0.1–50
Copper	0.2–10	0.005–0.1	0.1–50
Gold	0.5–20		
Iron	0.3–10	0.005–0.1	0.1–100
Lead	1–20	0.005–0.1	0.6–100
Lithium	0.1–2		0.06–100
Magnesium	0.02–2		0.45–100
Manganese	0.1–10	0.001–0.03	0.06–50
Molybdenum	1–20	0.003–0.06	0.12–100
Nickel	0.3–10	0.005–0.1	0.2–50
Platinum	5–75		
Potassium	0.1–2		1.5–100
Selenium		0.005–0.1	1.0–100
Silver	0.1–4	0.001–0.025	0.1–50
Sodium	0.03–1		
Strontium	0.3–5		0.03–50
Thallium			0.6–100
Tin	10–200	0.02–0.3	
Titanium	5–100		
Vanadium	2–100		0.1–50
Zinc	0.05–2		0.03–100

Data from *Standard Methods for the Examination of Water and Wastewater. 19th edition.* Water Environment Federation. Washington, D.C. 1995.

analytical work and an electrothermal AAS for individual samples for single elements at very low concentrations.

Table E.5 lists various operational and cost attributes of these metal analysis methods (Pitt et al. 1997). The trade-offs between the various types of equipment are obvious. The instruments with greater sensitivity cost more. Only an electrothermal AAS instrument can analyze many samples quickly (with an autosampler) with good sensitivity, but with only a few metals being analyzed at a time, at the most. The instruments that can analyze many metals at a time include the ICP units. However, only the ICP/MS units are capable of similar low sensitivities as the electrothermal AAS units. These units are mostly still being used in research environments and are not typically used in production laboratories, as they require well-trained specialized operators and are the most costly alternative shown.

In flame AAS, a sample is aspirated directly into a flame (typically air–acetylene) and is atomized. A light beam (from a hollow cathode lamp designed for a specific wave length) is directed through the flame and into a monochromator, and finally into a detector. The detector measures the amount of light absorbed by the atomized element. The lamp operating at the specific wavelength of the metal makes the method relatively free from spectral and radiation interferences. However, different schemes (continuum-source, Zeeman, or Smith-Hieftje) to correct for molecular absorption and light scattering interferences are typically used.

Table E.5 Attributes of Metal Analysis Methods

	Flame AAS	Electrothermal AAS	Plasma ICP	Plasma ICP/MS	Anodic Stripping Voltammetry	X-Ray Fluorescence
Capital cost ($US)	10,000–30,000	25,000–80,000	40,000–80,000	150,000–250,000	8000–25,000	25,000–60,000
Operational cost[a]	Low	Moderate	Moderate–high	High	Low to moderate	Low
Sensitivity	Good	Very good	Poor–good	Very good	Excellent	Poor (solid matrices only)
Operation (number of metals at a time)	Single	Single–few	Many	Many	Few	Few
Sample throughput	High	High	High	Low	Moderate	Moderate
Ease of use	Good	Moderate	Good–moderate	Poor	Moderate–poor	Moderate
External sample preparation	Acid digestion	Acid digestion	Acid digestion	Acid digestion	Filtration	Possibly grind and sieve to obtain uniform particles

[a] Approximate operational costs, including expendable supplies (gases, acids, filters, graphite tubes, etc.), but not labor ($/sample): low: 3–10; moderate: 10–25; high: >25.

From Pitt, R., S. Mirov, K. Parmer, and A. Dergachev. Laser applications for water quality analyses, in *ALT'96 International Symposium on Laser Methods for Biomedical Applications*. Edited by V. Pustovoy. SPIE — The International Society for Optical Engineering. Volume 2965, pp. 70–82. 1997.

Cold-vapor AAS is used for very sensitive determinations of mercury. In this scheme, the sample (modified with H_2SO_4, HNO_3, $KMnO_4$, and $SnCl_2$ to volatilize the mercury) is purged with air, which is then directed into an absorption cell placed in the light pathway where the flame unit is normally located.

Electrothermal (graphite furnace) AAS is much more sensitive than flame AAS because it can place a much greater density of atoms in the light pathway. Contamination is therefore much more critical than with flame units. Electrothermal AAS is subject to more interferences than flame AAS and is only recommended for very low concentrations of metals. However, because of the relatively low concentrations of many heavy metals found in stormwater, especially the dissolved fraction, graphite furnace AAS (Figure E.10) is the preferred method in this area of research (using a suitable background corrector to minimize most interferences).

Inductively coupled plasma atomic emission spectroscopy uses a controlled plasma from argon gas ionized by an applied radio frequency. A sample aerosol is directed into the plasma, which is at an extremely high temperature (6000 to 8000 K). This results in almost complete dissociation of the metal molecules and significantly reduced chemical interferences compared to most other metal analyses techniques. Another important advantage of the ICP is the extremely wide dynamic range of the instrument, as shown in Table E.4. An emission light emitted from the sample and plasma combination is focused in a monochromator and is detected using a series of photomultipliers set at specific wavelengths for the elements of interest.

The ICP/MS uses a mass spectrophotometer to separate the analyte ions emitted by the plasma and sample mixture according to their mass-to-charge ratios. This results in a much more sensitive unit (comparable to the electrothermal AAS), and it can detect multiple elements simultaneously.

Anodic stripping voltammetry is rarely used in a production laboratory, but it is a relatively common research instrument (Figure E.11). ASV is one of the most sensitive metal analysis techniques, even more sensitive than electrothermal AAS. Cyclic ASV is also capable of identifying

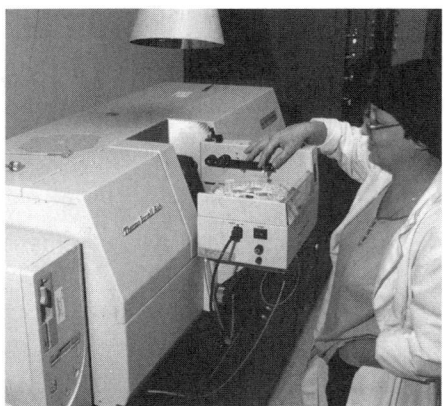

Figure E.10 Graphite furnace AAS used for storm-water analyses at the University of Alabama at Birmingham.

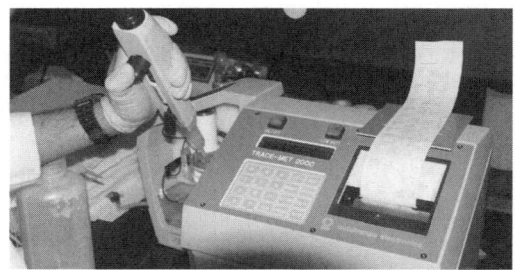

Figure E.11 Anodic stripping voltammeter (Outo-kompku) for heavy metal analyses.

different characteristics of the metals in the sample. The analyzer uses a three-step process. The first step typically plates a mercury film on a glossy carbon electrode. The second step plates the metals on the mercury film, and the third step strips the metals from the film as a function of increasing oxidizing potential. This last step allows the individual metals to be identified and quantified. Only metals that form an amalgam can be determined (such as cadmium, copper, lead, and zinc, metals of great interest in most environmental investigations). Because the instrument is so sensitive, great care must be taken to avoid contamination. Interferences may be caused by complexes that form between metals in the sample (such as between high concentrations of copper and zinc). ASV is especially well suited for analyzing heavy metals in saline waters (such as snowmelt) where graphite furnace procedures are subject to many interferences from the high salt concentrations.

X-ray fluorescence (Figure E.12) can also be used to detect heavy metals in solid samples, such as sediments and soils, including particulates trapped on filters (from water or air samples). The sample is irradiated with low-intensity X rays causing the elements in the sample to fluoresce. The emitted X rays from the irradiated sample are sorted by their energy level and are used to identify and quantify the metals of interest. Relatively little sample preparation is needed, especially for homogeneous samples. The technique is commonly used as a screening tool in the field to guide sampling for more accurate and sensitive laboratory analyses. Its relatively poor sensitivity limits its use for most environmental investigations, except for evaluating heavily contaminated sites.

Sample preparation is very critical for all of these metal analysis procedures. Typical sample preparation requires acid digestion using a combination of acids to reduce interferences by organic matter and to convert the metal associated with particulates (and colloids) to the free metal forms that can be detected. Nitric acid digestion with heat is adequate for most samples. However, hydrofluoric acid is also needed if the digestion is to completely release metals that may be tied up in a silica matrix. Unfortunately, hydrofluoric acid forms volatile compounds with some metals, resulting in their partial loss upon storage if not analyzed immediately. Almost all of the stormwater heavy metals can be released from the particulates using just nitric acid, especially considering metal losses from using a hydrofluoric acid digestion. A nitric acid and perchloric acid mixture may be needed to digest organic material in the samples. Microwave-assisted digestion (Figure E.13) has become more common recently because of improved metal recovery, much faster digestion, and better repeatability.

Figure E.12 X-ray fluorescence unit for analyses of heavy metals in solids.

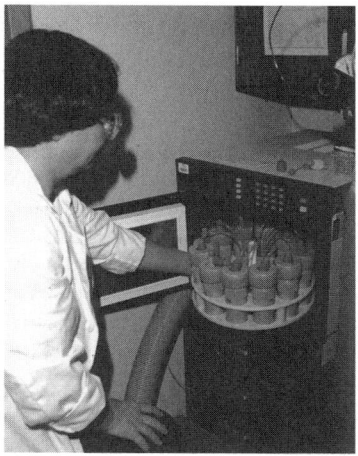

Figure E.13 Microwave digestion of stormwater samples for heavy metal analyses.

STORMWATER SAMPLE EXTRACTIONS FOR EPA METHODS 608 AND 625

The following paragraphs outline the modified organic extraction methods that have been used by UAB for the analysis of wet-weather flows (Pitt and Clark 1999). These modifications are necessary because of the large amount of particulates in the samples and the large particulate fraction of the organics of greatest interest. These particulates interfere with solid-phase extraction procedures, for example, resulting in very little recovery of organic toxicants using that method.

1. Samples are extracted using a liquid–liquid separatory funnel technique. This has been found to give the most reliable results, especially compared to solid phase extraction or critical fluid extraction methods, for stormwater samples (and most surface water samples). The problem with stormwater organics is that a substantial fraction of many of the organic compounds of interest are associated with particulates. This particulate fraction needs to be quantified, as stormwater has been shown to have significant effects on receiving water sediments. If emulsions prevent achieving acceptable solvent recovery with separatory funnel extraction, continuous extraction is used. The separatory funnel extraction scheme described below assumes a sample volume of 250 mL. Serial extraction of the base/neutrals uses 10 mL additions of methylene chloride, as does the serial extraction of the acids. Prior to the extraction, all glassware is oven baked at 300°C for 24 hours.

2. A sample volume of 250 mL is collected in a 400-mL beaker and poured into a 500-mL glass separation funnel. For every 12 samples extracted, an additional four samples are extracted for quality control and quality assurance. These include three 250-mL composite samples made of equal amounts of the 12 samples, and one 250-mL sample of reverse osmosis water. Standard solution additions consisting of 25 μL of 1000 μg/mL base/neutral spiking solution, 25 μL of 1000 μg/mL base/neutral surrogates, 12.5 μL of 2000 μg/mL acid spiking solution, and 12.5 μL of 2000 μg/mL acid surrogates are made to the separation funnels of two of the three composite samples and mixed well. Sample pH is measured with wide-range pH paper and adjusted to pH > 11 with sodium hydroxide solution.

3. A 10-mL volume of methylene chloride is added to the separatory funnel and sealed by capping. The separatory funnel is gently shaken by hand for 15 s and vented to release pressure (Figure E.14). The cap is removed from the separatory funnel and replaced with a vented snorkel stopper. The separatory funnel is then placed on a mechanical shaker and shaken for 2 min. After returning the separatory funnel to its stand and replacing the snorkel stopper with the cap, the organic layer is allowed to separate from the water phase for a minimum of 10 min, longer if an emulsion develops

(Figure E.15). The extract and any emulsion present is then collected into a 125-mL Erlenmeyer flask (Figure E.16).

4. A second 10-mL volume of methylene chloride is added to the separatory funnel, and the extraction method is repeated, combining the extract with the previously collected extract in the Erlenmeyer flask. For persistent emulsions, those with emulsion interface between layers more than one third the volume of the solvent layer, the extract including the emulsion is poured into a 50-mL centrifuge vial, capped, and centrifuged at 2000 rpm for 2 min to break the emulsion (Figures E.17 and E.18). Water phase separated by the centrifuge is collected from the vial and returned to the separatory funnel using a disposable pipette. The centrifuge vial with the extract is recapped before performing the extraction of the acid portion.

5. The pH of the remaining sample in the separatory funnel is adjusted to pH < 2 using sulfuric acid. The acidified aqueous phase is serially extracted twice with 10-mL aliquots of methylene chloride, as in the previous base/neutral extraction procedure. Extract and any emulsions are again collected in the 125-mL Erlenmeyer flask.

6. The base/neutral extract is poured from the centrifuge vial though a drying column of at least 10 cm of anhydrous sodium sulfate and is collected in a 50-mL beaker (Figure E.19). The Erlenmeyer flask is rinsed with 5 mL of methylene chloride, which is then used to rinse the centrifuge vial and then to rinse the drying column and complete the quantitative transfer.

Figure E.14 Initial hand shaking the separatory funnel and venting gas.

Figure E.15 Separation of organic solvent extract from water sample.

Figure E.16 Collecting solvent extract and emulsion after separation.

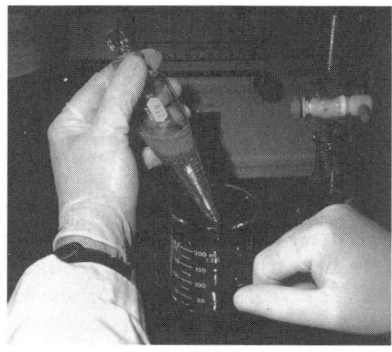

Figure E.17 Extract in centrifuge vial.

7. The base/neutral extract is transferred into a 50-mL concentration vial and is placed in an automatic vacuum/centrifuge concentrator from Savant (Figure E.20). (Vacuum concentration is used in place of the Kuderna–Danish method; Figure E.21.) Extract is concentrated to approximately 0.5 mL.

8. The acid extract collected in the 125-mL Erlenmeyer flask is placed in the 50-mL centrifuge vial. Again, if emulsions persist, the extract is centrifuged at 2000 rpm for 2 min. Water is drawn from the extract and discarded. Extract is poured through the 10 cm anhydrous sodium sulfate drying column and collected in the 50-mL beaker as before. The Erlenmeyer flask is then rinsed with 5 mL of methylene chloride, which is then poured into the centrifuge vial and finally through the drying column.

9. The acid extract is then poured into the 50-mL concentration vial combining it with the evaporated base/neutral extract. The combined extract is then concentrated to approximately 0.5 mL in the automatic vacuum/centrifuge concentrator.

10. Using a disposable pipette, extract is transferred to a graduated Kuderna–Danish concentrator. Approximately 1.5 mL of methylene chloride is placed in the concentration vial for rinsing. This rinse solvent is then used to adjust the volume of extract to 2.0 mL. Extract is then poured into a labeled Teflon-sealed screw-cap vial and freezer stored until analysis (Figure E.22).

Notes for method 608: under the alkaline conditions of the extraction step, α-BHC, γ-BHC, endosulfan I and II, and endrin are subject to partial decomposition. Florisil cleanup is not utilized unless the sample matrix creates excessive background interference.

When sediments are being analyzed for organic compounds, we use a semiautomated method in place of the traditional Soxlet extraction method. A Dionex ASE (accelerated solvent extractor) (Figure E.23) is used to extract organic compounds from the sediment, while an OI gel permeation chromatograph (Figure E.24) is used to clean up the extracts.

Figure E.18 Extract placed in centrifuge.

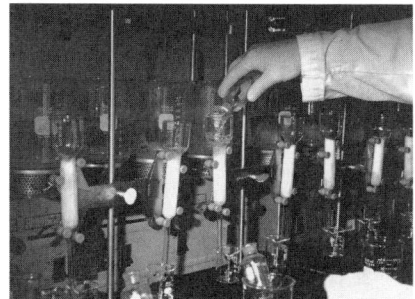

Figure E.19 Drying columns containing anhydrous sodium sulfate.

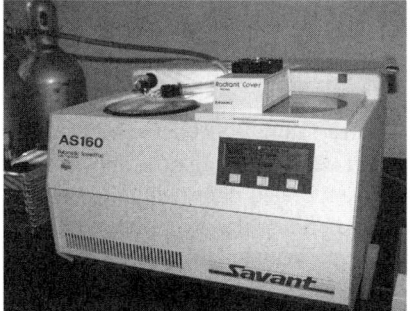

Figure E.20 Automatic vacuum/centrifuge concentrator (Savant AS 160).

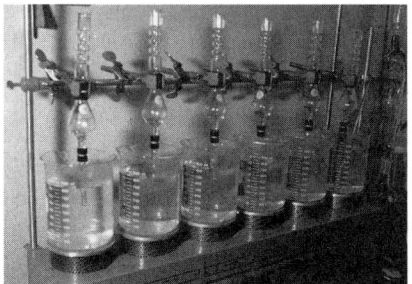

Figure E.21 Alternative micro Kuderna–Danish concentration method.

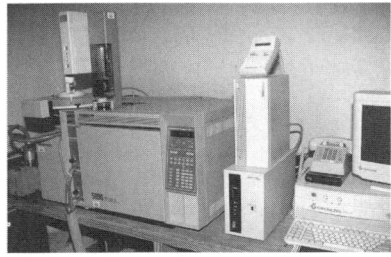

Figure E.22 GC/MSD used for organic analyses.

Figure E.23 Dionex ASE for automatic extractions of organics from sediment samples.

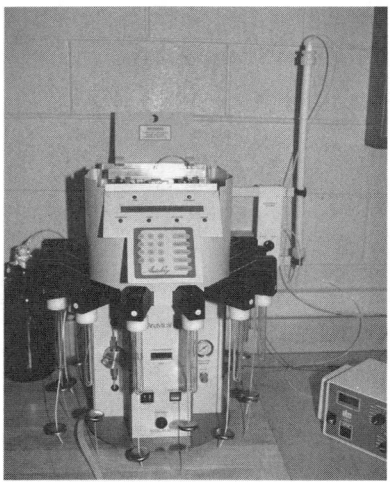

Figure E.24 OI GPC used to clean sediment extracts.

CALIBRATION AND DEPLOYMENT SETUP PROCEDURE FOR YSI 6000UPG WATER QUALITY MONITORING SONDE

This discussion on calibration and deployment setup procedures for the YSI 6000 is presented here due to the reliance on this water quality monitoring sonde for many different applications presented in this book. This discussion was prepared by John Easton, Ph.D. candidate, University of Alabama at Birmingham, who has used this equipment extensively during his research. These procedures are therefore a compilation of the instructions given by YSI, in addition to his field and lab experience with this equipment.

The YSI 6000upg Environmental Monitoring System is a multiparameter, water quality measurement and data logging system. It is intended for use in research, assessment, and regulatory compliance applications. This instrument, or sonde, is ideal for profiling and monitoring water conditions in lakes, rivers, wetlands, estuaries, coastal waters, and monitoring wells. It can be left unattended for weeks at a time with measurement parameters sampled at a user-defined setup interval and data saved securely in the unit's internal memory. The Model 6000upg is designed to house four field-replaceable probes (six sensors) and a depth sensor module in the sonde body. The 6000upg communicates with a computer with a terminal emulation program, or via the Ecowatch for Windows software. The data is easily exported to any spreadsheet program for sophisticated data analysis. The unit operates on eight C-size alkaline batteries. Depending upon the activated sensor configuration and frequency of data collection, the unit can provide up to 90 days of battery life.

The Environmental Research Area at UAB has four 6000upgs configured to collect the following measurement parameters: dissolved oxygen, conductivity, specific conductance, salinity, total dissolved solids, resistivity, temperature, pH, ORP, depth, level, and turbidity. Table E.6 gives the reported performance specifications for each sensor.

This method details how to calibrate the sonde for the following measurement parameters: specific conductivity, dissolved oxygen, depth, pH, and turbidity for freshwater monitoring, plus routine maintenance of the DO and conductivity probes. The temperature and ORP probes require no calibration, but should be checked against known standards. This method also describes how to configure the sonde for unattended deployment or sampling.

All calibration standards should be prepared fresh, and this procedure should be done at approximately 25°C. The following lists the materials and supplies needed for calibrations:

Materials
- One or more containers to hold calibration standards. YSI calibration cup or 800-mL beaker
- Large (5-gallon) bucket filled with tap water for rinsing the sonde between calibration solutions

Table E.6 Performance Specifications and Sensor Types in the YSI 6000 Sonde

Parameter	Sensor Type	Range	Accuracy	Resolution
Dissolved oxygen % saturation	Rapid Pulse – Clark-type, polarographic	0–200% air saturation	±2% air saturation	0.1% air saturation
Conductivity[a]	4 electrode cell with autoranging	0–100 mS/cm	±0.5% of reading + 0.001 mS/cm	0.01 mS/cm
Temperature	Thermistor	−5–45°C	±0.15°C	0.01°C
pH	Glass combination electrode	2–14 units	±0.2 units	0.01 units
ORP	Platinum ring	−999–999 mV	±20 mV	0.1 mV
Turbidity	Optical, 90° scatter, mechanical cleaning	0–1000 NTU	±5%	0.1 NTU
Depth — Medium	Stainless steel strain gauge	0–61 m	±0.12 m	0.001 m
Depth — Shallow	Stainless steel strain gauge	0–9.1 m	±0.06 m	0.001 m

[a] Report outputs of specific conductance (conductivity corrected to 25°C)

- Volumetric flasks, graduated cylinders, pipette, and pipette tips for preparation of calibration solutions
- Barometer. NOTE: *Remember that barometer readings which appear in meteorological reports are generally corrected to sea level and are not useful for your calibration procedure unless they are uncorrected and at the elevation and location of the sonde.*
- Dissolved oxygen probe maintenance kit, contains: O-rings, DO membranes, pencil eraser (or very fine sandpaper), electrode filling solution
- Several clean, absorbent paper towels or cotton cloths for drying the sonde between rinses and calibration solutions
- Computer (with Ecowatch software), connection cable for interfacing computer with sonde, AC power supply, and eight C-size alkaline batteries
- Allen wrench for removing sonde guard and battery compartment cover

Reagents
- Deionized water (diH$_2$O)
- pH buffers: 7.00, 4.01, and/or 10.01 (either 4.01 or 10.01, in addition to the 7.00 solution is suitable for two-point calibration)
- Conductivity standard, e.g., NaCl solution at 16,640 µS/cm @ 25°C
- Turbidity standard, e.g., Formazin solution at 4000 NTU

Initial Calibration Procedure
- Remove sonde guard
- Check to see if DO electrode is bright silver; if not, clean by gently rubbing with the pencil eraser. Clean eraser particles off probe completely. Fill probe well with filling solution and replace membrane. Put probe guard back onto sonde.
- Connect computer to sonde and connect sonde to external AC power supply

Conductivity Probe Calibration
- Prepare conductivity standard. Use a 1 mS/cm (1000 µS/cm) standard if the sonde is to be deployed in fresh water. For example, dilute typically available 16.640 mS/cm standard solution 1:16.64 with diH$_2$O (to prepare 500 mL, add 30 mL of 16.640 mS/cm standard and QS to 500 mL with diH$_2$O).
- Decant 1 mS/cm solution into calibration cup and immerse sonde into cup.
- Launch Ecowatch software. Open communications with sonde, and type "menu." From the sonde main menu select **2. Calibrate**. From the calibrate menu, select **1. Conductivity** to access the conductivity calibration procedure and then **1. SpCond** to access the specific conductance calibration procedure. Enter the calibration value of the standard you are using (1.000 mS/cm at 25°C) and press ENTER.
- The current values of all enabled sensors will appear on the screen and will change with time as they stabilize. Observe the readings under SpCond and when they show no significant change for approximately 30 s, press ENTER.
- The screen will indicate that the calibration has been accepted and prompt you to HIT ANY KEY to return to the **Calibrate** menu.

- If you receive an error message indicating that the calibration is out of range, assure yourself that the calibration solution was prepared correctly. If it was, remove sonde guard, and using small brush (located in pocket in user's manual) clean out the channel on the conductivity probe. BE GENTLE. Replace sonde guard and repeat calibration steps.
- Rinse the sonde in tap or purified water and dry.

DO Probe (and depth) Calibration

- Place approximately ⅛-in (3 mm) of water or a saturated sponge in the bottom of the calibration cup. Make sure the DO and temperature probes are *not* immersed in the water. Wait approximately 10 minutes for the air in the cup to become water saturated. NOTE: *if the transport cup is used, make certain that the cup is vented to the atmosphere by loosening the vent screw.*
- From the **Calibrate** menu, select **2. DO%** to access the DO% calibration procedure.
- Enter the current barometric pressure in mm Hg. (inches of Hg × 25.4 = mm Hg).
- Press ENTER and the computer will indicate that the calibration procedure is in progress.
- After approximately 1 min, the calibration will be complete. Press any key as instructed, and the screen will display the percent saturation value which corresponds to your local barometric pressure input. For example, if your local barometer reads 742 mm Hg, the screen will display 97.6% (742/760) at this point. If an error message is received, proceed to the diagnostics step; otherwise, press any key to return to the **Calibrate** menu (and skip the following diagnostic step).
- If an error message was received, conduct a diagnostics test. From the **Main** menu, chose **8. Diagnostics**. Check the DO charge. This value should read between 25 and 75 during calibration. If out of this range, then the probe needs to be cleaned (pencil eraser) or replaced. After cleaning, repeat the above DO calibration procedure.
- Following the DO calibration, leave the sonde in water-saturated air. From the **Calibrate** menu, select **3. Depth** to access the depth calibration procedure.
- Input 0.00 or some known sensor offset in feet. (The depth sensor is about 0.46 ft above the bottom of the probe compartment, and this offset value could be used if installing the unit vertically and depth in relation to the sonde bottom was desired.) Press ENTER and monitor the stabilization of the depth readings with time.
- When no significant change occurs for approximately 30 s, press ENTER to confirm the calibration and zero the sensor with regard to the current barometric pressure.
- Press any key to return to the **Calibrate** menu.

pH Probe Calibration

- Place approximately 400 mL of pH 7 buffer in a clean calibration cup. Allow at least 1 min for temperature equilibrium before proceeding.
- Immerse probe into solution. From the **Calibrate** menu, select **6. pH** to access the pH calibration choices and then **2. 2-Point**.
- Press ENTER and input the value of the buffer (7.00) at the prompt. Press ENTER, and observe the values under pH until the readings are stable for 30 s.
- Press ENTER. The display will indicate that the calibration is accepted. (If an error message is received, repeat with fresh buffer.)
- Press any key to continue.
- Rinse the sonde in water and dry before proceeding.
- Place approximately 400 mL of a second pH buffer solution in a clean calibration cup. The second buffer might be pH 4.01 if the monitored water is expected to be acidic, or pH 10.01 if the monitored water is expected to be basic. Allow at least 1 min for temperature equilibrium before proceeding.
- Press ENTER and input the value of the second buffer (4.01 or 10.01) at the prompt. Press ENTER, and observe the values under pH until the readings are stable for 30 s.
- Press ENTER. After the second value calibration is complete, press any key to return to the **Calibrate** menu.
- Rinse the sonde in water and dry before proceeding.

Turbidity Probe Calibration

- Prepare 100 NTU solution. Dilute 4000 NTU formazin solution 1:40 with diH$_2$O (pipette 25 mL of 4000 NTU formazin solution into 1-L volumetric flask and qs to 1 L). **Formazin is a hazardous material, and special care needs to be taken. Read and follow all precautions.**

- Select **9. Turbidity** from the **Calibrate** menu, and then **2. 2-Point**.
- To begin the calibration, immerse the sonde in approximately 300 mL of 0 NTU standard (clear, deionized water), and press ENTER.
- Input the value 0.00 NTU at the prompt, and press ENTER.
- After calibration of the mechanical wiper speed, the screen will display real-time readings, which will allow you to determine when turbidity values have stabilized. If the readings appear unusually high or low or are unstable, there are probably bubbles on the optical surface. Activate the mechanical wiper by pressing the "3" key to remove the bubbles.
- After stable readings are observed for approximately 40 s, press ENTER to confirm the first calibration. Press any key to continue.
- Dry the sonde and probes carefully and then place the sonde in approximately 300 mL of the second turbidity standard (100 NTU). Input the value 100.0 NTU, press ENTER, and view the stabilization of the values on the screen.
- As described previously, if the readings appear unusually high or low or are unstable, activate the wiper to remove bubbles and be sure to wait 40 s before confirming the calibration.
- After the readings have stabilized, press ENTER to confirm the calibration. Press any key to return to the **Calibrate** menu. Input "0" to return to the **Main** menu.
- Proceed to the deployment setup procedure.

Deployment Setup Procedure (for unattended monitoring)

- Unplug the AC power source, and continue this procedure using the sonde's internal (battery) power.
- Select **1. Run** from the sonde **Main** menu. The **Run** menu will be displayed.
- Select **3. Unattended** sample from the **Run** menu.
- The current time and date, all active sensors, battery voltage, and free flash disk space will be displayed.
- Note: if the current time and date are not correct, your unattended sampling study will not begin or end when you desire. To correct the time and date, see Section 2.5 in the instruction manual.
- You will be asked to enter the starting date. Use the following format: XX/XX/XX. For example to start on 1 January, 1999, enter "01/01/99."
- Enter the starting time. Use the following format: XX/XX/XX. You must include not only hours and minutes, but seconds. For example, if you want to start a study at 8 AM, you must enter "08:00:00."
- Enter the study duration in days. For example, for a 2-week study, enter "14."
- Enter interval in minutes. For example, to collect data every 15 minutes, enter "15."
- Enter the site description.
- You will be asked if all start-up information is correct. Check the information carefully (especially the estimated battery life) and, if you want to change something, press "N." If all information is correct, press "Y." The following message will be displayed briefly: *INSTRUMENT IS IN UNATTENDED MODE*.
- Continue to press "zero" until the Ecowatch software breaks communication with the sonde (after exit from the **Main** menu).
- Remove the communication cable from the sonde and screw on the waterproof connector cap. The sonde is now ready for deployment.

REFERENCES

Day, J. *Selection of Appropriate Analytical Procedures for Volunteer Field Monitoring of Water Quality.* MSCE thesis, Department of Civil and Environmental Engineering, University of Alabama at Birmingham. 1996.

Pitt, R.E., R.I. Field, M.M. Lalor, D.D. Adrian, and D. Barbe. *Investigation of Inappropriate Pollutant Entries into Storm Drainage Systems: A User's Guide.* Rep. No. EPA/600/R-92/238, NITS Rep. No. PB93-131472/AS, U.S. EPA, Storm and Combined Sewer Pollution Control Program (Edison, NJ), Risk Reduction Engineering Lab., Cincinnati, OH. 1993

Pitt, R., S. Mirov, K. Parmer, and A. Dergachev. Laser applications for water quality analyses, in *ALT'96 International Symposium on Laser Methods for Biomedical Applications.* Edited by V. Pustovoy. SPIE — The International Society for Optical Engineering. Vol. 2965, pp. 70–82. 1997.

Pitt, R. and S. Clark. *Communication Manhole Water Study: Characteristics of Water Found in Communications Manholes*. Final Draft. Office of Water, U.S. Environmental Protection Agency. Washington, D.C.
July 1999.

Standard Methods for the Examination of Water and Wastewater. 19th edition. Water Environment Federation.
Washington, D.C. 1995.

Sampling Requirements for Paired Tests*

* From R. Pitt and K. Parmer. *Quality Assurance Project Plan (QAPP) for EPA Sponsored Study on Control of Stormwater Toxicants*. Department of Civil and Environmental Engineering, University of Alabama at Birmingham. 1995.

Number of Sample Pairs Needed
(Power = 0.5 Difference = 10%)

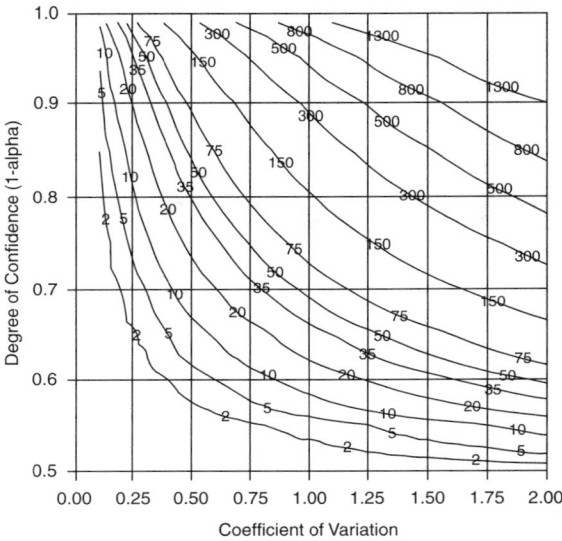

Number of Sample Pairs Needed
(Power = 0.8 Difference = 10%)

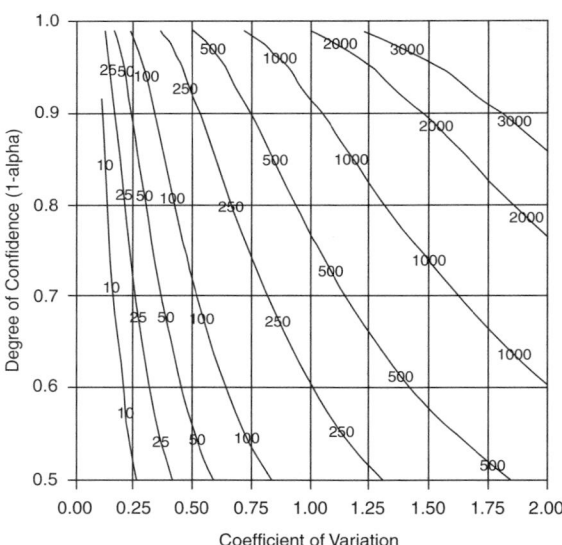

Figure F.1

Number of Sample Pairs Needed
(Power = 0.9 Difference = 10%)

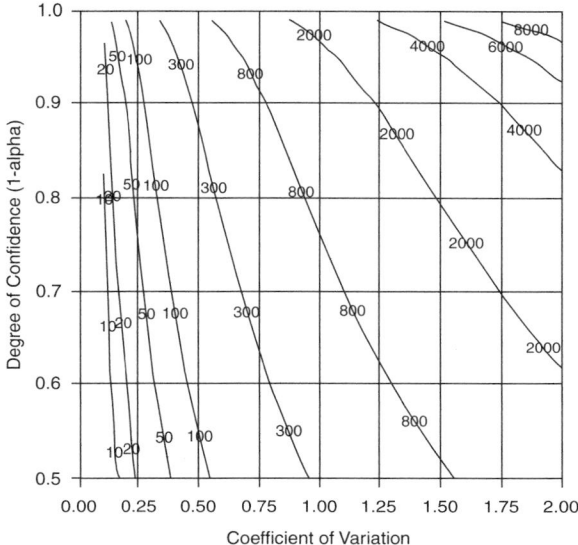

Number of Sample Pairs Needed
(Power = 0.5 Difference = 25%)

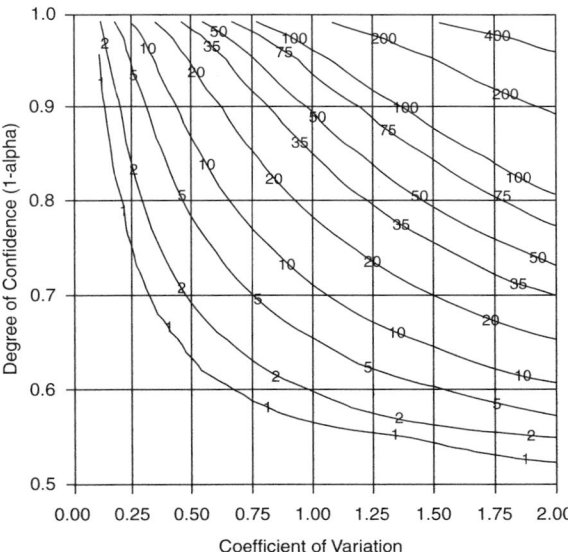

Figure F.2

Number of Sample Pairs Needed
(Power = 0.8 Difference = 25%)

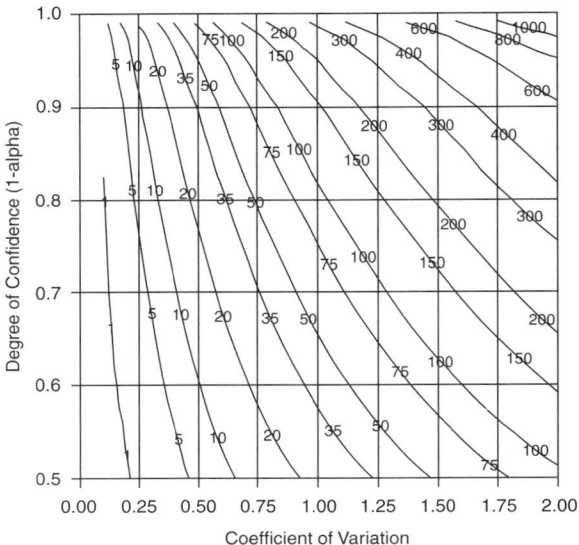

Number of Sample Pairs Needed
(Power = 0.9 Difference = 25%)

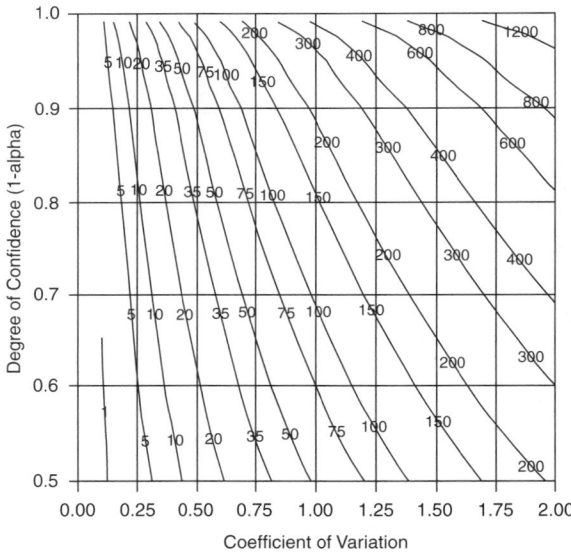

Figure F.3

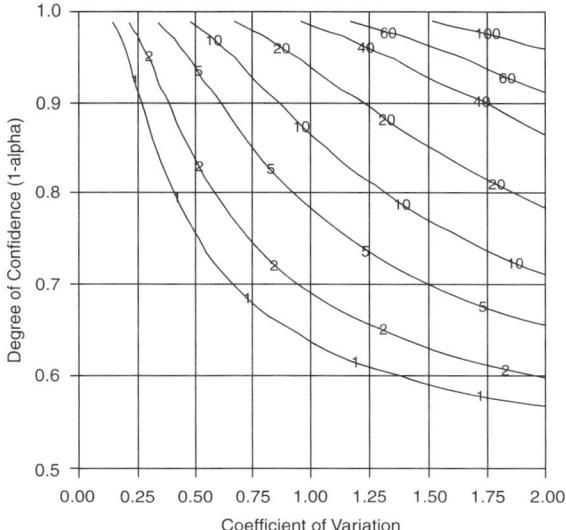

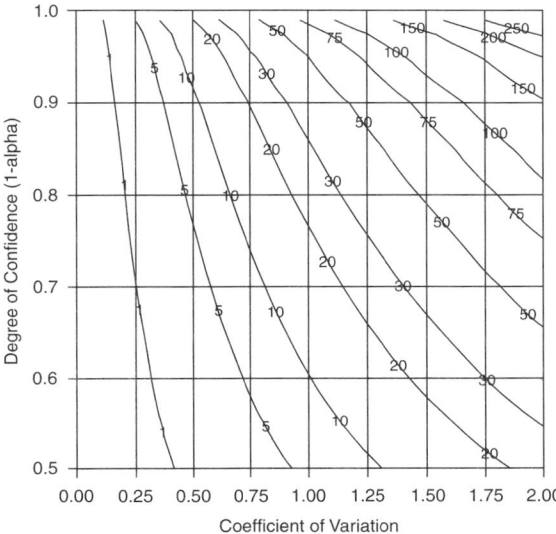

Figure F.4

**Number of Sample Pairs Needed
(Power = 0.9 Difference = 50%)**

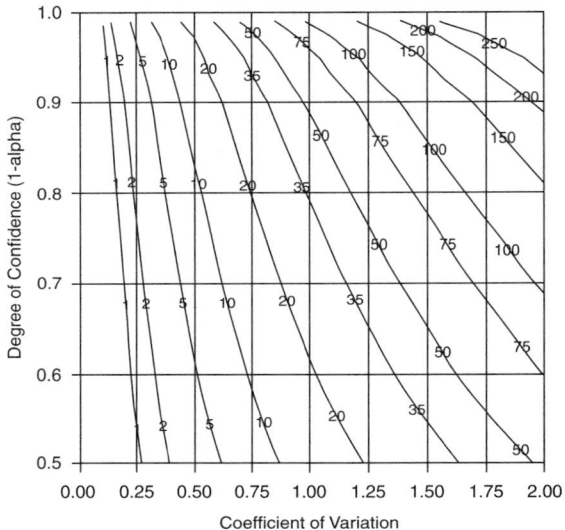

**Number of Sample Pairs Needed
(Power = 0.5 Difference = 75%)**

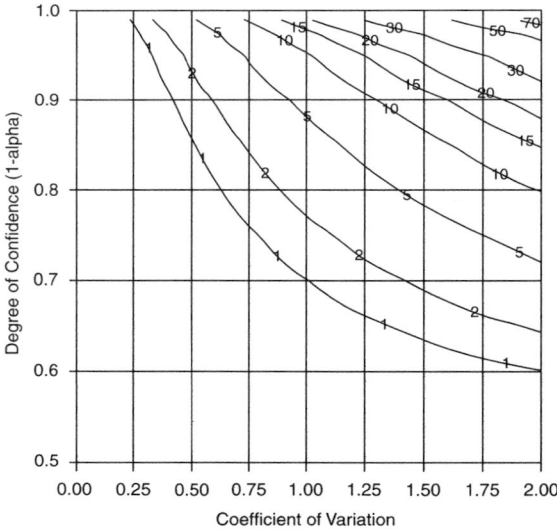

Figure F.5

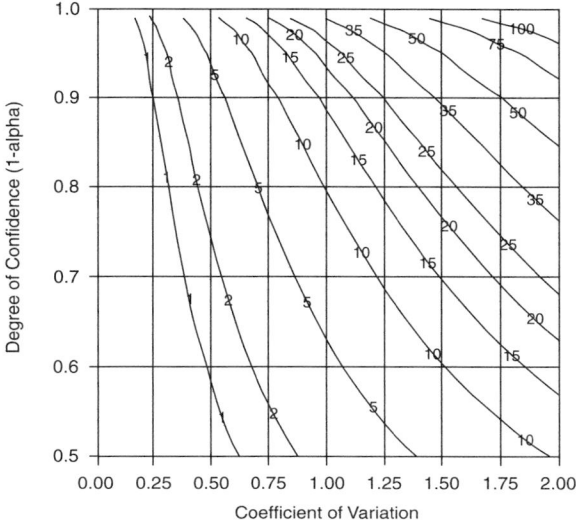

**Number of Sample Pairs Needed
(Power = 0.8 Difference = 75%)**

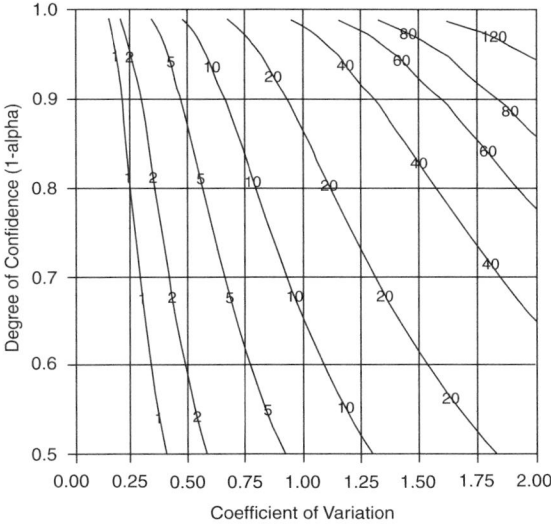

**Number of Sample Pairs Needed
(Power = 0.9 Difference = 75%)**

Figure F.6

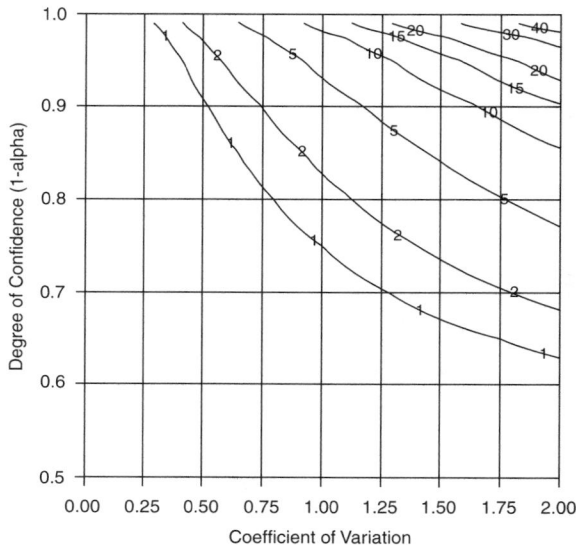

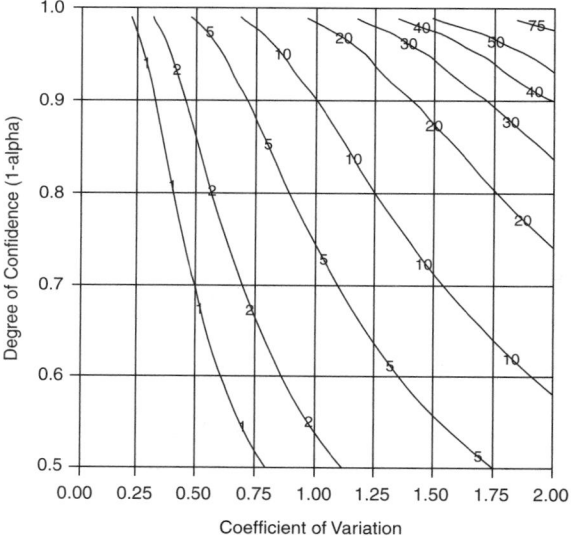

Figure F.7

Number of Sample Pairs Needed
(Power = 0.9 Difference = 95%)

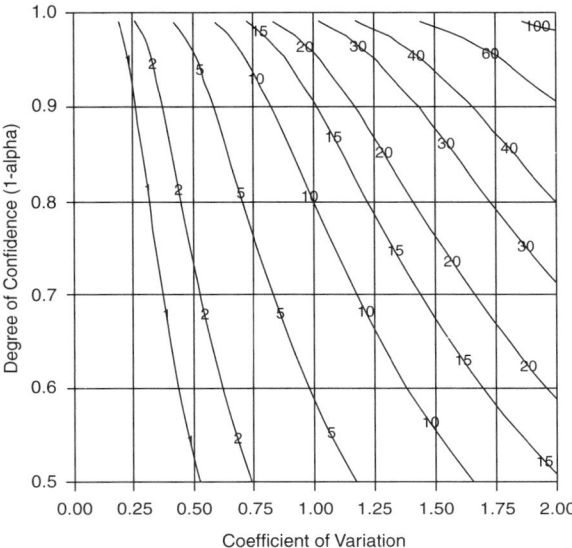

Number of Sample Pairs Needed
(Power = 90% Confidence = 95%)

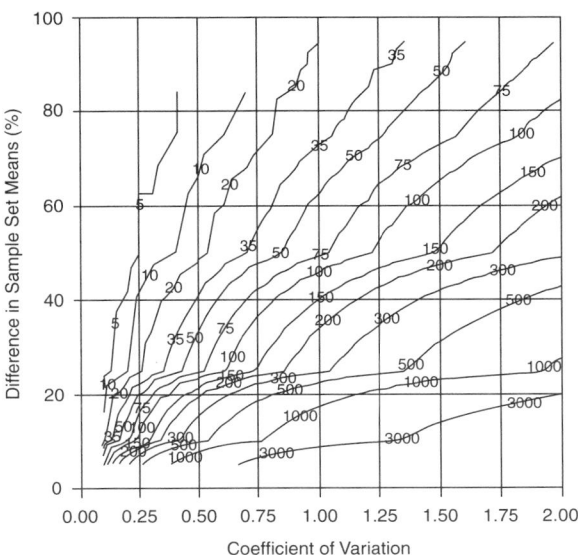

Figure F.8

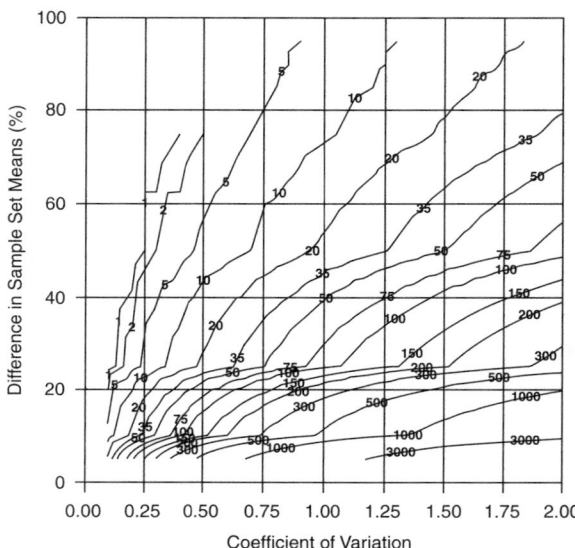

Figure F.9

Water Quality Criteria

CONTENTS

Introduction 798
EPA's Water Quality Criteria and Standards Plan — Priorities for the Future 798
Compilation of Recommended Water Quality Criteria and EPA's Process
    for Deriving New and Revised Criteria 799
Ammonia 813
    National Ammonia Water Quality Criteria 815
Bacteria 816
    Development of Bathing Beach Bacteriological Criteria 816
    Bacteria Criteria for Water Contact Recreation 820
Chloride, Conductivity, and Total Dissolved Solids 822
    Human Health Criteria for Dissolved Solids 822
    Aquatic Life Criteria for Dissolved Solids 823
Chromium 823
    Aquatic Life Effects of Cr^{3+} 823
    National Freshwater Aquatic Life Criteria for Cr^{3+} 824
    Human Health Criteria for Chromium 824
Copper 824
    Effects of Copper on Aquatic Life 824
    National Aquatic Life Criteria for Copper 825
    Human Health Criteria for Copper 825
Hardness 825
Hydrocarbons 826
Lead 827
    Aquatic Life Summary for Lead 827
    National Aquatic Life Criteria for Lead 828
    Human Health Criteria for Lead 828
Nitrate and Nitrite 828
    Human Health Nitrate and Nitrite Criteria 829
    Nitrate and Nitrite Aquatic Life Criteria 829
Phosphate 830
    Aquatic Life Summary for Phosphate 831
pH 832
    pH Aquatic Life Effects and Criteria 833

Suspended Solids and Turbidity 834
    Water Quality Criteria for Suspended Solids and Turbidity 835
Zinc...835
    Aquatic Life Criteria for Zinc..835
    Human Health Criteria for Zinc...836
Sediment Guidelines 836
References 839

INTRODUCTION

One of the most confusing aspects of conducting a receiving water study is attempting to compare acquired water quality data to appropriate standards and criteria. In many cases, available data have been obtained haphazardly without specific project objectives in mind. Inappropriate constituents also may have been measured, based more on convenience (and expense) than usefulness. The user is then left with trying to understand if a problem exists and determining the extent of the problem. This book has emphasized the need for careful experimental design (with clear objectives) and the need for a multidisciplinary approach in receiving water studies.

In all cases, the user will still need to compare acquired data with some type of objective. As stated in Chapter 8, however, care must be taken when comparing measured values with available criteria. In addition, many of the most commonly measured constituents (such as turbidity, Secchi disk transparency, and specific conductivity) are not directly comparable to water quality criteria, and are best evaluated through long experience at a monitoring location and through comparisons with observations obtained at reference sites. Finally, Chapter 8 (and elsewhere) lists reasons why water quality criteria are not directly applicable to stormwater-related conditions. Nevertheless, water quality criteria are important tools that cannot be overlooked. If measured conditions exceed established criteria, then problems may occur, requiring that the conditions be investigated further. However, the most serious problem associated with water quality criteria applied to stormwater is the likelihood of false negative conclusions, based on the observation of no, or few, exceedances. As noted elsewhere, problems caused by stormwater in receiving waters may more likely be associated with habitat disturbances and contaminated sediment than by elevated water quality concentrations. In addition, few receiving water studies include broad representations of toxicants and conventional pollutants, especially in sufficient numbers and sampling frequencies, to make statistically valid comparisons with the criteria.

The following sections of this appendix summarize U.S. Environmental Protection Agency water quality standards and criteria for selected constituents of concern when conducting a receiving water investigation. These criteria and standards are subject to periodic change, and it is important to review the most current listing from the EPA at: http://www.epa.gov/OST/standards. Much of the background discussion in this Appendix is summarized directly from EPA (1986b).

EPA'S WATER QUALITY CRITERIA AND STANDARDS PLAN
— PRIORITIES FOR THE FUTURE

In September 1998, the EPA announced a plan (URL: http://www.epa.gov/OST/standards/planfs.html) for working together with the states and tribes to enhance and improve the water quality criteria and standards program across the country. This plan describes new criteria and standards program initiatives that EPA and the states and tribes will take over the next decade. The development and implementation of criteria and standards will provide a basis for enhancements to the total maximum daily load (TMDL) program, National Pollutant Discharge Elimination System (NPDES) permitting, nonpoint source control, wetlands protection, and other water resources management efforts.

The EPA's Office of Water will emphasize and focus on the following priority areas for the Criteria and Standards Program over the next decade:

- Developing nutrient criteria and assessment methods to better protect aquatic life and human health
- Developing criteria for microbial pathogens to better protect human health during water recreation
- Completing the development of biocriteria as an improved basis for aquatic life protection
- Maintaining and strengthening the existing ambient water quality criteria for water and sediments
- Evaluating possible criteria initiatives for excessive sedimentation, flow alterations, and wildlife
- Developing improved water quality modeling tools to better translate water quality standards into implementable control strategies
- Ensuring implementation of these new initiatives and improvements by the states and tribes in partnership with EPA

Over the past two decades, state and tribal water quality standards and water quality-based management approaches have relied upon aquatic life use designations and protective criteria based primarily upon narrative, chemical-specific, and whole-effluent toxicity methodologies. Using these approaches, much progress has been made. However, not all of the nation's waters have achieved the Clean Water Act goal of "fishable and swimmable," and significant water pollution problems still exist. The EPA concludes that there is an essential need for improved water quality standards. Adding nutrient criteria and biological criteria to the water quality criteria and standards program ensures further improvements in maintaining and restoring aquatic life. Improved human health criteria will better protect against bioaccumulative pollutants, and new microbial pathogen controls will better protect human health (especially that of children) during water-related recreation. Better tools are also needed for controlling excessive sedimentation, flow alterations, and for protecting wildlife.

COMPILATION OF RECOMMENDED WATER QUALITY CRITERIA AND EPA'S PROCESS FOR DERIVING NEW AND REVISED CRITERIA

Section 304(a) of the Clean Water Act, 33 U.S.C. 1314(a)(1), requires the EPA to publish and periodically update ambient water quality criteria. These criteria are to "… accurately reflect the latest scientific knowledge … on the kind and extent of all identifiable effects on health and welfare including, but not limited to, plankton, fish, shellfish, wildlife, plant life … which may be expected from the presence of pollutants in any body of water. …" Water quality criteria developed under section 304(a) are based solely on data and scientific judgments on the relationship between pollutant concentrations and environmental and human health effects. These recommended criteria provide guidance for states and tribes in adopting water quality standards under section 303(c) of the CWA. The compilation was published in the *Federal Register* and can be accessed on the Office of Science and Technologies Home-page: http://www.epa.gov/OST/

The following tables are from the April 1999 compilation report (EPA 822-Z-99-001). In these tables, CMC refers to the "criterion maximum concentration" with an exposure period of 1 hour (generally corresponding to the earlier "acute" criterion), and CCC refers to the "criterion continuous concentration" with an averaging period of 4 days (generally corresponding to the earlier "chronic" criterion). "Freshwater" and "saltwater" refer to aquatic life uses in these waters.

Following these tables are discussions for many constituents of concern when conducting a receiving water investigation. These discussions, which briefly outline specific problems associated with different concentrations of the pollutants, are mostly from the 1986 EPA *Water Quality Criteria* report. Some of the criteria have been modified since that time, specifically for ammonia and bacteria, and those discussions have been modified to reflect these newer guidelines.

U.S. Recommended Water Quality Criteria for Priority Toxic Pollutants

#	Priority Pollutant	CAS Number	Freshwater CMC (µg/L)	Freshwater CCC (µg/L)	Saltwater CMC (µg/L)	Saltwater CCC (µg/L)	Human Health Water + Organism (µg/L)	Human Health Organism Only (µg/L)	Federal Register Cite/Source
1	Antimony	7440360					14[B,Z]	4300[B]	57FR60848
2	Arsenic	7440382	340[A,D,K]	150[A,D,K]	69[D,bb]	36[A,D,bb]	0.018[C,M,S]	0.14[C,M,S]	62FR42160 / 57FR60848
3	Beryllium	7440417					J,Z	J	62FR42160
4	Cadmium	7440439	4.3[D,E,K]	2.2[D,E,K]	42[D,bb]	9.3[D,bb]	J,Z	J	62FR42160
5a	Chromium III	16065831	570[D,E,K]	74[D,E,K]			J,Z Total	J	EPA820/B-96-001 / 62FR42160
5b	Chromium VI	18540299	16[D,K]	11[D,K]	1,100[D,bb]	50[D,bb]	J,Z Total	J	62FR42160
6	Copper	7440508	13[D,E,K,cc]	9.0[D,E,K,cc]	4.8[D,cc,ff]	3.1[D,cc,ff]	1300[U]		62FR42160
7	Lead	7439921	65[D,E,bb,gg]	2.5[D,E,bb,gg]	210[D,bb]	8.1[D,bb]	J	J	62FR42160
8	Mercury	7439976	1.4[D,K,hh]	0.77[D,K,hh]	1.8[D,ee,hh]	0.94[D,ee,hh]	0.050[B]	0.051[B]	62FR42160
9	Nickel	7440020	470[D,E,K]	52[D,E,K]	74[D,bb]	8.2[D,bb]	610[B]	4,600[B]	62FR42160
10	Selenium	7782492	L,R,T	5.0[T]	290[D,bb,dd]	71[D,bb,dd]	170[Z]	11,000	62FR42160
11	Silver	7440224	3.4[D,E,G]		1.9[G,G]				IRIS 09/01/91
12	Thallium	7440280					1.7[B]	6.3[B]	62FR42160
13	Zinc	7440666	120[D,E,K]	120[D,E,K]	90[D,bb]	81[D,bb]	9100[U]	69,000[U]	57FR60848
14	Cyanide	57125	22[K,Q]	5.2[K,Q]	1[D,bb]	1[D,bb]	700[B,Z]	220,000[B,H]	62FR42160 / IRIS 10/01/92
15	Asbestos	1332214					7 million fibers/L[L]		EPA820/B-96-001
16	2,3,7,8-TCDD dioxin	1746016					1.3E-8[C]	1.4E-8[C]	57FR60848
17	Acrolein	107028					320	780	62FR42160
18	Acrylonitrile	107131					0.059[B,C]	0.66[B,C]	57FR60848
19	Benzene	71432					1.2[B,C]	71[B,C]	62FR42160
20	Bromoform	75252					4.3[B,C]	360[B,C]	62FR42160
21	Carbon tetrachloride	56235					0.25[B,C]	4.4[B,C]	57FR60848
22	Chlorobenzene	108907					680[B,Z]	21,000[B,H]	57FR60848
23	Chlorodibromomethane	124481					0.41[B,C]	34[B,C]	62FR42160
24	Chloroethane	75003							
25	2-Chloroethylvinyl ether	110758							

No.	Compound	CAS							Publication
26	Chloroform	67663					5.7[B,C]	470[B,C]	62FR42160
27	Dichlorobromomethane	75274					0.56[B,C]	46[B,C]	62FR42160
28	1,1-Dichloroethane	75343							
29	1,2-Dichloroethane	107062					0.38[B,C]	99[B,C]	57FR60848
30	1,1-Dichloroethylene	75354					0.057[B,C]	3.2[B,C]	57FR60848
31	1,2-Dichloropropane	78875					0.52[B,C]	39[B,C]	62FR42160
32	1,3-Dichloropropene	542756					10[B]	1700[B]	57FR60848
33	Ethylbenzene	100414					3100[B,Z]	29,000[B]	62FR42160
34	Methyl bromide	74839					48[B]	4000[B]	62FR42160
35	Methyl chloride	74873					J	J	62FR42160
36	Methylene chloride	75092					4.7[B,C]	1600[B,C]	62FR42160
37	1,1,2,2-Tetrachloroethane	79345					0.17[B,C]	11[B,C]	57FR60848
38	Tetrachloroethylene	127184					0.8[C]	8.85[C]	57FR60848
39	Toluene	108883					6800[B,Z]	200,000[B]	62FR42160
40	1,2-*trans*-Dichloroethylene	156605					700[B,Z]	140,000[B]	62FR42160
41	1,1,1-Trichloroethane	71556					J[Z]	J	62FR42160
42	1,1,2-Trichloroethane	79005					0.60[B,C]	42[B,C]	57FR60848
43	Trichloroethane	79016					2.7[C]	81[C]	57FR60848
44	Vinyl chloride	75014					2.0[C]	525[C]	57FR60848
45	2-Chlorophenol	95578					120[B,U]	400[B,U]	62FR42160
46	2,4-Dichlorophenol	120832					93[B,U]	790[B,U]	57FR60848
47	2,4-Dimethylphenol	105679					540[B,U]	2300[B,U]	62FR42160
48	2-Methyl-4,6-dinitrophenol	534521					13.4	765	57FR60848
49	2,4-Dinitrophenol	51285					70[B]	14,000[B]	57FR60848
50	2-Nitrophenol	88755							
51	4-Nitrophenol	100027							
52	3-Methyl-4-Chlorophenol	59507							
53	Pentachlorophenol	87865	19[F,K]	15[F,K]	13[ab]	7.9[ab]	U	U	62FR42160
53							0.28[B,C]	8.2[B,C,H]	62FR42160, 57FR60848
54	Phenol	108952					21,000[B,U]	4,600,000[B,H,U]	62FR42160
55	2,4,6-Trichlorophenol	88062					2.1[B,C,U]	6.5[B,C]	62FR42160, 57FR60848
56	Acenaphthene	83329					1200[B,U]	2700[B,U]	62FR42160
57	Acenaphthylene	208968							
58	Anthracene	120127					9600[B]	110,000[B]	62FR42160
59	Benzidine	92875					0.00012[B,C]	0.00054[B,C]	57FR60848
60	Benzo(a)anthracene	56553					0.0044[B,C]	0.049[B,C]	62FR42160
61	Benzo(a)pyrene	50328					0.0044[B,C]	0.049[B,C]	62FR42160
62	Benzo(b)fluoranthene	205992					0.0044[B,C]	0.049[B,C]	62FR42160
63	Benzo(ghi)perylene	191242							

U.S. Recommended Water Quality Criteria for Priority Toxic Pollutants (continued)

	Priority Pollutant	CAS Number	Freshwater CMC (µg/L)	Freshwater CCC (µg/L)	Saltwater CMC (µg/L)	Saltwater CCC (µg/L)	Human Health For Consumption of: Water + Organism (µg/L)	Human Health For Consumption of: Organism Only (µg/L)	Federal Register Cite/Source
64	Benzokfluoranthene	207089					0.0044 B,C	0.049 B,C	62FR42160
65	Bis-2-chloroethoxymethane	111911							57FR60848
66	Bis-2-chloroethylether	111444					0.031 B,C	1.4 B,C	62FR42160
67	Bis-2-chloroisopropylether	39638329					1400 B		62FR42160
68	Bis-2-ethylhexylphthalate[x]	117817					1.8 B,C	170,000 B	57FR60848
69	4-Bromophenyl phenyl ether	101553						5.9 B,C	57FR60848
70	Butylbenzyl phthalate[w]	85687					3000 B	5200 B	62FR42160
71	2-Chloronaphthalene	91587					1700 B	4300 B	62FR42160
72	4-Chlorophenyl phenyl ether	7005723							
73	Chrysene	218019					0.0044 B,C	0.049 B,C	62FR42160
74	Dibenzoa,hanthracene	53703					0.0044 B,C	0.049 B,C	62FR42160
75	1,2-Dichlorobenzene	95501					2700 B,Z	17,000 B	62FR42160
76	1,3-Dichlorobenzene	541731					400	2600	62FR42160
77	1,4-Dichlorobenzene	106467					400 Z	2600	62FR42160
78	3,3'-Dichlorobenzidine	91941					0.04 B,C	0.077 B,C	57FR60848
79	Diethyl phthalate[w]	84662					23,000 B,C	120,000 B	57FR60848
80	Dimethyl phthalate[w]	131113					313,000	2,900,000	57FR60848
81	Di-n-Butyl phthalate[w]	84742					2700 B	12,000 B	57FR60848
82	2,4-Dinitrotoluene	121142					0.11 C	9.1 C	57FR60848
83	2,6-Dinitrotoluene	606202							
84	Di-n-octyl phthalate	117840							
85	1,2-Diphenylhydrazine	122667					0.040 B,C	0.54 B,C	57FR60848
86	Fluoranthene	206440					300 B	370 B	62FR42160
87	Fluorene	86737					1300 B	14,000 B	62FR42160
88	Hexachlorobenzene	118741					0.00075 B,C	0.00077 B,C	62FR42160
89	Hexachlorobutadiene	87683					0.44 B,C	50 B,C	62FR42160
90	Hexachlorocyclopentadiene	77474					240 B,U,Z	17,000 B,H,U	57FR60848
91	Hexachloroethane	67721					1.9 B,C	8.9 B,C	57FR60848
92	Idenol1,2,3-cdpyrene	193395					0.0044 B,C	0.049 B,C	57FR60848
93	Isophorone	78591					36 B,C	2600 B,C	62FR42160
94	Naphthalene	91203							IRIS 11/01/97

#	Compound	CAS							Reference
95	Nitrobenzene	98953					17^B	$1900^{B,H,U}$	57FR60848
96	N-Nitrosodimethylamine	62759					$0.00069^{B,C}$	$8.1^{B,C}$	57FR60848
97	N-Nitrosodi-n-propylamine	621647					$0.005^{B,C}$	$1.4^{B,C}$	62FR42160
98	N-Nitrosodiphenylamine	86306					$5.0^{B,C}$	$16^{B,C}$	57FR60848
99	Phenanthrene	85018							
100	Pyrene	129000					960^B	$11,000^B$	62FR42160
101	1,2,4-Trichlorobenzene	120821					260^Z	940	IRIS 11/01/96
102	Aldrin	309002	3.0^G		1.3^G		$0.00013^{B,C}$	$0.00014^{B,C}$	62FR42160
103	α-BHC	319846					$0.0039^{B,C}$	$0.013^{B,C}$	62FR42160
104	β-BHC	319857					$0.014^{B,C}$	$0.046^{B,C}$	62FR42160
105	γ-BHC (Lindane)	58899	0.95^K		0.16^G		0.019^C	0.063^C	62FR42160
106	δ-BHC	319868							
107	Chlordane	57749	2.4^G	$0.0043^{G,aa}$	0.09^G	$0.004^{G,aa}$	$0.0021^{B,C}$	$0.0022^{B,C}$	62FR42160 / IRIS 02/07/98
108	4,4′-DDT	50293	1.1^G	$0.001^{G,aa}$	0.13^G	$0.001^{G,aa}$	$0.00059^{B,C}$	$0.00059^{B,C}$	62FR42160
109	4,4′-DDE	72559					$0.00059^{B,C}$	$0.00059^{B,C}$	62FR42160
110	4,4′-DDD	72548					$0.00083^{B,C}$	$0.00084^{B,C}$	62FR42160
111	Dieldrin	60571	0.24^K	$0.056^{K,O}$	0.71^G	$0.0019^{G,aa}$	$0.00014^{B,C}$	$0.00014^{B,C}$	62FR42160
112	α-Endosulfan	959988	$0.22^{G,Y}$	$0.056^{K,O}$	$0.034^{G,Y}$	$0.0087^{G,Y}$	110^B	240^B	62FR42160
113	β-Endosulfan	33213659	$0.22^{G,Y}$	$0.056^{G,Y}$	$0.034^{G,Y}$	$0.0087^{G,Y}$	110^B	240^B	62FR42160
114	Endosulfan sulfate	1031078					110^B	240^B	62FR42160
115	Endrin	72208	0.086^K	$0.036^{K,O}$	0.037^G	$0.0023^{G,aa}$	0.76^B	$0.81^{B,H}$	62FR42160
116	Endrin aldehyde	7421934					0.76^B	$0.81^{B,H}$	62FR42160
117	Heptachlor	76448	0.52^G	$0.0038^{G,aa}$	0.053^G	$0.0036^{G,aa}$	$0.00021^{B,C}$	$0.00021^{B,C}$	62FR42160
118	Heptachlor epoxide	1024573	$0.52^{G,V}$	$0.0038^{G,Vaa}$	$0.053^{G,V}$	$0.0036^{G,Vaa}$	$0.00010^{B,C}$	$0.00011^{B,C}$	62FR42160
119	Polychlorinated biphenyls			$0.014^{N,aa}$		$0.03^{N,aa}$	$0.00017^{B,C,P}$	$0.00017^{B,C,P}$	63FR16182
120	Toxaphene	8001352	0.73	0.0002^{aa}	0.21	0.0002^{aa}	$0.00073^{B,C}$	$0.00075^{B,C}$	62FR42160

A This recommended water quality criterion was derived from data for arsenic (III), but is applied here to total arsenic, which might imply that arsenic (III) and arsenic (V) are equally toxic to aquatic life and that their toxicities are additive. In the arsenic criteria document (EPA 440/5-84-033, January 1985), Species Mean Acute Values are given for both arsenic (III) and arsenic (V) for five species and the ratios of the SMAVs for each species range from 0.6 to 1.7. Chronic values are available for both arsenic (III) and arsenic (V) for one species; for the fathead minnow, the chronic value for arsenic (V) is 0.29 times the chronic value for arsenic (III). No data are known to be available concerning whether the toxicities of the forms of arsenic to aquatic organisms are additive.

B This criterion has been revised to reflect the Environmental Protection Agency's q1* or RfD, as contained in the Integrated Risk Information System (IRIS) as of April 8, 1998. The fish tissue bioconcentration factor (BCF) from the 1980 Ambient Water Quality Criteria document was retained in each case.

C This criterion is based on carcinogenicity of 10^{-6} risk. Alternate risk levels may be obtained by moving the decimal point (e.g., for a risk level of 10^{-5}, move the decimal point in the recommended criterion one place to the right).

U.S. Recommended Water Quality Criteria for Priority Toxic Pollutants (continued)

D Freshwater and saltwater criteria for metals are expressed in terms of the dissolved metals in the water column. The recommended water quality criteria value was calculated by using the previous 304(a) aquatic life criteria expressed in terms of total recoverable metal, and multiplying it by a conversion factor (CF). The term "Conversion Factor" (CF) represents the recommended conversion factor for converting a metal criterion expressed as the total recoverable fraction in the water column to a criterion expressed as the dissolved fraction in the water column. (Conversion Factors for saltwater CMCs and CCCs are not currently available. Conversion factors derived for saltwater CMCs have been used for both saltwater CMCs and CCCs.) See "Office of Water Policy and Technical Guidance on Interpretation and Implementation of Aquatic Life Metals Criteria," October 1, 1993, by Martha G. Prothro, Acting Assistant Administrator for Water, available from the Water Resource center, USEPA, 401 M St., SW, mail code RC4100, Washington, DC 20460; and 40CFR§131.36(b)(1). Conversion Factors applied in the table can be found in Appendix A to the Preamble-Conversion Factors for Dissolved Metals.

E The freshwater criterion for this metal is expressed as a function of hardness (mg/L) in the water column. The value given here corresponds to a hardness of 100 mg/L. Criteria values for other hardness may be calculated from the following: CMC (dissolved) = exp{m$_A$[ln(hardness)] + b$_A$} (CF), or CCC (dissolved) = exp{m$_C$[ln(hardness)] + b$_C$} (CF) and the parameters specified in Appendix B to the Preamble- *Parameters for Calculating Freshwater Dissolved Metals Criteria That Are Hardness-Dependent.*

F Freshwater aquatic life values for pentachlorophenol are expressed as a function of pH, and are calculated as follows: CMC = exp(1.005(pH) − 4.869); CCC = exp(1.005(pH) − 5.134). Values displayed in table correspond to a pH of 7.8.

G This Criterion is based on 304(a) aquatic life criterion issued in 1980, and was issued in one of the following documents: Aldrin/Dieldrin (EPA 440/5-80-019), Chlordane (EPA 440/5-80-027), DDT (EPA 440/5-80-038), Endosulfan (EPA 440/5-80-046), Endrin (EPA 440/5-80-047), Heptachlor (EPA 440/5-80-019), Hexachlorocyclohexane (EPA 440/5-80-054), Silver (EPA 440/5-80-071). The Minimum Data Requirements and derivation procedures were different in the 1980 Guidelines than in the 1985 Guidelines. For example, a "CMC" derived using 1980 Guidelines was derived to be used as an instantaneous maximum. If assessment is to be done using an average period, the values given should be divided by 2 to obtain a value that is more comparable to a CMC derived using the 1985 Guidelines.

H No criterion for protection of human health from consumption of aquatic organisms excluding water was presented in the 1980 criteria document or in the *1986 Quality Criteria for Water.* Nevertheless, sufficient information was presented in the 1980 document to allow the calculation of a criterion, even though the results of such a calculation were not shown in the document.

I This criterion for asbestos is the Maximum Contaminant Level (MCL) developed under the Safe Drinking Water Act (SDWA).

J EPA has not calculated human health criterion for this contaminant. However, permit authorities should address this contaminant in NPDES permit actions using the State's existing narrative criteria for toxics.

K This recommended criterion is based on a 304(a) aquatic life criterion that was issued in the *1995 Updates: Water Quality Criteria Documents for the Protection of Aquatic Life in Ambient Water,* (EPA-820-B-96-001, September 1996). This value was derived using the GLI Guidelines (60FR15393-15399, March 23, 1995; 40CFR132 Appendix A); the difference between the 1985 Guidelines and the GLI Guidelines are explained on page iv of the 1995 Updates. None of the decisions concerning the derivation of this criterion were affected by any considerations that are specific to the Great Lakes.

L The CMC = 1/[(f1/CMC1) + (f2/CMC2)] where f1 and f2 are the fractions of total selenium that are treated as selenite and selenate, respectively, and CMC1 and CMC2 are 185.9 µg/l and 12.83 µg/l, respectively.

M EPA is currently reassessing the criteria for arsenic. Upon completion of the reassessment the Agency will publish revised criteria as appropriate.

N PCBs are a class of chemicals which include Aroclors, 1242, 1254, 1221, 1232, 1248, 1260, and 1016, CAS numbers 53469219, 11097691, 11104282, 11141165, 12672296, 11096825 and 12674112 respectively. The aquatic life criteria apply to this set of PCBs.

O The derivation of the CCC for this pollutant did not consider exposure through the diet, which is probably important for aquatic life occupying upper trophic levels.

P This criterion applies to total pcbs, i.e., the sum of all congener or all isomer analyses.

Q This recommended water criterion is expressed as µg free cyanide (as CN)/L.

R   This value was announced (61FR58444-58449, November 14, 1996) as a proposed GLI 303(c) aquatic life criterion. EPA is currently working on this criterion and so this value might change substantially in the near future.

S   This recommended water quality criterion refers to the inorganic form only.

T   This recommended water quality criterion is expressed in terms of total recoverable metal in the water column. It is scientifically acceptable to use the conversion factor of 0.922 that was used in the GLI to convert this to a value that is expressed in terms of dissolved metal.

U   The organoleptic effect criterion is more stringent than the value for priority toxic pollutants.

V   This value was derived from data for heptachlor and the criteria document provides insufficient data to estimate the relative toxicities of heptachlor and heptachlor epoxide.

W   Although EPA has not published a final criteria document for this compound it is EPA's understanding that sufficient data exist to allow calculation of aquatic criteria. It is anticipated that industry intends to publish in the peer reviewed literature draft aquatic life criteria generated in accordance with EPA Guidelines. EPA will review such criteria for possible issuance as national WQC.

X   There is a full set of aquatic life toxicity data that show that DEHP is not toxic to aquatic organisms at or below its solubility limit.

Y   This value was derived from data for endosulfan and is most appropriately applied to the sum of alpha-endosulfan and beta-endosulfan.

Z   A more stringent MCL has been issued by the EPA. Refer to drinking water regulations (40 CFR 141) or Safe Drinking Water Hotline (1-800-426-4791) for values.

aa This CCC is based on the Final Residue Value procedure in the 1985 Guidelines. Since the publication of the Great Lakes Aquatic Life Criteria Guidelines in 1995 (60FR15399, March 23, 1995), the Agency no longer uses the Final Residue Value procedure for deriving CCCs for new or revised 304(a) aquatic life criteria.

bb This water quality criterion is based on a 304(a) aquatic life criterion that was derived using the 1985 Guidelines (*Guidelines for Deriving Numerical National Water Quality Criteria for the Protection of Aquatic Organisms and Their Uses*, PB85-227049, January 1985) and was issued in one of the following criteria documents: Arsenic (EPA 440/5-84-033), Cadmium (EPA 440/5-84-032), Chromium (EPA 440/5-84-029), Copper (EPA 440/5-84-031), Cyanide (EPA 440/5-84-028), Lead (EPA 440/5-84-027), Nickel (EPA 440/5-86-004), Pentachlorophenol (EPA 440/5-86-009), Toxaphene (EPA 440/5-86-006), Zinc (EPA 440/5-87-003).

cc When the concentration of dissolved organic carbon is elevated, copper is substantially less toxic and use of Water-Effect Ratios might be appropriate.

dd The selenium criteria document (EPA 440/5-87-006, September 1987) provides that if selenium is as toxic to saltwater fishes in the field as it is to freshwater fishes in the field, the status of the fish community should be monitored whenever the concentration of selenium exceeds 5.0 µg/L in salt water because the saltwater CCC does not take into account uptake via the food chain.

ee This recommended water quality criterion was derived on page 43 of the mercury criteria document (EPA 440/5-84-026, January 1985). The saltwater CCC of 0.025 µg/L given on page 23 of the criteria document is based on Final Residue Value procedure in the 1985 Guidelines. Since the publication of the Great Lakes Aquatic Life Criteria Guidelines in 1995 (60FR15399-15399, March 23, 1995), the Agency no longer uses the Final Residue Value procedure for deriving CCCs for new or revised 304(a) aquatic life criteria.

ff This recommended water quality criterion was derived in *Ambient Water Quality Criteria Saltwater Copper Addendum* (Draft, April 14, 1995) and was promulgated in the Interim final National Toxics Rule (60FR22228-222237, May 4, 1995).

gg EPA is actively working on this criterion and so this recommended water quality criterion may change substantially in the near future.

hh This recommended water quality criterion was derived from data for inorganic mercury (II), but is applied here to total mercury. If a substantial portion of the mercury in the water column is methylmercury, this criterion will probably be under protective. In addition, even though inorganic mercury is converted to methylmercury and methylmercury bioaccumulates to a great extent, this criterion does not account for uptake via the food chain because sufficient data were not available when the criterion was derived.

U.S. Recommended Water Quality Criteria for Nonpriority Pollutants

	Nonpriority Pollutant	CAS Number	Freshwater CMC (μg/L)	Freshwater CCC (μg/L)	Saltwater CMC (μg/L)	Saltwater CCC (μg/L)	Human Health Water + Organism (μg/L)	Human Health Organism Only (μg/L)	Federal Register Cite/Source
1	Alkalinity	—		20000[F]					Gold Book
2	Aluminum pH 6.5–9.0	7429905	750[G,J]	87[G,I,L]					53FR3178
3	Ammonia	7664417	FRESHWATER CRITERIA ARE pH DEPENDENT — SEE DOCUMENT[D] / SALTWATER CRITERIA ARE pH AND TEMPERATURE DEPENDENT						EPA822-R-98-008 / EPA440/5-88-004
4	Aesthetic qualities	—	NARRATIVE STATEMENT — SEE DOCUMENT						Gold Book
5	Bacteria	—	FOR PRIMARY RECREATION AND SHELLFISH USES — SEE DOCUMENT						Gold Book
6	Barium	7440393					1000[A]		Gold Book
7	Boron	—	NARRATIVE STATEMENT — SEE DOCUMENT						Gold Book
8	Chloride	16887006	860000[G]	230000[G]					Gold Book
9	Chlorine	7782505	19	11	13	7.5	C		53FR19028
10	Chlorophenoxy herbicide 2,4,5,-TP	93721					10[A]		Gold Book
11	Chlorophenoxy herbicide 2,4-D	94757					100[A,C]		Gold Book
12	Chloropyrifos	2921882	0.083[G]	0.041[G]	0.011[G]	0.0056[G]			Gold Book
13	Color	—	NARRATIVE STATEMENT — SEE DOCUMENT[F]						Gold Book
14	Demeton	8065483	0.1[F]	0.1[F]					Gold Book
15	Ether, Bis Chloromethyl	542881					0.00013[E]	0.00078[E]	IRIS 01/01/91
16	Gases, Total Dissolved	—	NARRATIVE STATEMENT — SEE DOCUMENT[F]						Gold Book
17	Guthion	86500	0.01[F]	0.01[F]					Gold Book
18	Hardness	—	NARRATIVE STATEMENT — SEE DOCUMENT						Gold Book
19	Hexachlorocyclo-hexane-Technical	319868					0.0123	0.0414	Gold Book
20	Iron	7439896	1000[F]				300[A]		Gold Book
21	Malathion	121755	0.1[F]	0.1[F]					Gold Book
22	Manganese	7439965					50[A]	100[A]	Gold Book
23	Methoxychlor	72435	0.03[F]	0.03[F]			100[A,C]		Gold Book
24	Mirex	2385855	0.001[F]	0.001[F]					Gold Book
25	Nitrates	14797558					10,000[A]		Gold Book
26	Nitrosamines	—					0.0008	1.24	Gold Book
27	Dinitrophenols	25550587					70	14,000	Gold Book
28	Nitrosodibutylamine,N	924163					0.0064[A]	0.587[A]	Gold Book
29	Nitrosodiethylamine,N	55185					0.0008[A]	1.24[A]	Gold Book
30	Nitrosopyrrolidine,N	930552							Gold Book
31	Oil and grease	—	NARRATIVE STATEMENT — SEE DOCUMENT[F]						Gold Book
32	Oxygen, dissolved	7782447	NARRATIVE STATEMENT — SEE DOCUMENT[O]						Gold Book

#	Compound	CAS	Freshwater Acute	Freshwater Chronic	Saltwater Acute	Saltwater Chronic	Human Health (Water + Organism)	Human Health (Organism Only)	Publication
33	Parathion	56382	0.065ᴶ	0.013ᴶ					Gold Book
34	Pentachlorobenzene	608935					3.5ᴱ	4.1ᴱ	IRIS 03/01/88
35	pH	—		6.5–9ᶠ	6.5–8.5ᶠᐟᴷ			5–9	Gold Book
36	Phosphorus elemental	7723140			0.1ᶠᐟᴷ				Gold Book
37	Phosphate phosphorus	—	NARRATIVE STATEMENT — SEE DOCUMENT						Gold Book
38	Solids dissolved and salinity	—			250,000ᴬ				Gold Book
39	Solids suspended and turbidity	—	NARRATIVE STATEMENT — SEE DOCUMENTᶠ						Gold Book
40	Sulfide-hydrogen sulfide	7783064		2.0ᶠ	2.0ᶠ				Gold Book
41	Tainting substances	—	NARRATIVE STATEMENT — SEE DOCUMENT						Gold Book
42	Temperature	—	NARRATIVE STATEMENT — SEE DOCUMENTᴹ						Gold Book
43	Tetrachlorobenzene,1,2,4,5-	95943					2.3ᴱ	2.9ᴱ	IRIS 03/01/91
44	Tributyltin TBT	—	0.46ᴺ	0.063ᴺ	0.37ᴺ	0.010ᴺ			62FR42554
45	Trichlorophenol,2,4,5-	95954					2600ᴮᐟᴱ	9800ᴮᐟᴱ	IRIS 03/01/88

A This human health criterion is the same as originally published in the Red Book which predates the 1980 methodology and did not utilize the fish ingestion BCF approach. This same criterion value is now published in the Gold Book.

B The organoleptic effect criterion value is more stringent than the value presented in the non priority pollutants table.

C A more stringent Maximum Contaminant Level (MCL) has been issued by EPA under the Safe Drinking Water Act. Refer to drinking water regulations 40CFR141 or Safe Drinking Water Hotline (1-800-426-4791) for values.

D According to the procedures described in the *Guidelines for Deriving Numerical National Water Quality Criteria for the Protection of Aquatic Organisms and Their Uses*, except possibly where a very sensitive species is important at a site, freshwater aquatic life should be protected if both conditions specified in Appendix C to the Preamble- Calculation of Freshwater Ammonia Criterion are satisfied.

E This criterion has been revised to reflect the Environmental Protection Agency's q1* or RfD, as contained in the Integrated Risk Information System (IRIS) as of April 8, 1998. The fish tissue bioconcentration factor (BCF) used to derive the original criterion was retained in each case.

F The derivation of this value is presented in the Red Book (EPA 440/9-76-023, July, 1976).

G This value is based on a 304(a) aquatic life criterion that was derived using the 1985 Guidelines (*Guidelines for Deriving Numerical National Water Quality Criteria for the Protection of Aquatic Organisms and Their Uses*, PB85-227049, January 1985) and was issued in one of the following criteria documents: Aluminum (EPA 440/5-86-008); Chloride (EPA 440/5-88-001), Chloropyrifos (EPA 440/5-86-005).

I This value is expressed in terms of total recoverable metal in the water column.

J This value is based on a 304(a) aquatic life criterion that was issued in the *1995 Updates: Water Quality Criteria Documents for the Protection of Aquatic Life in Ambient Water* (EPA-820-B-96-001). This value was derived using the GLI Guidelines (60FR15393-15399, March 23, 1995; 40CFR132 Appendix A); the differences between the 1985 Guidelines and the GLI Guidelines are explained on page iv of the 1995 Updates. No decision concerning this criterion was affected by any considerations that are specific to the Great Lakes.

K According to page 181 of the Red Book:

 For open ocean waters where the depth is substantially greater than the euphotic zone, the pH should not be changed more than 0.2 units from the naturally occurring variation of any caes outside the range of 6.5 to 8.5. For shallow, highly productive coastal and estuarine areas where naturally occurring pH variations approach the lethal limits of some species, changes in pH should be avoided but in any case should not exceed the limits established for fresh water, i.e., 6.5–9.0.

U.S. Recommended Water Quality Criteria for Nonpriority Pollutants (continued)

L   There are three major reasons why the use of Water-Effect Ratios might be appropriate. (1) The value of 87 μg/l is based on a toxicity test with the striped bass in water with pH = 6.5–6.6 and hardness <10 mg/L. Data in "Aluminum Water-Effect Ratio for the 3M Plant Effluent Discharge, Middleway, West Virginia" (May 1994) indicate that aluminum is substantially less toxic at higher pH and hardness, but the effects of pH and hardness are not well quantified at this time. (2) In tests with the brook trout at low pH and hardness, effects increased with increasing concentrations of total aluminum even though the concentration of dissolved aluminum hydroxide particles. In surface waters, however, the total recoverable procedure might measure aluminum associated with clay particles, which might be less toxic than aluminum associated with aluminum hydroxide. (3) EPA is aware of field data indicating that many high quality waters in the U.S. contain more than 87 μg aluminum/L, when either total recoverable or dissolved is measured.

M   U.S. EPA. 1973. Water Quality Criteria 1972. EPA-R3-73-033. National Technical Information Service, Springfield, VA.; U.S. EPA. 1977. Temperature Criteria for Freshwater Fish: Protocol and Procedures. EPA-600/3-77-061. National Technical Information Service, Springfield, VA.

N   This value was announced (62FR42554, August 7, 1997) as a proposed 304(a) aquatic life criterion. Although EPA has not responded to public comment, EPA is publishing this as a 304(a) criterion in today's notice as guidance for States and Tribes to consider when adopting water quality criteria.

O   U.S. EPA. 1986. Ambient Water Quality Criteria for Dissolved Oxygen. EPA 440/5-86-003. National Technical Information Service, Springfield, VA.

U.S. Recommended Water Quality Criteria for Organoleptic Effects

Pollutant	CAS Number	Organoleptic Effect Criteria (μg/L)	Federal Register Cite/Source
1 Acenaphthene	83329	20	Gold Book
2 Monochlorobenzene	108907	20	Gold Book
3 3-Chlorophenol	—	0.1	Gold Book
4 4-Chlorophenol	106489	0.1	Gold Book
5 2,3-Dichlorophenol	—	0.04	Gold Book
6 2,5-Dichlorophenol	—	0.5	Gold Book
7 2,6-Dichlorophenol	—	0.2	Gold Book
8 3,4-Dichlorophenol	—	0.3	Gold Book
9 2,4,5-Trichlorophenol	95954	1	Gold Book
10 2,4,6-Trichlorophenol	88062	2	Gold Book
11 2,3,4,6-Tetrachlorophenol	—	1	Gold Book
12 2-Methyl-4-Chlorophenol	—	1800	Gold Book
13 3-Methyl-4-Chlorophenol	59507	3000	Gold Book
14 3-Methyl-6-Chlorophenol	—	20	Gold Book
15 2-Chlorophenol	95578	0.1	Gold Book
16 Copper	7440508	1000	Gold Book
17 2,4-Dichlorophenol	120832	0.3	Gold Book
18 2,4-Dimethylphenol	105679	400	Gold Book
19 Hexachlorocyclopentadiene	77474	1	Gold Book
20 Nitrobenzene	98953	30	Gold Book
21 Pentachlorophenol	87865	30	Gold Book
22 Phenol	108952	300	Gold Book
23 Zinc	7440666	5000	45FR79341

1. These criteria are based on organoleptic (taste and odor) effects. Because of variations in chemical nomenclature systems, this listing of pollutants does not duplicate the listing in Appendix A of 40 CFR Part 423. Also listed are the Chemical Abstracts Service (CAS) registry numbers, which provide a unique identification for each chemical.

U.S. Recommended Water Quality Criteria for Organoleptic Effects (continued)

U.S. RECOMMENDED WATER QUALITY CRITERIA

Additional Notes:

1. Criteria Maximum Concentration and Criterion Continuous Concentration

The Criteria Maximum Concentration (CMC) is an estimate of the highest concentration of a material in surface water to which an aquatic community can be exposed briefly without resulting in an unacceptable effect. The Criterion Continuous Concentration (CCC) is an estimate of the highest concentration of a material in surface water to which an aquatic community can be exposed indefinitely without resulting in an unacceptable effect. The CMC and CCC are just two of the six parts of aquatic life criterion; the other four parts are the acute averaging period, chronic averaging period, acute frequency of allowed exceedance, and chronic frequency of allowed exceedance. Because 304(a) aquatic life criteria are national guidance, they are intended to be protective of the vast majority of the aquatic communities in the United States.

2. Criteria Recommendations for Priority Pollutants, Nonpriority Pollutants, and Organoleptic Effects

This compilation lists all priority toxic pollutants and some non priority toxic pollutants, and both human health effect and organoleptic effect criteria issued pursuant to CWA §304(a). Blank spaces indicate that EPA has no CWA §304(a) criteria recommendations. For a number of nonpriority toxic pollutants not listed, CWA §304(a) "water + organism" human health criteria are not available, but, EPA has published MCLs under the SDWA that may be used in establishing water quality standards to protect water supply designated uses. Because of variations in chemical nomenclature systems, this listing of toxic pollutants does not duplicate the listing in Appendix A of 40 CFR Part 423. Also listed are the Chemical Abstracts Service CAS registry numbers, which provide a unique identification for each chemical.

3. Human Health Risk

The human health criteria for the priority and nonpriority pollutants are based on carcinogenicity of 10⁻⁶ risk. Alternate risk levels may be obtained by moving the decimal point (e.g., for a risk level of 10⁻⁵, move the decimal point in the recommended criterion one place to the right).

4. Water Quality Criteria published pursuant to Section 304(a) or Section 303(c) of the CWA

Many of the values in the compilation were published in the proposed California Toxics Rule (CTR, 62FR42160). Although such values were published pursuant to Section 303(c) of the CWA, they represent the Agency's most recent calculation of water quality criteria and thus are published today as the Agency's 304(a) criteria. Water quality criteria published in the proposed CTR may be revised when EPA takes final action on the CTR.

5. Calculation of Dissolved Metals Criteria

The 304(a) criteria for metals, shown as dissolved metals, are calculated in one of two ways. For freshwater metals criteria that are hardness-dependent, the dissolved metal criteria were calculated using a hardness of 100 mg/L as CaCO₃ for illustrative purposes only. Saltwater and freshwater metals' criteria that are not hardness-dependent are calculated by multiplying the total recoverable criteria before rounding by the appropriate conversion factors. The final dissolved metals' criteria in the table are rounded to two significant figures. Information regarding the calculation of hardness dependent conversion factors are included in the footnotes.

6. Correction of Chemical Abstract Services Number

The Chemical Abstract Services number (CAS) for Bis(2-Chloroisopryl) Ether, has been corrected in the table. The correct CAS number for this chemical is 39638-32-9. Previous publications listed 108-60-1 as the CAS number for this chemical.

7. Maximum Contaminant Levels

The compilation includes footnotes for pollutants with Maximum Contaminant Levels (MCLs) more stringent than the recommended water quality criteria in the compilation. MCLs for these pollutants are not included in the compilation, but can be found in the appropriate drinking water regulations (40 CFR 141.11-16 and 141.60-63), or can be accessed through the Safe Drinking Water Hotline (800-426-4791) or the Internet (http://www.epa.gov/ost/tools/dwstds-s.html).

8. Organoleptic Effects

The compilation contains 304(a) criteria for pollutants with toxicity-based criteria as well as non-toxicity based criteria. The basis for the non-toxicity based criteria are organoleptic effects for 23 pollutants. Pollutants with organoleptic effect criteria more stringent than the criteria based on toxicity (e.g., included in both the priority and non-priority pollutant tables) are footnoted as such.

9. Category Criteria

In the 1980 criteria documents, certain recommended water quality criteria were published for categories of pollutants rather than for individual pollutants within that category. Subsequently, in a series of separate actions, the Agency derived criteria for specific pollutants within a category. Therefore, in this compilation EPA is replacing criteria representing categories with individual pollutant criteria (e.g., 1,3-dichlorobenzene, 1,4-dichlorobenzene and 1,2-dichlorobenzene).

10. Specific Chemical Calculation

A. Selenium

(1) Human Health

In the 1980 Selenium document, a criterion for the protection of human health from consumption of water and organisms was calculated based on a BCF of 6.0 L/kg and a maximum water-related contribution of 35 µg Se/day. Subsequently, the EPA Office of Health and Environmental Assessment issued an errata notice (February 23, 1982), revising the BCF for selenium to 4.8 L/kg. In 1988, EPA issued an addendum (ECAO-CIN-668) revising the human health criteria for selenium. Later in the final National Toxic Rule (NTR, 57 FR 60848), EPA withdrew previously published selenium human health criteria, pending Agency review of new epidemiological data.

This compilation includes human health criteria for selenium, calculated using a BCF of 4.8 L/kg along with the current IRIS RfD of 0.005 mg/kg/day. EPA included these recommended water quality criteria in the compilation because the data necessary for calculating a criteria in accordance with EPA's 1980 human health methodology are available.

(2) Aquatic Life

This compilation contains aquatic life for selenium that are the same as those published in the proposed CTR. In the CTR, EPA proposed an acute criterion for selenium based on the criterion proposed for selenium in the Water Quality Guidance for the Great Lakes System (61 FR 58444). The GLI and CTR proposals take into account data showing that selenium's two most prevalent oxidation states, selenite and selenate, present differing potentials for aquatic toxicity, as well as new data indicating that various forms of selenium are additive. The new approach produces a different selenium acute criterion concentration, or CMC, depending upon the relative proportions of selenite, selenate, and other forms of selenium that are present.

EPA notes its currently undertaking a reassessment of selenium, and expects the 304(a) criteria for selenium will be revised based on the final reassessment (63FR26186). However, until such time as revised water quality criteria for selenium are published by the Agency, the recommended water quality criteria in this compilation are EPA's current 304(a) criteria.

U.S. Recommended Water Quality Criteria for Organoleptic Effects (continued)

B. 1,2,4-Trichlorobenzene and Zinc

Human health criteria for 1,2,4-trichlorobenzene and zinc have not been previously published. Sufficient information is now available for calculating water quality criteria for the protection of human health from the consumption of aquatic organisms and the consumption of aquatic organisms and water for both these compounds. Therefore, EPA is publishing criteria for these pollutants in this compilation.

C. Chromium (III)

The recommended aquatic life water quality criteria for chromium (III) included in the compilation are based on the values presented in the document titled: *1995 Updates: Water Quality Criteria Documents for the Protection of Aquatic Life in Ambient Water*, however, this document contains criteria based on the total recoverable fraction. The chromium (III) criteria in this compilation were calculated by applying the conversion factors used in the Final Water Quality Guidance for the Great Lakes System (60 FR 15366) to the 1995 Update document values.

D. Ether, Bis (Chloromethyl), Pentachlorobenzene, Tetrachlorobenzene 1,2,4,5-, Trichlorophenol

Human health criteria for these pollutants were last published in EPA's *Quality Criteria for Water 1986* or *"Gold Book."* Some of these criteria were calculated using Acceptable Daily Intake (ADIs) rather than RfDs. Updated q1*s and RfDs are now available in IRIS for ether, bis (chloromethyl), pentachlorobenzene, tetrachlorobenzene 1,2,4,5-, and trichlorophenol, and were used to revise the water quality criteria for these compounds. The recommended water quality criteria for ether, bis (chloromethyl) were revised using an updated q1*, while criteria for pentachlorobenzene, and tetrachlorobenzene 1,2,4,5-, and trichlorophenol were derived using an updated RfD value.

E. PCBs

In this compilation EPA is publishing aquatic life and human health criteria based on total PCBs rather than individual arochlors. These criteria replace the previous criteria for the seven individual arochlors. Thus, there are criteria for a total of 102 of the 126 priority pollutants.

Appendix A — Conversion Factors for Dissolved Metals

Metal	Conversion Factor (freshwater CMC)	Conversion Factor (freshwater CCC)	Conversion Factor (saltwater CMC)	Conversion Factor (saltwater CCC)[1]
Arsenic	1.000	1.000	1.000	1.000
Cadmium	1.136672–[(ln hardness) (0.041838)]	1.101672–[(ln hardness) (0.041838)]	0.994	0.994
Chromium III	0.316	0.860	—	—
Chromium VI	0.982	0.962	0.993	0.993
Copper	0.960	0.960	0.83	0.83
Lead	1.46203–[(ln hardness) (0.145712)]	1.46203–[(ln hardness) (0.145712)]	0.951	0.951
Mercury	0.85	0.85	0.85	0.85
Nickel	0.998	0.997	0.990	0.990
Selenium	—	—	0.998	0.998
Silver	0.85	—	0.85	—
Zinc	0.978	0.986	0.946	0.946

Appendix B — Parameters for Calculating Freshwater Dissolved Metals Criteria That Are Hardness Dependent

Chemical	m_A	b_A	m_C	b_C	Freshwater Conversion Factor (CF) Acute	Freshwater Conversion Factor (CF) Chronic
Cadmium	1.128	−3.6867	0.7852	−2.715	1.136672–[(ln hardness) (0.041838)]	1.101672–[(ln hardness) (0.041838)]
Chromium II	0.8190	3.7256	0.8190	0.6848	0.860	0.860
Copper	0.9422	−1.700	0.8545	−1.702	0.960	0.960
Lead	1.273	−1.460	1.273	−4.705	1.46203–[(ln hardness) (0.145712)]	1.46203–[(ln hardness) (0.145712)]
Nickel	0.8460	2.255	0.8460	0.0584	0.998	0.997
Silver	1.72	−6.52	—	—	0.85	—
Zinc	0.8473	0.884	0.8473	0.884	0.978	0.986

Appendix C — Calculation of Freshwater Ammonia Criterion

1. The one-hour average concentration of total ammonia nitrogen (in mg N/L) does not exceed, more than once every three years on the average, the CMC calculated using the following equation:

$$CMC = \frac{0.275}{1 + 10^{7.204 - pH}} + \frac{39.0}{1 + 10^{pH - 7.204}}$$

In situations where salmonids do not occur, the CMC may be calculated using the following equation:

$$CMC = \frac{0.411}{1 + 10^{7.204 - pH}} + \frac{58.4}{1 + 10^{pH - 7.204}}$$

2. The 30-day average concentration of total ammonia nitrogen (in mg N/L) does not exceed, more than once every 3 years on the average, the CCC calculated using the following equation:

$$CCC = \frac{0.0858}{1 + 10^{7.688 - pH}} + \frac{3.70}{1 + 10^{pH - 7.688}}$$

and the highest 4-day average within the 30-day period does not exceed twice the CCC.

AMMONIA

The ammonia criteria are only for the protection of aquatic life, as no criteria have been developed for the protection of human health (consumption of contaminated fish or drinking water). The water quality criteria are for general guidance only and do not constitute formal water quality standards. However, the criteria reflect the scientific knowledge concerning the effects of the pollutants and are recommended EPA acceptable limits for aquatic life.

The data used in deriving the EPA criteria are predominantly from flow-through tests in which ammonia concentrations were measured. Ammonia was reported to be acutely toxic to freshwater organisms at concentrations (uncorrected for pH) ranging from 0.53 to 22.8 mg/L NH_3 for 19 invertebrate species representing 14 families and 16 genera and from 0.083 to 4.60 mg/L NH_3 for 29 fish species from 9 families and 18 genera. Among fish species, reported 96-hour LC50 values ranged from 0.083 to 1.09 mg/L for salmonids and from 0.14 to 4.60 mg/L NH_3 for other fish. Reported data from chronic tests on ammonia with two freshwater invertebrate species, both daphnids, showed effects at concentrations (uncorrected for pH) ranging from 0.304 to 1.2 mg/L NH_3, and with nine freshwater fish species, from five families and seven genera, ranging from 0.0017 to 0.612 mg/L NH_3.

Concentrations of ammonia acutely toxic to fishes may cause loss of equilibrium, hyperexcitability, increased breathing, cardiac output and oxygen uptake, and, in extreme cases, convulsions, coma, and death. At lower concentrations, ammonia has many effects on fishes, including a reduction in hatching success, reduction in growth rate and morphological development, and pathologic changes in tissues of gills, livers, and kidneys.

Several factors have been shown to modify acute NH_3 toxicity in fresh water. Some factors alter the concentration of un-ionized ammonia in the water by affecting the aqueous ammonia equilibrium, and some factors affect the toxicity of un-ionized ammonia itself, either ameliorating or exacerbating the effects of ammonia. Factors that have been shown to affect ammonia toxicity include dissolved oxygen concentration, temperature, pH, previous acclimation to ammonia, fluctuating or intermittent exposures, carbon dioxide concentration, salinity, and the presence of other toxicants.

The most well studied of these is pH; the acute toxicity of NH_3 has been shown to increase as pH decreases. However, the percentage of the total ammonia that is un-ionized decreases with decreasing pH. Sufficient data exist from toxicity tests conducted at different pH values to formulate a relationship to describe the pH-dependent acute NH_3 toxicity. The very limited amount of data regarding effects of pH on chronic NH_3 toxicity also indicate increasing NH_3 toxicity with decreasing pH, but the data are insufficient to derive a broadly applicable toxicity/pH relationship. Data on temperature effects on acute NH_3 toxicity were limited and somewhat variable, but indications are that NH_3 toxicity to fish is greater as temperature decreases. There was no information available regarding temperature effects on chronic NH_3 toxicity. Examination of pH and temperature-corrected acute NH_3 toxicity values among species and genera of freshwater organisms showed that invertebrates are generally more tolerant than fishes, a notable exception being the fingernail clam. There is no clear trend among groups of fish; the several most sensitive tested species and genera include representatives from diverse families (Salmonidae, Cyprinidae, Percidae, and Centrarchidae). Available chronic toxicity data for freshwater organisms also indicate invertebrates (cladocerans, an insect species) to be more tolerant than fishes, again with the exception of the fingernail clam. When corrected for the presumed effects of temperature and pH, there was no clear trend among groups of fish for chronic toxicity values. The most sensitive species, including representatives from five families (Salmonidae, Cyprinidae, Ictaluridae, Centrarchidae, and Catostomidae), have chronic values ranging by not much more than a factor or two. Available data indicate that differences in sensitivities between warm- and cold-water families of aquatic organisms are inadequate to warrant discrimination in the national ammonia criterion between bodies of water with "warm-" and "cold-water" fishes; rather, effects of organism sensitivities on the criterion are most appropriately handled by site-specific criteria derivation procedures.

Data for concentrations of NH_3 toxic to freshwater phytoplankton and vascular plants, although limited, indicate that freshwater plant species are appreciably more tolerant to NH_3 than are invertebrates or fishes. The ammonia criterion appropriate for the protection of aquatic animals will therefore in all likelihood be sufficiently protective of plant life.

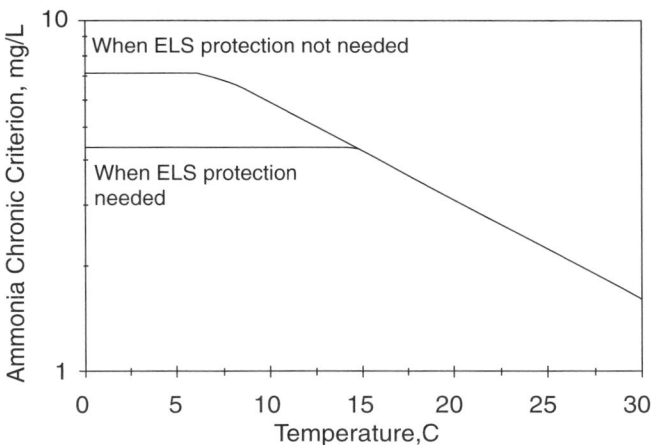

Figure G.1   Chronic criterion values for early life stages (ELS) of fish in the 1999 update; pH = 7.5.

National Ammonia Water Quality Criteria

The U.S. EPA has published a 1999 Update of Ambient Water Quality Criteria for Ammonia (1999 Ammonia Update). The 1999 Ammonia Update contains EPA's most recent freshwater aquatic life criteria for ammonia, superseding all previous EPA-recommended freshwater criteria for ammonia. The 1999 Ammonia Update pertains only to fresh waters and does not change or supersede the aquatic life criterion for ammonia in salt water, published in Ambient Water Quality Criteria for Ammonia (Saltwater) in 1989. The new criteria reflect recent research and data since 1984, and are a revision of several elements in the 1984 criteria, including the pH and temperature relationship of the acute and chronic criteria and the averaging period of the chronic criterion. As a result of these revisions, the acute criterion for ammonia is now dependent on pH and fish species, and the chronic criterion is dependent on pH and temperature. At lower temperatures, the dependency of chronic criterion is also dependent on the presence or absence of early life stages of fish (ELS). The effect of temperature and expected presence of early life stages of fish on the chronic criterion in the 1999 Update is shown in Figure G.1. The temperature dependency in the 1999 Update results in a gradual increase in the criterion as temperature decreases, and a criterion that is more stringent, at temperatures below 15°C, when early life stages of fish (ELS) are expected to be present.

EPA's recommendations in the 1999 Update represent a change from both the 1984 chronic criterion, which was dependent mainly on pH, and from the 1998 Update, in which the chronic criterion was dependent on pH and the presence of early life stages of fish. The temperature dependency of ammonia toxicity at temperatures below 20°C is incorporated directly into the criterion of the 1999 Update. The other significant revision in the 1999 Update is EPA's recommendation of 30 days as the averaging period for the ammonia chronic criterion. EPA recommends the 30B3 (the lowest 30-day average flow based on a 30-year return interval when flow records are analyzed using EPA's 1986 DFLOW procedure), the 30Q10 (the lowest 30-day average flow based on a 10-year return interval when flow records are analyzed using extreme-value statistics), or the 30Q5 as the appropriate design flows associated with the 30-day averaging period of the ammonia chronic criterion. In addition, EPA recommends that within the 30-day averaging period, no 4-day average concentration should exceed 2.5 times the chronic criterion, or criterion continuous concentration (CCC). Consequently, the design flow should also be protective of any 4-day average at 2.5 times the CCC. EPA believes that in the vast majority of cases, the 30Q10 is protective of both the CCC and any 4-day average at 2.5 times the CCC. However, if a state or tribe specifies the use of the 30Q5, then the state or tribe should demonstrate that a 7Q10 (the lowest average 7-day once-in-10-year flow using extreme-value statistics) is protective of 2.5 times the CCC, to ensure that any short-term (4-day) flow variability within the 30-day averaging period does not lead to shorter-term chronic toxicity.

BACTERIA

Development of Bathing Beach Bacteriological Criteria

Dufour (1984) presents an excellent overview of the history of bacterial standards and water contact recreation, summarized here. Total coliforms were initially used as indicators for monitoring outdoor bathing waters, based on a classification scheme presented by W.J. Scott in 1934. Scott had proposed four classes of water, with total coliform upper limits of 50, 500, 1000, and >1000 MPN/100 mL for each class. He had developed this classification based on an extensive survey of the Connecticut shoreline where he found that about 93% of the samples contained less than 1000 total coliforms per 100 mL. A sanitary survey classification also showed that only about 7% of the shoreline was designated as poor. He therefore concluded that total coliform counts of <1000 MPN/100 mL probably indicated acceptable waters for swimming. This standard was based on the principle of attainment, where very little control or intervention would be required to meet this standard. In 1943, the State of California independently adopted an arbitrary total coliform standard of 1000 MPN/100 mL for swimming areas. This California standard was not based on any evidence, but it was assumed to relate well with the drinking water standard at the time.

Dufour points out that a third method used to develop a standard for bathing water quality used an analytical approach adopted by H.W. Streeter in 1951. He used a ratio between *Salmonella* and total coliforms, the number of bathers exposed, the approximate volume of water ingested by bathers daily, and the average total coliform density. Streeter concluded that water containing <1000 MPN total coliforms/100 mL would pose no great *S. typhosa* health hazard. Dufour points out that it is interesting that all three approaches in developing a swimming water criterion resulted in the same numeric limit.

One of the earliest bathing beach studies to measure actual human health risks associated with swimming in contaminated water was directed by Stevenson (1953), of the U.S. Public Health Service's Environmental Health Center, in Cincinnati, OH, and was conducted in the late 1940s. They studied swimming at Lake Michigan at Chicago (91 and 190 MPN/100 mL median total coliform densities), the Ohio River at Dayton, KY (2700 MPN/100 mL), at Long Island Sound at New Rochelle and at Mamaroneck, NY (610 and 253 MPN/100 mL). They also studied a swimming pool in Dayton, KY. Two bathing areas were studied in each area, one with historically poorer water quality than the other. Individual home visits were made to participating families in each area to explain the research program and to review the calendar record form. Follow-up visits were made to each participating household to ensure completion of the forms. Total coliform densities were monitored at each bathing area during the study. More than 20,000 persons participated in the study in the three areas. Almost a million person-days of usable records were obtained. The percentage of the total person-days when swimming occurred ranged from about 5 to 10%. The number of illnesses of all types recorded per 1000 person-days varied from 5.3 to 8.8. They found an appreciably higher illness incidence rate for the swimming group, compared to the nonswimming group, regardless of the bathing water quality (based on total coliform densities). A significant increase in gastrointestinal illness was observed among the swimmers who used one of the Chicago beaches on 3 days when the average coliform count was 2300 MPN/100 mL. The second instance of positive correlation was observed in the Ohio River study where swimmers exposed to the median total coliform density of 2700 MPN/100 mL had a significant increase in gastrointestinal illness, although the illness rate was relatively low. They suggested that the strictest bacterial quality requirements that existed then (as indicated above, based on Scott's 1934 work) might be relaxed without significant detrimental effect on the health of bathers.

It is interesting to note that in 1959, the Committee on Bathing Beach Contamination of the Public Health Laboratory Service of the U.K. concluded that "bathing in sewage-polluted seawater carries only a negligible risk to health, even on beaches that are aesthetically very unsatisfactory" (Alexander et al. 1992).

Dufour (1984) pointed out that total coliforms were an integral element in establishing fecal coliform limits as an indicator for protecting swimming uses. As a result of the Stevenson (1953) study, reported above, a geometric mean fecal coliform level of 200 MPN per 100 mL was recommended by the National Technical Advisory Committee (NTAC) of the Federal Water Pollution Control Administration in 1968 and was adopted by the U.S. Environmental Protection Agency in 1976 as a criterion for direct water contact recreation (Cabelli et al. 1979). This criterion was adopted by almost all states by 1984. It was felt that fecal coliform levels were more specific to sewage contamination and had less seasonal variation than total coliform levels. Since fecal coliform exposures at swimming beaches had never been linked to disease, the NTAC reviewed the USPHS studies, as published by Stevenson (1953). The 2300 MPN/100 mL total coliform count association with gastrointestinal disease was used in conjunction with a measured ratio of fecal coliform to total coliform counts (18%) obtained at the Ohio River site studied earlier. It was therefore assumed that a health effect could be detected when the fecal coliform count was 400 MPN/100 mL (18% of 2300 = 414). Dufour (1984) notes that a detectable health effect was undesirable and that the NTAC therefore recommended a limit of 200 MPN/100 mL for fecal coliforms. Dufour (1984) mentions that, although likely coincidental, the 1968 proposed limit for fecal coliforms (200 MPN/100 mL) was very close to being theoretically equivalent to the total coliform limit of 1000 MPN/100 mL that was being replaced (200/0.18 = 1100).

The Cabelli et al. (1979) study was undertaken to address many remaining questions pertaining to bathing in contaminated waters. Their study examined conditions in New York (at a Coney Island beach, designated as barely acceptable, and at a Rockaway beach, designated as relatively unpolluted). About 8000 people participated in the study, approximately evenly divided between swimmers and nonswimmers at the two beaches. Total and fecal coliforms, *Escherichia*, *Klebsiella, Citrobacter–Enterobacter*, enterococci, *Pseudomonas aeruginosa*, and *Clostridium perfringens* were evaluated in water samples obtained from the beaches during the epidemiological study. The most striking findings were the increases in the rates of vomiting, diarrhea, and stomach ache among swimmers relative to nonswimmers at the barely acceptable beach, but not at the relatively unpolluted beach. Ear, eye, nose, and skin symptoms, as well as fever, were higher among swimmers compared to nonswimmers at both beaches. They concluded that measurable health effects do occur at swimming beaches that meet the existing health standards. Children, Hispanic Americans, and low-middle socioeconomic groups were identified as the most susceptible portions of the population.

Cabelli et al. (1982) presented data from the complete EPA-sponsored swimming beach study, conducted in New York, New Orleans, and Boston. The study was conducted to address issues from prior studies conducted in the 1950s (including Stevenson's 1953 study noted above) that were apparently contradictory. They observed a direct, linear relationship between highly credible gastrointestinal illness and enterococci. The frequency of gastrointestinal symptoms also had a high degree of association with distance from known sources of municipal wastewater. Table G.1 shows correlation coefficients for total gastrointestinal (GI) and highly credible gastrointestinal (HCGI) symptoms and mean indicator densities found at the New York beaches from 1970 to 1976. The best correlation coefficients were found for enterococci. In contrast, the correlation coefficients for fecal coliforms (the basis for most federal and state guidelines) were poor. Very low levels of enterococcus and *Escherichia coli* in the water (about 10 MPN/100 mL) were associated with appreciable attack rates (about 10/10,000 persons).

They concluded that swimming in even marginally polluted marine bathing water is a significant route of transmission for observed gastrointestinal illness. They felt that the gastrointestinal illness was likely associated with the Norwalk-like virus that had been confirmed in 2000 cases in a shellfish-associated outbreak in Australia and in several outbreaks associated with contaminated drinking water.

Table G.1 Correlation Coefficients between Gastrointestinal Symptoms and Bacterial Densities at New York City Beaches

Indicator	HCGI Correlation Coefficient	GI Correlation Coefficient	Number of Observations
Enterococci	0.96	0.81	9
Escherichia coli	0.58	0.51	9
Klebsiella	0.61	0.47	11
Enterobacter-Citrobacter	0.64	0.54	13
Total coliforms	0.65	0.46	11
Clostridium perfringens	0.01	−0.36	8
Pseudomonas aeruginosa	0.59	0.35	11
Fecal coliforms	0.51	0.36	12
Aeromonas hydriphila	0.60	0.27	11
Vibrio parahemolyticus	0.42	0.05	7

From Cabelli et al. 1982.

Table G.2 Correlation Coefficients for Bacterial Parameters and Gastrointestinal Disease (Freshwater Swimming Beaches)

	Highly Credible Gastrointestinal Illness	Total Gastrointestinal Illness	Number of Study Units
Enterococci	0.774	0.673	9
E. coli	0.804	0.528	9
Fecal coliforms	−0.081	0.249	7

From Dufour 1984.

Dufour (1984) also reviewed a series of studies conducted at freshwater swimming beaches from 1979 to 1982, at Tulsa, OK, and at Erie, PA. Only enterococci, E. coli, and fecal coliforms were monitored, based on the results of the earlier studies. Table G.2 shows the correlation coefficients for these three bacterial parameters and gastrointestinal disease.

These results are quite different from the results of the marine studies in that both enterococci and E. coli had high correlation coefficients between the bacterial levels and the incidence of gastrointestinal illness. However, the result was the same for fecal coliforms, in that there was no association between fecal coliform levels and gastrointestinal illness. Dufour (1984) concluded that enterococci would be the indicator of choice for gastrointestinal illness, based on scientific dependability. E. coli could also be used, if only fresh waters were being evaluated. Fecal coliforms would be a poor choice for monitoring the safety of bathing waters. However, he concluded that numeric standards should be different for fresh and saline waters because of different die-off rates for the bacteria and viruses for differing salinity conditions.

Other studies examined additional illness symptoms associated with swimming in contaminated water, besides gastrointestinal illness, and identified other potentially useful bacterial indicators. Seyfried et al. (1985), for example, examined users of swimming beaches in Toronto for respiratory illness, skin rashes, plus eye and ear problems, in addition to gastrointestinal illness. They found that total staphylococci correlated best with swimming-associated total illness, plus ear, eye, and skin illness. However, fecal streptococci and fecal coliforms also correlated (but not as well) with swimming-associated total illness. Ferley et al. (1989) examined illnesses among swimmers during the summer of 1986 in the French Ardèche river basin, during a time when untreated domestic sewage was entering the river. They examined total coliforms, fecal coliforms, fecal streptococci, *Pseudomonas aeruginosa,* and *Aeromonas* spp., but only two samples per week were available for each swimming area. The total morbidity ratio for swimmers compared to nonswimmers was 2.1 (with a 95% confidence interval of 1.8 to 2.4), with gastrointestinal illness the major illness observed. They found that fecal streptococci (FS) was the best indicator of gastrointestinal illness. A critical FS value of 20 MPN/100 mL indicated significant differences between the swimmers and nonswimmers. Skin ailments were also more common for swimmers than for nonswimmers and were well correlated with the concentrations of fecal coliforms, *Aeromonas* spp., and *P. aeruginosa*. They noted that a large fraction (about 60%) of the fecal coliforms corresponded to *E. coli,* and that their definition of fecal streptococci essentially was what North American researchers termed enterococci.

Many of the available epidemiological studies have been confined to healthy adult swimmers, in relatively uncontaminated waters. However, it is assumed that those most at risk would be children, the elderly, and those chronically ill, especially in waters known to be degraded. Obviously, children are the most likely of this most-at-risk group to play in, or by, water. Alexander et al. (1992) therefore specifically examined the risk of illness associated with swimming in contaminated sea water for children, aged 6 to 11 years old. This study was based on parental interviews for 703 child participants during the summer of 1990 at Blackpool beach, U.K. Overall, 80% of the samples at the Blackpool Tower site and 93% of the samples at the South Pier site failed to meet the European Community standards for recreational waters. All of the 11 designated beaches in Lancashire (including Blackpool beach), in the northwest region of England, continually failed the European directive imperative standards for recreational waters. During this study, statistically significant increases in disease were found in children who had water contact compared to those who did not. Diarrhea and loss of appetite had strong associations with the water contact group, while vomiting and itchy skin had moderate associations. No other variables examined (household income, sex of the child, sex of the respondent, general health, chronic or recurring illness in the child, age of the child, foods eaten, including ice cream, other dairy products, chicken, hamburgers, shellfish, or ice cubes, acute symptoms in other household members, presence of children under 5 in the household, and other swimming activities) could account for the significant increases in the reported symptoms for the children who had water contact.

Santa Monica Bay Project

This study was the first large-scale epidemiological study in the U.S. to investigate possible adverse health effects associated with swimming in ocean waters affected by discharges from separate storm drains (SMBRP 1996). This was a follow-up study after previous investigations found that human fecal waste was present in the stormwater collection systems (*Water Environment & Technology* 1996b; *Environmental Science & Technology* 1996b; Haile et al. 1996). This subsection was previously considered in Chapter 4 of this book, but is repeated here for comparison with the other discussions on the development of the standards for bacteria exposure from stormwater.

During a 4-month period in the summer of 1995, about 15,000 ocean swimmers were interviewed on the beach and during telephone interviews 1 to 2 weeks later. They were queried concerning illnesses since their beach outing. The incidence of illness (such as fever, chills, ear discharge, vomiting, coughing with phlegm, and credible gastrointestinal illness) was significantly greater (from 44 to 127% increased incidence) for oceangoers who swam directly off the outfalls, compared to those who swam 400 yards away, as shown on Table G.3. As an example, the rate ratio (RR) for fever was 1.6, while it was 2.3 for ear discharges, and 2.2 for highly credible gastrointestinal illness (HCGI) comprised of vomiting and fever. The approximated associations were weak for any of the symptoms, and moderate for the others listed. Disease incidence dropped significantly with distance from the storm drain. At 400 yards, and beyond, upcoast or downcoast, elevated disease risks were not found. The results did not change when adjusted for age, beach, gender, race, socioeconomic status, or worry about health risks associated with swimming at the beach.

These interviews were supplemented with indicator and pathogen bacteria and virus analyses in the waters. The greatest health problems were associated with times of highest concentrations (*E. coli* > 320 cfu/100 mL, enterococcus > 10^6 cfu/100 mL, total coliforms > 10,000 cfu/100 mL, and fecal coliforms > 400 cfu/100 mL). Bacteria populations greater than these are common in urban runoff and in urban receiving waters. Symptoms were found to be associated with swimming in areas where bacterial indicator levels were greater than these critical counts. Table G.4 shows the health outcomes associated with swimming in areas having bacterial counts greater than these

Table G.3 Comparative Health Outcomes for Swimming in Front of Storm Drain Outfalls, Compared to Swimming at Least 400 Yards Away

Health Outcome	Relative Risk	Rate Ratio	Estimated Association	Estimated No. of Excess Cases per 10,000 Swimmers (rate difference)
Fever	57%	1.57	Moderate	259
Chills	58%	1.58	Moderate	138
Ear discharge	127%	2.27	Moderate	88
Vomiting	61%	1.61	Moderate	115
Coughing with phlegm	59%	1.59	Moderate	175
Any of the above symptoms	44%	1.44	Weak	373
HCGI-2	111%	2.11	Moderate	95
SRD (significant respiratory disease)	66%	1.66	Moderate	303
HCGI-2 or SRD	53%	1.53	Moderate	314

From SMBRP 1996.

Table G.4 Heath Outcomes Associated with Swimming in Areas Having High Bacterial Counts

Indicator (and critical cutoff count)	Health Outcome	Increased Risk	Risk Ratio	Estimated Association	Excess Cases per 10,000 Swimmers
E. coli (>320 cfu/100 mL)	Earache and nasal congestion	46% 24%	1.46 1.24	Weak Weak	149 211
Enterococci (>106 cfu/100 mL)	Diarrhea w/blood and HCGI-1	323% 44%	4.23 1.44	Strong Weak	27 130
Total coliform bacteria (>10,000 cfu/100 mL)	Skin rash	200%	3.00	Moderate	165
Fecal coliform bacteria (>400 cfu/100 mL)	Skin rash	88%	1.88	Moderate	74

From SMBRP 1996.

critical values. The association for enterococci with bloody diarrhea was strong, and the association of total coliforms with skin rash was moderate, but nearly strong.

The ratio of total coliform to fecal coliform was found to be one of the better indicators for predicting health risks when swimming close to the storm drain. When the total coliforms were greater than 1000 cfu/100 mL, the strongest effects were generally observed when the total to fecal coliform ratio was 2. The risks decreased as the ratio increased. In addition, illnesses were more common on days when enteric viruses were found in the water.

The SMBRP (1996) concluded that less than 2 miles of Santa Monica Bay's 50-mile coastline had problematic health concerns due to the storm drains flowing into the bay. They also concluded that the bacterial indicators currently being monitored do help predict risk. In addition, the total to fecal coliform ratio was found to be a useful additional indicator of illness. As an outcome of this study, the Los Angeles County Department of Health Services will post new warning signs advising against swimming near the outfalls ("Warning! Storm drain water may cause illness. No swimming"). These signs will be posted on both sides of all flowing storm drains in Los Angeles County. In addition, county lifeguards will attempt to warn and advise swimmers to stay away from areas directly in front of storm drain outlets, especially in ponded areas. The county is also accelerating its studies on sources of pathogens in stormwater.

Bacteria Criteria for Water-Contact Recreation

A recreational water quality criterion can be defined as a "quantifiable relationship between the density of an indicator in the water and the potential human health risks involved in the water's recreational use." From such a definition, a criterion can be adopted which establishes upper limits for densities of indicator bacteria in waters that are associated with acceptable health risks for swimmers.

Table G.5 National Bacteria Criteria (Single Sample Maximum Allowable Density, counts per 100 mL)

		Designated Beach[a]	Moderate Full Body Contact Recreation[a]	Lightly Used Full Body Contact[a]	Infrequently Used Full Body Contact[a]	Drinking Water[b]
Freshwater	Enterococci	61	89	108	151	1
	E. coli	235	298	406	576	1
Marine water	Enterococci	104	124	276	500	1

[a] EPA 1986

The Environmental Protection Agency, in 1972, initiated a series of studies at marine and freshwater bathing beaches which were designed to determine if swimming in sewage-contaminated marine and fresh water carries a health risk for bathers, and, if so, to what type of illness. Additionally, the EPA wanted to determine which bacterial indicator is best correlated to swimming-associated health effects and if the relationship is strong enough to provide a criterion (EPA 1986a).

The quantitative relationships between the rates of swimming-associated health effects and bacterial indicator densities were determined using standard statistical procedures. The data for each summer season were analyzed by comparing the bacteria indicator density for a summer bathing season at each beach with the corresponding swimming-associated gastrointestinal illness rate for the same summer. The swimming-associated illness rate was determined by subtracting the gastrointestinal illness rate in nonswimmers from that for swimmers.

The EPA's evaluation of the bacteriological data indicated that using the fecal coliform indicator group at the maximum geometric mean of 200 organisms per 100 mL, as recommended in *Quality Criteria for Water*, would cause an estimated 8 illness per 1000 swimmers at freshwater beaches.

Additional criteria, using *E. coli* and enterococci bacteria analyses, were developed using these currently accepted illness rates. These bacteria are assumed to be more specifically related to poorly treated human sewage than the fecal coliform bacteria indicator. The equations developed by Dufour (1983) were used to calculate new indicator densities corresponding to the accepted gastrointestinal illness rates.

It should be noted that these indicators only relate to gastrointestinal illness, and not other problems associated with waters contaminated with other bacterial or viral pathogens. Common swimming beach problems associated with contamination by stormwater include skin and ear infections caused by *Pseudomonas aeruginosa* and *Shigella*.

National bacteria criteria have been established for contact with bacteria and are shown in Table G.5. State standards usually also exist for fecal coliform bacteria. Typical public water supply standards (Alabama's are shown) are as follows:

1. Bacteria of the fecal coliform group shall not exceed a geometric mean of 2000/100 mL; nor exceed a maximum of 4000/100 mL in any sample. The geometric mean shall be calculated from no less than five samples collected at a given station over a 30-day period at intervals not less than 24 hours. The membrane filter counting procedure will be preferred, but the multiple tube technique (five-tube) is acceptable.

2. For incidental water contact and recreation during June through September, the bacterial quality of water is acceptable when a sanitary survey by the controlling health authorities reveals no source of dangerous pollution and when the geometric mean fecal coliform organism density does not exceed 100/100 mL in coastal waters and 200/100 mL in other waters. When the geometric mean fecal coliform organism density exceeds these levels, the bacterial water quality shall be considered acceptable only if a second detailed sanitary survey and evaluation discloses no significant public health risk in the use of such waters. Waters in the immediate vicinity of discharges of sewage or other wastes likely to contain bacteria harmful to humans, regardless of the degree of treatment afforded these wastes, are not acceptable for swimming or other whole-body water-contact sports.

Standards for fish and wildlife waters are similar to the above standard for a public water supply, except Part 1 has different limits: "Bacteria of the fecal coliform group shall not exceed a geometric mean of 1000/100 mL on a monthly average value; nor exceed a maximum of 2000/100 mL in any sample." Part 2 is the same for both water beneficial uses.

CHLORIDE, CONDUCTIVITY, AND TOTAL DISSOLVED SOLIDS

Total dissolved solids, chlorides, and conductivity observations are typically used to indicate the magnitude of dissolved minerals in the water. The term *total dissolved solids* (or *dissolved solids*) is generally associated with fresh water and refers to the inorganic salts, small amounts of organic matter, and dissolved materials in the water. Salinity is an oceanographic term, and although not precisely equivalent to the total dissolved salt content, it is related (Capurro 1970). Chlorides (not chlorine) are directly related to salinity because of the constant relationship between the major salts in seawater. Conductivity is a measure of the electrical conductivity of water and is also generally related to total dissolved solids, chlorides, or salinity. The principal inorganic anions (negatively charged ions) dissolved in fresh water include the carbonates, chlorides, sulfates, and nitrates (principally in groundwaters); the principal cations (positively charged ions) are sodium, potassium, calcium, and magnesium.

Human Health Criteria for Dissolved Solids

Excess dissolved solids are objectionable in drinking water because of possible physiological effects, unpalatable mineral tastes, and higher costs because of corrosion or the necessity for additional treatment. The physiological effects directly related to dissolved solids include laxative effects principally from sodium sulfate and magnesium sulfate and the adverse effect of sodium on certain patients afflicted with cardiac disease and women with toxemia associated with pregnancy. One study was made using data collected from wells in North Dakota. Results from a questionnaire showed that with wells in which sulfates ranged from 1000 to 1500 mg/L, 62% of the respondents indicated laxative effects associated with consumption of the water. However, nearly one quarter of the respondents to the questionnaire reported difficulties when concentrations ranged from 200 to 500 mg/L (Moore 1952). To protect transients to an area, a sulfate level of 250 mg/L should afford reasonable protection from laxative effects.

As indicated, sodium frequently is the principal component of dissolved solids. Persons on restricted sodium diets may have an intake restricted from 500 to 1000 mg/day (National Research Council 1954). The portion ingested in water must be compensated by reduced levels in food ingested so that the total does not exceed the allowable intake. Using certain assumptions of water intake (e.g., 2 L of water consumed per day) and the sodium content of food, it has been calculated that for very restricted sodium diets, 20 mg/L sodium in water would be the maximum, while for moderately restricted diets, 270 mg/L sodium would be the maximum. Specific sodium levels for entire water supplies have not been recommended by the EPA, but various restricted sodium intakes are recommended because: (1) the general population is not adversely affected by sodium, but various restricted sodium intakes are recommended by physicians for a significant portion of the population, and (2) 270 mg/L of sodium is representative of mineralized waters that may be aesthetically unacceptable, but many domestic water supplies exceed this level. Treatment for removal of sodium in water supplies is also costly (NAS 1974).

A study based on consumer surveys in 29 California water systems was made to measure the taste threshold of dissolved salts in water (Bruvold et al. 1969). Systems were selected to eliminate possible interferences from other taste-causing substances besides dissolved salts. The study revealed that consumers rated waters with 320 to 400 mg/L dissolved solids as "excellent," while those with 1300 mg/L dissolved solids were "unacceptable." A "good" rating was registered for dissolved solids less than 650 to 750 mg/L. The 1962 U.S. Public Health Service Drinking Water

Standards recommended a maximum dissolved solids concentration of 500 mg/L, unless more suitable supplies were unavailable.

Specific constituents included in the dissolved solids in water may cause mineral tastes at lower concentrations than other constituents. Chloride ions have frequently been cited as having a low taste threshold in water. Data from Ricter and MacLean (1939) on a taste panel of 53 adults indicated that 61 mg/L NaCl was the median level for detecting a difference from distilled water. At a median concentration of 395 mg/L chloride, a salty taste was identified. Lockhart et al. (1955) when evaluating the effect of chlorides on water used for brewing coffee, found threshold taste concentrations for chloride ranging from 210 to 310 mg/L, depending on the associated cation. These data indicate that a level of 250 mg/L chlorides is a reasonable maximum level needed to protect consumers.

The causation of corrosion and encrustation of metallic surfaces by water containing dissolved solids is well known. By using water with 1750 mg/L dissolved solids as compared with 250 mg/L, service life was reduced from 70% for toilet flushing mechanisms to 30% for washing equipment. Such increased corrosion was calculated in 1968 to cost the consumer an additional $0.50 per 1000 gallons used.

The U.S. EPA has adopted secondary drinking water standards (40 CFR D143.3) and ambient water quality criteria. The National Secondary Drinking Water Maximum Contaminant Level (MCL) for chloride is 250 mg/L (40 CFR D 143.3). This corresponds roughly to a conductivity measurement of about 1200 $\mu S/cm^2$, but this is never exactly the case. However, the relationship between conductivity and chloride can be established on a site-specific basis. Chloride toxicity is increased when the counter ion of the chloride salt is not sodium.

Aquatic Life Criteria for Dissolved Solids

All species of fish and other aquatic life must tolerate a range of dissolved solids concentrations in order to survive under natural conditions. Studies in Saskatchewan found that several common freshwater species survived 10,000 mg/L dissolved solids, that whitefish and pikeperch survived 15,000 mg/L, but only the stickleback survived 20,000 mg/L dissolved solids. It was concluded that lakes with dissolved solids in excess of 15,000 mg/L were unsuitable for most freshwater fishes (Rawson and Moore 1944). The 1968 NTAC Report also recommended maintaining an osmotic pressure level of less than that caused by a 15,000 mg/L solution of sodium chloride.

Indirect effects of excess dissolved solids are primarily the elimination of desirable food plants and other habitat-forming plants. Rapid salinity changes cause plasmolysis of tender leaves and stems because of changes in osmotic pressure. The 1968 NTAC Report recommended the following limits in salinity variation from natural to protect wildlife habitats:

Natural Salinity (parts per thousand)	Variation Permitted (parts per thousand)
0 to 3.5 (fresh water)	1
3.5 to 13.5 (brackish water)	2
13.5 to 35 (seawater)	4

Alabama is an example of a state that has established a standard for chloride to protect aquatic life. A chloride criterion of 230 mg/L is used to protect aquatic life in the Cahaba River.

CHROMIUM

Aquatic Life Effects of Cr^{3+}

Acute values for Cr^{3+} are available for 20 freshwater animal species in 18 genera ranging from 2.2 mg/L for a mayfly to 71 mg/L for caddisfly. Hardness has a significant influence on toxicity, with Cr^{3+} being more toxic in soft water.

A life-cycle test with *Daphnia magna* in soft water gave a chronic value of 66 µg/L. In a comparable test in hard water, the lowest test concentration of 44 µg/L inhibited reproduction of *D. magna*, but this effect may have resulted from ingested precipitated chromium. In a life-cycle test with the fathead minnow in hard water, the chronic value was 1.0 mg/L. Toxicity data were available for only two freshwater plant species. A concentration of 9.9 mg/L inhibited growth of roots of Eurasian watermilfoil. A freshwater green alga was affected by a concentration of 397 µg/L in soft water. No bioconcentration factor was measured for Cr^{3+} with freshwater organisms.

National Freshwater Aquatic Life Criteria for Cr^{3+}

The procedures described in the guidelines indicate that, except possibly where a locally important species is very sensitive, freshwater aquatic organisms and their uses should not be affected unacceptably if the 4-day average (chronic) concentration (in µg/L) of Cr^{3+} does not exceed the numerical value given by:

$$e^{(0.8190(\ln(\text{hardness}))+1.561)}$$

more than once every 3 years on the average, and if the 1-hour average (acute) concentration (in µg/L) does not exceed the numerical value given by:

$$e^{(0.8190(\ln(\text{hardness}))+3.688)}$$

more than once every 3 years on the average. For example, at hardnesses of 50, 100, and 200 mg/L as $CaCO_3$ the 4-day average concentrations of Cr^{3+} are 120, 210, and 370 µg/L, respectively, and the 1-hour average concentrations are 980, 1700, and 3100 µg/L. Many states have adopted these equations to define the Cr^{3+} standards for freshwater aquatic life uses.

Human Health Criteria for Chromium

For the protection of human health from the toxic properties of Cr^{3+} ingested through water and contaminated aquatic organisms, the ambient water criterion is determined to be 170 mg/L. For the protection of human health from the toxic properties of Cr^{3+} ingested through contaminated aquatic organisms alone, the ambient water criterion is determined to be 3433 mg/L. In contrast, the ambient water quality criterion for total Cr^{6+} is recommended to be identical to the existing drinking water standard, which is 50 µg/L.

COPPER

Effects of Copper on Aquatic Life

Acute toxicity data are available for species in 41 genera of freshwater animals. At a hardness of 50 mg/L, the genera range in sensitivity from 17 µg/L for *Ptychocheilus* to 10 mg/L for *Acroneuria*. Data for eight species indicate that acute toxicity decreases as hardness increases. Additional data for several species indicate that toxicity also decreases with increased alkalinity and total organic carbon.

Chronic values are available for 15 freshwater species and range from 3.9 µg/L for brook trout to 60 µg/L for northern pike. Fish and invertebrate species seem to be about equally sensitive to the chronic toxicity of copper.

Toxicity tests have been conducted on copper with a wide range of freshwater plants and the sensitivities are similar to those of animals. Complexing effects of the test media and a lack of

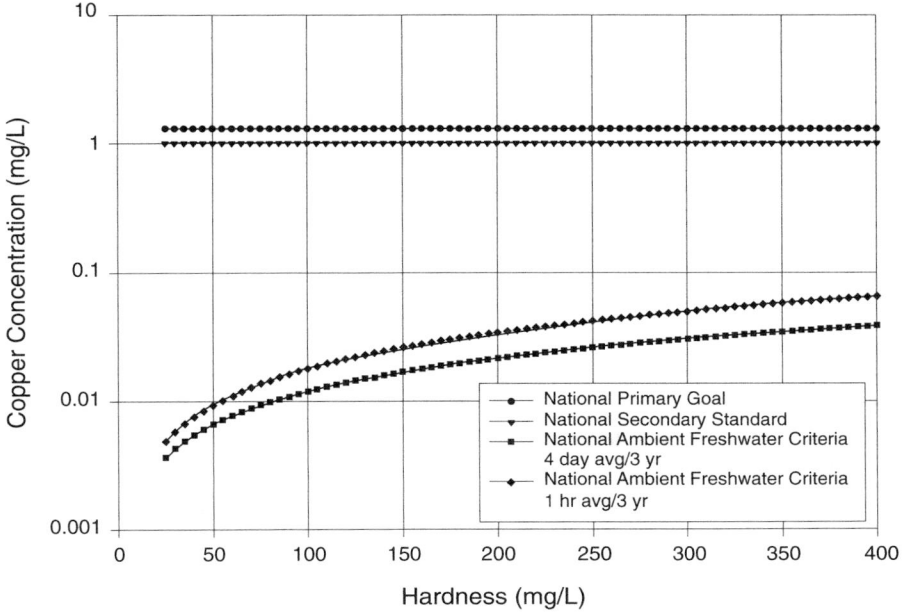

Figure G.2 National copper criteria.

good analytical data make interpretation and application of these results difficult. Protection of animal species, however, appears to offer adequate protection of plants. Copper does not appear to bioconcentrate very much in the edible portion of freshwater aquatic species.

National Aquatic Life Criteria for Copper

The U.S. EPA has established a national ambient water quality criteria for the protection of wildlife (EPA 1986b). The wildlife protection criteria are a function of hardness and are shown in Figure G.2.

Human Health Criteria for Copper

The U.S. EPA has established a primary drinking water goal (40 CFR D Subpart F 141.51) of 1.3 mg/L, a secondary drinking water quality MCL of 1.0 mg/L (40 CFR 143.3).

HARDNESS

Water hardness is caused by the divalent metallic ions (having charges of +2) dissolved in water. In fresh water, these are primarily calcium and magnesium, although other metals such as iron, strontium, and manganese also contribute to the hardness content, but usually to a much lesser degree. Hardness commonly is reported as an equivalent concentration of calcium carbonate ($CaCO_3$).

Concerns about water hardness originated because hard water requires more soap to form a lather and because hard water causes scale in hot water systems. Modern use of synthetic detergents has eliminated the concern of hard water in laundries, but it is still of primary concern for many industrial water users. Many households use water softeners to reduce scale formation in hot water systems and for water taste reasons.

There are no national standards for hardness, but water hardness has a dramatic effect on criteria for a number of heavy metals. "The affects of hardness on freshwater fish and other aquatic life appear

Table G.6 USGS Hardness Scale

Hardness (mg/L as CaCO₃)	Classification
<60	Soft
61–120	Moderately hard
121–180	Hard
>180	Very hard

From Leeden et al. 1990.

to be related to the ions causing hardness rather than hardness (EPA 1986b)." The USGS classifies the hardness of waters using the scale in Table G.6.

Natural sources of hardness principally are limestones which are dissolved by percolating rainwater. Groundwaters are therefore generally harder than surface waters. Industrial sources include the inorganic chemical industry and discharges from operating and abandoned mines.

Hardness in fresh water is frequently distinguished in carbonate and noncarbonate fractions. The carbonate fraction is chemically equivalent to the bicarbonates present in water. Since bicarbonates are generally measured as alkalinity, the carbonate hardness is equal to the alkalinity.

The determination of hardness in raw waters subsequently treated and used for domestic water supplies is useful as a parameter to characterize the total dissolved solids present and for calculating chemical dosages for water softening. Because hardness concentrations in water have not been proven to be health related, the final level of hardness to be achieved by water treatment principally is a function of economics. Since water hardness can be removed with treatment by such processes as lime-soda softening and ion exchange systems, a water quality criterion for raw waters used as a public water supply is not given by the EPA.

The effects of hardness on freshwater fish and other aquatic life appear to be related to the ions causing the hardness rather than by hardness as a general indicator. Both the NTAC (1968) and NAS (1974) panels have recommended against the use of the term *hardness* and suggested the use of the concentrations of the specific ions instead. For most existing data, it is difficult to determine whether toxicity of various metal ions is reduced because of the formation of metallic hydroxides and carbonates caused by the associated increases in alkalinity, or because of an antagonistic effect of one of the principal cations contributing to hardness, e.g., calcium, or a combination of both effects. Stiff (1971) presented an example showing that if cupric ions were the toxic form of copper, whereas copper carbonate complexes were relatively nontoxic, then the observed difference in toxicity of copper between hard and soft waters can be explained by the difference in alkalinity rather than hardness. Recent laboratory work has also shown that alkalinity may be more related to heavy metal toxicity than water hardness. As noted previously, however, carbonate hardness and alkalinity are the same.

Doudoroff and Katz (1953), in their review of the literature on toxicity, presented data showing that increasing calcium in particular reduced the toxicity of other heavy metals. Under usual conditions in fresh water and assuming that other bivalent metals behave like copper, it is reasonable to assume that both effects occur simultaneously and explain the observed reduction of toxicity of metals in waters containing carbonate hardness. The amount of reduced toxicity related to hardness, as measured by a 40-hour LC50 for rainbow trout, has been estimated to be about four times for copper and zinc when the hardness was increased from 10 to 100 mg/L as $CaCO_3$ (NAS 1974). As shown in other discussions for specific heavy metals, many of the heavy metal criteria depend on water hardness. The allowable concentrations of cadmium, chromium, lead, and zinc to protect fish and other aquatic life, are much less in soft waters than in hard waters, for example.

HYDROCARBONS

The U.S. EPA has promulgated criteria for several of the organic toxicants that can be found in stormwater or in urban receiving waters. In addition, the EPA has specific criteria for the detection of individual organic molecules. The MCLs (maximum concentration limits) for the individual chemicals are mostly all well below 0.1 mg/L (40 CFR D Subpart F 141.50 and Subpart G 141.61). The following table summarizes several of the criteria for toxic organics:

aldrin+dieldrin	0.002 µg/L (acute freshwater aquatic life)
	0.007 ng/L (human health)
chlorodane	2.4 µg/L (maximum conc. for acute freshwater aquatic life)
	0.046–4.6 µg/L (human health)
DDT and metabolites	1.1 µg/L (maximum concentration for acute freshwater aquatic life)
DDE	1.05 mg/L (acute freshwater aquatic life)
2,4-dichlorophenol	2.02 mg/L (acute freshwater aquatic life)
2,4-dimethylphenol	2.1 mg/L (acute freshwater aquatic life)
endosulfan	0.05 µg/L (acute freshwater aquatic life)
endrin	0.0023 µg/L (acute freshwater aquatic life)
pentachlorophenol	55 µg/L (acute freshwater aquatic life)
phthalate esters	940 µg/L (acute freshwater aquatic life)
polycyclic aromatic hydrocarbons	0.28–28 ng/L (human health)

Several of the compounds periodically found in urban runoff also have state and/or national standards for the protection of human health, including some that are recognized carcinogens. The following table lists typical limits (for Alabama, at 10^{-5} risk level):

	Water and Fish Consumption	Fish Consumption Only
Noncarcinogens		
2-Chlorophenol	0.12 mg/L	0.40 mg/L
Diethyl phthalate	23	118
Dimethyl phthalate	313	2900
Di-*n*-butyl phthalate	3	12
Isophorone	7	490
Carcinogens		
Benzo(ghi)perylene	0.03 µg/L	0.31 µg/L
Benzo(k)fluoranthene	0.03	0.31
3,3-Dichloro-benzidine	0.39	0.77
Hexachlorobutadiene	4.5	500
N-Nitrosodiphenylamine	50	160

Florida water quality criteria for organic toxicants include the following pesticide limits:

2,4-D	0.1 µg/L(potable water supply)
andrin+dieldrin	0.003 µg/L (potable water supply, recreation, fish and wildlife)
chlordane	0.01 µg/L (potable water supply)
	0.01 µg/L (recreation, fish and wildlife)
endosulfan	0.003 µg/L (potable water supply, recreation, fish and wildlife)
endrin	0.004 µg/L (potable water supply, recreation, fish and wildlife)
heptachlor	0.001 µg/L (potable water supply, recreation, fish and wildlife)
lindane	0.01 µg/L (potable water supply, recreation, fish and wildlife)
malathion	0.1 µg/L (potable water supply, recreation, fish and wildlife)
methoxychlor	0.03 µg/L (potable water supply, recreation, fish and wildlife)
mirex	0.001 µg/L (potable water supply, recreation, fish and wildlife)
parathion	0.04 µg/L (potable water supply, recreation, fish and wildlife)
toxaphene	0.005 µg/L (potable water supply, recreation, fish and wildlife)

LEAD

Aquatic Life Summary for Lead

The acute toxicity of lead to several species of freshwater animals has been shown to decrease as the hardness of water increases. At a hardness of 50 mg/L, the acute sensitivities of 10 species

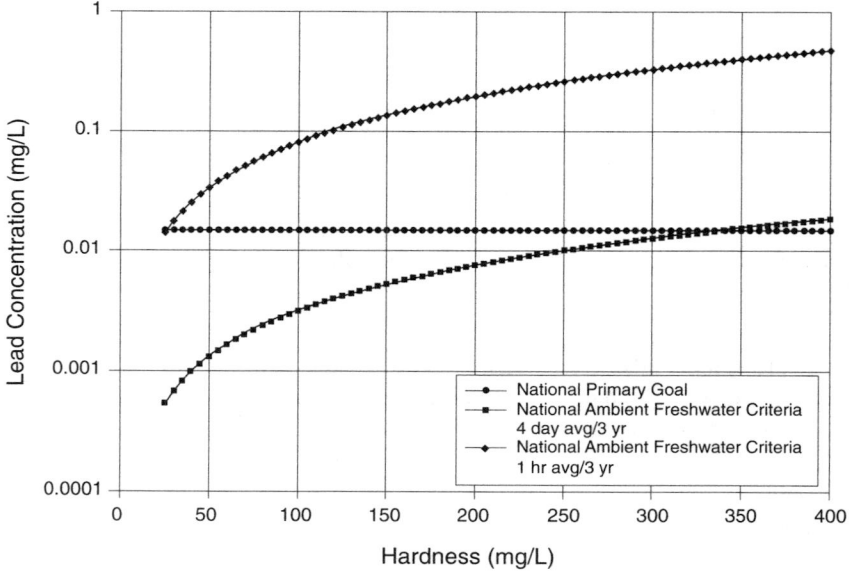

Figure G.3   National lead criteria.

range from 142 µg/L for an amphipod to 236 mg/L for a midge. Data on the chronic effects of lead on freshwater animals are available for two fish and two invertebrate species. The chronic toxicity of lead also decreases as hardness increases and the lowest and highest available chronic values (12.3 and 128 µg/L) are both for a cladoceran, but in soft and hard water, respectively. Freshwater algae are affected by concentrations of lead above 500 µg/L, based on data for four species. Bioconcentration factors are available for four invertebrate and two fish species and range from 42 to 1700.

National Aquatic Life Criteria for Lead

For the protection of wildlife, U.S. EPA has set a national freshwater criteria for lead that is a function of hardness. Figure G.3 shows these standards.

Human Health Criteria for Lead

The U.S. EPA has set the lead National Drinking Water MCL goal at 0 mg/L (40 CFR D Subpart F 141.51) and the National Drinking Action Level at 0.015 mg/L (40 CFR D Subpart I 141.80 (2) (c)).

NITRATE AND NITRITE

Two gases (molecular nitrogen and nitrous oxide) and five forms of nongaseous, combined nitrogen (amino and amide groups, ammonium, nitrite, and nitrate) are important in the nitrogen cycle. The amino and amide groups are found in soil organic matter and as constituents of plant and animal protein. The ammonium ion either is released from proteinaceous organic matter and urea or is synthesized in industrial processes involving atmospheric nitrogen fixation. The nitrite ion is formed from the nitrate or the ammonium ions by certain microorganisms found in soil, water, sewage, and the digestive tract. The nitrate ion is formed by the complete oxidation of ammonium ions by soil or water microorganisms; nitrite is an intermediate product of this nitrification process. In oxygenated natural water systems, nitrite is rapidly oxidized to nitrate. Growing plants assimilate nitrate or ammonium ions and convert them to protein. A process known as denitrification takes place when nitrate containing soils become anaerobic and the conversion to nitrite, molecular nitrogen, or nitrous oxide occurs. Ammonium ions may also be produced in some circumstances.

Among the major point sources of nitrogen entering water bodies are municipal and industrial wastewaters, septic tanks, and feed lot discharges. Nonpoint sources of nitrogen include farm-site fertilizer and animal wastes, lawn fertilizer, sanitary landfill leachate, atmospheric fallout, nitric oxide and nitrite discharges from automobile exhausts and other combustion processes, and losses from natural sources such as mineralization of soil organic matter. Water reuse systems in some fish hatcheries employ a nitrification process for ammonia reduction; this may result in exposure of the hatchery fish to elevated levels of nitrite (Russo et al. 1974).

Human Health Nitrate and Nitrite Criteria

In quantities normally found in food or feed, nitrates become toxic only under conditions in which they are, or may be, reduced to nitrites. Otherwise, at "reasonable" concentrations, nitrates are rapidly excreted in the urine. High intake of nitrates constitutes a hazard primarily to warm-blooded animals under conditions that are favorable to reduction to nitrite. Under certain circumstances, nitrate can be reduced to nitrite in the gastrointestinal tract. It then reaches the bloodstream and reacts directly with hemoglobin to produce methemoglobin, consequently impairing oxygen transport.

The reaction of nitrite with hemoglobin can be hazardous in infants under 3 months of age. Serious and occasionally fatal poisonings in infants have occurred following ingestion of untreated well waters shown to contain nitrate at concentrations greater than 10 mg/L nitrate nitrogen (as N) (NAS 1974). High nitrate concentrations are frequently found in shallow farm and rural community wells, often as the result of inadequate protection from barnyard drainage or from septic tanks (USPHS 1961; Stewart et al. 1967). Increased concentrations of nitrates also have been found in streams from farm tile drainage in areas of intense fertilization and farm crop production (Harmeson et al. 1971). Approximately 2000 cases of infant methemoglobinemia have been reported in Europe and North America between 1945 and 1950; 7 to 8% of the affected infants died (Walton 1951). Many infants have drunk water in which the nitrate nitrogen content was greater than 10 mg/L without developing methemoglobinemia. The differences in susceptibility to methemoglobinemia are not yet understood, but appear to be related to a combination of factors including nitrate concentration, enteric bacteria, and the lower acidity characteristic of the digestive systems of baby mammals. Methemoglobinemia systems and other toxic effects were observed when high nitrate well waters containing pathogenic bacteria were fed to laboratory mammals (Wolff et al. 1972). Conventional water treatment has no significant effect on nitrate removal from water (NAS 1974).

Because of the potential risk of methemoglobinemia to bottle-fed infants, and in view of the absence of substantiated physiological effects at nitrate concentrations below 10 mg/L nitrate nitrogen, this level is the criterion for domestic water supplies. Waters with nitrite nitrogen concentrations over 1 mg/L should not be used for infant feeding. Waters with a significant nitrite concentration usually would be heavily polluted and probably bacteriologically unacceptable.

The only national criterion for nitrate is 10 mg/L as N (40 CFR D Subpart F 141.51). The criterion applies to domestic water supplies. As noted above, the real danger from nitrate occurs when nitrate occurs in a reducing environment and converts to nitrite. The U.S. EPA set a National Primary Drinking Water MCL for nitrite at 1 mg/L as N (40 CFR D Subpart F 141.51).

Nitrate and Nitrite Aquatic Life Criteria

For fingerling rainbow trout, *Salmo gairdneri*, the respective 96-hour and 7-day LC50 toxicity values were 1360 and 1060 mg/L nitrate nitrogen in fresh water (Westin 1974). Knepp and Arkin (1973) observed that largemouth bass, *Micropterus salmoides*, and channel catfish, *Ictalurus punc-*

tatus, could be maintained at concentrations up to 400 mg/L nitrate without significant effect on their growth and feeding activities.

Nitrite forms of nitrogen were found to be much more toxic than nitrate forms. As an example, the 96-hour and 7-day LC50 values for chinook salmon were found to be 0.9 and 0.7 mg/L nitrite nitrogen in fresh water (Westin 1974). The effects of nitrite nitrogen on yearling rainbow trout, *Oncorhynchus mykiss*, showed that they suffered a 55% mortality after 24 hours at 0.55 mg/L; fingerling rainbow trout suffered a 50% mortality after 24 hours of exposure at 1.6 mg/L; and chinook salmon, *Oncorhynchus tshawytscha*, suffered a 40% mortality within 24 hours at 0.5 mg/L. There were no mortalities among rainbow trout exposed to 0.15 mg/L nitrite nitrogen for 48 hours. These data indicate that salmonids are more sensitive to nitrite toxicity than are other fish species, e.g., minnows, *Phoxinus laevis*, which suffered a 50% mortality within 1.5 hours of exposure to 2030 mg/L nitrite nitrogen, but required 14 days of exposure for mortality to occur at 10 mg/L (Klinger 1957), and carp, *Cyprinus carpio*, when raised in a water reuse system, tolerated up to 1.8 mg/L nitrite nitrogen (Saeki 1965).

The EPA concluded that (1) levels of nitrate nitrogen at or below 90 mg/L would have no adverse effects on warm-water fish (Knepp and Arkin 1973); (2) nitrite nitrogen at or below 5 mg/L should be protective of most warm-water fish (McCoy, 1972); and (3) nitrite nitrogen at or below 0.06 mg/L should be protective of salmonid fishes (Russo et al. 1974; Russo and Thurston 1975). These levels either are not known to occur or would be unlikely to occur in natural surface waters.

Recognizing that concentrations of nitrate or nitrite that would exhibit toxic effects on warm- or cold-water fish could rarely occur in nature, restrictive criteria were not recommended by the EPA.

PHOSPHATE

Phosphorus in the elemental form is very toxic (having an EPA marine life criteria of 0.10 µg/L) and is subject to bioaccumulation in much the same way as mercury. Phosphate forms of phosphorus are a major nutrient required for plant nutrition. In excessive concentrations, phosphates can stimulate plant growth. Excessive growths of aquatic plants (eutrophication) often interfere with water uses and are nuisances.

Generally, phosphates are not the only cause of eutrophication, but frequently it is the key of all the elements required by freshwater plants (generally, it is present in the least amount relative to need). Therefore, an increase in phosphorus allows use of other already present nutrients for plant growth. In addition, of all the elements required for plant growth in the water environment, phosphorus is the most easily controlled by man. In some aquatic systems, however, nitrogen compounds may be the most critical nutrients because of relatively large amounts of treated sewage (which is especially high in phosphates) in relation to other pollution sources, such as agricultural and urban runoff (which are high in nitrogen).

Phosphates enter waterways from several different sources. The human body excretes about one pound per year of phosphorus compounds. The use of phosphate detergents increases the per capita contribution to about 3.5 lb per year of phosphorus compounds. Some industries, such as potato processing, have wastewaters high in phosphates. Many nonpoint sources (crop, forest, and urban lands) contribute varying amounts of phosphorus compounds to watercourses. This drainage may be surface runoff of rainfall, effluent from agricultural tile lines, or return flow from irrigation. Cattle feedlots, birds, tree leaves, and fallout from the atmosphere all are contributing sources.

Evidence indicates that (1) high phosphorus compound concentrations are associated with accelerated eutrophication of waters, when other growth-promoting factors are present; (2) aquatic plant problems develop in reservoirs and other standing waters at phosphorus values lower than those critical in flowing streams; (3) reservoirs and lakes collect phosphates from influent streams and store a portion of them within consolidated sediments, thus serving as a phosphate sink; and (4) phosphorus concentrations critical to noxious plant growth vary and nuisance growths may result from a particular concentration of phosphate in one geographical area but not in another. The amount or percentage of inflowing nutrients that may be retained by a lake or reservoir is variable and will depend upon: (1) the nutrient loading to the lake or reservoir; (2) the volume of the euphotic zone; (3) the extent of biological activities; (4) the detention time within a lake basin or the time available for biological activities; and (5) the discharge from the lake.

Once nutrients are discharged into an aquatic ecosystem, their removal is tedious and expensive. Phosphates are used by algae and higher aquatic plants and may be stored in excess of use within the plant cells. With decomposition of the plant cell, some phosphorus may be released immediately through bacterial action for recycling within the biotic community, while the remainder may be deposited with sediments. Much of the material that combines with the consolidated sediments within the lake bottom is bound permanently and will not be recycled into the system, but some can be released in harmful quantities.

Aquatic Life Summary for Phosphate

Total phosphate concentrations in excess of 100 µg/L (expressed as total phosphorus) may interfere with coagulation in water treatment plants. When such concentrations exceed 25 µg/L at the time of the spring turnover on a volume-weighted basis in lakes or reservoirs, they may occasionally stimulate excessive or nuisance growths of algae and other aquatic plants. Algal growths cause undesirable tastes and odors in water, interfere with water treatment, become aesthetically unpleasant, and alter the chemistry of the water supply. They contribute to eutrophication.

To prevent the development of biological nuisances and to control accelerated or cultural eutrophication, total phosphates as phosphorus (P) should not exceed 50 µg/L in any stream at the point where it enters any lake or reservoir, nor 25 µg/L within the lake or reservoir. A desired goal for the prevention of plant nuisances in streams or other flowing waters not discharging directly to lakes or impoundments is 100 µg/L total P (Mackenthun 1973). Most relatively uncontaminated lake districts are known to have surface waters that contain from 10 to 30 µg/L total phosphorus as P (Hutchinson 1957).

The majority of the nation's eutrophication problems are associated with lakes or reservoirs, and currently there are more data to support the establishment of a limiting phosphorus level in those waters than in streams or rivers that do not directly impact such water. There are natural conditions, also, that would dictate the consideration of either a more or less stringent phosphorus level. Eutrophication problems may occur in waters where the phosphorus concentration is less than that indicated above and, obviously, such waters would need more stringent nutrient limits. Likewise, there are those waters within the United States where phosphorus is not now a limiting nutrient and where the need for phosphorus limits is substantially diminished.

There are two basic needs in establishing a phosphorus criterion for flowing waters: one is to control the development of plant nuisances within the flowing water and, in turn, to control and prevent animal pests that may become associated with such plants. The other is to protect the downstream receiving waterway, regardless of its proximity in linear distance. It is evident that a portion of that phosphorus that enters a stream or other flowing waterway eventually will reach a receiving lake or estuary either as a component of the fluid mass, as bedload sediments that are carried downstream, or as floating organic materials that may drift just above the stream's bed or float on its water's surface. Superimposed on the loading from the inflowing waterway, a lake or estuary may receive additional phosphorus as fallout from the atmosphere or as a direct introduction from shoreline areas.

Another method to control the inflow of nutrients, particularly phosphates, into a lake is that of prescribing an annual loading to the receiving water. Vollenweider (1973) suggests total phosphorus (P) loadings, in grams per square meter of surface area per year, that will be a critical level for eutrophic conditions within the receiving waterway for a particular water volume. The mean depth of the lake in meters is divided by the hydraulic detention time in years. Vollenweider's data suggest a range of loading values that should result in oligotrophic lake water quality:

Mean Depth/Hydraulic Detention Time (m/y)	Oligotrophic or Permissible Loading (g/m/yr)	Eutrophic or Critical Loading (g/m/yr)
0.5	0.07	0.14
1.0	0.10	0.20
2.5	0.16	0.32
5.0	0.22	0.45
7.5	0.27	0.55
10.0	0.32	0.63
25.0	0.50	1.00
50.0	0.71	1.41
75.0	0.87	1.73
100.0	1.00	2.00

There may be waterways where higher concentrations, or loadings, of total phosphorus do not produce eutrophication, as well as those waterways where lower concentrations or loadings of total phosphorus may be associated with populations of nuisance organisms. Waters now containing less than the specified amounts of phosphorus should not be degraded by the introduction of additional phosphates.

pH

pH is a measure of the hydrogen ion activity in a water sample. It is mathematically related to hydrogen ion activity according to the expression: $pH = -\log_{10} H^+$, where H^+ the hydrogen ion activity, expressed in moles/L. The pH of natural waters is a measure of the acid–base equilibrium achieved by the various dissolved compounds, salts, and gases. The principal chemical system controlling pH in natural waters is the carbonate system, which is composed of atmospheric carbon dioxide (CO_2) and resulting carbonic acid (H_2CO_3), bicarbonate ions (HCO_3^-) and carbonate ions (CO_3^{2-}) The interactions and kinetics of this system have been described by Stumm and Morgan (1970).

pH is an important factor in the chemical and biological reactions in natural waters. The degree of dissociation of weak acids or bases is affected by changes in pH. This effect is important because the toxicity of many compounds is affected by the degree of dissociation. One such example is for hydrogen cyanide. Cyanide toxicity to fish increases as the pH is lowered because the chemical equilibrium is shifted toward an increased concentration of a more toxic form of cyanide. Similar results have also been shown for hydrogen sulfide (H_2S) (Jones 1964). Conversely, rapid increases in pH can cause increased NH_3 concentrations that are also toxic. Ammonia has been shown to be 10 times as toxic at pH 8.0 as at pH 7.0 (EIFAC 1969).

The solubility of metal compounds contained in bottom sediments, or as suspended material, is also affected by pH. For example, laboratory equilibrium studies under anaerobic conditions indicated that pH was an important parameter involved in releasing manganese from bottom sediments (Delfino and Lee 1971).

Coagulation, used for removal of colloidal color and turbidity through the use of aluminum or iron salts, generally has an optimum pH range of 5.0 to 6.5. The effect of pH on chlorine in water principally concerns the equilibrium between hypochlorous acid (HOCl) and the hypochlorite ion (OCl⁻) according to the reaction:

$$HOCl = H^+ + OCl^-$$

High hydrogen ion concentrations (low pH) would therefore cause much more HOCl to be present, than at high pH values. Chlorine disinfection is more effective at values less than pH 7 (favoring HOCl, the more effective disinfectant). Water is therefore adjusted to a pH of between 6.5 and 7 before most water treatment processes. Corrosion of plant equipment and piping in the distribution system can lead to expensive replacement as well as the introduction of metal ions such as copper, lead, zinc, and cadmium. Langelier (1936) developed a method to calculate and control water corrosive activity that employs calcium carbonate saturation theory and predicts whether the water would tend to dissolve metal piping, or deposit a protective layer of calcium carbonate on the metal. Generally, this level is above pH 7 and frequently approaches pH 8.3, the point of maximum bicarbonate/carbonate buffering.

Since pH is relatively easily adjusted before, and during, water treatment, a rather wide range is acceptable for water serving as a source of public water supply. A range of pH from 5.0 to 9.0 would provide a water treatable by typical (coagulation, sedimentation, filtration, and chlorination) treatment plant processes. As the range is extended, the cost of pH-adjusting chemicals increases.

pH Aquatic Life Effects and Criteria

A review of the effects of pH on freshwater fish has been published by the European Inland Fisheries Advisory Commission (1969). The commission concluded:

> There is no definite pH range within which a fishery is unharmed and outside which it is damaged, but rather, there is a gradual deterioration as the pH values are further removed from the normal range. The pH range which is not directly lethal to fish is 5 to 9; however, the toxicity of several common pollutants is markedly affected by pH changes within this range, and increasing acidity or alkalinity may make these poisons more toxic. Also, an acid discharge may liberate sufficient CO_2 from bicarbonate in the water either to be directly toxic, or to cause the pH range of 5 to 6 to become lethal.

Mount (1973) performed bioassays on the fathead minnow, *Pimephales promelas*, for a 13-month, one-generation time period to determine chronic pH effects. Tests were run at pH levels of 4.5, 5.2, 5.9, 6.6, and a control of 7.5. At the two lowest pH values (4.5 and 5.2), behavior was abnormal and the fish were deformed. At pH values less than 6.6, egg production and hatchability were reduced when compared with the control. It was concluded that a pH of 6.6 was marginal for vital life functions.

Bell (1971) performed bioassays with nymphs of caddisflies (two species), stoneflies (four species), dragonflies (two species), and mayflies (one species). All are important fish food organisms. The 30-day TL50 pH values ranged from 2.5 to 5.4, with the caddisflies being the most tolerant and the mayflies being the least tolerant. The pH values at which 50% of the organisms emerged ranged from 4.0 to 6.6 with increasing percentage emergence occurring with the increasing pH values.

Based on present evidence, a pH range of 6.5 to 9.0 appears to provide adequate protection for the life of freshwater fish and bottom-dwelling invertebrates. Outside of this range, fish suffer adverse physiological effects, increasing in severity as the degree of deviation increases until lethal levels are reached:

pH Range	Effect on Fish
5.0–6.0	Unlikely to be harmful to any species unless either the concentration of free CO_2 is greater than 20 ppm, or the water contains iron salts which are precipitated as ferric hydroxide, the toxicity of which is not known
6.0–6.5	Unlikely to be harmful to fish unless free CO_2 is present in excess of 100 ppm
6.5–9.0	Harmless to fish, although the toxicity of other poisons may be affected by changes within this range

From EIFAC 1969

The U.S. EPA set a national drinking water secondary standard limiting pH ranges of domestic water supplies to 6.5 to 8.5 (40 CFR D 143.3). For the protection of fish and bottom-dwelling invertebrates, the U.S. EPA recommends that pH values should be less than 9 and greater than 6.5 (EPA 1986b).

SUSPENDED SOLIDS AND TURBIDITY

Suspended solids (sometimes referred to as nonfilterable residue) and turbidity are related to the solids content that is not dissolved. Turbidity refers to the blockage of light penetration and is measured by examining the backscatter from an intense light beam, while suspended solids are measured by weighing the amount of dried sediment that is trapped on a 0.45-μm filter, after filtering a known sample volume. The suspended solids test therefore measures a broad variety of solids that are contained in the water, including floatable material and settleable matter, in addition to the suspended solids. An Imhoff cone can be used to qualitatively estimate the settleable solids content of water. Subjecting the filter to a high temperature will burn off the more combustible solids. The remaining solids are usually referred to as the nonvolatile solids. The amount burned is assumed to be related to the organic fraction of the wastewater.

Turbidity (and color) can be caused mostly by very small particles (less than 1 μm), while the suspended solids content is usually associated with more moderate-sized particles (10 to 100 μm). Suspended solids can cause water quality problems directly, as discussed in the following paragraphs from *Water Quality Criteria* (EPA 1986b). They may also have other pollutants (such as organics and toxicants) associated with them that would cause additional problems. The control of suspended solids is required in most discharge permits because of potential sedimentation problems downstream of the discharge and the desire to control associated other pollutants.

Turbid water interferes with recreational use and aesthetic enjoyment of water. Turbid waters can be dangerous for swimming, especially if diving facilities are provided, because of the possibility of unseen submerged hazards and the difficulty in locating swimmers in danger of drowning (NAS 1974). The less turbid the water, the more desirable it becomes for swimming and other water contact sports. Other recreational pursuits, such as boating and fishing, will be adequately protected by suspended solids criteria developed for protection of fish and other aquatic life.

Fish and other aquatic life requirements concerning suspended solids can be divided into those whose effect occurs in the water column and those whose effect occurs following sedimentation to the bottom of the water body. Noted effects are similar for both fresh and marine waters. The effects of suspended solids on fish have been reviewed by the European Inland Fisheries Advisory Commission (EIFAC 1969). This review in 1965 identified four effects on the fish and fish food populations.

1. By acting directly on the fish swimming in water in which solids are suspended, and either killing them or reducing their growth rate, resistance to disease, etc.
2. By preventing the successful development of fish eggs and larvae
3. By modifying natural movements and migrations of fish
4. By reducing the abundance of food available to the fish

Settleable materials which blanket the bottom of water bodies damage the invertebrate populations, block gravel spawning beds, and if organic, remove dissolved oxygen from overlying waters (EIFAC 1969; Edberg and Hofstan 1973). In a study downstream from the discharge of a rock quarry where inert suspended solids were increased to 80 mg/L, the density of macroinvertebrates decreased by 60%, while in areas of sediment accumulation, benthic invertebrate populations also decreased by 60% regardless of the suspended solid concentrations (Gammon 1970). Similar effects have been reported downstream from an area which was intensively logged. Major increases in stream suspended solids (25 mg/L upstream vs. 390 mg/L downstream) caused smothering of bottom invertebrates, reducing organism density to only 7.3 vs. 25.5/ft^2 upstream (Tebo 1955).

When settleable solids block gravel spawning beds which contain eggs, high mortalities result. There is also evidence that some species of salmonids will not spawn in such areas (EIFAC 1969). It has been postulated that silt attached to the eggs prevents sufficient exchange of oxygen and carbon dioxide between the egg and the overlying water. The important variables are particle size, stream velocity, and degree of turbulence (EIFAC 1969). Deposition of organic materials to the bottom sediments can cause imbalances in stream biota by increasing bottom animal density (principally worms), and diversity is reduced as pollution-sensitive forms disappear (Mackenthun 1973). Algae, likewise, flourish in such nutrient-rich areas, although forms may become less desirable (Tarzwell and Gaufin 1953).

Plankton and inorganic suspended materials reduce light penetration into the water body, reducing the depth of the photic zone. This reduces primary production and decreases fish food. The NAS committee in 1974 recommended that the depth of light penetration not be reduced by more than 10% (NAS 1974). Additionally, the near-surface waters are heated because of the greater heat absorbency of the particulate material which tends to stabilize the water column and prevents vertical mixing (NAS 1974). Such mixing reductions decrease the dispersion of dissolved oxygen and nutrients to lower portions of the water body. Increased temperatures also reduce the capacity of the stream to contain dissolved oxygen.

Suspended inorganic material in water also sorbs organic materials, such as pesticides. Following this sorption process, subsequent sedimentation may remove these materials from the water column into the sediments (NAS 1974). However, the sedimentation of these polluted sediments can cause dramatic changes in the benthic microorganism populations, which in turn affect other aquatic life forms. Recent research associated with the effects of polluted sediments in urban streams is summarized in earlier chapters of this book.

Water Quality Criteria for Suspended Solids and Turbidity

The EPA water quality criterion for freshwater fish and other aquatic life is essentially that proposed by the National Academy of Sciences and the Great Lakes Water Quality Board: "Settleable and suspended solids should not reduce the depth of the compensation point for photosynthetic activity by more than 10 percent from the seasonally established norm for aquatic life."

States have selected numeric values for turbidity. Alabama, for example, uses the same standard for all designated uses: "There shall be no turbidity of other than natural origin that will cause substantial visible contrast with the natural appearance of waters or interfere with any beneficial uses which they serve. Furthermore, in no case shall turbidity exceed 50 Nephelometric units (NTU) above background. Background will be interpreted as the natural condition of the receiving waters, without the influence of man-made or man-induced causes. Turbidity levels caused by natural runoff will be included in establishing background levels." In addition, the state of Alabama has minimum conditions applicable to all state waters that includes: "State waters shall be free from substances attributable to sewage, industrial wastes, or other wastes that will settle to form bottom deposits which are unsightly, putrescent, or interfere directly or indirectly with any classified water use."

ZINC

Aquatic Life Criteria for Zinc

The U.S. EPA has set a national ambient water quality for the protection of wildlife as a function of hardness (EPA 1986b), and ambient water quality for the Great Lakes as a function of hardness (40 CFR 132.3 (b)). Figure G.4 shows these criteria.

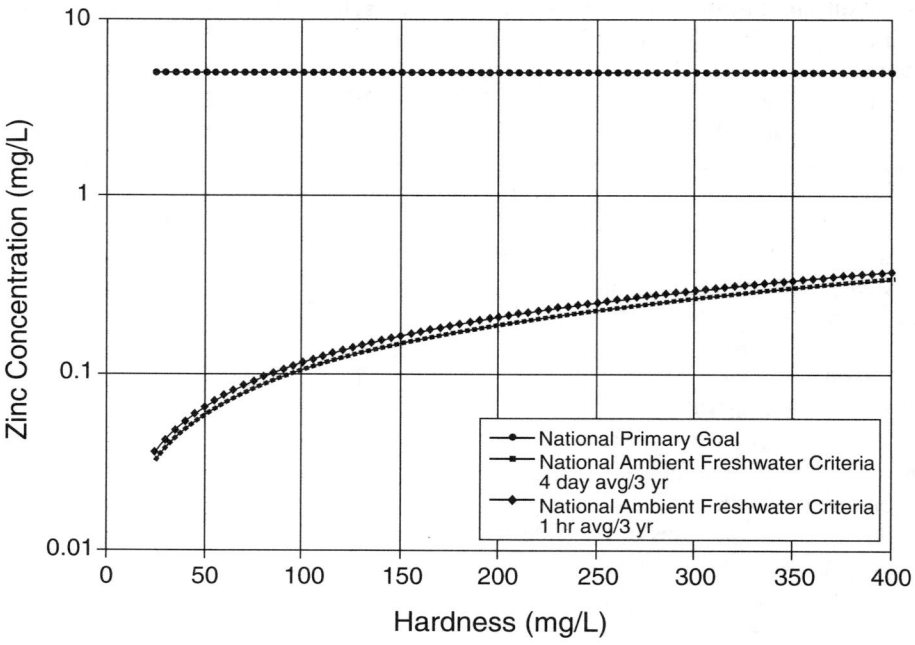

Figure G.4 Zinc criteria.

Human Health Criteria for Zinc

The U.S. EPA has set a national secondary MCL for zinc at 5 mg/L (40 CFR D 143.3), based on available organoleptic data, and to control undesirable taste and odor quality of ambient water. It should be recognized that organoleptic data have limitations as a basis for establishing water quality criteria, and have no demonstrated relationship to potential adverse human health effects.

SEDIMENT GUIDELINES

Water quality criteria and standards are proven to be useful tools for helping to assess receiving water quality and beneficial use attainment. For these reasons, it is logical that sediment quality criteria would also be a useful tool. However, the complexity of sediments has impeded establishing guidelines because of the lack of clear relationships between sediment characteristics and the bioavailability of associated contaminants. Nonetheless, several useful approaches have been proposed for establishing sediment guidelines (also called criteria, standards, guidelines, objectives, or assessment values). In recent years, there has been a tremendous increase in sediment contaminant research and monitoring, which has resulted in improved sediment quality guidelines. The U.S. EPA has proposed guidelines using a theoretical approach known as *equilibrium partitioning guidelines*. Concentrations of contaminants are predicted in interstitial water and compared to the chronic water quality criteria to establish whether the sediments are toxic. Currently there are only criteria for acenaphthene, phenanthrene, fluoranthene, dieldrin, and endrin. This approach normalizes nonpolar organic compounds to the sediment total organic carbon content and metals to the acid volatile sulfide content. Both these sediment parameters have been shown to strongly control bioavailability

Table G.7 Sediment Quality Guidelines for Freshwater Ecosystems

Substance	TEL	PEL	LEL	SEL	MET	TET	ERL	ERM	SQAL
Metals (in mg/kg DW)									
Arsenic	5.9	17	6	33	7	17	33	85	NG
Cadmium	0.596	3.53	0.6	10	0.9	3	5	9	NG
Chromium	37.3	90	26	110	55	100	80	145	NG
Copper	35.7	197	16	110	28	86	70	390	NG
Lead	35	91.3	31	250	42	170	35	110	NG
Mercury	0.174	0.486	0.2	2	0.2	1	0.15	1.3	NG
Nickel	18	36	16	75	35	61	30	50	NG
Zinc	123	315	120	820	150	540	120	270	NG
Polycyclic Aromatic Hydrocarbons (in µg/kg DW)									
Anthracene	NG	NG	220	3700	NG	NG	85	960	NG
Fluorene	NG	NG	190	1600	NG	NG	35	640	540
Naphthalene	NG	NG	NG	NG	400	600	340	2100	470
Phenanthrene	41.9	515	560	9500	400	800	225	1380	1800
Benz[a]anthracene	31.7	385	320	14800	400	500	230	1600	NG
Benzo(a)pyrene	31.9	782	370	14400	500	700	400	2500	NG
Chrysene	57.1	862	340	4600	600	800	400	2800	NG
Dibenz[a,h]anthracene	NG		60		NG		60		NG
Fluoranthene	111	2355	750	10200	600	2000	600	3600	6200
Pyrene	53	875	490	8500	700	1000	350	2200	NG
Total PAHs	NG	NG	4000	100000	NG	NG	4000	35000	NG
Polychlorinated Biphenyls (in µg/kg DW)									
Total PCBs	34.1	277	70	5300	200	1000	50	400	NG
Organochlorine Pesticides (in µg/kg DW)									
Chlordane	4.5	8.9	7	60	7	30	0.5	6	NG
Dieldrin	2.85	6.67	2	910	2	300	0.02	8	110
Sum DDD	3.54	8.51	8	60	10	60	2	20	NG
Sum DDE	1.42	6.75	5	190	7	50	2	15	NG
Sum DDT	NG	NG	8	710	9	50	1	7	NG
Total DDTs	7	4450	7	120	NG	NG	3	350	NG
Endrin	2.67	62.4	3	1300	8	500	0.02	45	42
Heptachlor epoxide	0.6	2.74	5	50	5	30	NG	NG	NG
Lindane (gamma-BHC)	0.94	1.38	3	10	3	9	NG	NG	3.7

PEL = Probable effect level; dry weight (Smith et al. 1996).

SEL = Severe effect level, dry weight (Persaud et al. 1993).

TET = Toxic effect threshold; dry weight (EC and MENVIQ 1992).

ERM = Effects range median; dry weight (Long and Morgan 1991).

NG = No guideline.

(e.g., Ingersoll et al. 1997). It does not appear that the U.S. EPA approach will result in additional guidelines in the near future. There have been several empirical approaches that are based on co-occurrence of adverse biological effects observed in the field or laboratory related to sediment contaminant concentrations. Tables G.7 and G.8 list some of the most reliable sediment quality guidelines available. Included in these are some "consensus" approaches that may be a first priority if one chooses to use a sediment guideline in their assessment. It is interesting to note that the majority of the approaches produce guidelines that are relatively similar; therefore, the consensus approach has added credibility.

Table G.8 Sediment Quality Guidelines for Polycyclic Aromatic Hydrocarbons (μg/g organic carbon)[a]

PAH	ERL[b]	ERM[b]	TEL[b]	PEL[b]	SLC[b]	LAET[b]	HAET[b]	EqP	TEC Mean	MEC Mean	Consensus EEC
Naphthalene	16	210	3	39	41	210	270				
Acenaphthylene	4	64	1	13	5	>56	130				
Acenaphthene	2	50	1	9	6[c]	50	200	230			
Fluorene	2	54	2	14	10	54	360				
Phenanthrene	24	150	9	54	37	150	690	240			
Anthracene	9	110	5	24	16	96	1300				
Low-molecular-weight PAH	57	638	21	153	115	616	2950				
Fluoranthene	60	510	11	149	64	170	3000	300			
Pyrene	66	260	15	140	66	260	1,600				
Benz[a]anthracene	26	160	7	69	26	130	510				
Chrysene	38	280	11	85	38	140	920				
Benzo[b]fluoranthene	32[c]	188[c]	7[c]	71[c]	32[c]	160	445				
Benzo[k]fluoranthene	28[c]	162[c]	6[c]	61[c]	28[c]	160	445				
Benzo[a]pyrene	43	160	9	76	40	160	360				
High-molecular-weight PAH	293	1720	66	651	294	1180	7280				
Total PAH	350	2358	87	804	409	1796	10,230	211	290 (119–461)	1800 (682–2,854)	10,000

[a] ERL = effects range-low;
ERM = effects range-median;
TEL = threshold effects level;
PEL = probable effects level;
SLC = screening level concentration;
LAET = low apparent effects threshold;
HAET = high apparent effects threshold;
EqP = U.S. Environmental Protection Agency criteria derived from equilibrium partitioning theory;
TEC = Threshold effect concentration;
MEC = Median effects concentration;
EEC = Extreme effects concentration.
[b] SQG at 1% OC.
[c] No SQG. Estimate assuming mean ratio to PAH mixture LC50 for other high-molecular-weight PAHs.

REFERENCES

Alexander, L.M., A. Heaven, A. Tennant, and R. Morris. Symptomatology of children in contact with sea water contaminated with sewage. *J. Epidem. Comm. Health*, 46, 340-344, 1992.

Bell, H.L. Effect of low pH on the survival and emergence of aquatic insects. *Water Res.*, 5:313. 1971.

Bruvold, W.H. et al. Consumer assessment of mineral taste in domestic water. *J. Am. Water Works Assn.*, 61:575, 1969.

Cabelli, V.J., A.P. Dufour, M.A. Levin, L.J. McCabe, and P.W. Haberman. Relationship of microbial indicators to health effects at marine bathing beaches. *Am. J. Pub. Health*, 69, 7, 690-696, July 1979.

Cabelli, V.J., A.P. Dufour, L.J. McCabe, and M.A. Levin. Swimming-associated gastroenteritis and water quality. *Am. J. Epidem.*, 115, 4, 606-616, 1982.

Capurro, L.R.A. *Oceanography for Practicing Engineers* . Barnes and Noble, Inc., New York, 1970.

Delfino, J.J. and G.F. Lee. Variation of manganese, dissolved oxygen and related chemical parameters in the bottom waters of Lake Mendota, Wisconsin. *Water Res.*, 5:1207, 1971.

Doudoroff, P. and Katz, M. Critical review of literature on the toxicity of industrial wastes and their components to fish. II. The metals and salts. *Sewage and Industrial Wastes*, 25, 7, p. 802, 1953.

Dufour, A.P. *Health Effects Criteria for Fresh Recreational Waters*, EPA-600/1-84-004. U.S. Environmental Protection Agency, Cincinnati, OH, 1983.

Dufour, A.P. Bacterial indicators of recreational water quality. *Can. J. Pub. Health*, 75, 49-56, January/February 1984.

EC and MENVIQ (Environmental Canada and Ministere de l'Environnement du Quebec) *Interim Criteria for Quality Assessment of St. Lawrence River Sediment*. Environment Canada. Ottawa, Ontario, 1992.

Edberg, N. and B.V. Hofstan. Oxygen uptake of bottom sediment studied *in-situ* and in the laboratory. *Water Res.*, 7:1285, 1973.

EIFAC (European Inland Fisheries Advisory Commission). Water quality criteria for European freshwater fish—extreme pH values and inland fisheries. Prepared by EIFAC Working Party on Water Quality Criteria for European Freshwater Fish. *Water Res.*, 3:593. 1969.

Environmental Science & Technology. News Briefs. 30, 7, pp. 290a, July 1996.

EPA (U.S. Environmental Protection Agency). *Suspended and Dissolved Solids Effects on Freshwater Biota: A Review*, Environmental Research Laboratory, U. S. Environmental Protection Agency, Corvallis, OR, EPA 600/3-77/042. 1977.

EPA (U.S. Environmental Protection Agency). Guidelines establishing test procedures for the analysis of pollutants under the Clean Water Act: final rule and interim final rule and proposed rule, 40 CFR. *Fed. Reg.*, 136:1-210, 1984.

EPA (U.S. Environmental Protection Agency). *Ambient Water Quality Criteria for Bacteria—- 1986*, EPA 440/5-84-002, U.S. Environmental Protection Agency, Office of Water Regulations and Standards, Washington, D.C., NTIS access #: PB 86-158-045. 1986a.

EPA (U.S. Environmental Protection Agency). *Quality Criteria for Water*. EPA 440/5-86-001. Washington, D.C., U.S. Environmental Protection Agency, May 1986b.

Ferley, J.P., D. Zmirou, F. Balducci, B. Baleux, P. Fera, G. Larbaigt, E. Jacq, B. Moissonnier, A. Blineau, and J. Boudot. Epidemiological significance of microbiological pollution criteria for river recreational waters. *Int. J. Epidem.*, 18, 1, 198-205, January 1989.

Gammon, J.R. *The Effect of Inorganic Sediment on Stream Biota*. Environmental Protection Agency. Water Poll. Cont. Res. Series, 18050 DWC 12/70, USGPO, Washington, D.C., 1970.

Haile, R.W., J. Alamillo, K. Barrett, R. Cressey, J. Dermond, C. Ervin, A. Glasser, N. Harawa, P. Harmon, J. Harper, C. McGee, R.C. Millikan, M. Nides, and J.S. Witte. *An Epidemiological Study of Possible Health Effects of Swimming in Santa Monica Bay*. Santa Monica Bay Restoration Project. Monterey Park, CA. May 1996.

Harmeson, R.H., et al. The nitrate situation in Illinois. *J. Am. Water Works Assn.*, 63:303, 1971.

Hutchinson, G.E. *A Treatise on Limnology*. John Wiley & Sons, New York, 1957.

Ingersoll C.G., T. Dillon, and G.R. Biddinger. Ecological risk assessment of contaminated sediments. Society of Environmental Toxicology and Chemistry, *SETAC Press*, Pensacola, FL, 1997.

Jones, J.R.E. *Fish and River Pollution*. Butterworth and Co., Ltd., London, 1964.

Klinger, K. Sodium nitrite, a slow acting fish poison. *Schweiz, Z. Hydrol.*, 19(2):565, 1957.

Knepp, G.L. and G.F. Arkin, Ammonia toxicity levels and nitrate tolerance of channel catfish. *The Progressive Fish Culturist*, 35:221, 1973.

Langelier, W.F. The analytical control of anti-corrosion water treatment. *J. Am. Water Works Assn.*, 28:150, 1936.

Lockhart, E.E., et al. The effect of water impurities on the flavor of brewed coffee. *Food Res.*, 20:598, 1955.

Long, E.R. and L.G. Morgan. *The Potential for Biological Effects of Sediment Sorbed Contaminants Tested in the National Status and Trends Program*. NOAA Technical Memorandum NOS OMA 52. *National Oceanic and Atmospheric Administration*. Seattle, WA, 175 pp+ appendices, 1991.

MacDonald, D.D. Freshwater sediment quality guidelines. *Arch. Environ. Contam. Toxicol.*, 1999.

Mackenthun, K.M. *Toward a Cleaner Aquatic Environment*. U.S. Government Printing Office, Washington, D.C., 1973.

McCoy, E.F. *Role of Bacteria in the Nitrogen Cycle in Lakes*. Environmental Protection Agency, Water Pollution Control Research Series, U.S. Government Printing Office (EP 2.10:16010 EHR 03/72). Washington, D.C., 1972.

Moore, E.W. Physiological effects of the consumption of saline drinking water. National Res. Council, Div. of Medical Sciences, *Bull. San. Engr., and Environment*, Appendix E. 1952.

Mount, D.I. Chronic effect of low pH on fathead minnow survival, growth and reproduction. *Water Res.*, 7:987, 1973.

NAS (National Academy of Sciences, National Academy of Engineering), *Water Quality Criteria, 1972*, U.S. Government Printing Office, Washington, D.C., 1974.

NRC (National Research Council). *Sodium Restricted Diets*. Publication 325, Food and Nutrition Board, Washington, D.C., 1954.

NTAC (National Technical Advisory Committee to the Secretary of the Interior), *Water Quality Criteria*, U.S. Government Printing Office, Washington, D.C., 1968.

Persaud, D., R. Jaagumagi, and A. Hayton. *Guidelines for the Protection and Management of Aquatic Sediment Quality in Ontario*. Water Resources Branch, Ontario Ministry of the Environment, Toronto, Ontario, 27 pp. 1993.

Rawson, D.S. and J.E. Moore. The saline lakes of Saskatchewan. *Can J. Res.*, 22:141, 1944.

Ricter, C.O. and A. MacLean. Salt taste threshold of humans. *Am. J. Physiol.*, 126:1, 1939.

Russo, R.C. et al. Acute toxicity of nitrite to rainbow trout. *J. Fish. Res. Bd.Can.*, 31:1653, 1974.

Russo, R.C. and R.V. Thurston. *Acute Toxicity of Nitrite to Cutthroat Trout*. Fisheries Bioassay Laboratory Tech. Report No. 75-3, Montana State University, 1975.

Saeki, A. Studies on fish culture in filtered closed circulating aquaria. II. On the carp culture experiments in the systems. *Bull. Jap. Soc. Sci. Fish.*, 31:916. 1965.

Seyfried, P.L., R.S. Tobin, N.E. Brown, and P.F. Ness. A prospective study of swimming-related illness, II Morbidity and the microbiological quality of water. *Am. J. Pub. Health*, 75, 9, 1071-1075, September 1985.

SMBRP (Santa Monica Bay Restoration Project). *An Epidemiological Study of Possible Adverse Health Effects of Swimming in Santa Monica Bay*. Santa Monica Bay Restoration Project. Monterey Park, CA, October 1996.

Smith, S.L., D.D. MacDonald, K.A. Keenleyside, C.G. Ingersoll, and J. Field. A preliminary evaluation of sediment quality assessment values for freshwater ecosystems. *J. Great Lakes Res.*, 22:624-638, 1996.

Stevenson, A.H. Studies of bathing water quality and health. *Am. J. Pub. Health*, 43, 529-538, May 1953.

Stewart, B.A., et al. Nitrate and other pollutants under fields and feedlots. *Envir. Sci Tech.*, 1:73, 1967.

Stiff, M.J. Copper/bicarbonate equilibria in solutions of bicarbonate ion at concentrations similar to those found in natural water. *Water Res.*, 5:171, 1971.

Stumm, W. and J.J. Morgan. *Aquatic Chemistry*. Chapter 4. John Wiley & Sons, Inc., New York, 1970.

Swartz, RC. Consensus sediment quality guidelines for polycyclic aromatic hydrocarbon mixtures. *Environ. Toxicol. Chem.*, 18(4):780-787, 1999.

Tarzwell, C.M. and A.R. Gaufin. Some important biological effects of pollution often disregarded in stream surveys. Proceedings of the *8th Purdue Industrial Waste Conference*. Reprinted in *Biology of Water Pollution*, 1967. Dept. of Interior, Washington, D.C. 1953.

USPHS (U.S. Public Health Service). *Public Health Service Drinking Water Standards*, rev. 1964. PHS Publication 95. Washington, D.C. 1961.

Vollenweider, R.A. Input output models. *Schweiz. Z. Hydrol.*, 1973.

Walton, G., Survey of literature relating to infant methemoglobinemia due to nitrate-contaminated water. *American J. Pub. Health*, 41:986. 1951.

Water Environment & Technology. Research Notes: Beachgoers at Risk from Urban Runoff. Vol. 8, no. 11, pg. 65. Nov. 1996.

Westin, D.T. Nitrate and nitrite toxicity to salmonid fishes. *The Progressive Fish Culturist*, 36:86. 1974.

Wolff, I.A. and Wasserman, Nitrates, nitrites, and nitrosamines. *Science*, 711:15, 1972.

Watershed and Receiving Water Modeling

CONTENTS

Introduction 843
Modeling Stormwater Effects and the Need for Local Data for Calibration and Verification 845
    Unit Area Loadings 845
    Simple Models..846
    Complex Models 852
    Receiving Water Models 855
    Geographical Information Systems (GIS) 857
Summary 860
References 866

INTRODUCTION

Models are important tools for watershed and receiving water analyses because they enable comprehensive evaluations of large systems and can predict future conditions. Models always have errors, but these can be reduced through good calibration and verification using locally obtained data, as described in this book.

For stormwater issues, most models can be separated into watershed models and receiving water models. Both are briefly addressed in this appendix. Many (and constantly increasing in numbers) public domain water quality models are available. Periodically, these are available on a CD-ROM from the EPA (*Exposure Models Library and Integrated Model Evaluation System,* EPA Office of Research and Development CD-ROM. EPA-600-C-92-002. Revised March 1996). Numerous specialized Internet sites also have download sites or links to the EPA download sites for acquiring these models and documentation. The main EPA source is through the EPA's Athens, GA, Center for Exposure Assessment Modeling (CEAM), where much of the EPA's water quality modeling support is available (downloads, short courses, etc.). One especially interesting reference available from Athens is *Rates, Constants, and Kinetics Formulations in Surface Water Quality Modeling* (second edition), EPA/600/3-85/040, prepared by Tetra Tech in 1985, but still highly useful. This report is available in PDF format from: http://www.epa.gov/ORD/WebPubs/surfaceH2O/surface.html. Not only does this report contain summaries of the model processes and lab and field data for the different fate processes, it also summarizes many field techniques that can be used to collect the needed local data.

The CEAM Internet site is at: http://www.epa.gov/CEAM/. The major models available at this web site are shown in Table H.1 (as of February 2000). These are all DOS-based, Fortran-coded programs. Very few Windows or Macintosh programs are available, but they will operate in the

Table H.1 DOS Programs Available to Download from the EPA's Center for Exposure Assessment Modeling (CEAMS) Group

File Name/Size (MB)	Description/Abstract/Release Notes	Version Number	Release Date
INSTALAN.EXE / 1.28	ANNIE-IDE tool kit	1.14	Sep 91
INSTALCI.EXE / .5	CEAM information system	3.21	May 95
INSTALCM.EXE / 1.63	CORMIX model / documentation	3.20	Dec 96
INSTALEX.EXE / 1.00	EXAMS model / documentation	2.97.5	Jun 97
INSTALFG.EXE / 1.07	FGETS model system	3.0.18	Sep 94
INSTALFW.EXE / 1.05	FEMWATER model / documentation	1.00	Jul 93
INSTALGC.EXE / 1.16	GCSOLAR model / documentation	1.20	Jul 99
INSTALHC.EXE / 8.44	HSCTM2D model / documentation	1.01	Nov 98
INSTALHS.EXE / 8.66	HSPF model / documentation	11.00	Apr 97
HSP11Y2K.EXE / .84	HSPF model / documentation / Year 2000 (Y2K) Patch	11.00	Dec 99
INSTALLC.EXE / .71	LC50 model / documentation	1.00	Jan 99
INSTALMT.EXE / 2.49	MINTEQ model / documentation	4.01	Dec 99
INSTALMS.EXE / 6.52	MMSOILS model / documentation	4.00	Jun 97
INSTALMM.EXE / 3.49	MULTIMED model / documentation	1.01	Dec 92
INSTALM2.EXE / 4.79	MULTIMED model / documentation	2.00 Beta	Oct 96
INSTALDP.EXE / 3.34	MULTIMDP model / documentation	1.00	Oct 96
INSTALOF.EXE / .34	Sample ANNIE-IDE application	1.61	Sep 91
INSTALOX.EXE / .40	OXYREF data base / documentation	1.00	Dec 98
INSTALP2.EXE / 2.76	PRZM2 model / documentation	2.00	Oct 94
INSTALP3.EXE / 5.15	PRZM3 model / documentation	3.12 Beta	Mar 98
INSTALPL.EXE / 1.44	PLUMES model / documentation	3.00	Dec 94
INSTALPT.EXE / 5.43	PATRIOT model / documentation	1.20	Nov 94
INSTALQ2.EXE / 2.21	QUAL2EU model system / Documentation	3.22	May 96
INSTALSW.EXE / 1.6	SWMM model system	4.30	May 94
INSTALSX.EXE / .39	SMPTOX3 model / documentation	2.01	Feb 93
INSTALWP.EXE / 3.14	WASP model / documentation	5.10	Oct 93

DOS shell of the Windows operating systems. Most of these programs were originally developed many years ago (with the processes reasonably well described in the Tetra Tech "rate" report of 1985, noted above).

There are numerous proprietary Windows "front-ends" for selected programs, along with proprietary versions that have substantial changes in the code. In addition, many private Internet sites also provide downloadable public domain water quality models, or "test" versions of commercial programs. Obviously, it is impossible to develop a complete list of available water quality models, and it is difficult for the user to select which model may be most appropriate for his or her specific use. Excellent model reviews are periodically prepared, such as *Compendium of Tools for Watershed Assessment and TMDL Development,* EPA-841-B-97-006, 1997. In addition, numerous listservers are available to provide excellent user support for specific models. A representative listing of list servers and other water quality modeling support is provided by Dr. Bill James at the University of Guelph at http://www.eos.uoguelph.ca/webfiles/james/homepage/Research/ListServers.html.

A major surface water quality modeling effort at EPA is directed toward supporting the Total Maximum Daily Load (TMDL) program. As part of this support, the BASINS model (*Better Assessment Science Integrating Point and Nonpoint Sources*), a Windows-based structure of several interconnected programs and a geographical information system (GIS), described later, has been developed. The main report is available as EPA-823, R-96001, May, 1996. Extensive Internet support, including downloads of the main program, and regional data, is available at http://www.epa.gov/OST/BASINS/. The structure of BASINS will allow additional models to be added to this framework. The extremely powerful aspect of BASINS is the GIS capabilities where local data can be easily integrated for model use. Individual CD-ROMS are available for each of the 10 EPA regions containing much local data. BASINS has six main components: nationally derived databases with Data Extraction Tools; assessment tools; utilities to facilitate organizing and evaluating the data, including land use data; Watershed Characterization Reports; water quality models; and the Nonpoint Source Model. It currently uses portions of HSPF for the land-based modeling component (NPSM, the Nonpoint Source Model), and QUAL2E and TOXIROUTE for the stream water quality models. Even though many of model components are older Fortran-coded modules, the Windows and GIS interfaces makes the model relatively easy to use.

BASINS is a large-scale model and may be too complex for focusing on specific smaller areas, or when detailed evaluations are needed. Figure H.1 is an overview of the individual environmental models commonly used (and evaluated in *Compendium of Tools for Watershed Assessment and TMDL Development*). Obviously, BASINS, although extremely powerful and needed for some applications, currently does not offer the flexibility that the wide range of individual models can.

MODELING STORMWATER EFFECTS AND THE NEED FOR LOCAL DATA FOR CALIBRATION AND VERIFICATION

A typical use of stormwater monitoring data is to calibrate and verify a model that will be used to examine many questions. Common uses of models are to determine the major sources of pollutants and to design control programs to effectively reduce the problem discharges. There are three general classes of stormwater models:

- Unit area loadings
- Simple models
- Complex models

Unit Area Loadings

Table 2.5 included unit area loading estimates for stormwater, based on numerous observations from throughout North America (mostly from the EPA's NURP projects, EPA 1983, and from other selected North American studies). Most of the available NURP data are from monitoring medium-density residential areas, but data from Wisconsin and Toronto included data from various land uses. These estimates are most useful when making preliminary assessments on a large scale, especially in preparing an experimental design for site-specific monitoring. As an example, these values can be used to identify the most significant land uses in a watershed and help direct the monitoring effort, as shown in Table 5.4 (repeated below as Table H.2) and Figure 5.7, a marginal benefit analysis. Obviously, the variations of unit area loadings can be very large, depending on specific conditions, but the basic rankings of land use related discharges are still useful for preliminary evaluations.

For most constituents, manufacturing industrial and commercial areas have the largest unit area loadings, while parks and low-density residential areas have the smallest unit area loadings. The importance of the areas in a watershed is obviously dependent on the size of the area. Medium-density residential areas comprise the majority of the land area for most cities, and therefore also for most large urban watersheds. These large areas increase the significance of this land use. However, relatively small amounts of industrial or commercial activity can overwhelm the residential contributions in small and moderate-sized drainages. Chapter 2 presented information showing the relative importance of industrial and residential areas in Toronto (Pitt and McLean 1986), based on a comprehensive monitoring program and measured unit area loadings for the major land uses.

The earlier Toronto discussion in Chapter 2 also showed how dry-weather flows and snowmelt contributions can be very important. That example stresses the need to consider all phases of flows that may be discharged from separate storm drainage systems. Few published unit area stormwater loading values include these other contributions that can have major effects on receiving water conditions.

Unit area loadings for a local area can be determined based on local monitoring data using one of the other modeling methods described below. Unit area loadings are a convenient method to summarize extensive monitoring data and to highlight potential problem areas, especially if integrated

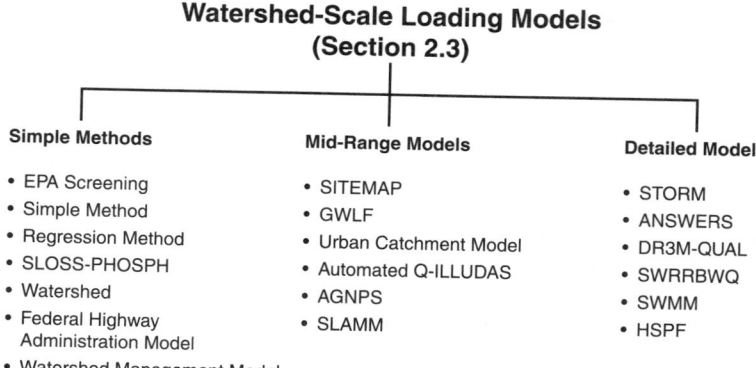

**Watershed-Scale Loading Models
(Section 2.3)**

Simple Methods
- EPA Screening
- Simple Method
- Regression Method
- SLOSS-PHOSPH
- Watershed
- Federal Highway
  Administration Model
- Watershed Management Model

Mid-Range Models
- SITEMAP
- GWLF
- Urban Catchment Model
- Automated Q-ILLUDAS
- AGNPS
- SLAMM

Detailed Models
- STORM
- ANSWERS
- DR3M-QUAL
- SWRRBWQ
- SWMM
- HSPF

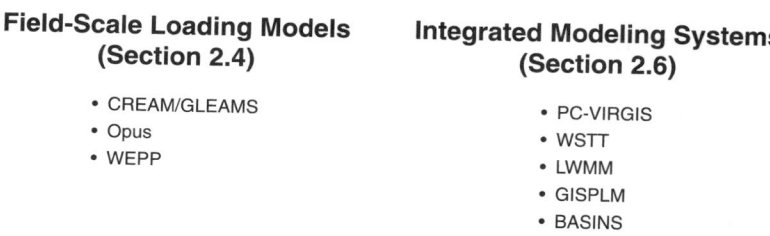

**Field-Scale Loading Models
(Section 2.4)**
- CREAM/GLEAMS
- Opus
- WEPP

**Integrated Modeling Systems
(Section 2.6)**
- PC-VIRGIS
- WSTT
- LWMM
- GISPLM
- BASINS

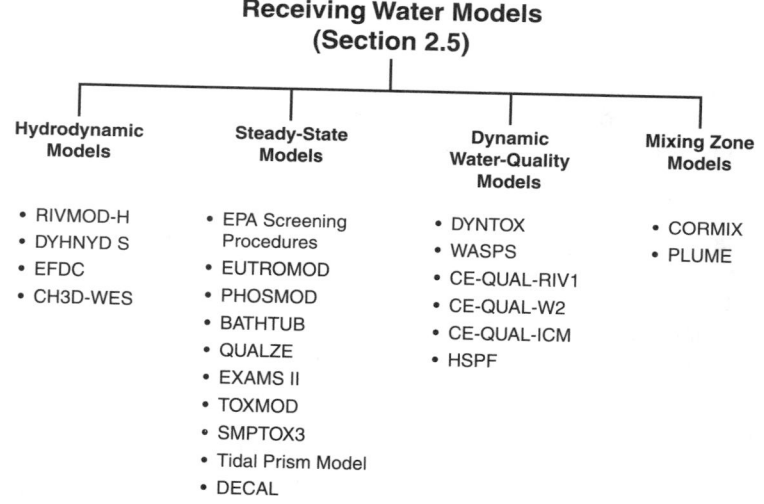

**Receiving Water Models
(Section 2.5)**

Hydrodynamic Models
- RIVMOD-H
- DYHNYD S
- EFDC
- CH3D-WES

Steady-State Models
- EPA Screening Procedures
- EUTROMOD
- PHOSMOD
- BATHTUB
- QUALZE
- EXAMS II
- TOXMOD
- SMPTOX3
- Tidal Prism Model
- DECAL

Dynamic Water-Quality Models
- DYNTOX
- WASPS
- CE-QUAL-RIV1
- CE-QUAL-W2
- CE-QUAL-ICM
- HSPF

Mixing Zone Models
- CORMIX
- PLUME

Figure H.1 Environmental models commonly in use. (From EPA. *Compendium of Tools for Watershed Assessment and TMDL Development.* EPA-841-B-97-006. U.S. Environmental Protection Agency. 1997.)

with a GIS. GIS has been successfully used with nonpoint source modeling activities to display the unit area loadings predicted from monitoring and modeling programs for many alternatives. Otherwise, the massive amounts of data generated is difficult to summarize in an easily presentable manner.

Simple Models

Simplified stormwater models usually take the general form:

$$\text{Unit Area Loading} = (\text{EMC}) \times (\text{Rv}) \times (\text{Rain})$$

Table H.2 Example Marginal Benefit Analysis

	Land Use (ranked by % mass per category)	% of Area	Critical Unit Area Loading	Relative Mass	% Mass per Category	Accum. (% mass)	Straight-line Model	Marginal Benefit
1	Older medium-density resid.	24	200	4800	22.8	22.8	6.25	16.5
2	High-density resid.	7	300	2100	10.0	32.7	12.5	20.2
3	Office	7	300	2100	10.0	42.7	18.8	24.0
4	Strip commercial	8	250	2000	9.5	52.2	25.0	27.2
5	Multiple family	8	200	1600	7.6	59.8	31.3	28.5
6	Manufacturing industrial	3	500	1500	7.1	66.9	37.5	29.4
7	Warehousing	5	300	1500	7.1	74.0	43.8	30.3
8	New medium-density resid.	5	250	1250	5.9	80.0	50.0	30.0
9	Light industrial	5	200	1000	4.7	84.7	56.3	28.4
10	Major roadways	5	200	1000	4.7	89.4	62.5	26.9
11	Civic/educational	10	100	1000	4.7	94.2	68.8	25.4
12	Shopping malls	3	250	750	3.6	97.7	75.0	22.7
13	Utilities	1	150	150	0.7	98.5	81.3	17.2
14	Low-density resid. with swales	5	25	125	0.6	99.1	87.5	11.6
15	Vacant	2	50	100	0.5	99.5	93.8	5.8
16	Park	2	50	100	0.5	100.0	100.0	0.0
	Total	100		21075	100			

Table H.3 Median EMCs and COVs for All Sites Monitored during NURP

Pollutant		Residential Median	Residential COV	Mixed Median	Mixed COV	Commercial Median	Commercial COV	Open/Nonurban Median	Open/Nonurban COV
BOD_5	mg/L	10.0	0.41	7.8	0.52	9.3	0.31	—	—
COD	mg/L	73	0.55	65	0.58	57	0.39	40	0.78
TSS	mg/L	101	0.96	67	1.14	69	0.85	70	2.92
Total lead	μg/L	144	0.75	114	1.35	104	0.68	30	1.52
Total copper	μg/L	33	0.99	27	1.32	29	0.81	—	—
Total zinc	μg/L	135	0.84	154	0.78	226	1.07	195	0.66
Total Kjeldahl nitrogen	μg/L	1900	0.73	1288	0.50	1179	0.43	965	1.00
NO_2-N + NO_3-N	μg/L	736	0.83	558	0.67	572	0.48	543	0.91
Total P	μg/L	383	0.69	263	0.75	201	0.67	121	1.66
Soluble P	μg/L	143	0.46	56	0.75	80	0.71	26	2.1

From EPA. *Results of the Nationwide Urban Runoff Program.* Water Planning Division, PB 84-185552, Washington, D.C. December 1983.

where EMC is the event mean concentration, Rv is the volumetric runoff coefficient (or the effective impervious area, EIA), and Rain is the total rain depth for the period of concern (usually a year). With the appropriate conversions, this simple equation predicts the unit area loadings for the monitored area. This is the method used in the stormwater permit applications for the EPA's NPDES (Nationwide Pollutant Discharge Elimination System) permit program.

The problems with this simplified model include: typically poor estimates of EMC, the Rv value varies for different rain depths, and the procedure cannot easily distinguish seasonal effects (unless EMC values are available for each season), and it cannot be used to evaluate the effectiveness of stormwater control practices.

The main problem with using this simplified model is obtaining an adequate estimate for the EMC. Table H.3 contains the basic concentration information from the EPA's NURP studies (EPA 1983) that are generally used for these analyses. The coefficient of variation (COV) values for these median values are seen to vary from 0.5 to more than 1.0. Figure H.2, also from the EPA's NURP studies (EPA 1983), illustrates the wide variations in observed concentrations for the common stormwater constituents. Wide concentration variations make it more difficult to distinguish between

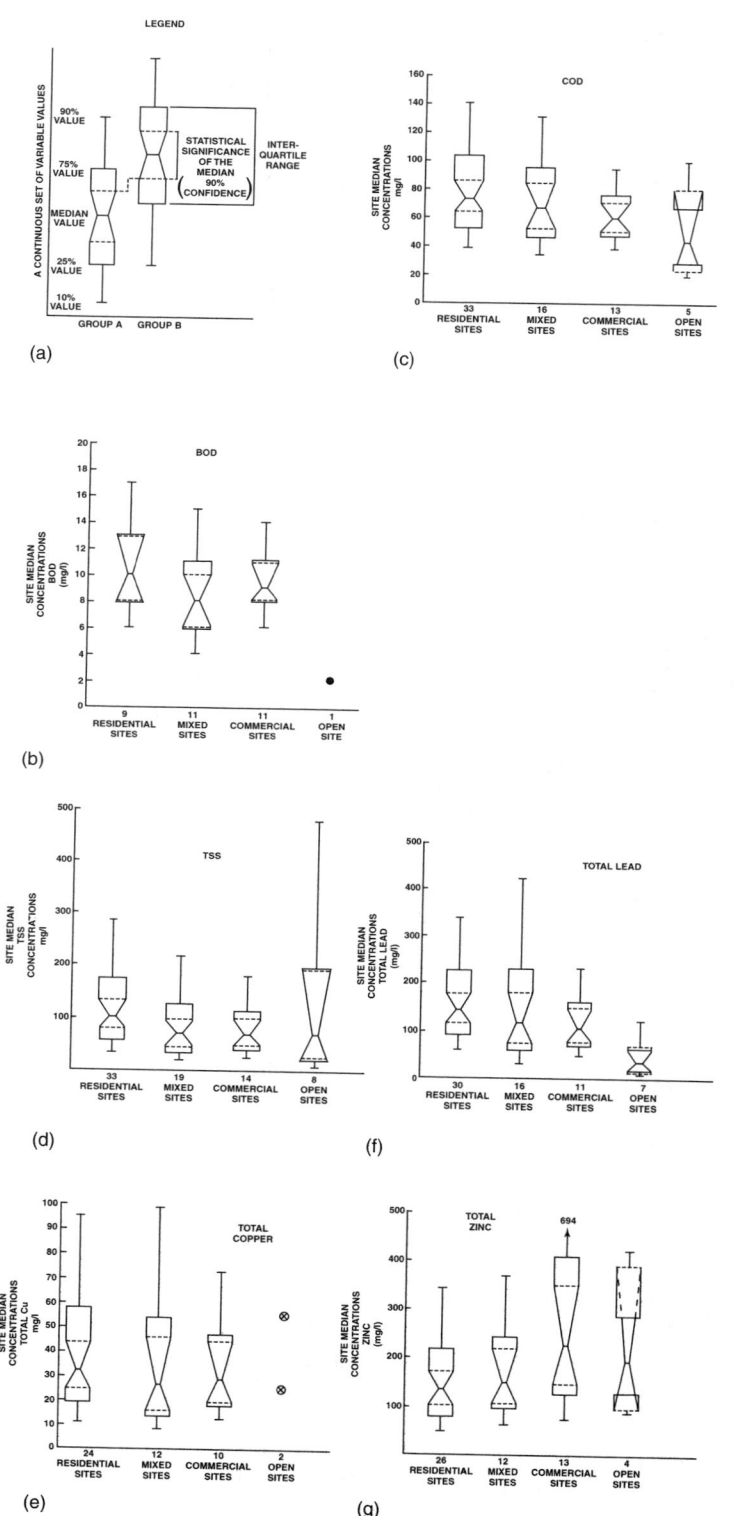

Figure H.2 Box plots of pollutant EMCs for different land uses. (From EPA. *Results of the Nationwide Urban Runoff Program.* Water Planning Division, PB 84-185552, Washington, D.C. December 1983.)

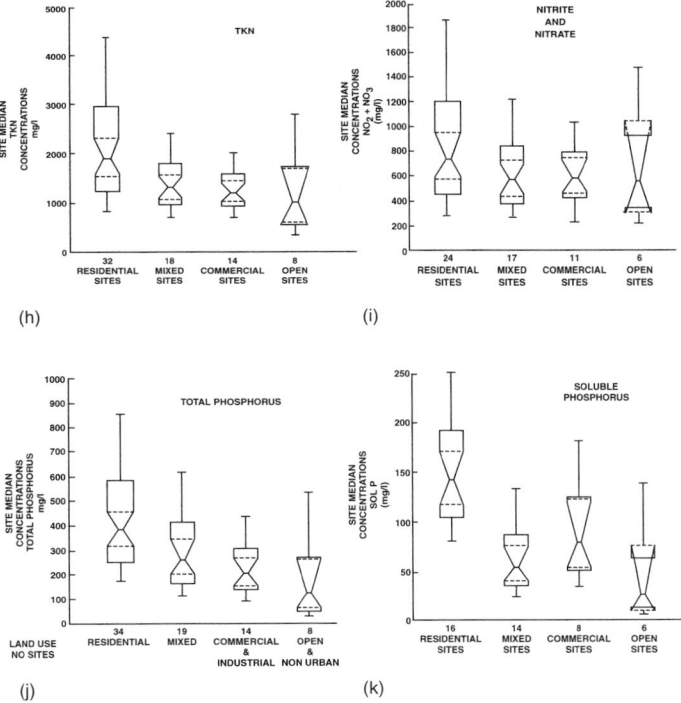

Figure H.2 (continued)

different land uses. As an example, Figure H.2 indicates that suspended solids, BOD, copper, zinc, and nitrite plus nitrate median values are not likely significantly different for any of the four land use categories shown. However, open site COD, phosphorus, and lead median concentrations are likely significantly less than for the other three land uses.

The stormwater permit program typically requires three events to be sampled to determine the EMC value. This small sampling effort likely results in inaccurate EMC estimates because of the relatively large variation in stormwater quality from the same sampling location. As seen in Figure H.3 (a duplicate of Figure 5.3), about 25 samples are required to estimate the EMC with an estimated error of 25% or less, if the COV is 0.5. Most of the time, the COV is even larger, requiring even more samples. The use of only three samples to determine the EMC value would likely result in errors of several hundred percent (using typical levels for confidence of 95% and power of 80%). Such large EMC errors would be reflected in similar errors in the calculated unit area loading values. This could result in incorrect conclusions concerning the relative pollutant sources and inappropriate expenditures of resources for stormwater control.

Errors also occur when selecting the volumetric runoff coefficient (Rv) value. For drainage design, the Rv value is assumed to be equal to the amount of directly connected impervious area. This is sometimes modified to be equal to the "effective" impervious area, as it is obvious that paved areas (and roofs) that drain to pervious areas contribute some runoff, but less than if the paved areas were directly connected to the drainage system. In addition, the Rv is different for different rain depths at the same area. Small rain depths are associated with relatively small Rv values, while larger rains produce larger Rv values, as shown in Figure H.4 (Pitt 1987). Table H.4 (Pitt 1987) illustrates how different urban surfaces contribute increasing fractions of rainfall as runoff. Therefore, if constant Rv values are used for all rains, large errors may occur for individual rains (overpredict for small rains and underpredict for large rains), although the annual average, or annual total, may be acceptable, assuming the monitored rains represent the complete set of annual

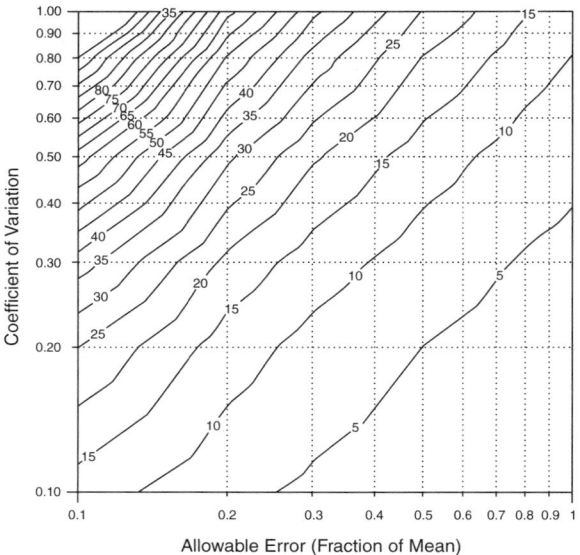

Figure H.3 Sampling requirements for different error goals, alpha of 0.05 and beta of 0.20 (duplicate of Figure 5.3).

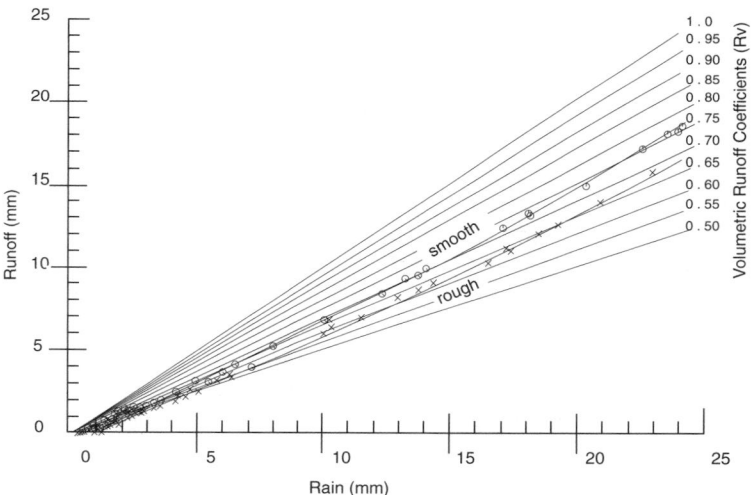

Figure H.4 Rainfall-runoff responses for pavement tests. (From Pitt, R. *Small Storm Urban Flow and Particulate Washoff Contributions to Outfall Discharges*. Ph.D. dissertation submitted to the Department of Civil and Environmental Engineering, University of Wisconsin, Madison. 1987. With permission.)

rains. If only moderate to large rains are monitored (a typical goal), then the averaged Rv for the monitored rains would be larger than the true annual averaged Rv.

Typical estimation methods used for runoff volume were developed for large drainage design storms (several inches in depth) and are not appropriate for the smaller events that are most significant in water quality studies. Table H.4 (Pitt 1987) shows how these runoff coefficients (the fraction of rain that occurs as direct runoff) for impervious areas vary greatly for different rain depths. After several inches of rain (in the range for drainage design studies), the Rv values for all paved and roof areas are between 0.9 and 0.99, resulting in little error if a constant 0.9 value is used. However, at 0.1 to 0.4 in of rain (the rain range where the water pollutants are becoming important), the Rv values for the different paved and roof areas vary greatly (from 0.25 to 0.95).

Table H.4 Observed Directly Connected Runoff Coefficients for Impervious Areas

	0.1 in	0.4 in	1.7 in	Depth When Coefficient Is about 0.9, in
Roads and other small impervious areas	0.4	0.6	0.8	3
Pitched roofs	0.7	0.9	0.98	0.25
Flat roofs	0.25	0.7	0.85	2
Large paved areas and freeways	0.95	0.97	0.99	0.05

From Pitt, R. *Small Storm Urban Flow and Particulate Washoff Contributions to Outfall Discharges*. Ph.D. dissertation submitted to the Department of Civil and Environmental Engineering, University of Wisconsin, Madison. 1987. With permission.

This would result in very large runoff prediction errors if a constant Rv value was assumed for all areas, especially when trying to predict where the runoff water originated.

Most of the annual rainfall is associated with many small individual events and not with the few rarer large rains. Figure H.5a shows measured rain and runoff distributions for Milwaukee during the 1983 NURP monitored rain year. Rains between 0.05 and 5 in were monitored during this period. Two large events (greater than 3 in) occurred during this monitoring period, which greatly bias these curves, compared to typical rain years. During this period:

- The median rain depth was about 0.3 in.
- 66% of all Milwaukee rains were less than 0.5 in in depth.
- For medium-density residential areas, 50% of runoff was associated with rains less than 0.75 in for Milwaukee.
- A 100-year, 24-hour rain of 5.6 in for Milwaukee could produce about 15% of the typical annual runoff volume, but only contributes about 0.15% of the average annual runoff volume when amortized over 100 years.
- Typical 25-year drainage design storms (4.4 in in Milwaukee) produce about 12.5% of the typical annual runoff volume but only about 0.5% of the average runoff volume.

Figure H.5b shows measured Milwaukee pollutant discharges associated with different rain depths for a medium-density residential area. Suspended solids, COD, lead, and phosphates discharges are seen to closely follow the general shape of the runoff distribution shown in Figure H.5a.

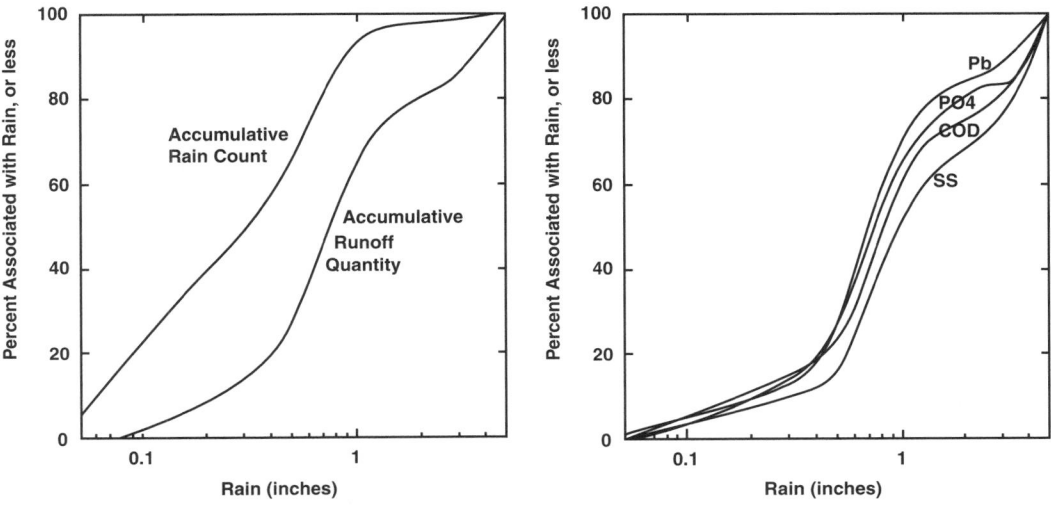

Figure H.5 Accumulative distributions of Milwaukee rain, runoff, and pollutant loadings for medium-density residential areas monitored during 1981 to 1983 (duplicate of Figures 6.1 and 6.2).

Table H.5 Observed Disturbed Urban Soil Volumetric Runoff Coefficients (RV) for Different Rain Depths

	0.1 in	0.4 in	1.7 in	Depth When RV Coefficient Is about 0.1
Clayey soils	0	0.15	0.25	0.2
Sandy soils	0	0	0.05	2.5

From Pitt, R. *Small Storm Urban Flow and Particulate Washoff Contributions to Outfall Discharges.* Ph.D. dissertation submitted to the Department of Civil and Environmental Engineering, University of Wisconsin, Madison. 1987. With permission.

Being able to accurately predict runoff volume is very important in order to reasonable predict runoff pollutant discharges. The shape of the runoff and pollutant runoff curves in Figure H.5 show three distinct regions (values given for Milwaukee):

- Common rains having relatively low pollutant discharges are associated with rains less than about 0.5 in in depth. These are key rains when runoff associated water quality violations, especially for bacteria, are of most concern.
- Rains between 0.5 and 1.5 in are responsible for about 75% of the runoff pollutant discharges and are key rains when addressing mass discharges.
- Rains greater than 1.5 in are associated with drainage design and are only responsible for relatively small portions of the annual pollutant discharges, even with the two unusually large rains that are included in these observations.

Similar relationships are observed for other regions in the country, but the specific rain depths associated with the three specific regions vary. In the southeast, the rain depths separating these three regions are about twice as large as observed for Milwaukee, for example.

Of course, the coefficients shown in Table H.4 can decrease substantially if the paved areas are not directly connected to the drainage system (especially important for roofs and parking areas), or if roadside grass swales are used. It should also be noted that disturbed urban soils contribute much more runoff for moderate rains than the typically expected values (Table H.5).

Complex Models

There are numerous models that fall in the mid-range and detailed model categories that are considered complex. These models all require the use of computers and varying amounts of input data, and they all require calibration and verification for local conditions. These models are constantly changing and new models are continually being developed. The selection of the most appropriate model for a specific situation is therefore important. A good source for model reviews that is periodically updated is the EPA's *Compendium of Watershed-Scale Models for TMDL Development* (EPA 1997). This document was developed for watershed planners and regulators who are responsible for preparing "Total Maximum Daily Load" (TMDL) discharge limitations for receiving waters that are affected by many pollutant sources, including stormwater.

Tables H.6 and H.7 are model summaries from the TMDL report (EPA 1997), while Tables H.8 and H.9 list some of the attributes of many models (including data requirements and overall model complexity). The main distinctions between the mid-range models and the detailed models are that most of the mid-range models are considered "planning" models (for evaluations), while the detailed models are more oriented toward specific design (including greater time-resolution in predicted flows and concentrations). As an example, the mid-range models typically do not require nearly as many details pertaining to specific drainage system layouts as do the detailed models: the mid-range models can operate with more lumped parameters (larger-scale average conditions), while many of the detailed models require detailed drainage system layout information. More of the detailed models can also

Table H.6 Evaluation of Model Capabilities — Mid-Range Models

	Criteria	NPSMAP	GWLF	P8-UCM	SIMPTM	Auto-QI	AGNPS	SLAMM
Land use	Urban	H	H	H	H	H	—	H
	Rural	H	H	—	—	—	H	—
	Point sources	M	M	H	—	—	H	H
Time scale	Annual	—	—	—	—	—	—	—
	Single event	L	—	H	—	—	H	—
	Continuous	H	N	—	—	H	—	H
Hydrology	Runoff	H	H	H	H	H	H	H
	Baseflow	L	H	—	L	L	—	L
Pollutant loading	Sediment	—	H	—	H	H	H	H
	Nutrients	H	H	H	H	H	H	H
	Others	—	—	H	H	H	H	H
Pollutant routing	Transport	L	L	L	L	M	—	M
	Transformation	—	—	—	—	—	H	—
Model output	Statistics	M	L	—	L	—	—	L
	Graphics	M	M	H	—	—	H	L
	Format options	H	H	H	H	L	H	H
Input data	Requirements	M	M	M	M	M	M	M
	Calibration	L	L	L	L	M	L	M
	Default data	H	H	M	M	L	M	M
	User interface	H	H	H	M	M	M	H
BMPS	Evaluation	L	L	H	M	M	M	M
	Design criteria	—	—	H	L	M	M	L
Documentation		H	H	H	L	M	H	M

From EPA. *Compendium of Tools for Watershed Assessment and TMDL Development.* EPA-841-B-97-006. U.S. Environmental Protection Agency. 1997.

include pollutant transformations and nonconservative pollutant behavior, especially for receiving water effects, than the mid-range models. Obviously, there are places where models of each type are needed. In some cases, it is useful to use a mid-range model to predict drainage area runoff conditions and a detailed model to evaluate specific issues pertaining to the drainage system and receiving waters. As an example, the Toronto Area Watershed Management Strategy program (TAWMS) used SLAMM to predict drainage area pollutant and flow discharges, SWMM to predict CSO discharges from the older sections of the city, and HSPF to evaluate receiving water conditions resulting from these discharges (OME 1986). In another example of multiple model use, engineers used SIMPTM in conjunction with SWMM to better predict Portland CSO overflow conditions (Roger Sutherland, personal communication, *Columbia Slough Management Plan*, prepared for the City of Portland's Bureau of Environmental Services). In another example of multiple model use, several cities in Wisconsin have used SLAMM in conjunction with geographical information systems to better prepare the input files required by the program and to display the model results (Thum et al. 1990; Ventura and Kim 1993). The use of a GIS is an especially powerful tool to summarize massive amounts of information, especially when making presentations to the community and to politicians.

Most of the mid-range models were originally developed on personal computers and some have relatively easy-to-use interfaces. The use of "default" values is also common for these models, sometimes restricting the use of locally obtained calibration data. The mid-range models included on Table H.6 are:

- NPSMAP, the Nonpoint Source Model for Analysis and Planning model is a spreadsheet template developed by Omicron Assoc. that predicts nutrient loadings for urban and agricultural areas.
- GWLF, the Generalized Loading Functions model was developed at Cornell University to assess point and nonpoint loadings of nitrogen and phosphorus from relatively large agricultural and urban watersheds. It includes rainfall/runoff processes and erosion predictions. Most of the processes are controlled by default values.

- P8-UCM, the Urban Catchment Model was developed by John Walker for the Narragansett Bay Project to simulate stormwater pollutants in small urban catchments. Evaluations of various management practices are possible with P8, including help in their sizing for specific control objectives. It incorporates many default values from the EPA's Nationwide Urban Runoff Program (EPA 1983).
- SIMPTM, the Simplified Particle Transport Model was developed by Roger Sutherland, of Pacific Water Resources, to simulate runoff, sediment, and yield of other pollutants from urban watersheds, including the evaluation of some control practices. Detailed particulate buildup and washoff processes are included, based on northwest regional data.
- Auto-QI, the Automated Qual-Illudas model was developed by Mike Trestriep at the Illinois State Water Survey to perform continuous simulations of runoff from impervious and pervious urban areas and to evaluate the effectiveness of selected control practices. It also includes components to examine receiving water impacts. A version of the model is linked to the ARC/INFO GIS program.
- AGNPS, the Agricultural Nonpoint Source pollution model was developed by the USDA Agricultural Research Service. It addresses potential impacts from point and nonpoint source pollution on surface and groundwater in agricultural watersheds. Alternative management programs are also evaluated. The spatial (grid) design of the model allows it to be interfaced to GIS and digital terrain models to simplify inputting the model parameters.
- SLAMM the Source Loading and Management Model was developed by Robert Pitt of the University of Alabama at Birmingham to evaluate the effects of urban development characteristics and source and outfall controls on pollutant discharges. It examines runoff from separate drainage areas that may include a wide variety of land uses and control practices. The outfall discharge estimates can then be evaluated in a separate model to evaluate receiving water impacts. Unique small storm hydrology and particulate washoff procedures, based on extensive field measurements, are incorporated in the model to more accurately predict the role of different source areas in generating stormwater pollutant discharges.

Detailed models were all originally developed on mainframe computers, but most have been ported to personal computers over the past several years. Most still have awkward user interfaces and require a group of skilled users to take advantage of most of their comprehensive capabilities, although proprietary Windows-based user interfaces and proprietary modifications of some of the more popular models (especially SWMM) are becoming common. The detailed models shown on Table H.7 include:

- STORM, the Storage, Treatment, Overflow Runoff Model was developed by the U.S. Army Corps of Engineers to continuously simulate urban runoff quantity, sediment, and several conservative pollutants. It has most commonly been used to evaluate the trade-offs between treatment and storage options for the control of CSOs.
- ANSWERS, the Areal Nonpoint Source Watershed Environment Response Simulation Model was developed by the University of Georgia to evaluate the effects of land use, management schemes, and conservation practices on the quantity and quality of watershed runoff. It stresses erosion and sediment transport processes.
- DR3M, the Distributed Routing Rainfall Runoff Model is supported by the U.S. Geological Survey and was developed to study conventional pollutants in predominantly urban areas. It produces detailed hydrographs and pollutant transport plots.
- SWRRBQ, the Simulation for Water Resources in Rural Basins model was developed by the USDA to simulate hydrologic, sedimentation, nutrient, and pesticide movement in large, complex, rural watersheds.
- SWMM the Storm Water Management Model was developed by the EPA to derive design criteria for structural stormwater controls. SWMM is likely the most commonly used detailed stormwater model, especially when evaluating sewerage issues and combined sewer overflows.
- HSPF, the Hydrological Simulation Program–FORTRAN was developed by the U.S. EPA to simulate water quantity and quality for a wide range of organic and inorganic pollutants from agricultural and urban watersheds. It is probably the most comprehensive model available, especially considering receiving water impacts. Chemical, biological, and physical processes are included to account for pollutant transport and transformations. However, much calibration information is required to effectively use all of HSPF's capabilities.

Table H.7 Evaluation of Model Capabilities — Detailed Models

	Criteria	STORM	ANSWERS	DR3M	SWRRBWQ	SWMM	HSPF
Land use	Urban	H	—	H	L	H	H
	Rural	—	H	—	H	L	M
	Point sources	H	—	H	H	H	H
Time scale	Annual	—	—	—	—	—	—
	Single event	L	H	L	L	H	H
	Continuous	H	—	H	H	H	H
Hydrology	Runoff	H	H	H	H	H	H
	Baseflow	L	—	L	H	H	H
Pollutant loading	Sediment	H	H	H	H	H	H
	Nutrients	H	H	H	H	H	H
	Others	H	—	—	H	H	H
Pollutant routing	Transport	—	M	H	H	L	H
	Transformation	—	—	—	—	L	H
Model output	Statistics	L	—	H	H	H	H
	Graphics	—	H	M	M	L	L
	Formal options	H	H	H	H	H	H
Input data	Requirements	M	H	H	M	H	H
	Calibration	L	L	M	M	H	H
	Default data	M	L	H	H	M	M
	User interface	—	—	M	M	—	—
BMPs	Evaluation	M	M	H	M	H	H
	Design criteria	M	M	M	—	H	H
Documentation		H	M	M	H	H	H

From EPA. *Compendium of Tools for Watershed Assessment and TMDL Development*. EPA-841-B-97-006. U.S. Environmental Protection Agency. 1997.

The mid-range models are probably the most commonly used because they are perceived as being easier to use and require less input information. That was certainly true when most of these detailed models were developed, but some of them, most notably SWMM, have a growing industry supporting their use, including the availability of much improved user interfaces. However, the cost of obtaining and using (entering and verifying) detailed information that may be required may not be justified by the intended use of the data generated by the model. The application of detailed models is more cost-effective when applied to address complex situations or objectives.

Dr. Bill James of the University of Guelph has long been an advocate of long-term continuous simulations. The cost of the required computer time has also decreased to the point where the use of long-term continuous simulations is no longer prohibitive, and is strongly recommended in order to obtain a much better understanding of watershed responses under a wide variety of conditions. The modeling of a few "design" storms may be satisfactory for simple drainage design considerations, where the only parameter of interest is peak flow rate. However, limiting simulations to only a few storms falls far short when a wide variety of water quality questions are important. The behavior of different stormwater quality control practices is also dependent on many different hydraulic parameters, not just peak flow rate. The use of several years of rainfall data in continuous simulations should therefore be considered the norm. If a model is not capable of continuous simulations, its usefulness is probably severely restricted to only the most rudimentary preliminary evaluations.

Receiving Water Models

Some of the models listed above include receiving water components (such as HSPF), but there are many additional models available that are specific for receiving waters and are in the public

Table H.8 Listing of Attributes of Commonly Used Urban Models

Attribute	DR3M-QUAL	HSFP	Statistical[a]	STORM	SWMM
Sponsoring agency	USGS	EPA	EPA	HEC	EPA
Simulation type[b]	C,SE	C,SE	N/A	C	C,SE
No pollutants	4	10	Any	6	10
Rainfall/runoff analysis	Y	Y	N[c]	Y	Y
Sewer system flow routing	Y	Y	N/A	N	Y
Full, dynamic flow routing equations	N	N	N/A	N	Y[d]
Surcharge	Y[e]	N	N/A	N	Y[d]
Regulators, overflow structures, e.g., weirs, orifices, etc.	N	N	N/A	Y	Y
Special solids routine	Y	Y	N	N	Y
Storage analysis	Y	Y	Y[f]	Y	Y
Treatment analysis	Y	Y	Y[f]	Y	Y
Suitable for screening (S), design (D)	S,D	S,D	S	S	S,D
Available on microcomputer	N	Y	Y[g]	N	Y
Data and personnel requirements[h]	Medium	High	Medium	Low	High
Overall model complexity[i]	Medium	High	Medium	Medium	High

[a] EPA procedure.
[b] C = continuous simulation, SE = single event simulation.
[c] Runoff coefficient used to obtain runoff volumes.
[d] Full dynamic equations and surcharge calculations only in Extran Block of SWMM.
[e] Surcharge simulated by storing excess inflow at upstream end of pipe. Pressure flow not simulated.
[f] Storage and treatment analyzed analytically.
[g] FHWA study, Driscoll et al. (1989).
[h] General requirements for model installation, familiarization, data requirement, etc. To be interpreted only very generally.
[i] Reflection of general size and overall model capabilities. Note that complex models may still be used to simulate very simple systems with attendant minimal data requirements.

From EPA. *Modeling of Nonpoint Source Water Quality in Urban and Non-urban Areas.* Office of Research and Development. U.S. Environmental Protection Agency. Washington, D.C. EPA 600/3-91/039. 1991.

Table H.9 Listing of Commonly used Non-Urban Runoff Models

Attribute	AGNPS	ANSWERS	CREAMS	HSPF	PRZM	SWRRB	UTM-TOX
Sponsoring agency	USDA	Purdue	USDA	EPA	EPA	USDA	ORNL & EPA
Simulation type	C,SE	SE	C,SE	C,SE	C	C	C, SE
Rainfall/runoff analysis	Y	Y	Y	Y	Y	Y	Y
Erosion modeling	Y	Y	Y	Y	Y	Y	Y
Pesticides	Y	N	Y	Y	Y	Y	N
Nutrients	Y	Y	Y	Y	N	Y	N
User-defined constituents	N	N	N	Y	N	N	Y
Soil processes							
Pesticides	N	N	Y	Y	Y	Y	N
Nutrients	N	N	Y	Y	N	Y	N
Multiple land type capability	Y	Y	N	Y	N	Y	Y
In-stream water quality simulation	N	N	N	Y	N	N	Y
Available on microcomputer	Y	Y	Y	Y	Y	Y	N
Data and personnel requirements	M	M/H	H	H	M	M	H
Overall model complexity	M	M	H	H	M	M/H	H

Y = yes, N = no, M = Moderate, H = High, C = Continuous, SE = Storm Event.

From EPA. *Modeling of Nonpoint Source Water Quality in Urban and Non-urban Areas.* Office of Research and Development. U.S. Environmental Protection Agency. Washington, D.C. EPA 600/3-91/039. 1991.

domain and available through the EPA's CEAM. Some included on the 1996 version of the *Exposure Models Library and Integrated Model Evaluation System* are listed below and in Figure H.6:

Surface Water Models	
Selected for 1st and 2nd Level Reviews:	**Selected for 1st Level Review Only:**
CEQUALRIV1	EXAMS:
CEQUALW2	FATE: fate of organics
CTAP: chemical transport & analysis program	GCSOLAR
DYNTOX: dynamic toxicity model	HEC-5Q & 6
EUTRO4	MICHRIV: transport in water & sediments
GEMS-EXAMS: geographical exposure modeling systems - EXAMs	PCPROUTE-PC: pollutant routing model
HSPF: hydrologic simulation program - fortran	PLUMES:
QUAL2E: enhanced stream water quality model	RESTMP: water temperature model
REACHSCAN	RIVMOD: sediment transport
SERATRA: in-stream sediment-contaminant transport	SEDDEP: settling of wastewater particulates
TOX 14	SMPTOX: stream toxic model
WQRRS: water quality (ecological cycling) in rivers and reservoirs	TERMS: thermal simulation of lakes
WASP5: water quality assessment program	TWQM: downstream transformation of problem constituents

Figure H.6 Surface water models included on the CEAM CD-ROM.

- SWAT (contains the GLEANS pesticide fate model as a component)
- PREWET (predicts fates of pollutants in wetlands)
- GWLF (a simple transport model)
- CREAMS (transport of soluble and sediment-attached chemicals)
- WASP5 (especially the TOXI5 and EUTRO5 components)
- MINTEQA2 (chemical equilibrium model)
- TWQM (in-stream effects of reduced species that may be discharged from dams)
- SMPTOX (toxicant interactions with stream bed sediments)
- WATEQF (chemical equilibrium model)
- VLEACH (chemical fate model)

Geographical Information Systems (GIS)

As indicated above, the use of GIS has become very important when modeling large areas. The main advantages of GIS include an ability to effectively display large amounts of information (relatively easy to incorporate with model output), and in some cases, to organize and automate the data input requirements for the models (requiring a much greater level of integration with a model). It has been especially important when working with nontechnical community groups and when summarizing modeling options. The visual presentation of the massive amounts of output results, or results from monitoring programs, is much more effective for communicating with diverse groups of people.

PUBLIC LAND RECORDS USED IN DIGITAL DATABASE

Data Item	Custodian	Document
Parcel	Dane Co. Land Records & Regulation	Section parcel maps
Soils	U.S.D.A. Soil Conservation Service	Dane County Soil Survey
Slope	U.S.D.A. Soil Conservation Service	Dane County Soil Survey
Slope	U.S. Geological Survey	Quadrangle maps
Wetlands	Wis. Dept. of Natural Resources	Wetlands inventory
Hydrology	U.S. Geological Survey	Quadrangle maps
Farm Tracts & Fields	Dane County, A.S.C.S.	NHAP aerial photo prints
Woodlots	Dane County, A.S.C.S.	NHAP aerial photo prints
Existing Land Uses	Dane County, A.S.C.S.	NHAP aerial photo prints
Planned Land Uses	Dane Co. Regional Planning Comm.	Town of Burke Land Use Maps
Planned Land Uses	Dane Co. Regional Planning Comm.	Hwy 151 Corridor Study
Planned Land Uses	City of Madison, Dept. of Planning	Burke Heights Dev't Plan
Land Use Zoning	Dane Co. Land Records & Regulation	Section zoning maps
Land Use Zoning	City of Sun Prairie, Dept. of Planning	City of Sun Pr. zoning maps
Land Use Zoning	City of Madison, Dept. of Planning	City of Madison zoning maps
Floodplain	Federal Emergency Mgmt. Agency	Flood boundary map
Existing Parks	Dane Co. Land Records & Regulation	Section parcel maps
Planned Parks	Dane County Parks Division	Cherokee Marsh Owner. Map
Existing Sewers	Madison Metro Sewerage District	Sewer. Dist. Interseptor Map
Existing Sewers	City of Sun Prairie, Dept. of Engineer.	City interseptor map
Planned Sewers	Madison Metro Sewerage District	Collection System Design Rep.
Urban Service Areas	Dane Co. Land Records & Regulation	Town of Burke Land Use Plan
Traffic Counts	Wis. Department of Transportation	1987 Highway Traffic maps
Roads/Hwys	U.S. Geological Survey	Quadrangle maps
Farm Tenure	Dane County, A.S.C.S.	Farm operator file
Farm Tenure	Dane Co. Land Records & Regulation	Section Parcel maps
Historic Buildings	Wisconsin Historic Society	Coded quadrangle maps
Archeologic Sites	Wisconsin Historic Society	Coded quadrangle maps
Watershed Boundary	Dane Co. Regional Planning Comm.	Watershed boundary map
Watershed Boundary	U.S. Geological Survey	Quadrangle maps

Figure H.7 Availability of data used in early GIS and stormwater modeling studies conducted in Dane County, WI. (From Pickett, S.R., O.G. Thum, and B.J. Hiemann. Using a land information system to integrate nonpoint source pollution modeling and land use development planning. *Land Information and Computer Graphics Facility,* The University of Wisconsin, Madison. 1989.)

GIS has been used for many years, but has recently become much more accessible with improvements in software and significant cost reductions in suitable computer equipment. Various communities in the State of Wisconsin, for example, have used GIS systems integrated with the Source Loading and Management Model (SLAMM) to graphically illustrate development and control options associated with urbanization (Haubner and Joeres 1996; Kim et al. 1993; Kim and Ventura 1993; Ventura and Kim 1993). Figure H.7 (Pickett et al. 1989) shows the availability of data used in some of the early studies conducted in Dane County, WI, while Figure H.8 (Kim and Ventura 1993) shows how the information is integrated with SLAMM to identify critical source areas. Figure H.9 (Pickett et al. 1989) is an example map showing expected changes in suspended solids discharges resulting from development options. SLAMM is currently available from www.winslamm.com.

As noted above, the current development and use of the BASINS model, especially for TMDL evaluations, relies heavily of a GIS framework (Lahlou et al. 1998). Tables H.10 through H.13, from the BASINS *User's Manual* (Lahlou et al. 1998) describe the information contained on the CD-ROMS specific for each EPA region. This wealth of information is available to initial analyses for a specific area, but users are encouraged to incorporate high-resolution information and locally derived data sets for more accurate use. The cartographic data (Table H.10) includes hydrologic boundaries and major roadways, plus census areas and various political boundaries. The environmental data (Table H.11) are to support watershed characterization and environ-

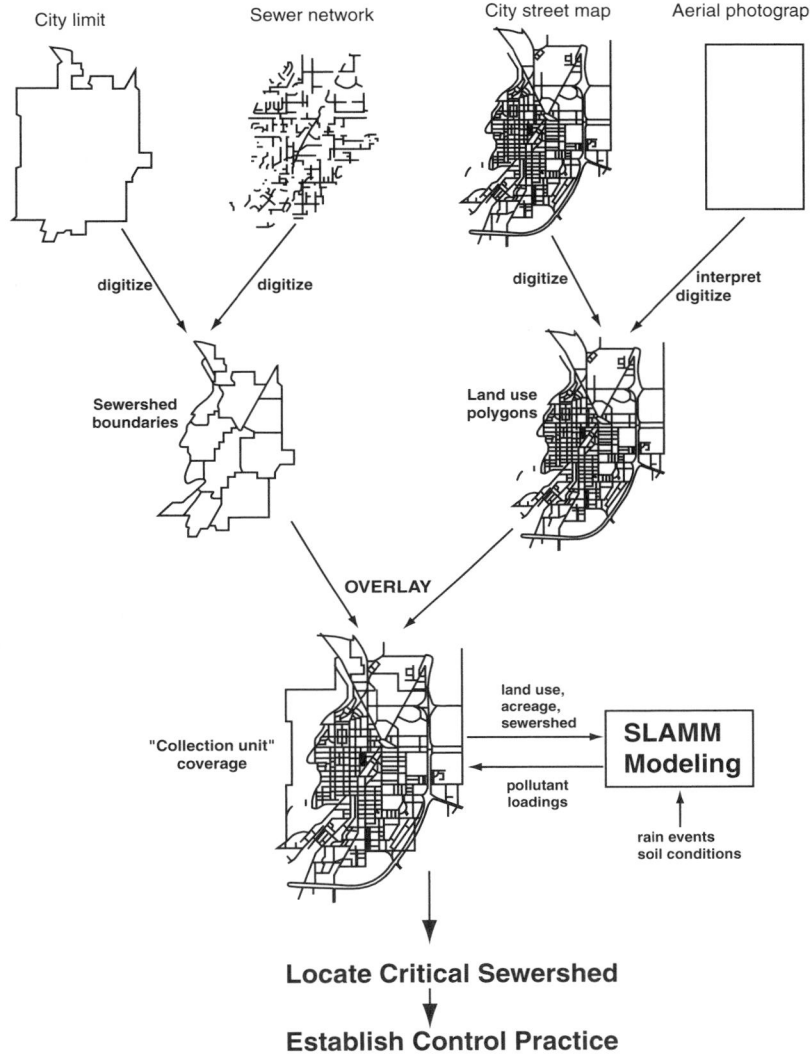

City limit   Sewer network   City street map   Aerial photograph

digitize   digitize   digitize   interpret digitize

Sewershed boundaries   Land use polygons

OVERLAY

"Collection unit" coverage

land use, acreage, sewershed

SLAMM Modeling

pollutant loadings

rain events soil conditions

Locate Critical Sewershed

Establish Control Practice

Figure H.8 Integration of information and modeling to identify critical source areas. (From Kim, K. and S. Ventura. Large-scale modeling of urban nonpoint source pollution using a geographical information system. *Photogrammetric Eng. Remote Sensing*, 59(10): 1539–1544. October 1993.)

mental analyses, and include data on soils, topography, land uses, and stream hydrography. This is the most important set of information for modeling local conditions. The environmental monitoring data (Table H.12) include statistical summaries of monitoring results, rainfall records, and limited biological conditions. The point source data (Table H.13) provide information on pollutant loadings from permitted facilities, plus locations of hazardous waste sites. The BASINS assessment tools allow users to make evaluations of water quality, while the available data management utilities delineate watershed boundaries, and can be used to modify the data, or to import new data into the system. The nonpoint source and stream models integrate these data to provide initial evaluations of watershed water quality conditions. BASINS can therefore be a very useful tool to focus specific monitoring efforts to investigate likely water quality problems and use impairments.

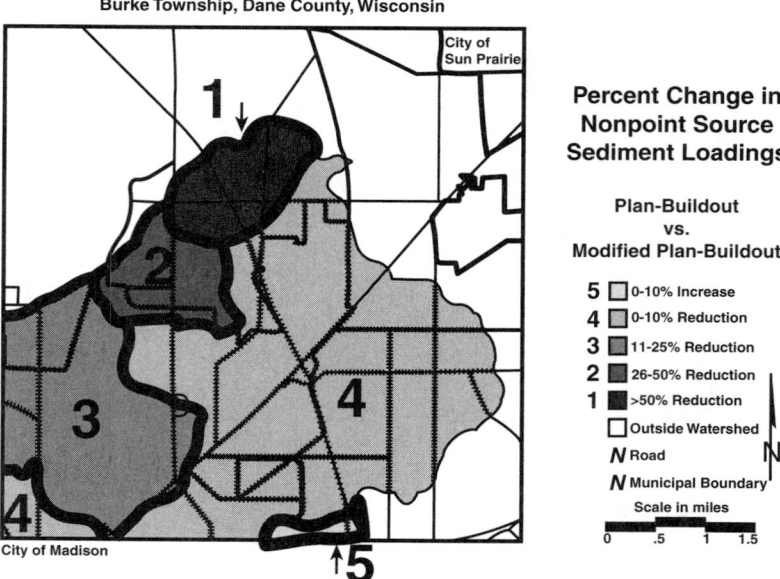

Figure H.9 Example of mapped results showing changes in suspended solids with different development options. (From Pickett, S.R., O.G. Thum, and B.J. Hiemann. Using a land information system to integrate nonpoint source pollution modeling and land use development planning. *Land Information and Computer Graphics Facility,* The University of Wisconsin, Madison. 1989.)

Table H.10 BASINS Base Cartographic Data

BASINS Data Product	Source	Description
Hydrologic unit boundaries	U.S. Geological Survey	Nationally consistent delineations of the hydrographic boundaries associated with major U.S. river basins
Major roads	Federal Highway Administration	Interstate and state highway network
Populated place locations	USGS	Location and names of populated locations
Urbanized areas	Bureau of the Census	Delineations of major urbanized areas used in 1990 Census
State and county boundaries	USGS	Administrative boundaries
EPA regions	USGS	Administrative boundaries

From Lahlou, M., L. Shoemaker, S. Choudhury, R. Elmer, A. Hu, H. Manguerra, and A. Parker. *BASINS, Better Assessment Science Integrating Point and Nonpoint Sources.* Version 2.0. EPA-823-B-98-006. Exposure Assessment Branch, U.S. Environmental Protection Agency. Washington, D.C. November 1998.

SUMMARY

The amount of data required to use these models can be very large. Tables H.14 and H.15 list some of these data needs for watershed-scale models (EPA 1991). Much of the information can be obtained from locally available sources and data summaries, but much will have to be extracted from detailed maps or the basic data to obtain the information in the necessary formats, accuracies, or time scales. In addition, the models need to be calibrated for site-specific conditions (especially pollutant characteristics and rainfall runoff relationships) and verified (comparing monitored outfall quality and quantity with modeled values). Receiving water models also require much local information for efficient use. Besides the watershed-scale information listed in these tables, specific stream processes (such as described in the *Rates,*

Table H.11 BASINS Environmental Background Data

BASINS Data Product	Source	Description
Ecoregions	U.S. Environmental Protection Agency (USEPA)	Ecoregions and associated delineations
National Water Quality Assessment (NAWQA) study unit boundaries	USGS	Delineations of study areas
1996 Clean Water needs survey	USEPA	Results of the wastewater control needs assessment by state
State soil and geographic (STATSGO) database	U.S. Department of Agriculture, Natural Resources Conservation Service (USDA-NRCS)	Soils information including soil component data and soils
Managed area database	University of California, Santa Barbara	Data layer including federal and Indian lands
Reach file version 1 (RF1)	USEPA	Provides stream network for major rivers and supports development of stream routing for modeling purposes (1:500k)
Reach file version 3 (RF3) alpha	USEPA	Alpha version of Reach File 3; provides detailed stream network and supports development of stream routing for modeling purposes (1:100K)
Digital elevation model (DEM)	USGS	Topographic relief mapping; supports watershed delineations and modeling
Land use and land cover	USGS	Boundaries associated with land use classifications including Anderson Level 1 and Level 2

From Lahlou, M. et al., *BASINS, Better Assessment Science Integrating Point and Nonpoint Sources.* Version 2.0. EPA-823-B-98-006. Exposure Assessment Branch, U.S. Environmental Protection Agency. Washington, D.C. November 1998.

Table H.12 BASINS Environmental Monitoring Data

BASINS Data Product	Source	Description
Water quality monitoring stations and data summaries	USEPA	Statistical summaries of water quality monitoring for physical and chemical-related parameters; parameter-specific statistics computed by station for 5-year intervals from 1970 to 1994 and 3-year interval from 1995 to 1997
Bacteria monitoring stations and data summaries	USEPA	Statistical summaries of bacteria monitoring; parameter-specific statistics computed by station for 5-year intervals form 1970 to 1994 and 3-year interval from 1995 to 1997
Water quality stations and observation data	USEPA	Observation-level water quality monitoring data for selected locations and parameters
National sediment inventory (NSI) stations and database	USEPA	Sediment chemistry, tissue residue, and benthic abundance monitoring data for fishing, including type of impairment
Listing of fish and wildlife advisories	USEPA	State reporting of locations with advisories for fishing, including 7Q10 low and monthly mean stream flow
Gauge sites	USGS	Inventory of surface water gaging station data including 7Q10 low and monthly mean stream
Weather station sites	National Oceanic and Atmospheric Administration (NOAA)	Location of selected first-order NOAA weather stations
Drinking water supply (DWS) sites	USEPA	Location of public water supplies, their intakes, and sources of surface water supply
Watershed data stations and database	NOAA	Location of selected meteorologic stations and associated monitoring information used to support modeling
Classified shellfish areas	NOAA	Location and extent of shellfish closure areas

From Lahlou, M. et al., *BASINS, Better Assessment Science Integrating Point and Nonpoint Sources.* Version 2.0. EPA-823-B-98-006. Exposure Assessment Branch, U.S. Environmental Protection Agency. Washington, D.C. November 1998.

Table H.13 BASINS Point Source/Loading Data

BASINS Data Product	Source	Description
Permit compliance system (PCS) sites and computed annual loadings	USEPA	NPDES permit-holding facility information; contains parameter-specific loadings to surface waters computed using the EPA Effluent Decision Support System (EDSS) for 1991–1996
Industrial facilities discharge (IFD)	USPEA	Facility information on industrial point source discharges to surface waters
Toxic release inventory (TRI) sites and pollutant releases data	USEPA	Facility information for 1987–1995 TRI public data; contains Y/N flags for each facility indicating media-specific reported releases
Superfund national priority list site	USEPA	Location of Superfund National Priority List sites from CERCLIS (Comprehensive Environmental Response, Compensation and Liability Information System)
Resource conservation and recovery information system (RCRIS) sites	USEPA	Location of transfer, storage, and disposal facilities for solid and hazardous waste
Minerals availability systems/mineral industry location system (MAS/MILS)	U.S. Bureau of Mines	Location and characteristics of mining sites

From Lahlou, M. et al., *BASINS, Better Assessment Science Integrating Point and Nonpoint Sources*. Version 2.0. EPA-823-B-98-006. Exposure Assessment Branch, U.S. Environmental Protection Agency. Washington, D.C. November 1998.

Table H.14 Typical Input Data Needs for Nonpoint Source Models

1. System parameters
   a. Watershed size
   b. Subdivision of the watershed into homogeneous subareas
   c. Imperviousness of each subarea
   d. Slopes
   e. Fraction of impervious areas directly connected to a channel
   f. Maximum surface storage (depression plus interception storage)
   g. Soil characteristics including texture, permeability, erodibility, and composition
   h. Crop and vegetation cover
   i. Curb density or street gutter length
   j. Sewer system or natural drainage characteristics
   k. Land use
2. State variables
   a. Ambient temperature
   b. Reaction rate coefficients
   c. Adsorption/desorption coefficients
   d. Growth stage of crops
   e. Daily accumulation rates of litter
   f. Traffic density and speed
   g. Potency factors for pollutants (pollutant strength on sediment)
   h. Solar radiation (for some models)
3. Input variables
   a. Precipitation
   b. Atmospheric fallout
   c. Evaporation rates

Adapted from EPA. *Modeling of Nonpoint Source Water Quality in Urban and Nonurban Areas*. Office of Research and Development. U.S. Environmental Protection Agency. Washington, D.C. EPA 600/3-91/039. 1991.

Table H.15 Data Needs for Various Quality Prediction Methods

Method	Data	Potential Source[a]
Unit load	Mass per time per unit tributary area	Derive from constant concentration and runoff, literature values
Constant concentration	Runoff prediction mechanism (simple to complex)	Existing model; runoff coefficient or simple method
	Constant concentration for each constituent	NURP; local monitoring
Spreadsheet	Simple runoff prediction mechanism	e.g., runoff coefficient, perhaps as function of land use
	Constant concentration or concentration range	NURP; local monitoring
	Removal fractions for controls	NURP; Schueler (1987); local and state publications
Statistical	Rainfall statistics	NURP; Driscoll, et.al. (1989); Woodward Clyde (1989); EPA SYNOP model
	Area, imperviousness. Pollutant median and CV	NURP; Driscoll (1986); Driscoll, et al. (1989); local monitoring
	Receiving water characteristics and statistics	Local or generalized data
Regression	Storm rainfall, area, imperviousness, land use	Local data
Rating curve	Measured flow rates/volumes and quality EMCs/loads	NURP; local data
Buildup	Loading rates and rate constants	Literature values
Washoff	Power relationship with runoff	Literature values

[a] Must be calibrated and verified using local monitoring.

From EPA. *Modeling of Nonpoint Source Water Quality in Urban and Non-urban Areas*. Office of Research and Development. U.S. Environmental Protection Agency. Washington, D.C. EPA 600/3-91/039. 1991.

Constants, and Kinetics Formulations in Surface Water Quality Modeling report prepared by Tetra Tech 1985) require calibration and verification. Tables H.14 through H.17 (EPA 1991) list some of the water quality variables modeled and the processes simulated by representative receiving water models. The techniques presented in this book, supplemented by the noted references, will enable the user to effectively collect the needed local data for model calibration and verification. Few models attempt to address in-stream biological process (beyond photosynthesis/respiration for DO evaluations and bacteria die-off). Biological beneficial uses are best compared to actual measurements and comparisons with reference streams. However, models are needed to predict likely future chemical and physical conditions that currently do not exist. The information in this book should enable reasonable evaluations of these predicted conditions for biological use impairments, at least by identifying potential areas of concern. The ability to model biological responses to chemical and physical changes (such as responses to habitat destruction and contaminated sediments that are likely the most serious issues in urban streams) is very uncertain. However, numerous site-specific investigations, especially in the Pacific Northwest and in Canada, are encouraging.

It is therefore important to consider the appropriate uses of models, especially in receiving water investigations. Models are important and critical tools in that they enable us to design experiments and monitoring activities effectively, and to look into the future and examine alternatives. However, there can be substantial error in their predictions, due to incorrectly described processes, lack of data and the natural variability of conditions that simply cannot be adequately explained. This error, coupled with our lack of understanding of cause and effect relationships between the more easily predicted physical/chemical parameters and biological conditions, warrants continued caution. With local experience associated with a commitment to long-term investigations in local waters, our understanding will improve along with our ability to make reasonable conclusions using modeling results.

Table H.16 Non-Toxic Constituents Included in Stream Models

Model Name	DO	CBOD or total BOD	NBOD	SOD	Temp.	Total P	Organic P	PO$_4$	Total N	Organic N
WQAM	X	X	X	X	X	X		X	X	
DOSAG1	X	X	X	X	X**					
DOSAG3	X	X	X*	X	X**			X		
SNSIM	X	X	X	X	X**					
QUAL-II	X	X	X*	X	X			X		
QUAL-IIe	X	X	X*	X	X			X		
RECEIV-II	X	X	X*	X	X**	X		X	X	X
WASP	X	X	X*	X	X**		X	X		X
AESPO	X	X	X*	X	X**		X	X		X
HSPF	X	X	X*	X	X			X		
HAR03	X	X			X**					
FEDBAK03	X	X			X**					
MIT-DNM	X	X			X**					
EXPLORE-1	X	X	X*	X	X**		X	X		X
WQRRS	X	X	X*	X	X			X		

Model Name	NH$_3$	NO$_2$	NO$_3$	Carbon	Algae or Chl-A	Zooplankton	pH	Alkalinity	TDS	Coliform Bacteria
WQAM									X	X
DOSAG1										
DOSAG3	X	X	X		X				X	X
SNSIM										
QUAL-II	X	X	X		X				X	X
QUAL-IIe	X	X	X		X				X	X
RECEIV-II	X	X	X		X				X	X
WASP	X		X	X	X	X	X	X	X	
AESPO	X	X	X	X	X	X	X	X	X	
HSPF	X	X	X	X	X		X	X	X	X
HAR03										
FEDBAK03										
MIT-DNM										
EXPLORE-1	X	X	X	X	X	X				
WQRRS	X	X	X	X	X	X	X	X	X	X

X* NBOD simulated as nitrification of ammonia.

X** Temperature specified by model users.

From EPA. *Modeling of Nonpoint Source Water Quality in Urban and Non-urban Areas.* Office of Research and Development. U.S. Environmental Protection Agency. Washington, D.C. EPA 600/3-91/039. 1991.

Table H.17 Conventional Pollutants Model Comparison as Used in Waste Load Allocations

Model	Water Quality Temporal Variability	Hydraulic Temporal Variability	Variable Loading Rates	Types of Loads	Spatial Dimensions	Water Body	Water Quality Parameters Modeled	Chemical/Biological Processes Simulated	Physical Processes Simulated
DOSAG-1	Steady-state	Steady-state	No	Multiple point sources	1-D	Stream network	DO, CBOD, NBOD, conservative	1st-order decay of NOBD, CBOD, coupled DO	Dilution, advection, reaeration
SNSIM	Steady-state	Steady-state	No	Multiple point sources and nonpoint sources	1-D	Stream network	DO, CBOD, NBOD, CONSERVATIVE	1ST-ORDER DECAY OF NBOD, CBOD, coupled DO, benthic demand(s), photosynthesis(s)	Dilution, advection, reaeration
QUAL-II	Steady-state or dynamic	Steady-state	No	Multiple point sources and nonpoint sources	1-D	Stream network	DO, CBOD temperature, ammonia, nitrate, nitrite, algae, phosphate, coliforms, nonconservative substances, three conservative substances	1st-order decay of NBOD, CBOD, coupled DO, benthic demand, CBOD setting, nutrient-algal cycle	Dilution, advection, reaeration, heat balance
RECEIV-II	Dynamic	Dynamic	Yes	Multiple point sources	1-D or 2-D	Stream network or well-mixed estuary	DO, CBOD, ammonia, nitrite, nitrite, total nitrogen, phosphate, coliforms, algae, salinity, one metal ion	1st-order decay of CBOD, coupled DO, benthic demand, CBOD settlings, nutrient-algal cycle	Dilution, advection, reaeration

(s) = specified.

From EPA. *Exposure Models Library and Integrated Model Evaluation System.* EPA Office of Research and Development CD-ROM. EPA-600-C-92-002. Revised March 1996.

REFERENCES

EPA. *Results of the Nationwide Urban Runoff Program.* Water Planning Division, PB 84-185552, Washington, D.C. December 1983.

EPA. *Modeling of Nonpoint Source Water Quality in Urban and Non-urban Areas.* Office of Research and Development. U.S. Environmental Protection Agency. Washington, D.C. EPA 600/3-91/039. 1991.

EPA. *Exposure Models Library and Integrated Model Evaluation System.* EPA Office of Research and Development CD-ROM. EPA-600-C-92-002. Revised March 1996.

EPA. *Compendium of Tools for Watershed Assessment and TMDL Development.* EPA-841-B-97-006. U.S. Environmental Protection Agency. 1997.

Haubner, S.M. and E.F. Joeres. Using a GIS for estimating input parameters in urban stormwater quality modeling. *Water Resour. Bull.,* 32(6): 1341–1351. December 1996.

Kim, K., P.G. Thum, and J. Prey. Urban non-point source pollution assessment using a geographical information system. *J. Environ. Manage.,* 39(39): 157–170. 1993.

Kim, K. and S. Ventura. Large-scale modeling of urban nonpoint source pollution using a geographical information system. *Photogrammetric Eng. Remote Sensing,* 59(10): 1539–1544. October 1993.

Lahlou, M., L. Shoemaker, S. Choudhury, R. Elmer, A. Hu, H. Manguerra, and A. Parker. *BASINS, Better Assessment Science Integrating Point and Nonpoint Sources.* Version 2.0. EPA-823-B-98-006. Exposure Assessment Branch, U.S. Environmental Protection Agency, Washington, D.C. November 1998.

OME (Ontario Ministry of the Environment). *Humber River Water Quality Management Plan.* Toronto Area Watershed Management Strategy, Toronto, Ontario, 1986.

Pickett, S.R., O.G. Thum, and B.J. Hiemann. *Using a Land Information System to Integrate Nonpoint Source Pollution Modeling and Land Use Development Planning.* Land Information and Computer Graphics Facility, The University of Wisconsin, Madison. 1989

Pitt, R. and J. McLean. *Toronto Area Watershed Management Strategy Study: Humber River Pilot Watershed Project.* Ontario Ministry of the Environment, Toronto, Ontario. 486 pp. 1986.

Pitt, R. *Small Storm Urban Flow and Particulate Washoff Contributions to Outfall Discharges.* Ph.D. dissertation submitted to the Department of Civil and Environmental Engineering, University of Wisconsin, Madison. 1987.

Tetra Tech, Inc. *Rates, Constants, and Kinetics Formulations in Surface Water Quality Modeling, 2nd edition,* EPA/600/3-85/040. Center for Exposure Assessment Modeling, U.S. Environmental Protection Agency. Athens, GA. 1985.

Thum, P.G., S.R. Pickett, B.J. Niemann, Jr., and S.J. Ventura. LIS/GIS: Integrating nonpoint pollution assessment with land development planning. *Wisconsin Land Information Newsletter,* University of Wisconsin, Madison. No. 2, pp. 1–11. 1990.

Ventura, S.J. and K. Kim. Modeling urban nonpoint source pollution with a geographical information system. *Water Resour. Bull.,* 29(2): 189–198. April 1993.

APPENDIX **I**

Glossary

Acclimation. (1) Steady-state compensatory adjustments by an organism to the alteration of environmental conditions. Adjustments can be behavioral, physiological, or biochemical. (2) Referring to the time period prior to the initiation of a toxicity test in which organisms are maintained in untreated, toxicant-free dilution water or soil with physical and chemical characteristics, e.g., temperature, pH, hardness, similar to those to be used during the toxicity test.

Acute. Involving a stimulus severe enough to rapidly induce a response; in toxicity tests, a response observed in 96 hours or less typically is considered an acute one. An acute effect is not always measured in the terms of lethality; it can measure a variety of effects. Note that acute means short, not mortality.

Acute-Chronic Ratio (ACR). The ratio of the acute toxicity (expressed as an LC50) of an effluent or a toxicant to its chronic toxicity (expressed as an NOEL). Used as a factor for estimating chronic toxicity on the basis of acute toxicity data.

Additivity. The characteristic property of a mixture of toxicants that exhibits a cumulative toxic effect equal to the arithmetic sum of the effects of the individual toxicants.

Anoxic. Without oxygen.

Antagonism. The property of a mixture of toxicants that exhibits a less-than-additive cumulative toxic effect.

Aquatic Community. An association of interacting populations of aquatic organisms in a given water body or habitat.

Bioaccumulation. Uptake and retention of environmental substances by an organism from all sources.

Bioavailability. The property of a toxicant that governs its effect on exposed organisms. A reduced bioavailability would have a reduced toxic effect.

Bioconcentration. Uptake and retention of environmental substances by an organism from water. A bioconcentration factor (BCF) can be calculated as the quotient of the concentration of chemical in the tissue (or whole) of an aquatic organism divided by the concentration in the water in which the organism resides.

Biological Assessment. An evaluation of the biological condition of a water body using biological surveys and other direct measurements of resident biota in surface waters.

Biological Criteria (biocriteria). Numerical values of narrative expressions that describe the reference biological integrity of aquatic communities inhabiting waters of a given designated aquatic life use.

Biological Integrity. The condition of the aquatic community inhabiting unimpaired water bodies of a specified habitat as measured by community structure and function.

Biological Monitoring. The use of a biological entity as a detector and its response as a measure to determine environmental conditions. Toxicity tests and biological surveys are common biomonitoring methods.

Biological Survey (biosurvey). Consists of collecting, processing, and analyzing representative portions of a resident aquatic community structure and function.

Chronic. Involving a stimulus that lingers or continues for a relatively long period of time, often 1/10 the life span or more. Chronic should be considered a relative term depending on the life span of an organism. A chronic effect can be lethality, growth, reduced reproduction, etc. Chronic means long term.

Community Component. Any portion of a biological community. The community component may pertain to the taxonomic group (fish, invertebrates, algae), the taxonomic category (phylum, order, family, genus, species), the feeding strategy (herbivore, omnivore, carnivore), or organizational level (individual, population, community association) of a biological entity within the aquatic community.

Conservative Pollutant. A pollutant that is persistent and not subject to decay or transformation.

Control. A treatment in a toxicity test that duplicates all the conditions of the exposure treatments but contains no test material. The control is used to determine the absence of toxicity of basic test conditions, e.g., health of test organisms, quality of dilution water.

Criteria (water quality). An estimate of the concentration of a chemical or other constituent in water which if not exceeded, will protect an organism, an organismal community, or a prescribed water use or quality with an adequate degree of safety.

Criteria Continuous Concentration (CCC). The U.S. EPA national water quality criteria recommendation for the highest in-stream concentration of a toxicant or an effluent to which organisms can be exposed indefinitely without causing unacceptable effect.

Criteria Maximum Concentration (CMC). The U.S. EPA national water quality criteria recommendation for the highest in-stream concentration of a toxicant or an effluent to which organisms can be exposed for a brief period of time without causing mortality.

Critical Life Stage. The period of time in an organism's life span when it is the most susceptible to adverse effect caused by exposure to toxicants, usually during early development (egg, embryo, larvae). Chronic toxicity tests are often run on critical life stages to replace long-duration, life cycle tests since the toxic effect occurs during the critical life stage.

Designated Uses. Those uses specified in water quality standards for each water body or segment whether or not they are being attained.

Dilution of Water. Water used to dilute the test material in an aquatic toxicity test in order to prepare either different concentrations of a test chemical or different percentages of an aqueous sample for the various test treatments. The water (negative) control in a test is prepared with dilution water only.

Disturbance. An event that causes a significant change from the "normal pattern" in an ecological system.

Diversity. The number and abundance of species in a specified location.

Ecological Assessment. An evaluation of the condition of a water body using water quality and physical habitat assessment methods.

Ecotone. A zone of transition between adjacent ecological systems having a set of characteristics uniquely defined by space and time scales and by the strength of interaction between adjacent ecological systems.

Effluent. A complex waste material, e.g., liquid industrial discharge or sewage, which is discharged into the environment.

Elutriate (extract). A sample of water obtained by mixing a solid sample with a specified weight ratio of solvent, usually water, for a specified time and then separating from the solid phase by setting, centrifugation, and/or filtration.

Impact. A change in the chemical, physical, or biological quality or condition of a water body caused by external sources.

Impairment. A detrimental effect on the biological integrity of a water body caused by an impact that prevents attainment of the designated use.

Macroinvertebrates. Large invertebrate organisms, sometimes arbitrarily defined as those retained by sieves with 0.425- to 1.0-mm mesh screens.

Median Lethal Concentration (LC50). The concentration of material to which test organisms are exposed that is estimated to be lethal to 50% of the test organisms. The LC50 is usually expressed as a time-dependent value, e.g., 24-hour or 96-hour LC50; the concentration estimated to be lethal to 50% of the test organisms after 24 or 96 hours of exposure. the LC50 may be derived by observation (50% of the test organisms may be observed to be dead in one test material concentration), by interpolation (mortality of more than 50% of the test organisms occurred at one test concentration and mortality of fewer than 50% of the test organisms died at a lower test concentration; the LC50 is estimated by interpolation between these two data points), or by calculation (the LC50 is statistically derived by analysis of mortality data from all test concentrations).

No Observed Effect Level (NOEL). The highest measured continuous concentration of an effluent or a toxicant which causes no observed effect on a test organism.

Patches (adjacent to ecotones in fluvial systems). Spatial units (e.g., biological communities and ecosystems) determined by patch characteristics and their interactions over various scales. Topography, substrate conditions, organisms, and disturbance influence patch composition, size, location, and shape.

Persistence. That property of a toxicant or an effluent which is a measurement of the duration of its effect. A persistent toxicant or toxicity maintains effects after mixing, degrading slowly. A nonpersistent toxicant or toxicity may have a quickly reduced effect after mixing, as degradation processes such as volatilization, photolysis, etc. transform the chemical.

Population. An aggregate of interbreeding individuals of a biological species within a specified location.

Quality Assurance (QA). A program organized and designed to provide accurate and precise results. Included are selection of proper technical methods, tests, or laboratory procedures; sample collection and preservation; selection of limits; evaluation of data; quality control; and qualifications and training of personnel.

Quality Control (QC). specific actions required to provide information for the quality assurance program. Included are standardizations, calibrations, replicates, and control and check samples suitable for statistical estimates of the confidence of the data.

Reference Controls. Tests using natural water or sediment samples collected from unimpacted areas of the site environs.

Regions of Ecological Similarity. Describe a relatively homogeneous area by similarity of climate, landform, soil, potential natural vegetation, hydrology, or other ecologically relevant variable. Regions of ecological similarity help define the potential for designated use classifications of specific water bodies.

7Q10. The discharge at the 10-year recurrence interval taken from a frequency curve of annual values of the lowest mean discharge for 7 consecutive days.

Static. Describing toxicity tests in which test materials are not renewed.

Sublethal. Involving a stimulus below the level that causes death.

Synergism. The characteristic property of a mixture of toxicants which exhibits a greater than additive cumulative toxic effect.

Total Maximum Daily Load (TMDL). The total allowable pollutant load to a receiving water such that any additional loading will produce a violation of water quality standards.

Toxic Acute Chronic (TC$_c$). The reciprocal of the effluent dilution that causes no unacceptable effect on the test organisms by the end of the acute exposure period.

Toxic Endpoints. Measurements of an acute or chronic toxicity for toxic substances, including exposure duration, concentration, and observed effects.

Toxic Unit Acute (TU$_a$). The reciprocal of the effluent dilution that causes 50% of the test organisms to die by the end of the acute exposure period.

Toxicant. An agent or material capable of producing an adverse response (effect) in a biological system, adversely impacting structure or function or producing death.

Toxicity. The inherent potential or capacity of a material to cause adverse effects in a living organism.

Uncertainty Factors. Factors used in the adjustment of toxicity data to account for unknown variations. Where toxicity is measured on only one test species, other species may exhibit more sensitivity to that effluent. An uncertainty factor would adjust measured toxicity upward and downward to cover the sensitivity range of other, potentially more or less sensitive species.

Water Quality Assessment. An evaluation of the condition of a water body using biological surveys, chemical-specific analyses of pollutants in water bodies, and toxicity tests.

Vendors of Supplies and Equipment Used in Receiving–Water Monitoring

CONTENTS

General Field and Laboratory Equipment...871
Automatic Samplers...872
Basic Field Test Kits...873
Specialized Field Test Kits 873
Parts and Supplies for Custom Equipment 873
Toxicity Test Organisms 874
Laboratory Chemical Supplies (and Other Equipment)..874

GENERAL FIELD AND LABORATORY EQUIPMENT

The following vendors and manufacturers supply a large variety of equipment and supplies typically needed for field environmental investigations:

- Cole-Parmer, 625 East Bunker Court, Vernon Hills, IL 60061-1844, USA. Phone: 800-323-4340, Fax: 847-247-2929. Internet: coleparmer.com

Cole-Parmer is also a comprehensive laboratory supply distributor and carries many field and laboratory items including injection pumps and pump samplers, dredge samplers, and field test kits.

- Cabela's, One Cabela Drive, Sidney, NE 69160. Phone: 800-237-4444, Fax: 800-496-6329. Internet: www.cabelas.com

Cabela's is a comprehensive hunting, fishing, and outdoor gear supplier. It carries low-cost GPS units, other navigation aids, waders, and other general outdoor equipment that is necessary when carrying out a receiving water investigation.

- Fisher Scientific, PO Box 4829, Norcross, GA 30091. Phone: 800-766-7000.

Fisher is a complete scientific equipment and supply distributor and handles a wide variety of laboratory equipment. It also has sample bottles and selected field test kits.

- Forestry Suppliers, 205 W. Rankin Street, Jackson, MS 39201. Phone: 800-647-5368.

Forestry Suppliers carries a selection of field supplies and equipment, including GPS receivers and differential correction units, manual water samplers, pump samplers, depth-integrated samplers, dredge samplers, core samplers, and field test kits.

- Halltech Environmental, Inc., #4-503 Imperial Road N., Guelph, Ontario, CANADA N1H 6T9. Phone: 519-766-4568, Fax: 519-766-0729. E-mail: sales@htex.com; Internet: www.htex.com.

Halltech sells many unique sampling supplies, including depth-integrated samplers and bedload samplers, GIS receivers and satellite telephones, limnology sampling equipment, many types of manual water samplers, cartography and survey equipment, and soil sampling equipment.

- Markson Scientific, Phone: 800-858-2243.

Markson carries a good variety of field equipment, especially its sample splitter and dipper sampler.

- Spectrum Technologies, Inc., 23839 W. Andrew Rd., Plainfield, IL 60544. Phone: 800-248-8873, Fax: 815-436-4460. E-mail: specmeters@aol.com.

Spectrum carries mostly agricultural sampling tools and soil analysis equipment. it carries the excellent line of Horiba dry sensors and the Sentron pH meter, along with inexpensive recording rain gauges, complete recording weather stations, continuous water temperature recorders, and soil moisture and compaction meters, for example.

- Yellow Springs Instruments (YSI), 1700/1725 Brannum Lane, Yellow Springs, OH 45387. Phone: 800-765-4974 or 937-767-7241; Fax: 937-767-1058, E-mail: info@ysi.com; Internet: ysi.com/ysi/envweb.nsf.

YSI is a long-time supplier of rugged field meters, especially for DO and conductivity. Its line of water quality sondes is also very comprehensive and the sondes are capable of long-term deployment and continuous data logging.

- Ben Meadows Company, 3589 Broad St., Atlanta, GA 30341. Phone: 800-241-6401. E-mail: mail@benmeadows.com; Internet: Web: benmeadows.com.

Ben Meadows is a supplier of field research equipment including such items as portable instrumentation and waders.

AUTOMATIC SAMPLERS

The following are selected distributors of automatic water sampling equipment:

- American Sigma (800-635-4567) automatic water samplers
- ISCO (800-228-4373) automatic water samplers
- Campbell Scientific of Logan, UT (801-753-2342) telemetry
- Hazco (800-332-0435) also sells (and rents) pump samplers and many other items
- Vortox Company (909-621-3843) source area samplers

BASIC FIELD TEST KITS

The following are vendors of field test kits and numerous other field equipment and laboratory supplies:

- CHEMetrics, Inc., Route 28, Calverton, VA 20138. Phone: 800-356-3072
- EM Science, 480 S Democrat Road, Gibbstown, NJ 08027. Phone: 800-222-0342
- HACH Company, PO Box 389, Loveland, CO 80539. Phone: 800-227-4224
- La Motte Company, PO Box 329, Chesterfield, MD 21620. Phone: 800-344-3100
- Sentron Integrated Sensor Technology, 33320 1st Way S, Federal Way, WA 98003. Phone: 206-838-7933

SPECIALIZED FIELD TEST KITS

The following vendors supply more specialized field test kits:

- Dexsil (PetroFlag for soil hydrocarbon screening), 1 Hamden Park Drive, Hamden, CT. Phone: 800-4-DEXSIL
- DTECH Environmental Detection Systems (immunoassay test kits), 480 Democrat Road, Gibbstown, NJ 08027. Phone: 800-222-0342
- Strategic Diagnostics, Inc. (SDI) (Water quality testing RaPID Assays test kits), 111 Pencader Dr., Newark, DE 19702-3322. Phone: 800-544-8881; Fax: 302-456-6782, E-mail: techservice@sdix.com; Internet: sdix.com
- Environmental Technologies Group (Metalyzer), 1400 Taylor Avenue, Baltimore, MD 21284. Phone: 800-635-4598
- FCI Environmental Inc. (PetroSense), 1181 Grier Drive, Building B, Las Vegas, NV 89119. Phone: 800-510-3627
- IDEXX (bacteria analysis equipment) 1 IDEXX Drive, Westbrook, MN 04092. Phone: 800-248-2483
- Industrial Municipal Equipment ('KoolKount Bacteria Assayer), PO Box 335, Bohemia, NY 11716. Phone: 800-858-4857
- Palintest USA (Palintest metal analyzer) (now distributed by AZUR Environmental), 21 Kenton Lands Road, PO Box 18733, Erlanger, KY 41018. Phone: 800-835-9629
- Tuner Designs (Fluorometers), 845 W. Maude Avenue, Sunnyvale, CA 94086. Phone: 408-749-0994
- Wilks Enterprise, Inc. (Infracal Oil in Water Analyzer), 140 Water Street, Norwalk, CT 06856. Phone: 203-855-9136

PARTS AND SUPPLIES FOR CUSTOM EQUIPMENT

The following sell interesting and hard-to-obtain supplies needed for custom construction of samplers and test units:

- Small Parts (stainless steel and nylon screens of many apertures, polypropylene mesh, etc.), 13980 NW 58th Court, PO Box 4650, Miami Lakes, FL 33014-0650. Phone: 800-220-4242, Fax: 800-423-9009, E-mail: smlparts@smallparts.com, Internet: smallparts.com
- Aquatic Ecosystems, Inc. (culture supplies, e.g., tanks, heaters, food, flowmeters, Lifeguard filters, activated carbon, tanks, pumps, fittings, and pipes made of many materials and sizes), 1767 Benbow Court, Apopka, FL 32703. Phone: 877-347-4788, Fax: 407-886-6787, Internet: aquaticeco.com
- Consolidated Plastics Company, Inc. (in situ chamber supplies [e.g., mailing tubing and end caps]), 8181 Darrow Road, Twinsburg, OH 44087. Phone: 800-362-1000, Fax: 330-425-3333, Internet: consolidatedplastics.com

TOXICITY TEST ORGANISMS

The vendors listed below supply toxicity test organisms and culture supplies:

- Aquatic Bio Systems, Inc. (ABS) (Toxicity test organisms), 1300 Blue Spruce Drive, Suite C, Fort Collins, CO 80524. Phone: 800-331-5916 or 970-484-5091, Fax: 970-484-2514, E-mail: absinfo@riverside.com
- Aquatic Research Organisms (ARO), PO Box 1271, Hampton, NH 03842-1271. Phone: 800-927-1650, Fax: 603-926-5278, E-mail: arofish@aol.com, Internet: holidayjunction.com/aro/
- Aquaculture Supply (culture foods and equipment, airstones, Spirulina, etc.), 33418 Old Saint Joe Road, Dade City, FL 33525. Phone: 352-567-8540, Fax: 352-567-3742, E-mail: ASUSA@Aquaculture-Supply.com, Internet: aquaculture-supply.com
- Argent Chemical Laboratories (Nitex mesh for *in situ* chambers, brine shrimp cysts), 8702 152nd Ave. NE, Redmond, WA 98052. Phone: 800-426-6258 or 206-885-3777, Fax: 206-885-2112, E-mail: email@argent-labs.com, Internet: argent-labs.com
- Azur Environmental (Microtox equipment and supplies), 2232 Rutherford Road, Carlsbad, CA 92008-8883. Phone: 760-438-8282, Fax: 760-438-2980, E-mail: maketing@azurenv.com, Internet: azurenv.com
- Pet Warehouse (culture foods and equipment: food, activated carbon, brine shrimp cysts, air pumps, air tubing, etc.) Dept. C93F, PO Box 752138, Dayton, OH 45475. Phone: 800-433-1160 or 937-428-6500, Fax: 800-513-1913 or 937-428-6505, E-mail: service@petwhse.com, Internet: petwhse.com
- Xpedx (Saalfeld Paper) (small, plastic cladocean toxicity testing cups), 4510 Reading Road, Cincinnati, OH 45229. Phone: 800-669-7101 or 513-641-5000, Fax: 800-880-5312 or 513-641-5003, Internet: xpedx.com

LABORATORY CHEMICAL SUPPLIES (AND OTHER EQUIPMENT)

The following vendors supply general laboratory supplies and equipment, plus many field supplies, such as test kits, meters, and sample bottles:

- Fisher Scientific (chemicals, reagents, and laboratory equipment and supplies [e.g., plastic centrifuge tubes]), 2000 Park Lane Dr., Pittsburgh, PA 15275-9952. Phone: 800-766-7000 or 412-490-8300, Fax: 800-926-1166, E-mail: fishersupport@plpit.fishersci.com, Internet: fishersci.com
- Millipore (Milli-Q system supplies and field bacteriological sampling equipment), 80 Ashby Road, Bedford, MA 01730. Phone: 800-645-5476, Fax: 617-275-5550, E-mail: order@millipore.com, Internet: millipore.com
- Supelco (glass, amber vials, standard solutions), PO Box B Bellfonte, PA 16823. Phone: 800-247-6628 or 814-359-3441, Fax: 814-359-5459, E-mail: supelco@sial.com, Internet: sigma-aldrich.com
- Sigma Chemicals (chemicals, dialysis tubing for air lines), PO Box 14508, St. Louis, MO 63178. Phone: 800-325-3010 or 314-771-5765, Fax: 800-325-5052, E-mail: sigma@sial.com. Internet: sigma-aldrich.com
- VWR (glass, amber vials and general equipment and supplies), 1310 Goshen Parkway, West Chester, PA 19380. Orders: 1-800-932-5000. Phone: 800-932-5000 or 610-431-1700, Fax: 610-429-9340, E-mail: solutions@vwrsp.com, Internet: vwrsp.com

Index

A

AAS, *see* Atomic absorption spectrometry
Acanthes lanceolata, 138
Acenaphthylene, 60, 448
Acetaldehyde, 740
Acetanilides, 436
Acetic acid, 740
Acetone, 261, 740
Acid
    extractable organics, 249
    handling, 271
    mine drainage, 74
    precipitation, 74
    storage cabinets, 738
    volatile sulfides (AVS), 108, 117, 198, 254,
        275, 325
Acoustic flowmeters, 358
Acoustic velocity meters, 377
Acrylic acid, 744
Acrylonitrile, 740
Acylating agents, 746
Adenosine triphosphate (ATP), 418
Aerococcus viridans, 487
Aeromonas
    hydriphila, 818
    spp., 819
Agriculture, 4, 5, 47
AHs, *see* Aliphatic hydrocarbons
AIDS patients, 625
Air pollution
    monitoring equipment, 312
    sources, 475
Air transportation, 187
Alabama hog sucker, 706
Alachlor, 60
Alder, 667
Alderflies, 52
Aldicarb, 436
Aldrin, 448
Alewife, 414, 702

Algae
    attached, 128, 133
    blooms, 114, 211
    communities, nutrient availability and, 406
    filamentous, 129, 133, 677
    freshwater, 519
    green, 518, 520
    growth,(s)
        effect of outfall on, 205
        excessive, 27
    mats, 27
    planktonic, 493
    survival, 151
Algicides, 5, 6
Aliphatic hydrocarbons (AHs), 78, 480
Alkalinity, 424, 445, 773
Alkylating agents, 745
Allergens, working with, 743
Alligator gar, 702
Allowable error, 232
Alosa pseudoharengus, 414
Ambient toxicity testing, 665
Ambloplites
    cavifrons, 707
    constellatus, 707
    rupestris, 414, 707
Amebiasis, 621
American eel, 704
Ammocrypta sp., 707
Ammonia, 806
    criteria, 813
    nitrogen, 146, 158, 397
    /potassium ratios, 474
Ammonification, 326
Ammonium hydroxide, 740
Ampelisca abdita, 527
Amphipods, 135, 151, 165, 520, 529
Aniline, 740
Animal husbandry, 185
Anion–cation balance, 252
Anionic surfactants, 471, *see also* Detergents
Anisoptera, 688

Annelida, 688
Anodic stripping voltammetric (ASV), 459, 778
ANOVA test, 580, 599
Anthracene, 61, 448, 480, 801
Antibody, *Giardia*-specific, 489
Anticake compound, 7
Ants, bioturbation by, 394
Apartment complex, stormwater pond adding value to, 25
Aphanizomenon flos aquae f. gracile, 175
Aplodinotus grunniens, 415
Applied statistics, 576
Aquarium air stone, suction using, 329
Aquatic assessments, most commonly used biological groups in, 116
Aquatic biota, flow requirements for, 350
Aquatic ecosystem assessment parameters, 108
Aquatic insects, food types of, 495
Aquatic life
    effects of copper on, 824
    habitat, maintaining quality, 618
    use impairments, receiving water investigation assessing, 445–446
Aquatic macroinvertebrates, effects of suspended solids on, 73
Aquatic microfauna, 73
Aquatic organism food availability, 143, 149
Aquatic toxicity testing, 712
Aqueous phase testing, useful species and life stages for, 520
Arachnoidea, 688
Arbacua punctulata, 523
Arctic char, 706
Arctic cisco, 705
Arctic grayling, 706
Arizona trout, 705
Aromatic samplers, 224
Arsenic, 132
Arthropoda, 688
Artificial drainage systems, 351
Artificial sediments, 534
Artificial substrate
    analysis of, 683
    macroinvertebrate colonization tests, 121
Artificial tracer, 362, 364
Asphalt degradation, 457
Assessment problem formulation, 101–221
    assessment tools, 107
    beginning of assessment, 108–119
        data quality objectives and quality assurance issues, 118–119
        formulation of conceptual framework, 113
        historical site data, 112–113
        initial site assessment and problem identification, 110–112
        selection of optimal assessment parameters, 113–118
        specific study objectives and goals, 110
    case studies of previous receiving water evaluations, 123–213
        current, ongoing, stormwater projects, 181–205
        longitudinal experimental design, 124–139
        long-term trend experimental design, 169–181
        outlines of hypothetical case studies, 205–213
        parallel creeks experimental design, 139–168
    example outline of comprehensive runoff effect study, 119–123
        confirmatory assessment, 122
        data evaluation, 122
        decision on problem formulation, 119–120
        project conclusions, 123
        project design, 120–121
        project implementation, 121–122
        question, 119
    rationale for integrated approach to assessing receiving water problems, 102–103
    study design, 107–108
    typical recommended study plans, 213–218
        components of typical receiving water investigations, 213
        example receiving water investigations, 213–218
    watershed indicators of biological receiving water problems, 103–107
Assessment score sheet, preliminary, 672
Asterionella, 175
ASTM standards on toxicity testing, 712
ASV, *see* Anodic stripping voltammetric
Atlantic salmon, 414, 706
Atmospheric contributions, sampling of, 310
    cold-vapor, 777
    electrothermal, 777
    graphite furnace, 778
Atomic absorption spectrometry (AAS), 774
ATP, *see* Adenosine triphosphate
Atrazine, 6, 60, 436
Automatic sampler(s), 259
    flow-weighted, 288
    line flushing, 282
    refrigerated, 280
Automatic sampling, advantages of manual sampling compared to, 260
Automatic source area samplers, 299
Automobile
    dealers, 187
    emissions, particulate lead from, 312
    exhaust, 457
    repair, 187
    service areas, 5

Autosampler, XYZ, 449
AVS, *see* Acid volatile sulfides
Awaous stamineus, 707
Aziridines, 745

B

Bacteria, 63, 188, 491
    analysis, 433
    biotypes, isolations of, 86
    coliform, 83, 277
    criteria, water-contact recreation, 820
    die-off, 84, 205
    fecal coliform, as indicators of inappropriate
            discharges of sanitary, 464
    Gram-negative, 90
    older, 84
    populations, interstitial water, 203
    presence of in stormwater runoff, 465
    protozoan cropping of, 494
    reagent, 731
    reproduction, 73
    respiration, 523
    sampling, 281
    sources
        dry-weather, 466
        urban, 82
    tests, 433
    wet-weather flow, 203
Bacteriological criteria, development of bathing
            beach, 816
Bandfin shiner, 706
Bank(s)
    erosion, 648, 649
    false, 649
    instability, 28
    soils, clayey, 618
    stability, 7, 661
    unstable, 55
    vegetative stability, 662
Banzoghiperylene, 801
Barium, 806
Baseflow water quality, 160
BASINS
    assessment tools, 859
    base cartographic data, 860
    environmental background data, 861
    environmental monitoring data, 861
    point source/loading data, 862
Basswood, 667
Bathing beach bacteriological criteria, 816
Beach debris, land-based sources of, 68
Bedded solids, 71

Bedload
    samplers, 295, 296, 410
    sediment, 409
Benchmarks, 612, 613
Benthic community assessment, 665–692
    agencies that have developed tolerance
            classifications and/or biotic indices, 687
    Ohio EPA invertebrate community index
            approach, 681–687
        field methods, 681–682
        laboratory methods, 682–683
        macroinvertebrate data analysis, 683–687
    Rapid Bioassessment Protocol, 665–681
        data analysis techniques, 669–681
        sample collection, 666–667
        sample sorting and identification, 667–669
Benthic invertebrates, 63, 348
Benthic macroinvertebrate(s), 116
    equipment and supplies, 685–686
    field data sheet, 668
    laboratory bench sheet, 670, 671
    sample log-in sheet, 669
Benthic organisms, 151
Benzaldehyde, 740
Benzene, 6, 38, 61, 740
Benzidine, 801
Benzo(a)anthracene, 424, 480
Benzo(b)fluoroanthenene, 61
Benzylbutyl phthalate, 262
Bering cisco, 705
Best management practice (BMP), 111, 462
Bias, 233
Bigeye chub, 702, 706
Bigeye shiner, 703, 706
Bigmouth buffalo, 703
Bigmouth shiner, 703
Bioaccumulation testing, *see* Toxicity and
            bioaccumulation testing
Bioassessment approach, flowchart of, 674
Biochemical oxygen demand (BOD), 73, 108, 325
    analyses, stormwater, 75
    decomposition rate, 18
    point source discharges of, 11
Bioconcentration factors, 49
Biofiltration, in parking area, 59
Biological degradation, 617
Biological endpoints, selection of for monitoring,
            115
Biological impairment benchmarks, categories of,
            613
Biological integrity, definition of, 347
Biological life objectives, 610
Biological toxicity fractionations, 537
Biosurveys, 337
Biotic Condition Index, 679

Biotic indices, agencies having developed, 687
Biotransformation, 78, 79, 80, 81
Bioturbation
    by ants, 394
    benthic invertebrate, 348
Birch, 667
Bird droppings, 83
Bis(2-ethylhexyl) phthalate, 6, 424
Bivalve tissue residues, 114
Blackberry vines, as shade for stream aquatic life,
        145
Black buffalo, 703
Black bullhead, 703
Black crappie, 415, 704
Black jumprock, 706
Black madtom, 707
Black redhorse, 70, 703
Blackchin shiner, 703, 706
Blacknose dace, 702, 706
Blacknose shiner, 703, 706
Blackside dace, 706
Blackstripe topminnow, 704
Blacktail redhorse, 706
Blank(s), 447
    calibration, 248
    equipment, 248
    instrument, 248
    method, 248
    reagent, 250
    trip, 248
    use of to minimize and identify errors, 248
Bleeding shiner, 706
Bloater, 705
Bluebreast darter, 705
Blue catfish, 703
Bluegill, 135, 415, 520, 536, 704
Blue-green algal blooms, 114
Blue sucker, 703
Bluntnose darter, 707
Bluntnose minnow, 703
BMP, see Best management practice
Boat electrofishing unit, 504
BOD, see Biochemical oxygen demand
Bonferroni t-test, 591
Bottom
    -dwelling organisms, 491
    sediments, scour of, 408
    sourcing and deposition, 661
    substrate, 660
Bowfin, 702
Box plots, 375, 848
Brassy minnow, 703
Bridge construction, 4
Brighteners, optical, 440
Brindled madtom, 704, 707

Bromine, 740
Brook silverside, 704
Brook stickleback, 705
Brook trout, 74, 414, 520, 536, 702, 706
Brown bullhead, 134, 414, 703
Brown madtom, 707
Brown trout, 414, 536, 702, 706
Bryozoa, 683, 688
Bubble sensor depth indicators, 377
Budget restrictions, 181
Buffalo fish, 414
Bug picking, from substrate samples, 498
Building evacuation procedures, 759
Bulk density, estimation of, 394
Bullhead minnow, 703
Bull trout, 706
Burbot, 704
Butane, 740
Butyl benzyl phthalate, 6, 38, 80, 424
Butyraldehyde, 740

C

Caddisflies, 52, 152
Calcium hypochlorite, 740
Calibration blank, 248
Campylobacter, 88
Campylobacterosis, 621
Cancer, 625, 745
Candidate critical sources, ranking of, 187
Canopy cover, 660
Capital costs, 431
Carbamates, 249, 436
Carbazole, 448
Carbon
    disulfide, 740
    fixation, 149
    sulfide, 744
    tetrachloride, 740, 800
Carcinogen(s), 747, 765, 827
Carcinogenicity, 507
Carcinogenic RaPID Assay, 435
Carp, 74, 414, 536, 701, 702
Catchbasin(s)
    cleaning, 71, 630, 631
    floatable material in, 70
    sediment accumulations, 162
    use, 632
Cation exchange capacity (CEC), 325
Catostomus commersoni, 414
Cattail plant segments, 129
Cattle feces contamination, 89
CCC, see Criterion continuous concentration

CDC, *see* Centers for Disease Control
CEC, *see* Cation exchange capacity
Cedar swamps, 52
Cell from hell, 89
Centers for Disease Control (CDC), 620
Central mudminnow, 702
Central silvery minnow, 703
Central stoneroller, 703
Centrifugation, 317
Ceriodaphnia dubia, 108, 208, 335, 502, 513, 515,
       546, 714, 718
Chain-of-custody
    forms, 273, 274
    seal, 271
Chain pickerel, 702
Channel
    alteration, 661
    banks, erosion of, 56
    bottoms, shifting of, 56
    conditions, stream flow-altering, 405
    geomorphology, 144
    lined, 55
    morphology, 64, 648, 661
Channel catfish, 134, 414, 520, 536, 703
Channel darter, 704
Channelization, 4, 7, 142, 349, 404, 406
CHEMetrics copper test kit, 773
Chemical(s)
    deactivation, 761
    endpoints, selection of for monitoring, 116
    exposure hazards, 423
    fingerprinting, 483
    manufacturing, 187
    mass balance equations, 476, 478
    neutralization, 761
    oxygen demand (COD), 73, 191, 325, 587, 851
    speciation, 78
    storage
        laboratory, 737
        requirements, 423
    waste
        disposal of down sink, 761
        disposal program, 760
        inorganic, 762
        organic, 762
        removal, 763
        water-reactive, 755
Chemical Response Unit, 758
Chewers, 495
Chinook salmon, 152, 414, 701, 702, 705
Chironomids, 165, 491
Chironomus
    riparius, 268, 520, 527
    tentans, 212, 268, 335, 520, 605,718, 723, 727,
       728
Chiselmouth, 701

Chi-square goodness of fit test, 586
Chlordane, 6, 38, 60, 122, 424, 436, 803
Chlorine, 740
Chloroacetone, 740
Chloroethane, 800
Chloroform, 6, 38, 61, 261, 440, 740, 801
3-Chlorophenol, 809
Chlorophyll *a* observations, 174
Chlorpyrifos, 6, 20, 436
Cholera, 621
Cholinesterases, 436
Chromic acid, 740
Chromium, 6, 823, 824
Chrysene, 6, 61, 448, 480, 802
Chrysochromulina, 175
Chum salmon, 414, 705
Churn splitter, 272
Cisco, 414
Citrobacter, 485
Cladophora
    dubia, 48
    glomerata, 48
    sp., 133
Clean Water Act (CWA), 3, 8
Clear View rain gauge, 378
Clinostomus elongatus, 706
Club moss, 362
Cluster sampling, 225
CMC, *see* Criterion maximum concentration
Coal mining, 4
Coarse particulate organic matter (CPOM), 666
    component, sampling of, 666
    sample, 669, 679
Cocconeis
    pediculus, 138
    placentula, 138
COD, *see* Chemical oxygen demand
Coefficient of variation (COV), 232, 244–245, 466
    control sample, 732
    values, 467
Coelenterata, 688
Coho embryo salmon, 154
Coho salmon, 414, 520, 702
Coincidence, 455
Cold-vapor atomic absorption spectrometry, 777
Coleoptera, 495, 689
Coliform bacteria, 83, 277
Collection methods, *see* Sampling effort and
       collection methods
Collectors, 410
Color comparator, 442
Combined sewer overflows (CSOs), 5, 15, 34, 68
    capture and control device, 363
    controls, effectiveness of, 106
    EPA-required, 69

Combustion products, 157
    diesel fuel, 167
    obsolete versions of enhanced, 577
Common carp, 414, 701, 702
Common shiner, 703
Community Loss Index, 673, 678
Community Similarity Index, 678, 684
Community structure, 411
Comparison tests, 580
Component–ecosystem interactions, 348
Compost
    -amended soils, 397
    ion-exchange capacity of, 398
Compressed gas cylinders, 757
Computer
    models, 106
    simulation, 351
Concentration-addition model, 516
Condition quality indicators, 106
Conductivity
    meters, 430
    probe calibration, 783
Confidence intervals, need for, 245
Confirmatory assessment, 193
Confirmatory studies, 616
Conifers, 667
Construction site(s)
    erosion
        characterization, 32
        controls, 28, 628
        rate of, 32
    inspections, 277
    runoff water quality, monitoring study of, 33
    soil erosion from, 31
Consumptive fisheries, 22, 124
Contact recreation areas, human health
        considerations associated with
        potentially contaminated, 124
Contaminant
    bioavailability, 254, 314, 327
    peaks, 315
    sources, characterizing, 539
Contamination
    data evaluation methods to indicate sources of,
        468
    detergents as indicators of, 470
    negative indicators implying, 468
    sources, distance-dependent association between
        health effects and, 624
    use of fecal sterol compounds as tracers of,
        477
Control
    charts, 250
    programs, effectiveness of, 12–13
Coosa shiner, 706

Copper, 6
    effects of on aquatic life, 824
    human health criteria for, 825
    national aquatic life criteria for, 825
Coprostanol, use of as tracers of contamination by
        sanitary sewage, 477
Coregonus
    artedii, 414
    clupeaformis, 414
Core-port suction, 327
Corer samplers, 289, 323
Correlation
    matrices, 592
    tests, 470
Corrosives, 752
    examples of, 752
    first aid for, 754
    health effects associated with, 753
    use and storage of, 753
Cottus sp., 707
Coulter Multisizer method, 455
COV, see Coefficient of variation
CPOM, see Coarse particulate organic matter
Crane flies, 52
Crayfish, 129, 133, 134, 481
Creek
    blowout, 55
    effects of erosion on, 156
    flows, salmon fisheries affected by, 155
    interstitial water quality, 147
    sedimentation, 143, 156
    sediment quality, 155
    system, dry-weather pollutants from, 56
    tributary flow rates, 174
Creek chub, 702
Creek chubsucker, 703
Creosote, 7
Cricotopus, 677
Criterion continuous concentration (CCC), 815
Criterion maximum concentration (CMC), 799
Croplands, erosion rate of, 32
Crop production, 4
Cryptosporidium, 88, 89, 197, 486
Crystal darter, 705
CSOs, see Combined sewer overflows
Curb-and-gutter drainage systems, 35
Current
    measurements, example calculation for, 360
    meter, 361
        flow monitoring, 360
        method, 357
Cutlips minnow, 706
Cutthroat trout, 74, 153, 154, 414, 701, 705
CWA, see Clean Water Act
Cyanazine, 60, 436

Cyclodienes, 436
Cyclohexane, 740
Cyclohexanone, 261
Cymatopleura solea, 138
Cymbella sp., 138
Cypress darter, 707
Cyprinus carpio, 414, 830

D

2,4-D, 118, 159
Dam construction, 4
Daphnia
   magna, 208, 268, 502, 528, 824
   pulex, 268, 502, 715
Darters, 508, 697, 707
Data
   analysis
      exploratory, 606
      techniques, 669, 694
   associations, 591
   dendogram of, 596
   mining, 576
   plots, basic, 583
   quality objectives (DQO), 109, 118, 119, 247
      descriptions of, 337
      quantification of habitat effects useful to meet,
         401
      sample integrity and, 314
   survival, 604
Data interpretation, 609–640
   evaluating biological stream impairments using
      weight-of-evidence approach, 611–619
   benchmarks, 612–615
   comments pertaining to habitat problems and
      increases in stream flow, 617–619
   process, 611–612
   ranking and confirmatory studies, 615–617
   evaluating human health impairments using risk
      assessment approach, 619–626
   deterministic approach, 619
   example risk assessment for human exposure
      to stormwater pathogens, 620–626
   probabilistic approach, 619–620
   identifying and prioritizing critical stormwater
      sources, 626–636
   case study, 628–629
   sources of urban stormwater contaminants,
      626–628
   use of SLAMM to identify pollutant sources
      and to evaluate control programs,
      629–636
   problem, 609–610

DCA, *see* Detrended correspondence analysis
DDT, 20, 436
Debris piles, 403
Decision making, errors in, 233
Deep sea sewage sludge disposal areas, 482
Deepwater sculpin, 705
Degraded cysts, 490
Deionized water, 783
Demeton, 806
Denitrification, 326
Denticula elegans, 138
Deoxygenation, 328
Deployment setup procedure, 785
Depth
   -integrated sediment sampler, 292, 294
   sensor, 428
Dermatitis, 621
Design rainfall, 245
Detection limits, reporting results affected by, 253
Detention pond(s), 26, 630
   dry, 209
   outfall, dry, 211
   side stream, 210
   wet, effect of, 211
Detergent(s), 471
   analyses, 472
   compounds, 484
   concentration, 442
   as indicators of contamination, 470
   test kit, 441
   whitener filter sets, 483
Detrended correspondence analysis (DCA), 694
Diatoma vulgare, 138
Diazinon, 6, 20, 60, 159
Dibenzo(a,h)anthracene, 448
Dibenzyl ether, 740
Dibutyl phthalate, 740
Di-N-butyl phthalate, 6, 38
Dicamba, 60
1,3-Dichlorobenzene, 424
1,2-Dichloroethane, 38
Dieldrin, 6, 38
Diesel fuel combustion products, 167
Diethanolamine, 740
Diethyl ether, 740
Diethyl phthalate, 38, 80, 262, 448
Digidot plot, 586
Dilution water, 717, 718, 719
2,4-Dimethylphenol, 262
Dimethyl sulfoxide, 740, 744
Dinitrophenols, 806
Diploneis sp., 138
Dip net sampling, 498
Dipper samplers, 290, 453
Diptera, 495, 689

Discharge(s)
  hillside, 199
  inappropriate, 461
  industrial wastewater, 598
  litter, 297, 399
  point source, 278
  pollutant, 353, 356
  sources, identifying inappropriate, 464
  stream, 349, 660
Disinfection by-products, 62
Dissolved metals, conversion factors for, 813
Dissolved organic carbon (DOC), 326
Dissolved organic matter (DOM), 677
Dissolved oxygen (DO), 17, 20, 417, 657, 720
  conditions, calculation of, 418
  curve, 201
  data, 420
  deficits, 49, 75
  levels, wet-weather, 75
  meters, 118, 430, 440
  probe calibration, 784
  problems, 428
  reading, elevated, 291
  receiving water levels, 85
Dissolved solids, 71, 158
  aquatic life criteria for, 823
  classification of, 72
  human health criteria for, 822
Distributed Routing Rainfall Runoff Model, 854
Disturbance, definition of, 347
Diuron, 118
DNA
  fingerprinting, 537
  profiling, 484
DO, see Dissolved oxygen
DOC, see Dissolved organic carbon
Dolly varden, 706
DOM, see Dissolved organic matter
Doppler flowmeter, 374
Doppler velocity sensors, 375
Dorosoma petenense, 414
DOS, 578
Dose–response restrictions, 604
Double-ring infiltration tests, 229
Downstream sampling stations, 361
DQO, see Data quality objectives
Drainage
  design studies, 850
  grass swale, 240, 632
  paths, 240
  systems
    artificial, 351
    man-made, 463
Dredge
  Ekman, 321

  Ponar, 320, 321
  sampler, 315
Dredging, 4, 7, 323, 403, 521
Drift
  method, 357
  organisms, 498
Drill auger, mixing sediment with, 325
Drinking water supply, 104
Drought, 125
Drowning, 67, 193
Dry detention basins, use of in controlling urban
        runoff discharges, 164
Dry detention pond, 209, 211
Dry sampling, 301
Dry-weather bacteria sources, 466
Dry-weather base flows, 34
Dry-weather flows, 10
  continuous, 460
  pollutants in, 463
Dry-weather outfall flow rates, highly irregular, 470
Dtech Immunoassay test kit, 435
Duckweed, 518
Duncan's multiple range test, 591
Dunner's test, 591
Durbin–Watson test, 598
Dusky darter, 704
Duskystripe shiner, 706
Dustfall, 310, 311
Dye
  continuous release rates of, 367
  dilution ratio of, 368
  injection current measurements, notation for
        mass balance calculations for, 367
  testing, 376
  tracers, 362, 364

E

Early life stages of fish (ELS), 815
Earthworm test, 519
Eastern banded killifish, 704
Eastern sand darter, 705
Ecoregions, 402
Ecosystem(s)
  characterization, 113
  complexities, 103
  degradation, assessment of, 4
  energetics, 346
  enhancement, 123
  quality
    degradation, 404
    stressors of, 400
  running water, 492

Ecosystem component characterization, 345–573
    aesthetics, litter, and safety, 398–400
        aesthetics, litter/floatables, and other debris, 398–400
        safety characteristics, 398
    benthos sampling and evaluation in urban streams, 491–501
        macroinvertebrate sampling, 494–501
        periphyton sampling, 493–494
        protozoan sampling, 494
    ecosystem structure and integrity, chaos and disturbance, 346–349
    fish sampling, 502–506
    flow and rainfall monitoring, 349–388
        flow monitoring methods, 357–377
        flow requirements for aquatic biota, 350–351
        pollutant transport, 356–357
        rainfall monitoring, 377–388
        urban hydrology, 351–356
    habitat, 400–423
        channelization, 404–406
        dissolved oxygen, 417–423
        factors affecting habitat quality, 403–404
        field habitat assessments, 410
        riparian habitats, 409–410
        substrate, 406–409
        temperature, 410–413
        turbidity, 413–417
    microorganisms in stormwater and urban receiving waters, 485–491
    soil evaluations, 388–398
        case study to measure infiltration rates in disturbed urban soils, 389–394
        observations of infiltration rates in disturbed urban soils, 394–397
        water quality and quantity effects of amending soils with compost, 397–398
    toxicity and bioaccumulation, 507–546
        bioaccumulation, 534–536
        emerging tools for toxicity testing, 536–546
        in situ toxicity testing, 530–534
        measuring effects of toxicant mixtures in organisms, 515–517
        pulse exposures, 514–515
        reason to evaluation toxicity, 507–512
        standard sediment testing protocols, 527–530
        standard water testing protocols, 517–527
        stormwater toxicity, 513–514
    water and sediment analytes and methods, 423–485
        conventional laboratory analyses, 447–459
        hydrocarbon fingerprinting for investigating sources of hydrocarbons, 483–485
        selection of analytical methods, 423–425
        use of field methods for water quality evaluations, 425–447
        use of tracers to identify sources of inappropriate discharges to storm drainage and receiving waters, 459–483
    zooplankton sampling, 502
Ecotones, definition of, 347
Ecotoxicological endpoints, 115
Ecotoxicology, 347
Ecowatch for Windows, 370
EDA, see Exploratory data analysis
Edge habitat, 405
Effects characterization, 11
Ekman dredge, 321
Electrical safety, 743
Electrofishing, 106, 503, 505
Electrophilic alkenes, 745
Electroshocking, 129
Electrothermal atomic absorption spectrometry, 777
Elegant madtom, 707
ELISA, see Enzyme-linked immunosorbent assay
ELS, see Early life stages of fish
Elutriate testing, 275
Embankment, lined, 55
Embeddedness, 646, 650, 660
Embryotoxins, working with, 744
EMC, see Event mean concentration
Emerald shiner, 414, 702
Emergency procedures, 758
Empirical model, 227
Endocrine disruption, 507
Endosulfan, 159, 827
Endosulfan sulfate, 38
Endrin, 38, 803
Endron ketone, 262
Enterobacter, 485
Enterococcus
    faecalis, 487
    faecium, 487
Environmental studies, principles for designing successful, 110
Environment Canada Biological Test Method Development Program, 711
Enzyme-linked immunosorbent assay (ELISA), 436, 479
Eohaustorius estuarius, 527
EPA, see U.S. Environmental Protection Agency
Ephemeroptera, 491, 495
Ephemeroptera, Plecoptera, and Trichoptera (EPT), 669, 673
Epibenthic invertebrates, bioturbation by, 348
Epoxides, 745
EPT, see Ephemeroptera, Plecoptera, and Trichoptera
Equilibrium partitioning guidelines, 836

Equipment blank, 248
Erosion, 311
    bank, 648, 649
    channel bank, 56
    control(s)
        construction site, 28, 628
        on-site, 452
        practices, 246
    effects of on creek, 156
    stream bed, 143, 155
    watershed, 659
Error(s)
    allowable, 232
    decision making, 233
    particle sampling, 282
    rainfall monitoring, 381
    runoff volume, 383
    sampling, 233, 251
    use of blanks to minimize and identify, 248
    watershed rain depth, 383
    wind-induced, during rainfall monitoring, 386
Escherichia coli, 82, 194, 195, 333, 433, 622, 817,
        821
Esox
    americanus vermiculatus, 414
    masquinongy, 414
Estradiol, 744
Estuary(ies)
    biological integrity of, 104
    eutrophication conditions, 104
    pollutants and sources affecting, 19
    pristine, 123
Etheostoma
    chlorosomun, 707
    fusiforme, 707
    gracile, 707
    nigrum, 707
    proeliare, 707
    sp., 707
    spectabile, 707
Ethyl acetate, 740
Ethylene dichloride, 740
Ethylene glycol, 740
Ethylene trichloride, 740
Eutrophication, 49
    accelerated, 830
    conditions, estuarine, 104
    problems, majority of nation's, 831
    processes, role of elevated turbidity levels in, 415
    transparency-associated, 180
Event
    mean concentration (EMC), 198, 213, 255
    plots, 371
Exoglossum maxillingua, 706
Expendable costs, 431

Experimental design, 109, 237
Exploratory data analysis (EDA), 583
Explosion hazard data, 765
Exposure
    assessment, 113, 623
    characterization, 11
    -effects interactions, 613
    –response relationship, 539
Extraction, solid-phase, 537
Extremely hazardous chemicals, 746

F

Fabricated metal products, 187
Factorial design, fractional, 231
Factorial experimental designs, major advantage of,
        227
False banks, 649
Family Biotic Index (FBI), 680
Family-level index, 675
Family-level tolerance classification, 680
Farming district, 247
Fathead minnow, 135, 520, 529, 703
FBI, *see* Family Biotic Index
FBM, *see* Flow Balancing Method
Fecal coliform(s), 39, 42, 114, 191, 195
    bacteria, as indicators of inappropriate discharges
        of sanitary sewage, 464
    concentrations of at highway runoff site, 83
    to fecal streptococci bacteria ratios, 83
Fecal indicators, in stormwater runoff, 82
Fecal pathogens, 120
Fecal sterol compounds, use of as tracers of
        contamination by sanitary sewage, 477
Fecal streptococci (FS), 39, 84, 818
Feeding measures, 676
Feedlot drainage, 83
Fenarimol, 436
Ferric chloride precipitation, removal of phosphorus
        by, 180
Fertilizer(s)
    application, roadside, 7
    lawn, 206
    nitrate leached from, 59
Fiberglass
    -reinforced epoxy material, 261
    window screening, 262
Field
    analytical methods, comparisons of laboratory
        and, 425
    classification, 668
    habitat assessments, 410
    manometer, 335

methods
    heavy metal, 437
    use of for water quality evaluations, 425
observation sheet, 392
sampling crew, 644
test kits, 429, 767
    analysis of organic compounds using, 434
    assembling appropriate set of, 432
    biggest difficulty with, 430
    evaluation of, 768–772
    selection of appropriate, 443
titration equipment, 444
Filamentous algae, 129, 133, 677
Fine particulate organic carbon, 146, 147
Fine particulate organic material (FPOM), 677
Fingerprinting, indication of contamination sources
        through, 475
Fire hazard data, 765
First-flush phenomenon, 285, 356
Fish, 128, 133, 152, *see also* specific species
    abundance, 699
    advisory, 30
    ammonia acutely toxic to, 814
    bioassay tests, side-stream, 54
    biomass
        seasonal trends of, 154
        total, 700
    community, 611
    consumption advisories, 51, 104
    death of from toxic material spills, 166
    disease surveys, 48
    -eating organisms, 116
    effects of suspended solids on, 74
    field collection data sheet, 695
    freshwater, 519
    gill damage, 154
    kills, 49, 211
        elevated nutrient loading and, 49
        massive, 53, 165
        sources associated with, 50
    populations
        characterizing, 122
        indices of, 504
    sampling, 502, 505
    seining, 504
    spawning conditions, deteriorating, 148
    species
        diversity, 144
        preferred temperature of some, 414–415
        total number of, 697
        trophic guilds used to categorize, 503
    surveys, 107
    tissue
        residues, 114
        sampling of, 122

Fish community assessment, 693–708
    data analysis techniques, 694–707
        fish abundance and condition metrics,
            699–707
        species richness and diversity, 697–698
        trophic composition metrics, 698–699
    sample processing, 694
Fisher's LDS, 591
Fishery(ies)
    consumptive, 22
    warm-water, 27
Fitted regression model, 597
Flagfin shiner, 706
Flammables, health effects associated with, 750
Flammable solvents, 749, 760
Flash point, 749
Flathead catfish, 704
Floatable litter
    characteristics, 71, 399
    sampling, 296
    wet-weather flows and, 68
Floatable(s), 398
    material sampling, 224
    matter, 469
    pollution, 70
Floating booms, litter controlled behind, 27
Flood
    control, 126
    -and-drought conditions, 64
    potential, 411
    prevention, 10, 26, 610
Floodplain, 652
    change factor, 111
    condition of, 652
    quality, 649
Florisil cleanup, 781
Floatable trash, 29
Flow
    Balancing Method (FBM), 169, 170, 180
        flow pattern in, 172
        in-lake tanks, 171
        system, 178
    calibration, 373
    component identification, 471
    measurement(s)
        equipment vendors, 374
        example calculation for, 360
        instruments, comparisons of available, 377
        methods for, 358
        subsurface, 394
    -metering equipment, calibration of, 366
    monitoring
        current meter, 360
        methods, 357, 376
        use of tracers in, 361
    recurrence interval, 246

Flowmeters
    acoustic, 358
    Doppler, 374
    magnetic, 358
Flow-weighted automatic samplers, 288
Flow-weighted composite sampling, 283, 285
Fluoranthene, 61, 161, 448, 480
Fluorene, 448, 802
Fluorescein, 363
Fluorescent dyes, water tracing using, 364
Fluorescent measurement instrumentation, 365
Fluorescent whitening agents (FWAs), 480
Fluoride, 441, 740
Fluorometers, calibration of, 366
Food web
    contamination, 123
    models, 620
Forest management, 4
Formaldehyde, 740
Formic acid, 740
Fortran-coded programs, 843
Fossil fuel combustion, 36
FPOM, *see* Fine particulate organic material
Fractional factorial design, 231
Frecklebelly madtom, 707
Freckled madtom, 704
Freeware, 575
Freeze core sampler, 316
Freeze-dried reagent, 487
Freezing core samplers, 322
Freshwater, 799
    algae, 519
    aquatic communities, biotic integrity of, 497
    drum, 415
    ecosystems, sediment quality guidelines for, 837
    fish, 519
    organisms, chronic toxicity data for, 814
Friction slope, 359
FS, *see* Fecal streptococci
Fuel leakages, 167
Fume hoods, 741
Fumigant, 6
FWAs, *see* Fluorescent whitening agents

G

Gambusia affinis, 133
Gammarus sp., 531
Garden store rain gauges, 388
Gas analyzer, portable, 468
Gas chromatograph with electron capture detector
        (GC/ECD), 523
Gas chromatograph with mass selective detector
        (GC/MSD), 459, 523

Gasoline, 167
Gastroenteritis, 621
Gastrointestinal illness, 816, 817
Gaussian distribution, 238
GC/ECD, *see* Gas chromatograph with electron
        capture detector
GC/MSD, *see* Gas chromatograph with mass
        selective detector
Geographical information system (GIS), 844, 857
Geomorphology, 411
GFAA, *see* Graphite furnace-equipped atomic
        absorption spectrophotometer
Ghost shiner, 703
Giardia, 197, 486
    cysts, degradation plot of, 489
    lamblia, 622
    -specific antibody, 489
Giardiasis, 621
Gilt darter, 704
GIS, *see* Geographical information system
Gizzard shad, 702
Glazed tile, 25
Glide habitats, 646
Glove materials, chemical resistance of, 740
Glycerol, 740
Golden redhorse, 703
Golden shiner, 74, 702
Golden trout, 705
Goldeye, 702
Goldfish, 701, 702
Golf courses, 185
Gomphonema sp., 138
Gomphosphaeria, 175
Gophers, bioturbation by, 394
Grab samplers, 315, 496, 685
Graphite furnace atomic absorption spectrometry,
        778
Graphite furnace-equipped atomic absorption
        spectrophotometer (GFAA), 523
Grass carp, 703
Grass pickerel, 414, 702
Grass swales, 58, 629, 630, 631
Gravel chub, 702
Gravity corers, 317
Gray redhorse, 706
Great redhorse, 703
Green algae, 518, 520
Greenside darter, 705
Green sunfish, 135, 414, 508, 698, 704
Groundwater(s)
    -associated biota, 326
    contamination
        phosphorus, 59
        potential sources of, 460
        problems, 57
    detection of viruses in, 61

flows, tipping bucket flow measurement device
        for measuring, 390
    hardness of, 826
    movement, piezometer measures of, 616
    MTBE contamination, 58
    pesticide contamination of, 60
    recharge, 126
        basin, 57
        decreases in, 28
        –surface water interactions, 326
        table, decreases in, 64
        upwelling, 326
        urbanization affecting, 31
Group comparison tests, comparing multiple sets of
        data with, 588
Guthion, 806

H

Habitat
    alterations, 16
    aquatic life, 618
    assessment(s), 277
        approach, generic, 653
        field, 410
        matrix, 653
        procedure for performing, 662
    characteristics, 405, 645
    definition of, 400
    degradation, 165
    designations, modified warm water, 112
    destruction, 28, 114, 214
    diversity, 405
    edge, 405
    evaluation index, qualitative, 404
    glide, 646
    goals, 549
    modifications, 63
    pool, 646
    problems, 617
    protection, major component of, 65
    quality, 103, 402, 403
    Quality Index, 401
    quantification, 114
    relationship between biological condition and,
        400
    restoration, 276
    riffle, 645
    riparian, 409
    run, 645
    Suitability Indices (HSI), 401
    surveys, 106, 107, 121
Habitat characterization, 643–663

Qualitative Habitat Evaluation Index, 643–652
    computing total QHEI score, 650–652
    geographical information, 643–645
    pool and glide habitats, 646–650
    riffle and run habitats, 645
    stream map, 652
USEPA habitat assessment for rapid
        bioassessment protocols, 652–662
    physical characterization, 658–662
    procedure for performing habitat assessment,
        662
    quality assurance procedures, 662
    water quality, 656–658
HACH
    color test kit, 773
    detergents test, 471
Halogenated aliphatics, 6, 161
Halogenated solvents, 760
Haphazard sampling, 225
Harbor facilities, 185
Harelip sucker, 703
Hazard
    assessments, 348
    identification, 11, 113, 620
    primary, 737
Hazardous waste
    elimination of nonhazardous waste from, 761
    sites, toxicity evaluation categories for, 519
HCGI, see Highly credible gastrointestinal
Headwater streams, removal of riparian vegetation
        in, 403
Health hazard data, 765
Heat sealing unit, 434
Heavy metal(s), 189
    analyses, 774
    emerging analytical methods for, 438
    field methods, 437
    in stormwater runoff, 76
    urban runoff, 79
Hemiptera, 495, 688
Heptachlor epoxide, 803
Herbicides, 118, 159, 249, 267
Herbivores, 495
Hester–Dendy samplers, 129
Hexachlorobenzene, 448, 744, 802
Hexachloroethane, 262
Hexagenia
    bilineata, 520
    limbata, 520
Hexane, 740
Hierarchical cluster analyses, 592
Highfin carpsucker, 703
Highly credible gastrointestinal (HCGI), 817
Highway runoff
    constituents, 7
    site, concentrations of fecal coliforms at, 83

Hillside discharge, 199
HIS, *see* Habitat Suitability Indices
Hitch, 53, 134, 135
Homogeneity tests, 177
Honest significant difference (HSD) test, 604
Horizontal water sampler, 291
Horneyhead chub, 702
Household garbage, 142
HPLC technology, organic analyses using, 459
HSD test, *see* Honest significant difference test
Human health
     criteria, 614
     impairments, evaluation of using risk assessment
          approach, 619
     problems, inappropriate discharges and, 461
     protection, 102
Humpback whitefish, 705
Hyalella azteca, 52, 135, 210, 335, 527, 546, 605,
     722, 725, 726
Hybobsis amblops, 706
Hydrobromic acid, 740
Hydrocarbons, 483, 826
Hydrofluoric acid, 740, 753, 778
Hydrogen peroxide, 740
Hydrological Simulation Program-FORTRAN, 854
Hydrologic change factor, 111
Hydrology, 349, 351, 411
Hydromodification, 4, 5
Hypentelium
     etowanum, 706
     nigricans, 706
     roanokense, 706
Hyporheic sampling, 326

I

IAI, *see* Indicator Assemblage Index
IBDU, *see* Isobutyldiene diurea
IBI, *see* Index of Biotic Integrity
ICI, *see* Invertebrate Community Index
ICP, *see* Inductively coupled plasma emission
          spectrometry
Ictalurus
     nebulosus, 414
     punctatus, 414, 520, 536
Ictiobus sp., 414
IDL, *see* Instrument detection limit
IFIM, *see* Instream Flow Incremental Methodology
Imidachloprid, 436
Immunoassay kits, 444
Impaction, 311
Inconnu, 706
Incubator, 434

Index of Biotic Integrity (IBI), 112, 402, 683,
     694
     metrics, regional variations of, 508–509
     steps in calculating, 696
Index of clumping, 235
Index of Well Being (IWB), 506, 694
Indicator Assemblage Index (IAI), 678
Inductively coupled plasma emission spectrometry
     (ICP), 774
Industrial wastewater discharge, 598
Infectious hepatitis, 621
Infiltration
     devices, 630
     rate measurements, 391
     test(s)
          apparatus, 389
          double-ring, 229
     trench, stormwater infiltration through, 59
In-lake flow balancing method, 180
In-lake tanks, FBM, 171
Inorganic chemicals, 762
In-place pollutants, 4
Insecta, 688
Insecticides, 6, 159
Insectivorous cyprinids, 509, 699
Insects, food types of aquatic, 495
In situ peepers, 327, 331
In situ testing, advantages of, 531
In-stream cover, 647
In-stream embryo bioassays, 154
Instream Flow Incremental Methodology (IFIM),
     350
In-stream temperature, 75
In-stream toxicity tests, 168
Instrument detection limit (IDL), 253, 583
Internal to external isomer ratio (I/E), 480
Interstitial water
     bacteria populations, 203
     chemistry, 203
     collection methods, 318
     degradation of, 202
     immediate collection and analysis of, 328
     isolation of, 329
     measurements, 202, 327
     quality, 371
     sampler selection guidelines, 317
     sampling, 326
Intolerant species, 508, 698
Invertebrate Community Index (ICI), 112, 683
Iodine, 740
Ion chromatograph, 449
Ion selective electrode (ISE), 426, 443
Iowa darter, 705
Iprodione, 436
I/R, *see* Internal to external isomer ratio

Irrigation
    return flows, 464
    water, 474
ISE, *see* Ion selective electrode
Isobutyldiene diurea (IBDU), 59
Isohyetal(s)
    method, 379
    preparation of for single rainfall, 380
Isophorone, 38, 448
Isoproturon, 436
IWB, *see* Index of Well Being

J

Jaccard Coefficient of Community Similarity, 678
Johnny darter, 705, 707
Judgment sampling, 225
Jussiaea sp., 136

K

Karst geology, 57
Kiyi, 705
Klebsiella, 485, 487
Kolmogorov–Smirnov one-sample test, 586
Kuderna–Danish method, 781
Kurskal–Wallis ANOVA on ranks test, 588, 591

L

La Motte Potassium Reagent Set, 442
LAB, *see* Linear alkylbenzenes
Laboratory
    analyses, conventional, 447
    analytical methods, comparisons of field and, 425
    chemical storage, 737
    information management systems (LIMs), 250
    personnel, selection of, 275
Laboratory safety, waste disposal, and chemical
        analyses methods, 735–786
    basic rules and procedures for working with
            chemicals, 738–743
        avoidance of routine exposure, 741
        choice of chemicals, 742
        electrical safety, 743
        equipment and glassware, 742
        fume hoods, 741–742
        housekeeping, 739
        labels and signs, 742–743
    laboratory protocol, 738
    personal safety practices, 738–739
    protective eyewear, 739
    protective gloves, 739–741
    protective clothing, 741
    unattended operations, 743
calibration and deployment setup procedure for
        YSI 6000upg water quality monitoring
        sonde, 782–785
chemical waste disposal program, 760–763
    chemical substitution, 761
    chemical waste containers, 760
    disposal of chemicals down the sink or
            sanitary sewer system, 761
    elimination of nonhazardous waste from
            hazardous waste, 761–762
    neutralization and deactivation, 761
    waste disposal, 762–763
    waste minimization, 760–761
comments pertaining to heavy metal analyses,
        774–778
emergency procedures, 758–759
    building evacuation procedures, 759
    mercury spills, 759
    minor spills, 759
    primary emergency procedures for fires,
            spills, and accidents, 758–759
field test kits, 767–774
fundamentals of laboratory safety, 737–738
    distribution of chemicals, 737
    laboratory chemical storage, 737
    procurement of chemicals, 737
    storage cabinets, 738
Material Safety Data Sheets, 763–767
    fire and explosion hazard data, 765
    hazardous ingredients/identity information,
            764
    health hazard data, 765–766
    physical/chemical characteristics, 764
    product name and identification, 764
    reactivity data, 765
    specific HACH MSDS information, 766–767
procedures for specific classes of hazardous
        materials, 748–758
    compressed gas cylinders, 757–758
    corrosives, 752–754
    flammable solvents, 749–750
    oxidizers, 750–752
    reactives, 754–757
stormwater sample extractions for EPA methods
        608 and 625, 779–781
use and storage of chemicals in laboratory,
        743–748
    chemical storage, 747–748
    procurement of chemicals, 743

transportation, 748
working with allergens, 743
working with chemicals of moderate or high
    acute toxicity or high chronic toxicity,
    744–747
working with embryotoxins, 744
LAF, *see* Laser atomic fluorescence
Lagoon runoff, 7
Lag plots, 598
Lake(s)
    hydraulic detention time of, 174
    hydraulic flushing rates, 178
    low elevation, 134
    phosphorus concentrations, 179
    pollutants and sources affecting U.S., 18
    sediment sampling in ice-covered, 336
    swimming restriction in urban, 28
Lake chubsucker, 703
Lake herring, 702
Lake sturgeon, 702
Lake trout, 414, 702, 706
Lake whitefish, 414, 702
Land
    disposal, 4, 5
    use
        category, 241
        monitoring, 239, 601
        predominant surrounding, 659
    waste disposal sources, 7
Landfill(s)
    runoff, 7
    sanitary, 4
Largemouth bass, 74, 415, 536, 704
Large organic debris (LOD), 618
Largescale stoneroller, 706
Largescale sucker, 701
Large woody debris (LWD), 408, 618
LAS, *see* Linear alkylbenzene sulfonates
Laser atomic fluorescence (LAF), 438
Laundry
    detergent samples, 483
    wastewaters, 461
Lawn fertilizers, 206
LC50, *see* Lethal concentration 50
LD50, *see* Lethal dose 50
LDV, *see* Less than detection values
Lead, 6, 132
    aquatic life summary for, 827
    bioaccumulation of, 133
    concentrations, dissolved, 145
    human health criteria for, 828
    reduction benefits, 636
Leaded gas, 6
Leaf
    core catcher, 323

packs, 666
shredding organisms, 150
Least brook lamprey, 701
Least madtom, 707
Leather products, 187
Lentipes concolor, 707
Leopard dace, 701
Lepidoptera, 495, 688
Lepomis
    cyanellus, 414
    gibbosus, 415, 536
    macrochirus, 415, 520, 536
    megalotis
Leptocheirus plumulosus, 527
Less than detection values (LDV), 583
Lethal concentration 50 (LC50), 764
Lethal dose 50 (LD50), 764
Life cycle measures, 676
Light–dark bottle method, 72
Light transmissivity, 188
Limestone quarry, 73
LIMs, *see* Laboratory information management
        systems
Lindane, 6, 60, 159, 424, 803, 827
Linear alkylbenzenes (LAB), 479
Linear alkylbenzene sulfonates (LAS), 479
Liquid
    flash point of, 749
    –liquid separatory funnel technique, 779
Litter, 398
    characteristics of floatable, 399
    control, 24, 27
    discharges, 297
    fast-processing, 667
    loose, 399
    material categories, discharged, 399
    slow-processing, 667
Livestock
    production, 4
    trampling, 649
    trucks, feces debris falling from, 83
LOD, *see* Large organic debris
LOEC, *see* Lowest Observed Effect Concentration
Log dragging, 73
Log-normal probability distribution, 253, 584
Logperch, 704
Longear sunfish, 704, 707
Longnose dace, 701, 702
Longnose gar, 702
Longnose sucker, 703
Lowest Observed Effect Concentration (LOEC), 604
Lumbriculus variegatus, 114, 120, 535, 724
LWD, *see* Large woody debris
Lycopodium, 362

M

Macrofaunal toxicity tests, 523
Macroinvertebrate(s)
    colonization tests, artificial substrate, 121
    counts, 682
    diversities, reduction of in urban streams, 51
    listing, phylogenetic order for, 688
    qualitative samples of, 681
    sampling, 494, 497
    surveys, 107
    taxonomy, 689
Macrophytes, 63, 257
Magnetic flowmeters, 358
Mailing lists, 578
Malathion, 60, 159, 806
Manholes, 162
Man-made drainage systems, 463
Manning's equation, 12, 361, 373
Manning's roughness coefficient, 359
Mann–Kendall test, 175, 176, 582, 602
Mann–Whitney signed rank test, 590
Mann–Whitney U tests, 470, 475
Manual pump samplers, 292
Manual samplers, selection of materials for, 261
Manual sampling
    advantages of compared to automatic sampling,
        260
    procedures, 289
Manual sheetflow samplers, 298
Manure-laden runoff, 89
Map gradient, 650
Maples, 667
Marginal benefit analysis, 242, 847
Margined madtom, 707
Marina facilities, 185
Marine debris, 104
Mass emission
    drainage monitoring stations, 184
    stations, 183
Material Safety Data Sheets (MSDS), 743, 763
Maximum Contaminant Level (MCL), 823
Mayflies, 52, 520
MBAS, see Methylene blue active substance
MCL, see Maximum Contaminant Level
MCTT, see Multichambered treatment train
MDL, see Method detection limit
Means quality control chart, 251
Meat packing wastes, 83
Mechanistic model, 227
Megaloptera, 688
Meio–microfaunal interactions, 531
Melosira, 138, 175
Mercury
    compounds, 744
    spills, 759

Metal(s)
    analysis methods, attributes of, 777
    conversion factors for dissolved, 813
    corrosion, 5
    optimal concentration ranges of in samples, 776
    plating, 7
    sample preparation procedures for identifying,
        458
    speciation, 314
    screening approach for in rivers, 544–545
Metalaxyl, 436
Metallic priority pollutants, 140
Methanogenesis, 326
Methemoglobinemia, 829
Method blank, 248
Method detection limit (MDL), 189, 249, 423, 583
Methoprene acid, 436
Methoxychlor, 38, 262
Methyl bromide, 801
Methyl cellosolve, 740
Methyl chloride, 6, 740, 801
Methylene blue active substance (MBAS), 479
Methylene chloride, 6, 38, 61, 740
Methylene urea, 59
Methylphenanthrene, 480
Metolachlor, 60
MFO, see Mixed function oxidase
Microbial activity tests, 523
Microbial-meiofaunal communities, 314
Microbiological sampling, 289
Microorganism(s)
    evaluations, sampling for, 487
    measurements, 549
    urban receiving water, 485
Micropterus
    dolomieui, 74, 415, 536
    punctulatus, 415
    salmoides, 74, 415, 536
Microtox
    osmotic adjusting solution (MOAS), 733
    screening test, 121, 445, 513, 730
MID, see Minimal infective dose
Midges, 520
Mimic shiner, 703, 706
Mineral scrapers, 495
Miners, 495
Minimal infective dose (MID), 621
Mini-piezometers, 334
Minytrema melanops, 706
Mirex, 806
Mississippi silverside, 134
Mississippi silvery minnow, 703
mIWB, see Modified Index of Well-Being
Mixed function oxidase (MFO), 515
MOAS, see Microtox osmotic adjusting solution
Mobile homes, 185

Model building, data associations and, 582
Model(s)
    calibration and validation, 11–12
    capabilities, evaluation of, 855
    concentration-addition, 516
    Distributed Routing Rainfall Runoff
    empirical, 227
    fitted regression, 597
    food web, 620
    mechanistic, 227
    Monte Carlo, 363, 476
    nonpoint source, 862
    non-urban runoff, 856
    pulse exposure, 614
    rainfall–runoff, 382
    receiving water, 843, 846, 855
    receptor, 475
    regression, 599
    Source Loading and Management, 854
    Storage, Treatment, Overflow Runoff, 854
    straight-line, 241
    stream
        non-toxic constituents in, 864
        predicting pollutant fates using, 544
    Urban Catchment, 854
    urban, 856
    washoff equation used in stormwater, 307
    watershed, 843, 846
Modified Family Biotic Index, 675
Modified Index of Well-Being (mIWB), 403
Mollusca, 689
Mollusks, 128
Monitoring
    initiation, 243
    program, personnel needed to carry out, 275
Monocyclic aromatics, 161
Monoethanolamine, 740
Monte Carlo analyses, 253
Monte Carlo mixing model, 363
Monte Carlo model, 476
Monte Carlo sampling routines, 577
Mooneye, 702
Morisita's Index, 679
Morone
    americana, 414
    chrysops, 414
    saxatilis, 414, 536
Morpholine, 740
Mosquito control, 6
Mosquitofish, 53, 135, 137, 704
Motor freight, 187
Motor vehicle activity, 167
Mountain brook lamprey, 701
Mountain madtom, 704, 707
Mountain sucker, 701

Mountain whitefish, 701, 705
Moxostoma
    anisurum, 706
    cervinum, 706
    congestum, 706
    duquesnei, 706
    hamiltoni, 706
    lachneri, 706
    poecilurum, 706
    rhothoecum, 707
    rupiscartes, 707
    valenciennesi, 707
MSDS, see Material Safety Data Sheets
Multichambered treatment train (MCTT), 525
Multidimensional scaling, 612
Multistage sampling, 225
Municipal point sources, 16
Municipal wastewater, 624
Muskellunge, 414, 702
Mussel populations, characterizing, 122
Mutagenicity, 507

N

Naphthalene, 740, 802
National Ambient Water Quality Monitoring
        Network (NAWQMN), 681
National Institute of Standards and Technology
        (NIST), 249
National Pollutant Discharge Elimination System
        (NPDES), 8, 513, 798
    permit compliance, 429
    stormwater permit program, 9
National Technical Advisory Committee (NTAC),
        817
National Water Quality Assessment (NAWQA), 118
Nationwide Urban Runoff Program (NURP), 34,
        307, 615, 629
Navicula spp., 138
NAWQA, see National Water Quality Assessment
NAWQMN, see National Ambient Water Quality
        Monitoring Network
Net sampling devices, 501
NEXRAD, 388
NIST, see National Institute of Standards and
        Technology
Nitrate, 828
Nitrite, 828
    aquatic life criteria, 829
    criteria, human health, 829
Nitrobenzene, 809
Nitrogen cycling, 115
N-Nitrosodimethylamine, 38

Nocomis micropogon, 706
NOEC, *see* No Observed Effect Concentration
NOEL, *see* No-observable-effects level
No exposure incentive, 9
Noncarcinogens, 827
Nonhazardous waste, 762
Nonmetallic minerals, mining of, 187
Nonparametric tests, 581
Nonpoint runoff receiving water impact research
          program, 51
Nonpoint sources (NPS), 3–4
     -affected streams, stream assessment factors for,
          111
     assessment, 276
     hydromodification category of, 7
     models, input data needs for, 862
     pollution
          categories, 4
          sources of, 4–8
Nonpollutant factors, 115
Nonspecified chemical waste, 763
Non-urban runoff models, 856
No-observable-effects level (NOEL), 546
No Observed Effect Concentration (NOEC), 604
Northern hog sucker, 703, 706
Northern madtom, 704
Northern pike, 702
Northern redbelly dace, 706
Northern squawfish, 701
Norwalk virus, 621
Notropis
     amnis, 706
     anogenus, 706
     ardens, 706
     atherinoides, 414
     boops, 706
     emiliae, 706
     galacturus, 706
     heterlepis, 706
     heterodon, 706
     hudsonius, 706
     hypselopterus, 706
     leuciodus, 706
     lutipinnis, 706
     nubilus, 706
     ozarcanus, 706
     photogenis, 706
     pilsbryi, 706
     rubellus, 706
     rubricroceus, 706
     signipinnis, 706
     telescopus, 706
     topeka, 706
     volucellus, 706
     whipplei, 706

     zaenocephalus, 706
     zonatus, 706
     zonistius, 706
Noturus
     albater, 707
     elegans, 707
     eleutherus, 707
     exilis, 707
     flavus, 707
     funebris, 707
     hildebrandi, 707
     insignis, 707
     laptacanthus, 707
     minitus, 707
     phaeus, 707
NPDES, *see* National Pollutant Discharge
          Elimination System
NPS, *see* Nonpoint sources
NTAC, *see* National Technical Advisory Committee
NURP, *see* Nationwide Urban Runoff Program
Nutrient
     availability, 406
     cycling, 347
     loads, 411
     tests, most common, 443

O

Oaks, 667
Ocean shorelines, pollutants and sources affecting
          U.S., 20
Odonata, 495, 688
Ohio lamprey, 701
Oil/gas
     extraction, 187
     production, 4
Oil in water optics, 426
Oligochaetes, 72, 165, 724
Omnivores, 108, 509
Oncorhynchus
     gorbuscha, 414
     keta, 414
     kisutch, 414, 520
     mykiss, 717, 830
     nerka, 414
     tshawytscha, 414, 830
On-site erosion controls, 452
On-site wastewater treatment, 4
O'opu alamoo, 707
O'opu nakea, 707
O'opu nopili, 707
Open vertical water sampler, 291
Optical brighteners, 440, 441

Orange-Peel sampler, 316
Orangespotted sunfish, 704
Orangethroat darter, 705, 707
Organic chemicals, 762
Organic compounds
    analysis of using field test kits, 434
    toxic, 77
Organic contaminants, 266
Organic-inorganic chelators, 77
Organic matter processing, 115
Organic scrapers, 495
Organic solvent extract, separation of from water
        sample, 780
Organic substrate components, 660
Organism(s)
    availability, 522
    photosynthetic, 518
Organochlorine(s)
    bioaccumulation of, 212
    pesticides, 249, 837
Organohalogen compounds, 746
Organophosphates, 267, 436, 535
Organ transplants, 625
ORP, *see* Oxidation-reduction potential
Orthophosphates, 130, 131, 132
Oscillatoria sp., 175
Osmerus mordax, 414
Outfall
    flow monitoring, 373
    structures, damage to, 469
Overexposure, signs and symptoms of, 766
Over-the-glasses safety glasses, 739
Overland flow sampling site, 197
Oxidation-reduction potential (ORP), 263, 418, 419
Oxidizers, 750
    common, 751
    first aid for, 752
    health effects associated with, 751
    use and storage of, 751
Oxygen
    -depleting substances, 18
    depletion, 73, 147, 166
    production, photosynthetic, 421
Ozark madtom, 707
Ozark minnow, 706
Ozark rockbass, 707
Ozark shiner, 706
Ozonated bromides, 62

P

Pacifastacus leniusculus, 135
Paddlefish, 701
PAHs, *see* Polycyclic aromatic hydrocarbons

Paired observations, 581
Paiute sculpin, 701
Pallid shiner, 706
Parallel stream analyses, 168
Parametric tests, 581
Paraquat, 436
Parasites, 103, 623
Parathion, 436, 807, 827
Particle
    characteristics, visual observations of, 457
    sampling errors, 282
    size(s)
        analysis, automated, 455
        distribution, 266, 407, 451
        methods to measure stormwater, 454
        settling velocity and, 451
Particulate(s)
    -associated toxicity, 19
    dry-fall, 310
    lead, automobile emission, 312
    removal process, 311
    residue, 39, 310
    sampling procedures, street surface, 301
Patch dynamics, 400
Pathogen(s), 78, 103, 610
    -contaminated waters, 85
    die-off tests, 200
    fecal, 120
    microorganisms, 61
    monitoring, in stormwater, 29
    from raw or poorly treated, 86
    risk assessment for human exposure to
        stormwater, 620
    in stormwater, 82
    survival, 254, 487
Paved area(s)
    drainage, 184
    sources of pollutants on, 627
Pavement
    temperature monitoring, 412
    tests, rainfall-runoff responses for, 850
    wear, 7
PCA, *see* Principal component analyses
PCB-1260, 38
PCP, 267
Peamouth, 701
Pearl dace, 706
Pearson correlation matrix, 593
Peeper(s)
    devices, 532
    disassembled, 330
    in situ, 327, 331
    large-volume, 331
    small-volume, high-resolution, 327
    wells, 331

PEL, *see* Permissible exposure limit
Pentachlorophenol, 6, 7, 36, 38, 81, 262, 424, 448, 614, 801
Perca flavescens, 415
Perchloric acid, 751
Percina
   shumardi, 707
   sp., 707
Periphyton, 254, 257, 258, 500
   populations, characterizing, 122
   sampling, 493, 501
Permissible exposure limit (PEL), 764
Peroxide-forming materials, 755
Personal protective equipment, 750
Perturbation, metric response to increasing, 675, 676
Pesticide(s), 189, 277, 526
   carbamate, 249
   contamination, of groundwater, 60
   cross-contamination, 60
   decomposition, 60
   detection, 436
   leaching, 60
   mobility, 60
   organochlorine, 249, 837
   organophosphate, 60, 249
Peterson sampler, 316
Petite Ponar dredge, 320
Petroleum refining, 187
Petrosense hydrocarbon probe, 427
Pfiesteria, 29
   monitoring program, 90
   piscicida, 89
pH, 832
   aquatic life effects, 833
   meters, 430, 439
   probe calibration, 784
PHABSIM, *see* Physical Habitat Simulation Model
Pharmaceuticals, 484
Phenanthrene, 6, 38, 161, 424, 480
Phenol, 38
Phenolics, 39, 161
Phenoxy acid herbicides, 267
Phosphate reduction benefits, 636
Phosphorus, 38
   budgets, treatment system, 172
   discharges, 179
   removal of by ferric chloride precipitation, 180
   removal rate, 173
   removal of from stormwater, 169
   soluble reactive, 146, 147
   treatment mass balance, 173
   trends, 177
Photobacterium phosphoreum, 520, 730
Photochemical decay, 364
Photodegradation times, 525

Photoinduced toxicity, PAH, 91
Photolysis, 78, 79, 80, 81
Photosynthesis, 420
   organisms, 518
   oxygen production rate, 422
   rates, 201, 411, 421
   test chambers having occurrence of, 419
Photosynthesis and respiration (P/R), 196, 200
   rates, 417, 420
   tests, *in situ*, 204, 440
Phototoxicity evaluations, 616
Phoxinus
   cumberlandensis, 706
   eos, 706
   erythrogaster, 706
   laevis, 830
Phthalate esters, 6, 36, 78, 161, 526
Physical characterization parameters, 658
Physical Habitat Simulation Model (PHABSIM), 401
Phytoplankton, 63
Piercers, 495
Pimephales promelas, 108, 210, 515, 546, 605, 710, 716, 719
Pink salmon, 414
Pirate perch, 704
Plankton, 122, 257
Plant equipment, corrosion of, 833
Plasticizers, 6, 78
Plastic samplers, 333
Plecoptera, 491, 495, 688
Plot(s)
   box, 375, 848
   data, 583
   Digidot, 586
   event, 371
   lag, 598
   QA/QC control, 577
   score, 595
   soil infiltration test, 390
   whisker, 587
Plywood, 262
Point source discharge, 278
Poison ivy, 193
Poisson distribution, 235
Pollutant(s)
   discharge(s)
      changes in from surface runoff and subsurface flows, 397
      ranking, 185
   generation, 240
   in-place, 4
   loading, 853
   mass discharges, 356
   potential sources of, 122

reduction, 588
sensitivity, 116
sources of on paved areas, 627
surface water, 104
-tolerant organisms, 51, 52
Pollution
floatable, 70
impacts, 651
-sensitive species, 116
-tolerant benthic macroinvertebrates, 314
-tolerant organisms, 137
Polychlorinated biphenyls, 837
Polycyclic aromatic hydrocarbons (PAHs), 6, 7, 36,
78, 197, 208, 435, 535, 837
-contaminated sediments, 48, 540
detection of in soil samples, 167
-photoinduced toxicity, 91, 209, 540
sediment quality guidelines for, 838
street dirt samples containing, 161
Polymerization reactions, 754
Polytetrafluoroethylene (PTF), 269, 328
Polyvinyl (PVC) samplers, 255
Pomoxis
annularis, 415
nigromaculatus, 415, 536
Ponar dredge, 320, 321
Ponar sampler, 316
Pool habitats, 646
Popeye shiner, 703
Population distribution characteristics, 237
Pore water
conditions, mini-piezometer measurements of,
334
sampling, 313
sediment sampling for interstitial, 336
squeezer, 329
toxicity test, 267
type of container and conditions recommended
for storing samples of, 266–267
Porifera, 688
Porosity, calculation of total, 394
Porous pavements, 630
Potable water, treated, 472
Potamogeton pectinatus, 136
PQL, *see* Practical quantification limit
P/R, *see* Photosynthesis and respiration
Practical quantification limit (PQL), 253, 583
Precipitation, 310, 313
Predator–prey
effects, 534
relationships, 22
Predators, 103
Price meter, 359
Prickly sculpin, 53, 701
Primary hazard, 737

Primary metals, 187
Principal component(s)
analyses (PCA), 591, 592
loadings of, 595
score plots of, 595
Printing and publishing, 187
Priority pollutants, metallic, 140
Pristine estuary, 123
Probability
information, need for, 245
plots, 584
sampling, 225
Procambarus clarkii, 133, 135, 136
Professional organizations, 578
Project field staff, 275
Prometon, 118
Protective eyewear, 739
Protozoa, 63, 88, 487
parasites, 623
sampling, 494
Pseudomonas aeruginosa, 42, 62, 86, 87, 88, 118,
197, 486, 487, 625, 817, 819, 821
PTF, *see* Polytetrafluoroethylene
Public land records, use of in digital database, 858
Public water supplies, 28
Pugnose minnow, 702, 706
Pugnose shiner, 703, 706
Pulse exposure model, 614
Pumpkinseed, 134, 415, 704
PVC samplers, *see* Polyvinyl samplers
Pyrene, 6, 38, 61, 81, 424, 448
Pyrethroids, synthetic, 436

Q

QA, *see* Quality assurance
QAPP, *see* Quality assurance project plans
QA/QC, *see* Quality control/quality assurance
QC, *see* Quality control
QHEI, *see* Quantitative Habitat Evaluation Index
Quality assurance (QA), 4, 118
objectives, quantitative, 424
procedures, 662
project plans (QAPP), 121
Quality control (QC), 4
Quality control/quality assurance (QA/QC), 224, 247
control plots, 577
officer, 274
problems, visual indications of, 250
procedures needed for during sample collection,
109
program, 251
requirements, 337

Quality prediction methods, data needs for, 863
Quantistrip method, for alkalinity, 773
Quantitative Habitat Evaluation Index (QHEI), 120, 404, 643, 645
Quillback, 703

R

Radar rainfall measurements, 388
Railroad transportation, 187
Rain, *see also* Rainfall
   characteristics, 452
   depth(s), 849
     errors in watershed, 383
     probability plots of, 190
   duration, 307, 308
   gauge(s), 121, 379
     calibration, 387
     Clear View, 378
     density, 381
     exposure, 386
     garden store, 388
     location, 309
     network, 380, 386
     proper placement of, 386
     recalibrated, 385
     sampler, 286
     spacing, 381
     stormwater monitoring, 378
     Thiessen polygons for, 380
     tipping bucket, 387, 388
   intensity, 308, 309
   temperature monitoring, 412
   volume, 308
   washoff of debris and soil during, 627
Rainbow smelt, 414, 702
Rainbow trout, 53, 74, 414, 520, 701, 702, 706
Rainfall
   depths, 353
   design, 245
   distribution(s), 385
     characteristics, 354–355
     urban watershed, 382
   energy, 33
   measurements, radar, 388
   monitoring, 349, 377
     errors, 381
     extreme in, 378
     methods, advantages and disadvantages of, 388
     wind-induced errors during, 386
   –runoff
     modeling, 382

   pattern, 381
   responses, pavement test, 850
  variability, 384
Randomly amplified polymorphic DNA (RAPD) markers, 537
Random sampling, 226
Range ratio, 253
Rank correlation coefficient, 684
RAPD markers, *see* Randomly amplified polymorphic DNA markers
Rapid Bioassessment Protocol (RBP), 654, 655, 656, 665
Rapid Bioassessment Protocol V (RBP V), 693
RBP, *see* Rapid Bioassessment Protocol
RBP V, *see* Rapid Bioassessment Protocol V
Reactives
   first aid for, 757
   health hazards associated with, 756
   use and storage of, 756
Reactivity, 763, 765
Reagent
   bacterial, 731
   blanks, analysis of, 250
   freeze-dried, 487
   SPADNS, 441
   waste, 443
Receiving water(s)
   aquatic organisms, effects of urban runoff on, 91
   assessment parameters, 114
   characterization, 187
   conditions, cause-and-effect relationships between urban runoff and, 30
   detrimental effects of urban and agricultural runoff on, 47
   effects, wet weather-related, 548
   impact(s)
     monitoring activities to assess, 609
     studies, 51
   investigation, 124, 213, 445–446
   levels, DO, 85
   microorganisms in urban, 485
   modeling, *see* Watershed and receiving water modeling
   models, 843, 846, 855
   nutrients entering, 76
   pH of, 364
   problems
     important pollutant causing, 416
     watershed indicators of biological, 103
   quality, components of integrated approach to assess, 102
   segment of interest, monitoring cost estimate for single outfall in single, 215
   swimming areas in urban, 28
   target factor, 111

typical urban, 216
ultra-urban area affecting local, 182
wet-weather flow impacts on, 15
Receiving water data, statistical analyses of,
        575–607
    comments on selected statistical analyses
            frequently applied to receiving water
            data, 582–605
        analysis of trends in receiving water
            investigations, 601–603
        comparing multiple sets of data with group
            comparison tests, 588–591
        data associations, 591–596
        determination of outliers, 582–583
        exploratory data analyses, 583–588
        regression analyses, 596–600
        specific methods commonly used for
            evaluation of biological data, 603–605
    selection of appropriate statistical analysis tools
            and procedures, 575–582
        computer software and recommended
            statistical references to assist in data
            analysis, 576–580
        selection of statistical procedures, 580–582
    statistical elements of concern when conducting
            receiving water investigation, 605–606
Receiving water uses, impairments, and sources of
        stormwater pollutants, 15–45
    beneficial use impairments, 22–29
        biological uses, 27–28
        human health-related uses, 28–29
        recognized value of human-dominated
            waterways, 22–26
        recreation uses, 26–27
        stormwater conveyance, 26
    likely causes of receiving water use impairments,
        30
    major urban runoff sources, 31–42
        construction site erosion characterization,
            32–34
        urban runoff contaminants, 34–42
Receptor model, 475
Reconnaissance surveys, 257
Redear sunfish, 134, 704
Redfin shiner, 703
Redox potential, 77
Red shiner, 703
Redside dace, 702, 706
Redside shiner, 701
Reference watershed, 111
Reformed seining, 505
Refrigerant, 6
Refuge areas, 150, 404
Regression
    analyses, 596, 599
    equation, 600

methods, 602
model, verifying of, 599
Regulatory agencies, fines imposed by, 233
Regulatory program, 8–10
Relative standard deviation (RSD), 432
Relative toxicity, calculation of, 732
Replicate sampling, 665
Representative qualitative sample, 693
Reservoirs, man-made, 126
Residuals, graphical analyses of, 597
Residue management, 4
Resource extraction, 4, 5, 7, 16
Resuspension
    effects, 521
    events, 348
    velocity, 545
Reverse osmosis (RO), 203, 430
Rhepoxynius abronius, 527
Rhinichthys atratulus, 706
Rhodamine B, 363
Rhodamine WT, 364, 365, 369
Rhododendrons, 667
Rhoicosphenia curvata, 138
Rhopalodia spp., 138
Riffle
    habitats, 645
    –pool boundary, 409
    /run
        quality, 649, 650
        sample, 666, 674
Riffle beetles, 52
Riffle sculpin, 53
Riparian areas, debris in, 27
Riparian cover factor, 111
Riparian habitats, 409
Riparian vegetation, 150, 618
    removal of, 75
    stabilization of stream banks by, 63
Riparian zone, 648
Riprapping, 142
Risk characterization, 11, 113, 619
River(s)
    classification system, 652
    concentration profiles of toxicants in, 547
    mouth, 409
    pollutants and sources impairing U.S., 16
    screening approach for metals in, 544–545
    sluggish, 134
    swimming beaches, 84
    temperature profiles in, 410
River carpsucker, 703
River chub, 702, 706
River darter, 704, 707
River redhorse, 703, 706
River shiner, 703
RO, *see* Reverse osmosis

Road construction, 4
Roadside fertilizer application, 7
Roanoke bass, 707
Roanoke hog sucker, 706
Rock bass, 414, 704, 707
Roof
    disconnections, 631
    drainage, 184
    runoff, 23
Rooftop temperature data logging, 412
Rosefin shiner, 703, 706
Rosyface shiner, 703, 706
Rotenoning, 505
Round whitefish, 705
Rovers, biological integrity of, 104
RSD, *see* Relative standard deviation
Run habitats, 645
Running water ecosystems, characteristics of, 492
Runoff, *see also* Urban runoff
    adverse aquatic life effects caused by, 50
    agricultural, 47
    calculated total, 283
    construction site, 43
    distribution characteristics, 354–355
    effect assessments, 402
    events, duration of, 371
    fecal indicators in stormwater, 82
    habitat problems caused by, 54
    heavy metals in stormwater, 76
    highway, 83
    long-term aquatic life effects of, 92
    manure-laden, 89
    monitoring projects, urban, 36
    on-site, effect of from industry, 207
    pollutants, urban, sources of, 157
    presence of bacteria in stormwater, 465
    sources, major urban, 31
    volume, 397
        errors, 383
        predicted, 850
        reduction benefits, cost-effectiveness data for, 635
    water
        matrix, 431
        sources, 157
    yields, stormwater, 159
Run–riffle–pool sequence, 404
Rural Clean Water Program, 601

S

Sacramento squawfish, 53
Sacramento sucker, 53, 134, 135

Safety glasses, over-the-glasses, 739
Saffron shiner, 706
Sailfin shiner, 706
Salmo
    clarki, 74, 414
    gairdneri, 74, 414, 520, 536, 706
    salar, 414, 706
    trutta, 74, 414, 536, 706
Salmon
    density, 153
    effects of sedimentation on stream-living, 154
    embryos, survival of, 165
    fishery, 155
Salmonella, 83, 86
    thompson, 86
    typhi, 622
    typhimurium var. *copenhagen*, 86
    typhosa, 816
Salmonellosis, 621
Salt applications, for winter traffic safety, 62
Salt dilution, 358
Saltwater, 799
Salvelinus
    alpinus, 706
    confluentus, 706
    fontinalis, 74, 414, 520, 536, 706, 717
    malma, 706
    namaycush, 414, 706
Sample(s)
    analysis of, 732
    bottle(s)
        cleaning, 269
        options, American Sigma, 279
    collection, 254, 273
    concentration variations, determining, 236
    containers, 269
    CPOM, 669, 679
    field processing of, 271
    fraction of rated as toxic, 524
    handling and preservation, 730
    laundry detergent, 483
    number of needed for comparisons between
        different sites, 244
    number of needed to characterize conditions, 231
    number of needed to identify unusual conditions, 243
    preservation, 274
    processing, 694
    representative qualitative, 693
    riffle/run, 666, 674
    setup options, 224
    shipping of, 272
    size equations, environmental research, 235
    sorting, 667
    transfer, 304

transportation of to laboratory, 273
volumes, 263
Sampler(s)
    aromatic, 224
    automatic, 259
        flow-weighted, 288
        line flushing, 282
        refrigerated, 280
        source area, 299
    bedload, 295, 296, 410
    choosing appropriate sediment, 315
    cleaning of, 270
    comparison of substrate, 500
    corer, 289, 323
    cycle time, 283
    dipper, 290, 453
    dredge, 315
    freezing core, 316, 322
    grab, 315, 496, 685
    horizontal water, 291
    -induced pressure waves, 315
    manual pump, 292
    manual sheetflow, 298
    modifications, 224
    Orange-Peel, 316
    periphyton, 501
    Peterson, 316
    plastic, 333
    polyvinyl, 255
    Ponar, 316
    precipitation, 313
    rain gauge, 286
    retrieval of, 681
    sediment
        depth-integrated, 294
        popular, 316
    semiautomatic, 259, 299
    settleable solids, 295
    sheetflow
        manual, 298
        semiautomatic, 299
    Shipek, 316
    siphon, 287, 288
    Smith–McIntyre, 317
    sticky paper fugitive, 312
    stream-net, 499
    submerged water, 290
    suspended particulate, 312
    tripped vertical water, 291
    tube, 292
    Van Veen, 316, 317
    Vortox, 299
Sampling
    artifacts associated with, 531
    atmospheric contribution, 310

automatic, advantages of manual sampling
        compared to, 260
bacteria, 281
benthos, 491, 549
cluster, 225
dip net, 498
dry, 301
dry-weather, 188
effort, marginal benefit associated with
        increasing, 243
error, 233, 251
first-flush, 285
fish, 502
floatable litter, 296
floatable material, 224
flow-weighted composite, 283, 285
haphazard, 225
hyporheic, 326
judgment, 225
lines, losses of particles in, 282
locations
    number of needed to be represented in
            monitoring program, 238
    selection of, 256
macroinvertebrate, 494
manual, 259, 260
methods
    fish, 505
    macroinvetebrate, 497
microbiological, 289
for microorganism evaluations, 487
multistage, 225
path, 652
paired, 258
periphyton, 493
plans, 225
pore water, 313
probability, 225
problems, 347
procedures
    manual, 289
    street surface particulate, 301
program, street dirt, 306
protozoan, 494
random, 226
replicate, 665
routines, Monte Carlo, 577
safety considerations, 255
search, 226
sediment, 313
soil, source area, 300
source area, 278, 297
station(s), 284
    downstream, 361
    lengths, 693

stratified random, 225
strategies, 356
system, multilocation, 193
systematic, 225
time-discrete, 299
time-weighted composite, 285
Sampling effort and collection methods, 223–344
  basic sample collection methods, 336–338
  data quality objectives and associated QA/QC
      requirements, 254
    identifying needed detection limits and
      selecting appropriate analytical method,
      252–254
    quality control and quality assurance to
      identify sampling and analysis problems,
      247–252
  experimental design, 224–247
    determining number of samples needed to
      identify unusual conditions, 243–244
    factorial experimental designs, 227–231
    need for probability information and
      confidence intervals, 245–247
    number of samples needed to characterize
      conditions, 231–243
    number of samples needed for comparisons
      between different sites or times, 244–245
    sampling plans, 225–227
  general considerations for sample collection,
    254–277
    basic safety considerations when sampling,
      255–256
    personnel requirements, 275–277
    sampler and other test apparatus materials,
      260–263
    selecting sampling locations, 256–260
    volumes to be collected, container types,
      preservatives to be used, and shipping of
      samples, 263–275
  receiving water, point source discharge, and
    source area sampling, 278–313
    automatic water sampling equipment,
      278–289
    manual sampling procedures, 289–297
    source area sampling, 297–313
  sediment and pore water sampling, 313–336
    interstitial water and hyporheic zone
      sampling, 326–336
    sediment sampling procedures, 313–324
Sand roller, 701
Sand shiner, 703
Sandy soil conditions, infiltration rates of, 395
Sanitary landfills, 4
Sanitary sewer overflows (SSOs), 195
SAS Institute, 579
Sauger, 415, 704

Scatterplots, 586
Scioto madtom, 704
Scorecard Litter Rating (SLR) Program, 399
Score plots, of principal components, 595
Sculpins, 707
Sea lamprey, 701
Search sampling, 226
Seasonal Kendall test, 603
Seattle tests, 398
Secchi disk, 416
  transparency data, 175
  transparency observations, 176
Sediment
  artificial, 534
  bacteria conditions, 17
  bedload, 409
  bioaccumulation studies, chambers for
    conducting, 541
  bioassay tests, 155
  characterization, 325
  chemical analyses, 117
  collection methods, 327
  cores, 324
  criteria, 521
  deposition, effect of erosion on, 156
  deposits, 660
  depth-integrated samples for suspended, 292
  devices for collecting, 319
  exposure chamber units, 532
  feeders, 495
  guidelines, 837
  integrity, 323
  mixing of with drill auger, 325
  oils, 660
  oxygen demand (SOD), 117, 196, 204, 417
  PAH-contaminated, 540
  particle size, 103
  phases, used in toxicity tests, 521
  profiles, devices for obtaining, 320
  properties, 128
  quality, 131
    analyses, 107
    criterion, 234
    guidelines, 837, 838
    triad, 91, 611
  receiving water problems caused by, 416
  sampler(s)
    choosing appropriate, 315
    depth-integrated, 294
    guidelines, 317
    popular, 316
  scour of bottom, 408
  shallow stream with contaminated, 336
  standard testing protocols, 527
  suspended, impacts associated with, 413

toxicity tasks, 729
transport, 143, 408, 618
traps, 295
Seed
germination, 519
pretreatment, 6
Selenastrum capricornutum, 108, 114, 206, 268,
517, 520, 546, 721
Semiautomatic samplers, 259
Semiautomatic sheetflow samplers, 299
Semipermeable membrane devices (SPMDs), 212,
333, 536
Semiquantitative survey, routine initial, 191
Semivolatile organic compounds, 189
Semotilus margarita, 706
Sen's nonparametric estimator of slope, 603
Separate sewer overflows (SSOs), 68
discharge point, 199
evaluation project, 333
Septic systems, 20
Septic tank(s)
discharge, 472
failures, 460
suspected failing, 481
Sequential extraction procedures, 458
Serratia marcescens, 487
Settleable solids samplers, 295
Settling column tests, 456
Settling velocity(ies), 545
methods to measure stormwater, 454
particle size and, 451
settling column tests for, 456
Sewage
-contaminated waters, 196, 200
discharges, into urban streams, 85
disposal
overboard, 625
systems, on-site, 7
fecal coliform bacteria as indicators of
inappropriate discharges of sanitary, 464
pathogens from raw or poorly treated, 86
raw, 202
treatment of sanitary, 25
treatment plant, 419
Sewerage
inlet cleaning, 164
maintenance, 12
Shallow water vibratory core collection, 321
Shannon index calculations, 506
Sheetflow sampler, 298
Shellfish harvesting, 114, 610
Shigella, 62, 87, 197, 486, 821
Shigellosis, 621
Shipek sampler, 316
Shipping containers, 272

Shock-sensitive materials, 756
Shore zones, 666
Shorthead redhorse, 703
Shortnose gar, 702
Short-term exposure limit (STEL), 764
Shredder abundance, 667
Shredders, 410
Shrubbery, as shade for stream aquatic life, 145
Sicydium stimpsoni, 707
Side stream
detention ponds, 210
fish bioassay tests, 54
SIE, *see* Stressor Identification Evaluation
Sieve analyses, 407, 454
SigmaPlot, 578
SigmaStat, 578
Silver chub, 702
Silverjaw minnow, 703
Silver lamprey, 701
Silver redhorse, 703, 706
Silver shiner, 706
Silvex, 159
Silviculture, 4, 5, 7
Simazine, 118
Siphon samplers, 287, 288
Site
assessment, initial, 110
topography, 33
Skimmer boats, 70
Skipjack herring, 702
SLAMM, *see* Source Loading and Management
Model
Slaughterhouse wastes, 89
Slender madtom, 704, 707
Slimy sculpin, 705
Slope, Sen's nonparametric estimator of, 603
Slough darter, 707
SLR Program, *see* Scorecard Litter Rating Program
Sludge farm runoff, 7
Slug discharge test, 368
Smallmouth bass, 74, 415, 536, 704
Smallmouth buffalo, 703
Smith–McIntyre sampler, 317
Snakes, urban stream corridors as habitat for, 67
Snowmelt, 34, 626
Sockeye salmon, 152, 414, 705
SOD, *see* Sediment oxygen demand
Sodium
adsorption ratio, 526
hypochlorite, 740
Soil
age, 391
bank, 618
characteristics, 300
clayey, 396

column extraction method, 394
compact sandy, 396
compaction, 391, 397
compost-amended, 397
conditions, infiltration rates of sandy, 395
erodibility, 33
erosion, 31, 311
evaluations, 388
extraction kits, 436
infiltration
    characteristics, importance of field tests of,
       396
    rates, 229
    test plot, 390
insects, 390
measurement of infiltration rates in disturbed
    urban, 389
moisture measurements, 392
noncompact sandy, 396
samples, detection of PAHs in, 167
sampling, source area, 300
studies of depth of pollutant penetration in, 63
surveys, 549
texture measurements, 393
triangle, 393
type, 452
urban, infiltration rates in disturbed, 394
washoff of during rain, 627
Solar panel, exposure of to vandalism, 374
Solid(s)
bedded, 71
dissolved, 71, 158
    aquatic life criteria for, 823
    classification of, 72
    human health criteria for, 822
    total, 189, 191
-solution reactions, 329
suspended, 71, 103, 131, 352, 446, 612, 834
    classification of, 72
    effects of on aquatic macroinvertebrates, 73
    effects of on fish, 74
    water quality criteria for, 835
Soluble reactive phosphorus, 146, 147
Sorption, 79, 81
Source area sampling, 278, 297, 300
Source Loading and Management Model (SLAMM),
    353, 629, 854, 858
Southern redbelly dace, 702, 706
Soybean farming, 5
SPADNS reagent, 441
Spearman's Rank Correlation, 679
Species
intolerant, 698
-level identifications, 683
population number, 612
richness, 115, 697

Speckled chub, 702
Speckled dace, 701
Speckled madtom, 707
Spectrophotometer, 429
SPMDs, *see* Semipermeable membrane devices
Spotfin shiner, 703
Spottail shiner, 703, 706
Spotted bass, 415, 704
Spotted gar, 702
Spotted sucker, 703, 706
SS, *see* Suspended solids
SSOs, *see* Separate sewer overflows
Standards and regulations, compliance with, 13
Staphylococcus aureus, 87
State–discharge curve, 359
Statistical procedures, selection of, 580
Statistical reference books, 576
Statistical software programs, 575, 577
Statistics, 579
StatSoft, 579
StatXact-Turbo, 590
Steelcolor shiner, 703, 706
STEL, *see* Short-term exposure limit
Stenodus leucichthys, 706
Sticky paper fugitive dust samplers, 312
Stizostedion
canadense, 415
vitreum, 415
Stonecat, 704, 707
Stoneflies, 152
Storage, Treatment, Overflow Runoff Model, 854
Storm drainage
identifying inappropriate discharges into, 463,
    484
sources of inappropriate discharges into, 459
systems, 33, 142
Storm drain outfalls, 127
health outcomes for swimming in front of, 194
swimming in front of, 820
Storm event(s)
hydrodynamics, 349
influence of on chemical element dynamics, 348
levels of organic nutrients during, 73
Storm samples, event-mean concentrations for series
    of, 239
Stormwater
aesthetic use of, 25
assessments, strengths and weaknesses of toxicity
    tests in, 511
BOD analyses, 75
characteristics, 43, 190
chronic toxicity associated with, 19
contamination, potential for, 207
conveyance, 26, 166, 610
direct pathogen monitoring in, 29
event, attributes of, 20

hardness, 159
human health effects of, 85
hydrometer analyses of, 456
indicator categories, 104
infiltration, 25
    devices, 62
    groundwater impacts from, 56
inlet sediment volumes, 162
management
    planning, 10–11
    practices, 139, 183
microorganisms in, 485
models, 12, 307
monitoring, rain gauges suitable for, 378
organic matter in urban, 76
outfall, swimming near, 28
pathogens, 82, 620
permit program, 849
pollutants, sources of, *see* Receiving water uses,
        impairments, and sources of stormwater
        pollutants
pond, advertising benefits of, 25
potentials for extreme heterogeneity in, 22
receiving water problems associated with, 22
removal of phosphorus from, 169
runoff
    characteristics of, 35
    effects, 256
    heavy metals in, 76
    problem of, 3–4
    warm weather, 42
    yields, 159
safety concerns with, 66
sample(s)
    analyses priority for automatically collected,
        189
    extractions, 779
sources, identifying and prioritizing critical, 626
toxicants, potential sources of, 6
toxicity, 513
treatment system operating cost breakdown, 172
typical microscopic view of particles in, 457
Straight-line model, 241
Stratified random sampling, 225
Stream(s)
    alteration, beneficial effect of, 151
    aquatic life, shade provided for, 145
    assessment factors, for nonpoint source-affected
        streams, 111
    bank(s)
        characteristics, 411
        modification, 4
        stabilization of by riparian vegetation, 63
    bed
        erosion, 143, 155
        /sediment monitoring, 410

canopy, 206
channelized urban, 26
characterization, 120, 125, 142
depth
    estimated, 660
    gauges, 121
diagram, 652
discharge, 349, 369, 660
fishing in urban, 29
flow, 617
    analyses, 107
    monitoring in, 357
friction slope, 350
hydraulics, 618
hydrologic balance, permanent change of, 65
improvement projects, 56
inappropriate sanitary sewage discharges into
        urban, 85
map, 652
measurements, 651
models
    non-toxic constituents in, 864
    predicting pollutant fates using, 544
monitoring, intermittent, 280
parameters, potential effects of sources of
        alteration on, 21
reach factor, 111
recovery program, 25
stability, protection of, 65
staff gauges, 349
temperature profiles in, 410
type, 657
velocity, 360, 660
Streamline chub, 702
Stream-net samplers, 499
Streamside cover, 662
Street
    cleaning, 631, 632, 633
        effects of in controlling urban runoff pollutant
            discharges, 163
        equipment, 163
        subsample, 304
    construction material, 303
    pavement condition, 305
    surface particulate sampling procedures, 301
    texture, 308
Street dirt
    accumulation
        measurements of, 305
        rates, 305
    contributions of to urban runoff discharge, 161
    loading, 308
    samples, PAHs found in, 161
    sampling program, 306
    subsample collection, 302
    washoff, 307

Streptococcus
    bovis, 486
    faecalis, 87, 486
Stress
    demonstration, 616
    –productivity–predation relationships, 346
Stressor(s)
    cause-and-effect relationships between biological
        impairments and, 637
    class identification, 616
    combinations, 103
    examples of identifying, 538
    exposures, 533
    Identification Evaluation (SIE), 91
    loadings, 207
    potential sources of, 122
    reduction, 123
Stressor categories, effects on humans and
        ecosystems and, 47–98
    effects of runoff on receiving waters, 47–63
        adverse aquatic life effects caused by runoff,
            50–54
        fish kills and advisories, 49–50
        groundwater impacts from stormwater
            infiltration, 56–63
        indicators of receiving water biological effects
            and analysis methodologies, 48–49
        observed habitat problems caused by runoff,
            54–56
    receiving water effect summary, 90–92
    stressor categories and effects, 63–90
        aesthetics, litter/floatables, and other debris
            associated with stormwater, 68–71
        dissolved oxygen, 73–75
        nutrients, 76
        pathogens, 78–90
        safety concerns with stormwater, 66–68
        solids, 71–72
        stream flow effects and associated habitat
            modifications, 63–66
        temperature, 75–76
        toxicants, 76–78
Striped bass, 414, 536, 704
Striped jumprock, 707
Striped shiner, 703
Student current meter, 359
Student–Newman–Keuls test, 591
Student's *t*-tests, 589
Subsample
    classification of organisms in, 668
    collection of, 303
    street cleaning, 304
    street dirt, 302
Substrate
    artificial, 683
    characterization, 407

quality factor, 111
samplers, comparison of, 500
samples, bug picking from, 498
Subsurface
    coal mining, 4
    flow measurements, 394
Suburban transit, 187
Suckermouth minnow, 702
Sucker species, 508, 698
Sulfate, 132
Sulfide-bound metals, 327
Sulfuric acid, 740
Sump pump discharges, 461
Sunfish species, 508, 697
Superfund sites, 619
Surface
    coal mining, 4
    cover, 33
    –groundwater interaction, 362
Surface water
    pollutants, 104
    quality, 146
    sampling locations, constituents monitored at,
        192
Surfactant(s)
    analyses, 472
    LAS from synthetic, 479
Surrogate species, laboratory-derived toxicity values
        for, 614
Survival data, 604
Suspended sediment
    depth-integrated samplers for, 292
    impacts associated with, 413
    samples, 224
Suspended solids (SS), 71, 103, 118, 198, 446, 612,
        834
    classification of, 72
    effects of on aquatic macroinvertebrates, 73
    effects of on fish, 74
    reduction benefits, 635
    water quality criteria for, 835
Suspension feeders, 495
Swallowers, 495
Swamp darter, 707
Swamps
    cedar, 52
    urban runoff-affected, 52
Swimmer's ear, 87
Swimming, 114
    areas, in urban receiving waters, 28
    beaches, river, 84
    in front of storm drain outfalls, 194, 820
Synedra, 138, 175
Synthetic organics, 7
Synthetic pyrethroids, 436, 535

SYSTAT, 379, 578
Systematic sampling, 225

T

Tadpole madtom, 704
Taxa richness, 671, 674
TDS, *see* Total dissolved solids
Tebuthiuron, 118
Telecommunications industry, 736
Telemetry equipment, 286
Telescope shiner, 706
Tennessee shiner, 706
Teratogenicity, 507
Termite control, 6
Test(s)
    ANOVA, 580, 599
    apparatus materials, 260
    aqueous phase, 520
    artificial substrate macroinvertebrate
            colonization, 121
    bacteria, 433
    bioaccumulation, 269, 616, 724
    Bonferroni *t*-, 591
    Chi-square goodness of fit, 586
    comparing multiple sets of data with group
            comparison, 588
    comparison, 580
    correlation, 470
    double-ring infiltration, 229
    Duncan's multiple range, 591
    Dunner's 591
    Durbin–Watson, 598
    earthworm, 519
    end of, 714, 722
    field, 396
    fish bioassay, 54
    HACH detergents, 471
    homogeneity, 177
    honest significant difference, 604
    infiltration, apparatus, 389
    in situ, advantages of, 531
    in-stream toxicity, 168
    kit performance, 430
    Kolmogorov–Smirnov one-sample, 586
    Kurskal–Wallis ANOVA on ranks, 588, 591
    Mann–Kendall, 175, 176, 582, 602
    Mann–Whitney signed rank, 590
    Mann–Whitney U, 470, 475
    microbial activity, 523
    Microtox screening, 445, 513, 730
    nonparametric, 581
    outfall, for optical brighteners, 440

    parametric, 581
    pathogen die-off, 200
    pavement, rainfall-runoff responses for, 850
    peeper, 204
    photosynthesis/respiration, 440
    pore water toxicity, 267
    Seasonal Kendall, 603
    Seattle, 398
    sediment bioassay, 155
    settling column, 456
    slug discharge, 368
    SOD, 122
    soil infiltration, 390
    Student's *t*-, 589
    Student–Newman–Keuls, 591
    toxicant reduction, 525
    toxicity
        ambient, 665
        approaches, 512
        ASTM standards on, 712
        emerging tools for, 536
        in situ, 530
        macrofaunal, 523
        sediment phases used in, 521
        strengths and weaknesses of in stormwater
                assessments, 511
    washoff, 306
    waters, fluorescence of, 471
    whole effluent toxicity, 48, 507, 514
    Wilcoxon rank sum, 590
Tetrachloroethylene, 38, 61, 801
Tetrahydrofuran, 261
Textile mills products, 187
Thalidomide, 744
Threadfin shad, 135, 414, 702
Three spine stickleback, 53, 134, 135, 701
Threshold limit value (TLV), 764
Thymallus oligolepis, 706
TIE, *see* Toxicity identification evaluation
Time-discrete sampling, 299
Time-of-travel velocity meters, 377
Time-weighted composite sampling, 285
Tippecanoe darter, 705
Tipping bucket rain gauges, 387, 388
Tire wear, 7
TLV, *see* Threshold limit value
TMDL, *see* Total maximum daily load
TOC, *see* Total organic carbon
Tolerance classification(s), 506
    agencies having developed, 687
    family-level, 680
Toluene, 6, 38, 61, 740, 744, 801
Tonguetied minnow, 702
Topeka shiner, 706
Torrent sculpin, 701

Torrent sucker, 707
Total dissolved solids (TDS), 118, 189, 191, 252
Total Kjeldahl nitrogen, 42, 158, 160
Total-load stations, 293
Total maximum daily load (TMDL), 10, 11, 798,
        844, 852
Total organic carbon (TOC), 130, 191, 325
Total petroleum hydrocarbons (TPH), 191, 435
Total suspended solids (TSS), 184
Toxaphene, 803, 827
Toxicant(s)
    concentration profiles of in rivers, 547
    food as source of, 620
    mixtures, measuring effects of in organisms, 515
    reduction tests, laboratory-scale, 525
    sources, 629
Toxicity, 763
    assay considerations, 522
    assessment, 113
    identification evaluation (TIE), 116, 122, 208,
            514, 729
        protocol, 538
        scheme, 458
        screening, 526
    particulate-associated, 19
    testing, 107
        approaches, 512
        ASTM standards on, 712
        emerging tools for, 536
        in situ, 530
        in-stream, 168
        sediment phases used in, 521
        strengths and weaknesses of in stormwater
            assessments, 511
Toxicity and bioaccumulation testing, 709–734
    general toxicity testing methods, 710
    in situ testing using confined organisms, 724–729
    methods for conducting long-term sediment
            toxicity tests with *Chironomus tentans*,
            718–724
        collection of egg cases, 719
        dissolved oxygen, 720–721
        ending of test, 722–723
        feeding, 720
        hatching of eggs, 719–720
        interpretation of results, 723–724
        monitoring emergence, 722
        monitoring survival and growth, 721
        placing organisms in test chambers, 720
    methods for conducting long-term sediment
            toxicity tests with *Hyalella azteca*,
            710–718
        acclimation, 713
        ending of test, 714–716

        feeding, 713
        interpretation of results, 716–717
        monitoring of test, 713–714
        placement of sediment into test chambers, 710
        placing organisms in test chambers, 713
    Microtox screening test, 730–733
        apparatus, 731
        calculations, 732–733
        health and safety information, 733
        interferences, 730
        precision and accuracy, 733
        procedure, 731–732
        reagents, 731
        sample handling and preservation, 730
        scope and application, 730
        summary of method, 730
        toxicity identification evaluations, 729
Toxic unit (TU), 516
TPH, *see* Total petroleum hydrocarbons
TRAACS 2000 continuous-flow analyzer, 449
Tracer(s), 377, 463
    artificial, 362, 364
    characteristics, of local source flows, 466
    dye, 362
    flow monitoring using, 361
    naturally occurring, 363
Trailer parks, 185
Transition zones (TZ), 326
Transparency, measurement of, 416
Transportation equipment, 187
Trash
    boom, 70
    floatable, 29
    racks, 67
Trend(s)
    analyses
        examples of, 109
        preliminary evaluations before use of, 601
    statistical identification of, 601
Triazines, 436
Tributyltin, 807
Trichloroethylene, 61, 740
2,4,6-Trichlorophenol, 448
3,5,6-Trichloropyridonol, 436
Trichoptera, 491, 495, 688
Trihalomethanes, 62
Trinitrotoluene, 741
Trip blank, 248
Tripped vertical water sampler, 291
Trophic composition metrics, 698
Trout-perch, 704
TSS, *see* Total suspended solids
TU, *see* Toxic unit
Tube sampler, 292

Tubificids, 52, 137
Turbidity, 131, 424, 446, 469, 658, 834
    excessive, 694
    probes, 439, 784
    relationship between *Daphnia magna* toxicity
        and, 543
    sensors, 371
    values, 204
TV surveys, 484
Typhoid fever, 621
TZ, *see* Transition zones

U

UF, Urea formaldehyde
Underground parking garages, 23
Underground storage tanks, leaks from, 461
Unit area loadings, 845
UNIX, 578
Upstream–downstream sampling design, 209
Upstream flow rate, 367
Urban aquatic environments, metal
        bioaccumulations in, 133
Urban area pollutant yields, 37
Urban bacteria sources, 82
Urban Catchment Model, 854
Urban drainage elements, 23
Urban fishing, 29
Urban hydrology, 351
Urbanization, 65
Urban models, attributes of, 856
Urban planning, initiated, 628
Urban runoff
    assessment of priority pollutant concentration in,
        131
    cause-and-effect relationships between receiving
        water conditions and, 30
    controls, 163
    discharge,(s)
        street dirt contributions to, 161
        use of dry detention basins in controlling, 164
    effects of on receiving water aquatic organisms,
        91
    hazardous substances observed in, 38
    heavy metals, 79
    monitoring, 36, 137
    pollutant(s)
        concentrations, 39
        sources of, 36, 143, 157
    sources, major, 31
    total solids in, 167
    yields, sewerage inlet cleaning effects in
        reducing, 164

Urban soils
    infiltration rates in disturbed, 394
    measurement of infiltration rates in disturbed, 389
Urban stormwater
    contaminants, sources of, 626
    organic matter in, 76
Urban watersheds, rainfall distribution in, 382
Urban waterways, stressed, 27
Urea formaldehyde (UF), 59
U.S. Environmental Protection Agency (EPA), 3
    CSOs required by, 69
    multimetric approach used by, 116
    National Pollutant Discharge Elimination
        System, 8, 513, 847
    National Water Quality Inventory released by,
        78
    Nationwide Urban Runoff Program, 34, 65, 615,
        629
    -sponsored research, on stormwater indicators, 30
U.S. Geological Survey (USGS), 139
USGS, *see* U.S. Geological Survey

V

Vacuum
    cans, reassembling of, 305
    pump, 742
    units, 304
Vandalism, 374
Van Veen sampler, 316, 317
Vapor
    density, 749
    pressure, 749
Variegate darter, 705
Vegetation
    overhanging, 647
    riparian, 150, 618
    surrounding outfall, 469
Velocity meters, 377
Vibratory corers, 316, 324
Vibrio
    cholerae, 622
    parahemolyticus, 818
Vinyl chloride, 801
Viral adsorption, 61
Viral gastroenteritis, 621
Viruses, 90, 490
VOCs, *see* Volatile organic compounds
Volatile organic compounds (VOCs), 61, 78
Volatile suspended solids (VSS), 118, 131, 198
Volatilization, 78, 79, 80
Vortox sampler, 299
VSS, *see* Volatile suspended solids

W

Walleye, 415, 704
Warm-water fishery, 27
Washoff
    equation, 307
    street dirt, 307
    tests, small-scale, 306
Waste
    assimilation capacity, 412
    containers, guidelines for, 760
    disposal methods, *see* Laboratory safety, waste
        disposal, and chemical analyses methods
    load allocations, conventional pollutants model
        comparison as used in, 865
Wastewater(s)
    effluent, treated, 88
    samples, sampling and handling requirements for,
        264–265
    sanitary, 472, 473
    treatment, on-site, 4
Water(s)
    accelerated eutrophication of, 830
    anion chromatographic conditions in, 450
    background fluorescence in, 365
    balance, treatment system, 173
    cation chromatographic conditions in, 450
    chemical analyses, 117
    clarity, 651
    collection methods, 533
    column
        pollutants, interaction of contaminated
            sediments and, 201
        quality, 51
        surrogate, 528
    contact recreation, 22, 624, 820
    creek interstitial, 148
    deionized, 783
    dilution, 717, 718, 719
    fluorescence of test, 471
    hardness, 825
    -holding capacity, 394
    interstitial
        bacteria populations, 203
        chemistry, 203
        collection methods, 318
        degradation of, 202
        immediate collection and analysis of, 328
        isolation of, 329
        measurements, 202, 327
        quality, 371
        sampler selection guidelines, 317
    irrigation, 474
    moccasins, 256
    odors, 658
    pathogen-contaminated, 85
    pore
        sampling, 313
        squeezer, 329
        toxicity test, 267
        type of container and conditions
            recommended for storing samples of,
            266–267
    quality, 128, 216, 656
        analyses, 107
        baseflow, 160
        characteristics, measurement of, 713
        construction site runoff, 33
        evaluations, 425, 618
        indicators, 105
        in-stream, 143
        monitoring sonde, 782
        observations, 585
        overlying, 724, 725, 727
        parameters, long-term *in situ* measurements
            of, 427
        pollutant constituent monitoring, 105
        probes, 281
        riparian zone components possibly affecting,
            411
        violations, 356
    -reactive chemicals, 755
    receiving, pH of, 364
    Reverse Osmosis quality, 203
    runoff, sources, 157
    sampler(s), 293
        horizontal, 291
        open vertical, 291
        tripped horizontal, 291
        tripped vertical, 291
    samples
        blank, 309
        sampling and handling requirements for,
            264–265
    sewage-contaminated, 196, 200
    stable isotope methods for identifying sources of,
        481
    stage, 651
    standard testing protocols, 517
    supplies, human health considerations associated
        with potentially contaminated, 124
    surface oils, 658
    tracing, using fluorescent dyes, 364
    treated potable, 472
Waterborne diseases, 90
Waterfront areas, 25
Water quality criteria, 797–841
    ammonia, 813–816
    bacteria, 816–822

bacteria criteria for water-contact recreation, 820–822
development of bathing beach bacteriological criteria, 816–820
chloride, conductivity, and total dissolved solids, 822–823
  aquatic life criteria for dissolved solids, 823
  human health criteria for dissolved solids, 822–823
chromium, 823–824
  aquatic life effects of Cr^{+3}, 823–824
  human health criteria for chromium, 824
  national freshwater aquatic life criteria for CR^{+3}, 824
compilation of recommended water quality criteria and EPA's process for deriving new and revised criteria, 799–813
copper, 824–825
  effects of copper on aquatic life, 824–825
  human health criteria for copper, 825
  national aquatic life criteria for copper, 825
EPA's water quality criteria and standards plans, 798–799
hardness, 825–826
hydrocarbons, 826–827
lead, 827–828
  aquatic life summary for lead, 827–828
  human health criteria for lead, 828
  national aquatic life criteria for lead, 828
nitrate and nitrite, 828–830
  human health nitrate and nitrite criteria, 829
  nitrate and nitrite aquatic life criteria, 829–830
nonpriority pollutant, 806–808
organoleptic effects, 809
pH, 832–834
phosphate, 830–832
priority toxic pollutant, 800–803
sediment guidelines, 836–838
suspended solids and turbidity, 834–835
zinc, 835–836
  aquatic life criteria for zinc, 835
  human health criteria for zinc, 836
Watershed(s)
  ammonia in, 192
  areal rainfall accuracies for fast-responding, 384
  areas, topographical maps used to determine, 120
  assessment projects, 346
  average rainfall depth, measurement of, 382
  characterization, 125, 141, 183, 348, 351
  chloride in, 192
  complexity matrix, 123
  development factor, 111
  erosion, 659
  heavily urbanized city, 17
  illicit problems in typical, 462
  increasing urbanization in, 179
  indicators of biological receiving water problems, 103
  investigation of parallel, 141
  land uses of, 258
  lead concentrations in, 192
  models, 843
  multi-, 123
  nitrate in, 192
  planning, 276
  rain depth errors, 383
  rainfall distribution in urban, 382
  reference, 111
  -scale loading models, 846
  sensitive species lost from, 507
  surveys, 398
  test sites, monitored annual pollutant discharges for, 40
  total coliforms in, 192
Watershed and receiving water modeling, 843–874
  complex models, 852–855
  geographical information systems, 857–860
  receiving water models, 855–857
  simple models, 846–852
  unit area loadings, 845–846
Waterways, human-dominated, 22
Weather station, 390
Weight-of-evidence (WOE), 610, 611
Western banded killifish, 704
Western sand darter, 705
WET, see Whole effluent toxicity
Wet detention pond(s), 66, 631, 633
  effect of, 211
  people living near, 25
  recommendations to maximize safety near, 67
Wetland
  acreage loss, 104
  health indicators, 116
Wet-weather
  discharge characteristics, land use monitoring for, 239
  quality, 214
Wet-weather flow(s)
  analyses, standard and modified methods for, 448
  bacteria, 203
  floatable litter associated with, 68
  use of multiparameter probe to indicate presence of, 370
Whisker plots, 587
White bass, 414, 704
White catfish, 703
White crappie, 415, 701, 704
White perch, 414, 704
White sucker, 414, 703
Whitetail shiner, 706

Whole effluent toxicity (WET), 507, 514
    calculations, 517
    tests, 48, 507
Whole-sediment manipulations, 542
Wilcoxon rank sum test, 590
Wind
    error, 386
    -transported materials, 311
Windows front-ends, 844
Winkler titration, 440
WOE, *see* Weight-of-evidence
Wood
    preservatives, 5, 6
    products, 187

X

X-ray fluorescence, 778
Xylene, 61, 436, 744
XYZ autosampler, 449

Y

Yard wastes, 67
Yellow bass, 704
Yellow bullhead, 701
Yellowfin shiner, 706
Yellow perch, 415, 701, 704
Yersiniosis, 621

Z

Zero runoff increase (ZRI), 64, 65
Zinc, 6, 132, 809
    aquatic life criteria for, 835
    bioaccumulation of, 133
    human health criteria for, 837
Zone of common flooding, 374
Zoogeography, 653
Zooplankton, 63, 116, 254, 268, 338, 491, 501
ZRI, *see* Zero runoff increase